ANTARCTIC RESOLUTION

edited by GIULIA FOSCARI / UNLESS

Lars Müller Publishers

FOUR ELEMENTS

SURVIVING IN THE CRYOSPHERE

ARCHIVE OF ANTARCTIC ARCHITECTURE

APPENDIX

"We travelled for Science [...] in order that the world may have a little more knowledge, that it may build on what it knows instead of on what it thinks."

Apsley Cherry-Garrard (Explorer, Zoologist), *The Worst Journey in the World*, 1922

Antarctic Resolution

Giulia Foscari

Accounting for approximately 10% of the land mass of Planet Earth,[1] Antarctica – first theorised by the ancient Greeks yet only "discovered" after the Industrial Revolution – is a realm we collectively neglect. Far from being a pristine natural landscape, the continent is a contested territory which conceals not only resources that might prove irresistible in a world with ever-increasing population growth but also scientific data crucial to inform future environmental policies. The kilometres-thick stratum of ice that has accumulated on its bedrock for millions of years, equivalent to around 70% of the freshwater on our planet,[2] represents both an indispensable resource for human survival and the greatest menace to global coastal settlements threatened by the rise in sea levels induced by anthropogenic global warming.

1 Planet Earth was first immortalised in its entirety on the 7th of December 1972 by Harrison Schmitt aboard the command module of Apollo 17. Known as NASA Image AS17-148-22727, the photograph (shown on the opposite page) is the first to capture the illuminated globe. Contrary to all previous lunar missions, during which the celestial bodies of the Earth and the Moon were obscured by complementary shadows, the trajectory of Apollo 17 allowed humankind to see the so called "Blue Marble" from afar. Prophetically, the crew's line of sight fell onto Africa and Antarctica, putting the latter centre stage. 2 See Tabatha Thompson and Ed Campion, "Joint NASA Study Reveals Leaks in Antarctic 'Plumbing System,'" NASA press release, 15 February 2007, nasa.gov [online].

Antarctic Resolution advocates the urgent removal of the goggles of Heroic Age explorers[3] and the rejection of the pixelated view of Antarctica offered to us by Big Data companies.[4] In their place it urges the construction of a high-resolution image focusing on the continent's unique geography, experimental governance system, contemporary geopolitical significance, unparalleled scientific potential, and extreme inhabitation model. Looking beyond what is visible, *Antarctic Resolution* also aims to unveil the intricate web of growing economic and strategic interests, tensions, and international rivalries that are deliberately enveloped in total darkness, as is the continent for six months per year.

Only a concerted effort by a transnational network of multidisciplinary polar experts could construct an image sufficiently focused to properly address the urgent issues concealed in 26 quadrillion tons of ice.[5] No single discipline can synthesise the complexity of the natural and political forces at play in what is unquestionably a Global Commons, i.e. an open-access site, like the high seas, the deep seabed, the atmosphere, and Outer Space, whose governance should equitably benefit humankind at large.

With this in mind, on the eve of the 200th anniversary of the first recorded sighting of Antarctica,[6] UNLESS[7] has invited over 200 specialists from the fields of aeronautics, anthropology, architecture, biology, chemistry, climate change, economics, engineering, geoscience, glaciology, history, law, literature, logistics, medicine, physics, political science, science, sociology, technology, and the visual arts to actively engage in dynamic dialogues, theoretical seminars, and radical workshops,[8] in an attempt to jointly close a paradoxical knowledge gap of the only continent devoted, on paper at least, purely to research.

3 The wooden snow googles with cross-shaped eye slits, like those worn by Captain Robert Falcon Scott during the National Antarctic Expedition in 1901–1904 (shown above), offered a limited field of vision to explorers who ventured into the unforgiving environment of Antarctica during the Heroic Age (1897–1922).
4 The ultimate metaphor for the collective neglect of Antarctica is the pixelated and fragmented view of the southernmost continent offered by Google Earth (shown unprocessed on the opposite page and, in detail, on the cover), which is reminiscent of the "blank spaces" on imperial maps.

5 See Peter Fretwell, *Antarctic Atlas: New Maps and Graphics That Tell the Story of a Continent* (London: Particular Books, 2020).
6 The first official sightings were recorded on the 27th of January 1820 by Admiral Fabian Gottlieb von Bellingshausen, on the 30th of January 1820 by Edward Bransfield, and on the 17th of November 1820 by Captain Nathaniel Brown Palmer. The first to land on the continent is believed to have been John Davis, on the 17th of February 1821.
7 UNLESS is a non-profit agency for change devoted to interdisciplinary research on extreme environments.

8 The workshops and seminars were held at the Architectural Association School of Architecture in London, at the Scott Polar Research Institute and the United Kingdom Antarctic Heritage Trust in Cambridge (United Kingdom), at the Pontifical Catholic University of Chile in Santiago (Chile), at the University of Hong Kong (Hong Kong), and at –Ness in Buenos Aires (Argentina).

Presented here in the form of single pixels, data to be analysed, and ideas to be built upon, the collaborative project offers no presumption of completeness and no authoritative finale. Alongside the authored texts, a vast portfolio including visualisations[9] produced by UNLESS and its Polar Lab[10] testifies to architecture's disciplinary aptitude for visualising phenomena that pertain to the realms of geopolitics, science, and sociology and to the designer's responsibility to critically analyse the contemporary condition and redirect collective behaviours. With subjects ranging in scale from neutrinos[11] and microscopic diatoms[12] to the cosmos, and in timescales from past glacial-era atmospheric concentrations[13] to projected post-human conditions, the publication aspires to broadcast knowledge of the continent and the Southern Ocean that surrounds it (which together constitute the "Antarctic" as a region), provoke change, and ultimately mobilise younger generations to undertake a true Antarctic resolution.

Equally significant to the content presented here are the blanks. I assume responsibility for any omission, which is hopefully excusable in the context of a work that attempts to examine a continent twice the size of Australia and a maritime area of tens of millions of square kilometres; any perceived scarcity of mention of certain nations and their accomplishments should not be interpreted as an editorial decision nor as a lack of effort. More problematic,

9 All the visualisations produced by UNLESS and the Polar Lab rely on primary sources that were often unveiled during deep dives into deserted archives and long conversations with Antarctic specialists. Evidence of these data-mining efforts is offered in the appendix, where all the sources are duly recorded. The cartography shown opposite offers a snapshot of the Antarctic's drifting borders. The ever-changing trajectory of the Convergence circumpolar current, the pulsating ice pack (which almost doubles the overall surface area of ice during the winter months), and the shifting Magnetic and Geomagnetic South Poles offer examples of the impossibility and inadequacy of trying to dominate the southernmost continent.
10 Launched by UNLESS, the Polar Lab is an experimental, transnational academic platform that brings together architecture students from universities around the globe with interdisciplinary Antarctic experts. With research hubs in Argentina, Brazil, Chile, Hong Kong, and the United Kingdom – directed respectively by Florencia Rodriguez, Sol Camacho, Arturo Lyon, Juan Du, Francesco Bandarin

and myself – the Polar Lab has brought to light unique Antarctic narratives concealed in regional archives and produced the "Antarctic Atlas" presented in the following pages.
11 Scientifically studied within the 1,000,000,000-cubic-metre block of ice of the IceCube Neutrino Observatory located at the South Pole, the almost massless and chargeless subatomic particles known as neutrinos encapsulate information on supernova explosions, cosmic rays, and the universe.
12 Coccolithophores are microscopic marine diatoms which are gravely affected by the excessive acidification of the warming Southern Ocean, the world's largest carbon sink. On its own, this body of water is responsible for the absorption of 40% of anthropogenic carbon dioxide and for activating the global oceanic circulation. The altered pH value of the oceanic water inhibits the calcification process of organisms such as the coccolithophores, endangering the entire marine ecosystem and aggravating biodiversity loss.
13 Antarctica is the largest planetary archive – an archive in which it is the very substance of the ice

that contains what is most evanescent in life: air and time. Ice cores drilled on the Antarctic Plateau allow palaeoclimatologists to extract scientific values on historic temperature fluctuations and reconstruct trends in atmospheric greenhouse-gas concentrations from past ice ages and "warm" interglacial eras. As shown on the following page and expanded upon in an article in *The Economist* dated 21st of September 2019, "in interglacials the carbon-dioxide level is 1.45 times higher than it is in the depths of an ice age. Today's level is 1.45 times higher than that of a typical interglacial". While relativising the infinitesimal history of humankind within deep time and heightening our awareness of the assured extinction of our own species, such time capsules of our planet's climate history are the foundation on which future sustainable environmental policies must, as a matter of urgency, be constructed, to reduce worldwide carbon-dioxide production.

Anne Noble, Bung in Antarctica on a blow-up globe, 2005

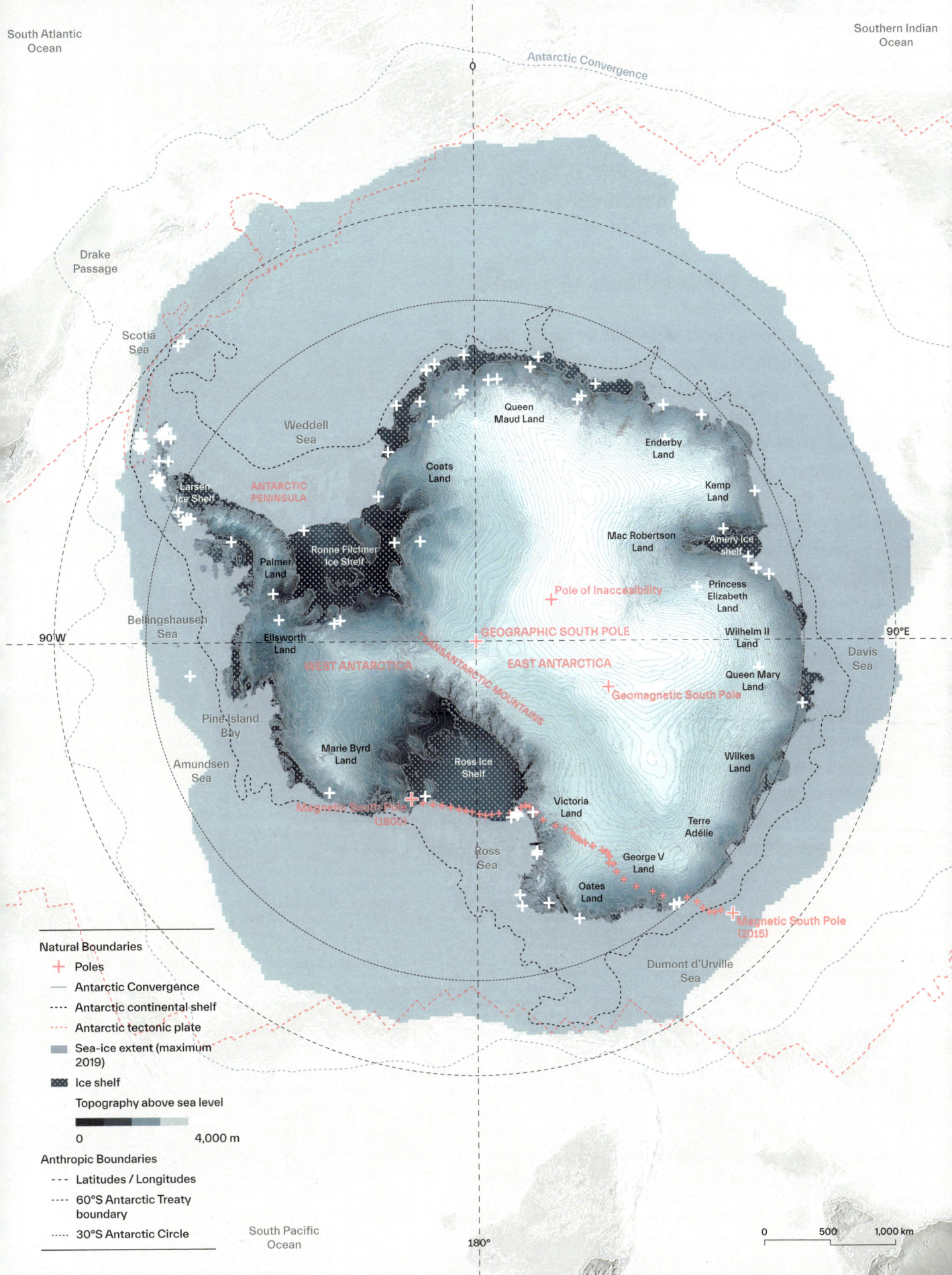

South Atlantic
Ocean

Southern Indian
Ocean

Antarctic Convergence

0

Drake
Passage

Scotia
Sea

Weddell
Sea

Queen
Maud Land

Enderby
Land

Coats
Land

Kemp
Land

Larsen
Ice Shelf

ANTARCTIC
PENINSULA

Mac Robertson
Land

Amery Ice
shelf

Palmer
Land

Ronne Filchner
Ice Shelf

Princess
Elizabeth
Land

Bellingshausen
Sea

Pole of Inaccessibility

90°W

Ellsworth
Land

WEST ANTARCTICA

GEOGRAPHIC SOUTH POLE

EAST ANTARCTICA

Wilhelm II
Land

90°E

Davis
Sea

Geomagnetic South Pole

Queen Mary
Land

Pine Island
Bay

TRANSANTARCTIC MOUNTAINS

Amundsen
Sea

Marie Byrd
Land

Ross Ice
Shelf

Wilkes
Land

Magnetic South Pole
(1800)

Victoria
Land

Terre
Adélie

Ross
Sea

George V
Land

Oates
Land

Magnetic South Pole
(2015)

Dumont d'Urville
Sea

Natural Boundaries

+ Poles

Antarctic Convergence

Antarctic continental shelf

Antarctic tectonic plate

Sea-ice extent (maximum
2019)

Ice shelf

Topography above sea level

0   4,000 m

Anthropic Boundaries

Latitudes / Longitudes

60°S Antarctic Treaty
boundary

30°S Antarctic Circle

South Pacific
Ocean

180°

0   500   1,000 km

though, is the blank page or, better put, the programmed "white-out"[14] of exploitable Antarctic resources.

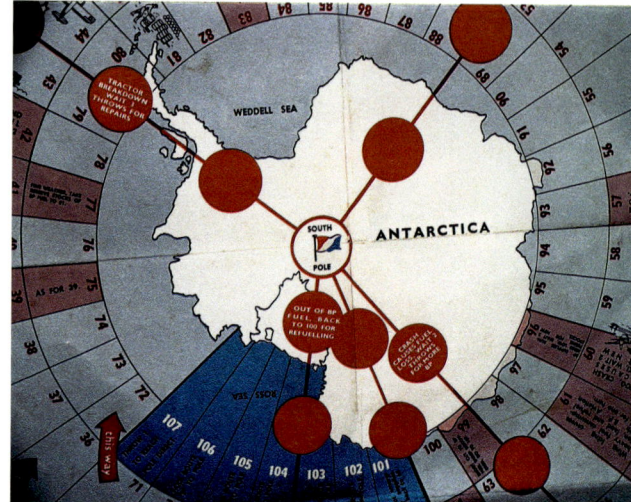

Antarctica's governance model relies on the Antarctic Treaty, an international Cold War–era agreement signed in 1959[15] as a legacy of the International Geophysical Year to promote the demilitarisation and denuclearisation of Antarctica and ensure that "international cooperation in scientific investigation" conducted in the continent would lead to the free exchange of "scientific observations and results".[16] Six decades after the agreement came into force, allowing for the progressive building of what is now called the Antarctic Treaty System, commercial and strategic interests in the vast untapped reserves of natural resources, in the marine species found in abundance in the Southern Ocean,[17] in Antarctic bioprospecting,[18] in polar tourism,[19] and in unperturbed access to global satellite navigation systems[20] are exerting new pressures on the already fragile Antarctic system of governance.

In this contemporary scenario, it appears as if scholars are no longer granted the cultural freedom to share their discoveries on Antarctic natural resources in the way that they were up until the signing of the Protocol on Environmental Protection,[21] and their subjection to the rules of conduct enforced by national programmes is expressed in

14 "White-out" is a term commonly used to identify an atmospheric phenomenon whereby the refraction of light onto the totalising whiteness of the Antarctic landscape blurs the horizon line, obliterates shadows, and overwhelms the human visual capacity by hindering depth detection. Alongside this disorienting weather-centric definition, "white-outs" can also refer to the conscious act of blanking out information. The use of the term in this context is non-binary and alludes to both definitions.
15 The Antarctic Treaty was signed on the 1st of December 1959 by the twelve nations that had been active during the International Geophysical Year: namely, Argentina, Australia, Belgium, Chile, France, Japan, New Zealand, Norway, South Africa, the Soviet Union, the United Kingdom, and the United States of America. To date the number of signatories has risen to include fifty-two states, twenty-eight of which are Consultative Parties and retain the right to vote on Antarctic affairs.
16 The Antarctic Treaty, opened for signature on the 1st of December 1959.
17 The marine species include krill, small crustaceans best known as representatives of the dominant animal species on Earth. Krill are overexploited

by national fishing enterprises, which vehemently object to the creation of much-needed Marine Protected Areas (MPAs).
18 Bioprospecting, which entails the systematic search for products and genetic information derived from bioresources, leverages the commercial value of unique biological compounds, predominantly for pharmaceutical applications.
19 According to data documented by the International Association of Antarctica Tour Operators' (IAATO) official website, retrieved on the 3rd of March 2021, Antarctic tourism has been growing exponentially over the decades, reaching the unprecedented number of 74,401 tourists in the austral summer season of 2019–2020.
20 While the Antarctic Treaty states, in its first article, that "Antarctica shall be used for peaceful purposes only", scholars argue that Antarctic satellite receiving stations have tacit dual civil-military capabilities which would enable new hybrid warfare strategies that surpass the traditional kinetic forms of warfare implied in the Treaty itself.
21 Geoscientific discoveries on Antarctic natural resources were freely accessible until the signing of the utopic Protocol on Environmental Protection

in 1991, when assumptions on natural resources were still based purely on "geological association", i.e. on memories of Antarctica's continental drift some 160 million years ago, from lands rich in oil, gas, coal, platinum-group metals, iron, uranium, copper, and chromium. (Doaa Abdel-Motaal, "Antarctic Resources and the Protocol for Environmental Protection," see pp. 136–37). Such transparency on Antarctic resources led Shell Oil BP to celebrate the 1957–1958 Commonwealth Transantarctic Expedition by producing a game that enacted "an exciting race to the South Pole by land, sea and air", whereby fuel depots (represented by red dots in the photograph above) are connected to the coastline via a road to the South Pole, and a "crash caus[ing] fuel loss" would force players to "wait 3 throws for more BP".

Anne Noble, board game produced by Shell Oil to commemorate the Hillary/Fuchs Antarctic expedition of 1957–1958, 2006

2021
Carbon-dioxide levels
reach 416.67 ppm

400
375
350
325
300
275
250
225
200
CO₂ (ppm)

6°
4°
2°
0°
-2°
-4°
-6°
-8°
-10°
-12°
Temperature (°C)

| 400,000 B.C.E | interglacial period | glacial era | 300,000 B.C.E | interglacial period | glacial era | 200,000 B.C.E | interglacial period | glacial era | 100,000 B.C.E | interglacial period | O |

CO₂ billion tons
34
32
30
28
26
24
22
20
18
16
14
12
10
8
6
4
2

Flaring
Cement
Gas
Oil
Coal

1997
The Kyoto Protocol
is signed.

2005
The Kyoto Protocol
enters into force.

2016
The Paris Agreement
enters into force.

1992
The United Nations
Framework
Convention on Climate
Change (UNFCCC)
is open for signature.

1994
The United Nations
Framework
Convention on Climate
Change (UNFCCC)
enters into force.

1755 1760 1765 1770 1775 1780 1785 1790 1795 1800 1805 1810 1815 1820 1825 1830 1835 1840 1845 1850 1855 1860 1865 1870 1875 1880 1885 1890 1895 1900 1905 1910 1915 1920 1925 1930 1935 1940 1945 1950 1955 1960 1965 1970 1975 1980 1985 1990 1995 2000 2005 2010 2015 2020

an eloquent silence. With the Antarctic Environmental Protocol – and thus the provisions that ban mining and resource exploitation – subject to review as of 2048, this black hole, which is no less haunting than the ozone hole that looms over Antarctica, preludes conflicts that will be difficult to govern and, if unresolved, potentially catastrophic to our ecosystem.

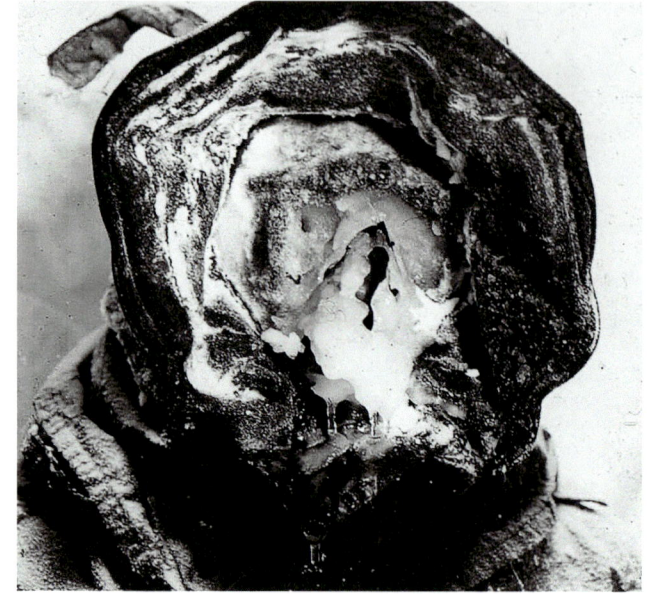

The paradoxes induced by the current science-centred governance model are accompanied by a legislative vacuum, by the unwillingness of the Consultative Parties to the Antarctic Treaty (who manage the continent "in the interest of all mankind"[22] – in other words, on behalf of the remaining 85% of nations) to formulate a pertinent legal framework for the southernmost continent. The normative void born out of the treaty's deferral of national claims, and the consequent absence of sanctions, limits the enforcement of regulations on all operations in Antarctica, including the building sector, which is not regulated by binding building standards either for new constructions or for the preservation of its historical artefacts. Consequently, the continent's intrinsic value – i.e. the moral standing of Antarctica's astonishing natural world, wilderness, and wonders, and the rights accorded to them – is challenged, just like anywhere else on our planet, not only by the manifest risks posed by global warming but also by environmental pressures and harmful interference arising from human activity.

Humans who devote their lives to science so that "the world may have a little more knowledge, that it may build on what it knows instead of what it thinks"[23] inhabit this un-normed geography in scientific research stations that shield them from the harsh, deserted environment, which has recorded temperatures of –89.2°C and wind speeds of 259 kilometres per hour.[24] The stations themselves (whose combined surface area equals that of the

22 The Antarctic Treaty, opened for signature on the 1st of December 1959.
23 Apsley Cherry-Garrard, *The Worst Journey in the World* (London: Constable, 1922).
24 The fragility of humankind and its uncertain survival within the unforgiving Antarctic environment is epitomised by the photograph taken during the 1911–1914 Australasian Antarctic Expedition by Frank Hurley (shown above). Upon his return from taking scientific measurements during a blizzard, meteorologist Cecil Madigan found himself wearing a so-called ice mask.

Pentagon[25]) reveal an under-theorised, accelerated history of architecture from primordial huts[26] to hyper-technological mobile architectures reminiscent of provocative 1960s projections of utopian cities.[27] With national iconography morphing into façades proudly projecting national identities against the white backdrop, embassy-like stations act as political devices, reinforcing national agendas on the territory and resisting true forms of international cooperation.[28]

Within these bases, the transient population of the only continent that has neither indigenous peoples nor other long-term resident populations lives in confinement, experiencing an isolation far greater than the one imposed on astronauts living in the International Space Station. Beyond the walls of their huts and stations, a white, iconoclast landscape – predominantly lifeless yet "living"[29] – stands in contrast to the world these people left behind. The distorted day cycle with its weeks-long sunset, the sensory deprivation, and the unexpected high-density cohabitation which obliterates privacy challenge the well-being of Antarctic inhabitants, who are scrupulously selected for their innate physical and mental stability. While the arts historically played a "medicinal role",[30] allowing for intellectual escape in simulated lives, today digital technologies and virtual realities[31] alleviate the frequent overwintering syndromes. Acting as a prototyping ground for extreme inhabitation, Antarctica – the ultimate "Space analogue"[32] – may prove essential for cosmic endeavours which contemplate (as yet unearned[33]) life beyond Planet Earth.

While we attempt to optimise extreme inhabitation strategies, Antarctica rejects and literally ejects the structures we build on its ice. The relentless movement of glaciers towards the ocean as they calve,[34] carrying with them all

25 Council of Managers of National Antarctic Programs, *Antarctic Station Catalogue* (Christchurch: COMNAP, 2017), comnap.aq [online].
26 Astonishingly, the first hut ever built on the continent – constructed at Cape Adare by Carsten Egeberg Borchgrevink and his party during the British Antarctic Southern Cross Expedition (1898–1900) – is still standing.
27 While architecture on all other continents could be understood as the final act of a gradual metamorphosis of the landscape, in the absence of matter and raw materials, all Antarctic buildings (including the radical station by Hugh Broughton Architects, which builds upon Archigram's "Walking City" project) consist of an assembly of imported prefabricated elements.
28 See Alan D. Hemmings, "International Antarctic Stations," see pp. 582–83.
29 Ursula K. Le Guin, "Sur: A Summary Report of the Yelcho Expedition to the Antarctic, 1909–1910," *The New Yorker,* 1 February 1982.
30 Elizabeth Leane, "The Medicinal Role of Reading in the Heroic Age," see p. 356.
31 Admittedly not yet exploited to the fullest within polar architectures.
32 Scott Parazynski, "Inhabiting Space," see pp. 580–81.
33 Neri Oxman in conversation with Paola Antonelli, "Design Emergency," Instagram live interview, 29th of January 2021.
34 Calving is a natural phenomenon whereby sections of glaciers floating on the ocean break off and fall into the water. Charged with symbolic value as markers of the retreating cryosphere and rising global sea levels, it is hard to imagine that calving glaciers could conceal even greater evidence of humanity's impact on Planet Earth. Yet, as proven by the photograph of Halley I Station's cross section taken by the crew of the *Polarsirkel* shown on the previous page, temporary apparitions of an

the captive evidence of humanity in the form of polluted frozen air bubbles or buildings, is a prelude to a future in which disembodied technologies[35] and forms of surveillance will allow for the reduction of the anthropic footprint on the continent in favour of automated unmanned scientific research.

The current imbalance between ice loss through melting and iceberg formation via snowfall – which occurs six times faster than forty years ago, with ice melting at a rate of 6,300 tons per second[36] – is provoking an alarming global rise in sea levels and calls for immediate action.

The conviction with which polar experts, practitioners, and thinkers have joined forces to author this book is evidence that the specialist community recognises the need to act together, driven by the determination that, at the very least, the knowledge of Antarctica ought to be shared as a Global Commons.

Learning from Antarctica's spirit of cooperation, built upon the total reliance on one another to survive, *Antarctic Resolution* aspires to launch a platform, an agency for change, in which planetary citizens can engage in a coordinated and unanimous effort – independent of nation – to shape the future of Antarctica, and, in turn, of our "Spaceship Earth".[37]

archaeology of debris are not unusual on the Antarctic Barrier.
35 As explored in Elena Glasberg, *Antarctica as Cultural Critique: The Gendered Politics of Scientific Exploration and Climate Change* (New York: Palgrave Macmillan, 2012).
36 David Vaughan, research data shared with UNLESS, 2020.
37 Richard Buckminster Fuller, *Operating Manual for Spaceship Earth* (1969; repr., Zurich: Lars Müller Publishers, 2008).

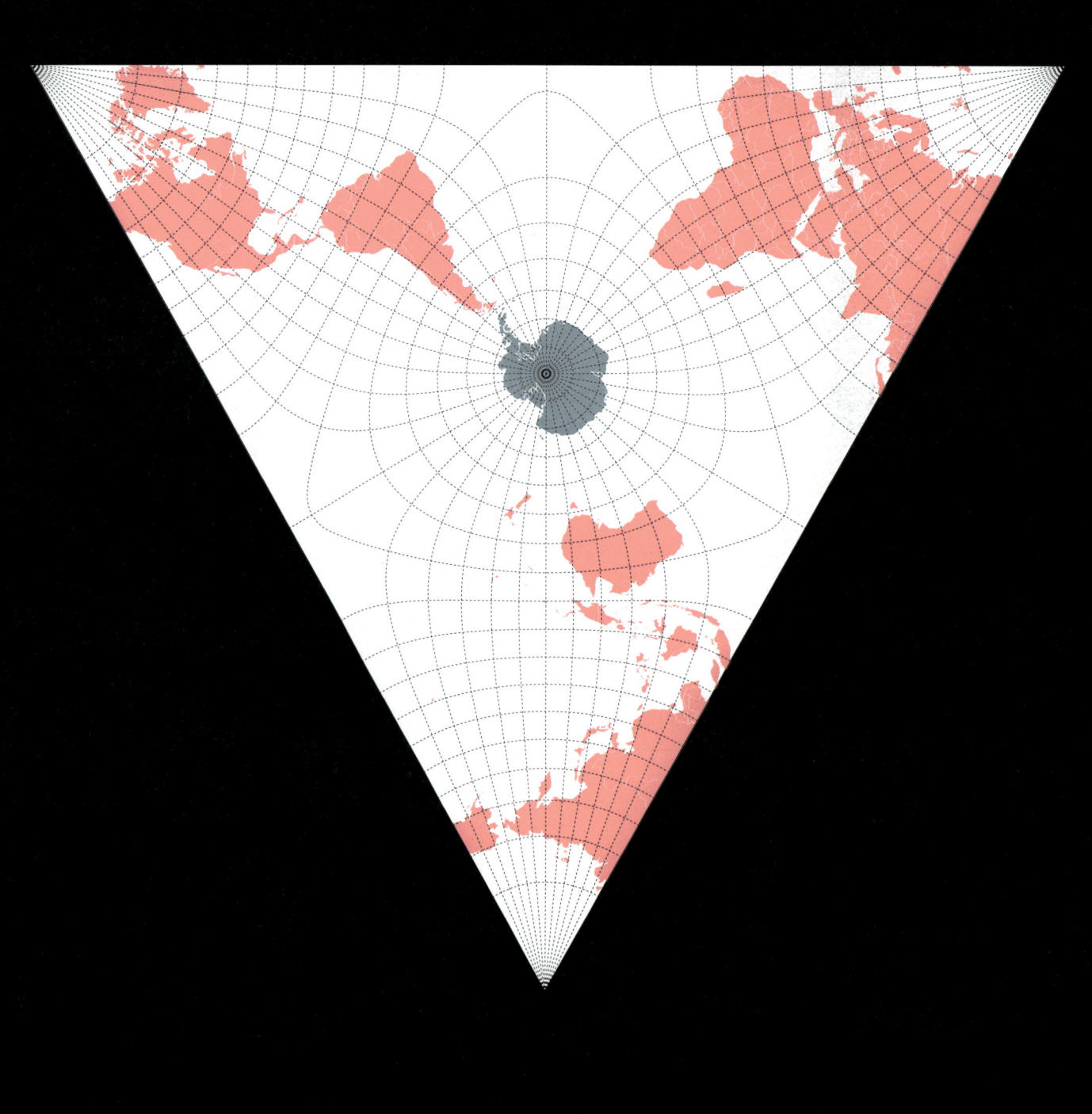

"Our little Spaceship Earth is only eight thousand miles in diameter, which is almost a negligible dimension in the great vastness of space. Our nearest star – our energy-supplying mother-ship, the Sun – is ninety-two million miles away, and the next nearest star is one hundred thousand times further away. It takes two and one-half years for light to get to us from the next nearest energy supply ship star. That is the kind of space-distanced pattern we are flying. Our little Spaceship Earth is right now traveling at sixty thousand miles an hour around the sun and is also spinning axially, which, at the latitude of Washington, D.C., adds approximately one thousand miles per hour to our motion. Each minute we both spin at one hundred miles and zip in orbit at one thousand miles. That is a whole lot of spin and zip. […]

Quite clearly we have vast amounts of income wealth as Sun radiation and Moon gravity to implement our forward success. Wherefore living only on our energy savings by burning up the fossil fuels which took billions of years to impound from the Sun or living on our capital by burning up our Earth's atoms is lethally ignorant and also utterly irresponsible to our coming generations and their forward days. Our children and their children are our future days. If we do not comprehend and realize our potential ability to support all life forever we are cosmicly bankrupt. […]
This all brings us to a realization of the enormous educational task which must be successfully accomplished right now in a hurry in order to convert man's spin-dive toward oblivion into an intellectual mastered power pullout into safe and level flight of physical and metaphysical success […]."

Richard Buckminster Fuller (Architect, Engineer, Inventor), *Operating Manual for Spaceship Earth*, 1969

"In the vicinity of the Poles one must adopt a new conception of time and direction. To try to think in terms of north and south, noon or midnight, or even to-day or to-morrow, is to become hopelessly involved in meaningless, contradictory phrases."

Rory Sweet, Henry Cookson, and Rupert Longden (Explorers) "Novo to Inaccessibility Antarctic Expedition", 2007

A transnational team of specialists
from the fields of aeronautics, anthropology,
architecture, biology, chemistry, climate
change, economics, engineering, geoscience,
glaciology, history, law, literature, logistics, medicine, physics, political science,
science, sociology, technology, and
visual arts authored *Antarctic Resolution.*
Their voices are featured in black.

All drawings, cartographies, data visualisations, images, captions, and quotes were
produced and/or curated by Giulia Foscari /
UNLESS. The editorial voice is featured in
fluorescent orange.

Rear Admiral Richard Byrd's
"Must soon turn north" note is a powerful
reminder that, close to the South Pole, all
directions are North. Hence, all the architectural drawings published in the "Archive
of Antarctic Architecture" are oriented
towards South. Within the authored texts,
reference to such drawings is indicated
by ↗ page number.

CE OR
H

01
Ant-Arctica
In-Cognita

The Discovery of Antarctica

Robert Headland

ROBERT HEADLAND is a fellow of the Royal Geographical Society who has served as an archivist and curator for the Scott Polar Research Institute, University of Cambridge (United Kingdom). Headland became involved with the Antarctic Heritage Trust, authored *A Chronology of Antarctic Exploration* (Cambridge University Press, 2009) and is the recipient of the Polar Medal (United Kingdom, 1984).

Early concepts of Antarctica, hypotheses of an unknown Southern Continent, were originated by ancient Greek philosophers. Distant exploratory voyages became more frequent after the 1500s and many penetrated far southern regions, rounding southern Africa and South America. As the extent of the oceans became known, the ancient concept of *Terra Australis* progressively disappeared. Charts of Antarctic regions steadily showed more ocean as speculative lands were disproved, resulting in the diminution of the theoretical continent into Australasia and unknown regions closer to the Antarctic Circle.

Major voyages of exploration began in 1768 by Captain James Cook of the British Navy, who was instructed "to search for the postulated Great Southern Continent". Among his results was an extensive survey of the oceanic region south of Australasia. Cook's second voyage, from 1772 to 1775, made the first crossing of the Antarctic Circle in January 1773 and two more while he circumnavigated the yet unknown continent. Cook's expedition discovered South Georgia and some of the South Sandwich Islands, making the earliest landing in Southern Ocean regions. Following his explorations, the southern continent was reduced further to only the frigid regions, and maps progressively started incorporating new discoveries rather than perpetuating speculation.

Captain Cook's report of abundant seals, particularly on South Georgia, with a lucrative market for their pelts in Canton, led sealers to rapidly extend their range to these areas. For more than a century, sealers discovered many and visited nearly all the peri-Antarctic islands. Details of their discoveries are sparse, as sealers' interests were limited to the industry, and secrecy was an important part of its capital. The peak of the sealing industry was during the early 1820s.

The discovery of lands near the Antarctic Peninsula was partly a consequence of the independence of South American states. The end of Spanish monopoly led many vessels to venture to South America. Inevitably some trading with Pacific Ocean states were blown off course while rounding the notorious Cape Horn. These included a voyage from London aboard the *Williams,* commanded by William Smith, which left Río de la Plata in January 1819, bound for Valparaíso. The *Williams* was blown far southwards in the Drake Passage, eventually discovering the South Shetland Islands on the 19th of February 1819. Smith then reached Valparaíso in March and during his return voyage endeavoured to find the islands again, but had sailed too far west; it was only on the third voyage from Montevideo to Valparaíso that he relocated them, landing on King George Island in October, where he took formal possession for King George III. A fourth voyage was accompanied by Edward Bransfield, who made a survey of the discoveries, including the first sighting and charting of the Antarctic Peninsula on the 30th of January 1820. News of these discoveries circulated quickly, and sealers soon became very active in the islands.

Contemporaneously, a Spanish vessel, the *San Telmo,* bound for Peru with a complement of 644, was lost in the Drake Passage in a severe storm. There is evidence that some survivors landed on the South Shetland Islands in September 1819, several months before Smith, but they perished.

A Russian expedition, led by Fabian Bellingshausen, was circumnavigating the Antarctic aboard the *Vostok* and *Mirny* from 1819 to 1821. On the 27th and 31st of January 1820, near the 0° meridian, two coastal ice features of Queen Maud Land were sighted and recognised as land; this was the earliest sighting of Antarctica. The expedition extended James Cook's surveys and explored south of the Antarctic Circle on six occasions.

From the 1820s, almost 100 sealing vessels working around the South Shetland Islands also had possibilities of sighting the Antarctic Peninsula. The

"It may have been about our year 750 that the astonishing Hui-Te-Rangiora, in his canoe Te Iwi-o-Atea, sailed from Rarotonga on a voyage of wonders in that direction (South): he saw the bare white rocks that towered into the sky from out the monstrous seas, the long tresses of the woman that dwelt therein, which waved about under the waters and on their surface, the frozen sea covered with pia or arrowroot, the deceitful animal that dived to great depths – 'a foggy, misty dark place not shone on by the sun'. Icebergs, the fifty-foot-long leaves of bullkelp, the walrus or sea-elephant, the snowy ice fields of a clime very different from Hui-Te-Rangiora's own warm islands – all these he had seen."

John Cawte Beaglehole (Historian), *The Discovery of New Zealand,* 1939

"What is the nature of the snow- and ice-covered land observed at so many points towards the South Pole? Is there a sixth continent within the Antarctic Circle or merely a range of lofty volcanic hills?"

James Murray (Biologist, Explorer), Royal Geographic Society, November 1893

ALAS, ANTARCTICA!

Heroic Age explorers who had signed up "for hazardous journey. Low wages, bitter cold, long hours of complete darkness. Safe return doubtful. Honour and recognition in event of success" (as promised by Shackleton in his famous advertisement) sighted the continent from their vessels. Captured by Herbert Ponting in a photograph taken aboard the *Terra Nova* – whose crew left Cardiff on the 15th of June 1910 and landed on the continent only on the 4th of January 1911 – their view from high up the mast would most often focus on the barrier, "a sweeping crescent of ice about 150 *shaku* high [...] [which resembled] a series of pure white folding screens, or perhaps a gigantic white snake at rest",[1] as described by Nobu Shirase in 1913. The sight "struck [them] to the very heart by a feeling of awe".[2] Upon arrival, celebrations were due. Notoriously, the French, led by Jean-Baptiste Charcot, who redirected their expedition south to save Otto Nordenskjöld and his ship, the *Antarctic,* knew how to do that best, as shown in the lower image, in which Ernest Gourdon and Paul Pléneau enjoy a glass of Mumm champagne in Antarctica on Bastille Day in 1904.

British Sealing Voyage, J. Weddell. Ships *Jane* and *Bufoy*. Britain

Smith and Bransfield expedition, E. Bransfield and W. Smith. Ship *Williams*. Britain

Russian exploring expedition, T. von Bellingshausen. Ships *Vostok* and *Mirny*. Russia[i]

Fanning-Pendleton Sealing Expedition, N.B. Palmer. Ships *Hero, Fredrick, Hersilia, Free Gift, Express, Essex*. United States

American sealing expedition, J. Davis. Ships *Huron* and *Huntress*. United States

British Naval expedition, H. Foster. Ship *Chanticleer*. Britain

Enderby-Bros sealing expedition, J. Biscoe. Ships *Tula* and *Lively*. Britain

French exploring expedition, J-S. Dumont d'Urville. Ships *Astrolabe* and *Zélée*. France

Enderby Bros sealing expedition, J. Balleny. Ships *Eliza Scott* and *Sabrina*. Britain

United States Exploring Expedition, C. Wilkes. Ships *Vincennes, Peacock, Porpoise, Sea Gull, Flying Fish, Relief*. United States

British Naval expedition, J.C. Ross. Ships *Erebus* and *Terror*. United Kingdom

Sealing and whaling voyage, M. Cooper. Ship *Levant*. United States

British naval voyage, Sir C. Wyville Thompson. Ship *Challenger*. Britain.

German sealing and exploring expedition, E. Dallmann. Ship *Groenland*. Germany

Dundee whaling expedition, A. Fairweather. Ships *Balaena, Active, Diana, Polar Star*. Scotland

Norwegian whaling expedition, C.A. Larsen. Ship *Jason*. Norway

Norwegian whaling expedition, C.A. Larsen. Ships *Castor, Hertha, Jason*. Norway

Norwegian sealing and whaling expedition, H.J. Bull. Ship *Antarctic*. Norway

Belgian Antarctic Expedition, A. de Gerlache. Ship *Belgica*. Belgium

British Antarctic Expedition, C. Borchgrevink. Ship *Southern Cross*. Britain[ii]

German South Polar Expedition, E. von Drygalski. Ship *Gauss*. Germany

Swedish South Polar Expedition, N.O.G. Nordenskjöld. Ship *Antarctic*. Sweden

British National Antarctic Expedition, R.F. Scott. Ship *Discovery*. Britain

Scottish National Antarctic Expedition, W.S. Bruce. Ship *Scotia*. Scotland

French Antarctic Expedition, J-B.E.A. Charcot. Ship *Le Français*. France

British Antarctic Expedition, E. Shackleton. Ship *Nimrod*. Britain

French Antarctic Expedition, J-B.E.A. Charcot. Ship *Pourquoi-Pas?* France

Norwegian Antarctic Expedition, R.E.G. Amundsen. Ship *Fram*. Norway[iii]

British Antarctic Expedition, R.F. Scott. Ship *Terra Nova*. Britain

Japanese Antarctic Expedition, C. Shirase. Ship *Kainan Maru*. Japan

German South Polar Expedition, W. Filchner. Ship *Deutschland*. Germany

Australasian Antarctic Expedition, D. Mawson. Ship *Aurora*. Australia

British Imperial Trans-Antarctic Expedition, E. Shackleton. Ship *Endurance*. Britain

British Expedition to Graham Land, J.L. Cope. Ship *Svend Foyn*. Britain

Shackleton-Rowett Antarctic Expedition, E. Shackleton. Ship *Quest*. Britain

Wilkins-Hearst Antarctic Expedition, H. Wilkins. Ships *Hektoria, William Scoresby*. USA and Britian.

United States Antarctic Expedition (BAE I), R.E. Byrd. Ships *City of New York, Eleanor Bolling, Sir James Clark Ross, C.A. Larsen*. United States

British, Australian, New Zealand Research Expedition (BANZARE), D. Mawson. Ship *Discovery*. United Kingdom, Australia, New Zealand

First Ellsworth Antarctic Expedition, L. Ellsworth. Ship *Wyatt Earp*. United States

Second Byrd Expedition (BAE II), R.E. Byrd. Ships *Bear of Oakland and Jacob Ruppert*. United States

Second Ellsworth Antarctic Expedition, L. Ellsworth. Ship *Wyatt Earp*. United States

British Graham Land Expedition, J. Rymill. Ship *Penola*. United Kingdom

Third Ellsworth Antarctic Expedition, L. Ellsworth. Ship *Wyatt Earp*. United States

Norwegian Antarctic Expedition, L. Christensen. Ships *Thorshavn* and *Firen*. Norway

German Schwabenland Expedition, A. Ritscher. Ship *Schwabenland*. Germany

Fourth Ellsworth Antarctic Expedition, L. Ellsworth. Ship *Wyatt Earp*. United States

United States Antarctic Service Expedition (BAE III), R.E. Byrd. Ships *North Star* and *Bear*. United States[iv]

US Navy Antarctic Developments Project (Operation Highjump), R.E. Byrd. 13 Ships. United States

Ronne Antarctic Research Expedition, F. Ronne. Ship *Port of Beaumont*. United States

US Navy Second Antarctic Developments Project (Operation Windmill), Cdr G. Ketchup. Ships *Burton Island* and *Edisto*. United States

Norwegian-British-Swedish expedition, J. Giaver, Norway, United Kingdom. Ship *Norsel*. Sweden[v]

1772–1775
James Cook is the "first" to cross the Antarctic Circle.

[i]27 January 1820
Thaddeus von Bellingshausen is the "first" to sight Antarctica.

[ii]1898–1900
The British Antarctic Expedition is the first to overwinter in Antarctica.

[iii]1911
Roald Amundsen is the first to reach the South Pole.

[iv]1939
Rear Admiral Richard Byrd is the first to fly over the South Pole.

[v]1949
The first International Scientific Expedition is launched.

1958
The first successful land traverse of Antarctica is completed.

1820 1825 1830 1835 1840 1845 1850 1855 1860 1865 1870 1875 1880 1885 1890 1895 1900 1905 1910 1915 1920 1925 1930 1935 1940 1945 1950

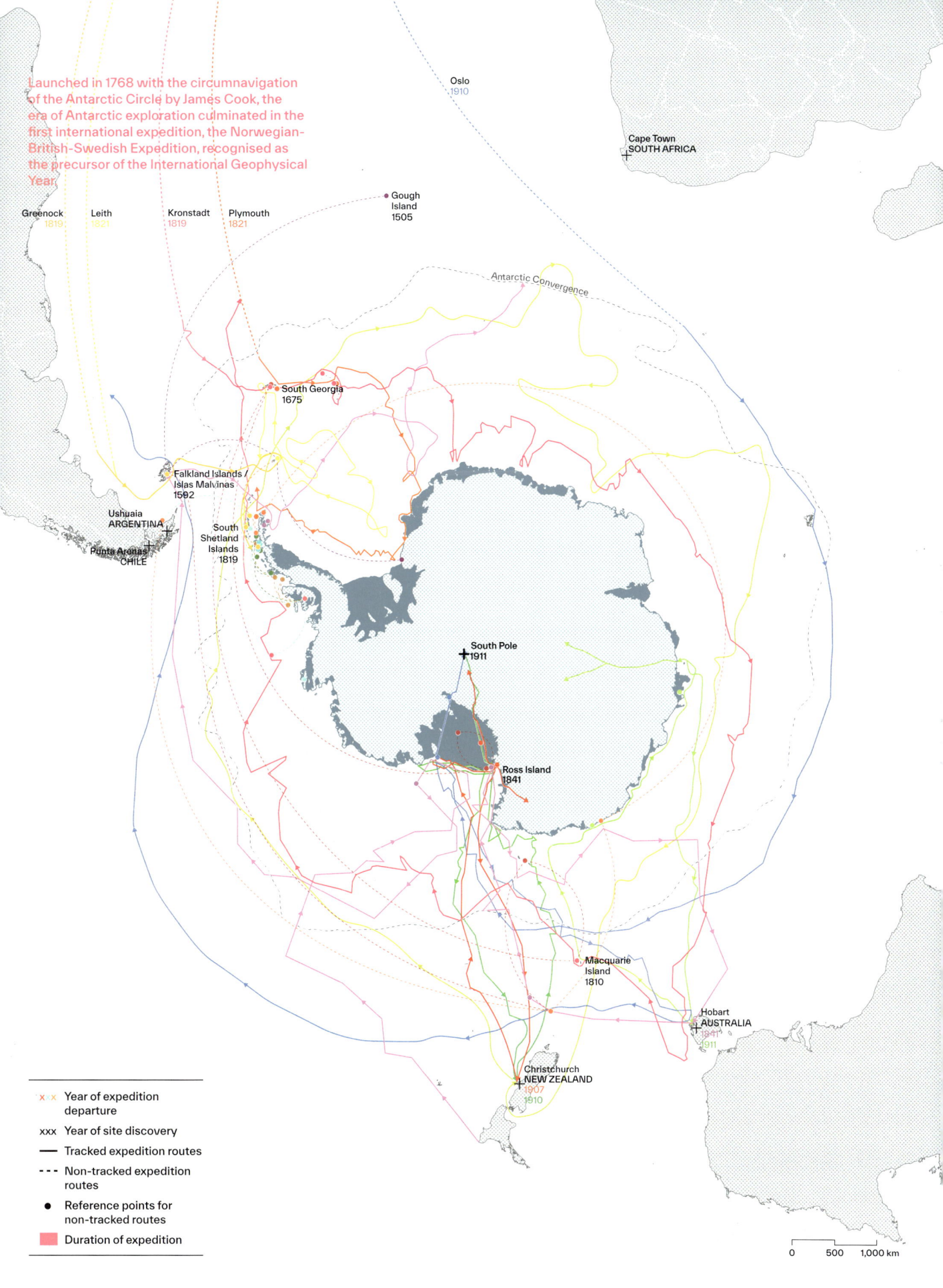

Launched in 1768 with the circumnavigation of the Antarctic Circle by James Cook, the era of Antarctic exploration culminated in the first international expedition, the Norwegian-British-Swedish Expedition, recognised as the precursor of the International Geophysical Year.

Oslo
1910

Cape Town
SOUTH AFRICA

Greenock
1819
Leith
1821
Kronstadt
1819
Plymouth
1821

Gough
Island
1505

Antarctic Convergence

South Georgia
1675

Falkland Islands /
Islas Malvinas
1592

Ushuaia
ARGENTINA

Punta Arenas
CHILE

South
Shetland
Islands
1819

South Pole
1911

Ross Island
1841

Macquarie
Island
1810

Hobart
AUSTRALIA
1841
1911

Christchurch
NEW ZEALAND
1907
1910

x x x Year of expedition
       departure

xxx Year of site discovery

—— Tracked expedition routes

- - - Non-tracked expedition
       routes

● Reference points for
   non-tracked routes

▮ Duration of expedition

0   500   1,000 km

best-known of these is a United States voyage by Nathaniel Palmer aboard the *Hero* which saw the coast on the 17th of November 1820. Reports of several sealers making early landings in Antarctica are also known from this period. Of the five known accounts, the most detailed was by John Davis, from New Haven, Connecticut, in the United States, who landed near Hughes Bay from the *Cecilia* on the 7th of February 1821. Sealers' explorations included the farthest southern voyage of James Weddell from London, with the *Jane* and *Beaufoy,* who reached 74°15′ S in the sea later named after him.

During the sealing period, when approximately 1,200 sealing voyages were recorded, fewer than 20 exploring and scientific expeditions were dispatched. French, American and British expeditions were associated with the determination of the magnetic poles from 1837 to 1843. These expeditions extended coastal mapping and entered the Ross Sea. In 1874, coordinated expeditions from Britain, France, Germany and the United States had observatories on several of the islands for the transit of Venus. As discoveries accumulated, charts of the Antarctic progressively showed more land, especially around the northern Antarctic Peninsula, the Ross Sea and Weddell Sea. The first International Polar Year of 1882–1883 had the earliest winter Antarctic scientific party based on South Georgia.

Exploration of the interior of Antarctica began during the brief but intense Heroic Age from the 1890s to the First World War. Whaling reconnaissance was a practical beginning, but the theoretical inception was in 1895 with an International Geographical Conference adopting a resolution for Antarctic exploration. During these decades, the earliest winters were spent south of the Antarctic Circle (1898, led by Adrien de Gerlache and the complement of the *Belgica* while beset) and on Antarctica (1899, 10 men led by Carsten Borchgrevink on Cape Adare [p. 614]). Scott Island, the last of the peri-Antarctic islands, was discovered, and the general limits of Antarctica became known. Twice in 1902, aircraft (hydrogen balloons) were used for aerial observations, and in March 1903 the first permanent meteorological station was opened (on the South Orkney Islands). The South Pole was reached twice in the 1911–12 summer, first by Roald Amundsen of Norway, then by Robert Falcon Scott from Britain only 33 days later. Technological developments resulted in the earliest Antarctic sound recordings in 1902, ciné films in 1903 and radio communications with Antarctica in 1913. Expeditions attained the close vicinity of the South Magnetic Pole, then well inland, in 1909 and 1912. Eleven of the extant historic huts of the Antarctic date from this time. During this period, coastal maps and charts were vastly extended and improved, although most of the continental interior remained unknown. It was also the beginning of the formal definition of sovereign claims.

Commercial whaling in the Southern Ocean began on South Georgia in 1904 and continued until 1987. Its major period was between the two World Wars, when the majority of Southern Ocean vessels belonged to Norwegian whaling fleets and to associated scientific investigations. Whale oil was a profitable commodity and most other whale products had commercial use. Whalers were responsible for discovering many coastal regions of Antarctica and for the establishment of Norwegian territories. Scientific and exploratory expeditions of several nationalities were sporadically active during the whaling period; many of these were assisted by the whalers.

"After that most people forgot it for a while, and when all the other white spaces on the map had been coloured in, they came back to it. The British were especially keen on Antarctica, as they had done Africa and spent much of the nineteenth century fretting over the Arctic. By the time the twentieth century rolled around they were fully engaged in the great quest for the south. For these British people the quest culminated in the central Antarctic myth, that of Captain Scott, a man woven into the fabric of our national culture as tightly as the pattern in a carpet."

Sara Wheeler (Writer), *Terra Incognita: Travels in Antarctica,* 1996

"At about a quarter past 11 o'clock we cross'd the Antarctic Circle, for at Noon we were by observation four Miles and a half South of it and are undoubtedly the first and only Ship that ever cross'd that line. […] Lands doomed by nature to everlasting frigidness and never once to feel the warmth of the Sun's rays, whose horrible and savage aspect I have no words to describe; such are the lands we have discovered, what may we expect those to be which lie more to the South, for we may reasonably suppose that we have seen the best as lying most to the North, whoever has resolution and perseverance to clear up this point by proceeding farther than I have done, I shall not envy him the honour of the discovery but I will be bold to say that the world will not be benefited by it."

James Cook (Captain, Explorer, Cartographer), Personal Logbook, 1770

The timeline of Antarctic explorers reveals a selection of expedition leaders who have successfully led their parties to secure documented primacy of discovery of the Antarctic coastline and plateau.

1773

British Captain James Cook, aboard the *HMS Resolution,* is the "first" to cross the Antarctic Circle on the 17th of January 1773 and to circumnavigate Antarctica.

27th of January 1820

Russian Admiral Fabian Gottlieb von Bellingshausen (with Mikhail Lazarev) is the "first" to report sighting Antarctica, by discovering an ice shelf at Princess Martha Coast – now known as Fimbul Ice Shelf – at 69°21'28" S, 2°14'50" W.

30th of January 1820

Irish sailor Edward Bransfield, an officer in the British Royal Navy, is the "second" person to sight Antarctica, reporting the existence of the Trinity Peninsula, the northernmost point of the Antarctic continent.

17th of November 1820

American seal hunter Captain Nathaniel Brown Palmer is the "third" person to sight the Antarctic continent, from the Orleans Channel.

7th of February 1821

American sealer John Davis is thought to be "first" person to land on Antarctica, by setting foot at Hughes Bay, near the northernmost tip of the Peninsula.

1898–1900

Anglo-Norwegian polar explorer Carsten Borchgrevink and the party of the British-financed *British Antarctic Expedition* aboard the *Southern Cross* are the first to over-winter in Antarctica. Their arrival at Cape Adare on the 24th of January 1895 marks the first undisputed landing on the continent.

17th of January 1909

Led by Irish explorer Ernest Shackleton, the Northern Party of the *British Antarctic Expedition* aboard the *Nimrod* is the first to reach the South Magnetic Pole. The team members accompanying him in this endeavour are Douglas Mawson, Alistair Mackay and Edgeworth David.

14th of December 1911

Led by Norwegian explorer Roald Amundsen, the *Norwegian Antarctic Expedition* party, including Helmer Hanssen, Sverre Hassel, Oscar Wisting and Olav Bjaaland, are the first to reach the Geographic South Pole.

17th of January 1912

British Captain Robert Falcon Scott, leader of the *British Antarctic Expedition,* and his party, including Lawrence Oates, Henry Robertson Bowers, Edward Adrian Wilson and Edgar Evans, are the second expedition team to reach the Geographic South Pole.

1910–1912

Japanese explorer Nobu Shirase joins the race to the South Pole by leading the *Japanese Antarctic Expedition* at the same time as Roald Amundsen and Robert Falcon Scott. The Japanese party reaches a latitude of 80°05′ S.

28–29th of November 1929

American Rear Admiral Richard Byrd is the first to fly over the South Pole. Departing at 3:29 p.m. on the 28th of November 1929, the flight aboard *Floyd Bennett* from Little America Station on the Ross Ice Shelf to the pole and back takes 18 hours and 41 minutes.

1958

Sir Edmund Percival Hillary is the to first to complete the overland crossing of Antarctica via the South Pole, during the *Commonwealth Trans-Antarctic Expedition.* Hillary is the first person ever to have reached both poles and the summit of Everest.

The earliest use of powered aircraft in the Antarctic was made by Hubert Wilkins in 1928, a development which greatly facilitated the exploration and mapping of the interior. United States expeditions, led by Richard Byrd and Lincoln Ellsworth, vastly extended discovery and surveying with aircraft, including a flight to the vicinity of the South Pole. This was the beginning of what has been termed the mechanical age of Antarctic exploration. Scientifically, it was also when knowledge of the Southern Ocean, especially biological oceanography, greatly increased. Maps and charts continued to improve, although areas such as the "Phantom Coast" remained almost unknown. A major cartographical development was the publication of an Australian Antarctic map in 1939 with assiduously compiled data from all national expeditions.

After the Second World War, regular annual expeditions from an increasing number of countries were the principal Antarctic activity; most were governmental. Permanent occupation of Antarctica began in 1944 at Base A ↗ p. 644 and Hope Bay on the Antarctic Peninsula. Many assertions of national sovereignty over Antarctic territories were reinforced during this stage. International law became involved and some national politics became passionate, resulting in one instance of failure of diplomacy with resort to military force. At the end of this period was the International Geophysical Year (1957–1958), a major event in the development of science throughout the world. It included a cooperative and coordinated Antarctic research programme undertaken by twelve countries, some with existing stations and others specially established. In total, 54 stations were open during the 1957 winter, which remains the largest number in Antarctic history. Maps and surveys continued to improve, especially with comprehensive aerial reconnaissance of the Antarctic Peninsula and South Shetland Islands. The regions around many national stations became well charted.

One of the consequences of the International Geophysical Year was a general appreciation of the efficiency of scientific cooperation in Antarctic exploration and research, which resulted in the establishment of the Special (later Scientific) Committee on Antarctic Research. This, with several other factors, notably the accommodation of distinct national interests, promoted negotiations which culminated in the Antarctic Treaty, signed by all the countries active in the Antarctic in 1959. This was the beginning of a *pax Antarctica* and Antarctica becoming known as a "continent for science". Charting and mapping became more coordinated as the remoter coasts were plotted from aircraft. It was the beginning of a survey of bedrock beneath the ice sheet developed with seismic, radar and radio-echo soundings from surface traverses and aircraft.

From 1959 the current, or Treaty, period of Antarctic history developed after the pattern of informal cooperation gave way to formalised methods of administration. Subsequently, concepts of the sensitivity of Antarctic biota and associated environmental factors were advanced. Internationally, this resulted in various conventions supplementing the Antarctic Treaty with regulations endeavouring to manage most human activity in the far south. This period was one when tourism and similar non-governmental visits increased rapidly in popularity, such that tourists formed the largest proportion of Antarctic visitors.

From the 1970s, a third exploitative period of the Southern Ocean involving fishing and associated industries developed rapidly, but largely under scientific control to catch less than the maximum sustainable yield. It is notable that during this period, seal and whale populations have vastly recovered from earlier overly exploitative periods. Administrative organisations regulating biological exploitation, logistics, tourism and eventually an Antarctic Treaty Secretariat were established. Coordination of mapping was improved by an attempt to regulate toponymy. Surface elevation and subglacial maps were thus published, combining surveys from several nationalities with remote sensing techniques, allowing for the chartering of subglacial lakes and mountain ranges. What at first was a mere abstract Greek theory became, in two centuries of exploration, a known and measurable three-dimensional land mass – one that occupies a tenth of the surface of the Earth.

"Polar exploration is at once the cleanest and most isolated way of having a bad time which has been devised. […] the desire for knowledge for its own sake is the one which really counts and there is no field for the collection of knowledge which at the present time can be compared to the Antarctic. Exploration is the physical expression of the Intellectual Passion. And I tell you, if you have the desire for knowledge and the power to give it physical expression, go out and explore. […] If you march your Winter Journeys you will have your reward, so long as all you want is a penguin's egg."

Apsley Cherry-Garrard (Explorer, Zoologist), *The Worst Journey in the World*, 1922

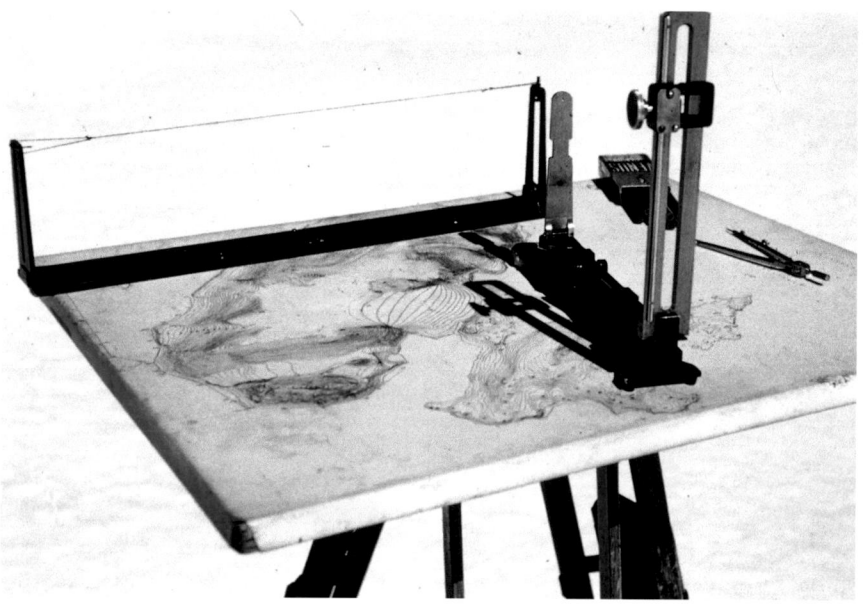

Asymptotic Cartography

William L. Fox

WILLIAM L. FOX is the founding director of the Center for Art + Environment at the Nevada Museum of Art in Reno, Nevada (United States). He is an art critic, science writer and cultural geographer, and has published 16 books on cognition and landscape, including *Terra Antarctica: Looking into the Emptiest Continent* (Counterpoint, 2007). Fox spent two and a half months in Antarctica with the US Antarctic Artists and Writers Program.

Mapping is an asymptotic endeavour, because we get increasingly close to accurately reducing the Earth to a data set expressed in two dimensions – but only close. Among the reasons are the fractal nature of topography and the ever-changing shape of the ground. The premier example of these dynamics is found in Antarctica, a continent that doubles in size every austral winter, when the surrounding ocean freezes. Approximately 97% of Antarctica is covered in ice so thick it depresses the bedrock below sea level. The ice itself relentlessly moves outward from the interior of the continent; the ice sheets and glaciers change thickness, elevation and rate of flow according to climate conditions.

For most of history, Antarctic cartography has been as much a product of imagination as measurement, starting with Aristotle (384–322 BCE) postulating an unknown southern continent to balance the known northern ones. Claudius Ptolemy published a map around 150 BCE that connected a *Terra Incognita* to the southern ends of Africa and India. The original map was lost, but the coordinates in his *Geographia* were rediscovered in the late 1400s, and subsequent Renaissance maps used them until Vasco da Gama rounded the Cape of Good Hope and sailed across the Indian Ocean in 1497, thus disproving the connection. Nonetheless, the idea of a great southern continent that was inhabited and fecund stubbornly resisted evidence to the contrary. The first modern atlas, compiled by Abraham Ortelius in 1570, included the famous world map *Types Orbis Terrarum.* Based on dozens of previous works by other cartographers, most prominently Gerardus Mercator, it included an enormous *Terra Australis Nondum Cognita* that stretched as far north as the Tropic of Capricorn, which bisects South America and Australia.

The idea of a fertile great southern continent, *Terra Australis Incognita,* was whittled down by subsequent navigators and map-makers, but not fully eliminated until Captain James Cook circumnavigated the Pacific Ocean from 1772 to 1775, which took him below the Antarctic Circle three times. Cook proved that no continent existed north of the Antarctic Circle and deduced there must be a frigid southern continent generating the "ice islands" they encountered. Nineteenth-century explorers Thaddeus von Bellingshausen, John Biscoe, Jules Dumont d'Urville and James Ross probed progressively farther south to finally encounter the continent and begin mapping the actual coastlines. By 1875 the *Süd-Polar-Karte* from Adolf Stieler's extraordinarily detailed atlas approximated the actual coastline and size of Antarctica.

Yet it was not until the Heroic Age of Antarctic exploration in the early twentieth century that land surveys were conducted by theodolite, which, somewhat like those made earlier from the sea by sextant, triangulated positions by measuring angles between features. Scott's 1901–1904 expedition mapped glaciers in the Ross Sea region, and Shackleton's journey of 1907–1909 forayed up onto the East Antarctic Ice Sheet. The next major advance for Antarctic cartographers was the use of aerial images beginning in 1928, when pictures were made with handheld still cameras poked out of airplane windows. This effort matured into trimetrogon photography, developed by the American military in the early 1940s, which used three wide-angle cameras at the bottom of a plane to capture a photographic survey from horizon to horizon. While flying at approximately 6 km height with one camera pointed vertically downward and the other two positioned to each side at 60° from the vertical, an overlapping set of images made it possible to interpret Antarctic topography by using stereography. Under the direction of Admiral Richard Byrd, who in 1929 had been the first person to fly over the South Pole, *Operation Highjump* (1946–1947) charted most of the Antarctic coastline, and his *Operation Deep Freeze I* (1955–1956) established

On "a beautiful sunny day" in February 1956, Derek Searle had "finished drawing the fair copy of the plan of the base site [located on Horseshoe Island] and harbour."[1] Little could he know that plane-table maps (shown on the left) would be surpassed by cartographies derived from aerial images. First performed as early as 1902, when the crew of the *Discovery* boarded a "captive balloon" and "Ernest Shackleton, Third Lieutenant with *Discovery*, with childish glee, ascended in the balloon, [...] brought a camera with him and took the first photographs of the bleakness of the ice shelf in his vision",[2] aerial campaigns were conducted systematically on the continent starting in 1928. Promptly replaced by satellite images, such as those captured by LIMA (shown above), then by REMA, a Reference Elevation Model of Antarctica that collated 187,585 images collected over six years, the recently developed mapping system offers a detailed view of the continent's topography. Comprising images captured by a constellation of polar-orbiting satellites licensed by the National Geospatial-Intelligence Agency (part of a department of the US Department of Defense), the 150-terabyte data set allows scientists to obtain a detailed understanding of the continent's geography down to a few metres, enabling the close monitoring of ice and stress fractures that occur between mountains and large ice shelves – such as that pictured at the side of Larsen C. Due to the seasonal absence of sunlight in Antarctica, REMA images can thus far only be captured from December to March.

the first permanent bases in the Antarctic in preparation for the 1957–1958 International Geophysical Year (IGY). The mid-1950s saw most of the continent mapped from the air.

In 1972, *Landsat 1* began the era of remote sensing in the Antarctic by surveying the continent in large swathes using multispectral cameras and scanning radiometers. The first Antarctic space-based map was useful, but the process was still relatively slow and the optics were unable to see through the Antarctic cloud cover. In 1997, the first *RADARSAT* mapping mission pierced Antarctic clouds with synthetic aperture radar (SAR) and made a high-resolution image of the entire continent in 18 days. Its limitations were imprecision regarding the elevation of the surface, and that much of the South Pole region remained unmapped. *ICESat* deployed high-precision laser altimetry in 1999, but that technology couldn't gather data through cloud layers.

By the early 2000s, as the capacities and ubiquity of computers dramatically increased, the United States Geological Survey (USGS) ceased printing paper maps of the Antarctic and elsewhere. Antarctic cartographic information became both digitally borne and viewed, and its source was remote sensing. Maps were read on computers and only when necessary printed out by the user.

In a collaboration among the National Aeronautics and Space Administration (NASA), the British Antarctic Survey (BAS), the United States Geological Survey (USGS) and the National Science Foundation (NSF), 1,100 images taken by *Landsat 7* between 1999 and 2001 were collated into the Landsat Image Mosaic of Antarctica (LIMA) for the 2007–2008 International Polar Year (IPY). It offered a resolution ten times greater than the earlier *Landsat* image. The latest digital satellite map was released in September 2018. The Reference Elevation Model of Antarctica (REMA) was compiled from 187,585 digital stereoscopic satellite images acquired from 2009 to 2017, which meant cloud-free images. The data set was so large that a supercomputer was used to process it, and the resulting topographic depictions were accurate to variations in elevation of less than a metre.

The asymptotic approach continues, as the actual shape and terrain under the ice are not mapped to a great degree of resolution or accuracy, and surface mapping data is not available in real time. As glaciers and ice sheets disappear during the current process of global warming, new lands will emerge and be mapped. The deployment of a much larger number of satellites may enable the creation of on-demand real-time data. Exploration in the Antarctic has mostly been replaced by recreational adventuring, but Antarctic mapping will continue to be of strategic interest to national governments as Antarctic resource exploitation and related military actions, despite the Antarctic Treaty, increase in feasibility.

"The Colleges of Cartographers set up a Map of the Empire which had the size of the Empire itself and coincided with it point by point. Less Addicted to the Study of Cartography, Succeeding Generations understood that this Widespread Map was Useless and not without Impiety they abandoned it to the Inclemencies of the Sun and of the Winters. In the deserts of the West some mangled Ruins of the Map lasted on, inhabited by Animals and Beggars; in the whole Country there are no other relics of the Disciplines of Geography."

Jorge Luis Borges (Writer, Poet), *Dreamtigers,* 2004

Since its conceptualisation in ancient Greece, Antarctica's discovery and representation induced paradigm shifts in the understanding of the world as a whole – not only by introducing the revolutionary notion that the planet is a globe, but also by triggering constant revisions of the notation of all other continents whose territory was relational to the vast unknown *Terra Australis Incognita*. An analysis of Antarctic mapping unveils at once the history of the discovery of the continent and the evolution of cartography from hemispheric Greek theories, via the Ptolemaic World Map, to the familiar latitude- and longitude-informed Renaissance's perspectival globe introduced by Ortelius.

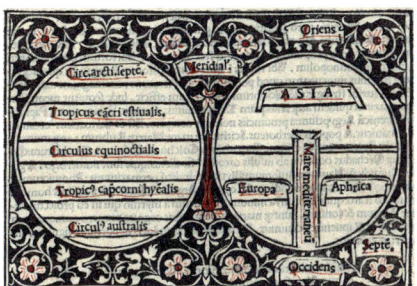

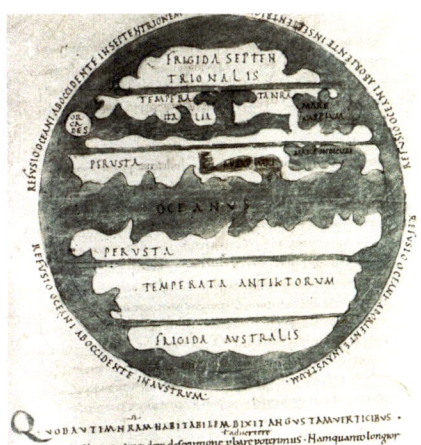

Seventh century

Tenth century

1459

Zonal and T-O map from *Novissime Hystoria,* 1503

This zonal and T-O map presents two medieval interpretations of the world. These include, on the left, a visualisation of the primitive zonal Pythagorean concept by which the world, intended to be spherical, is articulated in parallel zones, guaranteeing a balancing act between the known northern Arcticus and an unknown, yet theorised, Ant-Arcticus. On the right, the T-O tripartite map, introduced in the seventh century by Isidore, Archbishop of Seville, in his *Etymologies,* proposes a reading of the world by identifying Europe, Africa and Asia, separated by the Mediterranean Sea, the Nile and the Don rivers, and surrounded by the sea – the "O".

De Diversis Generibus Musicorum
Macrobius, Pseudo-Jerome, southern Germany

Authored by Alfred Hiatt
First appearing in the late Roman author Macrobius' commentary on Cicero's *Dream of Scipio,* this abstract world map was copied by a south German monastery around 1000 CE. Writing in the early fifth century CE, Macrobius sought to explain references in Cicero's text to areas beyond the Roman Empire. Macrobius outlined a theory according to which the spherical world was divided into five zones: two cold zones in the far north and south (marked "Frigida Septentrionalis" and "Frigida Australis" on this image); a central zone of extreme heat ("Perusta" – burned up), through which an equatorial ocean ("Oceanus") ran; and two temperate zones, one in the northern hemisphere ("Temperata Nostra" – our temperate zone) and one in the southern hemisphere ("Temperata Antoecorum" – the temperate zone of the antichthones, i.e. those living on the opposite side of the Earth). A second, complementary theory envisaged the division of the Earth into four land masses, two in each hemisphere, only one of which comprised the known world of Europe, Asia and Africa. The ocean was believed to encircle the entire world: the inscriptions on the image explain that water flows from the equatorial sea towards the North and South Poles. Macrobius' text became a standard source of scientific and philosophical information during the Middle Ages.

Mappa Mundi
Fra Mauro, Venice

Compiled in Venice in the mid-fifteenth century, the 2.4 × 2.4 m, detailed *Mappa Mundi* is considered a masterpiece of medieval cartography. Combining knowledge provided by the newly rediscovered Ptolemaic theories and by Marco Polo's recent explorations, the south-oriented, circular planisphere is surrounded by four smaller spheres which represent: a cosmologic diagram of the solar system following Ptolemaic precepts; the four elements; the Garden of Eden; and finally, the North and South Poles. On the latter, Fra Mauro writes on his map, "I do not think it derogatory to Ptolemy if I do not follow his *Cosmographia,* because to have observed his meridians or parallels or degrees, it would be necessary in respect to the setting out of the known parts of this circumference to leave out many provinces not mentioned by Ptolemy. But principally in latitude, that is from south to north, he has much *terra incognita,* because in his time it was unknown." The accuracy and conceptual strength of Fra Mauro's map has led even the National Aeronautics and Space Administration (NASA), whose famous *Apollo 17* image of the Earth strikingly resembles Fra Mauro's cartography, to name a lunar crater after the Venetian monk.[1]

1482

1508

World (untitled), Ulm edition of Ptolemy

Claudius Ptolemy, one of the greatest
geographers who lived in Alexandria in the
second century BCE, was the first to reduce
a spherical world to a map based on a longitude and latitude system. Reinforcing Greek
theories of the presence of a counterbalanc-
ing land mass in the south, the Ulm edition
(second only to the 1477 Italian edition) uses
Ptolemy's second projection and includes
both the Ptolemaic Antarctica and the Arctic.

World Map
Francesco Rosselli, Florence

Authored by Alfred Hiatt
Rosselli's planisphere was one of the first
European maps to attempt to assimilate
news of Columbus' discoveries in the New
World. Showing a land mass in South
America identified as "Terra S. Crucis sive
Mundus Novus" (Land of the Holy Cross
or New World), the map also includes the
Caribbean islands and a North American
land mass, apparently connected to Asia.
Place names deriving from Columbus'
voyages appear on the coast of southeast
Asia, in accordance with the explorer's
belief that he had reached Asia. Rosselli's
map shows also an unidentified land
mass in the southern hemisphere, directly
beneath the Cape of Good Hope. The
word "Antarcticus" on the land mass refers
to the Antarctic Circle, which Rosselli has
marked in black, just above his signature,
"F. Rosello Florentino fecit". Five red marks
on the northern coast of the land suggest
habitation of some kind, and it is likely that
this depiction of a southern land derives
from a misreading of Florentine explorer
Amerigo Vespucci's account of his
voyages to the New World.

"The story of Antarctica as a
unique global village can only
be appreciated by an analysis
over time of its discovery and
exploration, and the behaviour
of its guests. The constant
record of these events has been
the maps drawn over many
hundreds of years, initially to
predict its presence, then to
validate its detection, and most
recently to judge its place in
the wider world."

Robert Clancy, John Manning and Henk Brolsma,
*Mapping Antarctica: A Five Hundred Year Record of
Discovery*, 2013

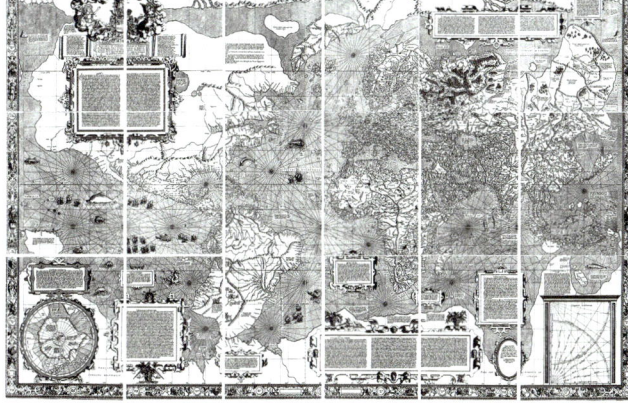

1531

1569

Nova, et Integra Universi Orbis Descriptio
Joannes Janssonius

*Nova et Aucta Orbis Terrae Descriptio ad
Usum Navigantium Emendate Accommodata*
Gerardus Mercator

Authored by Alfred Hiatt
Produced by the Parisian mathematician
Oronce Fine as a revision of his World Map
of 1519, this cartographic work marks the
emergence of a prominent southern land in
the European cartographic imagination.
Fine's map shows the northern and southern
hemispheres side by side, with comparable
land masses and ocean surface. The Antarctic land mass, labelled "Terra Australis
recenter inventa sed nondum plene cognita"
(Southern Land, recently discovered, but
not yet fully known), appears with a realistic
coastline, mountain ranges, and a sprin-
kling of toponyms ("Brasielie Regio"; "Regio
Patalis") and presents an annotation of the
Antarctic Circle. The toponyms (which derive
from garbled early modern voyage narratives in the case of "Brasielie", and from Pliny
the Elder's *Natural History* in the case of
"Regio Patalis") offer verisimilitude to the
hypothesis of the southern land, rooted
in classical and medieval science, supported
by the evidence of early modern exploration. *Terra Australis Incognita* (or "Nondum
Cognita") endured through the rest of the
sixteenth century as a standard feature of
world maps and was only fully erased from
maps following James Cook's explorations
of the South Seas in the third quarter of
the eighteenth century.

As the title itself implies, this map intends to
offer "a new and more complete representation of the terrestrial globe properly adapted
for use in navigation". The ambition of facilitating accurate sailing routes led the Flemish-
German geographer to develop an innovative
cartography based on a cylindrical projection which allows all parallels and meridians
to be straight and perpendicular to each
other, allowing for linear constant bearings,
the rhumb lines. The innovative conformal
map projection has the side effect of increasing the size of any area proportionally to
the distance to the equator; it derives that
Antarctica, still *terra incognita* at that time,
appears very large, elongated at the bottom
of the map. Indebted to the Portuguese and
Spanish portolan tradition, the Mercator map
remains to date one of the pivotal world
representations.

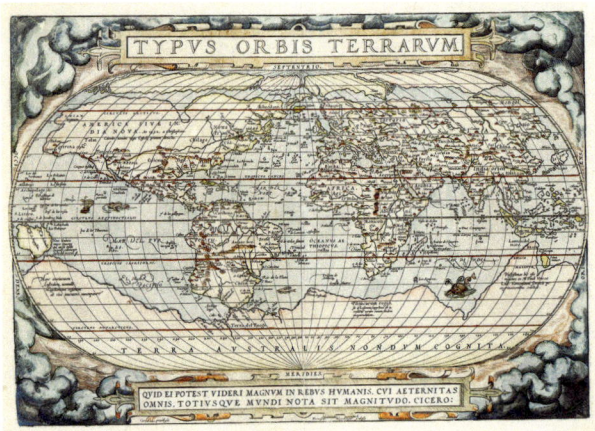

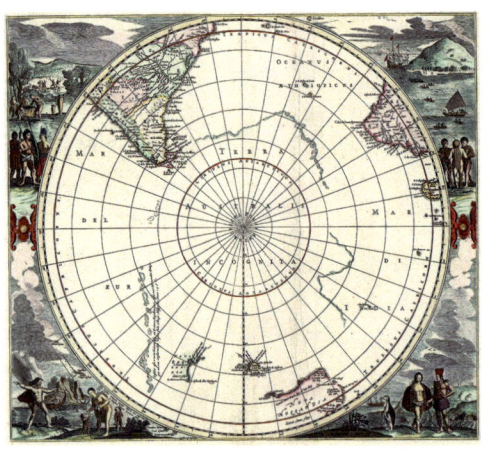

1570

1650

Typus Orbis Terrarum / Franciscus Hogenbergus sculpsit
Abraham Ortelius, Antwerp

Considered to be the "first 'modern' standardised atlas", the *Theatrum Orbis Terrarum* by Abraham Ortelius visualises the newly theorised Renaissance understanding "that the Earth's continents moved in relation to each other" and "that the Americas were 'torn away from Europe and Africa [...] by earthquakes and floods'" and that the "vestiges of the ruptures reveal themselves, if someone brings forward a map of the world and considers carefully the coasts of the three [continents]". The cartography thus reflects the contemporary idea of the continents' formation, according to which the supercontinent of Pangaea separated into Laurasia and Gondwana 200 million years ago.[2]

Polus Antarcticus Terra Australis Incognita
Joannes Janssonius

This early map of *Terra Australis* centred on the South Pole is considered one of the earliest realistic maps of the continent, at once revealing parts of South America, Africa and Madagascar as well as the coastlines of Australia, New Zealand and Tasmania. The fictional outline of *Terra Australis Incognita* included in previous centuries of world idealisations and representations was removed, as its conceptualised colossal scale had been disproved by many explorers while its presence had not been confirmed.

"Maps not only plot the surface of the land but, seen in historical sequence, also show us how our attitudes have imposed changing conceptions of landscape on that same land. Sometimes those landscapes are more mental than physical – a national territorial claim is just a line on a piece of paper – but sometimes maps are prelude to and evidence of our actual manipulation of space into place."

William L. Fox (Art Critic, Science Writer),
Terra Antarctica, 2005

1658

Orbis Terrarum Nova et Accuratissima
Tabula, Auctore Nicolao Visscher
Nicolaes Visscher, Amsterdam

Nicolaes Visscher's world map is informed
by the successful circumnavigation of
Australia by Abel Tasman in 1642 and 1643.
Reflecting the Franco-Dutch cartographic
paradigm shift which matured during
the period between the establishment of
the Dutch East Indies Company in 1602
and James Cook's circumnavigation, and
distancing itself from earlier intuitional
representational practices, this is known
to be one of the first world maps to be
based on factual discoveries rather than
a hypothesis.

1660–1672

Haemisphaerium Stellatum Australe
Antiquum
Andreas Cellarius

Focusing on the southern hemisphere,
this map by Cellarius shows many of
the classical zodiac constellations, including
the Southern Cross, first described by
Amerigo Vespucci in *Mundus Novus* (1504).
Aside from its undisputed beauty, the
map is historically relevant as it manifests
an awareness of the complexity of locating
one's position on the chart while navigating
in open seas in proximity to the magnetic
poles. In the absence of steady magnetic
bearings due to the failings of the compass in the proximity of the South Magnetic
Pole (a phenomenon which was yet to be
fully understood, despite the early invention
of the compass needle in the twelfth century and the deeper understanding of magnetic variations formulated by Christopher
Columbus in 1492), sailors in the Southern
Ocean could rely only on the elevation of
the sun above the horizon at noon (for the
latitude), star charts and astronomical tables.

1870

Ice Chart of Southern Hemisphere
United Kingdom Hydrographic Office

Issued by the British Admiralty's Hydrographic Office, this series of cartographies
were the first official charts commissioned
by a government to ensure a safe journey
across the Southern Ocean. Compiled by
overlaying information extracted from the
voyages of Cook (1772–1775), Bellingshausen
(1819–1821), Weddell (1822–1824), Foster
(1828–1829), Biscoe (1830–1832), Balleny
(1839), D'Urville (1839), Wilkes (1839) and
Ross (1841–1843), and later updated using
information from the voyages of Scott
(1901–1904) and Shackleton (1908–1909),
the relevance of this illustration derives
not from its accuracy (which remains debat-
able), but from its attempt to create a precursor to the Sailing Directions used by future
explorers of the Southern Ocean.

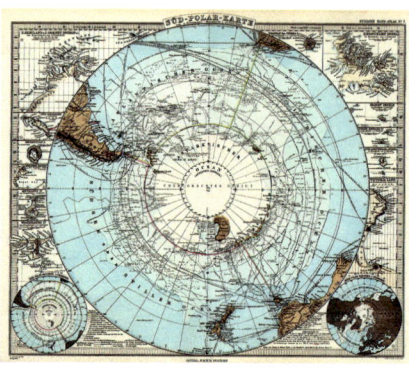

1850

Untitled World Map

Illustrated to accompany Ivar Hamre's
account of the largely overlooked Japanese
attempt to compete with Amundsen and
Scott in the race to the South Pole, this
cartographic work depicts the world, and
Antarctica, from the perspective of Nobu
Shirase's 1911–1912 expedition. Challenging
the policy of seclusion imposed upon Japan
from 1603 to 1868 during the Tokugawa
period, Shirase's expedition to the Southern
Ocean relied on cartographic information
from the Renaissance.[3]

1891

South Polar Map
August Heinrich Petermann, Adolf Stieler,
Carl Vogel, Hermann Berghaus, Hermann
Habenicht

Published in a time in which "cartographic
knowledge of the territories of the earth [was]
far less than generally supposed" as Petermann himself stated in 1866, the "South Polar
Map", drawn to include recent discoveries
by Wilkes, still reflects a rather undefined Antarctic coastline. Quoting Petermann, who
led the Geographische Mitteilungen Institut
in Gotha and was a great promoter of explorations, in late nineteenth-century cartography
"the African and Australian *terrae incognitae*
shrink more and more, and there remain [only]
a few blank spots, maybe 'wild territories',
where there is 'nothing'". Ultimately, he states,
"everything we see on our maps is just the
first step, the beginning of a more accurate
knowledge of the earth's surface."[4]

"The prodigious depths of snow
above, and the endless expanse
of ensnaring sea around are
mostly impregnable to man. He
who contemplates an attack
on this heatless undersurface
of the globe will find many
tempting allurements and many
disheartening rebuffs. [...]
The battle, however, should be fought, though it promises
to be the fiercest of all human
engagements. Science demands
it, modern progress calls for
it, for in this age a blank upon
our chart is a blur upon our
prided enlightenment."

Frederick Albert Cook (Explorer, Physician,
Ethnographer), *Through the First Antarctic Night*, 1900

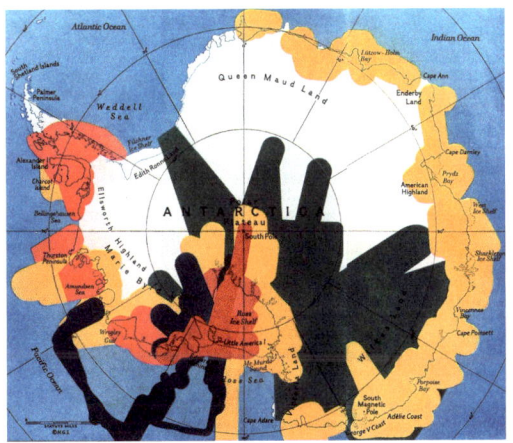

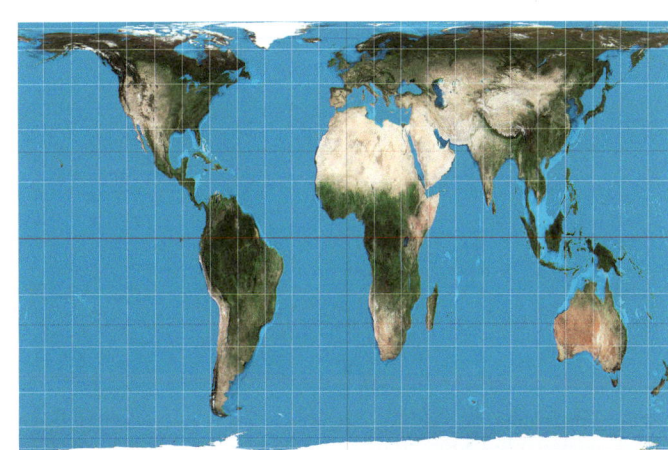

1956

1973

Admiral Richard Evelyn Byrd's map
of Antarctica
National Geographic Society

"This map, based on official Navy logs,
shows by colors how the Admiral's five
expeditions, spanning three decades,
successively put out exploring fingers
to bare the profile of the vast white continent. Byrd's conquest of the South Pole
by air and the discovery of Marie Byrd Land
climaxed his 1928–1929 venture. By ship
and plane, more than 450,000 square miles
of land and surrounding waters were explored and mapped by the 1933–1935 party.
Delineation of the Pacific coastline highlighted the 1939–1941 journey. Flights during
Operation Highjump covered an area more
than half as large as the United States and
recorded 10 new mountain ranges. Aerial
surveyors with *Operation Deep Freeze I*
swept across Wilkes Land and from the
Ross Sea to the Weddell Sea – a 3,200-mile
flight, the longest yet made in Antarctica.
Altitude and visibility determined the area
seen by observers aboard each flight. In
good weather crew men at 7,000 feet could
scan the ice sheath for 100 miles on either
side."[5]

The colours red, blue, pink, yellow and green
represent respectively: the first Byrd Antarc-
tic Expedition (1928–1930), the second Byrd
Antarctic Expedition (1933–1935), the United
States Antarctic Service (1939–1941), the
United States Navy Antarctic Expedition
Operation Highjump (1946–1947) and the
United States Navy Operation Deep Freeze I
(1955–1956).[6]

"The major figure in the shift
away from human-centered
Antarctic knowledge was an
American, Rear Admiral
Richard Byrd, whose aerial
mapping of the continent
extended the scope of the
human eye beyond the limits
of the terrestrially situated
body. It was this very notion
[…] [that] powered tech-
nologies into the robotic
surveillance devices […] The
difficulties of imposing human-
based technologies of seeing,
sensing, and knowing highlight
the ways the materiality of
Antarctica has shifted with and
under technological changes
and epistemological arrival.
By the twentieth century
Antarctica, the long-deferred
completion of the globe, has
itself become global."

Elena Glasberg (Professor at New York University, Essayist,
Speaker), *Antarctica as Cultural Critique: The Gendered
Politics of Scientific Exploration and Climate Change*, 2012

Peters World Map

Building upon Mercator's map, yet fundamentally questioning its Eurocentricity as
a means of promoting synchronoptic theories
of world history, German scholar Arno
Peters developed a much-contested (almost)
equal-area projection. Referencing the
prior work of Scottish clergyman James Gall,
the "fair" projection seems to inevitably
still neglect Antarctica, which remains an
underrepresented silver lining in this
renowned cartographic work.

WILLIAM KENTRIDGE is an internationally acclaimed artist based in Johannesburg, South Africa. His work spans a variety of media which include drawing, writing, film, performance, music, theatre and collaborative practices to create works of art that are grounded in politics, science, literature and history, while maintaining a space for contradiction and uncertainty. His aesthetics are drawn from the medium of film's own history, from stop-motion animation to early special effects.

SOUTH POLAR REGIONS
William Kentridge

The tapestries that I have been making over the last fifteen to twenty years have largely used found images and found colour. The colour often comes from maps; there is a particular colour palette of maps from different eras and different places. In this tapestry I wanted the range of blues, which was there to represent the kind of sky, so I used the map of the polar regions where the white of the ice becomes the white of the clouds, and the blue of the sea becomes the blue of the sky, and the ground and landscape is the black paper or the torn shapes with green pencil crayon on top. The figures moving across the landscape are based on figures that I used in a mural along the banks of the Tiber river for a project called "Triumphs and Laments" in 2016. They were drawings of different historical figures and historical events and in this case the movement of refugees from Africa to Europe around that era. The figures started as ink drawings to then become sten-

cils which were then cut out, and the images were then washed onto the banks of the Tiber river in Rome. These same figures, reconstituted, cut up, rearranged and changed in scale, become the figures walking across the landscape of the South Polar region. The objects on their heads are their own worlds that they carry, with some reference to all those polar explorers carrying their worlds on their sledges behind them as they tried to find new unexplored regions. The South Polar regions also refer to the most inhospitable places in the world, and so it is the difficulty of moving from an inhospitable place to what will likely be another inhospitable place. The text at the bottom "and when he returned" comes from a parable from West Africa that says "he that fled his fate, a journey of sixty years, and when he returned, who was waiting for him by the roadside? That is fate that said to him, 'Come, my friend. Let us go and eat.'" The pleasure of the tapestries is their transformation by Magritte Stephens and her weavers at the studio outside Johannesburg.

The mohair from which the tapestry is made is grown on goats in Lesotho, a country surrounded by South Africa. The wool is then spun and dyed in Swaziland, where Stephens has a property with a family weaving activity, and then the weaving itself is done at the studio outside Johannesburg, forty minutes from my own studio. The work is always a collaboration between the drawing that I give to Magritte to work with and the way this gets translated into the individual stitches of colour of the tapestry. Four to five weavers make it from a full-scale drawing behind the loom, and over the years the weavers have acquired extraordinary skills, so that some are doing the lettering while others are doing the fine lines. The tapestry itself will be a six-month project to weave.

Drawing the Antarctic Resolution Atlas

Sol Camacho and Carolina Passos

SOL CAMACHO is the director of the Polar Lab Brazil. She is an architect, urban designer, researcher and curator based in São Paulo, where she directs RADDAR, a studio working on multi-scalar architecture and landscape heritage conservation and adaptive reuse projects. CAROLINA PASSOS is an architect, urban designer and data scientist. She is the founder and urban data expert of Mapping Lab.

All the cartographies in this book, specifically produced for it, were drawn by the Polar Lab as part of the global network dedicated to celebrating the 200th anniversary of the discovery of the southernmost part of our planet, a celebration that speaks out loud of the pressing environmental issues that Antarctica faces, which are inevitably interconnected with the rest of the world.

The Antarctic Resolution Atlas, consisting of more than 80 carefully drawn and curated cartographies permeating the book, is one of the major outcomes of the collective transnational effort embraced by the Polar Lab since 2018.

Eight different world projections were chosen to support this effort: the Antarctic Polar Stereographic, the Azimuthal Equidistant, the Dymaxion, the Gauss-Krüger, the Lee Conformal, the Spilhaus, the South Pole Orthographic and, finally, a customized Azimuthal projection.

The deliberate action of drawing on various bases holds the argument that our world can be seen from a wide range of perspectives that totally depend on socio-political interests. This asserts that the most popular view of the world is a cultural construct that has consistently relegated Antarctica to the bottom of the map and often deformed it beyond recognition. This set of maps contests this stance and brings Antarctica – a continent fundamental to the global ecosystem – to the foreground.

The eight world projections, available in open-source GIS, were the canvas for laying down information on subjects as varied as topography, ice, wind, and precipitation dynamics, magnetism, temperature, climate change, traces of ancient vegetation, penguin and krill flows, historic travel routes, locations of scientific research bases, the identification of built heritage and historic sites, tourism, time zones, and governance models on the frigid continent – none of which, at first, most of us knew virtually anything about.

Still, the complex cartographic endeavour was pursued naturally; a multiscalar approach to design has permeated our architecture practice for years. Gathering information and translating it into drawings to convey a message is one of the most relevant ways in which architects, urban planners and landscape designers have contributed to the fields of planning, conservation and development.

All of the maps in the Antarctic Resolution Atlas are fully sourced with technical data received from the authors of the respective texts and other consolidated publications, to recent information directly sourced from active research centres. The gathering of this information was a collective effort led by the Polar Lab network located in Argentina, Brazil, Chile, Hong Kong and the United Kingdom. Its product is a testament to yet another characteristic of the cartographic building process: multitudinal and multidisciplinary collaboration.

Nonetheless, we can't ignore the fact that all drawings have an intrinsic problem, in addition to the reduction of the Earth into two dimensions: being static is a problematic condition. Without seeing the dynamism of the constant iterations and continuously evolving relationships beneath, on, and above land, ice and water, the reader is unable to fully grasp the forces at play. In that sense, responsive platforms like Bedmap, the visualization tool developed by the British Antarctic Survey, remain important sources for in-depth studies of the Antarctic continent. Nevertheless, despite their limitations and flaws, the cartographies have a clarity that seeks to raise awareness of the continent with precision and immediacy.

The Antarctic Resolution Atlas is a call to action. It is a tool developed to disseminate knowledge of the pivotal role Antarctica has in the global ecosystem to the widest audience possible. Data was translated into lines, texts, and areas plotted onto the very latest cartographies to show with blatant evidence the status of the planetary crisis and the forces at play, to encourage reflection and action.

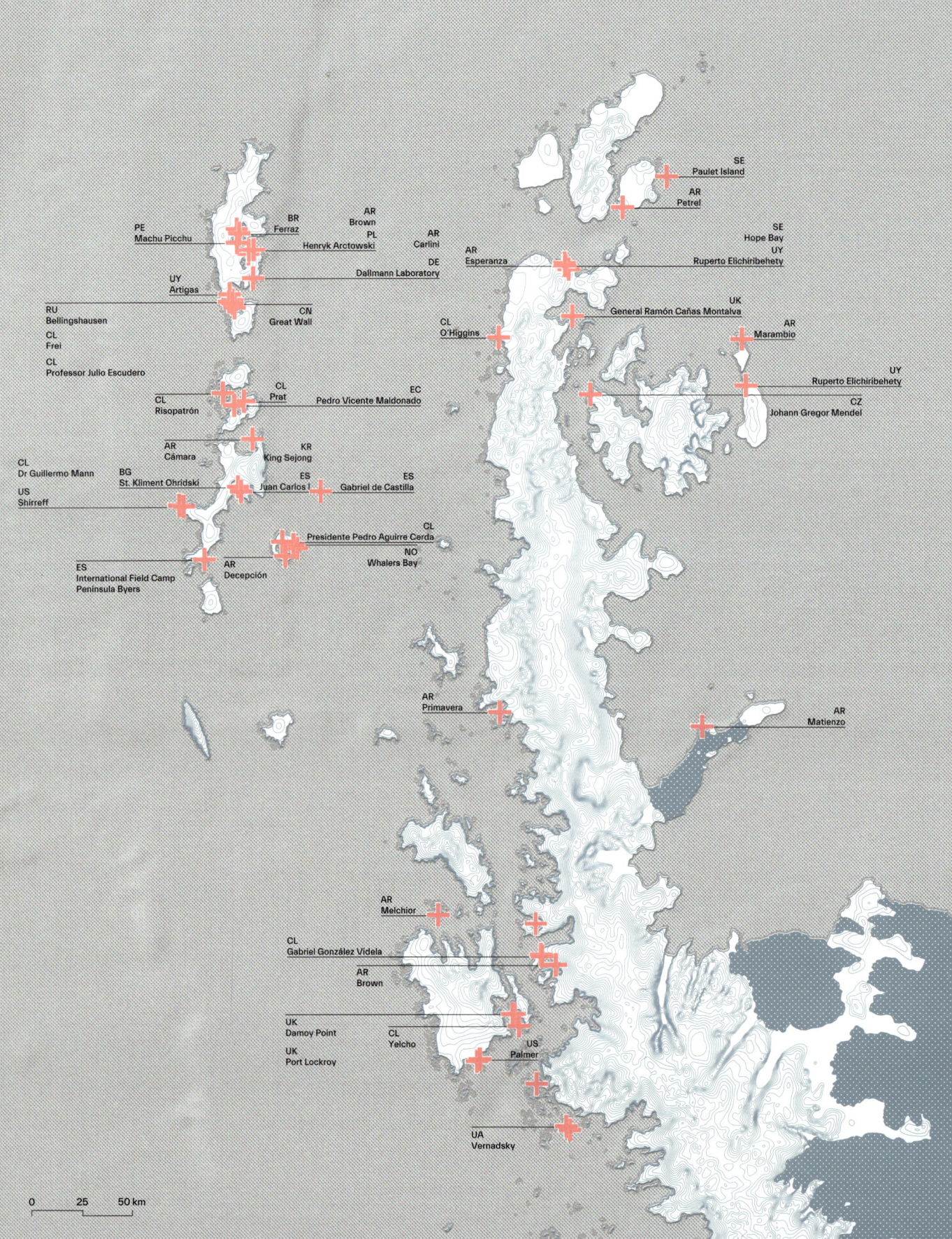

PE
Machu Picchu

BR
Ferraz

AR
Brown

PL
Henryk Arctowski

AR
Carlini

DE
Dallmann Laboratory

AR
Esperanza

SE
Paulet Island

AR
Petrel

SE
Hope Bay

UY
Ruperto Elichiribehety

UY
Artigas

RU
Bellingshausen

CN
Great Wall

CL
Frei

CL
O'Higgins

UK
General Ramón Cañas Montalva

AR
Marambio

CL
Professor Julio Escudero

CL
Risopatrón

CL
Prat

EC
Pedro Vicente Maldonado

UY
Ruperto Elichiribehety

CZ
Johann Gregor Mendel

AR
Cámara

KR
King Sejong

CL
Dr Guillermo Mann

BG
St. Kliment Ohridski

ES
Juan Carlos I

ES
Gabriel de Castilla

US
Shirreff

CL
Presidente Pedro Aguirre Cerda

ES
International Field Camp
Peninsula Byers

AR
Decepción

NO
Whalers Bay

AR
Primavera

AR
Matienzo

AR
Melchior

CL
Gabriel González Videla

AR
Brown

UK
Damoy Point

CL
Yelcho

US
Palmer

UK
Port Lockroy

UA
Vernadsky

0   25   50 km

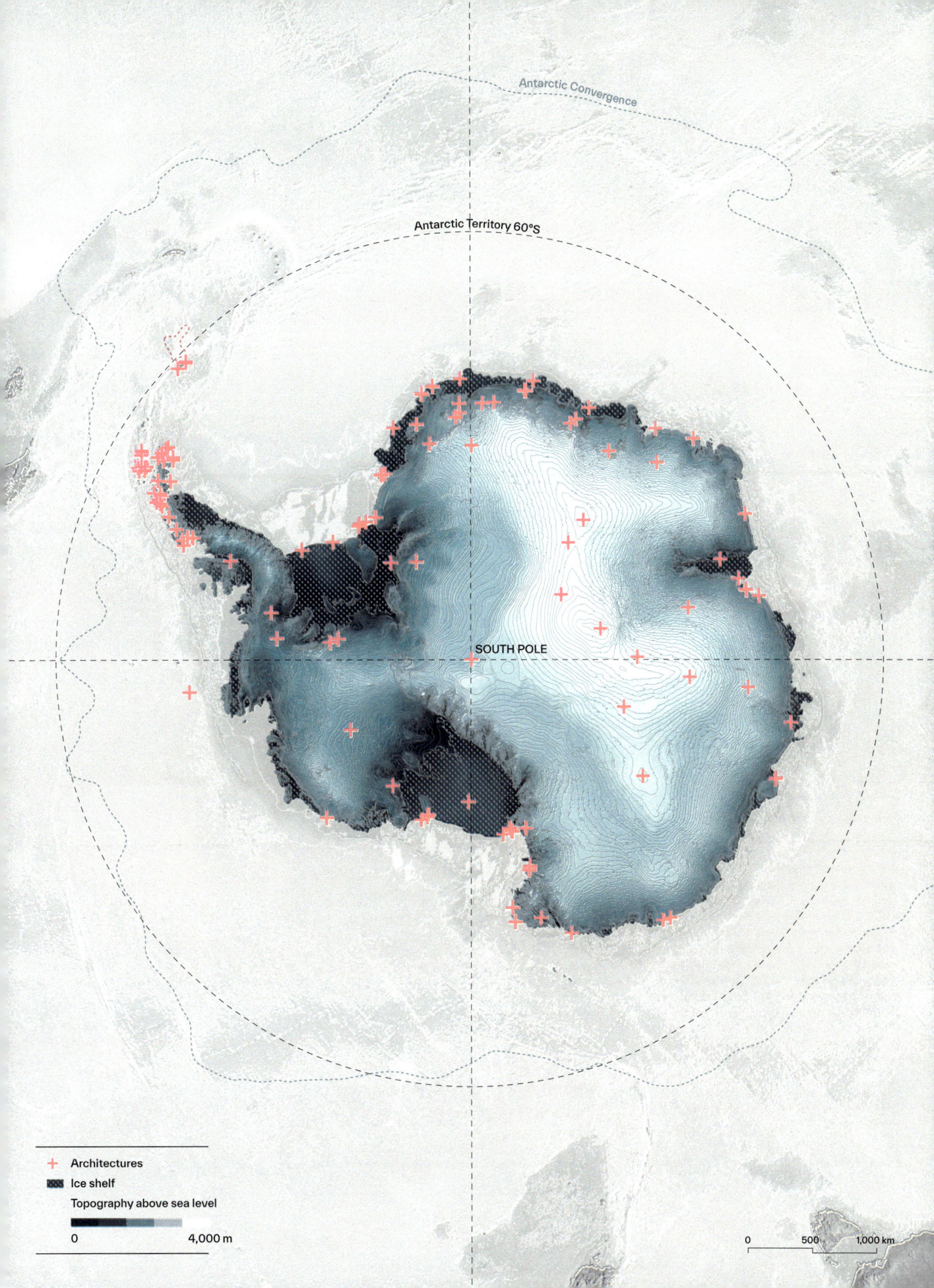

Antarctic Convergence

Antarctic Territory 60°S

SOUTH POLE

Architectures
Ice shelf
Topography above sea level

0   4,000 m

0   500   1,000 km

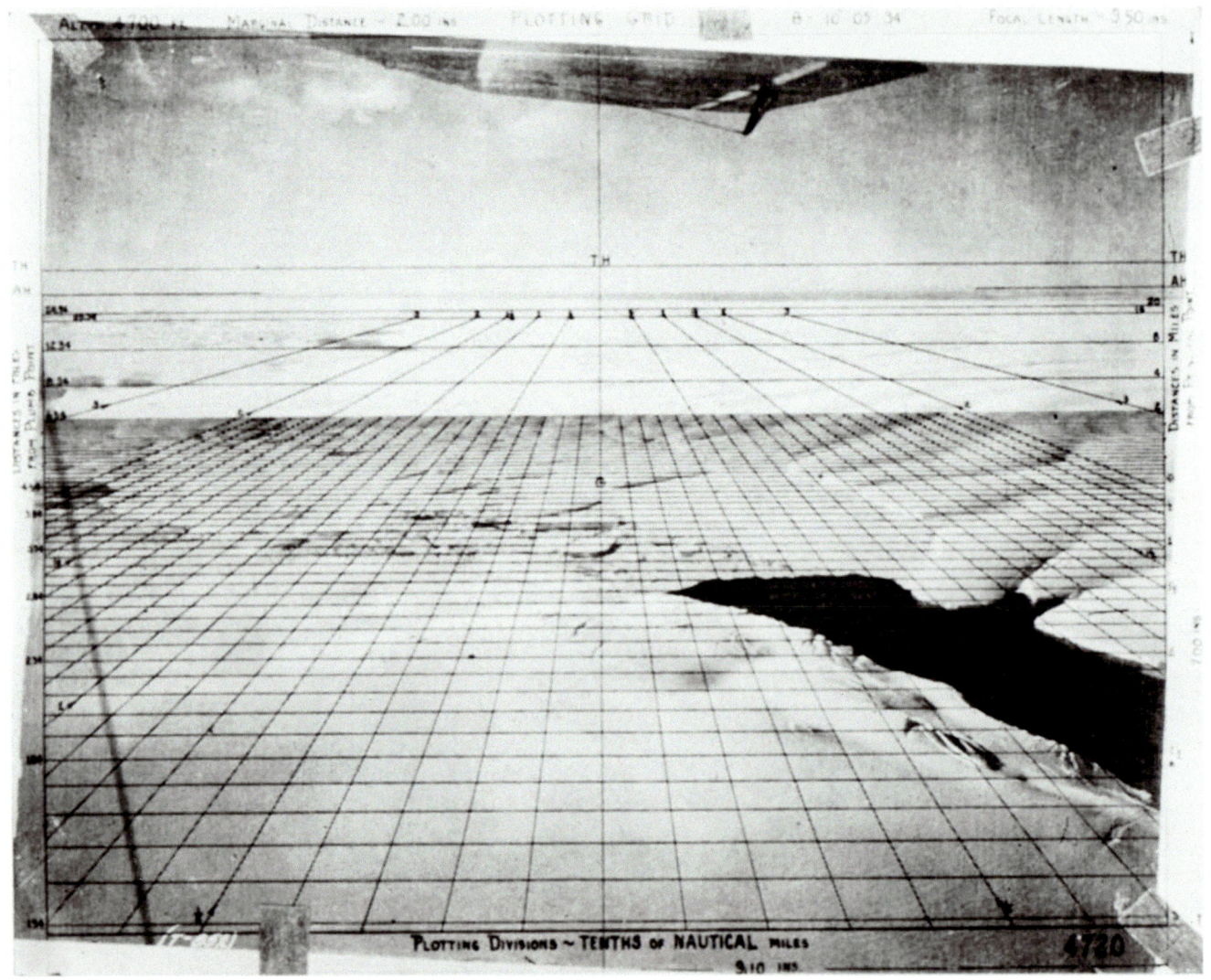

The production of accurate cartographies derived from early aerial surveys of Antarctica, such as those led by Rear Admiral Richard Byrd in King Edward VII Land and the northern part of Marie Byrd Land in December 1929, was no easy task, especially when mapping the seemingly scaleless environment of the plateau. To overcome the challenge, an experimental mapping strategy was defined whereby bearing grids were drawn on celluloid sheets placed over a paper protractor and subsequently overlaid on photographs, such as those shown in the picture at the side taken by Ashley C. McKinley. Due to the "complete absence of ground control points", the orientation of the bearing grid was informed by shadows and reflections and verified by comparing inbound and outbound flight trajectories.

"The problem of making a map from a series of photographs taken at altitudes not accurately known, depicting a terrain composed of hills and mountains of unknown elevations, and regulated by ground control points that were conspicuous by their almost complete absence, seemed at first well-nigh hopeless. […] In making the map, one of the biggest problems was the establishment of scale."

Richard Evelyn Byrd (Admiral, Explorer, Aviator) and Harold Eugene Saunders (Captain, Naval Engineer), "The Flight to Marie Byrd Land", *Geographical Review*, 1933.

Climate Models. Antarctica and the Politics of Digital Representations

Juan Francisco Salazar

JUAN FRANCISCO SALAZAR is a professor of communications and media studies at Western Sydney University (Australia) and is currently an Australian Research Council Future Fellow. He is an environmental anthropologist and documentary film-maker with academic and artistic interests in social-ecological change, environmental futures and indigenous politics. Supported by the Chilean Antarctic Institute, Salazar has undertaken ethnographic and film work in Antarctica, where he directed *Nightfall on Gaia* (2015).

From the realm of the imaginary to direct scientific observation, and now increasingly via remote sensing, the Antarctic region has been exposed to a very particular logic of representation. This has occurred through an array of lenses, sensing devices, and imaging technologies. Over the last decade, the Antarctic has been increasingly sampled using a range of technologies for calculation and measurement, while at the same time being imagined via distinctive observational *loci*: a diversity of gazes – and gazing bodies – that cut across a range of practices and modes of knowledge production. Here, models act as engines of enquiry within new "algorithmic architectures", thus becoming central to the capturing of experience[1] rather than merely acting as digital technologies of representation that index and reproduce empirical facts. Examining critically the dominance of scientific modelling as a means of generating environmental narratives of Antarctica, one could argue that such techniques of visualisation pervade visions of Antarctic futures and enact a politics that involves anticipating nature through "an aesthetics of prediction".[2]

In an age defined by digital culture, in which notions of the social emphasise the fluid, assembled, and multilayered nature of "societies", and the local and the global are irrevocably entangled with the boundaries of "the national" becoming inescapably porous, contemporary representational politics of the Antarctic have become profoundly problematic.

Across the Antarctic and Southern Ocean regions, an intricate system of different sensors continually generate data that is recorded, interpreted, applied, and preserved over a long period, on a scale spanning over 14 million square kilometres. Digital techniques of landscape visualisation are becoming ever more pervasive in presenting and conveying transformations to Antarctic landscapes and oceanscapes, providing examples of the ways in which our experience of space and place is digitally mediated.

This points to the importance of critically examining how our contemporary scientific imagination of the Antarctic (and its futures), centred around practices of geo-visualisation, offers prime examples of multimodal frameworks that integrate different forms of representation – visual, textual, and numerical.

It is not only the political force of Antarctic images that is being shaped, but also the social life of scientific models. This refers specifically to the circulation of anticipatory scenarios produced by Antarctic scientists that rely on algorithmic modelling as devices for anticipating the continent's socio-ecological futures. Whether it is the ice-sheet modelling used by glaciologists to trace the flows of ice and the movement of glaciers in an attempt to understand glacial cycles and the evolution and future of ice-stream dynamics, or the climate- and Earth-system modelling used by atmospheric scientists and climatologists to try to define the global reach of the Antarctic atmosphere and Southern Ocean, scientific visualisations of and in Antarctica are becoming future-making practices oriented by the endeavour to detect, project – and hopefully manage – future changes in Antarctic ecosystems.

Much of Antarctic climate science relies on model-based forecasting and predictions, which often become justifications for future-oriented action. Computer modelling and simulation aggregate data to build scenarios of probable and possible future events. As integral practices of contemporary digital networked culture, they enable scientific representations to gain and exercise a degree of authority that often shapes strategic geopolitical narratives of the Antarctic. At the same time, they can animate global environmentalist sentiments towards the Antarctic as the last wilderness in need of protection and care. The very production of Google Earth software, Lisa Parks argues, "is symptomatic of a global economy in which most nation-states are unable to control the production and circulation of representations of their own territories",[3] where nothing seems out of reach or out of bounds for the globalising vision of satellites.

Autonomous underwater vehicles (AUVs) used by an expedition to the Larsen region of the Antarctic Peninsula led by the Scott Polar Research Institute (SPRI) in 2019 reveal a high-resolution digital visualisation of ridges located at the bottom of the western Weddell Sea. Interpreted by scientists in Cambridge as features that are generated at the ice grounding zone, i.e. the point where "the ice flowing off Antarctica into the ocean becomes buoyant and starts to float", the images focus on "the rungs [which] are created as the ice at this location repeatedly pats the sediments as the tides rise and fall."[1]

One Map, One Ocean.
The Athelstan Spilhaus Projection

Bojan Šavrič, John Nelson, and David Burrows

BOJAN ŠAVRIČ is a geodesist and software developer at the Environmental Systems Research Institute (United States). JOHN NELSON is a cartographer and user experience designer at the Environmental Systems Research Institute (United States). DAVID BURROWS is a geographer and senior principal software developer at the Environmental Systems Research Institute (United States).

When we dip our toe into the ocean, we are connected to a vast and deep entity that encircles our globe. The coldest, deepest drops of water are contained within the same basin – flowing, sinking, mixing, swirling, and rising to connect our skin with every creature that glides within its saline medium. Certainly, as a default mode of thinking, it is easiest to consider a named body of water as a discrete, segregated presence, cut off from other named bodies of water, other coasts, other depths, other ecosystems. And this false picture of a limited reach with finite consequences drives how we think of our actions and their impact. The reality, however, is that Earth's oceans are, regardless of their designated name, a single body of water. One ocean wraps itself around our home – narrowing and widening, rising to shallows, and plunging to depths. A single ocean.

Cartographers have the unavoidable task of flattening our home into something that can be represented on a sheet of paper or a screen. There is just no way around this truncating, slicing, and cutting of the Earth's surface if they are to build a cartographic model of the planet. It is the nature of cartographers to lie in this way; they use mathematics – a projection – to do it. But make no mistake: creating or choosing a map projection is a carefully considered process of choosing which version of the truth to preserve and which to skew. Historically, when they have taken pen to parchment, or data to pixels, they bring our terrestrial biases to that process and flatten this Earth of ours into maps that focus on dry land. Continents are invariably preserved, while oceans are a convenient *terra incognita* to be split, compressed, trimmed or otherwise distorted.

While a handful of map projections were created to focus on the marine environment, they invariably depicted the world's oceans as segregated lobes or vastly distorted areas that end abruptly at the edge of the map, failing to portray the oceans holistically, as a unified entity. Athelstan Spilhaus, an inventor, academic, dreamer and futurist active throughout the twentieth century, noted this opportunity and set his considerable imagination to resolving a view of the world's oceans in an uninterrupted perspective. Born in Cape Town, South Africa, in 1911, he would call many places in the world his home while researching, inventing, organising and communicating in an array of scientific fields. John F. Kennedy once remarked to Spilhaus, "The only science I ever learned was from your comic strip",[1] referring to his weekly *Boston Globe* cartoon feature illustrating the possibilities and future of science.

Among the sprawling interests of Athelstan Spilhaus was communicating the urgency of exploring and understanding the sea. To reveal the world's oceans as a truly singular entity, Spilhaus worked with geodesists to derive a map whose maths would manifest the vision of an uninterrupted portrait of one world ocean. This map was first presented graphically in a *Smithsonian* article in 1979[2] and in a handful of publications in the years since.[3] Athelstan Spilhaus passed away in 1998, leaving behind an important legacy of imaginative science. The complex underlying mathematics required to reproduce and share his map were unfortunately never published, however, and access to this powerful perspective remained out of reach.

Committed to the agenda pursued by Spilhaus, a team of geographers and geodesists from the Environmental Systems Research Institute (ESRI) based in Redlands, California, undertook an investigative effort to determine the mathematical properties of the map and its projection by combing through historical articles for clues and attempting several iterations of its positional configuration. The team discovered the equations and offered Spilhaus's geographic perspective into the hands of thousands of map-makers, geologists, conservationists, oceanographers and other scientists.[4]

In this map, 70% of the Earth's surface seems more contained. This large expanse of seawater becomes distilled into a unified ocean entity revolving around the often-discounted Antarctic continent, which was consciously positioned at the centre. Something tangible and connected, a vulnerable body whose waters mix and flow to all depths and whose coasts take centre stage. Land masses are perched along its meandering shore; humankind is positioned about its perimeter. With this perspective, we have, in a renewed sense, a picture of our responsibility and shared stewardship of the medium that connects us. Enabling the map-makers of today to pour their data into Spilhaus's uniquely projected view of our watery world, one which emphasises the interconnectedness of the world's ecosystems, will, hopefully, provide a new tool to catalase collective awareness and calls for action.

> "Even though Antarctica appears to be at the end of the earth on traditional maps, it is at the centre of the world's oceans. The strongest current in the world, the Antarctic Circumpolar Current, runs clockwise around the continent. It drives the global 'conveyor belt' that circulates water through the oceans, a beating heart that pumps life through the seas."

Leslie Hook (Environment and Clean Energy Correspondent) "Why Penguins May Help Us Predict the Impact of Climate Change", *Financial Times,* 2020

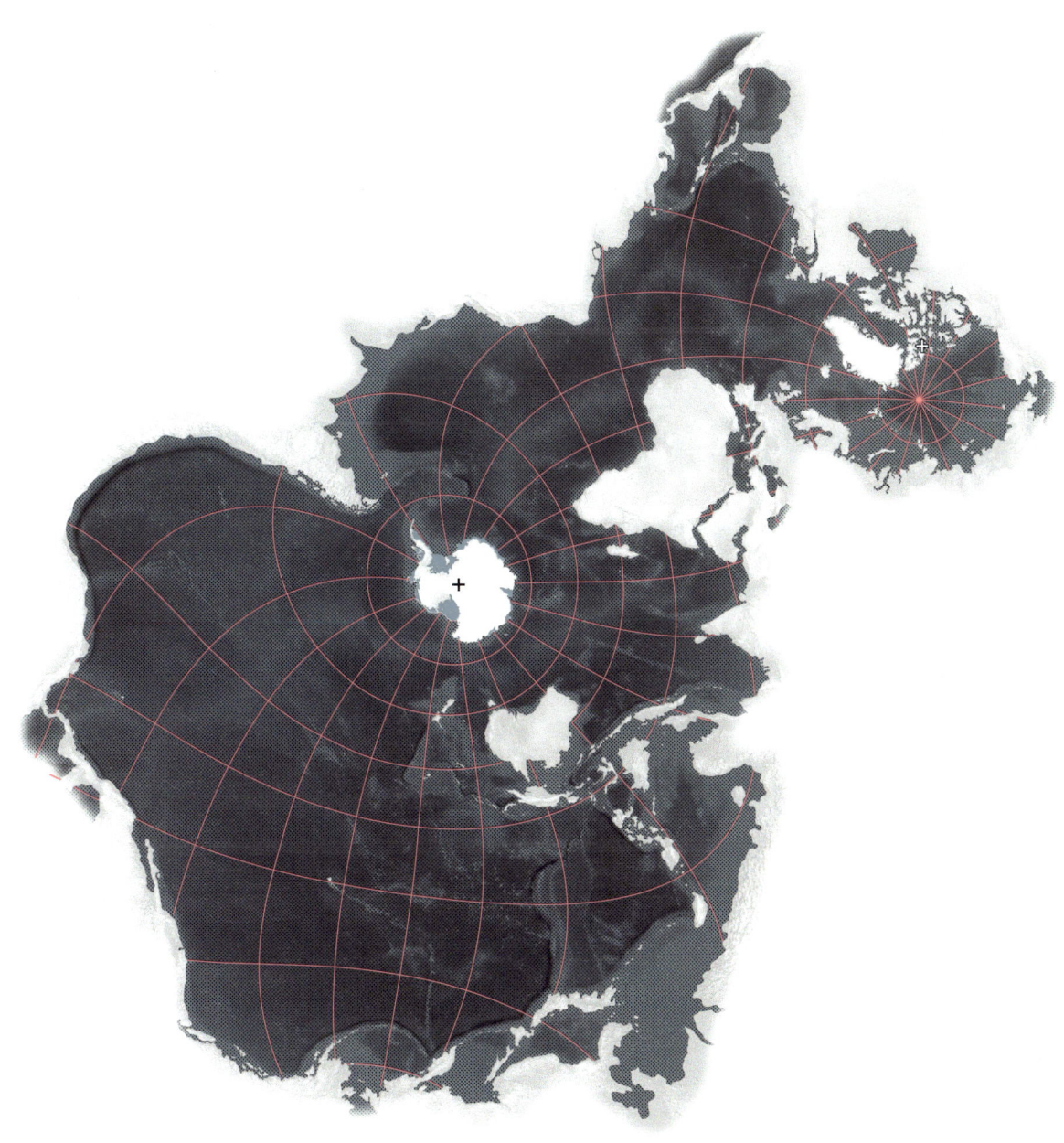

Constructing a Place through Literature. The Image of Antarctica

Elizabeth Leane

ELIZABETH LEANE is Associate Dean Research in the College of Arts, Law and Education at the University of Tasmania (Australia), chief officer of the Standing Committee on Humanities and Social Sciences within the Scientific Committee on Antarctic Research (SCAR), and arts editor of *The Polar Journal*. With a degrees in physics and literary studies, Prof. Leane has published seven books, including *Antarctica in Fiction* (Cambridge University Press, 2012).

The idea of an imaginative literature of Antarctica comes as a surprise to many people. This is, after all, a continent with no indigenous people, no permanent inhabitants, a relatively short period of human interaction, and a strong association, since the mid-twentieth century, with scientific investigation. Where does imagination fit into this "laboratory for science" that nobody can really call home?

The answer is everywhere. Humanity's late encounter with the far south means that imagination has played a unique role in the way humans have constructed Antarctica as a *place.* Before anyone had physically encountered Antarctica, diverse cultures told stories about a frozen land to the south. Indigenous people living in relatively high southern latitudes, such as the Selk'nam of Tierra del Fuego, have long incorporated knowledge of a cold region further south into their cultural traditions. In the northern hemisphere, ancient Greek philosophers hypothesised the existence of a land mass in the south balancing that of the north, and also believed that this region, like the far north, would be frigid. Myths of a great southern land fuelled European voyages of exploration during the Renaissance and began to generate fictional accounts of possible civilisations at the pole. Probably the earliest is the dystopia satire *Mundus Alter et Idem* (1605) by Joseph Hall, which describes an icy, dark but inhabited polar land lying at the extreme of a sprawling southern continent.

Over the following centuries, as commercial and exploratory voyages pushed further south, these speculations were tempered by direct experience. Despite the continent having been circumnavigated (by a British expedition led by James Cook) in the 1770s, officially sighted about fifty years later and landed upon not long after that, Antarctica remained mysterious. Its coast was beginning to be mapped, but nothing was known of its interior or whether the land encountered was part of an archipelago or a continental land mass stretching to the pole. Theories of open polar seas and even holes to the interior of the planet were circulating at the time. Creative writers drew upon this mixture of knowledge and speculation as they created their own Antarctic narratives.

During this period – the late eighteenth and early nineteenth centuries – two of the most enduring images of the far south were generated: *gothic Antarctica* and *utopian Antarctica.* The first represents the far south as an enigmatic, deathly region that promises the traveller both unprecedented insight and possible destruction. Its progenitors in English are Samuel Taylor Coleridge, whose narrative poem *The Rime of the Ancient Mariner* (1798) centres on a violent act against the non-human committed in an ice-entrapped ship, and Edgar Allan Poe, whose *MS Found in a Bottle* (1833) and *The Narrative of Arthur Gordon Pym of Nantucket* (1838) both end with terrifying but awe-inspiring encounters with phenomena at or near the South Pole. This tradition is well captured in a line from Joseph Conrad's novel *Lord Jim* (1900): "... the chilly Antarctic can keep a secret, but ... in the manner of a grave".[1] Conrad, perhaps the only pre-Heroic Age novelist to have direct experience of high southern latitudes, used the region as a setting for his short story *Falk* (1901) – another tale in which an Antarctic mariner undergoes a horrific and life-changing experience. These gothic associations have continued to pervade cultural responses to the far south, most prominently in science fiction and horror texts, including H. P. Lovecraft's *At the Mountains of Madness* (1936) and Joseph W. Campbell, Jr's *Who Goes There?* (1938; adapted multiple times for film, mostly famously as *The Thing* in 1982), but also in literary fiction such as Rebecca Hunt's *Everland* (2014), in which the icescape of an Antarctic island is imbued with a vague but persistent sense of menace.

The counterpart to this gothic tradition is the Antarctic utopia. Here, the seeming blankness and remoteness of the South Polar region, geographically

"And now there came both mist and snow,
And it grew wondrous cold:
And ice, mast-high, came floating by,
As green as emerald.

And through the drifts the snowy cliffs
Did send a dismal sheen:
Nor shapes of men nor beasts we ken –
The ice was all between.

The ice was here, the ice was there,
The ice was all around:
It cracked and growled, and roared and howled,
Like noises in a swound!"

Samuel Taylor Coleridge (Poet, Literary Critic, Philosopher, Theologian), *The Rime of the Ancient Mariner*, 1798

Illustration by Gustave Doré for *The Rime of the Ancient Mariner* by Samuel Taylor Coleridge (1798).

Constructing a Place through Literature. The Image of Antarctica

58

opposite the large population centres of the north, make it an ideal region for writers to imagine new ways of living, either as models to which to aspire or as satirical comments on their own society's shortcomings. James Fenimore Cooper's *The Monikins* (1835) and James De Mille's *Manuscript Found in a Copper Cylinder* (1888) are perhaps the best-known examples, although there are many more obscure texts in this genre. The utopian tradition (along with its dystopian variant) has been a strong element of Antarctic literature throughout the twentieth century, acting as a kind of barometer of public attitude toward the far south at different times. US science fiction of the 1930s, for example, produced techno-optimistic visions of the continent geoengineered to enable resource extraction and comfortable inhabitation. After the signing of the Antarctic Treaty in 1959, and particularly the Protocol on Environmental Protection in 1991, speculations began to focus on the Antarctic as a space for international cooperation and/or environmental protection, in texts such as David Poyer's *White Continent* (1980) and Kim Stanley Robinson's *Antarctica* (1997). More recently, the centrality of the Antarctic to planetary warming – as a record of past climate (through ice cores), a region impacted by warmer ocean and air temperatures, and a source of sea-level rise – has produced Antarctic climate fiction, or "cli-fi", narratives such as Witi Ihimaera's *The Purity of Ice* (2012), Ilija Trojanow's *The Lamentations of Zeno* (2011) and Paul McCauley's *Austral* (2017).

The gothic and utopian traditions do not exhaust Antarctic fiction, which extends to almost every genre – there are even category romances set in the far south, such as the Mills & Boon novel *Frozen Heart* (1980) by Daphne Clair. Unsurprisingly, the promise of excitement and danger in the extreme climate has produced a long line of popular texts, from boys' adventure fiction of the late nineteenth century to the many Antarctic thrillers of the late twentieth. Famous Heroic Age expeditions led by Scott, Shackleton, Amundsen, Mawson and others have also generated many imaginative responses, including Kåre Holt's *The Race* (1976; translated from the 1974 Norwegian-language original) and Beryl Bainbridge's *The Birthday Boys* (1991). The domination of Antarctic stories by white men – both as characters and authors – has been met in recent years by a number of narratives writing against this tradition, including Ursula K. Le Guin's short story *Sur* (1982), Mojisola Adebayo's play *Moj of the Antarctic* (2008) and Mat Johnson's novel *Pym* (2010).

Nonetheless, the twin visions of a place of hidden secrets and unknown threats, and a place of future hope and new ways of coexisting with the natural environment, still dominate Antarctic cultural response, given new impetus by the climate crisis. In the same way that the imagination enabled civilisations many millennia in the past to speculate about the far southern region, contemporary literature allows us to explore possible meanings and futures of Antarctica.

"The darkness had materially increased, relieved only by the glare of the water thrown back from the white curtain before us. Many gigantic and pallidly white birds flew continuously now from beyond the veil […] But there arose in our pathway a shrouded human figure, very far larger in its proportions than any dweller among men. And the hue of the skin of the figure was of the perfect whiteness of the snow."

Edgar Allen Poe (Writer), *The Narrative of Arthur Gordon Pym of Nantucket*, 1838

The Sphinx of the Ice Fields by George Roux, as featured on the cover of *An Antarctic Mystery* by Jules Verne (1897), was sighted by UNLESS in the Antarctic Peninsula in the austral summer of 2018–2019.

"But I was under a spell which drew me towards the unknown, that unknown of the polar world whose secrets so many daring pioneers had in vain essayed to penetrate. And this time, who could tell but that the sphinx of the Antarctic regions would speak for the first time to human ears! […] The Antarctic Sphinx was simply a colossal magnet.

Under the influence of that magnet the iron bands of the Halbrane's boat had been torn out and projected as though by the action of a catapult. This was the occult force that had irresistibly attracted everything made of iron on the Paracuta. […] Was it, then, the proximity of the magnetic pole that produced such effects?"

Jules Verne (Writer, Poet and Playwright), *An Antarctic Mystery*, 1899

Photography's Trace on Ice

El Glasberg

EL GLASBERG is a senior lecturer in expository writing for the Tisch School of the Arts at New York University (United States). Glasberg first discovered Antarctica through Edgar Allan Poe's 1838 sea journey *The Narrative of Arthur Gordon Pym*, and authored *Antarctica as Cultural Critique: The Gendered Politics of Scientific Exploration and Climate Change* (Palgrave, 2012).

Icy. Remote. Empty … Penguins? When asked about Antarctica, most people draw a blank. Filling in for this lack of knowledge, hardy clichés are repeated: Antarctica is the last place on Earth, a frozen laboratory for science and the only place on Earth never to have known war. All this against a backdrop of images depicting the endless array of ice and sky. But what people really don't know about Antarctica is that it is in many ways as mediated, surveilled and controlled as a bank lobby.

Two historical accidents help to account for this strange mix of Antarctic ignorance and its hyper-mediation. One factor is that humans did not gain footing on ice until around 1895, initiating the Heroic Age of Antarctic exploration that ended with the First World War. And because those explorers carried cameras among their supplies, Antarctica's belated discovery coincided with its photographic representation. Secondly, as European empire gave way, an extraordinary post-Second World War international science treaty emerged, one that continues today to manage the ice. Every step of the way, photography has provided documentation, evidence, orientation and, for its best practitioners, artistic expression. And every step of the way, the vast iciness pushed to the limits the photographer's tools, body and vision.

The Antarctic as a subject fits, often awkwardly, into the traditions of landscape photography, travel photography, war photography, nature photography and the fine arts. Stephen Pyne labels the prominent feature of Antarctica, its ice, an "aesthetic sink" that sucks in energy, money and artistic effort, giving nothing in return.[1] Yet if Antarctica has been a disappointment under capitalism, its very challenges to the body and to traditional culture have induced ever-greater artistic efforts – and rewards.

Travel photographer and accidental explorer Herbert Ponting, a self-proclaimed "camera artist" of the Heroic Age, strained to create drama and depth of field by showing the scale of human, animal and geographic features. Insisting on an already obsolete daguerreotype method that was both laborious and clumsy, Ponting went beyond documenting the establishment of a British base. Highlighting exotic bergs and penguins, he paradoxically normalised Antarctic ice by fitting it into the orientalist tropes of travel photography. Ponting's images provided much more than ancillary proofs or stand-ins for territorial claims; they provoke in the viewer Antarctica's existential challenge to human being, seeing and knowing.

At first glance, "Adelie Penguin Tracks and Sledge Tracks Crossing, 8 December 1911" seems to show only an empty ice field topped by a sliver of sky, a dark land mass on the left and a string of clouds edging off to the right. The foreshortened field is composed of tracks trailing away towards opposite edges of the horizon, forming an X. The tracks are a found object of sorts, man-hauled sledge tracks overwriting those of a penguin. Their layering suggests human imposition over nature consonant with a broadly conceived modernity. Ponting's crossroads thus reflects a shift from empire based on territorial claim and direct occupation to one based on representations such as photography and film. No longer a direct index of national territorial claim, photography in a science-managed Antarctica nevertheless bears the traces of heroic nationalism, empire and military occupation. Most ominously, Ponting's crossroads depicts the Anthropocene, the era in which human activity has indelibly marked the Earth.

Photographers continue along Ponting's crossed tracks, reflecting both Antarctica's existential challenges to humanity and humanity's fateful imprint upon nature. New Zealand's Anne Noble has produced the most sustained and significant body of work in and about Antarctica. From the lush disorientation of

Herbert George Ponting, *Adelie Penguin Tracks and Sledge Track Crossing* (1911)

"Whiteout" (2001/2008) to her re-photography of Heroic Age explorers and Antarctic tourism sites to her documentation of the aesthetics of support work on ice – "Bitch in Slippers" (2008) – Noble takes photography itself to its limits while pressuring its role under the management of science by nations and private corporations and its less acknowledged cultures of labour, consumerism and museumification.

Connie Samaras and An-My Lê both mine the traces of the US military in Antarctica's ever-expanding built environment. Samaras' "Buried Fifties Station" (2006), a low-altitude aerial view of a seemingly empty polar plateau, slowly reveals a tiny aperture, the entrance to the now abandoned and off-limits pre-Treaty station. Noble's experiential whiteout meets Samaras' epistemological blackout, and the result is a "black site".[2] It seems the more Antarctica is photographed, the less it can be known. "Buried Fifties Station" undermines the mapping and documentary function of Antarctic landscape photography, opening up Antarctica's science-managed surfaces to eerie sci-fi possibilities and architectures of absence and suppressed information.

An-My Lê includes her deadpan depictions of US installations at McMurdo Sound and the South Pole in her series "At Home Abroad" (2004–2007) of US overseas military bases. US military bases, like black sites and other extraterritorial spaces, are not quite fully national; they are strange and uncanny, often by definition hidden or unacknowledged – a status that Lê extends to the scientific stations of Antarctica. She further suggests the affinity of Antarctic built environments to a wasteful and dispiriting suburban sprawl, to question the impact of national and corporate presence on the continent and to pressure Antarctica's exceptional status as a continent devoted to peace.

Yet, amid the troubling traces of the past and impending environmental doom, life goes on in an Antarctica become prosaic, just another remote workplace or overrun tourist destination. Antarctica's defining geographic remoteness has been virtually breached as science support workers and tourists using smartphone cameras instantly upload and circulate countless images to the Internet. Unmanned cameras such as Galileo, a spider-like robot built to descend into an active volcano, and the Extreme CCTV remote surveillance device, which can withstand prison riots and lava bombs, suggest science management's deep connection to exploration history as well as the incomplete demilitarisation of its infrastructure.[3] Data based on the satellite mapping of glaciers support projections of a disastrous global rise in sea level, while ever-increasing numbers inhabit new scientific stations in an environment as inhospitable as that of the Moon but for a breathable atmosphere. Far from replacing human presence, remote sensing goes step by step with unsustainable growth in Antarctica. These cameras – if not their images – may outlast both humanity and the ice.

"The 'Whiteout' photographs aim to inspire the search for an image and the struggle to see. They also aim to evoke both the wonder of the ice and the irony that through looking at white (and all the constants and subtleties of that colour, tone, and hue) we can examine the act of seeing itself – the processes of creating imaginatively what is not there – and our human capacity for making that absence a presence for which we long."

Anne Noble (Photographer, Artist), *Ice Blink: An Antarctic Imaginary*, 2011

Anne Noble, *Whiteout #15* (2008)

Connie Samaras, *Buried Fifties Station* (2006)

Imagined, Recreated, and Rephotographed. Ice Blink

Anne Noble

ANNE NOBLE is an artist, a professor of fine arts at Massey University (New Zealand), and an NZ Arts laureate. She has visited Antarctica as an NZ Arts fellow and a US National Science Foundation (NSF) fellow. Her work examines the imagination and representation of Antarctica. She has exhibited and published widely on the photographic construction of the Antarctic imaginary. Her books include *Ice Blink: An Antarctic Imaginary* (Clouds, 2011), *These Rough Notes* (VUW, 2012), and *The Last Road* (Clouds, 2013).

When we look at Antarctica, from the deck of a ship, or in a make-believe *tableau,* and fix it in our gaze, what we see is a figment of the imagination. The sight we encounter is a sight already seen, image upon image fixed in the shadow of our dreaming by the medium of photography itself. First seen and drawn by artists and cartographers (who, because mirages were frequent, mapped whole coastlines that did not exist), then photographed by the great Heroic Age photographers such as Herbert Ponting and Frank Hurley, the Antarctica of our dreams is a visual domain cast in a pattern already set – as a picturesque wilderness (but we are there) and as a photographer's paradise (ice cliffs, and penguins floating on bergs). Photography confirms a "having been there" that is desirable, touchable and ultimately purchasable. It becomes the pretext for travel and loads a geographic imaginary that renders the traveller blind.[1]

Antarctica, Discovery Museum, Dundee, Scotland, 2005 (opposite page)

The Polar Sea #5, Antarctic Centre, Tasmania, Australia, 2006

Storm, Christchurch International Antarctic Centre, New Zealand, 2004

Christchurch Antarctic Centre, New Zealand, 2005

Paradise Harbour, Eleven Spectacular Days Antarctic Tour, 2005

Eleven Spectacular Days Antarctic Tour, 2005 (opposite page)

02
Antarctic Pie. The Menacing Geometry of Power

National Claims in Antarctica

Robert Headland

ROBERT HEADLAND is a fellow of the Royal Geographical Society who has served as an archivist and curator for the Scott Polar Research Institute, University of Cambridge (United Kingdom). Headland became involved with the Antarctic Heritage Trust, authored *A Chronology of Antarctic Exploration* (Cambridge University Press, 2009) and is the recipient of the Polar Medal (United Kingdom, 1984).

In the Antarctic, as elsewhere, there is an accepted convention that no person should be beyond the reach of law. On land, the necessary jurisdiction is based on territorial sovereignty. At sea, it is based on nationality – flag state jurisdiction. In the Antarctic, without indigenous peoples, there is no agreement as to whether jurisdiction is to be based on either of these. Although a variety of national claims have been asserted, none has found general acceptance, although a consensus has been reached under the Antarctic Treaty of 1959. The nature of the territorial boundaries in Antarctica is entirely artificial: geographic coordinates alone are used; no monuments or natural boundaries, such as mountain ranges or glaciers (in the absence of rivers), are involved, as would be the norm elsewhere on planet Earth.

The earliest sovereign claim within the Antarctic Treaty region was made when the South Shetland Islands were discovered in 1819, while claims to continental territory began in 1829 and continued though to 1843. Formal claiming resumed during the Heroic Age and became relatively frequent after 1928, when aerial exploration became practicable. Several expeditions were dispatched with the object of making claims, although for many others this was incidental. Such claims, generally involving a landing, the ceremonial raising of a flag and promulgation of a formal declaration, have been made by individuals of various nationalities throughout the Antarctic region.

Such acquisition of territorial sovereignty – by discovery, then proclamation – was considered to confer an inchoate title on the discovering state. The entire Antarctic continent is subject to such claims. To perfect such a title, it was necessary for a proclamation of geographically defined territory to be followed by administrative acts responding to governmental needs. In the Antarctic Treaty region, claims have been made for Argentina, Australia, Britain, Chile, France, Germany, Japan, New Zealand, Norway and the United States. Some of these have resulted in a formal claim to a specific geographic region; others have not, although such claims have been neither cancelled nor abandoned, except for German and Japanese claims renounced by treaty in 1919 and 1951, respectively. The German claim to Neuschwabenland in 1939 was a formal ceremony that involved the swastika flag; it was conducted only five days after Norway had proclaimed Queen Maud Land on the 14th of January 1939. Subsequent circumstances were not propitious for the German claim to progress, although it remains of historical interest. The Japanese claim, to Yamato Yukihara in the Ross Sea region, was made in 1912 by Nobu Shirase's expedition.

The countries whose citizens have been active in Antarctica and whose governments have made defined territorial claims are Argentina, Australia, Britain, Chile, France, New Zealand and Norway. Australia, Britain, France, New Zealand and Norway mutually recognise each other's claims, but none of the formally defined territorial claims are universally recognised. Ecuador, exceptionally, claims territory for geographic reasons.

Some Antarctic territories have involved negotiation; for example, Britain and Australia (for the Australian Antarctic Territory) and France (for Terre Adélie) settled the boundaries diplomatically. In contrast, the Antarctic Peninsula regions are subject to unresolved claims put forward by Britain in 1908, Chile in 1940 and Argentina in 1943. British attempts to remedy consequent counter-claims by legally binding means, and later before the International Court of Justice, did not succeed. In 1959 the problem passed into abeyance under the Antarctic Treaty.

Prior to the Antarctic Treaty, seven existing claims to territorial sovereignty in Antarctica had been defined. However, there was no commonly accepted

"We hoisted Her Majesty's flag and the other Union Jack afterwards and took possession of the plateau in the name of His Majesty. While the Union Jack blew out stiffly in the icy gale that cut us to the bone, we looked south with our powerful glasses, but could see nothing but the dead white snow plain."

Sir Ernest Henry Shackleton (Explorer), *The Heart of the Antarctic: Being the Story of the British Antarctic Expedition 1907–1909*

"So we plant you, dear flag, on the South Pole, and give the plain on which it lies the name King Haakon VII's Plateau."

Roald Amundsen (Explorer), quoted in Roland Huntford's *Scott and Amundsen: The Last Place on Earth, 1910–1912*

means of exercising jurisdiction on land other than on territorial sovereignty, and no forum existed in which any other possibility might have been considered. Two factors have largely protected the region from the worst consequences of this situation. First, the Antarctic regions, with their very small populations, have comparatively little need for government; second, the provisions of Article IV of the 1959 Antarctic Treaty suspended, but did not abolish, the territorial claims.

In February 1967, two decades before Ecuador acceded to the Antarctic Treaty, the National Constituent Assembly in Quito formally declared the Territorio Ecuatoriano Antártico in the part of the Antarctic between 84°30′ W and 96°30′ W. This was based on a geopolitical theory of confrontation; the sector approximated the territorial limits of Islas Galápagos. A formal protest was made by Chile, as the Ecuadorian territory intruded upon that claimed by Chile. Perhaps inadvertently, it also included Peter I Øy of Norway. Twenty years later, in June 1987, the Assembly confirmed the sovereign claim. A few months later, in September of that year, Ecuador acceded to the Antarctic Treaty, thus becoming bound by Article IV concerning sovereignty and territorial claims. Nothing in the Treaty, however, makes this, or any of its other articles, retroactive. Thus Ecuador has the eighth, and most recently, defined territory in Antarctica.

There are no formally proclaimed territories in the sector between the Ross Dependency and the Ecuadorian limits. Although some authorities assert that these are "unclaimed", they have been the subject of several historical national claims (including by Britain, Japan, Norway and the United States). However, no defined territory includes the sector. Acknowledgement of claims varies; for example, South Africa's recognition of Queen Maud Land was in return for the transfer of Norway Station to South Africa in 1960. The policy of several other states, including Belgium, Japan, Russia and United States, is neither to assert nor to acknowledge any Antarctic sovereign claim. The United States, owing to four decades of extensive exploration and claims starting in 1929, retains a strong basis for a sovereign territory, particularly in the Marie Byrd Land region. Russian claims date from Bellingshausen's pioneering expedition of 1819–1821, which included very early sightings of Antarctica and six crossings of the Antarctic Circle.

Few published Antarctic territorial maps are entirely complete and correct, even when the Ecuadorean claim is excluded. Argentine, Australian, French and New Zealand territories are usually represented correctly, including their northern limit (60° S). Those of Chile and Norway are often shown with this limit, although neither specifies it. The British Antarctic Territory is often shown in its pre-1962 form – even in maps published more than half a century later; and the Norwegian territory is commonly shown as not extending to the South Pole, even in most post-2015 maps. Peter I Øy, the earliest Norwegian Antarctic claim, is frequently omitted. Ecuadorian territory rarely appears on maps and similarly has no northern limit defined.

Article IV of the Antarctic Treaty, to which all states active on the continent have adhered, practically solves disputed sovereignty. Thus no acts or activities shall constitute a basis for asserting, supporting or denying an existing claim, and no new claim or enlargement of an existing claim shall be asserted while the Treaty is in force.

The Australian flag is raised by Dr Phillip Law at the naming ceremony of Mawson Station in February 1954 (top), while soldiers salute the Argentine flag during *Operación 90,* the first Argentine ground expedition to the South Pole, conducted in 1965 by ten soldiers of the Argentine army (bottom).

ESTABLISHING THE FORMAL DEFINITIONS OF THE SOVEREIGN TERRITORIES SOUTH OF 60° S
Robert Headland

The formal definitions of the sovereign territories south of 60° S (the Antarctic Treaty latitude with geographical statistics) were made, in the order of their establishment, by Britain, New Zealand, France, Norway, Australia, Chile, Argentina and Ecuador.

Britain established national claims in 1908, when royal letters patent consolidated claims to South Georgia, the South Orkney Islands, South Shetland Islands, South Sandwich Islands and Graham Land as the Falkland Islands Dependencies. These five places are within a region bounded by 50° S between 20° W and 80° W. In 1917, the definition was amended to include all islands and territories whatsoever in a smaller area bounded by 20° W to 50° W south of 50° S and 50° W to 80° W south of 58° S. After the Antarctic Treaty came into force, the British Antarctic Territory was designated as all territories between 20° W and 80° W south of 60° S. This territory has a surface area of 1,709 × 103 sqkm (12.3% of Antarctica, 82% of which is contra-claimed by Argentina and 45% by Chile). It includes the South Shetland Islands and South Orkney Islands.

Following Britain's claims, the Ross Dependency, established in 1923 by a British Order in Council and situated between 160° E and 150° W south of 60° S, was administered by a governor who was also the Governor-General of New Zealand. This was the earliest adoption of the 60° S parallel in Antarctic territories. In 1983 the territory was incorporated into the Realm of New Zealand. The Ross Dependency territory has a surface area of 450 × 103 sqkm (3.2% of Antarctica), including the Balleny Islands and Scott Island.

A presidential decree of 1924 reserved French rights in Terre Adélie, but did not specify the extent. Negotiations involving Britain and Australia resulted, in 1938, of a decree fixing its boundaries as 136° E to 142° E south of 60° S. The territory has an area of 432 × 103 sqkm (3.1% of Antarctica).

In 1931 a royal proclamation placed the small Peter I Øy (68°51′ S, 90°37′ W) under Norwegian sovereignty. The annexation of the much larger Queen Maud Land in 1939 was as the coast between the Falkland Islands Dependencies and Australian Antarctic Territory with the land lying within, and the environing sea. Neither a northern nor a southern extent was specified until 2015, when the Storting (parliament) defined the southern limit as extending to the South Pole. The Norwegian territory has an area of 2,700 × 103 sqkm (19.5% of Antarctica).

In 1933 a British Order in Council established the Australian Antarctic Territory as the islands and territories other than Adélie Land between 160° E and 45° E south of 60° S. The 1938 French specification of Terre Adélie divided the Australian territory into two sectors, which have a total area of 5,896 × 103 sqkm (42.5% of Antarctica).

In 1940 a presidential decree defined Territorio Chileno Antártico as all lands, islands, reefs of rocks, and glaciers (pack ice) in the sector between longitudes 53° W and 90° W. This had no northern boundary. Chile is the only country to claim territory over the annually variable extent of maritime pack ice. The Chilean territory has an area of 1,251 × 103 sqkm (9.0% of Antarctica, 30% of which is counter-claimed by Argentina and 45% by Britain). It includes the South Shetland Islands.

A formal memorandum from the Argentine Foreign Minister to the British Ambassador in Buenos Aires in 1943 reaffirmed Argentine sovereignty over all Antarctic lands south of 60° S between 25° W and 68°34′ W. In 1947 the western boundary was extended to 74° W. Antártida Argentina has an area of 1,462 × 103 sqkm (10.5% of Antarctica, 11% of which is counter-claimed by Chile and 100% by Britain). It includes the South Shetland Islands and South Orkney Islands.

In 1967 the Constituent Assembly specified Territorio Antártico Ecuatoriano as lying between 84°30′ W and 96°30′ W with no northern limit. This sector has an area of 323 × 103 sqkm (2.3% of Antarctica). It contains Peter I Øy of Norway and a 65% counter-claim by Chile.

ROBERT HEADLAND is a fellow of the Royal Geographical Society who has served as an archivist and curator for the Scott Polar Research Institute, University of Cambridge (United Kingdom). Headland became involved with the Antarctic Heritage Trust, authored *A Chronology of Antarctic Exploration* (Cambridge University Press, 2009) and is the recipient of the Polar Medal (United Kingdom, 1984).

National territorial claims in the Antarctic region, in order of inception

- British Antarctic Territory (20° W to 80° W, south of 60° S) [BAT]; inception 1908, present form 1962
- Ross Dependency, New Zealand (160° E to 150° W, south of 60° S) [RD]; inception and present form 1923
- Terre Adélie, France (136° E to 142° E, south of 60° S) [TA]; inception 1924, present form 1938 Peter I Øy, Norway (68°51′ S, 90°37′ W) [PIØ]; inception and present form 1931
- Australian Antarctic Territory (45° E to 136° E and 142° E to 160° E, both south of 60° S) [AAT(E) and AAT(W)]; inception 1933, present form 1938
- Queen Maud Land, Norway (25° W to 45° E, no northern limit) [DML]; inception 1939, present form 2015
- Territorio Antártico Chileno (53° W to 90° W, no northern limit) [TAC]; inception and present form 1940
- Antártida Argentina (25° W to 74° W, south of 60° S) [AA]; inception 1943, present form 1947
- Territorio Antártico Ecuatoriano (83°30′ W to 96°30′ W, no northern limit) [TAE]; inception and present form 1967

Reserved right to claim:
United States of America
Soviet Union

+ Stations operating during the 1957–58 IGY

▨ Sea-ice extent (2019)

"Chief among its historical discontinuities, Antarctica is not now and has never been a national space. While the globe was coming under the reign of sovereigns, and later of nation-states, Antarctica was remote as myth until the turn of the twentieth century."

Elena Glasberg (Professor at New York University, Essayist, Speaker), *Antarctica as Cultural Critique: The Gendered Politics of Scientific Exploration and Climate Change*, 2012

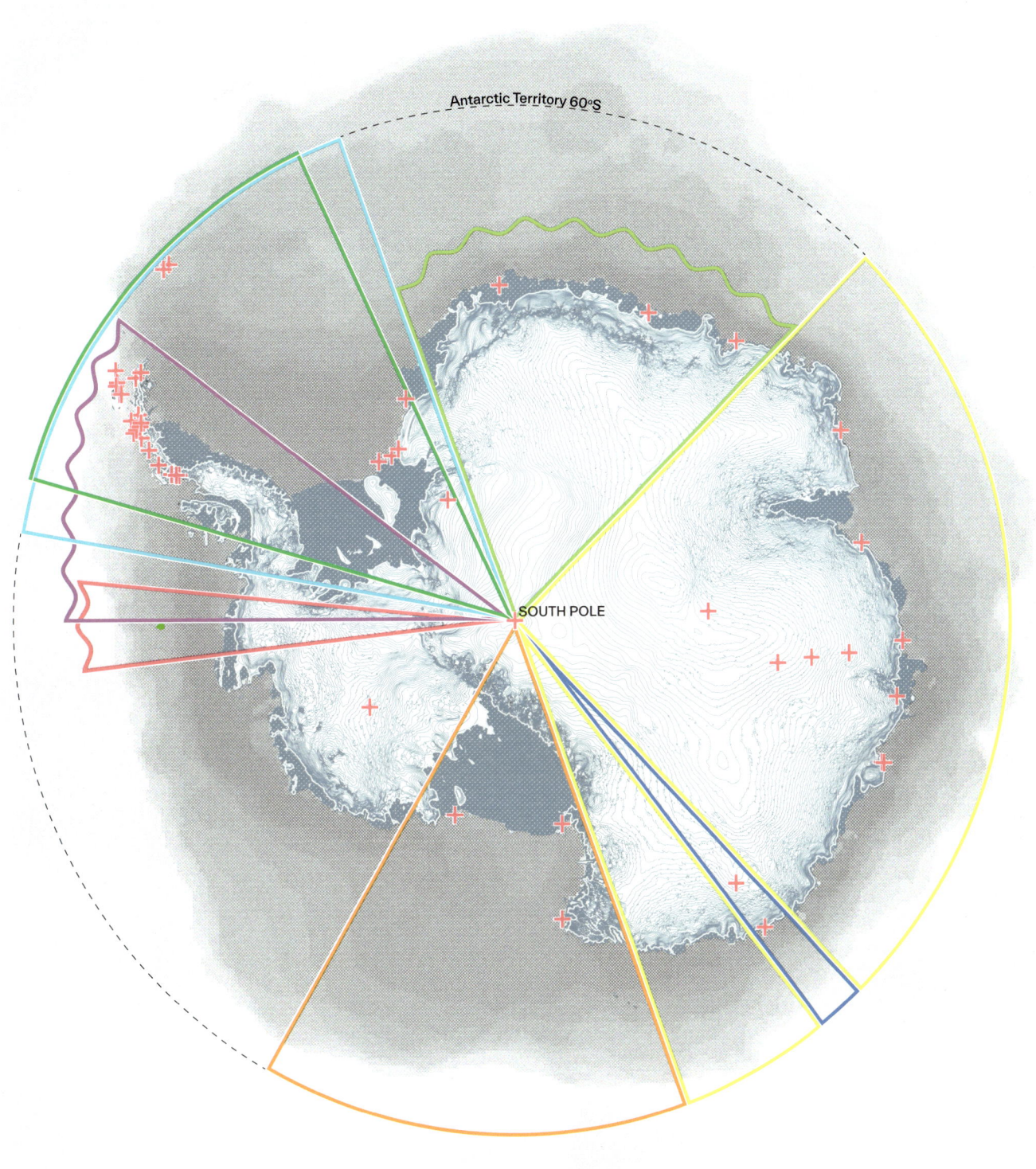

Antarctic Territory 60°S

SOUTH POLE

0   500   1,000 km

Neuschwabenland

Cornelia Lüdecke

CORNELIA LÜDECKE is the chief officer and founder of the Expert Group on the History of Antarctic Research within the Scientific Committee on Antarctic Research (SCAR). She is a professor of the history of natural sciences at the University of Hamburg (Germany), focusing on the history of polar research and meteorology. Prof. Lüdecke is a member of the editorial boards of scientific journals, including *Polar-forschung, Polar Record* and *The Polar Journal*.

The catapult ship *Schwabenland,* equipped with two Dornier Wal flying boats, was sent to Antarctica under the Third Reich's Four-Year Plan in 1938 with the aim of securing German whaling in Antarctic waters. Germany was preparing for war and could no longer afford to import whale oil from Norway to produce margarine, detergents, soap or glycerine. To cope with this situation, the construction of a national whaling fleet was begun in Hamburg and Bremen in 1936. During the season of 1938–1939, it comprised seven factory ships and sixty-eight whale catchers. It became soon evident that to support the fleet, it was necessary to construct a base on the Antarctic coast, which meant occupying a piece of land for it. However, the latter – according to international law at the time – required evidence of legal bases which would grant Germany the right to occupy that land. To that end, German ministries united to plan what is known as the *Schwabenland Expedition,* a short summer Antarctic campaign aimed at producing a map, which would lead to the occupation of the unknown area between 5° W and 20° E. The government's ambition was to present such a map at the International Whaling Conference in London in July 1939. Though Norway forestalled Germany by asserting a claim to this same region just before the Germans reached the site, the regime's photographic survey flights began according to the original plan on the 19th of January 1939. Alfred Ritscher, the expedition leader, developed a special flight routine which is still used on aircraft expeditions all over the world. The daily planning started with a report at an evening meeting on the readiness of the aircraft and on the weather forecast for the next day, based on information provided by the German whaling fleet and radiosonde measurements of upper-air wind and temperatures taken from "Schwabenland". If the weather forecast was favourable, an aeroplane was prepared on the catapult overnight and the photographic flight would start the next morning (provided the forecast two hours before take-off was still the same). A second plane was always ready in case the first machine was faulty, and repairs were scheduled after its return. Between the 19th of January and the 3rd of February 1939, six flights were performed by one plane along the mountain ranges of Neuschwabenland, while the second aeroplane was used to explore the coast and to make a landing on the ice shelf, where the crew later raised the flag of Nazi Germany. During the final reconnaissance flight, the expedition team discovered an ice-free area and called it Schirmacher Oasis, after the pilot.

During the expedition, the Germans took photogrammetric images of an area of about 250,000 square kilometres, but they neither set foot on Antarctica nor performed any geodetic measurements on land. Upon their return to Germany, a map was published only after the International Whaling Conference. A second expedition, which aimed to make accurate ground measurements and adjust the coordinates of the map, was cancelled due to the Second World War. The official names of the German discoveries were published only after 1952, while efforts to correct the map continued until 1986. The expedition remained an unsuccessful historical attempt at granting sovereignty rights to Germany in Antarctica, as no official claim of Neuschwabenland was ever made.

This timeline juxtaposes a comparative analysis of the Antarctic surface area claimed by the eight claimant nations (seen in relation to their own territory) and the prolific construction and flag-hoisting activities undertaken (among other reasons) to reinforce their declarations of territorial sovereignty.

+ Antarctic Stations

● Size of national claim

Argentina 1,461,597 sqkm

Australia 5,896,500 sqkm

Chile 1,250,257.6 sqkm

France 432,000 sqkm

New Zealand 450,000 sqkm

Norway 2,700,000 sqkm

United Kingdom 1,709,400 sqkm

* Reserved right to claim: United States of America Soviet Union

Size of country

● Argentina 2,740,000 sqkm

● Australia 7,680,000 sqkm

● Chile 743,800 sqkm

● France 547,660 sqkm

● New Zealand 267,710 sqkm

● Norway 304,280 sqkm

● United Kingdom 241,930 sqkm

Hoisting of national flag

1912
Nobu Shirase,
Ross Ice Shelf
Japan

1929
Nils Larsen,
Peter I Øy
Norway

1935
Klarius Mikkelsen,
Ingrid Christensen Coast
Norway

1904
William Bruce,
South Orkney Islands
Argentina

1935
Lincoln Ellsworth,
Ellsworth Land
United States of America

1942
Antonio Oddera,
Deception Island
Argentina

1898
Carsten Borchgrevink,
Cape Adare
United Kingdom

1911
Kristian Prestrud,
Cape Colbeck
Norway

1929
Lawrence Gould,
Marie Byrd Land
United States of America

1931
Hjalmar Riiser-Larsen,
Prinsesse Ragnhild Coast
Norway

1909
Ernest Shackleton,
Polar Plateau
United Kingdom

1911
Roald Amundsen,
South Pole
Norway

1931
Douglas Mawson,
Mac Robertson
United Kingdom (for Australia)

1957–1958
International
Geophysical Year

1819, William Smith, South Shetland Islands, United Kingdom
1820, Edward Bransfield, South Shetland Islands, United Kingdom
1821, George Powell, South Orkney Islands, United Kingdom
1840, Dumont d'Urville, Terre Adélie, France
1841, James Ross, Victoria Land, United Kingdom

1939
Paul Ritscher,
Neuschwabenland
Germany

1959
Stipulation of the
Antarctic Treaty

Paradoxically, for a continent whose sovereignty claims were suspended in time by an international treaty and which has no indigenous peoples, the "tattered flag – like the incomplete map and the battered bodies and minds of the early heroes such as Byrd and Scott"[1] indeed embodies the Antarctic continent to this day, as Elena Glasberg argues. Proposals to absorb all national flags into a black canvas or to design an Antarctic flag that advocates collaborative transnational governance remain utopic, while traces of anthropic contamination at the biological and building scale leave tangible evidence of human presence on the continent.

"In 1983 a staff doctor burnt the Argentinean Almirante Brown station to the ground when the setting sun announced the onset of winter. His act has been interpreted as such: 'The thought of spending the next seventy days without light was unbearable and so he decided to evacuate himself and other residents'. While it is tempting to view his action as a literal manifestation of lunacy, perhaps it should be viewed differently: as the most categorical form of architectural critique. There was, after all, light in the station – so one imagines that its built environment was missing other important qualities."

Nadim Samman (Curator, Art Historian), "Other Antarctica", *Antarctic Pavilion* (online)

ARCHITECTURAL FLAG

The rhetoric of the flag and the ritual of hoisting national flags to demarcate claims of sovereignty perpetuated religiously by all Heroic Age explorers were taken to another level on the southernmost continent with the amplification of national symbolism at architectural scale. With the deployment of architecture as a political device, embassy-like stations display national flags on their façades, supposedly to facilitate their visibility on the white reflective landscape. This practice, explicit mainly at the Argentine and Chilean bases from the 1950s and 1960s that punctuate the Peninsula, also resonates in contemporary stations that introduce the colours of their respective flags in the cladding of their façades to reaffirm the national identity of the commissioning national programme and make less explicit, yet recognisable, statements on the intended permanence of the respective nation in Antarctica.

LUCY + JORGE ORTA are artists. They operate a collaborative visual arts practice and employ a diversity of media including painting, sculpture, film and performance to realise bodies of work that address key social and ecological challenges. Aided by the logistical crew and scientists stationed at Marambio Base in the Antarctic, the pair realised the ephemeral art piece *Antarctic Village No Borders* and raised the *Antarctic Flag*.

SANTIAGO SIERRA is an artist currently based in Madrid (Spain). After graduating in fine arts at the Complutense University (Spain), he completed his artistic training in Hamburg (Germany). Sierra's career unfolded primarily in Mexico and Italy.

FLAG FOR ANTARCTICA
Lucy + Jorge Orta

Antarctica embodies Utopia. The extreme climate imposes mutual aid and solidarity, freedom of research, sharing and collaboration for the well-being of Planet Earth. The continent is a metaphor, embodied by an immaculate whiteness that contains all the wishes of humanity, spreading a message of hope for future generations.

To this end, the Antarctica Flag project, developed by artists Lucy and Jorge Orta since 1995,[2] is a supranational emblem of human rights. The flag represents a kaleidoscope of different nations. As if through a prism, it focuses all the national colours into the sum of light, echoing the snowy whiteness of purity. Through the flag, all identities coexist, side by side, hand in hand. The boundaries of each nation blend, symbolising that they belong to a larger common identity.

Alongside the Flag, *Antarctica World Passports* have been offered since 2008 to all those who pledge to become a member of the Antarctic world community, either virtually or at live events programmed in cultural institutions and international NGO forums. The travelling *Antarctica World Passport* project provides a platform for discussing issues relating to the environment, politics, autonomy, habitat, mobility and relationships among people.

THE BLACK FLAG
Santiago Sierra

On the 14th of April 2015, Santiago Sierra hoisted the *Black Flag* at the geographic North Pole, at a latitude of 90° N. Eight months later, on the 14th of December 2015, exactly 104 years after Roald Amundsen's successful Norwegian expedition to the South Pole, Sierra planted a second *Black Flag* at the geographic South Pole, at latitude 90° S.

Standing in stark contrast to the colourful flags of all nations and those of all Antarctic Treaty member states symbolically located at the South Pole, quoting Howard Ehrlich: "The black flag is the negation of all flags. It is a negation of nationhood which puts the human race against itself and denies the unity of all humankind. Black is a mood of anger and outrage at all the hideous crimes against humanity perpetrated in the name of allegiance to one state or another. It is anger and outrage at the insult to human intelligence implied in the pretences, hypocrisies, and cheap chicaneries of governments [...] But black is also beautiful. It is a colour of determination, of resolve, of strength, a colour by which all others are clarified and defined."

ANNE NOBLE is an artist, a professor of fine arts at Massey University (New Zealand), and an NZ Arts laureate. She has visited Antarctica as an NZ Arts fellow and a US National Science Foundation (NSF) fellow. Her work examines the imagination and representation of Antarctica. She has exhibited and published widely on the photographic construction of the Antarctic imaginary. Her books include *Ice Blink: An Antarctic Imaginary* (Clouds, 2011), *These Rough Notes* (VUW, 2012), and *The Last Road* (Clouds, 2013).

PISS POLES
Anne Noble

Flags on bamboo poles are ubiquitous Antarctic markers of the Heroic Age journey. Usually coloured red, blue, green or black, they point the way south, mark the presence of a hazard, a buried store of food, a fuel line or (in the case of yellow flags) an approved site for pissing outdoors. Piss poles must be hard to use. In Antarctica, women are issued a piece of equipment called a feminine urinary device (FUD). I used to imagine mustering the courage to point one at a piss pole. From a distance, I would watch men peeing at these poles and every now and again would register the sight of an odd body shake and wonder if indeed it was a woman peeing through her FUD.[1]

"The Amundsen-Scott research station [was built] where all the sectorial claims of the claimant countries meet – thus symbolically setting the US above all of them. In this sense the pole station is an unmistakeable concretization of US policy not to recognize any existing territorial claims, and to reserve the right to lay its own claims in the future if and whenever this might prove advantageous. […] [This is an] example of science in action as symbolic capital in the international political arena."

Aant Elzinga (Emeritus Professor, University of Gothenburg), "The Continent for Science" in *Handbook on the Politics of Antarctica*, 2017

"At the counter I browsed the Antarctica pins that I never buy, and scrutinized one that depicted the Antarctic continent flanked by American flags. The clerk told me that last summer the NSF, which usually has little to do with the running of the store, instructed her to remove the 'Made in Taiwan' sticker from the back of the pins before displaying them."

Nicholas Johnson (Writer), *Big Dead Place: Inside the Strange and Menacing World of Antarctica,* 2005

Naming the Antarctic

Carlo Baroni, Adrian Fox, Ursula Harris, and Jean-Yves Pirlot

CARLO BARONI is a delegate to the Standing Committee on Antarctic Geographic Information of SCAR. ADRIAN FOX is the head of the Mapping and Geographic Information Centre at the British Antarctic Survey (United Kingdom). URSULA HARRIS is a mapping and spatial data manager at the Australian Antarctic Data Centre. JEAN-YVES PIRLOT is a former deputy director general of the Belgian Mapping Agency for the National Geographic Institute (Belgium).

Physically isolated by the vast Southern Ocean, extensive sea ice and frequent fog, and completely without an aboriginal population, Antarctica was the last continent to be discovered and explored. The Antarctic continent was first sighted in 1820, but a *Terra Australis Incognita* or *Nondum Cognita* was hypothesised long before its discovery. Deriving its name from the Greek *Antarktikos,* the name Antarctica literally means "opposite the Arctic (*Arktos*)".

According to Aristotle, the lands of the northern hemisphere, under the constellation of Arktos, the Bear, had to be counterbalanced, for symmetry, by a southern continent; thus the idea, or noumenon, of the Antarctic continent received a name well before its discovery.

Areas with a long human presence usually have a panoply of place names, and every feature that needs a name has a name. The toponyms can be rich in meaning, reflecting the palimpsest of recent and historical events and past environments in the landscape.

Contrastingly, in Antarctica, explorers and cartographers were presented with a blank canvas, with the consequence that place names in the continent are *de facto* a shorthand for the history of exploration over only the last two centuries.

During the late nineteenth and early twentieth centuries, successive national Antarctic expeditions, supplemented by the discoveries of sealers and whalers, slowly assembled the jigsaw puzzle of the geography of Antarctica. The initial blank canvas progressively evolved into detailed nautical charts and topographic maps. This gradual process can be traced in the clusters of related place names that appear on maps today.

Exploration of the continent started in the Antarctic Peninsula region. Early explorers (such as William Smith and Edward Bransfield aboard the Royal Navy ship *Williams,* and Thaddeus von Bellingshausen aboard the Russian vessels *Vostok* and *Mirny,* first sighted and explored the South Shetland Islands and northern Antarctic Peninsula in 1819 and 1820–1821, respectively. They gave descriptive names to about 20 major features, including King George Island (after George III); Peter I Island (Tsar Peter the Great); Alexander Coast, now Island (Tsar Alexander I); Smith Island (after William Smith); Cornwallis and Clarence Islands (Royal Navy admirals); and Tower and Desolation Islands. These categories of place names denote important figures and sponsors, expedition members, and descriptive features, setting the pattern for place-naming in Antarctica.

First explored in the 1820s, the northern Antarctic Peninsula and the Weddell Sea sector present names of sealing ships (along with names of their owners and captains), and evocative names capturing the dangers of navigating uncharted stormy waters in small wooden ships: Neck or Nothing Passage is a perilous channel used by desperate sealers to escape storm winds; Neptune's Bellows captures unpredictable gusts at the narrow entrance to Deception Island. The hostile and impenetrable environment of the ice became emblematic of the sense of fear, mystery and perdition, outside time and space, representing the allegorical limit of rational knowledge.

Antarctica's exploration history is clearly reflected in regional differences in the source of geographical names, and in the overlapping toponyms attributed by different expeditions to the same feature. Since toponymy has long been seen as a way to "fly the flag" and leave a lasting imprint of national interests in an area, there are many place names from national expeditions in the first half of the nineteenth century, among them from the expeditions led by Thaddeus von Bellingshausen (Russia), 1819–1821; Jules-Sébastien Dumont d'Urville (France), 1838–1840; Charles Wilkes (USA), 1838–1842; and James Ross (Britain), 1839–1843.

"Captain Nemo had brought a spyglass with a reticular eyepiece […] and he used it to observe the orb sinking little by little along a very extended diagonal that reached below the horizon. […] If the lower half of the sun's disk disappeared just as the chromometer said noon, we were right at the pole. 'Noon!' I called. 'The South Pole!' Captain Nemo replied in a solemn voice, handing me the spyglass, which showed the orb of day cut into two exactly equal parts by the horizon. […] Just then, resting his hand on my shoulder, Captain Nemo said to me: […] 'Well now! In 1868, on this 21st day of March, I myself, Captain Nemo, have reached the South Pole at 90°, and I hereby claim this entire part of the globe, equal to one-sixth of the known continents.' 'In the name of which sovereign, Captain?' 'In my own name, sir!' So saying, Captain Nemo unfurled a black flag bearing a gold 'N' on its quartered bunting. Then, turning toward the orb of day, whose last rays were licking at the sea's horizon: 'Farewell, O sun!' he called. 'Disappear, O radiant orb! Retire beneath this open sea, and let six months of night spread their shadows over my new domains!'"

Jules Verne (Writer, Poet and Playwright), *Twenty Thousand Leagues Under the Sea,* 1870

The alluvial diagram visualises the Scientific Committee on Antarctic Research Composite Gazetteer of Antarctica (SCAR CGA) identifying which type of nomenclature was given to topographic features and anthropic settlements.

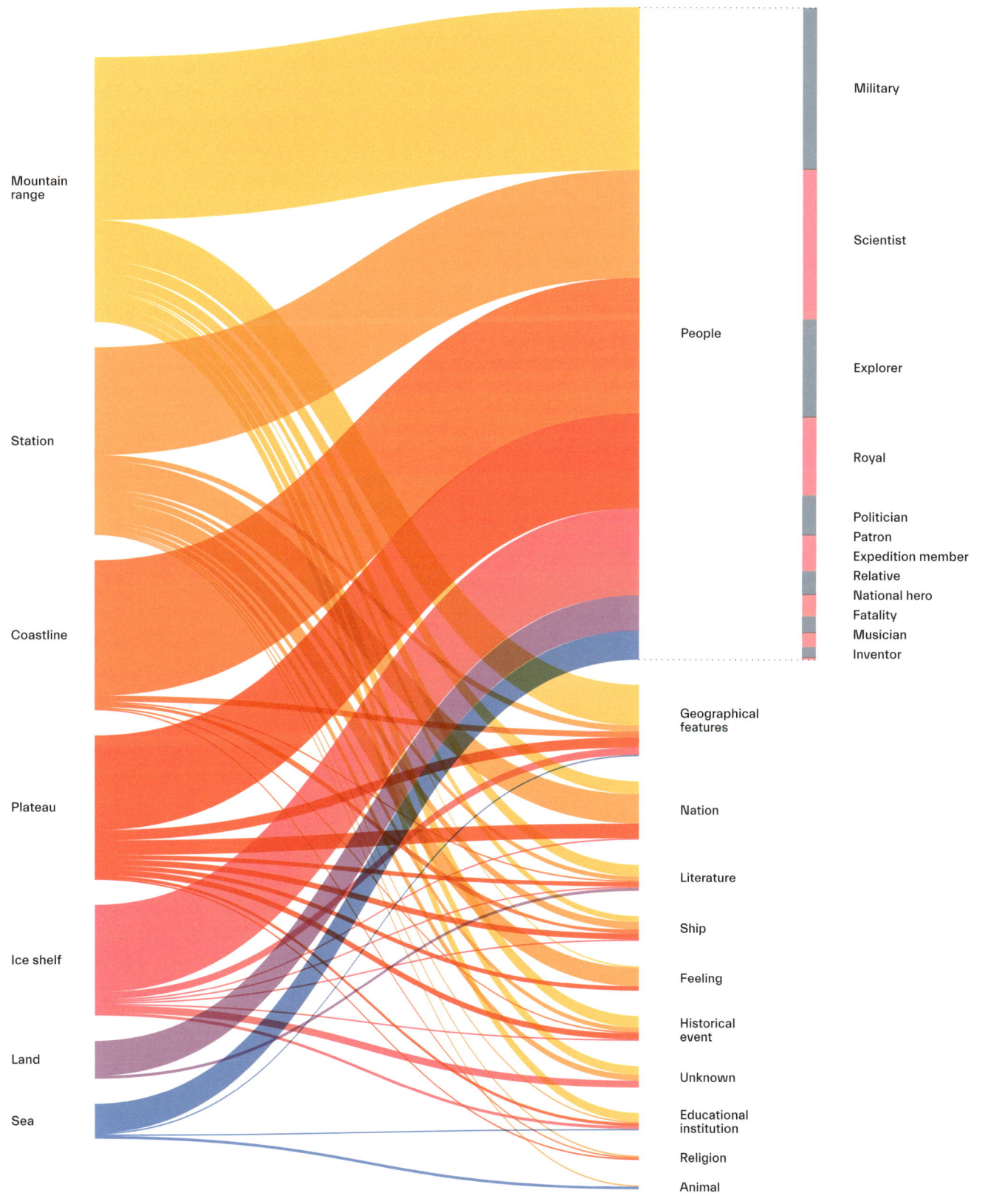

From the early twentieth century until the Second World War, expeditions began to explore the interior: Robert Scott (Britain), Roald Amundsen (Norway), Ernest Shackleton (Britain), Adrien de Gerlache (Belgium), Otto von Nordenskjöld (Sweden) and Douglas Mawson (Australia) by land; Hubert Wilkins, Richard Byrd and Lincoln Ellsworth (United States) by air in the late 1920s and 1930s. Echoes of difficulties endured; risks taken and the dramatic losses suffered during this Heroic Age remain indelible in such place names as Hope Bay (Nordenskjöld), Inexpressible Island (Scott's Northern Party, *Terra Nova* expedition), and Memorial Cross on Observation Hill at Ross Island – named to commemorate the death of Captain Scott and his companions.

Programmes of systematic surveying and mapping, and scientific research, became central after the Second World War. Excluding the first Antarctic flights between the late 1920s and 1930s, previous expeditions had been exclusively seaborne and had mapped only the coastal regions and inland features observable from the sea. The era of regional aerial photography, starting in the late 1940s, and overland programmes of travel for surveying and science systematically explored the largely unknown interior of the Antarctic continent. This led to a surge in place-naming; whole mountain ranges, prominent peaks, large glaciers, ice shelves and other topographic features were seen for the first time.

The International Geophysical Year (1957–1958) represents the divide between the old world of Antarctic exploration and the modern approach which considers Antarctica as a land protected for peace and science, the heritage of humanity, and a natural reserve, as the Antarctic Treaty established in 1959.

The increasing number of national programmes operating in Antarctica translated into a multiplication of national gazetteers, leading to overlapping naming, mainly from a simple lack of coordination, but in part consciously promoted to guarantee national interests.

This resulted in a naming redundancy that to date counts about 38,000 gazetted names for only 19,900 features. This phenomenon is especially acute in the Antarctic Peninsula and in the South Shetland Islands. To establish a means of accurately identifying existing names and preventing the multiple naming or duplication of names on maps, scientific publications, and databases, the Scientific Committee on Antarctic Research Composite Gazetteer of Antarctica (SCAR CGA) was founded as a compilation of place names extracted from individual national gazetteers.

Including topographic, under-ice and near-shore features south of 60° S, the Gazetteer assigns a unique identifier and lists all applicable names for each feature. Since 2008, Italy and Australia have jointly managed the Gazetteer, with Italy responsible for the editing and Australia for maintaining the database and the associated website. The Scientific Committee on Antarctic Research (SCAR) Standing Committee on Antarctic Geographic Information coordinates the project and seeks to develop "international principles and procedures for Antarctic place names".

Today, Antarctica's pivotal role in the Earth's system is well-recognised. International scientific research is both venturing into relatively unknown areas to answer emerging scientific questions, for example the International Thwaites Glacier Collaboration 2018–2023, and working in more detail elsewhere, for example at Little Dome C with the Beyond Epica Oldest Ice project. It is thus of paramount importance to both scientific research and the supporting operational infrastructure to refer unequivocally to places using names that offer clarity and precision, at once resolving conflicting nomenclature and allowing for the everlasting need to name the unnamed as the resolution of the research deepens and more features require a name.

The 38,000 gazetted names for only 19,900 features included in the Scientific Committee on Antarctic Research Composite Gazetteer of Antarctica (SCAR CGA) are overlaid below as a manifesto to the cacophony of geopolitical ambitions at play on the continent.

"The South Polar regions, in respect to their nomenclature, are different from every other part of the world. As there are no inhabitants, there are no native names. All the names applied to all places beyond 60° S have been given either by explorers or by stay-at-home geographers. […] A philosophic nomenclature of the Antarctic regions must, in the nature of things, be based on geographic and historic facts. Certain names may well be descriptive and others should be commemorative. This is tantamount to saying that Antarctic nomenclature must be founded on truth and justice, since geography and history are simply records of things as they are or of events which have happened. A proper Antarctic nomenclature must have its source in loyalty to humanity and to science, and not spring from servile obedience to national prejudices and national greed. The golden rule of doing unto others as you would like done unto yourself should ever be borne in mind, and international fair play should insist upon justice being awarded to all."

Edwin Swift Balch (Lawyer, Artist, Mountaineer), *Bulletin of the American Geographical Society*, 1912

Data from the Scientific Committee on Antarctic Research Composite Gazetteer of Antarctica (SCAR CGA) is mapped below to highlight the sheer volume of names attributed to Antarctic topographic features by country, offering insight into the relation between the naming strategy and sovereignty claims.

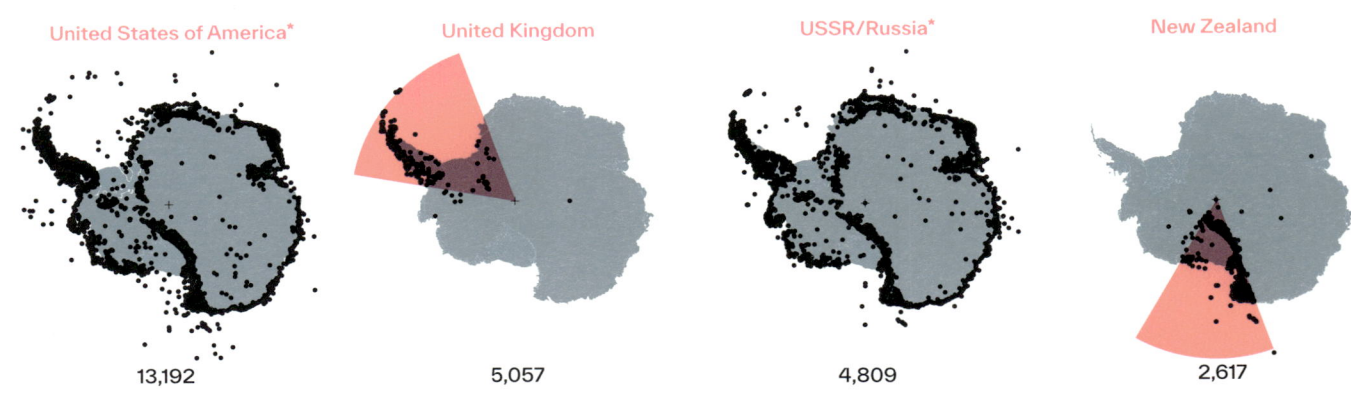

United States of America*	United Kingdom	USSR/Russia*	New Zealand
13,192	5,057	4,809	2,617

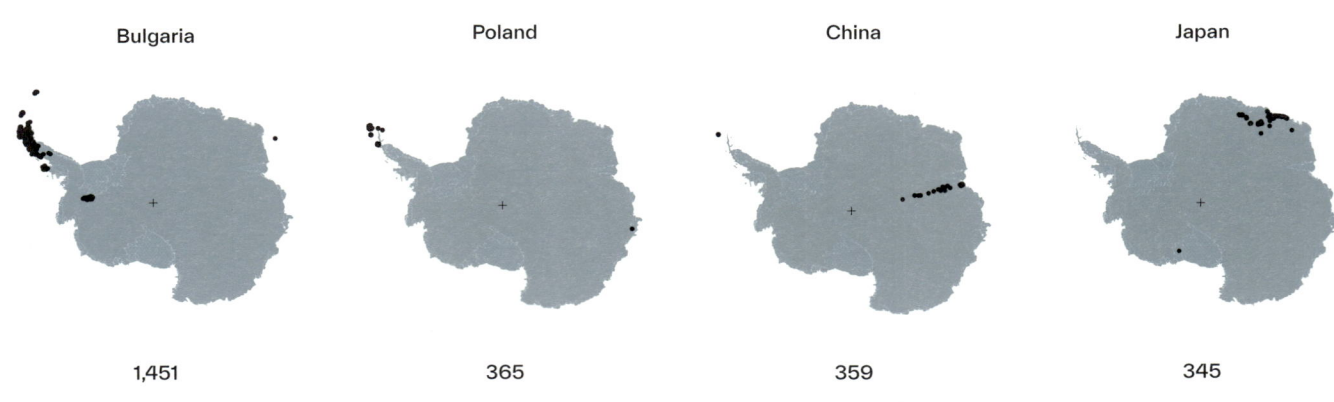

Bulgaria	Poland	China	Japan
1,451	365	359	345

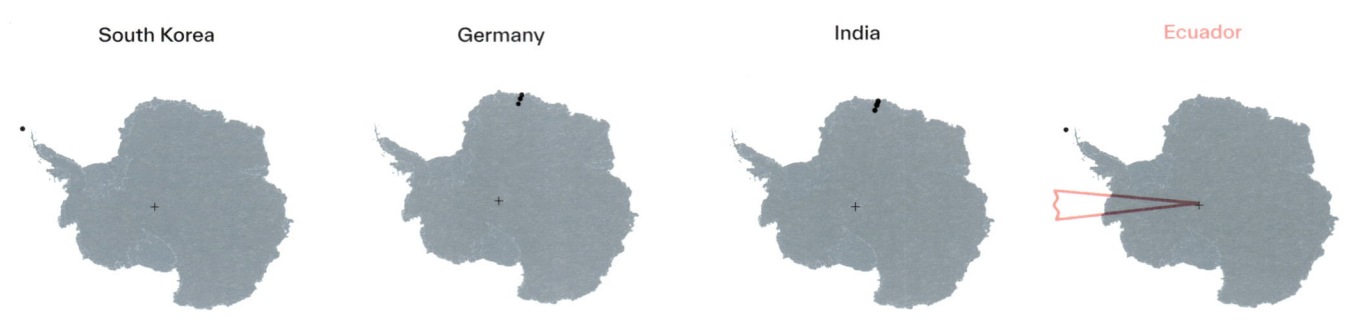

South Korea	Germany	India	Ecuador
27	21	21	9

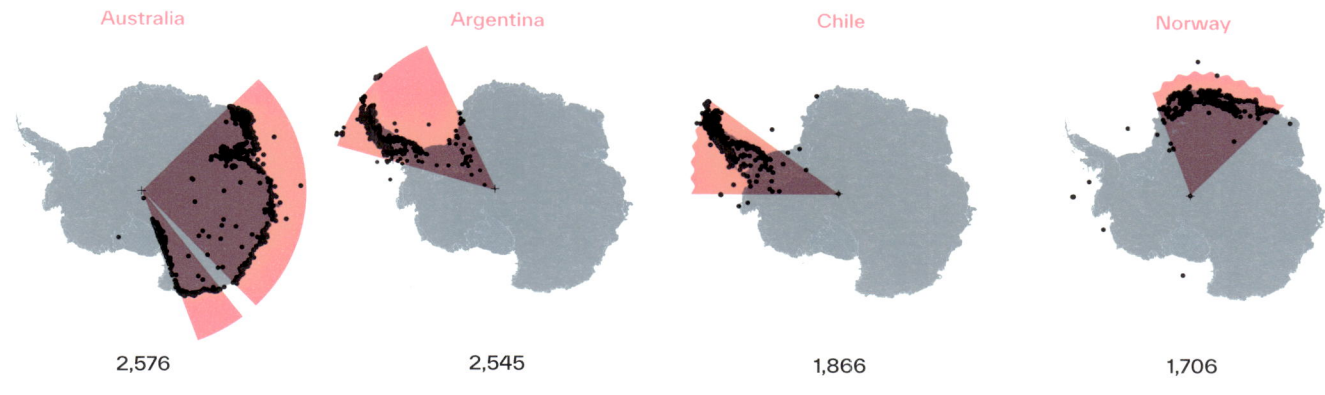

Australia

2,576

Argentina

2,545

Chile

1,866

Norway

1,706

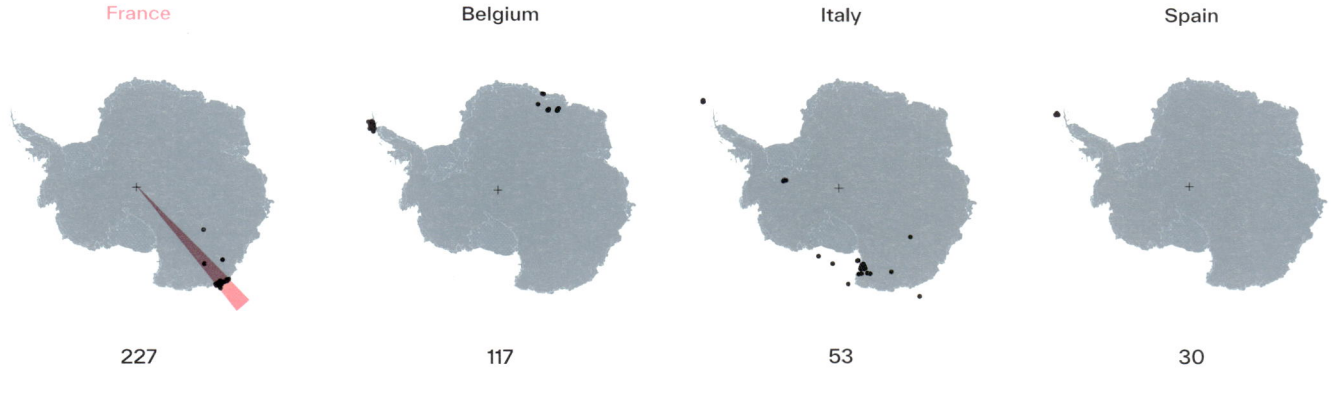

France

227

Belgium

117

Italy

53

Spain

30

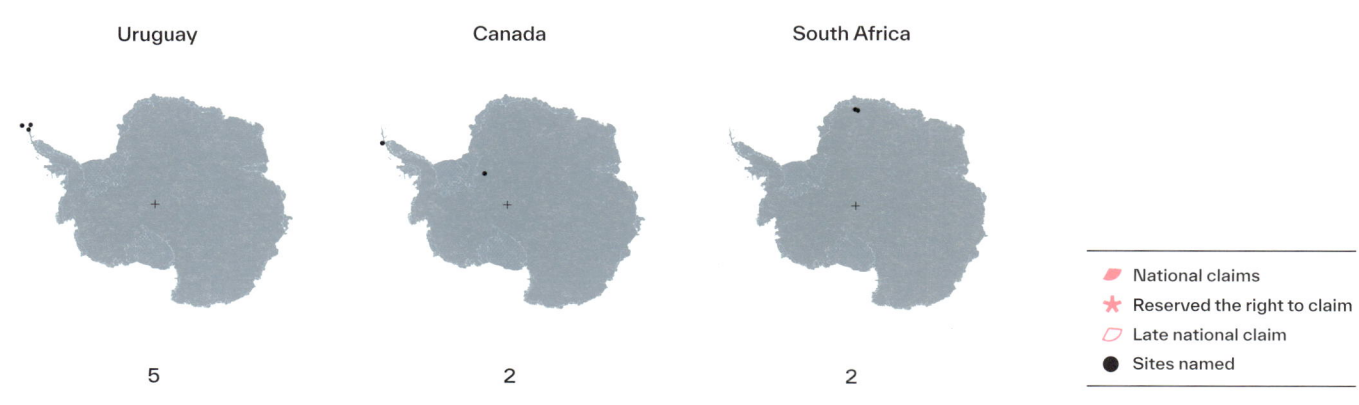

Uruguay

5

Canada

2

South Africa

2

National claims

Reserved the right to claim

Late national claim

Sites named

Political Conflict on the Antarctic Peninsula

Adrian Howkins

ADRIAN HOWKINS is co-principal investigator at the United States' National Science Foundation (NSF)-funded McMurdo Dry Valleys Long-Term Ecological Research project in Antarctica, and a reader of environmental history at the University of Bristol (United Kingdom). Focusing on the environmental history of the polar regions, Dr Howkins' publications include *Frozen Empires: An Environmental History of the Antarctic Peninsula Region* (Oxford University Press, 2017).

Tourists visiting the Antarctic Peninsula frequently find themselves mildly baffled by the overt displays of nationalism and sovereignty claims they encounter. There are various reasons why tourists come to this remote part of the world – to experience wilderness and isolation, to photograph the wildlife, to visit their "final" continent – but the search for political tension is rarely one of them. It can come as something of a surprise, therefore, when they find that they've stepped into the middle of an active and occasionally tetchy three-way sovereignty dispute among Argentina, Britain and Chile. Large national flags proudly fly from ships and scientific stations, Argentine and Chilean military personnel encountered on visits ashore smilingly proclaim ownership for their respective country, and tourists are tacitly encouraged to acknowledge British sovereignty by sending postcards using the Royal Mail Post Office at the Port Lockroy ↗ p. 644 Museum. "What is going on?" they ask. "Why are these three countries contesting the ownership of this desolate, ice-covered landscape?"

The sovereignty dispute in the Antarctic Peninsula offers one of the most obvious examples of Antarctica's unusual, and still largely unresolved, political and legal status.[1] It offers a case study of the ongoing conflict between imperialism and nationalism in Antarctica, which might be labelled an environmental history of decolonisation.[2] In the first half of the twentieth century, Argentina, Britain and Chile each made claims to sovereignty over the Antarctic Peninsula region, arguing that these claims were simply a formalisation of ownership that was already in place.[3] In making its case for sovereignty, Britain drew upon assertions of environmental authority in presenting the Antarctic Peninsula as a useful but fragile environment in need of scientifically informed stewardship. In contrast, at least initially, Argentina and Chile rejected this scientific vision and promoted instead a form of environmental nationalism which had at its heart a more visceral sense of connection based on commonalities such as shared geology, shared weather and shared history. On several occasions, the conflict threatened to turn violent, as happened most dramatically on the 1st of February 1952, when Argentine naval personnel fired machine guns over the heads of British scientists during a confrontation at Hope Bay.[4]

At the centre of this conflict was Juan Domingo Perón, Argentina's president from 1946 to 1955.[5] Perón wanted an end to Britain's "informal empire" of trade and influence in Argentina; he used British claims to the Islas Malvinas or Falkland Islands and the Antarctic Peninsula region as a way of demonstrating Britain's imperial intentions. While the British population living on the Islas Malvinas or Falkland Islands limited the scope of what Argentina could do there without provoking an armed confrontation, the unpopulated Antarctic Peninsula region offered an ideal stage for Perón to perform Argentine nationalism and seek to realise his "Antarctic dream" of Argentine sovereignty over Antarctica. While Perón was happy to draw upon direct connections to the Antarctic environment and make a case for Argentine sovereignty based on environmental nationalism, he also believed that Argentina could defeat Britain at its own imperial game of using science as a justification for ownership of the Antarctic Peninsula region. Such an approach highlights the rhetorical power of science within Antarctic politics, while the contradictions between environmental authority and environmental nationalism remain largely unresolved.[6]

The Cold War between the United States and the Soviet Union further complicated the contested political situation in the Antarctic Peninsula region. Neither superpower wished to give the other any advantage in Antarctica, and without making specific claims of their own, both reserved their right to the whole continent.[7] As the most accessible part of Antarctica, the Peninsula region

Chilean President Gabriel González Videla was the first head of state to visit Antarctica, in February 1948. In his speech inaugurating O'Higgins Station, he stated that Chile had "to defend the sovereignty and unity of [the Chilean] territory, from Arica to the South Pole".[1] González Videla expanded on this a few weeks later by adding that they "would deny [their] glorious history [...] [and] their past if [they] were to renounce a single piece of [their] territory, only because there are those [the British] who believe that acts of imperialism [...] constitute a title of sovereignty."[2] Informed by the 1908 Chilean theories of "Antarctandes"[3] (i.e. an understanding that Chile and Antarctica are intrinsically one territory, as the "geological extension of the Andes mountains underneath the Drake Passage [reaches] Antarctica")[4] and later "geographical imperatives" of "the new science of geopolitics",[5] González Videla insisted on the right of Chile to the territory of the Antarctic Peninsula. The latter, also known as "Antartida Chilena", is represented in the 1945 cartography shown at the side.

had clear geopolitical value, and the United States in particular was quite active in the Peninsula region during the most active period of the dispute in the 1930s, 1940s and 1950s. The American activity witnessed the further militarisation of Antarctica, as occurred most dramatically with the 4,700 military personnel and thirteen naval ships that sailed to the continent during the 1946–47 season. While the American presence threatened to add another layer of instability to the conflict, the broader geopolitics of the Cold War in fact had an overall restraining influence: as allies of the United States against the Soviet Union, Argentina, Britain and Chile were unwilling to upset this alliance through direct military confrontation.[8]

        In 1957 and 1958, twelve nations were involved in scientific research in Antarctica as part of the global enterprise known as the International Geophysical Year (IGY).[9] These countries included Argentina, Britain and Chile as well as the United States and the Soviet Union. Shortly after the end of the IGY, the signing of the Antarctic Treaty in 1959 by all twelve countries involved in scientific research on the continent brought to an end the most active period of the sovereignty dispute.

        The role played by the IGY in creating a "continent dedicated to peace and science" through the Antarctic Treaty continues to be debated.[10] Rather than as the simple idealism of science triumphing over political discord, the close connections between science and politics might better be viewed as a continu-

Allí termina todo
y no termina:
allí comienza todo:
se despiden los ríos en el hielo,
el aire se ha casado con la nieve,
no hay calles ni caballos
y el único edificio
lo construyó la piedra.
Nadie habita el castillo
ni las almas perdidas
que frío y viento frío
amedrentaron:
es sola allí la soledad del mundo,
y por eso la piedra
se hizo música,
elevó sus delgadas estaturas,
se levantó para gritar o cantar,
pero se quedó muda.
Sólo el viento,
el látigo
del Polo Sur que silba,
sólo el vacío blanco
y un sonido de pájaro de lluvia
sobre el castillo de la soledad.

Pablo Neruda (Poet, Diplomat, Nobel Prize Laureate for Literature), *Piedras Antárticas*, 1938

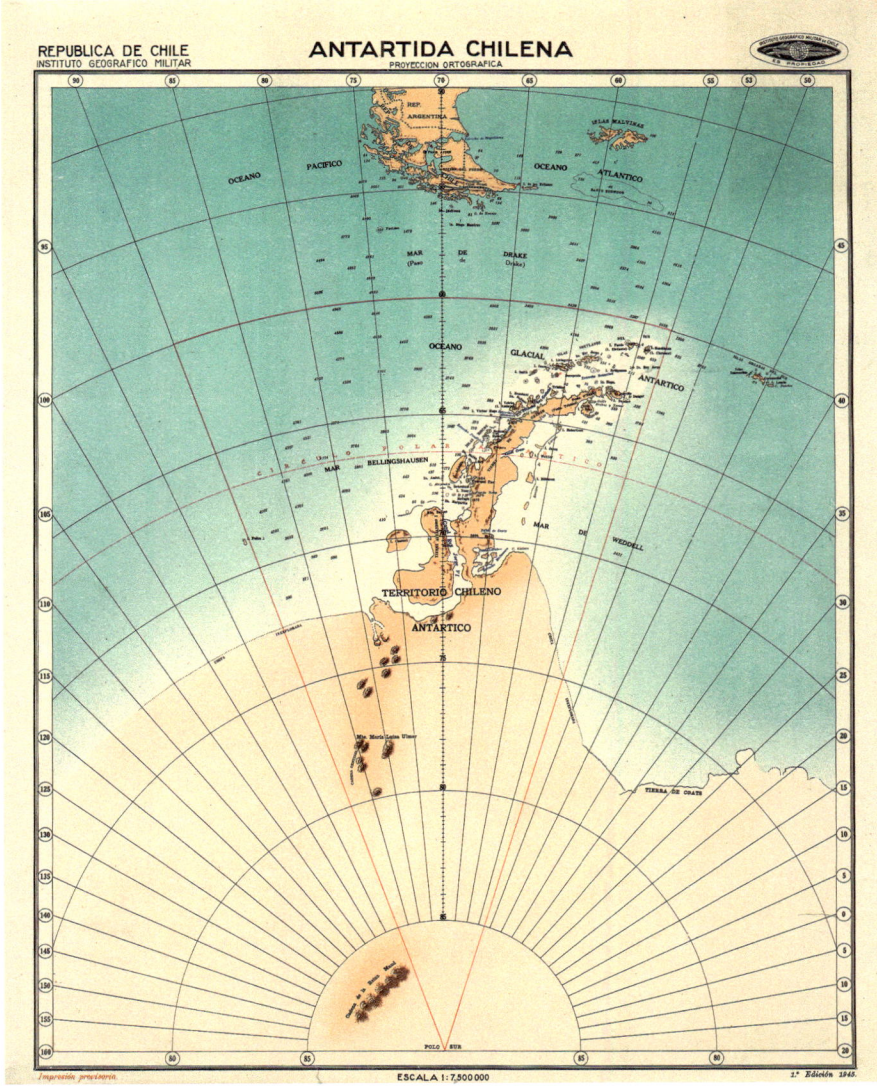

ation of the British imperial policy of using science to promote sovereignty claims. By suspending all sovereignty claims and reservations of rights rather than recognising them or rejecting them, the Antarctic Treaty left the underlying question of ownership deliberately unresolved.

Today it is tempting to interpret the "banal nationalism" of the ongoing territorial dispute in the Antarctic Peninsula in stark contrast to the enlightened science taking place in the research stations. At a time when the Antarctic Peninsula is on the global front line of climate heating and glacier melting, and scientists working in the region are rightly seen as working "for the good of humanity", this sense of contrast is perhaps stronger than ever.[11] But even a short examination of the contested political history of the Antarctic Peninsula region suggests that this is a flawed distinction. Antarctic science itself is highly politicised. In much the same way that British imperial officials used science to support their claims to the Falkland Islands Dependencies, the consultative members of the Antarctic Treaty might be seen as collectively using science to justify the existence of this international system. The close connection between science and politics in Antarctica is highlighted by Article IX of the Antarctic Treaty, which states that a country must be conducting "substantial scientific research" in Antarctica before it can become a consultative member of the Antarctic Treaty. Instead of bringing about a genuine decolonisation of Antarctica, the history of the sovereignty dispute in the Antarctic Peninsula helps to demonstrate that the 1959 Antarctic Treaty represented a continuation of imperial assertions of environmental authority and the creation of a new form of "frozen empire".[12]

The last three words of *Sur,* Ursula K. Le Guin's 1981 fictional feminist critique of Euro-American exploration narratives, expand upon the metaphor of the footprint as the quintessential symbol of imperial Eurocentric male heroic history. In consciously leaving no trace of their *Yelcho* expedition (neither physically on the snow, nor verbally), Le Guin reinstates in her postcolonial analysis the moral superiority of the supposedly "subaltern speakers" represented by Latin American women.

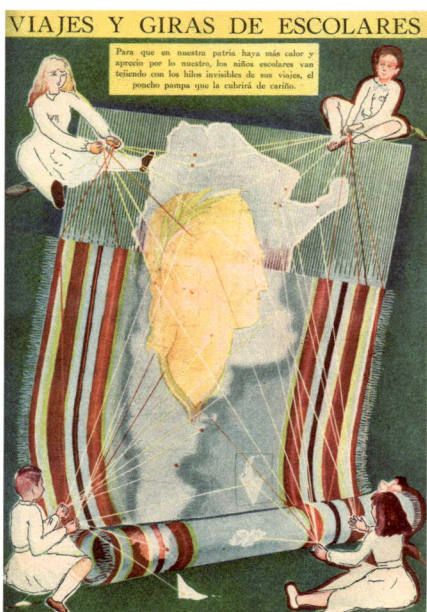

The Argentine ambition of Antarctic sovereignty – sustained by arguments of geological continuity and proximity, and reinforced by tactical civilian occupation in permanent settlements – was stated openly in the Peronist manifesto *La Nación Argentina: Justa, Libre, Soberana,* published in 1950 by the State Control Office of the Presidency of the Nation. In the publication, Antarctica is included proudly in all national cartography, emphasising the role that so-called Antártida Argentina had (and arguably still has) on Argentine culture. This is exemplified by the two images extracted from the pioneering propaganda publication, shown above, in which Antarctica

features in school educational weaving looms as part of the national territory, and is depicted even while addressing industrial productivity, despite the evident absence of any plants on the continent itself.[6]

"On the twenty-second of December, 1909, we reached the South Pole. [...] We discussed leaving some kind of mark or monument, a snow cairn, a tent pole and flag; but there seemed no particular reason to do so. Anything we could do, anything we were, was insignificant, in that awful place. [...] We are old women now, with old husbands, and grown children, and grandchildren who might some day like to read about the Expedition. Even if they are rather ashamed of having such a crazy grandmother, they may enjoy sharing in the secret. But they must not let Mr. Amundsen know! He would be terribly embarrassed and disappointed. There is now need for him or anyone else outside the family to know. We left no footprints."

Ursula Kroeber Le Guin (Writer), *Sur: A Summary report of the Yelcho Expedition to the Antarctic 1909–1910,* 1982

38

SUR

A SUMMARY REPORT OF THE YELCHO EXPEDITION TO THE ANTARCTIC, 1909-10

EL MAPA EN EL DESVÁN

ALTHOUGH I have no intention of publishing this report, I think it would be nice if a grandchild of mine, or somebody's grandchild, happened to find it some day; so I shall keep it in the leather trunk in the attic, along with Rosita's christening dress and Juanito's silver rattle and my wedding shoes and finneskos.

The first requisite for mounting an expedition—money—is normally the hardest to come by. I grieve that even in a report destined for a trunk in the attic of a house in a very quiet suburb of Lima I dare not write the name of the generous benefactor, the great soul without whose unstinting liberality the Yelcho Expedition would never have been more than the idlest excursion into daydream. That our equipment was the best and most modern— that our provisions were plentiful and fine—that a ship of the Chilean government, with her brave officers and gallant crew, was twice sent halfway round the world for our convenience: all this is due to that benefactor whose name, alas!, I must not say, but whose happiest debtor I shall be till death.

When I was little more than a child, my imagination was caught by a newspaper account of the voyage of the Belgica, which, sailing south from Tierra del Fuego, was beset by ice in the Bellingshausen Sea and drifted a whole year with the floe, the men aboard her suffering a great deal from want of food and from the terror of the unending winter darkness. I read and reread that account, and later followed with excitement the reports of the rescue of Dr. Nordenskjöld from the South Shetland Islands by the dashing Captain Irizar of the Uruguay, and the adventures of the Scotia in the Weddell Sea. But all these exploits were to me but forerunners of the British National Antarctic Expedition of 1901-04, in the Discovery, and the wonderful account of that expedition by Captain Scott. This book, which I ordered from London and reread a thousand times, filled me with longing to see with my own eyes that strange continent, last Thule of the South, which lies on our maps and globes like a white cloud, a void, fringed here and there with scraps of coastline, dubious capes, supposititious islands, headlands that may or may not be there: Antarctica. And the desire was as pure as the polar snows: to go, to see—no more, no less. I deeply respect the scientific accomplishments of Captain Scott's expedition, and have read with passionate interest the findings of physicists, meteorologists, biologists, etc.; but having had no training in any science, nor any opportunity for such training, my ignorance obliged me to forgo any thought of adding to the body of scientific knowledge concerning Antarctica, and the same is true for all the members of my expedition. It seems a pity; but there was nothing we could do about it. Our goal was limited to observation and exploration. We hoped to go a little farther, perhaps, and see a little more; if not, simply to go and to see. A simple ambition, I think, and essentially a modest one.

Yet it would have remained less than an ambition, no more than a longing, but for the support and encouragement of my dear cousin and friend Juana ——. (I use no surnames, lest this report fall into strangers' hands at last, and embarrassment or unpleasant notoriety thus be brought upon unsuspecting husbands, sons, etc.) I had lent Juana my copy of "The Voyage of the 'Discovery,'" and it was she who, as we strolled beneath our parasols across the Plaza de Armas after Mass one Sunday in 1908, said, "Well, if Captain Scott can do it, why can't we?"

It was Juana who proposed that we write Carlota —— in Valparaíso. Through Carlota we met our benefactor, and so obtained our money, our ship, and even the plausible pretext of going on retreat in a Bolivian convent, which some of us were forced to employ (while the rest of us said we were going to Paris for the winter season). And it was my Juana who in the darkest moments remained resolute,

How to Claim a Continent.
Operation Tabarin and the Post Office

Camilla Nichol

CAMILLA NICHOL is the chief executive of the United Kingdom Antarctic Heritage Trust, a British charity in charge of the conservation of six heritage sites on the Antarctic Peninsula and promoting Antarctic heritage in the United Kingdom. Nichol holds an AMA from the Museums Association and is a fellow of the Royal Geographical Society, a trustee of the Burton Constable Foundation, and chair of the Cromwell Museum Trust (United Kingdom).

Ever since Antarctica was first sighted, there has been a political component to Antarctic exploration. In 1908, the first sovereign claim was made by the United Kingdom, in order to control whaling revenue and taxation. Other nations followed suit, but later claims were made as political manoeuvres to assert dominance in the region. Several nations competed to lay territorial claims to portions of Antarctica, some of which overlap and compete, most notably on the Antarctic Peninsula. This came to a head during the Second World War, when the United Kingdom, Chile and Argentina began establishing permanent bases, stationing personnel on the continent throughout the year as an effective means of asserting sovereignty.

In 1943, the British Admiralty and Colonial Office established a secret operation to build such wintering bases. It was code-named *Operation Tabarin,* purportedly after the famous Parisian nightclub. The cover story was that it was set up to monitor German U-boat activity in the Southern Ocean, while its real purpose was to counter Argentina's encroachment of the British Antarctic Territory. In 1944, two permanent bases were established on the Antarctic Peninsula, at Port Lockroy and at Whaler's Bay on Deception Island; in 1945, a third was established at Hope Bay. These were strategic locations on the Peninsula from where they could monitor Argentine activity and be a visible British presence. The expedition was led by Lieutenant James Marr, who recruited educated men, mostly scientists, who would keep themselves occupied with scientific

work during their long season. Base A at Port Lockroy ↗ p. 644 represented the start of the British, state-funded scientific programme in Antarctica; it is the birthplace of what is now the British Antarctic Survey.

At the end of the war, *Operation Tabarin* was disbanded and turned into the Falkland Islands Dependencies Survey (FIDS). Port Lockroy continued to operate. During the fifteen years that followed, several more British bases were erected, until in 1958–1959 there were thirteen bases and three refuges, all simultaneously occupied. Each carried out intensive scientific and mapping programmes, the data from which is the foundation of modern Antarctic science.

Argentina, Chile and many other nations were also establishing bases, largely on the Peninsula given its proximity to South America, and tensions were rising. Added to this was the Cold War, which was also gently playing out on the ice of Antarctica as the Soviet Union and the United States both reserved the right to claim the continent.

In 1962, after the signing of the Antarctic Treaty, Port Lockroy finally closed, and with it the FIDS became the British Antarctic Survey. Stations continued to be established up and down the Peninsula. Today, more than thirty nations operate scientific programmes in Antarctica, with stations all over the continent. These are both creating vitally important scientific data for the world and asserting a territorial stake in this unique continent.

THE POST OFFICE AS GEOPOLITICAL DEVICE

Post Offices have played a role in Antarctic operations since the times of Scott and Shackleton – both of whom realised the promotional potential of having special stamps and post offices at their huts. But it wasn't until *Operation Tabarin* (1943–1945) that the emphasis shifted to focus on the political power of the Great British Post Office.

Stamps have been issued in the Southern Ocean by the United Kingdom since the late nineteenth century, when stamps were first issued for the Falkland Islands Dependencies, and have been avidly collected since. But it wasn't until *Operation Tabarin* that stamp issues were seen to be both a tool for sovereignty and a valuable income stream. Prior to this period, tensions had been mounting between Argentina, Chile and the United Kingdom in the race for sovereignty over the Antarctic Peninsula. Stamp issues were a convenient propaganda tool exploited by Argentina in their depictions of their Antarctic territory, including the Falkland Islands or Islas Malvinas and South Georgia – both considered British territories by the United Kingdom.

Operation Tabarin was a government-sponsored operation launched to establish permanent, wintering bases on Antarctica to counter the claims of Argentina and Chile, with permanent occupancy being the strongest means to resists others' efforts. The establishment of a post office, with the sending and receiving of mail, reinforced this sovereignty over the territory and was an essential activity. At first, the United Kingdom claimed the Antarctic Peninsula region as a Dependency of the Falkland Islands, meaning it came under the administrative and governance of the Governor of the Falkland Islands or Islas Malvinas. It made sense, then, that the postal system of the same controlling nation was extended to the enlarged territory.

Special stamps were not printed, but large volumes of stamps were overprinted with the names of the Dependencies to distinguish them by their postal location. Graham Land (the Antarctic Peninsula), the South Shetlands, the South Orkneys and South Georgia were the principal locations. Almost five million of these stamps were overprinted and put into circulation at stations all over the area. As bases were established at Port Lockroy, Hope Bay, Whalers Bay (Base B ↗ p.702), the Argentine Islands and at many other locations along the Peninsula, each

opened post offices and were issued with the Graham Land stamps. This continued until 1962, when the Falkland Islands Dependencies, including Graham Land and the Antarctic Islands, became the British Antarctic Territory under its own governance, and the Falkland Island Dependencies Survey became the British Antarctic Survey. Port Lockroy closed as scientific operations were concentrated at other bases. This meant that from 1963 onwards, new stamps for the British Antarctic Territory could be issued.

Stamps continue to be issued today for the British Antarctic Territory – around six sets per year – and are still sold to British personnel working in Antarctica and to visitors. Some 80,000 items of mail are handled each year at the post office at Port Lockroy and are sent around the world. This tradition is a much-cherished service by all those who find themselves in Antarctica, and one that is upheld for its geopolitical importance.

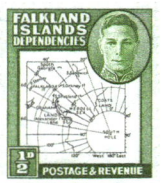

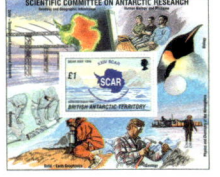

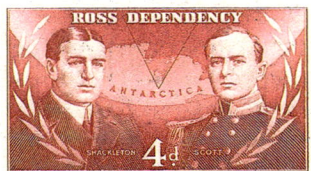

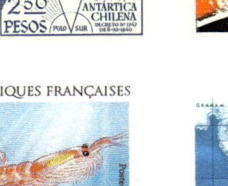

On the 15th of January 1908, the New Zealand Post Office issued the first stamps ever produced for the Antarctic. Used by members of the British Antarctic Expedition led by Ernest Shackleton, these represent the first of many stamps issued for the white continent and originally sent from makeshift post offices like the one shown on the left at Port Lockroy. The role of the post office in asserting sovereignty claims was soon recognised by the United States as well during the 1947–1948 *Ronne Antarctic Research Expedition* (the last privately sponsored American expedition and the first to include women in the overwintering party), and later attested to by Scott Smith, who personally cancelled approximately 10,000 pieces of mail during 10 years at the pole. The latter stated that "having a post office and issuing stamps for an entity [was] considered one of the important aspects in a case for a geographic claim [and] the United States made him a postmaster down there to cancel envelopes to counteract the British claim to that part of Antarctica."[1]

A Concise Military History of Antarctica

Joseph Micallef

JOSEPH MICALLEF is a historian, author, keynote speaker, syndicated columnist and war correspondent. He has been an advisor to the American government and is currently the honorary consul to Oregon from the Republic of Malta. Micallef has also appeared as a commentator on a variety of broadcast venues including CNN, Fox News and Fox News Radio. He is an opinion columnist for Military.com and a contributor to *Forbes*.

It seems a misnomer to speak of Antarctica's military history. It is the one continent on Earth on which a war has never been fought. Indeed, it is impossible to find a conflict there that rises above a minor skirmish. The 1982 Falklands War briefly touched the subantarctic island of South Georgia in the South Sandwich Islands. Likewise, the First and Second World Wars on occasion brushed the Antarctic region, though the area was of little military significance.

Nonetheless, the projection of military power to achieve broader policy goals is a theme that weaves in and out of Antarctica's history, a point underscored in the Trump administration's 2020 memorandum announcing a plan to launch a fleet of polar icebreakers to both the Arctic and the Antarctic. The competition for the region's resources (both real and theoretical) over the last two centuries has often had a military dimension, albeit covertly. Moreover, disagreements over the legal basis of Antarctic claims has meant that demonstrations of military power have often been used as a means for strengthening claims or warding off counterclaimants.

The military history of Antarctica begins with early nineteenth-century whaling, limited to an occasional showing of the flag by a frigate, typically British or American, to keep at bay foreign whaling ships or protect those of nationals. By mid-century, however, whaling and sealing activity had declined as wildlife dwindled. Industrial-scale whaling emerged as a large, profitable industry at the end of the nineteenth century.

The military role in Antarctica, however, was still *sub rosa.* Leadership roles in scientific expeditions were sometimes held by serving or reserve military officers. A country's navy might bring supplies or render assistance, but the military's involvement was minor. Projecting military power openly in Antarctica was difficult, save for the occasional visiting navy ship deployed there.

The First World War largely sidestepped the Antarctic region, although the Falkland Islands were the scene of a naval battle where a British squadron destroyed a flotilla of German heavy cruisers. Little else happened. Trade routes through the South Atlantic ran far north of 60° S with little reason for submarines or the naval forces hunting them to venture further south. Pursuant to the Versailles Treaty, Germany renounced any claim to the Antarctic continent and many other territories. After the First World War, aeroplanes began to be used for exploring and mapping the Antarctic interior, increasing the competition over land claims.

There were two additional developments over the course of the 1930s that would significantly transform the geopolitics of the Antarctic continent. In 1939, the United States government, spurred by Antarctic explorer Richard Byrd, began organising the US Antarctic Service (USAS). This organisation was tasked with establishing and maintaining three permanent US bases on the Antarctic continent.

A second development was the growing role of Nazi Germany and Japan in the region. During the 1930s, both Japan and Germany had looked to the Antarctic whaling industry as a source of animal fat for their food and munitions industry. In 1938, German Air Marshal Hermann Goering organised an expedition to Antarctica. Germany wanted to reduce its dependence on Norway for whale oil and potentially create a base for military operations in the South Atlantic.

The expedition to Queen Maud Land, a territory later renamed Neuschwabenland, was widely seen as the first step towards an eventual German Antarctic claim. During the Second World War, Kriegsmarine ships sometimes used old whaling stations to resupply German submarines in the South Atlantic. Despite persistent rumours, however, no evidence of a secret Nazi base in Antarctica was ever found.

Rear Admiral Richard Byrd (in the picture, on the left) takes a look around the site of Little America I and II, where his 1928 and 1934–1935 expeditions camped.

- ● Military
- ⊘ Scientists
- ◌ Supporting staff
- ● Tourists

Washington's growing interest in Antarctica precipitated countermoves by Chile and Argentina, triggering a rivalry between Argentina and Great Britain over the setting up and manning of bases in the region. In July 1941, Argentina announced it would deploy naval personnel to its base on Laurie Island in the South Orkneys. Further missions followed, between 1941 and 1943, to various subantarctic islands and the Antarctic Peninsula. In each instance, Argentine military personnel raised the Argentine flag and deposited chests containing the Argentine Act of Possession.

In response, the Admiralty dispatched an armed merchant ship, the HMS *Carnarvon Castle,* to Deception Island to raise the British flag. The War Cabinet also authorised *Operation Tabarin,* designed to strengthen British claims in the region.

In 1946, with the end of the Second World War in sight and spurred on by Rear Admiral Byrd's proposals, the United States Navy launched *Operation Highjump.*

Commencing on the 26th of August 1946, *Operation Highjump* was the first major expedition to Antarctica that didn't disguise its primary mission as scientific research. Its goal was to "train members of the Navy and to test ships, aircraft and other military equipment under frigid conditions."[1] It also surveyed the Antarctic interior to create more accurate maps and identified American claims by dropping markers displaying the American flag and containing declarations of sovereignty at the limits of each flight.

Designated Task Force 68, it included 4,700 men, 13 ships, including the aircraft carrier USS *Philippine Sea,* 33 aircraft and six helicopters. It was the largest show of military force that had ever been deployed in Antarctic waters. It also established the Little America IV base on the Ross Ice Shelf. *Operation Highjump* ended in late February 1947.

The US Navy returned the next year with *Operation Windmill* (designated Task Force 39). Also known as the Second Antarctica Developments Project, this was intended as an exploration and training mission. The icebreaker USS *Burton Island* was the smaller flotilla's flagship. In addition, in 1947, the United

"Byrd's fourth expedition, called 'Operation Highjump,' in the summer of 1946–47, was the most massive sea and air operation theretofore attempted in Antarctica. It involved 13 ships, including two seaplane tenders and an aircraft carrier, and a total of 25 airplanes. Ship-based aircraft returned with 49,000 photographs that, together with those taken by land-based aircraft, covered about 60 percent of the Antarctic coast, nearly one-fourth of which had been previously unseen."

Lize-Marié van der Watt (Researcher, KTH Royal Institute of Technology), *Encyclopædia Britannica,* 2020

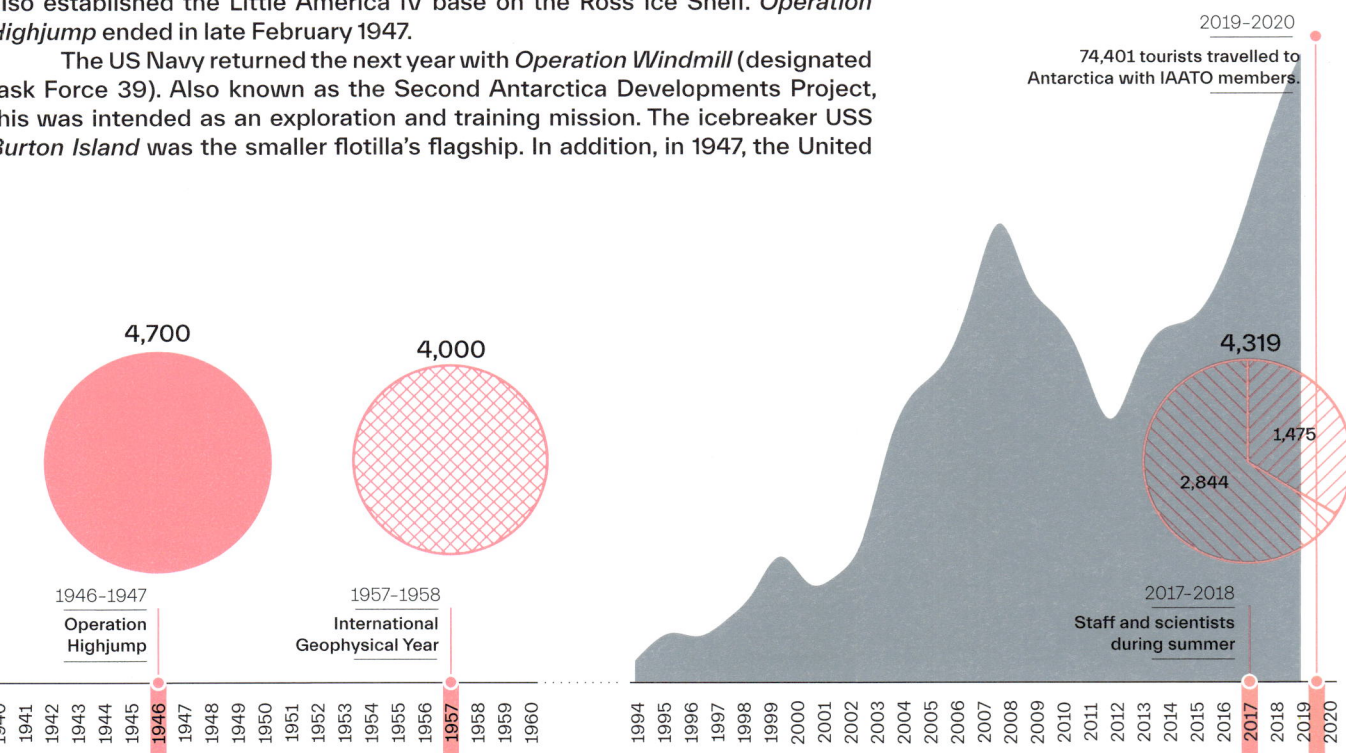

2019–2020
74,401 tourists travelled to Antarctica with IAATO members.

4,319

1,475

2,844

2017–2018
Staff and scientists during summer

4,700

1946–1947
Operation Highjump

4,000

1957–1958
International Geophysical Year

1940 1941 1942 1943 1944 1945 1946 1947 1948 1949 1950 1951 1952 1953 1954 1955 1956 1957 1958 1959 1960

1994 1995 1996 1997 1998 1999 2000 2001 2002 2003 2004 2005 2006 2007 2008 2009 2010 2011 2012 2013 2014 2015 2016 2017 2018 2019 2020

US Navy Douglas R4D Skytrain aircraft are lined up on the ice at the Little America IV Station in Antarctica during *Operation Highjump* (top), while USS *Burton Island* (AGB-1), USS *Atka* (AGB-3) and USS *Glacier* (AGB-4), shown at the bottom, push an iceberg out of the channel in the "Silent Land" near McMurdo Station in December 1965.

"I have just attended a Staff Meeting at the Pentagon. I have stated fully my discovery and the message from the Master. All is duly recorded. The President has been advised. I am now detained for several hours (six hours, thirty-nine minutes, to be exact). I am interviewed intently by Top Security Forces and a Medical Team. It was an ordeal!!!! I am placed under strict control via the National Security provisions of this United States of America. I am ordered to remain silent in regard to all that I have learned, on the behalf of humanity!!! Incredible! I am reminded that I am a Military Man and I must obey orders."

Richard Evelyn Byrd (Admiral, Explorer, Aviator), 11 March 1947

States successfully lobbied for the inclusion of what Washington called the "American Quadrant", 24° W to 90° W, as part of the security zone covered by the Inter-American Treaty of Reciprocal Assistance.

In 1948, the Chilean Antarctic Expedition established the General Bernardo O'Higgins Riquelme Base on the Antarctic Peninsula at Cape Legoupil. Still operated by the Chilean military, the base has one of the longest records of continuous habitation among Antarctic bases.

The following year, Argentina dispatched a flotilla of eight warships into Antarctic waters, establishing a manned base at Hope Bay on the tip of the Antarctic Peninsula, and sought to create other bases. Several minor skirmishes occurred between 1948 and 1953 between Argentine and British military forces at Hope Bay and Deception Island. In 1949, Argentina, Chile and Great Britain signed the Tripartite Naval Declaration, committing the signatories to refrain from deploying warships south of 60° S. This agreement remained in effect until 1961.

The 1957 International Geophysical Year (IGY) provided the impetus for increased international activity in Antarctica. The US Navy's *Operation Deep Freeze I,* Task Force 43, supported American scientists working in Antarctica as part of the IGY. During the IGY, the United States landed an R4D Skytrain, a modified Douglas DC-3, on the South Pole: this established the first permanent American base there. The Deep Freeze operation would evolve into US Naval Support Antarctica, tasked with the yearly resupply of US bases there. Successive Deep Freeze operations continued under Navy auspices until 1999, when they were transferred to the 109th Airlift Wing of the New York Air National Guard. Other countries, including Great Britain and the Soviet Union, also began to rely primarily on their naval forces to supply their new Antarctic facilities. Many countries still rely on their military to operate their Antarctic bases, even though no military activities are conducted there.

The 1959 Antarctic Treaty was the first arms-control agreement signed during the Cold War. It prohibited the militarisation of the Antarctic and any testing of weapons, either nuclear or conventional, on the continent, although it did permit the use of military forces for logistics. Occasionally, military personnel have rotated through the bases either in administrative, security, scientific or logistical roles, but no military forces are stationed there. Beyond an occasional demonstration of military power, there has not been a significant permanent deployment of military forces in Antarctica. Displays of military force have, however, woven in and out of Antarctica's history, inevitably playing out in the shadow-boxing that often occurred around competing national claims.

The recent Memorandum on Safeguarding US National Interests in the Arctic and Antarctic Regions issued by United States President Donald Trump on the 9th of June 2020 called for the development of "a ready, capable, and available fleet of polar security icebreakers that is operationally tested and fully deployable by fiscal year 2029."[2] In addition, the memorandum advocates that the US secure "associated assets and resources capable of ensuring a persistent United States presence in the Arctic and Antarctic."[3] The main thrust of the Trump administration's polar strategy, however, has been the Arctic, not Antarctica.

There is a risk that a stepped-up American presence in Antarctica and an increase in the deployment of dual-use equipment could trigger a corresponding response from Russia and China, and possibly even Chile and Argentina, producing a sort of arms race around dual-use equipment. While such a development would not be in violation of the ATS, it is clearly a step in the wrong direction.

The Protocol on Environmental Protection to the Antarctic Treaty bans resource development on the Antarctic continent until 2049. From a practical standpoint, the exploitation of Antarctic mineral resources outside of marine fisheries is not currently economically viable.

Notwithstanding advances in military technology, the difficulty of projecting military power in Antarctica remains a significant constraint to its use. The Antarctic Treaty System has avoided the potential of military confrontation on the Antarctic continent. Continued adherence to its provisions is the surest way of avoiding such conflicts in future and the best way to ensure that the military history of Antarctica remains a misnomer.

In 1928, Admiral Richard Byrd's first expedition to Antarctica was the largest and best-equipped human endeavour that had been organised by any country on the white continent to that date. At an estimated cost of $400,000, the expedition secured the United States the primacy of the first flight over the South Pole, on the 29th of November 1929, leading to the naming of numerous geographic features in honour of John D. Rockefeller Jr. and Byrd's own wife (Marie Byrd), as well as to a promotion. The ticker-tape parade held in New York upon the return of triumphant newly appointed "Rear Admiral" Byrd in 1929 (featured in the image above) offers an insight into the value attributed to this achievement by the American nation at large. Subsequently, Byrd travelled to the continent four more times, the third of which as commander-in-chief of the US Navy's *Operation Highjump* (1946–1947), and subsequently as the officer in charge of the United States' Antarctic programmes and senior authority for government Antarctic matters during *Operation Deep Freeze.*

Strategic Interests in Antarctica. The Example of China

Anne-Marie Brady

ANNE-MARIE BRADY is a professor of political science and international relations and a research associate at Gateway Antarctica at the University of Canterbury, Christchurch (New Zealand). Prof. Brady is a global fellow at the Wilson Center in Washington, DC (United States), and executive editor of *The Polar Journal*.

China has strong geostrategic, political, military, economic and scientific interests in Antarctica. Chinese polar analysts divide these interests into three categories: security (both traditional and non-traditional); resources (namely Antarctic minerals and hydrocarbons, fishing, tourism, transport routes, water and bioprospecting); and science and technology. The last of these relies on access to Antarctica for the roll-out of the BeiDou navigational system and China's space science programme.

The Antarctic Treaty permits the orderly exploitation of certain Antarctic resources: free access to the continent for scientific research, free access to the continent for individual exploration and adventure, managed fishing, and unlimited tourism and bioprospecting. However, since the 1991 Protocol on Environmental Protection entered into force in 1998, mineral exploitation and exploration have been banned, although scientific research into Antarctic minerals has not.

Seen from the People's Republic of China (PRC) government's point of view, the protocol simply postponed what Chinese polar policymakers believe is the inevitable opening up of Antarctic resources. In 2000, a PRC Ministry of Land and Resources report assessing Antarctic mineral resources concluded that the Antarctic Treaty and its various instruments establish the political prerequisites for both "protecting the Antarctic environment and the future utilisation of Antarctic resources".[1] At present, any of the original signatories of the protocol can request a review at any time, though any change would require consensus. After 2048, however, any modification to the terms of the protocol is allowed to pass with a three-quarters majority vote of the parties – though all thirty-three of the original signatories must also agree.

Since the same signatories were party to the original Convention on the Regulation of Antarctic Mineral Resource Activities (CRAMRA), which would have permitted mineral exploitation in Antarctica, it does not seem impossible to believe that by mid-century these same states will be ready to sign a similar agreement, especially since eight states have already publicly stated their interest in exploiting Antarctic minerals.[2]

Polar Research Institute of China (PRIC) researchers estimate that there are 500 billion tons of oil and 300 to 500 billion tons of natural gas on the Antarctic continent, plus a potential 135 billion tons of oil in the Southern Ocean.[3] In 2009, PRIC staff produced a book-length study investigating the full range of Antarctic mineral resources and their legal status, stating that "when all the world's resources have been depleted, Antarctica will be a global treasure house of resources".[4] Access to the considerable natural resources in Antarctica is seen as essential for the continued growth of the Chinese economy.[5] A 2013 report com-

mented that "regardless of how the spoils are divided up, China must have a share of Antarctic mineral resources to ensure the survival and development of its one billion population".[6] To that end the country recently produced a government report that included chapters on the "preliminary exploration of mineral resources in Antarctic waters" and "surveyed coal reserves".[7]

A further important Antarctic resource of interest to China is fishing. Antarctic fishing is managed by the Commission for the Conservation of Antarctic Marine Living Resources (CCAMLR). Fishing in the Southern Ocean sees three main players, of which China has the third-largest catch of Antarctic krill (after Norway and Korea), taking 12% of the total catch in 2018.[8] In 2015, China announced that it planned to double or even quadruple its existing krill catch to between one and two million tons per year,[9] and current levels of fishing have already quadrupled the level of krill taken at the time that statement was made. To support this goal, China launched the world's largest krill boat in 2019 and plans a second.[10]

Chinese tourism into Antarctica has also grown exponentially in the last five years. In 2010, Chinese Antarctic diplomat Wu Yilin recommended that China encourage Chinese tourism operators to become active in Antarctica in order to take advantage of a legitimate Antarctic "resource" – the pristine environment – and gain market share before restrictions on tourism numbers are introduced.[11] China is now the second-largest source of Antarctic tourists after the United States, and the Chinese government wants to make China "a major tourism nation in Antarctica."[12] Becoming a significant market for Antarctic tourists adds weight to China's Antarctic authority, influence and presence.

China dramatically expanded its Antarctic activities in the years from 2005 to 2015. Due to budgetary constraints, growth on some projects such as base-building has now slowed. Yet projects such as the roll-out of the BeiDou global navigation system have not slowed their pace. For China, as for other great powers, Antarctica's strategic importance continues to grow.

Antarctica as a Hybrid Warfare Environment?

Anne-Marie Brady

Article I of the Antarctic Treaty affirms that states may not engage in any measure of a military nature in Antarctica. Essentially banning the establishment of military bases, the carrying out of military manoeuvres, or the testing of any type of weapon, the Treaty restricts military activities in Antarctica and the surrounding seas to "peaceful purposes" only. As permitted by the Treaty, most countries adhere to the article and use their militaries in Antarctica solely for logistics and scientific support, and report duly on details of military personnel or equipment.

China stands out on this matter, as the country is steadily expanding the level of involvement of the People's Liberation Army (PLA) in Antarctica to enhance China's Antarctic operating capacity and enable PLA personnel to gain experience operating in extreme environments. Furthermore, the Chinese navy is rapidly expanding its capabilities and reach in the polar regions (citing significant global shipping interests as the official justification)[1] and is setting up an intercontinental Antarctic air route and permanent airfields).

The subtleties in the definition of military activities become all the more critical if one considers that the unresolved territorial status of Antarctica has the potential to enable military powers to place space tracking and ground receiving stations for polar satellites with global coverage on the continent that would be unwelcome on the sovereign territory of other states. Satellite receiving stations and telescopes, housed at Antarctic bases, have unspoken dual civil-military capabilities, as their infrared telescopes can be used to search for enemy satellites, drones and missile launches, and identify whether they have been shot when targeted, enhancing (in times of war) defensive capabilities in air-sea battles. Furthermore, some countries (including China, Russia and the United States) are conducting research on Antarctic high-frequency active auroras to exploit the defence-related potential uses of the ionosphere to use electromagnetic pulses to upset or even destroy enemy electronics, and use solar flares to interfere with military and civilian communication. Antarctic satellite receiving stations thus play an important role in helping militaries enhance Command, Control, Communications, Computers, Intelligence, Surveillance and Reconnaissance (C4ISR) systems capabilities, missile timing and missile positioning, enhancing situational awareness in a tactical environment, improving interoperability and providing surveillance and intelligence capacity.

The United States established Global Positioning System (GPS) ground stations in Antarctica in 1995, the same year full operational capability for GPS was achieved.[2] China installed BeiDou ground satellite receiving and processing stations at Changcheng and Zhongshan Stations in 2010,[3] at Kunlun Station in early 2013,[4] and completed further upgrades to the Zhongshan Station ↗ p.760 facilities in early 2015. Russia installed three GLONASS ground stations in Antarctica in 2009 with plans to expand to seven ground stations there by 2020.[5] In December 2018, BeiDou-3 (with five open channels and five closed military channels) achieved full global coverage.[6]

Unquestionably problematic, there is a conundrum as to how to deal with this new situation. While the recent declaration passed at the 2019 Antarctic Treaty Consultative Meeting restates that all military activities in Antarctica shall continue to be peaceful,[7] such a declaration is dependent on thinking appropriate to traditional kinetic forms of warfare, but it does not seem to take into consideration the new hybrid warfare environment that is at play in Antarctica today.

"China's focus on becoming a polar great power represents a fundamental reorientation – a completely new way of looking at the world. Nothing symbolizes this shift more than China's vertical world map [...]. The PLA [People's Liberation Army] has been using this vertical world map to help determine the location of satellites and satellite receiving stations for BeiDou-2, China's strategic weapons navigating system, in order to chart China's new direction in the most literal sense. The map is the visual representation of China's new global *Realpolitik:* pragmatic, assertive of China's national interests, cooperative where it is possible to be cooperative, and yet ready to face up to conflict."

Anne-Marie Brady (Professor, University of Canterbury), *China as a Polar Great Power,* 2017

03
The Governance System of a Citizenless Continent

Science, Politics, and Governance in the Antarctic Treaty System. A Symbiotic Relationship

Karen N. Scott and Donald R. Rothwell

KAREN N. SCOTT is a professor of law at the University of Canterbury (New Zealand) and the president of the Australian and New Zealand Society of International Law. DONALD R. ROTHWELL is a professor of international law at the Australian National University College of Law.

Antarctica has uniquely been "constructed *through* science"[1] as a geopolitical and legal space.[2] Science has played and continues to play a fundamental role in supporting and, to an extent, defining states' political engagement with the continent.[3] Science has undoubtedly been used as a surrogate for occupation, facilitating a form of "colonisation" of Antarctica;[4] this is most visible through the operation of research stations by the seven claimant states within "their" zones and through science-related infrastructure on the continent more generally. Equally, non-claimant states have used science as a means to deny territorial claims; this is perhaps no better illustrated than by the US station Amundsen-Scott↗ p. 670 at the South Pole, "a location which manages to both physically straddle, and allow the US to philosophically set itself above, all sovereign claims".[5] Yet science also provided both the motive and the means to negotiate the 1959 Antarctic Treaty,[6] under the auspices of which a successful, effective and enduring management regime for Antarctica has been developed. Thus science and its rhetoric have been described as providing a "necessary lubricant for a conflict-free Antarctica",[7] but rather than supplanting political rivalry, it has emerged as a "new form of geopolitics"[8] simultaneously serving both national and international interests.

The "gentlemen's agreement" that underpinned the 1957–1958 International Geophysical Year (IGY) stipulated that scientific activities taking place during the IGY could not be used to support or deny territorial claims to Antarctica and that scientific activities could take place in any location on the continent irrespective of existing claims.[9] While the characterisation of the IGY and the Antarctic Treaty as cause and effect is overly simplistic,[10] this principle was subsequently developed into the ultimate "agree to disagree" clause, which provides the foundation of the Antarctic Treaty System (ATS) through Article IV of the Antarctic Treaty. The prioritisation of science in Antarctica is evident from the preamble to the Antarctic Treaty, which also enshrines the freedom of scientific research (Article II), requires the free dissemination of scientific information and promotes the exchange of research personnel (Article III). Antarctic science is similarly prioritised by other parts of the ATS, notably the 1991 Environmental Protocol to the Antarctic Treaty.[11] Article III of the protocol asserts that Antarctica's "value as an area for the conduct of scientific research, in particular research essential to understanding the global environment, shall be fundamental considerations in the planning and conduct of all activities in the Antarctic Treaty area". The protocol permits exemptions to its strict rules prohibiting interference with native Antarctic fauna and flora and limiting access to vulnerable areas where such interference or entry is necessary for scientific purposes,[12] as well as providing for the designation of protected areas for the specific purpose of protecting scientific research activities.[13]

Science also plays a unique role in determining the balance of power between states within the Antarctic Treaty decision-making forum. While any state may accede to the Treaty, only those states that demonstrate their "interest in Antarctica by conducting substantial scientific research activity there, such as the establishment of a scientific station or the despatch of a scientific expedition" may participate in decision-making at annual (and other) Antarctic Treaty meetings.[14] Thus scientific research is the "currency"[15] through which entry to the Antarctic management club can be bought, and it is research that constitutes "symbolic capital"[16] within the Antarctic Treaty System. A consequence of the deliberate institutionalisation of a co-dependent relationship between science and political participation under the Treaty, however, is that science is sometimes not "just the means through which politics is mediated

"In the national policy sphere, there is great affection for Antarctic science. It is the celebrated 'right stuff' because the impeccable credentials of 'science' (who can be against that – unless of course it tells you something inconvenient?) provide the perfect alibi for country X being there. National science programmes do all sorts of valuable science for us, but until our governments really get serious about climate change, the thing they do *par excellence* is fly the national flag. And weather your state is a claimant (and remains rather keen to maintain its options, profile and influence, despite non-recognition by others and the present constraints of the Antarctic Treaty's Article IV) or a semi-claimant, let alone a non-claimant (but none the less keen on maximising national opportunities, status and future options), it feels it ought to be there, ensuring it has a place at the table. This, surely, sounds political."

Alan D. Hemmings (Professor, University of Canterbury), *Handbook on the Politics of Antarctica*, 2017

but political in itself".[17] Furthermore, making science a criterion for participation by necessity privileges wealthy states and opens the ATS to criticism that Antarctica is managed on *behalf of* rather than *by* the international community.[18]

This notwithstanding, the symbiotic relationship between science, governance and politics in the Antarctic has overall been beneficial not only to the region and the security of the ATS, but also to the international community more generally. By providing the means to channel the rhetoric of sovereignty and political influence, science has obviated the need for states to engage in other, potentially more damaging activities, such as demonstrations of military might or the extraction of resources. Moreover, the investment in Antarctic science made by the majority of Antarctic Treaty Consultative Parties – irrespective of motivation – over more than an 80-year period has returned a wealth of scientific information of global benefit, not least in relation to climate change.

Nevertheless, the constructive role of science within the ATS should not be taken for granted. At times, scientific values have come into conflict with other ATS values, notably environmental and wilderness values. A high-profile illustration of this conflict was Russia's controversial project to drill into Lake Vostok over more than a twenty-year period, demonstrating the limits of Antarctic environmental governance, in particular, the environmental impact assessment process under the Environmental Protocol.[19] Dual-use science, especially where research has a military application, has also been subject to concern, with criticism that it conflicts with the principle of peaceful purposes under Article I of the Antarctic Treaty.[20] The most recent incarnation of this concern is the installation of ground-based receiving stations in Antarctica for satellite navigation systems. The United States, China, Norway and Russia all operate ground stations in Antarctica associated with GPS, BeiDou, TrollSat and GLONASS, respectively, and the extent to which this technology can be classified for scientific research and/or compatible with the peaceful purposes provisions of the Antarctic Treaty is questionable.[21] The relationship between science and commercial gain, particularly in the context of "applied science", is equally divisive, and the question of who should benefit from the profits derived from scientific research has been assiduously avoided by the parties to the Antarctic Treaty, anxious to eschew issues that indirectly raise questions relating to sovereignty and jurisdiction.

"Antarctica, though preserved from legal claims until 2049, is in fact increasingly occupied by national science programs and other nonnational entities [...] Legal and scientific regimes purport to control and manage access to Antarctica in the name of environmentalism and global equity for non-first-world nations."

Elena Glasberg (Professor at New York University, Essayist, Speaker), *Antarctica as Cultural Critique: The Gendered Politics of Scientific Exploration and Climate Change*, 2012

The Council of Managers of National Antarctic Programs (COMNAP) 2017 Antarctic Stations Catalogue reveals that, on average, only 1 out of 9 individuals deployed in the Antarctic during the austral summer are scientists, and only 13.8% of the total square metres of the station are reserved for scientific laboratories.

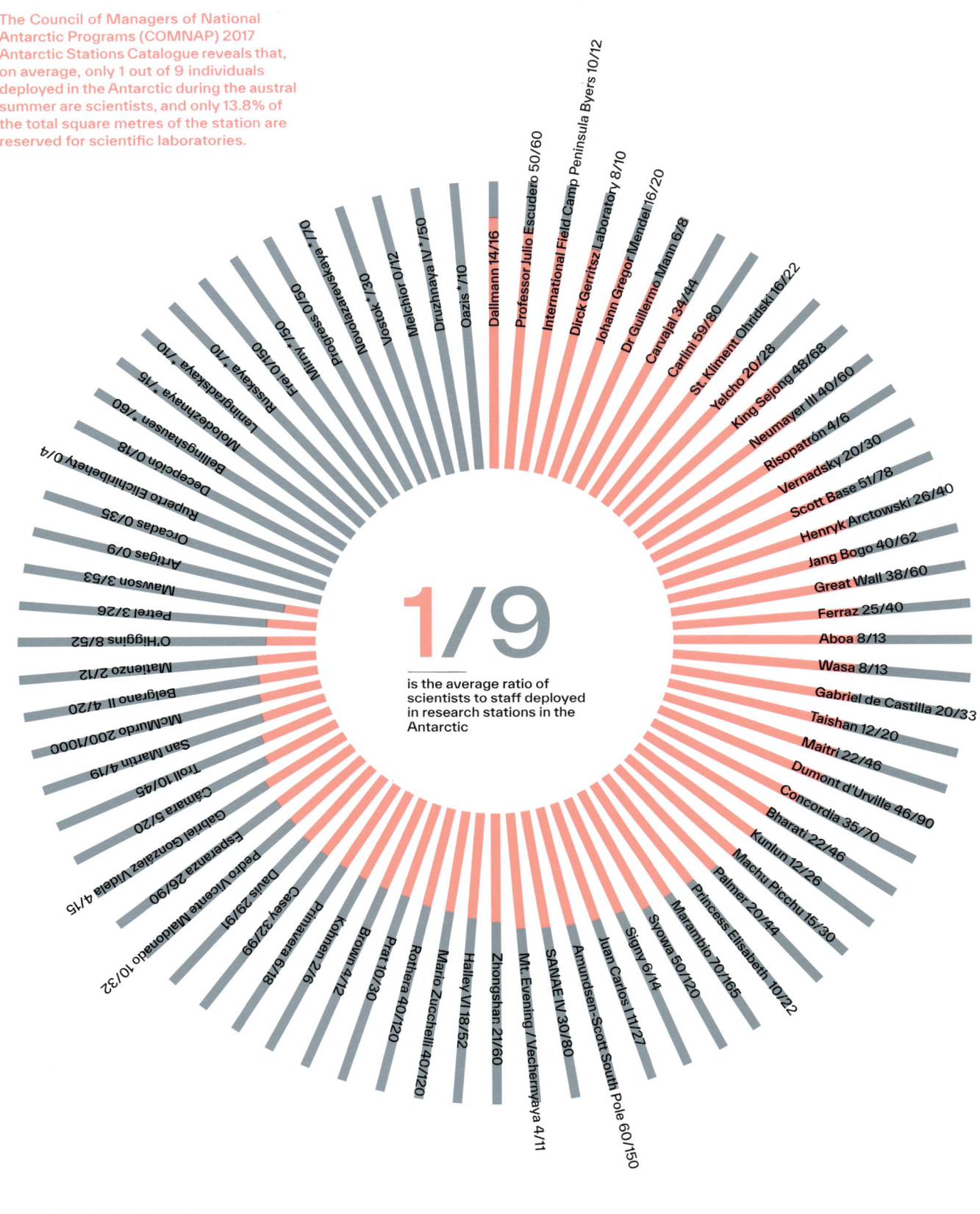

1/9

is the average ratio of scientists to staff deployed in research stations in the Antarctic

Dallmann 14/16
Professor Julio Escudero 50/60
International Field Camp Peninsula Byers 10/12
Dirck Gerritsz Laboratory 8/10
Johann Gregor Mendel 16/20
Dr Guillermo Mann 6/8
Carvajal 34/44
St. Kliment Ohridski 16/22
Yelcho 20/28
King Sejong 48/68
Neumayer III 40/60
Risopatrón 4/6
Vernadsky 20/30
Scott Base 51/78
Henryk Arctowski 26/40
Jang Bogo 40/62
Great Wall 38/60
Ferraz 25/40
Aboa 8/13
Wasa 8/13
Gabriel de Castilla 20/33
Taishan 12/20
Maitri 22/46
Dumont d'Urville 46/90
Concordia 35/70
Bharati 22/46
Kunlun 12/26
Machu Picchu 15/30
Palmer 20/44
Princess Elisabeth 10/22
Marambio 70/165
Syowa 50/120
Signy 6/14
Juan Carlos I 11/27
Amundsen-Scott South Pole 60/150
SANAE IV 30/80
Mt. Evening Vechernyaya 4/11
Zhongshan 21/60
Halley VI 18/52
Mario Zucchelli 40/120
Rothera 40/120
Pratt 10/30
Brown 4/12
Kohnen 2/6
Primavera 6/18
Casey 32/99
Davis 29/91
Pedro Vicente Maldonado 10/32
Esperanza 26/90
Gabriel González Videla 4/15
Cámara 5/20
Troll 10/45
San Martín 4/19
McMurdo 200/1000
Belgrano II 4/20
Matienzo 2/12
O'Higgins 8/52
Petrel 3/26
Mawson 3/53
Artigas 0/9
Orcadas 0/35
Ruperto Elichiribehety 0/4
Deception 0/18
Bellingshausen *0/60
Molodezhnaya *0/15
Leningradskaya *0/15
Russkaya *0/10
Frei 0/150
Mirny */50
Progress 0/50
Novolazarevskaya */70
Vostok */30
Molchior */20
Druzhnaya */50
Oasis */10

Legend
- Number of scientists
- Total number of staff

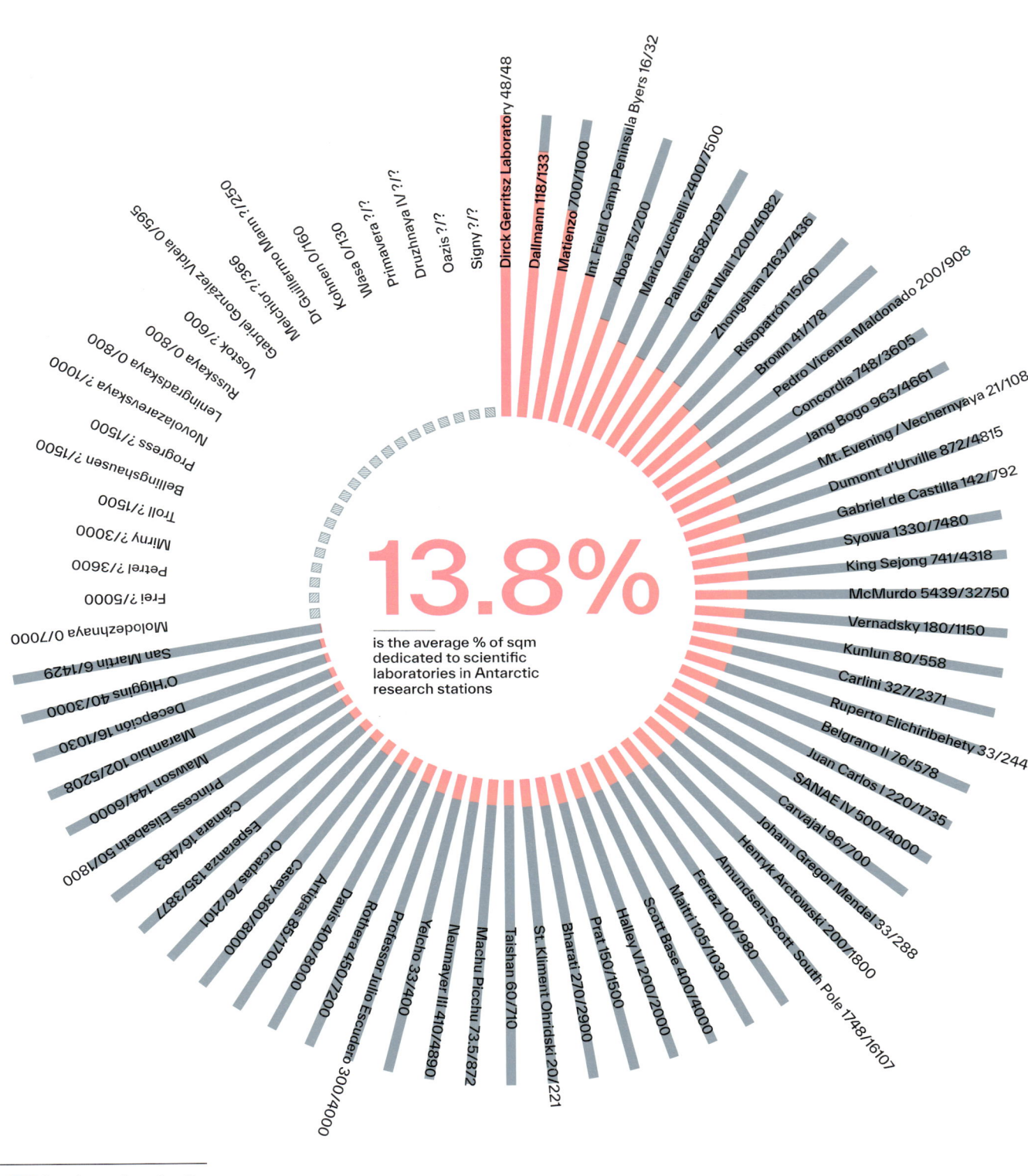

13.8%

is the average % of sqm dedicated to scientific laboratories in Antarctic research stations

Dirck Gerritsz Laboratory 48/48
Dallmann 118/133
Mattienzo 700/1000
Int. Field Camp Peninsula Byers 16/32
Aboa 75/200
Mario Zucchelli 2400/7500
Palmer 658/2197
Great Wall 1200/4082
Zhongshan 2163/7436
Risopatrón 15/60
Brown 41/178
Pedro Vicente Maldonado 200/908
Concordia 748/3605
Jang Bogo 963/4661
Mt. Evening / Vechernyaya 21/108
Dumont d'Urville 872/4815
Gabriel de Castilla 142/792
Syowa 1330/7480
King Sejong 741/4318
McMurdo 5439/32750
Vernadsky 180/1150
Kunlun 80/558
Carlini 327/2371
Ruperto Elichiribehety 33/244
Belgrano II 76/578
Juan Carlos I 220/1735
SANAE IV 500/4000
Carvajal 96/700
Johann Gregor Mendel 33/288
Henryk Arctowski 200/1800
Amundsen-Scott - South Pole 1748/16107
Ferraz 100/980
Maitri 105/1030
Scott Base 400/4000
Halley VI 200/2000
Prat 150/1500
Bharati 270/2900
St. Kliment Ohridski 201/221
Taishan 60/710
Machu Picchu 73.5/872
Neumayer III 410/4890
Yelcho 33/400
Professor Julio Escudero 300/4000
Rothera 450/7200
Davis 400/8000
Artigas 85/700
Casey 360/8000
Orcadas 135/387
Esperanza 16/483
Cámara 16/483
Princess Elisabeth 50/1800
Mawson 144/6000
Marambio 102/5208
Decepción 16/1030
O'Higgins 40/3000
San Martín 6/1429
Molodezhnaya 0/7000
Frei ?/5000
Petrel ?/3600
Mirny ?/3000
Troll ?/1500
Bellingshausen ?/1500
Progress ?/1500
Novolazarevskaya ?/1000
Leningradskaya 0/800
Russkaya 0/800
Vostok ?/600
Gabriel González Videla 0/595
Dr Guillermo Mann ?/250
Melchior ?/366
Kohnen 0/160
Wasa 0/130
Primavera IV ?/?
Druzhnaya IV ?/?
Oazis ?/?
Signy ?/?

Sqm devoted to scientific laboratories

Station's total sqm

Missing data

International Polar and Geophysical Years

Susan Barr

SUSAN BARR was the founding president of, and is now Arctic advisor to, the International Council on Monuments and Sites (ICOMOS) Polar Heritage Committee, which she founded in 2000. She has worked since 1979 solely with polar heritage and history, with extensive field work in both polar areas, authoring many publications. In 2019 Dr Barr was appointed a life member of ICOMOS Norway in recognition of her work for polar cultural heritage.

The Heroic Ages of polar exploration, the heroes themselves and the stories of achievement and suffering are gripping pieces of the long and eventful history of man's gradual understanding of the polar areas and their impact on the rest of the world. Yet in the cause of the advancement of scientific knowledge, these individual and national feats were not necessarily the most important. The patient, systematic collection of comparable observations and measurements and the exchange of data between fellow and even rival scientists are what in the end bring global society forward.

Both exploration and commercial expeditions routinely collected meteorological and other data from their travels, and in the early nineteenth century both magnetic and meteorological networks were established around the world which decided on standardised methods of observation and analysis. Information from the Arctic was considered to be important for a deeper understanding of the Earth's magnetic and meteorological conditions, and international cooperation was the answer for an effective approach. The Austro-Hungarian naval officer and Arctic explorer Carl Weyprecht (1838–1881) returned from a national expedition to Franz Josef Land in 1872–1874 and raised the issue of the need for systematic and synchronous observations from a network of land stations spaced as far as possible all around the Arctic.

Weyprecht's initiative resulted in what became known as the First International Polar Year (IPY), which involved eleven nations establishing and operating twelve stations in the Arctic and subarctic and two in the subantarctic from the 1st of August 1882 into September of the following year. Already existing magnetic and meteorological observatories around the world also took part, and the methods, means and aims were coordinated by an international committee. Each of the fourteen expeditions that established dedicated Polar Year stations had an exciting story to tell, with varying degrees of success and adventure. One of the two American expeditions was rescued after two years, and only six of the twenty-five participants survived, while the Dutch expedition spent ten months working from a camp on the sea ice after their ship was destroyed in the ice.

The participating countries were the Austro-Hungarian Empire, Denmark, Finland, France, Germany, the Netherlands, Norway, Russia, Sweden, the United Kingdom and the United States. The main disciplines involved were meteorology, geomagnetism and auroral phenomena, but each expedition also carried out a variety of other observations wherever possible.

The first IPY was a success in that it provided the impulse that led to the following IPYs and International Geophysical Years (IGYs) and also collected a large amount of systematic data. It did not, however, result in the international cooperation that Weyprecht had envisaged; the accumulated observations and measurements were mostly left to national initiatives to analyse and publish. The published general accounts of the expeditions are a rich source of information to historians about such expeditions of the time as well as the national emphases and differences.

By the early 1930s, the meteorological and electrical conditions of the upper atmosphere were in focus, and registering instruments such as radiosondes could be connected to kites, balloons, dirigibles and other aircraft. Geophysical observations that could improve weather forecasts for the developing area of air transport as well as global seaborne transport were sought after. In 1926, Professor Leonid Breitfuss in Berlin proposed a second Polar Year, fifty years after the first, particularly for this reason, and the International Meteorological Organization supported the initiative. Investigation of the polar

"All over the world men of science are mobilizing for the most intensive survey of man's physical environment ever attempted, the International Geophysical Year."

Hugh Latimer Dryden (Aeronautical Scientist, Civil Servant, NASA Deputy Administrator), "The International Geophysical Year", *National Geographic*, 1956

atmosphere, terrestrial magnetism and aurorae were the main objectives, while other disciplines such as glaciology were pursued by some participating expeditions.

Finally, in 1932–1933, forty-four countries participated, with twelve European states, Canada and the United States establishing twenty-seven stations around the Arctic. Five additional magnetic stations lay near or south of the equator, while ship-based observations were made in the southern hemisphere, since means were not available at this time to operate stations in Antarctica. Other high-elevation stations participated as well, as did Soviet marine expeditions in their Arctic waters, thus giving the second IPY a far more global extent than the first.

Data compilation and analysis were restricted by the economy of the early 1930s and then by the political run-up to the Second World War, but many of the observation series were collected and stored by the Danish Meteorological Institute and became available in this way.

The International Geophysical Year launched in 1957–1958 and known as the "Third IPY" is discussed elsewhere in this publication. The IGY was again much larger than its predecessor twenty-five years earlier. This time, it introduced space technology and brought Antarctic research stations to the forefront. The international cooperation established there was to lead directly to the formation of the Antarctic Treaty System, which still serves the continent today.

These international years of scientific cooperation, first and foremost in the polar areas but with a further geographical extent (including space exploration), became increasingly comprehensive and difficult to summarise. Although the main scientific activity had been concentrated on the geophysical sciences, many other disciplines also played a part. Not least, the material legacy is a source of study for today's scientists from the humanities and social sciences.

The fourth IPY, launched in 2007–2008, ran for 24 months and was not formally concluded until June 2010. Concentrated again on the polar regions, this edition of the IPY introduced traditional knowledge from local people and societies for the first time. It also coincided with the start of general awareness of forms of climate change that were more than natural cycles, and which particularly affected the polar areas, especially the Arctic. Inclusion and inspiration of the younger generation of upcoming polar scientists was also prioritised, with "outreach" in all forms being required of the scientific projects. International cooperation was by now well-established and was sponsored by the International Council for Science (ICSU) and the World Meteorological Organization (WMO), while each participating country was expected to provide funding for research projects. The Scientific Committee on Antarctic Research (SCAR) assumed responsibility for coordinating all IPY-related Antarctic research, while the International Arctic Science Committee (IASC) helped in the planning of Arctic research.

This extensive international effort in the field of polar science was extremely inspiring to the more than 10,000 scientists and 50,000 students and supporters from 63 countries who participated in some way. 228 international IPY projects were recorded, and many other contributions added to the wealth of knowledge collected. The aims of inspiring young people, local residents and the general public were met and great efforts continue to keep the legacy of collected data and established observational systems operational. Cooperation between scientists who had previously focused on either the Arctic or the Antarctic, but now on both, has also been a long-lasting result.

On the occasion of the International Geophysical Year, in 1957–1958, the International Geophysical Year National Academy of Sciences (IGY NAS) Committee compiled educational booklets to offer an introduction on environmental sciences. The images above represent two of the six original posters designed by the artist Herbert Danska.

The Precursor of the International Geophysical Year. The Norwegian-British-Swedish Antarctic Expedition

CHRISTEL MISUND DOMAAS is an advisor at the University Library in Tromsø (Norway). Misund Domaas holds a master's degree in history from the University of Tromsø and a postgraduate certificate in Antarctic studies from the University of Canterbury (New Zealand).

Christel Misund Domaas

On the 11th of February 1950, the Maudheim ↗ p. 648 wintering base was established on the Quar Ice Shelf at 71°02.6' S 10°55.5' W in Queen Maud Land, the territory claimed by Norway in 1939. This base would serve as the home of the Norwegian-British-Swedish Antarctic expedition.[1]

The idea of a Norwegian-British-Swedish expedition was conceived during the Second World War by the Swedish geographer Hans Wilhelmsson Ahlmann (1889–1974) to research global melting and glacial retreat in the southern hemisphere and to seek the causes and effects of the *klimatförbettring* – the "climate improvement" – he had registered in the northern hemisphere.[2] The possible global effects and his belief in a Scandinavian obligation to participate in the quest for knowledge led him to focus on international cooperation to realise his plans. Ahlmann sought to continue the Swedish polar tradition and planned the expedition as a continuation of Otto Nordenskjöld's prior

expedition to Graham Land.[3] He expected to attract British collaboration, leveraging on the fact that the country was in territorial dispute with Argentina and Chile in this region.[4] The Royal Geographical Society accepted his offer of collaboration in September 1945.

Photographs from the *Schwabenland Expedition* (1938–1939) of snow-free mountains with interspersed glacier tongues showed the area around Queen Maud Land to be ideal for geological and glacial science.[5] Driven by the conviction that "to determine whether the contemporary climate changes are of regional or universal character, it is of the utmost importance to study the glaciers in Antarctica",[6] Ahlmann also approached Norway as a potential partner, owing to that country's territorial claim of the area.[7] Instrumental to successfully gaining their participation was the growing Soviet initiative on polar research and exploration[8] and the argument Ahlmann made regarding the

consequent threat that the Soviets posed to Norwegian territorial interests as he urged an increase in Norwegian polar activity. The British agreed to this cooperation on the condition of Norwegian leadership of the expedition, underscoring the importance of strengthening the Norwegian claim in Queen Maud Land.[9] In April 1946, the Norwegian Geographical Society took responsibility for preparing the expedition, while the Royal Geographic Society and the Scott Polar Research Institute formed a committee in Britain and Ahlmann led the Swedish committee. After the Norwegian Geographical Society received an initial grant in October 1946, they invited the head of the Norges Svalbard- og Ishavsundersøkelser, Anders Orvin, to participate in the planning and assist in the logistics of the proposed joint expedition (as was desired by the Ministry of Trade and Norwegian Foreign Minister Lange). However, Orvin announced that the office was too busy with its own Arctic expedition preparations,

and the Norwegian Geographic Society was left with the full burden of planning the Antarctic expedition.[10]

Unexpected problems in acquiring a ship, along with conflicting internal objectives, postponed the start of the expedition until the director of the newly founded Norwegian Polar Institute, Harald Ulrik Sverdrup, took on full responsibility for the expedition in March 1948. The expedition finally left Norway in mid-November 1949 and reached Queen Maud Land on the 11th of February 1950. Comprised of personnel from Norway, Sweden and the British Commonwealth (United Kingdom, Canada, Australia and an observer from South Africa), the expedition addressed at once ambitions of scientific discovery and other motives, including access to weather data, whaling grounds and resources, as well as military training and territorial control. All these factors contributed to the realisation (and funding) of the expedition. Unlike previous Norwegian ventures to the Antarctic, the expedition was government-funded and, even though it was not promoted as a political expedition in order not to stir smouldering conflict in the politically fraught post-war period, it was operated as a means of reaffirming an official statement of national presence and consolidation.

During the expedition, the scientific activity and sheer presence validated Norwegian sovereignty. Set beforehand, the scientific programme was adapted on site to respond to challenges arising from the context and the available expertise. Designed to support national interests with cartography as its main objective, the scientific programme offered an opportunity to test new equipment and working methodologies as well as unveil groundbreaking information about the continent. Charting unknown areas and naming land translated into the production of printed and published maps that could be used as hard political currency in the disputes over sovereignty,[11] and the possibility of testing new equipment and training personnel undercover on a "civilian" expedition proved to be of great military interest.[12] To that end, all three countries had military representatives on the committees and among the expedition crew; nevertheless, the use of military air crews was technically also a necessity for the execution of the expedition and for the scientific surveys.

Although operating at the end of the world, the reverberations of changes in the geopolitical situation caught up with the expedition. The high expectations set for the newly established Norwegian Polar Institute did not persist, and the envisioned ambitions of fruitful cooperation resulted in a hefty administration. Despite the turbulence following the expedition – especially the disputes over settlement with Britain – the projected public image of harmony prevailed[13] and the Norwegian-British-Swedish expedition has a legacy of being a scientifically motivated expedition that placed scientific investigation before geographical discovery in the name of planetary climate concerns. This legacy is what granted the expedition the title of being a worthy forerunner of the upcoming International Geophysical Year (1957–1958).[14]

The Norwegian-British-Swedish expedition thus became the first multilateral expedition to the Antarctic region. The logistical groundwork, the scientific programme and the output provided valuable knowledge and marked the beginning of an era of collaboration in Antarctic policy. While unsuccessful in the search for causes of global warming, the *symbolic value* of science remained and the continued emphasis on research motivated further cooperation, leading to the IGY and to the established reading of the continent for science.

The Pivotal Role of the International Geophysical Year (IGY) 1957–1958

James Fleming and Cara Seitchek

JAMES FLEMING is the Charles A. Dana Professor of Science, Technology, and Society at Colby College and a research associate at the Smithsonian Institution (United States). CARA SEITCHEK works for the Smithsonian Institution and is an adjunct instructor at the University of California, Los Angeles and American University (United States).

On the 5th of April 1950, at a dinner party honouring British geophysicist Sidney Chapman – hosted by James and Abigail Van Allen – atmospheric physicist Lloyd V. Berkner proposed a worldwide study of the oceans, atmosphere, cryosphere and the role of the sun in Earth's geophysical systems. His proposal evolved into the International Geophysical Year, which complemented the previous two International Polar Years.

The International Geophysical Year (1st of July 1957 to 31st of December 1958) involved 67 nations in a coordinated global effort that took advantage of recent advances in the geophysical sciences and developments in rocketry, computing and instrumentation. The International Council of Scientific Unions (ICSU) sponsored it in commemoration of the 75th and 25th anniversaries of the first two International Polar Years and appointed a special organising committee led by Chapman to oversee the many activities. The IGY placed special emphasis on Antarctica, space science, and Earth-sun connections. Approximately 80,000 scientists and volunteers worked together to investigate aurorae and airglow, cosmic rays, geomagnetism, glaciology, gravity, ionospheric physics, longitude and latitude determination, meteorology, oceanography, rocketry, seismology and solar activity.

IGY architect Sydney Chapman promoted the rhetoric of scientific internationalism, simultaneously endorsing the national interests of the participating nations, while fostering a high level of international cooperation during the Cold War. All nations received an invitation to participate in the IGY. While the IGY Special Committee provided uniform instructions, each nation established its own national programme, research teams and scientific plans. Nations also advocated for government funding and logistical support, crucial elements for success. Eventually, states created research stations concentrated around, but not limited to, Antarctica, the Arctic, the equator and meridians 10° E, 70° W, 110° E and 140° W.

On the eve of the IGY, in a radio and television broadcast, US President Dwight D. Eisenhower shared his hopes that the International Geophysical Year might demonstrate the ability of people of all nations to work together harmoniously for the common good, and that cooperation could become common practice in other fields of human endeavour.

Despite these aspirations and the Special Committee's support for cooperative efforts, international tensions were inevitable. The United States and the Soviet Union engaged in a heated space competition that resonated with the general public. The Soviet Union launched the first of its artificial satellites, *Sputnik 1,* in October 1957, followed by the United States with *Explorer 1* three months later. These dramatic events of the early space age struck a public nerve in the depths of the Cold War and overshadowed, at the time, the larger and more lasting scientific accomplishments of the International Geophysical Year, which was already underway. When Taiwan joined the IGY, the People's Republic of China dropped out and conducted geophysical research on its own.

The IGY left two major diplomatic legacies of broad significance: International Space Law and the Antarctic Treaty. The latter, signed by 12 nations in 1959, specifies that Antarctica shall be used for peaceful purposes only and prohibits all military activities, including nuclear testing and the disposal of nuclear waste. It sets Antarctica aside as a scientific preserve, establishes freedom of investigation, preserves and conserves the living resources of Antarctica, and encourages the free international exchange of scientific results, personnel and observations. It requires open inspection and advance notice by signatory nations of all activities on the continent. It also provides for the peaceful resolution of any

"The legal resolution of national claims on the continent, and its enduring designation as *terra nullius* – free of sovereignty and citizenship – would emerge, not incidentally, alongside a similar legal order for the entire non-terrestrial (i.e. "celestial") cosmos. This grand new legal order was the result of an informal dinner party held in a compact suburban home […] on the evening of Wednesday, April 5, 1950. […] It is illustrative of both the intimate scale of the post-war scientific community and the influence it had forged for itself in the fires of wartime science that an idea hatched by the group of the tails of this discovery and end of the dinner party ('while sipping brandy') would quickly become global scientific and diplomatic policy."

Nicholas de Monchaux (Architect, Dean at Massachusetts Institute of Technology), *Kosmos* (online essay), 2020

"And of those probing Antarctica's ice and seas and rock and air, the United States had mounted the most ambitious program – an effort amounting to about $ 250,000,000. Only two percent of that was required for the scientific studies as such. We had to spend some $ 245,000,000 just to set science up in business on the inhospitable ice."

Rear Admiral George John Dufek (Commander, US Naval Support Force Antarctica), "What We've Accomplished in Antarctica", *National Geographic,* 1959

Sputnik 1, the world's first satellite, launched into orbit by the Soviet Union, exacerbated American fears that "unless [American] defense policies are promptly changed, the Soviets will move from superiority to supremacy,"[1] fuelling the so-called Space race. After the launch of *Sputnik 2* on the 3rd of November 1957 with Laika aboard (the first living being to enter orbit), the confrontation became untenable, and the United States revived a joint US Army and US Navy proposal called Project Orbiter, to launch its first scientific satellite into orbit from Cape Canaveral. Two hours after the launch on the 31st of January 1958, in a crowded press conference in which a scale model of the satellite was hoisted by team members as shown in the historic photograph, President Eisenhower announced, "The United States has successfully placed a scientific satellite in orbit around the Earth. This is part of our participation in the International Geophysical Year." Thanks to the Geiger tube radiation detector designed by James Van Allen, *Explorer 1* contributed to the discovery of a radiation belt (the Van Allen belt) of energetic charged particles that are captured by the planet's magnetic field.

The ambitions outlined by the IGY led to a far-reaching wave of optimism beyond the scientific community. As stated by the Duke of Edinburgh during a television interview on the eve of the programme launch, the IGY offered a unique opportunity, as "it is seldom that this world of ours acts together [...] Yet, for the next 18 months, east and west, north and south will unite in the greatest assault in history on the secrets of the Earth [...] At the same time, it may well help to solve the real problem – the conflict of ideas."[2]

disputes. In a similar vein, the Committee on the Peaceful Uses of Outer Space (COPUOS) was formally established by the United Nations in 1959 to promote cooperation in the peaceful use of Outer Space and to share information regarding its exploration.

Beyond the space race itself, and in the depths of the Cold War, the IGY facilitated new discoveries about Earth and new ways of working together internationally. Some of the most notable discoveries during the IGY include submarine mid-oceanic mountain ridges, the Van Allen radiation belts, the south polar vortex and a mathematical understanding of the pear-like shape of the Earth. Accomplishments include ice coring, the monitoring of atmospheric trace gases, and improved global mapping. The overall goal, paraphrasing a report from the United States National Academy of Sciences, was to observe geophysical phenomena and to secure data from all parts of the world in a coordinated manner so that results could be archived, collated and shared in a meaningful manner.

The IGY Special Committee established three World Data Centres: one in Washington, DC, one in Moscow, and one jointly managed by a consortium of western European nations, Australia, and Japan. These serve as data archives and ongoing sites for international research.

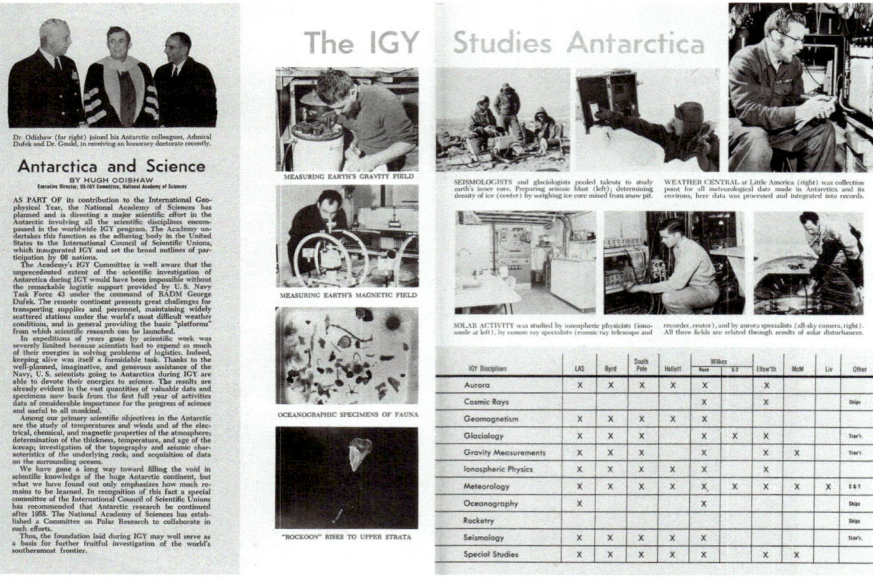

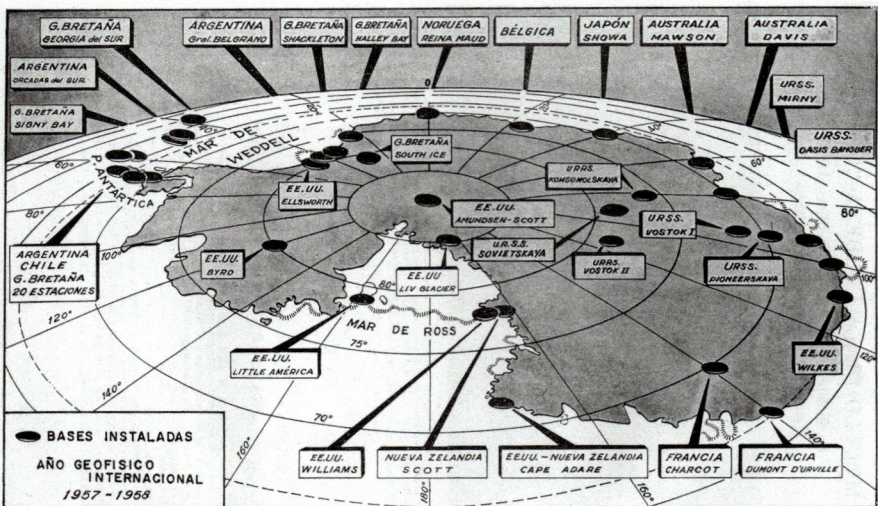

"When the United States built 'the seven cities of Antarctica' as part of the International Geophysical Year, those stations were regarded as temporary bases for 18 months of intensive study. But by early 1959, when I took over the Antarctic command from Rear Adm. George J. Dufek, it was already abundantly clear that our scientific investigation of the White Continent must continue. The years since then have witnessed the transition from the IGY's program of temporary occupancy to a long-term program of scientific effort. Without question, we are in Antarctica to stay."

Rear Admiral David M. Tyree (Commander, US Naval Force Antarctica, 1959–1963), "New Era in the Loneliest Continent", *National Geographic*, 1963

Argentina
Australia
Austria
Belgium
Bolivia
Brazil
Bulgaria
Burma
Canada
Ceylon
Chile
Nationalist China
Colombia
Cuba
Czechoslovakia
Denmark
Dominican Republic

East Africa
Ecuador
Egypt
Ethiopia
Finland
France
German Democratic Republic
German Federal Republic
Ghana
Greece
Guatemala
Hungary
Iceland
India
Indonesia
Iran
Ireland

Israel
Italy
Japan
Korea, Democratic Republic
Malaya
Mexico
Mongolian People's Republic
Morocco
Netherlands
New Zealand
Norway
Pakistan
Panama
Peru
Philippines
Poland
Portugal

Southern Rhodesia
Romania
Spain
Sweden
Switzerland
Thailand
Tunisia
Union of South Africa
Soviet Union
United Kingdom
United States of America
Uruguay
Venezuela
Vietnam, Democratic Republic (north)
Vietnam, Republic (south)
Yugoslavia

433

2,069

The idea of the International Geophysical Year – conceived by James Van Allen, Lloyd Berkner and Sydney Chapman during a conversation "while sipping brandy"[3] at a casual dinner – was formally accepted by the Comité Speciale de l'Année Geophysique Internationale (CSAGI) in 1951. The 18-month programme (scheduled to run from the 1st of July 1957 until the 21st of December 1958) was detailed in four meetings held between 1953 and 1958 in major European cities. With more than 20 nations agreeing to participate by May 1954, the IGY led to the publication of 2,069 papers, 433 of which address Antarctica specifically.

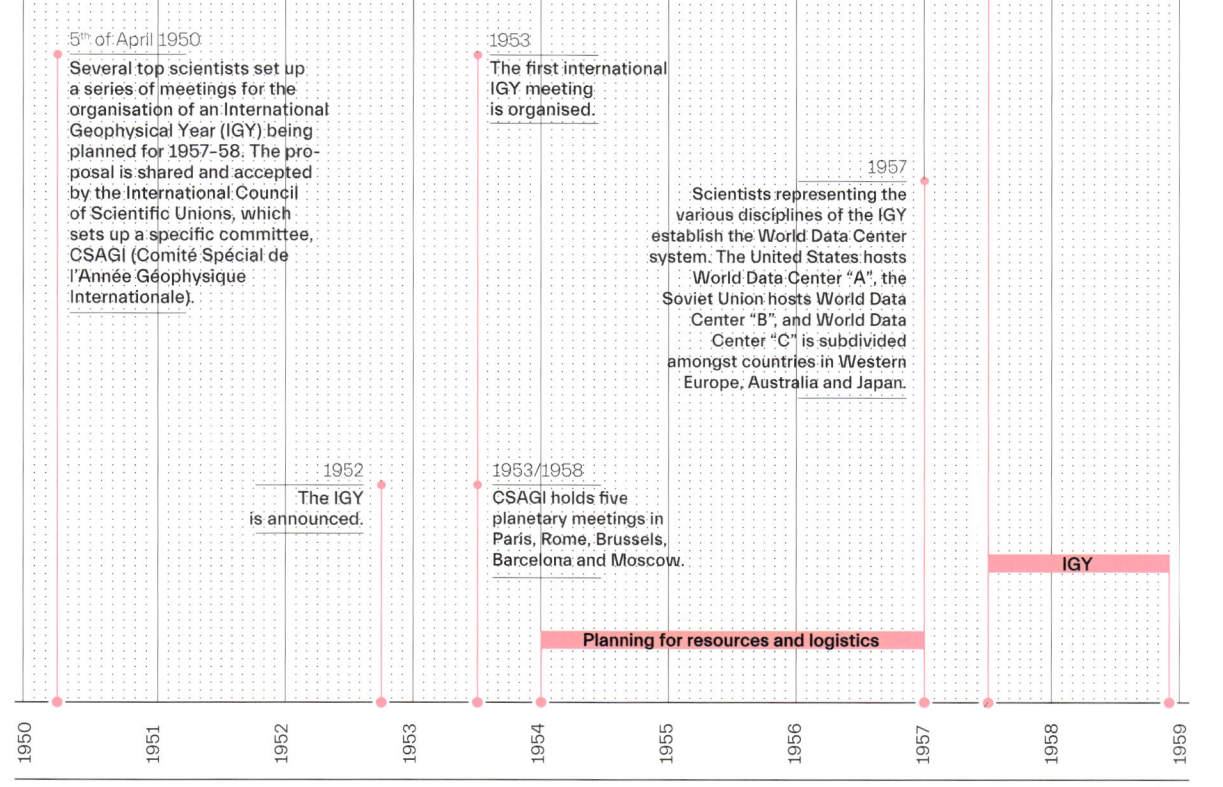

5th of April 1950
Several top scientists set up a series of meetings for the organisation of an International Geophysical Year (IGY) being planned for 1957-58. The proposal is shared and accepted by the International Council of Scientific Unions, which sets up a specific committee, CSAGI (Comité Spécial de l'Année Géophysique Internationale).

1953
The first international IGY meeting is organised.

1957
Scientists representing the various disciplines of the IGY establish the World Data Center system. The United States hosts World Data Center "A", the Soviet Union hosts World Data Center "B", and World Data Center "C" is subdivided amongst countries in Western Europe, Australia and Japan.

1952
The IGY is announced.

1953/1958
CSAGI holds five planetary meetings in Paris, Rome, Brussels, Barcelona and Moscow.

IGY

Planning for resources and logistics

1950 | 1951 | 1952 | 1953 | 1954 | 1955 | 1956 | 1957 | 1958 | 1959

The Antarctic Treaty

Klaus Dodds

KLAUS DODDS is a professor of geopolitics and the director of research and knowledge exchange at Royal Holloway, University of London (United Kingdom). He is co-editor of *Handbook on the Politics of Antarctica* (Edward Elgar, 2017), co-author of *The Scramble for the Poles* (Polity, 2016) and the editor-in-chief of *Territory, Politics, Governance* (Routledge). Prof. Dodds is an honorary fellow of the British Antarctic Survey and a trustee of the Royal Geographical Society (United Kingdom) and Regional Studies Association (United Kingdom).

In December 1959, twelve parties led by the United States signed the Antarctic Treaty. After six weeks of negotiation, a treaty emerged which was genuinely transformative in the midst of the Cold War. It is a highly notable multilateral international agreement, which made the case for demilitarisation, denuclearisation and international cooperation based on scientific endeavour and confidence-building measures such as the right to inspect one another's activities in the Antarctic. As an exercise in arms control, the Antarctic to this day remains the only continent never to have experienced the curse of war and conflict.

The genesis of the Antarctic Treaty lies at the intersection of three factors: the contested geopolitics of the polar continent, the politics of the Cold War and the more immediate legacy of the 1957–1958 International Geophysical Year.

By the mid-1940s, the Antarctic was a contested piece of real estate. A vast, poorly mapped and barely understood continent was subject to rival territorial claims. Argentina, the United Kingdom and Chile were locked in a struggle to assert their territorial dominance in the Antarctic Peninsula, the area closest to the southern edge of South America. Although not in conflict, the three parties were quite serious in their determination to seek primacy. They all claimed the same area, extending across hundreds of thousands of square kilometres of ice, rock and water. The three states established their own research stations and funded rival mapping and surveying projects. They produced their own maps, proclamations and resource evaluations. All three parties sought allies to strengthen their positions in the region. The United Kingdom turned to its Commonwealth partners, Australia and New Zealand, as well as to Norway, an old ally in the Antarctic. Argentina and Chile turned to the United States and appealed for hemispheric solidarity. All three counterclaimants were open to the possibility that the United States might join the claimant club (Argentina, Australia, Chile, France, New Zealand, Norway and the United Kingdom).

The politics of the Cold War, however, held the United States' interests in the Antarctic in abeyance. Up until the late 1940s and early 1950s, there was interest in the United States in making a claim to the Antarctic. As with the established members of the claimant club, the United States could reasonably claim that it had been at the forefront of activities related to discovery, exploration and scientific research. Since the 1920s, private and state-sponsored expeditions, including by the United States Navy, had been highly active in traversing the Antarctic continent. There was even an unclaimed sector adjacent to the Pacific Ocean. The United Kingdom hoped the United States might be persuaded to assert a claim to that unclaimed sector and thus join the special club of entitlement to Antarctic territory. What cemented that hope was the emergence of the Soviet Union as a state determined not to be excluded from any future plans to manage the Antarctic. By the early 1950s, it became clear that the two superpowers were intimately tied to Antarctica; their collective presence proved decisive when it came to later negotiations.

Finally, the political and scientific legacy of the International Geophysical Year (IGY) is notable. First discussed in 1950, the IGY was a special period (lasting some eighteen months) designed to encourage all countries to investigate Planet Earth and its relationship to the Moon, Outer Space and the global environment. Investigating Antarctica was integral to this endeavour, and twelve parties were committed to carrying out research in the region. In order to secure the spirit of scientific internationalism, all the parties agreed that the contested politics of Antarctica would have to be put aside. The so-called "Gentlemen's Agreement" (1955) enabled all the parties to carry out their scientific activities regardless of territorial claims and counterclaims. The United Kingdom was free

"The [Antarctic Treaty] AT, a disarmament treaty that came into force in 1961, […] proved instrumental in reducing political tensions and sublimating these instead in scientific rivalry and cooperation. The Treaty set forth a viable governance regime for Antarctica outside the UN system and put control in the hands of a limited number of nation states. […] Performance of substantial research activity in Antarctica became […] the criterion that qualifies new nations for full-fledged membership of the 'club' responsible for managing Antarctic affairs. Thus science acquired symbolic value as political capital […] At present the Treaty remains effectively a select club dominated by the claimant nations and the Cold War warriors (USA and Russia), and […] the return on the investment in Antarctic activities in terms of significant science or political initiatives seems lacking for several countries. […] Science in its own right is never enough to motivate a mobilization of resources and efforts on the scale seen in Antarctica."

Aant Elzinga (Emeritus Professor, University of Gothenburg), "The Continent for Science" in *Handbook on the Politics of Antarctica*, 2017

to locate all its IGY activities in its own territorial sector, while the United States and the Soviet Union could establish their own stations and research programmes anywhere on the continent and offshore. While never entirely free from tension and intrigue, the IGY did cement a view that scientific cooperation could coexist with political compromise. Throughout all of this, the United States and the Soviet Union retained their own rights to make a territorial claim to the polar continent in future. Antarctic scholars call both countries semi-claimants.

Mindful of this legacy and concerned about the aftermath, the United States invited the eleven other IGY Antarctic parties to participate in a series of preparatory meetings to further consider the governance of Antarctica. The meetings established a road map for the eventual Antarctic Treaty Conference, which opened in early October 1959. Under the leadership of the United States, the conference delegates were able to negotiate a new framework of governance. Remarkably, the parties were able to avoid the public scrutiny of the international community. Attempts by India to raise the "Question of Antarctica" at the United Nations (UN) General Assembly in 1956 and 1958 were rejected by those conference parties. India was persuaded by the United States to drop its public questioning and was reassured that any treaty would be attentive to the need to ensure that the continent was free from conflict and strife.

After six weeks of negotiation, the 1959 Antarctic Treaty was opened for signature on the 1st of December. This short treaty, consisting of fourteen articles, posited the Antarctic as a zone of peace and cooperation and a place for scientific endeavour. Emphasis was placed on consensus; Article IV famously declares that the territorial claims affecting Antarctica should be held in abeyance for the duration of the treaty. The eventual ratification of the Treaty in June 1961 was never entirely free of controversy or high drama. Article IV was a highly divisive issue to some of the claimant states such as Argentina, Australia and France. Argentine and Australian commentators were very critical of demands that their countries should have their territorial claims suspended. Australia was concerned about the current and future intentions of the Soviet Union. Cold War paranoia was part and parcel of Antarctic geopolitics.

While the Antarctic Treaty is often cited as securing the Antarctic continent and the surrounding ocean (extending up to 60 degrees latitude), it was a bitter pill for the claimants to swallow. Non-claimants and the two semi-claimants, the United States and the Soviet Union, had to keep their territorial and resource interests in check. Over the next three decades, the signatories had to invest further in the spirit of the 1959 Treaty and develop additional legal instruments to address resource exploitation and environmental protection. With no definitive agreement on sovereignty, negotiations were always going to be sensitive. By the 1980s, new powers such as China and India ensured that the Antarctic became more globalised. Environmental groups such as Greenpeace challenged the signatories to be less secretive and more attentive to environmental protection. Mineral resource exploitation was banned by the 1991 Protocol on Environmental Protection after some six years of negotiations on a possible mineral exploitation regime.

Buffeted by ongoing climate change and rising Asian powers, the polar continent is part and parcel of a global geopolitical landscape. Claimant states such as Australia and New Zealand have expressed concerns about China's logistical, resource and scientific plans. Antarctic tourism continues to prove popular, with some 50,000-plus visitors making the journey to "the ice" each year. Environmental protection sits uneasily with pressure on resources (such as fishing conservation and protected marine areas), political cooperation in an era

The photograph shown at the top captures the instant in which United States Ambassador Herman Phleger, who chaired the 1959 Conference on Antarctica, signed the Antarctic Treaty on behalf of the United States on the 1st of December 1959 in Washington, DC. The inscription on the photograph is a dedication from Herman Phleger to geologist and Antarctic explorer Laurence McKinley Gould, chief scientist during Richard Byrd's first Antarctic expedition, "without whom [reads the dedication] there would be no Antarctic Treaty". The first meeting of Antarctic Treaty members, captured in the image at the bottom, was held more than a year later, on the 1st of July 1961.

of worsening relations with Russia and suspicion of China, and the resilience of the treaty spirit grounded in consensus, compromise and the free exchange of knowledge. For now, the continent remains demilitarised.

An anthropogenic Antarctic is a very different situation to the one encountered in the late 1950s. Sixty years earlier, there was speculation aplenty about the depth and extent of the Antarctic ice sheet. No one was talking about climate change. Now we worry about the fate of those glaciers and ice sheets. It is now unthinkable that only twelve countries should determine the fate of an entire continent. Antarctica's ecology and geopolitics are now becoming ever less exceptional. The continent is warming and melting. It is attracting ever more international political attention. By the end of the current century, we might even witness a world in which the milder Antarctic Peninsula is permanently inhabited by communities displaced by rising sea levels elsewhere.

How long will the Antarctic Treaty endure? For now, signatories inhabit a system of governance that accommodates their complementary and competing interests and wishes. One way to prepare for the future is not to mythologise it, but to accept that the science, infrastructure and fishing are proxies for territorialisation, settler colonialism and mineral extraction, respectively.

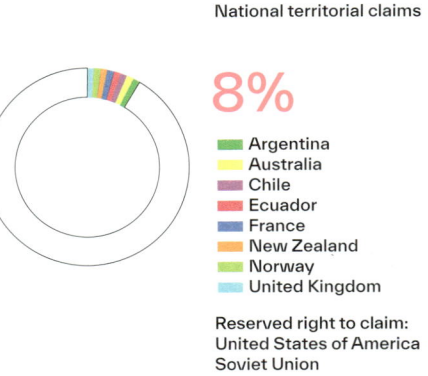

1957
National territorial claims

8%

- 🟩 Argentina
- 🟨 Australia
- 🟪 Chile
- 🟥 Ecuador
- 🟦 France
- 🟧 New Zealand
- 🟩 Norway
- 🟦 United Kingdom

Reserved right to claim:
United States of America
Soviet Union

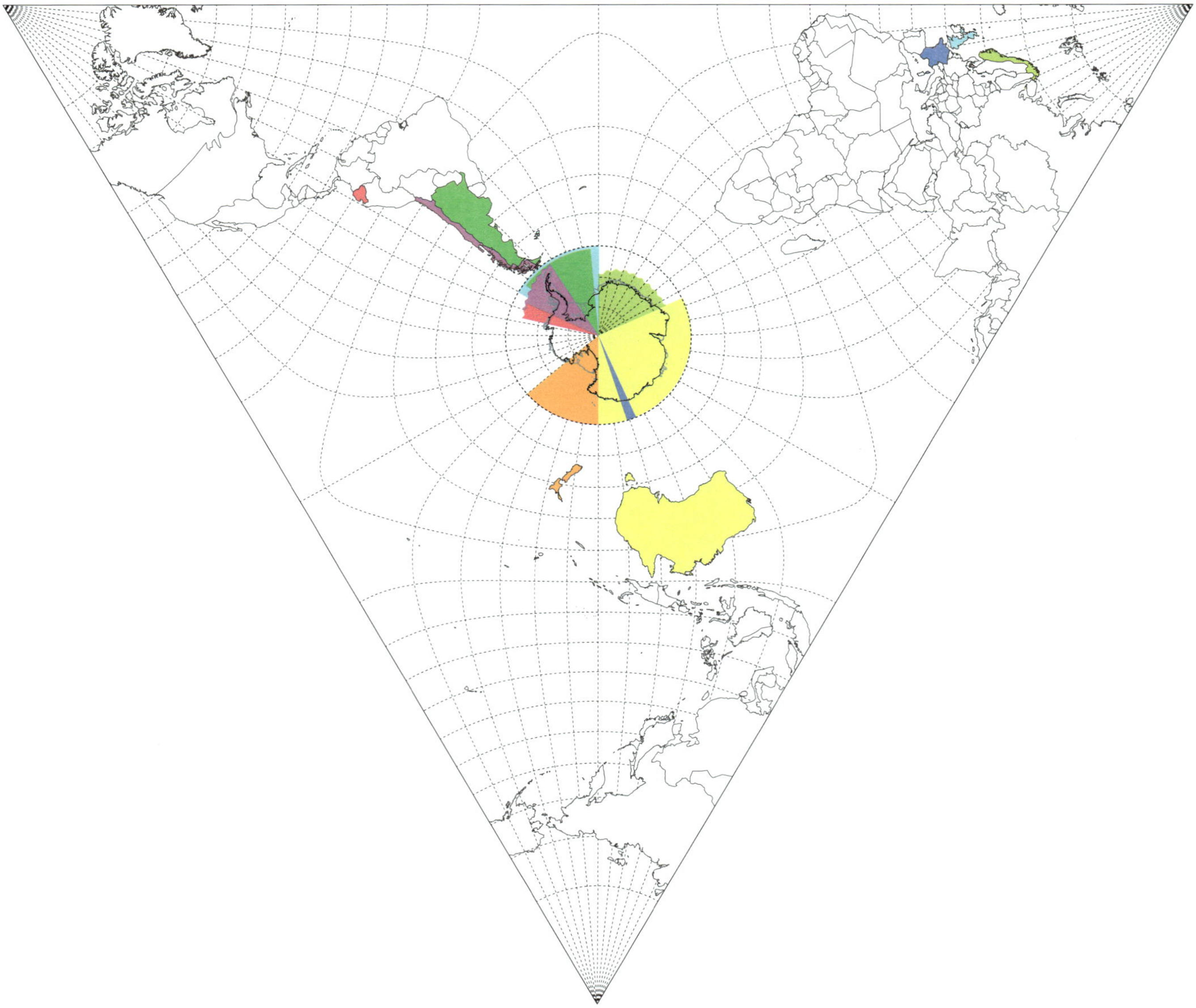

2020
Antarctic Treaty Members

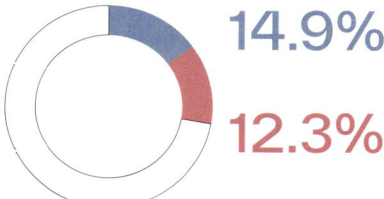

14.9% 29 out of 195 countries in the world are Consultative Parties to the Antarctic Treaty.

12.3% 25 out of 195 countries in the world are Non-Consultative Parties to the Antarctic Treaty.

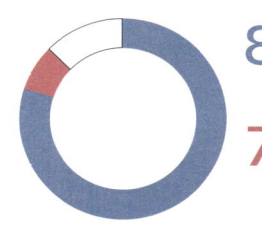

80% The cumulative GDP of Consultative Parties to the Antarctic Treaty accounts for 80% of worldwide GDP with 68,226.539 billion dollars.

7.6% The cumulative GDP of Non-Consultative Parties to the Antarctic Treaty accounts for 7.6% of worldwide GDP with 6,464.051 billion dollars.

13% The cumulative GDP of Non-Parties to the Antarctic Treaty accounts for 13% of worldwide GDP with 10,734.199 billion dollars.

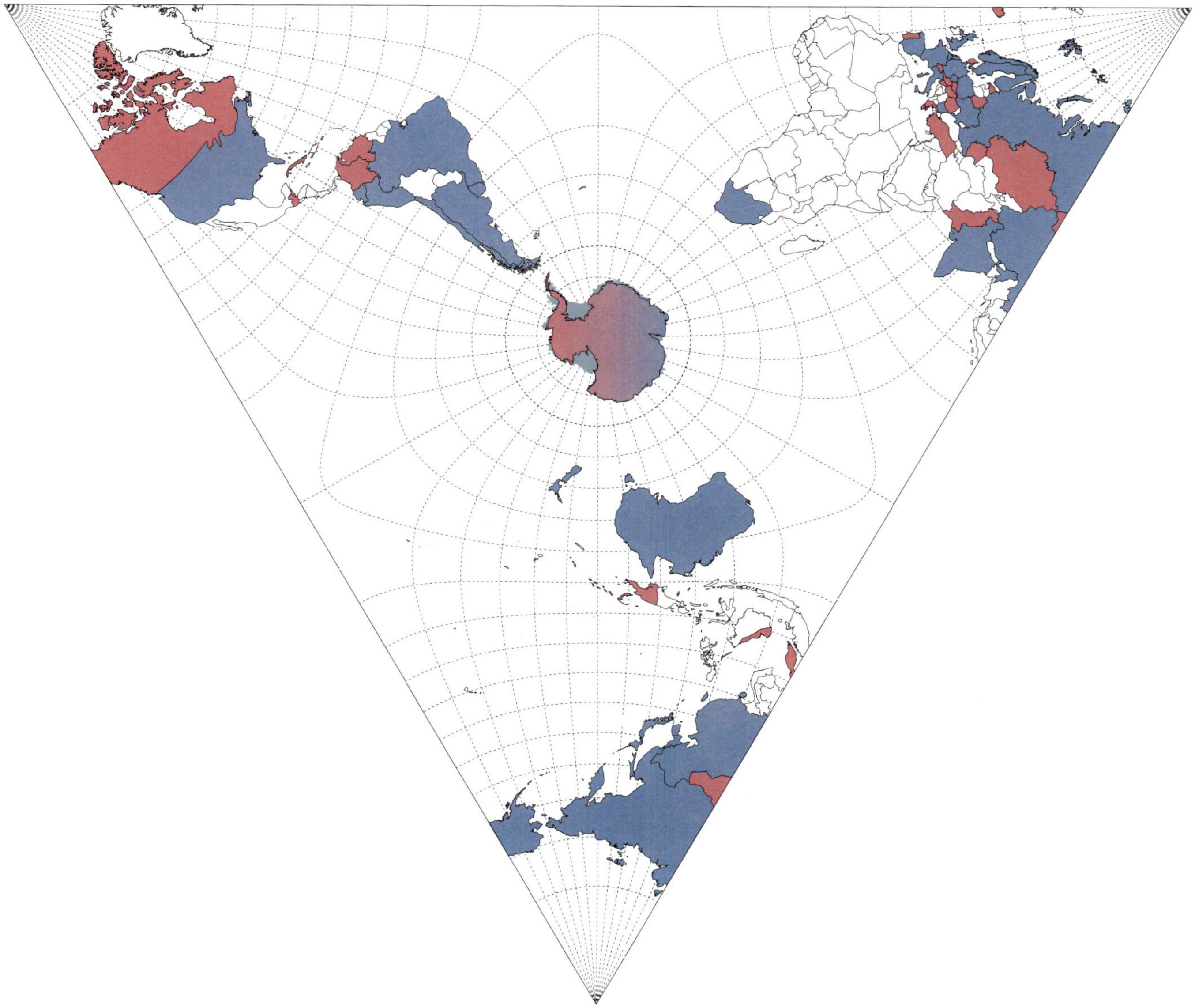

Twelve states signed the Antarctic Treaty in December 1959. The original signatories retain a special status among the Consultative Parties, which now comprise 17 additional nations that have proved their commitment to the continent by "conducting substantial research activity [...], such as the establishment of a scientific station or the dispatch of a scientific expedition" (Art. IX). Alongside the Consultative Parties, 25 Non-Consultative Parties have also committed to the Treaty, raising the overall number of countries involved in the management of the continent to 28% of the world's countries, of which technically only 15% have recognised voting power.

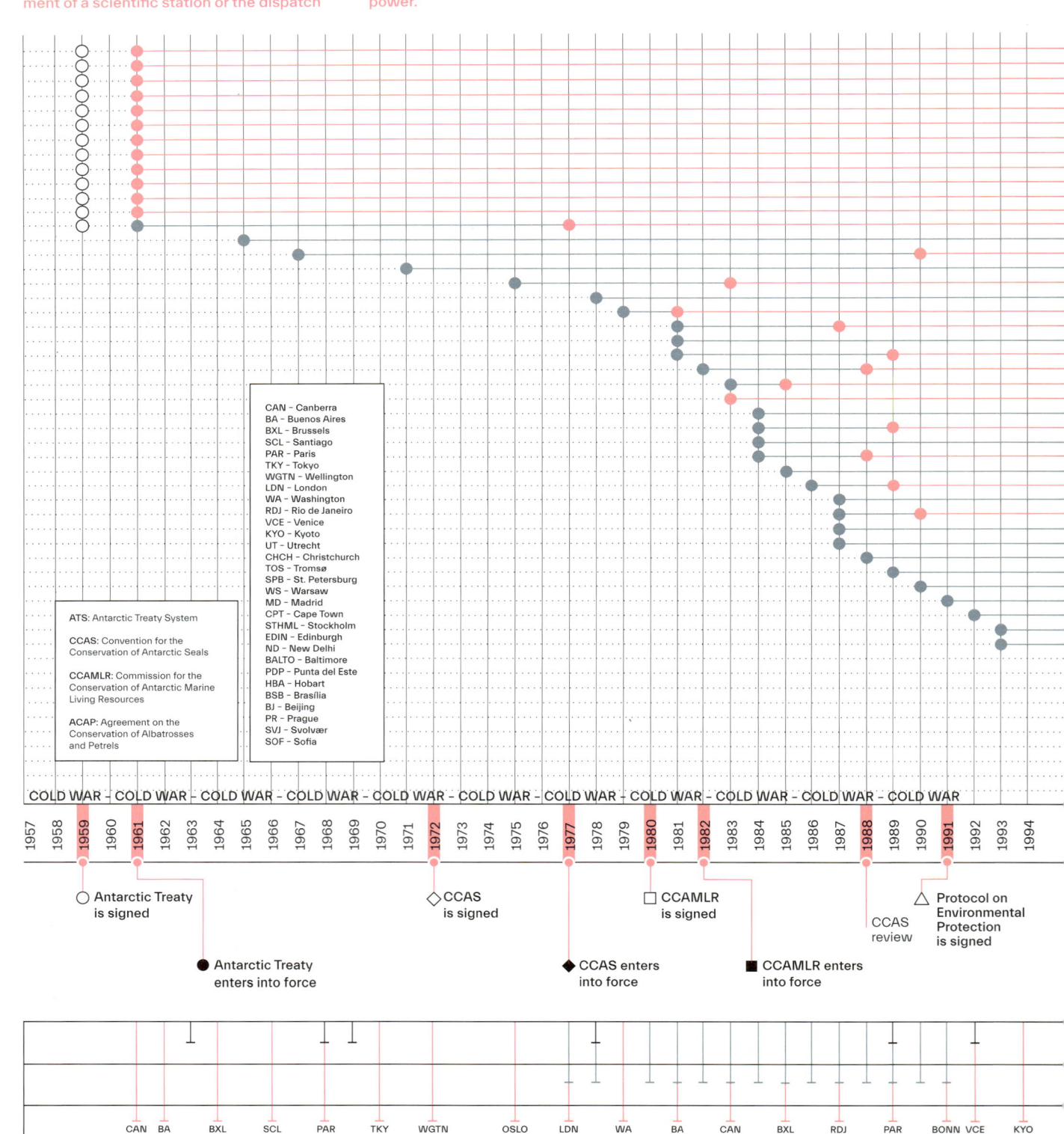

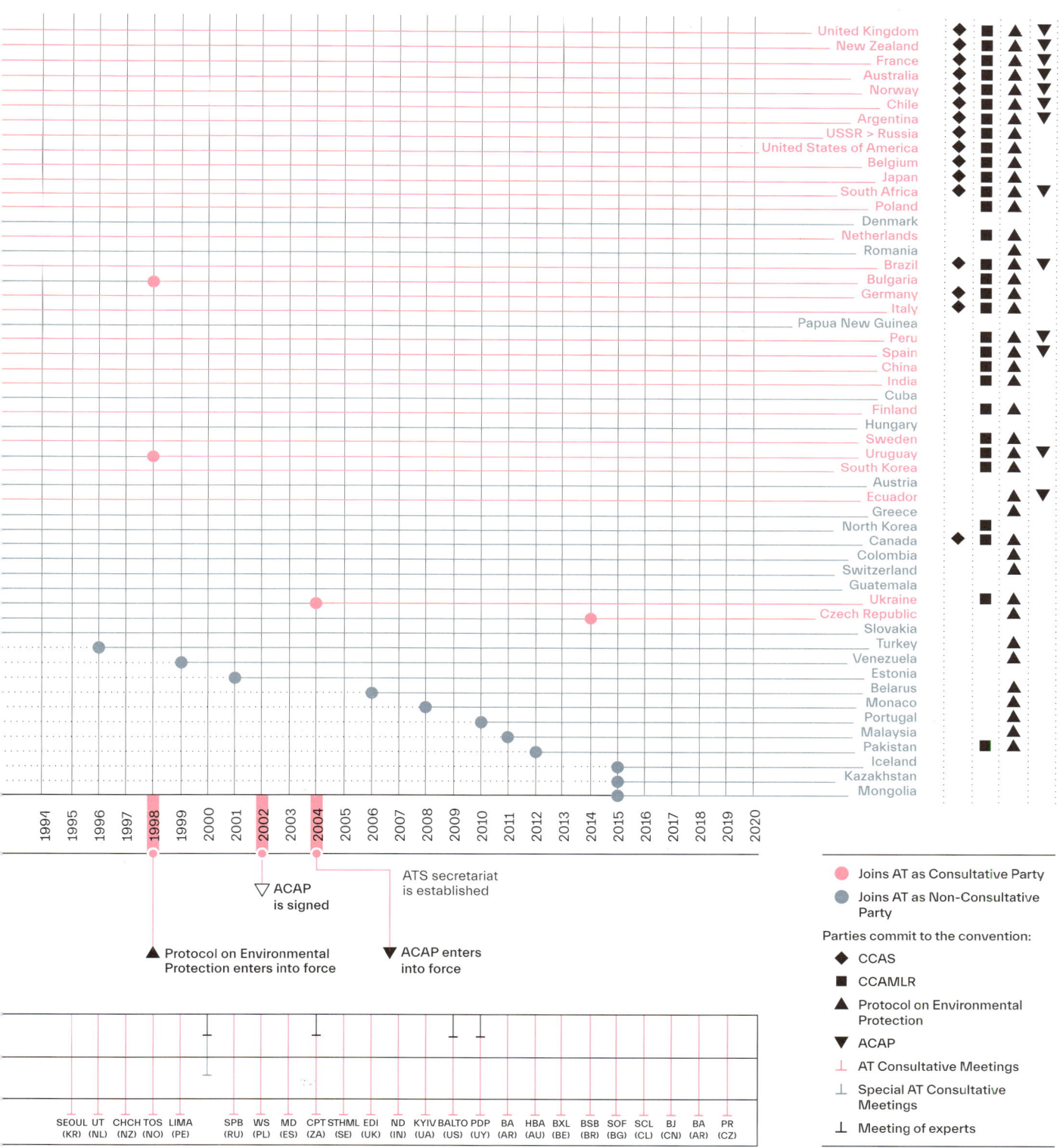

Joins AT as Consultative Party

Joins AT as Non-Consultative Party

Parties commit to the convention:

◆ CCAS

■ CCAMLR

▲ Protocol on Environmental Protection

▼ ACAP

⊥ AT Consultative Meetings

⊥ Special AT Consultative Meetings

⊥ Meeting of experts

▽ ACAP is signed

ATS secretariat is established

▲ Protocol on Environmental Protection enters into force

▼ ACAP enters into force

The Antarctic Treaty and the Protocol on Environmental Protection are the two key agreements around which the governance of the Antarctic is centred. Their instruments revolve around yearly meetings: the Antarctic Treaty Consultative Meeting (ATCM) and the Committee for Environmental Protection (CEP). The first sees Consultative and Non-Consultative Parties, alongside observers and invited experts, convene "for the purpose of exchanging information, consulting together on matters of common interest pertaining to Antarctica, and formulating and considering and recommending to their governments measures in furtherance of the principles and objectives of the Treaty" (Art. IX). The ambition of the ATCM is for the Consultative Parties to formulate (in collaboration with all other attendees) and adopt by consensus (binding) measures as well as (non-legally binding) decisions and resolutions that provide guidelines for the management of the Antarctic Treaty area. CEP meetings are held concurrently with the ATCM and focus on environmental protection matters. Alongside Consultative and Non-Consultative Parties, observers such as the Scientific Committee on Antarctic Research (SCAR), the Commission for the Conservation of Antarctic Marine Living Resources (CCAMLR) and the Council of Managers of National Antarctic Programs (COMNAP), as well as experts including the Antarctic and Southern Ocean Coalition (ASOC) and the International Association of Antarctica Tour Operators (IAATO), are invited to attend the meetings. Founded in 2003, the Buenos Aires-based Antarctic Treaty Secretariat supports the ATCM and CEP annual meetings – hosted by the Consultative Parties by rotation in alphabetical order of their English names – and facilitates the exchange of information between party members. The cancellation of the 2020 ATCM meeting scheduled for June in Finland due to COVID-19 raises questions about the future of the decision-making process on Antarctica and its reliance on a single yearly event.

"The geopolitics of Antarctica – complex, potentially explosive, deadly serious and ice cold – were played out not on the snow-fields but in the ring of the international circus of conferences they engendered and the carpeted corridors of capital cities."

Sara Wheeler (Writer), *Terra Incognita: Travels in Antarctica*, 1999

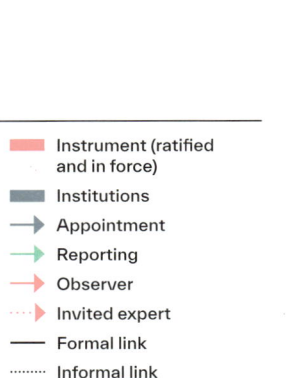

Instrument (ratified and in force)

Institutions

Appointment

Reporting

Observer

Invited expert

Formal link

Informal link

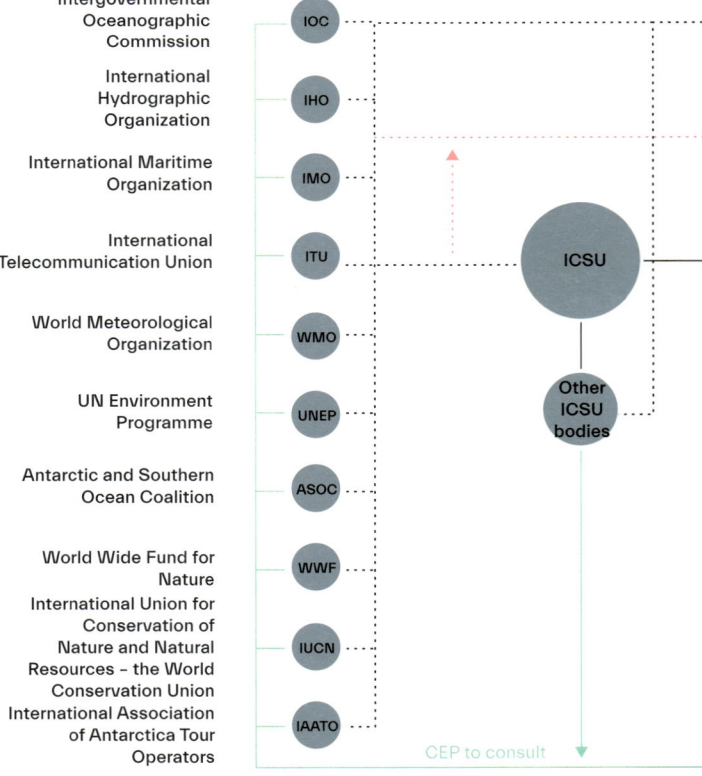

Intergovernmental Oceanographic Commission — IOC

International Hydrographic Organization — IHO

International Maritime Organization — IMO

International Telecommunication Union — ITU

World Meteorological Organization — WMO

UN Environment Programme — UNEP

Antarctic and Southern Ocean Coalition — ASOC

World Wide Fund for Nature — WWF

International Union for Conservation of Nature and Natural Resources – the World Conservation Union — IUCN

International Association of Antarctica Tour Operators — IAATO

ICSU

Other ICSU bodies

CEP to consult

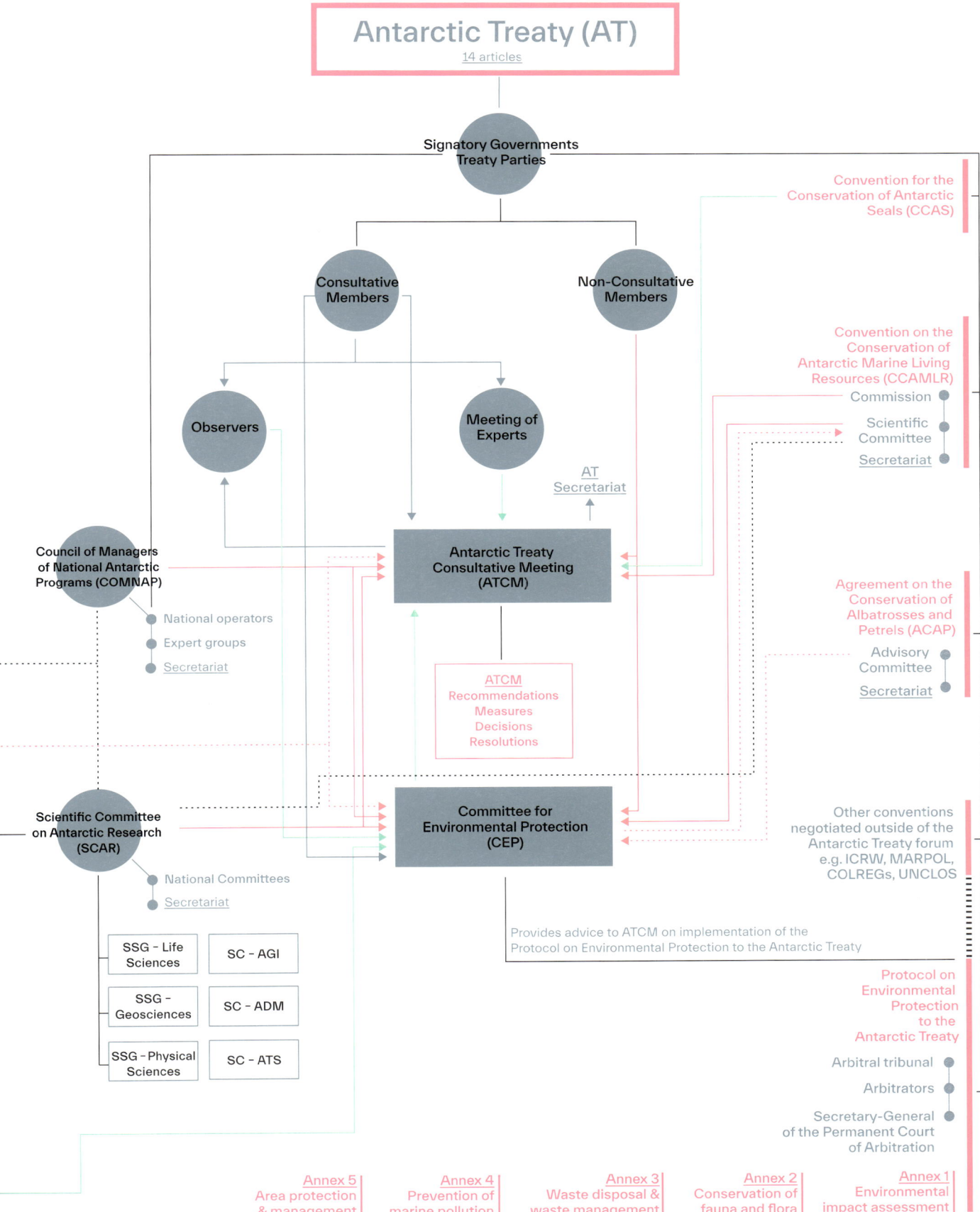

Antarctic Treaty (AT)
14 articles

Signatory Governments
Treaty Parties

Consultative
Members

Non-Consultative
Members

Observers

Meeting of
Experts

AT
Secretariat

Convention for the
Conservation of Antarctic
Seals (CCAS)

Convention on the
Conservation of
Antarctic Marine Living
Resources (CCAMLR)

Commission
Scientific
Committee
Secretariat

Council of Managers
of National Antarctic
Programs (COMNAP)

National operators
Expert groups
Secretariat

Antarctic Treaty
Consultative Meeting
(ATCM)

ATCM
Recommendations
Measures
Decisions
Resolutions

Agreement on the
Conservation of
Albatrosses and
Petrels (ACAP)

Advisory
Committee
Secretariat

Scientific Committee
on Antarctic Research
(SCAR)

National Committees
Secretariat

Committee for
Environmental Protection
(CEP)

Other conventions
negotiated outside of the
Antarctic Treaty forum
e.g. ICRW, MARPOL,
COLREGs, UNCLOS

Provides advice to ATCM on implementation of the
Protocol on Environmental Protection to the Antarctic Treaty

SSG – Life Sciences	SC – AGI
SSG – Geosciences	SC – ADM
SSG – Physical Sciences	SC – ATS

Protocol on
Environmental
Protection
to the
Antarctic Treaty

Arbitral tribunal
Arbitrators
Secretary-General
of the Permanent Court
of Arbitration

Annex 5
Area protection
& management

Annex 4
Prevention of
marine pollution

Annex 3
Waste disposal &
waste management

Annex 2
Conservation of
fauna and flora

Annex 1
Environmental
impact assessment

THE ANTARCTIC TREATY

The Governments of Argentina, Australia, Belgium, Chile, the French Republic, Japan, New Zealand, Norway, the Union of South Africa, the Union of Soviet Socialist Republics, the United Kingdom of Great Britain and Northern Ireland, and the United States of America,

Recognizing that it is in the interest of all mankind that Antarctica shall continue forever to be used exclusively for peaceful purposes and shall not become the scene or object of international discord;

Acknowledging the substantial contributions to scientific knowledge resulting from international cooperation in scientific investigation in Antarctica;

Convinced that the establishment of a firm foundation for the continuation and development of such cooperation on the basis of freedom of scientific investigation in Antarctica as applied during the International Geophysical Year accords with the interests of science and the progress of all mankind;

Convinced also that a treaty ensuring the use of Antarctica for peaceful purposes only and the continuance of international harmony in Antarctica will further the purposes and principles embodied in the Charter of the United Nations;

Have agreed as follows:

Information on how to access the complete version of the Antarctic Treaty can be found on page 910.

ARTICLE I

1.   Antarctica shall be used for peaceful purposes only. There shall be prohibited, _inter alia_, any measures of a military nature, such as the establishment of military bases and fortifications, the carrying out of military maneuvers, as well as the testing of any type of weapons.

2.   The present Treaty shall not prevent the use of military personnel or equipment for scientific research or for any other peaceful purpose.

ARTICLE II

Freedom of scientific investigation in Antarctica and cooperation toward that end, as applied during the International Geophysical Year, shall continue, subject to the provisions of the present Treaty.

ARTICLE III

1.   In order to promote international cooperation in scientific investigation in Antarctica, as provided for in Article II of the present Treaty, the Contracting Parties agree that, to the greatest extent feasible and practicable:

(a)   information regarding plans for scientific programs in Antarctica shall be exchanged to permit maximum economy and efficiency of operations;

(b)   scientific personnel shall be exchanged in Antarctica between expeditions and stations;

(c)   scientific observations and results from Antarctica shall be exchanged and made freely available.

The Scientific Committee on Antarctic Research

Gary Wilson

GARY WILSON is the chief scientist at the Institute of Geological and Nuclear Sciences and a professor of geology and marine science at the University of Otago (New Zealand). He is the vice-president of the Scientific Committee on Antarctic Research (SCAR) and has undertaken more than 30 research expeditions to Antarctica. He holds a BMus and a PhD. Dr Wilson's research has helped to uncover the dynamic history of Antarctica's climate and ice sheets.

The Scientific Committee on Antarctic Research, otherwise known as SCAR, is a thematic organisation of the International Council for Science[1] charged with facilitating and coordinating international Antarctic research and bringing issues emerging from Antarctic research to the attention of policymakers. SCAR, now formally in its sixth decade, has its roots in the 1957–1958 International Geophysical Year (IGY). During the IGY, a dozen nations maintained 40 geophysical stations in Antarctica; some were already operational, and others were established specifically to contribute to IGY.[2] The main focus of the IGY scientific programme in Antarctica was to develop an Antarctic-wide view of the geophysical properties of the continent. Studying Antarctic meteorology, the magnetic field and auroras, along with performing seismic soundings to determine ice thickness, were some of the major undertakings that saw scientists overwinter through the polar night across the continent including, for the first time, at the South Pole.

At the close of the IGY, the twelve nations involved decided to continue their Antarctic programmes and see Antarctica preserved as a continent for peace and science by means of the Antarctic Treaty.[3] At its first Consultative Meeting in 1961, Treaty parties adopted SCAR in an advisory capacity, partly to coordinate the international science effort in Antarctica, but also to progress and direct future scientific effort evaluating the continent's role within the global Earth system. From its inception, SCAR's purview has included not just the Antarctic continent, but also the surrounding Southern Ocean and subantarctic islands.

In its early years, SCAR was organised into a series of working groups focused around the major scientific disciplines and activities.[4] Some of the working groups, such as Cartography (later changed to Geodesy and Cartography), Meteorology, Upper Atmosphere Physics, and Crustal Geophysics, grew out of IGY activities. Others, such as Biology and Geology, were quickly added as SCAR turned its attention to defining the Antarctic continent and the life it supported. SCAR promptly also added conservation to the activities of the Biology working group. Logistics (later to become the Council of Managers of National Antarctic Programs, or COMNAP)[5] was also included in the early suite of working groups.

Among the various working groups, SCAR researchers set about documenting the seventh continent and unearthing the secrets it held. The first ice cores were collected from Byrd[6] and Vostok[7] stations, and Antarctica's place in continental drift was confirmed with the discovery of common fossils between the strata of the Beacon Supergroup exposed in the Transantarctic Mountains and similar sediments in Australia, South Africa and India.[8] The first few decades also saw SCAR set up a number of groups of specialists and working parties including groups working on seals, fishing and birds. The importance of krill in the food chain was also becoming apparent.[9] The cycle of specialist symposia in the geological, biological and physical sciences was also established and continues today on a four-year rotating cycle.

In the late 1970s and 1980s, membership of SCAR began to grow beyond the small group of members that had established SCAR twenty years previously. Scientific programmes also expanded with coordinated observations around the continent from different national programmes. Conservation practices progressed with the establishment of Antarctic Specially Protected Areas (ASPA)[10] and codes of conduct for Antarctic research and logistics activities. Expeditions reached most of the exposed land to document the geology of Antarctica, and traversed the continent making soundings through the ice sheet and collecting a wide range of ice cores. The later part of the 1980s also saw the discovery of the ozone hole above Antarctica.[11]

By the 1990s, SCAR membership had grown significantly and, along with the expanding science effort, the number of specialist science groups also grew into an increasingly complicated web of technical parties and working groups.[12] This led to a change in structure in the 2000s, with the main coordination of scientific activities occurring within three standing scientific groups – Life Sciences, Geosciences, and Physical Sciences, coordinating a number of evolving expert and action groups with particular focuses. By the early 2000s, SCAR was also taking a more global view in its scientific research programmes, which are time-limited (eight-year) internationally coordinated approaches to scientific challenges, including Antarctica and the Global Climate System, Antarctic Climate Evolution,[13] Evolution and Biodiversity in the Antarctic, Subglacial Lake Exploration[14] and Interhemispheric Conjugacy Effects in Solar-Terrestrial and Aeronomy Research. These new collaborative programmes resulted in some landmark assessments of the continent, such as the report by Turner and colleagues on the state of climate change in Antarctica.[15] SCAR also became more involved with collaborative programmes involving the Scientific Committee on Oceanic Research and the World Metrological Office, and they have joined with the International Arctic Science Council to take a bipolar approach to more global issues, including a joint effort in the 2007–2008 International Polar Year.[16]

The restructuring of SCAR also recognised its important role in providing scientific advice through its Standing Committee on the Antarctic Treaty System and nurturing future Antarctic researchers through capacity-building, education and training. A further offshoot of the International Polar Year is the Association of Polar Early Career Scientists[17] which now provides important input on the development of research programmes. Recently, SCAR has also expanded its Social Science Research through the Standing Committee on Humanities and Social Sciences.

SCAR research is now firmly focused on some of the more significant challenges facing humanity: the impact of the changing climate on Antarctica's ice sheets and surrounding oceans and the threat to biodiversity that comes with changing terrestrial and ocean conditions in the Antarctic. It is through major international collaborative programmes that SCAR researchers have not only responded to the global challenge of climate change but have directly influenced our understanding of how the global climate system works. Long ice cores from the European Project for Ice Coring in Antarctica (EPICA)[18] and the geological drilling beneath the Ross Ice Shelf (ANDRILL project)[19] taught us how important changing levels of carbon dioxide in the atmosphere are and how they are linked to changing global temperatures and changing sea levels from melting ice sheets. Attention is now turning to how resilient the Antarctic system will be to the current changes in global temperature to identify – and ideally help to prevent – future ecological tipping points.

At present, the most prominent collaborative Scientific Research Programmes (SRPs) of SCAR include AntClim21, AntEco, AnT-ERA, PAIS and SERCE. Antarctic Climate Change in the 21ˢᵗ Century (AntClim21), State of the Antarctic Ecosystem (AntEco), Antarctic Thresholds – Ecosystem Resilience and Adaptation (AnT-ERA), Past Antarctic Ice Sheet Dynamics (PAIS), and Solid Earth Response and Influence on Cryosphere Evolution (SERCE). With 44 member countries, and 9 International Science Council unions, SCAR's pivotal role in facilitating international scientific collaboration in and for Antarctica remains unquestioned.

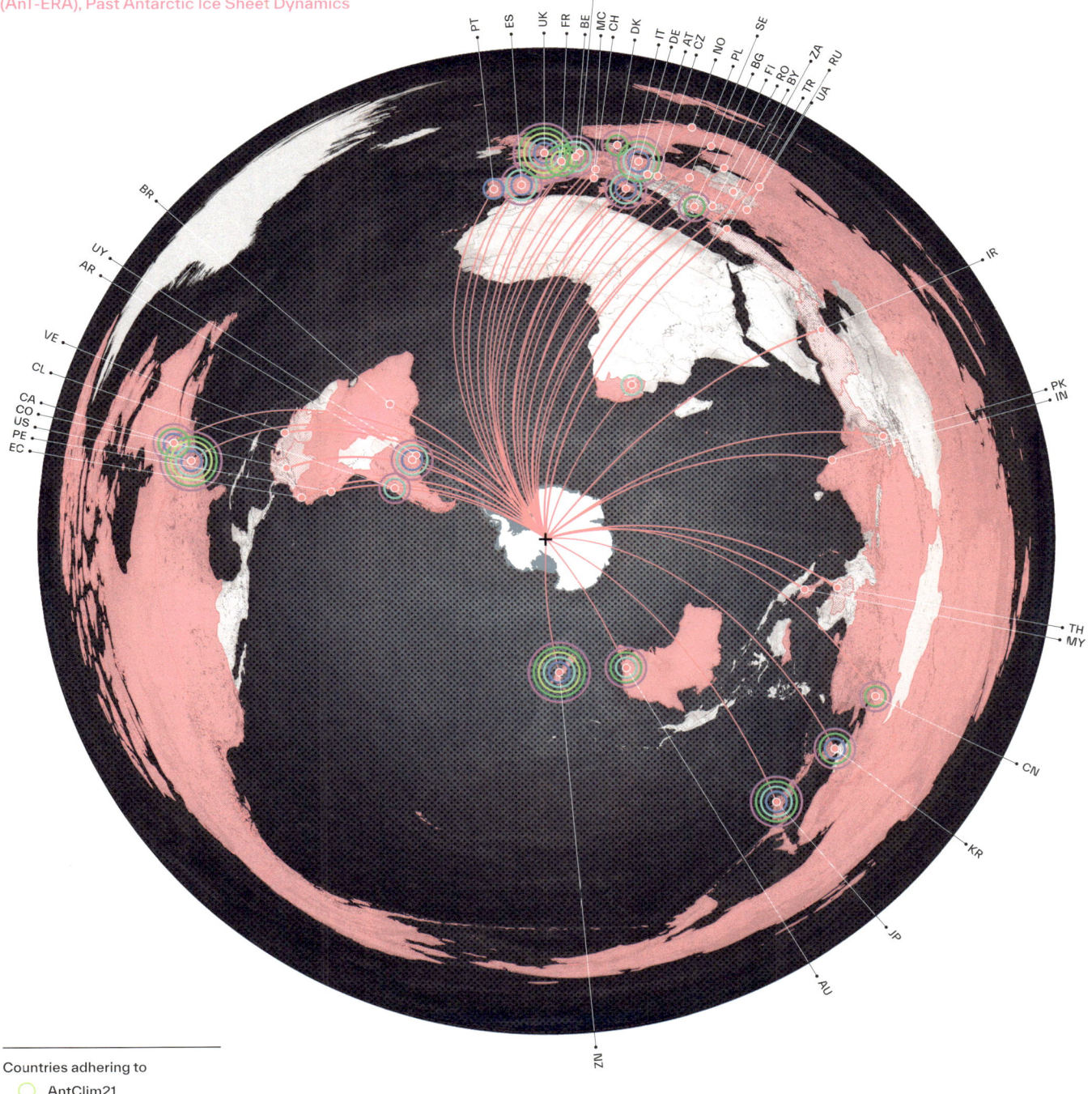

Countries adhering to

○ AntClim21
○ AntEco
○ AnT-ERA
○ PAIS
○ SERCE
▨ SCAR (full member)
▨ SCAR (associate member)

Facilitating International Collaboration in Antarctic Science. The Role of COMNAP

Gillian Wratt

GILLIAN WRATT is a past chief executive of Antarctic New Zealand, the chair of the Council of Managers of National Antarctic Programs (COMNAP), and the vice-chair of the Committee for Environmental Protection (CEP). She is the author of *A Story of Antarctic Co-operation: 25 Years of the Council of Managers of National Antarctic Programs* (COMNAP, 2013). Wratt now chairs the Steering Group for the NZ Antarctic Science Platform.

The Council of Managers of National Antarctic Programs (COMNAP) was created in 1988 by the managers of the Antarctic programmes of the then 22 Consultative Parties to the Antarctic Treaty. It had, and still has, a pragmatic rationale of sharing knowledge and expertise, facilitating cooperation, and providing practical, technical and non-political advice to the Antarctic Treaty Consultative Meetings (ATCMs). COMNAP's purpose is "to develop and promote best practice in managing the support of scientific research in Antarctica".[1]

The members of COMNAP are the organisations responsible for the operation of Antarctic stations and the infrastructure that supports science across the continent. They put the Antarctic Treaty System principles of "a natural reserve devoted to peace and science" into effect on a day-to-day basis and have unparalleled first-hand knowledge of Antarctica, enabling the international cooperation on which Antarctic activities rely.

Today, COMNAP brings together the national Antarctic programmes of all 29 countries that are the current Consultative Parties to the Antarctic Treaty and one from a Non-Consultative Party (Belarus). In addition, six observer programmes participate in COMNAP. Between them, the COMNAP members manage and operate 86 Antarctic stations, 10 intercontinental runways and approximately 50 research and resupply vessels. Many of the COMNAP members have operated in the Antarctic since the International Geophysical Year (IGY) in 1957–1958. As with other international organisations, its members operate using different languages, cultures and levels of resourcing for their activities, as well as markedly different organisational structures within their governments.

Despite its structure as an association comprised of government programmes, COMNAP was not set up at the instigation of governments nor as a formal mechanism of the Antarctic Treaty. It has no official intergovernmental status or funding and was initially looked upon with some trepidation by both the diplomatic community and the Scientific Committee on Antarctic Research (SCAR). Through its relations with the formal Antarctic Treaty mechanisms, it has to tread carefully between formal positions of its members' governments, providing practical advice based on the realities of managing national programme activities in the uncompromising Antarctic environment. In this context, it has offered advice to the ATCM on wide range of topics, including waste management, oil spills, emergency response and contingency planning, environmental monitoring, visitor management, air and ship operations, and search and rescue.

COMNAP is a recognised ATCM observer organisation, holding the same status as SCAR and the Commission for the Conservation of Antarctic Marine Living Resources (CCAMLR). This status provides for COMNAP to contribute advice and recommendations to the ATCM in the form of working papers on a range of topics within its purview. Thus far, COMNAP has submitted 157 papers to the ATCM. From 1992 to 2010, it was the thirteenth most frequent contributor of working papers to the meetings.[2]

Science is the currency of the Antarctic Treaty. Antarctic science is facilitated, supported and delivered through COMNAP members. In addition to providing advice to the ATCM and providing a means for communication between national programme managers, it provides forums and guidance for its members to facilitate collaboration in supporting Antarctic science. It has held regular COMNAP and logistics symposiums, run workshops on a wide range of topics, developed guidelines and manuals, set up online systems, and engaged effectively on Antarctic science and operational issues with SCAR and other organisations. One of these initiatives is the Antarctic Station Catalogue.[3] This project began as a collaboration with the EU-PolarNet on their European Polar Infrastructures Project. Information on the Antarctic stations of all 30 COMNAP member programmes can be found in the Catalogue, which was produced as a tool for supporting international cooperation in science. The Catalogue is also an informative tool for the broader Antarctic community and for the general public.

In 2014, the SCAR Horizon Scan exercise identified the 80 most critical Antarctic research questions for the next 20 to 30 years.[4] COMNAP followed this with an Antarctic Roadmap Challenges Project to provide a framework for government investment in the enabling technologies, essential access, infrastructure and logistical support and people needed to answer these critical questions.[5] Achieving this collective vision of a possible path to the future will not be possible without international collaboration through COMNAP.

COMNAP's accumulated expertise and practical solutions to working in a challenging environment are essential to meeting the demanding international and scientific objectives for Antarctica in the twenty-first century.

> "Since […] active performance must mostly be an independent or even unaided effort on the part of an individual nation, the criterion fosters nationalism and at first effectively discouraged the introduction of genuinely multi- or international research platforms of the kind one finds in other areas of Big Science, for example physics (CERN in Geneva) […] The advent of real 'international' stations simply under a SCAR flag or the AT's own emblem still seems a long way off."
>
> Aant Elzinga (Emeritus Professor, University of Gothenburg), "The Continent for Science" in *Handbook on the Politics of Antarctica*, 2017

Founded in 1988, COMNAP brings together
30 National Antarctic Programs of Consultative
and Non-Consultative Parties to the Antarctic
Treaty, and 6 observer National Antarctic
Programs. The COMNAP Secretariat, originally
located in Washington, DC and then Tasmania,
is presently situated in Christchurch.[1]

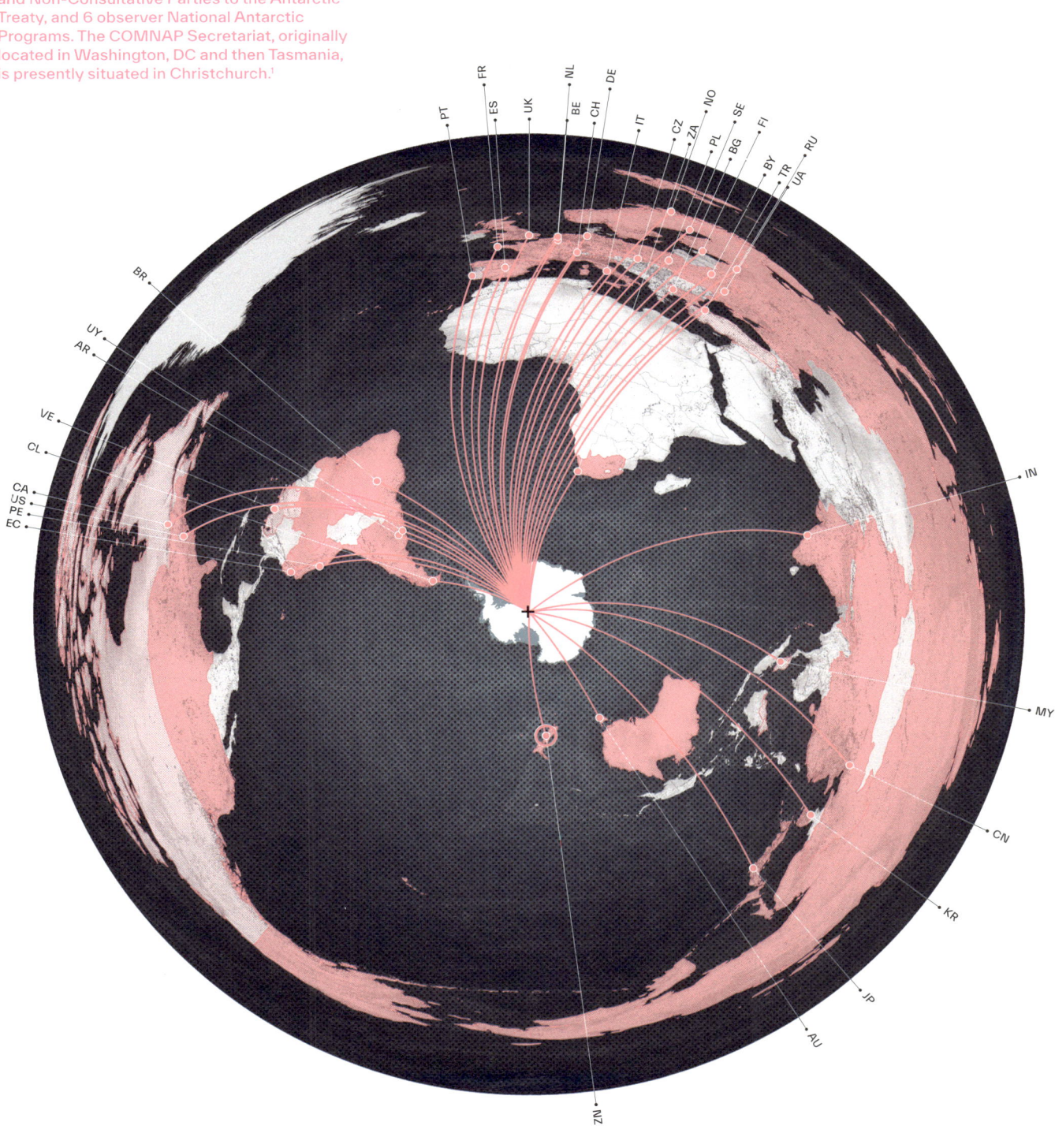

COMNAP members
COMNAP observers
Program headquarters
COMNAP headquarters
South Pole

The National Antarctic Programs, active since 1928 as shown in the timeline below, differ substantially in their organisational structures. Ranging from being rather independent bodies devoted to science and preservation (as in the United Kingdom model, whereby all Historic Sites and Monuments and research stations are owned by UK Research and Innovation, UKRI), to military-supported scientific institutions (as exemplified in the Chilean case study), each nation has an intricate Antarctic management system.[2]

UK Government
FCO
Department for Business, Energy, Industry and Strategy
UKRI
NERC
Historic Sites and Monuments
UKAHT
Detaille Hut
Stonington Island
Damoy Hut
Port Lockroy
Wordie House
Horseshoe Island
Subantarctic Research Stations
Antarctic Research Stations
Antarctic Support Stations
BAS
Bird Island
King Edward Point
Fossil Bluff
Sky Blu
Halley VI
Signy
Rothera

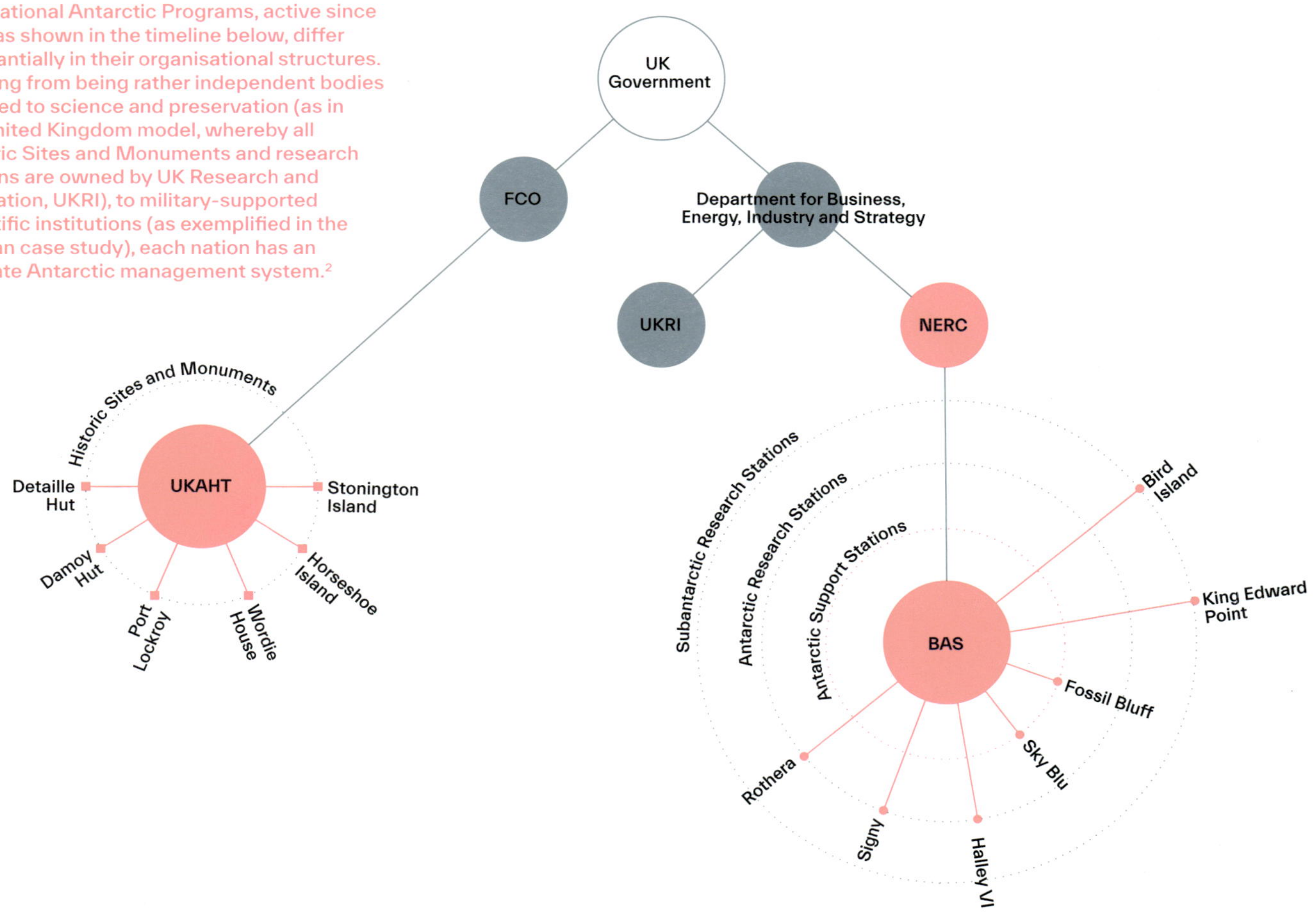

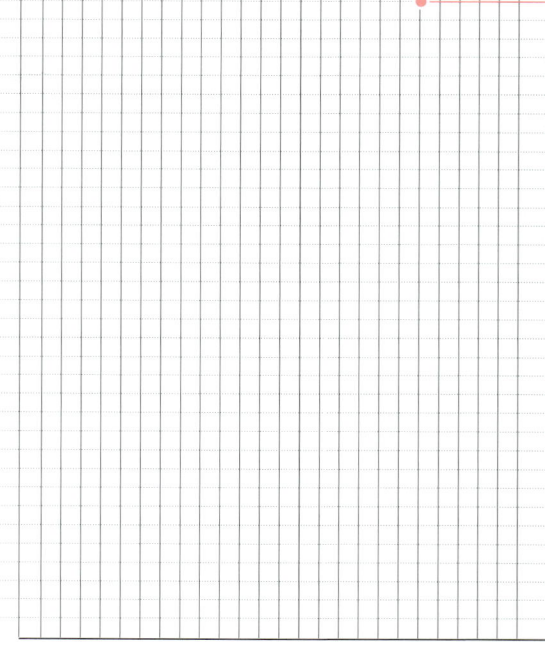

Norway – Norwegian Polar Institute
Australia – Australian Antarctic Program (AAD)
Peru – Division of Antarctic Affairs
Japan – National Institute of Polar Research (NPRI)
Russia – Arctic and Antarctic Research Institute (AARI/RAE)
United States – National Science Foundation (NSF) Office of Polar Programs
South Africa – South African National Antarctic Programme (SANAP)
United Kingdom – British Antarctic Survey (BAS)
Chile – Chilean Antarctic Institute (INACH)
Argentina – Argentine Antarctic Program (DNA)
Poland – Institute of Biochemistry and Biophysics (PAS) Department of Antarctic Biology
Uruguay – Uruguayan Antarctic Institute (IAU)
Germany – Alfred Wegener Institute for Marine and Antarctic Research (AWI)
China – Chinese Arctic and Antarctic Administration (CAA); Polar Research Institute of China (PRIC)
India – National Centre for Polar & Ocean Research (NCPOR)
Brazil – Brazilian Antarctic Program (PROANTAR)
Sweden – Swedish Polar Research Secretariat
Belgium – Belgian Federal Science Policy and Polar Secretariat
Italy – Italian National Antarctic Research Program (PNRA)
Republic of Korea – Korean Polar Research Institute (KOPRI)
Ecuador – Ecuadorian Antarctic Institute (INAE)
Finland – Finnish Antarctic Research Program at the Finnish Meteorological Institute (FINNARP)
Canada – Polar Knowledge Canada (POLAR)
France – French Polar Institute Paul-Émile-Victor (IPEV)
Bulgaria – Bulgarian Antarctic Institute (BAI)
Ukraine – National Antarctic Scientific Center of Ukraine
Czech Republic – Masaryk University
New Zealand – Antarctica New Zealand
Spain – Spanish Polar Committee (CPE)
The Netherlands – Netherlands Organization for Scientific Research (NWO)
Portugal – Portuguese Polar Program (PROPOLAR)
Venezuela – Venezuelan Institute for Scientific Research
Malaysia – Sultan Mizan Antarctic Research Foundation
Belarus – Belarus National Academy of Sciences
Turkey – Polar Research Institute (PRI)
Switzerland – Swiss Committee on Polar and High Altitude Research
1928
1929
1930
1931
1932
1933
1934
1935
1936
1937
1938
1939
1940
1941
1942
1943
1944
1945
1946
1947
1948
1949
1950
1951
1952
1953

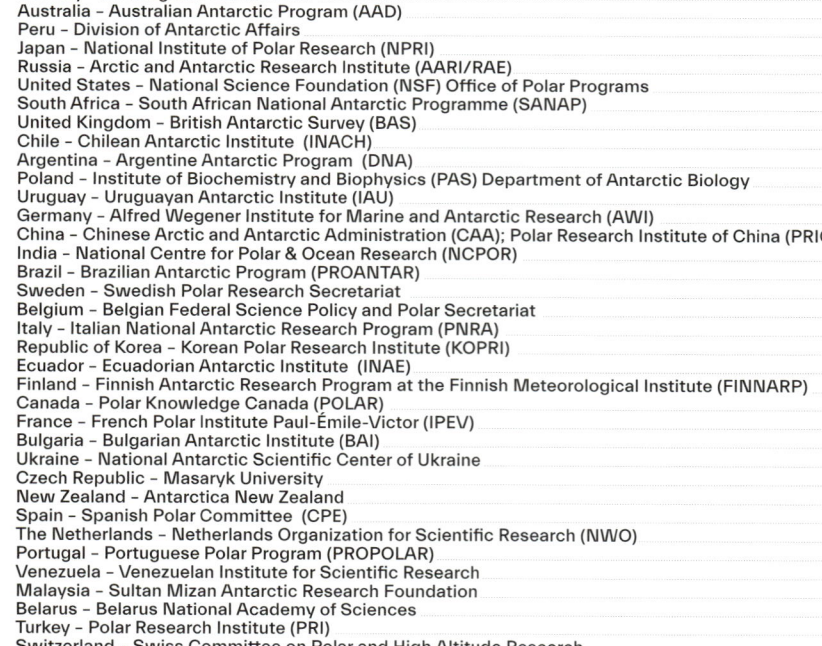

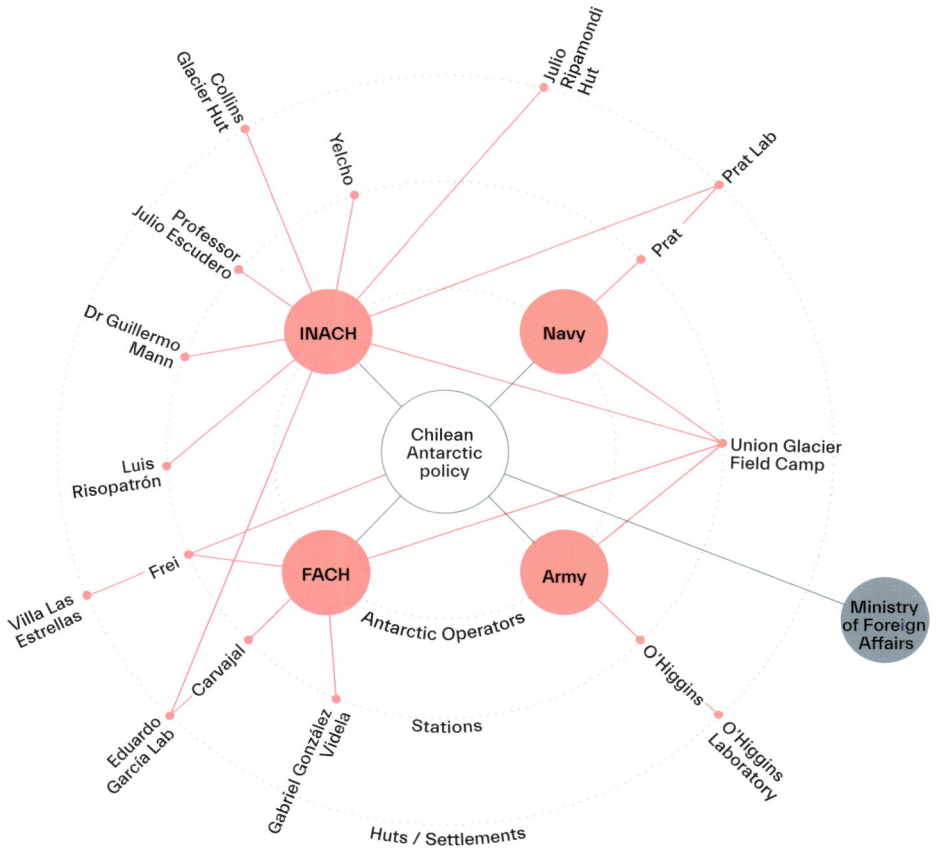

Chilean Antarctic policy
INACH
Navy
FACH
Army
Ministry of Foreign Affairs
Antarctic Operators
Stations
Huts / Settlements
Collins Glacier Hut
Yelcho
Julio Ripamondi Hut
Prat Lab
Prat
Professor Julio Escudero
Dr Guillermo Mann
Luis Risopatrón
Frei
Villa Las Estrellas
Carvajal
Eduardo García Lab
Gabriel González Videla
Union Glacier Field Camp
O'Higgins
O'Higgins Laboratory

Organisation operating in the Antarctic

Governmental Organisation

Historic Site and Monument

Antarctic Station

United Kingdom management

FCO: Foreign & Commonwealth Office Polar Regions Department

BAS: British Antarctic Survey

International Association of Antarctica Tour Operation

NERC: Natural Environment Research Council

UKAHT: United Kingdom Antarctic Heritage Trust

UKRI: United Kingdom Research and Innovation

Chile management

FACH: Chilean Air Force

INACH: Chilean Antarctic Institute

Establishment of National Programme

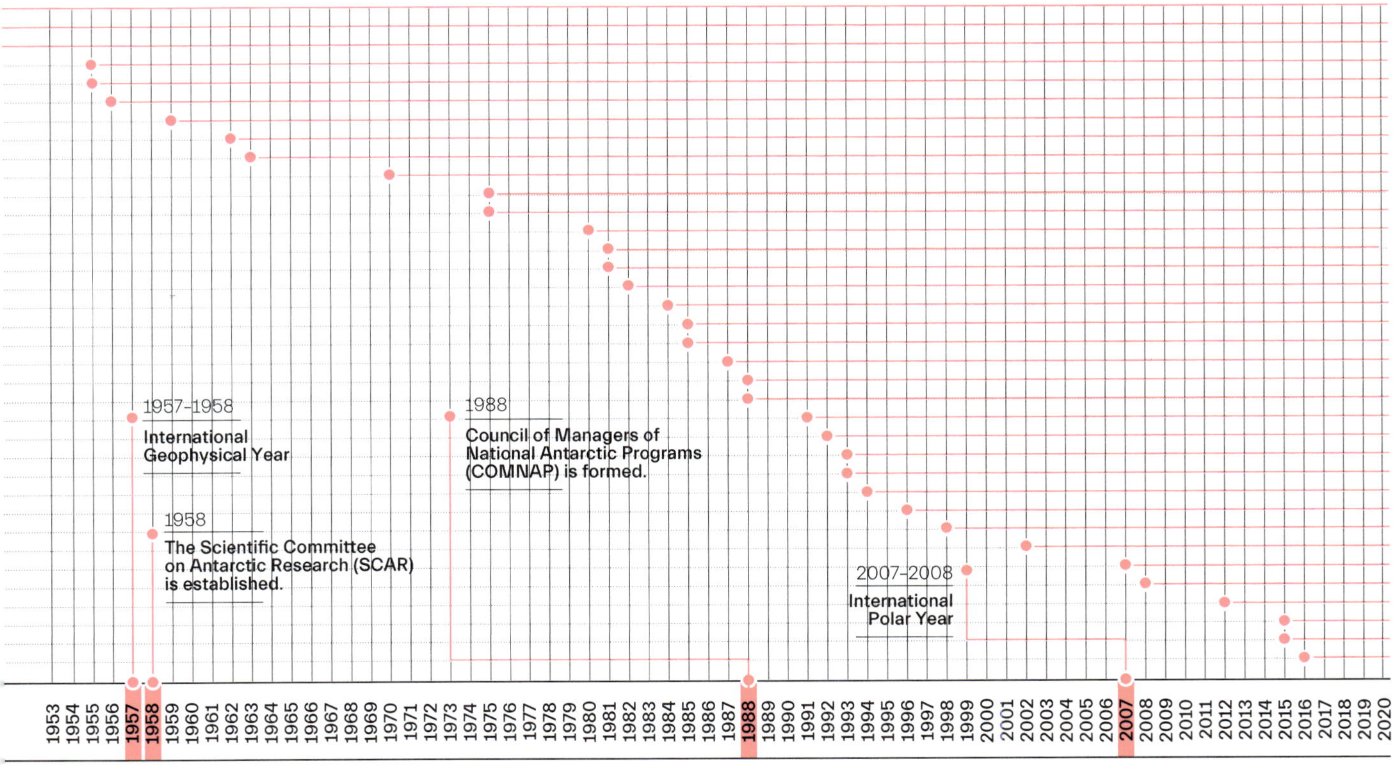

1957–1958
International Geophysical Year
1958
The Scientific Committee on Antarctic Research (SCAR) is established.
1988
Council of Managers of National Antarctic Programs (COMNAP) is formed.
2007–2008
International Polar Year
1953 1954 1955 1956 1957 1958 1959 1960 1961 1962 1963 1964 1965 1966 1967 1968 1969 1970 1971 1972 1973 1974 1975 1976 1977 1978 1979 1980 1981 1982 1983 1984 1985 1986 1987 1988 1989 1990 1991 1992 1993 1994 1995 1996 1997 1998 1999 2000 2001 2002 2003 2004 2005 2006 2007 2008 2009 2010 2011 2012 2013 2014 2015 2016 2017 2018 2019 2020

Antarctica's Latest Governance Challenge. Coronavirus

Alan D. Hemmings and Bob Frame

ALAN D. HEMMINGS is professor at the Gateway Antarctica Centre for Antarctic Studies and Research at the University of Canterbury (New Zealand). BOB FRAME is an adjunct associate professor at the Gateway Antarctica Centre for Antarctic Studies and Research at the University of Canterbury (New Zealand).

In the first quarter of 2020, in the year of the bicentenary of the discovery of the continent, amidst the blizzard of news stories around the coronavirus pandemic (COVID-19), unsurprisingly, little attention was given to Antarctica. Then, Antarctic coverage began to cast Antarctica itself as a place untouched while reporting the nightmare of tourism vessels coming out of Antarctica with suspected or actually infected people aboard being denied port entry. Through the second half of 2020, attention shifted to the challenges the pandemic posed to the normal operation of Antarctic science programmes and tourism operation for the forthcoming Antarctic summer season of 2020–2021. The broader implications of COVID-19 in and around Antarctica and for the governance and management of Antarctic affairs largely escaped consideration.

Yet COVID-19 poses governance issues for Antarctica; its effects on the continent, either directly or tangentially (i.e. manifesting elsewhere), are here examined across three notional time periods: the period from eruption of the pandemic in early 2020 until the commencement of the 2020–2021 Antarctic summer season (say, November 2020); the period of the 2020–2021 Antarctic summer season (November 2020 to April 2021); and the longer term (from mid-2021 out some years ahead).

Antarctic stations in East Antarctica were largely isolated by March 2020. In the Antarctic Peninsula, states shut down vessel access to their stations from mid-March in order to reduce their risk exposure as the seriousness of the pandemic became clear. The challenges of 2020 were to ensure the safe return of vessels, aircraft, expedition staff, crews and passengers from Antarctica after the 2019–2020 operating season. Most national Antarctic programmes ordinarily end their operating season, change over personnel, and return outgoing people, aircraft and ships to their home bases by the end of April. This was achieved in 2020 as well, albeit with obvious complications around transit through Antarctic gateway states which were introducing more stringent border controls. Sadly, in late December 2020, Chilean authorities reported that up to 36 people associated with its O'Higgins Station on the northern Antarctic Peninsula had tested positive for the virus. Later reports (sourced to the Chilean military) took the number up to 58, with 21 of these aboard the Chilean Navy ship Sargento Aldea, which had serviced the O'Higgins Station. A further case was reported from Villa Las Estrellas, next to Chile's main facility on King George

Island (in the South Shetland Islands, quite a long way from O'Higgins, which is on the mainland of the Antarctic Peninsula), where Chile maintains an airstrip used by a number of other programmes. King George Island has the largest number of Antarctic stations of any location in the Antarctic, and so any breakout into that wider international community would be a serious matter. Happily, at the time this went to press (mid-January 2021), no further instances were reported.

Because tourism operates into April in the Antarctic Peninsula, this activity, routing home via South America, was where we saw most issues. A small number of vessels reported infection, and in at least one case very high levels, which confined conditions, geographical remoteness and the difficulties of accessing ports, medical care and repatriation exacerbated into frightening situations.

Antarctica is not just a place; it is also an international geopolitical and legal focus involving more than 50 states, and this is ordinarily given effect through the diplomatic meetings and associated technical meetings of the Antarctic Treaty System.

In 2020 virtually all global diplomatic meetings closed down. Some major meetings were postponed; others were switched to scaled-down online sessions. The contemporary Antarctic Treaty System (ATS) is organised around two annual diplomatic meetings: the Antarctic Treaty Consultative Meeting (ATCM) – which gives effect to the 1959 Antarctic Treaty and the 1991 Protocol on Environmental Protection; and the meeting of the Commission for the Conservation of Antarctic Marine Living Resources (CCAMLR) – which gives effect to the 1980 CAMLR Convention. Both depend on technical groups to provide critical input for the political decision-making process. Decision-making in both forums is through consensus of the parties in attendance. The ATS diplomatic programme was largely curtailed in 2020.

The ATCM was due to be held in Helsinki starting in late May 2020. It was cancelled due to COVID-19, and thus there was no capacity to formally and collectively agree to do anything in relation to the pandemic (e.g., prohibit ship visits to stations, share data on risk management, or adopt measures to mitigate risk to biota such as seals). Given the fact that COVID-19 did not reach Antarctica in 2020, and with reductions in human activity in Antarctica, some of this may be moot, but it is striking that at a time of international uncertainty the institution

responsible for Antarctic management appeared to have lost the capacity to act collectively.

The second diplomatic meeting of 2020 – the Commission of CCAMLR, scheduled between the 26th of October and the 6th of November – was significantly reduced and conducted entirely online. Some preparatory technical group meetings – in Tokyo and Kochi – were cancelled, and the advisory Scientific Committee meeting, like the Commission, was significantly scaled back and conducted online.

Cancelling the CCAMLR Commission meeting for 2020 would have had significant consequences for Antarctic activity, since, among other things, the annual Commission meeting allocates fish catches for the year ahead. Given that closing down the fishery was judged not to be acceptable (globally, states have treated fishing like agriculture, as a vital activity to maintain), once a meeting in person was ruled out (the usual venue is Hobart, Tasmania, so a very long way for most delegations and thus additionally risky for them and Hobart's residents in a pandemic), agreement had to be reached on conducting a decision-making meeting remotely for the first time in the history of the ATS.

If collective decision-making under the ATS is effectively suspended (as in the case of the ATCM) or significantly constrained (as in the case of the virtual CCAMLR Commission meeting), much more decision-making is left to individual states, residing most significantly, perhaps, with those southern "gateway" states whose cities provide final departure and reception ports for Antarctic expeditions – Ushuaia (Argentina), Punta Arenas (Chile), Cape Town (South Africa), Hobart (Australia), Christchurch and Lyttelton (New Zealand). Indeed, as the COVID-19 pandemic unfolded, the International Association of Antarctica Tour Operators (IAATO) advised that "all vessels scheduled to depart for the peninsula [sic] and/or the sub-Antarctic islands have ceased, or been postponed due to new regulations imposed at Antarctic Gateways".

For the 2020–2021 Antarctic summer season, and the years beyond, gateway states will be a factor in policymaking around national Antarctic programme activity and critical in any Antarctic tourism that survives and any fishing activity that necessitates port calls. They will be heavily invested, not only because of concerns about domestic exposure risk – and the fact that it is their hospitals to which injured or ill people are evacuated from Antarctica – but also because they are also

"A provisional and contested set of statements about how the world is cannot be used directly as a rule for what governments should do.

Ministers have to decide for themselves. They must take responsibility for these decisions and their own inevitable mistakes, rather than relying

on science as if it were an apolitical and indisputable tablet of stone."

Alex Stevens (Professor, University of Kent), "Governments Cannot Just 'Follow the Science' on COVID-19", *Nature Human Behaviour*, 2020

the states responsible for air and maritime rescue coordination in the sectors that extend down into the Antarctic. The Council of Managers of National Antarctic Programs (COMNAP), whose purpose is "to develop and promote best practice in managing the support of scientific research in Antarctica", will also be important in enabling coordination between national Antarctic programmes.

Worldwide, cruise tourism has been devastated by COVID-19, with consequences for polar tourism, too, for the 2020 Arctic summer season in which some ordinarily expect to operate and the 2020–2021 Antarctic summer season. Large global tourism companies appear to be mothballing ships and plans, so the global conveyor belt of large-vessel tourism will likely mostly shut down as well. States prohibited or otherwise discouraged visits both to their metropolitan territories and to Antarctic stations during the 2020–2021 Antarctic summer. If COVID-19 is still active (as seems probable), and vaccination is not yet widespread by the middle of 2021, we would expect this to continue. One has to say that the outlook for Antarctic tourism looks bleak over the next few years.

For most national Antarctic programmes, operational constraints due to domestic lockdowns, reduced finances (as in the case of Ecuador in 2018 due to internal austerity measures), and/or redirection of national efforts will be significant. Many states reduced the scale of their Antarctic science programmes in the 2020–2021 Antarctic summer season. Some may reduce programmes or defer station rebuilds or equipment acquisition, and some may be unable to sustain their programmes over the next several years. Among the effects, there was little significant Antarctic climate-change-related fieldwork in the 2020–2021 summer season, although remote-sensing monitoring from research stations and work on existing data sets continued.

To the extent that Antarctic operations continue through 2021 and beyond, there will be particular concern to minimise the risk of introducing COVID-19. Although the immediate focus will be on avoiding human infection, the finding that the virus has contaminated zoo animals (big cats) suggests that care should be taken to prevent any Antarctic infection of biota – particularly of seals. So, for some time yet, in addition to extraordinary precautions to prevent transmission through station resupply and personnel changeovers, perhaps requiring the screening of ingoing people, there may be significant reductions in multinational

projects and field operations – a mainstay of normal Antarctic science – and even of purely national research where this concerns potentially at-risk biota.

If some of the issues canvassed here prove correct – the likely collapse of the Antarctic tourism industry and a contraction in national Antarctic programme activities and Antarctic marine harvesting – one might expect the effects of these to resonate for some years, perhaps up to a decade from now. There may be structural changes in the nature of Antarctic activity and mechanisms of governance.

Building on the precedent and experience of conducting a virtual CCAMLR in 2020, if no ATS diplomatic meetings can be convened in person again in 2021, mechanisms will be needed to allow collective Antarctic governance and decision-making remotely. Technical advisory groups to the ATCM and CCAMLR already do much of their work between sessions and remotely, but they would need to be formally mandated, and hitherto this has been done at physical meetings. Remote decision-making mechanisms will need, *inter alia,* to be able to establish and mandate these groups. This is new ground for the ATS, but it is for the vast majority of diplomatic and technical fora in every other field as well.

The problems facing an Antarctic tourism industry hitherto largely based on ships and "expedition cruising" may accelerate the long-anticipated (but not manifested) shift from ship-borne to air-mediated tourism. If so, this has implications around mass-tourism growth trajectories, infrastructure ashore (airstrips and hotels), and in-area small aircraft and ship operations. Alternatively, after a hiatus, we may see a recovery to essentially the present model of tourism, given the recent commissioning of new polar vessels.

Might crises concerning the viability of Antarctic stations encourage station sharing or international stations? Do we face the possibility that some states will disappear from Antarctic engagement for shorter or longer periods? Will it stimulate the use of different ways of working and acquiring data such as remote platforms (robots, buoys and satellites) with a greater emphasis on predictive modelling?

If the states that are currently major drivers of that research (e.g. the US) are among the worst hit by the pandemic, we could witness an overall reduction in Antarctic research over the next few years. Will this catalyse the

displacement of the United States by another single state (China?) or a group of states as leaders of international research in Antarctica, or will everything essentially "snap back" to the *status quo ante* at some point? The pandemic had hardly got underway before *The Atlantic* magazine was reporting the suggestion of some international relations scholars that some states were "taking advantage" of the pandemic in Antarctica.

Beyond economics, much of course hinges on how "science" itself comes out of a global crisis whose understanding and resolution seem utterly dependent upon its effective deployment. If COVID-19 reinvigorates a rational, evidence-based and integrated approach to human challenges and aspiration, then a revisualisation of Antarctica as a place in which to conduct vital international scientific research – most obviously in relation to anthropogenic climate change – is in prospect. However, if in the meantime humanity as a whole loses one (or two) seasons of climatological field work, there might be unrecoverable gaps in the small number of long-term climate records that we depend upon at a time when attempts are being made to attribute changes to anthropogenic causes.

Finally, what about the environment? Media coverage has made much of apparent environmental recovery during the shutdown of significant swathes of human activity – whether tourism in Venice, air traffic over Europe, industrial emissions in China, or cetaceans popping up in previously congested harbours. Antarctica, however, is not a site of industrial activity nor is it subject to activity whose halting would stimulate comparable transformations. Even over (say) two years of appreciably reduced global human activity, it is unlikely that we could detect any transformation in relation to climate change or other environmental parameters. Only if the consequences of COVID-19 resulted in structural changes in human activity or behaviour over a longer period would we expect to see effects in the Antarctic. Perhaps, for reasons going beyond Antarctic futures, this is what we might hope for.

Antarctic Law Enforcement

Donald R. Rothwell and Karen N. Scott

DONALD R. ROTHWELL is a professor of international law at the Australian National University College of Law. KAREN N. SCOTT is a professor of law at the University of Canterbury (New Zealand) and the president of the Australian and New Zealand Society of International Law.

Antarctica is currently experiencing and will continue to experience ever greater numbers of visitors.[1] Whether they come as members of accredited scientific research expeditions, as fishermen exploiting the Southern Ocean, or as tourists engaging in air- or sea-based activities, the number of persons visiting and temporarily staying in Antarctica is increasing. For example, tourists from China travelling to Antarctica rose from fewer than 100 in 2008 to 8,000 by 2018.[2] When the greater accessibility of Antarctica, owing to more transportation options and the impacts of climate change, is also taken into account, all of the signs point to greater human activity in Antarctica into the future. More Antarctic visitors raise issues with respect to the application of law and ultimately law enforcement. Antarctic law applicable to all persons on the continent is diverse, and can extend from the prohibition of mining activities[3] to criminal law for common assault,[4] resulting in a range of different interests in law enforcement, including the interests of the global community and those of the territorial claimants.

Antarctic law enforcement in the 2020s also raises issues that affect Antarctic security. The ability of a state to apply and enforce its laws is one of the traditional elements of sovereignty. Effective law enforcement is generally taken as a given in the developed world. A failure on the part of one of the seven Antarctic claimants to apply and enforce law could therefore be used against that state if a territorial claim was contested. It would be argued that a territorial claim has not been perfected if law cannot be enforced.[5]

Law enforcement is also crucial to the success and security of the Antarctic Treaty System (ATS), as is the case with the effectiveness of most treaty regimes. At a base level, lawlessness among scientific research expeditions is clearly not conducive to the efficient conduct of Antarctic scientific research. Visitors such as tourists also need to be aware that they are subject to laws that regulate their activities, otherwise "codes of conduct" for the protection of the environment and especially wildlife will ring hollow.[6] Those Antarctic visitors who are, on the other hand, intent on engaging in illegal activities, such as illegal fishers, also need to be aware that they do not enjoy immunity from prosecution just because they are operating within a remote location governed by the structures of the ATS. Antarctic research stations in effect become small enclaves for the application of national law for the scientists and base staff. But laws have limited reach beyond the stations. The 1959 Antarctic Treaty[7] has relatively little to say on the question of enforcement, with the only express reference to jurisdiction – on the basis of nationality – articulated in relation to observers designated under Article VII(1) and research personnel on exchange pursuant to Article III(1)(b)(3) of the Treaty.[8] The further measures relating to the exercise of jurisdiction as envisaged by Article VIII(2) of the Treaty have never eventuated. While this approach was designed to minimise sovereignty tensions that could arise from law enforcement during the Cold War, the question that arises is whether such an approach can be justified in the 2020s.

These gaps are highlighted in the case of Antarctica's unclaimed sector. What law applies there, and how is it enforced? While the sector is not claimed by any state, it is not void of law. ATS instruments apply and national law will follow Antarctic expeditioners, but gaps will arise in the case of tourists whose home country has no interest in Antarctic affairs, including law enforcement. Likewise, there can be issues when a non-party to the Antarctic Treaty seeks to engage in Antarctic affairs, such as commencing a scientific expedition. While these activities will often be undertaken in cooperation with an existing Treaty party, that does not extend to the application of law or to law enforcement.

"Under the terms of the 53-nation Antarctic Treaty, workers accused of serious crimes at a research base are subject to the jurisdiction of their home country."[1] This citation from a 2016 article in *The New York Times* finds its visual counterpart in the 1950s Peronist propaganda image in which Argentine Justice hovers over the globe, called upon at once to judge over Argentina and Antártida Argentina, the sector of Antarctica claimed by the South American country. Expounding on his argument about "Crime and Punishment in Antarctica", journalist Bryant Rousseau wrote, "so when an American cook attacked a co-worker with a hammer in 1996 at McMurdo Station [...] the FBI sent agents to investigate and take the cook into custody. In the meantime, he was simply confined to a hut. Where was he going to run?"

The evolution of the ATS has created additional challenges, but also opportunities for law enforcement. As the regime has developed and expanded, greater emphasis has been given to the need for Treaty parties to take their obligations seriously and apply and enforce additional ATS provisions, not only to their nationals but also to others who may engage in Antarctic and Southern Ocean activities. This especially became a pressing issue following the adoption of the 1980 Convention on the Conservation of Antarctic Marine Living Resources (CCAMLR),[9] which sought to target Southern Ocean fishing, but also following the adoption of the 1991 Protocol on Environmental Protection to the Antarctic Treaty (the so-called Madrid Protocol).[10] CCAMLR raised pressing issues for law enforcement given that it had a predominant maritime focus, with the target being fishing and marine harvesting in which fishers had the ability to use vessels that were under the flag of a CCAMLR state party or of a non-state party. Subsequent adoption of the Madrid Protocol resulted in even greater emphasis on the need for effective law enforcement, as a result of the impact of Article 7 and its prohibition on mineral resource activities. The protocol also adopted very extensive environmental protection measures, which imposed obligations upon parties to ensure the adoption and implementation of a comprehensive range of initiatives. Nevertheless, a violation of the protocol by any state undertaking mining – irrespective of whether they are a party – would challenge the legitimacy of the ATS. Formal dispute resolution measures could be attempted, but these would in turn throw up complex legal challenges. All these examples of new Antarctic law have no significant consequence unless they are subject to enforcement.[11]

To date, these Antarctic law enforcement issues have by and large been contained within the ATS legal framework or by the application of national law. Loopholes have from time to time been identified, such as the prosecution of illegal Southern Ocean fishermen[12] or Japanese whaling under Australian law,[13] but generally Antarctica law enforcement has held firm despite the challenges.

Yet some other security challenges may not be so readily addressed within this framework. Terrorist activities would pose a particularly significant challenge. Much of the global criminal law framework which has developed in response to terrorism is based upon territorial jurisdiction in cases when a terrorist incident occurs within the territory of a state, upon nationality jurisdiction when a terrorist act is committed by a national or against a flagged ship or aircraft, and also upon grounds of universality based upon a growing number of counterterrorism conventions. If a terrorist attack were to take place in Antarctica, how would law be enforced? To begin with, there is no Antarctic police force. Beyond that, the actual application of law to an Antarctic terrorist act would be problematic, depending on where it took place. Other principles of international law could be called upon to fill any gap in the Antarctic legal regime, but terrorist acts on the high seas have highlighted previous challenges of law enforcement beyond national jurisdiction. As is often the case, Antarctica can highlight difficult law enforcement issues. While the scenarios may previously have been unthinkable, they are increasingly becoming more probable.

"In international law, a *condominium* is a political territory (state or border area) in or over which multiple sovereign powers formally agree to share equal dominium (in the sense of sovereignty) and exercise their rights jointly, without dividing it into national zones. Although a condominium, coined in the eighteenth century […], has always been recognised as a theoretical possibility, condominia have been rare in practice. A major problem, and the reason so few have existed, is the difficulty of ensuring co-operation between the sovereign powers; once the understanding fails, the status is likely to become untenable. Antarctica is a *de facto* condominium, governed by parties to the Antarctic Treaty System that have consulting status."

Louis de Gouyon Matignon (Writer, LGM Editions), "The Legal Status of Antarctica", *Space Legal Issues*, 2019

04
Antarctic Resources. Temptations and Accountability

Antarctic Resources and the Protocol on Environmental Protection

Doaa Abdel-Motaal

DOAA ABDEL-MOTAAL is the former executive director of the Rockefeller Foundation Economic Council on Planetary Health at the University of Oxford (United Kingdom) and has served in various high-level positions in the United Nations and other multilateral organisations. Dr Abdel-Motaal is the author of *Antarctica: The Battle for the Seventh Continent* (Praeger, 2016), nominated for the 2018 Mountbatten Award for Best Book.

Antarctica existed in the minds of the ancient Greeks long before it was found. They believed that the large masses of land to the north of the planet had to have a counterweight to the south. They were the first to name Antarctica, calling it *Terra Australis Incognita.* Definitive proof of the continent's existence, however, only emerged in the 1772–1775 voyage of Captain James Cook.[1] While Captain Cook dismissed Antarctica as a land too harsh to benefit humanity, his report of plentiful seals on the island of South Georgia ironically opened the way for the continent's commercial exploitation. Since his voyage, Antarctica has been charted mostly by successive sealing and whaling expeditions.[2]

Antarctica's geology is the outcome of the formation and subsequent break-up, millions of years ago, of the supercontinent known as Gondwanaland. Antarctica's mineral potential is therefore frequently assessed through "geological association" with the lands from which it spun.[3] There has been little to no actual mining or drilling in Antarctica because of its harsh climate and the adoption in 1991 of an Environmental Protocol by parties to the Antarctic Treaty which banned, in Article 7, mining on the continent.[4]

The information available to date points to a continent rich in mineral resources and rare earths. The continent is divided into two main geographical regions, East Antarctica and West Antarctica, which are separated by the Transantarctic Mountains. East Antarctica makes up two thirds of the continent.[5] One of its most promising areas is the Dufek Massif, which straddles the Transantarctic mountain chain and is thought to harbour valuable platinum-group metals.[6] East Antarctica is also known to harbour iron, uranium, other industrial minerals, and coal.[7] The Antarctic Peninsula is also known to hold copper, iron, chromium and other deposits.[8] Antarctica's best hydrocarbon prospects appear to be in the parts of the Ross and Weddell Sea basins that are free of permanent ice.[9]

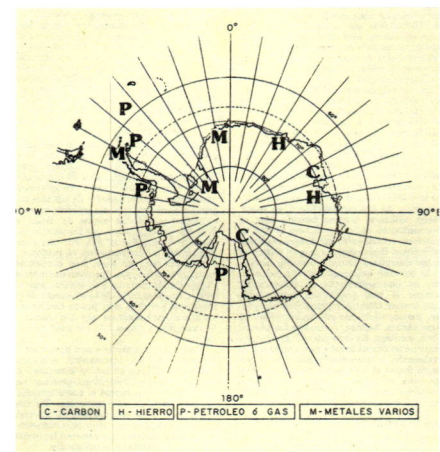

C - CARBON H - HIERRO P - PETROLEO ó GAS M - METALES VARIOS

The fact that Antarctica is rich in mineral deposits was known prior to the signing of the Antarctic Treaty, i.e. since IGY explorers found traces of a number of metals, including gold, copper, lead, chromium, molybdenum, antimony, zinc, and tin, on the continent. However, as stated at the time by Dr Laurence M. Gould, the director of the United States IGY Antarctic programme, while expounding on the Dirty Diamond Coal Co. mine in Antarctica shown in the image below, "the frozen continent [might] contain the largest coal deposits in the world, though probably of poor grade. But even if it were good, ice and cold, distance and expense would prohibit profitable mining."[1] To date, the average 2.5 km-thick ice sheet that lies above the continental bedrock still inhibits any attempt at accessing Antarctic resources. Their exploitation is prohibited (at least until 2048) by the Protocol on Environmental Protection which entered into force in 1998, replacing the much-contested Convention on the Regulation of Antarctic Minerals Resource Activities (CRAMRA), originally adopted by consensus of 20 Antarctic Treaty Consultative Parties in 1988. While the whole concept of Antarctic resource detection and extraction should be eliminated from national agendas, as Klaus Dodds stated in his recent interview with the *Financial Times,* it is evident that "once you get [...] explicit about resource exploitation, you [still] raise the troubling issue of who owns Antarctica. That's the issue that haunts the Antarctic Treaty, and the Treaty System more generally."[2]

DIRTY DIAMOND COAL CO. No. 1 Mine

The Southern Ocean contains a number of important fisheries, with four main types of fish currently exploited commercially: krill, mackerel icefish, and the Antarctic and Patagonian toothfish. Finally, Antarctica holds 70% of the world's fresh water, a commodity that may soon be in short supply as global warming advances.[10]

The twentieth-century politics of Antarctica have very much been dominated by its resource potential and strategic location. While the Consultative Parties dedicated Antarctica to peace and scientific research in the 1961 Antarctic Treaty, froze the seven territorial claims that had been made on the continent,[11] and postponed the question of mining through the Environmental Protocol,[12] competition for Antarctica has only become fiercer.

One of the main outcomes of the Antarctic Treaty has been an explosion in the number of scientific research stations established by the Consultative Parties across the continent. Today, eighty to one hundred different facilities belonging to the twenty-nine Consultative Parties dot the continent, with certain countries owning more than ten stations.[13] A large number of stations are concentrated on the Antarctic Peninsula,[14] which is the most accessible part of the continent and one of the fastest-warming parts of the globe.[15]

While valuable scientific research is indeed conducted in Antarctica, many of the stations have not produced sufficient science to justify their presence, prompting the suspicion that they were built mainly to create a foothold on the continent and fly their national flag. In fact, a prominent Antarctic scholar once asked why the world had succeeded in establishing an International Space Station (ISS) but not an international research station in Antarctica, arguing that the desire to assert national sovereignty in Antarctica is the main cause.[16]

Today, a simple look at the map demonstrates that Antarctica is no longer an unoccupied continent. Every corner of it has been given a name: it has 38,000 place names in total, with only twenty Consultative Parties having assigned the bulk of those names.[17] This has prompted some in the developing world to call for the "scientific decolonisation" of Antarctica.[18]

The rapid expansion of scientific activity in Antarctica, together with other economic activities such as tourism and bioprospecting, has not been without harm to its fragile environment.[19] While the Environmental Protocol bans mining on the continent, it does not regulate economic activities such as bioprospecting, and creates a jurisdictional vacuum that limits the enforcement of environmental regulations. This puts Antarctica's fragile environment at risk.[20]

Evidence of this lies in the following simple numbers. Whereas almost 15% of the Earth's land surface is designated as protected, only 1.5% of Antarctica's ice-free portions benefit from a similar status under the Antarctic Treaty. This makes it one of the world's least-protected continents.[21] Antarctica's waters are equally under-protected due to overfishing. In fact, the term "illegal, underreported and unregulated" fishing was coined specifically to describe the plight of the Southern Ocean.[22] What is increasingly clear is that Antarctica has been protected more by its ice cover than by its treaty system. As its ice continues to melt, a fresh conversation will need to be had on its future.

"Antarctica has always been a political space. But technology has made it a lot less remote, and as we have begun to be able to do things there beyond the shareable public goods of scientific research, new economic interests have emerged. Given certain sorts of futures, these economic interests might now become very grand indeed – after all, this region is some 10 per cent of the planet. This reality, the possibility that in one or a few generations the gloves are off and an Antarctic resources jamboree underway, has given new intensity to Antarctic politics in the past decade. [...] The Antarctic is now the object of international rivalry. Its future, who will determine the pathway to that future, and who will be the beneficiaries, are all in contest."

Alan D. Hemmings (Professor, University of Canterbury), *Handbook on the Politics of Antarctica*, 2017

"In order for companies to justify the massive costs of mining in Antarctica, supergiant reserves must be discovered there. With the exploitation of supergiant reserves there would be catastrophic damage to the Antarctic environment from facilities, towns, roads, airstrips, waste disposal facilities and spills. Not even the strictest regulations would prevent this damage."

Will Martin (Director, Wilderness Society's Antarctica project), in "U.S. Seeks Moratorium on Antarctic Minerals," *The New York Times*, 14 November 1990

The Role and Impact of Antarctic Specially Protected Areas and Antarctic Specially Managed Areas

Kevin A. Hughes

KEVIN A. HUGHES is vice-chair of the Committee for Environmental Protection (CEP), a member of the United Kingdom delegation to the Antarctic Treaty Consultative Meeting (ATCM), and an environmental researcher at the British Antarctic Survey (United Kingdom). His Antarctic interests include conservation, area protection, geological heritage, non-native species and environmental monitoring. Dr Hughes has visited the polar regions eleven times and overwintered once.

Antarctica is the most isolated and climatically extreme continent on the plant, yet climate change and human impact are increasingly putting its environments and ecosystems at risk. The continent is dominated by ice. Only approximately 0.3% of its area is free of permanent ice, and it is on these generally small and often remote "islands" of ice-free ground that the majority of Antarctica's terrestrial life, comprising a restricted range of biological groups, exists.[1] Coastal ice-free areas are important nesting sites for birds and haul-out sites for seals, while nearshore marine environments can support exceptionally rich benthic biodiversity.[2]

The arrival of humans in Antarctica has put pressure on Antarctic environments and habitats. The construction of research stations and other infrastructure to facilitate scientific research has displaced species and destroyed habitat. Rapidly increasing numbers of tourists (approximately 80,000 during the 2019/20 season), who predominantly visit the Antarctic Peninsula region, present a potential threat to biological communities, geological heritage and historic sites.[3] Protected area designation provides one mechanism by which the parties to the Antarctic Treaty (signed in 1959) can agree to protect sites that are vulnerable to human impact, as well as more broadly protect areas that are representative of the diverse range of scientific, environmental, historical and intrinsic values present on the continent.

The Antarctic Treaty System provides a framework for international agreements concerning Antarctica, including for the designation of protected areas. The Agreed Measures for the Conservation of Antarctic Fauna and Flora (1964) described Antarctica as a "special conservation area" and provided the framework for the designation in 1966 of the continent's first protected areas to protect "unique natural ecological systems".[4] In 1975, the scope of the protected areas system was expanded to include sites for scientific investigation and, in particular, areas vulnerable to harmful interference from human activities. In 1991, the Protocol on Environmental Protection to the Antarctic Treaty was agreed upon; it designated Antarctica as a "natural reserve, devoted to peace and science". The protocol largely superseded earlier agreements on conservation, including area protection, and in 2002, Annex V to the protocol, concerning "Area Protection and Management", came into effect, integrating existing protected areas into the new system. Annex V established two new types of protected area: Antarctic Specially Protected Areas (ASPAs) and Antarctic Specially Managed Areas (ASMAs).

ASPAs represent the highest level of protection within the Treaty system and are designated to protected outstanding environmental, scientific, historic, aesthetic or wilderness values, or ongoing or planned scientific research. Entry to an ASPA is prohibited without a permit from a national governmental authority, and all those accessing the area must comply with the associated management plan, which must be reviewed regularly. ASPAs should be designated within a "systematic environmental-geographic framework" to protect specific characteristics and values, for example, to keep an area inviolate from human interference or to protect representative examples of terrestrial and marine ecosystems, unusual or important assemblages of species, the type locality of any habitat or species, areas of importance for research, or sites of outstanding geological, glaciological, geomorphological, historic, aesthetic and wilderness value.

ASMAs are designated to assist in the planning and coordination of activities, avoid possible conflicts, improve coordination between parties or minimise environmental impacts. ASMAs can be further divided into Access, Historic, Scientific, Restricted, Visitor and Facilities and Operations Zones, and can also

> "At a national level, it is difficult to know the exact nature of the political motivations of environmental managers when making proposals for protected areas. Proposals to create protected areas, and then the ensuing management of these places, offer states (both claimants and non-claimants) an opportunity to behave like *de facto* sovereigns by taking a lead in protecting the environment. [...] However many environmental discussions may take place at a supposedly depoliticized technocratic level, the reality is that politics and environmental regulation have been inextricably connected since the first sighting of the Antarctic continent in the early nineteenth century."
>
> Adrian Howkins (University of Bristol), "Politics and Environmental Regulation in Antarctica: A Historical Perspective" in *Handbook on the Politics of Antarctica,* 2017

contain ASPAs and Historic Sites and Monuments (HSMs). No permit is required to enter an ASMA, but the management plan for the area should be adhered to.

Currently, seventy-two ASPAs exist.[5] Marine, glacial and terrestrial environments, totalling approximately 3,860 square kilometres, have been protected over the past fifty-four years, predominantly in an *ad hoc* manner. Only approximately 1.5% (760 square kilometres) of the continent's ice-free ground is protected, which falls far short of global conservation targets, including Aichi Target 11, which calls for the protection of at least 17% of terrestrial area.[6] ASPAs are often small, with 55% having an area of less than five square kilometres, and are frequently located close to research stations, mainly within the Antarctic Peninsula, the Ross Sea region and the ice-free oases of coastal East Antarctica. Consequently, almost a third of Antarctica's eco-regions (called Antarctic Conservation Biogeographic Regions)[7] lack any ASPAs to protect representative habitats, and many species are not afforded higher levels of protection.[8] Furthermore, the full range of values and characteristics the ASPA system aims to systematically protect is inadequately represented in the existing protected area network.

The Antarctic Treaty area currently contains six ASMAs, with up to six nations involved in the management of each area, sometimes with input from the International Association of Antarctica Tour Operators (IAATO) and the Antarctic and Southern Ocean Coalition (ASOC, representing non-governmental conservation organisations).[9] The spatial scale varies greatly, from circa 100 square kilometres at ASMA No. 4 Deception Island to 26,344 square kilometres for ASMA No. 5 Amundsen-Scott South Pole Station ↗ p. 790.

Currently, fifteen out of the twenty-nine Consultative Parties to the Antarctic Treaty act as the proponents for ASPAs, and have worked increasingly in partnership to propose and manage areas.[10] However, the rate of ASPA designation has slowed markedly: only four ASPAs were designated during 2010–2019, compared to twenty-five during 1980–1989, although three new protected areas are now under consideration by the Committee for Environmental Protection (CEP). It remains to be seen whether recent efforts by the Scientific Committee on Antarctic Research (SCAR) and the CEP to promote more strategic designation of ASPAs are effective.[11] As Antarctica becomes increasingly busy, ASMAs are likely to become more important for effective management of the Antarctic environment. However, some ASMA proposals have been unsuccessful due to a lack of international consensus (the Fildes Peninsula on King George Island and Dome A in East Antarctica, for example),[12] and it is currently unclear how the potentially conflicting geopolitical ambitions of some parties will affect the future use of the ASMA management tool.[13]

Overall, with climate change and human impact threatening the biodiversity, natural features and intrinsic values of the continent, the case for a strategic and rapid expansion of the protected areas system is clear.

The timeline at the side and the cartography shown on the following pages examine the Antarctic Specially Protected and Managed Areas (ASPA and ASMA) by revealing the total square kilometres of each and highlighting the respective managing countries. To date, Antarctica has 72 ASPAs and 6 ASMAs, of which more than 60% were proposed by at least one claimant party. These preservation measures safeguard a total of 760 sqkm, equivalent to 1.5% of Antarctica's ice-free area. This percentage is strikingly low if compared to what is being done elsewhere on our planet. According to UNESCO, on the other six continents, 149 million sqkm, equivalent to 15% of Earth's land mass, are protected.

The tropical time capsule that Paul Rosero Contreras installed at Paradise Bay provocatively introduced a foreign species to Antarctica for a short period. His action, theoretically prohibited under Annex V of the Protocol on Environmental Protection, which forbids the introduction of any alien flora and fauna to the continent, was possible because the small cocoa plant from the Ecuadorian Amazon rainforest was duly concealed in a temperature-controlled container that inhibited the possibility of contamination. The work, conceived in the context of the Antarctic Biennale curated by Alexander Ponomarev, was intended to reflect on Antarctica's tropical past (over 90 million years ago) as a reminder of the potential effects of climate change. Today, almost nothing remains of the lush vegetation. Only upon careful scrutiny one can detect, on the 0.3% of Antarctica's surface that is ice-free ground, unexpected life forms such as those shown in the image below taken by UNLESS. These include Antarctic hair grass, Antarctic pearlwort, Xanthoria, Usnea, Caloplaca, and Rhizocarpon, among others.

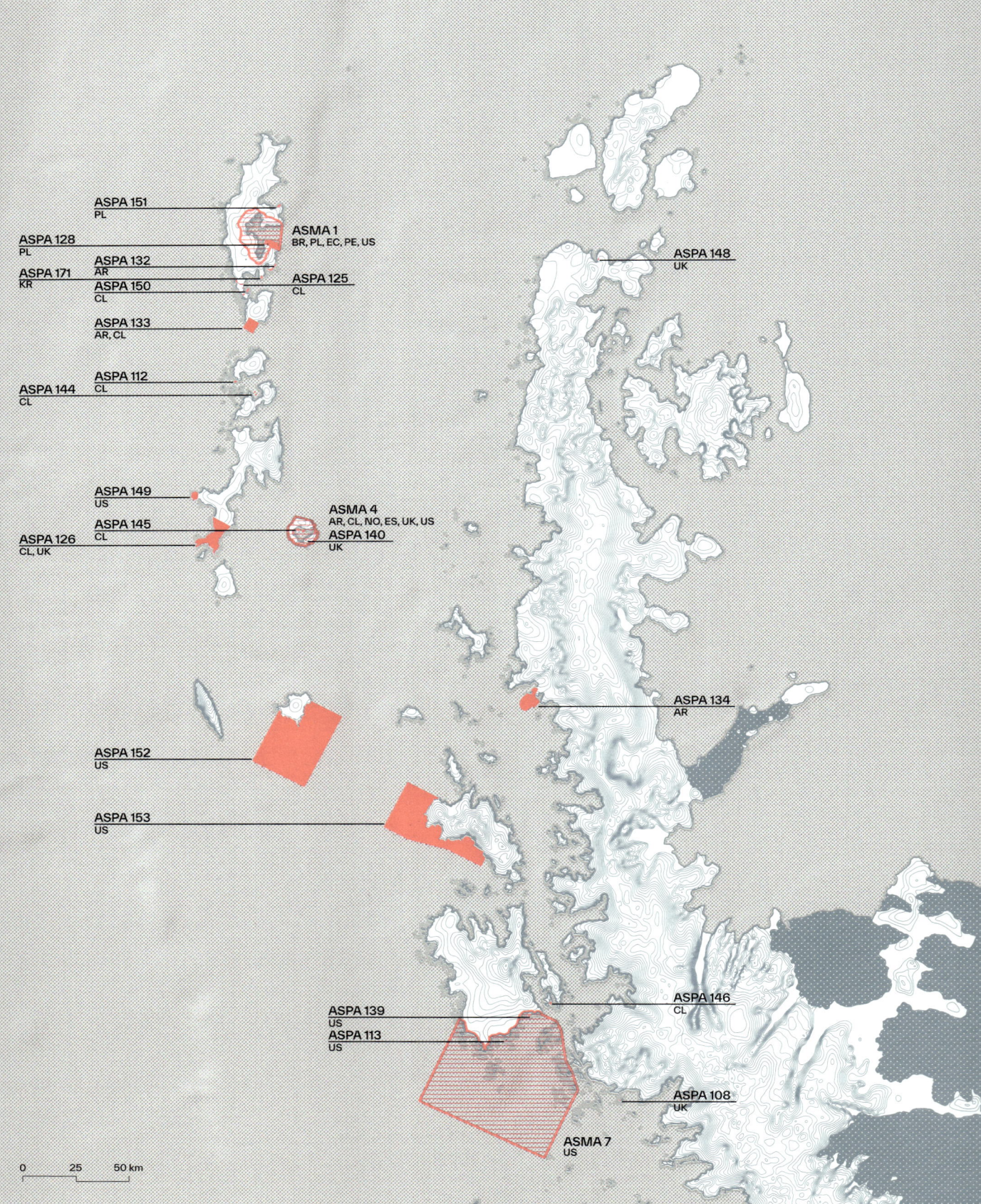

ASPA 151
PL

ASPA 128
PL

ASPA 171
KR

ASPA 132
AR

ASPA 150
CL

ASPA 133
AR, CL

ASPA 144
CL

ASPA 112
CL

ASPA 149
US

ASPA 145
CL

ASPA 126
CL, UK

ASMA 1
BR, PL, EC, PE, US

ASPA 148
UK

ASPA 125
CL

ASMA 4
AR, CL, NO, ES, UK, US

ASPA 140
UK

ASPA 134
AR

ASPA 152
US

ASPA 153
US

ASPA 146
CL

ASPA 139
US

ASPA 113
US

ASPA 108
UK

ASMA 7
US

0   25   50 km

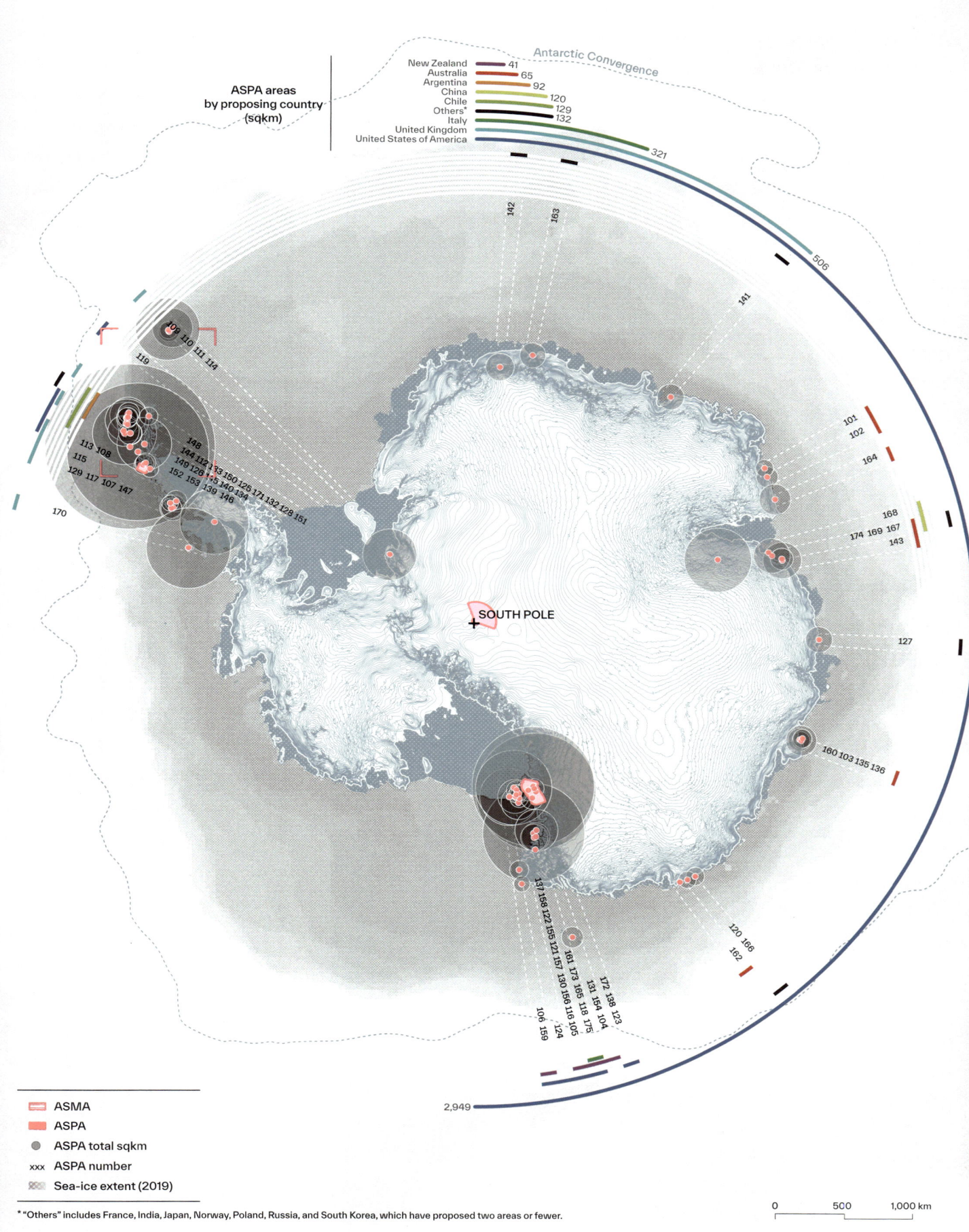

ASPA areas
by proposing country
(sqkm)

Antarctic Convergence

New Zealand	41
Australia	65
Argentina	92
China	120
Chile	129
Others*	132
Italy	
United Kingdom	321
United States of America	506

142
163
141
101
102
164
168
174 169 167
143

119
108 110 111 114

148
144 112 133 150 125 171 132 128 151
113 108 149 126 155 140 134
115 152 153 139 146
129 117 107 147
170

SOUTH POLE

127

160 103 135 136

120 166
162

137 158 122 155 121 157 130 156 116 105
161 173 165 118 175
172 138 123
131 154 104
106 159
124

2,949

ASMA

ASPA

ASPA total sqkm

xxx   ASPA number

Sea-ice extent (2019)

* "Others" includes France, India, Japan, Norway, Poland, Russia, and South Korea, which have proposed two areas or fewer.

0   500   1,000 km

Antarctic Biodiversity

Peter Convey

PETER CONVEY is a polar ecologist based at the British Antarctic Survey (United Kingdom), where he is currently the deputy leader of the Survey's Biodiversity, Evolution and Adaptation Team. He has over 31 years' experience in Antarctic and Arctic environments. Dr Convey's research focuses on the evolution of polar biodiversity and biogeography, analysing the polar regions to model past and future global consequences of climate change.

Although it has around twice the area of Australia, only a tiny fraction of Antarctica (0.2%–0.3%) is free of ice or snow today. The continent is surrounded by the vastness of the Southern Ocean, which comprises 10% of the world's oceans, and both have been isolated from lower latitudes by strong circumpolar atmospheric and oceanic circulations for many millions of years.[1] Much deeper than that of other continents (because of the vast volume of ice pressing down on the land mass), the Antarctic continental shelf is physically isolated from lower-latitude shelves, as are the much smaller shelf depths of associated remote islands in the Southern Ocean.

Both land and ocean face globally severe environmental conditions, exacerbated by the extreme seasonality of high latitudes.[2] The marine environment is among the most thermally stable on Earth, typically varying as little as a degree or two annually, but it experiences extreme seasonal variation in factors such as sea-ice cover (on average doubling the area of the continent in winter) and primary production (the biological fixation of carbon dioxide in photosynthesis). In contrast, temperatures on land, while on average colder than the sea, are much more variable, and the major driver of terrestrial biodiversity is liquid water availability.[3]

Antarctic terrestrial diversity is low,[4] although microbial and viral diversity may not conform to this pattern.[5] However, marine and particularly benthic (sea-floor) diversity may be second only to coral reefs globally.[6] In both realms, diversity estimates are limited by sampling effort and quality, including lack of taxonomic expertise. Molecular studies are increasingly confirming previously unknown cryptic diversity hidden through lack of morphological differentiation[7] and high proportions of endemic species typify many groups of Antarctic terrestrial and marine organisms.[8] This creates important conservation and management challenges for the continent and the Southern Ocean.[9]

There are various terrestrial ecosystems in Antarctica.[10] Inland, they are predominantly montane (with mountain ranges and isolated nunataks) and frigid deserts. In physically isolated coastal oases around the continent,

Inhabited by extraordinary marine species that have evolved to survive in extreme waters and temperatures, the Antarctic seabed is a colourful underwater landscape that sets itself apart from the white, featureless territory that lies above the water level. Salps as shown on the left, i.e. "tubular, gelatinous sea creatures" which travel the oceans "by pumping water through their barrel-shaped bodies, [and] constantly strain the ocean water for algae cells", coexist with more than 35 species of starfish, sea spiders, anemones and nudibranchs in an ever-changing landscape. A realm, to quote photographer Dhritiman Mukherjee, in which "everything is constantly moving: the icebergs move, the sea ice moves [...] [and] you have no landmarks." Such an environment and the incredible Antarctic biodiversity never cease to surprise biologists and palaeontologists who consider that "the predominance of filter feeders, sponges and sea squirts on the seabed [such as those captured by the Census of Antarctic Marine Life shown above] confer on the marine community an 'archaic' character, which for some scientists recalls the Paleozoic communities that were once widespread throughout all oceans".[1] According to Bruno David and Thomas Saucède this suggests that such species might have survived several cycles of ice-sheet advances and retreats. "It is nevertheless recognized today that Antarctic marine communities are more recent in origin, and that the absence of large predators and the predominance of sessile fauna is an evolutionary answer to the glaciomarine conditions progressively established during the Cenozoic Era [65.5 million years ago to present], rather than a legacy of the distant past."[2] Alongside such almost fantastical fauna, marine biodiversity counts numerous other species, among them majestic creatures such as the monumental jellyfish shown on the right, captured by Norbert Wu using HDTV technology, whose profound isolation enabled them to grow larger than their warm-water relatives.

and particularly on the Antarctic Peninsula and Scotia Arc archipelagos, the eco-systems are more developed, dominated by a diversity of mosses and lichens.

True land or freshwater vertebrates are not to be found on the Antarctic continent and are limited to a very small number of birds on the subantarctic islands. However, concentrations of marine vertebrates transfer large quantities of nutrients to terrestrial environments, acting as important local drivers of ter-restrial productivity and diversity.[11] Faunal diversity is therefore largely limited to invertebrates, including insects (only two species live on the continent itself), mites, springtails and micro-invertebrates such as nematodes, tardigrades and rotifers, with a limited number of crustaceans in freshwaters.[12]

Vegetation is similarly restricted, with only two flowering plant species on the continent, and relatively low diversities of mosses and lichens – com-pared, for instance, with northern high-latitude vegetation on Greenland or Svalbard. A feature of much of this biota is its antiquity within Antarctica. Rapidly increasing evidence from most groups of extant terrestrial biota, and most re-gions of Antarctica with ice-free ground, shows they have persisted within these regions for many millions of years throughout, at least, multiple Pleisto-cene (up to 2 my), Pliocene (~5 my) or Miocene (>14 my) glacial cycles and in some cases since before the final stages of Gondwana breakup (>30–40 million years ago).[13]

The majority of Antarctic biodiversity is marine. Diversity is high in some taxonomic groups but low in others. The relatively high benthic biodiversity is perhaps surprising given the depth of the continental shelf (elsewhere diversity typically decreases with depth) and the lack of ice-free intertidal and shallow subtidal habitats around the continent, which generally acts as a significant contributor to diversity at lower latitudes.[14] Among the bony fish, notothenioids account for more than 70% of fish in the Southern Ocean, being the only major family that remained in, and radiated in, the cold waters that developed after

"At planetary timescales, the Antarctic biophysical system is somewhat removed from the rest of the planet due to its objective distance from all other continents, to the degrees of separation offered by the Polar Front and to the presence – over the past 10 to 20 million years – of cold and stable ocean temperatures. In Antarctica's oceans this facilitated the evo-lution of remarkable unique biological biodiversity."

Craig Stevens (Scientist, National Institute of Water and Atmospheric Research), 2020

0   100   200 km

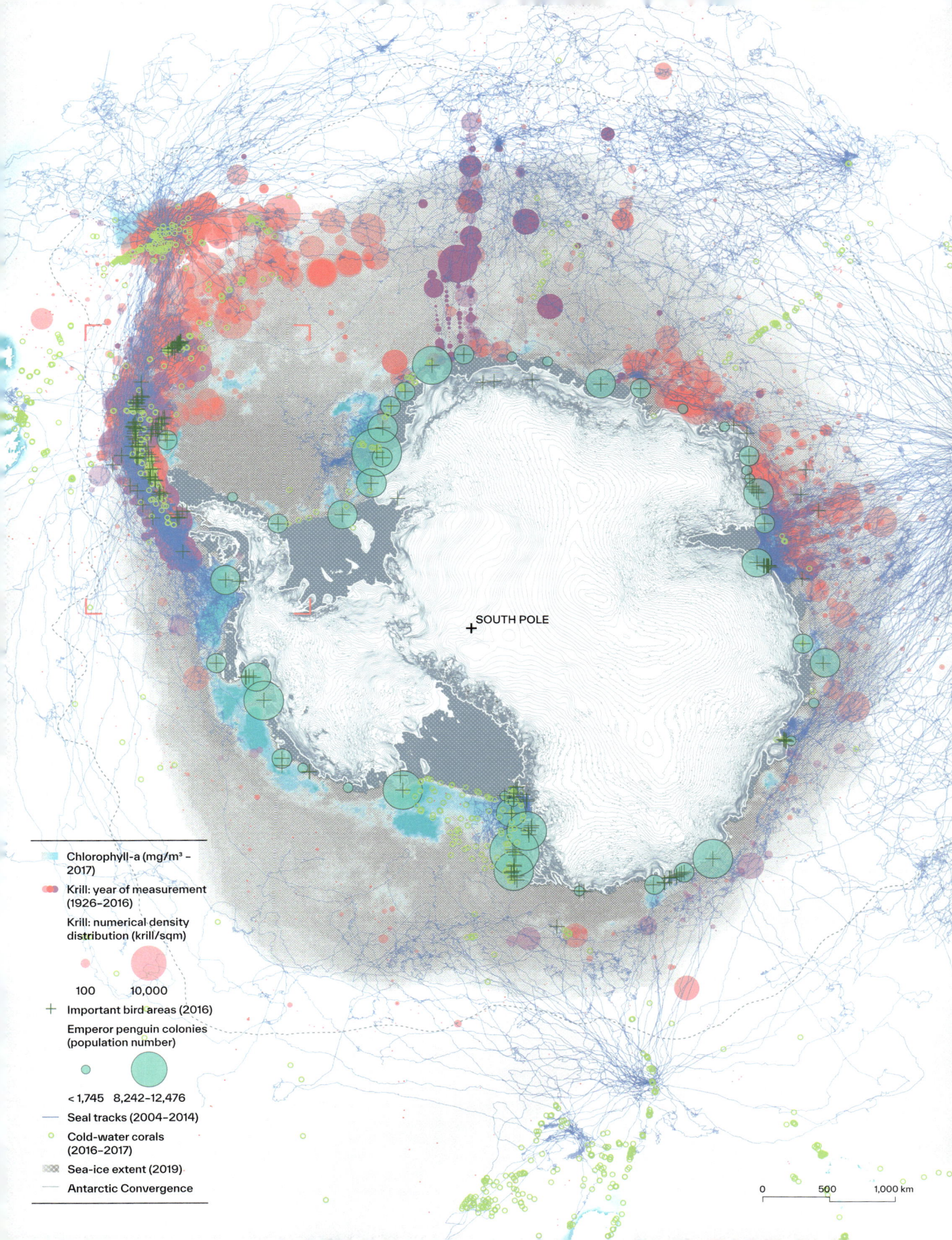

SOUTH POLE

Chlorophyll-a (mg/m³ – 2017)

Krill: year of measurement (1926–2016)

Krill: numerical density distribution (krill/sqm)

100   10,000

Important bird areas (2016)

Emperor penguin colonies (population number)

< 1,745   8,242–12,476

Seal tracks (2004–2014)

Cold-water corals (2016–2017)

Sea-ice extent (2019)

Antarctic Convergence

0   500   1,000 km

the formation of the isolating Antarctic Circumpolar Current between 15 and 20 million years ago.[15]

Pelagic primary production supports the Southern Ocean marine ecosystem, including spectacular concentrations around the subantarctic islands of iconic marine birds, penguins, seals and whales. There are also several important commercial fisheries, in particular targeting krill (a pelagic crustacean often described as a keystone species central to the Southern Ocean food web) and the lucrative toothfish. These fisheries are regulated in an ecosystem-based approach by the Convention for the Conservation of Antarctic Marine Living Resources.

The Southern Ocean is also thought to be the greatest sink globally for increasing atmospheric carbon dioxide, although this is in turn leading to ocean acidification and the consequential threat to elements of its biodiversity, especially those with exposed carbonate skeletons. High marine biodiversity is underlain by multiple scales in the region's environmental heterogeneity, its isolation and its age[16] – which has encouraged evolutionary radiation. Furthermore, during glacial cycles, life was restricted to isolated continental shelf or slope refugia with subsequent expansion during interglacial periods.[17]

Since the early days of terrestrial research on the continent, various terminology has been used to refer to different regions of the continent. However, most experts today recognise three large-scale biogeographic zones (the sub-, maritime, and continental Antarctic) based on broad similarities in overall climates and ecosystems within each.[18] More recently, sixteen distinct terrestrial Antarctic Biogeographic Conservation Regions (ACBRs) have been identified.[19]

An increased understanding of biogeographic complexity, combined with contemporary trends of environmental change and direct human impacts through increased connectivity between both terrestrial and marine regions, use of the limited extent of ice-free ground, and introduction of non-native species, poses multiple challenges to the governance system and the conservation of the Antarctic region.[20] Failing to adequately protect either terrestrial or marine biodiversity across the different regions of Antarctica and the Southern Ocean, the current system of environmental protection in Antarctica should be reviewed with a sense of urgency.[21]

Nobert Wu's photograph shown below, depicting orcas peacefully swimming between sea-ice floes, is a clear reference to the iconic picture taken by Herbert Ponting during the 1910–1913 British Antarctic Expedition when he found himself surrounded by "wolves of the sea"[3] which "almost toppled him into the ocean."[4]

"Since first seeing some of these wolves of the sea off Cape Crozier I had been anxious to secure photographs of them [...] The whales dived under the ice, so, hastily estimating where they would be likely to rise again, I ran to the spot adjusting the camera as I did so. I had got to within six feet of the edge of the ice – which was about a yard thick – when, to my consternation, it suddenly heaved up under my feet and split into fragments around me; whilst the eight whales, lined up side by side and almost touching each other, burst from under the ice and 'spouted'. [...] Fortunately the shock sent me backwards, instead of precipitating me into the sea, or my Antarctic experiences would have ended somewhat prematurely."

Herbert George Ponting (Photographer, Explorer), *The Great White South*, 1921

Fishing in the Southern Ocean and the Role of CCAMLR

Cassandra M. Brooks

CASSANDRA M. BROOKS is an assistant professor of environmental studies at the University of Colorado Boulder (United States). Her expertise lies at the intersection of marine science, environmental policy and science communication. With a PhD from Stanford University on international ocean policy, Brooks was awarded a Switzer Fellowship in Environmental Leadership. In the past 15 years she has published over 200 articles in scientific journals, books and popular outlets.

Antarctica is exceptional. The coldest, windiest, iciest, driest and remotest of continents is widely celebrated for its rich history of exploration, science and diplomacy and for its extraordinary beauty. It's also exceptionally important. Since its discovery, scientists have documented how the Antarctic is vital to Earth's systems.[1] Despite its extreme environment, life thrives in incredible diversity and abundance.[2] The freezing Southern Ocean that surrounds the Antarctic continent teems with whales, seals, penguins, toothfish and krill, and this frozen seascape harbours some of the last remaining great wildernesses on the planet.[3] However, fishing pressure combined with the cumulative impacts of climate change jeopardises the future of Antarctic life in the Southern Ocean.[4] The Commission for the Conservation of Antarctic Marine Living Resources (CCAMLR) faces the enormous challenge of regulating the use of economically valuable marine resources while protecting the integrity of the Antarctic marine ecosystem, all under conditions of rapid environmental change.

The pelagic-whaling factory ships of the 1920s, such as that shown in the image above with a stern ramp to allow for hauling the whales onto the bow, inherit their design concept from the Norwegian motorised catcher boats developed around 1865 by Svend Foyn, and mark the beginning of the modern whaling industry. The astounding growth of the latter in the 1930s (which increased from 17 floating factories, 61 catcher boats and the production of one million barrels of oil from 14,000 whales in the 1927–1928 season to 41 factories, 200 catchers and 3.6 million barrels from more than 40,000 whales)[1] led the newly established International Council for the Exploration of the Sea (ICES) to state "that if the expansion continues at the present rate there is a real risk of those stocks being so reduced as to cause serious detriment to the industry" and urged "governments of the countries interested in whaling [to], as a matter of urgency, give serious consideration to the question of taking immediately temporary measures." To address this crisis, an International Whaling Commission (IWC) was set up in the immediate aftermath of the Second World War "to provide for the proper conservation of whale stocks" and "make possible the orderly development of the whaling industry". The organisation successfully imposed a general moratorium on commercial whaling in 1982 and is still active. A century after the infamous whaling industry apex, which almost led to species extinction, and the International Court of Justice's 2014 warning to Japan to temporarily halt its whaling programme, the 2018 decision taken by Japan to leave the IWC and continue to fish whales commercially (albeit only in Japanese waters) has left the international community alarmed.[2]

Forty years ago, the Antarctic Treaty parties came together and negotiated the 1980 Convention on the Conservation of Antarctic Marine Living Resources (CAMLR Convention). This Convention was negotiated rapidly in response to expanding fisheries for Antarctic krill (*Euphausia superba*) – the key prey species of the Southern Ocean food web.[5] In accordance with the principles of peace, science and environmental preservation embodied in the Antarctic Treaty System, the Convention's explicit objective (expanded upon in Article II) is to conserve marine living resources. While conservation includes "rational use", scientific and commercial exploitation of living resources are subject to articulated conservation principles. These principles mandate that fishing activities not cause significant changes in exploited species nor to their predators and prey, while also avoiding adverse effects on the Southern Ocean ecosystems. This approach has been recognised as farsighted, as well as arguably the world's most successful international management body for marine living resources and one that best exemplifies an all-encompassing ecosystem approach. CCAMLR is a regime that operates through yearly meetings (held in Hobart, Tasmania) at which new policies, in line with the operational procedures of the Antarctic, are proposed and ratified by consensus by the Convention signatories – i.e. the twenty-five member states plus the European Union (1980 CAMLR Convention, Article XII).

Climate change coupled with increasing fishing pressure presents a new challenge that is testing the management integrity of CCAMLR. Regions of the Antarctic are among the most rapidly changing on the planet, facing global repercussions of sea level rise, ocean circulation and climate regulation.[6] Locally, climate change is driving fluctuations in sea ice cover, shifts in marine population distributions and decreases in primary productivity. Potential declines in ice-dependent Antarctic krill, the very foundation of the Southern Ocean food web, could lead to disruptions throughout the whole ecosystem.[7] As Southern Ocean biota adapt to their changing environment, pressure on commercial fisheries has amplified in recent years. Antarctic krill (increasingly harvested for omega-3 pills and fishmeal) reached the unprecedented high catch of more than 380,000 tons in the 2019 season,[8] while Patagonian and Antarctic toothfish (*Dissostichus eleginoides* and *D. mawsoni*) are sold as lucrative Chilean sea bass and are exploited across the region with rising demand.[9] Meanwhile, scientists still do not fully understand the potential ecosystem impacts of toothfish fisheries,[10] and the combined impact of fishing and climate change is likely to have greater effect than the impact of either alone.[11]

In 2002 CCAMLR joined the international movement to contribute to a global network of Marine Protected Areas (MPAs). Extensive research supports the idea that MPAs – areas where fishing and other human activities are restricted – can conserve biodiversity, and perhaps most importantly in the case of the Southern Ocean, can enhance resilience to the impact of climate change.[12] By 2005, the Convention began working towards identifying priority areas for protection and compiling the best available science to guide the development of an ecologically representative network of Southern Ocean MPAs.[13] In 2009, CCAMLR adopted its first MPA south of the South Orkney Islands to protect approximately 94,000 square kilometres as a no-take reserve (Conservation Measure 91-03). Subsequently, in 2011, CCAMLR adopted a framework to guide the MPA process (Conservation Measure 91-04), and individual member states started proposing MPAs in their historic regions of interest.[14]

In 2016, CCAMLR made history by establishing one of the world's largest MPAs in the Ross Sea, conserving approximately two million square kilometres, with more than 70% being a dedicated area fully off-limits to fishing (Conservation Measure 91-05). Three additional MPAs (located in the Weddell Sea, the East Antarctic and the Antarctic Peninsula, respectively) would contribute significantly to a representative network of Southern Ocean protected areas, but they are still under negotiation[15] largely due to political barriers, economic interests and some states' efforts to reinterpret the CAMLR Convention. Even efforts to incorporate climate change into CCAMLR's decision rules have been hindered.[16]

The CAMLR Convention explicitly defines conservation as including rational use, yet during discussions about the South Orkney Islands MPA in 2009, China stated that conservation should not compromise rational use. By 2015, some fishing states indicated that MPAs must not restrict rational use,[17] implying that MPAs should not interfere with current or future fishing. The scramble for resources in the Antarctic has been described as "delayed" due to the strength of the governance systems in place in the Antarctic,[18] yet fishing states largely outnumber non-fishing states[19] and a new resource frontier is emerging in the Southern Ocean. The securing of fishing access now and in future, as well as global and regional geopolitics, have increasingly challenged CCAMLR's ability to progress on any conservation initiatives, including MPAs.[20]

The Antarctic is historically a place of great diplomacy, science and conservation. It is rightly celebrated as such. The future of Antarctic ecosystems depends on CCAMLR rising swiftly to new challenges and employing more precautionary management, including adopting well-designed MPAs.

"Life here under the ice has remained unchanged for millennia, but in the last 200 years, much of Antarctica's wildlife has had to face new predators: human beings. We devised new hunting techniques and used them so mercilessly that we almost exterminated the great whales […] More than one and a half million whales were slaughtered in Antarctic waters. The blubber was stripped from their massive bodies and boiled down in vats to make margarine and soap."

Sir David Attenborough (Natural Historian, Broadcaster), BBC series "Seven Worlds, One Planet", 2019

"The Southern Ocean, which surrounds Antarctica, is becoming a significant fishing ground, as resources in other seas are depleted."

Leslie Hook (Environment and Clean Energy Correspondent) and Benedict Mander (Latin America Correspondent), "The Fight to Own Antarctica", *Financial Times*, 2018

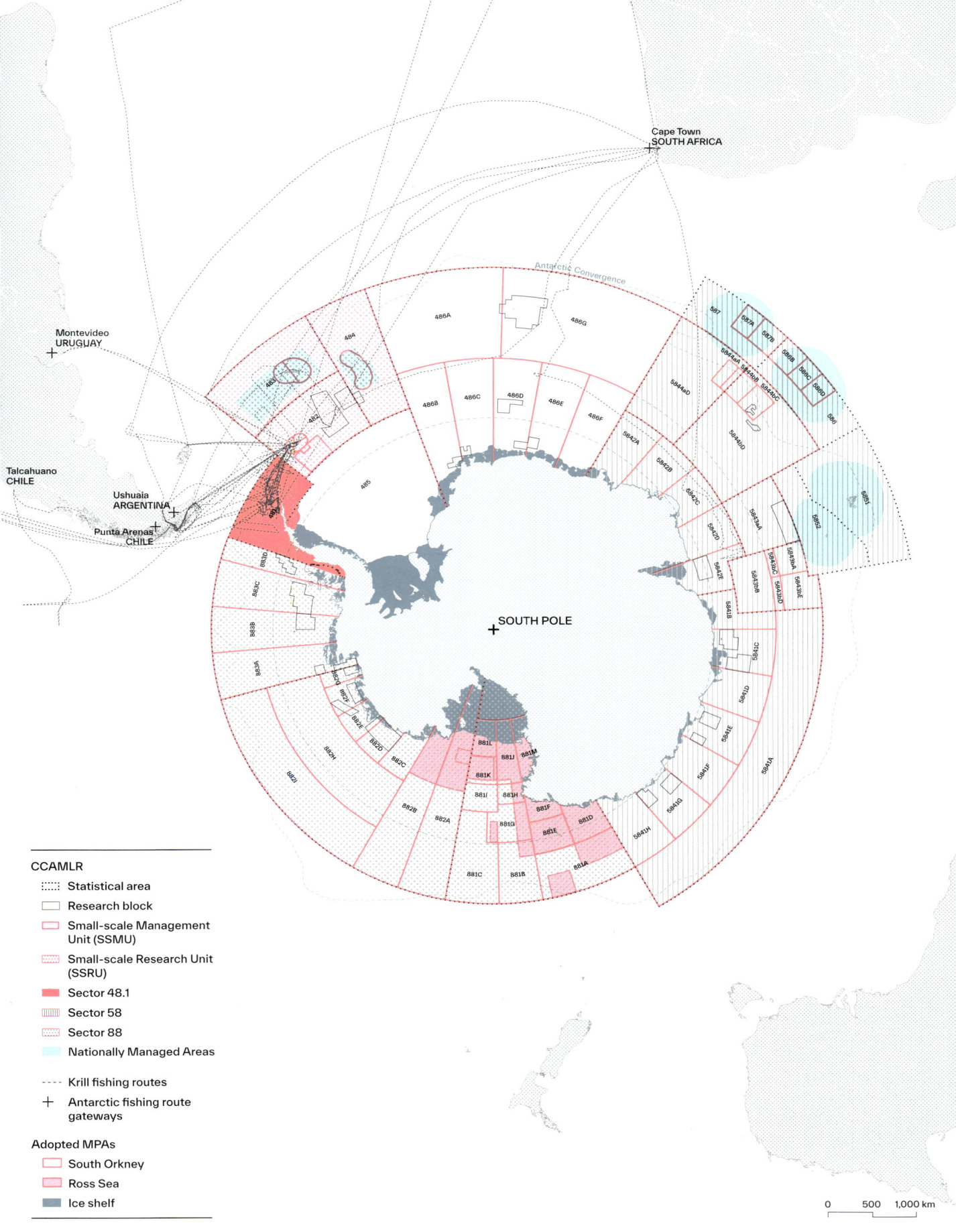

CCAMLR

┊┊┊┊ Statistical area

▢ Research block

▭ Small-scale Management
Unit (SSMU)

▦ Small-scale Research Unit
(SSRU)

■ Sector 48.1

▥ Sector 58

▦ Sector 88

▨ Nationally Managed Areas

---- Krill fishing routes

+ Antarctic fishing route
gateways

Adopted MPAs

▭ South Orkney

▭ Ross Sea

■ Ice shelf

SOUTH POLE

Antarctic Convergence

Cape Town
SOUTH AFRICA

Montevideo
URUGUAY

Talcahuano
CHILE

Ushuaia
ARGENTINA

Punta Arenas
CHILE

0   500   1,000 km

Marine Protected Areas

Andrea Kavanagh

ANDREA KAVANAGH is the director of The Pew Charitable Trusts. Her work promotes efforts to protect Antarctica's Southern Ocean, one of the world's last great wilderness areas. Kavanagh has worked for more than two decades to establish solid marine protections and precautionary toothfish and krill fishing regulations in the Southern Ocean.

The icy waters of Antarctica's Southern Ocean constitute one of the world's greatest wilderness, where thousands of species found nowhere else thrive, while also being home to numerous species of penguins, seals and whales that all depend on large swarms of Antarctic krill – the base of the region's delicate food web.

In a broader context, the Southern Ocean serves as the beating heart of the global ocean and the planet's health. Vital nutrients that well up from Antarctica's deep ocean are carried away by currents far beyond the Antarctic Convergence, their nutrient-rich waters breathing life into coastal fisheries north of the equator. The ocean also provides a critical climate mitigation service by removing and storing human-caused carbon emissions from the atmosphere.

Scientists believe this ecosystem is changing, with temperatures that are increasing faster than nearly anywhere else on Earth. A recent study[1] reported that the range of Antarctic krill has shifted more than 400 kilometres south since the 1970s due to warming ocean waters, a move that could threaten the many species that depend on krill, including whales. Another study revealed that concentrated krill fishing[2] in the Antarctic Peninsula is negatively affecting the breeding and nesting of the region's penguins – including chinstrap and emperor penguins, which are threatened with extinction. This recent news from the Southern Ocean echoes a United Nations report on the state of global biodiversity, which warned that species extinction rates are accelerating, with about one million species already threatened. A 2019 report from the Intergovernmental Panel on Climate Change concluded that the ocean and Earth's ice-covered regions, or cryosphere, are on the front line of the climate crisis, and recommended that global leaders act to increase the number and size of Marine Protected Areas (MPAs).

Marine scientists agree that establishing a network of large MPAs throughout the Southern Ocean is essential to protecting biodiversity and providing resilience to climate change. The latest science confirms that to regenerate ocean life, we need to establish Marine Protected Areas covering at least 30% of the ocean by 2030. Antarctic MPAs can contribute to this large percentage target, especially since only 13% of the ocean can be defined as wilderness and much of it is located in polar regions – yet, counterintuitively, these areas are less protected than all others. Preservation of Antarctica's unique biological resources is important and achievable. In 1982, the Commission for the Conservation of Antarctic Marine Living Resources (CCAMLR) was established under the umbrella of the Antarctic Treaty, and in 2002, CCAMLR became the first international body to commit to creating a network of MPAs, a decision based on a mission to protect, rather than exploit, life in the Southern Ocean. To date, CCAMLR has designated two MPAs in Antarctica and proposed three others. Progress has been made in these intervening years, including acknowledging that the science backing the currently proposed MPAs – in East Antarctica, the Weddell Sea and the Antarctic Peninsula – is sound; but the political will to designate them has been stymied. The lack of progress is at direct odds with the urgent need dictated by global climate change and biodiversity loss.

CCAMLR will hold its 40th meeting in October 2021, the same year that marks the sixty-year anniversary of the Antarctic Treaty going into force. To mark these two important occasions, member governments must make up for lost time by designating the three currently proposed MPAs. These areas, together with existing MPAs in the Southern Ocean and subantarctic waters, would protect more than 7 million square kilometres, significantly contributing to the goal of protecting 30% of the world's ocean by 2030 and realizing CCAMLR's vision to protect the extraordinary marine ecosystems of the Southern Ocean and the penguins, seals, krill and whales that call this place home.

"The next few years are absolutely essential for the future of our oceans and we are in desperate need for governments to come together and do what is best for these amazing ecosystems […] Now we want to go one better and create the world's largest protected area. We want to create that momentum that says this is not just possible, it is inevitable if we are to protect the wildlife that call the ocean home and crucially help mitigate the worst effects of climate change. […] World leaders shouldn't allow an ocean wilderness to be exploited by a handful of companies. In the 1980s it took a global movement to protect the Antarctic's land. Now we need to protect its oceans."

Will McCallum (Head of Oceans, Greenpeace), "World's Biggest Wildlife Reserve Planned for Antarctica in Global Campaign" in *The Guardian*, 2018

The timeline of the worldwide Marine Protected Areas evidently shows that it took almost thirty years after the signing of UNCLOS to include an MPA in Antarctic waters. Following the successful, yet small, South Orkney Islands MPA, a second Antarctic MPA was established in 2017 in the Ross Sea. Covering an area of 1.55 million sqkm (of which 1.12 million sqkm are fully protected), the latter is to date the world's largest MPA.

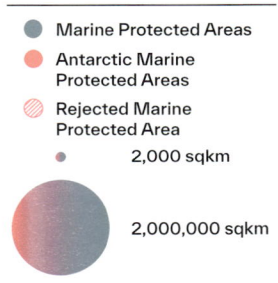

● Marine Protected Areas

● Antarctic Marine Protected Areas

◔ Rejected Marine Protected Area

• 2,000 sqkm

2,000,000 sqkm

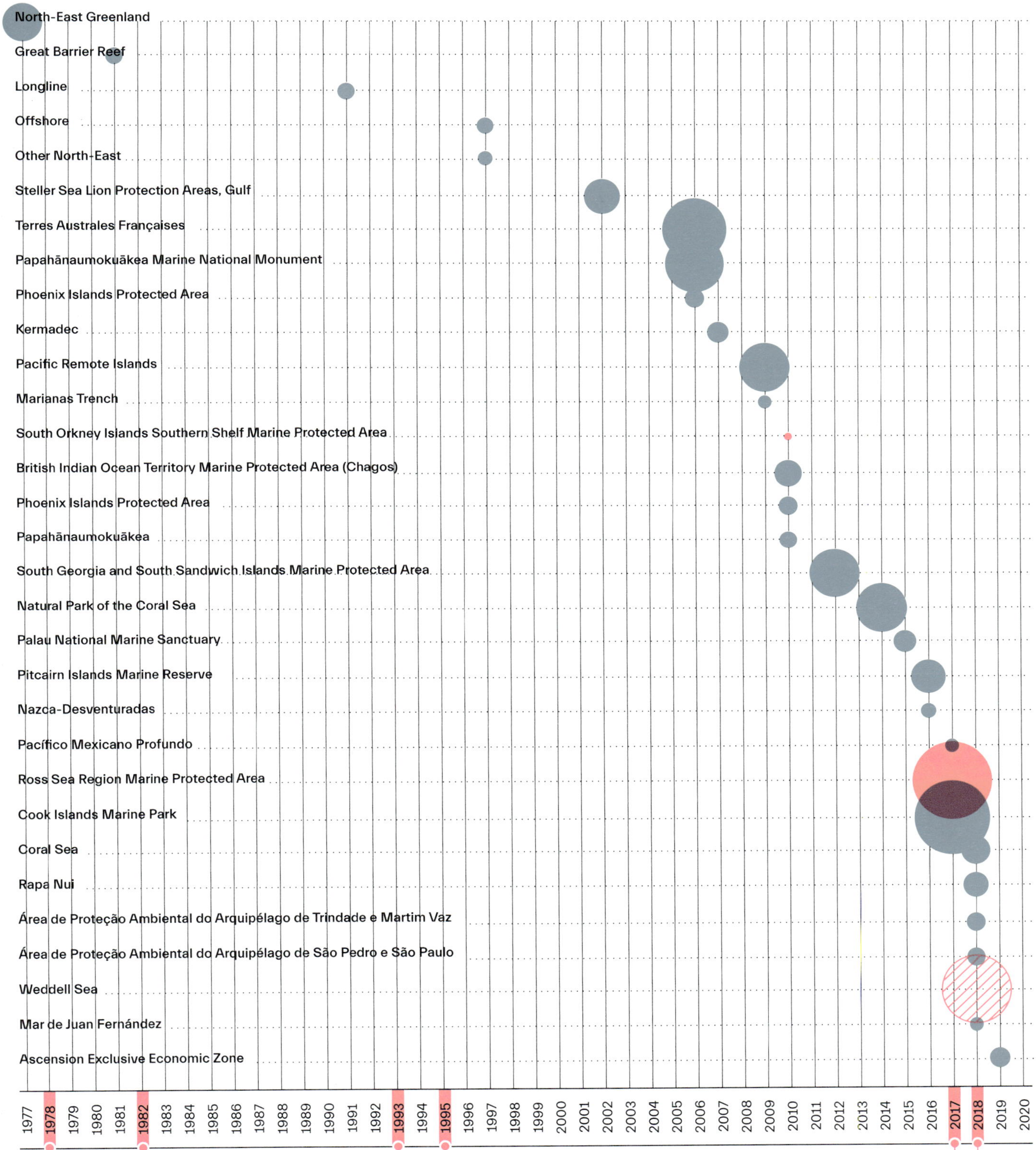

North-East Greenland
Great Barrier Reef
Longline
Offshore
Other North-East
Steller Sea Lion Protection Areas, Gulf
Terres Australes Françaises
Papahānaumokuākea Marine National Monument
Phoenix Islands Protected Area
Kermadec
Pacific Remote Islands
Marianas Trench
South Orkney Islands Southern Shelf Marine Protected Area
British Indian Ocean Territory Marine Protected Area (Chagos)
Phoenix Islands Protected Area
Papahānaumokuākea
South Georgia and South Sandwich Islands Marine Protected Area
Natural Park of the Coral Sea
Palau National Marine Sanctuary
Pitcairn Islands Marine Reserve
Nazca-Desventuradas
Pacífico Mexicano Profundo
Ross Sea Region Marine Protected Area
Cook Islands Marine Park
Coral Sea
Rapa Nui
Área de Proteção Ambiental do Arquipélago de Trindade e Martim Vaz
Área de Proteção Ambiental do Arquipélago de São Pedro e São Paulo
Weddell Sea
Mar de Juan Fernández
Ascension Exclusive Economic Zone

1977 1978 1979 1980 1981 1982 1983 1984 1985 1986 1987 1988 1989 1990 1991 1992 1993 1994 1995 1996 1997 1998 1999 2000 2001 2002 2003 2004 2005 2006 2007 2008 2009 2010 2011 2012 2013 2014 2015 2016 2017 2018 2019 2020

1978
ASOC is
established.

1982
UNCLOS enters
into force.

1993
The Convention on
Biological Diversity
enters into force.

1995
The Jakarta Mandate
on Marine and
Coastal Biological
Diversity is adopted.

2017
The Ross Sea MPA agreement enters into force
5 years after it was originally proposed in 2012.

2018
China, Norway and Russia block the proposal for the
world's biggest sanctuary in the Antarctic ocean.

Spatial Exceptions at Sea

Juan Du

JUAN DU is the director of the Polar Lab Hong Kong. She is an associate professor and associate dean of architecture at the University of Hong Kong, where she directs the Urban Ecologies Design Laboratory. The projects within her practice, IDU, range from community centres to informal settlements upgrades. Prof. Du's extensive writings include *The Shenzhen Experiment* (Harvard University Press, 2020).

While special zones around the world are predominately utilised to expedite industrial and economic development, large-scale spatial zoning, such as Marine Protected Areas (MPAs), have become an important environmental and political tool for marine conservation and management. MPAs are used by national and international organisations to designate sites of uniquely important environmental, and in some cases cultural, significance that are threatened by human activities such as fishing and mining. These demarcated areas are often also living research laboratories for scientists working on subjects from biodiversity to climate change. With increasing awareness of the impacts of oceanography and marine activity on the some of the world's most urgent environmental risks, there has been a significant increase in MPAs in recent decades. According to the *Atlas of Marine Protection,* there are currently more than 15,000 MPAs in the world, covering 5% of our global ocean.[1] There are only two MPAs in Antarctica, at the South Orkney Islands and in the Ross Sea region, and they hold unique historical, ecological and political importance for the scientific community as well as our international habitat at large.

The world's first MPA on international waters beyond national jurisdiction, the South Orkney MPA, was established in 2009 at the South Orkney Islands Southern Shelf in Antarctica by the Commission for the Conservation of Antarctic Marine Living Resources (CCAMLR).[2] Known as one of the first efforts in the world to recognise the importance of protecting entire ecosystems rather than isolated species, the South Orkney MPA was a major achievement for the CCAMLR and a milestone for marine conservation in establishing an international landmark to spatially protect ecosystems as opposed to isolated species. The South Orkney MPA was also groundbreaking in establishing protocols of international negotiation, spatial planning and scientific data requirements to support designation as well as serving as a demonstration case for a post-designation management framework for research activities, data sharing, monitoring and reporting.[3]

In 2010, following the momentum of the South Orkney MPA, the CCAMLR community began to discuss and gather support for a new MPA in Antarctica's Ross Sea region just off the coast of Victoria Land and the Ross Sea Ice Shelf. The Ross Sea shelf and slope is only 2% of the total Southern Ocean, yet it is home to the world's largest colonies of Adélie and emperor penguins, Antarctic petrels,

Ross Sea killer whales and South Pacific Weddell seals, as well as an abundance of their food source, the Antarctic krill and toothfish.[4] Complex oceanography contributes to the extraordinary biodiversity of the Ross Sea region. While it is a small percentage of Antarctica's continental shelf and Southern Oceans sea, the Ross Sea region hosts a variegated physical setting of Antarctic mountains and ice sheets, strong katabatic and cyclone winds, and diverse spatial ocean depth – all contributing to dynamic ice formations and unique biological processes, such as the highest production of phytoplankton biomass in Antarctica.[5] The Ross Sea region is also the source of the second-largest Antarctic bottom water, which has a significant impact on global fluctuations of ocean heat and sea level rise.[6]

The Ross Sea area also hosts a proportionally high level of human presence in Antarctica. The Ross Sea region is home to some of the largest and most active research stations: McMurdo ↗p. 866 of the United States, Scott Base ↗p. 852 of New Zealand, Mario Zucchelli ↗p. 752 of Italy, Gondwana of Germany and Jang Bogo ↗p. 820 of South Korea. During the austral summer months, research and touristic field camps, fishing vessels and cruise ships bring more human activity. It seems a matter of urgent necessity to establish a protected area for this globally significant and ecologically unique place on Earth.

In 2011, both New Zealand and the United States filed formal petitions to establish a Ross Sea region MPA and in 2012, the two petitions become a joint proposal. Compared to the 94,000 square kilometres of the South Orkney Islands MPA, the proposed Ross Sea MPA was on a massive scale at 1.55 million square kilometres, which would have made it the world's largest marine protected area. However, the designation process stalled over different national interests, with particular opposition from China and Russia over the terms of restrictions versus use, and questions over the permanence of the restrictions.[7] Negotiations, consensus-building and revisions to reach the eventual establishment took another five years, with China agreeing to a revised version in 2015, and Russia also giving its consent by 2017.[8] Both countries have expressed long-term interest in the area. In particular, China's fifth research station since 2018 was approved for construction, on the Ross Sea region's Inexpressible Island.

The eventual Ross Sea region MPA was set for a period of thirty-five years. It contained three different types of zones with different

restrictions and allowances through the CCAMLR's Conservation Measure 91-05 (2016).[9] Within the overall MPA, the largest part, a "General Protection Zone", is set as a "no-take" area where no marine life or minerals may be extracted. Set within the central protected area, just off the Ross Sea Ice Shelf, is a Special Research Zone (SRZ) that allows for Antarctic toothfish research fishing. And on the western side of the protection area is a Krill Research Zone (KRZ) that allows for the fishing of krill for research purposes. The Ross Sea region outside of the MPA would follow the general measures of the CCAMLR's ecosystem conservation approach, which allows for managed commercial fishing. While the SRZ and KRZ resulted from spatial and temporal negotiations and compromises, these zones of exception still enabled the CCAMLR to reach the difficult goal of unanimous consensus.

In 2010, the world missed a deadline set after the 1992 Earth Summit in Rio to establish Marine Protected Areas for 10% of the global ocean. It was replaced by "Aichi Target 11", which extended the deadline to 2020 – a target we have failed to meet yet again. Among the many challenges, the often complex nature of maritime sovereignty is a major obstacle, as is evidenced by the majority of current Marine Protected Areas set within national jurisdictions. However, most of the world's ocean consists of international waters beyond national jurisdiction, and thus 95% of the Earth's water volume could only be regulated through negotiated diplomacy. The international collaborative model for debating, planning, designing, managing and monitoring MPA in Antarctica offers lessons, and cautions, for global environmental action.

Ross Sea MPA
1.8 million sqkm

Iran
1.6 million sqkm

The Ross Sea MPA is divided into three zones (KRZ, GPZ and SRZ), each of which allows specific activities and restricts others. Fishing is allowed only in certain areas within it and must be conducted in accordance with conservation measures set out by CCAMLR.[1] In addition to the two approved Antarctic MPAs, numerous attempts to designate other areas in the Southern Ocean remained unsuccessful. Most notable among them is the 2018 proposal to designate a 1.8 square kilometre area in the Weddell Sea, which was boycotted by Russia, China and Norway.

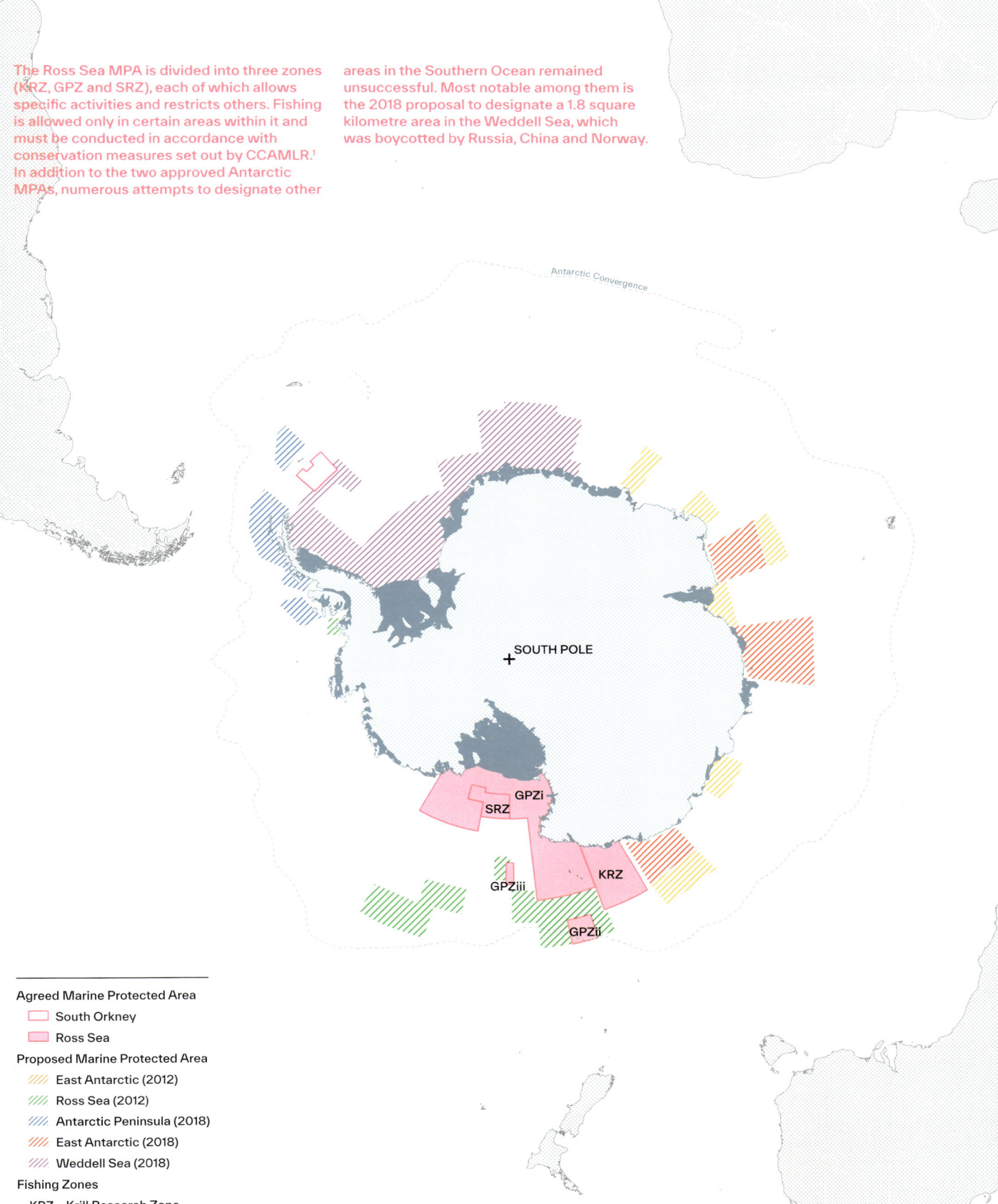

Antarctic Convergence

SOUTH POLE

GPZi
SRZ
GPZiii
KRZ
GPZii

Agreed Marine Protected Area

South Orkney

Ross Sea

Proposed Marine Protected Area

East Antarctic (2012)

Ross Sea (2012)

Antarctic Peninsula (2018)

East Antarctic (2018)

Weddell Sea (2018)

Fishing Zones

KRZ – Krill Research Zone

GPZ – General Protection Zone

SRZ – Special Research Zone

Ice shelf

0   500   1,000 km

The Krill Market

Daniel Kiss and Swadheet Chaturvedi

DANIEL KISS and SWADHEET CHATURVEDI hold M.Arch. degrees in landscape urbanism from the Architectural Association School of Architecture (United Kingdom), where they took an active part in the Polar Lab. Their thesis, "Dynamic Domains of Antarctica", envisions the dynamic management of commercial krill fishing activities through the models of regulatory systems of global common resources.

Euphausia superba, commonly known as Antarctic krill, is the most sought-after marine species in the Southern Ocean. Of interest to an ever-growing number of stakeholders, the krill fishing industry has grown exponentially due in part to the technological evolution of fishing equipment.

Tiny in scale, the Antarctic krill has a pivotal role in the marine ecosystem. Grazing directly upon the large amounts of phytoplankton – which breeds successfully in Antarctica due to various nutrients dissolved into the ocean through the sea ice and is responsible for the absorption capacity of the Southern Ocean which accounts for approximately 60% of the global atmospheric carbon[1] – the krill is a keystone species in the marine food web. In the South Georgia shelf, for example, predators depend on the krill for their dietary requirements to varying degrees: 25% for flying birds, 22% for seals, 19% for penguins and 3% for whales, respectively.[2]

Commercial harvesting of krill is facilitated by their tendency to aggregate into huge swarms that can stretch for kilometres.[3] The evolution of extractivist technologies in recent years has multiplied the efficiency of krill catch through modern techniques such as "eco-harvesting" – a new fishing system developed and applied to one of the Norwegian vessels, *Saga Sea* – reducing the amount of krill perishing during extraction and lowering the by-catch of other species.[4] Claiming the development of sustainable fishing techniques, companies receive Marine Stewardship Council (MSC) certification and define themselves as eco-sensitive organisations undermining but not justifying the ongoing conversations on the necessary heightened regulatory protection framework in Antarctica.

Freshly caught krill is processed to extract essential proteins and amino acids that make it marketable for the aquaculture industry and for the production of fish oil tablets widely recognised for their health benefits.[5] With the global value of krill oil set at US\$ 204.4 million in 2015, global revenues are expected to nearly double by 2021,[6] especially as it is proved that fish oil guarantees higher efficiency in fish farming.[7] Therefore, commercial krill fishing is extremely valuable to countries whose economy relies on aquaculture, such as Norway, where the largest source of foreign exchange after oil and gas is related to the salmon-farming industry.[8] Interest is also high in the other Antarctic fishing states, which include China, South Korea, Chile and Ukraine[9] – notably all permanent member states of the Antarctic Treaty System. Cumulatively, such states conduct Antarctic krill-fishing activities with a fleet of "only" eleven vessels.

Fishing activities – potentially in contrast with the 1998 Environmental Protocol, which prohibits "any activity relating to mineral resources, other than scientific research"[10] – are regulated by the Convention for the Conservation of Antarctic Marine Living Resources (CCAMLR), which attempts to limit annual biomass catch to a maximum of 150,000 tons per year, based on the "Krill Yield Model" (KYM) developed by CCAMLR's Ecosystem Monitoring Program (CEMP). To implement a precautionary approach and avoid irreversible anthropogenic damage, CCAMLR established that the Peninsular marine area is further subdivided into Smaller-Scale Management Units (SSMU).

Perhaps not surprisingly, given the relative ease of accessibility and calm weather conditions, almost all of the krill caught in the last forty years has been concentrated in a single SSMU called the Bransfield Strait West, i.e. in Area 48 (which accounts for 13,695,378 square kilometres), more than half of which is further condensed into Area 48.1, which has a surface of 697,000 square kilometres. As a result of inadequate static regulations of an otherwise dynamic marine system, the adjacent ecosystem has been under severe stress, since it corresponds to the chosen habitat of large colonies of Antarctic predator species that depend on krill, especially during the breeding season.

Given the steady growth in demand for the commodification of Antarctic krill, the global community, incentivised by non-governmental organisations such as Greenpeace, proposed a future in which at least 30% of oceans would be designated as geographically static Marine Protected Areas around the oceans.[11] Launching a much-needed discussion on how we collectively envision regulatory systems for resource extraction from global commons which adapt to the dynamism of natural systems in space and time, the adoption of such a proposal is key to ensuring the future of the fragile Antarctic ecosystem.[12]

> "The global omega-3 market is expected to reach US\$ 8.30 billion in 2024, growing at an annual growth rate of 10.63%, for the duration spanning 2020–2024. […] Krill oil is also emerging significantly in the market, owing to its EPA and DHA content, which are beneficial for healing various health problems."
>
> Omega 3 Market (Fish & Krill Oil), Report, 2020–2024

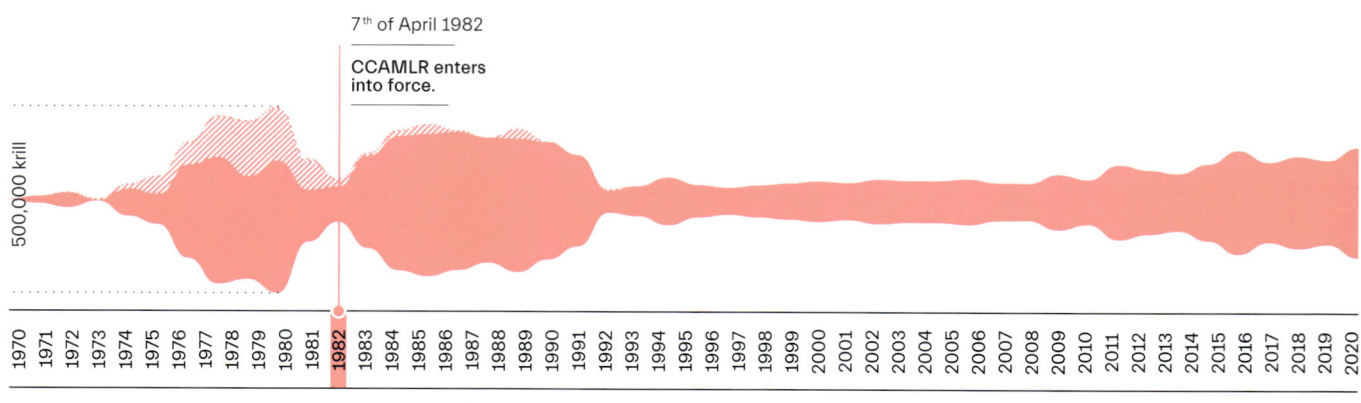

7th of April 1982

CCAMLR enters into force.

500,000 krill

1970 1971 1972 1973 1974 1975 1976 1977 1978 1979 1980 1981 1982 1983 1984 1985 1986 1987 1988 1989 1990 1991 1992 1993 1994 1995 1996 1997 1998 1999 2000 2001 2002 2003 2004 2005 2006 2007 2008 2009 2010 2011 2012 2013 2014 2015 2016 2017 2018 2019 2020

Aker BioMarine is one of the greatest krill-fishing stakeholders in the Southern Ocean. Based in Oslo, with warehouses in Montevideo, the company offloads "3,000–4,000 million tons of krill meal in only thirty-six hours"[1] and processes it to supply both fisheries with aquaculture feed and health-care retailers with omega-3 supplements. According to the company, "in 2016 alone, fish farmers using [their] krill meal produced an additional 175 million servings of salmon without increasing the volume of feed." The statement adds that "an independent study conducted by the research institute Nofima concluded that farmed salmon grows 10–25% faster due to QRILL Aqua krill meal."[2]

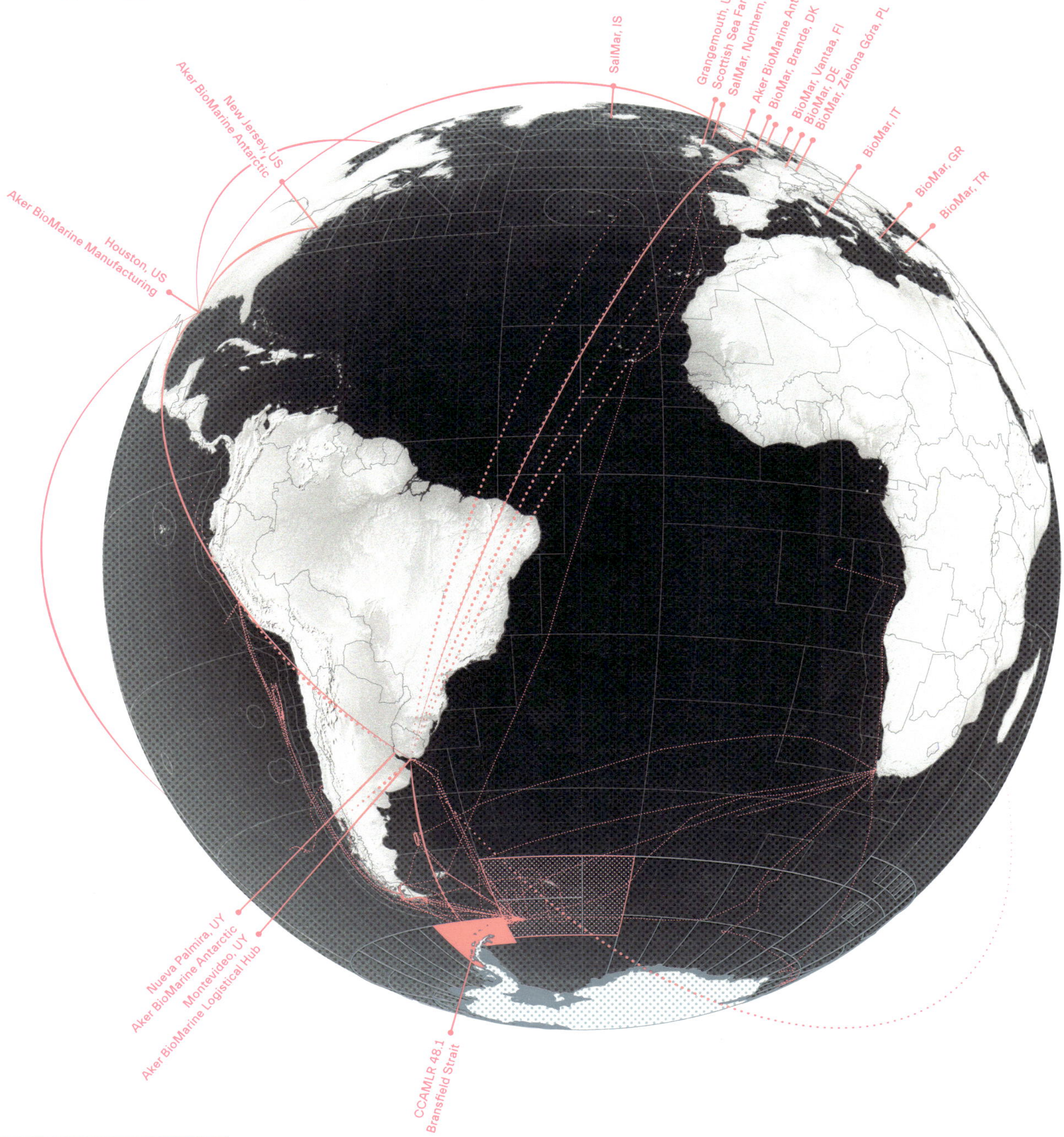

New Jersey, US

Aker BioMarine Antarctic

Aker BioMarine Manufacturing

Houston, US

SalMar, IS

Grangemouth, UK

Scottish Sea Farms

SalMar, Northern, Lysaker, NO

Aker BioMarine Antarctic Services, NO

BioMar, Brande, DK

BioMar, Vantaa, FI

BioMar, DE

BioMar, Zielona Góra, PL

BioMar, IT

BioMar, GR

BioMar, TR

Nueva Palmira, UY

Aker BioMarine Antarctic

Montevideo, UY

Aker BioMarine Logistical Hub

CCAMLR 48.1

Bransfield Strait

Fishing routes

FAO statistical zones

CCAMLR

Krill catch within Sector 48

Krill catch within Sector 48.1

Modern fishing vessels such as the *Antarctic Endurance,* the first purpose-built krill vessel commissioned by Aker BioMarine in 2019 (which cost the Norwegian company approximately US$ 118 million), are the technological successors to historic whaling factory ships. With a carrying capacity varying between 642 and 3,150 tons, they cumulatively secure 42,000 tons of fish-hold capacity for the five leading krill nations.

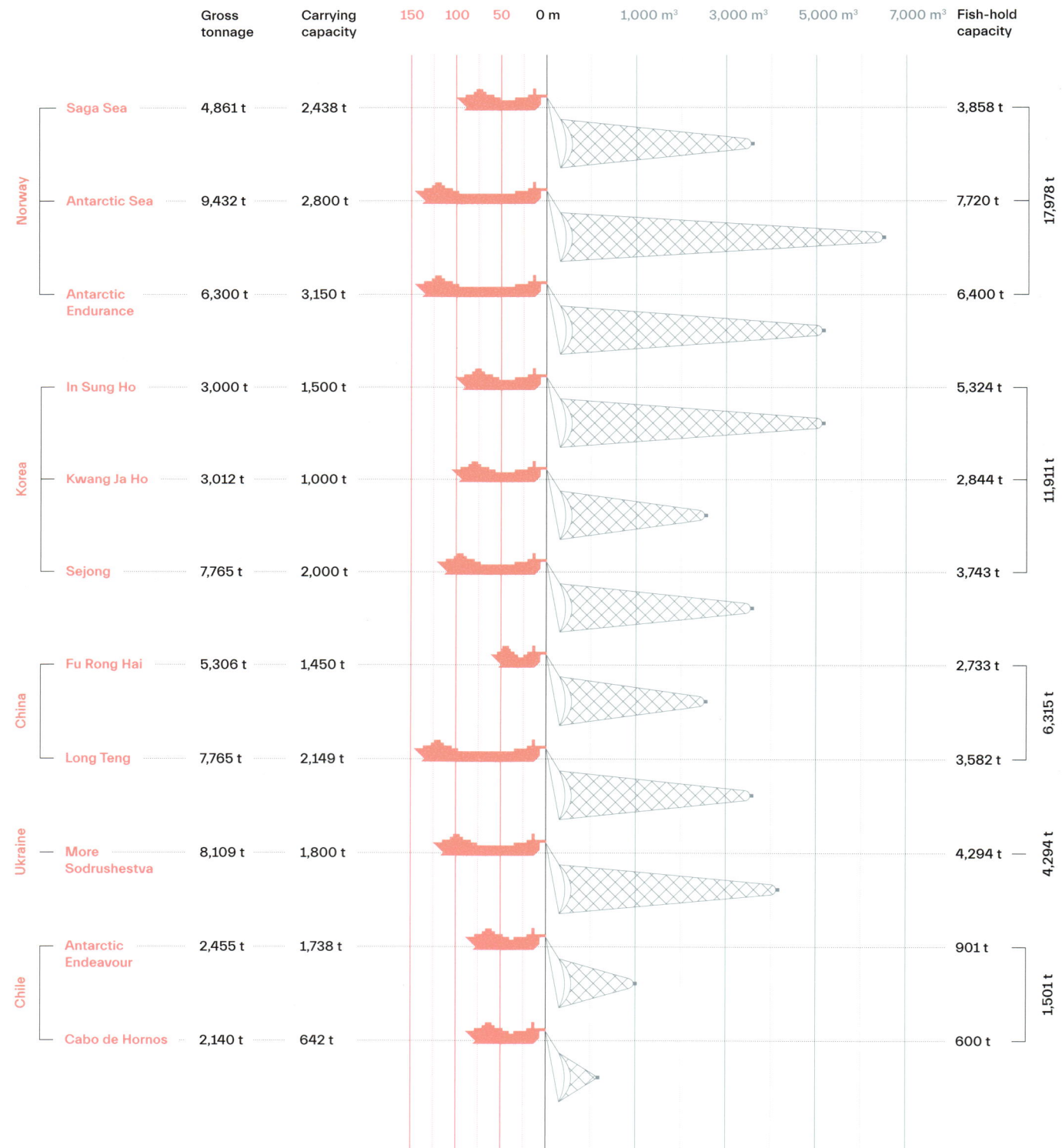

		Gross tonnage	Carrying capacity		Fish-hold capacity	
Norway	Saga Sea	4,861 t	2,438 t		3,858 t	17,978 t
	Antarctic Sea	9,432 t	2,800 t		7,720 t	
	Antarctic Endurance	6,300 t	3,150 t		6,400 t	
Korea	In Sung Ho	3,000 t	1,500 t		5,324 t	11,911 t
	Kwang Ja Ho	3,012 t	1,000 t		2,844 t	
	Sejong	7,765 t	2,000 t		3,743 t	
China	Fu Rong Hai	5,306 t	1,450 t		2,733 t	6,315 t
	Long Teng	7,765 t	2,149 t		3,582 t	
Ukraine	More Sodrushestva	8,109 t	1,800 t		4,294 t	4,294 t
Chile	Antarctic Endeavour	2,455 t	1,738 t		901 t	1,501 t
	Cabo de Hornos	2,140 t	642 t		600 t	

The Emerging Issue of Plastic Pollution in Antarctica

Elisa Bergami and Ilaria Corsi

ELISA BERGAMI and ILARIA CORSI are members of the Scientific Committee on Antarctic Research (SCAR)'s Steering Committee of the Plastic in Polar Environments Action Group. Ilaria Corsi is the co-ordinator of the Plastics in Antarctic Environment project (PLANET) funded by the Italian National Antarctic Research Programme.

Antarctica has long been depicted as a pristine environment, isolated by the Antarctic Circumpolar Current. However, anthropogenic impacts are increasingly putting polar ecosystems under pressure and new threats such as plastic pollution have emerged. Records of plastic litter on Antarctic islands date back to the 1980s, and its occurrence has now been confirmed in Antarctic surface waters, sea ice, sediments, and biota, although at low concentrations when compared with other regions worldwide.

Plastic debris may be brought to Antarctica by oceanic currents, favoured by strong winds and storm events, as shown by the latest high-resolution ocean circulation models.[1] Human influence is also likely to be responsible for the spread of plastic contamination: seventy-six base stations are operating below 60°S, managed by twenty-nine nations, in addition to tourism and fishing vessels, which are abundant in the austral summer. In this context, plastic pollution, in the form of fragments, fibres, and films, may derive from multiple local sources, such as logistics, field-based activities, waste water (synthetic textile fibres and personal care products) as well as from the fragmentation of large plastic materials. Once in the environment, plastics inevitably breaks down owing to the action of sunlight, water, temperature and biota, which alter their physico-chemical structure and cause embrittlement and erosion. Through weathering, plastics may take several decades or even centuries to degrade, depending on their intrinsic properties, such as size, shape, and polymeric composition as well as their interaction with the surrounding environment.

In Antarctica, the harsh environmental conditions caused by the low temperatures and high seasonal variations in UV radiation, together with the presence of sea ice and organisms, are critical factors that may significantly influence the behaviour and fate of plastics. For example, plastic particles not only tend to agglomerate in the presence of salts and exudates released from sea ice algae but may also adhere to ice crystals, thus becoming incorporated in the sea ice as it forms.[2] As a result, sea ice acts as a dynamic reservoir of these pollutants, which can be released in large quantities when it melts, a process that is accelerated by climate change.[3]

As evidenced by the wide spectrum of recent research papers on plastic pollution in Antarctica published by members of the SCAR Action Group "Plastic in Polar environments",[4] the potential impacts of plastics on Antarctic wildlife are manifold. Macroplastics (above 5 millimetres in size) are associated with physical damage, such as entanglement, ingestion, and starvation in Antarctic seabirds and marine mammals. Microplastics (below 5 millimetres in size) have been found ingested in Antarctic marine species belonging to different trophic levels, from benthic invertebrates up to penguins. Toxicity might be associated not only with the presence of these plastics but also with the release of harmful chemicals from the polymer structure (such as plasticisers, monomers), adsorbed contaminants (such as metals, persistent organic pollutants) and/or associated pathogens growing on their surface.

A pioneering bench-scale study conducted with the smallest plastic fraction, nanoplastics (under 1 micrometre in size), shows how Antarctic sea urchin (*Sterechinus neumayeri*) immune cells respond to this emerging threat.[5] Although their presence has not been quantified yet, the nanoplastics, which are comparable to viruses or even large proteins in size, may directly interact with cell membranes, leading to behavioural and physiological changes and triggering toxicity in the exposed organisms.

Antarctic krill (*Euphausia superba*), a keystone species of Antarctic marine ecosystems, can be particularly exposed to this contamination, since it has been shown to ingest and fragment microplastic beads, releasing nanoplastics.[6] Our recent study demonstrates that krill can directly feed on nanoplastics, which, once incorporated in krill faecal pellets, may thus redistribute in the water column, with detrimental effects on Antarctic pelagic ecosystems and biogeochemical cycles.[7]

The latest news from the South reveals that plastic pollution does not only affect Antarctic marine wildlife but has also reached terrestrial food webs, as shown by the polystyrene traces found in the digestive tract of the Antarctic soil collembolan *Cryptopygus antarcticus*,[8] which was feeding on a piece of polystyrene foam covered by algae, moss, and lichens on King George Island. In this region, stranded polystyrene foam has previously been associated with the spread of antibiotic resistance genes,[9] suggesting that microplastics may alter soil properties, microbial communities, and biodiversity within Antarctic terrestrial ecosystems.

Current estimates of plastics in the Antarctic Peninsula warrant urgent action to unravel the spread of this emerging issue and provide answers to the many unexplored questions regarding their potential harm to such fragile ecosystems. Antarctica is experiencing prominent changes in the biosphere due to the combined effects of climate change and human pressure, and it represents an open-air laboratory for scientists studying plastic contamination as an additional stressor to remote environments, as well as a window on the future of Antarctica in the Anthropocene.

"There was a massive increase in marine-life mortality. It is as if someone started cutting down a rainforest repeatedly every year so there was no hope of any seedling growing to become a mature tree. That's what has happened in this cove in the last decade as the duration of sea ice has reduced. It's one of the most startling natural changes in the seabed that we know of in the world."

Simon Morley (Marine Biologist, British Antarctic Survey), "Inside Antarctica: The Continent Whose Fate Will Affect Millions", *Financial Times*, 2018

Bioprospecting in Antarctica

Julia Jabour

JULIA JABOUR is a senior lecturer in law and policy at the University of Tasmania (Australia). She has served as an observer with the Australian delegation at Antarctic Treaty Consultative Meetings (ATCM) on a number of occasions. She has been researching, writing and lecturing on polar governance for more than 20 years and hosted the annual Polar Law Symposium in Hobart (Australia) in 2014 and 2019. Dr Jabour has visited Antarctica six times.

Bioprospectors look for novelty in living organisms with a view to commercialising it. The Antarctic has an extreme climate: it is cold, dry, windy, elevated and subject to dramatic fluctuations in seasonal ice and light. Its relatively undisturbed and isolated fauna and flora have made unique functional modifications to ensure their continued survival. According to Alex Rogers, "The Antarctic biota is highly endemic, and the diversity and abundance of taxonomic groups differ from elsewhere in the world".[1] These conditions lead to the potential for novelty being present, especially in marine organisms, which is what attracts bioprospectors. Since the Antarctic is governed by an international legal framework rather than by individual sovereign countries, determining who has the right to access and use the biological resources and who should profit from their development is problematic.

Although archaea (an extremophile) were originally regarded as nothing more than bacteria that existed in extreme environments, later discoveries have challenged this assumption to the extent that they are now recognised as prominent actors in the biosphere. An ongoing surge in exploration efforts is leading to an increase in utilisation of archaea in biotechnology, in fields as diverse as biocatalysis, biocomputing, bioplastics production, bioremediation, bioengineering, food, pharmaceuticals, and nutraceuticals. Archaea are used in liposomes for drug delivery and cosmetics, waste treatment, and food industry molecular biology, and are also being explored to screen cancer patients' sera. To date, bioprospecting for archaea is the main focus of bioprospecting on the southernmost continent. Novel extremophiles and their biochemical processes are likely to remain the most important commercial application of the genetic resources of the Antarctic.[1]

The governance provisions for the continent of Antarctica and its surrounding Southern Ocean are complex as they involve, at once, a dedicated system of Antarctic-specific laws and overlapping international legal measures. The first is legally defined as "The Antarctic Treaty, the measures in effect under the Treaty, its associated separate international instruments in force and the measures in effect under those instruments".[2] The instruments of relevance are the Antarctic Treaty itself, adopted in 1959, the Convention on the Conservation of Antarctic Marine Living Resources (CCAMLR, adopted in 1980 and put into force in 1982) and the Protocol on Environmental Protection. New Antarctic laws and regulations are discussed and defined in occasion of the annual meetings of the Consultative and Non-Consultative Parties to the Treaty. In addition to these, there is a suite of relevant international laws that overlap with the Antarctic framework; these include, but are not limited to, the United Nations Convention on the Law of the Sea and the Convention on Biological Diversity.[3]

The Antarctic Treaty contains an artful compromise on sovereignty, recognising claims that existed prior to 1959 but suspending any further assertions on this subject. This means that the "ownership" of Antarctic resources is controversial, as this technically falls into seven claims, three of which (by Chile, Argentina and the United Kingdom) partially overlap, with two additional reserved claims (the United States and the Russian Federation); but no sovereignty is *de facto* active. The Treaty's Environmental Protocol reaffirms this position on sovereignty and mandates environmental assessments for any and all human activities on the continent. These assessments must meet baseline principles, including the protection of biodiversity. The CAMLR Convention also reaffirms the Treaty's position on sovereignty, offering guidance to the conservation of marine life by regulating harvesting activities. It does not, however, expressly target marine genetic resources nor the activity of prospecting for them. An argument can therefore be made that Antarctica and the Southern Ocean are "areas beyond national jurisdiction"[4] with such categorisation finding justification under the Law of the Sea Convention and the Convention on Biodiversity.

"Bioprospecting" *per se* does not have any specific legal regulation. Rather, under Antarctic law, the activity of *in situ* sample collection[5] is classed either as "scientific research" or "fishing". Provided the activity does not breach the egalitarian ambitions of the Treaty, both the Environmental Protocol's laws and the CCAMLR Convention's conservation and fishing regulations permit such activity irrespective of any intent to commercialise the findings without sharing benefits – in other words, without regard for any rights the rest of the world might be owed from the downstream development of unowned resources.

For two decades, the Antarctic Treaty Consultative Parties (i.e. the decision-makers) have discussed bioprospecting at the annual meetings and dismissed it on the grounds that it is an activity in need of targeted regulation. Without even a commonly agreed-upon definition of "bioprospecting", they haven't agreed on any specific regulation and treat it as an authorised human activity such as scientific research or fishing. There is no consensus among the parties on how to proceed. Notwithstanding this, they are adamant that they will not accept external interference in their affairs. The United Nations General Assembly's decision to establish a working group to develop an international legally binding instrument under the United Nations Convention on the Law of the Sea for the conservation and sustainable use of the marine biological diversity of areas beyond national jurisdiction has been a challenge to this position. The Antarctic Treaty Consultative Parties have declined correspondence and cooperation with this UN working group, stating that "the ATS [Antarctic Treaty System] is the [only] competent framework within which to address the conservation and sustainable use of biodiversity in the Antarctic region".[6] The potential for conflict is thus significant,[7] and the matter is far from settled in the eyes of some parties to the Antarctic legal framework and some non-governmental organisations that represent civil society.

"Bioprospecting bridges the divide between conventional scientific research and a commercial harvesting activity. It occurs in Antarctic terrestrial, limnological and marine environments (and hence transcends any one regulatory instrument of the [Antarctic Treaty System] ATS). With its commercial sensitivity aspects, potentially it challenges the obligation to free exchange of scientific information under the [Antarctic Treaty] AT."

Alan D. Hemmings (Professor, University of Canterbury), "Antarctic Politics in a Transforming Global Geopolitics", in *Handbook on the Politics of Antarctica*, 2017

Antarctic Tourism

Daniela Liggett

DANIELA LIGGETT is the chief officer of the Standing Committee on the Humanities and Social Sciences at the Scientific Committee on Antarctic Research (SCAR). She is a social scientist with a background in environmental management, Antarctic politics and tourism. Dr Liggett is a senior lecturer at the University of Canterbury (New Zealand) and is on the editorial boards of *Polar Geography*, *The Polar Journal* and *Advances in Polar Science*.

Tourist visits to the Antarctic region date back to the nineteenth century, when a small number of paying passengers travelled to the New Zealand subantarctic islands on resupply missions or relief voyages.[1] However, regular fully commercial visits to the Antarctic only became a feature by the 1960s, when the Swedish-American entrepreneur Lars-Eric Lindblad first chartered Argentine and Chilean naval vessels to take tourists to the Antarctic Peninsula and then commissioned the building of an ice-strengthened vessel, the M/V *Lindblad Explorer,* specifically for Antarctic tourism.[2] The M/V *Lindblad Explorer* undertook its maiden voyage in 1961, which is now hailed as the year that modern era of Antarctic tourism began, as from that time onwards annual tourist cruises to the Antarctic Peninsula have taken place.[3]

The M/V *Lindblad Explorer,* which later operated under the name M/V *Explorer,* had the misfortune of also being the first vessel to sink in the Antarctic on a dedicated tourist cruise. Flying the Liberian flag, it encountered ice in Bransfield Strait in the Antarctic Peninsula and sank.[4] The event stimulated detailed discussions at Antarctic Treaty Consultative Meetings (ATCMs) about the safety of tourism operations in general but also about the risks involved in maritime operations in poorly charted Antarctic waters. This prompted the ATCM to call for an Antarctic Treaty Meeting of Experts (ATME) on Ship-Borne Tourism, which was held in Wellington, New Zealand, in 2009. Tourism had been discussed at ATCMs since the late 1960s, when the Treaty parties' attention focused on the interaction between tourism and research stations and centred around the concern that visits by tourists would interfere with research activities.[5] In response to growing numbers of tourists visiting Antarctica and a diversification of activities, Antarctic Treaty Consultative Parties (ATCPs) have given more attention

Emphasising the relentless growth of Antarctic tourism witnessed in the past decades, the timeline reveals significant events that have affected the industry trend, such as the short-lived recession informed by the quasi-concomitant sinking of the tourist vessel MV *Explorer* and the global economic crisis of 2007. A comparative analysis of the number of tourists by country highlights that while twentieth-century tourism was dominated by Americans and Europeans, the turn of the century has seen the advent of Chinese-based tourism, which (if reviewed in light of a recent announcement on investments in specialised polar fleets) is expected to keep rising.

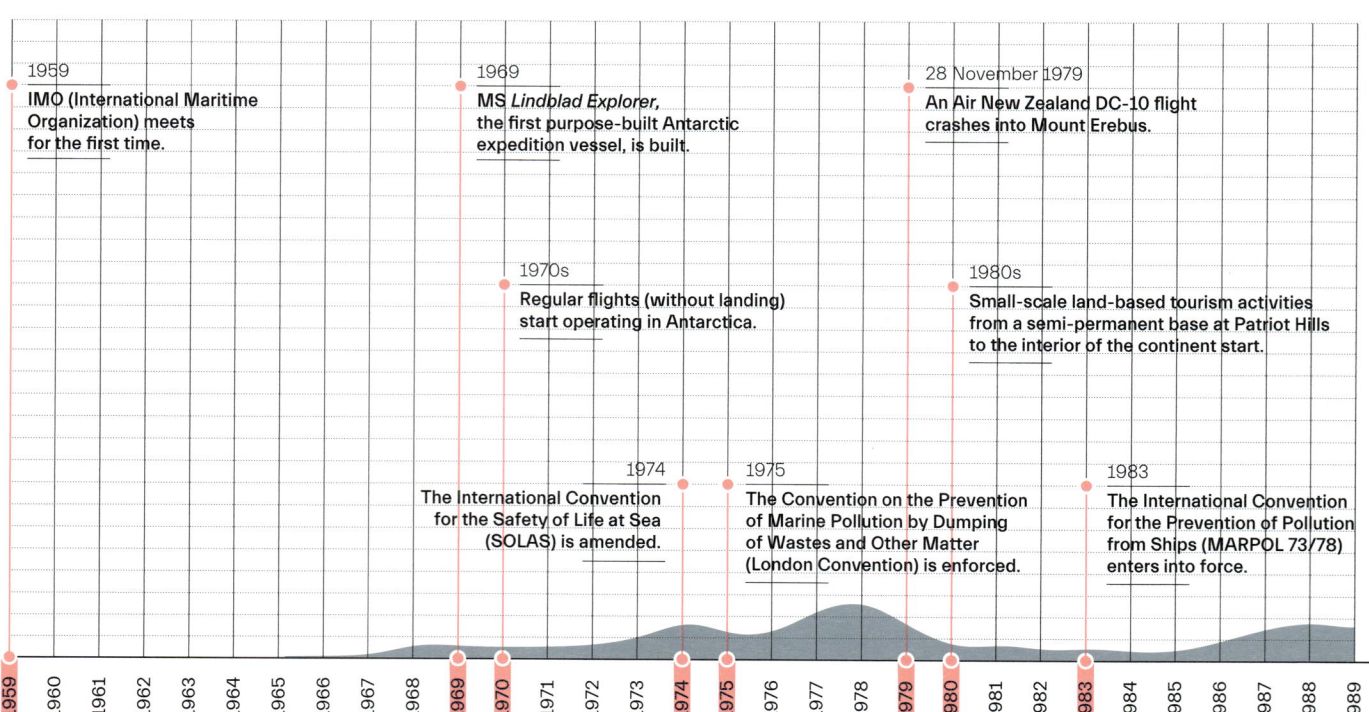

1959 — IMO (International Maritime Organization) meets for the first time.

1969 — MS *Lindblad Explorer,* the first purpose-built Antarctic expedition vessel, is built.

28 November 1979 — An Air New Zealand DC-10 flight crashes into Mount Erebus.

1970s — Regular flights (without landing) start operating in Antarctica.

1980s — Small-scale land-based tourism activities from a semi-permanent base at Patriot Hills to the interior of the continent start.

1974 — The International Convention for the Safety of Life at Sea (SOLAS) is amended.

1975 — The Convention on the Prevention of Marine Pollution by Dumping of Wastes and Other Matter (London Convention) is enforced.

1983 — The International Convention for the Prevention of Pollution from Ships (MARPOL 73/78) enters into force.

1959 1960 1961 1962 1963 1964 1965 1966 1967 1968 1969 1970 1971 1972 1973 1974 1975 1976 1977 1978 1979 1980 1981 1982 1983 1984 1985 1986 1987 1988 1989

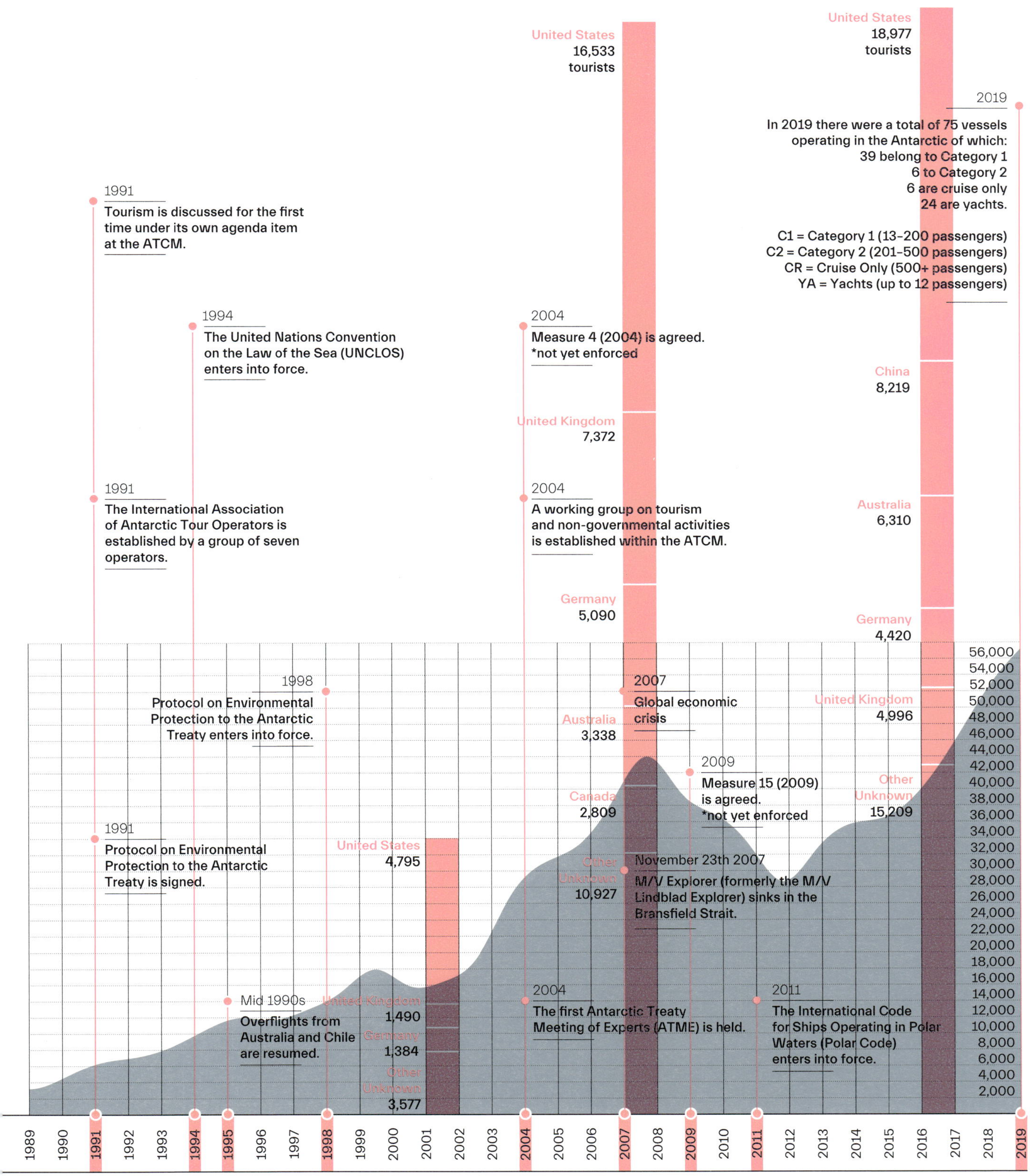

United States
16,533
tourists

United States
18,977
tourists

2019

In 2019 there were a total of 75 vessels operating in the Antarctic of which:
39 belong to Category 1
6 to Category 2
6 are cruise only
24 are yachts.

C1 = Category 1 (13–200 passengers)
C2 = Category 2 (201–500 passengers)
CR = Cruise Only (500+ passengers)
YA = Yachts (up to 12 passengers)

1991

Tourism is discussed for the first time under its own agenda item at the ATCM.

1994

The United Nations Convention on the Law of the Sea (UNCLOS) enters into force.

2004

Measure 4 (2004) is agreed.
*not yet enforced

China
8,219

United Kingdom
7,372

1991

The International Association of Antarctic Tour Operators is established by a group of seven operators.

2004

A working group on tourism and non-governmental activities is established within the ATCM.

Australia
6,310

Germany
5,090

Germany
4,420

1998

Protocol on Environmental Protection to the Antarctic Treaty enters into force.

Australia
3,338

2007

Global economic crisis

United Kingdom
4,996

2009

Measure 15 (2009) is agreed.
*not yet enforced

Other Unknown
15,209

Canada
2,809

1991

Protocol on Environmental Protection to the Antarctic Treaty is signed.

United States
4,795

Other Unknown
10,927

November 23th 2007

M/V Explorer (formerly the M/V Lindblad Explorer) sinks in the Bransfield Strait.

Mid 1990s

Overflights from Australia and Chile are resumed.

United Kingdom
1,490

2004

The first Antarctic Treaty Meeting of Experts (ATME) is held.

2011

The International Code for Ships Operating in Polar Waters (Polar Code) enters into force.

Germany
1,384

Other Unknown
3,577

56,000
54,000
52,000
50,000
48,000
46,000
44,000
42,000
40,000
38,000
36,000
34,000
32,000
30,000
28,000
26,000
24,000
22,000
20,000
18,000
16,000
14,000
12,000
10,000
8,000
6,000
4,000
2,000

1989 1990 1991 1992 1993 1994 1995 1996 1997 1998 1999 2000 2001 2002 2003 2004 2005 2006 2007 2008 2009 2010 2011 2012 2013 2014 2015 2016 2017 2018 2019

"What is to be expected in the future in this region? […] If as seems possible even warmer conditions supervene; we shall arrive at a topography like that of the Himalayas. We may imagine a high level plateau crossed by summer trails. […] At Cape Evans mountaineers will land to tackle Mount Erebus, and aeroplanists will descend for refreshment on their arrival from New Zealand. The less energetic will proceed in the comfortable steamers of the Antarctic Exploitation Company to the chalets of the Beardmore. Here start the summer motor trips to the South Pole. When?"

Thomas Griffith Taylor (Geographer, Geologist), "A Chapter of Antarctic History", *The South Polar Times*, 1911–1913

Far from being as perilous and adventurous as in the Heroic Age, crossings of the Drake Passage in contemporary vessels departing from the Antarctic gateways of South America – Ushuaia and Punta Arenas – have become a weekly occurrence for Antarctic tourism operators. As described in a recent *Financial Times* article, visitors to the seventh continent now reach the Antarctic Peninsula harmlessly while being entertained with "part[ies] around the heated swimming pool, hot dogs and waffles, glühwein and champagne, [all accompanied by a soundtrack provided by] a band on deck in parkas and beanies [that] plays hits from the 1960s." [1] This is staged only about 80 kilometres from where Shackleton's *Endurance* had been imprisoned in pack ice a century earlier.

"The Antarctic is a tourist paradise, and the day will come when the *Geographic* will be carrying ads like this: 'Cruise to unspoiled Antarctica. See the comical penguins, the great Ross Ice Shelf, the steaming volcano Mt. Erebus, and icy mountains untouched by man. Visit Little America and the actual huts of Scott and Shackleton. Enjoy winter sports in an unsurpassed setting. Landing craft and sightseeing helicopters will whisk you quickly ashore. Warm staterooms, all meals aboard ship. Make your reservations now."

Rear Admiral George John Dufek (Commander, US Naval Support Force Antarctica), "What We've Accomplished in Antarctica", *National Geographic*, 1959

to tourism issues since 1991, when tourism was discussed under its own agenda item (Tourism and Non-Governmental Activities) at ATCMs.[6]

As the number of tourists grew and as the collapse of the Soviet Union resulted in former Soviet ice-strengthened vessels of the academic fleet as well as icebreakers becoming available for long-term charter on the free market, the number of expedition cruises to the Antarctic increased exponentially.[7] At the same time, tourism activities increasingly diversified, as tour operators tried to distinguish themselves from their competitors; by the early twenty-first century, tourists could, for instance, camp, kayak, snowshoe and ski in the Antarctic, with scuba diving and marathons appearing on tourist itineraries more regularly as well. As tourism activities diversified and as the numbers of tourists grew, Antarctica witnessed a shift in the nationalities of tourists. While visitors from the United States are still dominating the market, China has discovered the continent as an exotic destination, and the second most prevalent nationality of Antarctic tourists is now Chinese.[8]

Cruise-based tourism is, by far, the most common form of Antarctic tourism. Presently, 99% of all tourists visit the Antarctic by ship and, of these, three-quarters are partaking in expedition cruises, which include landings in the Antarctic. Landings are offered only by vessels carrying up to five hundred passengers, and the number of passengers landed at any one time may not exceed one hundred. While this approach is encoded in Measure 15 (2009), a binding regulatory mechanism agreed by the Antarctic Treaty Consultative Parties,[9] it is also embedded in the International Association of Antarctica Tour Operators' (IAATO) bylaws,[10] which mandate that their member operators adhere to the spirit of this measure, even though it has not yet officially entered into force.

IAATO, a voluntary member organisation of Antarctic tourism organisers or other Antarctic tourism stakeholders, was founded by seven tour operators in 1991. With its suite of hortatory guidelines, safe operating procedures and mandatory bylaws, the organisation has since played a significant role in the *in situ* management of Antarctic tourism.[11] Currently, IAATO member operators dominate the Antarctic tourism market, with all commercial operations involving vessels other than yachts which carry up to twelve passengers organised by IAATO members. IAATO also successfully represents its member operators at ATCMs and has managed to establish a significant level of trust in its members' operational standards, to the point where ATCPs have not determinedly pursued a top-down approach to tourism regulation through the Antarctic Treaty System, but have relied on IAATO and its member operators to ensure environmentally responsible and safe tourism in the Antarctic.[12]

While still retaining the ultimate regulatory power (at least as far as operators and tourists in Antarctic Treaty signatory states are concerned), ATCPs rely largely on the goodwill of the operators themselves and the regulation put in place through the 1991 Protocol on Environmental Protection to the Antarctic Treaty which entered into force in 1998 and governs all human activities in the Antarctic, including tourism and non-governmental activities, which find specific mention in the protocol. The protocol, like any international treaty, is implemented through the national legislation of its signatory states, with certain differences in interpretation and enforcement being present.

In addition to the protocol and the aforementioned Measure 15 (2009), the ATCPs agreed upon another binding measure, Measure 4 (2004), which focuses on contingency planning and insurance coverage, as well as a host of non-binding guidelines aimed at managing tourism activities in the Antarctic. But, just like Measure 15 (2009), Measure 4 (2004) is not yet in force.[13] Some of these guidelines have grown out of IAATO's codes of conduct – Recommendation XVIII-1 (1994), for example – and all are formally supported by IAATO, which has shown itself to be considerably proactive in advancing and maturing their set of codes of conduct for operations in Antarctica.[14]

Beyond the regulation imposed through the ATCPs and IAATO, a number of instruments used by the International Maritime Organization (IMO) have particular relevance for Antarctic ship-borne tourism. First and foremost is its International Code for Ships Operating in Polar Waters, also called the Polar Code,[15]

"After a year of pandemic, travelers are looking for open spaces, safety, and sunshine – and are ready to put down serious cash to get it…. [One] online travel aggregator […] has seen a 25% increase in bookings to the southernmost continent compared to last year's pre-pandemic numbers. On Google, Antarctica has charted the most significant growth in interest from American travelers."

Nikki Ekstein (Travel Editor, Bloomberg), "Travel Industry Sees Glimmers of Recovery in Africa, Antarctica", *Bloomberg*, 22 February 2021

which entered into force on the 1st of January 2017. It requires tour operators to adhere to stricter standards on matters such as vessel design and equipment or staff training. In addition, the Polar Code mandates that any vessel certified under the International Convention for the Safety of Life at Sea (SOLAS)[16] needs to have a Polar Ship Certificate, which identifies the vessel's operational limits based on its ice class. Other important IMO conventions that are relevant to Antarctic ship-borne tourism include: the International Convention for the Prevention of Pollution from Ships (MARPOL 73/78),[17] which endeavours to minimise and prevent marine pollution; the United Nations Convention on the Law of the Sea (UNCLOS), which identifies the responsibilities and rights of flag states; and the Convention on the Prevention of Marine Pollution by Dumping of Wastes and Other Matter (London Convention) as well as the related Protocol to the Convention on the Prevention of Marine Pollution by Dumping of Wastes and Other Matter (London Protocol).[18]

The thirty-two new ice-strengthened vessels that have been commissioned over the last few years and that are now entering the market to replace the ageing vessels of the former Soviet academic fleet have all been built with the standards required by the Polar Code in mind and aim to reduce the carbon footprint of expedition cruising.[19]

Overall, since the founding of IAATO, the organisation and its member operators have aspired to keep the local and regional environmental impact of their visits as low as possible and have been successful in proactively managing tourist activities *in situ.* However, cruise tourism to the Antarctic has a considerable carbon footprint,[20] which the introduction of new and more environmentally sustainable vessels might, at least to a certain extent, address. Indirectly, the travel restrictions imposed by governments in response to the COVID-19 pandemic as well as the resulting global economic recession and reduced air travel capacity are similarly likely to reduce the number of Antarctic tourists in the upcoming season and thereby the environmental impact of Antarctic tourism, although nobody can yet foresee what Antarctic tourism in a post-COVID-19 world will look like.

Throughout the early Heroic Age, highly constructed iconic images, such as Herbert Ponting's 1911 double portrait of Thomas Clissold and an emperor penguin on a leash, shown at the side, and the notorious photo of piper Gilbert Kerr playing the bagpipes for a penguin secretly tied by a hidden rope during the Scottish National Antarctic Expedition, are reflective of the predominance of a human-centric understanding of Planet Earth. A century later, such behaviour is banned, and Antarctic visitors are required to pay the utmost respect to the continent's wildlife. Orange flags (shown on the opposite page) temporarily punctuate the white landscape at each tourist landing site, delineating areas of respect towards all penguin colonies and other wildlife habitats.

Ship-Based Antarctic Tourist Landing Procedures

Solan Jensen

SOLAN JENSEN is a professional expedition leader on the Peninsula of Western Antarctica and in the Arctic region. He received a degree in philosophy, leading him to a career in the polar territories. Jensen first visited Antarctica in 2002 as part of the construction team of the new Amundsen-Scott South Pole Station and has been working on the continent since.

The visitor experience of "protecting Antarctica" begins long before arrival. It takes the form of educational pre-departure information sent to the visitor by the outfitter, sharing protocols and objectives the industry has developed to meet and extend the regulations of the Antarctic Treaty System for the purpose of maintaining the natural integrity of Antarctica as we venture into it. In just over fifty years, the modern Antarctic tourism industry has seen a massive increase in numbers of visitors from hundreds to tens of thousands a year. Considering the potential environmental impact of increasing tourism, seven private tour operators organising excursions in Antarctica came together in 1991 to shape, practise and promote safe and environmentally responsible travel in this remote, wild region of the world.

Thirty years later, more than 100 companies are voluntary members of the International Association of Antarctica Tour Operators (IAATO) who work with the Treaty parties on the management of Antarctic tourism. The majority of Antarctic tourism is marine-based, venturing ashore for brief periods. Common safety challenges relate to the intensity and changeability of marine conditions as well as to snow and ice hazards ashore, while common environmental impact challenges relate essentially to the management of people, the visiting global public.

Industry measures established to ensure safe landing while leaving minimal impact range from mandatory biosecurity cleaning prior to and between landings, a restriction on the number of people ashore; guide-to-visitor ratios; site-specific and activity-specific guidelines; wildlife-watching guidelines, including minimum approach distances for certain species; pre- and post-visit activity reporting; mandatory instruction and education briefings for all visitors; guide training standards, including annual testing; contingency and emergency medical evacuation plans.

Day to day, prior to any landing, expedition guides consider the "theatre of operations" with a series of practices, questions, evaluations and analyses. Are we visiting a known site or a new site? What are the potential hazards to manage? What are the conditions underfoot at this site, at this time of the season? Is there a history of an incident or a near miss at this location? In what ways is the landing site exposed to adverse weather changes that could trigger an immediate emergency recall or potentially strand a party ashore?

All the equipment used daily for communications, boating, setting up and managing landing sites is constantly examined, maintained and cleaned in order to avoid any bacterial contamination on the continent. In fact, biosecurity disinfection is a ritual that precedes any landing and is mandatory for anyone leaving the ship. In the initial biosecurity and cleaning briefing, confirmed bio-security procedures are mandatory, the crew gives instructions on the procedures and educates the visitors on why such procedures are important for maintaining biological integrity and diversity. The aim is for each visitor to have an awareness of the risks and a commitment to self-responsibility. The procedure begins with the detailed physical examination of all items of gear intended to come into contact with the onshore environment. Critical points of contact are boot treads, Velcro, zippers, cuffs, tripods and backpack mesh. For significantly soiled items, a pre-wash with hot water and soap is directed. Then all foreign debris is removed by hand, vacuum and wire brush. The final step for any item that will physically touch the ground is immersion in a disinfectant, followed by time allowing the solution to dry on the item. This is why we use the disinfectant the day prior to the first landing and then at the end of the initial landing, and each subsequent landing. This allows time for the disinfectant solution to dry on the item and take effect before the next outing.

Prior to every landing, all guides review the IAATO and Antarctic Treaty site guidelines. An advanced party of guides proceeds ashore to set up the site for hazard management, minimal impact to wildlife, control (and ease) of movement and scenic reveal. Flags and ropes are used to determine off-limits areas and minimum approach distances to stationary wildlife. One example of this is the 5–10 m approach limit to nesting sea birds, including penguin colonies. If there is an observed change in bird behaviour, then we must move away to a greater distance. Some species, such as nesting giant petrels, need a greater distance due to their sensitivity, with an approach limit of 25–50 m.

To minimise the impact of crowding, 100 persons are allowed ashore at one place at any one time. Smaller and more sensitive landing areas can have limits of 50, 25 or even 10 persons ashore. These allowances correspond to a ratio of guides to visitors. The current minimum guide-to-visitor ratio when ashore is 1:20, and the norm is closer to 1:10. Some landing sites have even smaller ratios in order to heighten control and the ability for expected interpretation opportunities.

All activities have guidelines. For example, guidelines for small-boat operations in the vicinity of ice are used to avoid entrapment, striking or capsizing due to ice. There are guidelines for observing whales and other cetaceans from small vessels; these include slow zones and no-go zones in relation to an individual animal or group.

Experiencing the vast wild beauty of the Antarctic Peninsula is deeply personal. The Antarctic tourism industry resolves to maintain and grow operating standards in order to protect Antarctica. To this objective, expedition guides vehemently interpret, educate, instruct, facilitate and police, if necessary.

The ultimate goal of tourism in Antarctica is to facilitate the experience of an emotional connection to the landscape, the ocean and wild species for each visitor. Reverence, wonder and a sense of moral responsibility towards Planet Earth accompany Antarctic visitors returning to inhabited continents. This can be a source of enhanced personal and collective sustainability measures and behaviours. However, despite all the efforts made on site by the tourism industry and enlightened visitors to protect the southernmost continent, the question remains: What are we doing to protect Antarctica when we are thousands of miles away from it?

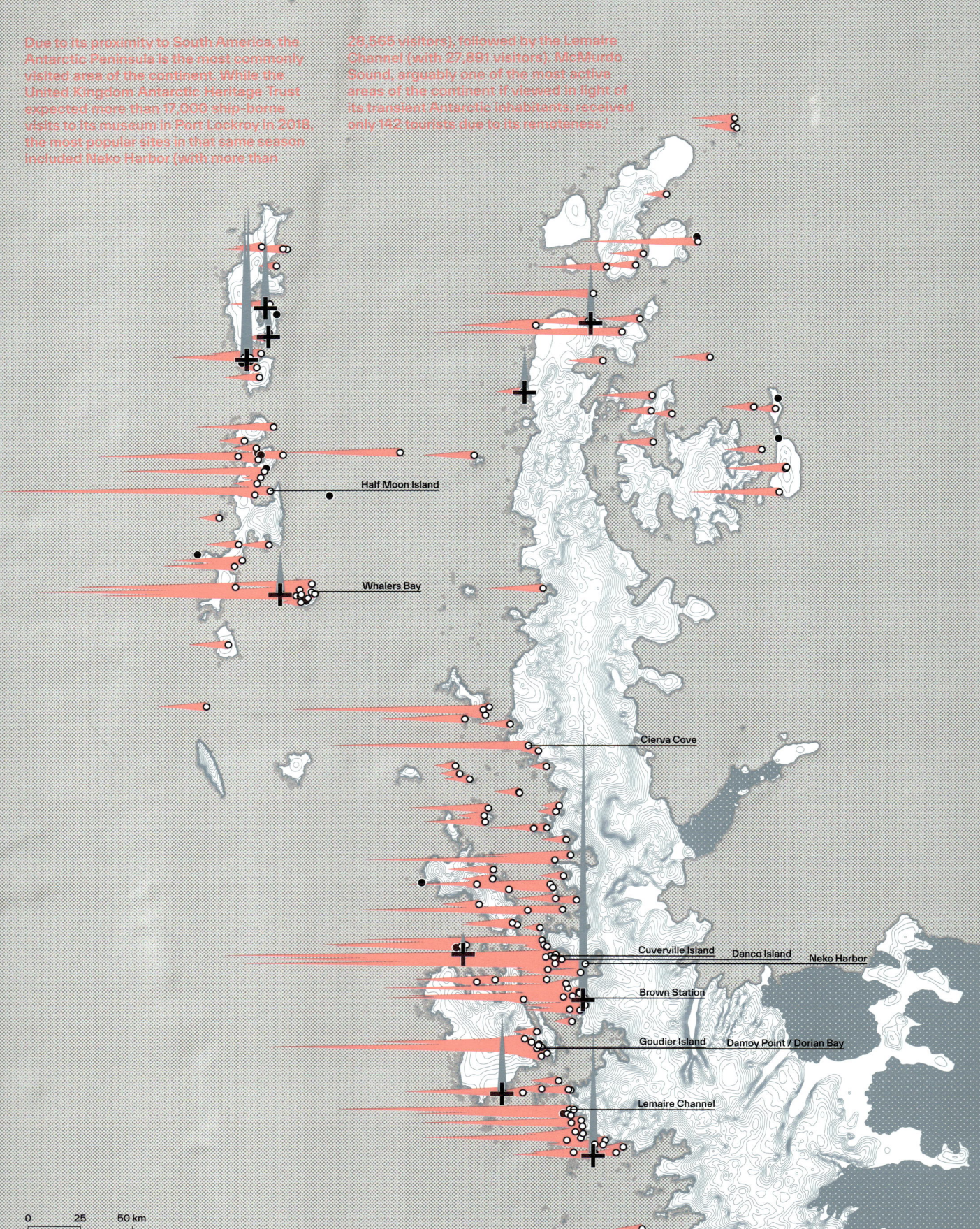

Due to its proximity to South America, the Antarctic Peninsula is the most commonly visited area of the continent. While the United Kingdom Antarctic Heritage Trust expected more than 17,000 ship-borne visits to its museum in Port Lockroy in 2018, the most popular sites in that same season included Neko Harbor (with more than 28,565 visitors), followed by the Lemaire Channel (with 27,891 visitors). McMurdo Sound, arguably one of the most active areas of the continent if viewed in light of its transient Antarctic inhabitants, received only 142 tourists due to its remoteness.[1]

Half Moon Island

Whalers Bay

Cierva Cove

Cuverville Island   Danco Island   Neko Harbor

Brown Station

Goudier Island   Damoy Point / Dorian Bay

Lemaire Channel

0   25   50 km

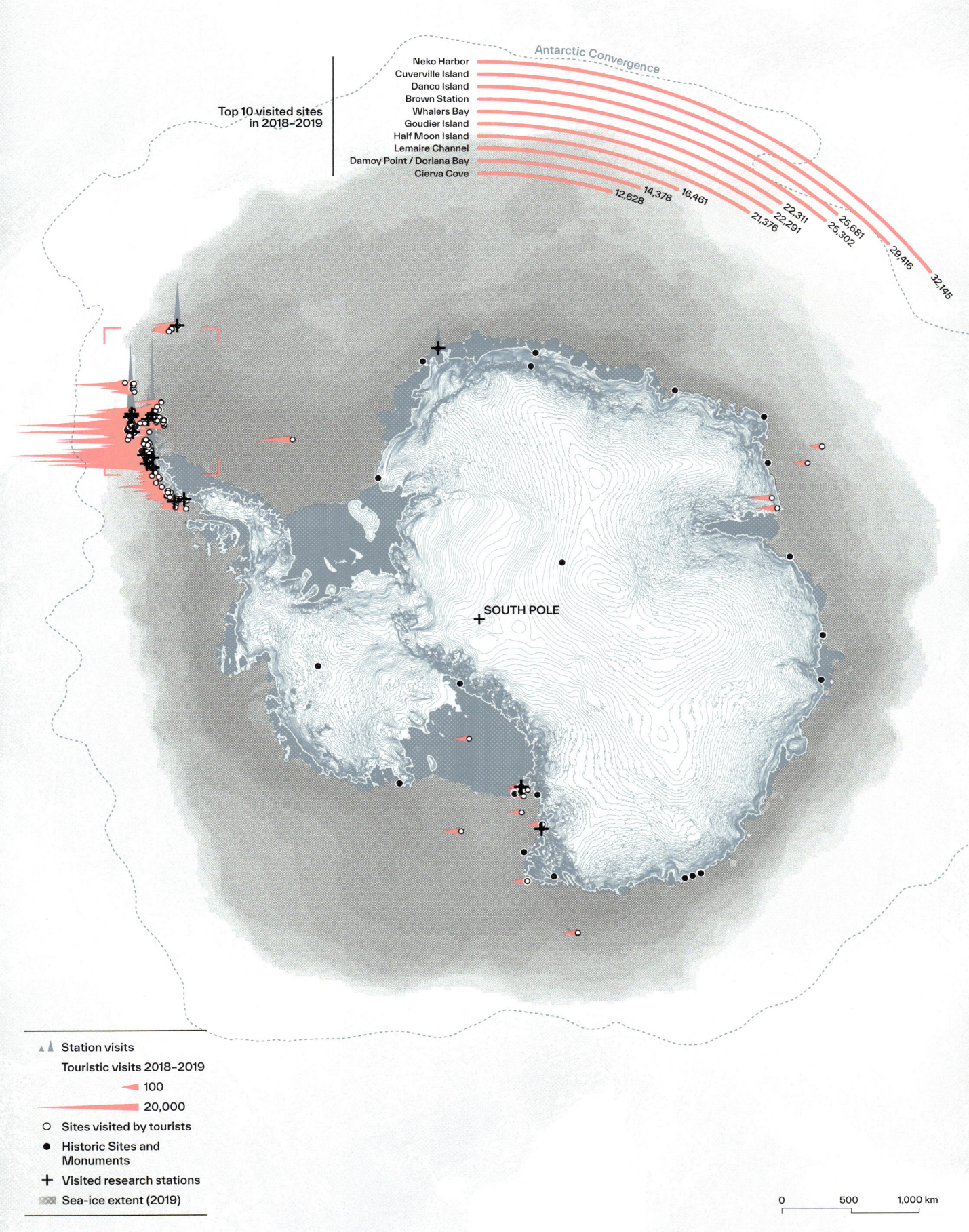

Top 10 visited sites in 2018–2019

Neko Harbor
Cuverville Island
Danco Island
Brown Station
Whalers Bay
Goudier Island
Half Moon Island
Lemaire Channel
Damoy Point / Doriana Bay
Cierva Cove

Antarctic Convergence

32,145
29,416
25,302
25,681
22,291
22,311
21,376
16,461
14,378
12,628

SOUTH POLE

▲ ▲ Station visits

Touristic visits 2018–2019
◄ 100
20,000

○ Sites visited by tourists

● Historic Sites and Monuments

+ Visited research stations

░ Sea-ice extent (2019)

0   500   1,000 km

05
"The Question of Antarctica"

Antarctica. Territory Beyond Possession?

Christy Collis

CHRISTY COLLIS is the associate director of the Office for Learning and Teaching at the University of Southern Queensland (Australia). She is a cultural geographer whose research focuses on the legal and cultural dynamics of territorial possession, specifically applied to Antarctica, outer space, and the deep seabed. Her writings include "Territories Beyond Possession? Antarctica and Outer Space" (*The Polar Journal*, 2017).

This essay begins with two – correct – statements: Antarctica is un-owned; Antarctica is owned.

This complexity is exemplified at a site: Taishan Station↗ p. 824, eastern Antarctica. Taishan is a Chinese station built in 2013; its practical function is to serve as a relay point between two of China's other bases. Geopolitically, however, Taishan is much more than a practical building. Taishan, a Chinese state installation, is built in the Australian Antarctic Territory (AAT), an area of almost six million square kilometres which envelops nearly 42% of the continent. To its Chinese owners, Taishan sits in un-owned space, part of an entire continent beyond the reaches of conventional state territorial sovereignty. To Australia, however, Taishan sits uncomfortably on Australian sovereign territory: a legally permitted but geopolitically unwelcome intrusion into Australian polar space.[1] "'We should have no illusions about [China's] deeper agenda' in Antarctica, warned Peter Jennings, executive director of the Australian Strategic Policy Institute and a former senior official in the Australian Department of Defence."[2] What makes Antarctica so complex is that both the Chinese and the Australian perspectives are entirely legally correct: Antarctica is un-owned, and Antarctica is owned; Taishan sits in unclaimed territory, and Taishan is located in the Australian Antarctic Territory.

As the example of Taishan Station signals, Antarctica is a complex and seemingly contradictory geopolitical space. To forty-seven of the fifty-four states which are parties to the Antarctic Treaty, Antarctica is a space that is not only un-owned, but also un-ownable: it is a continent beyond possession. But to seven of the Treaty's party states, Antarctica is ownable, and it is theirs. To Argentina, Australia, Chile, France, New Zealand, Norway and the United Kingdom, Antarctica is the same kind of space as Europe or South America: a continent available for, and subject to, state territorial claims. Between them, these seven states lay claim to 80% of the continent. So how did Antarctica come to be this contradictory space? How can it be both owned and unpossessable simultaneously?

In the late nineteenth and early twentieth centuries, Antarctica was following a familiar narrative of spatial transformation: as explorers arrived, Antarctica was being transformed from a vast, un-owned space into a series of state territorial possessions. Claimant states such as Australia and France relied on "discovery of *terra nullius*" as the basis of their claims. Under this international territorial law, un-owned land – *terra nullius* – can be claimed through a codified set of legal and geopolitical acts and processes: these include having an officially invested explorer touch the land and read out a proclamation of possession, filing evidence of the explorer's visit, and "effective occupation" of the land through occupation and administration. The early twentieth century was a particularly busy period for this form of territorial claim in Antarctica. Spain and Argentina, meanwhile, invoked a very different law for their Antarctic claims. According to Spain and Argentina, the papally decreed 1494 Treaty of Tordesillas divided up the globe neatly between Spain and Portugal along approximately 46° W longitude. The South American states argued that this line extended all the way to the South Pole, and thus included Antarctica. From the discovery of *terra nullius* and its inheritance from a Renaissance pope, to its inception as a possession, Antarctica has always been complex.

By the early 1950s, Antarctica was under increasing geopolitical pressure. The United States and Russia, deep in their Cold War enmity, had each established a significant presence in Antarctica, with both states arguing that they had the historical right to claim polar land. The seven claimant states and the

United States grouped together to try to form a condominium to govern Antarctica; this initiative collapsed when Russia rejected it. The United States sent 4,700 military men to construct Little America Station on the Ross Ice Shelf. The question of territorial possession of Antarctica was devolving into dispute.

But in 1957–1958, in a global display of the amicability of science, Antarctica became something other than a conventional space in which claimants and non-claimants squabbled for territorial possession. As part of the 1957 International Geophysical Year (IGY), twelve countries sent scientists and support workers to Antarctica, not to claim territory, but to conduct internationally collaborative science. This experiment in international collaboration in Antarctica proved so successful that when it was over, the states that had participated in the Antarctica IGY came together and wrote the Antarctic Treaty. Argentina, Australia, Belgium, Chile, France, Japan, New Zealand, Norway, South Africa, the Soviet Union, the United Kingdom and the United States signed the Treaty, putting it into force by 1961. And with that, Antarctica underwent a massive geopolitical transformation.

While its declaration of Antarctica as a site for "peace and science" is laudable, the Antarctic Treaty's unique approach to territory is of particular importance. It is important both because it has to date forestalled polar conflict over land, and because it created an entirely unique type of geopolitical and legal space. According to Article IV of the Treaty, the original seven claimant states may keep their polar claims (though not expand them), but no other state is legally compelled to recognise those claims.

This is the kind of space that Antarctica remains today. Under the terms of the Treaty, Antarctic land is dually configured as simultaneously un-owned and owned. No new claims may be made to the territory of Antarctica: the 1.5 million square kilometres of the continent between 90° W and 150° W remain the only unclaimed land on Earth.[3] More radically, the Treaty allows all states to reject the seven territorial claims that divided the continent in 1957. To all of the world's states other than this historically select handful, Antarctica is unowned, and unownable, space. It remains *terra nullius,* but not the type of *terra nullius* that can be claimed. To these states, Antarctica is the Earth's only truly international land. Yet at the same time, the seven claimant states retain their claims: in their geographies, Antarctica is no different from any other claimable – and claimed – site on Earth. Antarctica thus remains a space of both/and: both claimed and unclaimable. The Treaty holds these contradictory spaces in restless tension.

If you say that Taishan occupies Australian sovereign territory, you are correct; if you say that Taishan sits on an entirely unclaimed continent, you are also correct. To date, this bifurcated spatiality has successfully held in Antarctica. But how long will it last? Although the 1991 Protocol on Environmental Protection bans commercial mining in Antarctica until 2048, once mineral extraction becomes economically viable in the polar south, the Treaty's delicate balancing act will come under intense pressure. However, for now, Antarctica remains a geopolitically and legally unique space: an entire continent that is at once possessed and unpossessable.

"*Res* (an object in the legal sense, anything that can be owned) is not yet the object of rights of any specific subject. Such items are considered ownerless property and are free to be acquired by means of '*occupatio*'. In Roman law, *occupatio* was an original method of acquiring ownership of un-owned property (res nullius) by occupying with intent to own."

"*Res nullius* could be defined as 'nobody's thing'. It is a Latin phrase used in *ius privatum* (Latin for private law), based upon property and contract, concerned [with] relations between individuals. It means 'something without a master', that is to say which has no owner but which is nevertheless appropriable."

"*Res communis* could be defined as a 'common thing'. It is a Latin phrase used in *ius publicum* (Latin for public law): [in] the past, public law regulated the relationships of the government to its citizens, including taxation […] *Res communis* preceded today's concepts of the *commons* and *common heritage of mankind*."

Louis de Gouyon Matignon (Writer, LGM Editions), "The Res Communis Concept in Space Law", *Space Legal Issues*, 2019

Antarctica as a Global-Commons. Promises and Pitfalls

Sanjay Chaturvedi

SANJAY CHATURVEDI is a professor of international relations and dean of the Faculty of Social Sciences at South Asian University, New Delhi (India). He is the chairman of Indian Ocean Research Group, Inc. (IORG), chief editor of the *Journal of the Indian Ocean Region* (Routledge), and regional editor of *The Polar Journal*. Chaturvedi's publications include *The Polar Regions: A Political Geography* (John Wiley & Sons, 1996) and, co-authored, *Climate Terror: A Critical Geopolitics of Climate Change* (Palgrave Macmillan, 2015).

The generic term "Global Commons",[1] when viewed through the lens of state jurisdiction, invokes the idea of open-access spaces (i.e. Antarctica, the high seas, the deep seabed, the atmosphere, the radio-frequency spectrum, the internet, and Outer Space) and resources (i.e. biodiversity, carbon, genes, water or forests) beyond the sovereignty of any state. Normatively speaking, Global Commons are perceived as being held in common, or in trust, for both the present and future generations, and endowed with the capacity to provide public goods. The question of their governance, including proactive policy responses to planetary concerns such as biodiversity conservation, sustainable development and climate change, remains subject to the critical principles of equitable access, stewardship[2] and both intergenerational and intra-generational equity.

The deliberate hyphenation of the phrase "Global-Commons" in the case of Antarctica invites critical attention to the remarkably differentiated, complex and contested nature of both words, "Global" and "Commons". Further, it enables the relentless interrogation of a totalising narrative of "global regulation", of universally accessible common-pool resources, allegedly on behalf of humanity as a whole. How does one map and measure the extent to which commonality is achieved and sustained in common-pool resources as public goods? This can be done in legal (i.e. spaces outside national jurisdictions), geopolitical (i.e. shared cooperative sovereignty), or ethical and moral (i.e. equitable access and benefit) terms. Intermittently invoked in the course of Antarctica's transformation over the past seven decades as a "continent of science",[3] "treasure trove of marine and mineral resources"[4] and "natural reserve devoted to science and peace"[5] – the subliminal notion of Global-Commons poses some challenges. It continues to exhibit a serious mismatch, bordering on a fault line, between both the resilient imperial imprints and colonial legacies in territorial claims that are legally "frozen" but geopolitically active, and the promise of open access and the equitable sharing of benefits. These are enshrined in the recognition in the Preamble to the Antarctic Treaty of 1959 that "it is in the interest of all mankind that Antarctica shall continue forever to be used exclusively for peaceful purposes and shall not become the scene or object of international discord".[6]

Who speaks for Antarctica as Global-Commons, and for whom does Antarctica speak as Global-Commons? Despite several remarkable achievements in Antarctic science and regional governance – including the establishment of marine protected areas (MPAs) in the high seas – the intriguing question of representation in the entangled cobweb of nationalisms and universal commitment to sustainable commons remains opaque and unanswered. The limits to governing a space as a Global Commons with authority and multilateral collaboration, which have been appropriated in the past, is obvious in the ongoing attempts to regulate Antarctic commercial tourism as one of the peaceful, and thus legitimate, uses of the Antarctic alongside science and marine resources. Within the overarching context of the Anthropocene, how Antarctica should be approached and governed as a Global-Commons in the post-COVID-19 global geopolitical order remains to be seen. There are good reasons to believe, however, that having precariously evolved thus far on the intersection of polar exceptionalism, contested territoriality and the pursuit of "global knowledge commons" through national Antarctic science programmes,[7] it appears that the notion of an Antarctic Global-Commons will become more, and not less, compelling, both ethically and geopolitically, for the instruments of governance under the Antarctic Treaty System (ATS). This applies especially around the year 2048, when the current ban on mining and mineral resource activities in the Antarctic Treaty area, under the 1991 Protocol on Environmental Protection, might be subjected to review.

The Antarctic

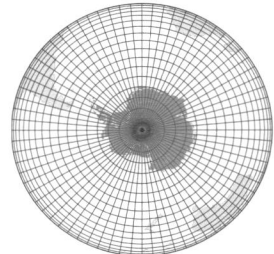

High Seas

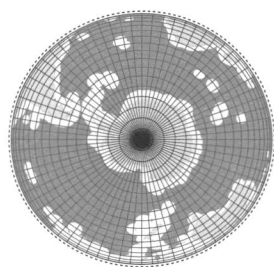

The Atmosphere

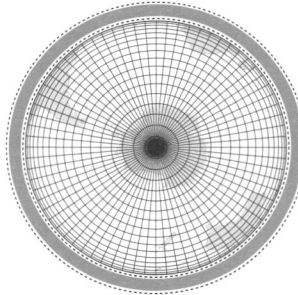

Outer Space

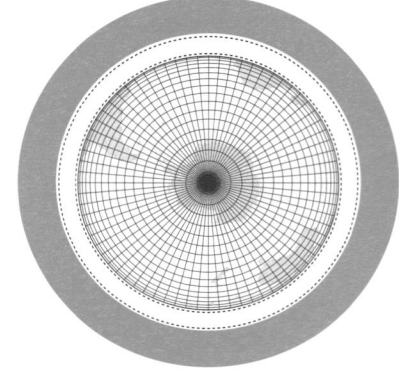

Where the Ice Meets the Waves. Antarctica and the Law of the Sea

Tim Stephens

TIM STEPHENS is a professor of international law at the University of Sydney Law School and deputy director at the University of Sydney Marine Studies Institute (Australia). He teaches public international law, with a particular focus on the law of the sea, the law of the polar regions and international environmental law. He is a Fellow of the Australian Academy of Law. Dr Stephens is the co-editor, with Ben Saul, of *Antarctica in International Law* (Hart Publishing, 2015).

The Antarctic continent and the Southern Ocean are not only physically distinctive; they are also the subject of a unique body of international law.[1] Antarctica was originally a domain of exploration and sovereign acquisition, with seven states laying claim to parts of the continent.[2] Later, during the International Geophysical Year (IGY, 1957–1958), there was intense international Antarctic scientific investigation, prompting proposals to formalise this cooperation. The resulting 1959 Antarctic Treaty achieved this and became the fulcrum of a collection of treaties and instruments now called the Antarctic Treaty System (ATS).

The Antarctic Treaty addressed the major concerns of the twelve original parties. It placed sovereign claims in abeyance, but did not question their legality. It promoted the freedom of scientific investigation and cooperation. It ensured that Antarctica was demilitarised, would be used only for peaceful purposes, and would not be used for nuclear testing or the disposal of nuclear waste. Moreover, the Treaty had a major maritime dimension, with the Antarctic Treaty Area (ATA) embracing not only the continent itself, but also the Southern Ocean to 60 degrees south latitude.

There have been two major ATS developments of relevance to the Antarctic marine environment. The first was the 1980 Convention on the Conservation of Antarctic Marine Living Resources (CCAMLR). A conservation agreement rather than a mere fisheries treaty, CCAMLR is a comprehensive regime for protecting the Antarctic marine ecosystem within the Antarctic convergence. The second was the 1991 Protocol on Environmental Protection (also known as the Madrid Protocol), which banned mining in the ATA indefinitely, designated Antarctica "as a natural reserve, devoted to peace and science" (Article 2) and established a wide-ranging mechanism for protecting the Antarctic environment, including its marine areas.

Because of the geographical reach of the ATA and the CCAMLR area, two legal orders coexist in Antarctic waters: the Law of the Sea and the ATS. The Law of the Sea is an ancient part of the international legal order; its core elements can be traced to European and Asian state practice and scholarship, particularly from the seventeenth century onwards.[3] From the middle of the twentieth century, a new ocean order began to emerge which coupled core elements of the ancient regime (such as the idea of the free seas) with demands from newly independent and developing states for a fairer share of ocean resources.[4] As with the Antarctic legal regime, the Law of the Sea was fundamentally transformed as a result of the IGY, which produced new knowledge about the oceans.[5]

These developments culminated in the adoption of the United Nations Convention on the Law of the Sea in 1982 (UNCLOS). UNCLOS has been described as the "constitution for the oceans",[6] and it commands almost universal support. Central to UNCLOS is its definition of maritime zones. These are the 12 nautical mile (M) territorial sea; the contiguous zone, which extends a further 12 M; the 200 M exclusive economic zone (EEZ); and the continental shelf, which is at least 200 M. The shelf may be wider still if a coastal state substantiates its entitlement to the satisfaction of the Commission on the Limits of the Continental Shelf (CLCS), an expert body established by UNCLOS. UNCLOS also preserves the high seas beyond national jurisdiction, and creates a new maritime zone, the deep seabed (the "Area"), the resources of which are placed under international management. The contemporary law of the sea is a highly detailed legal order with many aspects, particularly its environmental dimensions, of relevance to the polar seas (see, for example, the International Maritime Organization's International Code for Ships Operating in Polar Waters).

Territorial Sea: zone, not exceeding 12 nautical miles from the baselines along the coast, in which the coastal state has sovereignty over the seabed, subsoil, water column, water surface and airspace.[1]

Contiguous Zone: zone up to 24 nautical miles from the baselines that may be claimed by coastal states for the exclusive purpose of exercising jurisdiction over custom, fiscal, immigration and sanitary matters. It is not a zone of sovereignty, but only jurisdiction.

Exclusive Economic Zone (EEZ): zone that coastal states may claim up to a distance of 200 nautical miles from their territorial sea baseline. With the EEZ states enjoy exclusive sovereignty rights for the purpose of exploring and exploiting, conserving and managing the living and non-living resources of the seabed, subsoil and water column. In the EEZ some high-seas freedoms are preserved – i.e. all states enjoy the freedom of navigation, overflight, to lay submarine cables and pipelines, etc.

Continental Shelf: seabed and subsoil of the submarine areas that extend beyond the territorial sea to a distance of 200 nautical miles. Within the CS, states have exclusive sovereign rights for the purpose of exploring its natural resources. This is inherent, i.e. it automatically belongs to coastal states and is not dependent on occupation or proclamation (as is an EEZ).

Area: seafloor area beyond nautical jurisdiction; it is known as "the common heritage of mankind". No part of the Area or its mineral resources may be subject to sovereign claim or acquisition, and its mineral wealth is vested in the International Seabed Authority (ISA).

High Seas: the water column above the Area and above the continental shelf is designated as high seas. The high seas are an area not subject to acquisition. Within the high seas, all states enjoy extensive freedoms of navigation, overflight, to lay submarine cables, to fish and conduct scientific research.

Where the Ice Meets the Waves. Antarctica and the Law of the Sea

176

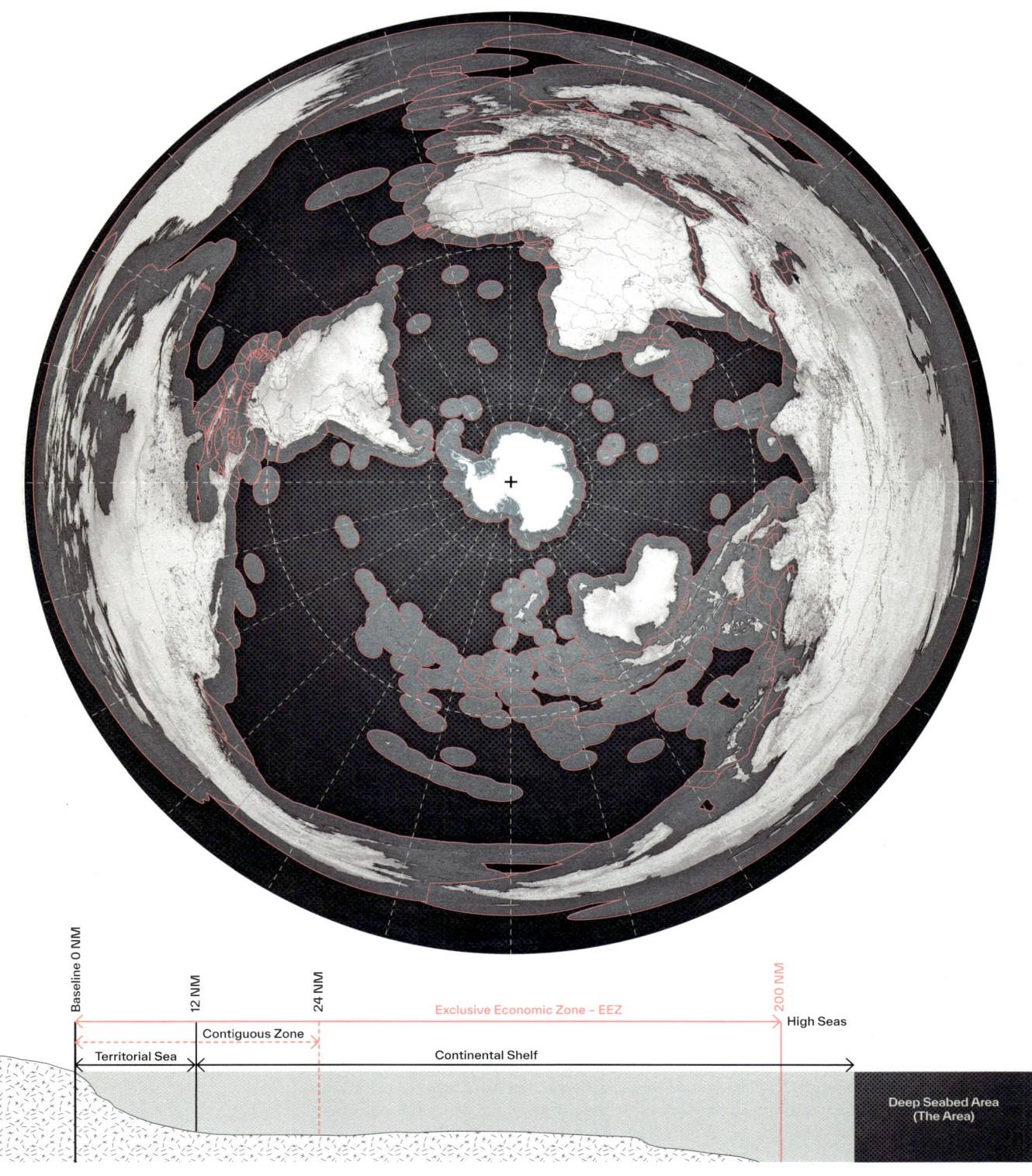

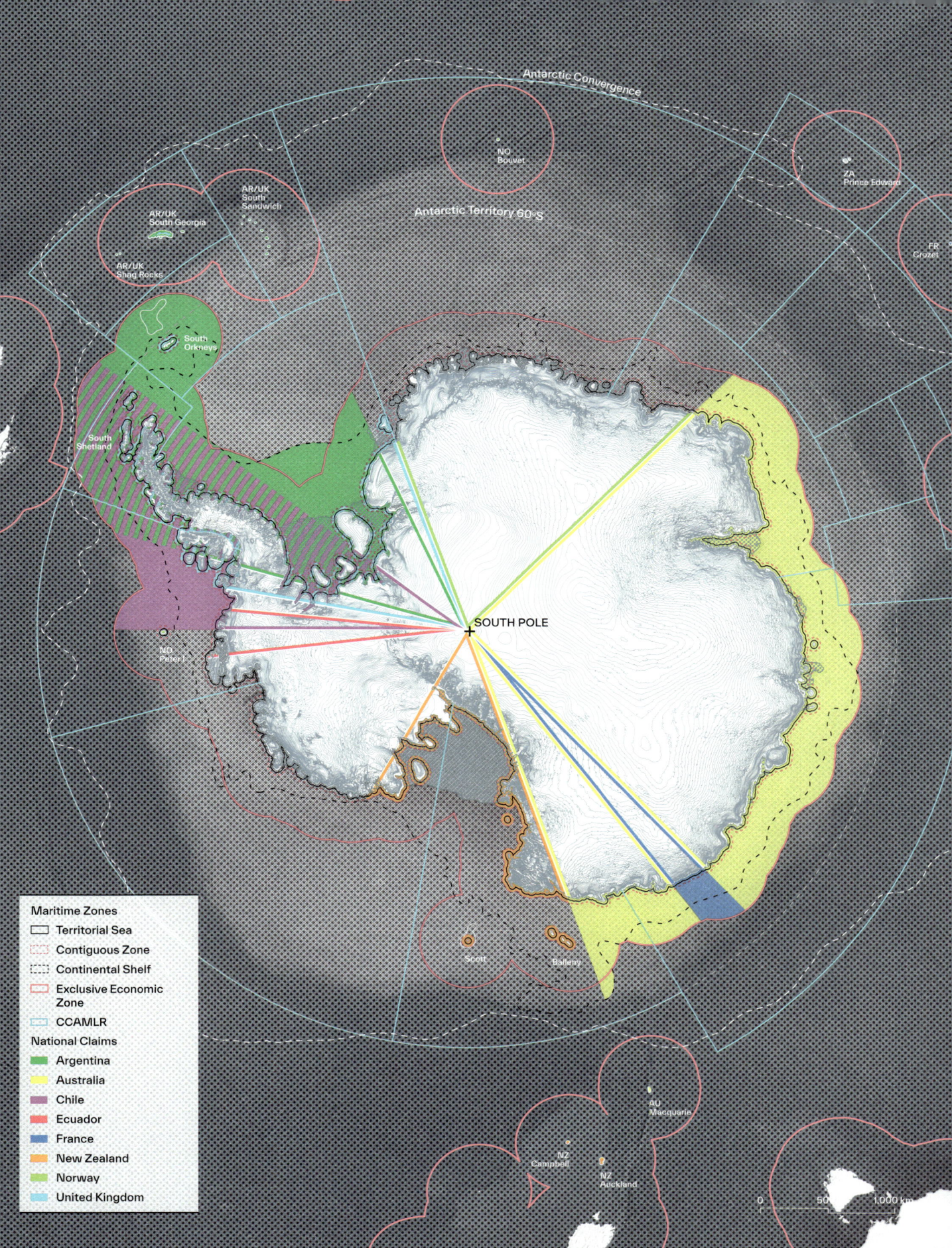

Antarctic Convergence

NO
Bouvet

ZA
Prince Edward

Antarctic Territory 60°S

FR
Crozet

AR/UK
South Sandwich

AR/UK
South Georgia

AR/UK
Shag Rocks

South
Orkneys

South
Shetland

South
Pole

SOUTH POLE

NO
Peter I

Scott

Balleny

Maritime Zones

- Territorial Sea
- Contiguous Zone
- Continental Shelf
- Exclusive Economic Zone
- CCAMLR

National Claims

- Argentina
- Australia
- Chile
- Ecuador
- France
- New Zealand
- Norway
- United Kingdom

AU
Macquarie

NZ
Campbell

NZ
Auckland

0   500   1,000 km

There has been much discussion of the relationship between the ATS and the Law of the Sea.[7] A central point of contestation is the interaction between the Antarctic Treaty and the assertions by Antarctic claimants to maritime zones.[8] Australia, Argentina, Chile, France, New Zealand and the United Kingdom have all stated that they possess a territorial sea adjacent to Antarctica, and Norway has reserved its right to a similar entitlement.[9] Australia, Argentina, France, Chile and New Zealand have also claimed contiguous zones. There have also been claimant assertions of resource zones offshore the Antarctic. Australia, Argentina, Chile and France have all proclaimed Antarctic EEZs.[10] The continental shelf does not require an active claim or proclamation, yet Chile, Australia and Argentina have all issued one.[11] Moreover, all seven claimants have either submitted data or preliminary information to the CLCS in respect of Antarctic extended continental shelves (Australia, Argentina, Chile and Norway) or reserved the right to do so in future (France, New Zealand, United Kingdom).[12]

In the Antarctic continental margin, questions arise over the interaction between the ATS, particularly the Madrid Protocol, and the Law of the Sea. If the continental shelf around Antarctica is not under coastal state control, then it could be regarded as part of the Area under the Law of the Sea, and its seabed resources the common heritage of mankind. Additionally, even if this shelf is not part of the Area, what of the seabed further north to the ATA boundary? Is the ATS a *lex specialis* that derogates from UNCLOS? These important legal questions remain unresolved for the reason that deep seabed mining in the Southern Ocean is currently impractical.[13]

Claimants contend that there is no incompatibility between their maritime assertions and the Antarctic Treaty prohibition on new territorial claims, or enlarging existing ones. Non-claimants, on the other hand, argue that any Antarctic coastal maritime zones should be treated like terrestrial claims and not be recognised. However, this disagreement has not undermined the ATS, principally because the claimants have never sought to exercise any jurisdiction in Antarctic waters. For example, Australia did not apply its laws prohibiting whaling to Japan's research whaling programme in the waters of the Australian Antarctic territory, but instead successfully challenged the programme as being unlawful under the 1946 International Convention for the Regulation of Whaling.[14]

The Southern Ocean is the meeting point of two great legal systems: the global Law of the Sea and the regionally focused ATS. Despite the obvious potential for tension and conflict, there has been pragmatic accommodation between these two orders.[15] While there are occasional points of friction (a recent one being how to manage the marine biodiversity of the high seas), this mostly harmonious meeting of the international law of the waves and the law of the ice appears likely to endure.

"The polar regions are losing ice, and their oceans are changing rapidly. The consequences of this polar transition extend to the whole planet […] yet the governance landscape is currently not sufficiently equipped to address cascading risks and uncertainty in an integrated and precautionary way within existing legal and policy frameworks."

Andrea Kavanagh (Director, Pew Charitable Trusts), *Intergovernmental Panel on Climate Change Special Report on the Ocean and Cryosphere in a Changing Climate*, 2019

Bounding Antarctica.
The Greater Southern Ocean

Alan D. Hemmings

ALAN D. HEMMINGS is professor at the Gateway Antarctica Centre for Antarctic Studies and Research at the University of Canterbury (New Zealand). BOB FRAME is an adjunct associate professor at the Gateway Antarctica Centre for Antarctic Studies and Research at the University of Canterbury (New Zealand).

What is a meaningful boundary for the place we call the Antarctic? Well, it has varied over time. First it was the continent and surrounding islands; then these plus some surrounding waters; and later it became the biophysical space bounded by the Antarctic convergence. That progression is complicated by a second consideration: that what makes a meaningful boundary depends upon the issue one seeks to bound.

Following a summary of the various historic "boundings" of Antarctica, it is argued that a much larger spatial area extending up to approximately 35° S better captures the *geopolitical* realities of the contemporary Antarctic. The latter provides a more coherent functional area within which to situate much that currently occurs in the Antarctic. Since this enormous area – a spatial construction that could be defined as "the Greater Southern Ocean" – is overwhelmingly beyond national jurisdiction, it is an area across which international management is required.[1]

Political boundaries appeared in Antarctica even before the final geographical boundaries were known. In asserting the first territorial claim in 1908, the United Kingdom also set 60° S as the northern political boundary.[2] Subsequent claimants followed this in their own delimitations. The Australian and New Zealand claims have a common imperial root with the British claim; the French and Norwegian claims were reconciled with the boundaries of these three, and Argentine and Chilean claims were cast as direct competitors to the British claim. Accordingly, for the Antarctic Treaty to secure a *modus vivendi* on claims positions, it also made its area of application the area south of 60° S. With high-seas freedoms reserved in the Antarctic Treaty, it became necessary to adopt new instruments to manage activities in the marine environment. The first, the Convention for the Conservation of Antarctic Seals, nonetheless also set its northern boundary at 60° S.[3] Similarly, the Protocol on Environmental Protection, which focuses on environmental protection and the prohibition of mineral resource activities, takes 60° S as its boundary, although it also embraces undefined "dependent and associated ecosystems", which presumably include areas north of 60° S. The Convention on the Conservation of Antarctic Marine Living Resources (CCAMLR) adopted a more northerly boundary, approximating to the position of the Antarctic Convergence or Polar Front. This reflects an ecological boundary between cold Antarctic surface waters and warmer subantarctic waters, is coincident with the Antarctic Treaty boundary at 60° S across the Pacific sector, while extending to 50° S in the Atlantic sector and as far north as 45° S in the Indian Ocean sector. The CCAMLR boundary is thus a proxy for the area of the Antarctic Treaty System (ATS) as a whole. The conventional "Antarctic" diplomatic space operationalised through the ATS is still focused on this area.

In the twentieth century, relatively little apart from transit occurred in the ocean space north of the CCAMLR boundary. However, in the early twenty-first century, substantial fishing activity immediately north of the Antarctic Convergence triggered a further phase of institutional development: a series of Regional Fisheries Management Organizations (RFMOs) based on a set of new geographically delimited fisheries conventions across the south Atlantic, Indian and Pacific oceans. Each took the northern boundary of CCAMLR as its southern boundary. These RFMOs, comprising the South East Atlantic Fisheries Organisation (SEAFO), South Pacific Regional Fisheries Management Organisation (SPRFMO) and Southern Indian Ocean Fisheries Agreement (SIOFA) – plus the contentious regional "Galápagos Agreement" – were adopted over a nine-year period (2001–2009) and (apart from the Galápagos Agreement) were in force by 2012.[4]

The regime of the ATS has been in what can be considered an institutional hiatus since its last instrument, the Protocol on Environmental Protection, was

"Interpretative flexibility is indicative of how Antarctic science has been and still is enmeshed in Antarctic politics and how the boundaries drawn in some respects are human constructions. SCAR as a science body chooses natural boundaries […] while 60° S is an expedient political boundary as are all political boundaries in the long run."

Aant Elzinga (Emeritus Professor, University of Gothenburg), "The Continent for Science", in *Handbook on the Politics of Antarctica*, 2017

adopted in 1991 and entered into force in 1998. It looks as if the effort shifted to the immediate north, with a decade of intense activity to create a new array of RFMOs. For large parts of these new RFMOs, the ecosystems and activities are a very long way from the Antarctic in every sense. But in their southern parts, they collect appreciable parts of the Antarctic ecosystem (the Convergence is permeable) through transboundary stocks of both target and non-target species and human activities. There is also a tight coupling between key members of these RFMOs and the key ATS fishing states.[5] In various ways, we see the emergence of a new maritime domain, significant in Antarctic terms but no longer adequately bounded by the Antarctic regime. This domain extends from the coastline of Antarctica to the mid-latitudes of the south Atlantic, Indian and Pacific oceans. These developments are largely driven by states which have territorial interests in the high Antarctic and subantarctic littoral and/or global fishing interests. The states of the global South are poorly represented within both the new RFMOs and (aside from the giants) the ATS.[6]

The proposition that the area of significance – the functional bounding of the Antarctic – has changed is not of itself a criticism of that change. There may be reasons for concern: it is manifestly driven by resource exploitation – is that sustainable? Is access and decision-making open to all or weighted in favour of the usual suspects? There may be benefits, if, for example, consistent (high) standards in relation to environmental performance are extended over a wider area. The point is that it seems no longer sensible to frame the totality of human engagement with Antarctica within the boundaries of the historic ATS. The new reality, consisting of human activity of increasing complexity and intensity occurring over a far larger area, is upon us. If earlier boundings of the Antarctic can be regarded as having been historically contingent, we should see no impediment to framing the boundary differently now. The notional bounding at 35° S is just that: notional. It is an ambit claim, a reflection of the sense that the totality of Antarctic-centric activity has now burst through the previous boundary of the Antarctic Convergence. States may take a while to accept that such a bounding has practical significance. This is not a reason for, nor an impediment to, scholars who consider the Antarctic more holistically, recasting their frames of reference.

- — The Greater Southern Ocean (35 degrees south)
- — 1959 Antarctic Treaty
  1972 CCAS
  1991 Madrid Protocol
- — 1980 CCAMLR
- — 2000 Galápagos Agreement
- — 2001 Convention on the Conservation and Management of Fishery Resources in the Southeast Atlantic
- — 2006 South Indian Ocean Fisheries Agreement
- — 2010 Convention on the Conservation and Management of High Seas Fishery Resources in the South Pacific Ocean
- + South Pole

Taken by UNLESS in the austral summer of 2018–2019, the two photographs depict the moment at which the expedition ship crossed the hazy, ever-shifting threshold of the Convergence (left) and the man-made political boundary of the Antarctic Treaty (right), respectively, at 60 degrees south.

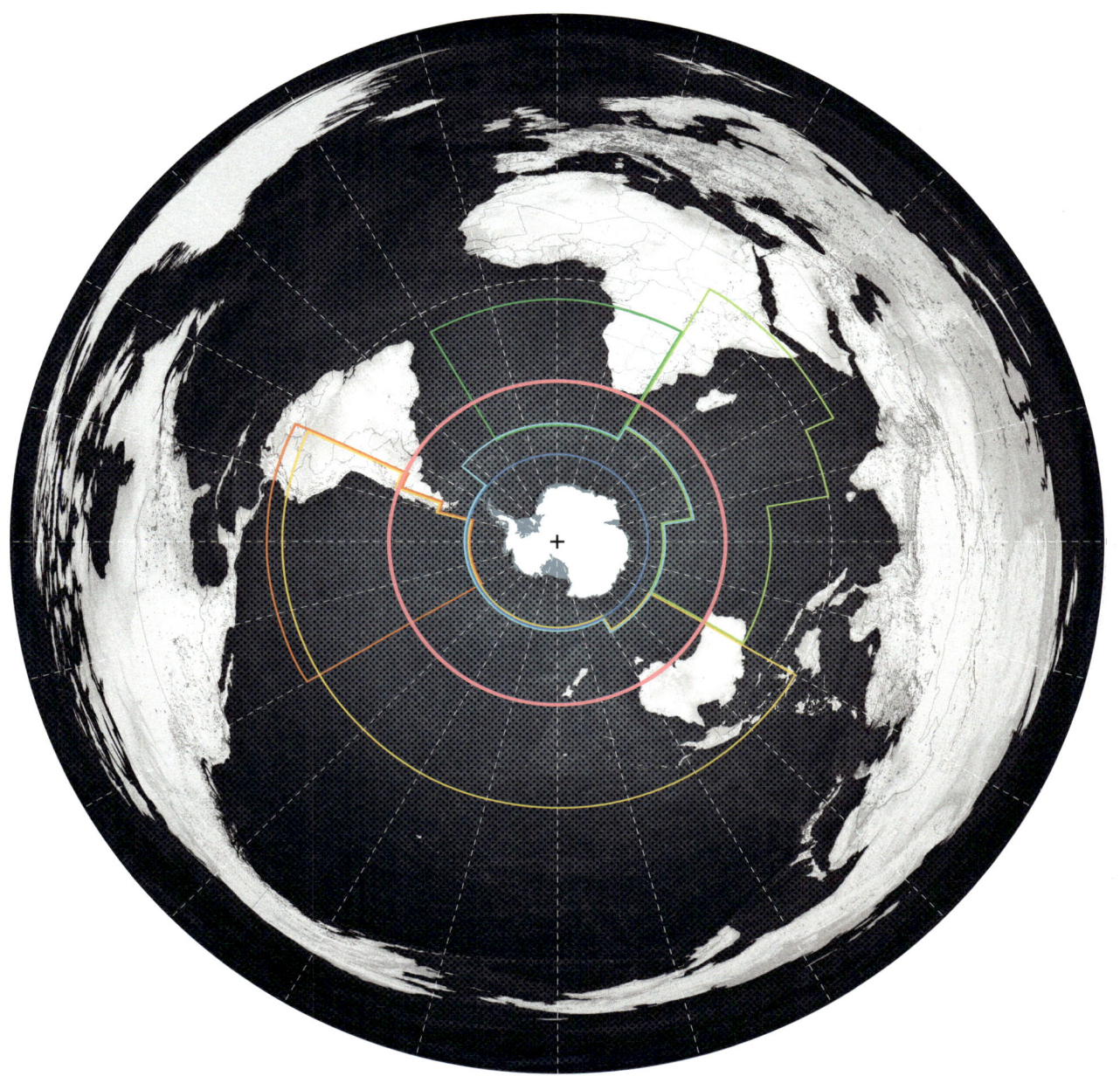

The "Question of Antarctica" at the United Nations

Doaa Abdel-Motaal

DOAA ABDEL-MOTAAL is the former executive director of the Rockefeller Foundation Economic Council on Planetary Health at the University of Oxford (United Kingdom) and has served in various high-level positions in the United Nations and other multilateral organisations. Dr Abdel-Motaal is the author of *Antarctica: The Battle for the Seventh Continent* (Praeger, 2016), nominated for the 2018 Mountbatten Award for Best Book.

In 1982, the prime minister of Malaysia, Dr Mahathir bin Mohamad, stood before the United Nations General Assembly to urge the international community to turn its attention to Antarctica. Congratulating its members on the recently concluded United Nations Law of the Sea, which declared the seabed and its resources to be the common heritage of mankind, he argued that the seventh continent belongs to all of humanity. "The United Nations must convene a meeting in order to define the problem of these uninhabited lands, whether claimed or unclaimed, and to determine the rights of all nations to these lands," were his exact words. His call – formalised in a letter sent on the 11th of August 1983 sent alongside the representatives of Antigua and Barbuda – resulted in the "Question of Antarctica" being placed on the agenda of the United Nations for more than two decades (1983–2005).

   Prompting Malaysia to make this request was the news that Consultative Parties to the Antarctic Treaty had started to negotiate a mining regime for the continent in secrecy. The world could not sit by and watch a handful of countries monopolise the resources of the last remaining continent. His call was echoed by the G-77 alliance of developing countries, which urged the United Nations Secretary-General to prepare a report on the "Question of Antarctica."[1] Issued in 1984, the report constituted the first comprehensive picture ever to be placed before the international community of the continent's mineral resources, the scope of the territorial claims put forward in the earlier part of the twentieth century, and the political settlement that followed.[2]

   Between 1908 and 1940, seven nations had made unilateral territorial claims in Antarctica. They included Australia, which claimed the lion's share of the continent, France, New Zealand, Chile, the United Kingdom, Argentina and Norway. These claims were never recognised by the international community. The two superpowers at the time, the United States and the Soviet Union, refrained from putting forward their own claims out of fear that Antarctica would become an extension of the Cold War battlefield.[3]

   The report by the Secretary-General on the "Question of Antarctica" described how the different claimants had tried to oust each other from the continent, tearing down each other's flags, destroying each other's scientific stations, and sending rival expeditions to the same locations. Fearing that these conflicts would spiral out of control, the United States had spearheaded the search for a permanent solution for Antarctica. This led to the Antarctic Treaty in 1961.

   The Secretary-General conveyed the international community's concern that the Treaty's twelve original signatories had established an exclusive club with a high "entrance fee". Consultative Party status, which confers decision-making powers upon its holder, was made obtainable only upon demonstration of "substantial scientific research activity", defined as the establishment of a scientific research station or the dispatch of a scientific expedition to the continent.[4] This led the developing world to characterise the treaty as a regime "advantageous only to the privileged few".[5] The Secretary-General also conveyed the concerns of many United Nations members regarding the mining negotiations that had started, the large number of scientific research stations that Treaty parties had placed on the continent, and the complete absence of oversight by the United Nations in the governance of Antarctica.

   In reality, while the Treaty had brought the territorial claims in Antarctica to a halt, it had not put an end to all forms of competition for the continent. Some twenty years after its signature, the Consultative Parties turned their attention to mining. They agreed that a regulatory regime needed to be negotiated before the discovery of important minerals, and started negotiating what became known as the Convention on the Regulation of Antarctic Mineral Resource Activities (CRAMRA).[6] In drawing attention to these negotiations, the Secretary-General's report led to calls by the United Nations to prevent the mining of Antarctica until the entire international community could be involved.[7]

In fact, the negotiations of a mining regime collapsed, in part, because of opposition by the United Nations. But they also subsided because of two other very significant developments. Environmental groups had started to protest against the opening of Antarctica to commercial exploitation; and politicians from claimant states had begun to express the fear that their territorial claims would be eroded were miners of all nationalities to be allowed into the area of their claim.[8] In other words, the mining regime died as much because of external pressure as because of internal back-pedalling on the need for such a regime by claimant states.

Discussions of the "Question of Antarctica" at the United Nations thus became less heated when the mining regime was debated. In 1991, the regime was replaced with an Environmental Protocol that banned mining in Antarctica. However, the "Question of Antarctica" was only taken off the agenda of the United Nations – with a note of caution that the General Assembly would nevertheless "remain seized of the matter" – when parties to the Antarctic Treaty recognised the need to expand the treaty's membership, and therefore its legitimacy. Malaysia, India and other developing countries were gradually brought on board.

Has the Question of Antarctica truly been settled, though, or has it merely been postponed? Realities on the ground reveal continued fierce competition for the continent in the form of new scientific research stations and the assignment of place names to the sections of Antarctica where stations are placed.[9] As Antarctica becomes warmer, the world will no doubt return to the "Question of Antarctica" once again.

"What I found was a place that would probably be unrecognizable to the likes of Robert Scott […] We see Antarctica's beauty and the danger global warming represents, and the urgency that we do something about it. […] All this may be gone, and not in the distant future, unless we act, together, now."

Ban Ki-moon (Former United Nations Secretary-General), "UNSG Visit to Antarctica", November 2007

"Henceforth all the unclaimed wealth of this earth must be regarded as the common heritage of all the nations […] Now that we have reached agreement on the Law of the Sea, the United Nations must convene a meeting in order to define the problem of uninhabited lands […] It is now time that the UN focus[ed] its attention on these areas, the largest of which is the continent of Antarctica."

Mahathir bin Mohamad (Former Prime Minister of Malaysia), UN General Assembly, 1982

1947

During *Operation Highjump,* Rear Admiral Richard Byrd flies over the South Pole with "a cardboard box containing small multi-coloured little flags of the United Nations" to be dropped on the continent to symbolise "the ideal of brotherhood among people".[1]

1947

The New York Times advocates a trusteeship for Antarctica, "which should be held in trust for the peoples of the world".[2] Petitions are submitted to the UN Trusteeship Council, proposing the creation of a Polar Trusteeship under the protection of the United Nations.

1948

UNESCO's director general, Julian Huxley, urges UNESCO to organise an International Antarctic Research Institute.

1949

Edward Shackleton, British member of parliament (MP) and son of the famous explorer, officially states that "the Antarctic [...] is being left to direct negotiations between the governments concerned. Yet there is no problem in the world that the United Nations is better suited to handle. [...] The Antarctic should be administered internationally, and the United Nations should be the body to do it."[3]

1956

Walter Nash, prime minister of New Zealand, proposes a trusteeship to turn Antarctica into "a world territory under the control of the UN".[5]

1956

At the UN General Assembly, Indian diplomat Arthur Lall raises the "Question of Antarctica" twice, recommending that the continent should be managed by the UN as Antarctica "is of great importance to the international community as a whole and not merely for certain countries [...]. The United States should call upon all States to utilise this territory solely for peaceful purposes [...]. The action proposed can only be taken by the world as a whole."[6]

1961

Brian Roberts, former head of the polar regions in the Foreign and Commonwealth Office and an active participant in the AT negotiations, points out that in future, the Consultative Parties to the Treaty (ATCPs) "will no longer be able to make effective decisions about the peaceful development of the Antarctic without steadily growing opposition. Thus, a marked change of approach would be required if they were to be recognised by the UN as responsible trustees acting on behalf of a much wider group of nations."[8]

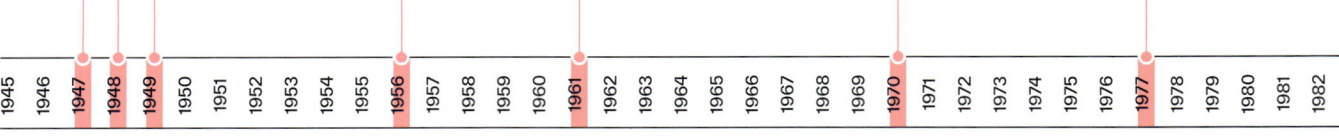

| 1945 | 1946 | 1947 | 1948 | 1949 | 1950 | 1951 | 1952 | 1953 | 1954 | 1955 | 1956 | 1957 | 1958 | 1959 | 1960 | 1961 | 1962 | 1963 | 1964 | 1965 | 1966 | 1967 | 1968 | 1969 | 1970 | 1971 | 1972 | 1973 | 1974 | 1975 | 1976 | 1977 | 1978 | 1979 | 1980 | 1981 | 1982 |

| 1945 Foundation of the United Nations | 1947 Calls for UN action on Antarctic matters | 1958 International Geophysical Year | 1959 The Antarctic Treaty (AT) is signed | 1961 The AT enters into force | 1964 Inclusion of Agreed Measures to the AT | 1967 Outer Space Treaty | 1972 CCAS | 1980 CCAMLR | 1982 UNCLOS | 1979 Moon Treaty |

COLD WAR PERIOD – COLD WAR PERIOD – COLD WAR PERIOD – COLD WAR PERIOD – COLD WAR PERIOD – COLD WAR PERIOD – COLD WAR PERIOD – COLD WAR PERIOD –

| 1945 | 1946 | 1947 | 1948 | 1949 | 1950 | 1951 | 1952 | 1953 | 1954 | 1955 | 1956 | 1957 | 1958 | 1959 | 1960 | 1961 | 1962 | 1963 | 1964 | 1965 | 1966 | 1967 | 1968 | 1969 | 1970 | 1971 | 1972 | 1973 | 1974 | 1975 | 1976 | 1977 | 1978 | 1979 | 1980 | 1981 | 1982 |

After the Second World War and the founding of the United Nations, the unresolved question of "who owns Antarctica" led to numerous proposals seeking UN involvement in the governance of the continent. These resurfaced in subsequent decades, and the "Question of Antarctica" was put annually to the UN General Assembly between 1983 and 2005.[4] Stalled by the development of the Antarctic Treaty System (ATS), the continent is to date governed by approximately one quarter of the world's countries, despite technically being regarded as a Global Commons.

1947

The UN Polar Trusteeship is not pursued in order to avoid unwelcome precedents for the Arctic regions in the midst of the Cold War and due to uncertainties regarding the legitimacy of applying the UN Charter's trusteeship articles 75–85 to a region without indigenous peoples.

1948

Demands for UN governance of the Antarctic as proposed by *The New York Times* and supporters fall upon deaf ears, since the governments actively involved in the continent – some of which had already announced sovereignty claims – prefer to retain direct control.

1956

The Indian request noted above is withdrawn, allegedly due to "a heavy UN agenda and the [fact that the] exploration of Antarctica was still proceeding".[7]

1958

The Indian proposal is finally rendered obsolete by AT negotiations.

1961

The AT places Antarctica's governance under the control of a limited number of states active on the continent, rather than granting this duty and right to the entire international community acting through the United Nations. The Treaty thus ultimately assigns the UN a virtual non-role in the affairs of Antarctica.

1964

The Antarctic Treaty Consultative Parties (ATCPs) reject the idea of acting upon a British proposal to clarify the ATS's relationship with international organisations to avoid reviving a debate about the UN's role in the region.

1972

The Antarctic Treaty Consultative Meeting (ATCM) opposes New Zealand's effort to revive the UN's role via a trusteeship proposal.[9]

1970

The UN proposes a New International Economic Order (NIEO), supported by the *Group of 77* developing nations, claiming that every state possesses the right to participate fully in discussions of global problems and should share equitably in the benefits resulting from their "Common Heritage".

1982

The prime minister of Malaysia, Dr Mahathir bin Mohamad, delivers a speech to the UN General Assembly as a leading advocate of a UN-based alternative to the AT: "Henceforth all the unclaimed wealth of this Earth must be regarded as the common heritage of all the nations [...] Now that we have reached agreement on the Law of the Sea [...] it is now time that the UN focus[ed] its attention on these [uninhabited] areas, the largest of which is the continent of Antarctica."[14]

2005

UN member states adopt *Resolution 60/47* in the General Assembly, without a vote, and agree that the UN will remain "seized" on the "Question of Antarctica" but will no longer discuss it regularly.[15]

1977

Ted Rowlands, British minister of state at the Foreign and Commonwealth Office, presses the case that the ATCPs should be acceptable to a wider international community and affirms that "the test of that acceptability will largely depend on the clarity with which we are seen to be serving the long-term interests of the Antarctic and the world community rather than short-term illusions of national advantage."[12]

1983

Antarctica officially appears on the UN General Assembly's agenda for the first time. The "Question of Antarctica" remains on the yearly agenda until 2005.

2007

Ban Ki-moon conducts the first-ever visit to Antarctica by a United Nations Secretary-General and states, "What I found was a place that would probably be unrecognizable to the likes of Robert Scott [...]. We see Antarctica's beauty and the danger global warming represents, and the urgency that we do something about it. [...] All this may be gone, and not in the distant future, unless we act, together, now."[16]

THE QUESTION OF ANTARCTICA – THE QUESTION OF ANTARCTICA – THE QUESTION OF ANTARCTICA

1982 | 1983 | 1984 | 1985 | 1986 | 1987 | 1988 | 1989 | 1990 | 1991 | 1992 | 1993 | 1994 | 1995 | 1996 | 1997 | 1998 | 1999 | 2000 | 2001 | 2002 | 2003 | 2004 | 2005 | 2006 | 2007 | 2008 | 2009 | 2010 | 2011 | 2012 | 2013 | 2014 | 2015 | 2016 | 2017 | 2018 | 2019 | 2020

1982
UN Convention on the Law of the Sea (UNCLOS)

1983
Anti-AT movement led by Malaysia and the UN

1991
Protocol on Environmental Protection of the AT is open for signature

2007
Fourth International Polar Year

■ events leaning towards Antarctica joining the UN

■ events discouraging Antarctica's inclusion in the UN

– COLD WAR PERIOD – COLD WAR – COLD WAR

1982 | 1983 | 1984 | 1985 | 1986 | 1987 | 1988 | 1989 | 1990 | 1991 | 1992 | 1993 | 1994 | 1995 | 1996 | 1997 | 1998 | 1999 | 2000 | 2001 | 2002 | 2003 | 2004 | 2005 | 2006 | 2007 | 2008 | 2009 | 2010 | 2011 | 2012 | 2013 | 2014 | 2015 | 2016 | 2017 | 2018 | 2019 | 2020

1975

The ATCM adopts *Recommendation VIII-8,* which states that "the Antarctic Treaty places a special responsibility on the Contracting Parties to exert appropriate efforts, consistent with the Charter of the United Nations, to the end that no one engages in any activity in the Antarctic Treaty Area contrary to the principles and purposes of the Treaty."[10]

1975

The ATCM rules out the need for any external intervention by the UN by adopting *Recommendation VIII-13* and reaffirming that, though welcoming the offer of cooperation by the UN Environmental Program (UNEP), the ATCPs possess "prime responsibility for Antarctic matters", including environmental protection.[11]

1977

A briefing paper from the British Foreign and Commonwealth Office reiterates the point: "If the Antarctic was brought under the control of a worldwide agency, possibly the United Nations, it would be far more difficult to achieve the level of cooperation that has been possible within the Antarctic Treaty framework. [...] An ill-functioning international agreement might tempt countries active in the Antarctic to act unilaterally with possible risks for the environment and with damaging effects for Antarctic scientific cooperation."[13]

1983

The ATCPs object unanimously to the Malaysian initiative on the "Question of Antarctica" and concede under Article III-2, acceding states, UN specialised agencies and other international and non-governmental organisations (such as the International Maritime Organization, the United Nations Environment Programme and the World Meteorological Organization) observer status at ATCMs for their capability of supporting and contributing to the ATS's scientific and technical ambitions.

2009

Hillary Clinton, then US secretary of state, says that "the [Antarctic] Treaty is a blueprint for the kind of international cooperation that will be needed more and more to address the challenges of the twenty-first century".[7]

1991

The objective of the ATS, an internationally agreed-upon regime for the governance of the region south of latitude 60° S, seeking to preserve Antarctica's status as a continent for peace and science and a special conservation area, is redefined by the Protocol on Environmental Protection. In effect, ATCPs, acting through the ATS, manage the Treaty area independently of the UN.

The Role of NGOs in the Antarctic Treaty System

James N. Barnes

JAMES N. BARNES is chair of the board of the Antarctic and Southern Ocean Coalition (ASOC), which he co-founded in 1978. He served on the board of Greenpeace USA from 1982 to 1985 and became Greenpeace International's UN representative in 1983, serving in that position for five years. Barnes has devoted more than 40 years of his work to protecting Antarctica.

From 1959 to 1978, the Antarctic Treaty Consultative Parties (ATCPs) maintained a closed, opaque system of governance and management. Second-tier states (Non-Consultative Parties) were not invited to meetings; reports of what occurred at Antarctic Treaty Consultative Meetings (ATCMs) were not published; no outside observers or NGOs were invited to attend; and the ATCPs chose not to inform the United Nations of their stewardship of the region.[1]

The Scientific Committee on Antarctic Research (SCAR) could be seen as an exception to this, depending on one's definition of "NGO". As the key scientific advisory body in the Antarctic Treaty System (ATS), SCAR played an important role for decades. Most of its members come from science organisations funded by governments, but it has maintained its independence and objectivity. In 1975, SCAR formed a group of specialists on marine living resources in the Southern Ocean and held an international workshop with John Gulland as rapporteur. Gulland's report noted the wide interest in Southern Ocean resources.[2] As an example of SCAR's initiative, following ATCM Recommendation IX-2 in 1977 (which laid out interim guidelines for krill harvesting and called for a special Consultative Meeting to develop what became the Commission for the Conservation of Antarctic Marine Living Resources, CCAMLR), SCAR launched BIOMASS, a complex scientific experiment focused on krill's place in the ecosystem and its lifecycle. International coordination was crucial to bringing together all the ships and equipment needed for this important project to be successful.[3] NGOs thus began requesting slots in national delegations – of which the most supportive were Australia, New Zealand, the United Kingdom and the United States – since 1978, when the Antarctic and Southern Ocean Coalition (ASOC) was founded.

On the 21st of April 1980, ASOC wrote to each ATCP on the draft of CCAMLR, suggesting a number of modifications, including stronger language in Article II about the "ecosystem-as-a-whole" principle and the protection of whales from excessive krill fishing. For the first time, the public, albeit indirectly, could see what was happening behind the closed doors.[4] Governments were led to declare the 1980s an "International Decade of Southern Ocean Research" in light of the many unknowns reflected in global conversation about the marine ecosystem, and NGOs requested that ASOC and other organisations be accredited as observers to CCAMLR.[5] Several of ASOC's suggestions were accepted by CCAMLR negotiators, including a better formulation of Article II,[6] but the Treaty parties refused to accredit NGOs as observers. ASOC member groups continued placing individuals on various national delegations while mounting demonstrations outside the venues of meetings. The international NGO newspaper *ECO* was published regularly, providing a record of what occurred at each negotiation and official meeting.[7]

Several other NGOs had important roles during the late 1970s and 1980s, in particular the International Institute for Environment and Development (IIED), which held international conferences and produced numerous reports based on its good access to decision-makers.[8] Richard Sandbrook and Barbara Mitchell in IIED's London office, and Lee Kimball in the US office, were the key people involved. Lee was also invited to serve on US delegations as an NGO advisor; she had a pre-eminent role in organising the Beardmore Glacier meeting in 1985, which was hosted by the US during the minerals negotiation. Improbably, a mixed group of scientists, diplomats, government experts and representatives of NGOs were flown there from New Zealand to spend a week talking day and night while experiencing the glories of Antarctica.[9] IIED was instrumental in spreading accurate information into mainstream media as well as to specialists about the ATS, and ASOC relied on its reports in preparing its campaigns. ASOC worked closely with the International Union for Conservation of Nature (IUCN), a hybrid conservation organisation that brings together scientists, legal and policy experts in and out of government, environmental organisations, government agencies and some states. Resolutions are agreed at its triennial General Assemblies. Since the late 1970s, ASOC submitted detailed Antarctic Resolutions at every General Assembly, which are useful diplomatically. IUCN's excellent international connections facilitated ASOC's advocacy activities, providing many allies and contacts. The IUCN was invited to participate in CCAMLR meetings as an observer in 1982 and was offered that status at the ATCM in 1987.[10]

ASOC continued to request accreditation to the ATCM and CCAMLR meetings every year, but the consensus decision-making rule allowed a single country to create a blockade until 1988, when ASOC finally obtained observer status at CCAMLR.[11] Negotiations of the Convention on the Regulation of Antarctic Mineral Resource Activities (CRAMRA) from 1983 to 1988 were also closed to observers, but a number of ASOC experts were invited to join their national delegations. After the Minerals Convention failed to achieve consensus in 1989, ASOC was invited to participate as an observer to the ATCM in 1990 for the two-year negotiation of what became the Protocol on Environmental Protection to the Antarctic Treaty. At the conclusion of the negotiations, ASOC's status was normalised along with relevant United Nations agencies, the tourism industry group IAATO, and IUCN. All have participated actively in the Antarctic Treaty System since then.[12] Even though NGOs have no role in decision-making, their participation has improved outcomes by the ATCPs and CCAMLR parties, provided detailed information to the media, and maintained public interest and support for Antarctic conservation.[13]

Without the sustained involvement of NGOs as ASOC, the ATS wouldn't have agreed to include a strong ecosystem-as-a-whole principle in CCAMLR, would have ratified a minerals treaty that opened the region to development, would not have put into force an Environmental Protocol to the Treaty, and wouldn't now be considering designating further large Marine Protected Areas. An active involvement of NGOs in Antarctic governance is thus pivotal to regulating anthropic actives in the fragile polar ecosystem and ultimately for the future of Planet Earth.

"The team [of Greenpeace] monitored pollution from neighboring bases and held other nations accountable for their actions. Greenpeace made headlines when 15 protesters blocked the French from building an airstrip at Dumont D'Urville [Station]. The construction work was controversial because it involved dynamiting habitats of nesting penguins. French scientists even admitted an airstrip violated terms of the Antarctic Treaty."

Greenpeace, *Creating the World Park Antarctica*

Antarctica as a World Park

James N. Barnes

The 1959 Antarctic Treaty had limited scope and membership, with only seven claimants (Argentina, Australia, Chile, France, Norway, New Zealand and the United Kingdom) and six non-claimants (Belgium, Germany, Japan, South Africa, Soviet Union and the United States). Together they controlled the fate of the region. Fishing wasn't regulated, even though several fish stocks had collapsed by 1975 and there was a rush to harvest krill, the foundation species of the marine ecosystem. Potential mineral development and exploitation were quietly being discussed among the twelve countries behind closed doors.

The phrase "World Park" was first introduced in a Resolution issued at the second World Parks Congress, held in the Grand Teton National Park, Wyoming, in 1972. Formed in 1978 by the Australian Conservation Foundation, Friends of the Earth, the Fund for Animals, Greenpeace, the Sierra Club, the World Wildlife Fund and ninety other organisations in thirteen countries, the Antarctic and Southern Ocean Coalition (ASOC) upheld the idea of a World Park or World Preserve as a means of protecting the Antarctic in order to maintain peace, promote science and protect wilderness values. A World Park in Antarctica would prohibit minerals activities and regulate all fishing based on scientific research and quotas that account for longer-term ecosystem impacts.

Negotiation of a fisheries agreement was moving forward among the Antarctic Treaty Consultative Parties (ATCPs), but faced difficulty as several countries wanted to continue fishing unhindered. Both fishing and minerals development were on the agenda of the tenth annual Antarctic Treaty Consultative Meeting (ATCM) in September 1979. On the 19th of April 1979, ASOC wrote to President Jimmy Carter urging the US to take the lead in protecting the Antarctic from commercial development by setting it aside as the first "World Preserve".[1] That request was renewed on the 10th of September 1980, prior to the eleventh ATCM.[2] While the initiative generated positive discussions with President Carter in the US and officials in Australia, Belgium, Chile, France, Italy, New Zealand and the United Kingdom, fishing and minerals interests blocked the proposal from being discussed formally by ATCPs. ASOC members persevered in promoting the concept of a World Park during the next two years, building a global movement to oppose minerals development.

Adopted on the 20th of May 1980 and entering into force on the 7th of April 1982, the Convention on the Conservation of Antarctic Marine Living Resources (CCAMLR) absorbed the "ecosystem-as-a-whole" principle that ASOC had promoted for managing fisheries[3] and provided the framework for large Marine Protected Areas (MPAs), crucial for Antarctica's future. Although fishing interests sometimes outweigh conservation in CCAMLR, two large MPAs have since been designated, with several more on the table, and many strong fisheries regulations have been adopted.[4]

By January 1983, when the ATCPs began negotiating a Minerals Convention in Wellington, New Zealand, a World Park campaign was launched.[5] With Greenpeace International playing a major role, millions of people around the world signed petitions against minerals development.[6] Greenpeace induced the UN General Assembly to begin holding annual debates on the "Question of Antarctica" in 1983, which brought the minerals negotiation to a global audience. Additionally, Greenpeace published detailed reports each year, including the text of successive drafts of the proposed Convention. Example of this include the Beeby Drafts (I and II), reprinted in Greenpeace International, *The Future of the Antarctic: Background for a U.N. Debate,* App. 8 (1983) and *The Future of the Antarctic: Background for a Second U.N. Debate,* App. 8 (1984).[7] Greenpeace's loud cry of urgency shattered Antarctica's widely disseminated reputation as a pristine continent managed without fault by its caretakers. Aiming to be invited to participate in Antarctic Treaty discussions (an honour bequeathed only upon those who have proven significant investment in the continent) in 1987, Greenpeace built a World Park Antarctic Base, a year-round base on Cape Evans, Ross Island.[8] From this newly acquired position, Greenpeace and ASOC campaigned actively against the adoption of the Convention on the Regulation of Antarctic Mineral Resource Activities (CRAMRA),[9] quoting Bernard Heber to argue that a World Park regime "would better maintain a long-run supply of the important public good and common property resource".[10] Finally, ASOC and its partner Jacques Cousteau convinced Australia[11] and France[12] to reject it one year later, in 1989. Given the consensus principle of the Antarctic Treaty, the Convention was put to a halt. ASOC thus urged the ATCPs to negotiate an environmental protection agreement with a ban on minerals activities. On the 4th of October 1991, the Protocol on Environmental Protection to the Antarctic Treaty was signed in Madrid, including a ban on minerals exploration and exploitation.[13] Entering into force in 1998, this "Madrid Protocol" was a major victory.[14] Although the ban could be reviewed in 2048 if any ATCP requests it, the rules for creating a new minerals treaty are strict. While management of the Antarctic under the Protocol on Environmental Protection and CCAMLR has gone reasonably well, wilderness values remain at risk, given the pace at which new scientific stations are being built and mass tourism and fishing are being developed.[15] Whereas the operative reality is far from the original World Park vision, the concept remains a motivating force for the public, informs ongoing campaigns, and underpins NGO hopes for Antarctica's future.

The Antarctic through the Lens of Herbert Ponting

Photographs taken between 1910 and 1912 as the official photographer of the British Antarctic Expedition led by Captain Robert Falcon Scott

"In all my travels in more than thirty lands I had seen nothing so simply magnificent as this stupendous work of Nature. The grandest and most beautiful monuments raised by human hands had not inspired me with such a feeling of awe as I experienced on meeting with this first Antarctic iceberg."

Herbert George Ponting (Photographer, Explorer),
The Great White South, 1921

FOUR ELE

MENTS

06
The Revolutionary Ocean

The Global Carbon Budget

Judith Hauck

JUDITH HAUCK is the head of the Helmholtz Young Investigator Group for Marine Carbon and Ecosystem Feedbacks in the Earth System (MarESys) and deputy head of Marine Biogeosciences at the Alfred Wegener Institute for Polar and Marine Research (Germany). Dr Hauck is responsible for the ocean carbon sink estimate in the annual Global Carbon Budget report (GCB).

Humans emit carbon dioxide (CO_2) through the burning of fossil fuels, cement production, and land-use changes – such as deforestation. Cumulative CO_2 emissions from the period between 1750 and 2019 amount to 685 gigatons of carbon with two-thirds of this stemming from fossil fuels and the remainder from land-use changes. If all this CO_2 had stayed in the atmosphere, we would already be observing atmospheric CO_2 concentrations of 600 ppm. That is roughly equivalent to what is expected in the middle of this century according to the latest IPCC projections and would be accompanied by a global temperature rise of 3–4°C. It is fortunate for human civilisation that atmospheric CO_2 concentrations are, in fact, much lower. The highest-ever directly measured daily atmospheric CO_2 concentration since 1958 was reached on the 25th of May 2020 with 418.04 ppm. We know this with certainty and with precision, thanks to the pioneering and persevering work of Charles Keeling, who started the first time series of atmospheric CO_2 concentration in Hawaii in 1958, a historical record that has been continued ever since. Given the typical scientific funding situation with grants for a few years, this is a remarkable achievement.

If we know with certainty that all emitted CO_2 would lead to atmospheric CO_2 of 600 ppm, yet we are at 418 ppm at most, this begs the question, where has all the carbon gone? The answer lies in the natural carbon cycle, which exchanges CO_2 between the three large carbon reservoirs of atmosphere, ocean, and terrestrial biosphere. While the atmospheric fraction, the part of the emitted CO_2 that remains in the atmosphere, has to date stayed astonishingly constant at around 45%, the other 55% is taken up by the ocean and the terrestrial biosphere. Ocean and land thereby provide a huge service to our society by slowing down the increase in atmospheric CO_2 and, by extension, slowing down the impact of climate change.

The Global Carbon Budget is a scientific publication that quantifies the CO_2 emissions and fluxes between ocean, land, and atmosphere caused by anthropogenic perturbations (CO_2 emissions, climate change) on an annual basis. The annual ocean carbon sink in the Global Carbon Budget is estimated from process models that simulate the circulation as well as chemical and biological processes in the ocean responsible for the CO_2 uptake. These models are forced on the basis of the observed atmospheric CO_2 concentration and weather patterns but have no *a priori* information relating to CO_2 concentrations in the surface water from measurements in the specific years. Observation-based estimates of the ocean carbon sink are discussed by comparison, but their strength lies in assessing the variations from year to year rather than the actual mean number.

The role of the ocean in the global carbon cycle cannot be overstated. The ocean is the largest carbon reservoir on Earth containing forty times as much carbon as the atmosphere and ten times as much carbon as the land. On long timescales of thousands to tens of thousands of years, the ocean will be the ultimate sink for anthropogenic CO_2. This timescale is set by two processes, which both enhance the capability of the ocean to take up CO_2. The first is the overturning circulation of the ocean, bringing carbon from the surface into the deep ocean, and the second is the reaction of the CO_2 with sedimentary minerals.

Over the last decade of 2009–2018, the ocean has taken up 235% and the land 295% of anthropogenic CO_2 emissions. Since pre-industrial times, the ocean has taken up 25% of all cumulative CO_2 emissions. Over the same period, the land carbon sink has taken up 30% of cumulative emissions but has also released a comparable amount of CO_2 through land-use change emissions.

The Southern Ocean is the main conduit by which this anthropogenic CO_2 enters the ocean. It stands out as a region with low temperatures, a vast surface

> "Human activities are estimated to have caused approximately 1.0°C of global warming above pre-industrial levels, with a *likely* range of 0.8°C to 1.2°C. Global warming is *likely* to reach 1.5°C between 2030 and 2052 if it continues to increase at the current rate."
>
> Intergovernmental Panel on Climate Change, *Special Report on the Ocean and Cryosphere in a Changing Climate*, 2018

The "explosion of fossil-fuel use is inseparable from everything else which made the 20th century unique in human history. As well as providing unprecedented access to energy for manufacturing, heating and transport, fossil fuels also made almost all the Earth's other resources vastly more accessible. The nitrogen-based explosives and fertilisers [...] transformed mining, warfare and farming. Oil refineries poured forth the raw materials for plastics. The forests met the chainsaw. In no previous century had the human population doubled. In the 20th century it came within a whisker of doubling twice. In no previous century had world GDP doubled. In the 20th century it doubled four times."[1] Such unprecedented growth – shown in the trend line at the side, which visualises the exponential rise in CO_2 emissions by nation overlaid on the rise in world population – comes at a high cost. As shown in the diagram (opposite), which builds upon work produced at the Milken Institute School of Public Health at the George Washington University, there is an inverse relationship between polluting countries (which tend to have greater climate adaptability) and nations who "lack resources to mitigate or recover from severe weather"[2] and who are thus most vulnerable to the effects of climate change.

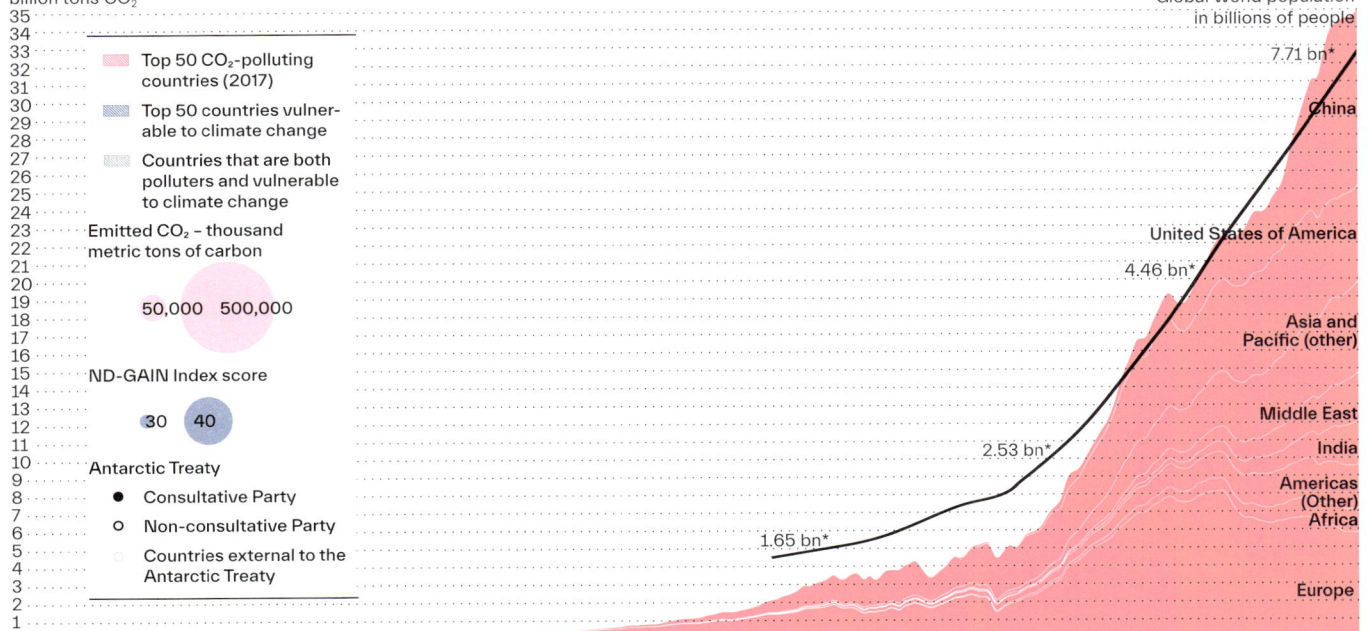

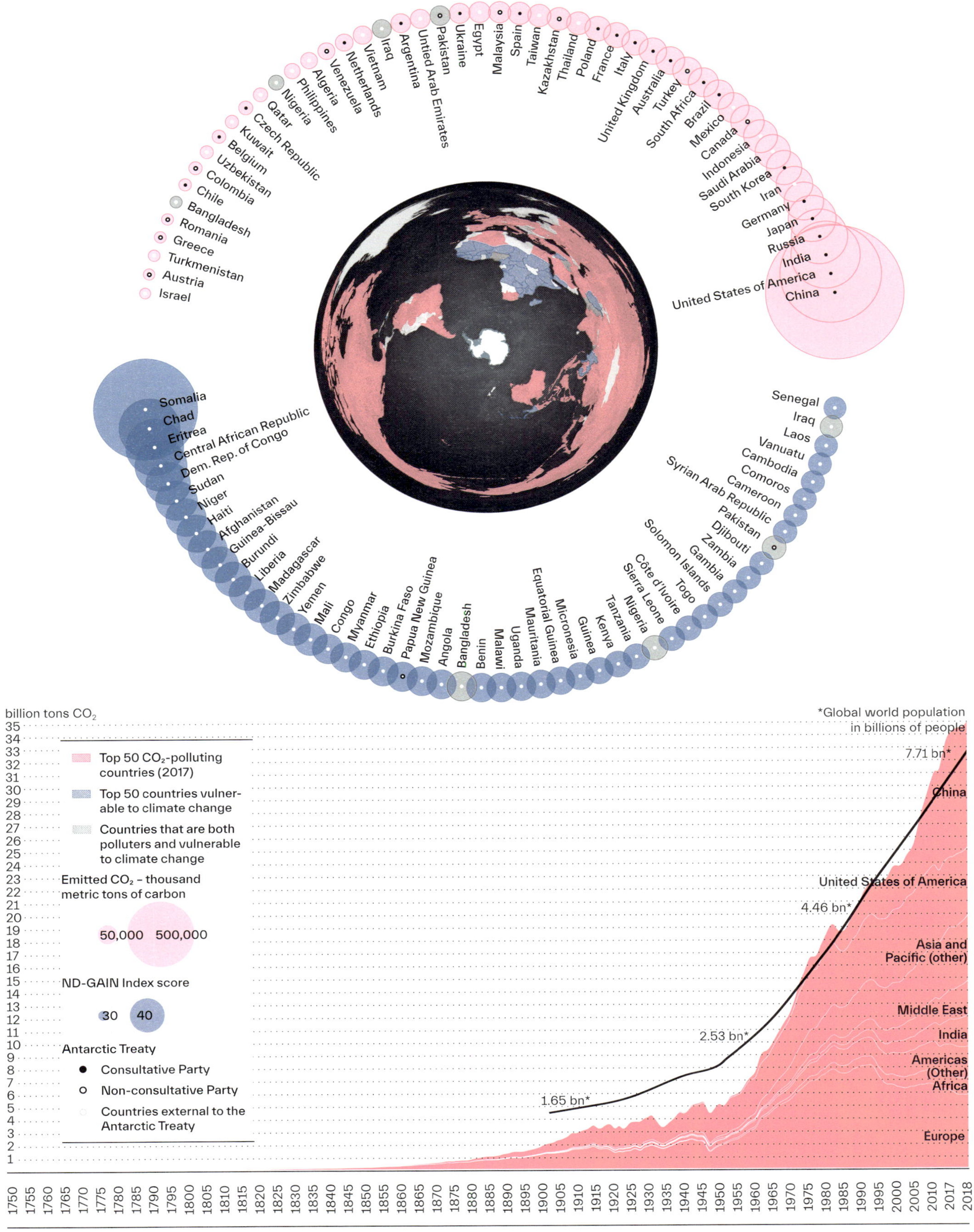

billion tons CO₂

Top 50 CO₂-polluting countries (2017)

Top 50 countries vulnerable to climate change

Countries that are both polluters and vulnerable to climate change

Emitted CO₂ – thousand metric tons of carbon

50,000   500,000

ND-GAIN Index score

30   40

Antarctic Treaty

● Consultative Party

○ Non-consultative Party

Countries external to the Antarctic Treaty

*Global world population in billions of people

7.71 bn*

China

United States of America

4.46 bn*

Asia and Pacific (other)

Middle East

India

Americas (Other)

Africa

2.53 bn*

1.65 bn*

Europe

area, and deep- and bottom-water formation regions. CO_2 is more soluble at low temperatures and the large surface area makes this high CO_2 flux into the seawater possible. As the surface ocean carbon content equilibrates quickly with the CO_2 concentration in the atmosphere, the main bottleneck for ocean carbon uptake is the transfer of CO_2 into the ocean interior. This transport of carbon away from the air-sea interface therefore facilitates further CO_2 uptake at the surface. Deep-water formation regions, such as the Southern Ocean and the North Atlantic, which connect the surface ocean with the ocean interior, are therefore the areas where most oceanic CO_2 uptake occurs. The subduction – or sinking – of surface water north of the Antarctic Circumpolar Current to intermediate water depths of around 1,000 metres, which also marks the formation of Antarctic Intermediate Water, is the key process that allows the surface Southern Ocean to continuously absorb more CO_2 from the atmosphere. Another process is the formation of Antarctic Bottom Water at the ice shelves close to the Antarctic coast, where the surface water cools and becomes saltier when sea ice forms. Cold and saline water is dense enough to sink to the bottom of the ocean and spread through the world's oceans at depth, sequestering the carbon for centuries. Although it plays a key role for the global carbon cycle, the Southern Ocean is one of the least sampled regions owing to its remoteness and hostile environment. Comparisons between process models and observation-based estimates in the Global Carbon Budget are made more difficult by a shortage of winter data in the Southern Ocean. This leaves us with unanswered scientific questions, such as how much the Southern Ocean carbon sink varies from year to year. Autonomous vehicles, such as floats and sail drones equipped with high-quality pCO_2 measuring instruments, promise to provide more clarity in future. It is crucial for the present-day Southern Ocean carbon sink to be quantified, in order to improve the process models so that they match the observations better. Only well-calibrated models can be used as a tool to investigate the service and surprises that the Southern Ocean might have for humankind in the future. The global ocean is interconnected in its circulation, and while the Southern Ocean is remote to us, any local change may have global implications.

Analysing the temporal evolution of the atmospheric CO_2 concentration in relation to the ocean and land carbon sinks, expressed in parts per million (ppm) between 1958 and 2018 (below), reveals that if all CO_2 emissions had stayed in the atmosphere, atmospheric CO_2 concentration would have reached 524 ppm by the end of 2018 – a number that would then rise to 600 ppm when all CO_2 emissions since industrialisation are taken into account.[3] The total emissions of 524 ppm were absorbed respectively by the atmosphere (409 ppm), the ocean (c. 50 ppm), and the land (c. 60 ppm).

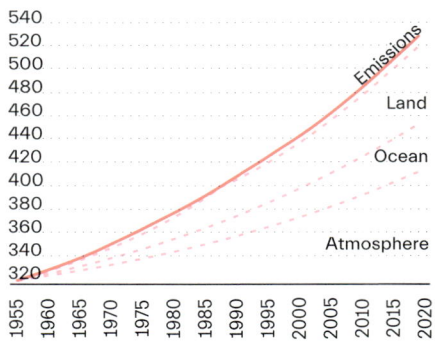

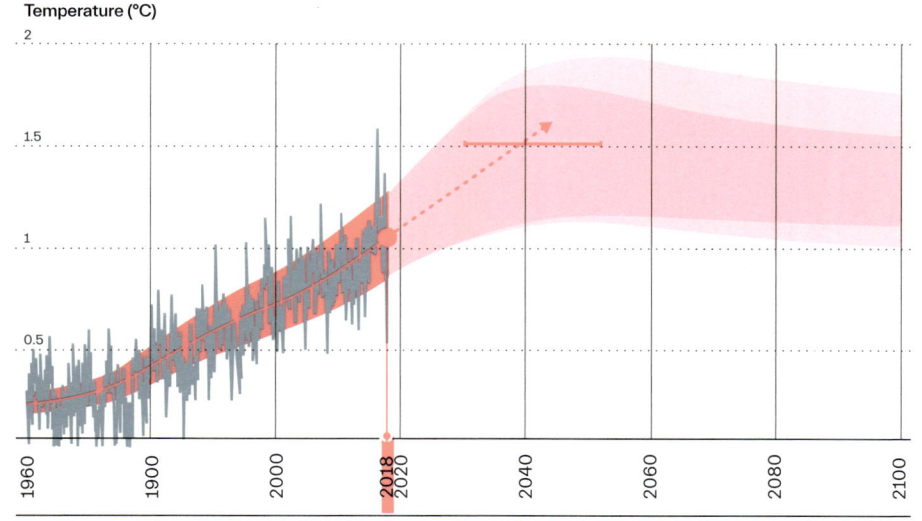

Temperature (°C)

Estimated anthropogenic warming to date and likely range

Observed monthly global surface temperature

Likely range of modelled response to stylised pathways

Global CO_2 emissions reach net zero in 2055, while net non-CO_2 radiative forcing is reduced after 2030.

Faster CO_2 reductions make it more likely that warming can be limited to 1.5°C.

No reduction in net non-CO_2 radiative forcing makes it less likely that warming can be limited to 1.5°C.

The Southern Ocean Overturning Circulation

Jean-Baptiste Sallée

JEAN-BAPTISTE SALLÉE is an oceanographer at the French National Centre for Scientific Research (CNRS) and researcher at the Pierre-Simon Laplace Institute (France). His work tackles questions of the dynamics of the ocean and climate with active research efforts on the study of the Southern Ocean. Dr Sallée is a lead author of the sixth report of the Intergovernmental Panel on Climate Change (IPCC).

Almost a third of the world's ocean surface is at our antipodes, circulating between the Antarctic continent and the southernmost inhabited places on Earth. This immense ocean is unique in many ways and a place of many extremes. The Southern Ocean is subjected to the strongest winds on the planet, the roaring forties and the howling fifties; it hosts the Antarctic Circumpolar Current, the most vigorous current of the planet, circulating clockwise without being obstructed by continental barriers and moving 170 million cubic metres of water per second (nearly 800 times the flow of the river Amazon); a large part of its surface is covered in winter with ice floes, a refuge for unique fauna and biodiversity; it is in direct contact with the immense Antarctic glaciers, which flow over its surface at the fringes of the Antarctic continent, creating huge subglacial cavity seas of hundreds or thousands of square kilometres.

The singular nature of the Southern Ocean geography and currents gives it a pivotal role in determining the climate. It literally acts as a natural sponge removing carbon and heat from the atmosphere to store them in the depths of the ocean, thus reducing the global warming that we are experiencing in the atmosphere. More than half of the increase in heat produced by global warming is absorbed in the Southern Ocean. In fact, half of the carbon emitted into the atmosphere by human activities is absorbed by the world's oceans, and it is in the Southern Ocean that the majority of this uptake takes place. The importance of this region is explained by an ocean circulation that connects the surface to the great depths below.

The fundamental vertical structure of the world ocean consists of a succession of different horizontal layers, with little vertical exchange. The three main ocean layers are the surface layer, which continually exchanges heat, freshwater, carbon, and other climatically important gases with the atmosphere; a layer called "pycnocline", which is characterised by an enhanced density contrast between shallower and deeper layers and inhibits circulation and mixing; and a deep ocean layer isolated from the atmosphere. At high latitudes, the vertical density contrast is reduced since the surface waters are very cold because of the temperature of the atmosphere above. The surface ocean is actually so cold that in can be colder than subsurface waters, and the stability of the water column is instead ensured by salinity (surface waters are fresher than subsurface waters). In addition, in the Southern Ocean, the presence of a very strong Antarctic Circumpolar Current, associated with the extreme winds, reduces the density contrast by tilting the density surfaces from horizontal to vertical, creating vertical pathways allowing a vertical circulation to develop and enhancing the propagation of the climate signal from the atmosphere all the way down to the ocean abysses.

The vertical circulation associated with the geographical connection that the Southern Ocean creates across the three main ocean basins, Atlantic, Indian, and Pacific, allows a three-dimensional global ocean circulation to exist, connecting all the regions of the world at all depths. Within this global circulation, the Southern Ocean feeds the world's oceans with dense oxygenated water that lies at the bottom and slowly moves northward into the northern hemisphere, in all basins. The production of these bottom waters is one of the main engines of global ocean circulation.

But if the Southern Ocean does us and the planet this service, it is partly at its own expense. The Southern Ocean water masses are warming faster than any other ocean depths on Earth. The upper kilometre within the Antarctic Circumpolar Current is warming at a rate of 0.1–0.2°C per decade, which is a very significant amount for the ocean. The ocean is harder to heat than the

"Could the waters of the Atlantic be drawn off, so as to expose to view this great sea-gash, which separates continents, and extends from the Arctic to the Antarctic, it would present a scene the most rugged, grand, and imposing. The very ribs of the solid earth, with the foundations of the sea, would be brought to light, and we should have presented to us at one view the empty cradle of the ocean."

Matthew Fontaine Maury (Oceanographer, Astronomer, Naval Officer), 1872–1876

"To date, the ocean has taken up more than 90% of the excess heat in the climate system. By 2100, the ocean will take up 2 to 4 times more heat than between 1970 and the present if global warming is limited to 2°C, and up to 5 to 7 times more at higher emissions. Ocean warming reduces mixing between water layers and, as a consequence, the supply of oxygen and nutrients for marine life. The ocean has taken up between 20 [and] 30% of human-induced carbon dioxide emissions since the 1980s, causing ocean acidification. Continued carbon uptake by the ocean by 2100 will exacerbate ocean acidification."

Intergovernmental Panel on Climate Change, *Special Report on the Ocean and Cryosphere in a Changing Climate*, 2018

atmosphere, but it is also very hard to reverse the process, so that deep ocean warming makes global climate change irreversible on human timescales. The significant warming of the Antarctic Circumpolar Current has recently been formally attributed to human-induced activities. Closer to the Antarctic margins, the warming is more subtle but is of great concern as it contributes to the melting of the Antarctic ice sheet, jeopardising its stability and threatening coasts all over the Earth with a potentially significant rise in sea level. The warming reaches all the way to the ocean floor at more than 4 to 5 kilometres from the surface, providing a clear warning signal that has been observed since the 1990s, even if deficiencies in the observation systems operating at those great depths currently still limit our understanding of the exact causes of this change. Associated with the process of warming, the Southern Ocean water masses are affected by change in the global hydrological cycle, which modifies the salt content of these masses, with important consequences for their circulation (salt content alters the density of the water, especially in a cold-water environment, which in turn impacts the circulation). For instance, as Antarctic ice melts, dense waters produced near the ocean surface become less dense and are therefore expected to sink less, thereby affecting the global ocean circulation. The uptake of carbon also causes an acidification of Southern Ocean water masses, endangering part of the biodiversity: with it comes the risk that a critical threshold will be passed, at which point it will become difficult for organisms to grow shells, with cascading impacts on the entire ecosystems.

Because of its remote location, it is only fairly recently that field observations of the Southern Ocean have been implemented on a regular basis. It is now clear that sustained observation of this part of the world is critical for monitoring climate change and to enable a better understanding of the underlying mechanisms at play, which have critical consequences worldwide. Large parts of this ocean are still unexplored, with some of them almost unreachable, forcing us to review our methods, adjust our equations, and invent new robots to measure its depths and tools to cope with these unique conditions. These efforts are essential. It is by going to the Southern Ocean that explorers and scientists have reported the clear and unique power of this very specific place on Earth, and it is only through extensive and repeated analysis of observations collected on-site that the fragility of this giant becomes apparent.

"There is this old and wrong idea of Antarctica as an island continent. It is not so isolated; it is really connected."

Marcelo Leppe (Director, Chilean Antarctic Institute) in "Why Penguins May Help Us Predict the Impact of Climate Change", *Financial Times*, 27 February 2020

"The subpolar ocean remoteness does not protect the continent from the impacts of climate change. The warmth of the atmosphere induced by human activity affects wind patterns over Antarctica too, and this in turn is changing sea ice patterns and ocean water flows around the continent. Enhanced ice melting affects the upper ocean stratification and changes in water column characteristics initiate a feedback loop that affects the production of sea ice, directly influences marine life behaviour, alters productivity patterns which drive the biogeochemical carbon pump and carbon sequestration to the deep ocean, a vital part of the climate system."

Craig Stevens, James Renwick, and Miles Lamare (Climate Scientists), 2020

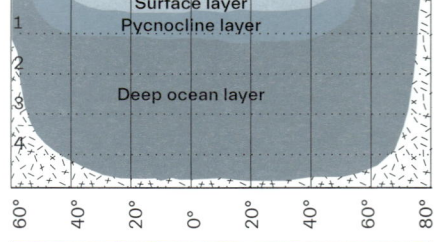

The ocean density structure is articulated in three layers, of which the pycnocline zone isolates the deep ocean from contact with the atmosphere in the mid-latitude areas and at the equator. As shown in the diagram, deep ocean waters are in contact with the atmosphere only at high latitudes, i.e. at the poles.

The Thermohaline Circulation, also known as the ocean conveyor belt, transports and mixes the planet's ocean waters. Driven by differences in temperature and seawater densities (i.e. salinity), the deep, abyssal water circulation influences regional weather patterns. The revolutionary Southern Ocean, with its large volumes of low-salinity cold water and the Antarctic Circumpolar Current (a deep ocean belt established some 25 million years ago around Antarctica), plays a central role in Earth's water dynamics.

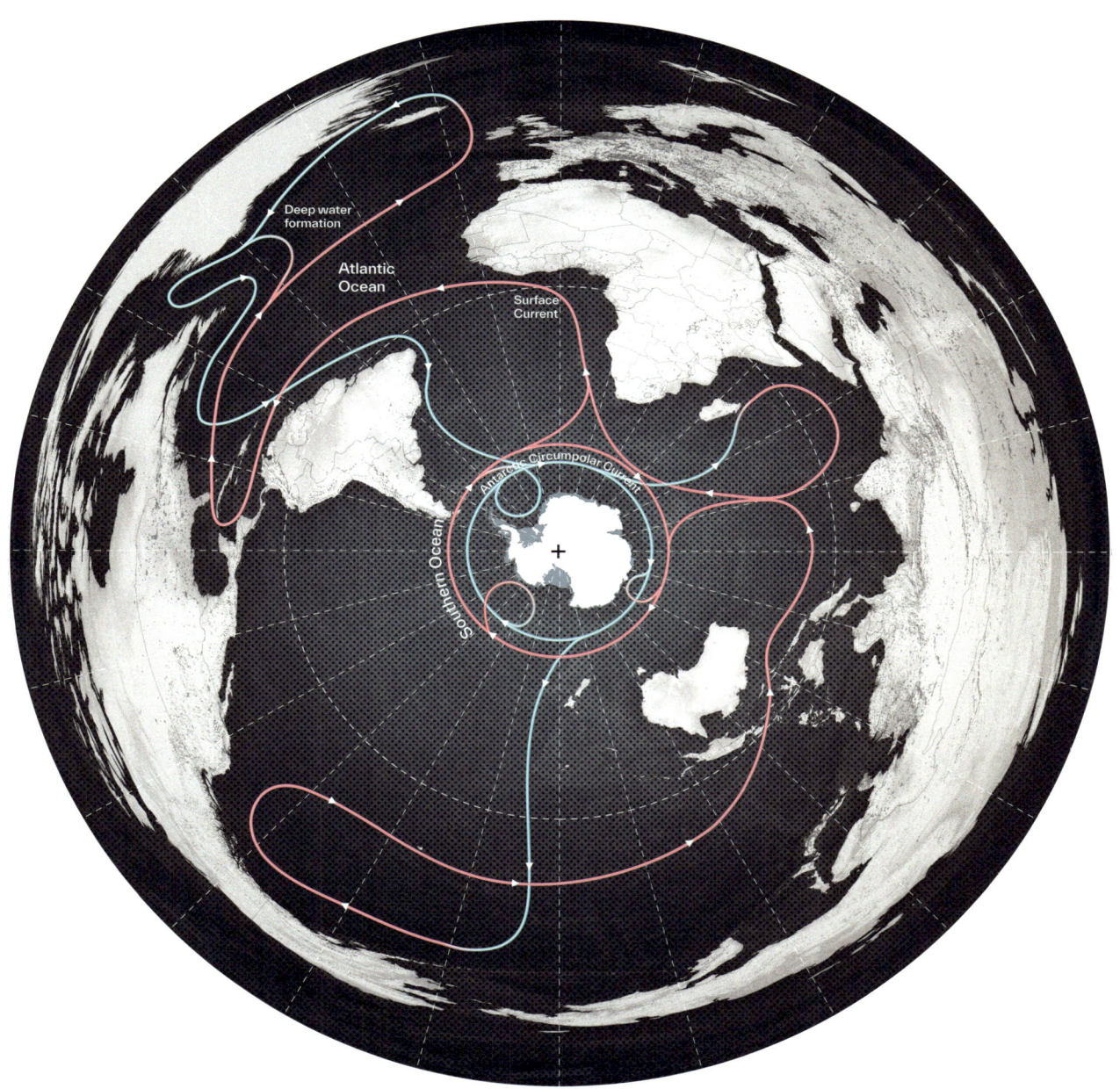

Ocean Currents

— Cold-water current

— Warm-water current

+ South Pole

The Variable Southern Ocean Carbon Sink

Peter Landschützer

PETER LANDSCHÜTZER is a marine biogeochemist at the Max Planck Institute for Meteorology (Germany), specialising in the exchange of carbon dioxide between the ocean and the atmosphere. He works with ocean observations and novel data interpolation techniques such as neural networks. Dr Landschützer is particularly interested in the Southern Ocean and the processes driving the air-sea carbon dioxide flux in high-latitude waters.

The Southern Ocean, defined by the waters surrounding the Antarctic continent, plays an important role in our global climate system. Despite its relatively small ocean area (which accounts for circa 20% of the global ocean) approximately 40% of the anthropogenic CO_2 produced since the beginning of industrialisation is stored in these waters. The driving force of the CO_2 flux is the concentration or partial pressure difference between the ocean and the atmosphere. If the ocean surface partial pressure exceeds the partial pressure of CO_2 in the overlying atmosphere, then the surface waters are supersaturated and will release CO_2 into the atmosphere. Likewise, if the partial pressure of CO_2 in the atmosphere exceeds the partial pressure at the sea surface, the ocean, being undersaturated, acts as a carbon sink for atmospheric CO_2.

The Southern Ocean's importance for the global carbon cycle can be directly linked to its unique circulation. The westerly wind circulation powers the Antarctic Circumpolar Current and creates a meridional overturning circulation. The westerly winds further create an upward transport of deep, aged water masses in the Southern Ocean that have not been in contact with the atmosphere in centuries. As a result, the dissolved carbon contained in these waters is often referred to as natural carbon. Oversaturated with CO_2 at the surface, these upwelled water masses release this natural carbon into the atmosphere.

Owing to the solubility pump, further north of 44°S, the Southern Ocean represents an intense sink for carbon from the atmosphere. Here, warm subtropical waters move southwards and cool, causing carbon to dissolve quickly and drawing down the surface CO_2 concentration. The resulting undersaturation of surface waters leads to a substantial uptake of carbon dioxide from the atmosphere. In pre-industrial times the outgassing of natural carbon dominated the Southern Ocean carbon balance, making the Southern Ocean a net source of carbon for the atmosphere; however, since industrial times this has shifted and atmospheric CO_2 concentrations have kept rising, creating a stronger undersaturation of marine surface waters in the Southern Ocean, which in turn results in the enhanced uptake of man-made carbon from the atmosphere. This has led to the exceptional gradual transformation of the Southern Ocean (as the only ocean basin on our planet) from a source of CO_2 to a carbon sink.

Despite the importance of the Southern Ocean to our climate system, it still remains under-sampled with respect to CO_2. Direct, high-accuracy measurements of the surface ocean partial pressure of CO_2 (i.e. the quantity that determines the flux of CO_2 between the ocean and the atmosphere) are sparse, as they historically required ships as measurement platforms. Owing to limited commercial interest, few ships sail the rough Southern Ocean waters, with fewer doing so in the even harsher winter season. Only a small group of regions, such as Drake Passage are well observed as a result of regular research cruises and Antarctic supply ships taking essential CO_2 measurements.

Therefore, Southern Ocean research today relies on autonomous measuring techniques: floats equipped with biogeochemistry sensors are now circling the Antarctic continent shedding new light on the Southern Ocean carbon cycle. Additionally, alternative autonomous platforms such as sail-drones and drifters support the CO_2 fleet in the Southern Ocean. Although these measurement platforms have only existed for a couple of years, they have already revolutionised our view of regional carbon dynamics in the Southern Ocean.

In the 1990s and early 2000s, the evolution of the Southern Ocean carbon sink was a source of great concern in the scientific world. Many researchers pointed out that the Southern Ocean was not taking up as much carbon as we would expect based on the increase in atmospheric CO_2 levels. The source for

"The Southern Ocean is the key region globally for the upwelling of interior ocean waters to the surface, enabling waters that were last ventilated in the pre-industrial era to interact with the industrial-era atmosphere and the cryosphere. New water masses are produced that sink back into the ocean interior. Such export of both extremely cold and dense Antarctic Bottom Water and the lighter mode and intermediate waters represents important pathways for surface properties to be sequestered from the atmosphere for decades to millennia. This upwelling and sinking constitutes a two-limbed overturning circulation, by which much of the global deep ocean is renewed."

Intergovernmental Panel on Climate Change, *Special Report on the Ocean and Cryosphere in a Changing Climate*, 2019

Decadal trends in the air–sea pCO_2 differences and air–sea CO_2 fluxes for the Southern Ocean (opposite) show a weakening sink trend in the 1990s followed by a reinvigoration past the turn of the millennium, resulting in an extended period during which the Southern Ocean carbon sink was substantially smaller than predicted on the basis of the long-term climatological uptake and the increase in atmospheric CO_2. The trendline depicts the air–sea CO_2 flux anomalies, i.e. the deviations from the mean over the period 1982 to 2016, from the Self-Organising Map – Feed-Forward Neural Network (SOM-FFN) estimate (silver[1]) and from the mixed-layer scheme (fluorescent[2]). Negative values indicate a stronger sink. Also shown are the air–sea CO_2 flux anomalies from seven global biogeochemical models that contributed to the Global Carbon Project (grey[3]), the multi-model mean (dashed silver), and the expected uptake based on the average of four of these simulations that did not consider climate variability. The shading indicates the uncertainty of the two observation-based products.

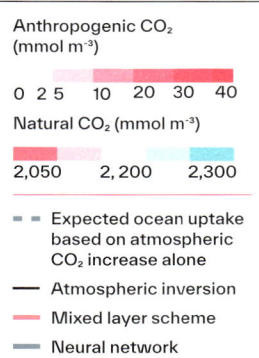

Anthropogenic CO₂
(mmol m⁻³)

0 2 5   10   20 30   40

Natural CO₂ (mmol m⁻³)

2,050   2,200   2,300

- - -   Expected ocean uptake
        based on atmospheric
        CO₂ increase alone
———   Atmospheric inversion
———   Mixed layer scheme
———   Neural network

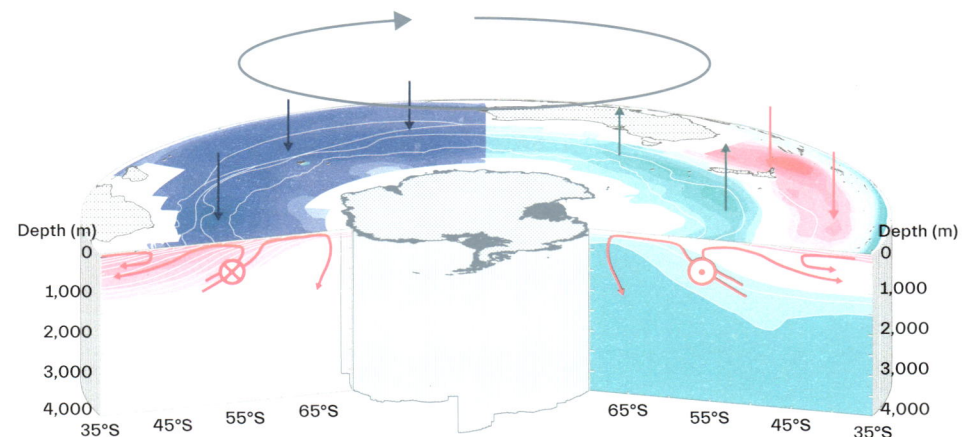

The schematic diagram of the zonal-mean overturning circulation of the Southern Ocean (above) and the associated fluxes of anthropogenic (left) and natural (right) CO₂, based on recent ocean inversion estimates.[4] The vertical sections depict the distribution of anthropogenic CO₂ (left) and dissolved inorganic carbon (right) for 1994,[5] with one indicating the Antarctic Circumpolar Current flowing into the plane (circle with X), and the other showing its movement out of the plane (circle with dot).

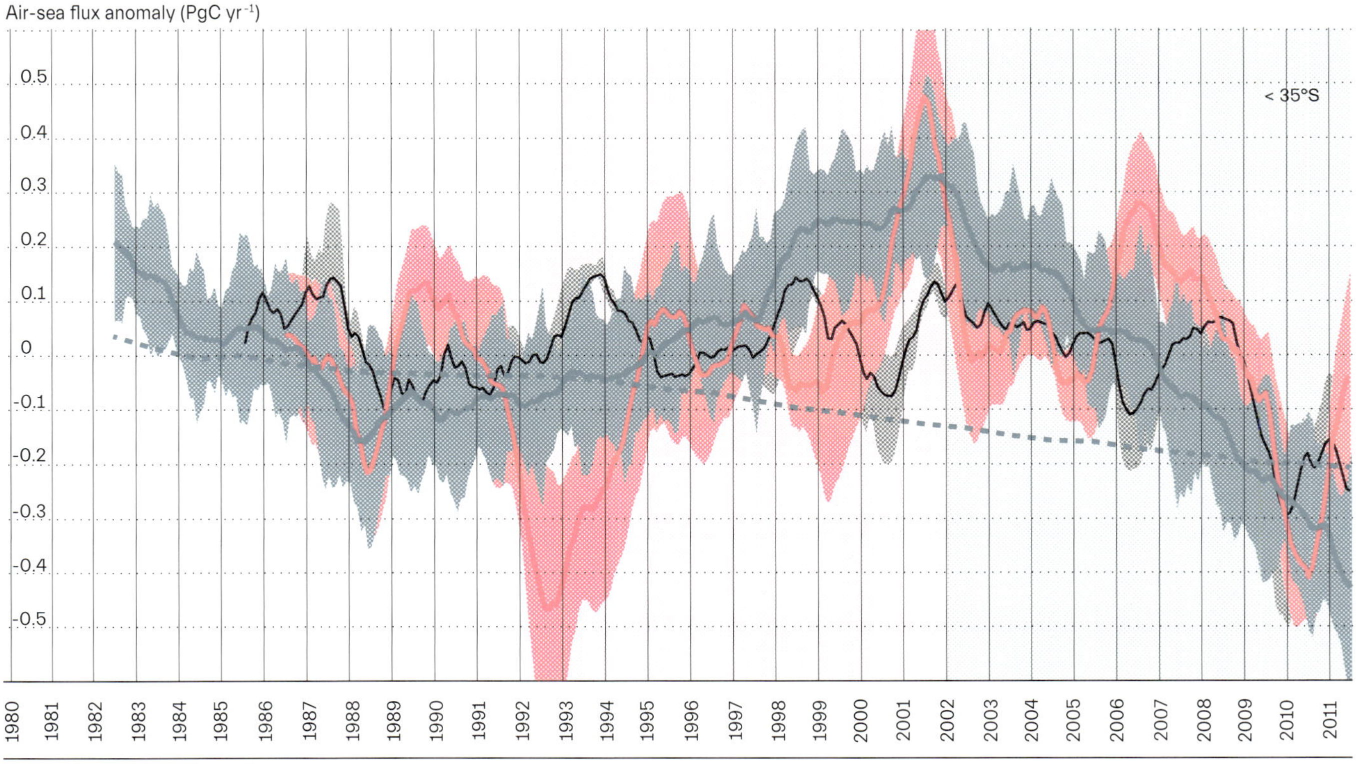

Air-sea flux anomaly (PgC yr⁻¹)

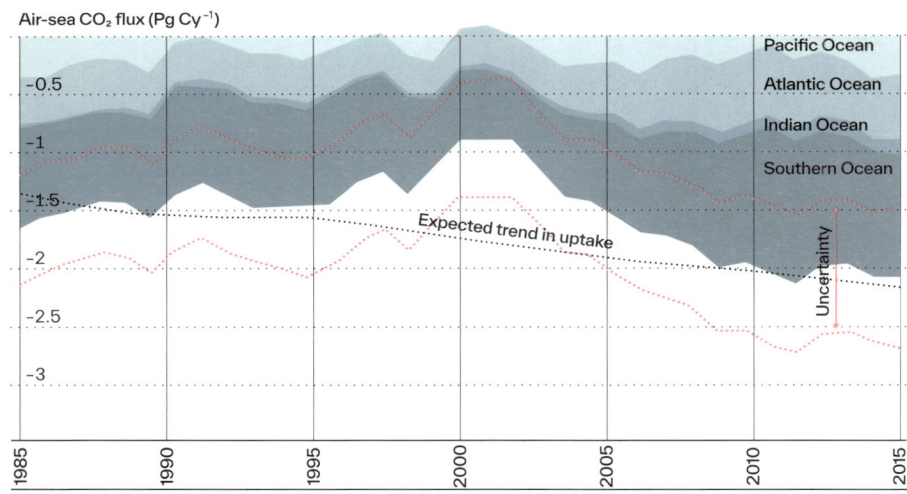

Air-sea CO_2 flux (Pg C y^{-1})

The timeline reveals the evolving capacity of the ocean to act as a global sink for atmospheric CO_2 and the respective contributions of each basin, separated at 35°S. The area between the dotted fluorescent lines indicates the uncertainty interval around the global mean, while the dotted black line represents the expected uptake based on the increase in atmospheric CO_2 estimated from a hindcast simulation with an ocean biogeochemistry model[6] scaled to the current global CO_2 flux.[7] Negative fluxes indicate a sink for atmospheric CO_2.[8]

this saturation of the Southern Ocean carbon sink was the fierce westerly winds around Antarctica. In our changing climate the westerly wind belt, which drives the Southern Ocean circulation and overturning, was observed to shift towards the South Pole and to intensify. As a result, more natural carbon-rich deep water was brought to the surface and ultimately released into the atmosphere. About a decade later in 2010, the strength of the Southern Ocean CO_2 uptake bounced back to levels we would expect, which was linked to circulation changes and local weather patterns. Most recently, the Southern Ocean has lost its sink strength again, indicating that the ability of the Southern Ocean to take up man-made carbon dioxide might be subject to significant decadal variations.

To date, the future evolution of the Southern Ocean still remains a mystery. The historical under-sampling of the harsh and ominous waters of the Southern Ocean leaves us with a limited understanding of the historical, present, and future variability of this important carbon sink. The wealth of new observations derived from autonomous devices will overcome this impasse in the future; however, it will take several years of deployment of autonomous devices to be able to resolve the question of whether the observed variations in the Southern Ocean carbon sink are recurring oscillations or whether the Southern Ocean has indeed lost its capacity to take up excess man-made carbon dioxide.

"The Southern Ocean, defined as the vast region south of 35°S that encircles Antarctica without any impeding continents […] covers only approximately 20% of the world's ocean's surface, [yet] it is responsible for approximately 40% of the global oceanic uptake of anthropogenic CO_2 and for approximately 75% of the excess heat generated in the Earth system."

Nicolas Gruber, Peter Landschützer, and Nicole Lovenduski, in "The Variable Ocean Carbon Sink", *Annual Review of Marine Science*, 2019

Southern Ocean Acidification and Its Impact on Microorganisms

Katherina Petrou

KATHERINA PETROU is a phytoplankton eco-physiologist. She is a senior lecturer at the School of Life Sciences, programme director at Marine Biology and leader of the Marine Microphycology Lab at the University of Technology Sydney (Australia). Specialising in Southern Ocean and Antarctic phytoplankton responses to climate change, Dr Petrou's interests include phytoplankton phenotypic plasticity and phytoplankton-bacteria interactions.

Since the Industrial Revolution, atmospheric carbon dioxide (CO_2) concentrations have increased by approximately 48% from pre-industrial levels of 280 parts per million (ppm) to over 415 ppm in 2020. Driven predominantly by the burning of fossil fuels and deforestation, this rate of increase is an order of magnitude faster than what has occurred in our geological past, with atmospheric levels currently higher than they have been in the past 800,000 years. Based on the current, rather bleak, outlook for any substantial reductions in our emissions in the coming decades, it is likely that the CO_2 in our atmosphere will rise another 70% (exceeding ~700 ppm) before the end of this century. This is a global-scale "experiment" in which we are changing our atmosphere at a pace that far exceeds many species' ability to adapt. There will, inevitably, be serious consequences for the biosphere, and particularly our oceans.

The oceans have helped lessen the impact of CO_2 emissions on our climate, having taken up a considerable fraction of the anthropogenic emissions produced since industrialisation. Without this ocean sink, CO_2 concentrations in our atmosphere would have already exceeded 470 ppm, making the present-day effects of climate change more intense. This oceanic uptake of CO_2, however, comes at a hidden cost; as oceans absorb CO_2, the chemistry of the water is altered and its pH lowered, resulting in what has become known as "ocean acidification". As the largest oceanic CO_2 sink, the Southern Ocean surface waters are expected to exceed 650 ppm by the year 2100, making the biology of the Southern Ocean especially vulnerable to the effects of acidification.

Although invisible, microscopic marine microorganisms dominate the Southern Ocean and are key players in the ocean's biogeochemical cycles. One of the most important groups of microorganisms are the phytoplankton that link the atmosphere with ocean biology. They inhabit the sunlit surface waters of the ocean, utilising the CO_2 and nutrients available in the seawater to convert energy from the Sun into organic carbon via photosynthesis. Across the globe, phytoplankton are responsible for half the planet's primary production (acting as Earth's partner lung alongside its forests) and are the base of oceanic food webs, sustaining all life in the ocean, from krill to whales. Marine bacteria are another essential group of microbes – without which the oceans would be empty of life as we know it. These are the nutrient recyclers, degrading and transforming organic nutrients for reuse and ultimately determining the fate of organic matter. Any changes to marine microbial physiology or ecology from ocean acidification will therefore have significant implications for the Southern Ocean carbon sink, pelagic food web, and nutrient cycling.

Over the years, accumulating evidence of the effects of ocean acidification have started to reveal clear responses in some important phytoplankton. For example, one major group are the coccolithophores, which build intricate plates (coccoliths) from calcium carbonate (chalk). These white chalky plates are highly reflective, and when coccolithophores bloom, the surface ocean turns milky white, forming large swirling patterns that can often be seen from space. However, production of the calcium carbonate required to build these plates is highly dependent on the pH of the seawater, and lower pH renders the seawater corrosive to shells and skeletons of many marine organisms, including coccolithophores. Laboratory studies have shown that ocean acidification may prevent the formation of these plates, creating uncertainty as to the fate of these ecologically important organisms in a high-CO_2 world.

Phytoplankton communities, however, are polyphyletic (made up of many different species), and therefore ecosystem-scale responses to change will be governed by the response of individual organisms within the community.

> "Ocean acidification is [...] predicted to have serious consequences for cold-water ecosystems. [...] The Southern Ocean will be impacted early by acidification due to its unique chemistry. [...] Current atmospheric CO_2 concentrations have resulted in a drop of about 0.1 pH units (a 30% increase in acidity), and if current trends continue, ocean pH could drop by an average of 0.5 units to about 7.8 around the year 2100 under the IS92a *'business as usual'* emissions scenario. The latter represents an ocean that is 320% more acidic than it was in pre-industrial times. Despite that change, the ocean will still be in a slightly alkaline state, the boundary between acid and alkali lying at a pH of 7."
>
> Antarctic and Southern Ocean Coalition (ASOC), "Ocean Acidification" (online)

As such, understanding how ecosystems will be affected by ocean acidification requires studies on natural mixed-species communities to account for species interactions and other influencing factors. In recent years, studies on community responses to acidification have been carried out using large-scale mesocosms – experiments in which large volumes (200–6,000 litres) of seawater with natural microbial communities are exposed to CO_2 enrichment. These studies have shown that predicted acidification affects different microbes in different ways, in many cases changing the productivity and species composition of the communities. Studies on natural Southern Ocean phytoplankton communities have even shown differing responses depending on the oceanic region. Communities from the east coast showed a shift towards smaller species, while those from the west shifted towards larger species, revealing strong regional variability and highlighting the complexity involved in making predictions from individual community studies.

In the Southern Ocean, diatoms are the dominant phytoplankton group, blooming en masse in the cold polar waters and readily inhabiting the sea ice. Like coccolithophores, diatoms build hard cell walls, but rather than using chalk, they build beautiful geometric armour out of a glass-like compound (hydrated silicon dioxide). This protective shell is heavy and therefore allows diatoms to sink readily to the ocean floor, taking carbon and silica with them. This process makes diatoms important conduits for ocean carbon burial (long-term removal of CO_2 from the atmosphere) and major players in the oceanic carbon pump. It has been theorised that non-calcifying phytoplankton, like diatoms, could benefit from rising oceanic CO_2 concentrations, leading to predictions of increased primary productivity in the Southern Ocean by the end of the century. However, in a recent study, it was shown that ocean acidification reduced the rate of silica production in several large Antarctic diatom species, showing that ocean acidification may in fact impede carbon and silica export from the surface ocean through its negative effect on diatoms.

Sensitivity to ocean acidification has also been shown in Southern Ocean bacterial communities, with acidification of the seawater resulting in an increase in the number of bacteria in East Antarctic waters, while in the Ross Sea, it reduced bacterial diversity but increased metabolic activity. In both cases, activity and production were enhanced, which could indicate increased nutrient recycling in surface waters, possibly reducing carbon export, but more work is needed to conclusively evaluate the impacts of acidification on marine bacterial communities.

Despite many questions remaining on the biological effects of lower pH on marine microorganisms, evidence has already mounted to provide little doubt that ocean acidification will severely impact the biology of the Southern Ocean. If we keep "experimenting" with our atmosphere, and human CO_2 emissions are left unchecked, the direct effect on ocean microorganisms may imperil the functioning of the Southern Ocean food webs, carbon storage, and nutrient cycling – processes that together play a major role in long-term global climate regulation.

0   1 μm

A micrograph of a diatom species belonging to the genus *Thalassiosira,* photographed with a Transmission Electron Microscope, is shown opposite, whilst the microscopic image of a complete coccosphere of *Emiliania Huxleyi* collected from surface waters in early spring at the Integrated Marine Observing System (IMOS) Southern Ocean Time Series site (SOTS: 47°S 142°E) is shown above. The two scale bars identify their dimensions whereby 1 micron is equivalent to one millionth of a metre, and the *Emiliania Huxleyi* is photographed at a 9,500× magnification.

"The universal existence of such an invisible vegetation as that of the Antarctic Ocean, is a truly wonderful fact. [...] I now class the *Diatomaceae* with plants, probably maintaining in the South Polar Ocean that balance between the animal and vegetable kingdoms, which prevails over the surface of our globe. [...] The end these plants serve in the great scheme of nature is apparent, on inspecting the stomachs of many sea-animals [...]. Owing to the indestructible nature of their shields, they tell their own tale."

Sir Joseph Dalton Hooker (Botanist, Explorer), in *Flora Antarctica* (best known as *The Botany of the Antarctic Voyage of H.M. Discovery Ships* Erebus *and* Terror *in the years 1839–1843, under the Command of Captain Sir James Clark Ross*) which offers the first account of the significance of phytoplankton, 1847

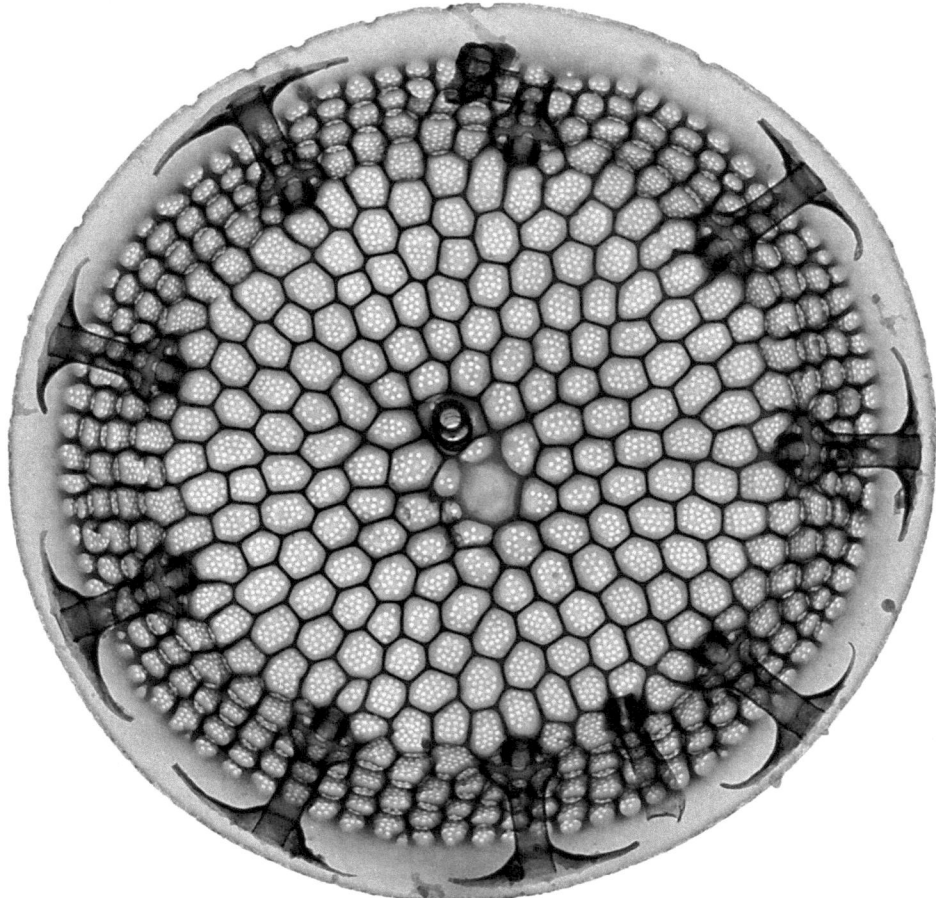

0   0.5   1 μm

"Phytoplankton are the grass of the sea. They are floating, drifting, plant-like organisms that harness the energy of the Sun, mix it with carbon dioxide that they take from the atmosphere, and turn it into carbohydrates and oxygen. Phytoplankton are critical to the marine food web, being the primary producers of food for the oceanic food web, from zooplankton to fish and shellfish to whales. Like plants and trees on land, phytoplankton give us a lot more than food. Scientists estimate that at least 50% of the oxygen in our atmosphere has been produced by phytoplankton. At the same time, they are responsible for drawing down significant portions of the carbon dioxide from the air. The tiniest of living organisms exert an outsized influence on the planet. […] The Visible Infrared Imaging Radiometer Suite (VIIRS) on the Suomi NPP satellite captured the extensive phytoplankton blooms stretching from the tip of South America across to the Antarctic Peninsula."

National Aeronautics and Space Administration (NASA) Earth Observatory, "Oxygen Factories in the Southern Ocean" (online), 2016

SHAUN O'BOYLE is a photographer and architectural designer. His work uses photography to explore architecture and built environments in high-latitude regions. He is a three-time grantee of the National Science Foundation (NSF)'s Antarctic Artists and Writers Program and the recipient of a Guggenheim fellowship in photography. O'Boyle's work has been featured in numerous galleries, museums, books, and magazines, including *Smithsonian* magazine.

THE FATE OF PTEROPODS IN AN ACIDIFIED OCEAN
Gretchen Hofmann, Juliet M. Wong, and Kevin Marquez Johnson

Ocean Acidification (OA) is the result of increasing levels of atmospheric carbon dioxide. Absorbed by the ocean and gradually transformed into carbonic acid, carbon dioxide eventually impacts the ocean ecosystem by increasing the ocean's acidity. Considered a climate-change-related stressor that can be a major threat to marine life, ocean acidification particularly affects shell-making animals because the increasing acidity dissolves away their shells. As such, future acidification threatens many ecologically important marine organisms, some of which, such as the somewhat uncharismatic pteropods (also known as sea butterflies), are important fish food in the Antarctic as they are key components in the food chains of seals and penguins: the peril of their extinction is thus amplified to the Antarctic waters and beyond.

Funded by the US National Science Foundation, the Bravo 134 (B-134) research team analysed the effects of climate change and ocean acidification on an important member of marine food webs: a small pelagic invertebrate called the pteropod. Collected using plankton nets, the pteropods were brought to McMurdo Station ↗ p. 866 and analysed at the Crary Lab in fresh seawater. While the research proved that ocean acidification did indeed negatively impact pteropods, mechanisms of rapid acclimation were detected in some specimens, which displayed the ability to rapidly change their gene expression to compensate for the stress of more acidic pH in their local seawater. While this aspect of the study was still underway, scientists were encouraged to see that these small but mighty creatures might have tools to adjust to moderate ocean acidification; however, the study confirmed that there is a limit to the pteropod's genetic flexibility and their responses failed at high levels of atmospheric carbon dioxide – i.e. those levels that reflect predicted "business as usual" levels, something we hope our global community can work to avoid.

FISH HUT 03
Shaun O'Boyle

Much of the research undertaken by the Bravo 134 scientists is conducted using fish hut FH03 – a wooden structure that can be pulled out on the sea ice by heavy equipment. The huts have a hole in the floor which enables access to McMurdo Sound via a hole drilled through the sea ice. FH03 contains an electric oceanographic winch for deploying the plankton net used to collect pteropods. In this image, Dr Umihiko Hoshijima takes a break from sampling pteropods while a visiting Weddell seal catches its breath before diving back down into the water below. Called "Messy" by the seasonal Bravo 134 team, the seal appeared at regular fifteen-minute intervals to recover from her dives. Between Weddell seal visits, the team used the winch to pull up the plankton net and collect pteropods, some of which are analysed immediately, while others are transported to the Crary Lab aquarium at McMurdo Station ↗ p. 866 to study how they might respond to future climate-change-related ocean acidification simulations.

"Climate change is the most likely factor behind the declines. [...] Some species are thriving, while others are at risk of extinction [...] Each of the five species of penguins in Antarctica [chinstrap, adélie, emperor, gentoo, and king penguins] face a different destiny as the continent warms. One of them is actually thriving – the hardy gentoo. [...] The outlook for the other species in not so good. The emperor penguin, which lives closest to the pole, is the most threatened. [...] Adélie penguins – the most plentiful species with four million pairs – appears to be increasing in the relatively cool East Antarctica but declining in the warming parts of the Antarctic Peninsula."

Leslie Hook (Environment and Clean Energy Correspondent), "Why Penguins May Help Us Predict the Impact of Climate Change", *The Financial Times*, 2020

Antarctica's Bio-indicators

Philip Trathan

PHILIP TRATHAN is head of conservation biology at the British Antarctic Survey (United Kingdom) and a UK delegate to the Commission for the Conservation of Antarctic Marine Living Resources (CCAMLR). With over 250 peer-reviewed publications and twenty trips to the Antarctic, PhD DSc Dr Trathan has been made an Officer of the Order of the British Empire, and a portion of Antarctica has been named after him: the Trathan Coast.

Without doubt, our planet is changing, including those wild, remote places about which we all care. One of the last remaining wild places is Antarctica – a continent of ice and rock, surrounded by the Antarctic Circumpolar Current, the largest wind-driven current on Earth and the only ocean current to travel the whole way around our planet. Though the continent is species poor, the ocean teems with the richness of life, including the largest species of all, the Antarctic blue whale. Since those earliest days when it was mainly sealers and whalers who ventured South, Antarctica's variability has been of interest. Annual cycles lead to enormous seasonal variation in physical properties, including in winds, ocean mixing, temperatures, and the changing extent of the sea ice. Each of these, in turn, has important consequences for biological communities, such as the seabirds and marine mammals that breed or feed in Antarctica's productive waters. During the last century, we have also learned that the Southern Ocean varies over longer timescales, including in response to natural atmospheric phenomena such as the El Niño-Southern Oscillation, which has an important influence on ecosystems throughout Antarctica.

Furthermore, and perhaps more importantly, we also now know that other factors can lead to variability in the Antarctic ecosystem structure and function and lead to directional changes, rather than simply influencing the observed levels of natural variation such as seabird productivity and breeding success or variation in diet. It is these changes that seriously challenge our understanding and which we must consider if we are to maintain and appreciate Antarctica and other such wild places.

Possibly the three most important drivers of ecological change are as follows: ongoing greenhouse gas emissions (driving recent, rapid, regional warming, changes in wind and in sea ice, glacial retreat, ocean acidification, and climate change); the recovery of populations of both fish and marine mammals following their historical over-exploitation; and other more recent human activities such as increasing tourism and modern-day fishing. The direct effects and possible interactions of these different factors mean that understanding the future consequences for ecosystems and for ecological change are challenging. Therefore, designing appropriate bio-indicators is vital if we want to understand and mitigate future change. The concept of bio-indicators has been in use for more than half a century and they commonly include information related to ecological processes, individual species, or communities that can be used to assess the status of the environment and how it might be changing over time. Bio-indicators are regularly used to assess environmental change, in relation not only to anthropogenic disturbances such as pollution or land-use change but also to natural phenomena such as air temperature, rainfall, or the date of spring migrations.

The initiation of many monitoring programmes in the Antarctic came about because of other related human activities. For example, the *Discovery Investigations* in the early twentieth century led to the first integrated marine ecological studies, directed at understanding the great whales, the krill they feed upon, and the oceanography of their habitat. Similarly, the *International Geophysical Year* in the mid-twentieth century led to the establishment of many national programmes and monitoring facilities, as part of a worldwide effort to understand the Earth and its environment. More recently in the late-twentieth century, the establishment of the *Commission for the Conservation of Antarctic Marine Living Resources* led to a monitoring programme to understand the ecosystem impacts of fishing.

Now, given climate change, there have been new international endeavours to identify coordinated sets of monitoring observations, or *ecosystem*

Far from the freedom enjoyed by the chinstrap penguins captured by Sebastiao Salgado between Zavodovski and Visokoi islands in the South Sandwich Islands (opposite page), the "gentoo penguin [shown below] wears a radio backpack that provides monitoring biologists with data on blood flow and pressure. The neck rig draws blood samples by remote control [taken as a 'bioindicator']. After a few days in the service of science, the bird [was] released in a nearby rookery, unencumbered and unharmed. The project [published by National Geographic in 1971, was conceived to help] man understand penguin physiology and adaptation to a harsh environment."[1]

Essential Ocean Variables – eEOVs. These long-term observations should enhance our capacity to understand ecosystem status. They will help identify key processes driving marine ecosystems, allow us to judge the long-term effects of people on marine resources, food webs, and biodiversity, and enable us to determine sustainable levels of activities, such as fisheries, and to assess the risk of passing tipping points in the face of change, so that we may identify the actions needed to mitigate this.

Over the past 250 years, the Southern Ocean has witnessed significant change. Human activities have driven many such changes, including the historical depletion and recent recovery of Antarctic fur seals, a species now classified by the International Union for Conservation of Nature (IUCN) as *least concern,* even though early sealers had driven populations to very low numbers. Similarly, the renaissance of some cetacean species is such that numbers are now increasing, with humpback whales now also classified by the IUCN as *least concern.* The recovery of these species demonstrates how enlightened management can reverse earlier human impacts.

Nevertheless, in the present day the status of albatross populations continues to demonstrate how humans are still having a detrimental effect upon marine ecosystems. Longline fisheries for high-value fish species set lines with baited hooks that are now one of the main threats to albatrosses, as they are attracted to the bait, become hooked on the lines, and drown. Of the twenty-two albatross species, the IUCN classifies seven as *vulnerable,* five as *endangered* and three as *critically endangered.* The remaining species are classified as *near-threatened.* By-catch mitigation remains a key management objective, as the most serious threat for wandering albatross populations remains by-catch in longline fisheries, where mitigation is absent or insufficient.

Wandering albatrosses also demonstrate how other threats can still be a problem, as adults regularly provision their offspring with items that include marine debris, highlighting the continuing presence of plastic in our oceans.

In the future, climate change is likely to lead to other, possibly more pervasive risks, especially for cold-adapted species that rely upon sea ice, such as emperor penguins. Using species as bio-indicators should assist managers in making informed decisions and ensure species do not reach the diminished population levels seen in the past.

Currently, many biological time series are relatively short when compared with physical data series. Moreover, coherent spatially and temporally matched data series that enable structured analyses are vanishingly rare. This means that separating signals of change from background noise is difficult and may remain challenging until long after ecosystems show long-term impacts.

Biological datasets also tend to reflect those species that are most amenable to study – land-based predators such as penguins, seals, and flying seabirds. Moreover, most are limited by the seasonal cycle of the Antarctic, only providing information during the summer. This ignores the vital winter period and probable carry-over effects into the spring that dictate individual and population status at the onset of breeding. Very few data sets reflect the dominant prey species, let alone the alternate prey of the monitored predators. Importantly, even for the better-studied species, monitoring sites are often conveniently located close to the hubs of logistic operation, leading to large biases in geographic distribution.

Despite their limitations, studies have already detected changes attributed to regional warming, including the loss of ice shelves and the retreat of glaciers, the reduction in the extent and duration of seasonal sea ice, and the warming of the ocean. Linking these changes to altered species distribution and abundance is more problematic, except for changes in phytoplankton, Antarctic krill, mesopelagic fish, and penguins, which are blatantly evident, particularly along the west Antarctic Peninsula, where rates of warming in the latter part of the twentieth century have been some of the highest anywhere on Earth. If krill populations decrease, for example, the implications for the wider marine system will be profound, given that krill forms the main dietary component for many species of fish, as well as iconic species of Antarctic seabirds and marine mammals.

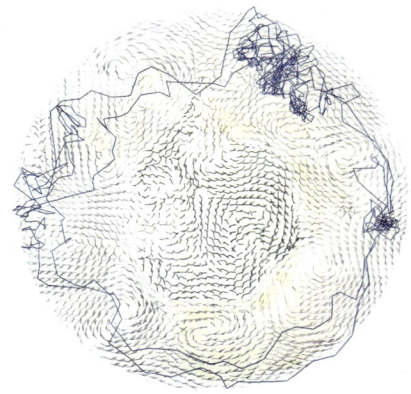

ALBATROSS
Chris Drury

The image is of one day of wind over Antarctica, with the line of the flight of a tagged albatross over eighteen months, drawn in blue crayon around the central circle. This bird lives on the wing, often just 2 metres above the waves of the southern oceans. The line of its flight circumnavigates the globe in a clockwise direction, which is also the direction of the winds and the circumpolar current. Where the cold waters of Antarctica meet the warmer currents – this is where all the krill is in the seas, and so this is where most of the life is, from fish to birds to mammals. The circumpolar current is also driving the climate of the planet.

Climate change will also bring with it other risks to the ecosystem, including increasing accessibility for fishing, science, and tourism, changes to commercial transport and shipping movements, increased pollution, and the introduction of alien species – all of which carry significant risks. Some of these risks (such as the introduction of alien species in combination with altered ecosystem properties) remain poorly understood, despite the fact that international agreements (namely the Protocol on Environmental Protection to the Antarctic Treaty) provide for the comprehensive protection of the Antarctic environment and its dependent and associated ecosystems by explicitly forbidding the introduction of any alien species.

Developing a suite of eEOVs with the statistical power to allow for the detection of change and the attribution of cause requires considerable investment in both hardware and human resources. Almost certainly, investment by national operators interested in science will be insufficient. What is needed now is a concerted effort to set up integrated international studies that recognise and document the drivers of change. It is only with objective scientific evidence, agreed internationally by world experts, that we will save our wild places. Acceptance of objective evidence inexorably leads to responsibility and action. Where the *realpolitik* of life gets in the way, those responsible have ample opportunity to ignore evidence, and as such, responsibility.

Our planet is changing rapidly, and we are unlikely to have the opportunity to spend time designing perfect monitoring systems. Evidence of change is accumulating already, so we need to decide what future we would like.

"Whales are a really important indicator of ecosystem health. By being able to gather information on the grandest scales afforded by satellite imagery [such as that provided by the WorldView-3 satellite on fin whales, humpback whales, southern right whales and grey whales], we can understand something more generally about the oceans' health and that's really important for marine conservation."

Jennifer Jackson (Whale Expert, British Antarctic Survey), in "Scientists Count Whales from Space", *BBC News*, 1 November 2018

"It is true there will be losers and winners with climate change, and this will depend on their capacity to adapt. [...] It will make a difference not only for the emperor penguin, it will make a difference for every species on earth, including us and our children."

Stéphanie Jenouvrier (Ecologist, Associate Scientist at the Woods Hole Oceanographic Institution), "Why Penguins May Help Us Predict the Impact of Climate Change", *The Financial Times*, 2020

Emperor penguins (at the side) are living indicators. Their population trends can illustrate the consequences of climate change since their life cycle is tied to sea ice. Living in such delicate balance with their rapidly changing environment, they have become our modern-day canaries. If climate change continues at its current rate, emperor penguins could virtually disappear by the year 2100 owing to the loss of Antarctic sea ice. However, a more aggressive global climate policy can halt their otherwise unavoidable march to extinction.

"The alarming part is not just that [chinstrap penguins] are declining but that we don't understand what is going on – and who knows what else is going on, what else is declining, under our noses, that we are unaware of. [...] The chinstraps are the canary in the coal mine for a whole host of changes that are happening on the Antarctic Peninsula [...]. Time might be running out to figure this out, before these changes are irreversible."

Heather Lynch (Associate Professor of Ecology and Evolution, Stony Brook University), "Why Penguins May Help Us Predict the Impact of Climate Change", *The Financial Times*, 2020

Emperor Penguins and the Impact of Climate Change

Stéphanie Jenouvrier

STÉPHANIE JENOUVRIER is an associate scientist at the Woods Hole Oceanographic Institution (United States). Her work as an ecologist focuses on predicting the effect of climate change on population dynamics, specifically on seabirds in the Southern Ocean. Dr Jenouvrier contributes to characterising species' responses to climate change, which informs the Intergovernmental Panel on Climate Change (IPCC) assessment.

Emperor penguins are the only species that breed during the Antarctic winter on top of the sea ice. Using the ice as a home base for breeding, feeding, and moulting, the penguins arrive at their colony from ocean waters in March or April after sea ice has formed. In mid-May the female lays a single egg. Throughout the winter, males incubate the eggs while females make a long foraging trip to open water to feed. When female penguins return to their newly hatched chicks with food, the males have fasted for four months and lost almost half their weight. After the egg hatches, both parents take turns feeding and protecting their chick. In September, the adults leave their young so that they can both forage to provide for their chick's growth. In December, everyone leaves the colony and returns to the ocean, but if the sea ice breaks up too early, the young will not have acquired the right plumage to survive in the Antarctic cold waters. Throughout this annual cycle, the penguins rely on a sea-ice period – nicknamed the "Goldilocks zone" of conditions – to thrive. They need openings in the ice that provide access to the water so they can feed, while also requiring a thick, stable platform of ice to raise their chicks until they fledge.

Sea-ice conditions have been shown to affect the birds' foraging behaviours, reproduction, survival, and hence population dynamics. In the 1970s, for example, the numbers of emperor penguins breeding in Terre Adélie, Antarctica, experienced a dramatic decline when several consecutive years of low sea-ice cover caused widespread deaths among male penguins. Based on the relationships between the sea ice and fluctuations in penguin life histories, climate-dependent-demographic models project that this population will decline toward extinction by 2100. However, sea-ice conditions show considerable regional variations, and climate change will also produce regionally varying trends. A global pan-Antarctic assessment found that all fifty-four known emperor penguin colonies would be in decline by 2100, and 80% of them would be quasi-extinct if greenhouse gas emissions continue on their present trajectory. Accordingly, the total population of emperor penguins will decline by 86% relative to its current size if nations fail to reduce their carbon dioxide emissions. Individuals may respond to climate change by moving permanently to other locations to track suitable climatic conditions, but for emperor penguins, dispersal behaviours have little capacity to offset climate-driven population declines. Notwithstanding, global climate policy has a larger ability to safeguard the future of emperor penguins than their intrinsic dispersal abilities. If the Paris Agreement succeeds in stabilising average global temperatures at 1.5°C above pre-industrial levels, emperor penguin numbers would decline by 31% – still drastic, but viable refuges would remain available to support some colonies.

Near-term global policy decisions over the next decade will thus have dramatic impacts on the viability of this iconic Antarctic predator and will likely shape the future of Antarctica's biota. Until the goals of the Paris Agreement on climate change are met, other conservation actions are necessary, including listing the emperor penguin as an Antarctic Specially Protected Species under the Antarctic Treaty, an endangered species under the United States Endangered Species Act, updating its status to Vulnerable on the International Union for Conservation of Nature Red List, and increasing spatial protection at breeding sites and foraging grounds by creating large-scale Marine Protected Areas.

The Impact of Climate Change on the Antarctic Krill

Andrea Piñones

ANDREA PIÑONES is an oceanographer at the Research Center Dynamics of High-Latitude Marine Ecosystems (IDEAL) at the Universidad Austral de Chile and at the Center for Oceanographic Research (COPAS Sur-Austral) at the Universidad de Concepción (Chile). Piñones specialises in biological-physical interactions in high-latitude systems, physical oceanography, and climate change.

Euphausia Superba – commonly known as the Antarctic Krill – is a small crustacean that has the distinction of making up the largest biomass on Earth (estimated at 379 million tons in 2009) after humans (which collectively account for roughly 350 million tons, calculated on the basis of 6.9 billion people averaging 50 kilograms each). Antarctic krill is also the most abundant, keystone species in the Antarctic marine food web. Irregularly located around Antarctica, its circumpolar distribution is rather patchy and consistent with the distribution of its food. Responsible for the efficient energy transfer between lower trophic levels (such as primary producers, i.e. organisms that perform photosynthesis) and higher trophic levels (such as cetaceans, pinnipeds, and birds), Antarctic krill is a species of great commercial interest and the fisheries for this crustacean are carried out in various regions of the Antarctic, mostly concentrated in the northern Antarctic Peninsula, predominantly in the regions of the Bransfield and Gerlache Straits.

The Antarctic krill begins its life cycle during the austral summer when females release krill eggs in regions far from the coast, in the open ocean above the continental shelf. Krill eggs are heavier than their surroundings and sink into the water column down to considerable depths (600–1,200 metres); as embryos descend, they develop and hatch at depth. Once the larvae hatch, they ascend to the surface and as they rise, they continue their development and growth until reaching the first larval stage that requires feeding, known as *calyptopis* 1 (C1). This early life cycle of krill is known as the descent-ascent cycle and during this period the embryos and larvae consume their own energy reserves, their development and growth closely linked to the temperature and density of the ocean. During the descent through the water column, the krill embryos encounter different bodies of water. One of them, the Circumpolar Deep Water (CDW), plays an important role in the development and hatching of the embryos. Transported around Antarctica by the Antarctic Circumpolar Current and led from the open ocean to coastal regions of the continental shelf through intrusions that are channelled by bathymetric depressions and submarine canyons, the CDW is a body of water that is characterised by having a maximum temperature below a depth of 200 metres. This relatively warmer water mass found at depth allows the krill to hatch at a shallower depth and, therefore, shortens the ascent. Thus, the krill reaches the surface in a better physiological condition to continue its development during the Critical Period 1 (CP1), i.e. during the rest of the summer, autumn, and winter. Once on the surface, krill larvae have a narrow window (of approximately ten days) in which to find food before reaching a starvation threshold, after which the larvae do not survive even if food becomes available. This period, known as Critical Period 2 (CP2), generally occurs when there is high primary productivity on the surface, i.e. during the austral summer extending to early autumn. This allows krill – which in its adult state measures between 5 and 7 centimetres, weighs up to 2 grams, and has a life expectancy of six years – to accumulate enough energy reserves to face the winter. A third critical period (CP3) occurs during the first winter, when the krill relies on sea ice biota (SIB) as a food resource and uses sea ice for shelter. Thus, during any of the three critical periods, the krill depends on the environmental conditions of temperature, chlorophyll, and sea ice for its survival.

During the last four decades, the Antarctic Peninsula has been one of the Antarctic areas most affected by changes in environmental conditions. The most notable changes have been an increase in westerly winds, an increase in air and deep ocean temperatures, and a decrease in the period of sea ice, in terms of both its time of formation and extent. Changes in the time required for the formation of sea ice, in the ocean temperature, and in the availability of krill food exert a huge influence over the survival of the larvae. In order to determine projections of how the habitat of the krill will be modified by further anticipated changes in environmental conditions caused by greenhouse gas emissions, biological and ocean model simulations have been implemented. The latter include environmental scenarios defined as "business as usual", i.e. a relentless emission of greenhouse gases into the atmosphere at the same rate as today without any mitigation policy – that is, the worst possible environmental scenario by the end of the century.

The simulations projected a scenario in which, if food availability decreases as expected by 50% from current values, krill

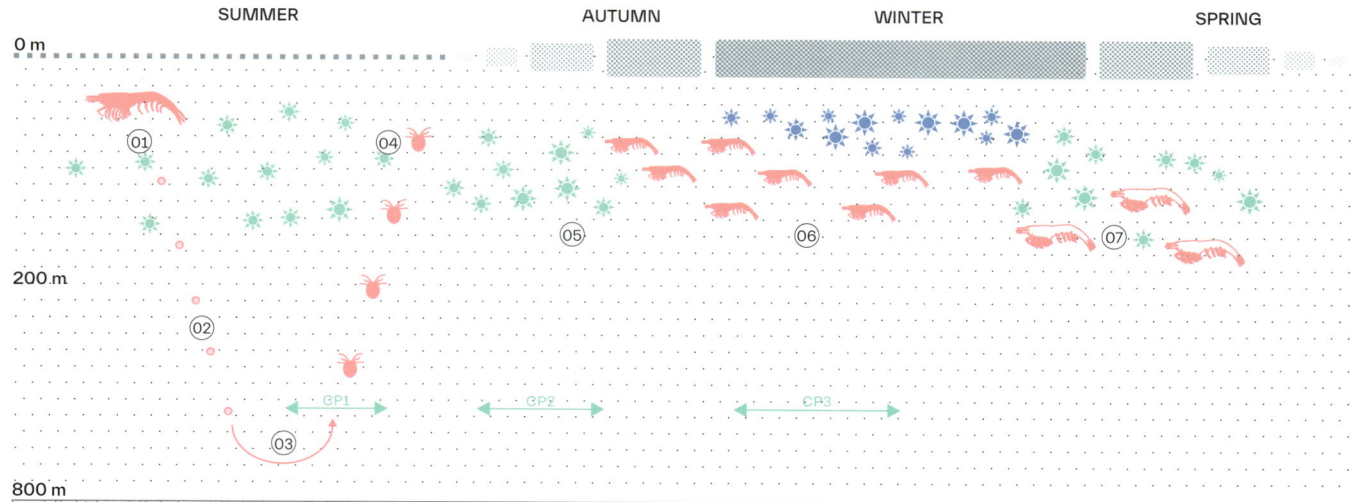

Underwater images by Laurent Ballesta and Andre Izzotti unveil the beauty of Antarctic marine life by capturing penguins ready to rise to the surface from the Antarctic abyss, and documenting a swarm of krill at night. The latter image was used by Greenpeace as the cover for *License to Krill,* an investigation published as part of their "Protect the Antarctic" campaign which unveils how, "despite the [krill fishing] industry's attempts to portray itself as one of the world's most sustainable fisheries, [... the] pattern of fishing activity [is] increasingly close to shore and in the immediate vicinity of penguin colonies and whale feeding grounds [...] in areas which have been put forward as ocean sanctuaries. [This] carries huge environmental risks."[1]

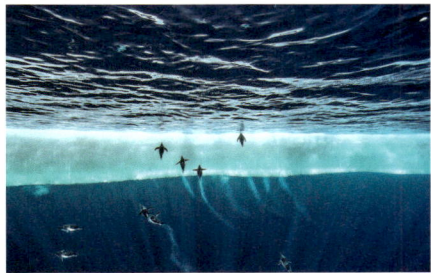

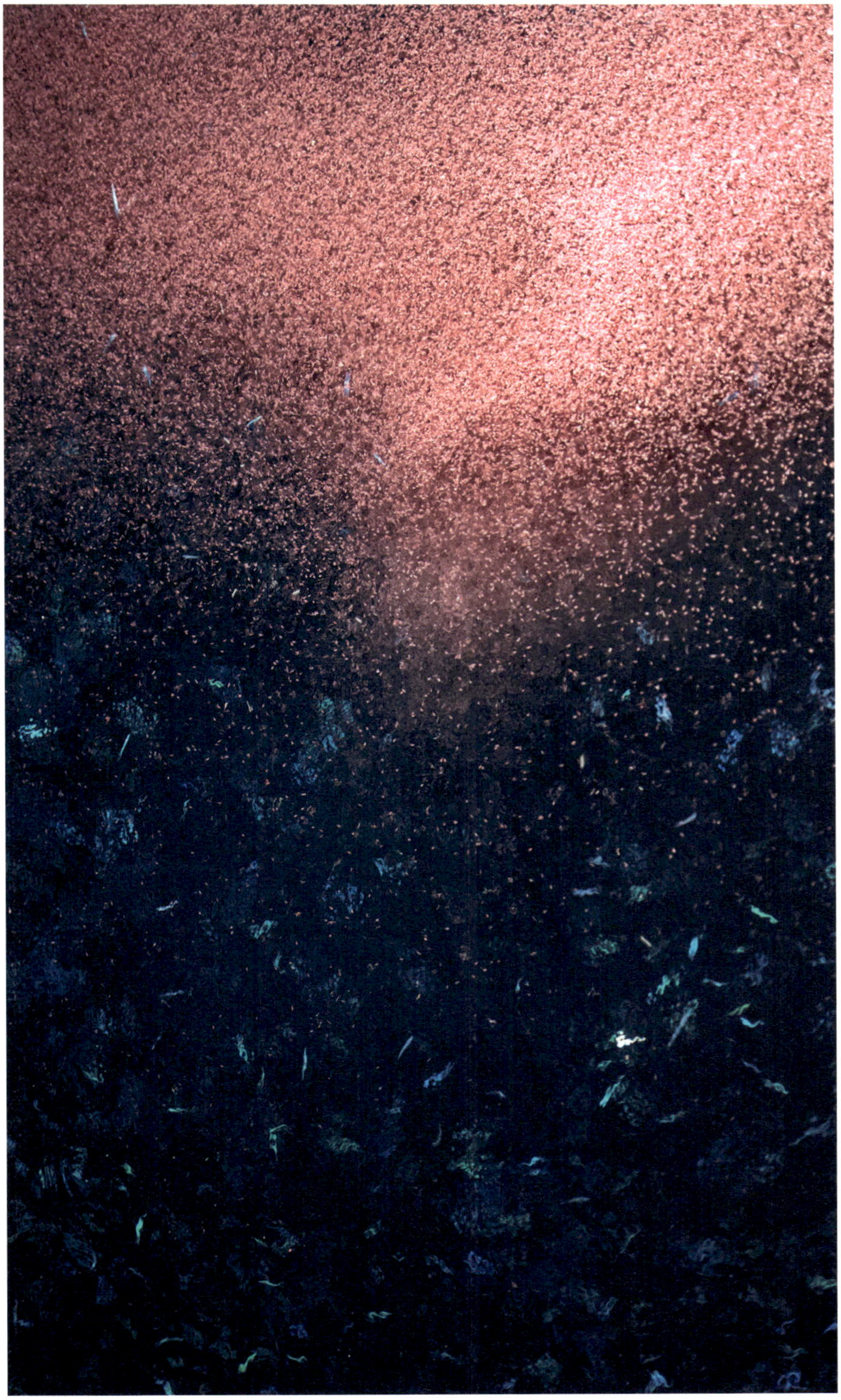

habitat could decrease by up to 80% and, by 2100, only localized regions along the western Weddell Sea, isolated areas of the Indian Antarctic sector, and the Amundsen/Bellingshausen Sea would support successful spawning habitats for krill. One of the most striking results of the model is the disappearance of the habitat along the Antarctic Peninsula. This implies that both populations throughout the peninsula and the populations of the Scotia Sea and South Georgia (which depend on the flow of larvae to those regions) will be greatly impacted if the peninsula's habitat disappears.

Furthermore, these results suggest that in the face of climate change scenarios, not only could krill populations be affected but the phenomena would extend to all the organisms that depend on it, such as whales, penguins, seals, and other seabirds.

01 Females
02 Embryos
03 Hatching
04 Calyptopis 1
05 Chlorophyll
06 Larvae
07 Juveniles
CP1 Critical period 1
CP2 Critical period 2
CP3 Critical period 1
Sea
Seafloor

Antarctic Sea Ice and Its Impact on the Global Climate System

Dirk Notz

DIRK NOTZ is a professor of cryosphere research at the University of Hamburg, group leader in the Ocean in the Earth System Department at the Max Planck Institute for Meteorology (Germany), and a lead author of the sixth report of the Intergovernmental Panel on Climate Change (IPCC). Dr Notz investigates the past and future evolution of Antarctic and Arctic sea ice, both in the field in polar regions and in large-scale model simulations.

The 24th of October 1915 was a Sunday. Unbeknownst to the world, a ship was about to be stuck fast and forever in the greatest sea-ice expanse on our planet. At a latitude of around 69°S, in the Weddell Sea off the coast of Antarctica, the sea ice closed around Sir Ernest Shackleton's ship, *Endurance,* with a force that caused the hull to splinter and the crew to abandon ship. She sank within a matter of days. Such is the power of polar sea ice.

However, the real power of Antarctic sea ice to affect our lives goes largely unseen by humans. As sea ice expands around the Antarctic continent in the harsh cold that prevails through the winter – covering at its maximum an area roughly the size of Europe and Australia combined – it expels enormous amounts of salt into the underlying ocean. This is because only the actual water contained in the sea freezes, while its salt content becomes more and more concentrated into a highly salty, liquid brine that creates a porous network within the sea ice. As time goes by, much of this salty brine trickles into the underlying ocean, making its water saltier than it would be without the formation of sea ice. This salt makes the water heavier, and it starts sinking towards the sea floor, spreading all around the world's ocean basins. Hence, the vast expanse of the world's ocean is filled with cold water from the Southern Ocean, pumped downwards through the formation of sea ice. In this way, the remote formation of sea ice in the Southern Ocean affects ocean circulation and ocean water masses all around our planet.

The melting of sea ice also affects the stratification of the ocean. Unlike the Arctic, where the ice is largely enclosed by the surrounding land masses and stays near the North Pole throughout the summer, the ice in the Southern Ocean is free to drift around. Much of this drift occurs in a slow northward motion, such that the ice encounters warmer and warmer water masses underneath. These warm water masses then slowly melt the ice from below, in contrast to Arctic sea ice where most melting occurs at the ice surface from warm air and sunshine. The melting ice, devoid of much of its earlier salt content, creates a layer of relatively fresh water at the ocean surface, which is then mixed through the upper ocean by wind and currents.

This growing and waning of Antarctic sea ice throughout the annual cycle is much more pronounced than the equivalent seasonal cycle in the Arctic. In the Arctic, a substantial amount of sea ice still survives the summer, even though the ice-covered area is decreasing rapidly with the ongoing effects of climate change and rising global temperature, whilst in the Antarctic, only little ice survives the summer owing to the substantial sea-ice drift towards the equator. The long-term evolution of Antarctic sea ice in our warming climate has so far also been quite different compared to its equivalent around the North Pole; the sea-ice area in the Southern Ocean has barely changed in the past few decades and has even expanded slightly over the period of reliable satellite observations since 1979. Until 2015, this expansion was quite substantial, only to be brought to a sudden halt with a decrease of the ice cover to the levels of the early record in the past few years.

The different responses of Arctic and Antarctic sea ice can be understood by the opposite geographical settings and the resulting major drivers of the ice coverage: around the North Pole, the ice is largely trapped by land. Its area develops, therefore, primarily in response to the prevailing air temperature, so that the ice coverage decreases rapidly in a warming climate. This decrease of the ice cover is so pronounced at this point that the complete loss of the Arctic sea ice in summer over the next few decades can barely be avoided. In contrast, the ice in the Southern Ocean is free to move around, and the air masses over the ice are comparably cold. Offshore winds from the Antarctic continent drive

"We are now again firmly stationed in a moving sea of ice, with no land and nothing stable on our horizon to warn us of our movements. Even the bergs, immense, mountainous masses, […] sail as we do, and with the same apparent ease. […] The entire horizon drifts with us. We are part of an endless frozen sea."

Frederick Albert Cook (Explorer, Physician, Ethnographer), *Through the First Antarctic Night,* 1900

Whilst it had taken the *Discovery* only four days to get through the pack ice on Scott's first Antarctic expedition, the *Terra Nova* remained "mired in the most of ice" for more than twenty days during his second expedition in 1911, when members of the western geological party led by geologist Griffith Taylor were captured by Herbert Ponting man-hauling on the ice floes (opposite page, top). However, the sea ice was never so treacherous as during the *Imperial Trans-Antarctic Expedition* of 1914–1916 when it crushed and sank the three-mast barquentine *Endurance,* which had carried Sir Ernest Shackleton and a crew of twenty-seven men from Sanderfjord in Norway to the Weddell Sea in the Antarctic.[1]

"We saw the blue sea lying in an almost flat calm, the white ice floes scattered on its surface, and the two ships, *Kainan-maru* and *Fram,* floating in lonely isolation alongside the expanse of sea ice which covered the entire bay. This was a *sumie* world painted in Indian ink on white paper. […] Turning to look in the other direction, we saw a boundless plain of white stretching undisturbed into infinity […]. The sun was reflected off the white snow with dazzling brightness, and we were all struck to the very heart by a feeling of awe."

Nobu Shirase (Explorer, Army Officer), *Nankyoku no kyoku,* 1913

the ice away from the coast. The resulting open water around the coast freezes rapidly, and vast expanses of new ice form throughout the polar winter. The overall area of the ice cover is hence primarily determined by the strength and direction of the prevailing winds rather than by the temperature patterns of the atmosphere. With the increase in offshore winds around Antarctica in recent decades, the ice has been distributed over a larger ocean area. This explains the slow increase of the sea-ice cover observed in the Southern Ocean.

Climate-model simulations often fail to capture these processes, because they have difficulties in correctly simulating the wind patterns around the Antarctic continent. In addition, the models lack the spatial resolution to represent the small ocean eddies that play a major role in the redistribution of water masses and sea ice in the Southern Ocean. Hence, we currently lack the tools to reliably quantify the future evolution of Antarctic sea ice in response to ongoing climate warming. Qualitatively, however, the future of sea ice in the Southern Ocean can be projected: in the long term, warming of the ocean water will cause additional melting of the ice cover. With this retreat, and with a decrease of ice formation, salt release into the Southern Ocean will diminish, with possibly far-reaching impacts on ocean circulation.

A reduction of the sea-ice area would also lessen the atmospheric cooling mechanism of the ice itself, which de facto acts as an extensive reflector of sunlight. The potential future loss of this cooling surface would contribute to a rapid rise in temperature over large coastal areas of Antarctica and a substantial sea-level rise from the melting of the ice sheets, revealing the invisible force that the retreating sea ice cover has to impact human and natural life all around the planet. How soon this will happen – and how far-reaching these consequences will be – is up to humankind, since the invisible exchange of power between humans and sea ice goes in both directions, and we are ultimately the authors of the changes that are amplified by the sea-ice cover.

Any form of frozen seawater is called "sea ice". However, the appearance of sea ice changes drastically throughout its growth and decay. Winds, waves, air temperature, and snowfall affect the ice cover throughout all stages of its development, giving rise to a myriad of different sea-ice shapes. The most common developments of sea ice under windy conditions in open water are frazil ice, grease ice, and pancakes. Under calm conditions (such as protected bays) nilas forms. After the development of the initial sea-ice cover, the ocean continues to lose its heat through the ice to the atmosphere, allowing for new ice to form at the bottom of the existing ice. The young ice cover gets thicker and thicker, effectively dampening wave activity. Still, the impact of storms, currents, and waves shifts the ice around, breaking the ice cover into individual ice floes, which are subsequently pushed into one another, sometimes forming high sea-ice ridges and allowing open water to form between the ice floes, where new ice starts to grow. These processes form the fascinating, varied landscape of sea ice, whose detailed functioning we have only just begun to understand.

FRAZIL
h = variable
Ø = 3–4 mm (crystals)
Small needle-like ice crystals suspended in water that merge under calm conditions to form thin sheets of ice on the surface

GREASE
h = 1–10 mm (above water)
h= up to 1,000 mm (below)
Coagulated crystals forming a soupy layer which makes the water surface resemble an oil slick. It behaves in a viscous fluid-like manner.

SLUSH
h = variable
Viscous floating mass of snow and grease ice floating in water after a heavy snowfall

SHUGA
Ø = 10–50 mm
Accumulation of spongy white ice lumps, formed from grease ice or slush and sometimes from anchor ice rising to the surface

NILAS
h = up to 100 mm
Thin elastic crust of ice, easily bending, thrusting in a pattern of interlocking "fingers"

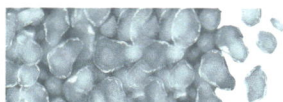

PANCAKE
h = up to 100 mm
Ø = 300–3,000 mm
Mainly circular pieces of ice with raised rims due to the pieces striking against one another. Its appearance may rapidly cover wide areas of water.

FLOEBIT
h = up to 2,000 mm
l = up to 10,000 mm
Relatively small piece of sea ice, composed of one or more hummocks or part of ridges frozen together and separated from any surroundings

FLOEBERG
h = up to 5,000 mm
l = up to 15,000 mm
Massive piece of sea ice composed of a hummock, or a group of hummocks frozen together, and separated from any ice surroundings

ICEFLOE
h = variable
l = variable
A cohesive sheet of ice floating in the water. Floes are subdivided according to horizontal extent.

Small ice cake	up to 2 m
Ice cake	2–20 m
Floe small	20–100 m
Floe medium	100–500 m
Floe big	500–2,000 m
Floe vast	2,000–10,000 m
Floe giant	over 10,000 m

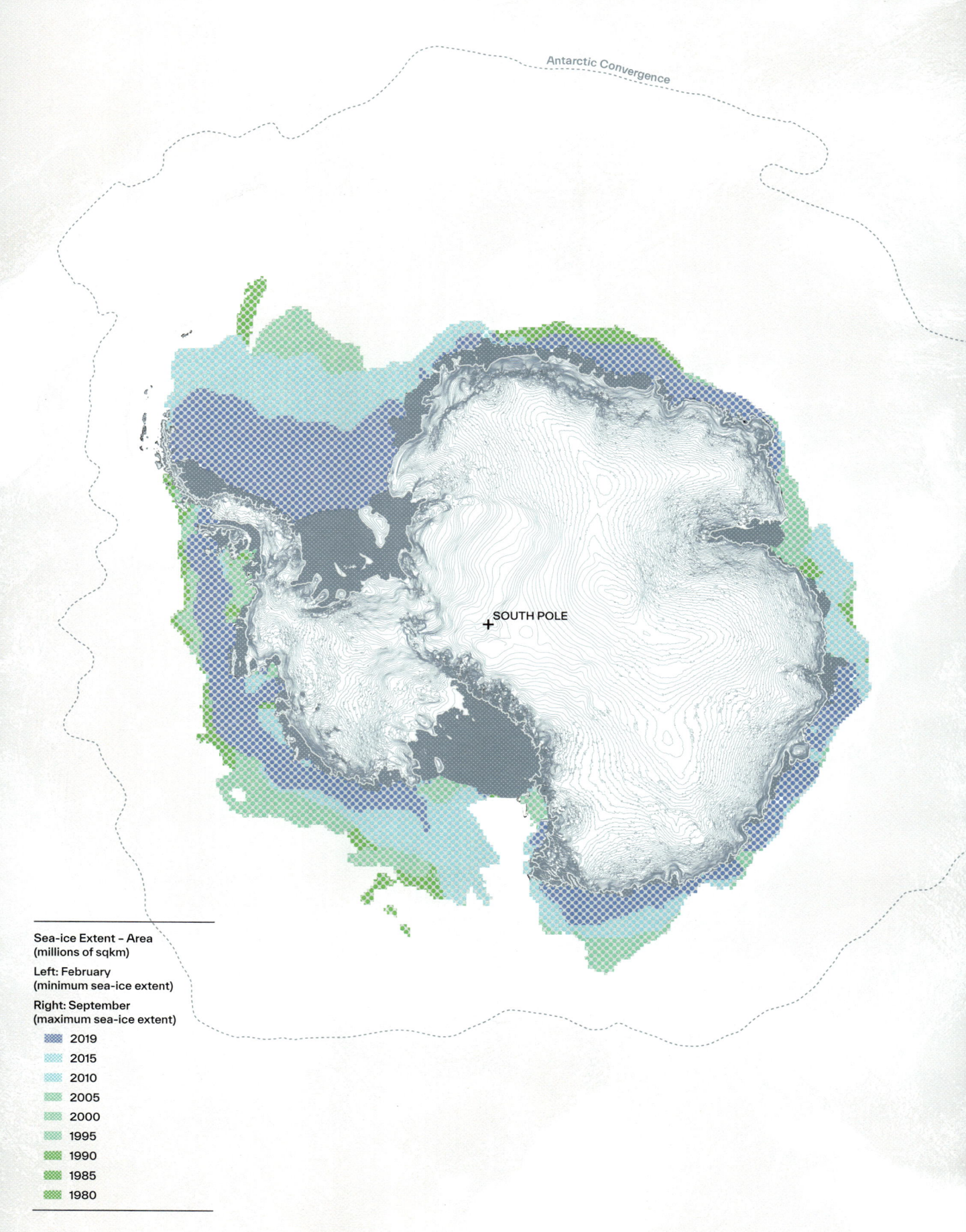

Antarctic Convergence

+ SOUTH POLE

Sea-ice Extent – Area
(millions of sqkm)

Left: February
(minimum sea-ice extent)

Right: September
(maximum sea-ice extent)

2019
2015
2010
2005
2000
1995
1990
1985
1980

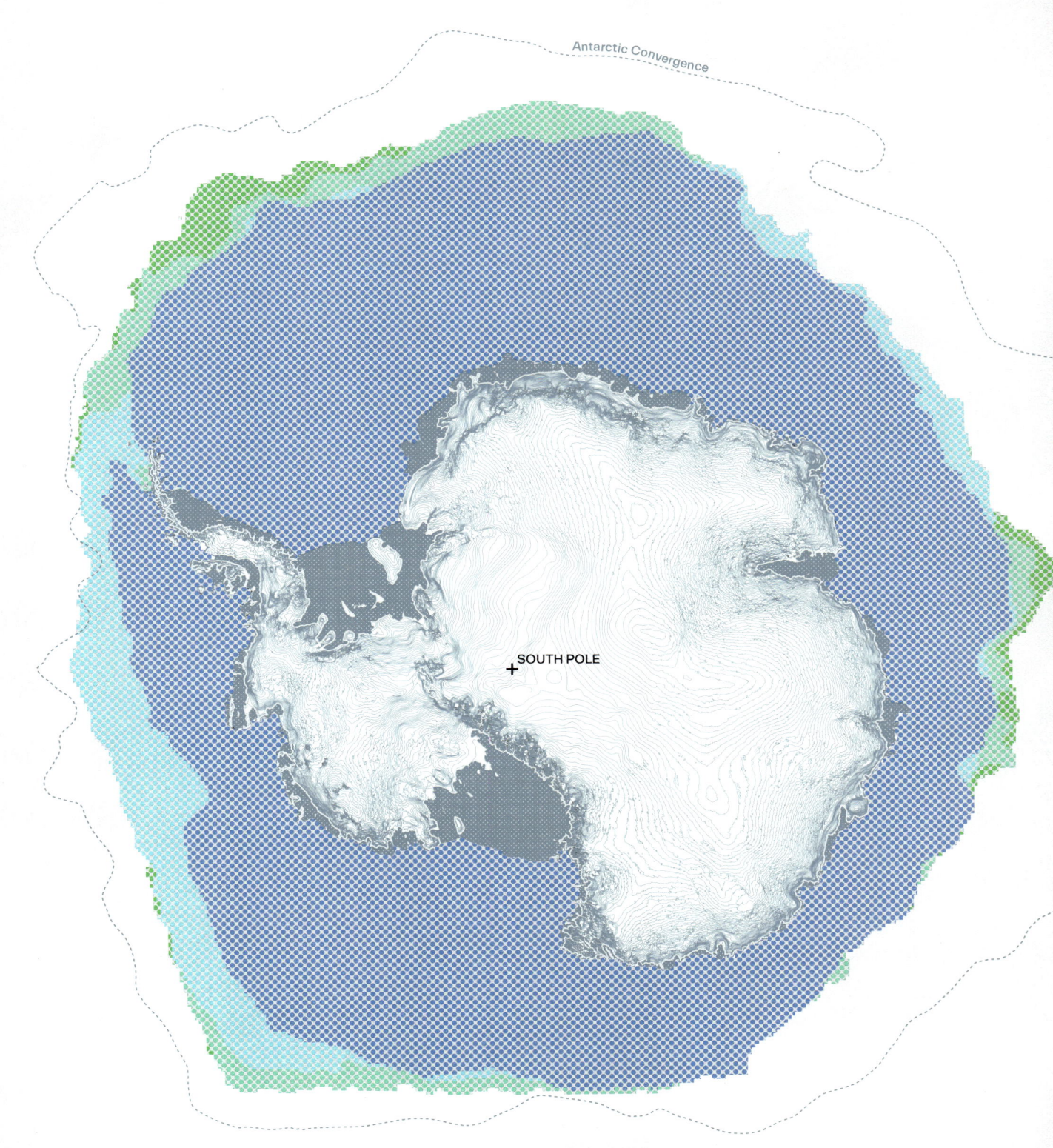

Antarctic Convergence

SOUTH POLE

The Antarctic is a living, pulsating continent.
Its surface almost doubles in winter as the
surrounding ocean freezes forming an ice pack.
Antarctic sea ice typically reaches its annual
maximum extent in mid- to late September and
its annual minimum in late February or early
March, varying in surface between approximately 2 million square kilometres in March
and 15 million square kilometres in October.[2]

0   500   1,000 km

07
Twenty-Six Quadrillion Tons of Ice. The Urgencies of Contemporaneity vs. the Truths of "Deep Time"

"We had discovered an accursed country. We had found the Weather Factory of the blizzard."

Sir Douglas Mawson (Geologist, Explorer, Academic),
The Home of the Blizzard, 1915

As suggested in an early *Argentine Antarctic Bulletin,* Antarctica and the Southern Ocean should be referred to as the "Climatic Factory" of the world, due to their undisputed role in the ecosystem of our planet at large.

Antarctica. The World's Coldest Desert

Ralf Brauner

RALF BRAUNER is a professor of marine meteorology at the Jade University of Applied Sciences (Germany) in the Department of Maritime and Logistic Studies. Prof. Brauner has participated in several expeditions to Antarctica over the past twenty years, forecasting weather on the continent for several months at a time at research stations and on scientific vessels.

Meteorology in the Southern Hemisphere is very different from its northern counterpart. The winds blowing from the mid-latitudes towards the Antarctic edge are stronger on average than those of the northern oceans – hence they are known as the Roaring Forties, Furious Fifties, and the Screaming Sixties. One reason for this is that the Antarctic continent is, on average, significantly colder than the Arctic region, by about 15°C (the world's coldest temperatures have been measured at the Russian overwintering Vostok Station↗ p. 862, with a record of −89.2°C), and its plateau is extremely dry. The Antarctic is thus known as the world's largest and driest desert.

The Earth's temperature is about 15°C on average near the surface. This is due to factors such as the atmosphere, the composition of the atmospheric gases, and their absorption of short- and long-wave radiation. Without the atmosphere and its special composition of regulating gases, the daily temperature would vary between −140°C at night and +80°C during the day. Other planets, such as Mars have different atmospheres and some, such as Mercury, have none at all. As a consequence of the Earth's shape and orientation, the incoming radiant energy from the Sun at the poles is different from its equivalent at mid-latitudes or the equator. This means that high latitudes tend to lose energy to space rather than receiving it from the Sun, while low latitudes and the equator gain energy.

Were there no compensation by the atmosphere and ocean, the North and South Pole would each get colder and the region around the equator would become warmer. Warm air and warm ocean currents transport heat energy away from the equator towards the poles while cold air moves in the other direction towards the equator. The wind system is the driver for ocean circulation near the surface, which is by no means the immediate surface, as we imagine it, since the impact of internal friction and the movement of the water is transferred to a depth of 100–200 metres. As the frictional drag in water is higher than that of air, the movement of water is much slower and the typical speeds of the currents near the surface are limited to 1 knot, which, on average, corresponds to a tenth of the velocities in the atmosphere. However, despite the reduced speed, the four-fold heat capacities of water versus air means that the ocean is able to transport more energy than the atmosphere by a ratio of, on average, eighty (ocean) to twenty (atmosphere).

While big gyres like the Brazil Current, Agulhas Current, and East Australian Current transport warm water masses southward, and the cold water masses of the Humboldt Current, Benguela Current, and West Australian Current head north, the Antarctic Circumpolar Current plays the largest role as it flows eastward around Antarctica and keeps the continent cool and frozen.

Meanwhile, in the atmosphere, three vertical atmospheric circulation cells on each hemisphere transport the warm and cold air masses from the poles to the equator and contrariwise. This takes place in the troposphere, which is the lowest layer in the atmosphere with a height ranging from 6 kilometres at the poles to 15 kilometres around the equator. The tropical regions around the equator receive the most heat, and warm air rises up to the top of the troposphere.

At a height of approximately 15 kilometres from the surface, the flow is directed south in the southern hemisphere, the air cools down and descends around the 30°S latitude, forming the Subtropical Highs – with clear skies and weak winds. (Only a portion of the descending air moves north near the surface; known as the trade winds, the air of both hemispheres eventually converges near the equator in the Intertropical Convergence Zone, also known as the "meteorological equator" or the "Doldrums"). The remaining larger portion of warm

The 1910–1913 British Antarctic Expedition had two main objectives: to continue scientific research and to secure the South Pole for the British Empire. As one of the many tools brought by the expedition party to the Antarctic, the sunshine recorder (top) was instrumental in registering the amount of sunshine at Cape Evans, data which was important when trying to understand the weather, climate, and temperature. Griffith Taylor was put in charge of such meteorological equipment, which was "a slow and painful job at −40°C",[1] while the head meteorologist, Dr Simpson, took meteorological observations on Vane Hill, as captured by Edward Atkinson in March 1911 (middle). Years later, in the 1930s, data on upper air temperatures, wind velocity and atmospheric pressure were taken with large kites (bottom), as the one held by William (Cyclone) Haines, official meteorologist in three of Rear Admiral Richard Byrd's Expeditions.

Planet Earth's climate and weather is driven, essentially, by the imbalance between the "incoming" electromagnetic radiation (i.e. solar radiation) of the Sun and the "outgoing" long-wave radiation emitted by all objects on Earth, or, in other words, between the amount of energy absorbed at the tropics and the energy lost at the poles. Solar radiation and the laws of thermodynamics are thus key to the planet's climate. Travelling through the vacuum of space, solar radiation penetrates the atmosphere (which is held around the planet due to gravitational force), reaching the Earth's surface. While part of it is reflected back into space (by clouds or extensive ice surfaces), the rest is captured by the ground. The long-wave radiation emitted by the Earth and its atmosphere follows an opposite trajectory, exiting the atmosphere and dissipating in space. As electromagnetic radiation from the Sun enters the various zones of the Earth's atmosphere, some ultraviolet radiation (UV) is captured in the stratosphere (where the temperature tends to be constant) by the gas ozone. The remainder of the solar radiation reaches the lower level of the atmosphere (troposphere) and is absorbed by the

ground. As the ground heats up, the air closer to it is warmed and only cools as it moves away from the surface, increasing in altitude. Since the Sun's rays do not reach the Antarctic in winter, there is a net radiation deficit in this region and a phenomenon called "inversion" occurs, as air in contact with the cold surface becomes incrementally colder – a phenomenon which, coupled with the high altitudes of the Antarctic topography, explains why Antarctica is even colder than the Arctic. Owing to the Earth's orbit around the Sun and the inclination of the orbital plane axis, greater amounts of solar radiation penetrate at the tropics than in the polar regions, where the solar rays hit the surface at a very oblique angle during the summer and are entirely absent for six months of the year. In the months in which solar radiation does reach the poles, the albedo (i.e. the refractive qualities of the ice surface) reflects approximately 80% of the energy, enhancing the temperature difference between the tropics and the poles. This temperature difference thus activates ocean currents and winds that seek to re-establish a balance between the radiation budgets.

Large-scale atmospheric circulation "cells" thus shift warmer air towards the poles in a relatively constant way that is determined by the Earth's size, rotation, and atmospheric depth. As the (denser) cold air sinks, it accelerates down the sloping terrain of the topography and is deflected left by the rotation of the planet, forming the so-called katabatic winds (which blow from 90°S to 65°S in the lowest part of the atmosphere), which in turn can generate blizzards, i.e. dense storms of drifting snow. Because of the low temperatures in the atmospheric column in Antarctica during winter, a strong bank of westerly winds forms what is called a polar vortex that drives temperatures down even further. As hot air rises and (denser) cold air sinks, there is an area of sinking motion at the pole that reduces the possibility of precipitation (which is correspondingly high at the equator) and dissipates the clouds. For this reason, Antarctica is a desert, with very low levels of precipitation. Though this might seem counter-intuitive for a continent that has a kilometres-thick ice sheet above the bedrock, the precipitation averages only 1 centimetre per year.

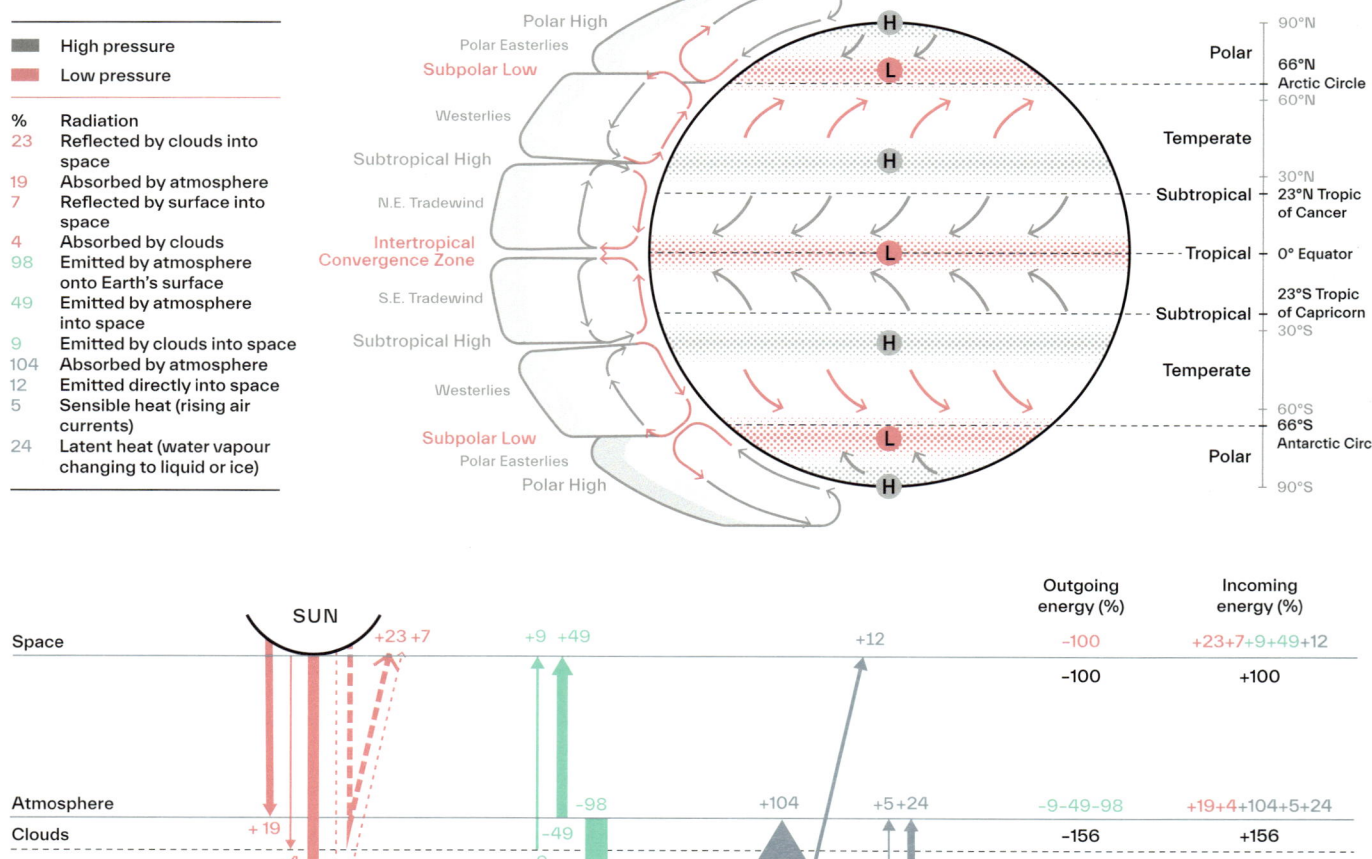

	High pressure
	Low pressure

%	Radiation
23	Reflected by clouds into space
19	Absorbed by atmosphere
7	Reflected by surface into space
4	Absorbed by clouds
98	Emitted by atmosphere onto Earth's surface
49	Emitted by atmosphere into space
9	Emitted by clouds into space
104	Absorbed by atmosphere
12	Emitted directly into space
5	Sensible heat (rising air currents)
24	Latent heat (water vapour changing to liquid or ice)

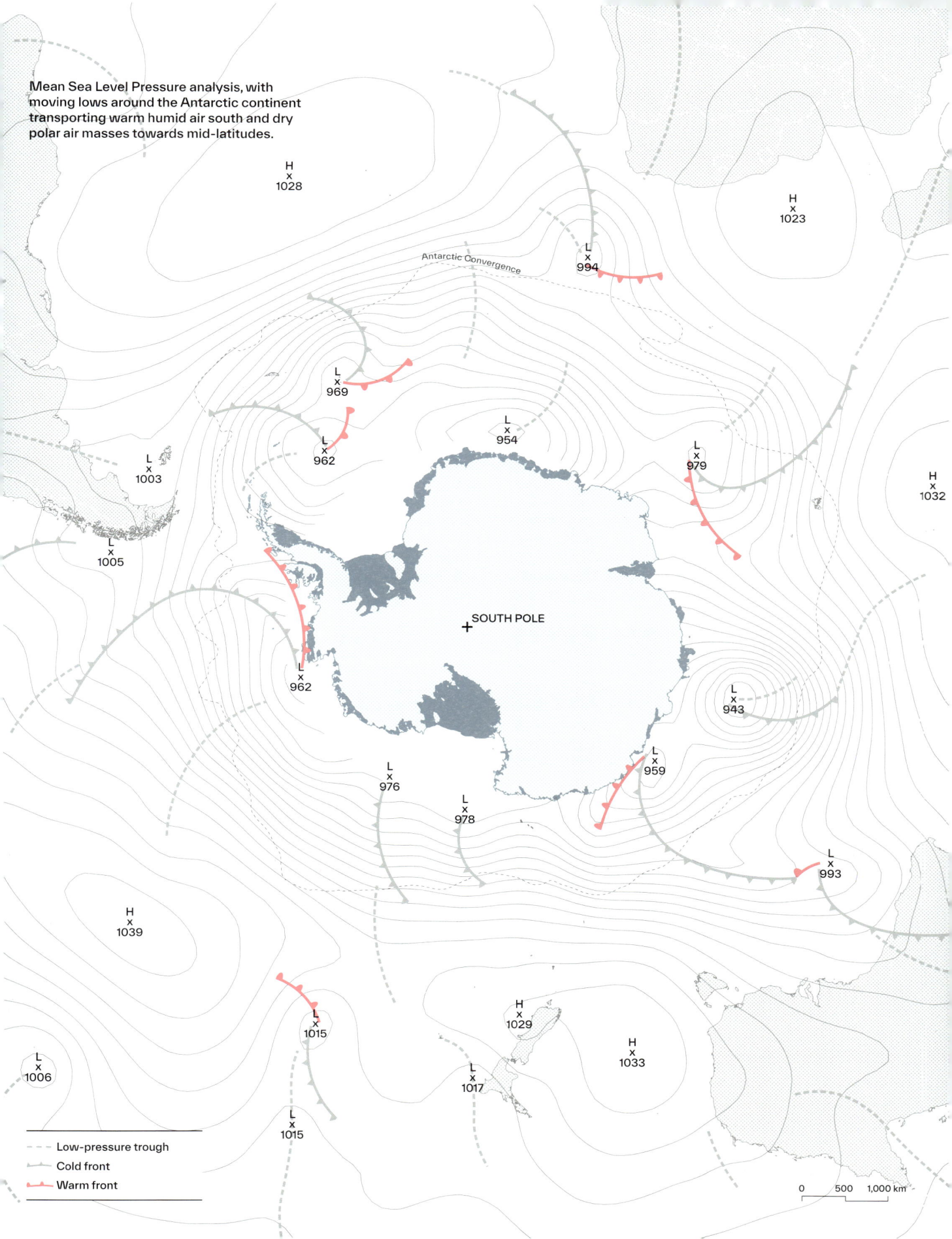

Mean Sea Level Pressure analysis, with moving lows around the Antarctic continent transporting warm humid air south and dry polar air masses towards mid-latitudes.

H
x
1028

H
x
1023

Antarctic Convergence

L
x
994

L
x
969

L
x
962

L
x
954

L
x
979

L
x
1003

H
x
1032

L
x
1005

+ SOUTH POLE

L
x
962

L
x
943

L
x
959

L
x
976

L
x
978

L
x
993

H
x
1039

L
x
1015

H
x
1029

H
x
1033

L
x
1006

L
x
1017

L
x
1015

- - - Low-pressure trough

▲▲▲ Cold front

▲▲▲ Warm front

0   500   1,000 km

air moves towards the south, saturated by moisture, and transforms into a westerly wind. Such subtropical air masses eventually reach the Polar Front around the mid-latitudes (between 40°S and 70°S) and meet the cold air coming from the South Pole which is deflected by the Coriolis Force into a south-easterly direction.

At the Polar Front, the low-pressure systems move mainly from west to east around the Antarctic continent. Acting as "mixing machines" for the warm and humid subtropical air and the cold and dry polar air masses, the lows inform a strong and permanent high-pressure system over the cold Antarctic continent and over the South Atlantic, South Pacific Ocean, and the Southern Indian Ocean.

Strong westerly winds up to hurricane force (> 64 knots/118 kilometres per hour) and gusts up to 140 kt/259 kmph produced by low-pressure systems can be found in the regions from latitude 40°S to 60°S, mainly influencing the Antarctic Peninsula and Sub-Antarctic Islands. The low-pressure systems, generally born over the Southern Ocean, move on a south-eastward course towards the Antarctic coast. Such wind speeds are calm over the inner Antarctic continent, while on the edge of the continent strong to stormy easterly winds prevail, accompanied by precipitation and blizzard activity.

Overall, the Antarctic air is extremely dry. The high plateau of Antarctica receives little precipitation and is considered the world's largest and driest desert, while some coastal areas (especially in the Antarctic Peninsula) experience rain precipitation at any time of the year caused by low-pressure frontal systems. Above approximately 15 kt/28 kmph, snow drifting can be observed with loose snow on the surface – it is only above wind speeds of 30 kt/55 kmph that gusting snow reduces visibility to a few hundred metres. Visibility may be reduced to a few metres by gale-force winds and stronger blizzards (often associated with deep low-pressure systems). The visibility is zero during so-called white-outs, one of two distinct meteorological Antarctic phenomena alongside katabatic winds.

The katabatic winds form on the elevated plateau of inland Antarctica owing to the cold temperatures and dense air and accelerate undisturbed from the high-elevation areas of the ice cap over more than 1,000 kilometres down to the coast. Moving, on average, at 20 knots, in some glacial valleys their speed reaches storm and hurricane force, while if the slope is gentle, the wind speeds may be more or less calm.

White-outs are a typical Antarctic hazard like katabatic winds, blizzards, and strong wind effects. This optical phenomenon occurs in uniformly overcast conditions over a snow-covered surface. Associated with diffuse shadow-less illumination that causes a lack of surface contrast and horizon, in Antarctica white-outs are exacerbated, as the perception of snow-covered orographic features depends solely on the shadows that they produce. The disorientation suffered by any person venturing out in white-out conditions – where they may not know if they are going up- or downhill and will not be able to keep to a straight path unassisted – is absolute.

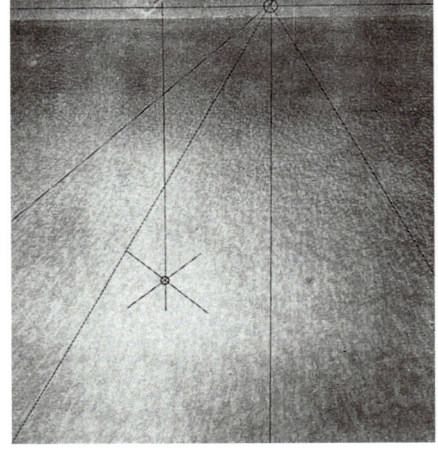

In their "The flight to Marie Byrd Land with a description of the map" issue of the *Geographical Review* in 1933, Rear Admiral Richard Byrd and Commander Harold E. Saunders, expand on the aerial survey and related cartographic challenges of mapping the "hitherto unpenetrated area to the northeast of Alexandra Mountains in King Edward VII Land". The image above is but one of the many photographs captured by Captain Ashley McKinley with overlaid lines tracing the parallel trajectory of drifts of snow used to define the "true horizon" within an otherwise featureless plane. The parallel lines, with "a featureless reflection of the sun" were in fact sasturgy similar to those depicted at the side by Sebastian Copeland.

"In some areas, as on the South Polar Plateau, there were found long drifts of snow in the lee of the ice ridges, or zastrugi, left by the last previous blizzard. In such open areas it was reasonable to assume that within the limits of the field of view of the camera in a series of successive photographs the blizzard wind had blown from a fixed direction and had left the drifts of snow all parallel to one another. The lines of these drifts, if prolonged, should then all meet at one point on the true horizon."

Richard Evelyn Byrd (Admiral, Explorer, Aviator) and Harold Eugene Saunders (Naval Engineer), "The Flight to Marie Byrd Land: With a Description of the Map", *Geographical Review*, 1933

"In Antarctica, relentless winds have created a unique type of snowdrift pattern known as megadunes. Formed by centuries of nearly continuous winds, megadunes are 1 to 8 meters high, and 2 to 6 kilometers from crest to crest. [...] Perpendicular to the megadunes is a corduroy-like pattern, [...] created by linear sets of sastrugi [shown below], ubiquitous snow sculptures like frozen waves on the ocean. Among the megadunes, sastrugi assume huge, meter-high forms."

National Aeronautics and Space Administration (NASA) Earth Observatory, "Antarctic Megadunes" (online), 2005

Antarctica as Environmental Archive

Carlo Barbante and Jacopo Gabrieli

CARLO BARBANTE is the director of the Institute of Polar Sciences of the National Research Council (Italy). JACOPO GABRIELI is a researcher at the Institute of Polar Sciences of the National Research Council (Italy).

Just as archaeologists dig deep into the ground in search of ancient tools and artefacts, palaeoclimatologists study the climate of the past, taking a similar scientific approach, which entails searching for clues to the history of our planet's climate by studying the stratigraphy of coral reefs, ocean and lake sediments, and glaciers. The large polar ice caps represent the best-preserved, naturally existing environmental archives. They were formed by the slow and constant accumulation of snowfall over millions of years and form a continuous succession of layers that can reach up to several kilometres in thickness. Each snowflake embodies a moment in history and reveals what our planet was like at the time of every single snowfall. The temperature is directly recorded in the water molecule through the different isotopic ratios of oxygen and hydrogen that compose it. Traces of dust, ash, metals, or pollen betray changes in the circulation of air masses or the occurrence of global events such as sudden volcanic eruptions. Even the air bubbles trapped between the ice crystals are considered tiny fossil air pockets, nothing less than a remnant of what the atmosphere was like when the ice formed. By analysing these bubbles, scientists are able to accurately measure atmospheric greenhouse gas concentrations, reconstructing their trends during the various glacial epochs and relating them to other climatic parameters.

The history of polar ice cores began in 1935, when a 15-metre-long section was extracted from the snows of Greenland. It was only in 1966 that American scientists concluded a six-year project by drilling through the entire Arctic ice sheet at the Camp Century site, reaching the bedrock at a depth of 1,388 metres. In Antarctica, ice coring started in 1956, when Russian scientists began drilling operations near the Vostok Antarctic Station. The scientific campaign was a real adventure that lasted more than thirty years and in the end, in 1998, after incredible technical struggles, they extracted what remains of the deepest ice core ever taken from the Antarctic ice cap reaching an astounding depth of 3,623 metres. The analysis of these samples represented a revolution for palaeoclimatology. For the first time, it was possible to associate temperature changes with concentrations of greenhouse gases in the atmosphere over the past 400,000 years, demonstrating their close correlation.

In January 2005, a team of European researchers reached the bedrock of the Dome C ice sheet close to the Italian French base at a depth of 3,260 metres. The project, called "European Project for Ice Coring in Antarctica", is popularly known as EPICA – a name that at once acts as an acronym and represents the epic impact deriving from the scientific results obtained. EPICA set a new record, not only for the depth it reached but also for having extracted 800,000-year-old ice – an ice repository representing eight glacial and interglacial cycles.[1]

"Beyond EPICA – Oldest ice" succeeded EPICA in 2020 with the aim of sampling the oldest ice on the Antarctic continent. Possibly representing the pinnacle of polar research in Antarctica, Beyond EPICA will keep hundreds of researchers busy for the next decade at least.[2]

After years of accurate geophysical measurements and computer simulations, the site chosen for the drilling is only 40 kilometres away from Concordia Station ↗p. 780 at Dome C. The proximity was necessary to extend the climatic record obtained with the EPICA project and reach the unprecedented date of 1,500,000 years ago. The ambitious scientific goal aims to reconstruct and study how the climate changed during the Mid-Pleistocene Transition (between 900,000 and 1,200,000 years ago), when the periodicity of the ice ages passed from around 41,000 to the current 100,000 years. The reason for this variation is still unknown to humankind and this is the mystery that the "Beyond EPICA – Oldest ice" project sets out to solve.

"The crucial milestone was that, one day, on the coast, drinking whisky in a caravan in which our small team had taken shelter for the night, as I dropped a small sample of ice into the whisky, I noticed bubbles escaping, and I said to myself, perhaps these bubbles bear witness to the composition of our atmosphere at the time when the ice formed. It took us a while to confirm it, but it was a real discovery. We were able to reconstruct not only the past temperature, but also the composition of the atmosphere and in particular the impact of human activity – for example, we found radioactive fallout at the South Pole, with nobody around, that had come from Europe via the sky. And with the composition of the atmosphere indeed we saw something which would later be important – greenhouse gasses … we were actually able to trace the history of these changes and link it to the rise in temperatures – there in that region as well as all over the planet."

Claude Lorius (Director Emeritus at CNRS, Glaciologist), in an interview, 2015

The study of ice cores in Antarctica dates back to the early twentieth century, with several cores being drilled throughout the International Geophysical Year reaching average depths of 400 metres. Since the record 2,164-metre-deep core drilled at Byrd Station in the 1960s, numerous drilling campaigns have been successfully completed. These have occurred at Vostok and Mirny Stations since the 1970s; on Dome C and Kohnen Station (as part of EPICA), with cores being extracted at depths of 3,260 metres and 2,760 metres respectively; at McMurdo Station; at Taylor Dome and Siple Dome (reaching 554 metres in 1994 and 1,004 metres in 1999 respectively); in the West Antarctic Ice Sheet (3,045 metres in 2011); and on Dome F (where the Japanese drilled to a depth of 3,035 metres in 2006, reaching ice estimated to be 720,000 years old), to name a few.[1]

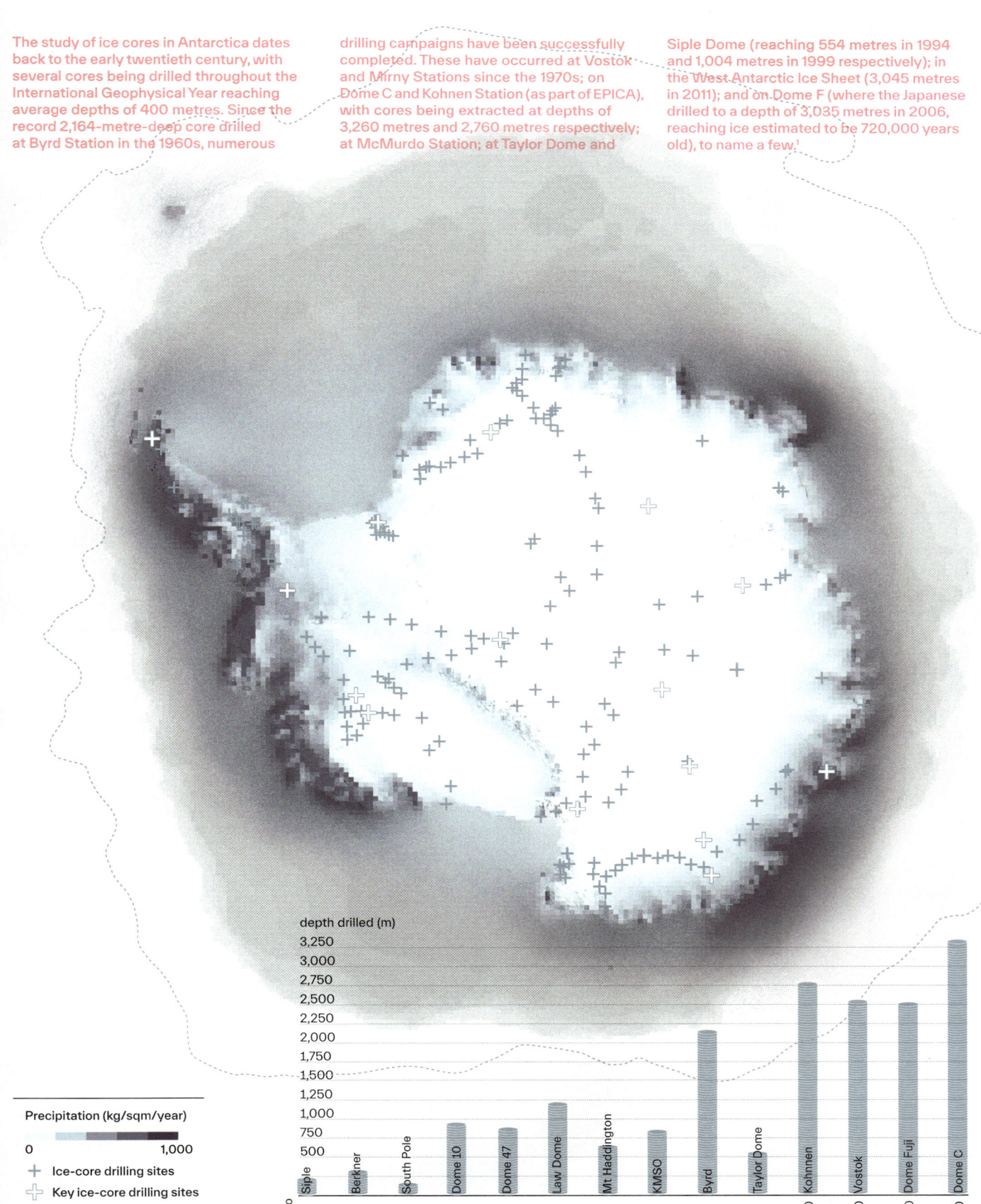

depth drilled (m)

3,250														
3,000														
2,750														
2,500														
2,250														
2,000														
1,750														
1,500														
1,250														
1,000														
750														
500														

Siple · Berkner · South Pole · Dome 10 · Dome 47 · Law Dome · Mt Haddington · KMSO · Byrd · Taylor Dome · Kohnen · Vostok · Dome Fuji · Dome C

years ago
250 · 1,000 · 1,250 · 2,000 · 2,000 · 10,000 · 14,000 · 14,000 · 70,000 · 130,00 · 190,000 · 220,000 · 340,000 · 740,000

Precipitation (kg/sqm/year)

0   1,000

+ Ice-core drilling sites
+ Key ice-core drilling sites
- - - Antarctic Convergence
Sea-ice extent (2019)

The drilling activities in the polar regions require a great deal of energy and financial outlay, both in terms of human and logistical resources. The sheer amount of resources required for each endeavour naturally brings up questions about the necessity of such operations: Is the study of past climate change really so important? What motivates scientists to work for years on end in the polar regions to collect these precious samples?

The reasons, as is often the case, are manifold. In part, they reflect human nature and our innate propensity to relentlessly research what remains unclear – this might suggest that scientists could be described as modern versions of Aeneas, who, leaving behind the ashes of Troy, set sail to discover the unknown. Alongside this noble instinct, there are important political reasons. The data on past temperatures and greenhouse gas concentrations obtained from the analysis of ice cores is necessary to validate the climate models that allow us to make predictions for the future. In a climate model, all the variables influencing the climate on Earth (the distance of the planet from the Sun, concentrations of greenhouse gases, movements of oceanic and atmospheric masses, the distribution of forests and sea ice, etc.) are thought to interact with each other according to the laws of classical physics, dating back to well before the twentieth century. To test the vigour of climate models in different future scenarios, however, it is necessary to validate them by simulating what has already happened in the past. The climatic and environmental records obtained from the analysis of ice cores represent a rich reservoir of essential information for the verification of these simulations, providing data both on the temperature and the variables that determine their progress.

A careful analysis of the layers of ice also allows the presence of polluting compounds to be detected: these include heavy metals, pesticides, or combustion residues deriving from human activities, which are distributed globally thanks to atmospheric currents. Thus, a study of the variations in the concentrations of these compounds at ultra-trace levels provides scientists with precise indications of humankind's impact on the environment and can help us to accurately measure the effectiveness of policies put in place to contain polluting emissions.

Ice cores are therefore formidable environmental archives that conceal, among their layers, the climatic history of our planet and unquestionably testify to the level of responsibility that humankind must bear for preserving or damaging the global ecosystem. Often described as an all-encompassing ancient book, written in a language that is not always accessible, the study of Antarctica's ice is an essential means to model and inform the future of Planet Earth.

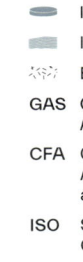

	Ice core
	Ice sheet
	Bedrock
GAS	Greenhouse Gas Analyses
CFA	Continuous Flow Analyses (chemical analyses)
ISO	Stable Isotope Characterisation
PP	Physical Properties
SC	Spare Core

French glaciologist Claude Lorius – shown below as he analyses ice cores in a laboratory carved in the snow at Charcot Station – was the first scientist to reveal, in his 1963 PhD thesis, the potential for measuring oxygen and hydrogen isotopes in ice and their role in helping us understand the planet's climatic history. Building upon his later hypothesis that air bubbles captive in the ice samples could represent the composition of atmospheric gases at the time of encapsulation, Lorius's twenty-year-long expedition fieldwork contributed significantly to the process of documenting the Earth's temperature trends and unequivocally determining what induced the current interglacial period. The ice-core sections shown below offer visual evidence of Charcot's prophecy by unveiling respectively prehistoric air bubbles and a strata of volcanic ash 22,000 years old.

Scientists leading the EPICA project identified Dome C on the Antarctic Plateau as the site for their ambitious deep-ice-core drilling project, which was deployed between 1996 and 2004. Developed in proximity to the Italo-French Concordia Station, EPICA reached a drilling depth of 3,270.2 metres, interrupting the exploration some 5 metres above the continental bedrock and retrieving ice cores which are estimated to be 900,000 years old.

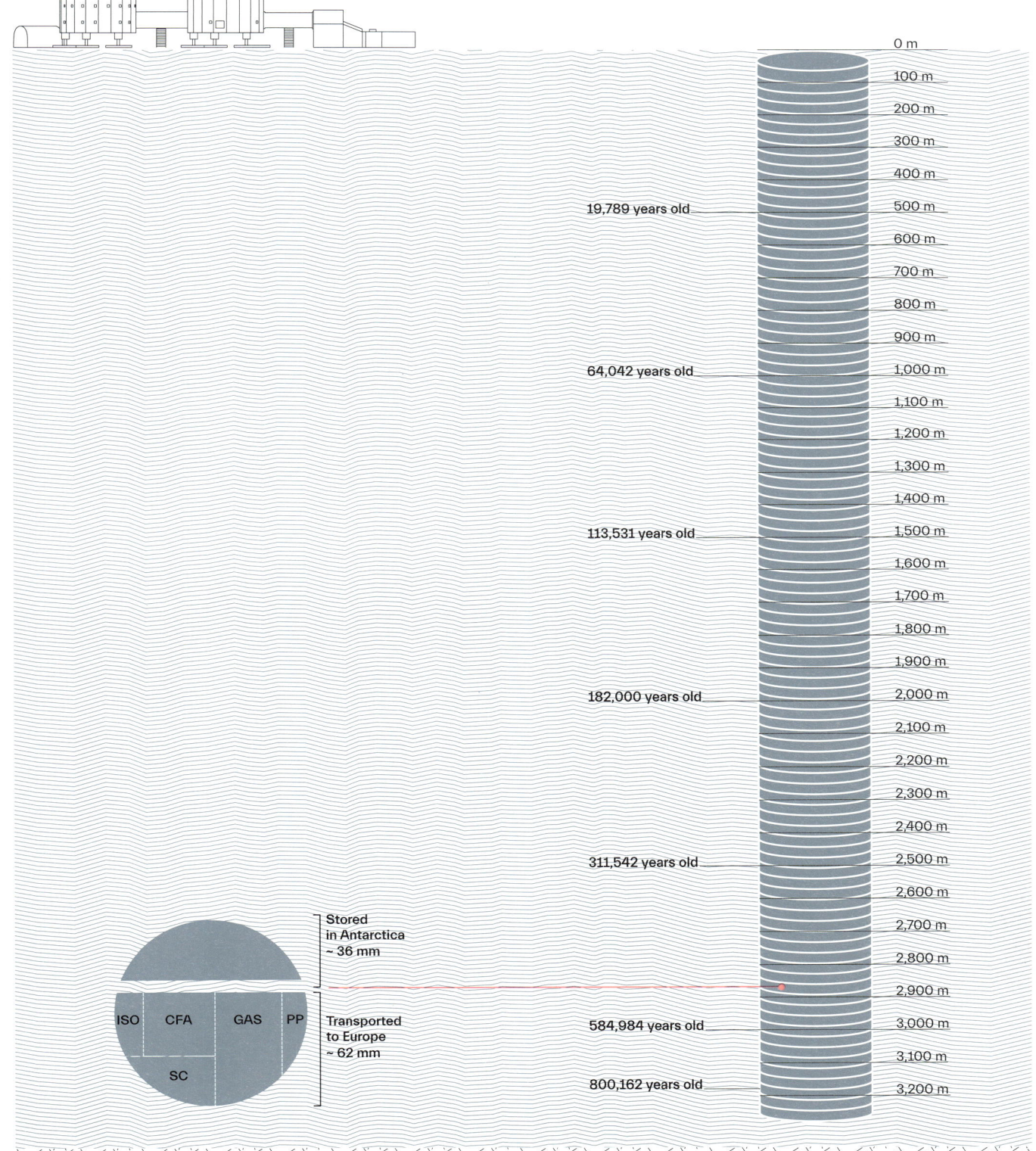

19,789 years old — 500 m

64,042 years old — 1,000 m

113,531 years old — 1,500 m

182,000 years old — 2,000 m

311,542 years old — 2,500 m

584,984 years old — 3,000 m

800,162 years old — 3,200 m

Stored in Antarctica ~ 36 mm

Transported to Europe ~ 62 mm

ISO   CFA   GAS   PP

SC

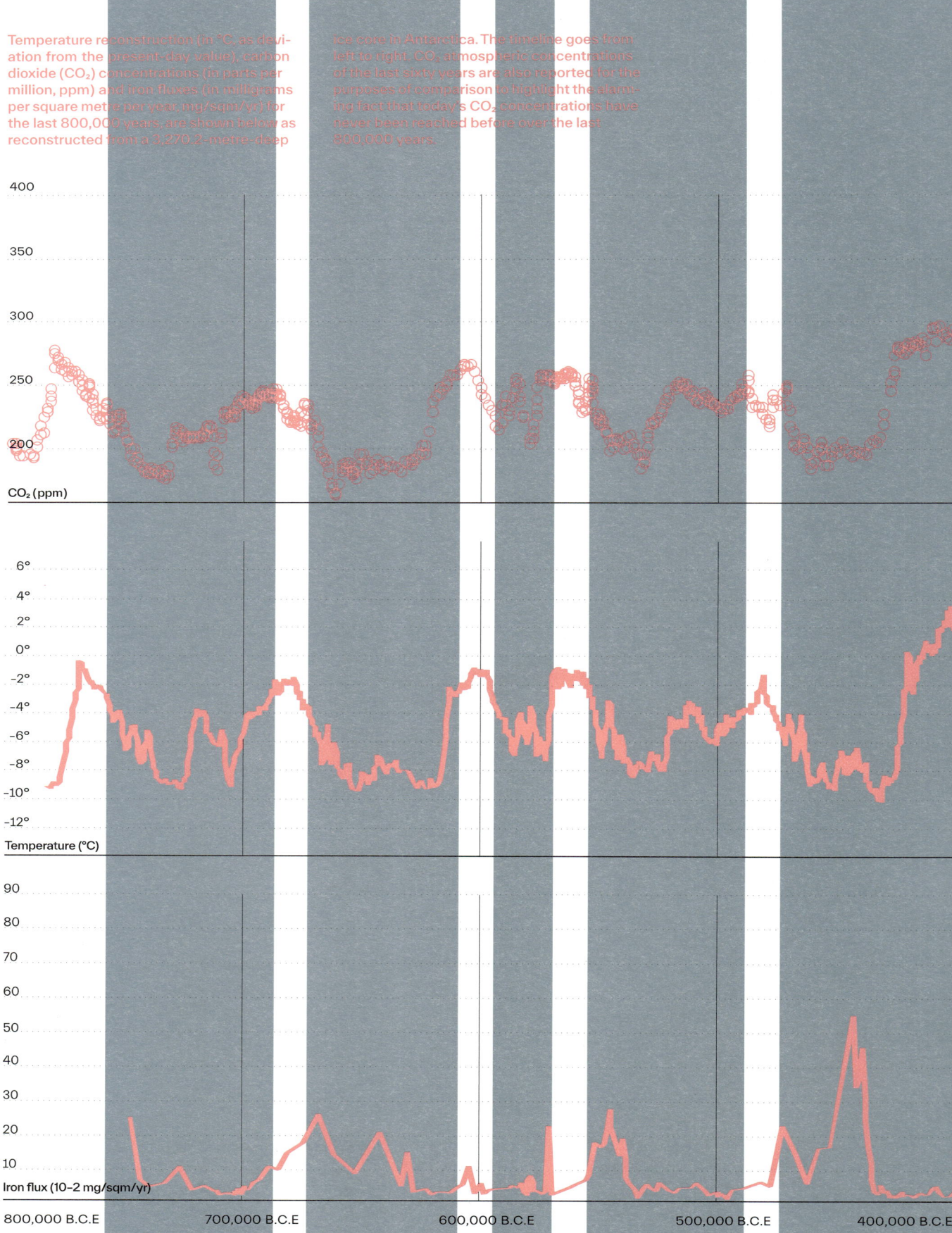

Temperature reconstruction (in °C, as deviation from the present-day value), carbon dioxide (CO₂) concentrations (in parts per million, ppm) and iron fluxes (in milligrams per square metre per year, mg/sqm/yr) for the last 800,000 years, are shown below as reconstructed from a 3,270.2-metre-deep ice core in Antarctica. The timeline goes from left to right. CO₂ atmospheric concentrations of the last sixty years are also reported for the purposes of comparison to highlight the alarming fact that today's CO₂ concentrations have never been reached before over the last 800,000 years.

400
350
300
250
200

CO₂ (ppm)

6°
4°
2°
0°
-2°
-4°
-6°
-8°
-10°
-12°

Temperature (°C)

90
80
70
60
50
40
30
20
10

Iron flux (10–2 mg/sqm/yr)

800,000 B.C.E   700,000 B.C.E   600,000 B.C.E   500,000 B.C.E   400,000 B.C.E

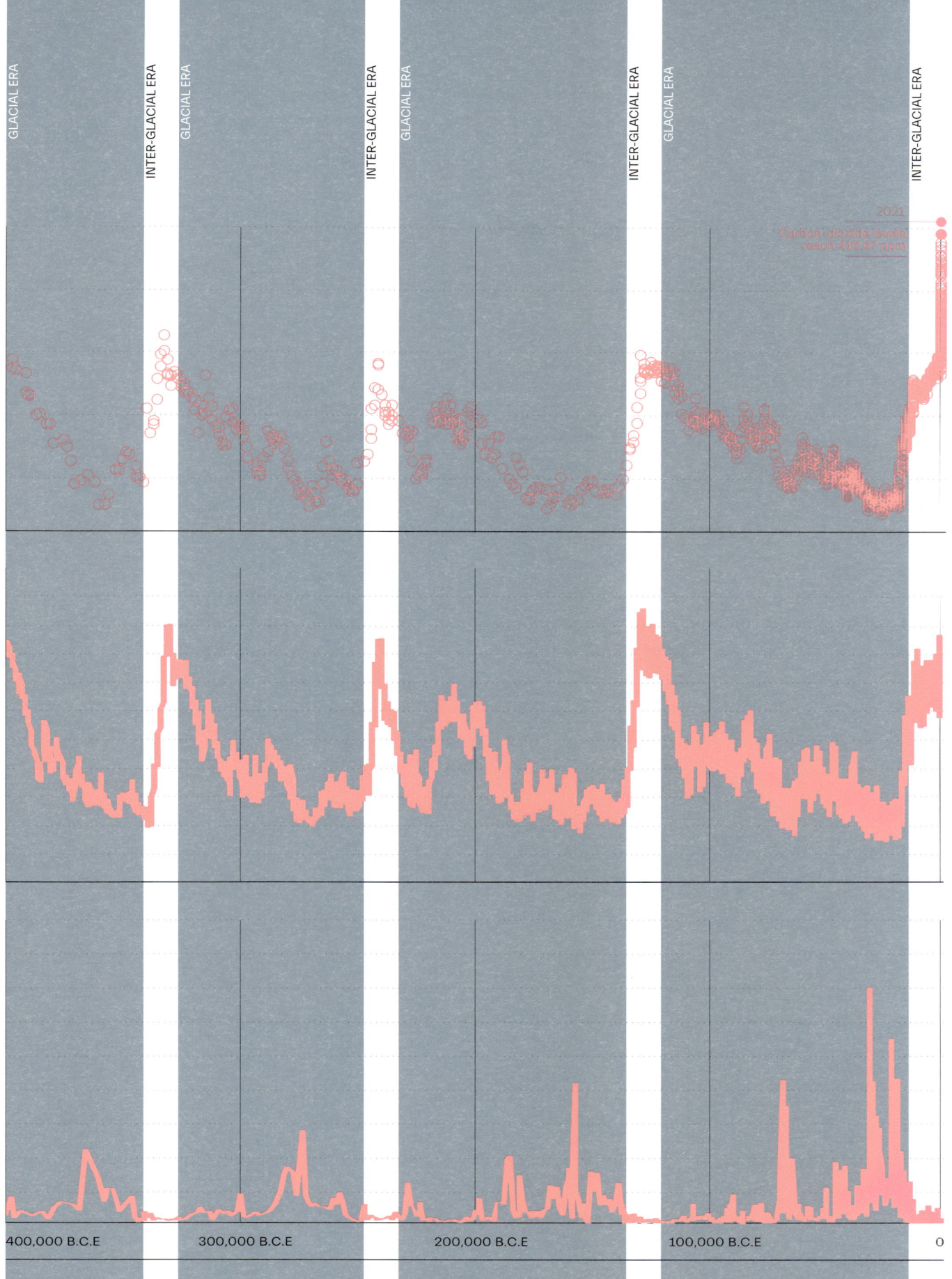

2021

Carbon-dioxide levels
reach 416.45 ppm

400,000 B.C.E   300,000 B.C.E   200,000 B.C.E   100,000 B.C.E   0

"Ice cores are long tubes of raw, blue, deep-frozen time."

Marie Darrieussecq (Writer), *White*, 2005

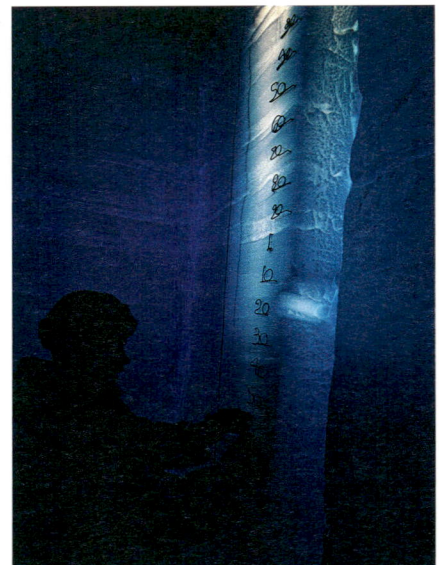

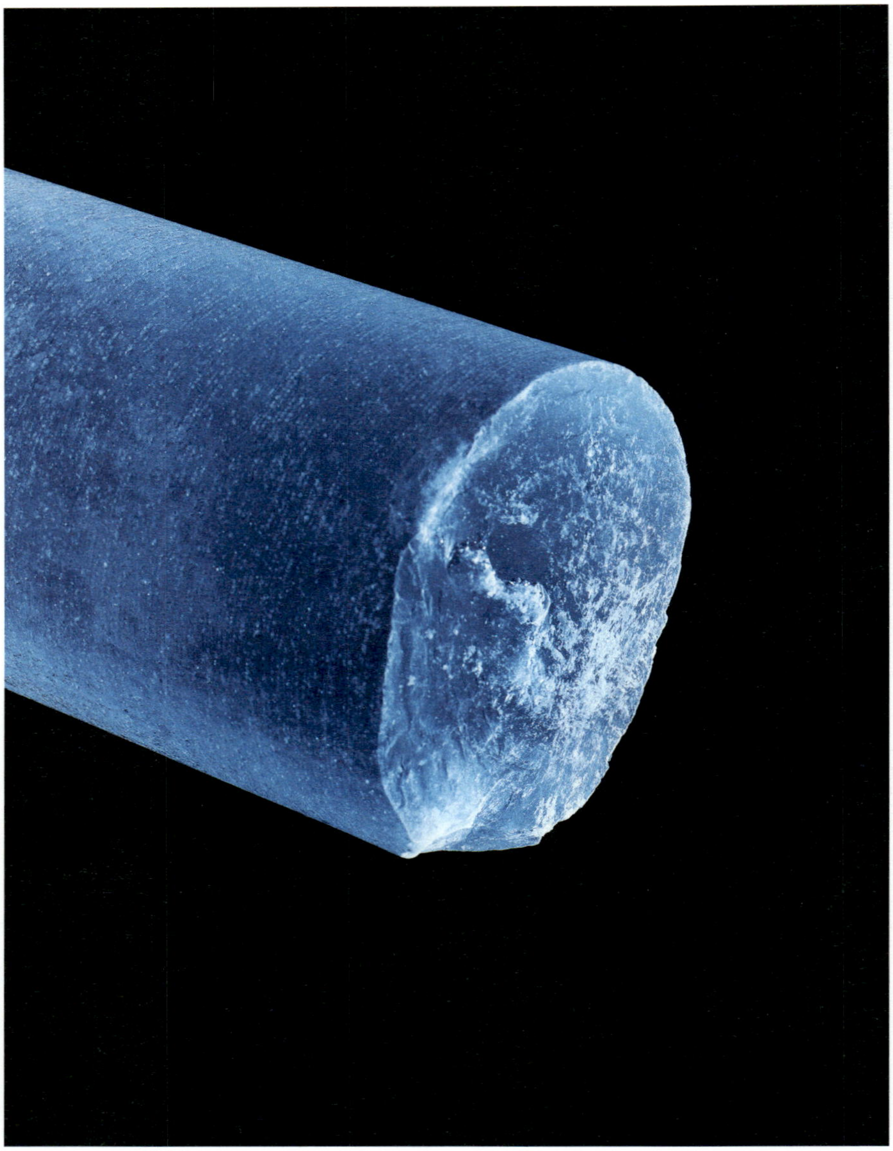

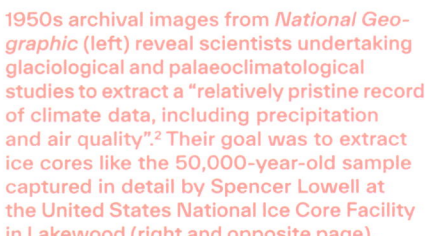

1950s archival images from *National Geographic* (left) reveal scientists undertaking glaciological and palaeoclimatological studies to extract a "relatively pristine record of climate data, including precipitation and air quality".[2] Their goal was to extract ice cores like the 50,000-year-old sample captured in detail by Spencer Lowell at the United States National Ice Core Facility in Lakewood (right and opposite page).

"The year is 2220. Everyone alive on Earth today is long dead, and the world looks very different. Desperate to study the atmospheres of the past, on a mission to understand how the hostile 23rd century climate could unravel further, a scientist has travelled to the Antarctic hinterland. Her gloved hand pulls open a door buried in the snow. Behind it lies the top of a staircase, which spirals down into the freezing darkness. […] Inside a steel vault that has been sealed for more than a century, frost sparkles from steel canisters. They hold ice scraped from mighty glaciers that once topped the highest mountains. She has seen the glaciers in pictures, but now – thanks to forward-thinking predecessors – she has the chance to analyse them and their valuable record of the past. […] Back in 2020, scientists are still racing to make this vision a reality. They are crisscrossing the globe, scaling peaks and drilling out lengthy sections of glacier ice while it still stands. Because time is dripping from the mountains. […] As the ice disappears, so does a unique archive."

David Adam (Writer), "The Rush to Sock Away Glacier Ice Before It All Melts", *The Atlantic*, 2020

Ice Memory. An International Salvage Program

Carlo Barbante and Jacopo Gabrieli

CARLO BARBANTE is the director of the Institute of Polar Sciences of the National Research Council (Italy). JACOPO GABRIELI is a researcher at the Institute of Polar Sciences of the National Research Council (Italy).

According to the latest Intergovernmental Panel on Climate Change (IPCC) reports,[1] the average global temperature has increased by about 1°C over the past century and, at the highest altitudes of the planet, the value is almost double that.[2] The mountainous areas of our planet are particularly sensitive to climate change, to the extent that they are considered true sentinels of current global warming.[3]

Starting from the second half of the nineteenth century, alpine glaciers have generally been retreating almost continuously, losing on average 60% of their mass at speeds that have increased year on year. In addition to the well-known consequences in terms of the environment, water resources, and mountain ecosystems, the melting of a glacier implies the destruction of a veritable natural archive that stores invaluable information on the climate and environment of the past, spanning periods from several centuries to tens of millennia, depending on the glacier.

The history of our planet and its people is contained and preserved in the ice, as if the ice crystals constitute a unique ancient manuscript recorded in a frozen library. And so, just as a fire could destroy the history of our past in a few hours, the disappearance of a glacier is also an incalculable loss of our heritage and knowledge.

The international Ice Memory project, initiated by a consortium of researchers well known for their contributions in the field of climatic reconstructions from ice cores (Carlo Barbante, Italy; Jerome Chapellaz, France; Margit Schwikowski, Switzerland), aims to recover ice cores from the most important endangered glaciers in order to preserve the information they contain and make it available to future generations, notably for the purposes of conducting scientific investigations that are not possible today due to limitations in analytical techniques and process understanding. From Africa to the Himalayas, from the Bolivian Andes to Mongolia, dozens of glaciers will be drilled as part of this project. For Italy, the Institute of Polar Sciences at the Italian National Research Council (CNR), in addition to taking part in expeditions on Mont Blanc and Kilimanjaro, will lead the activities on the Grand Combin and Monte Rosa glaciers, where long-term climate reconstructions are still possible. In addition, samples will be taken from some very particular glacial bodies that are unfortunately, at this point, in an advanced stage of melting: Calderone (Apennines, 2,900 m.a.s.l. ice, south of Europe), Montasio (Julian Alps, 1,900 m.a.s.l., the lowest-altitude glacier in Europe), and Marmolada (3,100 m.a.s.l., the largest glaciers in the Dolomites). These are real glacial legacies that are disappearing before our eyes.

At least two ice cores will be extracted for each glacier. Of these, one will be immediately analysed in order to obtain a general characterisation of the samples (stratigraphy, dating, profiles of stable isotopes, dust, black carbon, major ions, trace elements, etc.), while the other will be transferred to Antarctica, the coldest place on the planet. At the Italo-French base located on Dome C, Concordia Station ↗ p. 780 (75°S, 123°E; 3,220 m.a.s.l.), where the average temperatures are around –54°C, this unique and invaluable repository of ice cores will be created, dug in the millennial ice – a real sanctuary for future generations, similar to the Global Seed Vault hosted in the frozen ground of the Svalbard Islands.

Owing to climate change and the accelerated pace at which the glaciers are thinning, Ice Memory is a true race against time. This project, supported by UNESCO, which offered its patronage, will keep scientists busy for at least ten years.

Driven by the same sense of urgency as the Ice Memory programme and informed by an equal awareness that the climate crisis that humanity is witnessing today – unless addressed by means of prompt enforcement of sustainability policies such as the Paris Agreement – will only worsen in the future, the Svalbard Global Seed Vault represents the ultimate effort to salvage genetic information and data of all the world's crop collections. Located on the island of Spitsbergen in the remote Arctic Svalbard archipelago (and captured here by Spencer Lowell), the vault features duplicates of seed samples from the world's crop collections stored in gene banks distributed globally. To quote a 2012 article by *The Economist,* "The Svalbard vault is a backup for the world's 1,750 seed banks, storehouses of agricultural biodiversity."[1] Holding more than 980,000 samples, with the capacity to store 4.5 million varieties of crops containing on average 500 seeds each, the project can be thought of as one of the greatest archives of our planet. Situated at the poles, both the Seed Vault and Ice Memory serve as Earth's natural repositories of irreplaceable data and genes.[2]

"There are three vaults leading off from the chamber, but only one is currently in use, and its door is covered in a thick layer of ice, hinting at the temperatures inside. In here, the seeds are stored in vacuum-packed silver packets and test tubes in large boxes that are neatly stacked on floor-to-ceiling shelves. They have very little monetary value, but the boxes potentially hold the keys to the future of global food security. Over the past 50 years, agricultural practices have changed dramatically, with technological advances allowing large-scale crop production. But while crop yields have increased, biodiversity has decreased to the point that now only about 30 crops provide 95% of human food-energy needs. Only 10% of the rice varieties that China used in the 1950s are still used today, for example. The United States has lost over 90% of its fruit and vegetable varieties since the 1900s. This monoculture nature of agriculture leaves food supplies more susceptible to threats such as diseases and drought. The seeds lying in the deep freeze of the vault include wild and old varieties, many of which are not in general use anymore. And many don't exist outside of the seed collections they came from. But the genetic diversity contained in the vault could provide the DNA traits needed to develop new strains for whatever challenges the world or a particular region will face in the future. One of the 200,000 varieties of rice within the vault could have the trait needed to adapt rice to higher temperatures, for example, or to find resistance to a new pest or disease. This is particularly important with the challenges of climate change."

Jennifer Duggan (News Editor), *TIME Magazine,* 2020

The Technicalities of Ice Drilling

Pavel Grigorievich Talalay

PAVEL GRIGORIEVICH TALALAY is a professor and the director of the Polar Research Center at Jilin University (China). Prof. Talalay has taken part in six polar expeditions and was involved in drilling the deepest hole in ice, at Vostok Station, Antarctica.

Some may think that continued use of the planet's mineral resources will require the discovery of new deposits in unexplored Antarctic areas. However, this is not the case. The Protocol on Environmental Protection to the Antarctic Treaty includes a ban on all commercial drilling and mining until at least January 2048. Currently, drilling in Antarctica is restricted to research and scientific studies of physical, chemical, biological, and geological processes in polar regions, Earth's climate changes in the recent past, and the reasons for such changes, along with other natural science questions.

To operate in the extreme environmental conditions of Antarctica – which include low temperatures, glacier flows, an absence of roads and infrastructure, and intense winds and snowfall – conventional equipment needs to be modified or purpose-built to be able to drill holes in ice. Depending on the nature of the ice disintegration at the borehole's bottom, the techniques that have been developed can be divided into mechanical and thermal drilling methods. Mechanical drilling tools most commonly utilise cutting, while thermal drilling tools use heat to melt the ice.

Ice holes are created by full-face boring, which only produces cuttings or melted water, or core drilling, which (as the name suggests) produces cores – cylindrical samples of ice. Borehole depths range from quite shallow, at a depth of just a few metres (for the installation of ablation stakes, measuring temperatures at the bottom of the active layer and revealing anthropogenic pollution), to very deep – these can reach depths of more than 3,000 metres for fundamental studies of ice-sheet dynamics and of the subglacial environment. Drilling deeper than 300–400 metres in ice brings with it a high risk of open-hole closure: hence, when drilling at greater depth, the borehole is filled with a low-temperature drilling fluid.

The simplest ice drills are hand- or power-driven portable augers. These drills are small systems that can drill holes to maximum depths of approximately 50 metres. They are relatively lightweight and do not require a drilling fluid.

Different cable-suspended drills have been suggested for boring Antarctic ice that use an armoured cable with a winch instead of a pipe string to provide power and retrieve the down-hole unit. These purpose-built drills have different characteristics: electrically heated hot points are used to produce boreholes without a core by continuously melting ice; ice-coring electric thermal drills are used to retrieve ice-core samples; electromechanical "shallow drills" with an auger conveyer are used to remove drilling cuttings into a special down-hole chamber or deep electromechanical drills with near-bottom fluid circulation. The use of a cable allows a significant reduction in power and material consumption, decreases the time of round-trip operations, and simplifies the removal of cuttings or melted water from the hole.

Hot-water drills provide the fastest penetration in ice and are now actively used in Antarctica. During drilling, hot water is pumped at high pressure through a drill hose to a nozzle that jets hot water to melt the ice. Renewed interest in conventional rotary drill rigs has come from the current scientific need to study the base of the ice sheet and recover bedrock cores from ice-covered geologic provinces of Antarctica. To bore rapidly through the ice sheet, a conventional wire-line diamond drilling system remains a viable option.

The timeline explores the evolution of mechanical and thermal ice-core drilling technology deployed in the Antarctic since the beginning of the twentieth century.

1903

Trapped by ice for nearly fourteen months, the *Gauss* was used as a research station in the south of the Kerguelen Islands. Multiple 30-metre-deep holes were drilled by hand auger in the neighbouring iceberg to facilitate temperature measurements.

1940–1941

A drill tower stands above a 41-metre-deep hole drilled by hot point on the Ross Ice Shelf; in the background, a snow-covered Snow Cruiser expedites transportation in Antarctica.

1955–1956

The legendary SIPRE hand auger was used by the international scientific community to drill thousands of shallow holes (with a maximum depth of about 50 metres) in ice throughout the cold regions of the world.

1958–1959

A conventional drill rig was used to drill a hole at Little America V ↗ p. 654 on the Ross Ice Shelf. When it reached a depth of 254 metres, seawater began to leak into the hole.

1958–1959

A hot-point thermal drill is being prepared to drill the first ever borehole at the Pole of Cold at Vostok Station ↗ p. 862.

1967–1968

An electromechanical drill in service at Byrd Station reached the ice-sheet base at a record depth of 2,164 metres.

1977

A drilling shelter at Law Dome during the polar night. Drilled by a thermal core drill to a depth of 474 metres, the borehole shown in the photo neared the limit of dry-hole drilling.

1978–1979

The last core extracted by an antifreeze thermal drill located at J-9 Camp on the Ross Ice Shelf shows the ice formed by seawater freezing directly onto the bottom of the ice shelf.

1988–1989

The 22-metre-high tower at Site D47 in Adélie Land rises up above an 871-metre-deep hole made by a thermal drill.

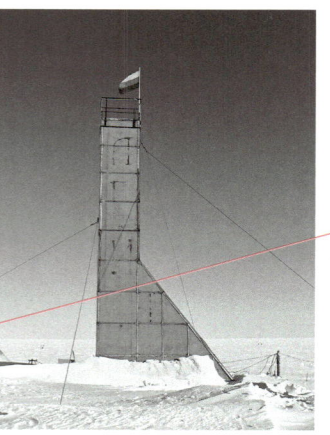

2011–2012

Nine-metre-high snow-covered drilling shelter above 5G borehole at Vostok Station. Here, after twenty-two years of hard work, the drillers made contact for the first time with the water of the subglacial Lake Vostok at a depth of 3,769 metres.

2012–2103

A hot-water drill reached Subglacial Lake Whillans and collected the first samples of pollution-free subglacial water and sediments.

1990

The first core is retrieved from the 5G borehole at Vostok Station that for two decades had the distinction of being the deepest hole drilled into ice.

2004–2005

The last ice core retrieved at Dome C from a depth of 3,270.2 metres, only 15 metres above the ice-sheet base. The core contains the oldest ice ever extracted, dating back some 800,000 years.

2005–2006

Drilling of the 2,500-metre-deep network of hot-water holes for the IceCube Neutrino Observatory at the South Pole

2014–2015

The thermal probe IceMole is ready to penetrate pressurised brine trapped in a crevasse at Blood Falls on the Taylor Glacier.

2015–2016

The Eclipse Drill reached the bedrock at the Allan Hills and recovered the oldest Antarctic ice, dating back 2.7 million years, from the near-base basal unit.

2018–2019

A core barrel shows a piece of bedrock core recovered from a 198-metre-deep borehole drilled at Zhongshan Station ↗ p. 760 with an electromechanical cable-suspended drill.

face of the glacier from satellites. The latter revealed that under the 3.5–4.2 km thick glacier of the Antarctic plateau lay a body of water the size of Lake Ontario.

In the 1998–1999 season, seismic soundings were supplemented by ground-based radar profiling of the glacier's body to help determine the composition of the lake's underlying surface. This helped to define the exact position and indentation of the lake's coastline, the presence of "islands" in the lake, and the thickness of the glacier, water column and sedimentary rocks at the bottom, while also making it possible to obtain a spatial picture of its distribution. It thus became evident that Lake Vostok has an average depth of 410 metres, housing a total volume of water of approximately 6,343 cubic kilometres. The coastline is 1,030 kilometres long, including 70 kilometres per island, and the surface area of the water is 15,500 square kilometres, excluding the 70 square kilometres of the islands. It is an asymmetrical lake divided into two parts of unequal size: the south side is deeper, covering an area of 2,100 square kilometres with a depth of 800 metres, while the north is larger and shallower, covering a total area of 10,800 square kilometres with an average depth of 300 metres. Later, it was established that geothermal heat from the Earth keeps the temperature of the lake water hovering at around –3°C, with the pressure caused by the weight of the overlying ice keeping the lake liquid despite its sub-zero temperature.

With a deep ice-core well already present at the site, it was very tempting to use the pre-existing structure to penetrate the subglacial lake. However, the filling fluid used in the well (consisting of a mixture of kerosene and freon) caused deep concern for environmentalists in connection with the possible pollution of the virgin waters of Lake Vostok, and drilling was thus suspended at a safe depth of several tens of metres. Drilling resumed in 2005, when AARI and the Mining Institute employees developed an environmentally friendly technology for penetrating the lake's waters. On the 5th of February 2012, at the Russian Antarctic station Vostok, the subglacial lake was finally unsealed through a deep-ice well at a depth of 3,769.3 metres, producing the longest ice core ever drilled to date. A second opening of the lake took place at a depth of 3,769.15 metres on the 25th of January 2015. Lake water rose into the well to a height of about 70 metres and was left to freeze to avoid the risk that any drilling fluid would enter the lake. Four days later, drilling operations were resumed until water poured into the well again on the 3rd of February 2015. In January 2009, the 5G-1 borehole was abandoned at a depth of 3,667 metres due to a drill accident, and a new borehole (5G-2) was launched from a depth of 3,598 metres.

Deuterium and oxygen-18 ratios measured in the 5G-1 core interval down to 3,650 metres allowed the isotopic properties of the 41-metre segment of Lake Ice 2 below 3,609 metres to be determined. This was helpful as scientists use isotopic data to study processes leading to the formation of lake ice and the hydrological regime of subglacial lakes. Studies of ice cores of frozen lake water have improved our understanding of the physical, chemical, and biological processes in the lake. Additionally, they have allowed the discovery of three phylotypes of bacteria that sample the true microbiota of the lake, an active circulation of water, several geothermal springs, and water oxygen saturation levels. Yet the main question remains unanswered: When and how was Lake Vostok formed?

"For this place could be no ordinary city. It must have formed the primary nucleus of some archaic and unbelievable chapter of earth's history whose outward ramifications, re-called only dimly in the most obscure and distorted myths, had vanished utterly amidst the chaos of terrene convulsions long before any human race we know had shambled out of apedom. Here sprawled a Palaeogaean megalopolis compared with which the fabled Atlantis and Lemuria, Commoriom and Uzuldaroum, and Olathoc in the land of Lomar, are recent things of today – not even of yesterday; a megalopolis ranking with such whispered prehuman blasphemies as Valusia, R'lyeh, Ib in the land of Mnar, and the Nameless city of Arabia Deserta."

Howard Phillips Lovecraft (Writer), *At the Mountains of Madness*, 1936

Ice-Shelf Retreat in Antarctica

David Vaughan

DAVID VAUGHAN is the director of science for the British Antarctic Survey (United Kingdom). He was a coordinating lead author for the 4th and 5th assessment of the Intergovernmental Panel on Climate Change (IPCC) and sits on the Science Board of the UK Natural Environment Research Council. Vaughan has participated in twelve Antarctic scientific field campaigns. He was awarded the Polar Medal and was made an Officer of the Order of the British Empire (United Kingdom).

An ice shelf is formed from ice that has accumulated on the continent of Antarctica over many thousands of years through snowfall, slowly flowing in glaciers towards the coast and on to the ocean. Here – unlike sea ice, which is formed by freezing seawater and is rarely more than a few metres thick – the ice shelf floats over the ocean, forming a layer of ice that can be several hundred metres thick. Since they are already floating on water, the calving ice shelves have little or no impact on sea level. However, their presence de facto affects sea level rise as they act as a "brake" of sorts, holding back the freshwater stored by glaciers that would otherwise melt into the ocean without restraint.

Over 80% of Antarctica's grounded ice drains through its fringing ice shelves, and glacier flow is sensitive to changes in ice-shelf extent and thickness. Vulnerable to changes in atmospheric conditions on their upper surfaces and exposed to heat transported at depth within the Southern Ocean at their bases, ice shelves are prone to collapse, which can lead to abrupt and potentially runaway ice-sheet retreat.

The process by which an ice shelf breaks away from a glacier is called calving. The causes of calving are complex, with no single reason attributed to each event. The ongoing and widespread retreat of ice shelves and floating glaciers around the Antarctic Peninsula in recent decades, for example, can be attributed to multiple factors such as changes in the surrounding ocean, the substantial atmospheric warming witnessed during the second half of the twentieth century, and the "föhn winds" that typically blow from west to east over the mountainous terrain, warming the surface on the eastern side of the peninsula. The combination of the two latter factors renders the long mountainous land mass known as the Antarctic Peninsula the warmest part of Antarctica and leads to the production, during the summer months, of significant amounts of meltwater. With an average temperature increase of 3°C over the last fifty years, the "limit of viability" of the ice shelves – i.e. the amount of meltwater each ice shelf can tolerate before it weakens and begins to retreat – has moved southwards and ice shelves that used to be stable are now retreating.

The first comprehensive study of glaciers around the coast of the Antarctic Peninsula was published in April 2005, revealing data that could inform a clear understanding of the real impact of climate change on the thinning of Antarctica. Results from the study produced by researchers at the British Antarctic Survey (BAS) and the United States Geological Survey (USGS) and published in the journal *Science,* showed that over the last fifty years, 87% of the 244 glaciers studied in the region had retreated, and that average retreat rates had accelerated. "The retreat began at the northern, warmer tip of the Antarctic Peninsula and, broadly speaking, moved southwards as atmospheric temperatures rose," commented lead author, Alison Cook. "This region has shown dramatic and localised warming but this is not the only factor causing the changes. It's a complex picture." BAS and USGS scientists have since analysed more than two thousand aerial photographs dating from 1940 and over one hundred satellite images from the 1960s onwards to calculate the position of glacier fronts along the coast of the Antarctic Peninsula, identifying previously unknown patterns of change.

Seven years later, in April 2012, an international team of scientists led by BAS developed new techniques, allowing them to differentiate, for the first time, between the two known causes of melting ice shelves – warm ocean currents attacking the underside and warm air melting the upper surface – and established that warm ocean currents are, in fact, the dominant cause of recent ice loss from Antarctica. Analysing over 4.5 million measurements made by

Trying to comprehend and model the process whereby ice shelves and glaciers are melting within the East Antarctic region, researchers monitor the Sørsdal glacier to understand whether "surface meltwater ponding is occurring more frequently on outlet glaciers, potentially reaching the bedrock and driving increased lubrication and acceleration of the glaciers."[1]

As shown in the diagrams on the following page, each year the equivalent of about 6 millimetres of global sea-level water is deposited in the form of snow on the Antarctic continent. By definition, for the ice sheets to be in equilibrium (i.e. "stable"), an equal amount of water should be released in the oceans by the relentless progression of the glaciers into the seas (and related calving of icebergs). The section of ice sheet that protrudes as a glacier tongue over the grounding line and floats on the ocean, known as an "ice shelf", partially supports the ice-sheet mass through buoyant force at the ice-shelf front. Defining the alignment which demarcates the transition from ice sheet to floating ice shelf, the "grounding line" acts as an invisible threshold beyond which the floating ice is thinned by the combined effect of the melting on its underside caused by warm ocean waters and the formation of fractures on the upper surface, induced by meltwater percolation, which results in iceberg calving. As the glacier's base recedes inland carved by warm water, the glacier retreats more rapidly, losing its grip on the sea floor, and the ice shelf becomes progressively less stable, ceasing to act as a brake to the continental ice's flow and thus accelerating the overall rate of melting. Monitoring this equilibrium and the exact location of the grounding zones (by measuring changes in surface elevation via GPS or satellite synthetic-aperture radar) is extremely important as a means to determine the speed at which Antarctic ice sheets are thinning.[2] In line with Archimedes' principle of buoyancy and informed by the density of pure ice (920 kg/m³) and that of seawater (1,025 kg/m³), icebergs – floating masses of frozen freshwater that broke off either from a glacier or from an ice shelf – reveal only one-tenth of their volume above water.

a laser instrument mounted on NASA's ICESat satellite, researchers mapped the changing thickness of almost all the floating ice shelves around Antarctica and revealed the pattern of ice-shelf melt across the continent. Their research, published in the journal *Nature,* confirmed that of the fifty-four ice shelves mapped, twenty (mainly on West Antarctica) are currently melting under the effects of warm ocean currents.

"The Antarctic glaciers are moving in our direction at a rate of three millimeters per year. Calculate when they'll reach us. Anticipate, in a film, what will happen."

Michelangelo Antonioni (Film Director, Screenwriter, Writer), *That Bowling Alley on the Tiber: Tales of a Director,* 1986

Phase 1

Snow accumulation over the ice sheet

Phase 2

Snow transformation into firn (granular snow)

Phase 3

Firn compressed in high density ice

Phase 4

Glacier moving due to snow and firn mass

Phase 5

Iceberg detaching from the glacier end (calving)

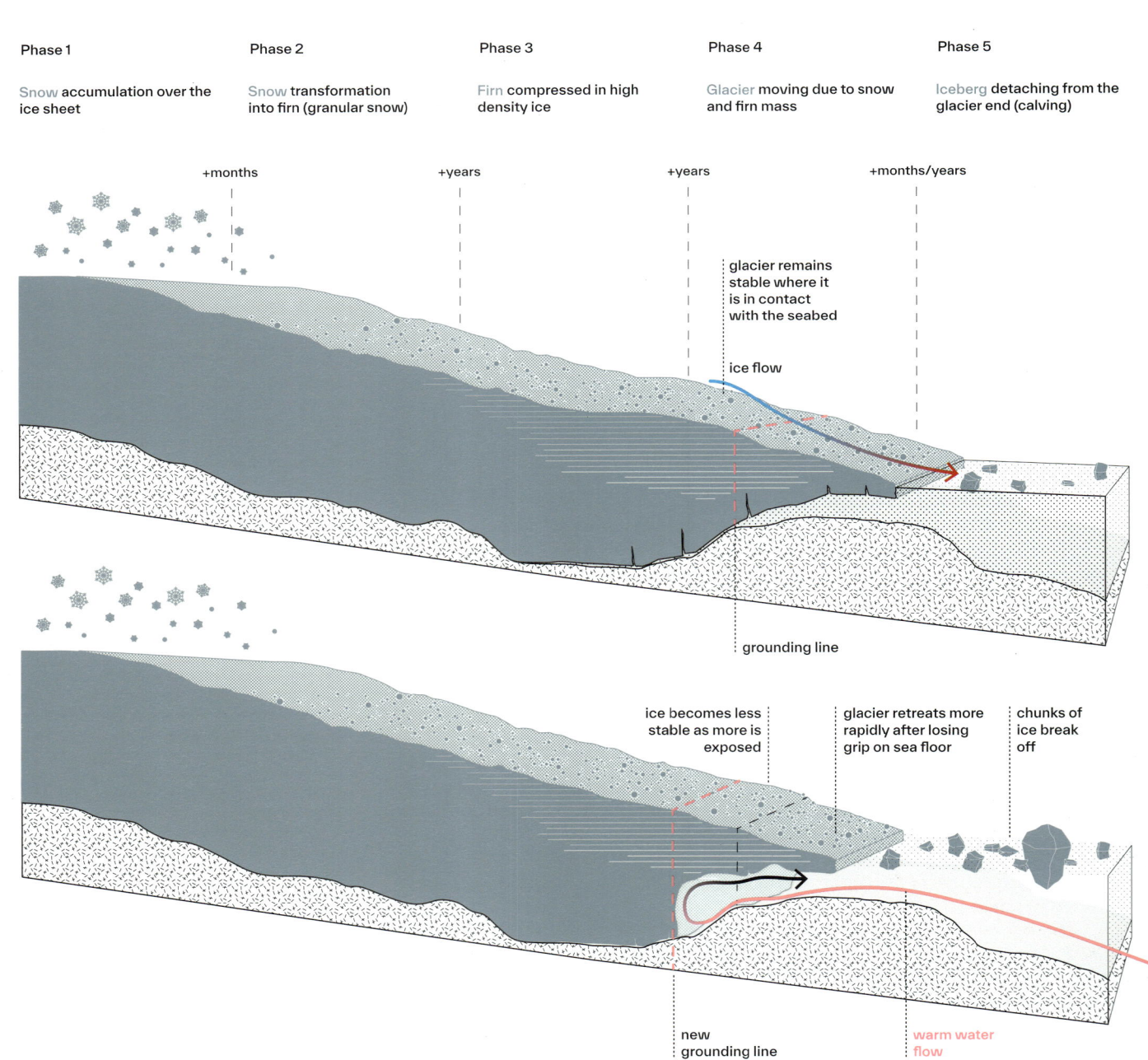

+months

+years

+years

+months/years

glacier remains stable where it is in contact with the seabed

ice flow

grounding line

ice becomes less stable as more is exposed

glacier retreats more rapidly after losing grip on sea floor

chunks of ice break off

new grounding line

warm water flow

The Larsen Ice Shelf

David Vaughan

DAVID VAUGHAN is the director of science for the British Antarctic Survey (United Kingdom). He was a coordinating lead author for the 4th and 5th assessment of the Intergovernmental Panel on Climate Change (IPCC) and sits on the Science Board of the UK Natural Environment Research Council. Vaughan has participated in twelve Antarctic scientific field campaigns. He was awarded the Polar Medal and was made an Officer of the Order of the British Empire (United Kingdom).

During the second half of the twentieth century the Antarctic Peninsula warmed up by 2.5°C, much faster than mean global warming. This warming had many impacts, but the most prominent were irreversible changes that led to the retreat of five ice shelves, floating extensions of the grounded ice sheet. In 1998, scientists of the British Antarctic Survey (BAS) used numerical models to predict the future of a specific ice shelf called Larsen B and said that if it "were to retreat by a further few kilometres, it too is likely to enter an irreversible retreat phase". Just a few years later this prediction was realised!

In 2002, a 200-metre-thick ice shelf with an area the size of Rhode Island collapsed into small icebergs and fragments, separating itself from the Antarctic continent and floating off to melt in the Weddell Sea. It was not the occurrence of the ice-shelf collapse itself that was alarming but rather the speed at which it happened – over a period of just forty days between January and March 2002. The section of ice-shelf lost was named "Larsen B". It covered an area of 3,250 square kilometres and was the latest drama in a region of Antarctica that had experienced unprecedented warming in the previous fifty years, the Antarctic Peninsula.

The Larsen Ice Shelf is located on the eastern (Weddell Sea) side of the Peninsula. Geographically, distinct sections were named alphabetically from north to south, with the largest section being Larsen C – the second largest ice shelf in the Weddell Sea sector of Antarctica. Since the mid-1990s, parts of the Larsen Ice Shelf had retreated and some collapsed: Larsen A collapsed in 1995, followed by Larsen B in 2002. Scientists continued to monitor Larsen C, located the furthest south towards the base of the Peninsula, while satellite images showed a crack forming along its surface. Over the years, what looked like a hairline crack grew, and an eventual break was foreseen.

Ice accumulated by snowfall on the plateau in the interior of the continent is naturally transported to coastal regions, where it is released in the oceans by the action of ice-shelf meltwater and icebergs. The outward motion of the ice and its flow velocity are reproduced at the side as originally processed by a NASA Research Team at UC Irvine on the basis of satellite data from CSA, JAXA, and ESA.[1]

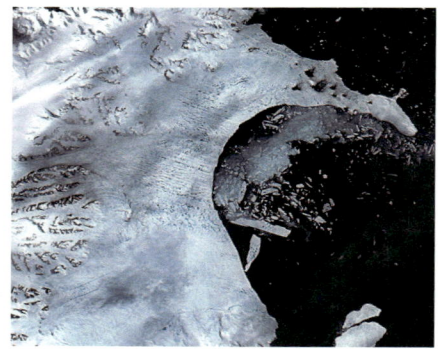

The Larsen B Ice Shelf – a 3,250 sqkm surface of ice floating on the ocean off the Antarctic Peninsula – collapsed before the eyes of scientists within the short time span of three months, between the 31st of January and the 13th of April 2002. Captured by the Moderate Resolution Imaging Spectroradiometer (MODIS) on NASA's Terra satellite, the sequence shows (from left to right): the ice shelf punctuated with parallel lines of blue pools of meltwater draining into crevasses; the retreated ice shelf splintering; and, finally, the shelf "disintegrated into a blue-tinged mixture (mélange) of slush and ice bergs. [By that stage], many of the bergs were too tall and narrow to float upright. They toppled over and spread out across the bay like a neat row of books that had been knocked off a shelf. When the bergs tipped over, the very pure ice on the bottom side of the ice shelf was exposed. The pale blue color is largely due to the reflection from this ice. Pure, thick ice absorbs a small amount of red light."[2]

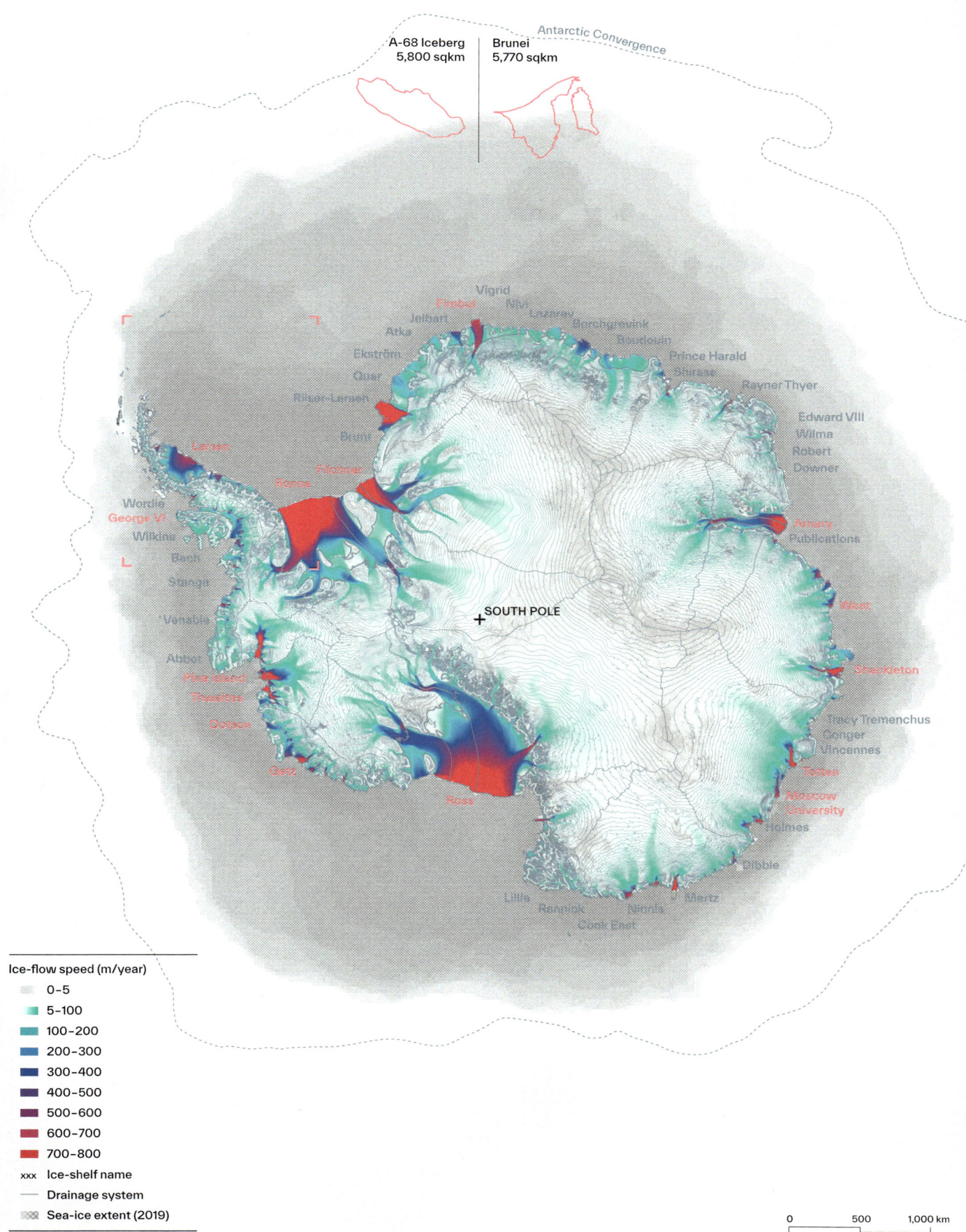

A-68 Iceberg
5,800 sqkm

Brunei
5,770 sqkm

Antarctic Convergence

Vigrid
Ekstra
Jelbart
Atka
Ekström
Quar
Riiser-Larsen

Nivl
Lazarev
Borchgrevink
Baudouin
Prince Harald
Shirase
Rayner Thyer

Brunt

Edward VIII
Wilma
Robert
Downer

Larsen
Filchner
Ronne

Amery
Publications

Wordie
George VI
Wilkins
Bach
Stange

West

Venable

SOUTH POLE

Shackleton

Abbot
Pine Island
Thwaites
Cosgrove

Tracy Tremenchus
Conger
Vincennes

Totten

Getz

Moscow
University
Holmes

Ross

Dibble

Lillie
Rennick
Cook East

Ninnis
Mertz

Ice-flow speed (m/year)

- 0–5
- 5–100
- 100–200
- 200–300
- 300–400
- 400–500
- 500–600
- 600–700
- 700–800

xxx   Ice-shelf name

—   Drainage system

▨   Sea-ice extent (2019)

0   500   1,000 km

The Antarctic Peninsula has witnessed an
unprecedented ice-shelf retreat in recent
decades. A comparative analysis of Antarctic
ice-shelf thinning along the coastline of
the Peninsula between 1950 and 2020 – with
a special focus on the infamous A-68 iceberg
– offers clear evidence of the accelerated
and alarming rate at which this phenomenon
is taking place.

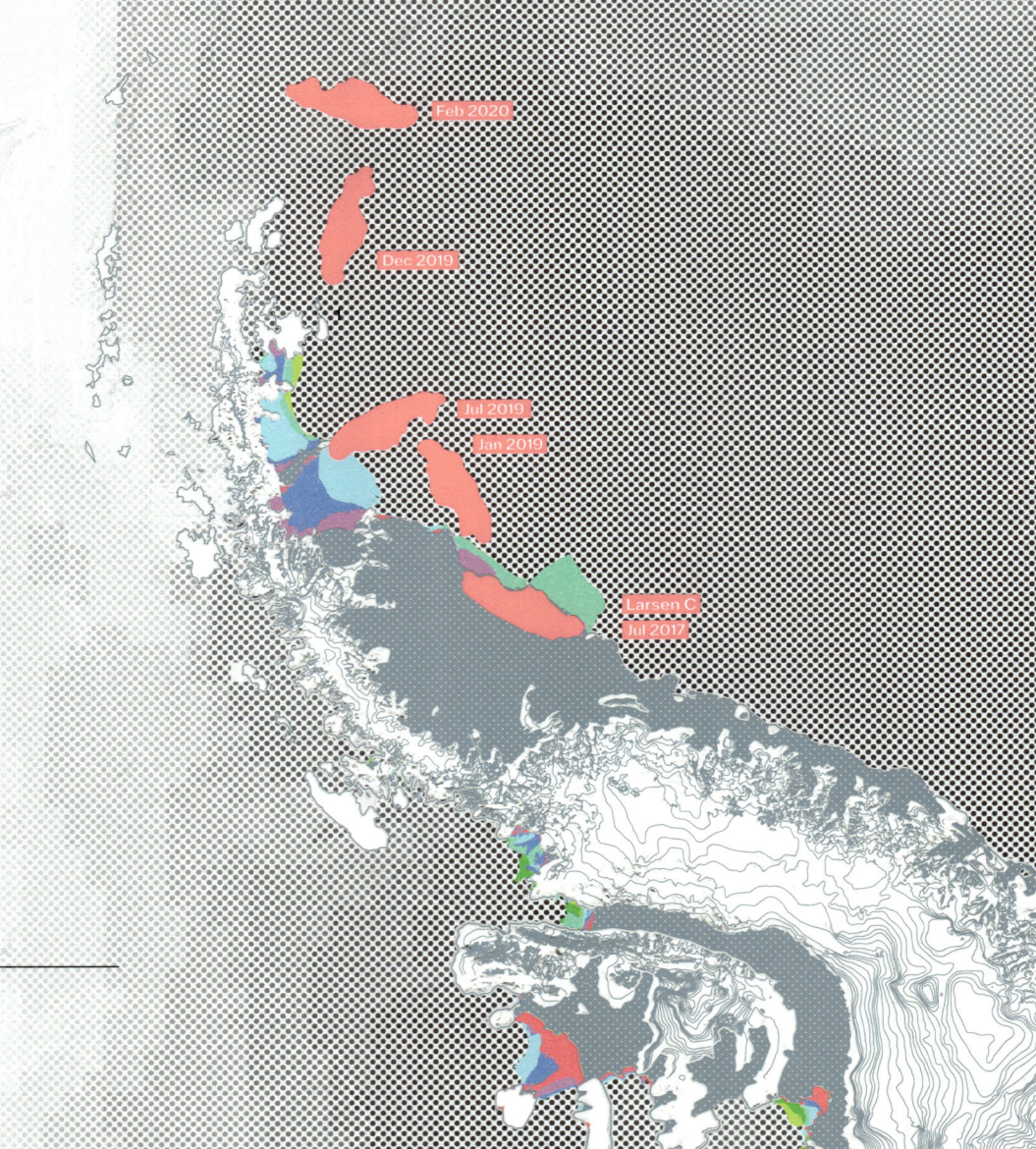

Feb 2020

Dec 2019

Jul 2019

Jan 2019

Larsen C
Jul 2017

Ice-shelf extent

- 1950
- 1960
- 1970
- 1980
- 1990
- 2000
- 2008
- 2020
- Sea-ice extent (2019)
- A68 Iceberg Route

0   100   200 km

In July 2017, after months of "hanging by a thread", a vast iceberg the size of Luxembourg, known as "A-68", broke off – or "calved" – Antarctica's Larsen C Ice Shelf. At more than 6,000 square kilometres in area, the new iceberg represents more than 10% of the total Larsen C ice shelf. Formed by a single massive crack, the iceberg was about 190 metres thick and has about 20 metres of "freeboard" sitting above the sea surface. As it broke away, it exposed the sea beneath and brought its marine life to light for the first time in perhaps 120,000 years. However, while the calving from Larsen C was significant, the formation of a single iceberg was not a collapse, and scientists have been left debating whether it was linked to climate change or just part of the natural life cycle of the ice shelf.

Importantly, the recent calving will not affect sea levels significantly because the ice shelf was already floating, and it is only the loss of ice that sits on land that contributes significantly to sea-level rise. However, if this calving forms part of a retreat that leads to further ice loss, then eventually the resistance it provides to the inland ice could be reduced. That would allow the inland glaciers to accelerate, contributing to sea-level rise. Larsen C ice shelf alone holds back the equivalent of about 1 centimetre of global sea-level rise, an amount equivalent to three years of current sea-level rise.

Evidence confirms that the Larsen C Ice Shelf has also been changing in other ways in recent years. Satellite data records show that the surface of Larsen C lowered between 1992 and 2010 and data from airborne surveys has indicated that this was due partly to the melting of ice from its base and partly to a loss of the air trapped in near-surface layers of the ice. The potential reasons for these changes include decreased snowfall, increased surface melting, changes in ice flow, and the melting of ice from below by ocean waters.

It is clear that human activities may have played a role in the atmospheric and ocean changes in the Antarctic Peninsula region through increasing greenhouse gases and ozone depletion, but natural fluctuations observed at this regional scale may also have been important.

The most recent computer simulations suggest that the A-68 portion was not critical to the stability of the rest of the ice shelf, and its loss alone will probably not have a significant long-term impact, so for the moment the ice shelf appears to be stable. However, the collapse of the neighbouring Larsen B Ice Shelf followed a period of successive calvings, which left the ice shelf vulnerable. If the A-68 calving of Larsen C is followed by similar events in the coming years and decades, then the ice shelf may be reduced to a point of instability, and collapse may follow.

"At around 5,100 sq km, the behemoth has been the largest free-floating block of ice in Antarctica since it broke away from the continent in July 2017. […] Having entered rougher, warmer waters – it is now riding currents that should take it towards the South Atlantic. […] A-68's name comes from a classification system run by the US National Ice Center, which divides the Antarctic into quadrants. Because the berg broke from the Larsen C Ice Shelf in the Weddell Sea, it got an 'A' designation. '68' was the latest number in the series of large calvings in that sector. […] When first calved in 2017, A-68 was close to 6,000 sq km, with an average thickness of about 190 metres."

Jonathan Amos (Science Correspondent), *BBC News*, 2020

"Mother of mighty ice bergs … the great grim giants you wean / Away from your broad white bosom"

Sir Ernest Henry Shackleton (Explorer), *The South Polar Times*, 1901–1903

West Antarctica Ice Shelf.
The Thwaites Glacier

David Vaughan

DAVID VAUGHAN is the director of science for the British Antarctic Survey (United Kingdom). He was a coordinating lead author for the 4th and 5th assessment of the Intergovernmental Panel on Climate Change (IPCC) and sits on the Science Board of the UK Natural Environment Research Council. Vaughan has participated in twelve Antarctic scientific field campaigns. He was awarded the Polar Medal and was made an Officer of the Order of the British Empire (United Kingdom).

Thwaites Glacier, located in West Antarctica, covers 192,000 square kilometres, a total area equal to that of Florida or the United Kingdom. Particularly susceptible to climate and ocean change, over the past thirty years the amount of ice flowing out of Thwaites and its neighbouring glaciers has nearly doubled, attracting scientific attention worldwide and adding to uncertainty in projections of sea-level rise. The current imbalance between ice loss through melting and iceberg formation and snowfall, accounts for about 4% of global sea-level rise. Further growth in this imbalance will increase this, and there remains the potential that a catastrophic and runaway collapse of the remaining glacier could, in coming centuries, lead to a significant increase in sea levels of around 65 centimetres.

It is important to note that while discussion often focuses on average global sea-level rise, for a variety of reasons this rise is not equally distributed around the world. One such reason relates specifically to ice sheets. As enormous quantities of ice are lost from the ice sheets, the pattern of gravitational anomalies around them changes such that sea levels in the area close to the glacier actually drop, while the sea level far from the ice sheet rises. As a result, sea levels in the northern hemisphere are actually more sensitive to ice loss in Antarctica than they are to ice loss from Greenland.

Established in 2018, the International Thwaites Glacier Collaboration (ITGC) aims to improve future predictions of global sea-level rise from Thwaites Glacier in West Antarctica through a better understanding of the present and past context of ice-sheet dynamics.

ITGC, which is jointly funded by the United States National Science Foundation and the United Kingdom Natural Environment Research Council, represents a global collaboration between the United States, United Kingdom, South Korea, Germany, and Sweden. It will last over six years and cost US$50 million. The United Kingdom and United States have invested heavily on Thwaites, as sea-level rise is a crucial issue for American and English coastlines and their protection is a national priority. However, the research will help improve our understanding of global sea-level projections in absolute terms.

Most researchers engaged in ITGC research travel through the United States McMurdo ↗ p. 866 Research Station and then to camps located on Thwaites Glacier. The logistics support in the field is provided jointly by the US Antarctic Program (USAP) and British Antarctic Survey. Two ambitious on-ice fieldwork seasons and several research cruises are required to support ITGC, and together they represent the largest, and most complex, field programme ever undertaken in Antarctica.

Five of ITGC's eight research projects (two are based on modelling that will not require field support) will work on and near the glacier to analyse different aspects of the ice, its bed, and how it interacts with the ocean.

Five dedicated teams of scientists and engineers have been working on Thwaites Glacier for the austral summer of 2019–2020 in below-freezing temperatures and extreme winds. The teams undertook work for several ITGC projects, some seeking to understand the interaction of the ice and the ocean beneath, others determining the history of the ice sheet over the past periods of glacial change. One project is focused on the boundary between fast-flowing ice and the slower-flowing ice that surrounds it. In coming seasons, other projects will extend this work to pursue an understanding of how the bed beneath the ice affects the ice flow.

Teams were deployed through an existing camp on the West Antarctic Ice Sheet (WAIS) Divide. Small aircraft were used to transport equipment and

"Thwaites Glacier is the size of Great Britain – or the US state of Idaho. Imagine a layer of ice about 1.5–2 km thick resting on top of that, then picture that vast mass, over the next few centuries, moving into the ocean."

Ted Scambos (Professor, Lead Principal Investigator, International Thwaites Glacier Collaboration), International Thwaites Glacier Collaboration (ITGC), 2020

"Glaciologists have described Thwaites as the 'most important' glacier in the world, the 'riskiest' glacier, even the 'doomsday glacier'. It is massive – roughly the size of Britain. It already accounts for 4% of world sea level rise each year – a huge figure for a single glacier – and satellite data show that it is melting increasingly rapidly. There is enough water locked up in it to raise world sea level by more than half a metre."

Justin Rowlatt (Journalist, News Reporter), BBC, January 2020

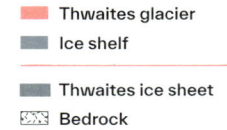

- Thwaites glacier
- Ice shelf
- Thwaites ice sheet
- Bedrock

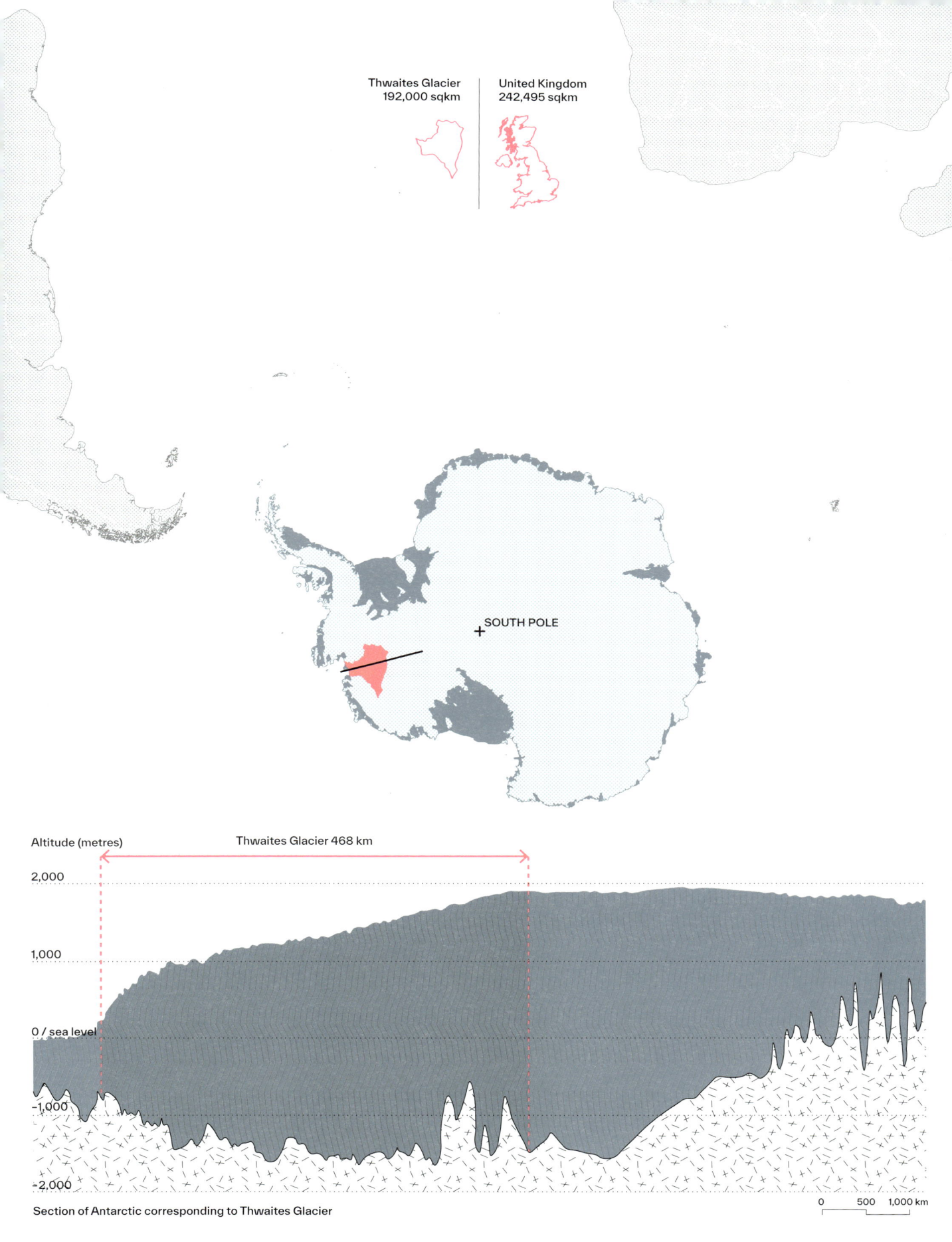

Thwaites Glacier
192,000 sqkm

United Kingdom
242,495 sqkm

+ SOUTH POLE

Altitude (metres)

Thwaites Glacier 468 km

2,000

1,000

0 / sea level

−1,000

−2,000

Section of Antarctic corresponding to Thwaites Glacier

0   500   1,000 km

people from the WAIS Divide to six smaller research camps, allowing scientists to use a range of methods to investigate the region. In particular, the MELT team, which comprises glaciologists and engineers, used hot water jets to drill through the ice at several sites, allowing them to deploy a suite of instruments – including the Icefin robot built by the Georgia Institute of Technology – beneath the floating ice shelf to observe how the ice interacts with the ocean and the underlying sediments.

A highlight of the 2019–2020 season were efforts to drill through Thwaites Glacier, with two teams using hot water to drill between 300 and 700 metres through the ice to the ocean and sediment beneath. The MELT team drilled two places beneath the glacier using hot water, including within two kilometres of the grounding zone, the area where the glacier meets the sea. The TARSAN team drilled at two locations about 30 kilometres further out on the floating shelf to explore the oceanographic conditions. Finally, the GHC team used the so-called Winkie Drill to extract four bedrock cores from beneath the ice, which will be used to understand the history of the ice sheet.

During this first ITGC field season, which marked the 200th anniversary of the first landings on the Antarctic continent, the ITGC research teams worked in Antarctica from November 2019 until March 2020. On Thwaites Glacier, over 100 scientists and staff participated and many more were needed to support them. With the field sites located up to 1,600 kilometres from either the English Rothera Station↗ p. 856 or the American McMurdo Station↗ p. 866, the logistical effort was enormous – however, the importance of the scientific questions that will be addressed by the ITGC programmes is truly global, and their success will benefit populations around the world.

At the time of writing, the impact of the coronavirus pandemic is already impacting the planning of upcoming ITGC fieldwork. While the timescales for completion of the remaining projects are uncertain, the commitment of the scientists, logistics experts, and funding agencies is not. For many of those involved, the goals of ITGC represent the realisation of career-long ambitions focused on work in Antarctica, which has always required patience and fortitude. The ITGC community is resolute and determined, and science will be delivered, just as soon as we can return to the ice.

"We designed Icefin [shown below] to be able to access the grounding zones of glaciers, places where observations have been nearly impossible but where rapid change is taking place. To have the chance to do this at Thwaites Glacier, which is such a critical hinge point in West Antarctica, is a dream come true. [...] We can definitely see it melting; there are a few places where you can see streams of particles coming off the glaciers, textures and particles that tell us it's melting pretty quickly and irregularly. [...] The grounding line is moving back – we know it's happening, but we don't know how quickly."

Britney Schmidt (Astrobiologist, Georgia Tech, NASA), 2020

Antarctica and Global Sea Level Rise

David Vaughan

DAVID VAUGHAN is the director of science for the British Antarctic Survey (United Kingdom). He was a coordinating lead author for the 4th and 5th assessment of the Intergovernmental Panel on Climate Change (IPCC) and sits on the Science Board of the UK Natural Environment Research Council. Vaughan has participated in twelve Antarctic scientific field campaigns. He was awarded the Polar Medal and was made an Officer of the Order of the British Empire (United Kingdom).

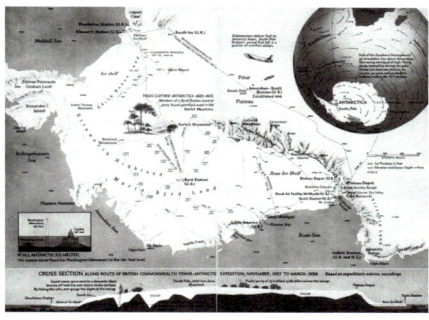

Together, the ice sheets of Antarctica and Greenland hold over 99% of all the ice and 90% of all the fresh water on Planet Earth. Today these massive ice sheets are changing, and as they lose their ice, sea levels around the world are rising. In future, as ice is lost at increasing rates, the fate of coastal communities depends on how these ice sheets – and in particular the West Antarctic ice sheet – feel the heat of global warming.

Globally, more than 100 million people live in coastal regions vulnerable to sea-level rise. Many of the world's largest cities are situated less than 10 metres above current sea level. While many of these cities have long histories of flooding, the effect of sea-level rise will be to increase the frequency and severity of the storm surges that damage economies and devastate lives. History has shown us that the chaos caused by the near-instant loss of critical infrastructure owing to coastal flooding can all but destroy even the most developed cities. As sea levels rise, economies will be damaged and many coastal communities displaced. Our maps will be redrawn, and precious coastal ecosystems will be drowned.

Since 1900, global mean sea level (GMSL) has risen by approximately 20 centimetres. The rate of sea-level rise has increased through the twentieth century and has reached its highest rate in both of the two decades of the twenty-first. Multiple satellite sensors confirm that the global average is around 3.2 centimetres per decade. Various processes contribute to this rise, but most, including the largest current contributors – the thermal expansion of warming ocean water and the melting of glaciers – are directly attributed to rising atmospheric temperatures.

It is only in the last two decades that scientists have been able to measure the last major contribution to global sea-level rise, that of the great ice sheets, Antarctica and Greenland. Even in this short period, however, they have observed an extraordinary increase in this contribution. This has tripled in the past two decades and now accounts for a third of total GMSL rise.

In Greenland, much of this acceleration can in large part be attributed to high air temperatures in summer, which have caused enhanced surface melting, but in Antarctica things are a little more complex.

The flow of the Antarctic Ice Sheet towards the ocean is to a large extent controlled by the floating ice shelves that fringe its coastline. These act to buttress, and effectively limit, the flow of the glaciers. However, measurements from satellites and on the ice shelves themselves show that many of Antarctica's ice shelves are thinning – a few are melting from above, but most are melting from below as warm ocean waters come in contact with the ice hundreds or thousands of metres below the surface.

A few ill-fated ice shelves, mostly around the Antarctic Peninsula, may have been thinning from above and below. Some, notably Larsen A in 1995 and Larsen B in 2003, became structurally unstable through warming and thinning and eventually collapsed in spectacular fashion.

The most concerning ice shelves are those along the coast of the West Antarctic Ice Sheet (WAIS). Here, recent changes in the ice shelves have led to some of the largest accelerations in ice-sheet flow seen in Antarctica. The vast Pine Island and Thwaites glaciers have been particularly hit, and this acceleration is particularly worrying for both of these. Inland on these glaciers, the beds on which the glacier rests slope downwards towards the centre of the ice sheet. Thinning at the point where the glacier goes afloat will drive this margin inland, reducing the amount of ice in contact with the bed and potentially leading to runaway ice flow. This particular vulnerability has led some scientists to predict a rapid future retreat of this portion of West Antarctica and, with it, rapid but

"Many explorers and scientists have guessed the depth of the ice beneath us. Estimates range far and deep. If the average guess is correct, the polar icecap, if melted, would raise the level of the seas by 40 feet – sufficient to require gondolas in the streets of London and to leave tourists stranded on the steps of Lincoln Memorial in Washington D.C."

Paul Allman Siple (Explorer, Geographer), "We are Living at the South Pole", *National Geographic*, 1958

uncertain rates of sea-level rise, beginning in a few decades but lasting many centuries.

Understanding the future of the West Antarctic Survey is the focus for many scientists, including the satellite and ice-sheet modelling communities, although field scientists such as those of the British Antarctic Survey and their partners involved in the recently launched International Thwaites Glacier Collaboration still have a vital role to play in understanding and quantifying the key processes in these most remote and harsh environments and collecting data that can be used to reduce uncertainty in projections.

To date, the most influential projections of global sea-level rise have been published by the Intergovernmental Panel on Climate Change (IPCC) Special Report on Global Warming of 1.5°C. These projections indicate that GMSL rise by 2100 is likely to lie in the range of 22–77 centimetres if global warming is limited to 1.5°C above pre-industrial levels, or 35–93 centimetres if warming reaches 2°C. It is the contribution of the West Antarctic ice sheet, which accounts for much of the uncertainty in such projections.

Beyond 2100, the sea level will continue to rise for many centuries, even if greenhouse gas (GHG) emissions are reduced to net zero in line with the 2016 Paris Agreement. However, the magnitude and the rate of this long-term sea-level rise is heavily dependent on near-term emissions reductions in coming decades. The sooner net-zero or net-negative GHG emissions are achieved, the higher the likelihood that long-term sea-level rise can be avoided. If GHG emissions are left unchecked, then the likelihood will increase that an irreversible collapse of the Pine Island and Thwaites glaciers and their neighbours in West Antarctica will be triggered. The complexity of the interactions between atmosphere, oceans, and ice, the relative short period in which changes have been observed, and the remoteness of these glaciers make this a uniquely difficult scientific question to address, but as the scale of what is at stake becomes apparent, our science communities are increasingly working together to provide coastal communities around the world the data they urgently need to embrace policies that can protect them.

ALEXANDER PONOMAREV is an artist and professional sailor. He has staged several art projects mostly in the oceans, the Arctic and Antarctica. He represented Russia in the National Pavilion at the Venice Biennale and conceived the first Contemporary Art Biennale in the Antarctic, held on board the research vessel *Akademik Sergey Vavilov*. Ponomarev is a member of the Russian Academy of Arts and an Officier des Arts et des Lettres awarded by France.

"The Antarctic ice sheet has an average thickness of around 1800 m, its maximum ice depth reaching more than 4770 m. […] The total volume of the ice sheet is estimated at 30 million km³, representing 80% of terrestrial freshwater."

David Winston Harris Walton (Emeritus Professor, British Antarctic Survey), *Global Science from a Frozen Continent*, 2013

"There is enough ice in Antarctica that if it all melted, or even just flowed into the ocean, sea levels [would] rise by 60 metres."

Martin Siegert (Professor, Imperial College London), *The Guardian*, 2017

ANTARCTIC BIENNALE
Alexander Ponomarev

Antarctica is our little moon, a polygon, a spaceport for a man's view and jump into space. Antarctica puts humanity in direct contact with the forces of planetary evolution, and it therefore creates the capacity for us to raise our ways of thinking/doing/relating/coexisting to a planetary scale.

Antarctica is also "the primordial ocean of zero culture" that does not have any "cultural heritage" associated with it (as it was never fully inhabited by humans). The Antarctic Biennale is our signal, another attempt to construct new forms of culture and art here at the end of the world, which are formed on new principles – Mobility, Interdisciplinarity and Multiculturalism, principles that allow creating self-sufficient communities of the *Homo faber!*

Alexis Anastasiou, Hopeless Utopia, 1st Antarctic Biennale, 2017

In the unforeseen scenario in which all the world's ice caps and valley glaciers should melt entirely, the global sea level would rise by 80.32 metres.[1] The unprecedented inundation would displace millions of people around the world, launching the greatest migration ever witnessed by mankind, and erase coastal settlements – highlighted in fluorescent orange in the cartography below. Should the sea level rise by "only" "0.74 meters by 2100 as is conservatively predicted, some 115 million people will likely be displaced and 420,000 km² of land will be lost to the encroaching seas."[2]

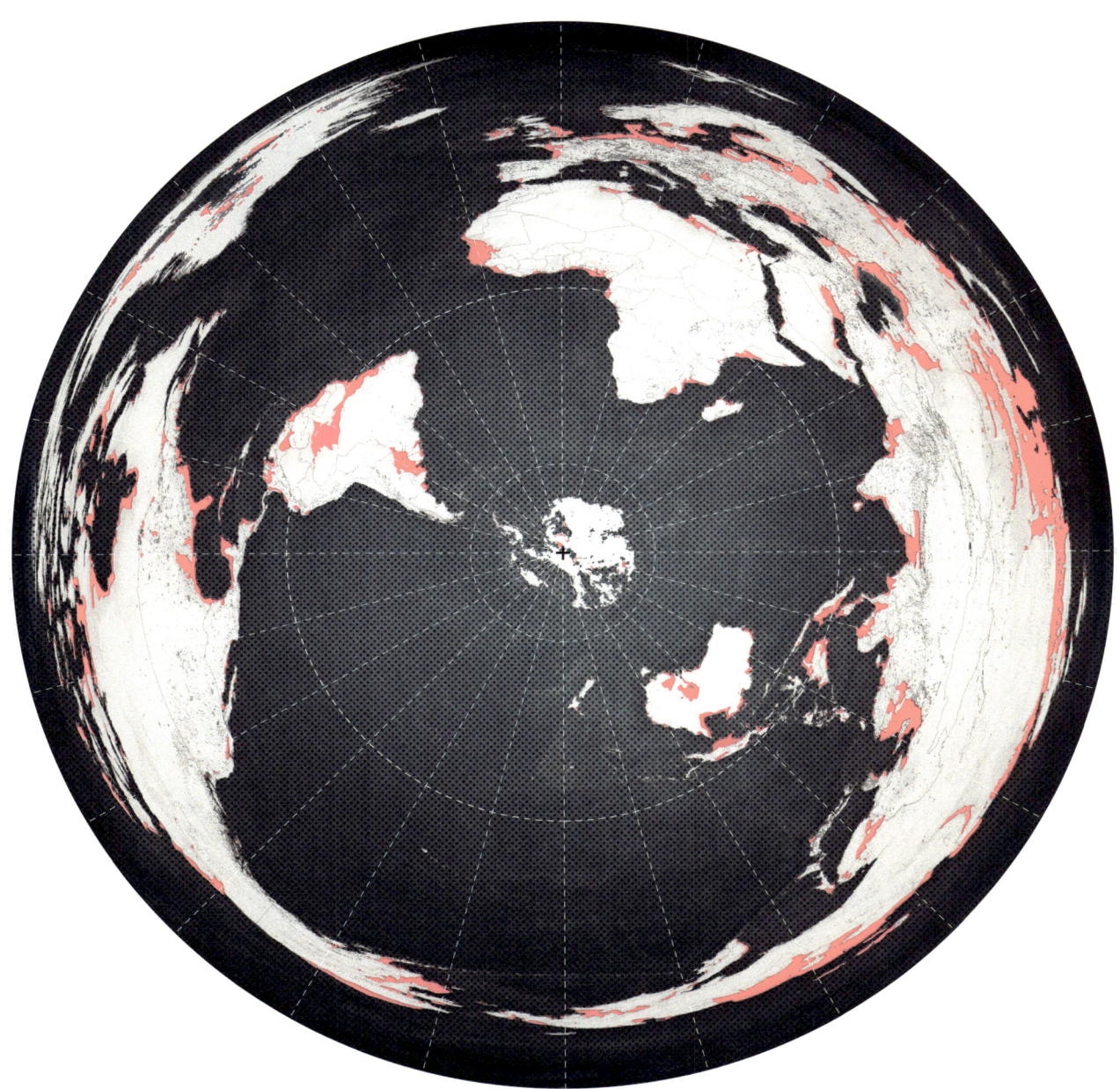

Topography above sea level (with ice caps melted)

Inundated areas as a result of all ice caps melted – Antarctica, Greenland, other ice caps, ice fields, and valley glaciers

+   South Pole

Greenland And Antarctica Are Melting Six Times Faster Than In The 1990s

Climate change: West Antarctica's Getz glaciers flowing faster

ears of satellite imagery shows receding of Thwaites Glacier

Teetering at the edge': Scientists warn of rapid melting of ntarctica's 'Doomsday glacier'

waites glacier is losing ice at an accelerating rate, threatening catastrophic sea-level rise

Antarctica collapse,

nland And Antarctica Are Melting
imes Faster Than In The 1990s

The Big Read World   + Add to myFT

The fight to own Antarctica

Competition for natural resources, research and tourism is putting pressure on the cold war-era treaty that guarantees order on the continent

CNN Weather Climate Storm Tracker Wildfire Tracker Video   Edition ⌄ 🔍 ⊙

Antarctica's colossal Thwaites Glacier is melting fast -- and scientists may have discovered why
By Emma Reynolds, CNN
⊙ Updated 1619 GMT (0019 HKT) September 9, 2020

Climate change is turning parts of Antarctica green, say scientists

Researchers map 'beginning of new ecosystem' as algae bloom across surface of melting snow

gest ice sheet on Earth more
nerable to melting than thought

ng evidence suggests that the last time the East Antarctic ice sheet collapsed,
d over 10 feet to sea level rise, and that it's likely to happen again.

Iceberg bigger than New York City broke off ice shelf in Antarctica

nt Ice Shelf in Antarctica calves

CNN By CNN | 3:21pm Mar 1, 2021

Radar images capture new Antarctic mega-iceberg

By Jonathan Amos
BBC Science Correspondent

imate change   + Add to myFT

Record Antarctic temperature met with the sound of cracking ice

igh of 18.3C this week underscores how rapidly climate change is
ffecting polar region

Melting Antarctic ice will r by 2.5 metres – even if Paris are met, study finds

Research says melting will continue even if tempe limited to 2C

South pole warming three times faste than rest of the world, our research sh

Dramatic change in Antarctica's interior in past three decades a
result of effects from tropical variability working together with
increasing greenhouse gases

Antarctica's colossal Thwaites Glacier is melt and scientists may have discovered why
By Emma Reynolds, CNN
⊙ Updated 1619 GMT (0019 HKT) September 9, 2020

Ancient Antarctic ice melt caused extreme sea level rise 129,000 years ago – and it could happen again
February 12, 2020 12:38pm GMT

This giant glacier in Antarctica is melting, and it cou raise sea levels by 5 feet, scientists say

By Leah Asmelash, CNN
⊙ Updated 2110 GMT (0510 HKT) March 25, 2020

Nasa detec

'I think it's called

A huge iceberg that's bigger than New York City broke off near a UK base in Antarctica

Big Read   Climate change

limate change: what Antarctica's loomsday glacier' means for the planet

waites Glacier is melting at an alarming rate,

Even the South Pole Is Warming, and Quickly, Scientists Say

Surface air temperatures at the bottom of the world have risen three times faster than the global average since the 1990s.

Reports   Events   Videos   🌐

giant ice shelves are vulnerable to dy warns

Giant crack frees a massive iceberg in Antarctica

Greenland's ice melting faster than any time in past 12,000 years

Increased loss of ice could trigger sea level rise of up to 10cm by end of century

CNN   **World**   Africa   Americas   Asia   Australia   China   Europe   India   Middle East   United Kingdom   Edition ⌄

Antarctic ice sheets capable of much faster melting than we thought

By Amy Woodyatt, CNN
🕐 Updated 1925 GMT (0325 HKT) May 28, 2020

ENVIRONMENT | NEWS

West Antarctica is melting—and it our fault

e sea level
mate goals

PLANET EARTH

Scientists find warm water beneath Antarctica's most at-risk glacier

Early Melting Detected Along the Antarctic
a

Thwaites Glacier is melting fast. But to understand how climate change is driving its decline, scientists need to send instrume through 2,000 feet of ice into the water below.

rises are

WORLD ECONOMIC FORUM

Biden returns US to Paris climate accor hours after becoming president

Global Agenda   Climate Change   The Ocean   Governance for Sustainability

The Washington Post

Global ice loss is catching up to worst-case scenario predictions

Capital Weather Gang

Temperature in Antarctica soars to near 70 degrees, appearing to topple continental record set days earlier

Climate graphic of the week: Icebe break from Brunt Ice Shelf

Glaciologists are monitoring whether the mass is at risk of causin damage as it is carried by the current

World's largest iceberg disintegrates into 'alphabet soup,' NASA photo shows
By Brandon Specktor · Senior Writer · 11 days ago

LATEST NEWS   To draw up road map ahead of UP polls, BJP state working committee to meet on March 15

Once the size of Delaware, iceberg A-68a is now a broken puzzle of ice.

Home / Technology / Science / Climate Change is causing Greenland, Antarctica to melt 6 times faster than in the 1990s

ntarctica melting

Climate Change is causing Greenland, Antarctica to melt 6 times faster than in the 1990s

08
A Continental Archipelago under Ice

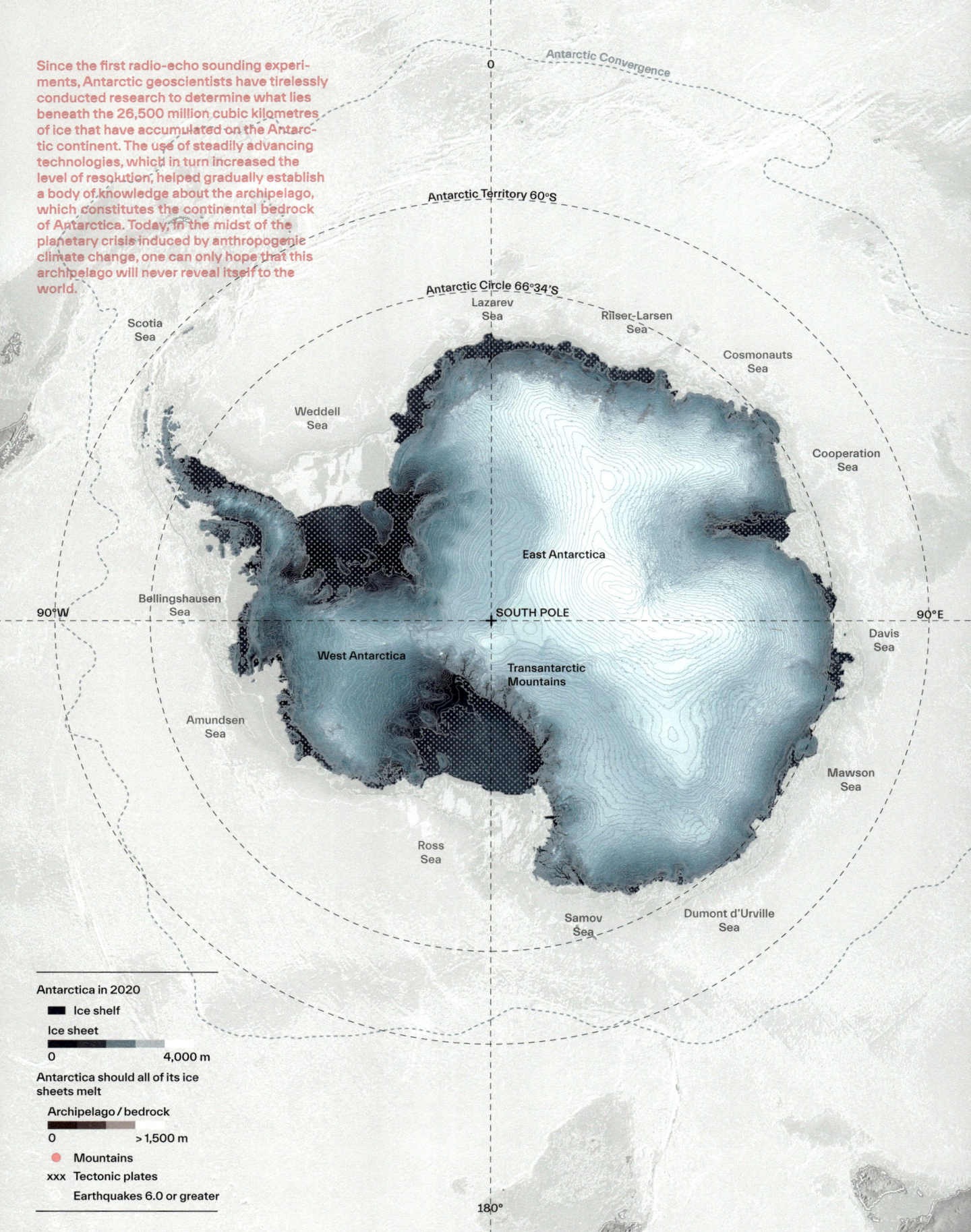

Since the first radio-echo sounding experiments, Antarctic geoscientists have tirelessly conducted research to determine what lies beneath the 26,500 million cubic kilometres of ice that have accumulated on the Antarctic continent. The use of steadily advancing technologies, which in turn increased the level of resolution, helped gradually establish a body of knowledge about the archipelago, which constitutes the continental bedrock of Antarctica. Today, in the midst of the planetary crisis induced by anthropogenic climate change, one can only hope that this archipelago will never reveal itself to the world.

Antarctic Convergence

0

Antarctic Territory 60°S

Antarctic Circle 66°34'S

Lazarev
Sea

Rïiser-Larsen
Sea

Scotia
Sea

Cosmonauts
Sea

Weddell
Sea

Cooperation
Sea

East Antarctica

90°W

SOUTH POLE

90°E

Bellingshausen
Sea

Davis
Sea

West Antarctica

Transantarctic
Mountains

Amundsen
Sea

Mawson
Sea

Ross
Sea

Samov
Sea

Dumont d'Urville
Sea

180°

Antarctica in 2020

Ice shelf

Ice sheet

0   4,000 m

Antarctica should all of its ice
sheets melt

Archipelago / bedrock

0   > 1,500 m

Mountains

xxx   Tectonic plates

Earthquakes 6.0 or greater

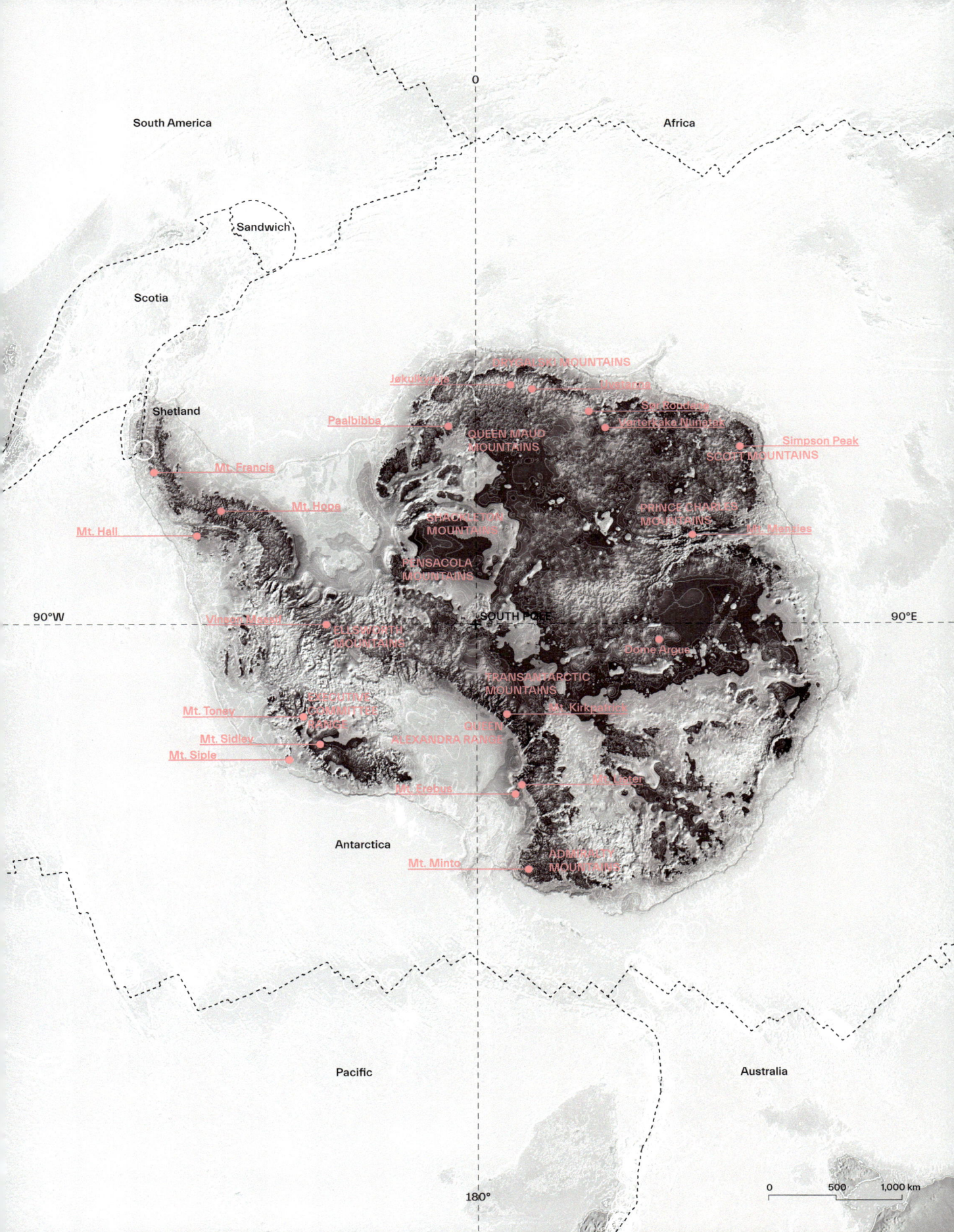

South America

Africa

Sandwich

Scotia

Shetland

0

MÜHLIG-HOFMANN MOUNTAINS

Jøkulkyrkje

Rüdtains

Sør Rondane

Vørterkake Nunatak

Simpson Peak

SCOTT MOUNTAINS

Paalbibba

QUEEN MAUD
MOUNTAINS

Mt. Francis

Mt. Hope

SHACKLETON
MOUNTAINS

PRINCE CHARLES
MOUNTAINS

Mt. Menzies

Mt. Hall

PENSACOLA
MOUNTAINS

90°W

Vinson Massif

ELLSWORTH
MOUNTAINS

SOUTH POLE

90°E

Dome Argus

TRANSANTARCTIC
MOUNTAINS

Mt. Toney

EXECUTIVE
COMMITTEE
RANGE

Mt. Kirkpatrick

Mt. Sidley

QUEEN
ALEXANDRA RANGE

Mt. Siple

Mt. Lister

Mt. Erebus

Antarctica

ADMIRALTY
MOUNTAINS

Mt. Minto

Pacific

Australia

0   500   1,000 km

180°

While Herbert Ponting was unable to appreciate the abstract quality of the kind of photographs he would have captured had he travelled with Scott's party to the South Pole, nor could he fathom what lay beneath the "featureless ice", over a decade later, on the 29th of November 1929, Ashley McKinley welcomed the opportunity to document the white landscape. During the epic first flight to the Pole aboard the Ford trimotor aircraft *Floyd Bennett,* McKinley captured Antarctica from above while performing radio-echo soundings, which were necessary to unveil the topography of the Antarctic continental bedrock.

"Anyway, after the party reached the Great Ice Barrier, there would be nothing to photograph but the level plain of boundless, featureless ice, or the long caravan stringing out toward the horizon."

Herbert George Ponting (Photographer, Explorer), *The Great White South,* 1921

"So far as we know, this polar plateau, as large as the United States, is a vast plain of snow and ice thousands of feet deep. Beneath may be mountains and valleys: we do not know. One of the significant contributions of the Anglo-New Zealand transantarctic expedition will be a series of dynamite explosions. Timed echoes will give us some idea of the thickness of the ice and conformation of the land below. Until then, we can only think of the interior of the Antarctic as a practically level whiteness going on almost forever, as if a snow-covered Kansas wheat field stretches from New York to San Francisco."

Paul Allman Siple (Explorer, Geographer), "We Are Living at the South Pole", *National Geographic,* 1958

Gondwana, Antarctica, and Continental Drift

Karsten Gohl and Georg Kleinschmidt

KARSTEN GOHL is a geophysicist and senior scientist at the Alfred Wegener Institute for Polar and Marine Research (Germany). GEORG KLEINSCHMIDT is a professor emeritus of geology at the Goethe University of Frankfurt (Germany).

The continent of Antarctica – larger than Australia or even Europe, with 97% of it covered in ice – is totally isolated and surrounded by the Southern Ocean. The distance to the southern tip of South America is nearly 1,000 kilometres, the distances to Australia and New Zealand are about 3,000 kilometres, the distances to the southern tips of Africa and of India are more than 4,000 and 9,000 kilometres respectively. Antarctica was not such an isolated continent in the past. Up to 160 million years ago, it was the central part of the supercontinent Gondwana – a huge continent formed around 500 to 600 million years ago through the collision of numerous smaller continental blocks.

Although Alfred Wegener first published his continental drift theory in his book *Die Entstehung der Kontinente und Ozeane* (The Origin of Continents and Oceans) in 1915, it took until the late 1960s for this theory to be widely accepted as plate tectonic principles based on increasingly convincing scientific evidence. Today, geophysicists and geologists have quite a thorough understanding of how the continents have drifted on the Earth's outer layer throughout Earth's history. Antarctica is an excellent example of how observed geophysical data and geological analysis of rock samples provide an integral picture of the continent's long geological history.

Dating back about 160 to 500 million years, Antarctica's history can be reconstructed by comparing similar rock formations in the Antarctic and the surrounding continents of the Southern Ocean. For instance, regions of South Australia are located on geological provinces with rock types and formation ages that show similarity with provinces in large areas of present-day East Antarctica, indicating that these regions were once juxtaposed. Large folded mountain ranges – geologically called "orogens" – are also used for reconstruction. Antarctica contains remnants of several of these orogens, some of which are about 500 million years old, examples being the Ross Orogen and the Pan-African Orogen, while others are much younger. Antarctic geologists have successfully identified geological structures from the Pan-African Orogen and the Ross Orogen that reposition Antarctica next to Africa and Australia in the context of Gondwana. The measurements of palaeomagnetic data from rock samples provide knowledge about the geographical location where this rock was formed, as the magnetic signal from that time remains preserved in the iron-bearing minerals. Therefore, this unique data can be used to re-establish the former positions of the continental fragments within the Gondwana supercontinent.

About 160 million years ago, the supercontinent Gondwana began to break apart, starting with Africa drifting away from Antarctica, later followed by India, Australia, New Zealand, and South America. The Earth's crust, forming the ocean basins between continents, preserves the magnetic signal at the time of formation, in much the same way as continental rocks do. But as oceanic crust forms steadily along mid-ocean ridges, these magnetic patterns can be identified with magnetic measurements from research ships and planes. These so-called magnetic-spreading anomalies are used to determine the ages of oceanic crust formation, providing the primary data set for plate tectonic reconstructions. According to the data, India broke off from Antarctica about 140 million years ago, after Africa, with Australia and New Zealand starting to separate from Antarctica 80–90 million years ago. The final tip of Australia, Tasmania and its South Tasman Rise, did not break apart from North Victoria Land in East Antarctica until 34 million years ago. Last came South America, with Patagonia separating from the Antarctic Peninsula in a complicated tectonic process 30–40 million years ago.

"Three hundred million years ago, Antarctica was located far to the north in the Gondwana supercontinent, surrounded by a tropical ocean. Progressively, plate tectonics drove Antarctica towards its modern south polar location, reached about 80 million years ago. […] The present East Antarctic ice sheet was probably formed 14 million years ago, while the West Antarctic ice sheet finished forming later (about 8 million years ago)."

David Winston Harris Walton (Emeritus Professor, British Antarctic Survey), *Global Science from a Frozen Continent*, 2013

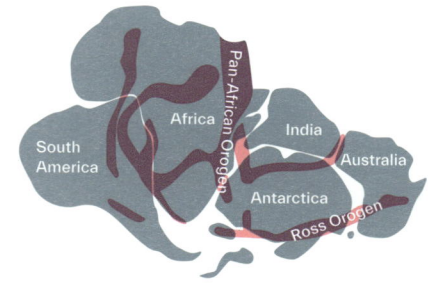

Antarctica was at the heart of the supercontinent Gondwana which existed from about 160 to 550 million years ago. The plate tectonics of continental fragments separated Antarctica from its surrounding land masses until it became completely isolated about 30 million years ago. The break-up of Gondwana, a slow process reconstructed at the side by retracing Bryan Storey's hypothesis, shows the supercontinent some 180 million years ago (1), its subsequent fragmentation when the South Atlantic opening occurred approximately 130 million years ago (2), the rift of Australia and New Zealand, dating to some 100 million years ago (3), and finally the isolation of Antarctica, which concluded approximately 35 million years ago (4).[1]

On long geological time scales, the break-up of continental pieces of the super-continent Gondwana, and particularly that of Antarctica, played a fundamental role in changing the global climate. The break-up of the last continental connections from Antarctica to Australia and South America enabled the establishment of a complete circum-Antarctic and circumpolar connection between the Atlantic Ocean, Indian Ocean, and Pacific Ocean. Strong southern hemispheric westerly winds, unobstructed by any land mass, power the largest ocean current system on Earth, the Antarctic Circumpolar Current, which diverts the warm currents coming from the tropics away from Antarctica. This geographical and thus thermal isolation of Antarctica initiated the beginning of the transition from a warm global climate to the relatively colder one of the last 30 million years.

From a subtropical land mass of the supercontinent Gondwana to a completely isolated and icy continent at the South Pole, the history of Antarctica is an exciting and unique story of long-term changes to the face of the Earth, to its environment and climate. Its present polar location and its gigantic ice masses (equivalent to a 60-metre rise in sea level), isolated by the surrounding Southern Ocean and the Antarctic Circumpolar Current, make this continent a key part of global climate processes.

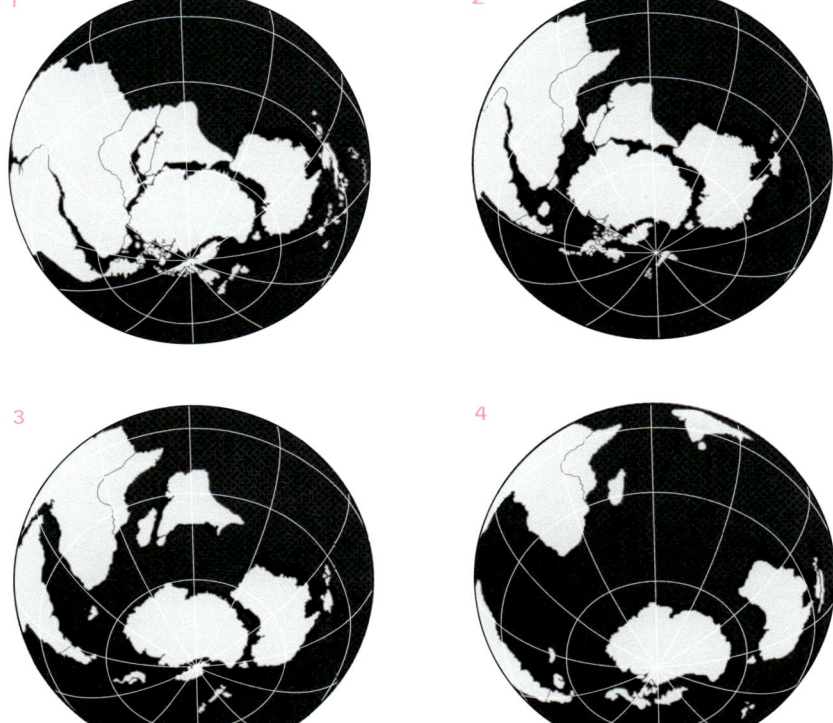

"And, yes, there probably were dinosaurs in the forests [...]. If you go to the tip of the Antarctic Peninsula, you'll find a whole range of fossils – things like hadrosaurs and sauropods, and primitive bird-like dinosaurs. The whole range of dinosaurs that lived in the rest of the world managed to get down to Antarctica during the Cretaceous."

Dame Jane Francis (Professor, British Antarctic Survey), *BBC*, 2020

Scientific Drilling on the Antarctic Margin

Gary Wilson

GARY WILSON is the chief scientist at the Institute of Geological and Nuclear Sciences and a professor of geology and marine science at the University of Otago (New Zealand). He is the vice-president of the Scientific Committee on Antarctic Research (SCAR) and has undertaken more than thirty research expeditions to Antarctica. He holds a BMus and a PhD. Dr Wilson's research has helped to uncover the dynamic history of Antarctica's climate and ice sheets.

Prior to the Dry Valley Drilling Project (DVDP) in the 1970s,[1] most of the geological attention had been focused on the Beacon Supergroup sequence that cropped out along the Transantarctic Mountains.[2] The Beacon Supergroup strata hold clues to Antarctica's central position in Gondwana and provide evidence for the Theory of Plate Tectonics and Continental Drift. However, the sequence ends with the Ferrar Dolerite injected across Antarctica around 177 million years ago,[3] as Gondwana began to break up. After that, there is limited evidence of Antarctica's history, as the continent moved into a mostly erosional phase and at some point became glaciated, although the record of that transition remains elusive. In order to decipher the last 100 million years of Antarctica's tectonic and climate history, geologists turned their attention to drilling around the margin of Antarctica into the sequences of sediment deposited offshore from the erosion of the continent.

The first of a sequence of drilling projects was the Dry Valley Drilling Project – a major collaborative undertaking between the United States, Japan, and New Zealand. It grew out of research conducted across the McMurdo Dry Valleys in the wake of the International Geophysical Year (IGY). This research had begun to alert geologists to the complexity of the Cenozoic history of glaciation and the fact that they could not unearth when or why Antarctica had become so heavily glaciated. Early drilling in the Valleys was associated with the Dry Valley lakes, but in 1974 two long cores (DVDP-10 and DVDP-11) were recovered from the thick sequence of sediments at the mouth of Taylor Valley revealing a complex glacial history. The lower intervals of the cores reveal multiple episodes of ice advance and retreat of the Taylor Glacier – sometimes extending beyond the coastline as a floating ice tongue and at times thick enough to ground on the sea floor. A major discovery, though, was a switch in the source of the sediments from the crystalline basement rocks of the valley to volcanic sediments out in McMurdo Sound[4] – implying a significant ice sheet on Antarctica and ice grounding across the Ross Sea. That transition was subsequently dated to the Plio-Pleistocene Boundary (2.59 million years ago).[5] DVDP culminated in an offshore drillhole (DVDP-15) recovered from the annual sea ice in the austral spring. The main aim for DVDP-15 was to recover the offshore Cenozoic glacigene sequence and get back to the pre-glacial sequence and warmer times in Antarctica. Unfortunately, the sequence turned out to be almost exclusively volcanic sand from the McMurdo Volcanoes, and the older glacial sequence was not penetrated before drilling terminated.

At the same time as drilling was taking place in the Dry Valleys, the *Glomar Challenger* drillship sailed into the Ross Sea on Leg 28 of the Deep Sea Drilling Project with the same aim of determining the long-term glacial, climatic, biostratigraphic, and geologic history of the Antarctic.[6] Cores from Leg 28 demonstrated that Antarctica was already extensively glaciated by the early Miocene (23 million years ago) and that sediments on the Antarctic Shelf were planed off in the Late Pliocene to Early Pleistocene (~2–4 million years ago) as grounded ice advanced out across the continental shelf. The early glaciation of Antarctica was confirmed by a core collected close to Macquarie Island by Leg 29 of the Deep Sea Drilling Project. Here, sedimentological and isotopic signals were used to identify early glaciation of Antarctica at the Eocene-Oligocene Boundary (~34 million years ago) coincident with the early development of the Antarctic Circumpolar Current.[7] Major glaciation followed in the Middle Miocene (~13 million years ago), when significant ice was predicted to have built up in East Antarctica.

By the end of the decade (to be precise, in 1979), the first core was recovered from the pre-Neogene (>24 million years ago) succession beneath McMurdo Sound – the McMurdo Sound Sediment and Tectonic Studies MSSTS-1 core.[8] The MSSTS-1 core was drilled in 195 metres of water from the floating sea ice about 15 kilometres offshore from Butter Point. Importantly, the drilling recovered a record of glacial advance and retreat throughout the Late Oligocene (between 24 and 29 million years ago) with coincident shoaling and deepening, respectively, implying that the glacial signal reflected significant ice growth and decay on East Antarctica. It wasn't for another seven years that the Cenozoic Investigations in the Ross Sea (CIROS-1) core penetrated the pre-glacial sediments. The CIROS-1 core was a major achievement – more than 700 metres of core were recovered in 197 metres of water only a few kilometres south of the MSST-1 site.[9] The lowermost part of the core is Eocene in age (>34 million years old), reflecting relatively warm conditions in Antarctica. That changed across the Eocene-Oligocene Boundary (~34 million years ago) with glacigene sediments beginning to be deposited. A major shoaling and switch to a dominantly glacial regime followed at the Mid-Oligocene (~29 million years ago).[10]

While the glacial history of Antarctica was slowly being pieced together from the cores recovered from the continental margin, the cause of glaciation was still unknown. Between 1997 and 1999, the Cape Roberts

Project drilled three deep cores from a dipping sequence 11–15 kilometres offshore from Cape Roberts, again from the annual sea ice but this time some 100 kilometres north of the CIROS and MSSTS sites. The three cores recovered a cumulative 1,500 metres of records spanning the Early Cenozoic (between 17 and 34 million years ago).[11] It was possible to date the Late Oligocene to Early Miocene (23–24 million years ago) interval with more precision than had been previously possible in Antarctic sequences, which arguably led to the discovery that the sequences of glacial advance and retreat were linked to the changing proximity of the poles to the Sun based on the obliquity cycle of the Earth's orbit through time.[12] Another decade later, the ANDRILL project recovered a 1,285-metre core (AND-1B) from beneath the floating McMurdo Ice Shelf.[13] The core sampled the succession that had accumulated in the flexural moat depressed into the crust from the building of the Erebus and earlier Ross Island volcanoes. The sequence recovered included a succession of more recent glacial and interglacial cycles and an extended snapshot of a warm interval in the Pliocene (~3.5 million years ago) when CO_2 levels peaked above 400 parts per million (ppm). The sediments indicated a complete loss of the West Antarctic ice sheet, and computer model simulations implied not just retreat of the ice shelf but loss of the West Antarctic Ice Sheet, contributing more than 3 metres to global sea-level rise.[14]

Recent drilling on the Amundsen Sea Shelf in Pine Island Bay used a remotely deployed sea-floor drill rig to recover sediments that predate those recovered from beneath the Ross Sea by the Cape Roberts Project.[15] The Amundsen Sea record reveals that, while still at the Pole, approximately 80 million years ago, and with atmospheric CO_2 levels greater than 1,000 parts per million, Antarctica supported a temperate rainforest of conifers and tree ferns.

Drilling on the Antarctic margin has revealed that Antarctica has supported a much more dynamic history than early exploration and initial observations implied. The state of the ice sheet and diversity of land vegetation seems to be strongly linked to the level of atmospheric CO_2. It will be very important for researchers to determine just how responsive the ice sheet is to middling levels of atmospheric CO_2 (500–1,000 parts per million), as the Earth looks set to be at these levels within the next fifty years.

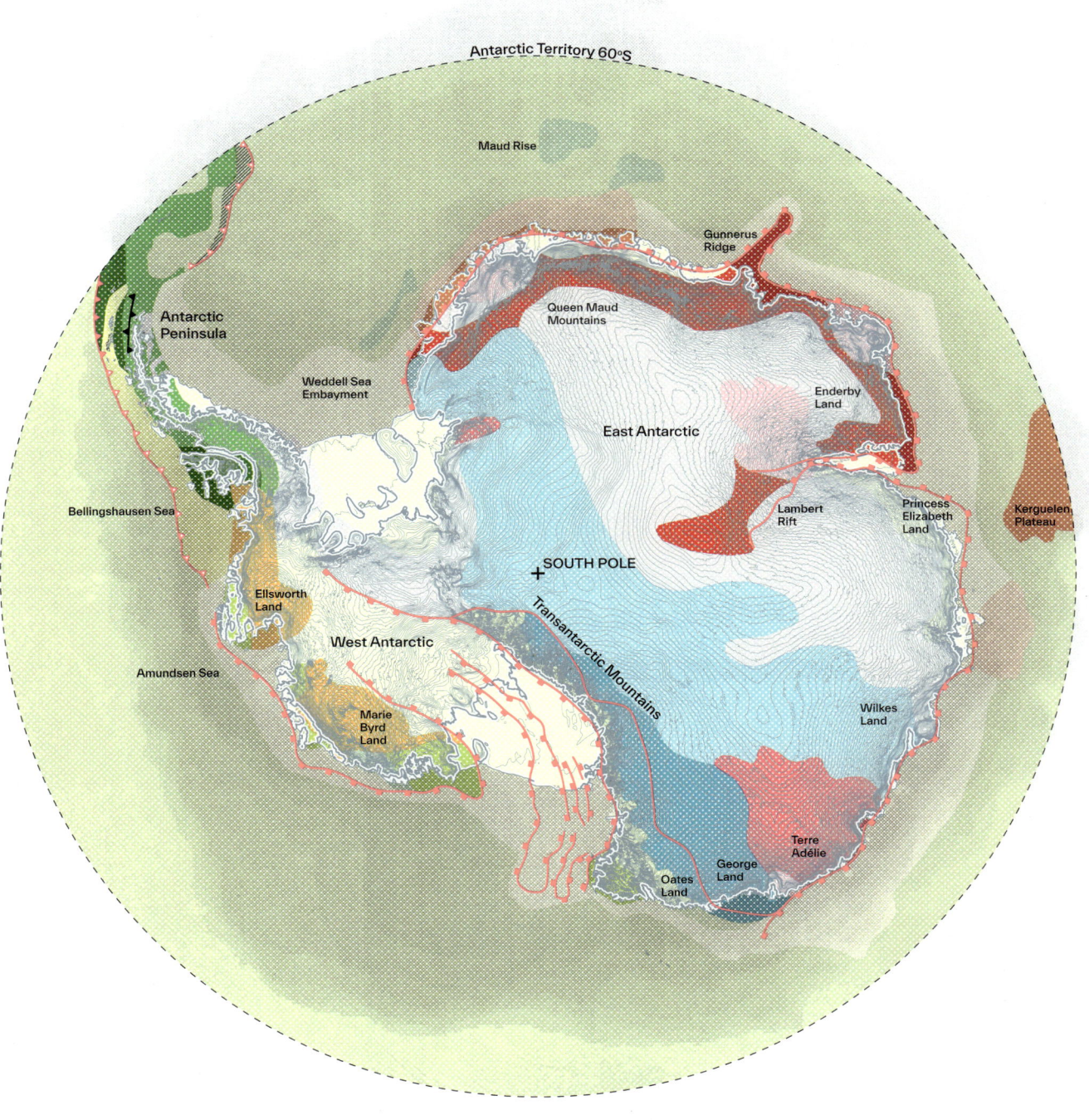

Antarctic Territory 60°S

Maud Rise

Gunnerus
Ridge

Antarctic
Peninsula

Queen Maud
Mountains

Enderby
Land

Weddell Sea
Embayment

East Antarctic

Bellingshausen Sea

Lambert
Rift

Princess
Elizabeth
Land

Kerguelen
Plateau

Ellsworth
Land

SOUTH POLE

West Antarctic

Transantarctic Mountains

Amundsen Sea

Wilkes
Land

Marie
Byrd
Land

Terre
Adélie

Oates
Land

George
Land

0   500   1,000 km

In April 2020 the BBC published the article *Dinosaurs Walked through Antarctic Rainforests,* exploring a recent discovery by scientists "drilling off the coast of West Antarctica [who] have found the fossil remains of forests that grew in the region 90 million years ago – in the time of the dinosaurs." The research was published in *Nature* by scientists from the Alfred Wegener Institute and British Antarctic Survey (BAS) who "used a novel cassette drill-mechanism called MeBo to extract core material some 30 metres under the seafloor" revealing "traces of ancient soils and pollen and even tree roots." According to Professor Jane Francis from BAS, the study (exemplified here by the X-ray of the sediment core, right) "represents the first evidence for Cretaceous forests so close to the South Pole – just 900 km away, at what would have been about 81–82 degrees South latitude."[1]

ARTURO LYON is the director of the Polar Lab Chile. He is an assistant professor at the School of Architecture of the Pontifical Catholic University of Chile. Lyon is the founder of Lyon Bosch Martic, where he developed architecture projects including the design of Las Majadas Hotel, the Chena Public Park, and the winning proposal for Nueva Alameda Providencia.

PLANT MIGRATION
Arturo Lyon

Antarctic fossil records have provided information that is key to understanding the evolution and biogeography of much of the flora and fauna of the southern hemisphere. Although currently only 3% of the continent's surface has exposed soils that might allow sites of palaeontological interest to be discovered, the continuous advance of ice melting is progressively revealing new petrified evidence of Antarctica's past.[1] The fossils have not only provided information about the various life forms that once inhabited Gondwana, they have also enabled an understanding of the different climate conditions in which they developed.[2] Evidence reveals that Antarctica's diverse ecosystems, when the continent was still part of Gondwana, included a great variety of plant species, such as the *Nothofagaceae* family, known as southern oak (*non-fagus* or false beech), currently represented by various species distributed in Australia, New Zealand, Chile, Argentina, New Guinea, and New Caledonia. Fossil remains of *Nothofagus* have been found in different parts of Antarctica, and the first record of it is attributed to Scott's *Terra Nova* expedition.[3]

Geological period

- 🟥 Mesoproterozoic mobile belt
- 🟥 Palaeoproterozoic massif
- 🟪 Archaean massif
- 🟫 Late Cenozoic
- 🟧 Late Cretaceous
- 🟨 Cretaceous
- 🟨 Forearc-related
- ⬜ Breakup-related
- 🟩 Late Mesozoic / Cenozoic fold belt
- 🟩 Early Mesozoic fold belt
- 🟩 Amundsen orogen
- 🟩 Borchgrevink orogen
- 🟩 Ross orogen
- ▦ Intra-plate folding beds
- 🟦 Gondwana system platforms
- 🟦 Early undivided platforms
- ⬜ Unknown geology

Major fault zones

- ⬩ Active subduction zone
- ⬩ Palaeosubduction zone
- ⬩ Rift boundaries
- — Assumed rift boundaries
- ⬟ Active spreading sea floor
- ▨ Neogene island arc

Geological time chart

CENOZOIC	MESOZOIC	PALAEOZOIC	PROTEROZOIC	ARCHAEAN

Neogene · Paleogene · Cretaceous · Jurassic · Triassic · Neo-proterozoic · Meso-proterozoic · Palaeo-proterozoic

Platforms

Mountain-building phases

Post-Gondwana sedimentary basin

Volcanic provinces

Precambrian massifs

Millions of years

0 · 20 · 40 · 60 · 80 · 100 · 150 · 200 · 250 · 300 · 400 · 500 · 600 · 1,000 · 1,500 · 2,000 · 2,500 · 3,000 · 3,500 · 4,000

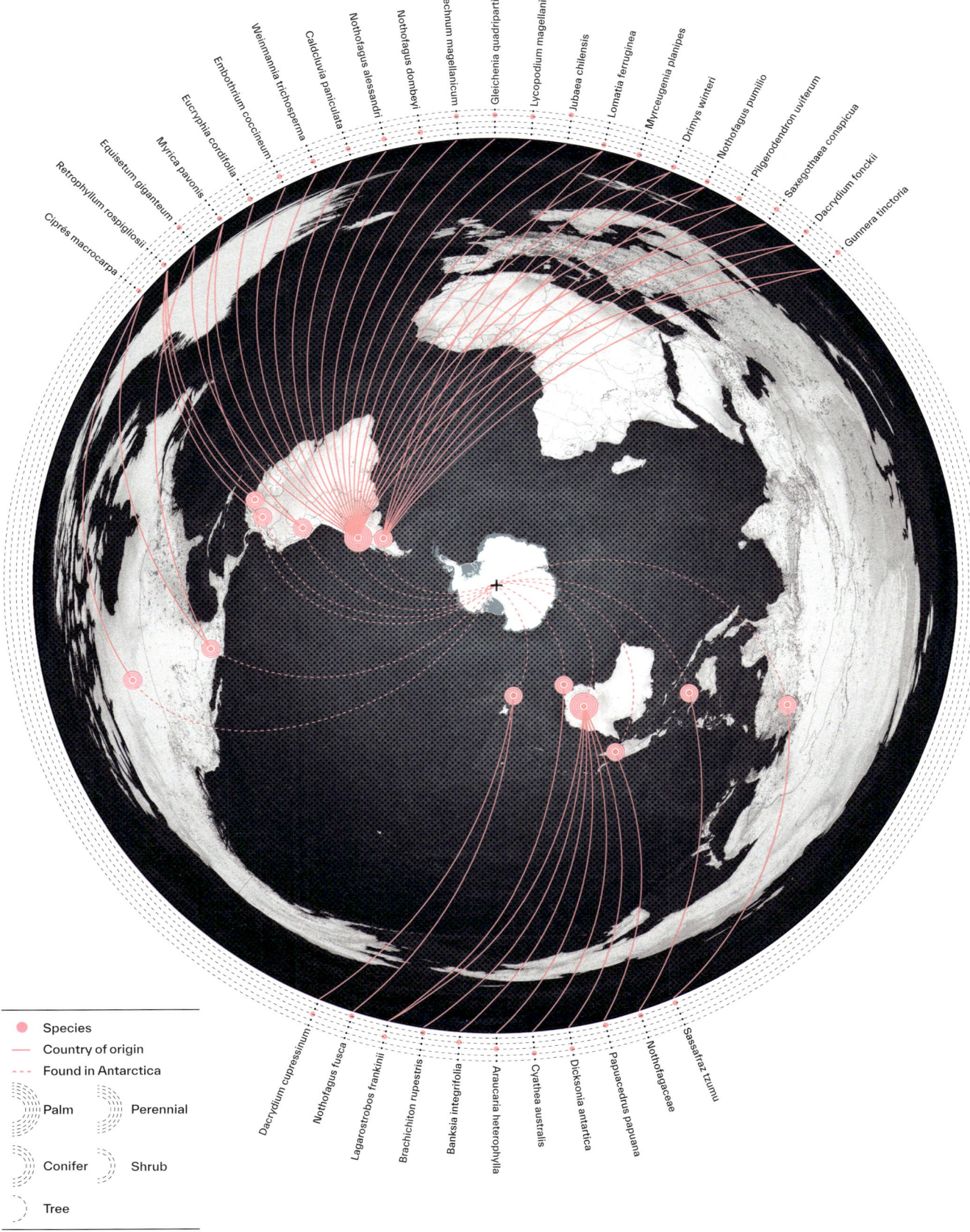

Cipres macrocarpa
Retrophyllum rospigliosii
Equisetum giganteum
Myrica pavonis
Eucryphia cordifolia
Embothrium coccineum
Weinmannia trichosperma
Caldcluvia paniculata
Nothofagus alessandri
Nothofagus dombeyi
Blechnum magellanicum
Gleichenia quadripartita
Lycopodium magellanicum
Jubaea chilensis
Lomatia ferruginea
Myrceugenia planipes
Drimys winteri
Nothofagus pumilio
Pilgerodendron uviferum
Saxegothaea conspicua
Dacrydium fonckii
Gunnera tinctoria

Dacrydium cupressinum
Nothofagus fusca
Lagarostrobos franklnii
Brachichiton rupestris
Banksia integrifolia
Araucaria heterophylla
Cyathea australis
Dicksonia antartica
Papuacedrus papuana
Nothofagaceae
Sassafraz tzumu

Legend

● Species
— Country of origin
-- Found in Antarctica

Palm   Perennial
Conifer   Shrub
Tree

Big Geo-Scientific Data in the Antarctic

Graeme Eagles and Mirko Scheinert

GRAEME EAGLES is a senior scientist and has been the head of Airborne Geophysics at the Alfred Wegener Institute for Polar and Marine Research in Bremerhaven (Germany). MIRKO SCHEINERT is a co-chair of the Geodetic Infrastructure of Antarctica (GIANT) group at the Scientific Committee on Antarctic Research (SCAR) and a senior scientist in the Group of Geodetic Earth System Research at the Department of Geosciences, Dresden University of Technology (Germany).

Antarctica is vast and remote, and covered by an ice sheet that is up to 4 kilometres thick. What lies beneath the ice? Well into the twentieth century, this question remained unanswered for geoscientists, whose maps showed at best a scattering of rock types that crop out of the ice at coastal nunataks, but otherwise merely a blank space. This started to change with the International Geophysical Year in 1957, which marked the onset of an era of near-surface remote sensing by sledges, manned aircraft, and, most recently, drones. Over this time, the areas of Antarctica yet to be visited by remote-sensing expeditions had become fewer and fewer so that, by around 2010, only relatively small parts of the continent could still be referred to as genuinely untouched. Millions of kilometres of survey, both systematic and opportunistic, had been conducted over the ice, delivering enormous quantities of data distributed along lines. This line data can be interpolated to form continuous grids using a combination of dedicated software and workflows developed either in the scientific setting, at universities and research institutes, or as a commercial undertaking for the resource-exploration industries. A tipping point in the use of such data and techniques was reached in the period between 2013 and 2018 with the completion and public availability of three large and complementary datasets: BEDMAP,[1] ADMAP,[2] and ANTGG.[3]

BEDMAP is a continent-wide compilation with a large quantity of radio-echo sounding (RES) data. RES illuminates the immediate underside of the ice sheet, enabling the rock surface of Antarctica, as well as large lakes, to be directly observed in radar reflections. Something of the geological history of Antarctica can be interpreted from these data sets by observing the variety of landforms produced by volcanism and tectonics, and by glacial and fluvial erosion. ADMAP, by contrast, compiles measurements of Antarctica's magnetic field variations. Earth's magnetic field is generated in its outer core by the convective motion of electrically conducting iron. These conditions produce a magnetic field, which in turn generates electrical currents as the iron continues to convect. The magnetic field created by this self-exciting dynamo will be maintained for as long as heat is lost by convection. It is enormously strong, and exits the planetary interior at the magnetic poles, curving around between them in colossal arcs that reach far out into space. The strong core field interacts everywhere with magnetically susceptible substances, including the wide variety of rock types in the upper part of Earth's crust. These interactions in turn produce smaller eddy fields, which locally deflect the core field, causing tiny yet measurable and predictable fluctuations in its strength and orientation. These are referred to as magnetic anomalies.

Measurements of Antarctica's gravity field are compiled in the ANTGG data set. The gravity field is simpler than the magnetic field in structure but can otherwise be viewed as similar. Any continuous mass concentration – or indeed any point mass – generates a gravitational field strength that is measurable by an appropriate device placed outside this mass. Everywhere on Earth, the dominant mass generating the gravitational attraction is the great bulk of the planet itself. Because of Earth's rotation, centrifugal acceleration is superimposed on this, and the resulting field is called the gravity field. Besides the gravitational acceleration, the field can be described by a scalar quantity, the gravitational potential. Its equipotential surface, close to the mean global ocean surface, is called the geoid. The geoid deviates by up to ±100 metres from a mean ellipsoid approximating the figure of the Earth. The gravitational acceleration vector is always perpendicular to the geoid and directed inwards. However, this acceleration is not exactly directed towards the centre of mass of the Earth but is deflected slightly by the effects of mass variations in the interior and at the

THE MAGNETIC CRUSADES

At the beginning of the nineteenth century, whilst it was extensively acknowledged that the Earth's magnetic field was constantly changing over time, the underlying reasons for such phenomena remained unknown. To establish a deeper understanding of the field, the United Kingdom, France, Germany, and Russia organised expeditions to expand the records of magnetic intensities around the world. One of the most prominent advocates for the so-called Magnetic Crusades was Sir Edward Sabine, who first made the case for "the next step in the advancement of knowledge", providing both theoretical and political justifications for new geomagnetic observations. In a meeting in 1837, he stated: "The earlier observations of terrestrial magnetism were made without reference to theory. As facts accumulated general conclusions arose. Their elaborate examination conducted to an hypothesis of four magnetic poles; and this, to the suggestion of new experiments to verify or disprove it. In the northern hemisphere the verification is complete, affording signal proof of the value of experiment directed by theory. A similar verification in the southern hemisphere is yet wanting; and the observations necessary for that purpose will also supply those elements of calculation whereby the hypothesis may be fitted for a detailed comparison with facts."[1] A year later, John Herschel proceeded with the planning of an Antarctic Expedition, which was officially approved in March 1839, leading to the appointment of James Clark Ross as its commander. Observatories were selected in both hemispheres, while an expedition was dispatched to the Southern Ocean to carry out a magnetic survey of the Antarctic. Whilst Ross's expedition to the Magnetic South Pole proved unsuccessful, along with attempts by Dumont D'Urville (1837–1840) and Charles Wilkes (1838–1842), observations from the Toronto records enabled Sabine to deduce that magnetic variations could be divided into a regular diurnal cycle, leading him to announce, in April 1852, that the Sun's eleven-year sunspot cycle was "absolutely identical" to the Earth's eleven-year geomagnetic cycle. Throughout the 1840s and 1850s, Sabine continued to superintend the operation of magnetic observatories which resulted in his magnum opus: the first magnetic survey of the globe. The South Magnetic Pole was to be claimed more than fifty years later, on the 16th of January 1909 by members of Sir Ernest Shackleton's *Nimrod* Expedition. There is now some doubt as to whether their location was correct, as the approximate position of the pole on the 16th of January 1909 was 72.25°S 155.15°E.[2]

surface of the Earth, which are related to the distribution of rock types of differing density, as well as water, ice, soil, and air. These variations can be expressed and referred to as gravitational anomalies.

Since the gravitational field is a global quantity, the ANTGG dataset was built upon satellite measurements that enable the long-wavelength signal to be resolved. In particular, observations of the Gravity Field and Steady-State Ocean Circulation Explorer (GOCE) were used to infer a global solution with a half-wavelength resolution of up to 80 kilometres. GOCE, a dedicated satellite mission exploiting the novel technique of gravity gradiometry, was launched in March 2009 and ceased operations in November 2013.[4] To consolidate the observational basis and increase resolution, terrestrial measurements need to be incorporated. In Antarctica this is a major challenge because of its vast expanse and the hostile environment. In this respect, airborne campaigns are a powerful means of surveying large areas and dealing with the complicated Antarctic conditions in the best way. As a result, gravimetry is mostly realised as part of a geoscientific airborne campaign comprising elements, or the entire suite, of geophysical-geodetic instrumentation. Additional techniques comprise shipborne measurements and relative gravity measurements directly carried out on the ice or bedrock surface. In this way, ANTGG was able to collect more than 13 million data points covering an area of about 10 million square kilometres (corresponding to about 73% of the entire area of the Antarctic continent). Coordinated by international science organisations – namely, the Expert Group on

"The Bedmap2 project is about more than making a map of the landscape. The data we've put together on the height and thickness of the ice and the shape of the landscape below are fundamental to modelling the behavior of the ice sheet in the future. This matters because in some places, ice along the edges of Antarctica is being lost rapidly to the sea, driving up sea level. Knowing how much the sea will rise is of global importance, and these maps are a step towards that goal."

Hamish Pritchard (Scientist, British Antarctic Survey), *The British Antarctic Survey,* 2013

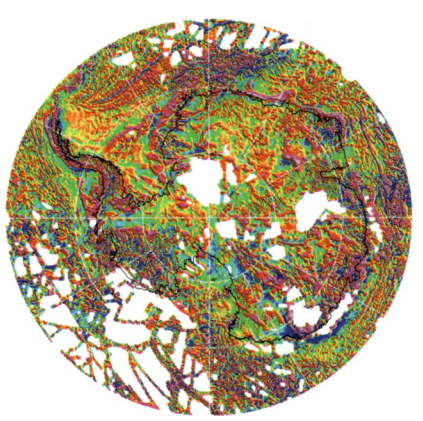

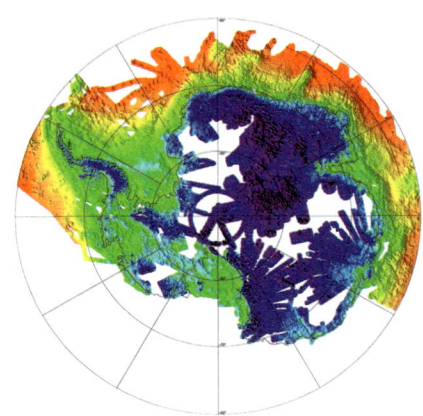

BEDMAP

ADMAP

AntGG

At the beginning of the millennium, a consortium of scientists worked together to collate and compile all the available geophysical data from approximately 120 previous surveys, providing an updated seamless digital elevation model of Antarctica's bedrock, known as Bedmap1. In 2013, Bedmap2 saw the addition of a further 27 million data points; a record overtaken by BedMachine Antarctica in 2019 (shown above), which includes modelled outputs and information on ice-flow motion derived from satellite data alongside survey data.[1]

The second version of ADMAP, shown above, represents the most up-to-date magnetic anomaly map for Antarctica: it offers a deeper understanding of the geological structure and tectonic history of the continent, while also shedding light on the lithospheric transition of Antarctica to its oceanic basins.

In 2003 the International Association of Geodesy (IAG) launched the first gravity anomaly grid at continental scale – known as Gravity and Geoid in Antarctica (AntGG) – for Antarctica. AntGG database compiles more than 13 million gravity data points and 1 million line-kilometres of aerogravimetry originating from terrestrial, airborne, and shipborne surveys. Altogether, the compilation covers 73% of the Antarctic continent including ice shelves.

PETER T. FRETWELL is a geographic information officer with the British Antarctic Survey (United Kingdom). He chairs several international Antarctic groups, including the Bedmap3 project and the Censusing Animal Populations from Space group. Dr Fretwell's interests include Antarctic satellite remote sensing of wildlife, mapping the terrestrial, ocean, and subglacial environment, polar geographic analysis, and GIS. He has recently published the *Antarctic Atlas* (Penguin, 2020).

Geodetic Infrastructure in Antarctica (GIANT) of the Scientific Committee on Antarctic Research (SCAR) and the Subcommission 2.4f Gravity and Geoid in Antarctica (AntGG) of the International Association of Geodesy (IAG) – the data provision was possible thanks to the collaboration of a dedicated international community.

In contrast to the geophysical applications, geodesy investigates the exterior gravity field of the Earth. The geoid serves as a reference surface for height systems, which is one of the main tasks of geodesy. The geoid also forms a reference surface when inferring the mean sea-surface topography, a major quantity in oceanography. Moreover, as the Southern Ocean is covered by sea ice, icebergs, or ice shelves, the geoid is designed to link the freeboard height (the height of the ice body above the mean ocean surface) and the surface ellipsoidal height. This relationship can be used to determine the thickness of the floating ice body; while the geoid – or adjacent equipotential surfaces – is used when studying Antarctic subglacial lakes.[5]

Magnetic and gravity anomalies can be presented as maps and profiles that allow interpretation of the distribution of rock types even if they are hidden beneath kilometres of ice or water. Magnetic anomalies are especially suitable for understanding the uppermost 15 kilometres or so of rocks in the crust, because the higher temperatures at greater depth reduce magnetic susceptibilities in nearly all rock types. Gravity anomalies are not similarly limited and so can be used to see deeper into the crust and mantle, albeit at the cost of having to deconvolve multiple signals from a variety of vertically stacked density contrasts.

The Antarctic big data trinity of BEDMAP, ADMAP, and ANTGG powerfully illustrates and confirms the expectation that Antarctica as a continent is only unusual in that it currently hosts perennial ice. Before this, like any other continent, Antarctica experienced a long history of plate tectonic activity. Famously, this included the break-up of the supercontinent Gondwana, which explains how some of the continent's geology can be understood by analogy to its now-distant Gondwanan neighbours, Africa, India, South America, Zealandia, and Australia.

East Antarctica – or broadly those parts of the continent that lie in the region swept out clockwise between the 30°W and 170°E meridians – consists of an amalgamation of very ancient continental cores, known as cratons, welded together by a network of so-called mobile belts, remnants of the deep roots of ancient mountain ranges and volcanic arcs that formed between the jostling cratons during the first 3.5 billion years of the planet's evolution. BEDMAP reflects this in the broad and relatively flat pattern of continental topography that existed as a consequence of hundreds of millions of years of erosion that preceded the build-up of the ice. ANTGG reveals the relatively weak gravitational attraction of the old thick crust, which is less dense than the mantle rocks it sits on, in terms of relatively weak gravitational attraction. The magnetic anomalies of ADMAP are subdued over the thick cratons but form intricate branching and anastomosing patterns over the diverse tectonically stretched and smeared rocks of the mobile belts.

The ANTGG gravity anomaly data set is of great value to geophysics, geodesy, and other geoscientific disciplines. Huge efforts have been and are still being made to improve the observational basis, mainly by carrying out aerogravimetric observations. The next generation ANTGG data set is already being processed with the aim of extending coverage and improving resolution.

By contrast, the big data trinity shows West Antarctica, between 170°E and 30°W, to be geologically much younger and, in places, still tectonically and

THE ANTARCTIC DIGITAL DATABASE
Peter T. Fretwell

The Antarctic Digital Database (ADD) is a dynamically updated, seamless data set of topographic data below 60°S. It is managed, updated, and maintained on behalf of the Scientific Committee on Antarctic Research (SCAR) by the Mapping and Geographic Information Centre at the British Antarctic Survey and is overseen by SCAR's Scientific Committee on Antarctic Geographic Information. Since it was first published in 1993, the ADD has become the primary source of topographic information for the continent and a crucial source of GIS data for a wide range of applications, including science, operations, planning, and mapping.

The ADD was first published in April 1993. The initial baseline data came from over two hundred regional scale maps from eleven national mapping agencies, digitised at the British Antarctic Survey in Cambridge. Most of these maps, produced after extensive aerial survey campaigns carried out in the decades following the International Geophysical Year of 1958, were in the scales of 1:200,000 and 1:250,000. Produced in light of the publication of the first standard global digital topographic data set, the Digital Chart of the World promoted in 1992 by the American Defence Mapping Agency, the database aimed at incorporating Antarctic data into future compilations of the world digital charts.

The input data was collected, edited, and harmonised into a seamless compilation at three differing scales: the most detailed scale varied considerably between regions; the second scale was a medium-resolution data set (planned to be applicable to a regional and continental scale of 1:1,000,000); and the third scale was the least detailed – a generalised set of data at a scale of 1:10,000,000. For each scale, both line and polygon data sets were provided along with data containing attributes of information such as coastline type (rock coastline, ice coastline, or grounding line), surface type for polygons (land or ice shelf), and the origin of the input data. Since 1993, the initial data has been supplemented and superseded by new vector data provided by national mapping agencies. At present, only around 5% of the original coastline data remains unchanged: however, some of the contour data and virtually all (96.4%) of the rock data remain from the original compilation.

The ADD's output provides topographic GIS vector data, with the main emphasis on

volcanically active. Starting over 500 million years ago, the crust of West Antarctica started to thicken rapidly with the addition to the Pacific margin of huge quantities of sediment and volcanic islands caused by the long-term action of plate convergence between the continent and the neighbouring ocean. This process, known as subduction, underlies the volcanism and mountain building that continues around much of the Pacific to the present day. West Antarctica was no exception, developing its own long-lived belt of coastal volcanoes, which are prominent features in BEDMAP, and whose deep roots of thick gabbro rocks produce strong magnetic anomalies in ADMAP. The crust came to be so thick that it developed a tendency to collapse under its own weight. The decay of radioactive elements in the sediments heated the crust, destabilising it further. On at least one occasion over the last 200 million years, a rising plume of convecting hot rocks in the mantle reached the underside of West Antarctica, giving rise to the presence of widespread volcanic and igneous rocks with characteristically strong magnetic anomalies visible in ADMAP. Perhaps in response to its own gravitational and thermal instability, West Antarctica started to spread out, its crust stretching in the process, starting around 85 million years ago and continuing until very recently. BEDMAP reveals the presence of multiple deep rift valleys formed during this stretching, whilst strong gravity anomalies in ANTGG testify to the presence of dense mantle rocks that reached up close to the surface beneath the thinning crust. The rise of this mantle material saw it decompress and partially melt, feeding a new population of inland volcanoes that appear as hundreds of spot-like magnetic anomalies scattered throughout the rift zone in West Antarctica.

The availability of these big data sets – and their future evolutions and additions, such as BEDMachine Antarctica – provides the raw materials not only for writing new chapters in the geological story of Antarctica but also for further deepening our understanding of Antarctica as the geological environment in which the present-day ice sheets formed and are maintained, and from which, in the future, they may be lost. Data sets similar to ADMAP have, for example, prompted efforts to determine the amount of heat reaching the base of the ice sheet from the deep earth below,[6] and data sets like those in ANTGG are used to determine the depth to the sea floor beneath the vast tongues of floating ice that extend out into the oceans from the continent.[7] Both are critical parameters in modelling how the ice might respond to ongoing climatic and oceanographic change, and thus, of direct consequence for understanding the coming rise in the global sea level and its potentially devastating consequences for coastal regions worldwide.

three primary topographic features: coastlines, contours, and the extent of rock outcrop. The coverage of these three primary data sets is seamless and continent-wide; they are each served out at a number of scales. Several other topographic types are also contained within the data set such as streams, lakes, and moraine, although the coverage of these is less complete. Now, updated on a regular basis, the database remains the primary Antarctic source of topographic data.

The nature of the Antarctic environment is dynamic. Ice coastlines, ice tongues, and ice shelves advance and calve back tens of kilometres over a natural cycle that could last decades, grounding lines shift, and the extent of rock outcrop changes over time. This means that the geography of the data included in the database is dynamic too and these changes are reflected in the GIS layers. Over the last twenty-five years, the database has documented both the evolution of knowledge about the topography of Antarctica and major changes in the physical landscape, such as the break-up of the Larsen Ice Shelf and the retreat of coastal glacier fronts. In addition to these natural changes, the availability of increasingly high-resolution satellite imagery and better geo-positioning of its features have revolutionised our knowledge of the continent and the quality of the geospatial data in the ADD. A number of continent-wide, remotely sensed data sets now make up important parts of the input data and this trend is likely to continue in the future.

The Land underneath the Ice Cap

David Vaughan

DAVID VAUGHAN is the director of science for the British Antarctic Survey (United Kingdom). He was a coordinating lead author for the 4th and 5th assessment of the Intergovernmental Panel on Climate Change (IPCC) and sits on the Science Board of the UK Natural Environment Research Council. Vaughan has participated in twelve Antarctic scientific field campaigns. He was awarded the Polar Medal and was made an Officer of the Order of the British Empire (United Kingdom).

Comprised of both land mass and ice, Antarctica is the least known and most poorly mapped of the continents. Mapping the topography and features of land beneath the ice in Antarctica relies on radio-echo sounding (RES). This technique normally requires an aircraft to fly over a region, using wing-mounted radar antennae to emit a radio signal that penetrates the ice, bouncing back from the point at which the ice meets the rock, sediment, or water beneath. Since its development in the 1960s, this technique has led to the discovery of subglacial lakes under the ice sheet, as well as the mapping of the land beneath the ice sheet.

Since the RES technique, at present, allows measurements to be made directly below the aircraft, glaciologists use techniques of interpolation to fill in the areas between the flight tracks. This is an unavoidably imprecise technique, especially where flight tracks are widely spaced, so significant sectors of Antarctica remain poorly resolved and critical spatial details are missing. The first "complete" digital subglacial map of Antarctica and its surrounding seas was named "Bedmap". It was delivered in 2001 with a second iteration – which incorporated ice thickness data from nineteen different research institutes dating back to 1967 and nearly a million line-miles of radar soundings – unveiled in 2013 (Bedmap2). The Bedmachine product was delivered in late 2019 and is based on a more advanced approach using an ice-sheet model to aid the interpolation. This marks a step away from normal cartographic practices but undoubtedly provides improvements to the subglacial topography in many data-poor areas.

Today, many researchers use RES measurements, seismic techniques, and cartographic data to map regional areas to support research in geology, glaciology, and many other more surprising areas of research, such as meteorite hunting. Most recently, NASA's Operation IceBridge campaigns have used long-range aircraft to collect many thousands of kilometres of RES data, without setting foot on the continent.

Several features of Antarctica's land mass were revealed for the first time in 2013 using Bedmap2, including a new deepest point: the bed under the Byrd Glacier in Victoria Land is 2,870 metres below sea level, making it the lowest point on any of the Earth's continental plates. A number of other key statistics emerged: the volume of ice in Antarctica is 4.6% greater than previously thought, with sufficient ice to raise global sea level by around 58 metres. The mean bed depth of Antarctica was located at 95 metres, i.e. 60 metres lower than previously estimated.

With the creation of the newly released Antarctic topography map "BedMachine" (published on the 12th of December 2019 in the journal *Nature Geoscience*), a team of glaciologists has unveiled the most accurate portrait yet of the contours of the land beneath Antarctica's ice sheet. This has helped identify which regions of the continent are going to be more, or less, vulnerable to future climate warming and provided ice-sheet modellers with the basic information they need to predict the future.

BedMachine improves on the previous cartographic approaches and produces an improved depiction of the bed under Antarctica's ice sheet by using a map of the ice-flow velocity and a model of ice flow. This can improve the process of interpolation significantly. The same methodology has been successfully employed in Greenland in recent years, transforming cryosphere researchers' understanding of ice dynamics and the mechanisms of glacier retreat.

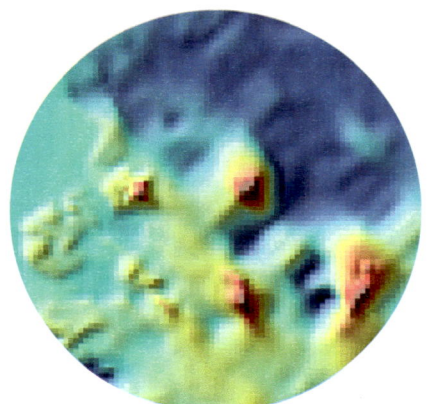

2013

Bedmap1, 2001
Resolution: 5 kilometres per pixel

2018

Bedmap2, 2013
Resolution: 1 kilometre per pixel

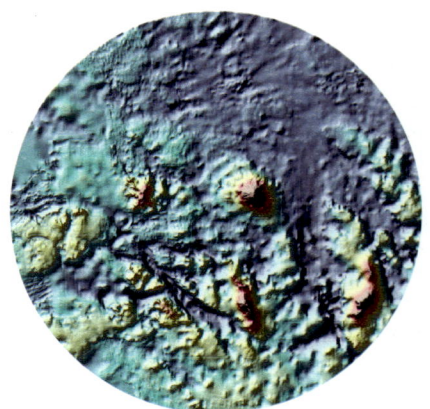

2020

BedMachine Antarctica, 2020
Resolution: 500 metres per pixel

The Geodynamic Evolution of the Continent and Magnetic Anomalies

Graeme Eagles

GRAEME EAGLES is a senior scientist at the Alfred Wegener Institute for Polar and Marine Research in Bremerhaven (Germany). He is the head of Airborne Geophysics at the Alfred Wegener Institute and has participated in numerous Antarctic marine and airborne expeditions. Dr Eagles completed a PhD on Antarctic plate tectonics at the British Antarctic Survey and the University of Leeds (United Kingdom).

Antarctica was first conceived in Western thought as an idea, *Terra Australis.* Whilst originally defined with respect to the Sun's transit, the association of the concepts of north and south, and thus Antarctica, with Earth's magnetic field has been culturally present since the eleventh-century invention of the magnetic compass in China and its adoption in the West in medieval times. However, the existence of *Terra Australis* as a separate southern continent was only proven centuries later, in the early nineteenth century with the first Russian, British, and American sightings of its coasts and ice shelves, and the continent remained as an empty space on maps for a further eighty years, until ship-based expeditions set out to explore it from Europe and Australia, returning the first geological and geographical details from the coastal regions. The first airborne surveys took place in the 1920s. Reaching the magnetic South Pole over land, and thus pinpointing it with laborious measurements of the inclination of the magnetic field, was a specific target throughout this early period.

The requirements of submarine warfare subsequently led to improvements in magnetic detection technology, such that so-called magnetic anomalies, the tiny effects of interactions between Earth's main magnetic field, generated deep in the outer core, and magnetically susceptible rock types in the upper 15 kilometres of the crust, could be measured quickly and reliably. Since the International Geophysical Year (IGY) in 1957, near-surface remote sensing by sledges and manned and unmanned aircraft in Antarctica has returned more than 3 million kilometres of magnetic survey data. These data sets have been interpolated to form a nearly continuous grid, ADMAP2, with a spatial resolution ranging between 500 metres and 50 kilometres. Consequently, it is now possible to interpret something of the geodynamic history of around 95% of the surface of Antarctica on the basis of the magnetic signals generated by rock contrasts buried by as much as 4 kilometres of ice.

The interior of Antarctica records the geodynamic history that it experienced over several supercontinent cycles, during which time the so-called Grunehogna, Napier, and Ruker cratons (blocks of earth's continental crust that have been stable for billions of years) repeatedly diverged from one another and collided again in various different configurations during the Earth's 4,500 million-year-long history. By 1,100 million years ago, the Ruker and Grunehogna cratons had collided in Queen Maud Land, in one of Antarctica's last cratonic collisions, trapping between them a large area of younger crust consisting of volcanic oceanic islands. Remnants of the island arcs appear as strong magnetic anomalies, sandwiched between more subdued signals from the less magnetically susceptible sediments covering the cratons and the surrounding sea floor.

Geodynamically, the Antarctic tectonic plate comprises not just the continent but also large areas of the sea floor that extend well into temperate southern latitudes. This deep sea floor is magnetically very distinct, revealing tell-tale patterns of stripes caused by the ongoing process of the sea floor spreading in the presence of periodic reversals in the polarity of the field generated in the outer core. One spectacular example of this pattern is in the Weddell Sea, where a strongly arcuate set of stripes records the 160-million-year-long history of South America's separation from Antarctica by a process of slow rotation around the Antarctic Peninsula.

The Antarctic continent itself today is defined by its continental margins, which are well defined by magnetic anomalies. The margin of West Antarctica has experienced ongoing (since approximately 500 million years ago) collision processes that do not involve cratons but instead are held between the oceanic crust of the Pacific Ocean (and its predecessors) and the continental crust of

"Simpson had also a small hut at the foot of his hill. It was divided into two compartments, in one of which he charted the form and movements of clouds with the aid of a 'camera obscura'; and in the other, which was heated with a small coal stove, he incarcerated himself for several hours each week to make 'absolute' calculations with a magnetometer, to check the instrument in the ice grotto."

Herbert George Ponting (Photographer, Explorer), *The Great White South*, 1921 western Antarctica. In this collision, the denser oceanic crust has been continually thrust beneath the continent, where it sinks into the planetary interior. Less dense sediments and seamounts are scraped off the ocean at the interface, thickening the continental crust there and giving rise to relatively subdued magnetic anomalies by virtue of their weak magnetic susceptibilities. Under pressure at depth below the continental margin, the oceanic crust loses water and volatile elements, which change the chemistry of the overlying mantle rocks so that they melt at lower-than-usual temperatures. The resulting magma finds its way to the surface to form a continuous chain of volcanoes, whose deeper roots of magnetically very susceptible gabbro give rise to a spectacular continuous magnetic anomaly known as the Pacific Margin Anomaly inland of the continental shelf edge.

The margins of the rest of Antarctica are geodynamically younger, all dating from the last 180 million years as the continent extracted itself, by extension and break-up, from its former neighbours in the supercontinent Gondwana. The geodynamic extension process sees continents become thinner from top to bottom and hot mantle rocks rise to fill the space created by thinning. As they rise, they decompress and melt, producing magma that is chemically distinct from – and at the surface less viscous than – that formed at collisional margins like West Antarctica's. The low viscosity sees the lavas generate broad, flat flows, rather than steep-sided volcanoes. In Antarctica, flows like this give rise to distinct broad, smooth, magnetic anomalies at the margins of the Weddell and Riiser-Larsen seas. Similar anomalies are also present in the continental interior – for example, all along the Transantarctic Mountains – and mark episodes of continental thinning that did not succeed in causing continental break-up.

With around 5% of the continent still never having been visited, basic magnetic surveying in Antarctica is still ongoing, while the results of some older surveys, which have very low resolution and large navigational uncertainties, are being replaced in order to keep pace with the requirements for future applications of magnetic anomaly data – for example, in estimating the magnitude of subglacial heat flow. The increased homogeneity of data quality will enable the depth of our understanding of Antarctica's geodynamic history to finally approach that of the inhabited continents.

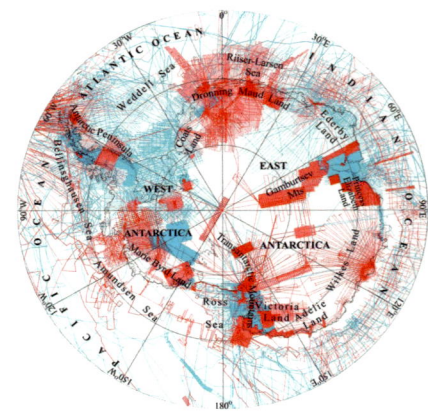

Compiled in 2018 in a project called ADMAP (Antarctic Digital Magnetic Anomaly Compilation), the cartography presents, in overlay, the total coverage of aeromagnetic and marine magnetic surveys acquired in Antarctica by the international magnetic community over a fifty-year period. The blue lines represent measurements taken during the first magnetic data compilation in 2001, while the red ones show those acquired since then; black lines represent all marine magnetic lines. In total, a staggering 3.5 million line-kilometres of data have been collected for the purpose of exploring the Antarctic continent and surrounding oceans.

Based on plate tectonic theories (exemplified in the diagram at the side), the Earth's geography and its lithographic section are in a continual state of evolution and the oceanic crust is constantly transforming at divergent plate boundaries as a result of mantle upwelling. The study of ancient planetary geography, also known as paleogeography, is informed by six fields of research. Of these, paleomagnetism makes it possible to measure the remnant magnetic field preserved in rocks containing iron-bearing minerals, which provide evidence about their original alignment to the Earth's magnetic field – which at the equator is parallel to the Earth's surface and at the poles almost perpendicular. The orientation captured by the solidifying rock offers insights into their original latitude of deposition prior to continental shifts. The periodic reverse polarity of Planet Earth (whereby the magnetic field of the North and South Poles flips) leads to anomalies in the magnetic field intensity that allows the age of fossil evidence to be determined.

Exploring Antarctic Subglacial Geology and Lithosphere

Fausto Ferraccioli

FAUSTO FERRACCIOLI is the science leader of the Geology and Geophysics Team and the head of Airborne Geophysics at the British Antarctic Survey (United Kingdom). He has led aerogeophysical studies of Antarctic subglacial geology over the last twenty-five years and has published 130 papers, including in *Nature* and *Science*. In 2010, Ferraccioli was awarded the Polar Medal by HRH Prince Philip for outstanding dedication and achievements in polar science.

Antarctica is the centrepiece of an ancient land mass that existed from about 180 to 550 million years ago called Gondwana, which also included Australia and New Zealand, India, Africa, and South America. Antarctica broke off from these continents, reaching the South Pole about 100 million years ago, where it became covered by vast ice sheets about 34 million years ago.

How does one explore this almost entirely ice-covered continent and reveal its jealously kept geological secrets? While the geology scattered along the coast provides tantalizing glimpses into over three billion years of history on Earth, one needs to expand the knowledge of what lies beneath the ice sheet (the so-called subglacial geology) and explore deeper down, peering into the rigid outer shell of Earth, which includes the crust and the upper mantle: the lithosphere. This is possible using airborne and satellite geophysical approaches. While radar observations allow for the mapping of the bedrock on which the Antarctic ice sheet flows, uncovering mountain ranges and deep subglacial basins and identifying subglacial lakes, magnetic and gravity data are required to reveal the subglacial geology and the deeper structure beneath the ice sheet.

By systematically conducting aeromagnetic and aerogravity surveys, the international community has collected a staggering 3.5 million line-kilometres of magnetic data, and gravity data for more than 70% of the continent. This data reveals major geological features, such as subduction zones, rift systems, cratons, orogens and suture zones (the places where continents or pieces of continents collided), and major sedimentary basins and volcanic provinces. By reconstructing, pixel by pixel, a large-scale picture of the white continent's "building blocks", this data enables fundamental tectonic processes to be studied. Using these magnetic and gravity data sets, we can now project the geology that we can see along the coast deep into the interior of the Antarctic ice sheet and formulate new geological hypotheses.

Despite the wealth of geophysical observations, many mysteries and enigmas remain. The Gamburtsev Subglacial Mountains are a prime example: they sit in the middle of East Antarctica and are completely ice-covered. Located in the middle of an ancient cratonic region, where one expects to see no mountains at all, these features resemble the rugged Alps. Analysis of radar, magnetic, and gravity data has made it possible for these phantom mountains to be mapped in detail and their existence explained on the basis of a combination

A	Bedrock data obtained via radar observations
B	Lithosphere data obtained via satellite gravity observations
01	Stationary or moving plate
02	Island arc
03	Trench at convergent boundary
04	Transform fault at transform boundary
05	Oceanic ridge at divergent boundary
06	Rising magma
07	Subduction zone
08	"Roll-back"
09	Hot spot
☐	Ocean
▨	Rock
	Asthenosphere
▬	Lithosphere
—	Oceanic crust
—	Continental crust

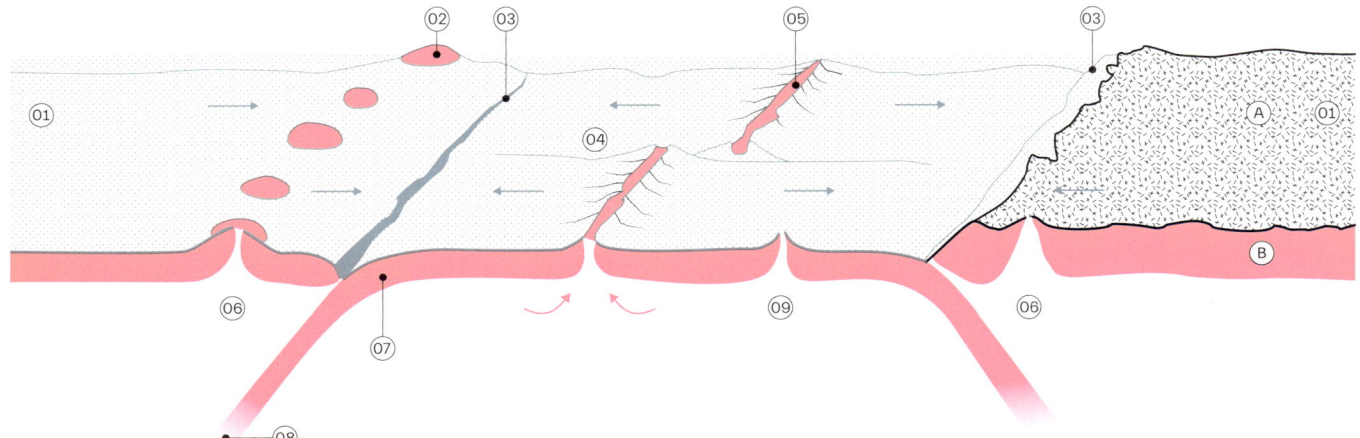

of a remarkably thick crust in interior East Antarctica and the superimposed effects of rifting and erosion caused by rivers and glaciers.

How can we peer even deeper into the continent's depths? In other continents, seismological approaches are often used, based on the analysis of seismic waves from distant earthquakes, but this is only partially possible in Antarctica since there are still large gaps in the coverage of seismic stations. This is where recent satellite gravity observations come into play. Using data from the GOCE satellites of the European Space Agency (ESA), scientists have been able to create the first 3D models of gravity anomalies measured from space and investigate how thick the crust and the lithosphere beneath are. West Antarctica, in particular, has a thin crust and lithosphere because it has been an area of extension for at least 100 million years. The West Antarctic Rift System has been compared in terms of its huge size to the Basin and Range Province that covers much of the inland Western United States and North-Western Mexico, as well as the East African Rift System, which is over 6,000 kilometres long. The extension of the lithosphere beneath the West Antarctic Rift System means that the warmer and softer asthenosphere lying beneath is in places shallower. As a result, the heat coming from the Earth (known as geothermal heat flux) is elevated and the mantle is softer and less viscous. Elevated geothermal heat flux beneath parts of the rift increases the amount of water beneath the ice sheet, affecting what is referred to as subglacial hydrology. And the lower viscosity affects how much and how fast the bedrock responds to a shrinking ice sheet – a process known as *Glacio Isostatic Adjustment* (GIA).

Two major European Space Agency projects are currently underway, *3D Earth* and *4D Antarctica.* These new projects are helping scientists study several other key aspects of Antarctica. While *3D Earth* is investigating the role of Antarctica in the global supercontinent cycle, focusing in particular on the early history of Antarctica, between about 1 and 1.8 billion years ago, *4D Antarctica* aims to quantify geothermal heat flux and determine its influence on subglacial hydrology, which in turn affects how the Antarctic ice sheets flow.

Satellite gravity anomalies are a useful tool for revealing the geoid (the mean height of the shape that the ocean surface would take under the influence of the gravity and rotation of Earth alone, in the absence of other influences such as winds and tides). The large-scale undulations in the geoid tell us about deep mantle processes. For example, in Antarctica a deep geoid low in the Ross Sea is thought to reflect ancient slabs of oceanic lithosphere and mantle flow.

"Universe is, inferentially, the biggest system. [...] We find no record as yet of man having successfully defined the universe – scientifically and comprehensively [...].
Einstein successfully equated the physical universe as E = Mc². [...] But the finite physical universe did not include the metaphysical weightless experiences of universe. All the unweighables, such as any and all our thoughts [...], are weightless.

I define universe [...] as follows:
The universe is the aggregate of all humanity's consciously-apprehended and communicated experience with the non-simultaneous, nonidentical, and only partially overlapping, always complementary, weighable and unweighable, ever omni-transforming, event sequences."

Richard Buckminster Fuller (Architect, Engineer, Inventor), *Operating Manual for Spaceship Earth,* 1969

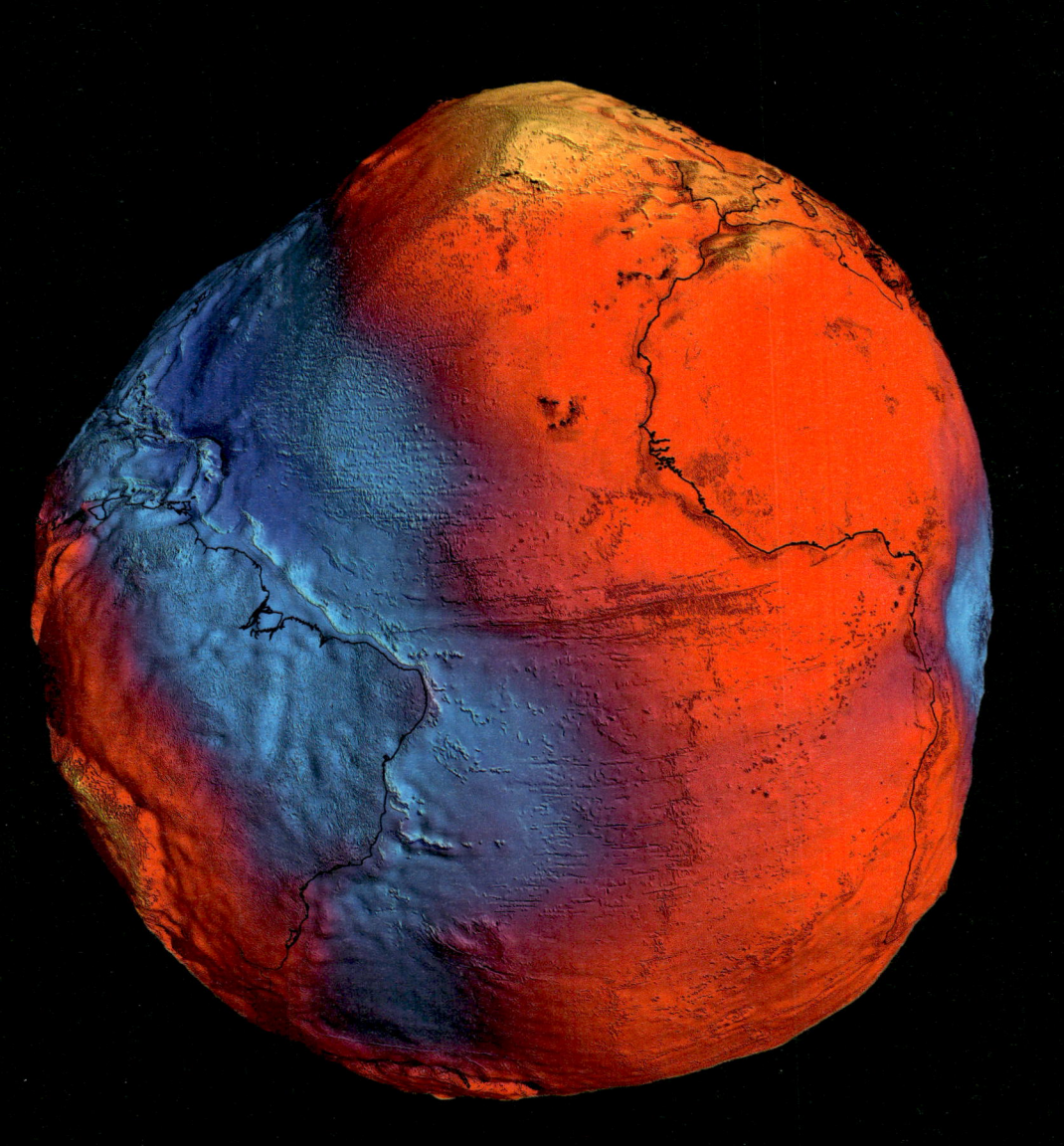

09
The White Desert. A Unique Viewing Platform into the Cosmos

The Magnetosphere

Louis J. Lanzerotti and Andrew J. Gerrard

LOUIS J. LANZEROTTI is a distinguished research professor of physics at the New Jersey Institute of Technology (United States). He served as principal investigator on several NASA interplanetary and planetary missions and has a long career of space- and ground-based research. ANDREW J. GERRARD is a professor and Physics Department chair at the New Jersey Institute of Technology and the director of the Center for Solar-Terrestrial Research (United States).

Planet Earth moves through the solar system enveloped in a huge comet-like structure: the magnetosphere. This structure, invisible to the eye, can only be determined by measurements made with sophisticated instrumentation flown on spacecraft. The magnetosphere is formed by the interaction of outflowing hot gas from the corona of the Sun (the solar wind, composed largely of protons and electrons) with Earth's nominally dipole magnetic field. On the sunward side of Earth, the front of Earth's magnetic field is compressed and forms a comet-like structure head, where the balance of the pressure from the flowing solar wind matches the pressure of the dipole field. This distance is nominally of the order of 10 Earth radii (1 Earth radius is 6,378 kilometres). Increases or decreases in solar activity – and thus the solar wind pressure – can cause this front boundary to move inward or outward respectively. When there is very high solar activity, measurements have shown that this front boundary is pushed by the solar wind deep inside the geosynchronous satellite orbit (about 36,000 kilometres). On the anti-sunward side, the magnetosphere is extended away from the Sun by a viscous-like interaction of the flowing solar wind with Earth's magnetic field, forming a comet-like tail. This magnetotail stretches to more than 100 Earth radii, reaching out beyond the orbit of the Moon (about 64 Earth radii).

Inside the magnetosphere, Earth's space environment can be considered to begin at an altitude of about 100 kilometres. This is where the neutral atmosphere begins to become partially ionised by ultraviolet and x-ray emissions from the Sun and by charged particles coming from higher altitudes in the magnetosphere proper. Earth's ionosphere, where the reflection of radio waves at

With a radius of almost 3,500 kilometres, the area outside the solid iron core of the Earth consists of a liquid mass of iron alloyed with nickel and other lighter components. The helical fluid motion of the rotating planet generates an electromagnetic dynamo effect that in turn produces a strong geomagnetic field, which has an almost ideal dipolar configuration at high latitudes (the poles). These fields permeate the planet and the space surrounding it, generating a large teardrop-shaped region known as the magnetosphere, which is formed by the opposing pressure between the particles escaping the Earth's gravity and the solar wind (i.e. the flux of protons and electrons escaping from the Sun's gravitational field). The two forces are balanced at a distance of some 65,000 kilometres outward toward the Sun on the Earth's dayside in an area approximately 100 kilometres thick called the magnetopause. Reconfigurations of the magnetosphere generate the so-called geomagnetic substorms that cause particles to penetrate the ionosphere, creating the auroral displays that are visible to the human eye. On the nightside, the terrestrial field presents a magnetotail that reaches past the orbit of the Moon.

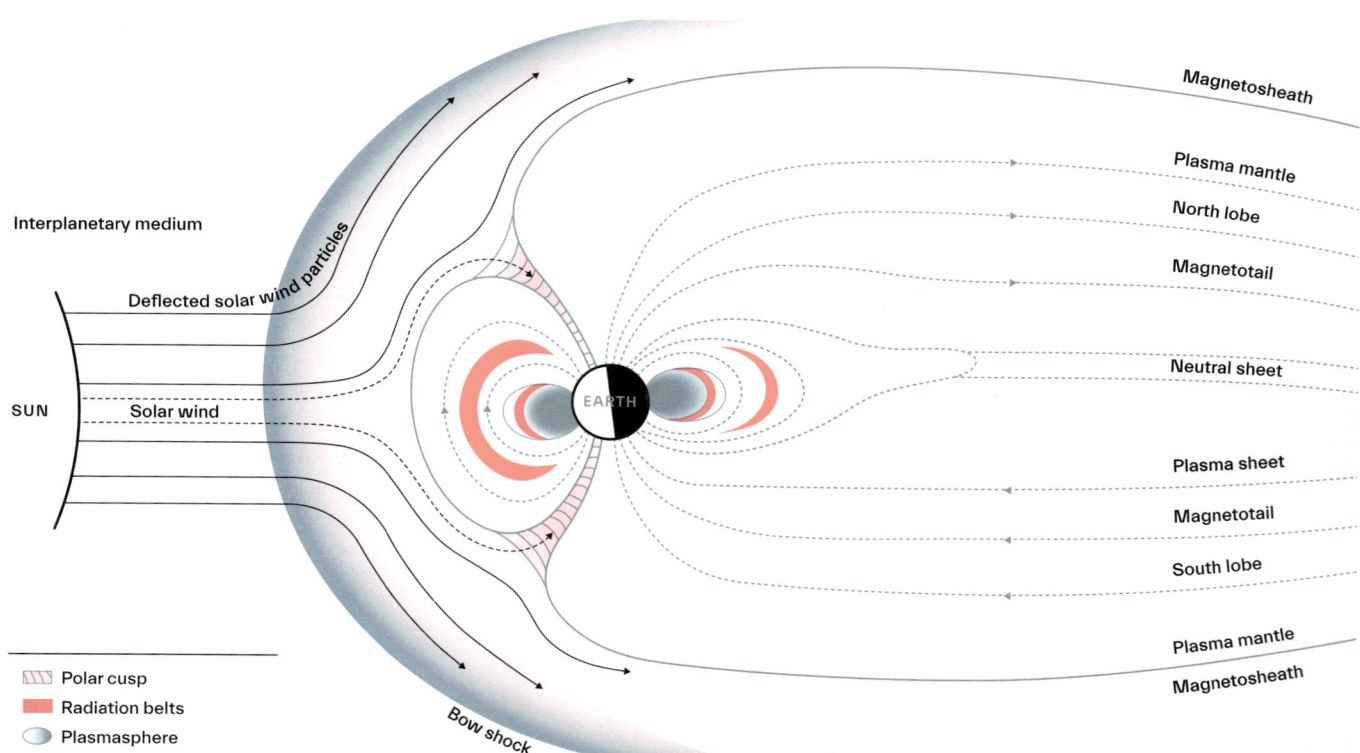

Interplanetary medium

Deflected solar wind particles

SUN

Solar wind

EARTH

Bow shock

Magnetosheath

Plasma mantle

North lobe

Magnetotail

Neutral sheet

Plasma sheet

Magnetotail

South lobe

Plasma mantle

Magnetosheath

Polar cusp
Radiation belts
Plasmasphere

altitudes of between 100 and 250 kilometres enables radio transmissions over substantial distances, exists between 100 and 1,000 kilometres; at about 1,000 kilometres and above, the Earth's "atmosphere" is almost completely ionised.

The magnetosphere proper is composed largely of protons (ionised hydrogen atoms) and electrons. There is also a small component of helium and oxygen ions, varying with time and coming both from Earth's atmosphere and from the Sun as carried by the solar wind. Such a population of charged particles is called a "plasma". The plasma particles in the magnetosphere are organized by the Earth's magnetic field. From the upper portion of the ionosphere and to an equatorial altitude of about 4 Earth radii, the protons and electrons are quite low energy, charged with less than a kilovolt or so, and are called a "cold plasma". This region, named the plasmasphere, can vary in radial distance with different levels of solar wind conditions.

Electrons of higher energies, from 10 kilovolts to many megavolts, tend to be organised into two main regions: the inner radiation belt and the outer radiation belt – the so-called Van Allen (radiation) Belts, named after their discoverer, Professor James A. Van Allen of the University of Iowa. The edge of the plasmasphere (the plasmapause) roughly forms a "boundary" between these two electron belt regions. This is because electromagnetic and plasma waves in the vicinity of the plasmapause can interact with the electrons, causing them to follow magnetic field lines into Earth's atmosphere (and thus contributing to the ionisation of the ionosphere). The inner electron belt is more stable in time than its outer belt, which can vary more in intensity and spatial extent under varying solar wind conditions. Under some quite disturbed conditions, a third electron belt has at times been measured. Higher-energy protons tend to be trapped by the more dipole-like magnetic field closer to Earth, in the range of 2 to 3 Earth radii. Large solar storms can often produce high-energy particles that propagate to Earth and access the magnetosphere, contributing at times to the radiation environment.

Since the time of the International Geophysical Year in 1957–1958, the Antarctic has played an important role in studies of the magnetosphere using

"Super-storms are rare events but estimating their chance of occurrence is an important part of planning the level of mitigation needed to protect critical national infrastructure."

Sandra Chapman (Professor, University of Warwick), *British Antarctic Survey*, 2020

01   Astronaut safety
02   Computer and memory
      upsets and failures
03   Micrometeorites
04   Surface and interior
      charging
05   Magnetic attitude control
06   Solar cell Damage
07   Atmospheric drag
08   Pipeline corrosion
09   Cellular disruption
10   GPS station navigation
11   Radar interference
12   Airline passenger radiation
13   Solar radio bursts
14   Telecommunication
      cable disruption
15   Electric grid disruption
16   Signal scintillation
17   Radio wave disturbance
      Rainfall water vapor
      Plasma bubble
      Earth currents
      Ionosphere currents

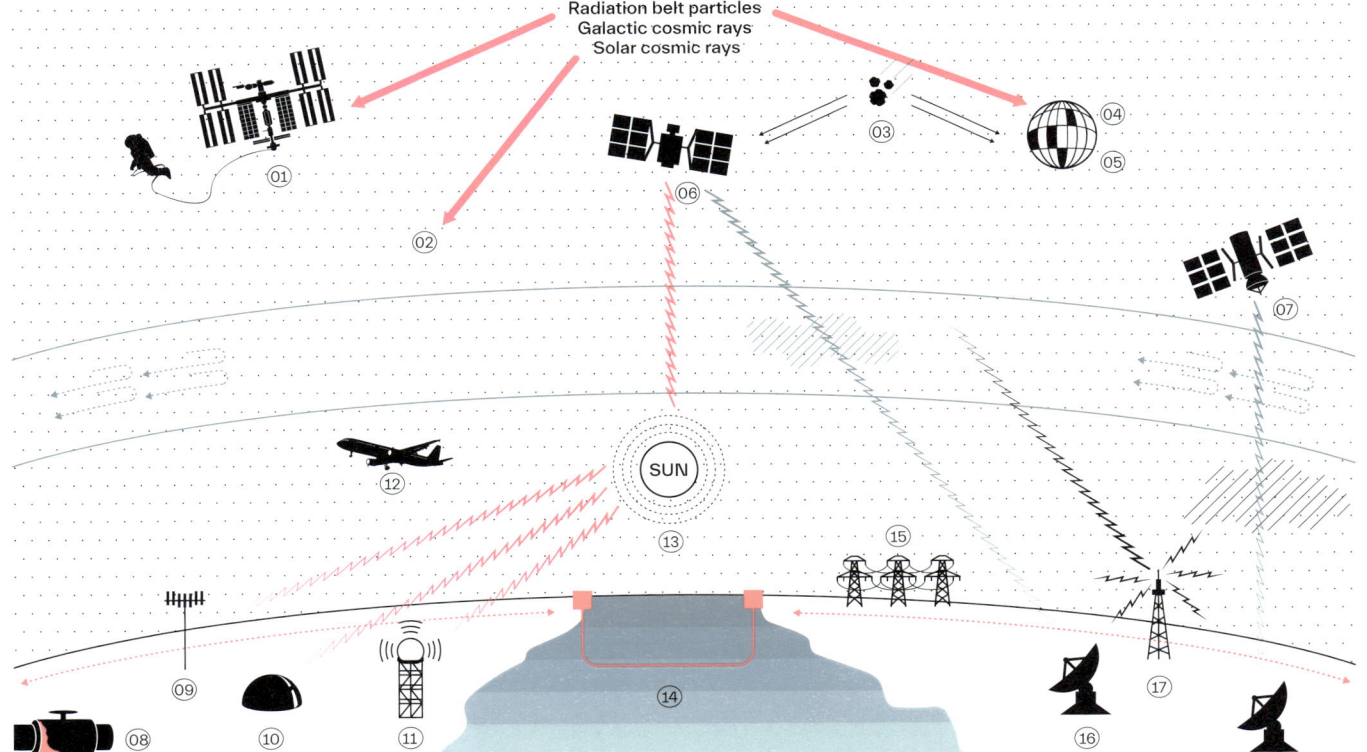

ground-based instruments, as well as instruments on high-altitude balloons and rockets. This is predominantly because the large land mass of the continent extends over a wide range of geomagnetic latitudes and includes magnetic field lines tied to the solar wind itself (i.e. the magnetosphere cusp). Magnetosphere studies continue to be undertaken from permanent and seasonal stations employing humans, as well as from many remote unmanned locations using solar and wind power.

Aurora borealis and aurora australis, circling the northern and southern geomagnetic poles respectively, are striking visual evidence of the processes of plasma physics in Earth's space environment. Aurora were also the striking visual accompaniments of spontaneous electrical currents that at times confused the earliest implementation of the telegraph – the first evidence of "space weather" effects on human technology. As William Henry Barlow, Superintendent Engineer of the Midland Counties Railway Company, wrote, "In every case which has come under my observation, the telegraph needles have been deflected whenever aurora has been visible."[1] It is now understood that moving electrical currents in the ionosphere (some of which are visually evidenced by the aurora) induce currents in the Earth. These "telluric" currents seek out the least-resistant path to flow: the high conductivity represented by the telegraph wires.

Research and adaptation to the Earth's magnetosphere has continued into current times: when long conductors such as telephone cables and pipelines are connected to the earth, unwelcome electrical currents can be induced in the conductors. These anomalous currents can disrupt and even damage electrical systems. Long wires in electric power grids can be severely susceptible to induced currents, with their connected electrical transformers at times damaged or completely destroyed.

We now recognise that human technologies have increased in number and diversity over the last 170-plus years: space weather effects have a significant impact on the operations of many of these technologies – particularly those used for communications, navigation, and electrical grids – for both civilian and national security objectives. In fact, the discovery of the radiation belts in the magnetosphere immediately meant that the space environment was not benign for human-built spacecraft or for human flight above the atmosphere. Thus, technology designs and operations must always take into account the space environment and its variability under changing solar conditions.

"The light is so wonderful; what causes this strange glow? It is clear as daylight, and yet the shortest day of the year is at hand. There are no shadows, so it cannot be the moon. No; it is one of the few really intense appearances of the aurora australis that receives us now. It looks as though Nature wished to honour our guests, and to show herself in her best attire. And it is a gorgeous dress she has chosen. Perfectly calm, clear with a starry sparkle."

Roald Amundsen (Explorer), *The South Pole: An Account of the Norwegian Antarctic Expedition in the Fram 1910–1912*

Aurora can be seen in the upper atmosphere in both hemispheres of planet Earth, at altitudes of about 100 to 300 kilometres – the auroral borealis in the north and the aurora australis in the south. Aurora are also found on other planets with intrinsic magnetic fields, especially the giant outer planets of the Solar System – Jupiter, Saturn, Uranus, and Neptune. The electrical phenomena of the aurora arise from the collision of charged particles (mainly electrons and protons) striking the high atmosphere as they propagate along magnetic field lines into Earth's polar cap regions. Because of the large land mass of the Antarctic, the aurora australis can be measured and studied at all local times with instruments placed on the Antarctic ice as Earth rotates on its axis under the aurora. Since there is an offset of Earth's spin axis and magnetic field axis, South Pole Station is generally almost directly under the aurora australis near midnight at the station. This situation provides an ideal opportunity to study the aurora with a wide spectrum of instruments that can be supported and maintained by the South Pole installation.

STELLAR AXIS
William L. Fox

Ninety-nine blue spheres, representing the brightest stars that would be visible during the long austral night, were deployed by Lita Albuquerque and her team on the Ross Ice Shelf near McMurdo Station, about 1,280 kilometres north of the South Pole. The ultramarine fibreglass spheres positioned on the ice ranged in size from 25 to 120 centimetres to represent the relative magnitude of the stars, or their apparent brightness to the naked eye. Working with astronomer Simon Balm, Albuquerque arranged the position of each sphere on the ice according to its place in the sky, and on the 22nd of December 2006, i.e. on the austral summer solstice, she invited fifty of the McMurdo staff and visiting scientists to create a human spiral inside the field of spheres. Envisioned as the first part of a larger land-art work (which later extended to the North Polar region), *Stellar Axis* was meant to link the stars through both poles as if a shaft of starlight were aligned with the rotational axis of the Earth, creating a spiral that is a metaphor for the conveyance of information, a kind of stellar DNA. *Stellar Axis* was the first large-scale art installation created in the Antarctic and was made possible by America's National Science Foundation Antarctic Artists and Writers Program, which, since its inception in the 1960s, has hosted more than 100 people on the continent.

The Earth's atmosphere features four distinct layers which include (from the planet towards space) the troposphere, the stratosphere, the mesosphere, and the thermosphere. Beyond that, the exosphere is a region 500 kilometres above the Earth's surface which is still contained within the planet's magnetosphere. The remoteness of the Antarctic continent allows for a thorough and diverse study of these layers via a wide variety of scientific devices ranging from weather balloons and telescopes to the remarkable IceCube Neutrino facility at the South Pole.

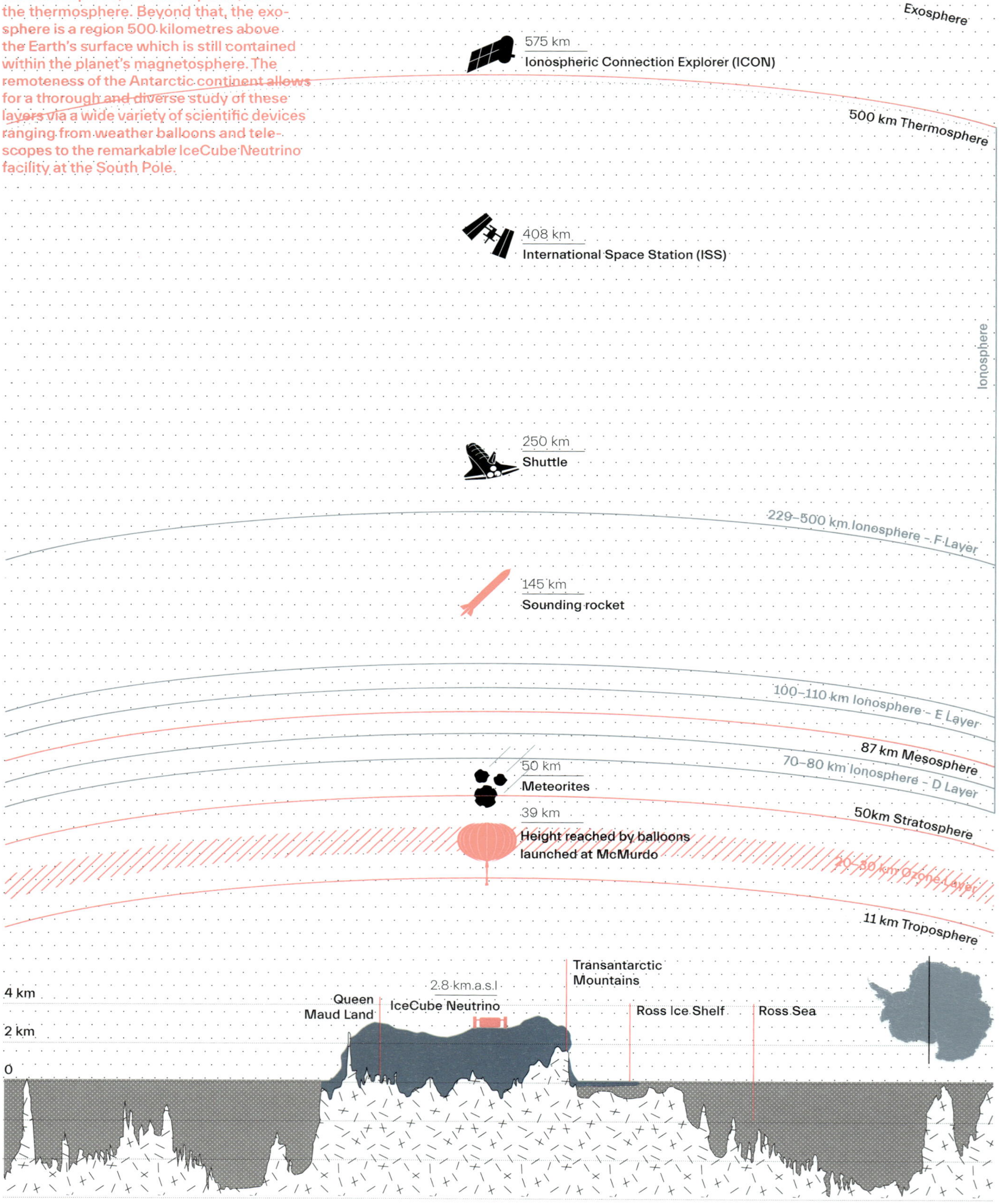

Exosphere

575 km
Ionospheric Connection Explorer (ICON)

500 km Thermosphere

Ionosphere

408 km
International Space Station (ISS)

250 km
Shuttle

229–500 km Ionosphere – F Layer

145 km
Sounding rocket

100–110 km Ionosphere – E Layer

87 km Mesosphere

70–80 km Ionosphere – D Layer

50 km
Meteorites

50km Stratosphere

39 km
Height reached by balloons
launched at McMurdo

20–30 km Ozone Layer

11 km Troposphere

Transantarctic
Mountains

4 km

2.8 km.a.s.l

Queen
Maud Land

IceCube Neutrino

Ross Ice Shelf

Ross Sea

2 km

0

09   The White Desert. A Unique Viewing Platform into the Cosmos

FOUR ELEMENTS   305

Atmospheric Chemistry.
The Discovery of the Ozone Hole

Jonathan Shanklin

JONATHAN SHANKLIN is one of the scientists who discovered the ozone hole, leading to a global shift in climate policy. He is an emeritus fellow at the British Antarctic Survey (United Kingdom). During his career, he has made twenty visits to the Antarctic, most recently visiting Halley, where he calibrated a new automated Dobson spectrophotometer. Shanklin's awards include the Polar Medal (United Kingdom) and the EPA Montreal Protocol Award (United States).

High up in the Earth's atmosphere, a thin layer of ozone protects humankind from the Sun's most harmful rays, safeguarding life on our planet and reducing the globe's temperature by absorbing some of the Sun's radiant energy. In 1985 a team of three scientists working for the British Antarctic Survey (BAS) – Brian Gardiner, Joe Farman, and Jonathan Shanklin – discovered a hole in the ozone layer over Antarctica. The paper they published in the scientific journal *Nature* in May 1985 provided an early warning of the damage being done to the ozone layer worldwide by chlorofluorocarbons and led to significant changes in the management of planet Earth, paving the way for international action and the signing of a successful agreement to counter the destruction – the 1987 Montreal Protocol.

The thinning of the ozone layer is caused by a chemical reaction in which man-made chemicals known as CFCs (chlorofluorocarbons) and halons are broken down into their constituent parts and react with the ozone in the upper atmosphere. Chemical reactions take place on the surface of stratospheric clouds in the ozone layer over Antarctica, converting the chlorine from the CFCs into an active form, which in turn breaks down the ozone when the Sun comes back in the spring. The ozone depletion thus only happens in places where the clouds are present, and during the Antarctic spring (September/October) ozone levels fall to less than half their normal amount and the ozone hole covers the entire continent.

Scientists from the United Kingdom began measuring the Antarctic ozone layer in 1956 at the first Halley Station [↗ p. 658], which was built for the International Geophysical Year (IGY) of 1957–1958. The ozone layer was originally studied because of its influence on the temperature of the atmosphere and was used to measure air circulation. It was not until the 1970s that ozone became the focus of attention as an indicator that long-term changes might be taking place in the atmosphere.

At the time, I was working for the British Antarctic Survey on quality control of meteorological data sent from the Antarctic, with a particular focus on analysing solar intensity measurements and data related to the ozone layer. Most of the quality control for the meteorological data was automated using computer programs – less powerful than the average mobile phone of today – whose verdicts were often questioned by the Antarctic observers.

Measurements from the Dobson Ozone Spectrophotometer at the Antarctic stations were written down on sheets of paper, and my first task was to digitise the records on the sheets. The digitisation revealed patterns of shortcomings: the Antarctic observers might have written down the wrong time or type of observation; the digitising staff might have jumped columns on the sheets; and occasionally there was an error in the measurements.

In those years, a growing concern was raised within scientific journals that exhaust gases from the Concorde aircraft or CFCs from spray cans might damage the ozone layer. My physics background led me to question such theories and I decided to present that years' data at an open day at the BAS Cambridge headquarters, comparing it with values computed from a decade earlier. I was assuming that the data would be the same: Concorde could keep flying and the public could keep using their spray cans. To my surprise, they weren't the same.

This discovery provided the impetus to continue working up the backlog, but progress was slow. Meanwhile, I had the opportunity to visit Antarctica for the first time with a twofold goal: to install a brand-new Dobson instrument at Halley Station and compare it against the existing one, and to understand the routine of meteorological work.
Upon my return, I continued work on reducing the ozone data until I had finished all the backlog. When I plotted the minimum spring-time ozone amount on a

ADVOCATING FOR CONSISTENT DATA SETS. THE DOBSON SPECTRO-PHOTOMETER
Jonathan Shanklin

For the past sixty years, the Dobson Spectrophotometer has been the instrument used to measure ozone concentrations at Halley Research Station in Antarctica. Designed in the 1920s, the undisputed standard device worldwide works by comparing the intensities of two wavelengths of ultraviolet light from the Sun. One wavelength is strongly absorbed by ozone and the other is only weakly absorbed. The ratio between the two tells scientists how much ozone is present in the atmosphere.

The amount of atmospheric ozone is measured in Dobson Units (DU). A normal measurement would be around 300 DU. This means that if all the ozone was collected in a vertical column above the instrument and brought down to sea level, it would form a layer just 3 mm thick. Since the discovery of the ozone hole, measurements of ozone have occasionally dropped below 100 DU. The exact calibration of the Spectrophotometer needs to be determined and this is found on sunny days by taking measurements at different times of the day. If the instrument is well calibrated, the amount of ozone measured during the course of an average day should be constant. If it appears not to be, then the calibration values may need adjustment. After an extended process of refinement, consistent ozone measurements become available.

Halley Research Station, in Antarctica, is renowned as one of the best places on Earth to measure the ozone layer and global pollution. All the buildings and vehicles at Halley are kept to the west of monitoring instruments so that when the normal easterly wind blows, the air has not been contaminated for many hundreds of kilometres. BAS scientists use sensitive instruments in a special atmospheric chemistry laboratory to monitor pollutants and aerosols in the air. As well as ozone data and paper filters, scientists collect weekly snow and air samples. The sampling equipment is so sensitive that researchers have to hold their breath when they use it.

sheet of graph paper, the conclusion was clear: there was a systematic decline in the amount of spring ozone and it certainly wasn't an anomaly. I wrote up a draft paper and placed it on my bosses' (and their bosses') desks. Along with listing suggestions for improvements to the paper, which included a possible theory on a chemical mechanism that could explain the observations, they wanted to publish the paper in *Nature.* Around Christmas 1984 the paper was ready, and *Nature* published it in May 1985.

The document gave the world a major shock. The prevailing theory had suggested that if ozone depletion was to happen because of CFCs, it would start at high altitude over the tropics. American satellite operators at NASA reviewed their data and it was confirmed. Very quickly the scientific community mobilised its resources and found that the ozone hole was forming over Antarctica because of CFCs, enhanced by the atmospheric conditions during the Antarctic polar winter. This discovery and the confirmation that the cause was indeed the CFCs quickly led to the Montreal Protocol, which sought to limit the introduction of ozone-depleting chemicals into the atmosphere. The protocol, signed by every UN Member State, has proven to be remarkably successful. Internationally, the United Nations Environment Programme launched an Ozone Secretariat, which will continue to oversee the complete phasing out of CFCs. Studies suggest that it will take until at least 2070–2080 for ozone levels to return to their natural levels.[1]

In many ways, the problem of the ozone hole was relatively easy to solve. The name "ozone hole" gave a graphic description of the phenomena and there is a perception that holes need to be filled in. With a thinning ozone layer there was the risk of greater exposure of people to ultraviolet light and hence a greater risk of skin cancer. These points were enough to make it a concern to voters. In parallel, industry was very willing to produce alternative chemicals and so there was no need for a change in lifestyle if CFCs were phased out. In addition, there

Juxtaposing the size of the ozone hole above Antarctica (as shown in the image to the side) with related pivotal events, from the discovery of its depletion in 1985 to the entry into force of the Montreal Protocol on Substances that Deplete the Ozone Layer in 1989, the timeline emphasises the impacts of climate targets and the consequent successful phasing out of the production of chlorofluorocarbons (CFCs).

Stratospheric ozone concentration in the Southern Hemisphere, Dobson units (DU)

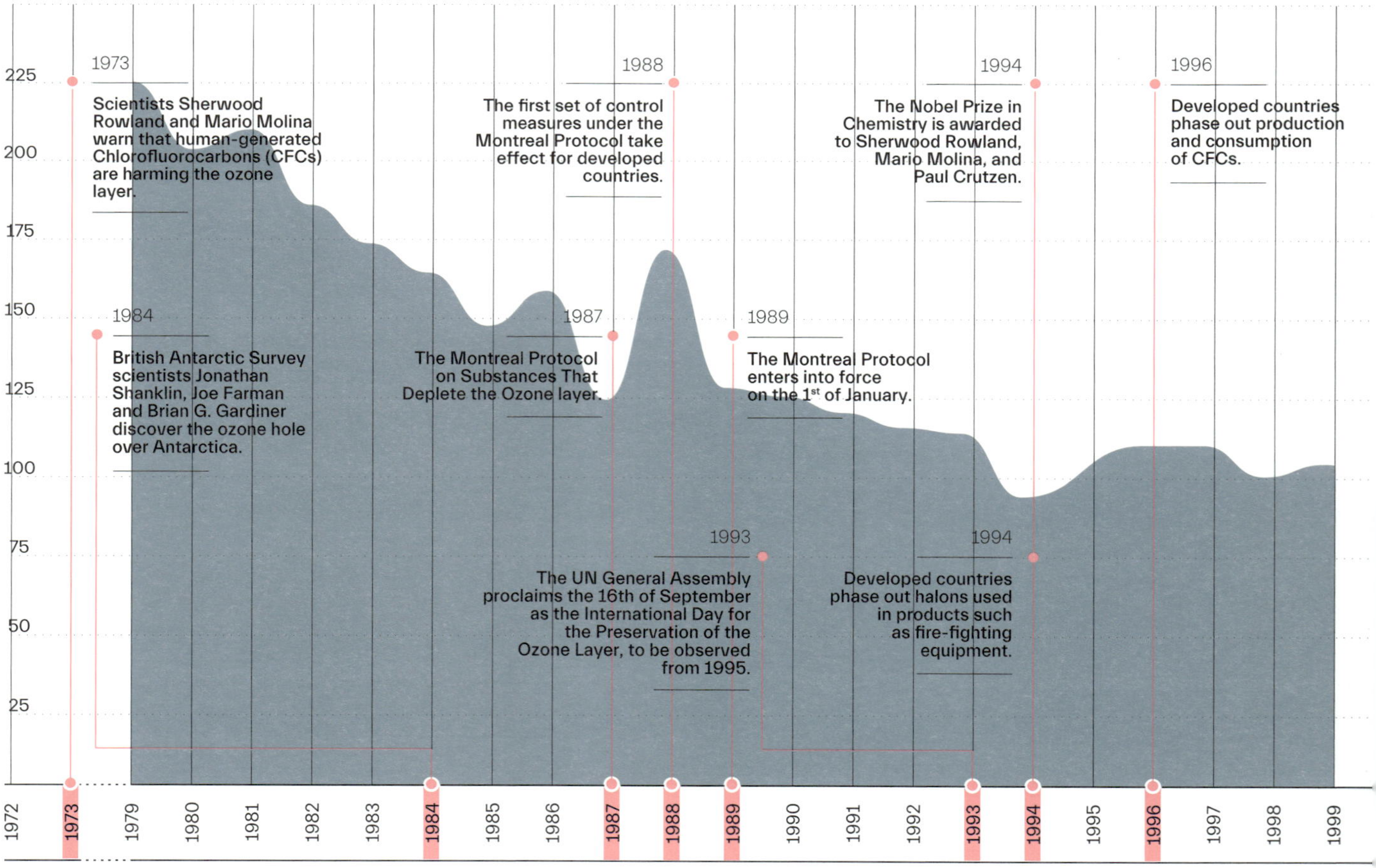

1973
Scientists Sherwood Rowland and Mario Molina warn that human-generated Chlorofluorocarbons (CFCs) are harming the ozone layer.

1988
The first set of control measures under the Montreal Protocol take effect for developed countries.

1994
The Nobel Prize in Chemistry is awarded to Sherwood Rowland, Mario Molina, and Paul Crutzen.

1996
Developed countries phase out production and consumption of CFCs.

1984
British Antarctic Survey scientists Jonathan Shanklin, Joe Farman and Brian G. Gardiner discover the ozone hole over Antarctica.

1987
The Montreal Protocol on Substances That Deplete the Ozone layer.

1989
The Montreal Protocol enters into force on the 1st of January.

1993
The UN General Assembly proclaims the 16th of September as the International Day for the Preservation of the Ozone Layer, to be observed from 1995.

1994
Developed countries phase out halons used in products such as fire-fighting equipment.

225
200
175
150
125
100
75
50
25

1972 1973 1979 1980 1981 1982 1983 1984 1985 1986 1987 1988 1989 1990 1991 1992 1993 1994 1995 1996 1997 1998 1999

was strong political support from the top in the 1980s, as the British Prime Minister Margaret Thatcher was trained as a chemist and so understood the science.

Every year since our discovery, an ozone hole has formed over Antarctica during the spring. Whilst it mostly stays over Antarctica and is circular, sometimes it becomes elliptical and can stretch north far enough to cross the tip of South America. As summer comes, the air warms, the stratospheric clouds disappear, and the ozone hole fills in. Recovery over Antarctica is likely to be slowed by what we are doing to the atmosphere with the introduction of gases such as carbon dioxide and methane. Whilst these gases act to warm the surface of the Earth, higher up in the ozone layer they have a cooling effect and hence increase the chance of cloud formation. Events outside our control may also play a part, whether it is a massive volcanic eruption or a giant meteor strike, either of which could affect the ozone layer. The hole has never extended as far as Australia and New Zealand, but the ozone layer over the Pacific does thin as a consequence of the ozone depletion over Antarctica.

The discovery of the ozone hole presents several lessons, which are often ignored in today's political climate. There is the need for continuous long-term monitoring of our planet, its atmosphere, land, oceans, and living things, in order for us to know what is changing. The rapid appearance of ozone depletion over Antarctica gave clear evidence of how quickly humankind can change our environment – it only took ten years for the ozone layer to go from normal to one-third depleted. Today, the focus is on climate change, but this is only one of many symptoms that affect our planet. Just as a responsible doctor considers all a patient's symptoms before diagnosing the underlying cause, so must we. The ozone hole, climate change, ocean plastics, soil degradation, water shortages, decline in biodiversity, and many other symptoms have one common factor at their root – human action. If we aspire to maintain our planet's environment, we must change; otherwise we are likely to be changed.

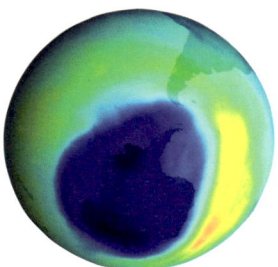

"Through the Montreal Protocol, a proven forum for solving environmental challenges like protecting the ozone layer, the world community has agreed to phase down the production and consumption of HFCs and avoid up to 0.5°C of warming by the end of the century – making a significant contribution towards achieving the goals we set in Paris." "

Barack Obama (44th President of the United States), "Statement by the President on the Montreal Protocol," The White House Press Release, 15 October 2016

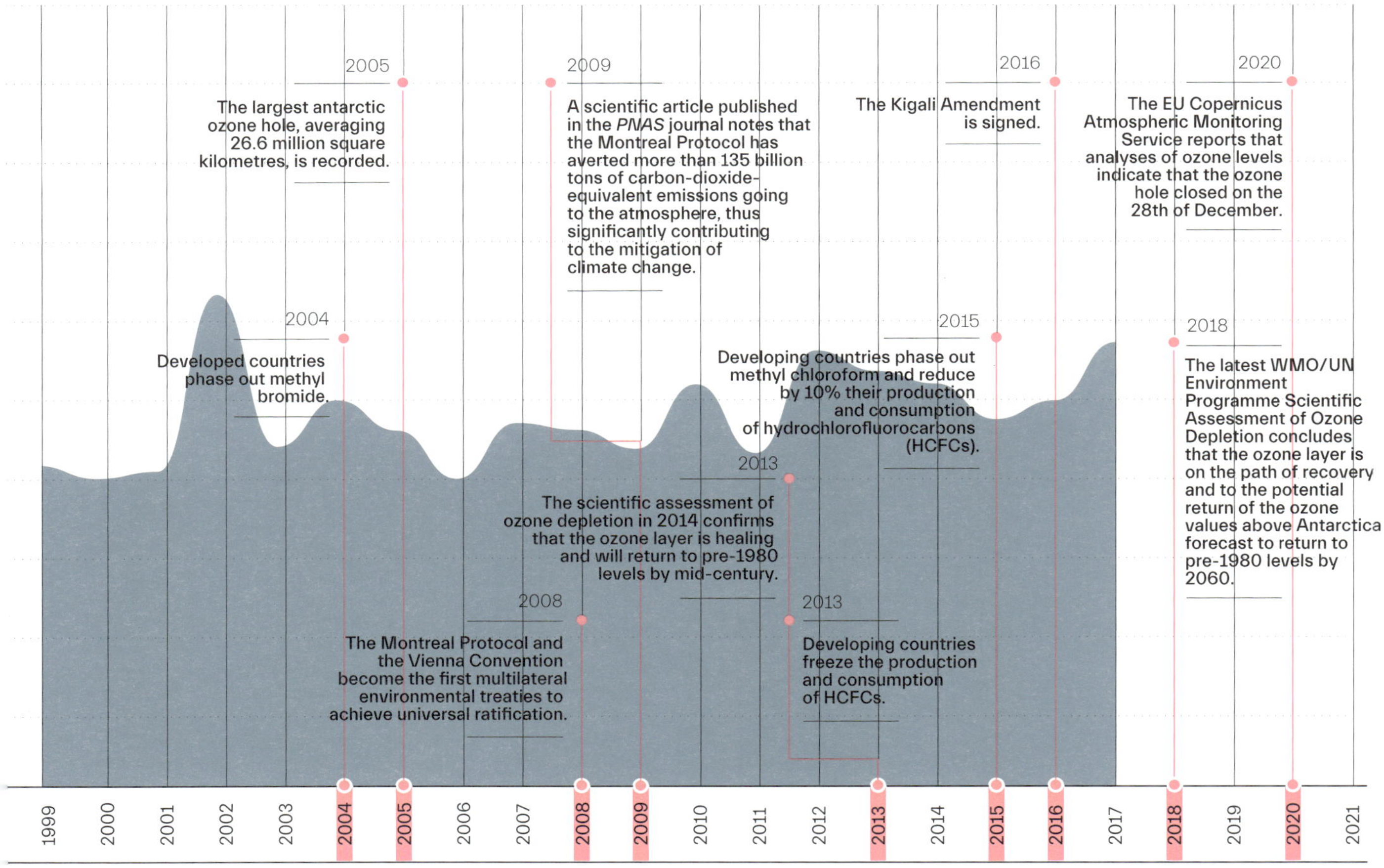

2005 The largest antarctic ozone hole, averaging 26.6 million square kilometres, is recorded.

2009 A scientific article published in the *PNAS* journal notes that the Montreal Protocol has averted more than 135 billion tons of carbon-dioxide-equivalent emissions going to the atmosphere, thus significantly contributing to the mitigation of climate change.

2016 The Kigali Amendment is signed.

2020 The EU Copernicus Atmospheric Monitoring Service reports that analyses of ozone levels indicate that the ozone hole closed on the 28th of December.

2004 Developed countries phase out methyl bromide.

2015 Developing countries phase out methyl chloroform and reduce by 10% their production and consumption of hydrochlorofluorocarbons (HCFCs).

2018 The latest WMO/UN Environment Programme Scientific Assessment of Ozone Depletion concludes that the ozone layer is on the path of recovery and to the potential return of the ozone values above Antarctica forecast to return to pre-1980 levels by 2060.

2013 The scientific assessment of ozone depletion in 2014 confirms that the ozone layer is healing and will return to pre-1980 levels by mid-century.

2008 The Montreal Protocol and the Vienna Convention become the first multilateral environmental treaties to achieve universal ratification.

2013 Developing countries freeze the production and consumption of HCFCs.

1999 2000 2001 2002 2003 2004 2005 2006 2007 2008 2009 2010 2011 2012 2013 2014 2015 2016 2017 2018 2019 2020 2021

Long-Duration Ballooning

Andrew T. Hynous

ANDREW T. HYNOUS is mission operations manager for the United States' National Aeronautics and Space Administration (NASA) Balloon Program Office, which conducts Long-Duration Balloon campaigns from McMurdo Station. Hynous directly supervises coordination efforts with the Columbia Scientific Balloon Facility, the National Science Foundation (NSF), the Antarctic Support Contract, and the United States Antarctic Program for campaign planning.

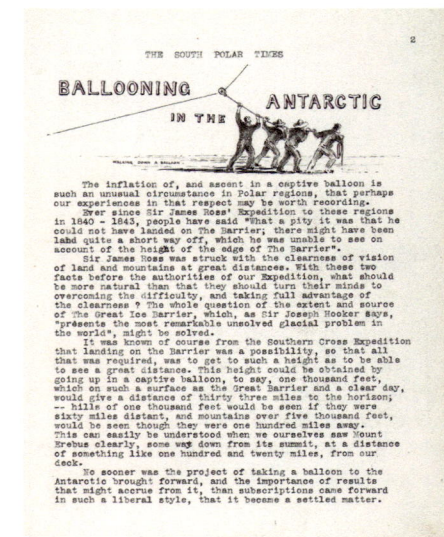

As documented in the *South Polar Times* shown above, early experimentation with "ballooning" was already underway at the beginning of the twentieth century as part of Heroic Age explorations. Ponting narrates, in fact, that "Dr Simpson had brought scientific equipment for observing air-currents, upper as well as lower. He had provided a number of small balloons made of gold-beater's skin, of a capacity of one cubic metre when fully inflated. Periodically, he would fill one of these balloons with hydrogen gas, which was made by immersing hydrate of calcium in water, in a generator of his own design. The ends of two strands of fine black silk (wound on reels containing ten miles) were tied to a small black parachute, to which a very light instrument was attached for recording the temperature and altitude. When all was ready, a slow-match fuse was ignited, the object of which was to burn through a short thread which held the parachute to the balloon, after a certain time had elapsed, he detach it, so that it fell. As the balloon ascended, it unwound the two silk threads from the conical reels, which were contained in a box with holes on the top through which the thread paid out. The balloon was then watched through a telescope, and its course recorded. When in due time the match detached the parachute, the little instrument was borne safely to the ground, and was easily discovered by following the trail of the black thread over the ice. By means of these balloons, the air-currents were investigated to a height of as much as five miles."[1]

For thirty years, the National Aeronautics and Space Administration (NASA) has been launching balloons the size of football stadiums into the stratosphere above McMurdo Station [↗ p. 866]. This truly unique location on the Ross Ice Shelf allows scientists to glimpse our universe in ways that cannot be duplicated anywhere else on our planet. Whether it be looking to see if there are any other Earth-like planets in our galaxy, measuring and counting the atomic particles flung from supernovas in far-off galaxies, or exposing bacteria to the harsh environment of the upper atmosphere to understand what would happen to a microorganism hitchhiker that caught a ride on the next rover mission to Mars: Antarctica presents a laboratory that is like no other.

Antarctica has very distinct advantages as a site for launching balloons when compared with other launch locations for stratospheric flights. Firstly, launches are conducted during the months of December and January in the Antarctic summertime. Stratospheric balloons are thermal vehicles and, as such, their float altitudes can vary depending on the heating and cooling that occurs over the day-night cycle. Since the latitude is below 77°S and launches occur during the austral summer, the Sun does not set, which provides for very good altitude stability. Secondly, at this time of year, the stratospheric winds above the pole become cyclonic in nature. This means that when a balloon is released and ascends to its flight altitude of around 40 kilometres, the prevailing winds will carry the balloon near that same latitude around the pole. The balloon will thus continue to circumnavigate the pole until the stratospheric winds begin to break down in the second half of January. These flights will generally last anywhere from ten to fourteen days for one circumnavigation of the pole and longer-duration flights are accomplished by allowing the balloon to conduct multiple circuits. The current record for a flight is over fifty-five days. Lastly, the Van Allen Radiation Belts are nearly non-existent above the South Pole. These belts are a kind of magnetic bottle around the Earth that protects it from solar and cosmic rays (i.e. those coming from the Sun and those from outside our solar system). These radiation belts can also interfere with some of the instruments used by scientists when they are conducting their experiments in different parts of the world. Anywhere else on the globe, the Van Allen Belts extend from between 640 and 58,000 kilometres above the planet but because they follow the magnetic field lines created by the Earth's core, at the poles they only extend several thousand metres. Therefore in Antarctica, stratospheric balloons can actually fly above the belts, allowing them – and the science instruments they carry – to be exposed to the harsh radiation that the planet is protected from elsewhere. Thus, when scientific instruments are launched from Antarctica, they are able to obtain images and measurements in greater detail than anywhere else.

These balloon-driven research missions allow scientists to make leaps of understanding that would be impossible elsewhere on our planet. One of the missions flown from Antarctica, for example, set out to determine the kind of elements (hydrogen, helium, iron, etc.) that are created when stars go supernova and to measure those elements that are heavier than iron. As the building blocks of our entire universe are continually being released by supernovas, the aim was to develop a scientific understanding of how our own solar system was made by taking specific measurements of such blocks. Another mission flown from Antarctica conducted a study on electron precipitation from the Van Allen Belts. By capturing the electrons that are quite literally falling from the sky, the scientists gained insight into how to better protect satellites and ground-based systems from the radiation found in space. Finally, in a future mission

The Long-Duration Balloon (LDB) facilities on the Ross Ice Shelf near McMurdo Station (shown at the side), known as the DB Hangar Buildings, are the tallest structures in Antarctica. Flanked by supplies of thousands of cubic metres of compressed helium in cylinders (below), used to fill balloons which carry LDB payloads into the stratosphere to altitudes of 40 kilometres, the buildings house scientific telescopes and balloon technology such as the Gamma-Ray Imager/Polarimeter for Solar flares (GRIPS) telescope and the Stratospheric Terahertz Observatory (STO) Telescope, both shown on the following spread.

from Antarctica, scientists plan to build telescopes into balloons to enable them to look for Earth-like planets. This proposed mission will hunt nearby solar systems and examine the planets to determine if any have an atmosphere that is similar to ours or contain moisture and might therefore be capable of sustaining life. Volumes could be (and have been) written on all the science that has been accomplished by the use of stratospheric balloons over the last three decades, but these examples are representative of the vast range of studies that can be carried out on this strange new world that is Antarctica.

Alongside the advantages of operating in Antarctica are the challenges that need to be overcome to support flight operations in one of the most inhospitable places on the planet. During a stratospheric balloon launch campaign from Antarctica, there can be close to a hundred people mobilised by NASA to directly support the launches over the length of the campaign. This includes teams from the universities that are conducting the scientific experiments during the balloon flight as well as NASA mission operations teams and representatives of the Wallops Flight Facility Safety Office. Most important is the crew from the Columbia Scientific Balloon Facility who actually conduct the launch operations. Along with the personnel who directly support flight operations, there are hundreds of people that are a part of the National Science Foundation (NSF), the Antarctic Support Contract, and the United States Antarctic Program (USAP) who live and work in McMurdo Station making stratospheric balloon launches possible. This includes all the personnel and professionals who make the station a thriving city.

Another set of challenges involved in conducting balloon launches from below the 77th parallel are the logistics of making it onto the continent in the first place. As much of the required equipment as possible is permanently stationed outside McMurdo, but a lot of material still needs to be brought in each year. Although the 3- to 4-ton scientific instruments are of primary importance here, other consumables must also be shipped in. These include, among things, helium gas, cryogens, and batteries. Other equipment that needs to be serviced or replaced may need to be taken off-continent at the end of each campaign. Thus logistics are organised via two very long and narrow channels – one by ship, and the other by air. Sea shipments require the sea ice to be broken and a dock to be built and can only be done once a year. There is a Herculean amount of planning and work that goes into this single undertaking. Air shipments rely on continual maintenance on a runway made out of ice, and, for heavy plane loads, there are two limited periods of the year when the ice is cold and hard enough to support multiple cargo flights. These are the only two ways of bringing in heavy equipment and they require months of planning, not just for stratospheric balloon launches but for all the other scientific experiments and research that can only be conducted from Antarctica.

The opportunities for conducting scientific research and exploration from McMurdo Station, Antarctica, are second to none and provide opportunities for new discoveries and paradigm-shifting breakthroughs. All of this is contingent on people from different backgrounds with a wide range of experience coming together from all over the world to support scientific research. This is the reason balloons are launched every year from this remote corner of the globe, with each launch guaranteeing that something new will be learned.

"There's a whole range of things you can do in astronomy at Antarctica that you can't do anywhere else on Earth. To beat Antarctica, you'd have to go to space. The good thing about Antarctica compared to space is that once you get something in space, you can't expect to change it, while in Antarctica, you can always build it bigger."

Michael Burton (Astronomer, Director at Armagh Observatory and Planetarium), 2011

The STO Telescope (opposite page, above), captured during the 2015–2016 flight-preparation phase, is a balloon-borne 80-centimetre telescope developed to explore the Milky Way in the far-infrared [CII] and [NII] lines. Funded by NASA and articulated in three main components (an 80-centimetre optical telescope, a terahertz instrument package, and a gondola), STO is designed to address a key problem in modern astrophysics: understanding the life cycle of the Interstellar Medium (ISM). The GRIPS telescope (opposite page, below), on the other hand, provides a near-optimal combination of high-resolution imaging, spectroscopy, and polarimetry of solar-flare gamma-ray / hard X-ray emissions from ~20 kiloelectron-volts to >~10 mega-electron-volts.

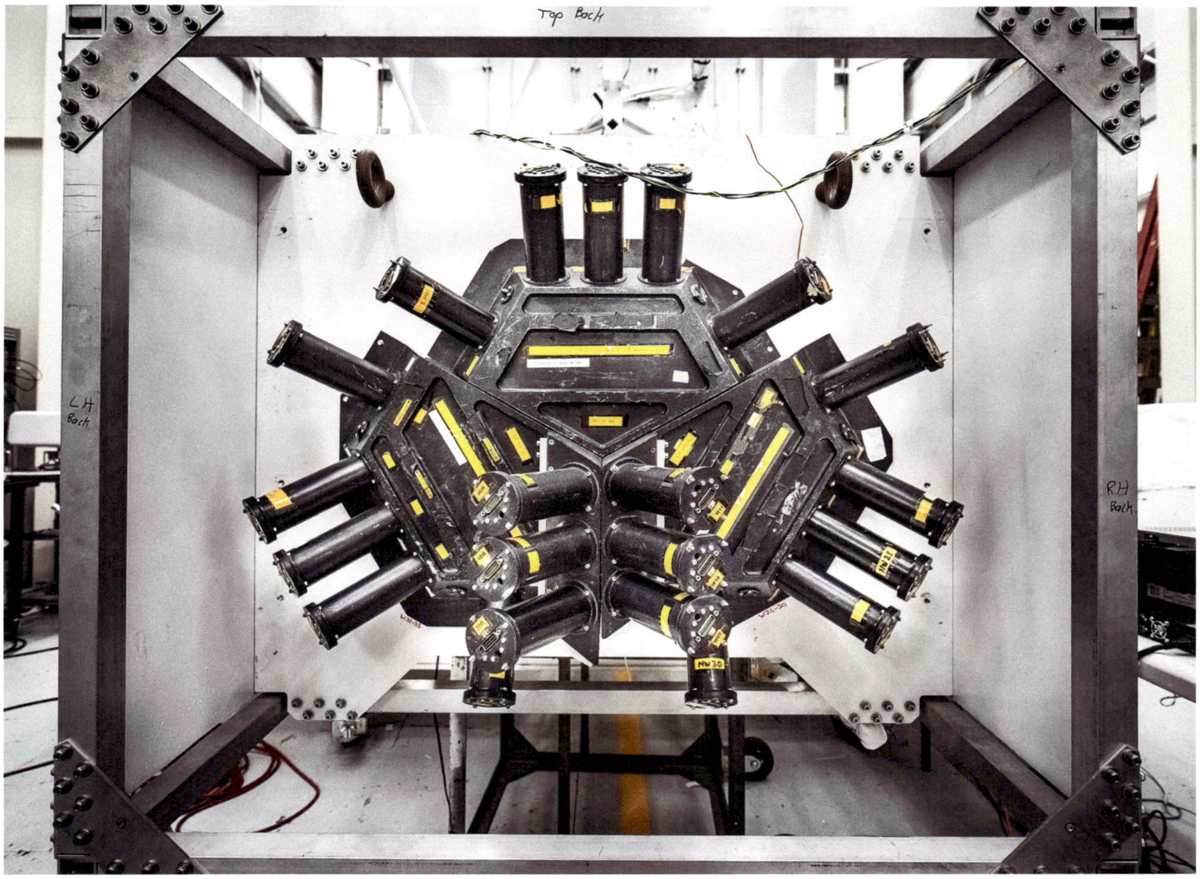
Top Back
LH Back
RH Back
NW30

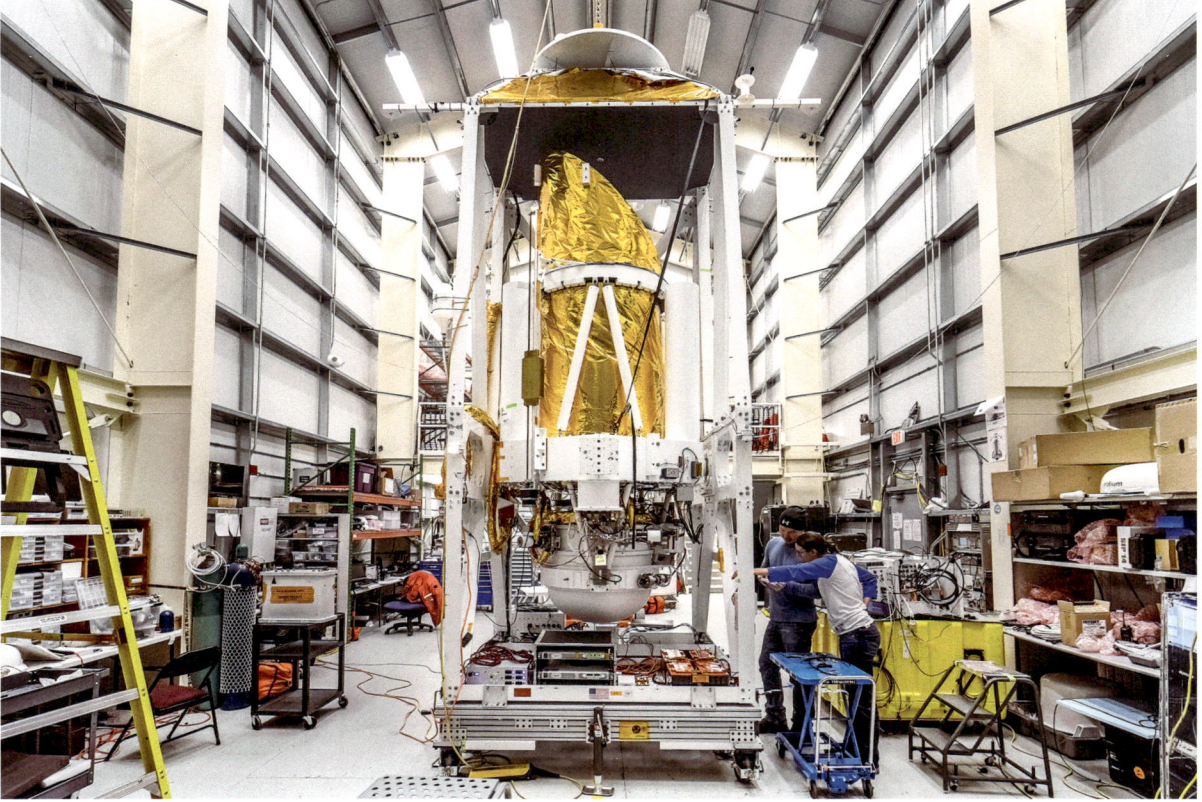

Through a selection of case studies, the timeline explores the evolution of the instruments deployed in the Antarctic continent to study the atmosphere and Outer Space since the late twentieth century.

1988

GRAD (Gamma Ray Advanced Detector) was designed to make observations of Supernova 1987A. Originally scheduled to fly on a Space Shuttle mission, the gamma-ray spectrometer was sidelined following the *Challenger* disaster in January 1987, before being adapted to a balloon platform by researchers from the University of Florida, DARPA, and the Goddard Space Flight Center. The experiment represents the first time a balloon of this size had been launched in such a remote location.[1]

1992–1998

HIREGS I (High Resolution Gamma-Ray and Hard X-Ray Spectrometer) was designed to study solar flare activity in an attempt to improve our understanding of how solar flares relate to the larger context of solar activity and the field/particle environment of the Earth. The project was developed as part of the NASA Space Physics Division's MAX 91 Programme.

2001–2003

TIGER (Trans-Iron Galactic Element Recorder) was designed to measure the elemental abundances of Galactic Cosmic Rays (GCRs), energetic atomic nuclei that originate from outside the solar system and are thought to be accelerated by supernovas to extremely high energies.

2006–2012

BLAST (Balloon-Borne Large Aperture Sub-millimeter Telescope) was designed to study the process of star formation in both local and cosmological galaxies. Solar panels mounted at the back of the telescope on the support structure for the sun shields provide power to the flight electronics by charging NiMH batteries. The panels face the Sun from only one side and are radiatively cooled by the sky from the other side.

2006–2020

ANITA (Antarctic Impulse Transient Antenna) was designed to detect ultra-high-energy cosmic-ray neutrinos and was the first NASA observatory for neutrinos of any kind. The antenna detects these ultra-high-energy neutrinos using the Askaryan effect.

2021

GUSTO (Galactic/Extragalactic Ultralong-Duration Balloon Spectroscopic Terahertz Observatory) is a NASA mission that will launch a high-altitude balloon including a 1-metre telescope. With a scheduled launch in 2021, the aim is to provide a comprehensive understanding of the inner workings of our galaxy and one of our companion galaxies, the Large Magellanic Cloud (LMC), by tracing all phases of the interstellar medium.

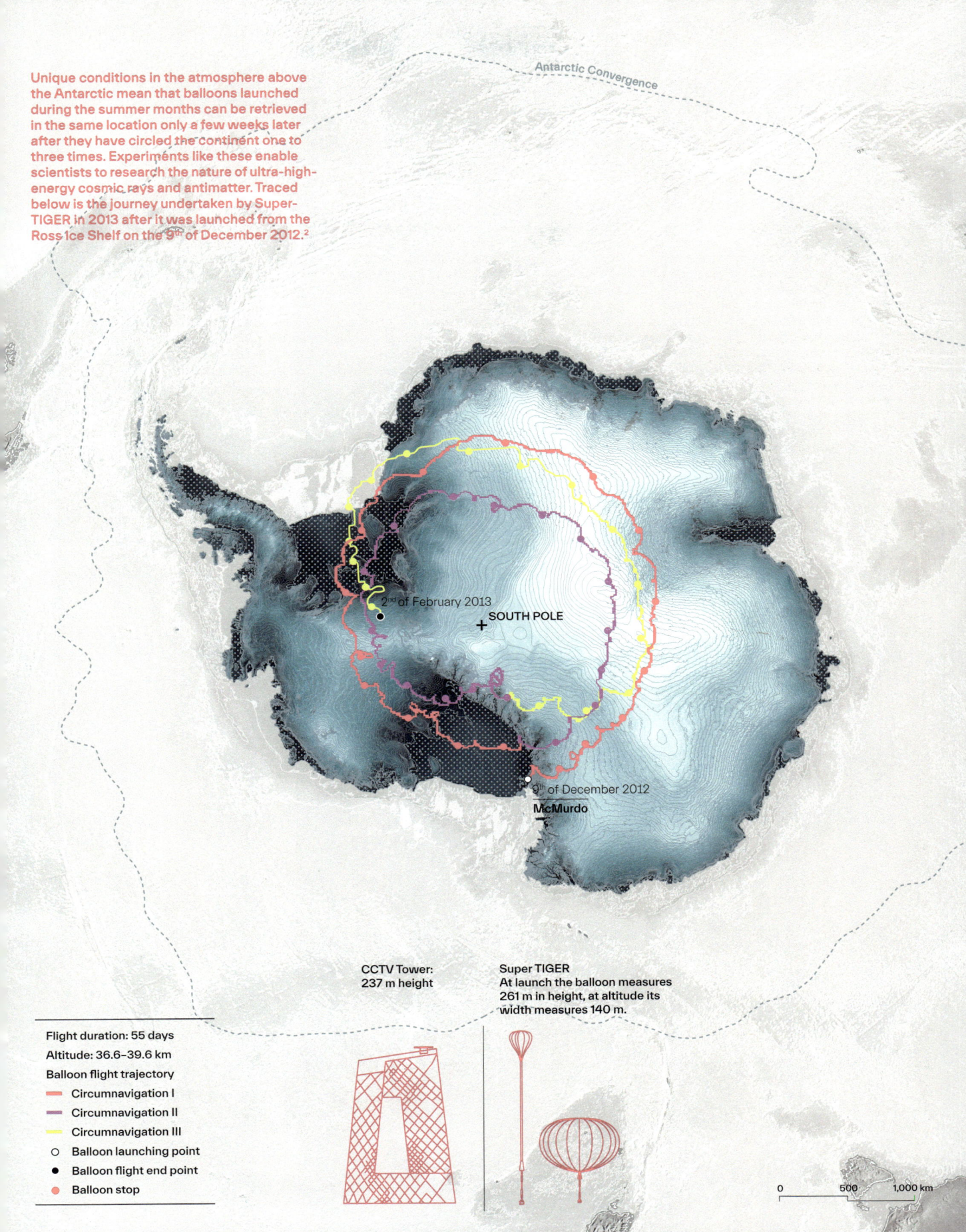

Unique conditions in the atmosphere above the Antarctic mean that balloons launched during the summer months can be retrieved in the same location only a few weeks later after they have circled the continent one to three times. Experiments like these enable scientists to research the nature of ultra-high-energy cosmic rays and antimatter. Traced below is the journey undertaken by Super-TIGER in 2013 after it was launched from the Ross Ice Shelf on the 9th of December 2012.[2]

Antarctic Convergence

2nd of February 2013

+ SOUTH POLE

9th of December 2012

McMurdo

CCTV Tower:
237 m height

Super TIGER
At launch the balloon measures 261 m in height, at altitude its width measures 140 m.

Flight duration: 55 days

Altitude: 36.6–39.6 km

Balloon flight trajectory

— Circumnavigation I
— Circumnavigation II
— Circumnavigation III
○ Balloon launching point
● Balloon flight end point
● Balloon stop

0   500   1,000 km

SuperTIGER

Brian Rauch

BRIAN RAUCH is principal investigator for the National Aeronautics and Space Administration (NASA) of the SuperTIGER stratospheric balloon-borne instrument flown from Antarctica and a co-investigator on the ANITA and X-Calibur balloons. Since he obtained his PhD for his work on the first two flights of the original TIGER instrument, Dr Rauch has been deployed to Antarctica four times.

The Super Trans-Iron Galactic Element Recorder (SuperTIGER) is a stratospheric balloon-borne instrument that flies above all but the last (roughly) 0.5% of the Earth's atmosphere to measure energetic particles from Outer Space called cosmic rays. SuperTIGER needs to fly at such a high altitude because the cosmic rays – rare bits of regular matter from outside the solar system – are stopped by the atmosphere. The cosmic rays SuperTIGER measures are the atoms that make up our everyday world, stripped of all their electrons and accelerated to almost the speed of light. All of the chemical matter we see on Earth and in the solar system is seen in the cosmic rays in much the same proportions, but the subtle differences provide clues as to where the cosmic rays come from, how they travel from their source to here, how the different types of atoms are made, and how the composition of the universe has changed over time.

Most of the regular matter in the Universe originated in the Big Bang, but over billions of years a small part of this has been trans-formed into heavier elements that make up much of our world. Most of this transformation occurs in stars, which support their bulk – and prevent it from collapsing – with the energy released in the fusion of smaller atoms into heavier ones. This fusion process can build heavier elements up to iron, but there it peters out, as elements formed beyond iron fusion require more energy than is produced. The largest stars die when their cores fuse into iron and they collapse and rebound in supernova explosions, which accelerate the cosmic rays. SuperTIGER measures the very rare cosmic rays heavier than iron that are made in other processes that take place slowly in massive stars or quickly in supernova explosions and massive stellar remnant mergers. These far rarer and heavier cosmic rays provide clues as to where they come from and how the heaviest elements are made.

TIGER cosmic ray abundance measurements of elements just heavier than iron support a model in which a significant fraction of the cosmic rays originate in groups of massive stars, where these heavy elements are apparently made. SuperTIGER confirmed these results, but in measuring even heavier elements, it has found evidence of different sources. The massive stellar remnant mergers responsible for recent gravitational wave measurements have also been observed to produce the heaviest chemical elements, and SuperTIGER may be able to help determine which elements are produced in these sources.

The Balloon-Borne Large Aperture Submillimeter Telescope

Gabriele Coppi

GABRIELE COPPI is a researcher at the University of Pennsylvania (United States), having been awarded a PhD from the University of Manchester (United Kingdom). He has managed the power system and the in-flight mapping software of the Antarctic Balloon-Borne Large-Aperture Submillimeter Telescope since 2018. Dr Coppi's writings include "The Balloon-Borne Large Aperture Submillimeter Telescope Observatory" and "In-Flight Performance of the BLAST-TNG Telescope Platform" (SPIE, 2020).

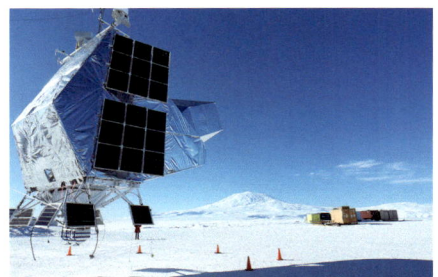

The Balloon-Borne Large Aperture Submillimetre Telescope, also known as BLAST-TNG, is the third iteration of the BLAST programme. The programme originated in the early 2000s with the goal of studying the star-formation rate throughout cosmic history. At the time, there was evidence that stars had mainly formed several billion years ago and that the formation process nowadays, by contrast, is very inefficient. The first flight in 2006 from Antarctica demonstrated this declining effect,[1] but the evidence was not deemed sufficient to explain the cause of the phenomena. In an effort to delve deeper into this, BLAST was relaunched from Antarctica in 2010 and 2012, introducing upgrades that would allow it to measure star-formation regions in the Milky Way, i.e. our own galaxy, and to study the universe nurseries in greater depth. The new version of BLAST – re-named BLASTPol as it was sensitive to the polarised radiation coming from star-forming regions – enabled a good understanding of the behaviour of the magnetic field in these regions. To date, magnetic fields are thought to be responsible for slowing down the star-formation process, and measuring them offers insights into the low star-formation rate that we observe today.

BLAST-TNG, like all previous BLAST iterations, was launched from Antarctica. The seventh continent is the ideal site for launching a balloon for several reasons. Firstly, because during the austral summer, the Sun is up for twenty-four hours, which means that the thermal environment does not vary significantly and a balloon filled with helium will have an almost constant volume since the gas will not expand or contract because of temperature changes. Secondly, the constant Sun also acts as a source of power for the experiment, keeping all the instrumentation active by providing up to 2 kW of power (similar to a hair dryer). Finally, in Antarctica there is a circumpolar stratospheric vortex in the summer – i.e. a low-pressure area in the stratosphere that generates circular winds around the poles – which keeps the balloon in a quasi-circular trajectory that is optimal for the experiment.

The high-altitude balloon flies at a height of almost 37,000 metres (37 km), higher than approximately 99.5% of the atmosphere. This area is an ideal laboratory for testing technology that may be used in future satellite projects, since the environment outside the atmosphere cannot be replicated in any laboratory in the world. As the atmosphere protects the Earth – as well as all the ground instrumentation – from dangerous radiation coming from space, the easiest way to test technology in a space-like environment is using a balloon.

Launching a balloon is a long process that begins with designing the instrument – a phase that lasts approximately two years, during which time all the instrumentation is designed to maximise the scientific output. After manufacturing, the instrument undergoes a rigorous testing period spanning several years until the team can be confident of its functionality. These tests are not only performed in the laboratories at the various collaborating institutions but also verified with a final integration test at the NASA facility in Palestine, TX, before going to Antarctica. When confirmation comes through that everything is working, the instrument is finally given the green light to be sent to Antarctica. There, the experiment is assembled one last time and is ready to be launched.

During the flight, the instrument's path cannot be modified – only a few subsystems can be controlled from the ground. Additionally, the connection between the science group and the payload is guaranteed only through a satellite link with a slow data rate (except for during the first twenty-four hours when there is also a direct connection through radio waves). This phase, called *line-of-sight,* is when the entire team stays awake to optimise the instrument's performance in-flight. BLAST-TNG generates almost 2 TB of data every day of observation, and it is impossible to download this volume of data and verify all the detectors owing to the limited communication bandwidth. Therefore, it is necessary to recover the experiment, and in particular the hard drives, to enable full data analysis. This is done with the help of the National Science Foundation (NSF), which organises special flights across the Antarctic continent to recover the experiments wherever they land.

The amount of data is related to the number of detectors included in the instrument. BLASTPol had 250 detectors to measure polarised radiation, and a natural evolution of this experiment was to increase both the number of detectors and their sensitivity. In BLAST-TNG the number of detectors was increased to almost 3,300 using a new type of detector technology that runs at lower temperatures. While BLASTPol used bolometer detectors at 315 mK (–272.835°C), BLAST-TNG uses KIDs (Kinetic Inductance Detectors) at 275 mK (–272.875°C).[2] Bolometers are detectors that measure the increase in temperature caused by the radiation hitting them, while KIDs are superconducting resonators that change their resonant frequency when the radiation hits them. By knowing the variation in temperature or the change in resonant frequency, it is possible to measure the incident power from the sky. BLAST-TNG is the first balloon to fly more than one thousand detectors using new KID technology. In addition, BLAST-TNG also has a bigger mirror than the previous iteration, with its diameter increased from 1.8 metres to 2.5 metres.[3]

Analysing the data collected from an experiment like BLAST-TNG is not an easy task. Due to the sheer volume of data collected, a supercomputer is required to perform any advanced data analysis – this process itself takes a couple of years. Moreover, BLAST-TNG is not simply taking pictures of the sky but rather constantly scanning areas of interest. This scan is then converted to a single image with polarisation data using reconstruction methods that know where the telescope is pointing at each moment and then converting the information to a value in a given pixel.

One particularity of BLAST-TNG is that all of those involved are professors, postdoc researchers, and PhD students, with the last group forming the majority. This means that this experiment and its enhancement of our knowledge of polar ballooning and star formation has in turn contributed to the development of a new generation of scientists.

The BLAST-TNG experiment, funded by NASA, is a truly collaborative effort. In terms of construction, its detectors were built by the National Institute of Standards and Technology (NIST) in Boulder, Colorado. "The instrument [depicted in the photograph above during the ascent phase just a few minutes after its 2020 release] features a cryogenically-cooled camera with extremely sensitive superconducting detectors, and a telescope with a 2.5-m-diameter carbon fiber composite mirror. The camera observes the thermal emission from interstellar dust around galactic molecular clouds, the locus of star formation in the galaxy, and is polarization-sensitive, enabling observations of the structure of magnetic fields in these regions."[1]

Situated in the South Pole Dark Sector, the South Pole Telescope (above) which includes the BICEP2 Telescope, and the Keck Array Telescope (shown in detail opposite) measures the Cosmic Microwave Background. The 10-metre South Pole Telescope was installed in 2011 and has been upgraded since, with the third-generation high-resolution 1500-square-degree camera SPT-3G installed in 2017,[2] while the detector of BICEP2 was created using micro-lithography and micro-machining. The Keck Array Telescope, captured by Shaun O'Boyle from inside the telescope mount with two receivers removed and three remaining, was deployed during the austral summers of 2010–2012 and underwent an upgrade during the 2019–2020 season.

"Using a telescope at the Pole can reduce noise not only from Earth's atmosphere, but from interstellar radiation and dust, too. That's because a telescope at Amundsen-Scott has perfect access to [...] the Southern Hole – a patch of unusually clean sky that allows for optimal viewing into very deep space and consequently to the very early universe."

Marc Kaufman (Writer), "The South Pole Is a Great Place to View Space", *National Geographic*, 2014

"The South Pole is the closest you can get to space and still be on the ground, it's one of the driest and clearest locations on Earth, perfect for observing the faint microwaves from the Big Bang."

John Kovac (Professor of Astronomy and Physics, Harvard-Smithsonian Center for Astrophysics in Massachusetts and Leader of the Background Imaging of Cosmic Extragalactic Polarization [BICEP] collaboration)

SHAUN O'BOYLE is a photographer and architectural designer. His work uses photography to explore architecture and built environments in high-latitude regions. He is a three-time grantee of the National Science Foundation (NSF)'s Antarctic Artists and Writers Program and the recipient of a Guggenheim fellowship in photography. O'Boyle's work has been featured in numerous galleries, museums, books, and magazines, including *Smithsonian* magazine.

ICE CUBE LABORATORY
Shaun O'Boyle

The IceCube Laboratory (framed by a "Sun halo" produced by sunlight streaming through ice crystals suspended in the air) is a ground-breaking international scientific infrastructure. A surface layout map placed on the Lab's roof deck (on which Wisconsin IceCube Particle Astrophysics Center director and IceCube operations director Kael Hanson stands in the photo to the left) visualises the complex layout of the cables that connect deep instrumentation placed below the ice to computers located in the laboratory's interior. Except for a densely instrumented region of eight cables near the centre of the array known as Deep-Core, such cables are placed 125 metres apart. Along these cables, DOM light sensors shown below were deployed into holes that were melted using a powerful 5-megawatt hot-water drill. Each hole took less than two days to drill, while the installation of the remaining instrumentation took another ten to twenty hours. The quality of the design and production of the light sensors is evident in the performance and robustness of these "Electronic Pearls", which convert light into an electrical pulse. On-board electronics, pro-tected within a thick protective glass sphere that allows the sensor to survive the extreme pressures of the ice as the water expands as it refreezes, digitise the light signal, register its intensity, and record the time to a billionth of a second. The digital optical module (DOM) is made up of numerous components, including a photomultiplier tube, a mainboard mounted into the board stack that digitally encodes the signal, transfers it to the surface, and performs timing calibration, and a penetrator cable that attaches to the mainboard. DOMs were assembled, calibrated, and tested by researchers at four sites: Stockholm University, DESY, UW–Madison's Physical Sciences Lab, and Uppsala University. Before shipping to the Amundsen-Scott Station ↗ p. 790 at the South Pole, they were tested in what is known as the Dark Freezer Laboratory, where an environment of darkness and extreme temperatures, ranging from room temperature to –55°C, is replicated. At the South Pole, they were first tested on the surface prior to actual deployment in the array deep in the ice, where they started recording data.[1]

IceCube Neutrino Observatory. Transforming Antarctic Ice into a Science Experiment

Francis Halzen, James Madsen, and Madeleine O'Keefe

FRANCIS HALZEN is principal investigator at the IceCube Neutrino Observatory, South Pole, and Hilldale, Gregory Breit Distinguished Professor of Physics and the director of the Institute for Elementary Particle Physics at the University of Wisconsin–Madison (United States). JAMES MADSEN is the executive director of the Wisconsin IceCube Particle Astrophysics Center. MADELEINE O'KEEFE is a communications specialist at the IceCube Neutrino Observatory (United States).

One might be inclined to think that the South Pole is a strange place to do science and, at first glance, this would be a correct assumption. It is a cold, dry, and desolate environment, cut off from most of civilization. For six months of the year, there is no sunlight – the entire continent lies in darkness. Why build an international science experiment in such a faraway place, such an inhospitable environment? The reason is that the South Pole has *ice,* and lots of it.

The IceCube Neutrino Observatory transformed a cubic kilometre of Antarctic ice into the largest and most innovative telescope in the world. By embedding 5,160 light sensors in the ice, scientists are able to look for the trails of light left by neutrinos: chargeless, nearly massless particles that are created in the Earth's atmosphere, in nuclear reactors, in the centre of the Sun, and during the most violent cosmic events, such as collapsing stars.

Neutrinos are famously elusive. They are "weakly interacting", which means they can fly through matter like bullets through rain. They are not deflected by galactic and extragalactic magnetic fields and are essentially never scattered or absorbed by radiation or dust. Neutrinos are the only known particles that can pass through Earth, which they accomplish more easily than light goes through glass.

It is precisely this "ghostly" behaviour that makes neutrinos so intriguing to study. Since they can travel such immense distances without being disturbed, neutrinos are the ideal astronomical messengers of powerful sources like the Sun, supernova explosions, and cosmic-ray accelerators. They come from the far reaches of the universe carrying information about their production intact.

Even though we cannot see neutrinos themselves, we can detect signals of their presence thanks to a phenomenon called Cherenkov radiation. This is blue light generated when charged particles move faster than the relative speed of light in a medium – in our case, ice. Cherenkov light is emitted by charged particles that are produced when a neutrino collides with a nucleus in the ice; the detailed light pattern subsequently observed by IceCube's network of light sensors – known as digital optical modules (DOMs) – reveals the direction of the original incident neutrino, making neutrino astronomy possible.

Due to neutrinos' extreme "shyness", large volumes of transparent matter are ideal sites in which to conduct this experiment; the estimated ideal size is 1 cubic kilometre.[1] Fortunately, nature offers deep natural Antarctic ice as well as deep ocean waters as mediums that can be successfully transformed into a Cherenkov detector.

Using Antarctic ice as a particle detector was a novel approach first proposed in 1989.[2] In the 1990s, the University of Wisconsin–Madison led the construction of an early version of a neutrino detector in Antarctica called AMANDA.[3] This project offered proof that Antarctic ice was suitable for detecting energetic neutrinos and acted as a testing ground for defining the technical and logistical challenges that would be faced while building IceCube (including flying twenty Hercules C-130 transport planes from Christchurch to the South Pole with over 45,000 kilograms of cargo and personnel each season).

There were multiple advantages to building IceCube in Antarctica. To begin with, sinking the photomultipliers into ice secured the fragile electronics of the sensors in a frozen, mechanically stable environment at constant and very cold temperatures, and it allowed investigators to walk around on top of the experiment. Better yet, the ice sheet is geologically stable (Antarctica almost never has earthquakes) and completely dark. Finally, the National Science Foundation (NSF) was already operating a research station at the geographic South Pole, with proven infrastructure designed to deliver, regardless of the weather conditions. Counterintuitively, this implied that installing light sensors up to

COSMIC RAYS AND NEUTRINO ASTRONOMY
Francis Halzen, Kael Hanson, and James Madsen

The most important finding associated with the discovery of cosmic neutrinos is that they overwhelmingly originate in extragalactic sources, with an energy density close to that in gamma rays and cosmic rays, suggesting that they may have sources in common.

The century-old problem of identifying the sources of cosmic rays was one of the main motivations for neutrino astronomy. Neutrinos are produced when cosmic rays (protons and heavier ions) interact with matter or radiation like the cosmic microwave background that permeates the universe. On the 22nd of September 2017, a multi-messenger campaign triggered by the coincident observation of a gamma-ray flare being watched by the Fermi satellite and a 290-TeV IceCube neutrino pinpointed the likely cosmic-ray accelerator as TXS 0506+056.[1] TXS 0506+056 is a *blazar,* a giant elliptical galaxy with two jets emitting light and particles moving close to the speed of light along the axis of the black hole's rotation. One of those blazingly bright jets – hence the name "blazar" – is pointing at Earth and can flare for minutes or for months. Subsequently, IceCube archival data covering nine and a half years of observations revealed a three-month burst of thirteen cosmic neutrinos in 2014–2015 from the same direction,[2] further supporting TXS 0506+056 as the source.

The ten-year IceCube neutrino sky map[3] has also produced evidence for neutrino emission from the active galaxy NGC 1068 (Messier 77), a nearby Seyfert galaxy undergoing a major process of accretion onto the black hole. A fraction of a few per cent of such special sources, now labelled by astronomers as gamma-ray blazars, is sufficient to accommodate the diffuse cosmic neutrino flux observed by IceCube. While rapid progress seems likely, the observations also convincingly make the case for the construction of more and larger neutrino telescopes with better angular resolution.

IceCube Neutrino Observatory. Transforming Antarctic Ice into a Science Experiment

320

ICECUBE NEUTRINO

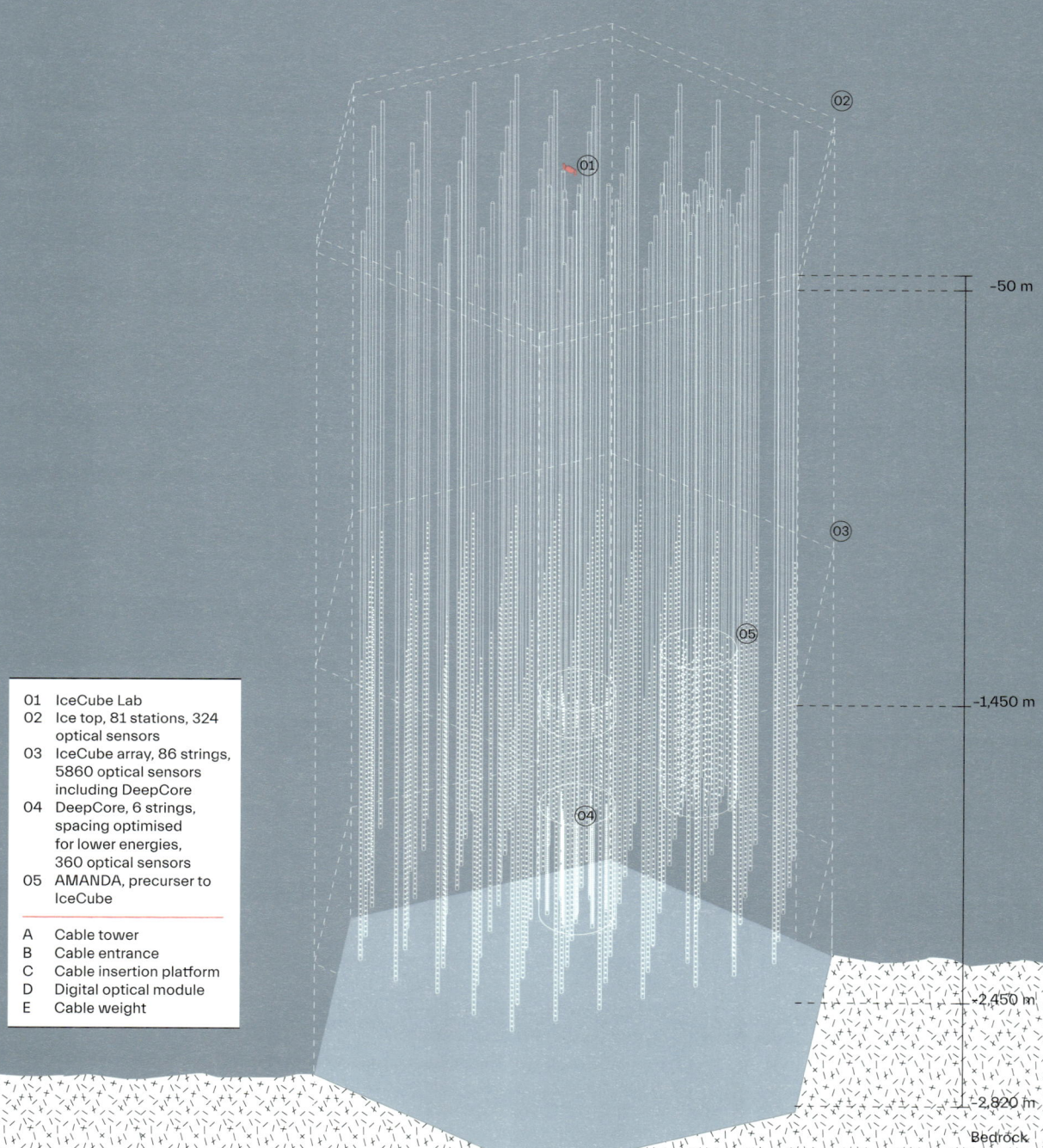

01 IceCube Lab
02 Ice top, 81 stations, 324 optical sensors
03 IceCube array, 86 strings, 5860 optical sensors including DeepCore
04 DeepCore, 6 strings, spacing optimised for lower energies, 360 optical sensors
05 AMANDA, precurser to IceCube

A Cable tower
B Cable entrance
C Cable insertion platform
D Digital optical module
E Cable weight

−50 m

−1,450 m

−2,450 m

−2,820 m

Bedrock

SOURCE
IceCube Collaboration

ICECUBE NEUTRINO LABORATORY AND DIGITAL OPTICAL MODULE

The IceCube Neutrino Observatory is a cubic-kilometre particle detector placed below the ice at the South Pole that reaches a staggering depth of about 2,500 metres below the snowpack. Here a network of 5,160 Digital Optical Modules (DOMs), each with a 25-centimetre photomultiplier tube and associated electronics, are attached to vertical "strings" frozen into eighty-six boreholes and arrayed over a cubic kilometre at depths ranging from 1,450 to 2,450 metres. Deployed on a hexagonal grid with 125 metres spacing, each string holds sixty DOMs, with the individual modules separated vertically by 17 metres.[1]

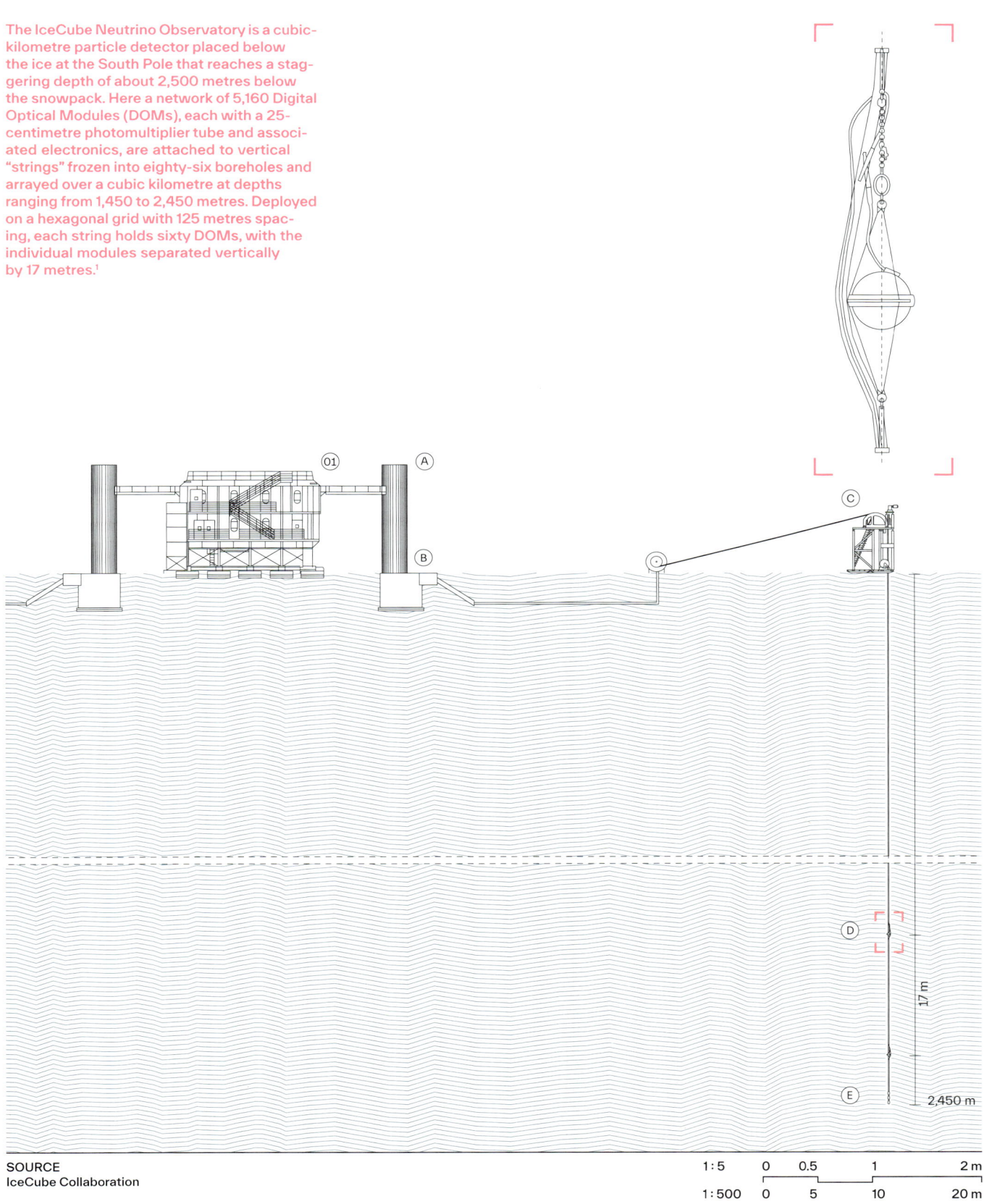

SOURCE
IceCube Collaboration

1:5   0   0.5   1   2 m

1:500   0   5   10   20 m

IceCube Neutrino Observatory. Transforming Antarctic Ice into a Science Experiment

322

2.5 kilometres beneath the geographic South Pole would be easier than building a detector in the sea.

IceCube drilling began in the austral summer of 2005. The drill that was used resembles a sleek rocket that dives, nose first, into the ice, gushing hot water out of its nose cone under pressure. It melts its way through the ice at a rate of 1.5 centimetres every second, to a depth of roughly 2.4 kilometres in less than two days. It leaves in its trail a hole of water that is kept liquid for a time by the highly insulating ice. The drilled path deviates from vertical by less than 1 metre over its 2 kilometres depth. The drill head is filled with electronics providing telemetry that is used, along with other methods, to determine the exact position of the instrumentation being deployed.

Once a hole had been completely drilled, the IceCube deployment crew lowered the cables carrying the light sensors into the hole, pulled by a 275 kilograms weight. For the next ten to twenty hours, the light sensors were attached to the sinking cable with carabiners of the kind used by rock climbers and plugged in at predetermined positions like beads on a rosary. Calibration devices to measure pressure and temperature, lasers, radio receivers, pulsing or steady light-emitting devices, and other useful gadgetry were also included. It takes several months for the hole to fully freeze and to find out whether each module survived the high pressure in the final moment of the freeze-in. If they made it, they are expected to live forever, protected a 2.5-centimetre glass casing.

On the 18th of December 2010, just after 6 p.m. New Zealand time, the last of IceCube's eighty-six cables was lowered into the Antarctic ice.[4] Seven austral summers of construction had come to an end, and IceCube was completed a decade after the proposal was first submitted. Two years later, we discovered neutrinos reaching us from the cosmos with energies exceeding – by a factor of ten thousand – those of the highest-energy neutrinos ever produced by particle accelerators.[5] For this discovery and for the construction of the novel instrument, IceCube was selected as the *Physics World* 2013 Breakthrough of the Year.[6]

Building IceCube was an extraordinary engineering achievement. Its success is a tribute to the perseverance and dedication of the international team of physicists, engineers, and technicians who overcame the many challenges of creating, building, and operating a detector in such an inhospitable environment. IceCube also stands as a testament to the importance of international collaboration in Antarctica, as the IceCube Collaboration involves 325 members from 52 institutions in 12 countries in North America, Australia, Europe, and Asia.

After a decade of operation, there is now an even more optimised and efficient design for the next-generation detector, for which an initial upgrade is already underway. With the IceCube Upgrade,[7] the team will deploy 750 advanced photodetectors and calibration devices near the centre of the existing IceCube detector. The additional instrumentation will improve our understanding of the optical properties of the ice, increase the detector's sensitivity to lower-energy neutrinos, and test ideas for new light sensors to be introduced in an even bigger, more ambitious expansion: IceCube-Gen2.[8] Gen2 will roughly double the instrumentation already deployed, increasing IceCube's volume tenfold and yielding an order-of-magnitude increase in neutrino detection rates.

The new advanced instrument will provide an unprecedented view of a high-energy universe of cosmic particle accelerators that is opaque to any form of electromagnetic radiation, moving from discovery to neutrino astronomy while also probing a wide range of fundamental physics. And Antarctica will be the hub for it all.

SHAUN O'BOYLE is a photographer and architectural designer. His work uses photography to explore architecture and built environments in high-latitude regions. He is a three-time grantee of the National Science Foundation (NSF)'s Antarctic Artists and Writers Program and the recipient of a Guggenheim fellowship in photography. O'Boyle's work has been featured in numerous galleries, museums, books, and magazines, including *Smithsonian* magazine.

COSMIC RAY OBSERVATORY
Shaun O'Boyle

The longest continuously running experiment at McMurdo Station was held at the Cosmic Ray Observatory or Cosray (shown in the image above), where low-energy cosmic rays were measured from 1960 to 2017. The observatory detected secondary sub-atomic neutral particles, neutrons, produced when primary cosmic rays interact in the Earth's atmosphere. One of a dozen sites around the world, the observatory is part of the neutron-monitoring network called Spaceship Earth, predominantly used by scientists to study how changing stellar magnetic fields, in this case on our Sun, accelerate particles to make the energetic cosmic rays. One of the three Cosray neutron monitor sections at McMurdo was removed in the 2014–2015 season and reinstalled at the Korean base Jang Bogo in December 2015. The other two sections were removed from McMurdo in the 2016–2017 season and have since been installed at Jang Bogo Station.

Involving more than three hundred researchers from fifty-two institutions in twelve countries, the IceCube Collaboration epitomises the importance of scientific collaboration. As stated by Professor Tom Gaisser, "The advantage of a collaboration is that working together makes it possible to accomplish what would be difficult or impossible for a few individuals or a single institution. In a successful collaboration the whole is more than the sum of its parts, and IceCube has already proved to be an outstandingly successful collaboration."[2]

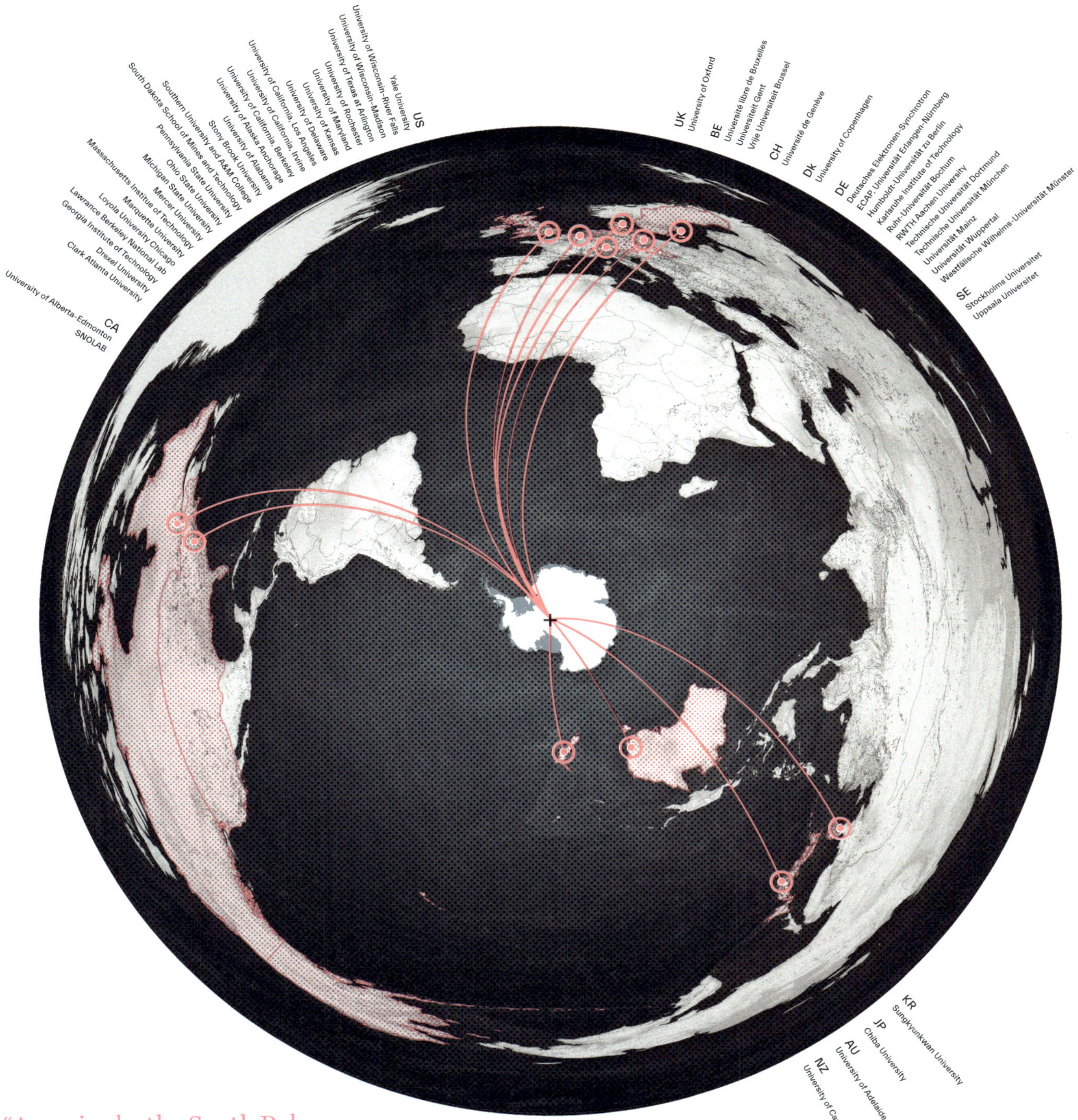

US
Yale University
University of Wisconsin-River Falls
University of Wisconsin-Madison
University of Texas at Arlington
University of Rochester
University of Maryland
University of Kansas
University of Delaware
University of California, Los Angeles
University of California, Berkeley
University of Alaska Anchorage
University of Alabama
Stony Brook University
Southern University and A&M College
South Dakota School of Mines and Technology
Pennsylvania State University
Ohio State University
Michigan State University
Mercer University
Massachusetts Institute of Technology
Marquette University
Loyola University Chicago
Lawrence Berkeley National Lab
Georgia Institute of Technology
Drexel University
Clark Atlanta University

CA
University of Alberta-Edmonton
SNOLAB

UK
University of Oxford

BE
Universiteit Gent
Université libre de Bruxelles
Vrije Universiteit Brussel

CH
Université de Genève

DK
University of Copenhagen

DE
Deutsches Elektronen-Synchrotron
ECAP: Universität Erlangen-Nürnberg
Humboldt-Universität zu Berlin
Karlsruhe Institute of Technology
Ruhr-Universität Bochum
RWTH Aachen University
Technische Universität Dortmund
Technische Universität München
Universität Mainz
Universität Wuppertal
Westfälische Wilhelms-Universität Münster

SE
Stockholms Universitet
Uppsala Universitet

KR
Sungkyunkwan University

JP
Chiba University

AU
University of Adelaide

NZ
University of Canterbury

"Amazingly, the South Pole
now ranks with the grand
research laboratories such
as Fermilab and CERN."

Francis Halzen (Principal Investigator at IceCube
Neutrino Observatory, South Pole, and Hilldale
and Gregory Breit Distinguished Professor of Physics
at the University of Wisconsin–Madison), 2011

The Antarctic through the Lens of Eliot Porter

Photographs taken between 1974 and 1976 as a guest of the United States National Science Foundation (NSF) Antarctic Program

"As world reserves of oil and gas go on shrinking, and as the richest mineral deposits approach exhaustion, international consortia will begin to exert pressure on governments to permit exploratory drilling in the unglaciated dry valleys […] and on the continental shelf of Antarctica."

Eliot Porter (Photographer), *Antarctica*, 1978

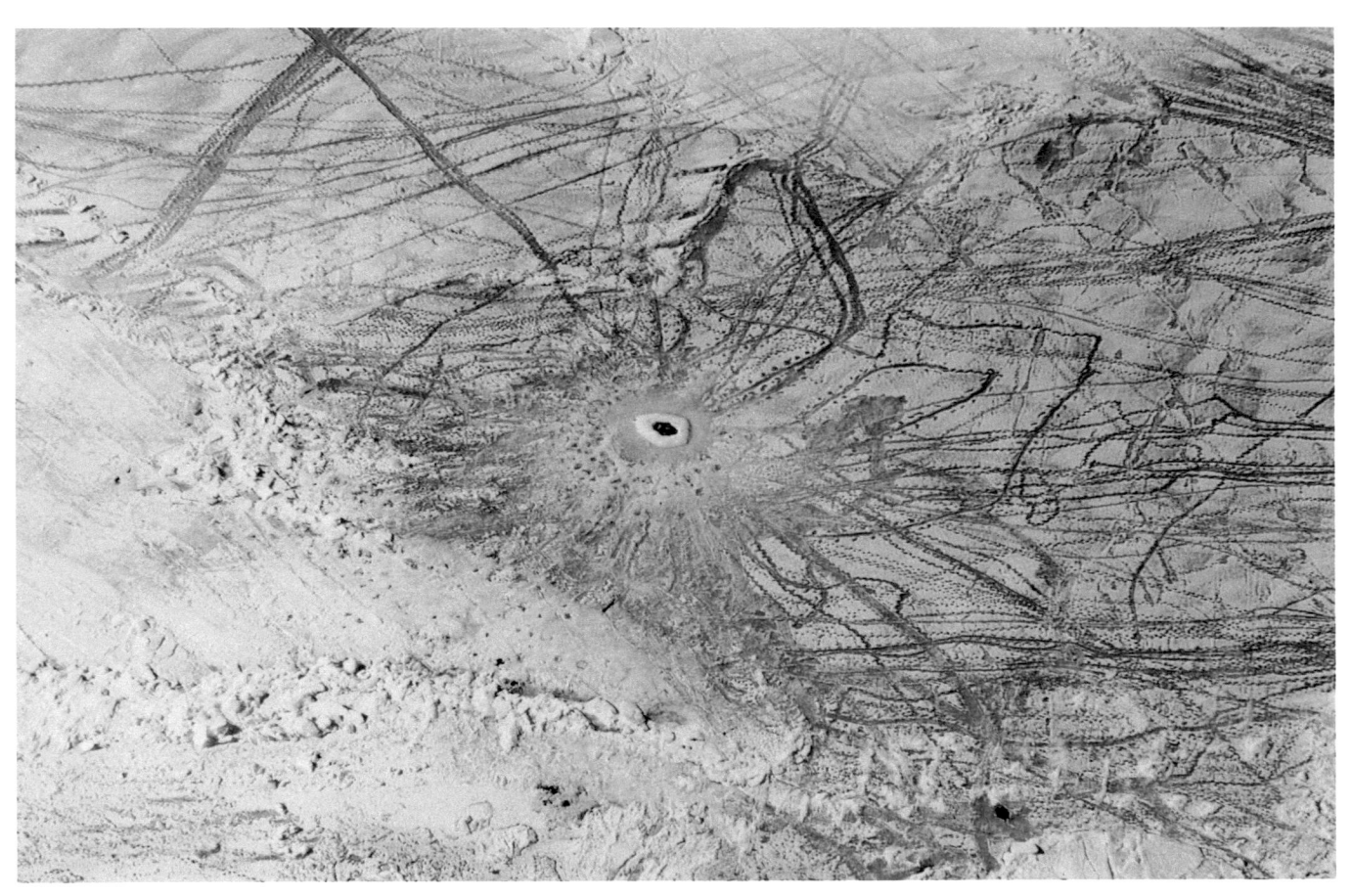

G IN THE
ERE

10
Extreme Habitation. Compressed Spaces and Dilated Times

The transient population of the only continent that has no indigenous peoples and no other long-term residents fluctuates in number in a way that is inversely proportional to the seasonal surface rhythms, whereby sea ice doubles the Antarctic surface in winter. As visualised in the diagram above, over 4,000 scientists and supporting staff are typically deployed in Antarctica during the summer months, while the number reduces drastically to approximately 1,000 during the austral winter. The visualisation reflects population data by station and by year as published in the COMNAP (Council of Managers of National Antarctic Programs) 2017 *Antarctic Station Catalogue*.

Inhabitation in the Extreme. The Winter-Over Syndrome

348

2020 in Venice, Italy. 45.4408° N, 12.3155° E

"Sunsight"[1] – 11:27
"Sunclipse"[2] – 13:20

"Sunsight" – 11:29
"Sunclipse" – 13:23

"It's as if even the simplest words are losing their meaning; as if 'this evening' refers to indefinite time, as if 'urgent' means 'later', as if a verb in the future describes a completed action."

Marie Darrieussecq (Writer), *White*, 2005

2020 in the Antarctic. 90.0000° S

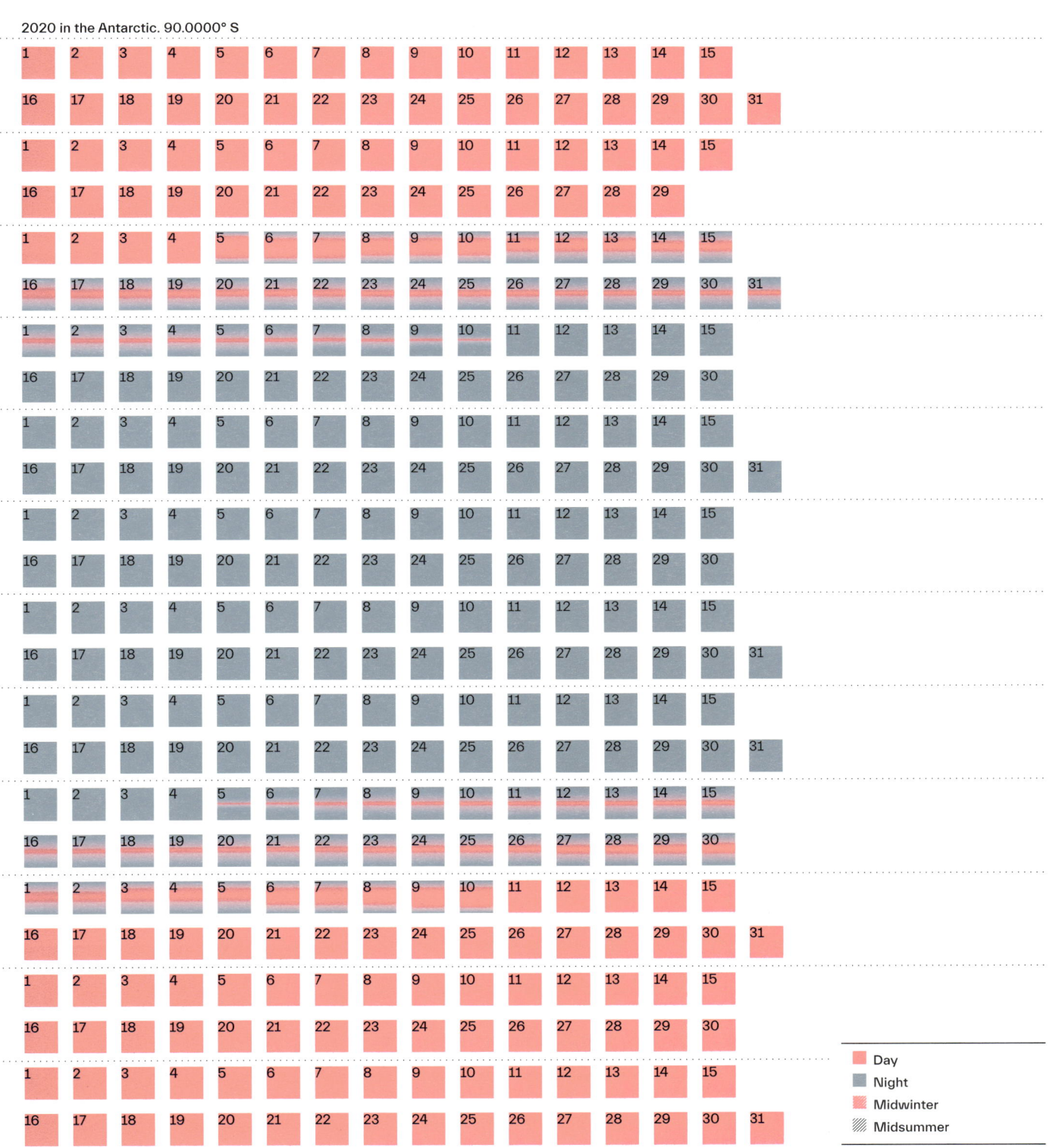

Day
Night
Midwinter
Midsummer

Interior Urbanism at the Pole.
Framheim vs. Cape Evans

Giulia Foscari

GIULIA FOSCARI is an architect, researcher, and writer who has been practising in Europe, Asia, and the Americas. She is the founder of UNA, an architecture studio focused on cultural projects, and of its alter ego UNLESS, a not-for-profit agency for change devoted to interdisciplinary research on extreme environments. Foscari taught at Hong Kong University and at the Architectural Association, where she ran a Diploma Unit and founded the Polar Lab.

While the legacy of the Space Race led to the launching of artificial satellites and human space flight, the Race to the South Pole – all aspects of which, apart from the tectonic one represented by the huts of the two contenders, have been scrutinised by exemplary scholars – defined for posterity models of inhabitation in the extreme that were unknown to humankind.

The absence of indigenous peoples and the hostility of the environment that confronted the British and Norwegian parties who endeavoured to secure primacy for their respective countries by being the first to hoist a flag on the South Pole were unprecedented. The physical and intellectual tabula rasa was absolute.

Faithful representatives of their homelands, whose respective cultures were deeply rooted in each filament of their DNA, the responses of Scott and Amundsen to the challenge of inhabiting Antarctica do not simply reflect the decisions they took prior to reaching the continent – namely which prefabricated wooden hut typology to acquire – but also the site in which they decided to establish their settlements. Scott erected Cape Evans ↗ p. 632 on the secure bedrock of Ross Island in proximity to Mount Erebus, whereas Amundsen (who anchored the *Fram* in the Bay of Whales) chose, for Framheim ↗ p. 628, the Ross Ice Shelf, which was seen to be "a whole degree farther south than Scott"[1] and ensured an abundant meat supply as well as optimal sighting of "The Barrier". "Of the opinion that the part of the Barrier on which the hut stood rested on land, so that any fear of a sea voyage was quite superfluous",[2] Amundsen and his team were blissfully unaware that the ice on which they lived was relentlessly shifting towards the ocean and could have calved at any time.

Approximately 460 kilometres further south, the *Terra Nova* expedition party settled in a location previously discovered by Scott during his first 1901–1904 expedition, which was seen as the "most admirable sandy flat […] with snow drift for the horses and easy access from the sea ice".[3] Here, Scott's team "made […] a truly seductive home, within the walls of which peace, quiet, and comfort reign supreme. […] A noble dwelling [that] transcends the word 'hut', […] a house of considerable size, in every respect the finest that has ever been erected in the Polar regions: 50 feet [15.2 metres] long by 25 feet wide [7.6 metres] and 9 feet [2.7 metres] to the eaves."[4] Here, in "the most comfortable dwelling-place imaginable", the party lived for two years, from 1911 to 1913.

Mirroring the layout of British vessel *Terra Nova*, Cape Evans was subdivided by a wall of stacked crates into "two large apartments", one known as the "Mess Deck" (which hosted the galley and the crew quarters), and the other – twice as big – known as the "Ward Room". The latter unfolded around a central area occupied by a large table and a stove, around which sleeping quarters were placed with varying degree of privacy (from Ponting's rather toxic darkroom and Scott's famous semi-enclosed cubicle to the unshielded bunk beds of the party members wryly labelled "the Tenements" recalling a notoriously dense multiple-dwelling building typology). Adorned with personal possessions ranging from photographs of loved ones to polar books, these "private" spaces were juxtaposed with more "public" areas devoted to research and leisure – the latter epitomised by the presence of a piano, a gramophone, and a magic lantern projector.

Often separated only by the millimetric presence of a curtain, the public and private areas were both inextricably contained within a 115-square-metre interior. There was no city outside the seaweed-insulated wooden walls of Cape Evans and for six months per year there was only darkness beyond the safe threshold of the door.

10 Extreme Habitation. Compressed Spaces and Dilated Times

SURVIVING IN THE CRYOSPHERE   351

With 25 people living in 105 square metres, seclusion was not an option at Cape Evans. Reflecting British social structures and military ranking, wooden crates (used as oversized wooden bricks to articulate the hut's interiors and visible in the photograph behind Petty Officers Evan and Creane) subdivided the hut into two main areas. Further partitioning allowed for greater degrees of privacy, from Herbert Ponting's enclosed darkroom, which was "eight feet long, six feet wide, and eight feet high"[1] and conveniently morphed into a (rather toxic) bedroom at night, and Robert Falcon Scott's semi-enclosed domestic cubicle to the open space in which all scientific and ordinary activities took place alongside party members' bunk beds and fold-out camp beds.

"The Hut had been entirely transformed from its customary appearance, by the draping of Union Jacks and sledging-flags and we sat down under festoons of bunting and coloured and embroidered silks to a feast, the bounteousness of which seemed almost incredible after our customary simple fare. [...] Then, after the company had been photographed, the table was moved aside, so that I might show about a hundred lantern-slides which I had prepared from my negatives of the Expedition. [...] Under the influence of rum punch the fun became fast and furious. [...] It was a marvellous inauguration of a season, which, as Scott wrote: 'For weal or woe must be numbered amongst the greatest in our lives.'"

Herbert George Ponting (Photographer, Explorer), *The Great White South*, 1921

The remoteness of Antarctica, and its extreme climatic conditions re-established the Semperian centrality of the hearth as "focus" of primordial architecture, a "site of all sites, as it were, the homestead pure and simple, toward which everything presences along and together with everything else and thus first is".[2] Alongside the Promethean element, a table competes as compositional fulcrum of the so-called wardroom, catalysing all social activities during the long Antarctic winter months. The table encapsulates the notion of "public space" in the internalised Antarctic inhabitation model. Reconfigured as needed, the table offers support for all studious duties, is used for daily prosaic rituals, and becomes the central stage for much-awaited celebrations such as midwinter and birthdays. A 90-degree rotation of the table, and its temporary erasure from the interior landscape, was an honour bestowed on rare occasions such as Herbert Ponting's lecture series (as captured on the 16th of October 1911 during a magic-lantern lecture on Japan), theatrical performances, and concerts by Cecil Meares and "His Master's Voice gramophone, which had been donated by the makers" to offer "pleasure [...] to those in exile".[3]

Living "captive" inside a hut – twenty-eight males confined in a big brother-like state without communication with the civilised continents – led the *Terra Nova* party to challenge the notion of the building and transfer to it qualities and programmes generally associated with the urban context, *de facto* internalising urbanism. As a consequence, spaces were endlessly reconfigured on a daily basis to allow sleeping quarters to become working areas, working areas to become entertainment spaces (where the party could "employ evenings in the most popular and profitable manner"[5]), entertainment spaces to morph into workshops, workshops (and darkrooms) to convert into bedrooms. The metamorphosis of Cape Evans – immortalised by the astounding pictures taken by the photographer in residence, Herbert Ponting – is a testament to the role that architecture has in extreme territories and to the discipline's role in mitigating physical and psychological disorders induced by the combined effects of isolation and hyper-dense cohabitation.

Meanwhile on the Ice Shelf, Framheim was assembled in fourteen days, following what might be defined as a well-executed master plan. The "camp", constructed upon levelled snow, featured the main hut as well as "sixteen-man tents" deployed for the storage of supplies (such as dried fish, fresh meat, coal, and wood) as well as to provide shelter for the dogs that would assist in the journey to the Pole. The advantages of building in snow became swiftly apparent to the Norwegian team as the floor of the dogs' tent was intentionally "sunk six feet [1.8 metres] below the surface of the Barrier [...] to create a draught-free space with air, light, and enough room".[6] Once complete, Amundsen himself remarked in his journals that the settlement "looked quite an important place" but did not expand too much on the actual 8- by 4-metre hut if not to share the "snug, cosy, and cleanly impression" it projected.

In an ode to ruthless efficiency, at Framheim all seven expedition members inhabited one unique space which was used for both sleeping and eating, transcending the social hierarchy segregation which characterised Cape Evans even during festivities such as Midwinter Day. Bunk beds were organised around a central table and privacy was achieved by pulling a curtain. The efficient Norwegian hut did not allow for much reconfiguration and the team had not had the luxury of carrying bulky instrumentation – be it scientific or musical. Yet surely Framheim's inhabitants must have suffered similar overwintering syndromes to those experienced at Cape Evans. Fortunately for Amundsen, nature came to their rescue, offering an alternative mode to address the individual needs of party members.

On the 21st of April the hut was buried up to its eaves by accumulated snow drift. The expedition members contemplated the possibility of alternative accommodation solutions, which briefly included a hypothesis that saw the living room transforming into a workshop, but this was immediately discarded as it did not seem a "good plan to give up the only room where [they] could sometimes find peace and comfort" while forcing expedition members to "be in each other's way all day long".[7] An enlightening realisation determined what occurred next: they needed to start "taking Nature in hand and working with her instead of against her".[8] Lacking a snow shovel, with which they could have potentially continued to excavate the entrance daily, they engaged in a "work of tunnelling which lasted a good while, for one excavation led to another, and [they] did not stop until [they] had a whole underground village – probably one of the most interesting works ever executed round a Polar station".[9] Referencing grand engineering works like the Simplon and Mont Cenis, and "affected" by a "building fever"[10] which saw them work tirelessly, the team at Framheim promptly transformed the settlement into an intricate system of under-snow tunnels which included a vapour bath, the "Carpenters Union" (where furniture elements were hollowed out of the walls), the "Smithy", the "Clothing Store", the "Crystal Palace", the "Depot", and a "deep hole to receive all the waste water from the kitchen".[11] By the time "all these undertakings were finished at the beginning of May" the expedition team had expanded the premises of the hut to such an extent that everyone had his own private space.[12]

By the time Roald Amundsen left for the South Pole on the 18th of October 1911, preceding Robert Falcon Scott and his team, his hut was almost fully buried under the snow. Accessible only via a wooden trapdoor which had been erected at an angle to ensure that the threshold would remain closed at all times preventing the risk that the Norwegian party would be involuntarily snowed in, Framheim's ordinary prefabricated hut became the extraordinary first exemplar of under-ice Antarctic architecture.

More than a century later, with hindsight as regards the outcome of the Race to the Pole, it is interesting to reflect on how the notion of "order" embedded in Amundsen's phrase "Victory awaits him who has everything in order [whilst] defeat is certain for him who has neglected to take the necessary precautions in time"[13] may be seen as referring not only to Amundsen's success but also to the architectural testament of the two precursors of Antarctic architecture, namely Cape Evans and Framheim.

Whilst Scott searched for comfort within the relatively palatial qualities of his hut (which was arguably constructed to serve as a home as well as scientific laboratory for what might be defined as a "gentlemen's expedition"), Amundsen rejected the allurements of comfort in favour of efficiency and optimisation. Within the Norwegian settlement, psychological relief from the enduring harsh climate, darkness, and isolation, appears to have been resolved via the excavation of individual spaces in which expedition members could exempt themselves throughout the day to convene for a more humble game of cards in the evening on the central table, which acted as a social catalyst. Whilst Captain Scott, in line with British imperial ideals sought to "terraform" Antarctica and colonise the space, Amundsen, who had been exposed to the Inuit culture, recognised the supremacy of the unforgiving environment and worked with it – within it.

The fate of these two huts is perfectly aligned with the sentiment that led to their different construction and inhabitation and at the same time ironic. While Framheim surrendered itself to the depths of the ice shelf that once shaped its architecture, and its remains are thought to have fallen to the bottom of the ocean bed with the calving of the Ross Ice Front (between 1957 and 1962),[14] Cape Evans stands proudly on the volcanic rock on which it was first erected, protected indefinitely and listed as Historic Site and Monument no. 16. Preserved initially by the freezing temperatures, and subsequently rehabilitated by a meticulous seven-year restoration campaign, conducted by the New Zealand Antarctic Heritage Trust, which crystallised one of the many configurations of the ever-morphing urban interior of the hut, Robert Falcon Scott's Cape Evans will remain for posterity, while the hut, tent, and flag hoisted by Roald Amundsen – who reached the South Pole on the 14th of December 1911, anticipating the British party by thirty-four days – are gone forever.

"We began to extend our premises. Along the axis of the house, facing west, a huge snowdrift had formed just in front of the entrance. It was our intention […] to remove it. At the last moment, it occurred to me that we could possibly use the drift and escape the work of shovelling and carting it away. We have now started to excavate it and arrange a direct connection with the house through an underground passage. By digging downwards, we can obtain all the space we need. Provisionally, we are digging out a carpenter's workshop for Bjaaland and Stubberud. […] I think the idea is good. […] Our excavation of […] workshops […] under the snow, has repaid itself many times. How would it look in our living space if we had dragged in sledges, ski, tents. Everybody always in each other's way. […] In addition to all these new rooms, we had gained an extra protection for our house."

Roald Amundsen (Explorer), *The South Pole: An Account of the Norwegian Antarctic Expedition in the Fram 1910–1912*

April 1911

A   Mess deck
01   Entrance
02   Stove
03   Workshop table
04   Sewing corner
05   Bulkhead (Venesta boxes)
B   Wardroom
06   Dining and working table
07   Ponting's darkroom
     (processing negatives)

October 1911

01   Entrance
02   Hair cutting
03   Horse stables
A   Mess deck
04   Beds
B   Wardroom
05   Lecture space
06   Ponting's darkroom
     (foldable bed)

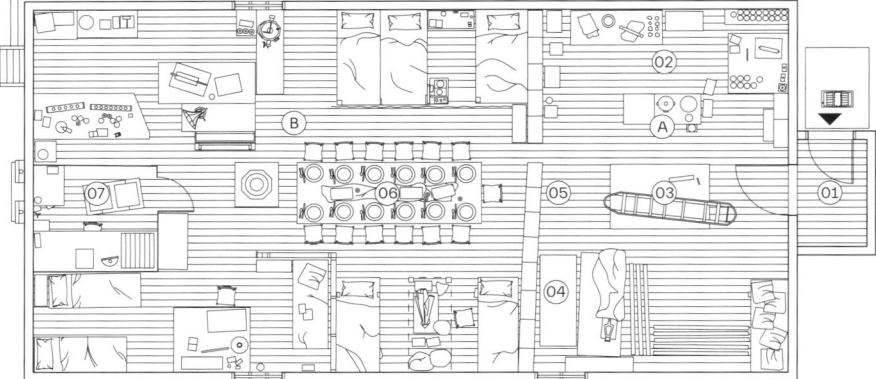

Cape Evans, Day, April 1911

Cape Evans, Evening, October 1911

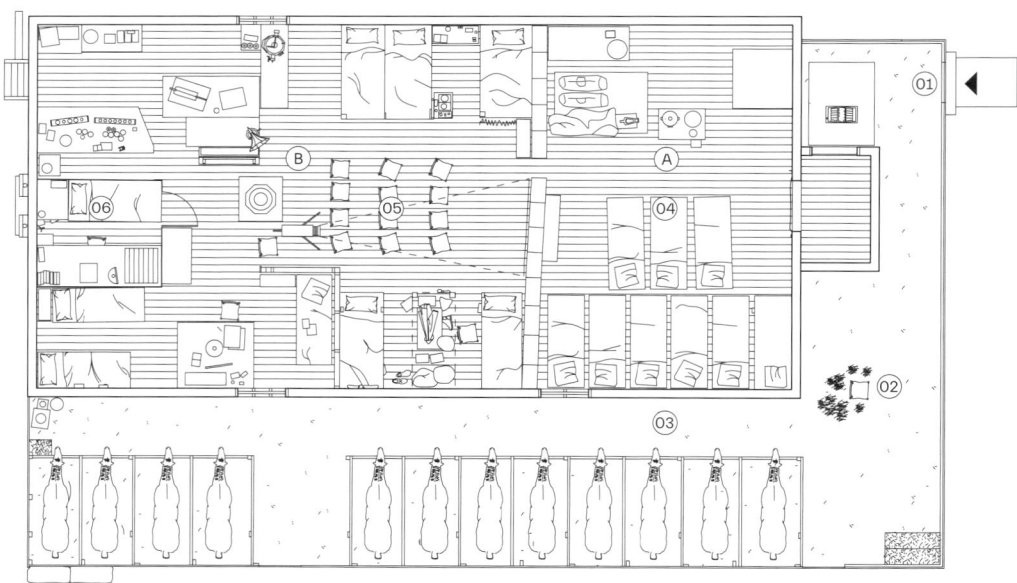

SOURCE
New Zealand Antarctic Heritage Trust, 2004

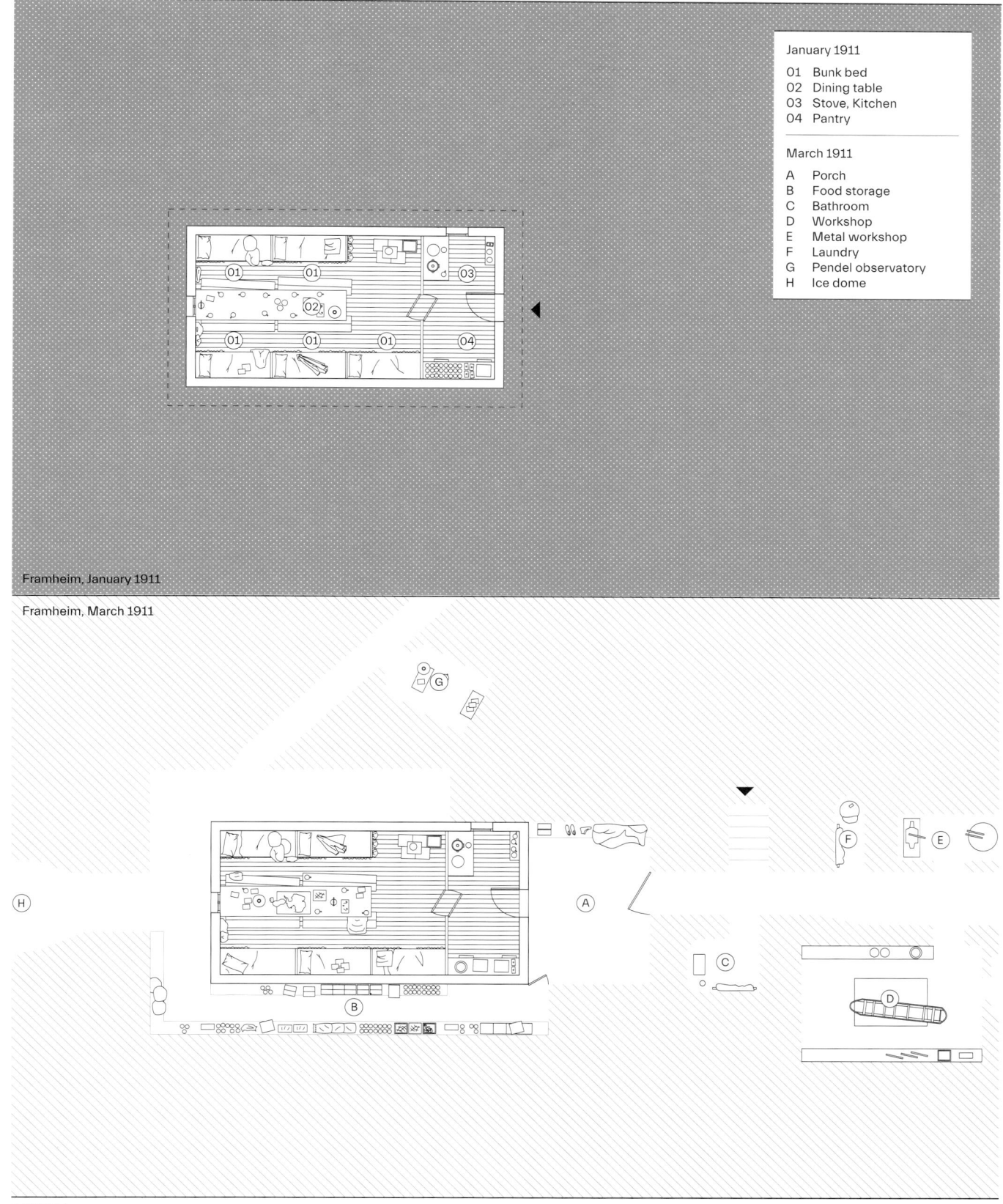

January 1911

01   Bunk bed
02   Dining table
03   Stove, Kitchen
04   Pantry

March 1911

A   Porch
B   Food storage
C   Bathroom
D   Workshop
E   Metal workshop
F   Laundry
G   Pendel observatory
H   Ice dome

Framheim, January 1911

Framheim, March 1911

SOURCE
Fram Museum, 1913

1 : 150   0   1.5   3   6 m

The Medicinal Role of Reading in the Heroic Age

Elizabeth Leane

ELIZABETH LEANE is Associate Dean Research in the College of Arts, Law and Education at the University of Tasmania (Australia), chief officer of the Standing Committee on Humanities and Social Sciences within the Scientific Committee on Antarctic Research (SCAR), and arts editor of *The Polar Journal*. With degrees in physics and literary studies, Prof. Leane has published seven books, including *Antarctica in Fiction* (Cambridge University Press, 2012).

In late 2018, reports appeared in the international press of a violent incident at Bellingshausen Station, a Russian Antarctic base. During a row in the canteen, an engineer stabbed his colleague – one of a group with whom he had lived over the previous winter – in the chest. The victim had to be evacuated to Chile, but the injury was not fatal, and it was only the Antarctic location that made the incident newsworthy. The initial news was followed by later coverage explaining why the attacker had been roused to such anger. The main provocation, according to these reports, was the victim's tendency to ruin his assailant's favourite leisure activity by giving away the endings of the books he liked to read.[1] While this unverified postscript to the story made only the more sensationalist press, to anyone who has studied the social life of early Antarctic expeditions, the account would ring very true, as reading was often the source of both intense pleasure and interpersonal conflict during the initial period of exploration of the continent.

Antarctic expedition leaders were certainly aware of the value of books and reading. In late 1911, Ernest Shackleton addressed the Poetry Society in London. Poetry had always played a large part in his life as an explorer. He wrote verse and had published it in an expedition "newspaper" that he also edited, the *South Polar Times;* he had a habit of reciting poems aloud, to the annoyance of his fellow explorers; and on his first visit to the continent he had taken part in a competition to determine the relative merits of famous poets (Browning versus Tennyson). Speaking to the Poetry Society, he noted that the men whom he had led got "real help" from studying poetry and were often inspired to write their own.[2] Five years later, when his Imperial Trans-Antarctic Expedition was in severe danger after its ship, the *Endurance* ↗ p. 640, was crushed by ice, he put his views into practice. Although, as his expeditioners prepared to live on ice floes, he had the dogs shot and commanded the men to discard most of their personal belongings, he allowed a number of books, including two volumes of poetry, to be retained.[3] In the ensuing months, as the men lived on the ice and then on the isolated Elephant Island, books became coveted items, with one man jealously guarding a secret volume in his sleeping-bag for six months.[4]

Reading was not just a luxury in the early years of land-based exploration of Antarctica (the so-called Heroic Age); it was a necessity. Where possible, expeditions brought with them extensive and wide-ranging libraries. Books, and particularly imaginative literature, served multiple functions on these expeditions: they provided a place to escape to in a crowded hut or tent but could also be the source of shared pleasure when read aloud, debated, and exchanged. If, as in the *Endurance* expedition, things went wrong, the value of books only increased. When the six men of the Northern Party of Robert F. Scott's *Terra Nova* expedition were forced to take refuge in an ice cave for many months, eating seal and penguin meat and burning blubber as fuel, the "worst" part of the tedious experience, according to their leader, was the absence of books.[5] They made constant use of the few titles they did have, reading a chapter of *David Copperfield* aloud each night, with the expedition doctor lifting the quota to two chapters when he noticed their spirits failing.[6]

Reading inspired writing: explorers would record their responses to the titles they read in their diaries, copying out favourite quotations and sometimes making extensive lists of their reading matter. The men of Douglas Mawson's Australasian Antarctic Expedition (1911–1914) wrote reviews of the books they read in their expedition newspaper written at Cape Denison ↗ p. 636. They developed strong feelings about novels that they read aloud as a group. They pretended a pillow was an unattractive male character from Owen Wister's Western novel *The Virginian,* and proceeded to beat it up; and they hugged each other at the conclusion of their favourite title, a girlish romance called *Lady Betty Across the Water*.[7] The imaginary worlds of literature became entangled with their lives in Antarctica – people and places were named after those featured in the novels they read. But fiction could also be provocative: in the middle of the dark polar winter, a fist fight broke out among Mawson's expeditioners when one man misinterpreted a comment another made about Arthur Conan Doyle's *The Hound of the Baskervilles*.[8] This was not a trivial incident; the man who took offence, Sidney Jeffryes, went on to suffer from a psychotic illness during and after his time in Antarctica, and the misinterpreted remark was an early indication of his growing paranoia.

Although books at times came second to other needs – Shackleton's men smoked some pages of the *Encyclopedia Britannica* – literature also kept men going, quite literally.[9] Apsley Cherry-Garrard recited verse to himself to match the rhythm of hauling a sledge; Douglas Mawson credited his escape from a crevasse – and hence his survival – to the poetry of Robert Service.[10] For these early expeditioners, books and literature were an indispensable form of mental medicine.

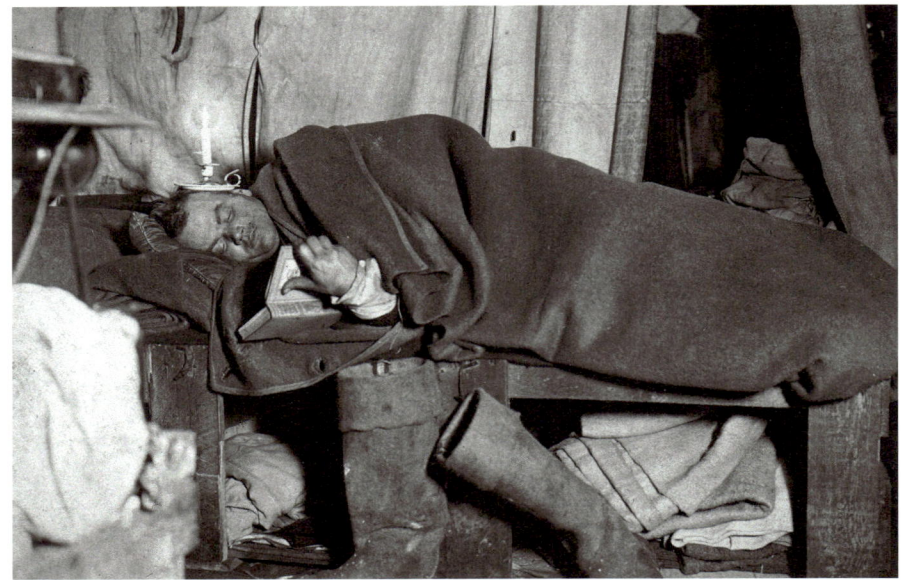

"We were in a hut the size of a garden shed and reading was the only form of escape I had."

The Economist, "An Expedition Reveals the Perils of Reading Dostoyevsky in Antarctica", 2019

10   Extreme Habitation. Compressed Spaces and Dilated Times

SURVIVING IN THE CRYOSPHERE   357

"[Music is] vital mental medicine, and we shall need it."

Sir Ernest Henry Shackleton (Explorer), in his diary during the *Nimrod* Expedition, 1907–1909

"It is necessary to be cut off from civilization [...] to realize fully the power music has to recall the past [...] to soothe the present and give hope for the future."[1] With these words, Apsley Cherry-Garrard, in his famous *The Worst Journey in the World,* reminds us of the role that music played during Heroic Age explorations in ensuring and preserving mental sanity for the expedition members. The presence of an official bagpiper in the 1902 Scottish National Expedition and the determination of meteorologist Leonard Hussey to save his six-pound banjo from a rapidly sinking *Endurance,* seen as a "mental medicine" that has absolute priority over other "vital" personal effects, offer ample evidence of the importance attributed to such art by expedition party members. Alongside music, which during Scott's British Antarctic Expedition (1910–1913) was famously enjoyed also by the dogs, literature played an important role in mitigating polar depression. Reading late at night in bed – as George Marston did during the British Antarctic Expedition in 1907 (shown on the opposite page) – was part of the daily routine until at least 11 pm. As stated later by Robert Falcon Scott in his diary entry on the 18th of June 1911 "at 11 pm, the acetylene lights are put out, and those who wish to remain up or to read in bed must depend on candlelight. The majority of candles are extinguished by midnight, and the night watchman alone remains awake to keep his vigil by the light of an oil lamp."[2]

THEATRE IN THE HEROIC AGE
Elizabeth Leane

The *Discovery* expedition (1901–1904), led by Robert F. Scott, took close to a thousand books to the Antarctic, including several play scripts. In the middle of their first winter, the men adapted and rehearsed an 1862 farce called *Ticket-of-Leave* (shown above, bottom right). Turning a storage hut into a theatre, with a two-foot-high stage, a drop scene and footlights, they performed the play in full costume and make-up in temperatures reaching well below –20°F. Ernest Shackleton and Edward Wilson created a programme. The actors were all drawn from the crew (Frank Wild is in the middle of the front row), while officers sat in the audience. Cross-dressing was clearly one of the pleasures of the event for the all-male expedition, and Gilbert Scott (no relation) as a maid was a general favourite. However, everyone was performing that night, with the attendees referring to the carriages they would take home, and oranges and nuts being served at the interval. The event represented a communal escape from the tedium and strangeness of a dark Antarctic winter, as the expedition hut was imaginatively transformed into the West End. The tradition of polar expedition theatricals dated back to 1819, when an overwintering British Arctic expedition inaugurated the practice, and continues through to the present day in the form of midwinter pantomimes.

THE *SOUTH POLAR TIMES*
Elizabeth Leane

Editing a house "newspaper" was an idea that Antarctic explorers adopted from their counterparts in the Arctic, where the tradition went back to the first deliberate shipboard overwintering in 1819–1920. The earliest and most prominent example of this unusual genre was the *South Polar Times*, produced on the two British expeditions led by Robert Falcon Scott. Edited in turn by Ernest Shackleton, Louis Bernacchi, and Apsley Cherry-Garrard, this newspaper benefitted from Edward Wilson's beautiful illustrations. Shown here is the cover of the first edition. Initially introduced as a way to keep expedition members occupied over the dark winter, the polar newspaper was also a bonding (and occasionally divisive) experience for these small communities and, as Shackleton suggests, a memorial of the more personal side of exploring. After the return of Scott's expeditions, limited editions of the newspaper (with the exception of the fourth volume) were published as a commemoration of the expedition. House newspapers became a common component of early Antarctic expeditions, sporting titles such as *The Antarctic Petrel, The Adélie Blizzard, The Glacier Tongue,* and *Das antarktische Intelligenzblatt* (Antarctic Intelligencer). When Shackleton led his own first expedition, he upped the stakes, taking a printing press to the Antarctic and producing the first book to come out of the continent, *Aurora Australis*.

Edited by Ernest Shackleton, illustrated by George Marston, printed by Ernest Joyce and Frank Wild, and bound with leather from horse harnesses on covers made of original Venesta wooden boards taken from the provision crates (one of which shows a stencil reading "Petit-Pois"), *Aurora Australis* was the first book ever published in the Antarctic. Produced at Cape Royds on Ross Island during the British Imperial Antarctic Expedition, the book was conceived as a means to ensure that "the spectre known as 'polar ennui' never made its appearance"[3] among Shackleton's party.

"It is too cold to keep the printer's ink fluid; it gets sticky and freezes. To cope with this a candle was set burning underneath the plate on which the ink was. This was all right, but it made the ink too fluid, and the temperature had to be regulated by moving the candle about."

James Murray (Biologist, Explorer) and George Marston (Artist, Explorer), *Antarctic Days: Sketches of the Homely Side of Polar Life by Two of Shackleton's Men*, 1913

10   Extreme Habitation. Compressed Spaces and Dilated Times

SURVIVING IN THE CRYOSPHERE   359

ANTARCTIC DIARIES
Elizabeth Leane

A surprising amount of information is compressed into the famous final page of the journal kept by Robert Falcon Scott during his second and last polar expedition. He writes of "stick[ing] it out to the end" – a phrase that makes sense only with knowledge of a prior argument about the possibility of using opium tablets to hasten this end. Having signed off, he adds a "Last Entry" requesting care for "our people" – presumably his own and his companions' families. Most remarkably, he locates the "pity" of the situation not in his own imminent demise in the frozen wastes of an ice continent, but in his inability to continue to write. For many Antarctic travellers, as for Scott, recording the experience of Antarctica in some form is just as important as the experience itself. While modern tourists and expeditioners often rely on the visual records enabled by digital cameras, go-pros, and selfie sticks, for early explorers the quintessential Antarctic textual product was the diary. Diaries served multiple functions: they were factual accounts of weather, distances marched, and goals achieved; personal memorials of never-to-be-repeated experiences; ways of marking time in a tedious, dark, and otherwise featureless winter; and a means of escaping into one's own thoughts – and privately venting one's frustrations – while living in a crowded tent or hut. In some expeditions, personal diaries were retained by the leader, so that he could write an official account based on multiple sources. When, as in Scott's case, the diary itself became the official account, it had a rawness and poignancy that no polished retrospective narrative could achieve.

"Wednesday, January 17 – Camp 69. THE POLE. Yes, but under very different circumstances from those expected. We have had a horrible day – add to our disappointment a head wind 4 to 5, with a temperature –22 degrees, and companions labouring on with cold feet and hands."

Robert Falcon Scott (Explorer, Royal Navy Officer), in his diary, 1912

"I am just going outside and may be some time."

Lawrence Oates (Explorer, Army Officer), in Robert Falcon Scott's diary, 1912

While personal journals allowed pioneering Antarctic explorers to record their data and experiences during the Heroic Age, for the 1910–1913 British Antarctic Expedition communication between huts and field stations was guaranteed by a telephone system supplied by the National Telephone Company Ltd – now British Telecom – in partnership with Ericsson. First installed at Discovery Hut, the "system [had] been arranged to work off one common battery of 12 accumulators giving 24 volts. [...] Provision [had] been made for one instrument with calling equipment [to be positioned] at each of the Aurora Stations, one at Hut Point, one in the instrument hut, and one in the living hut."[4] Early precursors of contemporary mobile phones, the instruments with Bakelite and aluminium handsets, were placed in wooden boxes to be easily transportable on sledges. Radio communications followed suit and were installed with great effort[5] at the Main Base at Cape Denison in 1912, during the Australasian Antarctic Expedition led by Mawson. Despite the introduction of this new technology, rightfully celebrated in the 1948 Peronist propaganda poster shown on the following spread, the kind of knowledge gap experienced by Shackleton (who famously was unaware of the advent of World War I and only discovered upon his arrival at the whaling station in South Georgia in 1914 that "millions are dying; Europe is mad; the world is mad"[6]) remained common, even in more recent times. To this end, John Dudeney, who lived at Britain's remote Faraday Base with twelve other men, wrote: "To this day, the years 1967 and 1968 are kind of a blank to me as far as world events, movies, and music goes." With the establishment of post offices along the coastline to reinforce geopolitical ambitions, delayed dialogues with people beyond the frozen continent were enabled by postal correspondence and "lengthy scrolls" – such as the comically long letter held by Seabee Earl Johnson in the image on the right on the opposite page.

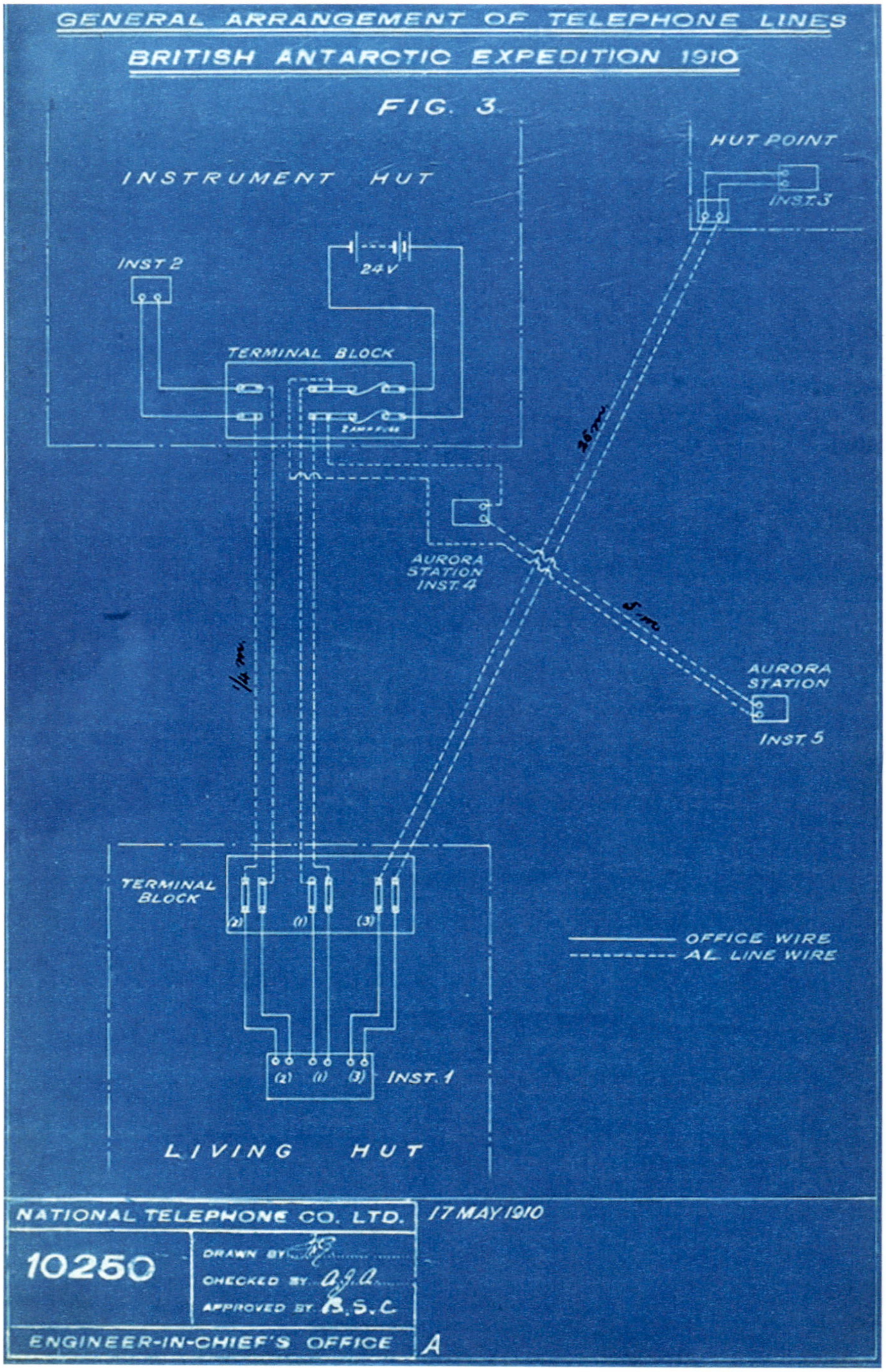

"At 5 o'clock the Hut Point telephone bell suddenly rang (the line was laid by Meares some time ago, but hitherto there has been no communication). In a minute or two we heard a voice, and behold! communication was established. [...] We were told it was blowing and drifting at Hut Point last night, whereas here it was calm and snowing; the wind only reached us this afternoon."

New Zealand Antarctic Heritage Trust, "History of Scott's Expedition" (online)

10   Extreme Habitation. Compressed Spaces and Dilated Times

SURVIVING IN THE CRYOSPHERE   361

Communication Abstinence

Arturo Lyon

ARTURO LYON is the director of the Polar Lab Chile.
He is an assistant professor at the School of Architecture of the Pontifical Catholic University of Chile.
Lyon is the founder of Lyon Bosch Martic, where he
developed architecture projects including the design
of Las Majadas Hotel, the Chena Public Park, and the
winning proposal for Nueva Alameda Providencia.

It is difficult to imagine human deployment in Antarctica over the last century without the existence of telecommunication technologies. In fact, this could arguably be considered the key element that has facilitated the overcoming of limitations imposed by climate, isolation, and long distances. Nevertheless, owing to their immaterial nature, which is almost impossible to depict in visual records, or to their advanced technologies, we usually do not acknowledge the fundamental role that telecommunications have played in Antarctic exploration.

While the famous 1915 Shackleton Imperial Trans-Antarctic expedition took months to ask for the rescue of its crew, the 1934 exploration to "Little America II" included a radio programme sponsorship that beamed via relays from Antarctica to CBS in New York, featuring a segment with Richard Byrd and two-way conversations with the explorers. The possibility of daily communication between Antarctica and other continents implied a huge increase in security and logistic capacities for the exploration and knowledge of this extreme territory. Together with this technology, radio towers appeared as a new feature in the Antarctic landscape, rising far above the flag poles that had previously dominated the skyline of the settlements.

Telecommunication technologies were incorporated as an essential element in the Antarctic expeditions and bases of the post-war period.[1] They required the deployment of hierarchical networks across different bases, which were to be strategically distributed in order to collect and transmit information in support of their operations and meteorological observation – crucial for airplane flights used to survey the Antarctic territory.

At the scale of the stations themselves, this technology required the installation of cable-tied antennas, a series of diesel engines to supply power, and fuel tanks for their long-term operation. Usually placed next to the sleeping areas, while the engines were in the service area and the tanks further away from the inhabited buildings, the telecommunications room played a fundamental role in the design, capacity, and resources of Antarctic bases. Consequently, many images of post-war bases show slender antennas located around them, establishing a set of light vertical structures in contrast with the solidity of the buildings and the strong horizontality of the Antarctic landscape.

Far from being a perfect system, communications in Antarctica were still described as one of the biggest problems posed for expeditions in the 1970s.[2] These problems related to difficulties in radio transmission caused by high noise, fading, and frequent signal blockages induced by various factors such as ionospheric disturbances in the auroral zone, the presence of magnetic storms, static energy produced by the snow suspended in contact with the antennas during storms, difficulties in establishing the equipment's ground connection in locations on hard rock, ice, or snow, and high levels of interference with electronic measurement equipment. After a number of substantial improvements, the use of radio has, to this day, remained a reliable communications technology, especially for missions conducted in the vicinity of the bases.

During the period in which several countries filed sovereignty claims and deployed their own infrastructure to sustain imperialistic ambitions, telecommunications, recognised as a necessary tool to guarantee safer operations and a better knowledge of the continent, fostered greater collaboration. The development of satellite technology (such as TDRS, Skynet, DSCS, or Iridium) enabled communications from the most extreme zones of Antarctica to any place in the world and became strategic in the transfer of scientific data collected on-site and in the collection of remotely sensed data for the observation of the earth, oceans, and atmosphere (along with spacecraft commanding support). Investments in telecommunications have also led to the provision of internet connections to the inhabitants of the bases, bringing about a noticeable change in the way of life, especially in conditions of confinement and extreme isolation.

Since many satellites move in polar orbits, there are zones of Antarctica in which their rotational trajectories tend to concentrate, rendering the respective sites ideal locations for satellite ground stations, while providing a frequent contact for telecommand and facilitating the unload of enormous volumes of information. These generally require substantial technological equipment, such as that deployed by the German Antarctic Receiving Station (GARS) built next to O'Higgins Base in 1990 (and dedicated to observations for long-term measurements for environmental research), the McMurdo Station ↗ p. 866 (1996) or the Troll Satellite Station built in Queen Maud Land in 2007. Frequently these stations take the form of a "radome" (radar + dome) in order to withstand the harsh climate conditions and the strong Antarctic winds, defining spherical volumes that populate the surroundings of the bases.

Within a continent in continuous flux and extreme isolation, the immaterial essence of telecommunications technology remains the only viable way of communication. In fact, Antarctica is the only continent not connected via fibre optic intercontinental cables. This is not due to the large distances, but to the instability of the ice, which is in a state of continuous flow. The high technology and great advances of the telecommunication networks leads us to question whether we should think of them as a support for field national operations in Antarctica or if, paradoxically, the Antarctic bases themselves have become a requirement for the deployment of telecommunication and global survey networks over the extreme south.

"A metal structure that rose above the otherwise nearly indistinguishable profile of rooms carved out beneath the ice and scattered boxes, oil drums, and equipment, the tower allowed direct communication to Washington, DC. Byrd's Antarctica was a futuristic amalgamation of sea, air, and disembodied technologies."

Elena Glasberg (Professor at New York University, Essayist, Speaker), *Antarctica as Cultural Critique: The Gendered Politics of Science Exploration & Climate Change*, 2012

LA RADIOFONÍA AL SERVICIO DE LA ENSEÑANZA

1948

CONCIERTOS

CONFERENCIAS

CANTOS ESCOLARES

CLASES MODELO

EFEMÉRIDES

MARCHAS PATRIÓTICAS

MÚSICA FOLKLÓRICA

LEYENDAS

DATOS ESTADÍSTICOS

CONSEJOS A LAS MADRES

POESÍA

DRAMATIZACIÓN DE HECHOS HISTÓRICOS

BIOGRAFÍAS CONFERENCIAS PARA LOS MAESTROS

On national programmes like the British Antarctic Survey, communications between Antarctic stations and field parties are ensured by installing Very High Frequency (VHF) and Single Side Band (SSB) radio systems on all vehicles and external buildings.[1] While during the summer months radio devices can be distributed one per group, during the dark winter, any individual exiting a module is required to have one with them at all times for safety.

In recent years, communication outside of the Antarctic is mostly achieved via satellite systems which act as relay stations. The satellites are used for the twofold purpose of transmitting scientific data and supplying the stations with Internet, telephone, and email services.[2] At the Amundsen-Scott Station, for example, the South Pole Tracking and Data Relay Satellite System (TDRSS) Relay 2, or SPTR2, was built to communicate with the NASA Tracking and Data Relay (TDRS) satellites. TDRS satellites, in particular, are a shared resource in the Space science community (including the Space Shuttle, International Space Station, and Hubble Space Telescope); as a result, their access schedules can be inconsistent or unpredictable.[3] Online times and durations via TDRS are also unreliable and vary daily; however, SPTR2 is typically online, providing South Pole Station with communications for between two and four hours each day. At numerous other stations, Internet reliability can be ensured twenty-four hours a day by using a geostationary satellite, but even in such fortunate circumstances streaming services are not allowed for private use as they take up too much bandwidth, and the downloading of large files is monitored by the communications manager. Researchers deployed in Antarctica are allowed to make personal calls but they have to bear the costs themselves and use calling cards; fortunately, "the cost of phoning home from Antarctica in 2016 is about 650 times less than it was [...] in 1986, 2p rather than £13 a minute."[4] The plurality of communications systems existing today on the continent effectively allows scientists and support teams deployed in Antarctica to have regular contact with home, greatly alleviating the symptoms induced by extreme isolation and communication abstinence.

"WYNAN YIGUM YIKYR WYSWO, John
WYNAN = Glad to hear you are better
YIGUM = I have grown a beard which is awful
YIKYR = This place gives you a pain at times – but it's worth it
WYSWO = Love and kisses"

Australian Antarctic Program, "ANARE Communications 1947–1985" (online), 2017

"YIKLA = This is the life."

Bernadette Hince (Honorary Lecturer, Australian National Dictionary Centre), *The Antarctic Dictionary: A Complete Guide to Antarctic English*, 2020

Extreme Habitation. Antarctica's "Architecture of Sleep"

Juan Du

JUAN DU is the director of the Polar Lab Hong Kong. She is an associate professor and associate dean of architecture at the University of Hong Kong, where she directs the Urban Ecologies Design Laboratory. The projects within her practice, IDU, range from community centres to informal settlements upgrades. Prof. Du's extensive writings include *The Shenzhen Experiment* (Harvard University Press, 2020).

Despite the enormous challenges and dangers presented by its natural environment, Antarctica's architecture has enabled human inhabitation to take place on the otherwise inhospitable continent. This feat of logistics, innovation, and design for extreme situations cannot be overstated. While over the past two centuries, building materials in Antarctica evolved from canvas and sealskin to Corten steel and reinforced fibreglass,[1] the interior spatial arrangement for sleep has not changed much. Similar to the explorers of the earliest expeditions at the beginning of the twentieth century, most inhabitants of Antarctica today still rest in sleeping arrangements similar to those of early settlements such as the Heroic Age's Cape Evans ↗ p. 632 "tenements" – that of shared dormitory-style beds, perhaps the most standard institutional "architecture of sleep".

A 2018 review of the existing sleep studies conducted in the continent found that "sleep disturbances are the main health complaints from personnel deployed in Antarctica".[2] And the various studies documented "consistent changes in sleep during the Antarctic winter, the common denominators being a circadian phase delay, poor subjective sleep quality, an increased sleep fragmentation, as well as a decrease in slow wave sleep".[3] The review called for multidisciplinary research "to elucidate the mechanisms behind these changes in *sleep architecture,* and to investigate interventions to improve the sleep quality of the men and women deployed in the Antarctic."[4]

Most of the existing sleep studies focused on the impact of Antarctica's extreme environment or work organisation on the subjects' biorhythms and sleep architecture.[5] *Sleep architecture* is a term used by medical professionals to describe the basic structural organisation of sleep, or the patterns of different stages between wakefulness and deep sleep over the course of a night. However, in the various sleep studies there are few descriptions or even mentions of the physical environment or spatial organisation of the bedrooms and bed spaces or of the "architecture of sleep".

An analysis of the spatial organisations of ten Antarctic research stations – namely, Scott Base ↗ p. 852 (New Zealand, 1957), McMurdo ↗ p. 866 (United States, 1969), The Great Wall (China, 1985), Zhongshan ↗ p. 760 (China, 1989), Concordia ↗ p. 780 (France & Italy, 2005), Amundsen-Scott ↗ p. 790 (United States, 2008), Princess Elisabeth ↗ p. 786 (Belgium, 2009), Bharati ↗ p. 804 (India, 2012), Halley VI ↗ p. 810 (United Kingdom, 2012), and Jang Bogo ↗ p. 820 (South Korea, 2014) – offers a better understanding of the current state of where people sleep in Antarctica. These stations differ considerably as regards building age, location, population, and settlement scale and represent a wide range of countries that signed the Antarctic Treaty at different dates.

McMurdo, Antarctica's largest and most populous research station, has the most spacious bedrooms, each averaging 26 square metres in floor area, generally shared by two residents, giving an occupation density of 13 square metres. The smallest bedrooms are at the Amundsen-Scott Station, with bedrooms of 5 square metres each, albeit with single occupancy. The stations' site locations contributed to these differences. Situated at the southern tip of Ross Island, McMurdo is one of Antarctica's most easily accessible stations, where vessels bringing supplied and materials arrive almost monthly. Amundsen-Scott, on the other hand, is located at the Earth's South Pole and is one of Antarctica's most remote locations. To reach it, building materials and equipment have to cross the South Pole Traverse, 1,600 kilometres through snow and ice. Furthermore, prior to the 2007 operation of the Traverse, all building materials had to be delivered by helicopter.

"As each of the cubicles has distinctive features in the furnishing and general design, especially as regards beds, it is worthwhile to describe them fully. This is not so trivial a matter as it may appear to some reader, for during the winter months the inside of the hut was the whole inhabited world to us. The wall of [meteorologist Lieutenant J. B.] Adams and [surgeon Dr Eric] Marshall's cubicle […] was fitted with shelves made out of Venesta cases, and there was so much neatness and order about this apartment that it was known by the address, 'No. 1 Park Lane'. In front of the shelves hung little gauze curtains, tied up with blue ribbon, and the literary tastes of the occupants could be seen at a glance from the bookshelves."

Sir Ernest Henry Shackleton (Explorer), in his diary during the Nimrod Expedition, 1907–1909

Even though the Amundsen-Scott Station has the smallest bedrooms, the spatial organisation of single occupancy is rare in Antarctica. Most of Antarctica's residents share bedrooms, most commonly with two to three persons in a room, with an average occupation density of 4 to 6 square metres. The bedrooms at Scott Base typically host five people, although the standard size of 25 square metres brings its average occupancy density into line with other stations. In terms of extreme density of space and occupation, Princess Elisabeth Station stands out in the analysis with the highest sleeping space occupation density of 2.2 square metres per person, where a bedroom of 8.9 square metres is shared between four individuals.

Princess Elisabeth is also the only summer station reviewed here. It is worth pointing out that in nearly all of the year-round stations, the over-winter population is reduced by 50% or more, which drastically alters the occupation density as well as the spatial and social qualities of the sleeping space. This is a particularly crucial aspect, as most of the Antarctic sleep studies are conducted during the over-winter months of June to September. The occupancy density of the station is much higher in the summer months (October to February), when the research stations reach a peak number of researchers and support staff, and the most pronounced difference is often the sleeping arrangements. It would thus be important to add this socio-spatial parameter, especially when conducting sleep studies that compare the winter and summer periods, which typically consider the differences of the natural environment and not the built environment within the stations themselves.

Apart from Concordia Station's green and brown bedroom interiors, white is the dominate colour in the bedrooms, across the various stations and presumably cultures. Most stations' sleeping spaces have additional accent colours such as blues or greys, and many are further accented by natural wood finishes to the furniture. Perhaps the use of white as a neutral tone is to be expected, but one wonders if this should be further studied along with the documented issues of sensory deprivation and overexposure to reflected light. Unlike typical natural or urban environments, the year-round snow results in extremely high luminance levels in the outdoor environment and a dramatic outdoor-indoor difference. The ratio of the outdoor to the indoor luminance in Antarctic is drastically higher than that in a typical urban area in other continents.

A 2014 study conducted at China's Great Wall Station found that the outdoor ground surface luminance level (697 candela per square metre [cd/sqm]) in Antarctica is 240 times that of the indoor dormitory floor (2.91 cd/sqm).[6] Furthermore, the recorded luminance level of the windows of the station (8,750 cd/sqm) was higher than that of the sky outside (6,490 cd/sqm). While most bedrooms in Antarctic stations have small windows to the outside, the "natural light" that enters could be mostly "unnatural" for the human senses. In the design of the 2012 Halley VI Station, in addition to collaborating with working colour psychologist Angela Wright, architect Hugh Broughton developed a special LED light next to the bed that can awaken the resident with simulated daylight at dawn rather than the window's night-less daylight during the summer months or the total darkness of the Antarctic winter months.

Built in the 1960s and 1970s, the two oldest research stations reviewed, Scott Base and McMurdo, are both undergoing major renovations. In designing the major overhaul of McMurdo Station, Denver-based OZ Architecture held numerous design consultation sessions with the 1,000+ inhabitants of the station, from scientists to administrators, as well as the majority of the station's population: from the support staff to the mechanics and chefs. Rick Petersen, principal of OZ Architecture, shared with the Polar Lab that of the hundreds of topics of discussion, the design of the bedrooms received the greatest input, some of it conflicting and, at times, heated. As perhaps the most intimate and overlooked space of the research stations, this architecture of sleep should be further examined for its potential ability to not only improve the quality of rest but also mitigate physiological and physical disorders for the inhabitants of Antarctica.

Curtains embody the quintessential threshold between private and communal areas in Antarctic architecture. The role of curtains exemplified above (top) by the textiles that demarcate the narrow area between the sleeping cubicles and the dining table (evidence of the extreme programmatic promiscuity of Framheim) has remained practically unchallenged since the Heroic Age. A photograph taken in the Korean Jang Bogo Station (bottom) offers an example of their contemporary use.

"The bunks were closed […] with a small opening, leaving yourself in an enclosure which can hold its own with our modern coffin; and, like this, it is private; for some minds it is absolutely necessary to be alone, out of sight and entirely undisturbed by others. It was by special recommendation from the doctor that I made this arrangement and found that it answered well."

Carsten Borchgrevink (Explorer), *First on the Antarctic Continent: Being an Account of the British Antarctic Expedition, 1898–1900*, 1901

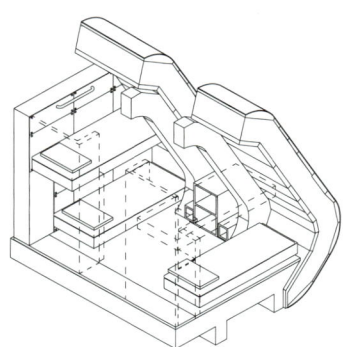

Princess Elisabeth, 2009

8.9 sqm, 4 beds

2.3 sqm per person

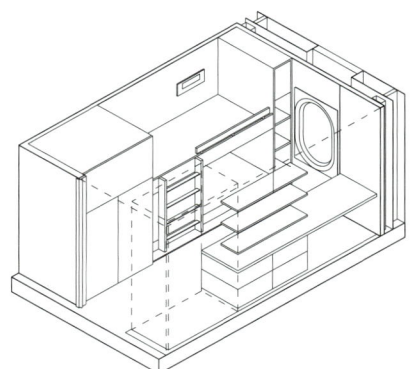

Halley VI, 2012

9 sqm, 2 beds

4.5 sqm per person

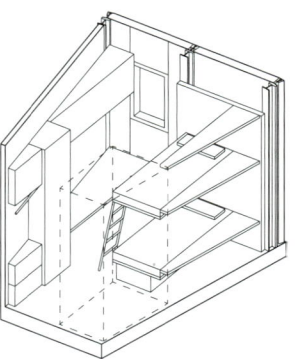

Concordia, 2005

9.3 sqm, 2 beds

4.7 sqm per person

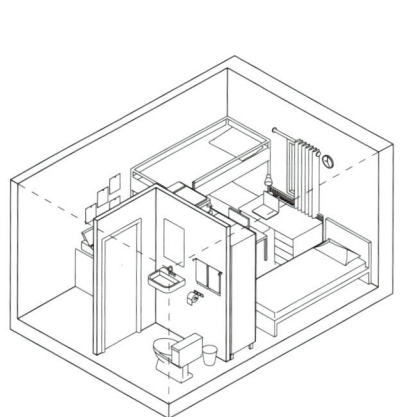

The Great Wall, 1985

10.5 sqm, 2 beds

5.3 sqm per person

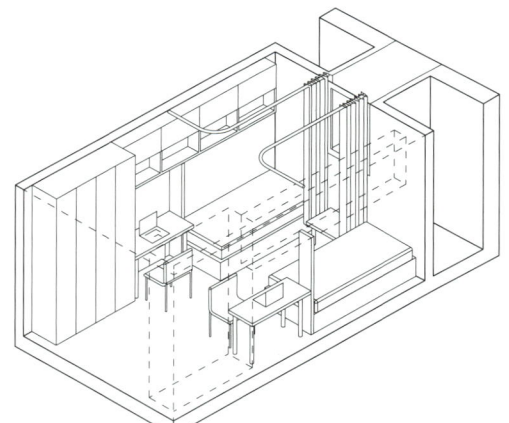

Jang Bogo, 2014

12 sqm, 2 beds

6 sqm per person

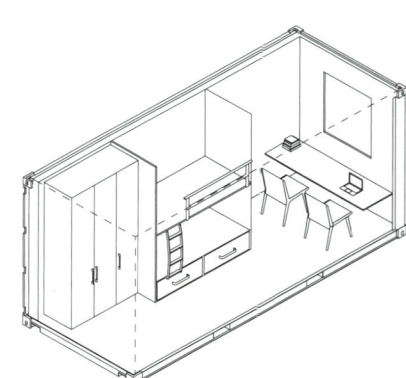

Bharati, 2012

14.6 sqm, 2 beds

7.3 sqm per person

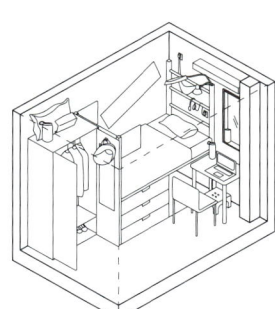

Amundsen Scott, 2008

5 sqm, 1 bed

5 sqm per person

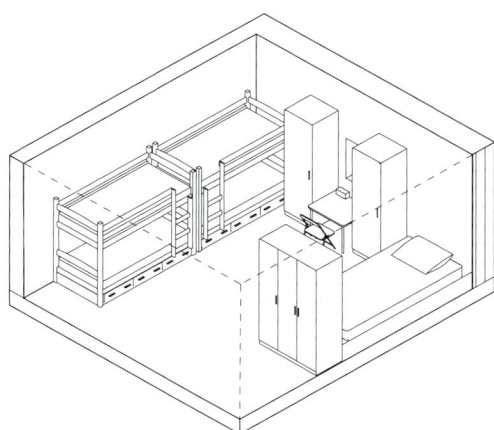

Scott Base, 1957

25 sqm, 5 beds

5 sqm per person

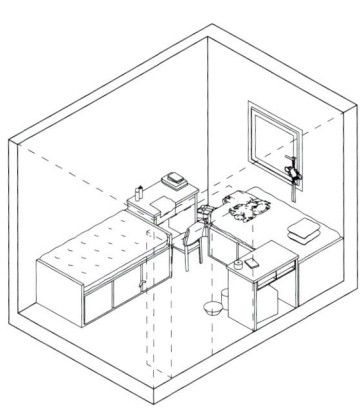

Zhongshan, 1989

11.4 sqm, 1 bed

11.4 sqm per person

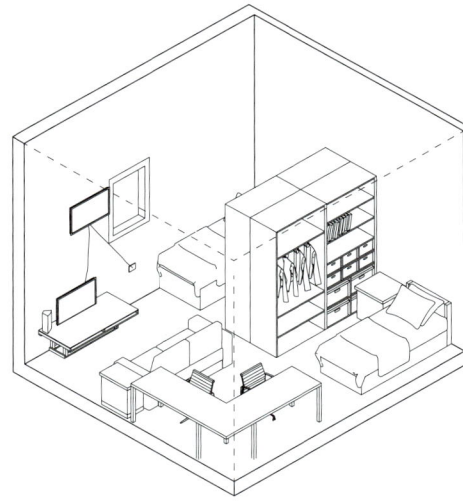

McMurdo, 1969

26.5 sqm, 2 beds

13.3 sqm per person

"Privacy in the ordinary meaning of the term did not exist."

Richard Evelyn Byrd (Admiral, Explorer, Aviator), *Alone: The Classic Polar Adventure*, 1938

"The first winter night I spent in the Antarctic, back in 1929, I was one of eight housed in barracks 10 × 10 square feet. At the Pole our space per man in the sleeping quarters alone was nearly this large. In fact, the total floor space of the Pole camp – 6,000 square feet – was about four times larger than at Little America 1929 base, which housed 42 men as compared to our 18. This in itself was a great step towards solving the psychological problems of a polar environment. The loosening up in living space gave a measure of privacy to our men. Also, the extra room required more servicing and left less spare time. Busy men are generally happy men."

Paul Allman Siple (Explorer, Geographer), "Man's First Winter at the South Pole", *National Geographic*, 1958

"In Halley VI bedrooms we have included a special alarm clock that we developed with Philips to help address seasonal affective disorder: it wakes you up with a daylight simulation light and balances serotonin and melatonin production to help with your cycle."

Hugh Broughton (Architect, Hugh Broughton Architects), in conversation with UNLESS, Architectural Association School of Architecture, 2019

"At the time I didn't know I was sleeping next to Sofia Loren as she was only discovered in 2010 along with all the other ladies of Port Lockroy. If only these walls could talk."

Sally Owen (Port Lockroy Staff, UKAHT), in conversation with UNLESS at Port Lockroy, Goudier Island, Antarctica, 2019

In 1914, when Shackleton published the famous advertising to recruit for his *Endurance* expedition, three women applied. Along with twenty-five further women who applied for the 1929 British, Australian, and New Zealand Antarctic Research Expedition, and the 1,300 women eager to join the 1937 British Antarctic Expedition, their request was rejected. It was not until 1939 that two Norwegian women first saw the continent (Ingrid Christensen and Mathilde Wegger) and it would take almost other two decades for a woman to work as a scientist aboard the Soviet Union ships *Ob* and *Lena*, along the Antarctic coastline. It is thus not surprising that in a male-dominated continent, in which even 1959–1973 British Antarctic Survey director Vivian Fuchs "firmly believed that the inclusion of women would disrupt the harmony and scientific productivity of Antarctic stations,"[1] the presence of women was relegated mainly as ornament on the walls. The painted pin-ups unveiled during recent restoration campaigns in the bedrooms of British Port Lockroy (on the left) and the ninety-two pornographic 1970s–1980s photos that covered the "Sistine Ceiling" (or better the Sex-tine Ceiling) of the Weddell Hut are a testament to the gender inequality that prevailed in the continent for over a century. According to an article published in 1983 by the *San Bernardino Country Sun* newspaper Antarctica was "one of the last macho redoubts, where men are men and women are superfluous."[2]

Sensory Deprivation and the Role of Colour in Antarctica

Angela Wright

ANGELA WRIGHT is the founder of Colour Affects (United Kingdom). She is an established expert on the unconscious effects of colour. In 1985, after studying both Freudian psychology and colour theory, she founded Colour Affects in order to test her colour theory empirically. Today, Wright advises blue chip companies across the world on colour psychology and she is an established spokesperson on the subject.

Colours are reflected wavelengths of light – the only section of the electromagnetic spectrum that is visible – and we are highly susceptible to these energies; every waking moment we are adjusting to the colours around us, whether consciously or unconsciously. Great minds throughout history have sought to understand the mysteries of colour, and dramatic progress was made in this field in the twentieth century. Today, we have a flourishing academic discipline called colour psychology. Many people simply enjoy the beauty of colour and give little thought to how it makes them feel, but generally most of us accept that there is more to colour than meets the eye.

Antarctica is a wondrous place offering unique opportunities for scientific research that would be impossible anywhere else on Earth. It could not, however, be described as hospitable. Its fluctuating inhabitants – especially those stationed on the plateau or ice shelves – experience extensive sensory and colour deprivation. If, as Johannes Itten would say, "Colour is life; for a world without colour appears to us as dead,"[1] the field of colour psychology is particularly important in Antarctic station design, as the calibrated use of colour in building interiors can offer a counterpart to the whiteness of the continent and help mitigate some of the symptoms of polar depression.

Halley VI ↗ p. 810, the British station located on a constantly shifting ice shelf that is essentially cut off from the rest of the world for about eight months a year, represents an interesting study case on the use of colour in Antarctic interiors. Given the extreme site, it was essential to the well-being of all concerned to create percep-

tions of variation and of light and warmth, ideally in an atmosphere of optimism and friendliness. The *Colour Affects System* of applied colour psychology was uniquely suited to this brief.

The system was developed from remarkable discoveries made in the 1980s, concerning links between patterns of colour and patterns of human behaviour in the natural world. It was found that every colour can be classified into one of just four colour groups and each colour group relates to one of four personality types. All the colours within each group have mathematical relationships that do not occur across the groups – suggesting a revolutionary new approach to objective colour harmony.

Loosely inspired by Nature's colour palette in springtime, the colour group identified as most effective for Halley VI was Group One – Type One. These colours are warm, light, clean, fresh, and bright; when this palette appears in the landscape, human beings instinctively recognise that their world is waking up after the long, dark winter, with the thought that "it is time to get busy – there is work to be done." Of course, spring is more or less apparent in different climates, but the underlying patterns appear to be universal.

The Group One colour palette created a lively and optimistic ambience that underpinned everything, drawing specific hues from Group One to emphasise the activities. Colour associations meant that colours were associated with specific programmatic requirements. Blue, for example, was used to enhance concentration, while red was adopted for spaces dedicated to physical activity.

"We hold approximately 100 gallons, 20 colours in all. […] It's suggested that in future only very pale colours and pastel shades should be ordered. […] Hopefully by using colours suggested as a guideline they will make Halley [III] a more pleasing place of work."

J. Scotcher (Builder, British Antarctic Survey), *British Antarctic Survey – Halley III Building Report*, 1979

"One has to think very carefully about sensory deprivation. […] At Halley there are very few natural features and in fact no smell because there are no trees, no plants, just ice. When we were designing the staircase [shown below] we introduced a very simple device of using Lebanese cedar panels because they have a gentle smell which would remind people of the woodland."

Hugh Broughton (Architect, Hugh Broughton Architects), in conversation with UNLESS, Architectural Association School of Architecture, 2019

The suffix -less is one that best encapsulates the true essence of the continent. The scale-less, apparently end-less territory not only denies humans a sense of scale and time but also deprives them of their senses. Colour-less and odour-less, the dry environment of the white desert challenges humans on all levels. While architects are starting to introduce colours and scented materials within the station design to overcome such abnegation, horticulture has been used since the early age of exploration to enrich taste (by adding fresh tomatoes and vegetables to an otherwise monotonous frozen or tinned meal) and bring a touch of colour and aroma to the station. The image on the right, taken by Emil Schulthess at Wilkes Station during the IGY, shows Dr Tressler's "Botanical Garden"[1] inserted within a Plexiglas dome originally conceived for aurora australis observations; the dome was built sideways to afford the garden constant sun exposure during the summer months.

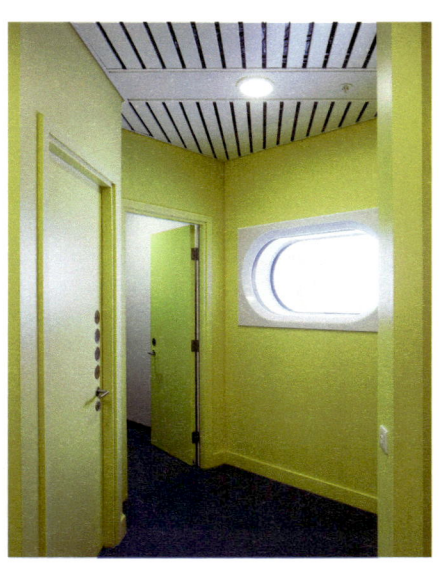

11
Importing and Transporting the Possibility of Life

From Polar Gateways to Antarctic Cities

Juan Francisco Salazar

JUAN FRANCISCO SALAZAR is a professor of communications and media studies at Western Sydney University (Australia) and is currently an Australian Research Council Future Fellow. He is an environmental anthropologist and documentary film-maker with academic and artistic interests in social-ecological change, environmental futures and indigenous politics. Supported by the Chilean Antarctic Institute, Salazar has undertaken ethnographic and film work in Antarctica, where he directed *Nightfall on Gaia* (2015).

Except as ports of entry and egress, cities are rarely considered in relation to Antarctica. Yet five cities positioned on the Southern Ocean Rim play a vital role in humanity's engagement with this fragile region. In recent years, the notion of polar gateways has gained expediency across national and city governments, polar-related businesses, and tourism operators to characterise both the geographical settings and historical conditions that a number of cities and ports along the Southern Ocean Rim share with the Antarctic region. Cape Town, Christchurch, Hobart, Punta Arenas and Ushuaia are regarded as the main Antarctic gateway cities in the polar community with a recognised Antarctic urban cultural heritage. These cities today have significant transport infrastructure and scientific logistics to and from Antarctica, as well as an increasing public engagement with the South Polar Region.

Located in zones with intense interconnectivity to the Antarctic, most travel to the Antarctic region is funnelled through these formally recognised international gateways. All significant engagement with the Southern Polar Region is coordinated through them, and the ensuing competition for the economic advantages that this traffic offers is not always positive. Each city is formed by long and complex histories of engagement with the Antarctic going back to the nineteenth century. There are, in fact, multiple ways through which these gateways are assembled, put to work, and acted upon by actors, materials, processes, and infrastructures. These gateways are linked to local political discourse, national security planning, frontier imaginaries, and ideational and commercial rivalries.[1] All five cities have historical, geopolitical, and economic connections with different regions of Antarctica, and each of them illustrates how memories of past events, narratives of polar exploration, literary fictions, and traces of material culture are mobilised both in the production of geopolitical and cultural imaginaries and in the promotion of ecopolar heritage tourism.[2]

There is an opportunity for these cities to see themselves as more than exit or entry points for tourists and base personnel travelling to Antarctica or five far-flung ports competing for the same northern hemisphere commerce, and rather as members of an interlinked Southern Rim network whose members can learn from and benefit each other. These Antarctic cities ought not to be just thoroughfares, but vibrant urban centres that embody the cosmopolitan values associated with Antarctic custodianship: international cooperation, scientific innovation, and ecological protection. In turn, there is the possibility that these cities could act collectively as global custodians of Antarctica, which has the potential to become either one of the most fiercely contested or one of the most positively collaborative zones in the world.

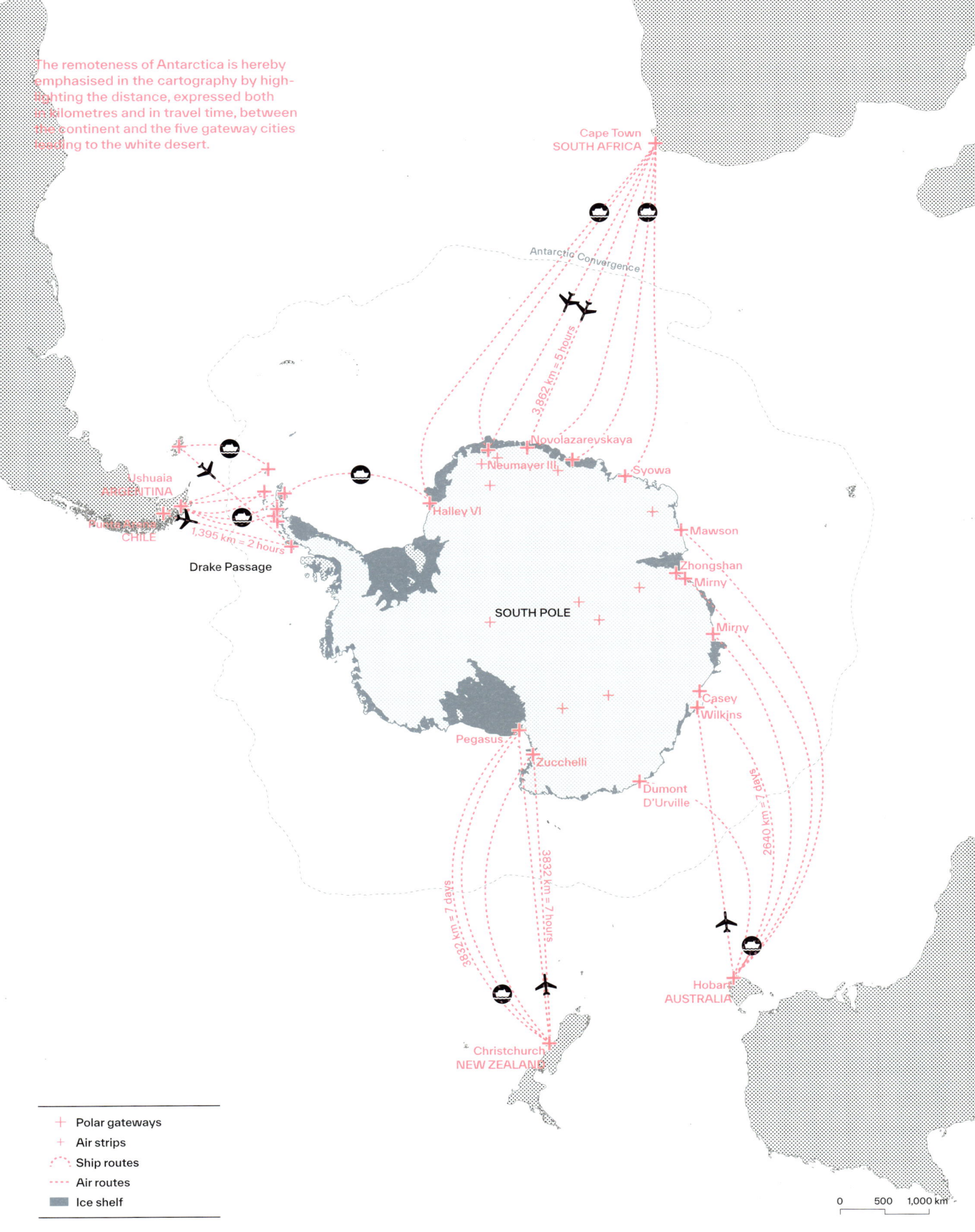

The remoteness of Antarctica is hereby emphasised in the cartography by high-lighting the distance, expressed both in kilometres and in travel time, between the continent and the five gateway cities leading to the white desert.

Cape Town
SOUTH AFRICA

Antarctic Convergence

3,862 km = 5 hours

Novolazarevskaya
Neumayer III
Syowa
Halley VI
Mawson
Ushuaia
ARGENTINA
Zhongshan
Mirny
Punta Arenas
CHILE
Mirny
1,395 km = 2 hours
Drake Passage
Casey
Wilkins
SOUTH POLE
Pegasus
Zucchelli
Dumont
D'Urville
2640 km = 7 days
3832 km = 7 days
3832 km = 7 hours
Hobart
AUSTRALIA
Christchurch
NEW ZEALAND

+ Polar gateways
+ Air strips
‑‑‑ Ship routes
‑‑‑ Air routes
Ice shelf

0   500   1,000 km

carefully, or million-dollar cranes could disappear beneath the ice within seconds. On this account heavy construction machinery is usually delivered in modular components that comply with the necessary weight restrictions and are assembled at the destination.

There are two main approaches to transporting supplies, materials, and equipment from a nation's homeland to Antarctica. The first is the use of regular freight vessels between the homeland and ports at Cape Town, Christchurch, Punta Arenas – i.e. the Antarctic gateways – with the cargo subsequently being transferred to ice-class vessels. The second is the use of ice-class vessels directly from the homeland to Antarctica. This latter option reduces the time-consuming task of transferring cargo between ships. Since the loading capacity of suitable ice-going vessels is limited, at least two journeys are required to construct a near-shore station; the number of ship journeys increases the further away from the coast the station is located, as more vehicles, sledges, worker accommodation and fuel will be required during the traverse. Generally, a minimum of two to three seasons are thus required to construct a new station.

Once unloaded, the cargo is usually placed on sledges, which are then pulled by tractor or skidoo across the sea ice to a safe location or staging point. When all materials and equipment have been offloaded, preparations are made to transport the cargo to the destination, which could range from a few hundred metres for coastal stations, to thousands of kilometres for inland stations. Mobile track cranes are required onshore to handle the cargo and assemble a new station.

If a new station is being constructed inland it is advantageous to pre-assemble parts of the station in the relative "comfort" of a coastal hub or existing station. This could include assembling steelwork, fixing containers together, mounting part of the façade on the structure, etc. Often such assemblies are pulled to inland stations by special track vehicles and custom-designed sledges.

A round trip from a coastal hub to an inland construction site, known as a traverse, can take several weeks. Loading and unloading cargo and setting up the construction camp is also time-consuming, limiting the construction window. In most cases it is unlikely that a station can be constructed in one austral winter and this creates the added complication of having to winterise a partly constructed station and ensure that all temperature-sensitive materials and components are protected from the elements during winter, ready for the next construction season.

Vessels such as the MV *Magga Dan,* constructed in Denmark by J. Lauritzen, and the helicopter *Sikorsky HRS-3*[2] (shown in the top image and the photograph opposite) – used, respectively, to resupply Australian Antarctic stations and to deliver goods in flight at Marble Point to the American field operation team who were inspecting the possibility of building an aeroplane runway during the International Geophysical Year – are representative of transport vehicles used in Antarctica to support building construction. When stations are located on the coastline, panels (such as those shown above, centre, from the Spanish Juan Carlos I Station designed by Hugh Broughton Architects) are discharged on barges and brought to shore; when the building site is situated above the Ice Barrier (as shown above, bottom), building material is hoisted with cranes directly onto sledges. The surreal scale of the buildings and logistic apparatus in Antarctica and their effects on the environment is epitomised by the photograph (in the next page) taken by Emil Schulthess in 1957 which depicts a 7-ton D-4 tractor being lowered to the ground at the South Pole by a gigantic parachute with a diameter of 30 metres and 80-metre-long cabling. The tractor – of which there were two at that time in the American station – was used to trace a single line on the polar plateau, the landing strip for US Air Force Hercules C-130.

Traverses and Intermodal Transportation

Paul Thur

PAUL THUR served as Traverse Operations Manager for the United States Antarctic Program for six years. Thur directed the development and refinement of the United States Antarctic Program traverse technologies in conjunction with the National Science Foundation (NSF), the Cold Regions Research and Engineering Laboratories, the United States Antarctic Program, Antarctic-support contract personnel, and other national programmes.

Although the golden age of Antarctic exploration ended long ago, and most overland hauling today is accomplished with motorised equipment instead of manpower, the logistics of moving supplies from place to place in Antarctica still present monumental challenges. Traversing can be broken down into three phases: planning, pre-departure preparation, and field operations – with planning taking the longest and being the most critical in determining success or failure.

Planning encompasses a range of tasks that include route selection, minimising environmental impacts, addressing safety concerns, depot planning, sled design, living facility design, crew selection, support and delivery of cargo quantities, contingency planning, and planning of preparatory pre-departure field operations. Heavy-duty traverse planners check routes using a combination of map-reading to avoid obvious obstacles, analysis of satellite imagery, which can show crevasse-prone areas, and on-the-ground assessment using ground-penetrating radar (GPR) to identify crevasses. Multi-year traverse routes are usually marked by bamboo poles with flags secured to the top, generally located 400 metres apart. This signage makes traveling the same surface year-after-year more reliable than using GPS coordinates, since snow and ice are constantly shifting and changing. The movements of waypoints along traverse routes vary on average from 2 metres a day on the Ross Ice Shelf to 6 millimetres a day on the Polar Plateau.

Apart from the shifting terrain, crevasses represent the greatest danger to traverses. Crevassed areas can be identified by assessing terrain characteristics and satellite imagery and confirmed on the ground by using GPR. Due to the high risk involved, crevasse crossing encompasses an extremely complex set of strategies. In broad terms, crossings are completed by establishing an extremely conservative safety ratio (bridge thickness to span length) through snow-strength analysis, identifying potential crevasses using GPR, profiling the crevasse by means of hot water drilling, and assessing the crevasse/bridge ratio. Operating with a mindset that prefers to open and fill a crevasse rather than take chances, crevasses are exposed with explosives set in the bridge and filled by pushing snow into the void.

The length of a traverse season varies greatly depending on the preparation time required at origination, route condition (comparing the new route versus the older established route), distance to be travelled, technological stages of the traverse (i.e. the level of experience amongst the traversers and planners), and traverse mission (delivery- or time-focused). The American traverse season lasts about 100 days maximum during the austral summer (late October to early February); for traverses led by experienced teams this could entail, for example, sixty days of pre-departure preparation and eighty-eight days on the trail.

A key function of traverses is to deliver cargo to the Antarctic Stations. To maximise cargo delivery, three strategies are deemed the most efficient: establishing supply depots, reducing the tare weight of hauling sleds, and, finally, reducing sled friction. Supply depots have been used since the beginning of polar exploration, including the very first expeditions to Antarctica led by Roald Amundsen and Robert Falcon Scott. Round-trip traverse plans should include staging fuel and supplies for the return, unless there is a valid reason or carrying all the supplies for the entire journey such as one-way journeys or different inbound and outbound routes.

Many nations have spent large amounts of research and development funds to ensure that new traverse technologies are durable and safe, and that their environmental impacts are negligible or controllable. The world of traversing was revolutionised by two high-impact innovations that were developed

Crevasses are deep fractures that form in glaciers in areas of high stress within the ice flow. Often extending for several kilometres, they can be tens of metres deep and wide. Mostly concealed by a thin veil of snow, crevasses have claimed many lives and are feared by all polar explorers, especially during whiteouts. Their detection – with ground-penetrating radar equipment attached to long booms on the front of tractors, as captured by Emil Schulthess in the image above, centre – is essential for safe traverses both by day and at night.

and refined concurrently: high-molecular-weight (HMW) polyethylene plastic sleds and rubberised fuel bladders. The former, developed to replace rigid cargo hauling platforms that conform to the surface as they are towed, are 20 millimetres thick and have tow plates installed at one end, while the latter, conceived to replace traditional fuel tanks, are almost a perfect fit for HMW sleds because they are flexible, durable, and easily secured. The payload efficiency, or the payload-weight-per-unit-towing-force values of black bladders (22.3), tan bladders (14.6), and steel skis under a reefer unit (2.6) shows the benefits of HMW and fuel bladders over steel skis.[1]

The great explorers carried only life-critical supplies, whereas today's national traverses focus on maximum cargo deliveries, which means minimising support cargo without endangering crews: traverses can thus carry an extremely large amount of weight. To illustrate this, a recent American-led traverse between McMurdo [↗ p. 866] and South Pole weighed more than 1.04 million kilograms on departure: 229,000 kilograms of fuel were burned to deliver 365,000 kilograms of fuel and 58,000 kilograms of cargo to the Amundsen-Scott Station [↗ p.790] and to return 80 kilograms of cargo to McMurdo, which equates to 2.2 kilograms delivered for every kilogram of fuel burned. The support cargo, including the fuel burned, weighed 630,000 kilograms. Thus, the ratio of support cargo to deliverable cargo was about 1.25:1.

Traverse tractors travel about 11 kilometres per hour and burn between 7 and 13 litres per kilometre depending on the weight of the cargo hauled, the resistance of the sled(s) towed, the terrain characteristics, the snow characteristics, and the operational speed/revolutions per minute. As traverses become more technologically advanced and operationally efficient, the ratio of support cargo to deliverable cargo is projected to decrease, with the ratio eventually approaching 1:1.[2]

Great efforts are taken in planning to reduce environmental impacts along the traverse route. Fuel spills are minimised by using a two-person fuelling team, a small containment sled for the mobile fuel pump, frequent inspections of rubberised bladders, bottles for urination (colloquially called "pee bottles" by the crew) and U-barrels (to bulk-capture waste designated for return to its point of origin), incinerator toilets (ash returned to point of origin), trash compactor (all trash returned to point of origin), and the reporting and documenting of any accidental releases to the environment.

Antarctic convoys generally feature a series of traverse vans mounted on sleds which are commonly used for a variety of purposes, ranging from living units to science laboratories, as well as generator vans. Since 2002, the National Science Foundation (NSF) has collaborated with ERDC's Cold Regions Research and Engineering Laboratory (CRREL) to develop "lightweight, high-efficiency fuel sleds that greatly increase the amount of fuel delivered compared to traditional sleds and enable large cost savings and emissions reductions compared with air transport."[1] As of 2008, bladder sleds have been implemented to transport fuel to all of NSF's polar research stations at the remote Amundsen-Scott South Pole Station, where three round trips are conducted each season to deliver over half of the station's total resupply needs. According to ERDC, "an economic analysis of three seasons' of deliveries to South Pole (2008–11) showed that traverse operations offset an average of 30 annual aircraft flights for cost savings of $2.0M/year. [...] Besides saving money, South Pole Traverse consumes one-fourth the fuel and emits less than 1% of the air pollutants compared with aircraft delivery of the same payload",[2] leading to all of the traverse-delivered payload to South Pole being fuel-transported in bladder sleds.

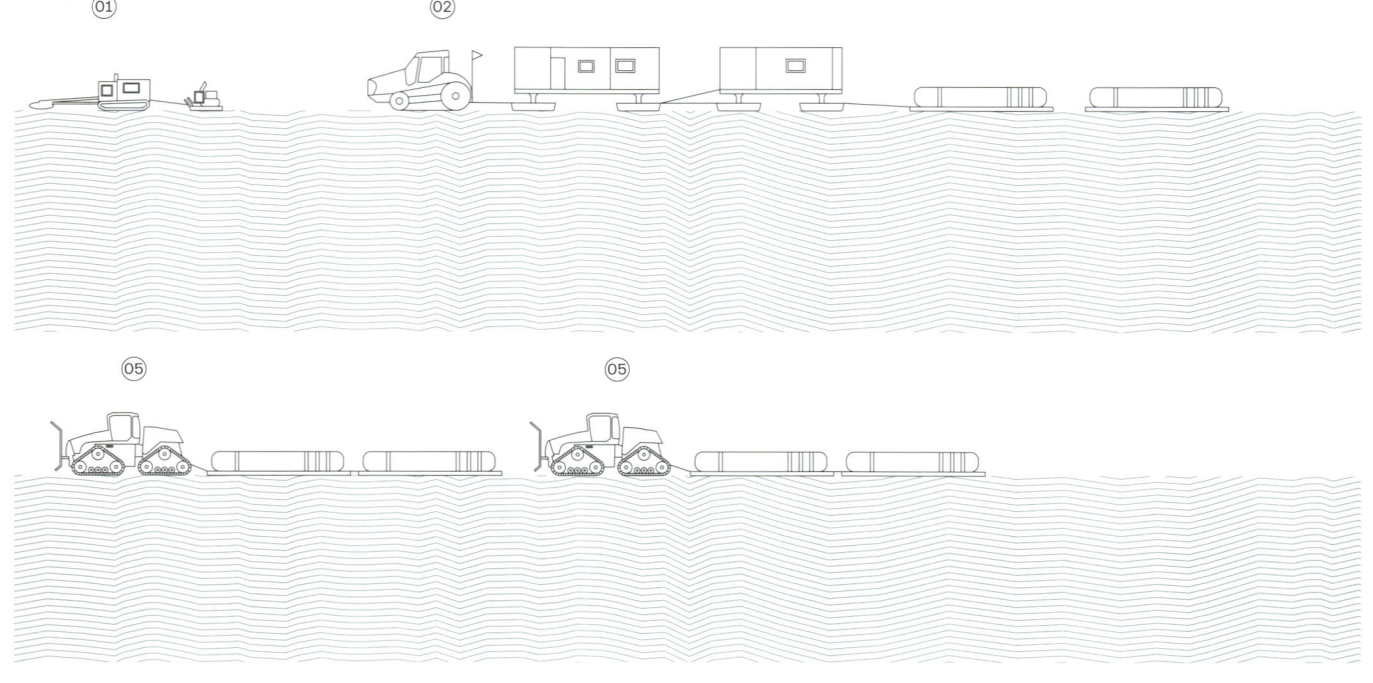

Two final critical aspects of a successful traverse are the crew selection and the organisation of the traverse accommodation and food supply. A good team leader is essential in a traverse environment because personnel issues and interpersonal problems are bound to arise and must be dealt with quickly. Sled design limits shared and personal spaces, but it is essential that crew members have individual private spaces. On average, the common areas for traverse units with ten members has a surface of approximately 15.8 square metres, while personal bunk spaces measure roughly 6 square metres. For Antarctic standards, these figures are not bad.

While traverses are used predominantly for supplying Antarctic Stations – i.e. to supply food, fuel, and medical and scientific equipment – they have a fundamental role in the logistics of building on the Antarctic Plateau. Building materials, mostly pre-assembled to facilitate construction on site, are transported entirely by traverses and the sleds represent an important constraint in determining the size and weight of prefabricated building units. Traverses are thus not only pivotal for the survival of the transient population of Antarctica but also inform the design, and enable the construction, of the Antarctic architectures that host them.

"Today's crevasses have been far more dangerous than any other we have crossed, as the soft snow hides all trace of them until we fall though. Constantly today one or another of the party has had to be pulled out of a chasm by means of his harness, which had alone saved him from death in the icy vault below."

Sir Ernest Henry Shackleton (Explorer), *The Heart of the Antarctic: Being the Story of the British Antarctic Expedition 1907–1909*

01   Radar vehicle
02   Tractor with living and generator modules and fuel
03   Tractor with tool shed and fuel
04   Tractor with freezer unit and fuel
05   Tractor hauling fuel

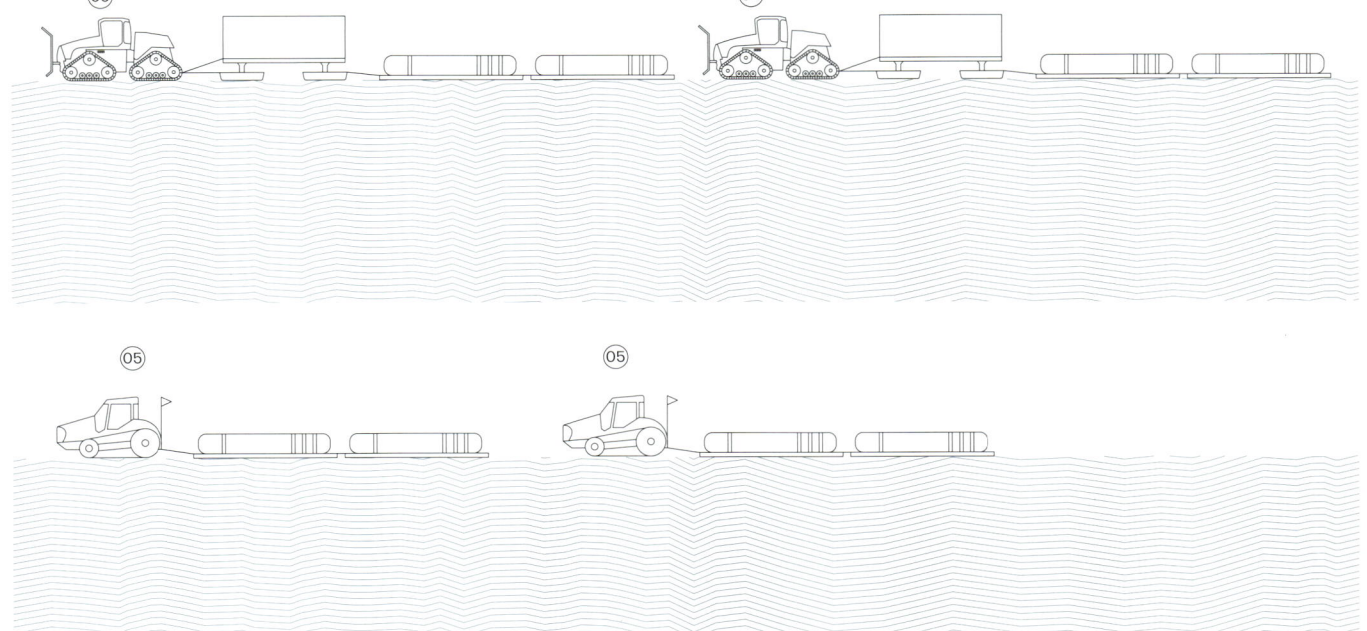

ANNE NOBLE is an artist, a professor of fine arts at Massey University (New Zealand), and an NZ Arts laureate. She has visited Antarctica as an NZ Arts fellow and a US National Science Foundation (NSF) fellow. Her work examines the imagination and representation of Antarctica. She has exhibited and published widely on the photographic construction of the Antarctic imaginary. Her books include *Ice Blink: An Antarctic Imaginary* (Clouds, 2011), *These Rough Notes* (VUW, 2012), and *The Last Road* (Clouds, 2013).

WHITE NOISE
Anne Noble

In Antarctica, the accumulation of drift eventually smothers everything that might momentarily poke itself above the vast white plains of the polar plateau. Over aeons and epochs, the ice piles up and, kilometres deep, it creeps its way off the polar plateau towards the coast, carrying anything buried within it. Across Antarctica there is an array of frozen human residue inching invisibly towards the coast. There is a D-78 tractor that was accidentally driven into a crevasse. Two almost-buried pink flags mark a large ice block of human waste that is now also slowly moving towards the sea. Even the bodies of Scott, Wilson, Evans, Bowers, and Oates still lying in their deep ice hummocks, will make their way back to the Ross Sea to join the bulldozer that was accidently dropped from a helicopter and crashed through the sea ice to rest on the bottom. This world of accumulating drift is continually shifted, carved, and reshaped for human habitation by machines. The biggest sites where ice is graded and crevasses blasted are human landscapes in creation, known by names such as "Mongo", "Hummer", "Strange Brew", and "Personal Space" – a litany of new road markers for twenty-first-century traverses across the Southern Continent.[1]

White Noise #5, 2008

White Noise #2, 2008

A Day in the Life of an Antarctic Traverser

Paul Thur

PAUL THUR served as Traverse Operations Manager for the United States Antarctic Program for six years. Thur directed the development and refinement of the United States Antarctic Program traverse technologies in conjunction with the National Science Foundation (NSF), the Cold Regions Research and Engineering Laboratories, the United States Antarctic Program, Antarctic-support contract personnel, and other national programmes.

The alarm clock's hushed beeps force me, scrambling, to silence them, lest I wake the other three souls sleeping in this 7.5-square-metre bunk room. I tried to find the quietest alarm clock, but the demand for such is apparently minimal. I rarely need a wake-up, but this alarm clock is an insurance policy against getting a delayed start. After all, would anyone really wake up the boss if he overslept?

My slippered footsteps fall loudly in the living module (LM) despite my intended near silence to respect others' sleep. I am usually the second or third person up, so I don't need to check, but I peer at the coffee maker for that tell-tale red glow of the power switch that means a glorious brew is on its way. Even "cheap, bulk" coffee is heaven in the Antarctic when you need it.

As per usual, I throw my fleece over my thermal sleeping attire and head out into the brisk morning air (darn, I forgot my sunglasses again). The face-slapping walk in between modules is only about thirty steps, but it goes a long way toward kick-starting the process of fully waking up. If it's not extra frigid, in the short distance I can visually check the weather, empty the pee bottle into the U-barrel, check the snow-melter level and continue into the generator/bathroom module (GM) to complete my morning routine. A quick check of the generator room completes my venture into the GM. The generators are checked by every crew member not because we all think we know what we're doing, but because it is the warmest room by far.

A quick check of email and recording of temperature and wind direction/speed and I am enjoying my oatmeal and coffee. Between slurps of coffee, I mention anything of note, review the daily plan, and mark the distance to the next of our fifty-four waypoints.

Carhartts, parkas, boots, hats, and well-seasoned work gloves are donned as everyone exits the LM to check fluids and unplug and start the equipment. As the equipment sputters to life and is left to warm up for fifteen minutes, we go silently and deliberately through our pre-departure tasks: electric draws shut off, generator stopped, all cabinets latched, bunk items secured, countertop items moved to the sink or floor, vehicle cords secured, snow melter closed, doors closed, and stairs raised and secured. Once all eight specialised tractors are backed up to the sled hitches and hitch-pins dropped in with a clank, we move slowly to position ourselves in our 0.8-kilometre-long travelling train. Once everyone has their music cued and ready, we head out around 8:00 am at a less-than-amazing 11 kilometres per hour.

I am lucky enough to get to stare at the ground-penetrating radar computer screen in the blacked-out rear box of the lead vehicle. I use my tunes and occasional conversation with the driver to keep my attention on the task at hand, crevasse detection. Although there are only two shear zones (areas of crevassing that occur between regions of slow- and fast-moving ice) along our trail, we always GPRed the entire route as a

precaution. We frequently break to stretch, for personal reasons, for inspections, or to re-secure cargo.

The lunch stop arrives about 11:45 am. With a flurry of activity and a well-choreographed dance to and from the microwave with mere seconds between cooking, we all fill our bellies. The pre-departure tasks are fewer, but it is just as impressive to watch the purposefulness with which we each go about our tasks. Breaks and inspections interrupt afternoon travel, but as 5:00 pm approaches, I determine the best location for our camp. Avoiding previous campsites and natural hazards, we stop the lead vehicle so I can direct the LM-GM sled into a flat and level spot for the night.

The flurry of activity at day's end is greater than the morning or lunch routine. Sleds are unhooked with strategic placement for morning departure, cargo and the 650,000 litres of fuel in bladders are inspected yet again, a generator roars to life, building heaters begin their task, fuelling operations coincide with cleaning snow and ice build-up from the tractors, and equipment is shut down and plugged in.

I take GPS coordinates and weather data as I head into the LM to prepare and send my daily check-in report to management. Spreadsheets beg to be filled in with data from the day (fuel usage, daily mileage, etc.). There are plenty of tasks remaining once the machines are fuelled and shut down. The only task that is scheduled is cooking dinner – the other jobs are done with little comment: flag the snow-gathering area, fill the snow melter, take out the trash. Even the non-social among us enjoy our nightly dinner, conversation, and common room clean-up. It is the chance to interact with others after spending all day in a mobile glass bowl by yourself.

Socialisation doesn't last very long before people begin vanishing to retire to their 6-square-metre personal bunk space separated from the world by a thick curtain with useless Velcro closures with peeling rubber

A Tucker Sno-Cat especially modified for the journey to the South Pole (by adding "multiple fuel pumps from a Cessna, multiple fuel valves, an escape hatch, insulated battery boxes, insulated cab interiors"[1] and by waterproofing the motor of the electrical system and the radar antenna) risked falling into the frozen abyss during the 1955–1959 Commonwealth Trans-Antarctic Expedition.

backing. I settle back to write in my journal. I review some of my previous journal entries to attempt to relive some of the most exciting and unique events of our trip or noteworthy revelations: how amazing it is that the most beautiful environment on the planet can become mundane when viewed at a snail's pace day after day, ice thunders that come out of nowhere when a loud crack comes from every direction resulting in the snow surface dropping some small distance, the sudden appearance of a snow petrel or skua, a plane dropping by to deliver a needed part and some mail, a bend on the trail instead of the monotonous arrow-straight drive, arrival or departure from Antarctic stations, and the awe-inspiring storms that force you into survival mode in the womb-like comfort of the modules.

I use my personal time at the end of the day to review our amazing journey and our team effort to overcome differing challenges. Specifically, the challenge of crevasses was always the most daunting. I revisit the thirteen days spent in the shear zone that we had to cross just 37 kilometres from the start of our route. We spent hours looking at the GPR screen to narrow down the location of those features of interest that our expert consultants had identified from GPR scans of the area a month earlier. Back and forth with the lightest traverse vehicle with its antenna mounted on a 6-metre boom, marking the likely extent of the crevasse on the surface with flags on bamboo poles. Once the crevasse was marked, we would tow the hot-water drill out and start drilling holes in the snow in a grid pattern. Soaked, mittened hands would gently lift and drop the copper drill head and hose down into the snow. All the while keeping an eye on the depth markings identified on the hose with peeling duct tape symbols. Once the hose and head began to free-fall, you knew you had just broken through the bottom of the bridge.

The best guess of bridge thickness was relayed to the scribe, who would dutifully mark it. This process would repeat itself anywhere from twenty to thirty times to attempt to determine the extent of the crevasse below the surface of our trail. If the crevasse was much wider than the bridge was thick, or if we couldn't determine what was going on below the surface, we would decide to open the crevasse. We installed a larger drill head on the hot-water drill that could accommodate sticks of explosives. In general, the entire thickness of the bridge could be loaded in holes on about a 2.4 metre pattern. Once the bridge was dropped, we would use the GPR to find a snow-gathering area among the maze of crevasses in this 6-kilometre section of our 1,656-kilometre trail that always was the most difficult. We would take hours to push snow into the void and then move onto the next feature of interest until we were on the Ross Ice Shelf, where we could begin our journey in earnest.

As the words blur and my head nods, I put down my journal, check my alarm clock, and ensure my iPod is charging for the following day. I drift off to sleep, hoping not to be woken by the rocking of the module due to wind from a sudden storm, the banging of a carelessly hung power cord against the outside of the module as it sways in the wind, or the sudden silence if the generator stops its constant, rhythmic hum.

The alarm clock's hushed beeps force me scrambling to silence them yet again on another of our blurred-together eighty-seven days in the field.

CHIARA MONTANARI has seventeen years of work experience in the polar environment, which saw her leading expeditions to the Italo-French Concordia Station and the Belgian Princess Elisabeth Station. An engineer and explorer by training, Montanari founded Complexity Aware, a consultancy for enterprises interested in developing an "Antarctic mindset", and authored *Chronicles from the Ice: 90 Days in Antarctica* (Mondadori, 2015).

ANTARCTIC MINDSET
Chiara Montanari

Since the so-called "Heroic Age of the Antarctic Exploration"[1] the lessons learned from Antarctica can offer valid testimony to how human beings have learned to adapt in conditions of uncertainty. Nowadays – despite the support of contemporary technologies – organising, running, and leading scientific missions in such extreme environments remains a challenge and this can provide enlightening insights for the current world we are living in, the so called VUCA[2] times. VUCA, an acronym denoting Volatility, Uncertainty, Complexity, and Ambiguity, describes contemporary situations at every level. Used to describe social dynamics like finance, technological innovation, cybersecurity and innovative design strategies, VUCA is also applied to global climate change challenges, particularly in the case of Antarctica. In extreme environments such as Antarctica, there is no room for mistakes and there is an urgency to manage quick changes, take wise decisions, and avoid major risks. For people operating solely in the field, resources to support errors become very limited. Leading a mission to Antarctica entails managing complex logistics as well as leading diverse teams (in Concordia Antarctic Station ↗p.780, for example, there are sixty to eighty people with very different backgrounds and diverse competences and nationalities). Teams have to deliver multiple projects with the result that several scenarios and risk analyses need to be prepared in advance, taking into account that any assumed situation may change at any minute. People working in polar environments must quickly learn how to perform at a high level while operating under the pressure of high risks, incomplete information, limited resources, and multiple unexpected events. This requires a variety of skills, both in terms of technical competencies and so-called "soft skills". It calls for a unique "Antarctic mindset".

FREIGHT TO VOSTOK STATION
Yuri A. Shibaev

Freight coming into Vostok Station ↗ p. 862 is delivered from Progress Station on six Pisten-Bully vehicles, each driven by two driver-mechanics who work in shifts around the clock for approximately eleven to fourteen days per trip. Along the 1,457 kilometres separating the two stations, two refuelling stations are located after 550 and 1,100 kilometres.

On a yearly base, there are two trips scheduled. The first one leaves Progress in early December to deliver diesel. Upon arrival, vehicles and staff help to clear the station area from snow. Once the necessary maintenance at Vostok is terminated, the vehicles return to Progress, in time for the arrival of the scientific vessel. The second trip resupplies the station with fuel, frozen food, groceries, and household goods (including clothing, underwear, spare parts, diesel engines, etc.). In total, about 200 cubic meters of diesel fuel, fifteen barrels of motor oil, fifty-five barrels of gasoline, thirty barrels of kerosene and 10 tons of products are imported to Vostok every year.

Upon its return to the vessel, the traverser carries back ice cores necessary for research purposes in the Arctic and Antarctic Research Institute (AARI) laboratory in Moscow and 5 to 10 tons of garbage. Some of the ice core's samples are subsequently returned to Vostok in the following season.

YURI A. SHIBAEV is the head of construction of the new wintering complex at Vostok Station on behalf of the Russian Arctic and Antarctic Research Institute (Russia). He acted as scientific director of the first Progress-East trip in the 53rd Russian Antarctic Expedition. Shibaev has participated in the implementation of the federal World Ocean Program, the creation of the Consortium of European Research Libraries, and various projects on palaeoclimate.

THE ANTARCTIC SNOW CRUISER

The infamous and gargantuan Antarctic Snow Cruiser was a vehicle intended for use in the United States Service Expedition. When it proved faulty, the extra-large automobile was first converted to a stationary crew quarters and subsequently abandoned (in 1941) as efforts shifted towards World War II. Developed under the direction of Admiral Richard Byrd and Thomas Poulter, in collaboration with the Research Foundation of the Armour Institute of Technology, the 37-ton, 17-metre-long vehicle took two years to design and construct, costing 150,000 US dollars (equivalent to almost 3 million US dollars today).[1] To justify the excessive costs, claims were made that the vehicle would have been able to explore 500,000 square miles (1,300,000 square kilometres) of unknown Antarctic territory in one Antarctic summer. Envisioned as "an unstoppable fortress for long-distance travel on the continent's endless swaths of snow and ice",[2] the cruiser was designed as a portable environment suitable for a crew of four to five individuals for up to a year. Once in Antarctica, the Snow Cruiser proved inapt: "the furthest it had managed to travel in a single shot was 92 miles [150 kilometres]"[3] as the diesel-electric hybrid powertrain was severely underpowered and the smooth tyres offered very little traction, sinking deep into the snow. Following its burial, in 1958 an international expedition managed to locate the vehicle by sighting a long bamboo pole known to mark its position and was able to uncover it using a bulldozer. Today, the Cruiser rests at the bottom of the Southern Ocean along with other Antarctic stations.[4]

PLAN AND ELEVATION

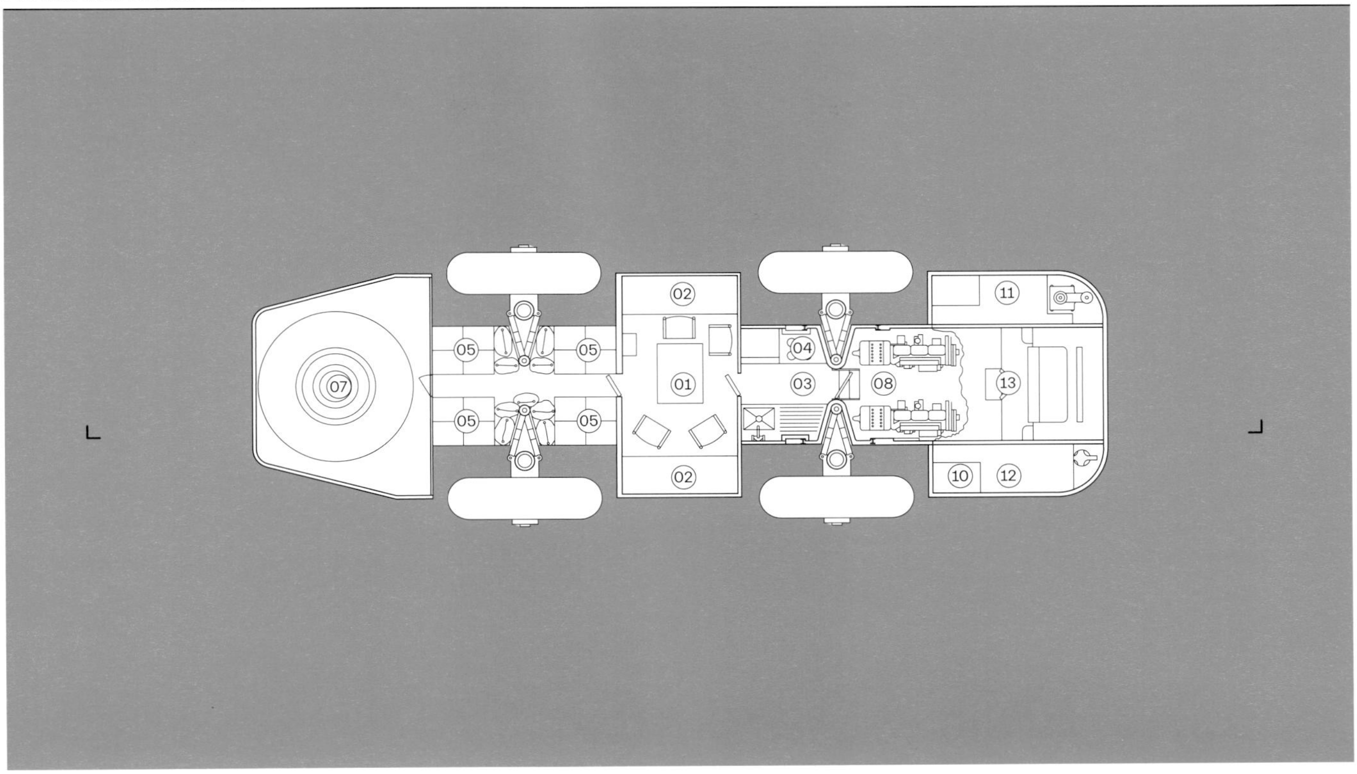

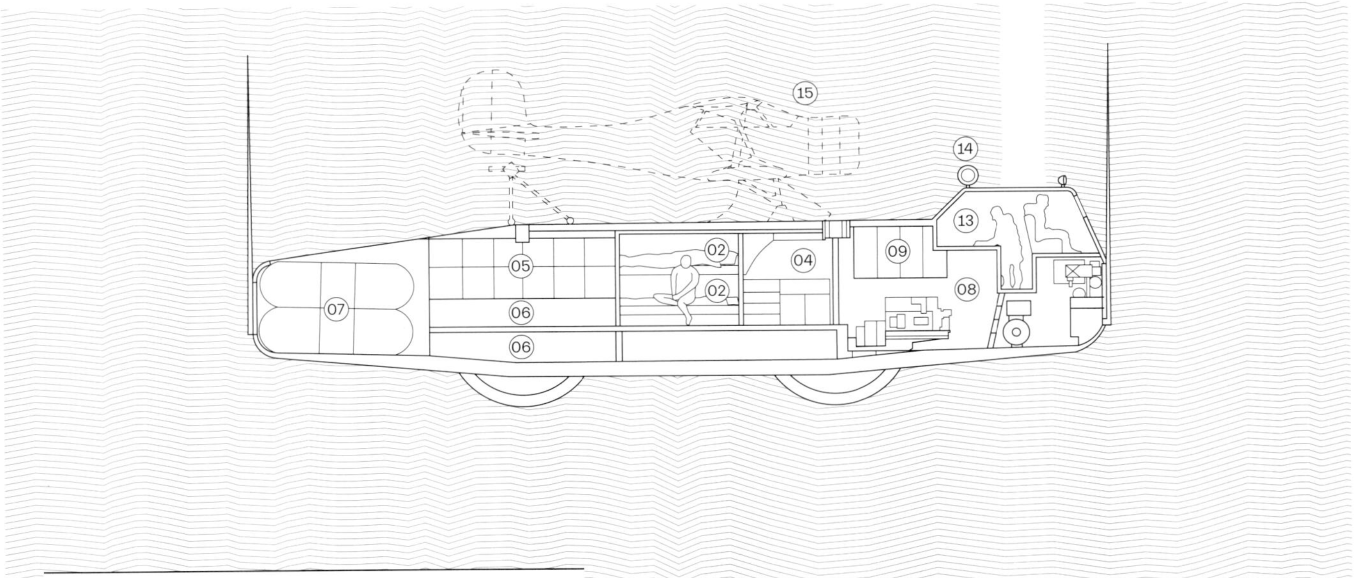

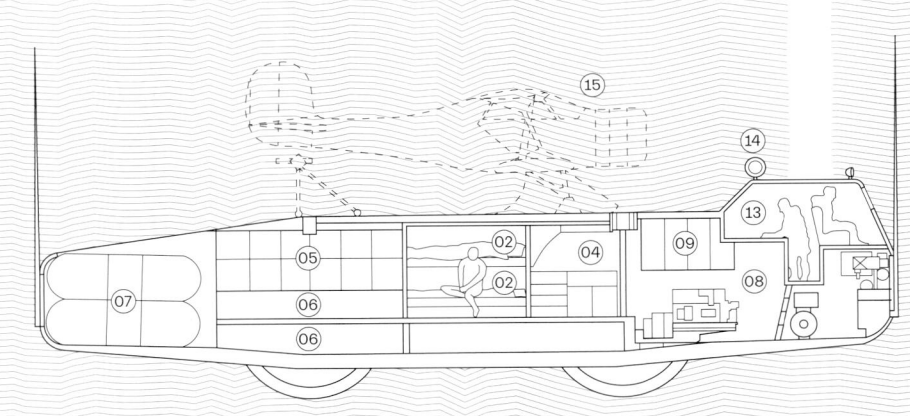

01	Living and sleeping quarters	08	Diesel engine room
02	Double-deck bunk	09	Tools, instruments
03	Darkroom and galley	10	Snow melter
04	Stove	11	Machine shop
05	Food store	12	Welding
06	Fuel tanks	13	Chart and control room
07	Spare tyres	14	Radio compass
		15	Plane moored on roof

SOURCE
Research Foundation of the Armour Institute of Technology in Chicago, 1939

1 : 100   0   1   2   4 m

1898–1900

Dogs were first brought to the Antarctic aboard the *Southern Cross* in 1898. Following the landing at Cape Adare ↗ p. 614, the seven-man shore crew experienced a four-day blizzard and owed their survival to their four-legged companions, who kept them warm within their fragile tents. Throughout the British expedition, the dogs proved of great assistance for travel purposes – a fact which was further epitomised by Roald Amundsen's party's reliance on them for their successful journey to the South Pole in 1911. In the years that followed, dogs played an important role in facilitating the exploration of the continent well into the "Scientific Age". Based on concerns that dogs might transfer diseases and disrupt local Antarctic fauna (and with an awareness of the intrinsic contradiction embedded in the Protocol on Environmental Protection, which imposed strict controls on the introduction of non-native species, while at the same time allowing huskies to be bred on the continent), Annex II to the protocol effectively banned dogs from the Antarctic by April 1994.

1907–1909

Bent on reaching the geographic South Pole during the British expedition of 1907–1909, Ernest Shackleton took with him dogs, ponies, and a purpose-built open two-seater "12/15hp New Arrol-Johnston", which "had a specially designed air-cooled, four-cylinder engine, used non-freezing oil, had a silencer that doubled as a foot-warmer, produced hot water by passing the exhaust pipe through a hopper that could be filled with snow, and could be fitted with a pair of ski runners on the front wheels."[1] Although, in an interview published in *The Car,* Shackleton computed that "under favourable circumstances [...] the machine [would] travel 150 miles in twenty four hours and [that] there would be a fair chance of sprinting to the Pole", the vehicle, parked in the garage at Cape Royds ↗ p. 624, was too heavy and the "wheels in duty turned violently round [...] burying themselves to such an extent that the car moved not an inch."[2]

1910–1913

During the "Race to the Pole" in 1912, Scott's party employed a combination of sled dogs, Manchurian ponies, and even a few motorised tractors, duly parked at Cape Evans ↗ p. 632. Both the motorised tractors and ponies were ill-suited for the job, with one of the sledges getting lost in the ice when the ship was being unloaded, and the ponies lacking appropriate snowshoes. Of the nineteen ponies brought south to aid in laying depots on the Ross Ice Shelf, nine were lost before the journey began. Scott and his team were forced to spend much of their journey man-hauling their heavy supply sledges on foot.

1957–1958

Snowcats were widely used during the International Geophysical Year as multipurpose vehicles; their truck cabs were equipped with benches which served at once as a surface for cooking and sleeping and as a staging area for the team's gear during fieldwork. Until the subsequent introduction of wannigans (movable single-purpose huts for cooking and sleeping), the multi-use device forced expedition members to constantly move, store, and reassemble items throughout the day. In addition to supporting scientific endeavours, the snowcats were also used by Vivian Fuchs – leader of the Commonwealth Trans-Antarctic Expedition (1955–1958) – in the first overland crossing of the Antarctic Continent.

Headed by the British explorer Sir Vivian Fuchs and the New Zealander Sir Edmund Hillary, the Commonwealth Trans-Antarctic Expedition was the first to reach the South Pole overland in forty-six years, since Amundsen and Scott's expeditions back in 1911 and 1912. Fuchs relied on six vehicles, including three Sno-Cats, two Weasel tractors, and a retrofitted Muskeg tractor, whilst Hillary and his team adopted the converted Ferguson TE20. Although Hillary's party was responsible for route-finding and for the laying of depots from Scott Base ↗ p. 678 towards the South Pole for Fuchs to use on the final leg of his journey from the Weddell Sea, once he had completed laying supply depots, he saw the opportunity to continue south, reaching the Pole on the 3rd of January 1958. Edmund Hillary was the first person to drive to the South Pole.

Faced with the challenge of creating an off-road vehicle that would be able to operate at –70°C, cover long distances on the snow and ice, and act as a temporary home, the Soviets developed the *Kharkovchanka* ("woman from Kharkov"). Measuring 8.5 meters in length, 3.5 meters in width, and 4 meters in height, and travelling at an average speed of 5 to 11 kilometres per hour the tractor was able to drag a cargo weight of 70 tons. On the 27th of September 1958 five fully equipped Kharkovchanka vehicles with trailer sleds left Mirny Station ↗ p. 668 heading west towards the South Pole, which they reached eighty-nine days later on the 26th of December 1958. Diverse iterations of the Kharkovchanka off-roaders were used in subsequent years to connect the six Soviet stations on the continent.[4]

1929

At around 1:00 am on the 29th of November 1929 Admiral Richard Byrd, Pilot Bernt Balchen, Harold June, and Captain Ashley McKinley were the first to fly over the South Pole aboard the *Floyd Bennett*. The explorers had left the day before at 3:29 pm from Little America and, as magnetic compasses were useless so near the pole, they had to rely on sun compasses and Byrd's skill as a navigator to reach 90°S. Once over the pole, the party dropped the American flag, weighted down by a stone from Floyd Bennett's grave at Arlington. In 1946–1947 Byrd's impressive aviator expertise led him to map the Antarctic continent extensively by air as part of Operation Highjump aboard the Douglas C-47 Skytrain. The combined aerial and ground survey covered an area of 8,900 kilometres of coastline and 3,900,000 square kilometres of the interior of the continent.[3]

1950s

UC-1 Otters were first introduced in support of Operation Deep Freeze for their ability to lift large amounts of cargo from makeshift airfields. The aircraft were pivotal for the construction of many of the American bases which were constructed on the remote Antarctic plateau, which was difficult to access at the time. Today Twin Otters are used for medium-length flights to supply deep-field parties travelling by snowmobile and sled.

1955

On the 20th of December 1955, the first air link between Antarctica and New Zealand was established when two Lockheed P2V-2 Neptunes and two R5D Skymasters flew from Christchurch to McMurdo Station ↗ p. 666. The following year, a ski-equipped R-4D Dakota was the first aircraft to land at the South Pole.

1964

The "Landing Snowplane" was an amphibious propellor-driven sledge with space for six people. It was sent to Base T (Adelaide) and Base Z (Halley) ↗ p. 658 for trials in 1964–1965 but was not widely used.

2018–2019

During the Austral Summer 2018–2019 the Clean2Antarctica expedition team aimed to reach the South Pole aboard the Solar Voyager, the first zero-emission electric vehicle powered uniquely by solar energy and built with 3D printed components made of recycled plastics.

2019

For over fifty years the LC-130A has been the main aircraft used for intercontinental flights from New Zealand to Antarctica. In recent years, the possibility of preparing more stable ice runways has led to the introduction, on the same route, of conventional passenger jets like the Boeing 757. On board these flights passengers must rely on webbing seats along the walls rather than conventional airplane seats.[5]

ANNE NOBLE is an artist, a professor of fine arts at Massey University (New Zealand), and an NZ Arts laureate. She has visited Antarctica as an NZ Arts fellow and a US National Science Foundation (NSF) fellow. Her work examines the imagination and representation of Antarctica. She has exhibited and published widely on the photographic construction of the Antarctic imaginary. Her books include *Ice Blink: An Antarctic Imaginary* (Clouds, 2011), *These Rough Notes* (VUW, 2012), and *The Last Road* (Clouds, 2013).

BITCH IN SLIPPERS
Anne Noble

In a vehicle yard at the back of McMurdo Station I found one of the old girls of the Ross Sea vehicle fleet, a giant pneumatic drill named *Bitch in Slippers*. The faded blue letters on her yellow body are testament to the Antarctic tradition of ascribing mostly girl's names to the 250 trucks, tractors, dozers, cranes, and excavators that move cargo, carve out the ice highways, maintain the ice runways, and work the roads on Ross Island. *Bitch in Slippers* was notorious for her name. When she first came out as a painted lady, there was an outcry, and the practice of naming vehicles was banned for a time before quietly re-emerging a few years on. *Bitch in Slippers* is an Antarctic inventory – a collection of photographs of a fleet of vehicles – and their names. *Brenda Hot Lips*, *Misty,* and *Hazel* inscribe the landscape, their colour palette and visual texts a contemporary marker of human presence and a reminder of the gendered nature of our relationship to Antarctica.[1]

12
Towards a
Temperate
"Existenzminimum".
From Seaweed
to Polystyrene

The Three Typologies of Heroic Age Huts

Michael Pearson

MICHAEL PEARSON is the president of the International Council on Monuments and Sites (ICOMOS) International Polar Heritage Committee. He is a historical archaeologist and heritage planner who has worked in Australia and internationally. Dr Pearson has taken part in ten research expeditions to Antarctica and has published extensively on polar history and heritage, industrial heritage, and world heritage.

During the Heroic Age of Antarctic exploration between 1895 and 1917, nine prefabricated huts were erected in Antarctica, intended to accommodate expedition members for at least one winter season as well as to provide spaces for use as laboratories, workshops, and darkrooms. Three of these huts were of Scandinavian, three of British, and three of Australian design, with each group of huts drawing on regional architectural influences and conscious innovation for polar conditions. Of these, six survive intact today. As well as their historical associations with the expeditions that occupied them, these huts are of interest because they were designed at a time when knowledge of the Antarctic environment was limited, and the science of efficient polar hut design was in its infancy.

The three groups of huts emerge as being very different in their concept and construction, yet within each group there were successes and failures in terms of providing adequate comfort for their occupants. The Scandinavian huts (Borchgrevink's Cape Adare hut ↗ p. 614 built in 1899, Nordenkjöld's hut erected at Snow Hill Island in 1902, and Amundsen's Framheim ↗ p. 628 hut constructed on the Ross Ice Shelf within the Bay of Whales in 1911) owe their design to Scandinavian building traditions and had many features in common, such as thick plank walls constructed on bearers with insulation between or over plank layers, and gabled roofs with lofts used as storage and insulation space. The British huts (Shackleton's hut at Cape Royds ↗ p. 624 built in 1908, Scott's hut erected in 1911 at Cape Evans ↗ p. 632, and Campbell's hut built in the same year at Cape Adare during Scott's expedition) were timber-framed on bearers or stumps with tongue-and-groove or weatherboard cladding and insulation material inserted between board layers. Their gabled roofs were open to the underside of the roof, and the additional internal volume (up to 50% more air space per unit of floor area than the Scandinavian designs) ameliorated their poor ventilation systems. The Australian hut designs (Scott's Hut Point hut ↗ p. 618 built in 1902, Mawson's Cape Denison hut ↗ p. 636 erected in 1912 in Commonwealth Bay, and Wild's hut erected on the Shackleton Ice Shelf in 1912 as part of Mawson's expedition) were timber-framed on stumps, clad with tongue-and-grooved boards, pyramidal roofs, and tarred paper or felt insulation. The Discovery Hut would appear to have been a stock-design prefabricated Australian homestead modified for Antarctic use and was a complete failure, whereas Mawson's and Wild's huts were the result of conscious design for Antarctic requirements, with fully enclosed verandas on three sides providing storage and insulation space. It may be that the Australians' ideas about building were shaped by the warm climate they were familiar with and led them to underestimate the need for additional insulation, but this was somewhat compensated for by the huts being almost totally covered by snow during winter.

There was no standardised design among the American, British, and Scandinavian Arctic exploration huts to guide Antarctic hut design. The Scandinavian design approach in both the Arctic and Antarctic was based strongly on traditional Scandinavian building techniques and materials, adapted as they were to a cold and snowy climate. The British cold-climate building tradition was largely masonry, but the manufacture of prefabricated timber buildings for use in the colonies was well established. Timber buildings were a common feature of Australian architecture, including prefabrication for use in rural areas (of which the Discovery Hut was an example), and to this tradition Douglas Mawson added design features with extreme Antarctic conditions in mind, based on his experience with Shackleton's expedition in 1907–1909.

Both the British and the Scandinavians were experimenting with insulation materials during this period, and this is reflected in the variety of such mate-

"The ship also carried several prefabricated huts. The main building, designed for the expedition's winter quarters, 15 metres by 8 metres in plan with a gabled roof rising to a central ridge 4.3 metres high, had been prefabricated in London. A trial erection of the hut took place at Officers' Point in Lyttelton; this revealed serious deficiencies in the sizes and quantities of some timbers, which were made good before the expedition sailed."

New Zealand Antarctic Heritage Trust, "History of Scott's Expedition" (online)

"There is easily seen evidence of the prefabrication of the building. Many of the panels have a letter and number, and a banding system was devised whereby stripes were painted (during a trial erection of the hut in Melbourne) from one panel to the next, and across the face of the connecting post. The prefix 'F' on the panels of the west elevation stands for 'front' and are in black; 'L' on the north elevation is for 'left' and is in blue; east elevation coding is 'B' for 'back' and is in white, and 'R' is 'right', the south elevation, and is in red. There are coding marks on the inside too, some clear and others less visible than those on the outside, being obscured by soot from the blubber stove."

Extract from New Zealand Antarctic Heritage Trust, *Conservation Report: Discovery Hut*, 2004

Prior to being erected in Antarctica, all of Heroic Age huts were pre-assembled by the manufacturers. The archival photographs at the side show (from top to bottom): hut at Hut Point, representative of the Australian design typology, preassembled in Melbourne prior to being constructed by Scott's expedition party in 1902 at Hut Point, Ross island; Cape Evans Hut, representative of the British design, preassembled in Lyttleton prior to being remounted in 1910 in the Antarctic, Ross Island; and Framheim, the Scandinavian hut, preassembled at Amundsen's home in Norway and subsequently deployed by Amundsen in the same year on the Bay of Whales, on the Ross Ice Shelf.

rials used in their huts, including papier-mâché, cellulose pulp, granulated cork, quilted seaweed, tarred paper, roofing felt, and linoleum. Some of these materials were well established and others were experimental. Other features also reflected technological developments, with the British and Australian huts using acetylene gas plants for lighting and compressed coal briquettes as fuel.

The factors which may have contributed to the success or failure of a particular hut include the thermal efficiency of the design, the effectiveness of the heating and ventilation systems installed, and the amount of floor and air space available to each man. Liveability appears to have depended on a combination of these factors rather than on any single one. Basic thermal efficiency calculations show the three Scandinavian huts to have been thermally efficient, Framheim outstandingly so. This is because they have multilayered walls and roofs with very thick timber layers and are all quite small huts with small exposed surface areas. At the other end of the scale are Mawson's hut and Scott's Cape

Evans hut both quite inefficient, due partly to their large size and partly to the incorporation of limited insulation. However, judging by the observations of the occupants, these were by no means the most uncomfortable huts to live in. The most obvious factor is heating. Two of the huts, Cape Royds and Discovery, were claimed to be uncomfortable because they were cold. In both cases only one stove was installed although they were designed to have two. At Borchgrevink's hut, which was thermally very efficient, too much heat was generated, and the hut overheated.

A factor counter-balancing heating is ventilation. At Framheim, perhaps the most comfortable Antarctic hut, an effective ventilation system, combined with a good stove and the most thermally efficient hut structure, made the hut warm and dry all winter. Both Borchgrevink's and Nordenskjöld's huts, while thermally efficient and well heated, had inadequate ventilation, which led to high humidity and foul air problems. The Mawson and Cape Evans huts, while at the bottom end of the scale in terms of thermal efficiency, had good heating and a large air space allowing proper ventilation. As a result, these two huts were reported to be quite comfortable to live in. Due to their size, the two smaller versions of these huts, Wild's and Campbell's, were more thermally efficient and even appear to have been quite comfortable.

Neither the Scandinavian, the British, nor the Australian huts can be said to be, as a class, better than the others. There were successes and failures within each group of huts, related to factors such as thermal efficiency, heating, ventilation, and living-space volume. One is left with the feeling that perhaps the discovery of a successful design solution also had a lot to do with serendipity, and that perhaps the Heroic Age was too short a period of time for the best combination of design factors to be recognised and integrated into a standard hut design that would be consistently effective in the Antarctic environment.

"It took supreme logistical skills to organise the acquisition of the building materials from building companies in four states. Mawson acquired prefabricated structural elements for two square-plan, pyramid-roofed accommodation huts, designed to Hodgeman's plans. The two main huts were clad and lined with tongue-and-groove Baltic pine boarding on an Oregon timber frame. Sheets of tar paper lining were used as additional wind protection under the external roof and wall boards. The largest structure (the one that became the main living quarters at Cape Denison) came from a Sydney timber merchant, George Hudson and Sons Ltd, which also sold prefabricated timber dwellings as 'Ready-Cut-Homes'. Hudsons provided a detailed breakdown of component members and an accompanying specification outlining the detailed erection process. A second pyramid-roofed hut (to be the Western Base) was supplied by Messrs Anthony, a Melbourne firm. A third smaller hip roof hut (originally meant for a third Antarctic base but finally used as a workshop at Cape Denison) was supplied by Walter and Morris Limited, Sarnia Timber Yards, Adelaide. Two more huts were purchased in Hobart. A small hut bought from Risby Brothers was used as the Magnetograph Hut at Cape Denison. A fifth hut, probably also purchased in Hobart from Risbys, was used for accommodation for the Macquarie Island party."

Australian Antarctic Division, "Shelters from the Stormy Blast" (online)

• Bed
♦ Stove

PLANS | HEROIC AGE HUTS

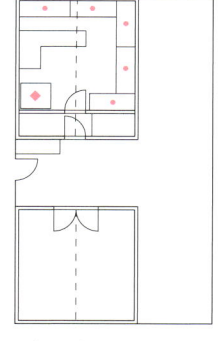

1899, Cape Adare huts
10 people, 54 sqm

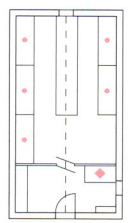

1902, Snow Hill hut
6 people, 25 sqm

1911, Framheim hut
9 people, 35 sqm

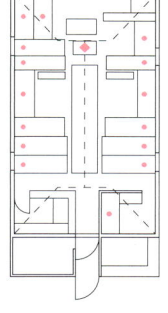

1908, Cape Royds hut
15 people, 64 sqm

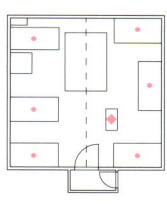

1911, Cape Adare hut (Campbell's)
6 people, 39 sqm

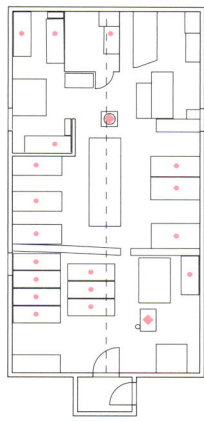

1911, Cape Evans hut
25 people, 105 sqm

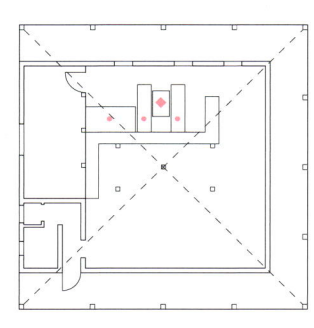

1902, Hut Point hut
16 people, 123 sqm

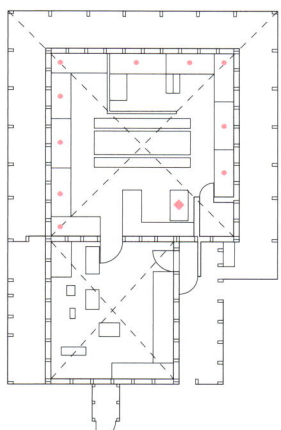

1912, Cape Denison huts
18 people, 143 sqm

1912, Wild's hut
8 people, 58 sqm

SCANDINAVIAN DESIGN

BRITISH DESIGN

AUSTRALIAN DESIGN

SOURCE
Expediton huts in Antarctica: 1899-1917, Michael Pearson

1 : 300   0   3   6   12 m

Borchgrevink's Cape Adare Hut. The First Antarctic Architecture

Michael Morrison

MICHAEL MORRISON is a conservation architect and the British representative at the International Polar Heritage Committee of the International Council on Monuments and Sites (ICOMOS). As the former chairman and senior partner of Purcell Architects, Prof. Morrison has worked, since 2002, on conservation projects devoted to Antarctic historic huts and whaling stations on the Peninsula, Ross Island, and South Georgia.

The word "unique" is frequently misused to describe something that, although special, is not the only one of its kind. However, this term can be correctly applied to the huts at Cape Adare that were constructed by Borchgrevink and his nine colleagues in February and March 1899. These are the first buildings to be erected in Antarctica. Where else in the world can one point to the first building ever erected on a continent that still survives in its original form?

Carsten Borchgrevink was a Norwegian living in Australia when he signed on as a member of the crew of the whaling vessel *Antarctic* in 1885. The ship visited Cape Adare where, according to his own words, he was the first man to set foot ashore. This first visit inspired him to try to make a return expedition with the intention of proving that it was possible to overwinter. Having failed to raise any funds in Australia, Borchgrevink went to Great Britain, where he persuaded the magazine publisher Sir George Newnes to back his endeavour to the tune of £40,000. This was much to the irritation of Sir Clements Markham (President of the Royal Geographical Society), who was busy trying to raise funds for the National Antarctic Expedition that was to be led by Captain Robert Falcon Scott. Borchgrevink's expedition was named "The British Antarctic Expedition 1898–1900", though in fact only two of the ten men who overwintered were British – the others being five Norwegians, two Laplanders, and one Australian.

The site where the *Southern Cross* landed Borchgrevink's party in February 1899 is composed of 45 hectares of shingle, named Ridley Beach by Borchgrevink, lying under the towering headland of Cape Adare ↗ p. 614, which rises abruptly out of the sea to a height of some 350 metres. The beach was then and still is the home to the world's largest colony of Adélie penguins with a reputed 750,000 birds in residence. Here, the expedition members erected two small timber huts

that still exist and are the focus of a conservation programme by the New Zealand Antarctic Heritage Trust.

There are, in fact, three huts at Cape Adare, the two Borchgrevink huts and the collapsed remains of the hut constructed by Scott's Northern Party of the *Terra Nova* Expedition in 1911. The Scott party hut was built with a conventional timber frame and clad in timber boarding internally and externally. Only the storm porch remains standing, with the floor and some remnants of walls lying on the shingle, thickly covered in penguin guano. By contrast the Borchgrevink huts are of solid Norwegian construction. One was used as a store hut and the other as a living hut. They are small, measuring 5 × 5 metres internally with a 1-metre extension to the living hut providing a cold porch and two small chambers, one used for taxidermy and the other as a darkroom. It is a very small space for the ten men who spent a year there. By contrast Scott's hut at Cape Evans ↗ p. 632, inhabited by twenty-five men, measures 15 × 7 metres (i.e. five times as big as the Borchgrevink huts), and Shackleton's hut at Cape Royds ↗ p. 624, occupied by fifteen men, measures 10 × 5.7 metres. It is not surprising that Bernacchi, in his book states that "the dimensions of the living room were too small to accommodate ten men comfortably" and "'with the accumulated dirt of months, the atmosphere of the interior [of the hut] became so foul as to be almost unbearable".

Known to have been made at a factory "Strommen Trevarefabrikk" near Kristiania (renamed Oslo in 1925), Borchgrevink's huts are of typical Norwegian construction, built of planks 7 centimetres thick by 14.5 centimetres high which are interlocked at the corners and are fitted together with a groove in the base and tongue in the top. They are further strengthened by iron tie rods which are threaded down though pre-bored holes from the wall plate to the underside of the foundation beams and tensioned with a nut at the base.

The store hut is left unclad internally, whereas the living hut has an internal lining of a layer of papier-mâché and internal match-boarding. The double floor is also interlined with the same material. Both huts had pitched roofs with substantial trusses at 1 metre centres covered in heavy boarding and then a skin of thick canvas. There is a boarded floor to a sizeable loft space accessed by a central trap door and used for storage. The store hut may well have had the same roof construction and attic – but the roof was taken off in 1900, when the expedition members were picked up by the *Southern Cross*. It is an astonishing

tribute to the strength of the construction that the four walls have survived for 120 years without the lateral stability provided by the roof and this in a location, which is one of the windiest places on Earth with major katabatic winds rolling down off the nearby Warning Glacier frequently reaching 11 or 12 on the Beaufort scale.

The huts were not revisited after Scott's party until 1956, when the icebreaker USS *Edisto,* which was part of the US Navy's Operation Deep Freeze, landed a party of sailors. Subsequently various landings were made but no attempt was made to repair the huts until the establishment in 1980 of a Historic Sites Management Committee of the Ross Dependency Research Committee, which has led to a series of major conservation campaigns.

Borchgrevink's expedition did achieve a notable number of firsts and could be said to be the beginning of the Heroic Age of Antarctic exploration. The huts were the first buildings to be erected on the continent, they proved that overwintering in the Antarctic was possible and they demonstrated that skis and dog teams were an effective means of transport. On their return to Britain the members of the expedition were largely ignored by the establishment – thus Scott and his expedition members did not benefit, as they might have done, from a lot of useful experience. The huts are currently undergoing a major conservation programme organised by the New Zealand Antarctic Heritage Trust.

"The ten men spent the last winter of the 19th century living in a tiny, cramped hut (with another hut for their supplies), perched on the edge of a wind-swept spit, surrounded by towering cliffs at remote Cape Adare in the northern reaches of the Ross Sea. Today, those huts still stand at Cape Adare, making Antarctica the only continent in the world where the first buildings to be constructed still exist."

New Zealand Antarctic Heritage Trust, *Conservation Plan: The Historic Huts at Cape Adare*, 2004

Antarctic Prefabrication. From Timber-Frame Structures to Panelised Construction

Geoff Cooper

GEOFF COOPER is Heritage Programme Manager for the United Kingdom Antarctic Heritage Trust, responsible for the conservation of six British Historic Sites on the Antarctic Peninsula. A conservation carpenter by trade, Cooper has spent five seasons working for the New Zealand and United Kingdom Antarctic Trusts in the Ross Sea and on the Antarctic Peninsula.

The British Antarctic Survey, together with its predecessors, the Falkland Islands Dependencies Survey (FIDS), have established over thirty bases or refuge huts on the Antarctic Peninsula since 1944, when the first bases were built. Like those of the Heroic Age some thirty years earlier, these early huts were primarily of a timber-frame design and construction, largely prefabricated in the United Kingdom and shipped to the Antarctic to be built by the scientists who were to occupy them.

The earliest British bases in the region were erected under the British government expedition codenamed *Operation Tabarin,* which aimed to establishing a permanent British presence on the Peninsula and, in doing so, assert Britain's territorial claims in the region. The first of these bases was Base B ↗ p.702 at Deception Island, which occupied pre-existing timber whaling huts that had been abandoned in the early 1930s. The first entirely new-built base was Base A ↗ p.644 at Port Lockroy, which was constructed in mid-February 1944. The core of this building was a "Spitzbergen" timber-frame structure prefabricated by Boulton and Paul, a large general manufacturing company located in Norwich, England, which specialised in prefabricated buildings.

Boulton and Paul manufactured and assembled their timber buildings in their workshops to ensure that all the components of the building fitted cohesively; they then numbered each component prior to dismantling the structure and packing it ready for transportation to Antarctica. Although these buildings used a familiar and relatively cheap form of construction, they had several shortcomings: they were complicated and slow to construct piece by piece in often sub-zero Antarctic conditions; the specific timbers they required were easy to lose in the snow (assuming that they managed to make it to site without being lost en route); the numbers on the timbers were often obscured, "lost", or difficult to read after their long and arduous journey to the bottom of the world, and the primary timber beams incorporated rudimentary mortice-and-tenon joinery, which often became difficult to assemble, requiring additional time for adjustments to make them fit, as the timbers had often warped during their passage through the tropics. Even Port Lockroy, which was constructed by Master

"The building of huts should not be considered a grim means to an end, the work of producing a really well-designed Antarctic hut should be considered experimental work in line with the general program of discovery work and important scientific studies."

Falkland Islands Dependencies Survey, *Horseshoe Island Building Report,* 1955

"A lot of these buildings were assembled by carpenters in the United Kingdom, shipped to Antarctica and often assembled by scientists. There weren't trained builders all the time."

Geoff Cooper (Heritage Programme Manager, United Kingdom Antarctic Heritage Trust), in conversation with UNLESS, Architectural Association School of Architecture, 2019

Carpenter Lewis "Chippy" Ashton, required numerous adjustments, resulting in him ignoring the pre-cut mortice and tenons to save time and resorting, it is said, to "plain carpentry".[1] Despite these numerous disadvantages, FIDS continued to use this traditional British form of construction for all of their numerous base buildings until the mid-1950s, only making minor changes to the layout of the rooms and the floor plan rather than altering the type of construction.

Those charged with actually constructing and living in these huts felt keenly the limitations of this form of construction. Kenneth Gaul, the Base Commander at the FIDS Base Y on Horseshoe Island, who constructed the kit sent by Boulton and Paul in 1955, observed that "the building of huts should not be considered a grim means to an end. The work of producing a really well-designed Antarctic Hut should be considered experimental work, in line with the general programme of discovery work comparable in importance with the various science studies."[2]

In 1957 a boat shed and store were erected at Base A. This building reflected an evolving and changing typology of construction at Boulton and Paul. Instead of coming as individual pieces to be assembled, it arrived on-site in the form of several pre-constructed and pre-clad timber-frame wall panels, which were simply bolted together upon the previously constructed floor deck. The floor itself comprised several 8- × 4-foot (1.2 × 2.4 m) floorboard panels, laid upon the floor joists. Similarly, the roof utilised panels of sarking boards which could be laid upon the bolted-together roof trusses. The continuing evolution of Boulton and Paul prefabricated construction for Falkland Dependencies was further advanced in 1958 with the construction of the new Generator Shed at Base A. Here, steel portal frames, formed by bolting together several short steel components, were connected by a few timber purlins. The steel-and-timber framework was then quickly infilled with an insulated roof and wall panels, joined together with pre-glazed windows.

Panellised construction was not new to Antarctic architecture, the first of the Heroic Age huts to be constructed on Ross Island at Hut Point [↗ p. 618] for the 1901–1904 British National Antarctic Expedition, used prefabricated panels to

"My first reaction was that they were asking the impossible and that they had better rethink their time scale. With some misgiving, as my reputation was on the line, I agreed to have a good look at the problem and see if there was any way it could be solved in time. [The base had to be] designed, transported, built and occupied (in less than a year). Antarctica might as well have been on the moon as far as I was concerned."

Frank Ponder (Architect), in New Zealand Antarctic Heritage Trust, *Conservation Plan: Hillary's Hut, Scott Base*, 2015

Scott Base was articulated in six interconnected buildings and three detached science huts. The dispersed massing organisation of the station was informed mainly by fire prevention strategies, as well as considerations relating to the ease and speed of assembly, and privacy (assuming that "with functions separated into different buildings the tensions of the long winter living in close quarters would be eased"[1]). In response to the unknown nature of the terrain, flexibility was equally important. Connected by "covered 'all-weather access' ways built of 20-gauge corrugated steel, specially rolled by Lysaght Ltd to allow for uneven ground conditions and different levels between the buildings to be accommodated, [and to facilitate] easy access to telephone cabling and other services that ran between the huts",[2] the structures were designed to host sleeping quarters, generators and snow-melters. "There were to be heavy asbestos curtains in the covered way at each junction box, to cut down draught in the tunnel between the buildings. As well, there was to be an automatic fire alarm system, and a manual alarm that could be activated before the automatic system was triggered by the heat; also, allowing for all possible scenarios, there were to be two hand-operated sirens outside. Dry powder fire extinguishers and bucket pumps for wood or paper fires were to be provided, with asbestos blankets in the generator buildings and the kitchen. Finally, all exposed timber was treated with a fire-retardant paint."[3]

infill the structural timber frame of posts, beams, and rafters. However, a good illustration of the contrasting construction methods is provided by the two bases constructed in support of the 1955–1958 Commonwealth Trans-Antarctic Expedition. Dr Vivian Fuchs, the Director of the Falkland Islands Dependencies Survey, led the expedition, building his crossing starting point base in the Weddell Sea region using a traditional FIDS-type hut supplied by Boulton and Paul. Due to numerous factors, not least the type of construction, the team was unable to complete their hut before winter set in and they were forced to overwinter in a combination of tents and an empty Sno-Cat packing case.

By contrast, the members of the support team in the Ross Sea, led by Sir Edmund Hillary, built their base, (Scott Base ↗ p. 678) using a prefabricated insulated panellised construction to establish the base quickly. The New Zealand architect Frank Ponder designed Scott Base using a large modular insulated panel construction for the floors, walls, and flat roof of each of the separate buildings, which comprised the base (following the lead set by the Australian Antarctic Programme, which built Mawson Station ↗ p. 652 in 1954 using a large insulated panel construction). These prefabricated panels were then simply butted up against each other and joined by means of long metal tie rods that passed through each elevation. When tightened, these rods compressed gaskets placed between the panels, providing a quick weatherproof structure. The speed of construction was demonstrated by the main structure for Hut A – the last remaining building from the original Scott Base – which was constructed in less than a day. While it is more expensive to build using panellised construction (it has been estimated that the cost of the Scott Base was double that of the Shackleton base), it clearly had many benefits and paved the way for more modern developments in Antarctica.

"The designing and building of the base huts and their fittings was probably the biggest problem that faced the expedition before its departure. Two basic types of hut were considered. The first […] was the traditional type of British hut that had been used for many years […]. In contrast was the second type which is widely used now by the Americans, the Russians, and the Australians. This uses large insulated panels which fit one into the other and are then bolted or wedged together. […] This method has the major advantage of being very quick to erect […]. The essential feature of this type of construction is that the more technical and fiddly work is done back in civilization under pleasant factory conditions, leaving only the minimum of work in the harsher climate out in the field."

New Zealand Antarctic Heritage Trust, *Conservation Plan: Hillary's Hut, Scott Base*, 2015

Containerisation in Polar Architecture

Hartwig Gernandt and Hans-Jürgen Meyer

HARTWIG GERNANDT is a scientist and polar researcher at the Alfred Wegener Institute for Polar and Marine Research (AWI) in Germany. As the head of AWI's polar infrastructure, Dr Gernandt oversaw the construction of Neumayer III Station. HANS-JÜRGEN MEYER is a physicist who teaches building physics at the University of Applied Science in Bremen (Germany).

Modern standard containers, as we know them today, were developed in the United States in 1955 by former trucking company owner Malcom McLean and engineer Keith Tantlinger, building upon knowledge accrued in years of international experimentation on intermodal transportation. The result were 8-foot- [2.44 m] tall by 8-foot- [2.44 m] wide units, 10-foot- [3.05 m] long, constructed from 0.1-inch- [2.5 mm] thick corrugated steel.

Freight containerisation grew apace, and in the 1970s the first 20-foot [6 m] transport containers appeared at piers for shipments. Reaching Germany via the Bremen Überseehafen with the MS *Fairland* on the 6th of May 1966, ideas on how to convert these "new boxes" into a wider range of applications emerged swiftly, and gradually gained importance as the primary method of choice for the logistic infrastructures of all national Antarctic research programs – which at the time represented the twelve Consultative Parties to the Antarctic Treaty that operated permanently occupied research stations on the Antarctic continent.

The first German application of such technology in the southernmost continent was commissioned on the 21st of April 1976 for a small research facility with the preliminary working title Container Station ↗ p. 730. Located at 70°46'39"S and 11°51'03"E (near the Russian station Novolazarevskaya, at the eastern end of the Schirmacher Oasis, Central Queen Maud Land), the East German annex station was assembled from mid-February until mid-April of the same year by converting 20-foot [6 m] transport containers (and units made of wood) into prefabricated and thermally insulated autonomous elements. Arranged on a steel substructure and subsequently lowered onto bedrock by a six-man expedition team, the six thermally insulated, autonomous prefabricated units were converted from simple shipping containers into prefabricated elements by the engineering team of the East German Academy of Sciences.

Although the internal design and installations were straightforward and relatively basic for a station, the research base provided decent living and working conditions, hosting manifold scientific activities for many years. Renamed Georg-Forster-Station when East Germany became a consultative member of the Antarctic Treaty in 1987, the station was closed in 1993 after the reunification of Germany. Subsequently all station facilities were dismantled and removed from the Treaty Area in accordance with the Protocol on Protection of the Antarctic Environment signed in 1991.

The small container assembly thus became representative of the first permanently occupied German research station in Antarctica, and presumably also the first utilization of the 20-foot transport containers converted into prefabricated autonomous modules for Antarctic stations. The site of the small container assembly is presently marked by a commemorative plaque and has been recognised as Historical Site and Monument (HSM) no. 187.

In 1978, the Federal Republic of Germany passed a law (in the West German parliament) on Germany's accession to the Antarctic Treaty. The newly established Alfred Wegener Institute (AWI) was tasked with coordinating and processing the development of all necessary polar infrastructure. Since shipping container conversions of newly available 20-foot ISO standard containers had improved significantly by that time, it seemed logical to focus on the development of containerisation not only for logistic transport chains into the Antarctic, but also for the construction of stations and elevated platforms built on snow or firn ground.

The first permanently occupied German research station on the Ekström Ice Shelf, Antarctica, Georg von Neumayer ↗ p. 736, was inaugurated on the the 3rd of March 1981 as a meteorological observatory committed to regularly reporting meteorological data to the global networks. Experimentation with building on ice

Prefabrication and containerisation are essential in Antarctica. The brief construction window available each year for building activities, the total absence of raw and processed materials on site, and the challenges of assembling a structure with "huge gloves at –40°C"[1] lead to the need to optimise building logistics (by exploiting the advantages of containerisation to the maximum and challenging prefabrication practices).[2] A short timeline of containerisation in the Antarctic is shown at the side.

THE LIBRARY ON ICE
Lutz Fritsch

The *Library on Ice* consists of a solitary 20-foot standard container drifting on Ekström Ice Shelf, while hosting a special artistic message to all overwintering crew members staying in isolation at Neumayer III Station. The library – an art project by the German sculptor Lutz Fritsch – includes a hand-picked collection of books in a specially furnished container module. Intended as a sanctuary on ice, a place of refuge, a room for contemplation to reflect upon existence in the Antarctic, the work is not simply a spatial entity on ice, but rather a real, usable library with 1,000 books, each donated by invited artists and scientists with the intention of enhancing the dialogue between the arts and science. Through this project, a singular library came into being at the end of the world. Without persistent effort on the part of station staff, the library will be slowly buried under the snow; the effort required from the Antarctic teams to clear it from snowfall serves as a poignant reminder to humankind to actively resist the loss of cultural knowledge.

1956

The first standard intermodal shipping container is invented and patented by Malcolm McLean in the United States of America.

1960s

The first prefabricated units, which resembled containers, were used at the US Plateau ↗p.698, Eights ↗p.694, and Brockton stations. Erected in 1965 at 3,624 m.a.s.l. to conduct solar and atmospheric studies, the five modules of Plateau Station were designed by the Alberta Trailer Company to fit in the US Air Force C-130 Hercules aircraft. The sledged-van modules, with minimal plumbing and electrical connections, were linked by interior "permawalk" areas.[1]

1976

Georg Forster Station ↗p.730, first called the "Container Station", was the very first Antarctic base to use the concept of deploying prefabricated standard container modules for inhabitation purposes in Antarctica. Featuring eight modules, which were carried on sledges for 120 kilometres from the unloading site to their destination on Queen Maud Land, the German station was assembled in six weeks.

1981

Built on the 200-metre-thick moving Ekström Ice Shelf, the Georg von Neumayer Station (GvN) ↗p.736 consisted of two parallel underground steel tubes which accommodated twenty-four standard container modules fitted for occupation by a wintering team of nine people.

1982

Built on the Filchner-Ronne Ice Shelf, Filchner Station was the first structure to be raised 4 metres off the ground on jackable *pilotis*. The station consisted of a steel open platform upon which eleven prefabricated container modules were assembled to host eight researchers throughout the summer season.

1986

Exploiting the flexibility offered by container modules, Mario Zucchelli Station ↗p.752, assembled in 1986 to accommodate twelve individuals, has been continuously expanded by means of simply adding containers. In its current form, the Italian station hosts up to 120 people and has a surface area of 7,500 square metres.

1993

Similar to its predecessor GvN, Neumayer II Station ↗p.768 consisted of a steel tube which accommodated sixty-seven prefabricated containers. Of these, fifty-six modules were devoted to accommodation and laboratories, while the rest were deployed for food storage and fuel tanks.

2009

In its third reincarnation, Neumayer III ↗p.796 abandoned the strategy of building under the snow, and positioned its 103 container modules, clad with 573 insulated panel elements, on two levels of a raised steel structure.

2012

In Bharati Station ↗p.804, the 134 inhabitable containers are not placed within a secondary steel structure (as in previous Antarctic buildings listed above), but are used themselves as structural elements, interconnected by customised corner castings. The simple structural system of the Indian station was subsequently clad with a single envelope.

led to the development of a special tubular structure consisting of steel trapezoidal profiles designed to provide mechanical protection against the permanently increasing weight of accumulated snow. This was later replaced by a second advanced tubular construction, with three times more space in two parallel tubes. Plugged into such tubes – each with a 26-foot [8 m] diameter – the prefabricated basic modules were placed in a row and were connected to each other by a continuous inner passage, about 4 feet [1.25 m] in width, which reduced the usable space of each module to approximately 32 square feet [3 sqm]. Programmes such as laboratories, the mess, and the hospital were constructed by grouping up to four basic modules. Safely operational between 1981 and 2009, despite being buried up to 49 foot [15 m] deep under accumulated snow, Neumayer II Station ↗ p. 768 marked the end of the tube concept and paved the way for a new generation of Antarctic buildings in compliance with the 1991 protocol requirements for sustainable research stations in the Treaty Area.

The next iteration, Neumayer III Station ↗ p. 796, consists of a very stable steel protected platform-trench construction which is carried by a hydraulic support system designed to keep the structure above the snow surface at all times through regularly raising and realigning the foundational snow level. Subject to no lifespan limitation, this system allows easy and complete future removal and disposal.

At Neumayer III Station, autonomous elements of converted transport containers are used to articulate the station itself, and for application-oriented projects in science and even art. Within the station, a total of 103 converted modules are arranged and installed on the platform on two levels along both sides of the corridor. The entire space of the modules is available for use and amounts to approximately 150 square feet [14 sqm]. Unlike within the autonomous modules, thermal insulation layers are not required for station units enabling a maximisation of the internal height and reaching 8-foot [2.5 m] clearance – a height almost identical to that of the original containers. The platform and its container assembly are clad with a special protective shell, consisting of 573 insulated panels mounted on a steel structure of rips and stringers. Between the protective cover and the thermal insulation of the container assembly itself, a new communal heated space is created: the so-called station gallery.

Known as one of the white continent's more modern research stations, Neumayer III marked an important technological advancement in inhabitation models on ice, allowing for life above the snow surface, as well as improved working and living modules. Offering generous internal mobility on several levels and reflecting the needs of the social structure of the station crew, the complex station construction represents a gigantic leap forward from its predecessors now buried deep in the firn, and marks an important milestone in Antarctic prototyping.

"The biggest challenge of building in Antarctica is that there is no blueprint. It has something to do with inventing: many things are completely new. You should be sure that they work, but you never know."

Bert Bücking (Architect, BoF Architekten), in conversation with UNLESS, Architectural Association School of Architecture, 2019

NEUMAYER - STATION
Neubau 1991/92
Bauherr: Alfred-Wegener-Institut, Bremerhaven
Planung: POLARMAR GmbH, Bremerhaven
Ausführung: Christiani & Nielsen GmbH, Hamburg
J. Heinr. Kramer GmbH & CoKG, Bremerhaven

polarmar

Recombining Architecture. The Role of the Container at Mario Zucchelli Station

Francesco Pellegrino

FRANCESCO PELLEGRINO is the technical manager of Mario Zucchelli Station in Antarctica. A mechanical engineer by training, he is a member of ENEA's Antarctic Unit (Italy), a construction manager of renewable energy plants, and a site supervisor of structural maintenance works. Having visited the continent six times, Eng. Pellegrino is an expert on renewable energy and energy-saving plants for application at extreme sites.

In 1985, a small group of Italian scientists arrived at Terra Nova Bay (Northern Victoria Land) in the Ross Sea area aboard the *Polar Queen* seeking an appropriate site on which to build a new station that could offer logistical support to scientific activities during the austral summer expeditions to Antarctica.

First built in 1986 on coastal granite rock, the initial facility of Mario Zucchelli Station ↗ p. 752 offered accommodation for twelve people and was assembled by combining ISO-20 steel containers transported by ship. In the following expeditions further containers were added to complete the living areas of the main building, and during the sixteenth expedition (2001–2002) a second floor was added to accommodate offices and an elevated "operating room" for the control of air operations.

Managed by the Italian National Antarctic Program (PNRA), to date the station's buildings, plants, storehouses, roads, and fuel tanks occupy an area of about 50,000 square metres. The T-shaped main building consists of 120 containers (44 for personnel accommodation, 44 for offices and 32 for laboratories) and covers an area of 2,000 square metres. The modules – standard ISO-20 steel container measuring 6.06 × 2.44 × 2.59 metres – are connected by removable steel bolts, supported by a steel frame raised by 1.5-metre-high plinths fixed to avoid snowdrifts. Silicone and metal covers are used to prevent water infiltration in the joints, and all structures are designed to guarantee a mechanical resistance to katabatic winds with a maximum speed of 250 km/h and insulate the interiors, assuming outdoor temperatures that reach a minimum of –40°C.

Clad with 100-millimetre-thick insulated panels, the external walls are further insulated by a safety gallery constructed around the building with landers for ordinary circulation and emergency. In recent years, to improve thermal insulation and eliminate water infiltration, a new roof was built over the main building consisting of 10-centimetre-thick insulation panels welded directly, with a slope, on the metal frame. Using aluminium frames, 400 square metres of solar panels were fixed on the new roof to produce renewable electricity for the station during the summer, when sunlight shines over the station for twenty-four hours a day.

Standard containers represent an economical modular construction solution, which has the twofold benefit of enabling building expansion through the addition of further containers to the original structure and

guaranteeing speedy dismantling if the station should be decommissioned.

In Mario Zucchelli Station, the lower floor of the main building hosts accommodation rooms, a canteen, a kitchen, food storage, the infirmary, and the laboratories dedicated to scientific fields such as Life, Earth, Atmospheric Sciences, Meteorology, and Climate Change research. The second floor houses offices, a library, and a meeting room, while the top floor contains the operation room, the communication centre, and a meteorological office. A modern aquarium laboratory with a total capacity of 40 cubic metres of seawater will be available in 2020–2021, allowing extensive in-house scientific experiments in marine biology on several Antarctic fish species – a well-known specialisation of the station.

Three wooden buildings were built between 1995 and 1999 to offer greater thermal and acoustic insulation to special accommodations (for pilots and guests, also known as "transiti" and "foresteria") as well as to leisure facilities such as the gym – known as the "pinguinattolo". The disadvantage of such wooden structures over the container assembly is that wood increases fire hazard risks and demands more maintenance.

Other containers, not structurally connected with the main building, enclose 750 square metres dedicated to power generation. These include electrical generators, a desalination water system, waste-water treatment, and an incinerator. During the 2017, 2018, and 2019 austral summer seasons, to reduce

fossil fuel consumption, a wind turbine farm was built with the assumption that it will produce energy even in winter, when the station is closed, providing electricity that is automated to satellite antennas, weather stations, and data recorders for scientific research.

Two hangars with a total area of 2,500 square metres are used as electrical and mechanical warehouses and carpentry workshops. The old hangars that served as warehouses and garages were built in 1988 and were enlarged in the 1990s for the construction of a helicopter depot and fire station. Construction of a gravel runway is underway in the Boulder Clay site, near the Station. This will become a permanent runway on the rock with a length of 2.2 kilometres that will increase the frequency of intercontinental links with New Zealand and Australia, securing a key logistic role for the Mario Zucchelli Station within the Terra Nova area.

Comandante Ferraz Station

Emerson Vidigal

EMERSON VIDIGAL is an architect and a co-founder, with Fabio Faria, Eron Costin, João Gabriel Rosa, and Martin Goic, of Estudio 41 (Brazil). Estudio 41 designed and built the Brazilian Antarctic station, Comandante Ferraz. Vidigal is also a professor at the Federal University of Paraná (Brazil).

On the 25th of February 2012, a fire destroyed 90% of the Brazilian station originally built in 1984. The event triggered a series of governmental actions to construct a new research facility that saw the Brazilian Navy, in charge of the Antarctic operation logistics section (PROANTAR), promptly remove the rubble of the former station and launch an international architecture competition for the project.

Located on King George Island, on the South Shetland archipelago some 125 kilometres from the Antarctic Peninsula, between Bransfield Strait and the Drake Passage, Comandante Ferraz ↗ p. 838 Research Station (Estação Antártica Brasileira Comandante Ferraz) is situated in a region that is a major hub for the scientific research facilities of several countries, such as Chile, Russia, China, Poland, and Peru.

Built on King George Island, specifically in the Keller Peninsula (a geological formation protected from strong ocean currents by Almirantado Bay), the immediate landscape of the Ferraz Station is the Morro da Cruz mountain range and the waterline of Martell Cove. Situated between meltwater lakes from which the drinking water was extracted, and occupying an area already characterised by intense human presence and a reduced biodiversity, its location was strategic.

The new building's footprint was informed by terrain studies (which favoured an orientation parallel to the Morro da Cruz to facilitate installation manoeuvres within the natural declivities), and by a careful study of the environmental zoning defined for the area in 2006 by the Almirantado Bay Environmental Management. Produced by Professors Rolf Roland Weber and Rosalinda Montone, the zoning report outlines and classifies the site in three key zones: a Restriction Zone, a Transition Zone, and a Use Zone. Standing for the most part in Use Zones 1 and 2 and avoiding occupying the Transition and Restriction Zones, the new Ferraz Station took advantage of the original fuel tank park and helipad that had survived the fire, by situating itself 3 metres above sea level, about 45 metres from the shoreline, on an imaginary line parallel to the shoreline.

Alongside topographic considerations, the linear configuration of the building reflects two key design parameters: transportation logistics and on-site assembly – two well-known challenges in Antarctica. Shipping the prefabricated systems from the preassembly lines to Antarctica is easiest when using 6-metre-long containers, which have the additional advantage of improving on-site assembly and accelerating construction speed – both essential if one considers that construction can only be executed during summertime, i.e. between the end of November and the beginning of April.

The logic of the modular construction is based on 20-foot containers which are raised above the ground by a metallic structure anchored on shallow foundations and are clad with a complex surface envelope. In morphologic terms, the building is configured as an extrusion of its cross-section, with a profile based on three main design premises: thermodynamics, aerodynamics, and landscape views. Raised for the most part above the ground to improve thermodynamic performance and to allow wind to sweep away snow, the main blocks (8 and 11 metres above sea level respectively) are clad with an aerodynamic wrapper that reduces horizontal forces at foundation level and on the steel structure. Given the topographic rise, the west block is elevated in relation to the east block to address issues of psychological well-being and ensure views of the Antarctic landscape from all the inhabitation units for the station's transient population.

Ajacent to the main building, Ferraz Remote Units are scattered across the site at a maximum distance to the main station of 3.5 kilometres. Comprising laboratories, logistic pavilions, safety units, and small hardware storages, they vary in floor area from 1.5 to 190 square kilometres. Conceived as remote laboratories, within their walls researchers perform weather analysis and ozone-layer monitoring. The VLF Laboratory, for instance, uses Very Low Frequency radio waves to obtain information about Space weather and its impacts on the Earth's ionosphere. Due to unstable weather conditions and the researchers' passage through remote areas of the Keller Peninsula, two refuges are stocked with water, food, and medicine and can each accommodate twelve people.

Through the juxtaposition of the small remote units and the newly built contemporary station, Comandante Ferraz at once stands out as an architectural exception and is subordinate to the unique geography and landscape of the southernmost continent.

The axonometric drawings highlight the assembly of fitted containers within the linear volume of the Brazilian station Comandante Ferraz.

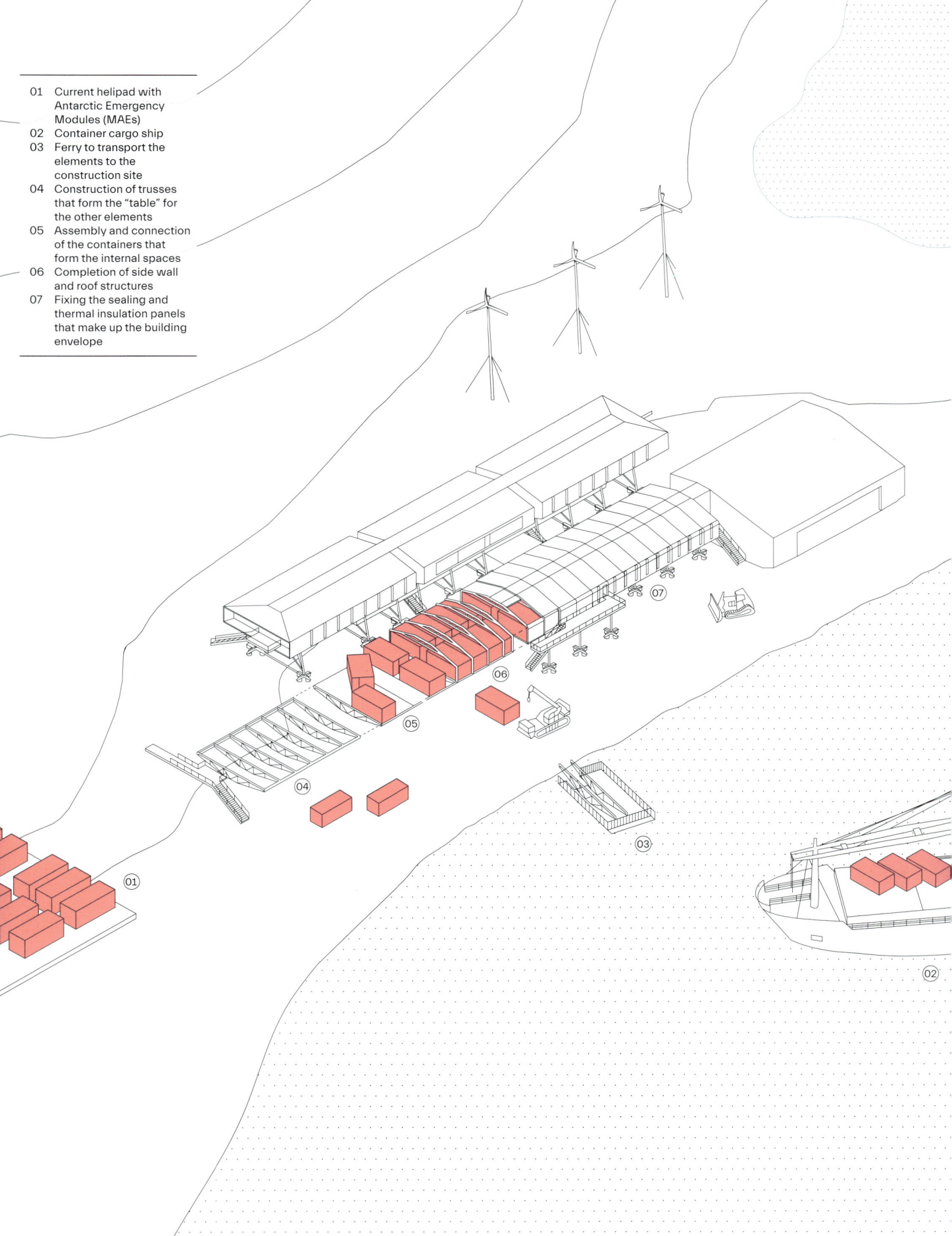

01 Current helipad with Antarctic Emergency Modules (MAEs)
02 Container cargo ship
03 Ferry to transport the elements to the construction site
04 Construction of trusses that form the "table" for the other elements
05 Assembly and connection of the containers that form the internal spaces
06 Completion of side wall and roof structures
07 Fixing the sealing and thermal insulation panels that make up the building envelope

Bharati Station

Bert Bücking

BERT BÜCKING is an architect and a co-founder of bof Architekten (Germany). His experience focuses on the interface between civil engineering and architecture. Bücking's interest in modular and sustainable building techniques led him to design and build, in collaboration with Ramboll Engineering, Bharati Station for the Indian Antarctic Program.

Located on a peninsula in the Larsemann Hills on the east of the white continent, Bharati ↗p.804 is India's third Antarctic research station. Unlike a number of other Antarctic stations, which are built on an ice shelf or on the Antarctic plateau, Bharati's site is located on solid rock. Similar to Russia's Progress Station or Bharati's neighbouring Chinese station, Zhongshan ↗p.760, the site's proximity to the Antarctic coast exposes it to extreme winds with gusts of up to 300 kilometres per hour and temperatures of −40°C, creating an atmosphere that is highly corrosive for metal surfaces. Even though the station does not require mobile foundations such as those deployed at Halley VI ↗p.810 or Neumayer III ↗p.796, Bharati remains an important study case for Antarctic architecture in terms of how it translates the awkward logistic challenges of building in the continent into a design opportunity, pushing the boundary of container-based architecture.

Because of its location, there are only four months in the year when milder temperatures allow the sea ice surrounding Bahrati Station to melt, rendering the coastline accessible. This constraint limits the annual construction period to a maximum of four months; a time in which the building components can be safely delivered and assembled on-site. This led to the definition of a "plug-and-play" design strategy, based on simple adaptations of 20-foot standard ISO high cube containers (which are 2.89 metres high rather than the standard 2.59 metres). A "container conglomerate" principle was subsequently developed to offer a structural system which would rely on the containers alone, by interconnecting them at their corner castings. Such a structure not only proved to be stiffer but also made it possible for the area conventionally devoted to the building to be limited to a girder framing system. The containers were prefabricated with the furniture already placed inside and were sealed off during shipping to protect their contents during the long journey.

Once assembled, an outer façade envelope wrapped the containers into a single surface to ensure thermal insulation. Designed to mediate the temperature difference between indoors and outdoors, particular attention was given to the cladding system. A "house in a house" strategy was thus introduced whereby the innermost layer (i.e. the station) was designed to retain a temperature of 22°C with 35% humidity (generated by the station's air conditioning system), while a non-conditioned secondary layer – technically an inhabited interstitial space – was devised to guarantee a temperature of 10°C with 10% humidity, with outside temperatures reaching −40°C with an average of 40% humidity. Standard off-the-shelf panels mounted on a light substructure were selected to construct such envelope, following the principle of using easily available and replaceable products, while minimising the number of parts and components.

Simulation models and wind tunnel tests were conducted to verify the façade resistance and resilience to Antarctic air pressure, and finally prefabrication was carried out in Duisberg, Germany, between March and August 2011. By November the containers were loaded in Antwerp, from where they headed south. Once the Russian icebreaker carrying the precious cargo on the final stretch from Cape Town to Antarctica reached the ice pack, helicopters transferred the containers (which were pre-emptively designed not to exceed 4–5 tons in weight) onto the coast. Here, during the second summer season (after all the infrastructural works, including the desalination plant had been safely installed), the station, which was designed to host forty-five inhabitants in the summer and fifteen overwintering crew, was assembled in four months.

01 The module is fitted out.
02 The modules are interlocked through corner castings.
03 The modules are transported.
04 The modules are loaded on a cargo ship.
05 The modules are given a special barcode to optmise unloading and costruction.
06 The modules are assembled on-site ready to be inhabited.

The diagram expands on the design process
that led to the realisation of the Indian station
Bharati, from the single container to the
finished building.

 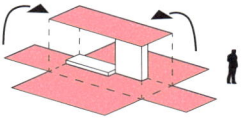

 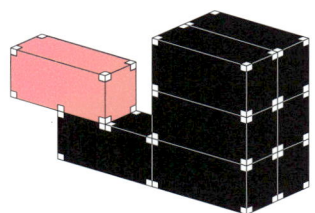

 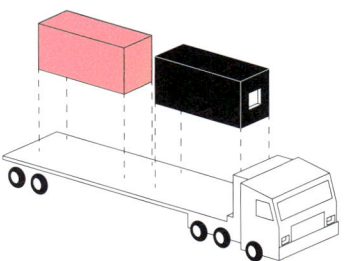 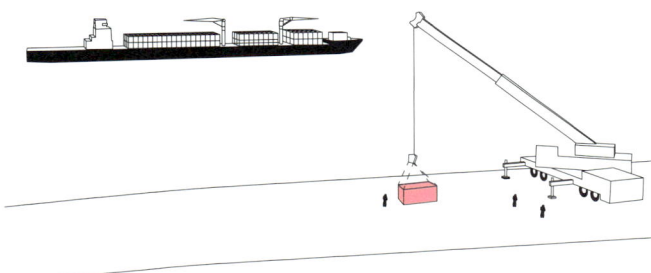

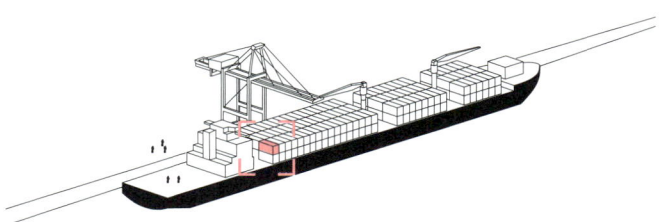 

The Role of the Window in Antarctic Architecture

Hugh Broughton

HUGH BROUGHTON is an architect and the founder of Hugh Broughton Architects (United Kingdom). He is the designer of the Halley VI Antarctic Research Station, the world's first relocatable polar base, and the Juan Carlos I Spanish Antarctic Base. Broughton is currently developing the designs for the Scott Base for Antarctica New Zealand, the Discovery Building at Rothera for the British Antarctic Survey (United Kingdom), and a master plan for Davis Station for the Australian Antarctic Division.

"A giant window fills the module with natural light. The glass incorporates nano gel – a special insulating material developed by NASA to reduce heat loss but maintain bright light levels."

Chris Thompson (Artist), "Inspired by Thunderbirds: Halley VI Antarctic Research Station", 2017 (online)

Antarctica is the coldest and windiest continent on the planet. Temperatures can drop to below –80° C in the coldest inland locations and katabatic winds at the coast can exceed 300 kilometres per hour, which is strong enough to carry airborne debris the size of small stones. These climatic extremes can place significant pressures on building fabric, requiring special technical solutions to ensure weathertightness.

As the most fragile component within a building envelope, windows considered for use in Antartica require special consideration. Not only do they need to resist the passage of cold temperatures and the impact of high winds, but they must also be able to control glare from light reflected off snow and ice during the 24-hour summer months and if they suffer damage, it must be relatively easy to replace them.

Despite the challenges of including glazing in the design of buildings in Antarctica, well-placed windows in buildings offer significant benefits. They introduce daylight into the building, reducing the need for artificial light and therefore lowering the use of energy, which is often created by generators consuming modified diesel. Glazing also offers views of the Antarctic landscape, helping to place residents within the context of their surroundings, enhancing their well-being and allowing them to see the weather outside. As the extent of glazing is, by necessity, limited, its positioning must be carefully considered. This, in turn, can lead the planners of new buildings to make the most out of the best views, as is shown at Bharati Station ↗ p. 804, designed by bof Arkitekten, in which the lounge is placed at first-floor level behind a large glazed screen overlooking the Larsemann Hills.

In designing windows for an extreme environment, a number of factors must be considered. The type of glass used in the window assembly will have a significant impact on the level of heat loss, the amount of glare, and the degree of transparency. Generally, glass will incorporate a low emissivity solar control layer, reducing the amount of ultra-violet and infrared light that can enter the building. The high levels of ultra-violet light in Antarctica make stringent demands on the glazing in terms of countering this glare. As a result, windows are often quite dark – similar to a pair of sunglasses. This is, however, not always possible. In some cases, maximum visibility is important, such as in the comms tower at the new Discovery Building at Rothera ↗ p. 856, which is used for air traffic control. In this case, the transparency required for the window to allow approaching airplanes to be clearly seen is offset by the angling of its glazing to minimise reflections.

The number of windowpanes is a key determinant of the insulating qualities of a glazed assembly. Triple-glazed windows are commonplace, and in the coldest locations quadruple glazing will be necessary. The voids between the sheets of glass are typically filled with an inert gas such as Argon to limit the transfer of cold through an air medium. To reduce heat loss, the sheets of glass must be separated by non-conductive thermoplastic spacers rather than the more traditional metal spacer.

In some instances, double glazing may be sufficient, although this is limited to special elements. An example of this is the cockpit-style rooflights on the upper level of the social module at Halley VI ↗ p. 810. In summer these 4.2-metre-long, 2.2-metre-wide skylights give views of the surrounding ice shelf and in winter offer a protected place from which to enjoy the spectacular displays of the aurora australis. Each of these rooflights is made up of three parts and is double glazed with high performance, solar control glass bonded with silicone to a stainless-steel frame. When the crew are watching TV, the cockpit rooflights can be blacked out with automatic blinds.

Where light transmission is more important than views, the glazing may incorporate a translucent insulating fill. Part of the giant window that illuminates the double-height space at the heart of the social module at Halley VI is fabricated with a white, double-glazed panel in which the gap between sheets of glass is filled with nanogel. This gives the glazing the same insulation value as a normal cavity wall, while still allowing 30% light transmission. As a result, the space has a pleasant glow throughout the summer with minimal heat loss.

The frame of the glazing is also a key consideration: if it is not insulated, it can be a significant source of heat loss. The outer element of the frame should be separated from the inner part by a non-conductive material such as a thermoplastic. This type of frame is described as thermally broken. The outer material may be different from the inner material: for example, colour-coated aluminium can be used externally and timber internally. In most cases, windows are not openable because if a window were accidentally left open, this would create the risk of winddriven snow being blown into the building. At Halley VI the windows do not have a traditional frame. Instead, the triple-glazed units are bonded into the fibreglass panels using structural silicone. This removes a thermal weak point from the building assembly to maximise the levels of insulation.

With constantly improving technology the amount of glazing in Antarctic buildings will increase. Some of the projects on the drawing board show extensive use of glazing which will envelop crews in the surrounding landscape. Examples include the extensive glass screens planned by Oz Architecture in the assembly space at McMurdo ↗ p. 866 and the full-width, full-height window to the dining space at the new Scott Base ↗ p. 852. This space is at first floor level and will enjoy extraordinary views across the Ross Ice Shelf towards the iconic Mount Erebus. Antarctica has majestic landscapes and making it possible to enjoy them from a warm environment thanks to carefully positioned glazing offers a remarkable opportunity for every Polar building designer.

The Evolution of the Architectural Envelope

Federica Sofia Zambeletti

FEDERICA SOFIA ZAMBELETTI is an architect at UNA, an international architectural office focusing on cultural projects based in Hamburg (Germany), and a researcher at UNLESS, a not-for-profit research agency devoted to interdisciplinary research on extreme environments, where she oversaw *Antarctic Resolution* as project manager. She is the founder of KooZA/rch, a digital platform researching the architectural imaginary.

Commonly referred to as the concealed technology enabling greater architectural sustainability, the Antarctic envelope has always been in a class of its own: a technological element unconcerned with stylistic architectural fetishisms and urban settlement conventions – a meticulously studied component designed to enable survival and habitation in Earth's coldest desert. The evolution of the Antarctic envelope is intriguing not only in terms of its independence from the urban context but also through its intimate relationship to the remote frozen environment and its complex set of logistics. An environment that seems immovable in its own right, yet fragile, as its ice shelves slowly and imperceptibly dissolve in a changing planet. What will the Antarctic envelope evolve to be in the next 100 years?

    Rather than a subjective design concept, the "U" value of a building envelope (developed by the French physicist Eugene Peclet in the 1850s) is a scientifically determined measure – a unit describing the flow of heat through a wall – a quantifiable matter of life or death within the coldest, harshest terrestrial climate of Antarctica,[1] where temperatures reach –80°C and wind travels at speeds of up to 327 kilometres per hour. From the early explorers' huts, which were clad with wooden tongue-and-groove boards and insulated with papier-mâché, seaweed, and wool to mitigate heat loss and wind, today's high-tech panelised structures feature cutting-edge insulation, whilst the exterior finishes are designed to resist discoloration, UV, and the abrasive impact of wind-driven snow ice.

    Whereas in the first half of the twentieth century, Antarctic structures did not diverge much from those of the early explorers, the advent of the International Geophysical Year (IGY) in 1957–1958 and the need for resilient architectures to assist the growing number of scientific programmes led to the optimisation and evolution of the architectural envelope. In line with the development and implementation of panelised systems worldwide (first in industry and later in housing from societies which were still reeling from the Great Depression and World War II[2]), the easier, more practical assembly provided by this system was quickly adopted in a continent where extreme conditions do not allow for great dexterity and movement. As the most experimental element of an Antarctic building, today the architectural envelope is carefully studied to define the optimum equilibrium between logistical optimisation, comfort, and a minimum footprint. Two architectures, both constructed on the occasion of the IGY, paved the way in terms of optimisation and comfort: Scott Base[↗ p. 678], constructed for the Commonwealth Trans-Antarctic Expedition, and Syowa[↗ p. 770] Station, erected by the first Japanese Antarctic expedition.

    Although the expedition led by Vivian Fuchs and Sir Edmund Hillary was to embark on the first overland crossing of the continent (a colossal endeavour), the expedition members regarded "the designing and building of the base huts and their fittings"[3] as the principal problem. The sentiment was shared by Frank Ponder, architect for the Ministry of Works (New Zealand), who, faced with the challenge of having to design, transport, build, and occupy the station in less than a year, remarked that "Antarctica might as well have been on the moon".[4] Rather than following in the footsteps of the British, which would have entailed the slow and tedious construction of a building assembled out of numbered and pre-cut timber elements, whilst "fiddling with innumerable nails and bolts in cold and unpleasant weather", Fuchs turned to large insulated panels.[5] Drawing on his experience of prefabricated buildings, Ponder merged this expertise with the design of "cool rooms" exploring the idea of a reverse refrigerator that would keep the cold out. The panel system resulted in the technical work being done

one of the many visitors to the site throughout the summer.

the last modules going up.

the now almost completed shell after an overnight snowfall.

The British erected two "Plastic Huts" in Antarctica, one at Signy Station (shown above), built in 1963 to contain living quarters and laboratories,[1] and one at Base B, Deception Island, erected in January 1966. The latter, which was notoriously painted green and named "Priestley House after Sir Raymond Priestley (Acting Director of FIDS, 1955–1959, and geologist on Scott's expedition, 1910–1913), was found to be missing on the 22nd of March 1985 when RRS *John Biscoe*"[2] sailed to the island.

ahead of time, leading to swift construction on-site in under a week for an architecture which lasted long beyond the expedition to become a prominent scientific station. In recognition of the importance of the hut for both exploration and scientific pursuits, today part of the station has been listed as a Historic Site and Monument under the Antarctic Treaty.

Defined as the precursor of the Japanese metabolist movement and of prefabricated housing,[6] the panel implemented in 1956 at the Japanese Syowa Station designed by Takashi Asada pioneered a construction system that prioritised the connection between the individual parts and rigorous modulation for easy assembly. Asada fully embraced the fundamentals of Metabolist architecture as one which could, similar to a living organism, metabolise and adapt to unpredictable changes in the environment – an approach deeply rooted in Japan's post-war history. First tested within the inhospitable landscape of the Antarctic continent, the architecture was intended as a tool for the purpose of survival,[7] with the architect looking to the practice of environmental engineering and climatological experiments undertaken at Hokkaido University's Low-Temperature Science Lab. In hindsight, one could argue that the pavilions presented at Osaka Expo '70 were in fact *evolutions* of the Antarctic unit with Antarctica playing the role of an experimental "test ground" for these early metabolism ideas, before they were reimported into the city and applied on an architectural and urban scale across everyday life. Beyond terrestrial applications, it is extraordinary that Misawa Homes, the producer of this first panelised structure, are today trialling the Mobile Polar Unit at Syowa for further usage on the Moon, putting to the test the notion of Antarctica as a space analogue.[8]

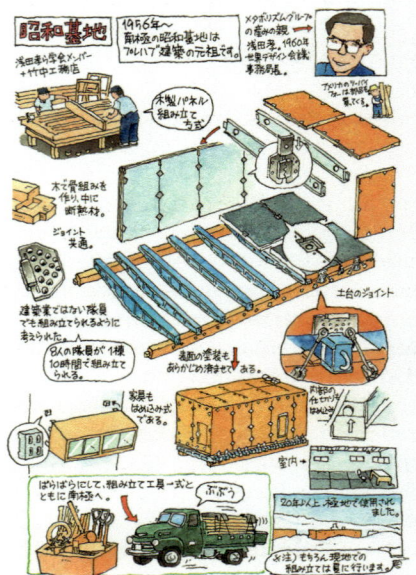

A different type of panel optimisation took place for buildings erected on snowfields. Due to the harsh environment, these buildings would rely on a duplication of their architectural envelopes, where their exterior layer would progressively detach itself from the interior layers of construction, acting as both a structural and a protective element. Modelled on Camp Century in Greenland, the American New Byrd Station↗ p. 690 (whose entire cargo had to be airlifted using LC-130 Hercules planes) consisted of a series of T-5 buildings constructed out of plywood and prefabricated fibreglass panels inserted into a second structural envelope of snow tunnels topped by a corrugated steel roof.[9] Similarly, in 1973, researchers at Halley Bay grappled with the construction of Halley III↗ p.716 as an under-snow facility built out of a series of prefabricated buildings erected within a second envelope of ARMCO tunnels, whose exact shape had been studied and tested for its ability to withstand the extreme forces exerted by the ice sheet.[10] The German National Programme also explored this method with the first Georg Von Neumayer Station↗ p.736, which was erected as a series of containers within an exterior steel structure. Perhaps the culmination of this approach is most evident in the geodesic dome↗ p.722 erected at the South Pole in the early 1970s. Rather than enclosing the urban settlement of Manhattan as famously collaged and explored conceptually by Buckminster Fuller, the dome at the South Pole acted as the envelope that enclosed the sprawling master plan of prefabricated structures in an attempt to shield them from accumulating snow drift. Nonetheless, like all the buildings erected on the plateau, the structure was not immune to the relentless accumulation of snow, which eventually lead to its demise and deconstruction during the 2007–2008 summer season.

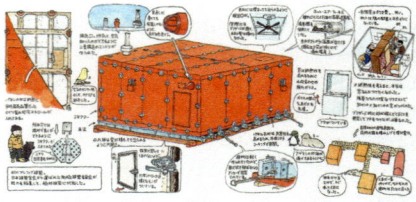

The progressive "raising" of stations on built stilts for structures erected both on the coast and inland (implemented on the basis of the Protocol on Environmental Protection to the Antarctic Treaty) has resulted in envelopes that are now concerned with their response to extreme temperatures and wind speeds. Recently built architectures like those of Bharati[p. 804] and Halley VI[p. 810] have been tested in wind tunnels to stress-test the design of the building envelope for optimisation, whilst Jang Bogo's[p. 820] envelope draws on the properties of a golf ball to maximise the deflection of wind.[11] Possibly most impressive in this regard is the equilibrium reached at Princess Elisabeth[p. 786], the first zero-emission summer-only station on the continent – an impressive achievement made possible by the meticulous design of the envelope and its integration with further sustainable technologies including solar panels and wind turbines. With greater attention being devoted to the psychological factors relating to living in such extreme environments, the building envelope had also come a long way in terms of apertures, from the windowless Halley II[p. 706] to the impressively large technological window in the central module of Halley VI, as developed by Hugh Broughton Architects with NASA.[12]

Although the implementation of the Protocol on Environmental Protection to the Antarctic Treaty (in particular Annex III) did have a significant impact on the architecture of Antarctic stations – namely, the possibility of removing them or raising them above ground – it is quite significant that the building envelope is still not legally subject to sustainability assessment methods such as BREEAM or LEED, protocols that are regularly applied in urban settings. The absence of a systematic set of regulations for building within this global commons means that national programmes and their chosen architects are free to define individual standards for efficiency. At present, it is easy to see how, in recent years, Rem Koolhaas's critique of the façade as "the element most invested with political and cultural meaning"[13] has gained particular momentum with the practice of using this as a canvas for more or less prominent national flags. The unique practice of extending this to the ground in order to enhance a station's architectural, and therefore national, presence within this global commons suggests we should question and consider the role of these buildings as political markers beyond their purpose as scientific stations.

1899

1909

1912

CAPE ADARE HUT ↗ p. 614
Mean annual temperature: –10.3 °C
Mean annual wind velocity: 104 km/h (max)
Cladding type: Wood match boarding
Envelope composition: Snow build-up (seasonal); exterior timber panelling; papier-mâché; timber framework; papier-mâché; interior timber panelling.

CAPE ROYDS HUT ↗ p. 624
Mean annual temperature: –17 °C
Mean annual wind velocity: 18 km/h
Cladding type: Wood match boarding
Envelope composition: Two to three tiers of casing; volcanic sand; flat galvanised iron; exterior weather boarding; felt insulation; timber framework; interior timber tongue-and-groove boarding.

CAPE EVANS HUT ↗ p. 632
Mean annual temperature: –17 °C
Mean annual wind velocity: 18 km/h
Cladding type: Wood match boarding
Envelope composition: Build-up of volcanic sand; exterior double boarding; quilted seaweed; timber framework; interior double boarding. The south and east walls also featured compressed forage bales.

"The floor in the dwelling hut was doubled with papier-mâché between the layers of planks. The walls were lined with wood, also with papier-mâché between the outer timber and the inner panelling."

Carsten Borchgrevink (Explorer), *First on the Antarctic Continent: Being an Account of the British Antarctic Expedition, 1898–1900*, 1901

"Some of the bunks have thick papier-mâché insulation on the wall side. […] There are in Savio's bunk a number of the dog coats nailed to the wall, presumably for extra insulation."

New Zealand Antarctic Heritage Trust, *Conservation Plan: The Historic Huts at Cape Adare*, 2004

"What with the double lining, double doors and windows, the want of ventilation and the sealskins (stitched over the roof) and snow outside, the living room, when a fire was burning, became at times too warm."

Louis Charles Bernacchi (Physicist and Astronomer) *To the South Polar Regions: Expedition of 1898–1900*, 1901

"The north elevation […] is clad with 61 timber tongue and groove boards (160 × 25 mm) fixed vertically over a layer of roofing felt insulation. Weatherboards on this elevation are covered with sheets of flat galvanised iron (some made of opened up tins). These metal sheets were nailed to the hut cladding to protect the timber from damage by the ponies."

New Zealand Antarctic Heritage Trust, *Conservation Plan: The Historic Huts at Cape Adare*, 2004

"To make certain that no air would penetrate from these sides, we built the first two or three tiers of cases a little distance out from the walls of the hut, pouring in volcanic earth until no gaps could be seen, and the earth was level with the cases. The rest of the stores were piled up to a height of six or seven feet. Before winter the windows on the south side were boarded up."

Sir Ernest Henry Shackleton (Explorer), *The Heart of the Antarctic: Being the Story of the British Antarctic Expedition 1907–1909*

"The sides have double boarding inside and outside the frames, with a layer of our excellent quilted seaweed insulation between each pair of boarding. The roof has a single match boarding inside, but on the outside is a match boarding, then a layer of 2-ply 'Ruberoid', then a layer of quilted seaweed, then a second match boarding, and finally a cover of 3-ply 'Ruberoid'. The first floor is laid, but over this there will be quilting, a felt layer, a second boarding, and finally linoleum; as the plenteous volcanic sand can be piled well up on every side it is impossible to imagine that draughts can penetrate into the hut from beneath. […] To add to the wall insulation the south and east sides of the hut are piled high with compressed-forage bales."

Robert Falcon Scott (Explorer, Royal Navy Officer), in his diary, 1912

1944

1954

1957

BASE A ↗ p. 644
Mean annual temperature: 1.8 °C
Mean annual wind velocity: 19.8 km/h
Cladding type: Wood match boarding
Envelope composition: 2.54 cm thick
tongue-and-groove exterior wood boarding;
air cavity; timber frame; aluminium foiled
tar paper lining; 2.1 cm thick tongue-and-
groove interior wood boarding.

MAWSON ↗ p. 648 (PTB MARK III STRUCTURE)
Mean annual temperature: –8.3 °C
Mean annual wind velocity: 41 km/h
Cladding type: Prefabricated modular panels
held together with threaded steel rod conduits running the length of the walls through
the panels forming a post-tensioned box.
(PTB Mark III)
Envelope composition: Aluminium and zincan-
nel sheeting; expanded bakelite (onazote)
insulation; timber frame; masonite hardboard
or plywood.

SCOTT BASE ↗ p. 678
Mean annual temperature: –19.8 °C
Mean annual wind velocity: 19.1 km/h
Cladding type: Prefabricated modular panels
held together with long tie rods.
Envelope composition: Aluminium sheets;
light timber frame; fire-resistant onazote
insulation; asbestolux lining.
Total panel thickness: 8.9 cm

"The entire roof and walls of the building were lined with aluminum foiled tar paper, over which was laid boards of one inch [2.54 cm] tongue and groove material, an air space being left in the walls by means of battens. The roof was covered with heavy rubberoid, well battened down, and tarred at the joints. The inner part of the building was completely lined with ⅚ inch [2.1 cm] tongue and groove material."

Andrew Taylor (Explorer, Surveyor), *Accomodation [sic] Construction at Port Lockroy*, 1944

"Due to contraction of the outside (& inside) T&G gaps have appeared. The entire north and east walls of the hut have been rubberoided, improving the warmth inside the hut. […] If it is to continue to serve for some years […] the entire inside of the hut should be lined with plywood or hardboard, to eliminate draughts caused by the contraction of the T&G."

Cameron, Henry Alan D. (Ionosphericist, Base Leader, Marine Geophysics Technician), *Base Building Report 1959, Base A*, 1959

"The design was a cooperative effort between ANARE, the Australian Construction Services (then the Commonwealth Department of Works), and a Melbourne cool room construction firm, Olympic General Products. The buildings were made up of standard cool room panels."

Australian Antarctic Program, "Australia's Antarctic Buildings: The DIY Era" (online)

"These buildings were well suited to the circumstance of the times being quickly erected – the panels could generally be all erected within a day – having efficient insulation and components which could be manhandled into place."

Philip Incoll (Architect), *AANBUS: The Creation of a Building System for Antarctica*, 1980

"The structural design that evolved for Scott base was for sandwich panels, 3½" [8.9 cm] thick and held together with long tie rods. They were to have a light timber frame and be sheathed in aluminium on the outside to avoid water vapour penetration and freezing within the walls, and for lightness in handling and transporting on a sledge from ship to site. The inner face of the panel was to be Asbestolux, a ¼" [0.6 cm] thick fibre cement material, which was in common use as a fire-resistant lining. The insulating core would be fire-resistant Onozote, a gas-expanded ebonite material used in the construction of cool stores. […] Huts A to D […] were built in Australia by Explastics Insulations Ltd of Melbourne, a company specialising in refrigerated cool stores."

New Zealand Antarctic Heritage Trust, *Conservation Plan: Hillary's Hut, Scott Base*, 2015

1962

NEW BYRD ↗ p. 690
Mean annual temperature: −25 °C
Mean annual wind velocity: N/A
Cladding type: Prefabricated modular panelled structure (T-5 type) inserted within snow trench below ice.
Envelope composition: Corrugated steel arch excavated in snow trench; exterior plywood skin; 6.35 cm glass wood fibre; interior plywood panel.

"The 'cut-and-cover' concept for undersnow construction […] initially evolved as an economical means of providing simple shelter from high winds and insulation from the extremes of surface temperature."

Malcolm Mellor (Glaciologist, Polar Explorer), *Methods of Building on Permanent Snowfield*, 1968

"Trenches were excavated […] to a depth of about 25 ft [7.6 m] […]. There were two principal types of trench cross section: wide trenches with vertical walls for roofing by 30- and 40-ft-span [9 and 12 m] Wonder arch, and narrower undercut trenches for roofing by 14-ft-span [4 m] corrugated steel arch. […] Snow from the excavations was blown over the arches as a backfill to give a finished snow cover of 2 to 3 ft [60 to 91 cm] on the arch crown. Inner buildings were prefabricated structures of the T-5 type. The modular panels were 4 ft [1.2 m] wide; they consisted of outer plywood skins with 2½ in. [6.35 cm] of glass fiber insulation between."

Malcolm Mellor (Glaciologist, Polar Explorer), *Methods of Building on Permanent Snowfield*, 1968

1976

MAWSON, DAVIS, AND CASEY ↗ p. 756 (AANBUS SYSTEM)
Cladding type: Prefabricated modular panels AANBUS type.
Envelope composition: Lysaght Zincalume-protected steel prefinished with silicone-modified polyester paint (continuously fabricated on a laminating machine as used in cool room manufacture); polystyrene; steel portal frame; 13 mm thick Australian Gypsum "Ultrawall" partition (1-hour fire rating). Where severe wear was likely (workshops), the lower part of the inner layer wall was replaced by 12 mm thick compressed asbestos cement.
Total panel thickness: 150 mm

"Timber framed panels had become expensive because of the labour required in framing up each individual panel. […] Frameless fibreglass panels had recently proven even worse in this regard. […] The current standard prefabricated cool room panels […] were then given detailed consideration. At this time, they had been in large scale use on commercial cool rooms for some years. […] This panel in contrast to all previously used panels was a standard production item manufactured in the same form by a number of firms throughout Australia. It was light in weight and strong. Panels were easily cut to length and shape. This would be possible even on site using simple equipment."

Philip Incoll (Architect), *AANBUS: The Creation of a Building System for Antarctica*, 1980

1984

HALLEY IV ↗ p. 740
Mean annual temperature: −20 °C
Mean annual wind velocity: 25.2 km/h
Cladding type: Interlocking prefabricated plywood-panelled structure constructed within an excavated tunnel of dense snow (maximum 2 m depth) and later backfilled to half the height of the tube.
Envelope composition: 9 mm Finnish (Keruing) birch plywood double external cladding; air cavity; triangular solid timber ribs; metal timber connectors; 50 mm thick insulation; 9 mm Finnish birch plywood single internal cladding with 94–97 mm thick redwood/whitewood fastening members – one at each either side.

"It was the intention during the design period to provide a consistent foundation level upon which the base was constructed which provided a density of 0.5 g/cm³. From the reports received to date it would appear that a density of this order was available at 5 m depth but not necessarily at surface level or at a 2 m depth. […] The tube deformation originally designed for 150 mm across the diameter with 5,000 mm ultimate stress. At the time of my survey there was a minimum deformation of 183 mm and a maximum of 260 mm. The design limit has been exceeded and the failure stage is being reached. […] Expected occupancy: 3 more years."

Alan Smith (Builder, Head of British Antarctic Survey Building Services), *Antarctic Station Halley Report No. 1 Structural Defects*, 1979

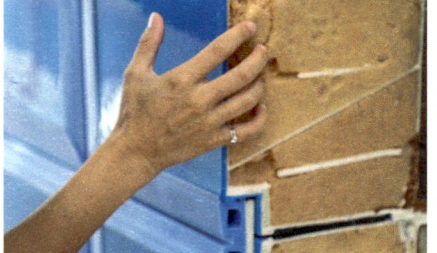

2008

PRINCESS ELISABETH ↗ p. 786
Mean annual temperature: –18 °C
Mean annual wind velocity: 50 km/h
Cladding type: Prefabricated modular panel
Envelope composition: 1.5 mm thick stainless steel sheet external cover (screwed into the multiply panel by Etanco 6.2 × 40 mm stainless steel screws fitted with watertight washers); 3 mm Ethafoam intermediate layer (protection and drainage for the EPDM waterproofing); 1.5 mm waterproofing layer with EPDM membrane Evalastic (Alwitra); 40 mm beechwood spacers holding two plyply panel layers – spacers are glued into the thickness of the plywood panels. The use of wooden spacers minimises the thermal bridges between the two supporting layers as much as possible); 42 mm thick multilayer laminated wood panel element; 400 mm thick graphite-coated expanded polystyrene insulation (Swisspor, lambda thermal conductivity value 0.029 W/mK); 80 mm multilayer laminated wood panels; Ampack Sisalex 514 aluminium vapour barrier glued on the entire surface with Ampacoll glue.
Total panel thickness: 60 cm

"The design of the envelope and the station at large was dictated by the extremely short time frame (under two months) within which it had to be built and sent to the Antarctic. The envelope specifically – and the prominent use of plywood following the principles of a 'cigar box' – was defined according to the skills and tools available to the craftsman. It is useless to dream something and then have no one able to build it."

Philippe Samyn (Architect, Philippe Samyn and Partners), in conversation with UNLESS, Architectural Association School of Architecture, 2019

2008

HALLEY VI ↗ p. 810
Mean annual temperature: –18 °C
Mean annual wind velocity: 25.2 km/h
Cladding type: Prefabricated modular panels.
Envelope composition: Colour coated aluminium architectural facing; 12 mm OSB board; SIP panel; 170 mm extruded polystyrene insulation core; 12 mm OSB board, field applied vapour barrier, steel super-structure, inner lining.
Total panel thickness: 44.8 cm[1]

"The construction of Halley VI will follow a production line approach based around five workstations. [...] Station 4: Cladding panels will be bolted to the steelwork. Cladding will be installed with mechanical fixings to enable easy removal when the station is decommissioned. [...] The cladding will be formed from relatively lightweight Glass Reinforced Plastic (GRP) panels fixed to the structure and joined with a silicone rubber gasket. The panels consist of closed cell polyisocyanurate foam insulation encapsulated within the GRP, finished with a layer designed to minimise discoloration, resist UV and the abrasive impact of wind driven snow and ice."

British Antarctic Survey and Natural Environment Research Council, *Proposed Construction and Operation of Halley VI Research Station, and Demolition and Removal of Halley V Research Station, Brunt Ice Shelf Antarctica: Final Comprehensive Environmental Evaluation*, 2007

2012

JANG BOGO ↗ p. 820
Mean annual temperature: –14 °C
Mean annual wind velocity: 21.6 km/h
Cladding type: Prefabricated modular panels
Envelope composition: 50 mm polymetal panel; 100 mm PIR panel; 150 mm intermediate space and structural frame; 12 mm wooden louvre; STACO panel.
Total panel thickness: 41 cm

"The dimple surface of a golf ball was applied on the outer surface of the building. It reduces the whirlpool phenomenon that occurs after the wind has blown away and thus prevents the snow from piling up. To utilize the dimple pattern as a design factor numerous patterns were revised with dimples that were diversified in terms of size, interval and density."

Space Group on Jang Bogo Antarctic Research Centre, 2011

Jang Bogo Station

Sang Leem Lee

SANG LEEM LEE is an architect and the chairman of the Space Group architectural practice (South Korea). He is the honorary president of the Korean Institute of Architects and was a UNESCO chair professor (Social Sustainability of Historic Districts). He has participated in numerous exhibitions, including at the Venice Biennale. Lee's architectural projects include the Jang Bogo Antarctic Research Station in Antarctica.

Antarctica has only two seasons: summer and winter. In the austral summer, the sun does not set, conversely in winter the continent is enveloped in constant darkness. Summer can be warm enough to allow outdoor activities, especially with the increasing temperatures (that are alarming in their numbers), yet this period only lasts for three months. For the rest of the year, the temperature is extremely low and outdoor activities can be very restricted. Researchers are forced to stay inside the station for long stretches of time, which induces feelings of confinement and stress. Reflecting on how architecture could act as a medium that allows for greater well-being (mental and physical) through spatial configurations, architects at Space Group analysed these local conditions and developed a design master plan for Jang Bogo ↗ p. 820, the Korean Antarctic station commissioned to counter the absence of the unique culture of South Korea in Antarctica.

Jang Bogo consists of a main building with research labs, bedrooms and living spaces, power plants, heavy equipment storage spaces, and independent labs. Typically, architects analyse a site's context for guidance, but without any urban context in sight, Antarctica's surface had to be regarded as a *tabula rasa*. The massing distribution of buildings was thus geared to ensuring ease of internal circulation, safety, construction feasibility, and mainte-

nance, while also relying on a deep understanding of the polar environment and a careful assessment of the prevailing wind direction, the skua zone, and, of course, the views.

After numerous design iterations, it was discovered that the minimisation of the building facility lines was guaranteed by positioning the buildings in a radial form. This concentric master-planning strategy had the additional advantage of allowing the research facilities to be independent of each other and oriented according to the wind direction. The compact design of the building pursues minimal surface area for energy efficiency and distinctive interior zoning for emergency cases, while its aerodynamic shape (with the innovative introduction of "dimples" in its outer cladding and seamless integrated form) allows the wind to flow over the structure, thereby mitigating the noise levels of the surrounding katabatic winds.

The form of "Samtaegeuk", or the "Triple Ultimate", recognizable in the shape of the station, represents the strongest symbol for harmony and balance among the Korean patterns and embodies the ultimate aim of the station, which is to harmonize with nature. The three resulting axes identify three different groups of facilities (Space Science, Earth Science, and Living & Maintenance), symbolising the trigram of the great ultimate, Heaven, Earth, and Human.

With the main building located at the centre, individual facilities (placed in a radial arrangement around the core, at a minimal distance from it) host research labs, bedroom areas, and living spaces. Promoting both independence and interconnectivity, these were planned as separate units and conceived as quiet places, at a distance from the invariably noisier common areas that include the kitchen, dining rooms, bath house, meeting areas, and station operation rooms.

Alongside the massing distribution, a variety of additional programmes and design decisions were introduced in the building to overcome and mitigate the stress of the researchers. Forced to live indoors for nine months per year, the life of the Antarctic crew revolves around a series of repeated activities, performed in indistinguishable living and working spaces. All of this is in addition to cold temperatures, low humidity, white nights, and no natural greenery in sight. Sensory stimulation therapies, scented matter and aromas, inner landscaping, natural lighting, and bath therapies were thus included within the building design to address such issues.

Stretching out in three directions, the station consists, in structural terms, of repetitive modules that were 100% prefabricated to overcome the challenges of building in Terra Nova Bay – where the conditions for construction are extremely difficult due to the presence of thick ice shelves that hinder access to the coastline and reduce the construction window to a maximum of sixty-five days per year. A modular construction system allowed for an accelerated construction process, a reduction in the number of on-site workers, and the minimisation of waste material. Each unit, prefabricated in Korea and assembled on-site, consisted of a modular structure with embedded inner finish, electric lines, ventilation ducts, and furniture. To improve the structural stability of the massing volumes, curved elements developed through Computational Fluid Dynamics (CFD) were introduced in the design and the outer surfaces became slanted.

Efficient energy management is the key issue in Antarctic stations. In order to achieve maximum efficiency, seasonal energy consumption patterns were carefully analysed and a strategy was chosen that saw a Combined Heat and Power (CHP) generation system as the main energy source of electrical power. The heat produced by the generator is used for heating and hot-water supply, while renewable energy sources, such as solar and wind, take up a minimum of 30% (96 kW) of the total capacity, minimising CO_2 emissions. While the double-layer walls of the building minimise energy loss by using state-of-the-art insulation, in the bedroom modules energy consumption is reduced by introducing shutting-out devices for electricity stand-by, installing LED light fixtures, and defining the orientation of the modules in such a way as to maximise the use of the all-day summer conditions (granting bedroom lighting an intensity up to 120,000 lux).

The concept of the dimpled surface of a golf ball was applied to the exterior skin of Jang Bogo Station. Compared to a smooth surface, dimple surfaces offer less resistance to wind and reduce the whirlpool phenomenon that might occur after the wind had blown away, thus preventing snow accumulation. Many iterations of dimple patterns were tested during the design process to solve the technical issues and make the building look more interesting by diversifying the size, interval, and densities of the dimples. By oversizing the dimple pattern, the openness ratio was maximised to let in as much natural light as possible, forcing a review of the window designs. Consequently, openings assumed a diamond shape, which was used throughout the station to ensure design coherence.

To make the station visible during a snow blizzard, Jan Bogo's cladding has a blue finish – a colour that was seen to have the twofold quality of maximising the aesthetics of the metal construction and harmonising with the surroundings, while standing out to ensure maximum safety.

"All the facilities are positioned in a radial shape with the main building at the center. All facilities are joined with the main building in a straight line and the map is simple."

Space Group on Jang Bogo Antarctic Research Centre, 2011

13
Geography of Science, "Domed Cities", and Mobile Architectures

September   October   November   Greenwich Meridian   December   January   February

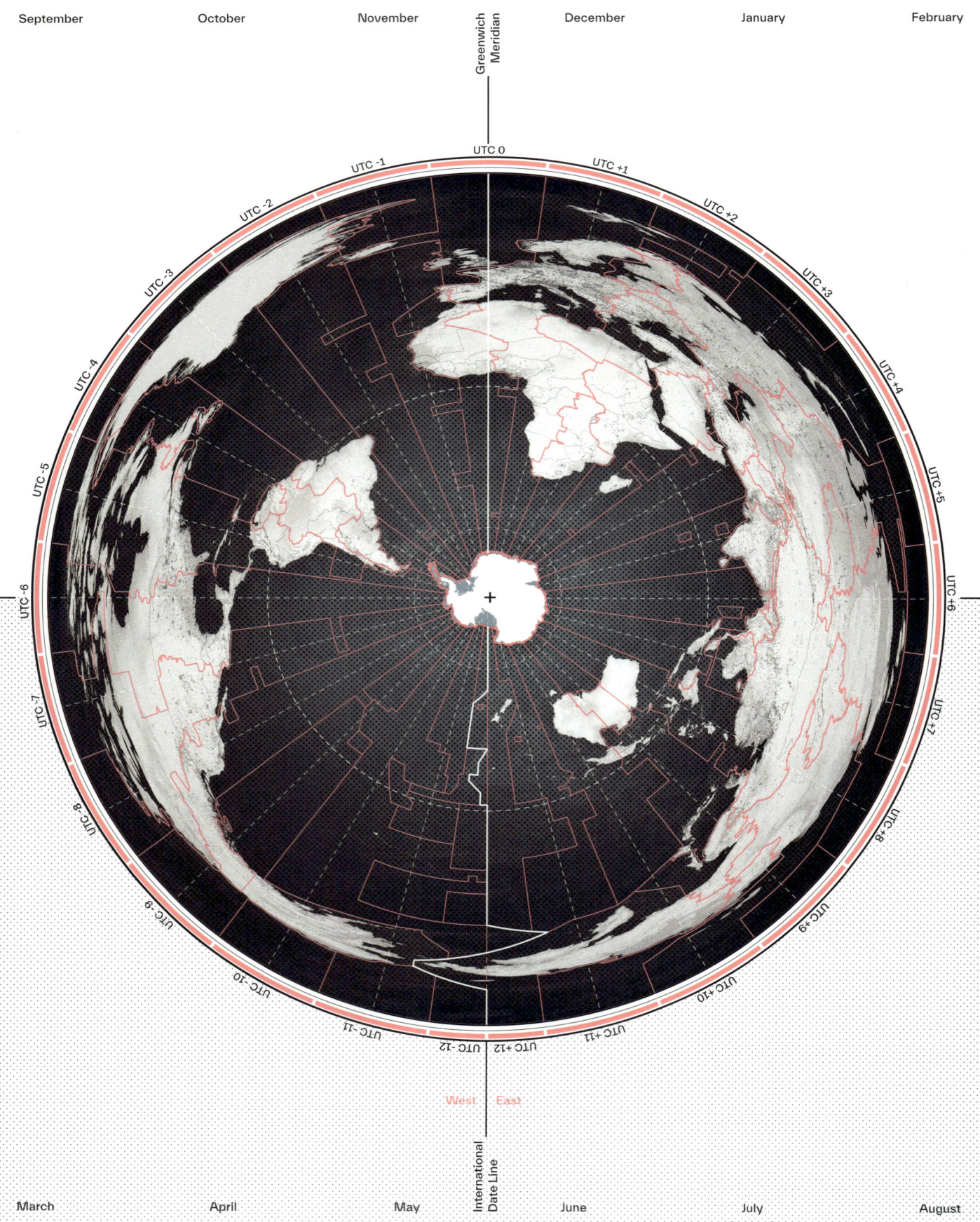

UTC 0
UTC -1
UTC -2
UTC -3
UTC -4
UTC -5
UTC -6
UTC -7
UTC -8
UTC -9
UTC -10
UTC -11
UTC -12
UTC +12
UTC +11
UTC +10
UTC +9
UTC +8
UTC +7
UTC +6
UTC +5
UTC +4
UTC +3
UTC +2
UTC +1

West   East

International Date Line

March   April   May   June   July   August

Abstract Master Plan

Giulia Foscari

GIULIA FOSCARI is an architect, researcher, and writer who has been practising in Europe, Asia, and the Americas. She is the founder of UNA, an architecture studio focused on cultural projects, and of its alter ego UNLESS, a not-for-profit agency for change devoted to interdisciplinary research on extreme environments. Foscari taught at Hong Kong University and at the Architectural Association, where she ran a Diploma Unit and founded the Polar Lab.

In the timeless, post-entropic landscape in which Smythson's crystalline sci-fi monument[1] landed, enacting a temporal collapse between prehistory and the industrial present,[2] the notion of a master plan traceable to ancient concepts of *cardo et decumanus maximus* is transformed beyond recognition.

Itself a monument to the immeasurability of time embodied by millions of years' worth of snow stratification that relentlessly archives data from Planet Earth's ecosystem, Antarctica rejects the two artificial notational systems that were conceived to "slice" the continent into temporal and geopolitical master-planned sectors: the conventional system of time zones and the twentieth-century sectoral map of Antarctic national claims.

Within the latter – opportunely "frozen in time" by the governance model introduced by the Antarctic Treaty System – embassy-like stations punctuate the seventh continent to reinforce latent geopolitical ambitions and enable scientists to conduct essential planetary research.

Ranging from single structures to proper Antarctic cities, the "Antarctic stations" have rarely been designed holistically up-front, and mostly grew organically, mirroring large suburban settlements which are far – both physically and conceptually – from what one might expect to find in the "pristine" territory of the white continent.

Only very few stations, amongst them the 1950s settlements of Little America V↗ p.654 and Ellsworth, for example, were truly master-planned. This was mostly out of necessity, since the structurally reinforced prefabricated standard units that were to be inhabited were installed on the Ross and Filchner ice shelves respectively and eventually buried under the snow. Consequently, it was of paramount importance for a master plan to be drawn up that determined the relational position of all buildings and for an appropriate system of enclosed pedestrian walkways to be designed. The two Antarctic stations retraced the model offered by Camp Century – built by the United States in Greenland in 1959 – which famously had an under-snow network of tunnels that extended or over 3 kilometres.

For master-planning to be successfully applied to stations above ground in Antarctica one would have to fast-forward to the late 1970s and analyse Philip Incoll's designs for Davis Station, in which the orientation and proximity of the buildings was informed by a thorough analysis of the prevailing winds and snow-drift accumulation,[3] paving the way for contemporary Antarctic design strategies.

Beyond these few examples, settlements in the Antarctic have mostly developed informally though time, constructing a rather dystopian anti-heroic landscape. This is the case, for example, with McMurdo Station↗ p.866, which institutionalises tropes that are familiar in suburban sprawl, offering a destabilising sense of permanence that is reinforced by the ever-expanding infrastructural systems that connect McMurdo's 105 buildings above ground.

The proposal developed by OZ Architects to condense the overall built footprint of the American station attempts to superimpose a master plan on the largest logistical hub in Antarctica in an operation that is paralleled (albeit at a different scale) by the neighbouring station, Scott Base↗ p.852, which is presently being redesigned by Hugh Broughton Architects. While the two stations share a sustainable wind farm and logistic infrastructure, it is evident that despite being both up for renewal at the same time, the United States and New Zealand have not considered the possibility of building a joint international station.

This is true almost everywhere in Antarctica. Despite the contiguity of many of the stations that punctuate the coastline, most nations prefer to build their own base and sustain substantial yearly logistical costs rather than join

The paradoxes of dispersed multinational cities in the Antarctic exemplified by the Fildes Peninsula are not restricted to the anthropological and social consequences of under-theorized inhabitation models in the extreme but extend to rather prosaic logistic issues like time zones. Accepted and adopted everywhere else on Planet Earth, the notion of time zones (shown in the diagram at the side) is dispensed with on a continent that is always either dark or light in favour of a rather arbitrary and surreal autonomy that allows each station to synchronise to a local time of its choice. This generally involves aligning the stations operational time zone with the local time of the country that manages the station itself, to the longitude on which it lies, or to the time zone of the countries that are in the closest proximity (and thus offer support) to the station in question. By virtue of this autonomy, for example, Australia's six stations operate according to their latitudes, while Russian stations synchronise with Moscow's time zone. In practice, this means, for instance, that Casey Station is three hours ahead of Mawson Station, even though they are only 3,000 kilometres apart on the coastline and both are Australian, while Vostok Station – which lies on the same longitude as Casey – uses Moscow time zone and is thus four hours ahead. The Amundsen-Scott South Pole Station is a special case. Although, theoretically the Earth's longitudes meet at the geographical South Pole, allowing the Americans to align themselves to all local times, the station observes New Zealand Standard Time to best coordinate with its supporting gateway country: New Zealand. Similarly, Palmer Station, also American, uses Chile Summer Time (CLST) simply because Chile is the closest gateway country to the continent.

forces with other countries, presumably to avoid any risk of lessening their geo-political power on the continent. An astounding example of this can be found in King George Island on the Fildes Peninsula, where stations established by four countries (Chile, China, Russia, and Uruguay) coexist in a symbiotic relation, each abiding to their own jurisdiction, synchronising to different time zones, and enacting a process of terraforming that challenges preconceived social and economic constructs promoting experimental forms of domesticity.

Shifting the attention further south, away from the coastline (where 84% of the active stations are located) and onto the Plateau, the density of buildings is much thinner, as is the oxygen level. On the featureless, almost scale-less land-scape, stations are isolated and self-reliant. This condition of absolute autonomy is best represented by the South Pole Station ↗ p.722. Designed and produced by Temcor of Torrance's co-founder and principal engineer Donald Richter – one of the 1,948 "disciples"[4] of Buckminster Fuller at Black Mountain College – the 50-metre-wide and 16-metre-high "polyframe dome" stood in Antarctica as tes-timony to Fuller's theory according to which "domed cities are going to be essen-tial to the occupation of the Arctic and the Antarctic."[5] Respectful of Fuller's "comprehensive anticipatory design science", which called for an "effective ap-plication of the principles of science to the conscious design of our total environ-ment in order to help make the Earth's finite resources meet the needs of all hu-manity without disrupting the ecological processes of the planet",[6] the polar geodesic dome contained eight buildings within an efficient interior environment. The project aimed to facilitate easy air circulation, guarantee a reduced loss of radiant heat, and maximise building material savings by virtue of its geometry. The latter was key, since all building components had to be flown to the South Pole in an LC-130 Hercules aircraft which had a maximum loading capacity of 9,000 kilograms.

Despite its scale and its utopian ambition of being a "city" rather than large-scale architecture, South Pole Station cannot be viewed as a true Antarc-tic master plan. Yet, the Dome (dismantled in 2008–2009) was located within what is the most interesting and unique example of polar master-planning – one that extends for thousands of kilometres describing (white on white) ab-stract forms on the snow which only Russian Avant Garde artists could have dared produce.

The "Quiet Sector", the "Dark Sector", the "Downwind Sector", and the "Clean Air Sector" all fall into the area defined by two intersecting semicircles with a radius of 20 and 150 kilometres, i.e. within the 26,344-square-kilometre site of the South Pole Antarctic Specially Managed Area (ASPA). Within the ASPA's "Scientific Zone" only regulated activities are allowed, as outlined in the twenty-one-page Management Plan set in place to enable "valuable scien-tific opportunities at the Pole, [while] protect[ing] the near-pristine environment and ensur[ing] that all activities […] can be conducted safely, environmentally responsibly and without disruption to scientific programs".[7] As urban zoning planning tools would do, the plan outlines the purpose of each sector whereby "the Clean Air Sector ensures a near-pristine air- and snow-sampling environ-ment for atmospheric and climate systems research; the Quiet Sector is an area where noise and equipment activities are limited to minimize vibration effects on seismological and other vibration-sensitive research; the Downwind Sector provides an area free from obstructions for balloon launches, aircraft operations, and other 'downwind' activities; [and] the Dark Sector aims to provide an area of reduced light pollution and low electromagnetic noise to help facilitate as-tronomy and astrophysical research."[8]

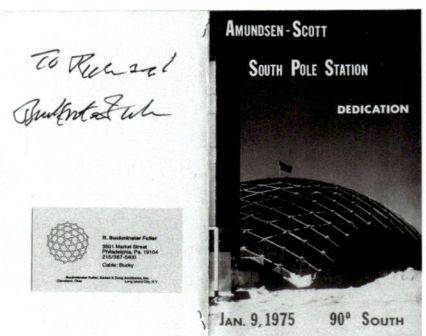

"Domed cities are going to be essential to the occupation of the Arctic and the Antarctic. The Russians are already experimenting with them in the Arctic. The Canadians are also studying them. Mining of the great resources of the Antarctic will require domed-over cities."

Richard Buckminster Fuller (Architect, Engineer, Inventor), *Utopia or Oblivion*, 1969

While the scientific master plan defines clear boundaries for the distinct research agendas, everything shifts relentlessly approximately 10 metres per year along with the ice sheet upon which it rests. To ensure a stable reference system (in contrast to the dynamic responsive GPS data adjustment of all features), a local grid bearing was established for the ASMA, zone, and sector boundaries. One particular staircase is central to this system. Deemed to be a permanent marker most readily recognisable in the landscape and on future maps, the circular aluminium tower staircase of the US Amundsen-Scott South Pole Station↗ p.790 provides the reference system for the most abstract and poetic master plan of the southern hemisphere.

Sublimating Smythson's much-sought-after entropy[9] whereby "time is still there [..] but it has lost its arrow",[10] the featureless landscape of the Antarctic plateau is almost impossible to navigate. In an environment prone to white-outs which obliterate any possibility of orienting oneself within a monochromatic, 360-degree light-refracted reality, in which even the compass arrow refuses to work due to the extreme perpendicular force of the Magnetic Pole, reaching the Dark, Quiet, and Clean sectors is an arduous task. Its master plan is invisible. No streets pave the way. The viability system, feebly demarcated by flags, offers only ropes – known to the Antarctic explorer as "hand lines" – to assist the scientists in making their way "home" to the station during blizzards and white-outs.

Orientation within the immense Antarctic tabula rasa is no easy task. To navigate between the Clean Air, Quiet, Downwind, and Dark Sectors, i.e. walking on a strata of ice 2,700 meters thick in a white and featureless landscape, scientists need markers. Flags and masts help one to find one's way on days with low visibility. During white-outs, when the light refraction renders the horizon line indiscernible, shadows are no longer cast, "visibility is pretty much zero, [and] you might not be able to see your hand in front of your own face".[1] Walking in such overwhelming whiteness is like walking with a bucket on your head; for this reason scientists deployed in Antarctic are trained to navigate holding onto ropes and perform a "bucket test", as captured in Werner Herzog's documentary *Encounters at the End of the World*.

Shaun O'Boyle, "Polar Plateau", South Pole Station, Antarctica

Robert Smithson, Proposal for a Monument
at Antarctica, 1966

Anne Noble, "The Quiet Sector", South Pole Station,
Antarctica, 2008

"For a few seconds we stood over the spot where Amundsen had stood, December 14, 1911 […]. There was nothing now to mark that scene: only a white desolation and solitude disturbed by the sound of our engines. The Pole lay in the centre of a limitless plain. […] And that, in brief, is all there is to tell about the South Pole. One gets there, and that is about all there is for the telling. It is the effort to get there that counts."

Richard Evelyn Byrd (Admiral, Explorer, Aviator),
Exploring with Byrd: Episodes of an Adventurous Life,
1938

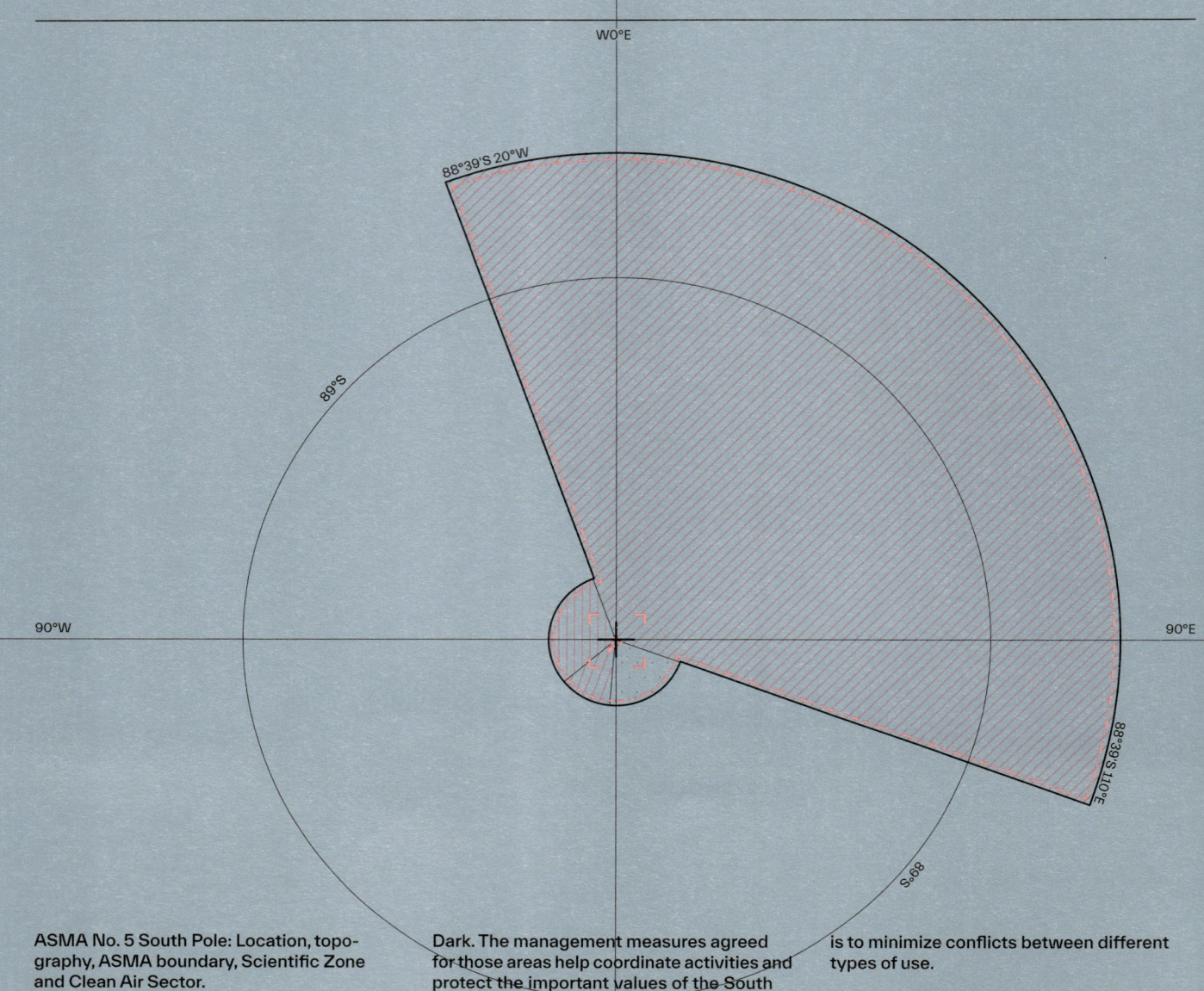

W0°E

88°39'S 20°W

89°S

90°W

90°E

88°39'S 110°E

89°S

ASMA No. 5 South Pole: Location, topography, ASMA boundary, Scientific Zone and Clean Air Sector.

"An area of ~26,344 square kilometres around the South Pole Station is designated as an Antarctic Specially Managed Area (hereafter referred to as 'the Area'). The Area has been designated in order to maximize the valuable scientific opportunities at the Pole, protect the near-pristine environment and ensure that all activities, including those to experience the extraordinary qualities of the South Pole, can be conducted safely, environmentally responsibly and without disruption to scientific programs. In order to help achieve the objectives of the Management Plan, the Area has been divided into Scientific, Operations, and Restricted zones. The Scientific Zone is further divided into four sectors: Clean Air, Quiet, Downwind and

Dark. The management measures agreed for those areas help coordinate activities and protect the important values of the South Pole. [...] This management plan establishes three types of zones within the Area: Operations, Scientific, and Restricted. [...]

Operations Zone: To ensure that science support facilities and related human activities within the Area are contained and managed within a designated area.

Scientific: To ensure those planning science or logistics within the Area, and all visitors to the Area, are aware of sites of current or long-term scientific investigation that may be sensitive to disturbance or have sensitive scientific equipment installed, so these may be taken into account during the planning and conduct of activities within the Area. A particular objective of the Scientific Zone

is to minimize conflicts between different types of use.

Restricted: To restrict access into a particular part of the Area and/or activities within it for a range of reasons, e.g. owing to special scientific values, because of sensitivity, presence of hazards, or to restrict emissions or constructions at a particular site. Access into Restricted Zones should normally be for compelling reasons that cannot be served elsewhere within the Area.

The Operations Zone has been established to contain primary human activity in the Area, including science support activities, main station services (e.g. living facilities), ski-way operations, and on-ground support facilities for Non-Governmental Visitors (NGVs)."[1]

W180°E

S

SOURCE
United States Antarctic Program

1 : 200,000,000   0   12.5   25   50 km

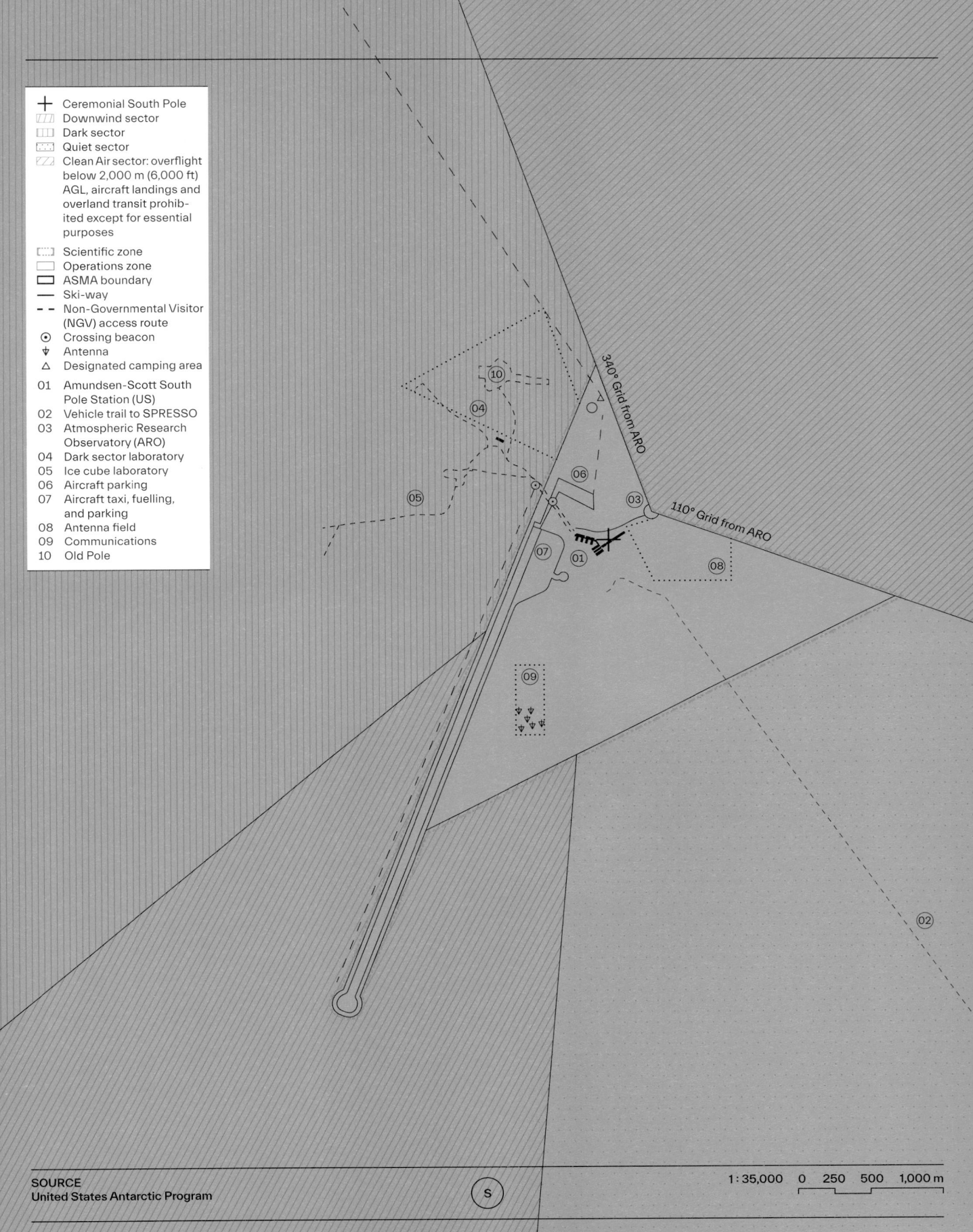

Legend:

+ Ceremonial South Pole
Downwind sector
Dark sector
Quiet sector
Clean Air sector: overflight below 2,000 m (6,000 ft) AGL, aircraft landings and overland transit prohibited except for essential purposes
Scientific zone
Operations zone
ASMA boundary
— Ski-way
- - Non-Governmental Visitor (NGV) access route
⊙ Crossing beacon
Ⴤ Antenna
△ Designated camping area

01 Amundsen-Scott South Pole Station (US)
02 Vehicle trail to SPRESSO
03 Atmospheric Research Observatory (ARO)
04 Dark sector laboratory
05 Ice cube laboratory
06 Aircraft parking
07 Aircraft taxi, fuelling, and parking
08 Antenna field
09 Communications
10 Old Pole

340° Grid from ARO
110° Grid from ARO

1 : 35,000   0   250   500   1,000 m

The Amundsen–Scott South Pole Station

Joe Ferraro

JOE FERRARO is an architect and a founding partner of Ferraro Choi and Associates Ltd (United States). He has been a lead designer of buildings and infrastructure in Antarctica since 1984. In addition to the Amundsen-Scott South Pole Station, Ferraro was the architect for the Albert P. Crary Science and Engineering Center at McMurdo Station and is currently the principal in charge of the design of the stations' Information Technology and Communications Primary Facility.

By 1991, the United States Amundsen–Scott Station ↗ p. 722 had reached the end of its design life span and the National Science Foundation (NSF) solicited architectural and engineering services for a major redevelopment of the station based on the following conceptual design directives: there had to be minimal environmental impact during construction and station operational life; the station should embed flexibility to adapt to unforeseen future scientific requirements; habitation and life safety facilities were to be improved; and, emerging technologies from the NSF/NASA Antarctica Space Analog programme should be incorporated in the design.

In 1992, Ferraro Choi and Associates Ltd. was selected to initiate architectural programming and engineering studies including all the building engineering as well as ice mechanics, waste management, and the analysis and design of alternative power systems. Their consultant, RWDI, was to provide snow studies that could guide the design team in defining the building in such way as to minimise snowdrift accumulation on and around the buildings. Since snowmelt does not occur at the site, snow accumulation has a dynamic impact on the station, diverting valuable energy and causing serious egress and structural problems.

Direct engagement with South Pole Station operations management staff and NSF representatives offered some initial inspiration for the design. This included recommen-

dations by the Director of NSF's Division of Polar Programs, Peter Wilkness, on practical brief requirements (such as the need for an enclosed area of approximately 12,952 square metres) and on the general design approach that should result in a station "so environmentally benign, that if it were lifted off the polar plateau, no trace of it would be found".

Modular elements were organised in a linear format perpendicular to the prevailing polar wind to allow the wind to scour the area below the buildings and to place a resulting drift in their leeward wind shadow. The drift was intuitively sized with a maximum height equal to the buildings' height above the snow surface (10 metres), tapering leeward for approximately thirty-five times the width of the buildings (1,200 metres). Above-surface buildings were also sited so that this permanent drift would not curtail on-surface functions or inundate existing below-surface buildings and therefore shorten their serviceable life. Conceived as a modular design capable of being periodically raised to extend their useful life and incrementally jacked to adjust for differential settlement, the station buildings' size, alignment, and shape were predicated on water flume and computer-based snowdrift studies conducted by RWDI.

Combined with the necessity for a phased design and construction approach, these prerequisites resulted in a linear, piano-key-like design plan of two-storey buildings connected to a segmented spine. This

approach allowed utilities to be distributed along the spine as new modules were constructed and brought into service and disconnected when building sections were raised. The design also allowed for some modules to be closed down to sub-operating temperatures to save energy during the reduced operations and staffing of the winter season.

A double-loaded corridor with the three upwind modules acting as an airfoil-shaped leading edge was conceived to accelerate wind speeds below the building. As they have views of the South Pole monument and arriving planes, the upwind buildings contain public, administrative, and research functions, while downwind buildings host the station's power plants as well as personnel berthing and recreational facilities. One of the building modules was designed specifically to cater for the needs of the reduced winter crew of fifty people and contains a separate backup power plant, galley, and emergency service connections.

The design intent was to construct a station that would be completely above surface and would measure 240 metres in length. However, funding, scheduling, and logistical considerations led to the establishment of a direct connection between the new building and its predecessor by extending the life span of the station areas for infrastructure located below surface and habitable spaces positioned above surface. Determining the optimum location for the new station required a quadrant analysis of the existing station complex rotated about the South Pole Station ↗ p. 790. Grounded in the station's distinct areas of scientific study and operation – which is articulated in four separate scientific study sectors, separated from each other and from station operations – the plan restricts unwanted activities that could disrupt, contaminate, or preclude ongoing scientific research projects.

The Amundsen-Scott South Pole Station is elevated on the Antarctic plateau. Opposite the station, the Ceremonial South Pole – consisting of a metal reflective sphere positioned over a barber's pole surrounded by national flags – stands some metres away from the actual geographic South Pole marker. The latter is redesigned each year and repositioned by the "Polies" on New Year's Eve to reflect the adjusted coordinates on the moving ice sheet.

"'The berms' in 'Polie' patois refers to the entire storage area on the near perimeter of the central encampment consisting of the Dome (now dismantled), new pole station, and outbuildings. An extensive sprawl of plywood platform covered with all sorts of off-loaded cargo forms an almost city-like network of streets. The berm area is like a warehouse without sides or a roof. […] If there can be a 'backside' to a circum-located pole, the berms are it. […] Lê's South Pole is a combination of (post) industrial office park and (de)militarized base. The support of science requires an infrastructure whose implications [secure] Antarctica within a map of globalization and capital flows at odds with depictions of the ice as pure, empty, or even as heroic and sublime."

Elena Glasberg (Professor at New York University, Essayist, Speaker), *Antarctica as Cultural Critique: The Gendered Politics of Science Exploration & Climate Change*, 2012

A Supra-regional Dispersed City. The Fildes Peninsula

Arturo Lyon

ARTURO LYON is the director of the Polar Lab Chile. He is an assistant professor at the School of Architecture of the Pontifical Catholic University of Chile. Lyon is the founder of Lyon Bosch Martic, where he developed architecture projects including the design of Las Majadas Hotel, the Chena Public Park, and the winning proposal for Nueva Alameda Providencia.

Located in the southern region of King George Island, the Fildes Peninsula is part of the archipelago of the South Shetland Islands. Due to its proximity to the American continent, the area is characterised by milder climatic conditions compared to the rest of Antarctica, making it one of the largest areas of ice-free ground in the polar region, with a high concentration of biodiversity. The retreat of the Collins Glacier, which formerly completely covered the Peninsula, in fact brought to light fossil-rich areas and a diffuse hydric system with more than sixty ponds.[1]

Due to the favourable conditions and the vicinity to the gateway cities of Punta Arenas and Ushuaia (1,240 kilometres), over the last fifty years the Fildes Peninsula has been the catalyst for the emergence of a supra-regional logistic hub. With six research stations, over 187 buildings, an aerodrome, a 13.4-kilometre road network, and a total population of three hundred people during the summer and one hundred during the winter, this is arguably one of the most urbanised areas in the whole white continent.[2]

Anthropic activities in the Fildes Peninsula started in 1968 with the construction of Bellingshausen, the Soviet station. One year later, to relocate the meteorological station of Pedro Aguirre Cerda Station which had been destroyed in 1967 by a volcanic eruption in Deception Island, the Chilean Air Force built President Frei Montalva Aerial Base ↗ p. 746 (also known as Frei Station) some 200 metres away from Bellingshausen. Alongside Frei Station, in 1984 Chile built the Teniente Rodolfo Marsh Martin Airport (with a 1,292-metre-long landing strip) and founded Villa Las Estrellas. One of only two civilian settlements in Antarctica (alongside Argentina's Esperanza Station ↗ p. 734), Villa Las Estrellas is a year-round inhabitation complex hosting families, which includes facilities such as an elementary school, a church, a supermarket, and a bank, as well as other public buildings. Occupied by families continuously from 1984 to 2018, the settlement now hosts researchers, civil and military personnel dedicated to the operation and logistics of the base.

Fildes's infrastructure was later complemented by the Artigas Station built by Uruguay in 1984, the Great Wall Station erected by China in 1985, and

> "King George Island pulls together over 20 nations into a small space, creating a working polar equivalent of utopian global fantasies typically only realised at World's Fairs and Disney's Epcot."
>
> Juan Francisco Salazar (Professor, Western Sydney University) and Jessica O'Reilly (Professor, Indiana University), *Inhabiting the Antarctic,* 2017

A careful analysis of the Fildes Peninsula master plan reveals a unique multinational city which revolves around the shared infrastructure of the Teniente Rodolfo Marsh Martin Airport. Operating approximately 2,000 flights per year, the Chilean airstrip is the only Antarctic airport (of the twenty airports that punctuate the continent) which has an IATA code and serves public flights for civilian and military to and from South America. All in close proximity to one another, the Argentine, Chilean, Chinese, and Russian stations share infrastructure, but not a jurisdiction and time zone, which are rigorously aligned to their respective countries of origin.

Stations
Artiga (UY)
Bellingshausen (RU)
Presidente Eduardo Frei Montalva (CL)
Professor Julio Escudero (CL)
Ripamonti Refuge (CL)
Great Wall (CN)

Teniente R. Marsh Airport (CL)

Historic Sites and Monuments (HSM)
❶ HSM 82
❷ HSM 52
❸ HSM 86
❹ HSM 50

Antarctic Specially Protected Area (ASPA)
❶ ASPA 125
❷ ASPA 150

▪ Building
▨ Landing strip
— Road infrastructure
⋯ Glacier
▥ Lagoon
— River
▨ Sea

COLLINS GLACIER

MAXWELL BAY

KING GEORGE ISLAND
FILDES PENINSULA

UY
Artigas Station

RU
Bellingshausen Station

ARDLEY
ISLAND

CL
Ripamonti
Refuge

Pier

Pier

CL
Teniente R. Marsh
Airport

CL
Professor Julio Escudero
Station

CL
Presidente Eduardo Frei
Montalva Station

CN
Great Wall Station

FILDES
STRAIT

NELSON
ISLAND

A Supra-regional Dispersed City. The Fildes Peninsula

446

the Antarctic Maritime Station constructed by Chile in 1987 (later destroyed by fire in 2018). In 1994, Chile completed its facilities by adding the Professor Julio Escudero Scientific Research Base, and some refuges in strategic outposts on the Peninsula.[3]

The developed infrastructure of the Fildes Peninsula, and the short 2.5-hour flight from South America, transformed this multinational cluster into a key stepping stone for accessing deeper areas in the continent. As underlined by the different national Antarctic programmes, the greatest deterrent for the human presence in Antarctica is the harsh climatic conditions of the region and, perhaps even more significantly, the extreme isolation of large parts of the continent and the long distances between consolidated infrastructures. From this perspective, Fildes (together with other supe super-regional rregional hubs such as McMurdo ↗ p. 866) will continue to play a critical role for human mobility in Antarctica over the next decades, acquiring increasing relevance proportional to the planned future explorations of the continent.

Due to its high population density, this infrastructural hub presents urgent challenges in terms of its compliance with the Antarctic Treaty System regulation on minimising the human footprint and safeguarding the fragile peninsular ecosystem; pressing issues such as the disturbance mitigation of flora and fauna, waste management, the use of drinking-water sources, waste-water treatment, oil contamination, noise and gaseous emissions, air and ship traffic, and the management of scientific and touristic activities must be urgently addressed and special attention must be given to the dedicated Antarctic Specially Protected Area – ASPA N.125 – which was listed to protect the palaeontological heritage present in 1.8 square kilometres of the Peninsula.[4]

The above is exacerbated by the pressure exerted by the expansion of the infrastructural system to service the constantly growing international logistic operations. The scale and relevance of the multinational infrastructure hubs, in compliance with stricter environmental protection policies, are questioning the current settlement model based on the accumulation of independent buildings and dispersed infrastructures. New typologies are being tested to prevent further spreading of the current informal dispersed structures, in favour of the construction of larger, multifunctional, and interconnected buildings that can achieve higher densities and better efficiency, while reducing the building footprint on Antarctic soil. This new tendency can be seen at McMurdo station, as discussed elsewhere in this publication, and at Frei.

The shift from dispersed settlements towards more compact and multifunctional complexes calls for a debate on the future of Antarctica that can guarantee innovative models for true international collaboration and foster a healthier relationship between infrastructure and landscape in lieu of the current informal utilitarian solutions that display the absence of strategic territorial planning. This relationship will be reassessed as large parts of the existing settlements are dismantled in the near future. In this context, the Fildes Peninsula should be regarded as a prototypical territory for growing anthropic occupation in Antarctica; a multinational city situated in a unique landscape that emerged from the retreat of the ice, where new modes of inhabitation and collaboration can be explored.

"The Fildes Peninsula […] can be described as a juxtaposition or meshwork of lively geographies: a territorial geography of networked national field stations; a non-human geography of bio-geophysical things, entities, processes, life-forms, events and phenomena; a material geography of international logistical cooperation; labour geographies of daily scientific practices and logistic personnel involved in field science support; and the leisure geographies of international tourists."

Juan Francisco Salazar (Professor, Western Sydney University), "Mediating Antarctica in Digital Culture: Politics of Representation and Visualisation in Art and Science", in *Handbook on the Politics of Antarctica*, 2017

Situated alongside Presidente Eduardo Frei Montalva military base, Villa Las Estrellas is one of two civilian settlements in Antarctica – the second being Base Esperanza. The Chilean town is designed to accommodate up to 150 civilians during the summer months, all of whom are required to remove their appendix[1] before moving to the continent.

Programmatically, the settlement includes a primary school (designed for approximately 15 students per year), a bank manned by a single banker that operates throughout the year, an air force hospital, a gymnasium, a Catholic chapel (Santa Maria de la Paz), a supermarket, and a souvenir shop.

Domesticity in the Extreme

Giulia Foscari

GIULIA FOSCARI is an architect, researcher, and writer who has been practising in Europe, Asia, and the Americas. She is the founder of UNA, an architecture studio focused on cultural projects, and of its alter ego UNLESS, a not-for-profit agency for change devoted to interdisciplinary research on extreme environments. Foscari taught at Hong Kong University and at the Architectural Association, where she ran a Diploma Unit and founded the Polar Lab.

In November 1977, Silvia Morella de Palma, then seven months pregnant, was airlifted along with her three children from their home in Argentina to join her husband, Captain Jorge de Palma, at Base Esperanza [p. 734] – a remote station located on the tip of the Antarctic Peninsula's Hope Bay. Two months later, on the 7th of January 1978, Emilio Marcos de Palma became the first person in history to be born in Antarctica. The Palmas were sent to Antarctica in a move orchestrated by the Argentinian government, spurred notably in response to Chilean President Augusto Pinochet's visit to Antarctica earlier that year, which marked his country's claims to dominance in the region. Between 1978 and 1983, eight Antarctic births were recorded – with ten families being sent to Base Esperanza for an annual residency to offer evidence of Argentina's colonization ambitions in the continent.[1] As soon as Emilio was born, a civil registry was set up at Esperanza, not dissimilar to that of a traditional village, to record his birth and those that followed,[2] as well as marriages. In a conscious effort to assert national claims, civilian presence in Base Esperanza, including families with children, has been continuous ever since.

The unbuilt precursor of Base Esperanza was conceived within the context of the geopolitically driven Antarctic settlement policy introduced in Argentina in the 1950s. At the time, General Hernán Pujato, founder and first director of the Argentine Antarctic Institute (IAA), promoted the idea of establishing a domestic settlement in the continent, launching a forestation project, introducing foreign mammal species (namely the husky) by building kennels, and encouraging production activities that included harvesting algae and whaling. The 1955 Argentine coup d'état and the subsequent signing of the Antarctic Treaty in 1959 interrupted General Hernán Pujato's project but Base Esperanza later embodied a number of its aims. Base Esperanza currently houses fifty-five overwintering inhabitants including ten families and two schoolteachers, who teach at its provincial school #38, *Presidente Raúl Ricardo Alfonsín,* which was inaugurated in 1978 – the year Emilio was born. Alongside the school, a church, a post office, and a radio station were constructed to allow the transplanted families to re-establish a somewhat normal existence in such remote territory, which is reachable only for a few months of the year by sea or by plane landing on the *Puerto Moro* landing platform. The name originally given to the base on the occasion of its inauguration (Fortín Sargento Cabral) was chosen in memory of the army's nineteenth-century process of colonising the deserted Patagonia region – a land that, conceptually, could be related to the "white desert" of Antarctica. This name highlights the geopolitical implications of promoting domestic settlements on the continent.

While the relevance of scientific findings produced in Hope Bay (which are limited to monitoring a large Adélie penguin rookery and conducting research on marine biology studies, palaeobotany, seismology, and meteorology) may raise questions regarding the station's true scientific ambitions, Base Esperanza's domestic nature certainly contributes to anthropological studies by offering insights into the medical and psychological consequences of growing up and residing in a remote territory.

Esperanza's modest wooden homes, school, and chapels may appear as small, quiet structures when compared with their experimental counterparts such as Hugh Broughton Architect's Halley VI or the Ice Cube Neutrino Observatory – renowned examples of Antarctic Polar science architecture – but they send ripples of wider statements about national claims and vested interests as examples not only of *dwelling* but *settling* on a continent that has no indigenous population of its own. Considered by some to be the last phase of human expansion on our planet, this small Antarctic town could indeed represent domestic architecture's key role in inhabiting other disputed, unclaimed, or extra-terrestrial frontiers.

"Permanencia, un acto de sacrificio."

Motto of Base Esperanza

An Unbuilt City in Antarctica

Claudio Williams

CLAUDIO WILLIAMS worked from a very young age with his father, Argentine architect Amancio Williams, until the end of his days. He faithfully directed, preserved and shared the work of the Williams Archive until 2020, when it was donated to the Canadian Centre for Architecture (CCA). He is the author of *Amancio Williams* (Gaglianone, 1990). Williams is also the owner of the company Lece Construcciones S.A. (Argentina).

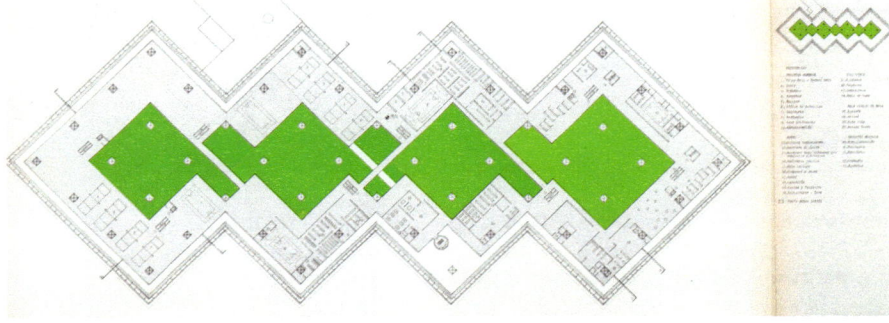

In 1980 the Argentinian Army's Antarctic Command launched an initiative to build a new project on the white continent. Until meeting Amancio Williams, the Antarctic Command Chief had no idea about the possibilities that would be presented to him and imagined a conventional city adapted to Antarctica's unique climate.

During that time, Argentina had several bases scattered throughout Antarctica. A large number of them were only operational during the summer, while others operated throughout the year. Amancio Williams's city was to be built in the Antarctic Peninsula, near the Esperanza and Marambio bases, which conveniently had a suitable runway for technical flights and eventually for future commercial flights. Transfers from Marambio to the new city would be by helicopter.

The project presented many challenges. One of these was Antarctica's lack of fresh water, and another was the difficulty of building on permafrost ground – a dynamic surface that cyclically freezes and thaws. Near Esperanza there is a freshwater lagoon that would have been the water source for the new city, though it was necessary to have a large reserve.

"The First City in Antarctica" was articulated in four prisms that were approximately 120 metres in length and width, to be constructed in phases. These modules were composed of two main parts. The half-buried base, which was to be constructed in concrete during the summer months, consisted of three levels and was 14 metres in height, whilst the lower part hosted a large freshwater reservoir, which would also serve as a fire extinguishing tank and an insulator against the permafrost ground. The city's mechanical equipment and warehouses were located in the upper levels.

Structurally, the main body of the city station was to be supported by the concrete base upon which a 22-metre metallic structure (manufactured and pre-assembled in Comodoro Rivadavia, Argentina) would support four levels, with additional suspended mezzanine floors.

A programmatic mix was addressed by including a sports and community infrastructure on the lower level. Here a 50 by 50-metre central flower bed was designed, which would allow natural plants to grow under illuminating systems that facilitate photosynthesis even during the permanently dark winter season. Laboratories and equipment for scientific activities were to be located on the higher levels, and the accommodation for the city's permanent population and potential tourists was placed on the highest floors. The city was crowned by a communications tower designed by Leonhardt, Andrä und Partners, from Stuttgart, Germany.

With four constructed modules, the city was to have a total capacity of nearly 2,000 inhabitants, of whom approximately 1,000 would form a permanent population dedicated to different scientific and technical disciplines, while the other 1,000 were assumed to be tourists, visitors, and participants in business conventions. "The permanent population [would] consist mostly of young married couples – mainly scientists and technicians – with young children who will go to daycare centers and primary schools. It [was seen to] be convenient if both men and women can perform professional duties to keep them active and simultaneously contribute to a smooth functioning of the city."

Amancio William's proposal thus advocated "a city created by its integration with the architecture and not by overlapping buildings", "a closed city that will have the same general conditions as in the layout of modern cities: Linear development, spatial architecture, [and] open spaces". Yet the city was also charged with the implicit sociopolitical aspiration to foster greater collaboration between nations – an issue that is still unresolved not only in inhabited continents but also in the "white desert", where national assertions latently survive. Amancio Williams "personally insist[ed …] on the need for a real union between Argentina and Chile. A union without hegemony. A union of brothers." He thought "that the way to proceed to reach a real union is neither through declamatory attitudes nor good wishes. [… The] union [could] be achieved by relying on common work between Argentinians and Chileans," and Antarctica's "First City" could have been the first testing ground for a new South American reality.

The "Informal Settlements" of Antarctica

Juan Du

JUAN DU is the director of the Polar Lab Hong Kong. She is an associate professor and associate dean of architecture at the University of Hong Kong, where she directs the Urban Ecologies Design Laboratory. The projects within her practice, IDU, range from community centres to informal settlements upgrades. Prof. Du's extensive writings include *The Shenzhen Experiment* (Harvard University Press, 2020).

Within the first few minutes of his documentary *Encounters at the End of the World,* Werner Herzog offered his initial impression of "Antarctica's largest settlement": "Of course I did not expect pristine landscapes and man living in blissful harmony with fluffy penguins, but I was still surprised to find McMurdo↗ p. 866 looking like an ugly mining town." Indeed, with over a hundred buildings supporting a population up to 1,200, McMurdo Station's spatial appearance and scale are much more urban than what one expects to find at the most remote place on Earth.

While Herzog's disappointment may have been influenced by the filmmaker's aesthetic quest, a 2003 review report commissioned by the US National Science Foundation (NSF) echoed the description of McMurdo Station as having a "remote outpost" feel that combines "aspects of a mining town, military base, and college campus" and has developed in a "haphazard manner."[1] In fact, most Antarctic research stations, while much smaller than McMurdo, also exhibit similar characteristics. However, to simply discount these visual and spatial qualities as unsightly and problematic would be a missed opportunity to examine the informal nature and dynamic processes of the research stations, as well as that of Antarctic spatial organisation in general.

Over the past decade, the established view of informal settlements around the world as problematic slums solely defined by deficiencies and poverty has been called into question.[2] International organisations, such as UN-Habitat, have started to recognise the complexities, and at times merits of informal processes and communities: "The 'informal' cannot be equated with 'illegal', since the 'informal' may be the only possible way of urbanisation in cities where no other options are available; the 'informal' cannot be equated with 'inferior' or 'marginal', since there are many examples where informal communities are superior to dysfunctional public housing projects. The 'informal' cannot be equated with 'poor', since middle-class families (even millionaires) can be found in non-formal neighbourhoods."[3] In certain aspects, the development of research stations in Antarctica exhibits many similar characteristics of informal settlements, from contested territorial and land ownership to the incremental, or "haphazard", nature of spatial growth; impermanence of built structures as well as formations of close-knit localised communities and collaborative operations.

The first century of human presence in the Antarctic was marked by resource extraction, by sealing and whaling industries, land sovereignty contestations, and the race to claim ownership over the Earth's last frontier. Many of the earliest physical stations established by the claimant states were motivated by their respective territorial claims. The 1957–1959 International Geophysical Year arose as an international effort to address, amongst other things, land rights conflicts and resulted in the development of the 1959 Antarctic Treaty. In this way, the first formal Antarctic governance planning could be viewed as a reaction to the informal actions of the various individual states. While the Treaty set regulations for the twelve signatory nations and suspended all territorial claims, it also incidentally encouraged the construction of more research stations by including a clause which states that conducting "substantial research activities" grants voting rights for Antarctica's collective governance.

Through this framework, the physical occupation of built structures placed on the international "free" land still symbolises a "temporary" territorial claim that always hints at the possibilities of a potential future revision of the Antarctic Treaty System (ATS), as of 2048. The very existence of the physical facilities, including approximately seventy active permanent research stations, could be viewed as an informal response to the formalised continental land regulation. While the first achieved goal of the ATS was to suspend territorial

"Principal port of entry, main air staging base, nerve centre, [...] no wonder that Naval Air Facility McMurdo has been called 'The New York of Antarctica.' Strictly speaking McMurdo is not an IGY [International Geophysical Year] station as it had no resident scientist aboard. But without it, there would be no IGY station at any point in Antarctica save possibly the granddaddy of them all, Little America. McMurdo is an air base and support center extarodinaire, communications center, and meteorological center with four or five support aerographers on full-time duty. To sum it up, 'McMudhole' is the air hub of the Antarctic universe."

United States Navy, *Operation Deep Freeze Task Force*, 1958

Ann My Lê's "Fuel Storage McMurdo" (below), taken from Observation Hill, and the images on the left by Shaun O'Boyle (focusing on the FEMC Trade shop, the Skua lost items storage, and the Dorm buildings) and Todd Anderson (bottom) offer a view of the "anti-heroic landscape"[1] of McMurdo.

contestation, the tension is always there, suspended. In some ways, one could argue that the back and forth between formal (international) regulations and informal (national/local) responses renders the ATS a dynamic governance system that keeps the constant vested interests of the now fifty-three supporting nations.

While the Antarctic Treaty created a continental framework for the stations' main purpose of international cooperation and scientific discovery, there were no clear planning or building regulations on how individual stations should be physically developed or operated. Each operating country had its own rules on spatial planning and building construction. Even for larger stations such as McMurdo, there was no effective master plan, and buildings were added, depending on needs and budget, from year to year.[4] However, while each station had a great deal of freedom in terms of organic growth, there were environmental consequences caused by unregulated activities, such as refuse incineration and sewage dumping. The 1991 Protocol on Environmental Protection to the Antarctic Treaty, formalised in 1998, aimed to regulate environmental protection by requiring all activities, including building construction, to undergo varying levels of Environmental Impact Assessment (EIA).[5] By that time, research stations such as McMurdo and the adjacent Scott Base ↗ p. 678, having been in operation with continuous expansion for twenty to thirty years, took efforts to remediate their buildings and facilities. The relative smaller scale and individualised nature of buildings and facilities added over time allowed for incremental upgrades and adaptive reuse.[6]

In informal settlements throughout the world, formal governance and informal processes often make up for each other's blind spots and deficiencies. Participation, trust, identity – some of the most important factors for long-term sustainability and resilience – are not results of regulatory governance but rather a product of informal activities, of open engagements and even contentious debates. While the Antarctic Treaty System established the overall reason for human presence as scientific research, international cooperation in science and environmental protection, national interests will always test the effectiveness of the ATS governance framework and stated goals. The requirement for consensus on the Treaty's environmental goals is achieved as much through policy setting as informally negotiated compromises and public perceptions.[7] Recognising and anticipating the interplay between formal regulatory planning and informal special interests could provide additional tools to further the goals of inclusivity and collaboration, as well as the long-term environmental well-being of the continent.

"As one might imagine, there are problems with adapting a transient military facility into a permanent scientific research base."

Alexis Madrigal (Journalist), "Designing a Better Antarctic Base for Science", *The Atlantic*, 2017

"Job definitions were loose. The engineer held yoga classes. Field guides turned into ski instructors on the slope behind the runway. A brisk tour of the station was led by the doctor, who also ran the post office. She later presented us with a fake foam bottom to teach us how to inject someone with morphine, should the need arise."

Pilita Clark (Business Columnist), "Inside Antarctica: The Continent Whose Fate Will Affect Millions", *The Financial Times*, 2018

It is hard to fathom how local species, such as the lonely penguin shown on the right, relate to what was described in Werner Herzog's 2007 documentary *Encounters at the End of the World* as "an ugly mining town, filled with Caterpillars and noisy construction sites", namely McMurdo. The American station – the largest logistics hub for scientific research on the continent – has grown organically since its inception without a master plan. The original thirty-seven structures known to have existed during the IGY in 1957–1958, have since risen to the astounding number of 116, spanning from "temporary" Nissen huts to three-storey-high accommodation buildings. Serviced by an intricate overground infrastructure system developed out of the impossibility of digging within the permafrost ground, McMurdo is difficult to navigate and pedestrians are forced to use purpose-built bridges to pass over the extensive exposed pipeline network captured on the left by Shaun O'Boyle.

Lessons in Temporal Permanence. McMurdo Station

Rick Petersen

RICK PETERSEN is an architect who is a partner at OZ Architecture (United States). Petersen led the master planning of McMurdo and Palmer Stations, as well as the building design of McMurdo Station under the Antarctic Infrastructure and Modernization for Science project, an initiative of the National Science Foundation (NSF)'s United States Antarctic Program.

As the oldest continually inhabited community on the Antarctic continent, McMurdo Station offers, paradoxically, a fascinating study of temporality and permanence. Located on volcanic Ross Island, McMurdo Station ↗ p. 866 fronts the relatively protected Winter Quarters Bay. Early explorers found this location – the closest to the South Pole to feature bare ground – to be an optimal point to leave their ships and begin their overland journeys.

One of these early explorers, Robert Falcon Scott, erected the first structure at the site of McMurdo Station, a simple wooden shelter at Hut Point ↗ p. 618 to support the British National Antarctic Expedition of 1901–1904. In order to minimise on-site labour in a remote and harsh climate, the structure represents an early example of prefabrication. Intended for inhabitation, it was quickly relegated to storage (with the expedition party residing aboard the ship) because its Australian design with overhanging verandas, conceived to increase shade in the hot Australian Outback, rendered the hut even colder inside than outdoors. Despite the fact that the hut was not used for its intended purpose, through its mere association with the "Golden Age" of Antarctic exploration, the wooden structure is unequivocally regarded as a historical artefact worthy of restoration and preservation as a Historical Site and Monument. This begs the question: What attributes distinguish temporary from permanent structures in Antarctica?

More modest prefabricated ancillary structures joined the Discovery Hut in 1955 when the United States Navy began erecting wood-framed "T-5", steel-framed "Robertson", and arched steel "Quonset Hut" buildings. Initially totalling twelve structures to support 130 men, what was then called the US Naval Air Facility McMurdo ↗ p. 666 was intended to be a temporary military field camp to support the United States' presence at the South Pole during the landmark cooperative scientific event of the International Geophysical Year in 1957–1958. These buildings, separated to minimise transmission of fire, were organised in a classic military orthogonal grid, carefully oriented according to the prevailing storm winds to minimise snow drift. But unlike gridded urban settlements that follow topography, the gridded grouping of McMurdo's early structures lay on one primary level created by the blasting of the sloping volcanic land. Today, many of these structures have been removed but it was once conceived as temporary, even though its foundations radically altered the site in an irreversible, permanent, fashion.

"Currently McMurdo is 10% efficient, which means that for every scientist there are nine to ten support staff. The primary ambition of the project was to improve that balance in favour of science."

Rick Petersen (Architect, OZ Architecture), in conversation with UNLESS, Architectural Association School of Architecture, 2019

"If McMurdo is the sprawling city, Scott Base [today] is the well-planned suburb."

Juan Francisco Salazar (Professor, Western Sydney University) and Jessica O'Reilly (Professor, Indiana University), *Inhabiting the Antarctic*, 2017

Some decades later, in 1962, another building housed in prefabricated modules sized specifically for shipment on standard 2.4- × 2.4-metre Air Force palettes was erected on site to power McMurdo Station. The structure housed a modest nuclear reactor, placed into service in 1962 and decommissioned eight years later due to cracked housings, and replaced by conventional diesel-fuel-driven generators, which serve as McMurdo's primary energy source today. (Meanwhile, between 2008 and 2010 three wind turbines were installed by the New Zealand government, reducing McMurdo's fossil fuel consumption by 11% and serving as a test case for an expanded and ever-lasting source of energy on Ross Island). The nuclear reactor's site has since been decontaminated and what remains is the structure's altered topography, heavy timber foundation members, and a commemorative plaque. Once again, while the facility was temporary, its traces will be long-lasting.

For the three decades between the 1960s and 1990s, structures were typically added to solve short-term needs without consideration for long-term efficiencies or strategic master planning.

Apart from McMurdo's first mixed-use building "N. 155" (built in 1968 to combine housing, food service, and recreation), most buildings continued to be single-purpose and separated from each other, with consequent reliance upon more and more vehicles to ferry goods between buildings. By 2012 McMurdo housed hundreds of vehicles. During this period McMurdo saw the addition of, amongst other things, the wood-shingled "Chalet" (1972), the clapboard-sided Chapel of the Snows (1989), and the metal-sided Science Support Center (1991). In addition to being of different colours, styles, scales,

and forms, none of these structures even shared the same orientation. The result is the haphazard appearance of McMurdo that does little to reflect the actual seriousness, investment, and stature of the United States Antarctic Research Program.

An exception to McMurdo's relative disorder and land manipulation is the Crary Lab from 1991, which represents McMurdo's first structure to be elevated. The building's simple massing of three connected parallel wings can easily expand or contract. Internally, Crary Lab can be easily reconfigured owing to its simple, repetitive floor plan and gabled volume. This flexibility allowed for a continued serviceability through the nearly hundred-year lifespan of its basic structural system; and when the building is eventually disassembled and removed from McMurdo, it will leave its site virtually untouched, embodying the perfect example of both permanence *and* temporality.

In response to the 2012 Blue Ribbon Panel Report,[1] which advocated "more and better science in Antarctica through increased logistical effectiveness", a master plan was created to consolidate McMurdo's 105 buildings into 18, reducing the station population from 1,200 to 850 and improving the "support-to-scientist" ratio by 40%, from 1:9 to 1:13. In addition to the significant logistical efficiencies resulting from consolidation, the new master plan aims to improve resource efficiencies, as the new, larger buildings will require nearly 30% less exterior envelope to contain the same volume, and the latter will be highly insulated, with a U-value of 0.013.

While efficiency was the primary focus, in the 2016 master plan the new buildings were also designed to improve wellness.

Physical wellness is achieved through strategies like the replacement of shared dormitory rooms (up to six people per room) with single-occupancy rooms to reduce the transmission of disease and guarantee privacy. In addition, "sky bridges" connecting buildings will provide for safe travel between functional units during inclement weather. Emotional wellness is enhanced through the privacy afforded by the personal living quarters and the inclusion of the new buildings' many interspersed social spaces.

The outward appearance of the planned new core building of McMurdo Station (approx. 28,000 square metres) reflects the United States' long-term commitment to scientific research on the Antarctic continent. By introducing a skirting wall which extends to the volcanic rock with the purpose of sealing off any accumulating ice and snow, even if the main structures are elevated, the station projects a perceived, not real, sense of permanence, which visually reaffirms the idea of a *settled* community.

McMurdo and Scott Base, the American and New Zealand stations on Ross Island, are only 3.2 km apart. Their proximity is such that if a short-stay visitor should dine at one station (paying by credit card) and withdraw cash from the ATM in the other station, his or her bank will most likely block the account due to the impossibility of personal transactions occurring in New Zealand and America in such a short timeframe.[1] Although the contiguity – which might naively lead one to formulate the unwelcome suggestion that the nations could combine forces (at least) during the winter months and keep only one station operational – does not translate into the design of a single shared station such as the Italo-French Concordia station, synergies between the two stations are significant, both in terms of logistics and power supply. The first relies primarily on the use of shared airport infrastructure which sees, for example, all American Air Force flights departing from New Zealand, while the latter translates into a shared energy grid which includes power produced by the joint Ross Island wind farm. Located on Crater Hill, the three-turbine wind farm is able to produce 1MW of power, equivalent to approximately half of the combined needs of the two bases, and allows a reduction of diesel-fuelled consumption by approximately 500,000 litres per annum, reducing significantly the carbon footprint of the two stations.

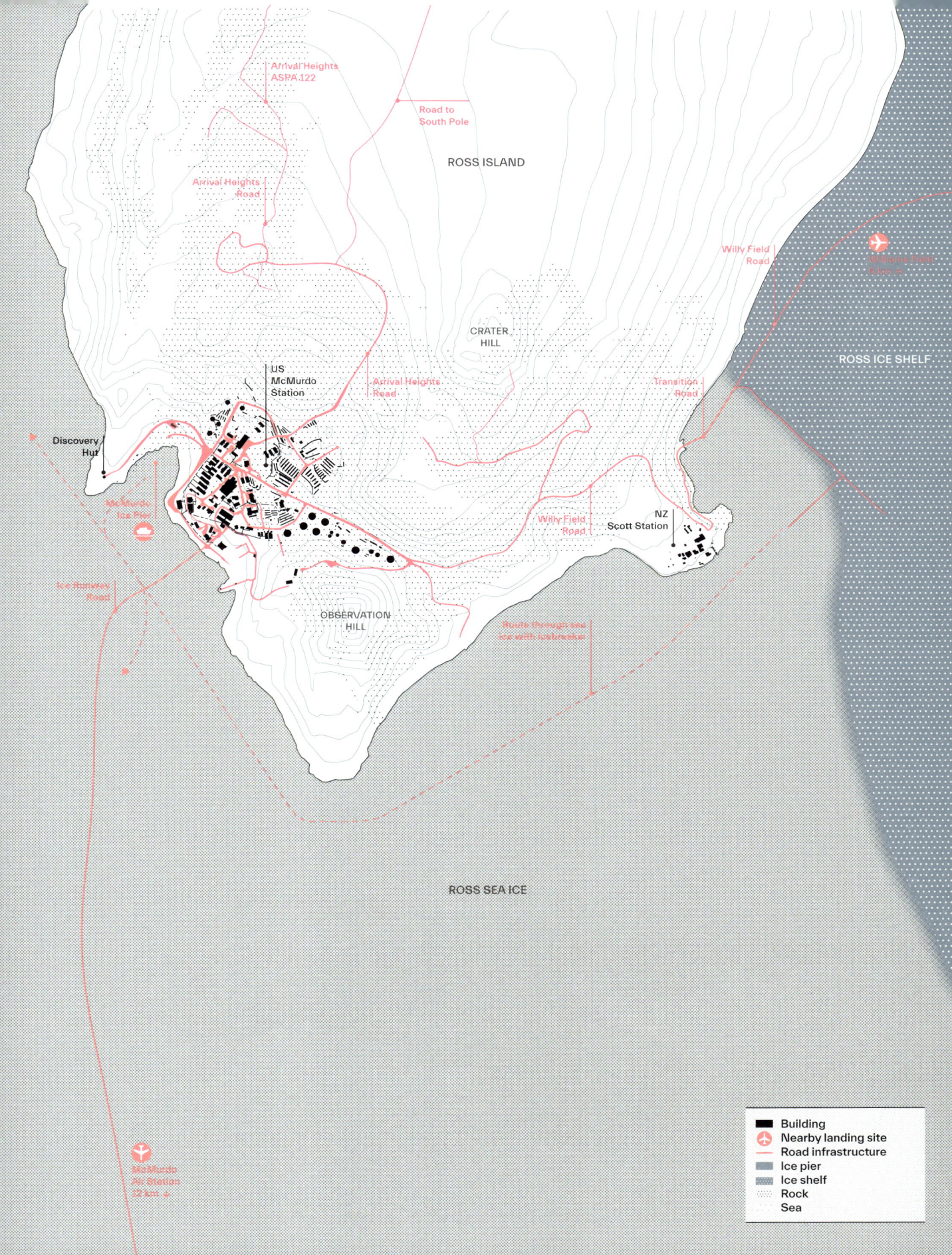

ROSS ISLAND

'Arrival Heights'
ASPA 122

Road to
South Pole

Arrival Heights
Road

Willy Field
Road

Williams Field
Skiway

ROSS ICE SHELF

CRATER
HILL

US McMurdo
Station

Arrival Heights
Road

Transition
Road

NZ
Scott Station

Discovery
Hut

McMurdo
Ice Pier

Willy Field
Road

Route through sea
ice with icebreaker

Ice Runway
Road

OBSERVATION
HILL

ROSS SEA ICE

McMurdo
Air Station
12 km →

■	Building
✈	Nearby landing site
—	Road infrastructure
▧	Ice pier
▨	Ice shelf
⣿	Rock
▨	Sea

Scott Base. Rethinking the Dispersed Master Plan Model

Hugh Broughton

HUGH BROUGHTON is an architect and the founder of Hugh Broughton Architects (United Kingdom). He is the designer of the Halley VI Antarctic Research Station, the world's first relocatable polar base, and the Juan Carlos I Spanish Antarctic Base. Broughton is currently developing the designs for the Scott Base for Antarctica New Zealand, the Discovery Building at Rothera for the British Antarctic Survey (United Kingdom), and a master plan for Davis Station for the Australian Antarctic Division.

Scott Base ↗ p. 678 is located on Pram Point on Ross Island at 77° South 166° East. Temperatures in the winter drop below –50°C degrees with 24-hour total darkness for four months. The first iteration of Scott Base was designed by Frank Ponder for Sir Edmund Hillary's Trans-Antarctic Expedition in 1957. A permanent presence, with year-round scientific observations, has been maintained in the Ross Dependency ever since and fosters the strong Antarctic bond which exists with other national Antarctic programmes operating out of the Ross Sea region.

By the late 1970s Hillary's base had reached the end of its effective life and was progressively demolished – although the original Hut "A" has been preserved and restored by the New Zealand Antarctic Heritage Trust (2017). The current base was constructed in phases from 1976 to 1983 and comprises a series of simple rectangular pitched roof buildings elevated on stub columns to allow the wind to scour snow from underneath. The buildings – many of which have reached the end of their effective life – are clad with coloured metal composite panels in Chelsea cucumber green and are interconnected with a corridor along their full length with levels arranged to suit the topography. In 2002 a large logistics building, the Hillary Field Centre, was added to the base.

The base is challenging to maintain, operational efficiency is reduced by the multiple levels and extended floor plan and, in some areas, buildings do not conform to the New Zealand Building Code. As a result, in 2015, Antarctica New Zealand embarked on the process of redeveloping Scott Base and providing a safe and fit-for-purpose facility to support science for the next fifty years.

The strategic objectives of the project are to provide an environment that keeps people safe and healthy, maintain New Zealand's continuous presence in the Ross Dependency, enable effective logistics support to maintain and enhance high-quality science that is relevant to New Zealand and the world. The new building design has to fulfil its role of protecting the Antarctic environment and maintaining New Zealand's credibility among Antarctic Treaty nations.

One of the project's biggest challenges is the limited space available for construction. The sloping site is contained by the sea to the south and east, sea cliffs to the west, the sea-ice transition road from McMurdo ↗ p. 866 station to the north, and areas of Antarctic flora to the north-west. There is a fall of 25 metres across the site.

The design comprises three interconnected, aerodynamically shaped two-storey buildings of matching width, which step down the hillside of Pram Point on Ross Island. There is a separate hangar serving twin helicopter pads. The three buildings are offset from each other to minimise the risk of snowdrift between them and are connected by enclosed links so that the lower level of the upper building connects to the upper level of the lower building. All the buildings are elevated above the ground to encourage the wind to flow beneath them, thereby minimising snow accumulation under the building footprint.

Future flexibility of the base is a key design driver. Each building is designed around a 6-metre repeating structural grid of primary

steel frames. This simplifies the process of reconfiguring the space when necessary because the structure is regularly spaced with large clear span zones between.

The upper building contains the living accommodation and is the primary point of entrance to Scott Base. The upper level contains a mix of single and twin prefabricated bedrooms, ablution blocks, and living spaces to support a summer population of one hundred people and a winter crew of fifteen. The bedrooms are split into four blocks. The single rooms for longer-stay residents take full advantage of the inspiring views. Twin rooms for those staying for shorter periods look into a double-height circulation zone illuminated by glazed roof lights. The dining room includes a glazed end wall with views towards Mount Erebus and Mount Terror. The lower level contains the medical suite, laundry, recreational spaces, food storage, shop, locker room, welcome lounge, and mechanical plant spaces.

The middle building contains laboratories and offices on the upper level. The lower level is dedicated to the preparation and staging of open-plan deep-field science expeditions with level access via a bridge link to the field stores in the lower building. The open-plan arrangement has been designed to provide safe and functional cargo flows in and out of the building. The lower building contains the vehicle workshop, inter-continental cargo handling area, waste management, and central storage.

A small roof deck will support science that requires unimpeded views of the horizon. A separate structure houses the helicopter hangar, drone workshops, and a meteorological balloon launch facility.

The interior design will foster a strong sense of well-being whilst minimising maintenance. Warm finishes are being selected for durability, comfort, economy, and style. Significant thought is given to ways in which the design can reflect New Zealand's cultural and natural landscape, capturing the essence of what it means to be a New Zealander, by conveying Māori values and reflecting New Zealand's history of involvement in Antarctica.

Central to Māori values is a sense of shared responsibility for the *mauri,* or life force, of the environment, and for the health and well-being of all people who depend upon it for their survival. This connectivity will be a key feature of the interior design. Windows are carefully placed to make the most of natural light and reinforce connections with the Antarctic landscape. The layout has been developed to include spaces for people to stop and chat as they walk from one place to another, fostering collaboration, and care for each other.

To minimise the environmental footprint of the base, most of the energy demand will be provided by the wind turbines of the Ross Island Wind Energy network. Heating will be provided by electric boilers. Only

when there is no wind and no energy available from the battery storage will the base be powered by fuel-driven generators. Waste heat from these generators will be collected and utilised to heat the base. Water will be produced using reverse osmosis, which converts seawater to drinking water. A vacuum drainage system will be used to dispose of waste water; this uses 1.5 litres per flush compared to 9 litres in a standard toilet, helping to save water and energy. The energy plant is distributed around the base with the duplication of key services such as water storage, power production, and communications to maximise the resilience of critical life-support in a remote and unforgiving environment.

"The original 1957 base took less than a year to design, to build and occupy, but ours will take something like twelve years so not everything is going in the right direction; some of these buildings are taking much longer than they used to."

Hugh Broughton (Architect, Hugh Broughton Architects), in conversation with UNLESS, Architectural Association School of Architecture, 2019

Calving Architecture.
Archaeology of the Debris

Giulia Foscari

GIULIA FOSCARI is an architect, researcher, and writer who has been practising in Europe, Asia, and the Americas. She is the founder of UNA, an architecture studio focused on cultural projects, and of its alter ego UNLESS, a not-for-profit agency for change devoted to interdisciplinary research on extreme environments. Foscari taught at Hong Kong University and at the Architectural Association, where she ran a Diploma Unit and founded the Polar Lab.

The Oxford dictionary tells us that cows are expected to calve, and so are glaciers.

Calving is a natural phenomenon whereby sections of glaciers floating over the ocean break and fall into the water. While the thinning process that sees ice shelves transforming into icebergs is normal in stable glaciers, the accelerated rate at which this is occurring in Antarctica today, a reflection of anthropogenic climate change, is a matter of grave concern to scientists and humanity. The "ice thunders" that accompany the formation of each fracture in the ice and violently shatter the absolute Antarctic silence awaken the consciousness of concerned citizens and fuel the spirited work of climate activists around the globe.

Charged with high symbolic value as physical evidence of retreating glaciers and rising global sea levels, icebergs calving off the "Barrier" rarely conceal further evidence of the human contamination of Planet Earth. Yet there are occasions when they do.

In January 1980 the crew and scientists aboard the *Polarsirkel* witnessed an unusual sight from the bow of their ship. The cross section of a building appeared captive within the ice shelf. Its contours against the white backdrop of the snow-ice resembled accurately drawn lines on a white piece of paper. The section belonged to Halley I ↗ p. 658 station, built by the United Kingdom on the Brunt Ice Shelf. Erected on the occasion of the International Geophysical Year in 1956, the prefabricated timber had a short life as it sunk in the depths of the snow in a matter of months. Replaced at first by Base Z – which was connected to its predecessor via a system of tunnels – Halley I was finally abandoned to its fate and supplanted in 1967 by Halley II ↗ p. 706.

The story of the Halleys is common to all architectures constructed on the shifting "ground" of the ice shelf – a "living ice" which at once drifts towards the ocean and abides by the mechanical and thermal properties of snow, which constantly undergoes complex deformations, especially if placed in contact with foreign surfaces. History has taught us that buildings erected on such malleable and cold substances will be engulfed within the crystalline matter.

The first building to disappear from the white horizon of the ice shelf was Framheim ↗ p. 628, the Heroic Age hut built by Roald Amundsen and his party in 1911. Followed swiftly by the tent they erected at the South Pole to signal to Robert Falcon Scott's southern party that they had been the first to reach the Pole, this fate was subsequently shared by all the other structures built on ice, including Rear Admiral Byrd's first Little America Station constructed on the coastline in close proximity to Framheim's original site.

It was only as a result of the wartime development of automated surface transport and large-scale aircraft that the supply of material within the polar plateau became possible, enabling expeditions to venture into the heart of the Antarctic. Driven by the exigencies of scientific exploration, on the occasion of the International Geophysical Year (1957–1958) the United States Navy built four surface stations on permanent snowfields: Little America V ↗ p. 654 and Ellsworth near the seaward margin of the ice shelves; Byrd and Amundsen-Scott on the Antarctic plateau. All stations featured a standard prefabricated building which "could be erected by hand labour in two hours using a four-man crew". The structures rested on foundations of "longitudinal timber sills laid on the snow surface parallel to the long walls of the building [...] which transferred the weight of the building to the footings" whilst creating an air gap to prevent the transmission of heat to the snow. Connected to one another by an intricate network of wooden corridors covered with chicken wire and burlap, the sunken stations were structurally damaged beyond repair in just a few years due to excessive

"As we rose we found that we were on a pressure ridge. We stopped, looked at one another, and then *bang* – right under our feet. More bangs, and creaks and groans; for that ice was moving and splitting like glass. [...] The deep booming of the ice continued, and it may be that the tide has something to do with this, though we were many miles from the ordinary coastal ice. [...] From under the tent came noises as though some giant was banging a big empty tank."

Apsley Cherry-Garrard (Explorer, Zoologist), *The Worst Journey in the World*, 1922

snowdrift accumulation, which reduced their lifespan to an embarrassing degree and forced them to be evacuated by 1959.

Relying solely on empirical knowledge gained through experience in Antarctica and the Arctic during the militarisation of Greenland in the 1950s,[5] the second generation of Antarctic stations built on the polar territory were no longer built *on* ice, but rather *in* ice. New Byrd Station[↗ p. 690], commissioned in 1962 to replace its buried predecessor and shelter a party of sixty men, was carved within the snow. Its floor, excavated some 7 metres below the surface, hosted prefabricated barracks, workshops, and offices, while its snow "walls" supported corrugated steel arches covered by a 60- to 100-centimetre-thick layer of snow. The building's typology closely followed that of Camp Century, as did its destiny. Crushed under the sheer weight of the ice, the cross-sectional area was reduced by 17% and the "Wonder Arches"[6] gradually collapsed, rendering the station uninhabitable and risking environmental contamination though fuel leakages.

The challenge of building within snow in Antarctica was subsequently met by both the Germans and the British with the establishment of the Georg Von Neumayer Stations I[↗ p. 736] and II[↗ p. 768] on the Ekström Ice Shelf and further iterations of Halley III[↗ p. 716] and Halley IV[↗ p. 740]. Experimenting with diverse materials and construction methods, both programmes explored the potential of under-snow facilities, developing structures that were progressively optimised to withstand multidirectional compression forces, such as Armco steel tubing.[7] Though undeniably interesting architecturally, these settlements exacerbated the "polar depression" symptoms experienced by scientists deployed on the seventh continent, who were not only living in isolation and darkness for most of the year but were doing so "underground".

Dante could not have foreseen that anyone would descend into the abyss of the cryosphere when, accompanied by Virgil, he ventured into the depths of the Earth, yet surely his verses must have echoed in the mind of scientists and logistic teams deployed in Antarctica as they undertook the long descent into

After a hard day's scientific work on the Antarctic Ice Sheet, scientists and staff descend back home, into the depths of the Antarctic snow, reaching the windowless and dark scientific stations of Halley III (below), Halley IV (opposite, top), and Byrd (opposite, bottom).

"As the cold and darkness increased, our whole existence changed. We became a family of moles, scuttling through glistening snow tunnels with lanterns and flashlights. We emerged only for short walks or tasks which had to be done."

Richard Evelyn Byrd (Admiral, Explorer, Aviator), "The Conquest of Antarctica by Air", *National Geographic*, 1930

the dark, claustrophobic, poorly ventilated below-ice stations they called "home". "Abandon all hope, ye who enter here."[8]

Despite being greeted with English humour by Edward Johnston's familiar London Underground logo from 1919, reaching the station never offered the comforts of the transit-system escalators. The circulation shafts were vertical, stairs were replaced by ladders cut into the snow, and the conduit was deep. Thus, it comes as no surprise that, as Professor David Walton shared with the Polar Lab, "researchers deployed in such stations never came back quite as they had left off."[9]

The ascent, however, must have been a spectacular sight. The Underground logo, lit by light refracted by the silvery landscape, must have appeared as a vision – as per Man Ray's 1938 reinterpretation of the famous symbol into a planet that "keeps London going".[10] Completing Dante's journey, the scientist "saw a point that sent forth so acute a light, that anyone who faced the force with which it blazed would have to shut his eyes."[11]

These underground stations – sunken unexpectedly or deliberately carved in ice – ceased to be built after the signing of the 1991 Protocol on Environmental Protection to the Antarctic Treaty, which enforced the obligation of party members to present an Environmental Impact Assessment for all intended activities on the continent (Annex I)[12] to the consensus-driven Antarctic governance body, declaring that all artefacts in disuse should be intended as waste and thus be removed from the continent to avoid contamination (Annex III).[13]

While this vital environmental policy led to the third – elevated – generation of mobile and responsive Antarctic architectures (with Halley VI ↗ p. 810 heralded as the first true architecture with a capital A on the continent), remains from the past are still imprisoned in the kilometre-thick ice sheet and move inexorably towards the ocean. As a temporary apparition within the shimmering Barrier, the archaeology of this debris offers a glimpse of neglected and under-theorised architectures, urging a collective paradigm shift.

"The temperature that night was –15.8 degrees, and I will not pretend that it did not convince me that Dante was right when he placed the circles of ice below the circles of fire."

Apsley Cherry-Garrard (Explorer, Zoologist), *The Worst Journey in the World*, 1922

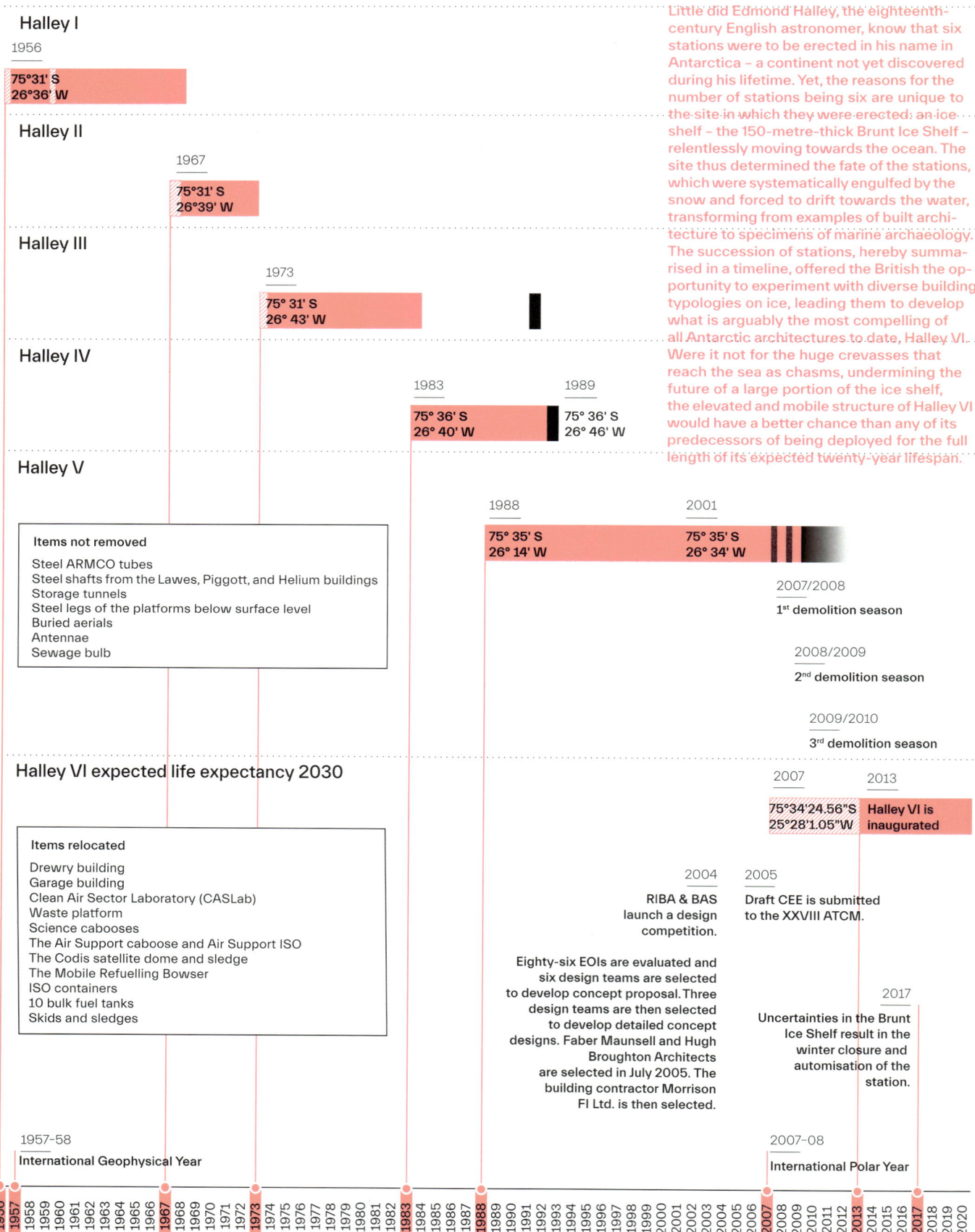

Halley I

1956

75°31' S
26°36' W

Halley II

1967

75°31' S
26°39' W

Halley III

1973

75° 31' S
26° 43' W

Little did Edmond Halley, the eighteenth-century English astronomer, know that six stations were to be erected in his name in Antarctica – a continent not yet discovered during his lifetime. Yet, the reasons for the number of stations being six are unique to the site in which they were erected: an ice shelf – the 150-metre-thick Brunt Ice Shelf – relentlessly moving towards the ocean. The site thus determined the fate of the stations, which were systematically engulfed by the snow and forced to drift towards the water, transforming from examples of built architecture to specimens of marine archaeology. The succession of stations, hereby summarised in a timeline, offered the British the opportunity to experiment with diverse building typologies on ice, leading them to develop what is arguably the most compelling of all Antarctic architectures to date, Halley VI. Were it not for the huge crevasses that reach the sea as chasms, undermining the future of a large portion of the ice shelf, the elevated and mobile structure of Halley VI would have a better chance than any of its predecessors of being deployed for the full length of its expected twenty-year lifespan.

Halley IV

1983

75° 36' S
26° 40' W

1989

75° 36' S
26° 46' W

Halley V

1988

75° 35' S
26° 14' W

2001

75° 35' S
26° 34' W

2007/2008

1st demolition season

2008/2009

2nd demolition season

2009/2010

3rd demolition season

Items not removed

Steel ARMCO tubes
Steel shafts from the Lawes, Piggott, and Helium buildings
Storage tunnels
Steel legs of the platforms below surface level
Buried aerials
Antennae
Sewage bulb

Halley VI expected life expectancy 2030

2007

75°34'24.56"S
25°28'1.05"W

2013

Halley VI is inaugurated

Items relocated

Drewry building
Garage building
Clean Air Sector Laboratory (CASLab)
Waste platform
Science cabooses
The Air Support caboose and Air Support ISO
The Codis satellite dome and sledge
The Mobile Refuelling Bowser
ISO containers
10 bulk fuel tanks
Skids and sledges

2004

RIBA & BAS
launch a design
competition.

2005

Draft CEE is submitted
to the XXVIII ATCM.

Eighty-six EOIs are evaluated and
six design teams are selected
to develop concept proposal. Three
design teams are then selected
to develop detailed concept
designs. Faber Maunsell and Hugh
Broughton Architects
are selected in July 2005. The
building contractor Morrison
FI Ltd. is then selected.

2017

Uncertainties in the Brunt
Ice Shelf result in the
winter closure and
automisation of the
station.

1957-58

International Geophysical Year

2007-08

International Polar Year

1956 1957 1958 1959 1960 1961 1962 1963 1964 1965 1966 1967 1968 1969 1970 1971 1972 1973 1974 1975 1976 1977 1978 1979 1980 1981 1982 1983 1984 1985 1986 1987 1988 1989 1990 1991 1992 1993 1994 1995 1996 1997 1998 1999 2000 2001 2002 2003 2004 2005 2006 2007 2008 2009 2010 2011 2012 2013 2014 2015 2016 2017 2018 2019 2020

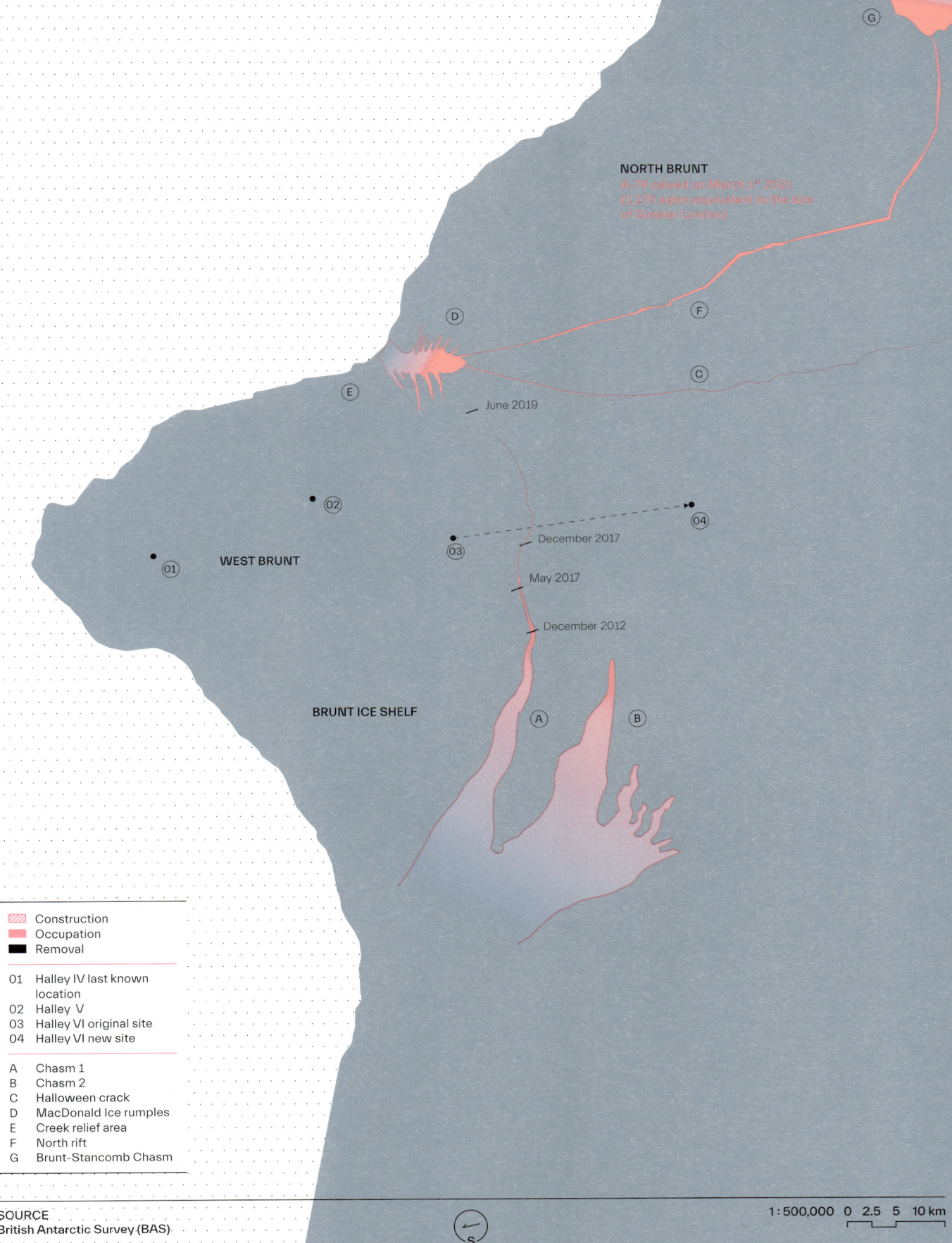

NORTH BRUNT
A-74 calved on March 2ⁿᵈ 2021
(1,270 sqkm equivalent to the size
of Greater London)

June 2019

WEST BRUNT

December 2017

May 2017

December 2012

BRUNT ICE SHELF

D
E
F
C
G
A
B
02
03
04
01

Construction
Occupation
Removal

01 Halley IV last known
    location
02 Halley V
03 Halley VI original site
04 Halley VI new site

A   Chasm 1
B   Chasm 2
C   Halloween crack
D   MacDonald Ice rumples
E   Creek relief area
F   North rift
G   Brunt-Stancomb Chasm

SOURCE
British Antarctic Survey (BAS)

1 : 500,000   0 2.5 5   10 km

S

HALLEY I	HALLEY II	HALLEY III

"Halley's history is indicative of trends across Antarctica. Stations have been built on grade, under the ice, on stilts, and using a combination of techniques. Such a trajectory is common for stations that are built on ice, as they must be rebuilt periodically."

Sam Jacob (Architect), *Ice Lab*, 2013

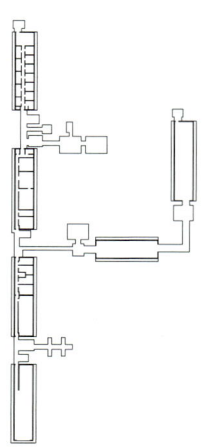

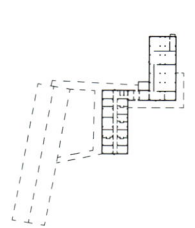

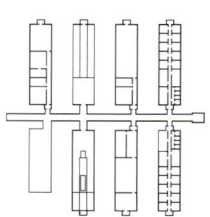

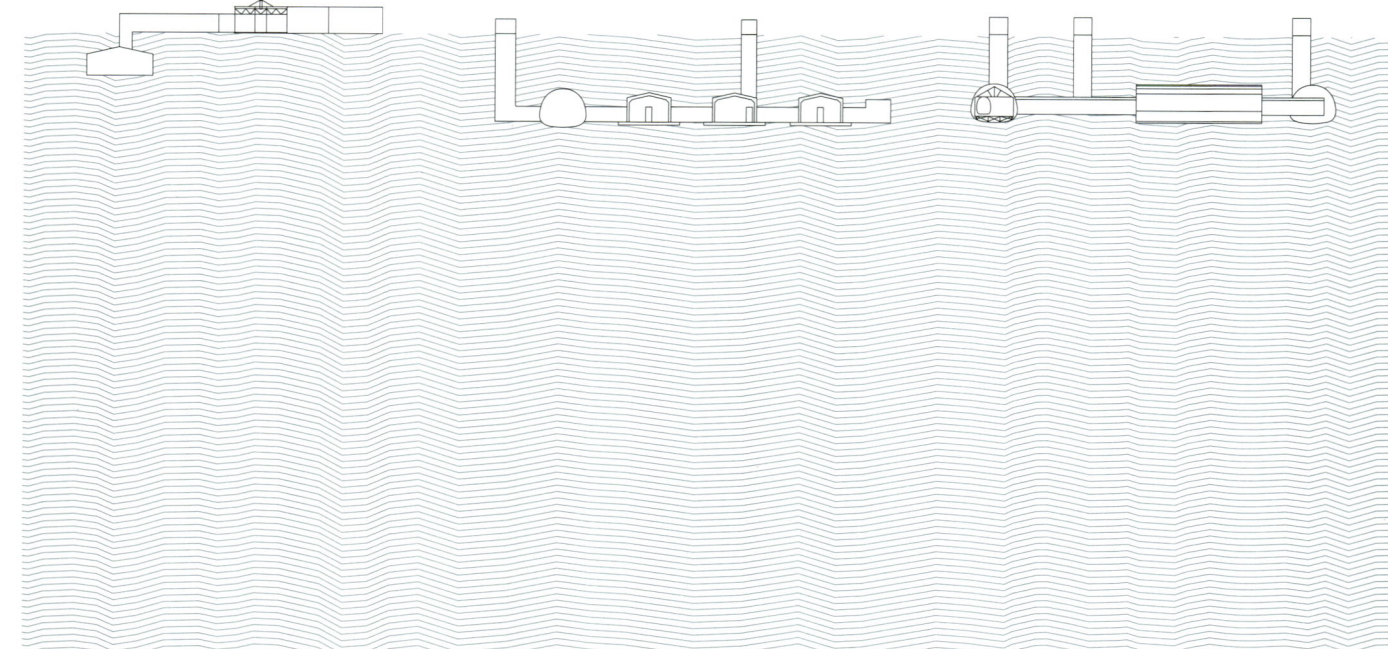

SOURCE
British Antarctic Survey (BAS)

HALLEY IV

HALLEY V

HALLEY VI

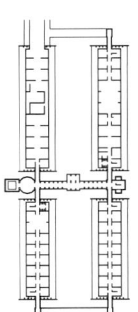
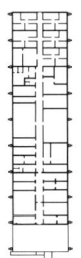
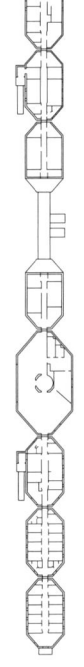

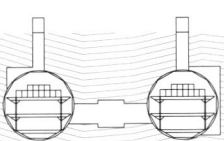
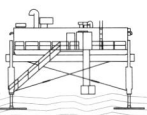

| 1 : 1,250 | 0 | 20 | 40 | 80 m |
| 1 : 1,000 | 0 | 10 | 20 | 50 m |

The Halleys

Hugh Broughton

HUGH BROUGHTON is an architect and the founder of Hugh Broughton Architects (United Kingdom). He is the designer of the Halley VI Antarctic Research Station, the world's first relocatable polar base, and the Juan Carlos I Spanish Antarctic Base. Broughton is currently developing the designs for the Scott Base for Antarctica New Zealand, the Discovery Building at Rothera for the British Antarctic Survey (United Kingdom), and a master plan for Davis Station for the Australian Antarctic Division.

1957

Halley ↗ p. 658 is located on the Brunt Ice Shelf at 75° South 25° West. The first station at Halley Bay was established by Sir Vivien Fuchs in 1956 in time for the International Geophysical Year (IGY) of 1957. It was a simple timber hut built on the ice surface in much the same way as the early huts erected by Borchgrevink, Scott, and Shackleton. As the snow levels rose around the building, it became buried. Access had to be redesigned via timber ladders and shafts but by 1968 life inside was no longer tenable as the structure was being crushed by the ice, forcing evacuation of the building.

1967

The second version of Halley, Halley II ↗ p. 706, was built in 1967. It comprised a series of timber buildings with steel supports to strengthen the structure against the ice. Despite this structural reinforcement, its life was short, and it was abandoned only six years after occupation in 1973.

1973

1983

Halley III ↗ p. 716 was completed in 1973 and comprised a series of timber structures built within Armco steel tubing to allow them to be buried. It is hard to imagine the quality of life within these buried structures, as the ice crushed the metal tubes and the space was filled with the fumes of petrol generators and sewage. Within ten years of its completion the base was buried up to 15 metres below the ice. Owing to the severe challenges of access and ventilation, it was abandoned in 1983 and left to drift within the ice shelf. Many years later, in 1995, the crew of the RRS *Sir Ernest Shackleton* sighted the ghostly remains of the compressed base emerging from the edge of the ice shelf about to drop into the sea.

Halley IV ↗ p. 740 was built in 1983 and was also designed to be buried. Two-storey buildings were constructed inside two parallel insulated, reinforced 9-metre-diameter timber tubes connected by a link. Access was provided by shafts containing ladders. The base continued to be used for ten years and was eventually abandoned in 1994, leaving it to a fate similar to its predecessor's.

1990

Halley V ↗ p.762, the precursor to the current base, introduced a significant design change. It comprised three single-storey insulated timber buildings elevated on steel legs around 4 metres above the ice. This allowed the wind to blow underneath, scouring the ice to create drifts on the leeward side. The station could be raised above the rising snow level using electric screw jacks. The steel legs were originally supported on timber pads but these were soon buried and the legs relied purely on the stiction of the ice to the steel for their support. Every summer a small crew of steel workers had to attend site to straighten up some of the legs before the buildings could be jacked – a process which involved everyone on base. The main platform – the Laws Building – contained accommodation, while the Piggott and Simpson Buildings provided space for science. A mobile garage and summer accommodation building were added to the base in 1998 and continued in operation. The base was supported by a series of small laboratory buildings built within elevated and modified shipping containers. The elevated platform was decommissioned following the completion of Halley VI and subsequently removed from Antarctica.

The Architectural Impact of Halley VI

Hugh Broughton

HUGH BROUGHTON is an architect and the founder of Hugh Broughton Architects (United Kingdom). He is the designer of the Halley VI Antarctic Research Station, the world's first relocatable polar base, and the Juan Carlos I Spanish Antarctic Base. Broughton is currently developing the designs for the Scott Base for Antarctica New Zealand, the Discovery Building at Rothera for the British Antarctic Survey (United Kingdom), and a master plan for Davis Station for the Australian Antarctic Division.

Halley is the most southerly science research station operated by the British Antarctic Survey (BAS) and is located on the 150-metre-thick floating Brunt Ice Shelf, which each year moves 400 metres towards the sea. Snow levels rise by 1 metre every year, and the sun does not rise for 105 days during winter. Temperatures drop to –56°C and winds blow in excess of 160 kph. Access by ship and plane is limited to a three-month summer window.

A research station has been occupied continuously at Halley since 1957 and in 1985 scientists working there first observed the hole in the ozone layer. Halley V↗ p. 762 was completed in 1992. Continuing occupation however seemed dangerous, as the station site had shifted so far from the mainland that it was at risk of calving as an iceberg. The station's legs were fixed in the ice and therefore it could not be moved, thus in 2004 BAS organised an international competition to select designers for a new station. The winning entry proposed a line of elevated ski-based modules.

Bedrooms, laboratories, office areas, and energy centres are housed in aerodynamic blue modules. A larger, two-storey, light-filled red module provides the social heart of the station and is used for living, dining, and recreation. Linked together, the modules create a self-supporting, infrastructure-free community. The station is arranged in a straight line perpendicular to the prevailing wind, so that snow drifts form only on the leeward side. This leaves the windward side free from drifts and creates a hard, icy surface across which vehicles can move. The base is split in two for life safety. Each half has its own energy centre and is self-sustaining in case of emergency. A bridge link allows the sharing of power, drainage and water. The modules are supported on giant steel skis and hydraulically driven legs that allow the station to mechanically "climb" up out of the snow every year. And, as the ice shelf moves out towards the ocean, the modules can be disconnected, lowered and towed further inland by bulldozers, as was done successfully for the first time in 2016.

The completion of Halley VI↗ p. 810 in 2013 marked a milestone in the development of a new architecture for Antarctica. On the one hand the station is an imaginatively pragmatic engineering-led response to a set of unusual criteria, in particular the need for relocation, which largely determined the modular, ski-based solution. On the other hand, it presents an architectural vision of a future of mobile research buildings which can quickly appear and then disappear leaving little trace of their presence once they have gone and therefore impacting as little as possible upon their environment.

Alongside the engineering achievement, the fleet-of-foot modules of Halley VI demonstrate a robust architectural response to the extreme environment,

EACH WALKING UNIT HOUSES NOT ONLY A KEY ELEMENT OF THE CAPITAL , BUT ALSO A LARGE POPULATION OF WORLD TRAVELLER-WORKERS.

A WALKING CITY

Although Charcot Station was one of the first Antarctic structures to be prefabricated on sleds and towed on-site, one could argue that the first example of mobile polar architecture was the *N.B. James Caird* lifeboat (top), shown while being hauled across the ice pack in 1917 before embarking on the 1,300-kilometre journey from Elephant Island to South Georgia to rescue the stranded Imperial Trans-Antarctic Expedition party led by Ernest Shackleton. Drawing on these principles, and some fifty years after the 24-square-metre semi-cylindrical French station was erected, Halley VI was assembled on hydraulically driven foundations resting on steel "skis". As shown in the image on the following spread, the latter allowed the mobile station to be relocated in 2017 upstream of a previously dormant ice chasm that threatened to cut the station off from the Brunt Ice Shelf. Reminiscent of provocative utopian projects from the 1960s, one could indeed argue that Halley VI builds upon paradigms first introduced by Archigram (left).

which manifests itself in the rigorous aerodynamic design, economic building form and resolved expression of parts making a whole. For example, the metal stairs, roof decks, and external balconies are all carefully considered bolt-ons which magnify the machine aesthetic of the station; the bold colours aid way-finding in poor weather and also provide a nationalist expression – a red, white, and blue marker in the midst of the cold, white landscape of Antarctica; and there is significant reference to other industrial sectors with the rounded windows and flexible silicone connectors between the modules both clearly reminiscent of high speed trains. In these many ways the external appearance is the epitome of British high-tech with its roots in the work of Archigram of the 1960s – there is little doubt that Halley VI is the physical embodiment of the Walking City.

The modules have a steel structure and are clad in highly insulated, air-tight composite glass-fibre panels. The prefabrication of structure, cladding, rooms, and services was maximised. Products were procured from around the world. The strategy was to select items which were tried and tested and then to use them in innovative ways. So, for example, the flexible silicone bellows which allow movement between modules whilst preserving weather tightness are adapted from the insulated connectors used in trains in Nordic countries; the skis which support the modules are giant versions of those used on standard sledges; and the translucent glazing to the big window in the red module is insulated with nanogel, which was first invented by NASA to insulate the nose cone of the space shuttle on re-entry to the Earth's atmosphere. The centre of pre-construction activities was in South Africa, where a full-scale trial erection of modules was undertaken prior to shipping to Antarctica by ice-strengthened cargo ship. The modules were erected on-site over four twelve-week summer seasons using a factory line approach.

To determine the optimum interior design, the architects talked extensively with people who had lived and worked at Halley to understand the rituals of an average day and uncover the psychological deprivations of Antarctic isolation. They then considered the architectural tools which could enhance the living experience and provide support through the long dark winters. This was done through the application of first principles to the design – considering people's response to light, space, volume, colour, and even smell. Where the exterior design finds its roots in British high-tech, the interior finds its inspiration in a more classical architectural language characterised by symmetry, enfilade layouts, strong axial views, and a restrained material palette. The rooflights which illuminate the main circulation are a feature of the architects' work. They provide unencumbered views of the sky free from physical context, whether in Antarctica or the middle of London. Within the central red module this approach reaches a high point with the top-lit veneer-lined spiral stair drum overlooking the double-height atrium with its dramatic inclined insulated glazing. This space fosters integration between people within a high-quality living environment which would be equally at home in a cultural or educational organisation in a more temperate setting.

The design of Halley VI works with its context to create a strong sense of place to support its crew within a cocoon developed around their needs. It provides a benchmark for the sustainability of future stations achieved through a highly insulated envelope, recycled heating, anaerobic sewage treatment, and low-water-use strategies. Simultaneously futuristic and classical in its design, it provides a sophisticated model for ergonomic living within an extreme environment, mediating successfully between its role as national symbol and home-from-home for the residents.

Halley VI harnesses and then expresses its technology in moves redolent of a set in a sci-fi movie to minimise its environmental impacts. The "kit-of-parts" approach helps to support an altogether quieter interior of carefully manipulated daylight playing softly on combinations of timber finishes and brighter colours to counteract sensory deprivation and support the crew through the loneliness of a polar existence. It is this combination which creates the architectural response, and which has shown the way to a future for Antarctic design which places the occupants at centre stage within containers that are immediately at home within their context.

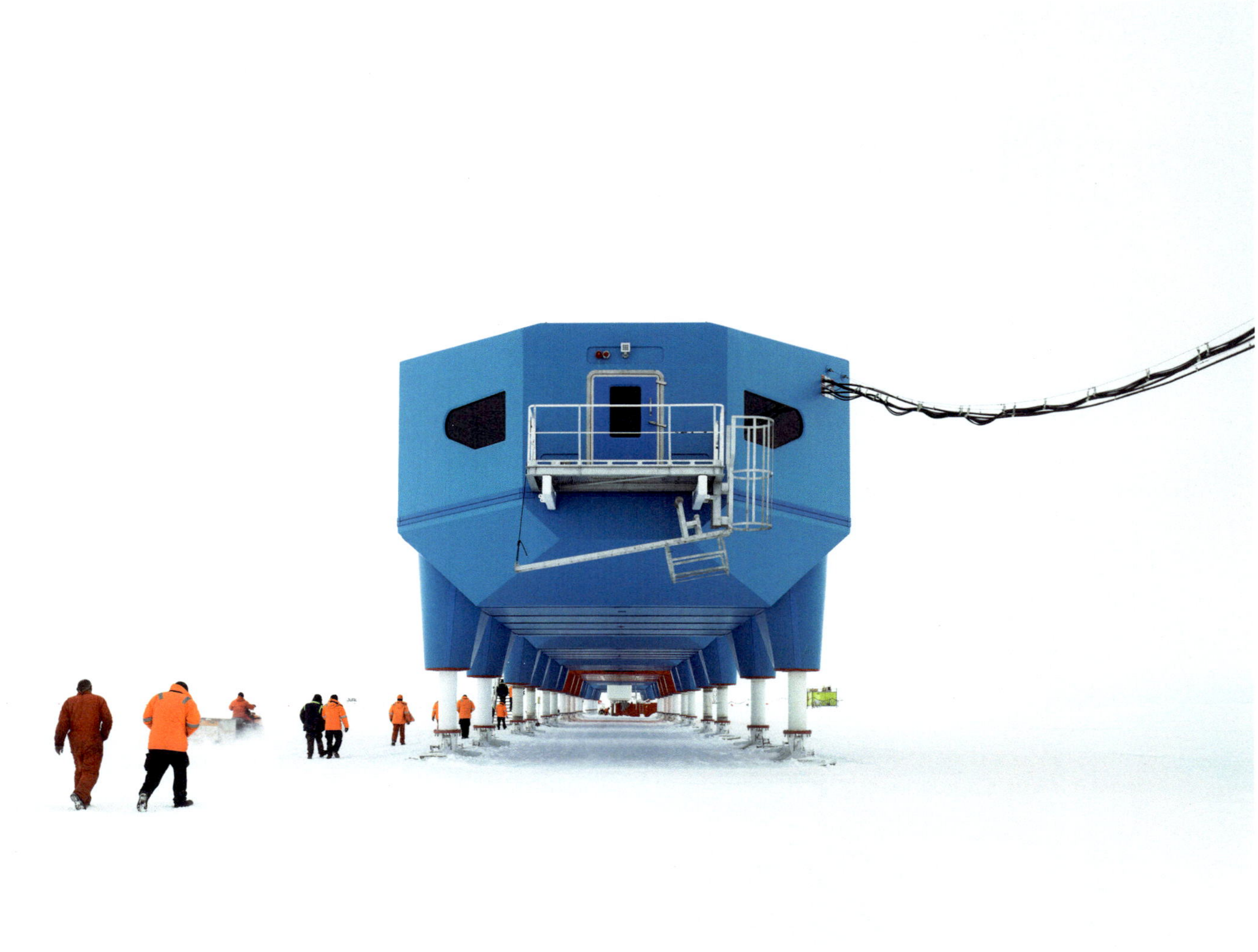

14
From Anthropic Footprints to Disembodied Technologies

Antarctic Building Design Standards

Neil Gilbert and Ceisha Poirot

NEIL GILBERT is a former environmental manager at Antarctica New Zealand who has worked on polar issues in various roles for over thirty years. CEISHA POIROT is New Zealand's representative to the Committee for Environmental Protection (CEP) and the expert group leader of the Council of Managers of National Antarctic Programs (COMNAP) Environmental Protection Group.

The first buildings on the Antarctic continent were erected in 1899 by the British Southern Cross Antarctic Expedition (1898–1900) led by Carsten Borchgrevink. Two prefabricated huts of Norwegian spruce, consisting of interlocking boards tightened by steel tie rods with the roof covered, in each case, with seal skins weighted down with bags of coal and boulders, were built at Cape Adare ↗ p. 614 in the Ross Sea region, one for accommodation and another as a small stores hut.[1]

Since the late nineteenth century, Antarctic stations, infrastructure and science support facilities have changed significantly, with modern buildings reflecting a range of designs depending upon their location (coastal or inland Antarctica) and functional requirements. Today, there are at least forty year-round stations and thirty-six seasonal stations in Antarctica,[2] and notably there are no internationally agreed building design standards for Antarctica – the only continent without such requirements. Each station has been individually designed to the specifications of the national Antarctic programme.

While the Antarctic Treaty System has not specifically addressed the topic of building design in Antarctica, following the adoption of the Protocol on Environmental Protection to the Antarctic Treaty in 1998, new stations built in the southernmost continent have been required to take account of the provisions of the protocol. These, however, are fairly minimal; for example the waste disposal and management provisions included in Annex III, require simply that sewage discharged into the sea is at least macerated for stations where occupancy exceeds thirty people.[3]

In Antarctica, all activities, including building a station, are subject to the Environmental Impact Assessment (EIA) requirements of the protocol, but this process primarily aims to assess and mitigate the impacts of the building process and operational phase of new stations on the local Antarctic environment. However, while the Antarctic EIA guidelines[4] do recommend that consideration be given to new technologies – such as the use of renewable energy sources, energy-efficient equipment, and building management systems that will help minimise atmospheric emissions – to waste-water treatment plants that may allow the reuse of treated water, to waste management practices, to collaboration opportunities, to the optimisation of the station footprint and the layout of the facilities and ancillary structures, and, finally, to the decommissioning of the old and new stations, they impose no formal obligations on parties to do so.

Founded in 1988 in order "to develop and promote best practice in managing the support of scientific research in Antarctica", the Council of Managers of National Antarctic Programs (COMNAP) regards the minimisation of environmental impacts from human activities as a key goal of their agenda. Since its inception, their environmental group has addressed practical environmental matters such as environmental impact assessments, environmental monitoring, waste management, and remediation, contingency planning and fuel management, and, most recently, alternative energy and energy management,[5] all of which are provided as guidelines when considering building design.

Other than the protocol and COMNAP guidelines, there are no known agreed standards that need to be met regarding durability, longevity, fabrication, insulation, sustainability, or emissions, although there is a general expectation that stations will be as sustainable as is practicable and many new stations have indeed incorporated sustainable design factors such as the use of renewable energy sources[6] and water recycling.[7]

Nonetheless, some national Antarctic programmes have recently attempted to construct new facilities and stations to recognised standards. In

"Navy Seabees work around the clock to complete the first Antarctic highway. Blasting, bulldozing full dirt, crushing rock and grading for the creation of 'Antarctic 6' was a full season chore completed with pride by the men of Mobile Construction Battalion SIX."

United States Navy, *Operation Deep Freeze '65 Report*, 1965

2018 the United Kingdom submitted its redevelopment of the wharf at its Rothera↗ p. 856 research station for a CEEQUAL award[8] – an international evidence-based sustainability assessment, ratings, and awards scheme,[9] and the redevelopment of New Zealand's Scott Base↗ p. 852 is currently considering a bespoke Green Star rating system[10] developed by the New Zealand Green Building Council. Other examples of building sustainability rating schemes that could be investigated for applicability include the United States Green Building Council LEED scheme[11] and the international BREEAM sustainability assessment method.[12]

While sustainable building schemes or standards are available, the initial upfront cost of building in Antarctica may still prevent the uptake of sustainable design. However, due to the complexity and difficulty of constructing in the southernmost continent coupled with the environmental sensitivities of Antarctica, all new structures would greatly benefit from consideration of the long-term benefits that ensue from analysing from the outset the building's whole-life cost.

"Approximately eighteen people will on average be permanently resident in the base. These people will produce approximately 20 kg of solid waste and 4,000 litres of sewage effluent daily. The design of the sewage treatment plan must be based on the peak sewage flow rate produced by the outfall from 80 people, amounting to approximately 16,000 litres per day."

SANAE IV Environmental, Health and Safety Impact Assessment, 1991

WASTE MANAGEMENT

In 1991 Annex III to the Environmental Protocol, commonly referred to as the Madrid Protocol, entered into force, revolutionising waste-management practices in the continent. According to the Australian Antarctic Government, prior to this Antarctic waste was either disposed of via open tips, landfills, and incinerators or via "sea-icing" – an approach which allowed for rubbish to be left floating on the sea ice in winter waiting for it to sink during the warmer months. Similarly, "sewage was burned or discharged with little or no treatment, whilst areas close to the stations and field camps became contaminated from oil and chemical spills."[1] Today, member states to the Treaty are requested to apply responsible waste-management principles and to designate a waste management official to develop and monitor (Article 10) waste-management plans (Article 8). Members of the expeditions receive waste-management training and are not permitted to bring hazardous materials, including polystyrene beads, PCBs and radioactive materials (Article 7). The waste that is produced is classified according to five main groups and management plans are to be reviewed in advance of the annual international meetings in accordance with the Antarctic Treaty (Article 9). In response to such regulations most stations, especially new ones, have invested heavily in sustainable waste-management solutions. Examples include Halley VI's bioreactor, which is able to separate dry incinerated sludge from clean water, reducing energy consumption. Waste is subsequently stored in Waste Barns, like the one captured above by Shaun O'Boyle, in which waste from McMurdo, Amundsen-Scott, and scattered field camps is processed and packaged prior to being transported to other continents.

Antarctic Environmental Impact Assessment

Neil Gilbert and Ceisha Poirot

NEIL GILBERT is a former environmental manager at Antarctica New Zealand who has worked on polar issues in various roles for over thirty years. CEISHA POIROT is New Zealand's representative to the Committee for Environmental Protection (CEP) and the expert group leader of the Council of Managers of National Antarctic Programs (COMNAP) Environmental Protection Group.

The idea of assessing the impact of human activity on the Antarctic environment was first discussed at the 6th Antarctic Treaty Consultative Meeting (ATCM) in 1970.[1] As early as 1975, the Antarctic Treaty Parties had agreed upon guidelines for Antarctic operators, which urged that in the planning of major operations in the Antarctic Treaty Area an evaluation of the environmental impact of the proposed activity be carried out.[2]

The Antarctic environmental impact assessment (EIA) system was further enhanced with the adoption of the Protocol on Environmental Protection to the Antarctic Treaty in 1991. A key principle of the 1991 protocol is that "activities in the Antarctic Treaty area shall be planned and conducted on the basis of information sufficient to allow prior assessments of, and informed judgments about, their possible impacts on the Antarctic environment and dependent and associated ecosystems."[3]

Article 8 of the protocol provides for environmental impact assessments to be undertaken for virtually all Antarctic activities, with procedural details set out in the protocol's Annex I.

The requirement to carry out an EIA is now well established within the Antarctic Treaty System, and virtually all activities that take place in Antarctica, both governmental and non-governmental, are subject to some form of EIA. The administering of this process across an entire continent is unprecedented.

The EIA framework set out in the protocol provides a flexible and scalable approach and establishes three levels of assessment (preliminary, initial, and comprehensive environmental evaluation) according to whether the activity in question is assessed as being likely to have less than, no more than, or more than a minor or transitory impact.

Preliminary and initial environmental evaluations are reviewed by national competent authorities within their own domestic administrative and legal systems. The requirements for Comprehensive Environmental Evaluations (CEEs) are more rigorous. CEEs must first be circulated in draft form to all parties to the protocol and to the Committee for Environmental Protection (CEP) for consideration. Further, all parties to the protocol must make a draft CEE publicly available for comment within their own country for a period of ninety days. A final version of the CEE that takes account of all comments received during the international review period must also be circulated to all parties,

along with notice of the final decision taken in relation to the activity. This international review mechanism is extremely rare in any EIA system and is a notable strength of the Antarctic EIA system. Such scrutiny also helps drive consistency of assessment and implementation for large-scale activities on the continent with the potential for greatest impact, such as the establishment of new stations or bases. The overarching objective of the protocol is that "the Parties commit themselves to the comprehensive protection of the Antarctic environment and dependent and associated ecosystems and hereby designate Antarctica as a natural reserve, devoted to peace and science."[4] Since the adoption of the protocol, the CEP has built up a considerable body of guidance material and other procedures to help protect the environment.[5] This body of work underpins and strengthens any Antarctic EIA.

Despite the strengths of the EIA system, concerns remain over a number of elements it contains. Many of these have been explored, though not necessarily resolved, by the CEP in recent meetings.[6]

A particular concern relates to the lack of any clear definition of the ambiguous term "minor or transitory impact", which results in differences of interpretation as to how they assess the level of EIA that is required for similar activities. There are examples where an EIA has arguably been undertaken at a lower level, with the consequence that these activities have not been subject to international assessment.[7]

There are also significant differences in the requirements for an Initial Environmental Evaluation (IEE) compared to those for a CEE, the requirements for the latter being far more detailed than the former. This has resulted in significant variability in the quality of IEEs, with some being extremely rigorous and possibly close to being a CEE, while other IEEs do little more than provide a very basic assessment of the potential impact of an activity.[8]

Further still, the requirements of what to include in an IEE make no explicit mention of the need to identify mitigation measures, which is regarded as a fundamental element of any EIA process.[9] The implication of this is that impacts that are judged to be no more than minor or transitory need not be further mitigated – even though such terms are open to broad interpretation.

The need to address cumulative impacts is a fundamental requirement for IEEs and CEEs,

though the adequacy with which Antarctic EIAs consider cumulative impacts has been the subject of considerable discussion within the CEP.[10] Concerns have also been raised with respect to parties assessing activities under their jurisdiction in complete isolation of separate, though potentially overlapping, activities being assessed by other parties[11].

Despite the efforts by the CEP to address this issue, the guidance available through the CEP's EIA guidelines[12] remains relatively modest, and there are opportunities to develop more comprehensive guidance for assessing cumulative impacts based on best-practice approaches elsewhere. Spatially oriented EIAs (as opposed to activity-oriented EIAs) have also been proposed as a means of better assessing cumulative impacts at key locations, such as tourist sites,[13] though none have yet been undertaken.

The fact that the majority of activities undertaken in Antarctica are subject to some form of prior environmental impact assessment is exceptional and establishes EIA as a core function in meeting the comprehensive environmental protection objectives of the protocol. Yet opportunities remain for further strengthening the Antarctic EIA system. With Antarctic environments under increasing pressure from expanding human activity and changing climate conditions,[14] there has never been a more important time to ensure that the EIA system is fit for purpose and up to date.

"The construction of China's fifth research base has also been controversial because preliminary building activities were started before the environmental impact assessment was complete, in violation of protocol. The lack of punishment for these – and similar infractions by other countries – is one of the weaknesses of the treaty system."

Leslie Hook (Environment and Clean Energy Correspondent) and Benedict Mander (Latin America Correspondent) "The Fight to Own Antarctica", *Financial Times*, 2018

Water Management in the Extreme

Tony McGlory

TONY MCGLORY is the director of Ramboll United Kingdom's Building Services Engineering team. Eng. McGlory has over thirty years' experience designing complex engineering projects. His responsibilities at Ramboll include leading Building Services Engineering design for projects in Antarctica and the subantarctic islands.

For human life to exist in any location a reliable supply of water must be available. Antarctica contains 70% of the Earth's fresh water, and is surrounded by the Southern Ocean, so at first glance water is certainly not a scarce resource on this continent. However, the water does not exist in a form suitable for human consumption even though snow, ice, and seawater are abundant. The methods Antarctic stations use to obtain drinking water vary depending on their location, the main factor being whether a station is positioned on the coast or inland.

Coastal stations with access to the sea convert seawater to drinking water using a process called reverse osmosis. This is a common water-purification process in which water is forced through a combination of filters and a partially permeable membrane. Although the treated water is safe to drink, it is devoid of nutrient-rich minerals including those that are beneficial for human consumption such as calcium and magnesium. To compensate for this deficiency, remineralisation is often applied by passing the treated water through a limestone filter bed allowing minerals to be absorbed and enhanced in conjunction with injections of very small quantities of carbon dioxide. Finally, to ensure that viruses and bacteria don't get into the potable water system, ultraviolet sterilisation is applied before the water is finally stored in a potable water tank.

To create 1 cubic metre of potable water typically requires 4 cubic metres of seawater and approximately 15 kilowatt-hours of electrical energy (consumed for the operation of the small-scale reverse osmosis system, excluding the power required to pump the seawater from the shoreline to the treatment area). Drawn from a well or coastal intake, the seawater is pumped into holding tanks. The well or seawater intake must be sufficiently deep to retain access to liquid seawater during winter and it must be held above its freezing point of around –2°C by trace-heating and insulating the pipework or by continually circulating the seawater within the pipelines.

Unlike their coastal counterparts, inland stations must melt snow or ice to create water. The most common way is to collect the snow or ice and place it in a melting tank. Collecting snow is labour intensive, and the amount that must be collected depends on the snow density; up to 10 cubic metres of snow could be required to create 1 cubic metre of water.

The American South Pole station, Amundsen-Scott ↗ p. 790, takes an alternative approach to sourcing fresh water, by adopting a technique developed by United States army engineer Raul Rodriguez in the 1960s. The Rodriguez Well, or Rodwell, entails creating a well by drilling through the ice using a hot water drill to form pools of water at the base of the well. A pump containing a heating element is then lowered into the well and used to maintain the water pool and pump fresh water up to the surface. The wells can be over 100 metres deep and generate huge quantities of water. As consumption increases the wells become deeper and wider; and once the well becomes too deep it is reused for depositing wastewater and a new well is drilled in its place to continue sourcing fresh water.

For any given snow or ice condition, the amount of energy that must be delivered to convert snow or ice into water will be the same irrespective of the method used. Assuming a snow temperature of –30°C, approximately 117 kilowatt-hours of energy is required to create 1 cubic metre of water at 5°C. However, this is only the starting point, as further energy will be expended to deliver water from the point of melting into the freshwater storage tank and this quantity varies from station to station. Energy is consumed: by the vehicles used to collect snow, through the inefficiency of heat exchangers, through the power required to pump water to the station, in the trace-heating of the external pipework to prevent freezing, and in generating electrical power for the chosen water treatment regime.

In theory, the water obtained from snow-melt is pure water and safe to drink. However, precipitation can absorb pollution from the atmosphere and snow and ice could become inadvertently contaminated by human activities around the station. Caution is therefore required and most stations run the melted snow through a filtration and treatment process similar to that described for seawater.

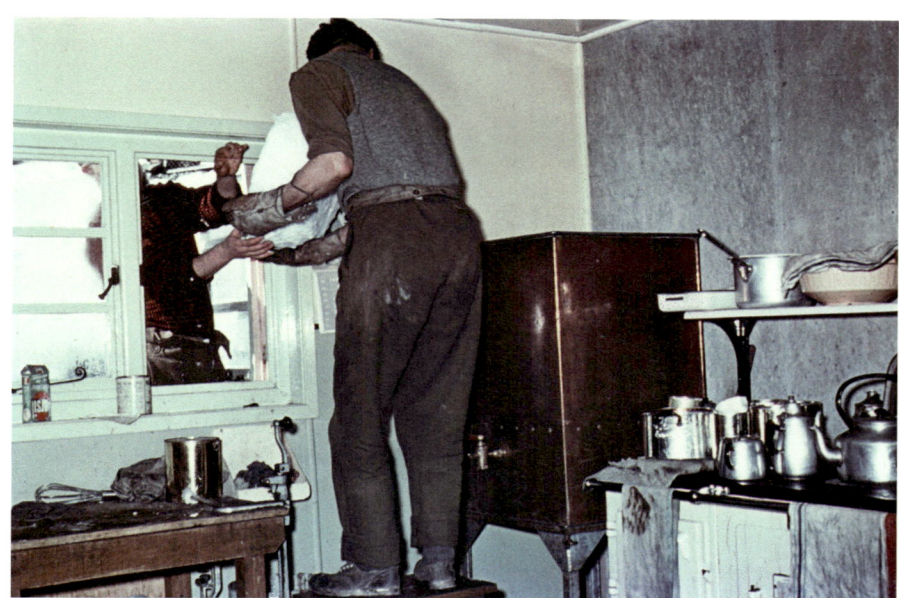

The size of the fresh water tanks for a station must be calculated carefully, they must be large enough to provide a suitable quantity of fresh water for daily use and must offer a level of resilience to cover the failure of the treatment system, but should not be over-sized. If there is not a sufficient regular consumption of water, the quality can deteriorate, creating conditions where bacteria can proliferate. It is therefore prudent to have at least two freshwater storage tanks; this allows one tank to be taken offline and cleaned without disrupting service and facilitates continued operation if one tank fails. Fresh water is finally pumped from these storage tanks into the pipe distribution system for the station.

Fuel is consumed creating, treating, heating, and pumping fresh water, along with collecting, pumping, treating, and disposing of waste water. Additional fuel is consumed transporting the fuel used on station to Antarctica, so even minor reductions in water consumption at the station can create a significant reduction in overall fuel consumption.

In locations subject to seasonal thawing where the topography can contain melt-water, or where the sun is capable of melting snow on metal building roofs, it is possible to recover meltwater and deposit it in the station meltwater storage tanks ready for treatment. Grey-water recovery systems can also reduce the need to create water:

waste water from washbasins, showers, washing machines, and even kitchens can be collected, treated using ultrafiltration and ultraviolet sterilisation, and reused. Although the water produced by this recycling process can be drinking quality, it is usually kept separate from drinking water and used for toilet flushing, laundry, and washing.

In the developed world, consumption of up to 100 litres of water per person per day is not uncommon, whereas water consumption in some Antarctic stations is below 30 litres per person per day. This substantial reduction in consumption levels is accomplished by using devices that reduce water consumption such as low-flow taps, low-flow showers, and vacuum toilets and is combined with a high standard of staff awareness, rationing, and discipline. In this respect, Antarctic water management could provide precious lessons for the more responsible use of water resources in the rest of the world.

Archival photographs on the opposite page taken at Base A on Goudier Island in 1956 depict the rather improvised solutions that were implemented in past decades to provide Antarctic scientists and expedition parties with drinking water. Cutting blocks of snow and passing them to a colleague in the kitchen through the window was routine in the southernmost continent for most stations. While nowadays desalination plants are implemented regularly on the coastline, water shortages on the plateau still lead to the absurdity of plastic water bottles being delivered to a continent which contains about 91% of all the ice in the world and approximately 70% of the planet's fresh water.

Halley VI

20 L

Water consumption per person per day

Achieved by implementing flow regulators, timed spray outlet controls for shower and taps, vacuum drainage systems (which consume only 1.2 litres per flush instead of the standard 9 litres), and retrofitted devices such as washing machines, dishwashers, and low-energy AAA-rated fridges and cookers.

United Kingdom

150 L

Water consumption per person per day

"There are a lot of useful technologies implemented within Antarctic stations, such as optimised water systems, which, if used more widely, would help us in the climate change battle".

Hugh Broughton (Architect, Hugh Broughton Architects), in conversation with UNLESS, Architectural Association School of Architecture, 2019

Global freshwater use in trillion cubic metres (m³)

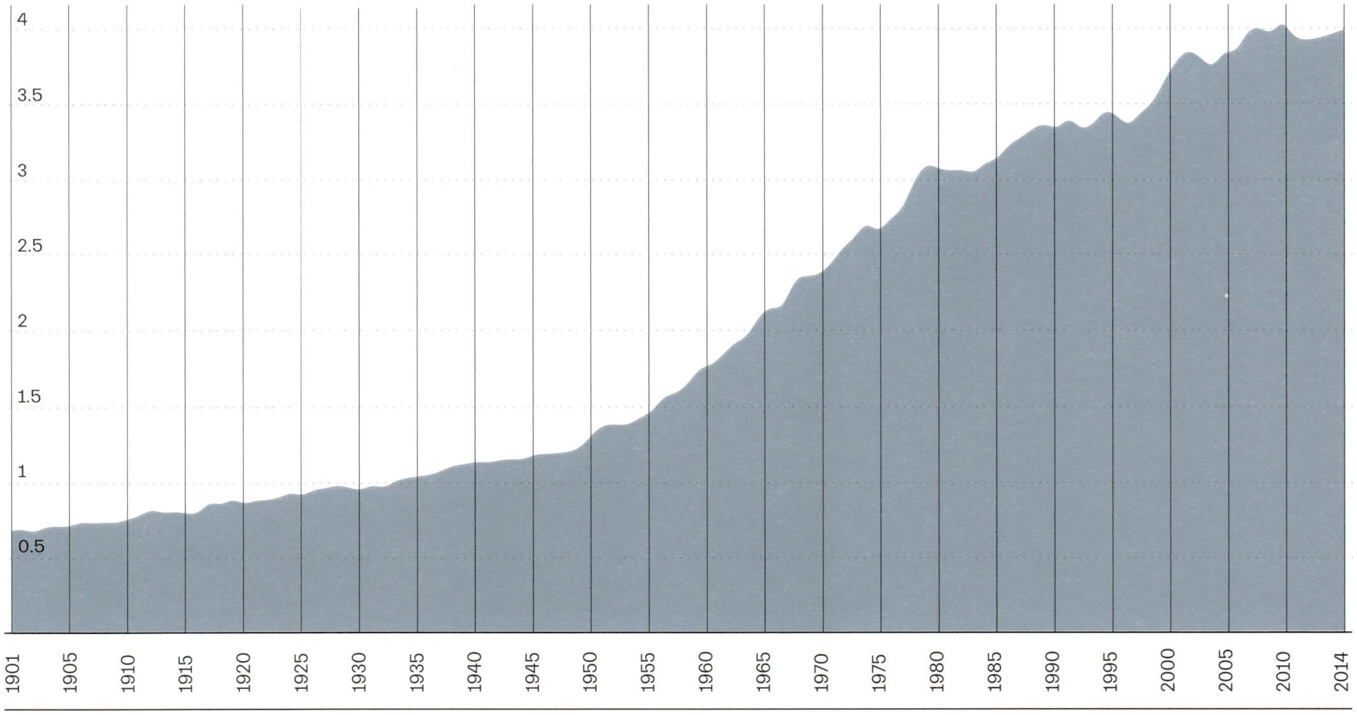

1901 1905 1910 1915 1920 1925 1930 1935 1940 1945 1950 1955 1960 1965 1970 1975 1980 1985 1990 1995 2000 2005 2010 2014

Rodriguez Wells and Water Management

Steve Theno

STEVE THENO is a professional engineer based in Alaska with over forty years' experience in the planning and design of facilities, energy systems, and infrastructure, predominantly in remote, cold regions. He is a former president of PDC Engineers (United States) who led the engineering design for the mechanical, electrical, and utilities systems for the third-generation Amundsen-Scott South Pole Station. Theno continues to be involved in Antarctic projects.

The Amundsen-Scott South Pole Station ↗ p.790 utilises Rodriguez Well technology for the station water supply. Developed by the Cold Regions Research and Engineering Laboratory (CRREL) in the mid-1990s, this innovative well typology is characterised by a thaw bulb in the snow pack with a diameter of approximately 30 metres and an average water depth of 25–30 metres, allowing a stored water volume of nominally 3.8 million litres – which is the equivalent of approximately fifteen months' supply at average station use.

The Rodriguez Well development at South Pole Station is founded on earlier research and field experience gained by CRREL in both Greenland and Antarctica and refined for the Amundsen-Scott Station operations. The thaw bulb for the well is initially formed using a submersible pump and a spray nozzle assembly attached to the bottom of the pump housing. Heated water is discharged from the spray nozzle into the snowpack beneath the pump, creating a borehole that penetrates down into the snowpack and forms an unlined well "casing". The development of the borehole continues until the pump assembly reaches a depth of approximately 60–70 metres, after which the snowpack transitions from an unconsolidated layer to firn, with a density approaching that of ice. At this point, the well pump elevation is held constant and water from the pump is heated and recirculated back into the well to create a large thaw bulb reservoir. The shape and dimensions of the thaw bulb are a function of the thermal energy input into the well, the amount of meltwater withdrawn and the characteristics of the snowpack. For the operating parameters of the well at the South Pole, the thaw bulb initially takes the shape of a tear drop. As water is drawn off for station use the thaw bulb continues to migrate down into the snowpack. The submersible pump elevation is adjusted downward to stay within the liquid water reservoir, so that as the well matures, the thaw bulb evolves to roughly the shape of a vertical cylinder with a spherical bottom.

The meltwater is very pure from a chemical and water quality standpoint, and the portion drawn off for station use is treated via filtration, pH adjustment, and disinfection. At South Pole Station, a treated water storage volume of approximately 19,000 litres is maintained, providing a buffer between the steady state output from the well and the fluctuating demands of the station. Waste heat from the power plant is used to heat water recirculated to the well, providing the necessary thermal energy input to maintain well operations.

The ultimate depth of the thaw bulb and capacity of the well is practically constrained by pumping energy requirements and the characteristics of the well pump. For South Pole operations a practical limit of 150–160 metres has been established, yielding a typical useful production life of five to ten years by melting ice that dates back to around 1000 BCE, depending upon the depth. Once a well has reached the end of useful water production, operations are transferred to a newly developed well and the closed well is reconfigured to be used for wastewater disposal. Traditionally, domestic waste water at the station has been disposed of by discharge into the snowpack, where it is eventually encapsulated as a frozen mass in the ice. The disposal thaw bulb was generated by the thermal energy in the waste water itself; the ultimate shape and capacity was a function of the mass inflow and thermal energy content coupled with the characteristics of the snowpack. Using the closed water wells now provides improved storage volumes, requires less development effort, and offers longer life.

With the new approach, the waste water is piped to the vacated well and discharged into the old well cavity. Over time, the accumulating waste-water mass freezes and after several years, when the well cavity is nearly full, it is sealed with an ice plug by filling the remaining volume with water. The sealed waste-water mass freezes solid, a state devoid of any biological activity and the waste water from the station is piped to the next available well cavity for continued operations. At South Pole Station, the underlying snow and ice pack is moving at a nominal rate of 10 metres per year in a direction 32.8° West of Grid North. The encapsulated frozen wastewater mass thus migrates with the pack movement, and, as the 20-centimetre annual snow accumulation at the surface slowly compresses the underlying pack, its depth below the surface gradually increases as it moves.

The implementation of the Rodriguez Well and its integration into the design and operation of Amundsen-Scott Station offered several advantages over traditional practices, including a reduction in operational costs. While the traditional approach to snowmelt supply is very labour intensive, once developed the Rodriguez wells operate with minimal maintenance and operational resource requirements and provide significant reserve capacity with an in situ storage volume of more than a year of normal station demand. Alongside labour and storage advantages offered by this well typology,

it allows for significant energy efficiency when compared to traditional snowmelt systems, which require substantial energy input, not only to achieve melting but also in terms of the fuel required to operate the front-end loaders performing snow mining operations. Furthermore, the nature of the Rodriguez Well thaw bulb allows the system to maximise the utilisation of waste heat from the power plant. During periods of excess waste-heat generation, the excess energy can be dumped into the Rodriguez Well, furthering the development of the station's water supply reserves, while during periods of limited waste-heat availability, the mass of the liquid reservoir in the well enables energy to be throttled. Ultimately, the Rodriguez Well offers a highly reliable water supply system with minimal operations and maintenance requirements once developed. And using deactivated well cavities for wastewater disposal provides additional benefits in terms of reduced development effort and increased disposal capacity per developed site for the waste-water system.

"The sewage waste disposal system consists of a 38-meter diameter 'bulb' that is installed 36 meters below the snow surface. After a hole is drilled in the snow, intrinsic heat in the wastewater stream melts the snow and forms a cavity. Over time, the depth of the bulb stabilizes. The sewage bulbs can typically accommodate up to approximately 7.6 million litres of waste. Historically, sewage bulbs at the main station have lasted five to six years before their use is discontinued and a new sewage disposal bulb developed."

National Science Foundation, *Final Environmental Impact Statement, Modernization of the Amundsen-Scott South Pole Station Antarctica*, 1998

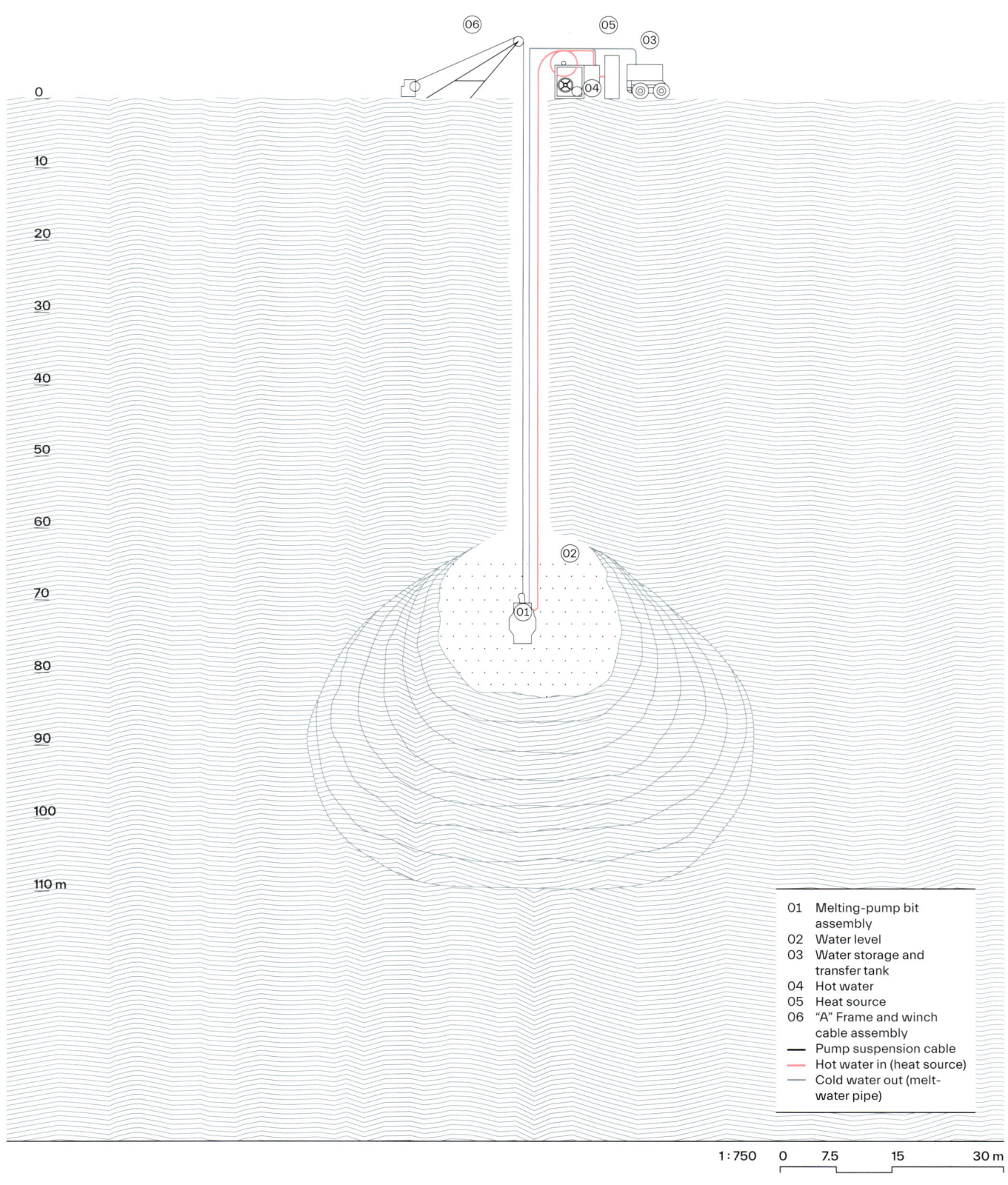

01	Melting-pump bit assembly
02	Water level
03	Water storage and transfer tank
04	Hot water
05	Heat source
06	"A" Frame and winch cable assembly
—	Pump suspension cable
—	Hot water in (heat source)
—	Cold water out (melt-water pipe)

1:750   0   7.5   15   30 m

The Challenges of Fire Safety in Antarctica

Tony McGlory

TONY MCGLORY is the director of Ramboll United Kingdom's Building Services Engineering team. Eng. McGlory has over thirty years' experience designing complex engineering projects. His responsibilities at Ramboll include leading Building Services Engineering design for projects in Antarctica and the subantarctic islands.

Fire has devastating consequences anywhere, but in Antarctica the consequences can be significantly more acute. Arguably, all Antarctic station buildings – be they used as laboratories, accommodation, food stores, or for power and heat generation, etc. – are essential, and potentially the loss of any building can endanger life, cause hardship, or ruin several years of scientific experimentation and advancement.

Despite the increased fire hazard due to the low moisture content of materials caused by the dry Antarctic atmosphere, the continent does not have specific fire regulations. Larger stations may have career firefighter representation on their teams, but most stations have staff with only a basic level of firefighting training. Each nation generally adopts its own domestic standards, or internationally recognised standards, enhancing them where possible to reflect the increased risks and consequences in Antarctica.

Typically, building regulations consider the safety of occupants by ensuring the inhabitants can evacuate the building quickly and safely. However, in Antarctica it is not always safe to go outside without preparation, and preparation can take precious time. It is therefore necessary to develop a holistic station fire strategy, rather than treating each building in isolation.

Both passive and active fire protection measures are recommended: the former is intended to contain a fire within the room of origin (thus restricting the spread of fire and smoke for a limited period of time) and to protect the integrity of escape routes; the latter entails automatic detection and protection measures that warn occupants of the existence of fire along with systems that attempt to suppress or extinguish the fire.

Essential passive fire protection measures include the following: achieving adequate spacing between buildings to prevent fire spread, using non-combustible construction materials where possible, and enhancing fire compartmentation and smoke-resistant enclosures within buildings beyond the normal ratings to contain or slow the spread of fire and increase the time available for escape. In addition, limitations on maximum travel distances must be in place to create escape routes to adjacent fire compartments. It is deemed imperative in Antarctic station design to include the provision of a refuge space where staff can congregate in the event of fire, particularly in locations where climate conditions may impede escape from a building and occupants have to wait for external assistance.

Increasing fire protection to load-bearing structural elements beyond the norm is also advantageous, as this will minimise, or ideally avoid, structural fire damage and expedite repair (time and cost), reinstatement, and reuse of the building. The first stage of active fire protection is the

automatic fire detection and alarm system, which typically comprises smoke detectors, heat detectors, sounders, and flashing beacons. Ideally this should be a station-wide system where all buildings are integrated into one fire alarm system. The benefit of this is that all occupants on station are alerted to a fire, no matter which building they may be located in. Often the fire alarm system is complemented by a voice alarm system which can provide verbal instructions, communicating the location of the fire, for example, or the point of refuge. Battery-powered emergency lighting is also required to illuminate escape routes in the event of a power failure.

The are many options available for fire suppression such as hydrant, hose reel, sprinkler, and mist and fog systems, as well as systems that utilise foam, powder, aerosols, and gaseous compounds. Hydrant, hose reel, and traditional sprinkler systems are familiar fire suppression measures used throughout

the world; however, these systems require large volumes of stored water to be readily available, which is difficult to achieve in Antarctic conditions. These systems also create a high risk of water damage post activation, particularly as the water can freeze and compromise the building's structural and thermal integrity. The other systems that are available, such as those using mist, fog, or gas, reduce or even obviate the need for firefighting water. Mist and fog systems will also reduce water damage and potential contamination resulting from run-off water. However, these systems have other limitations and complexities that need to be considered when selecting a suppression system for a building in Antarctica.

Perhaps counter-intuitively, fires (like the one shown on the left which affected two garages at McMurdo Station on the evening of the 1st of December 1981) are considered one of the greatest threats in Antarctica. Water, plentiful in its frozen state, is rare in liquid form, and survival outside an enclosed structure is, to say the least, uncertain. To overcome this risk, all inhabitable structures on the continent have a twin shelter that in case of emergency can house the inhabitants for as long as a whole winter. At Halley VI, the two accommodation units are connected by a pedestrian bridge running above two independent melt tanks that provide water to the station.

According to the 2017 Antarctic Station Catalogue published by the Council of Managers of National Antarctic Programs (COMNAP), the total built area of active Antarctic stations accounts for 211,062 square metres. The diagram below lists the stations by size from the smallest to the largest, ranging from Byers Field Camp (which accounts for 32 square meters) to McMurdo (which accounts for 32,750 square meters).

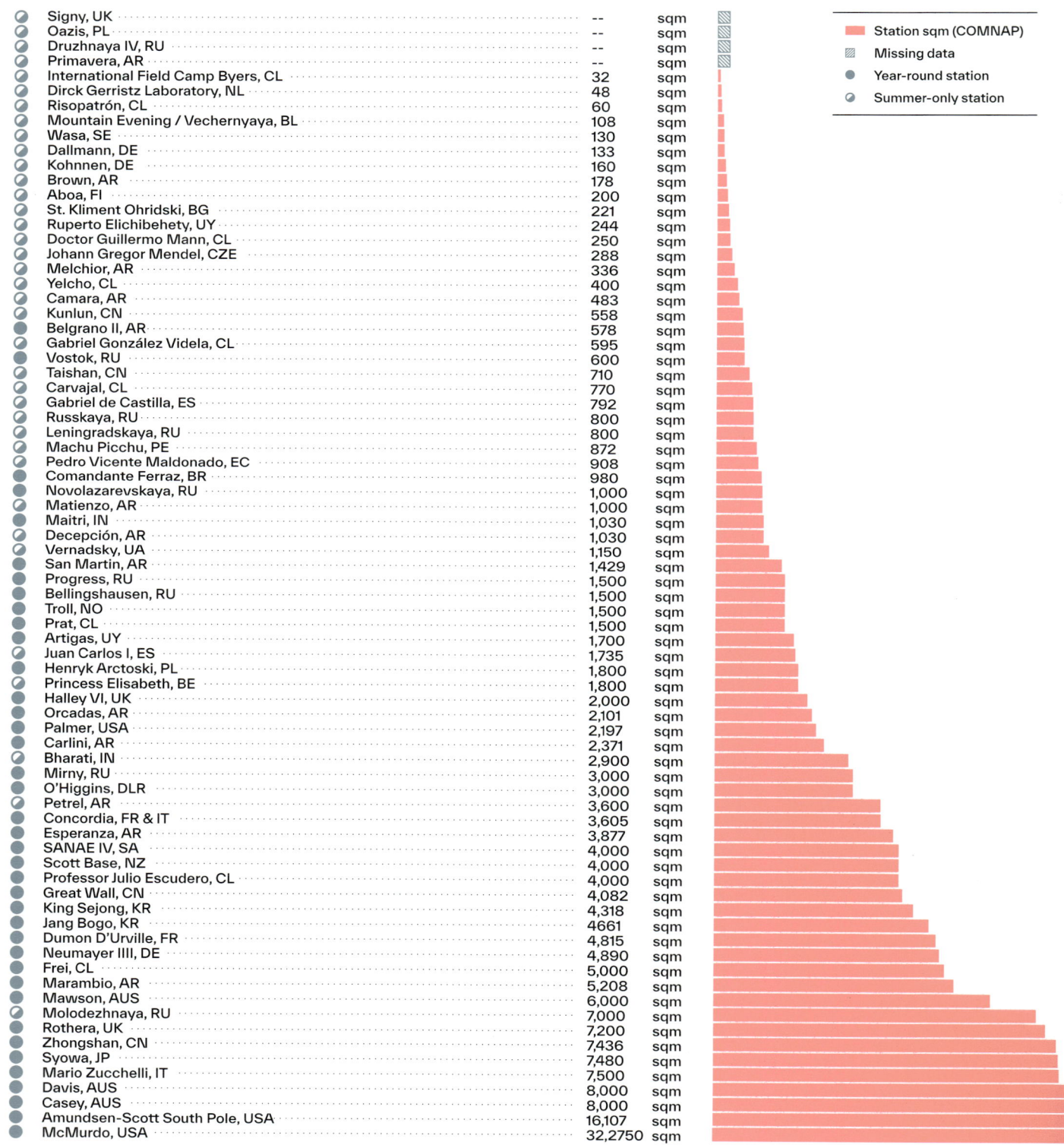

Legend:
- Station sqm (COMNAP)
- Missing data
- Year-round station
- Summer-only station

Station	sqm
Signy, UK	--
Oazis, PL	--
Druzhnaya IV, RU	--
Primavera, AR	--
International Field Camp Byers, CL	32
Dirck Gerristz Laboratory, NL	48
Risopatrón, CL	60
Mountain Evening / Vechernyaya, BL	108
Wasa, SE	130
Dallmann, DE	133
Kohnnen, DE	160
Brown, AR	178
Aboa, FI	200
St. Kliment Ohridski, BG	221
Ruperto Elichibehety, UY	244
Doctor Guillermo Mann, CL	250
Johann Gregor Mendel, CZE	288
Melchior, AR	336
Yelcho, CL	400
Camara, AR	483
Kunlun, CN	558
Belgrano II, AR	578
Gabriel González Videla, CL	595
Vostok, RU	600
Taishan, CN	710
Carvajal, CL	770
Gabriel de Castilla, ES	792
Russkaya, RU	800
Leningradskaya, RU	800
Machu Picchu, PE	872
Pedro Vicente Maldonado, EC	908
Comandante Ferraz, BR	980
Novolazarevskaya, RU	1,000
Matienzo, AR	1,000
Maitri, IN	1,030
Decepción, AR	1,030
Vernadsky, UA	1,150
San Martin, AR	1,429
Progress, RU	1,500
Bellingshausen, RU	1,500
Troll, NO	1,500
Prat, CL	1,500
Artigas, UY	1,700
Juan Carlos I, ES	1,735
Henryk Arctoski, PL	1,800
Princess Elisabeth, BE	1,800
Halley VI, UK	2,000
Orcadas, AR	2,101
Palmer, USA	2,197
Carlini, AR	2,371
Bharati, IN	2,900
Mirny, RU	3,000
O'Higgins, DLR	3,000
Petrel, AR	3,600
Concordia, FR & IT	3,605
Esperanza, AR	3,877
SANAE IV, SA	4,000
Scott Base, NZ	4,000
Professor Julio Escudero, CL	4,000
Great Wall, CN	4,082
King Sejong, KR	4,318
Jang Bogo, KR	4661
Dumon D'Urville, FR	4,815
Neumayer IIII, DE	4,890
Frei, CL	5,000
Marambio, AR	5,208
Mawson, AUS	6,000
Molodezhnaya, RU	7,000
Rothera, UK	7,200
Zhongshan, CN	7,436
Syowa, JP	7,480
Mario Zucchelli, IT	7,500
Davis, AUS	8,000
Casey, AUS	8,000
Amundsen-Scott South Pole, USA	16,107
McMurdo, USA	32,2750

The Footprint of Antarctic Stations

Shaun T. Brooks

SHAUN T. BROOKS is an associate at the University of Tasmania (Australia). He is an environmental scientist specialising in conservation with field experience across East Antarctica and the subantarctic. Dr Brooks' research focuses on the effects of the anthropic footprint on Antarctica as a means of promoting informed environmental protection policies.

The notion of *footprint* in Antarctica is important because it is the only continent where human impacts can be measured by where they are, as opposed to identifying wilderness areas where they aren't – as is the norm in the rest of the world. This reflects Antarctica's mandate as a continental "natural reserve, devoted to peace and science".[1]

There are many uses of the term footprint in Antarctica, but the focus on buildings and the disturbance they cause is singled out here as it represents the most evident, persistent, and widespread form of human impact on the continent. Disturbance to ice-free ground has been demonstrated as a proxy for further forms of environmental impact, and the hotspots of activity which buildings cause are inextricably linked with accidents, interference with wildlife, pollution, fuel spills, and introductions of non-native species. By measuring the footprint on the continent, understanding where it is, and reflecting on how planning can increase or reduce impact, progress can be made in protecting the vulnerable polar environment while still enabling access for science and tourism.

Up until the nineteenth century, Antarctica remained isolated from human activity and the impact that it inevitably has on the territory. Anthropic activities in the Antarctic can be traced back 200 years to the advent of the sealing industry, which developed rapidly within the Maritime Antarctic. Permanent infrastructure on the continent was first established in 1898 during the Heroic Age of exploration. The following fifty years saw a gradual increase in the level of development on the continent interrupted only by the two world wars. Although there were bursts and pauses of activity in Antarctica and territorial claims were made, the human footprint remained negligible. The 1950s, however, witnessed a turning point, culminating with the 1957–1958 International Geophysical Year, which shifted attention to Antarctica, and thus led to a massive footprint expansion. Although geopolitical tensions eased with the ratification of the Antarctic Treaty, the footprint has continued to grow ever since. In less than seventy years, research stations, and a handful of tourism facilities, spread across the continent, amounting to a building footprint, in 2019, of over 393,000 square metres – the equivalent of nearly 27,000 shipping containers. On ice-free land, the ground disturbance surrounding these buildings as detected by satellite reached 5,242,000 square metres – the equivalent of nearly 1,000 National Football League sports fields. These figures are important, as they show that infrastructure (primarily associated with research stations) has the most pronounced environmental impact on terrestrial Antarctica. This process is exacerbated by ongoing national investment in buildings that includes the construction of new stations, the modernisation of old stations, and the creation of new facilities such as runways.

Although huge wilderness areas still exist on the ice sheets and ice shelves of Antarctica, aligning with typical perceptions of the continent, ice-free land is under pressure. While geopolitical and research interests are common determining factors for the locations of research stations, the practicality and accessibility of coastal ice-free land has resulted in these areas being disproportionately targeted for construction. Ice-free areas make up less than 0.5% of the continent[2] but contain 81% of all station buildings. This small fraction of land also provides the essential habitat for most of Antarctica's biodiversity including all bryophytes, lichens, plants, terrestrial invertebrates, and most

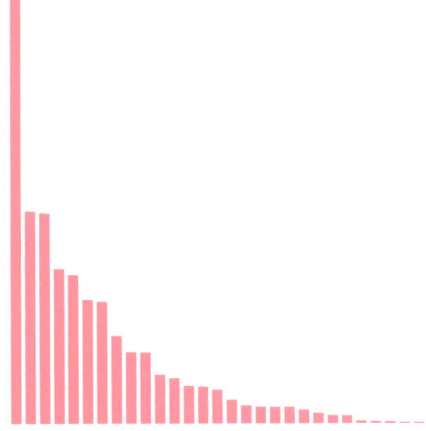

United States of America | 3 stations | 51,054 sqm
Argentina | 13 stations | 22,191 sqm
Australia | 3 stations | 22,000 sqm
Russia | 10 stations | 16,200 sqm
Chile | 9 stations | 15,575 sqm
China | 4 stations | 12,786 sqm
Italy | 1.5 stations | 9,300 sqm
United Kingdom | 3 stations | 9,200 sqm
Republic of Korea | 2 stations | 8,979 sqm
Japan | 1 station | 7,480 sqm
France | 1.5 stations | 6,615 sqm
Germany | 3 stations | 5,183 sqm
New Zealand | 1 station | 4,000 sqm
South Africa | 1 station | 4,000 sqm
India | 2 stations | 3,930 sqm
Spain | 3 stations | 2,559 sqm
Uruguay | 2 stations | 1,944 sqm
Belgium | 1 station | 1,800 sqm
Poland | 1 station | 1,800 sqm
Norway | 1 station | 1,500 sqm
Ukrain | 1 station | 1,150 sqm
Brazil | 1 station | 980 sqm
Ecuador | 1 station | 908 sqm
Peru | 1 station | 872 sqm
Czech Republic | 1 station | 288 sqm
Bulgaria | 1 station | 221 sqm
Finland | 1 station | 200 sqm
Sweden | 1 station | 130 sqm
Republic of Belarus | 1 station | 108 sqm
Netherlands | 1 station | 48 sqm

Please note that the measurements above are approximate values, as the COMNAP station catalogue does not account for all buildings of each station.

penguin rookeries and seabird nesting sites. The distribution of infrastructure is also significant, with more than half of all large ice-free areas around the coast of Antarctica showing signs of disturbance, which, it has been shown, impacts the biological and physical environment, leading to irrevocable loss of ancient landforms, permafrost destabilisation, contamination, and reduction of habitat suitability for polar ecosystems which are less able to cope with change than those in other climates.

What can looking at the footprint on Antarctica teach us about how to visit the continent, while minimising devastating human impacts on its fragile ecosystems? The site upon which stations and facilities are built is one of the most important factors: ice-covered land makes up over 99.5% of the continent yet it contains only 19% of the buildings. While establishing infrastructure on ice poses architectural and engineering design challenges, the magnitude of this landscape and its limited biodiversity mean that any impact on this environment has a far smaller footprint relative to those on ice-free sites. Any new proposal for infrastructure on the continent should thus first consider placing it on ice. For the majority of the existing stations already within ice-free areas, their exact location is key. Most facilities on ice-free areas are, in fact, built on soil and gravel sites, but these soft substrates are commonly vulnerable to impact, increasing the likelihood of creating a persistent disturbance footprint compared to those built on rock. The configuration of buildings and infrastructure has similar effects. Across all ice-free stations and tourist facilities, those that have their buildings within a contained "core" area have caused, on average, two and a half times less disturbance to the surrounding environment than those with a spread-out infrastructure. This also applies to multiple stations within close proximity that share infrastructure – avoiding the creation of unnecessary, environmentally damaging, duplicate facilities. But this reduces impact only if such research stations actually contribute a scientific benefit. As such, decisions on creating new infrastructure should be informed by careful evaluations of whether the resultant footprint is really necessary for science, if it could be avoided or simply minimised, and whether the scale of impact envisaged is consistent with Antarctica's designation as a continental natural reserve, rather than responding purely to convenience and cost considerations.

Superficially, the management of the footprint of stations in Antarctica – given its scale – may appear to be of little consequence in comparison to the scale of the environmental challenges faced globally, yet the state of the Antarctic environment is important, as it represents the only territory on planet Earth where true international cooperation and shared governance (not bound by active national claims) have the potential to preserve the region as a pristine natural reserve. Given the current situation and the exponential increase in the building footprint, if the construction of new facilities continues, strategic planning and design will be urgently needed to minimise the anthropogenic impact on nature and meet environmental obligations. Better planning of where and how stations are located, an incentive to share facilities, and greater use of previously disturbed sites have the potential to benefit access while simultaneously minimising the contamination impact on wildlife and preventing the irreversible loss of habitat and landscapes.

"Even if the purpose of a new station is primarily *bona fide* scientific research, current practice suggests that the need for a permanent station is relative. Of the 73 stations in the ATA [Antarctic Treaty Area] listed in the COMNAP website, 33 (45%) are active only in summer. Many summer scientific activities could be carried out of vessel or field camps of different sizes and characteristics. Even large-scale, multi-year scientific projects can, and have been, conducted from temporary camps."

Antarctic Southern Ocean Coalition in their report *Are More Antarctic Stations Justified?* presented at the XXVII ACTM, March 2004

Heroic huts and contemporary Antarctic stations alike leave a footprint on the Antarctic territory that cannot be overlooked. Whilst the storage crates that peppered the Cape Evans site during its fourteen-day period of construction (which occurred between the 4th and 18th of January 1911) and the plentiful instrument huts that surrounded Scott's famous settlement (including the magnetic hut, the meteorological screen, the magnetometer post, the instrument shelter, and the southern stores dump[1]) might project a sense of temporality that does not suggest high degrees of contamination, most contemporary stations, built on concrete cut-and-fill foundations, leave traces that are not negligible. Even the new Comandante Ferraz station, which was designed with great care after a fire erased its predecessor, is more dispersed in the landscape than is evident from studying the contained linear volume of the station itself.

"Approximately 40 percent of all tonnages flown to the inland stations consists of fuel. The Navy is constantly trying to find ways of cutting down on this requirement by reducing consumption on the spot, preventing leaks, by tightening inventory and by planning flights carefully to consume fewer flying hours in support of scientific programs."

Operation Deep Freeze 64, Task Force 43, 1964

To take off in conditions of "rarefied air and sand-like snow", airplanes required "JATO [expanding gas] boosters". The "seventy-seven pounds of fuel in each container" were connected to R4D at McMurdo Sound, propelling the planes forward "with a thousand-pound thrust". Planes then allowed "fuel drums [to] cascade plunging through the frosty air [with] cargo parachutes slow[ing] the descent."[1]

Built-In Redundancies or Infrastructure Resilience

Tony McGlory

TONY MCGLORY is the director of Ramboll United Kingdom's Building Services Engineering team. Eng. McGlory has over thirty years' experience designing complex engineering projects. His responsibilities at Ramboll include leading Building Services Engineering design for projects in Antarctica and the subantarctic islands.

Primary engineering systems are essential for preserving life in Antarctica. Electrical power generation, for example, must have provision for concurrent maintainability and it must be possible for any part of the essential systems, or subsystems, to be removed without impacting on operations, life safety, and mission equipment. One normal and one alternative distribution path should be created for all essential systems, and these must contain an element of redundancy so that standby equipment is available for immediate operation in the event of a failure of one component.

Commonly, engineers use the Uptime Institute Tier system[1] to define the level of redundancy built into engineering systems. This guidance was developed primarily for data centre infrastructure, but it can also be applied to any critical engineering system. The tier system operates across four levels of increasing resilience.

Essential systems should always be provided with a minimum of N+1 equipment redundancy, where N is the number of items needed to meet a given load, therefore N+1 retains

a spare unit in reserve. During maintenance or repair activities the risk of loss of a service may increase if no alternative path or equipment is available. If this risk of service loss during maintenance is not acceptable, a minimum N+2 configuration must be provided.

Systems cannot be considered in isolation, as availability can be affected by other systems or the buildings in which they are housed. For example, if essential systems and equipment are housed in the same building or plant room, there is a risk that a loss of the building or plant room, as a result of fire, for instance, could result in a loss of availability to the rest of the station. The optimum arrangement would thus be to house the duplicated equipment in separate independent buildings.

Since the cost of providing increased levels of resilience is high, up to three times the cost of a standard system depending on the chosen tier, a balance must be found between avoiding risk and meeting project budgets. In such cases it is not unusual to provide higher levels of resilience in the primary sections of the system, while allow-

ing resilience to be reduced in areas of the system where the consequences of failure are less critical and easier to manage.

The approach to resilience described in the electrical power example below is equally applicable to all essential systems in a station, for example: fuel, heating, water generation, waste treatment, sprinkler systems, etc. Renewable energy devices, when operating, would increase generating resilience, but since the generating power from these devices varies depending on climate, they are not factored into the resilience rating.

Due to its harsh climate and challenging accessibility, Antarctica is one of the most difficult places on Planet Earth for humans to live safely. Assistance from the rest of the world could take approximately six to nine months to reach a station, so all vital engineering systems need to rely on infrastructural redundancy to achieve the highest standards of resilience and thus life safety.

Using a power generation and electrical system as an example and assuming a 100 kW building peak electrical demand, the system from the generators up to the low voltage (LV) switchgear would have an overall plant redundancy of 2 (N+2) in normal operation. A loss of either building 1 or 2 would provide N+2 provision during the incident. If one of the remaining generators were in maintenance at the time the building loss occurred, resilience would be reduced to N+1 during this time. The generator resilience in the example exceeds N+1 and the distribution system possess alternative paths to the duplicate LV switchgear. This part of the system would be considered Tier III. The power supply between the distribution boards is a risk of single point of failure, as is the power connection from the renewable energy supplies, due to the single cable connections to the main distribution system. As a result, this part of the system would be considered Tier I, which often would be deemed a reasonable management of risk. As there are many distribution boards in a normal system, loss of one board would not normally be critical because power would be available at other boards in the system and such problems are usually easily and quickly repaired. If certain distribution boards are deemed essential, for example a medical area or server room, such boards could be fed by dual supplies and auto changeover switches.

Vostok Station

Yuri A. Shibaev

YURI A. SHIBAEV is the head of construction of the new wintering complex at Vostok Station on behalf of the Russian Arctic and Antarctic Research Institute (Russia). He acted as scientific director of the first Progress-East trip in the 53ʳᵈ Russian Antarctic Expedition. Shibaev has participated in the implementation of the federal World Ocean Program, the creation of the Consortium of European Research Libraries, and various projects on palaeoclimate.

Vostok Station ↗ p. 862 (the only Russian intra-continental Antarctic station) was founded on the 16ᵗʰ of December 1957 during the Second Complex Antarctic Expedition. The station was named after the *Vostok,* one of the two sloops led by Faddey Bellingshausen during the First Russian Antarctic expedition in 1819–1821, the expedition which famously led to one of the first sightings of the continent on the 28ᵗʰ of January 1820.

Located on the South Geomagnetic Pole (of the time), Vostok Station is well-known not only for having recorded the lowest temperature on our planet (–89.2°C), but also as the site in which the deepest ice-core well ever was drilled (3,769 metres deep) over the fifth largest freshwater lake in the world, also known as the largest subglacial lake in Antarctica (Lake Vostok). Thanks to advanced technology developed at the Climate and Environmental Research Laboratory (CERL) in the Arctic and Antarctic Research Institute, this allowed Russian scientists to extract climatic records that were over 1.3 million years old.

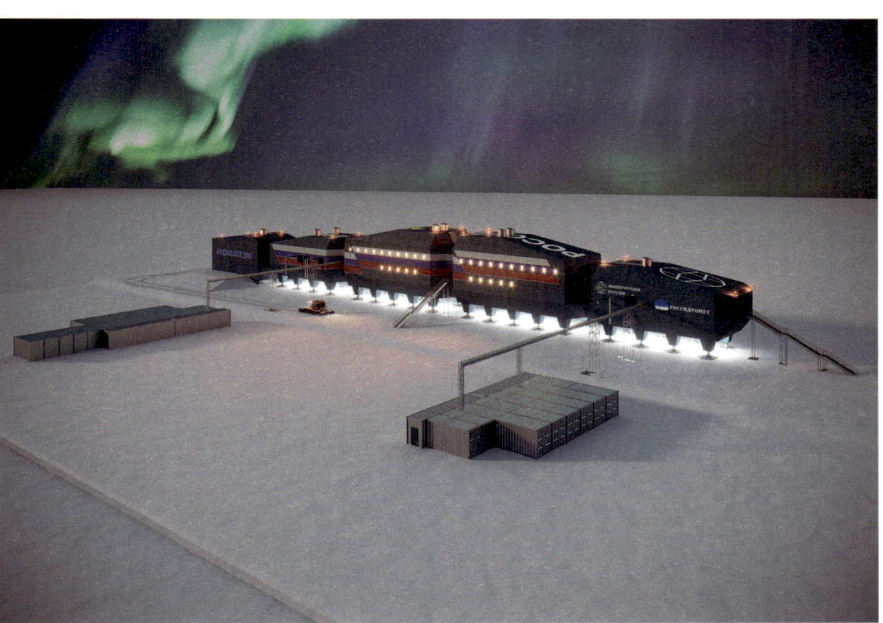

The location and climatic conditions at the Vostok Station are such that you can reach its ice dome either by plane or by land. For the latter, traverses may take from ten to forty days, forcing the parties and convoys to navigate first through a zone of crevasses (at a distance of up to 15 kilometers from the shore of the ocean), then though a *zastrug* zone (i.e. an area subjected to frequent storm winds and zero visibility, dense vertical snow ridges that reach 1.5 metres in height), and finally through a zone of very loose snow charged with severe snowdrift perils. Either way, the station can only be reached during the short window of austral summer between November and February, when the "milder" temperatures prevent the planes' skis from freezing and allow for the traverse. For the remaining months the station – located some 1,500 kilometres from the shores of the Southern Ocean – remains inhabited yet inaccessible.

Living at Vostok Station is especially hard since it is located at an altitude of 3,500 metres above sea level and thus presents reduced atmospheric air pressure, lower oxygen content (which, due to low temperatures corresponds to an oxygen level generally found at 5,000 metres above sea level), severe frosts, and low humidity. In addition, the few who call it home – also known as "Vostochnik", only drink distilled water, i.e. melted snow, deprived of the necessary minerals.

Over the course of sixty-three years Vostok Station was gradually covered with snow. It was only thanks to the enthusiasm of the "Vostochnik" that the entire complex could be maintained in a condition suitable for a small team to live and work in. A system of snow-covered crossings was built with passages dug under the snow. The main snow corridor was dug from the diesel power station side directly to the radio entrance, leading to the central area where the station personnel live. The road was named Solyanik Street, in honour of the Russian Antarctic expedition member who had wintered at Vostok for several years. Since it was first opened, only one reconstruction was completed at the station, in the 1970s.

In 2019 preparatory work began at the site for a new station. Designed on 3-metre-high supports to slow snow accumulation, the station will embody the following advanced measures: diesel and ventilation systems will be optimised to recycle heat from cooling systems to warm the air; the waste-water system is designed to recycle and purify 70% of the water consumed in the station, with solar panels being used to create meltwater in the summer months; LED-light systems will be used exclusively, with servers designed to have double/triple backups; and all products shipped from the mainland are designed to be used on-site with little to no waste. Given the extreme site conditions, a two-year stock of fuel and general supplies storage will grant a level of autonomy sufficient for two winters.

The work started in the 2019–2020 season by filling and compacting the snow base "slab" for the construction of the station. Given the low snow densities on the surface of the ice cap in this region (approx. 0.35 grams per cubic centimetre) the foundation slab was calculated to be 200 metres long and 120 metres wide, with a thickness of almost 3 metres.

Station modules are prefabricated and manufactured fully assembled to optimise the installation process in Antarctica. A control assembly and the station's main life-support systems will be tested in Russia beforehand and AARI personnel will be trained to use the complex engineering systems. The delivery of the Vostok Station modules deep into Antarctica will be carried out using a set of unprecedented measures of Antarctic intracontinental logistics operations carried out by the AARI Russian Antarctic Expedition. For its implementation more than thirty tractors and unique heavy transport platforms on skis will be used, allowing the transportation of goods weighing up to 60 tons. Over sixty driver-mechanics will be involved in conducting the sledge-caterpillar traverse.

The new wintering complex at Vostok Station will open up opportunities for the implementation of scientific projects, including the study of the water column and sediments of subglacial Lake Vostok and other frontier projects.

ANNE NOBLE is an artist, a professor of fine arts at Massey University (New Zealand), and an NZ Arts laureate. She has visited Antarctica as an NZ Arts fellow and a US National Science Foundation (NSF) fellow. Her work examines the imagination and representation of Antarctica. She has exhibited and published widely on the photographic construction of the Antarctic imaginary. Her books include *Ice Blink: An Antarctic Imaginary* (Clouds, 2011), *These Rough Notes* (VUW, 2012), and *The Last Road* (Clouds, 2013).

THE LAST ROAD
Anne Noble

"Almost one-hundred years after the explorer Robert Falcon Scott's fated Antarctic expedition, there was a quiet and unobserved traverse of the southern continent. Its mission was to lay a road – a traverse spanning 1,600 kilometres from McMurdo Station to the US base at the South Pole. I heard an interview on the radio with an engineer who described creating the road as a 'breakthrough project' that would enable all the fuel and heavy equipment, currently flown in hundreds of flights each year by military aircraft, to be hauled overland by tractors.

In 2008, the year I was in Antarctica as a grantee on the US National Science Foundation Artists and Writer's programme, the first delivery of 8,000 gallons [30,000 litres] of diesel in large rubber fuel bladders arrived at the South Pole station. Hailed as a success, the mission has now grown to haul annually as much as 900,000 litres of fuel across the continent. 'Each year the volume will increase [...] we have to work up to that,' the man on the radio said.

I noticed little diesel depots everywhere I went in Antarctica. Prosaic and colourful orange cans stacked up and signposted. To pass my skidoo license I had to know how to pull-start the motor, how to fill up, and how to clean up a fuel spill should I splash some diesel on the white carpet. Wherever there are people in the wide white terrain of Antarctica – from the coast to the Pole – there is the hum of machines supporting the march of curious individuals in their search for knowledge and experience."[1]

Fuel Depot, South Pole Station, 2008

Fuel Bladder, Siple Dome, The Polar Plateau, 2008

First car in Antarctica.

Could any car handle Antarctica?
Could it take the terrible battering of ice like broken glass?
Could it start, and travel, and keep travelling, in sub-zero temperatures?
Could it hold a course on a sea of ice?
In a place like this, could men entrust their lives to it?
If any car could do it, we felt the Volkswagen could.
So a VW1200 was shipped to Mawson to become the first sedan on the Antarctic continent.
How is it going?
Like the Volkswagen always does.

Australian Women's Weekly — July 10, 1963

Ice-hills and crevasses can't stop it, anymore than bush tracks or day-in, day-out, highway cruising can.
Its unique suspension takes the shock and shudder out of the ice as it does with the pot-holes around the corner.
The air-cooled engine can't freeze in Antarctica or boil in Bourke.
Because it's in the rear, the VW has traction through snow and ice (or on a slippery city street) that front-engine cars can't equal.
Four-vent heater.
Body work.
Gear change.

Baked enamel finish.
Everything helps the VW lick the toughest conditions in the world. But you don't have to take it to Antarctica to prove how good it is.
Just a five-mile test-drive with a VW Agent will do.
VW1200 sedan — from £849 — tax paid.

Page **37**

Official photograph Australian National Antarctic Research Expedition

"A much-circulated 1994 advertisement for Chrysler Corporation depicts the continent of Antarctica from the vantage point of space, zooming through the ozone hole directly onto the South Pole, from which a cartoon penguin gazes up plaintively as if to say, rescue me. The purpose of Chrysler's advertisement was not directly to promote its products, but rather its adherence to standards of pollution control, stating: 'The Ozone hole has protected us for 1.5 billion years. It's time we returned the favor.' […] Chrysler, the ozone-destroying avatar of fordist production, comes to the rescue of its own creation of Antarctica as an otherworldly, nonnationalized and nonindustrialized place through its zealous compliance with 'government guidelines' on the atmospheric pollutants, chlorofluorocarbons. […] Chrysler thus produces both problem and cure, along with the very place on which it stages its rescue drama."

Elena Glasberg (Professor at New York University, Essayist, Speaker), Antarctica as Cultural Critique: The Gendered Politics of Science Exploration & Climate Change, 2012

"In 1963, Volkswagen (VW) proclaimed itself the first car in the Antarctic. […] The license plate reading 'Antarctica 1' inserted the VW bug familiar to middle-class viewers all over the world [in]to the new terrain of Antarctica. That there was no corresponding national road '1' in the nonnationalized territory only strengthened the car maker's claim for its versatile and hardy bug. […] The self-promotion of the advertisement loops into the anticipatory maintenance of its own risk. This proto-disaster capitalism, ready to profit from even its own failures, characterizes the complexity of virtual and material scales of Antarctica's management under capital."

Elena Glasberg (Professor at New York University, Essayist, Speaker), Antarctica as Cultural Critique: The Gendered Politics of Science Exploration & Climate Change, 2012

Building Performance and Carbon Emissions

Tony McGlory

TONY MCGLORY is the director of Ramboll United Kingdom's Building Services Engineering team. Eng. McGlory has over thirty years' experience designing complex engineering projects. His responsibilities at Ramboll include leading Building Services Engineering design for projects in Antarctica and the subantarctic islands.

Whole-life carbon emissions are becoming the yardstick by which sustainable buildings are measured. Whole-life carbon includes not only the operational emissions resulting from energy consumption but also the carbon embodied in material manufacture and supply and construction processes throughout the entire building lifecycle, including demolition, disposal, and recycling. The focus on embodied carbon is gaining increasing recognition throughout the construction industry and a holistic approach to whole-life carbon is deemed necessary in responsible building design processes.[1]

The ideal aspiration for any building or infrastructure project is to achieve life-cycle net-zero carbon dioxide emissions; however, this is more difficult to achieve in Antarctica than in other areas of the planet due to the remoteness and extreme climate. Nevertheless, this should not hinder any attempt to get as close as possible to life-cycle net-zero carbon dioxide emissions.

All station buildings and infrastructure require energy to operate. For stations in Antarctica, this energy has traditionally come from fossil fuel sources, typically winter diesel or kerosene; both are expensive, difficult to transport, and a source of pollution. Consequently, any reduction in fuel consumption has an immediate positive effect on the cost, station logistics, and environment from a pollutant, health, and well-being perspective.[2]

The overarching hierarchal approach to achieving operational net-zero carbon dioxide emissions is the same, irrespective of project type or location, and is composed of three steps: minimising demand, maximising system efficiency, and introducing renewable energy sources.

The first step, minimising the demand for energy, often requires fundamental questions to be addressed: is a station or building actually required? Does the station size match the resource deployed and the quantity of science being undertaken? If a building is indeed required, can the building's thermal comfort parameters be reduced? Is it possible to avoid the need for a new building by adapting an existing building on a whole-life net-zero carbon basis?

Inhabiting Antarctica is one of the least sustainable activities humankind could choose to venture upon. Yet, the value of Antarctic scientific research is sufficient to justify the presence of humans and the related construction and logistic activities. These, in turn, require the supply of substantial amounts of power – to heat the stations in unforgiving weather conditions, desalinate seawater, melt snow, fuel transport vehicles, etc. Given the extremely large amounts of diesel burned simply to deliver the fuel supply to Antarctica, all such activities consume, on average, twice the amount of fuel that they consume on the other six continents. At the South Pole Station, for example, 750,000 litres of diesel fuel are stored annually to supply energy for the station, aircraft, vehicles, and heavy machinery, while the estimated annual fuel used at Halley VI is 240,000 litres (111 litres per square metre) – a 26% improvement compared to the 225,000 litres (150 litres per square metre) of its predecessor, Halley V.[1]

01 "Carbon-zero" station
Princess Elisabeth

47 Fossil fuel stations

Artigas	Great Wall	Molodezhnaya	Russkaya
Belgrano II	Halley VI	Novolazarevskaya	San Martin
Bellingshausen	International Field	Oazis	Sanae IV
Bharati	Camp Peninsula	Orcadas	Signy
Brown	Byers	Palmer	Vernadsky
Camara	King Sejong	Pedro Vicente	Vostok
Carlini	Kohnnen	Maldonado	
Concordia	Leningradskaya	Petrel	
Dallmann	Machu Picchu	Prat	
Davis	Maitri	Primavera	
Decepción	Marambio	Professor Julio	
Druzhnaya IV	Mario Zucchelli	Escudero	
Esperanza	Matienzo	Progress	
Ferraz	Melchior	Risopatrón	
Frei	Mirny	Ruperto Elichiribehety	

26 Fossil fuel/renewable stations

Aboa	Mawson
Amundsen-Scott South Pole	McMurdo
Carvajal	Mountain Evening/ Vechernyaya
Casey	Neumayer III
Dirck Gerritsz Laboratory	O'Higgins
Dr Guillermo Mann	Rothera
Dumont d'Urville	Scott Base
Gabriel de Castilla	St. Kliment Ohridski
Gabriel González Videla	Syowa
Henryk Arctowski	Taishan
Jang Bogo	Troll
Johann Gregor Mendel	Wasa
Juan Carlos I	Yelcho
Kunlun	Zhongshan

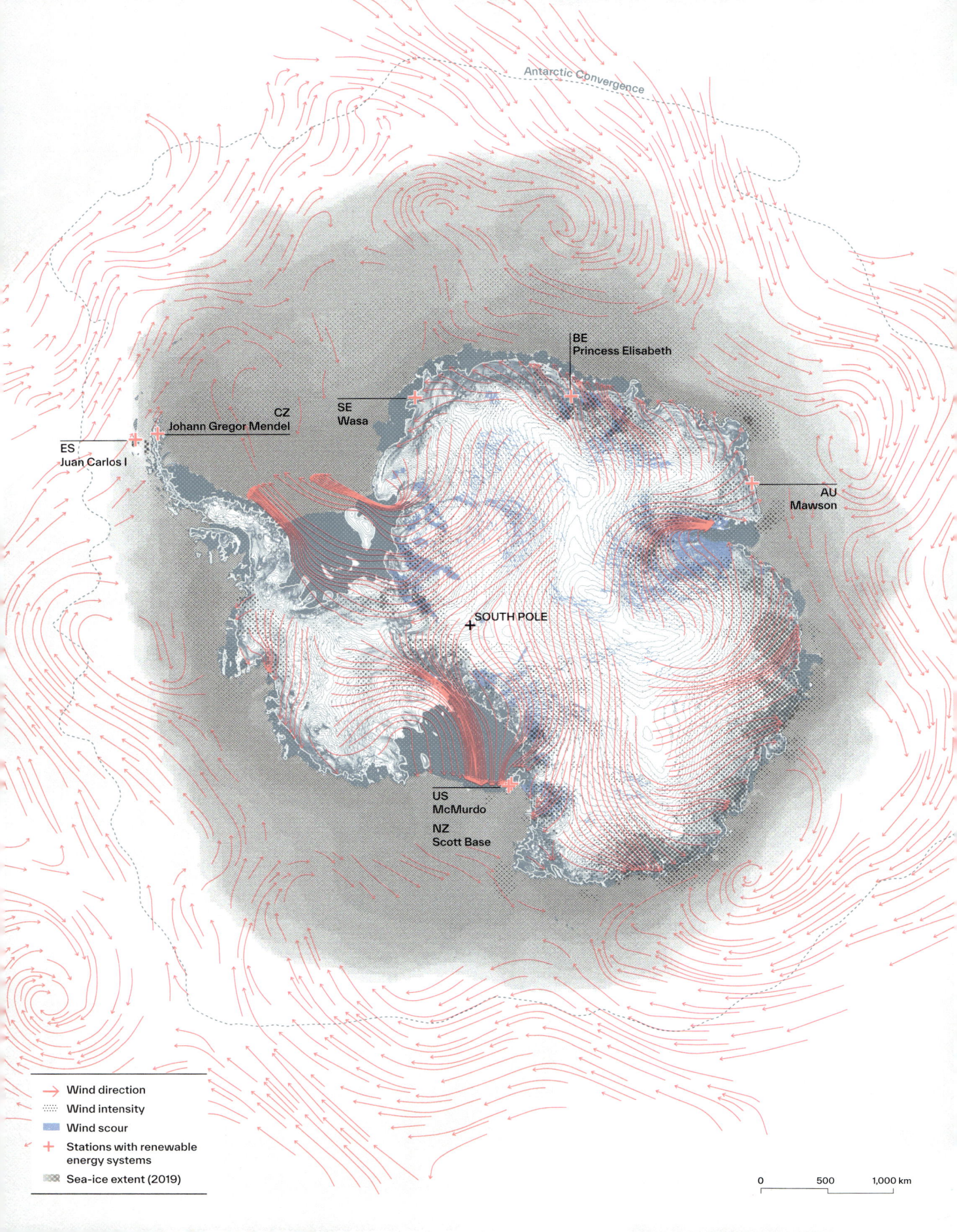

Antarctic Convergence

BE
Princess Elisabeth

SE
Wasa

CZ
Johann Gregor Mendel

ES
Juan Carlos I

AU
Mawson

SOUTH POLE

US
McMurdo

NZ
Scott Base

→ Wind direction

Wind intensity

Wind scour

+ Stations with renewable
energy systems

Sea-ice extent (2019)

0   500   1,000 km

In designing new buildings, passive design principles should be applied: for example, minimising building surface area to volume ratio, ensuring high standards of thermal insulation and airtightness, and maximising the use of beneficial solar radiation and natural daylight. Buildings should also be zoned to match occupancy patterns and allow flexibility to facilitate the winterising of building sections, as many stations run on a low occupancy through the winter.

The second step requires the maximising of the operational efficiency of the engineering systems needed to address the remaining demand identified from step one. Such measures might extend to capturing waste heat from electricity generators, including heat recovery ventilation systems, maximising boiler combustion efficiencies, demanding lead control systems and load management systems, using LED lighting, providing low-water-use sanitary accommodation, grey-water recycling, and natural snowmelt capture.

White goods and fixed electrical equipment with "A+"-rated energy performance should be selected and a comprehensive energy metering and targeting systems should be provided so that system performance can be monitored and optimised and unusual demand patterns identified, investigated, and corrected. The implementation of these measures reduces station energy demand whilst also minimising the scale of renewable energy installations required to achieve a net-zero carbon estate.

The final step in achieving operational net-zero carbon dioxide emissions is to meet the remaining energy demand using renewable energy sources alone. Although this is the ideal, most renewable energy sources are variable and seasonal, and thus there will certainly be a need for emergency standby power generation and it is therefore unlikely that the use of fossil fuels can be completely avoided in Antarctica in the short term.

Many Antarctic stations have successfully adopted low-carbon and renewable energy technologies, with a number of photovoltaic, wind power, and hydroelectric installations in operation. These installations provide a basis for minimising fuel consumption at their respective stations, whilst offering a platform to increase renewable energy generation capacity further.

Wind power – a system which normally yields the highest energy return per unit cost invested – has been implemented and proven in use at several stations. Although Antarctica benefits from a high wind resource in general, there are locations which have a relatively low wind resource, and others which are at such high altitude that the reduced density of the air significantly lessens the wind turbine output. Other constraints, such as proximity to bird life, air-

"Our decision-making process on the infrastructural system to be implemented on Antarctic stations is pretty much always driven by cost; I believe this will be soon replaced by a careful assessment of carbon emissions."

Oliver Darke (Head of Estates, British Antarctic Survey), in conversation with UNLESS, Architectural Association School of Architecture, 2019

Historically, the fuel supply for stations located on the plateau was delivered by aircraft, which released fuel-drum parachutes over the featureless landscape, whereas today it is mostly delivered on fuel bladders by traverse. These necessary activities are not without risk of contamination, as shown, for example, by the 1989 infamous 150,000-litre fuel leakage at South Pole,[2] and the seepage that occurred in Byrd station when the pressure exerted by the shifting ice sheet crumpled the oil drums buried in the snow. To address similar concerns, in August 1960 the United States Congress authorised the design and construction of a nuclear reactor for McMurdo, which resulted in the deployment of the "portable" PM-3A reactor in Ross Island. The reactor was assembled on a foundation of solid volcanic rock on the side of Observation Hill overlooking McMurdo Station and was operated by a twenty-five-person crew. During its ten-year lifetime, the PM-3A "Nukey- Poo" produced "over seventy-eight million kilowatt hours of electricity whilst the excess steam in the desalination plant generated thirteen million gallons [49 million litres] of fresh water." Although the implementation of the power station greatly reduced the station's consumption of fuel, its 438 malfunctioning cases, and the consequent lack of reliability of the system that was operative only 72% of the time between 1964 and 1972, led to the closure of the structure in 1972, when McMurdo returned to the sole use of diesel engine generators.[3] Today, wind turbines, which exploit the continent's strong katabatic winds, and solar panels, taking advantage of the six months of sunlight per year, are implemented at a handful of stations (including McMurdo) as a supplement to the diesel generators. Alas, to date, the stations implementing sustainable systems are only a few, and there is only one station which relies uniquely on renewable energy: the summer-only facility of Belgium's Princess Elisabeth Station located in Queen Maud Land, and shown on the opposite page.

craft movements, radar and radio masts, although not insurmountable, can further complicate the scale, positioning, and viability of installations in relation to the stations.

Photovoltaic power (PV) – whose electricity generation has a low power yield per unit area of panel and therefore requires large areas to generate significant quantities of power – has also been proven in use at several stations. Naturally available only during the summer season, this system can still provide a useful contribution to reductions in carbon emissions. Similarly, solar power, available for only approximately six months per year, is also in use in Antarctica and is deemed most suited to buildings that use large amounts of hot water, such as accommodation buildings.

Deep drilling would almost certainly be necessary in order to access a heat source suitable for generating electricity. Although this option is likely to be cost prohibitive in the short term it should not be completely ruled out, as future technological advances may overcome the challenges. Finally, generating power from wave- or tidal-powered devices seems unlikely during winter seasons. Even during summer seasons when the ocean is not frozen, sea ice poses a high risk of damage to floating or submerged devices.

Surplus renewable electrical energy can be used through the process of electrolysis to create hydrogen. Typically, it takes 50 kilowatt-hours of energy to create hydrogen. Hydrogen can be stored ready for reuse as either electricity or heat, using a fuel cell or hydrogen boiler, with the only by-product being water, which could also be put to use on station: for instance, to power station vehicles. The main disadvantages of hydrogen include the risks associated with the explosive nature of the gas and the large high-pressure storage capacities required to ensure a reliable supply. That said, the technology does warrant further investigation for use in Antarctica.

With a significant increase in renewable energy generation, there may come a point where the renewable electricity generated is greater than the electricity demand for the site; alternatively, climate or seasonal conditions may dictate that there is no renewable energy available. It is essential, therefore, to implement energy storage in order to capture renewable energy when generation exceeds demand so that it can be used when renewable energy generation is not available.

Energy storage could take several forms, as electrical energy can be stored in chemical batteries, fly wheels, and compressed air vaults. Surplus renewable electrical energy can also be used to heat water and create a thermal store that can be made available when heat demand exceeds the renewable heat source available. The capacity of thermal stores can be increased beyond that of water-based thermal stores using special phase-change materials – i.e. materials with a high heat of fusion which, when melting and solidifying at a certain temperature, can store and release large amounts of energy per unit volume.

In order to achieve a net-zero carbon station in Antarctica, it is certain that a combination of renewable energy generation and energy storage technologies will be required. Such a station energy configuration would open up opportunities to use renewable energy to power logistic operations on the station and perhaps deep field ventures. Engineers and architects have a great responsibility to design whole-life net-zero (or as near net-zero as practicable) carbon buildings and infrastructure to minimise their environmental impact on the Antarctic ecosystems, whilst still providing a safe, functional, and resilient place for scientists to live and work.[3]

"While wind power is used to supply the station with electricity year-round, solar power provides both electricity (photovoltaic panels) and hot water (solar thermal panels) during the austral summer. The station's systems are integrated and piloted by a programmable logic controller, which ensures optimal working and living conditions inside the station. This smart grid is three times more efficient than any conventional system and can be controlled remotely during winter. Princess Elisabeth Antarctica's Micro Smart Grid, the key feature that makes it a zero-emission station, is a unique system based on a Demand Power Management System."

International Polar Foundation, *Princess Elisabeth Antarctica: The First Zero Emission Polar Research Station*, 2013

Princess Elisabeth Station

Prefalux

PREFALUX is a general construction company specialising in wood construction, roofing and finishing, based in Luxembourg. Prefalux played an important role in the creation of Belgium's Princess Elisabeth Antarctica station, in particular in the development, production, and construction of the structure's wooden frame, its exterior walls, and its roof.

Located 200 kilometres from the Antarctic coast, Princess Elisabeth Station ↗ p. 786 is surrounded by an ocean of ice and snow. Anchored on a rocky spur (71°57' South, 23°20' East), the station is based at Utsteinen in Queen Maud Land (1,400 metres above sea level), and in the uninhabited 1,500-kilometre zone that separates the Japanese Syowa Station ↗ p. 770 from its Russian counterpart at Novolazarevskaya.

By an initiative of the International Polar Foundation and, in particular, the explorer Alain Hubert, a new scientific base was built not far from the former Roi Baudouin polar base (abandoned in 1968) as an experimental station powered mainly by renewable energy sources.

Designed by Philippe Samyn to withstand the extreme weather conditions of Antarctica, the station experiences significant temperature differences between exterior and interior. The exterior elements of the structure are subject to surface temperatures ranging from –40°C in winter to up to +30°C as the result of direct exposure to the sun in summer, while the temperature differences between inside and outside can reach

a delta of 60°C. In addition to the extreme temperature variation, the polar climate is also susceptible to violent winds, sometimes up to 200 kilometres per hour, carrying stones that can hit the walls with considerable violence and potentially damage the envelope.

Built on a piling system to avoid being buried under the snow and to allow snow drift underneath the structure, the front of the core component of the station finds itself 5 metres above the ground while the rear is only 2 metres above the ground. Weighing 200 tons, the main building has a total floor area of 400 square metres, while the technical areas located at the foot of the base occupy an additional 2,000 square metres.

Unique in its design and operation, Princess Elisabeth Station is the first in the continent that aims to use exclusively renewable energies. By design, the base's energy needs represent only 20% of those of a conventional Antarctic station of comparable size, rendering it – theoretically – the only summer station that could be powered solely by renewable energy. To achieve this goal, 20 square metres of thermal solar panels were

installed on the roof of the station and an additional 300 square metres of photovoltaic panels are placed on the main elevations, only partly occupied by windows. In proximity to the building, eight wind turbines, each with a power of 6 kilowatts, produce energy that in turn is stored in a series of batteries installed in the very heart of the station. Additional energy, unfortunately still very much necessary, is supplied by diesel generators.

To reduce the environmental impact of the base's implementation, special attention was given to the façade, with materials and technologies chosen that respect all the principles of eco-construction. Designed to minimise thermal bridges and to achieve the lowest possible U value (the wall's heat transmittance coefficient), the exterior wall has a complex system of layers with a total thickness of 528 millimetres and a thermal transmittance coefficient (U value) of only 0.06 watts per square metre per kelvin.

Striving for Zero Carbon Emission in Princess Elisabeth Antarctica Station and the Ellipsoid

Philippe Samyn

PHILIPPE SAMYN is a civil engineer, architect, and urban planner. He is the founder of the Samyn and Partners architectural and engineering practice (Belgium) and is involved in a range of international projects. Dr Samyn was responsible for the shell and core of the Belgian station, Princess Elisabeth, in Antarctica. For his technical research on sustainability, he was awarded the Global Award for Sustainable Architecture (2008).

The theoretical backgrounds first summarised here, which are valid for any construction in Antarctica, explain the shell and core as well as the energy characteristics of the PEA and lead me to conclude that an ellipsoid might be an appropriate shape.

THEORETICAL BACKGROUNDS

Limiting the carbon emissions of a construction and its operation requires that the three following items be limited:

1. The **total embodied energy** $E1$, which is the sum of the embodied energy in the structure Es, the envelope Ee, the finishes Ef, and the technics Et:

$$E1 = Es + Ee + Ef + Et$$

2. The **daily energy consumption**, composed mostly of thermal $E2$ and electrical $E3$ energy.

2.1 The thermal energy is the sum of the energy lost through the envelope (Eh) and the energy needed to heat the ventilation air (Ev) and to humidify this air (Ew), minus the human-released energy (Ep) – and possibly the solar gains (Eg):[7]

$E2 = Eh + Ev + Ew - Ep (-Eg)$, of which $(Ev + Ew)$ cannot be recuperated, this thermal energy being produced with electricity.

2.2 The electrical energy is the sum of the energy for lighting El, and operations Eo:

$E3 = El + Eo$,
this energy being dissipated in thermal energy and partially compensating for $E2$.

2.3 The global electrical energy consumption will thus be

$E = E2 - E3$, when $Eh \geq (E3 + Ep)$, or

$E = E3 + Ev + Ew$, when $Eh < (E3 + Ep)$

Healthy energy management should thus first aim at limiting $E3$, keeping in mind that it should compensate for $E2$ as much as possible, so as to be able to produce it locally, using only wind and solar energy.

The units here are as follows: metre (m), kilogram (kg), second (s), degrees Celsius or kelvin (C° or °K) and the derived units: hour (h = 3600 s), newton (N = kgm/s²), pascal (Pa or N/m²). All energy values are here expressed in joules (J) or (when mentioned in brackets) in kWh (1 kWh = 3.6 megajoules) per day (day) and per person (p).

As the carbon footprint is a consequence of energy production (kilograms of carbon dioxide per kilowatt-hour), its reduction means reducing the energy demand for building as well as for the use of a building. Energy is quantified in joules,[1] (or more commonly in kilowatt-hours, as used here) but is conveniently expressed in money. With a given level of performance, the least expensive construction will thus always be the most sustainable (the one with the lowest energy consumption and the smallest carbon footprint). Conversion tables for "money/energy" and "carbon dioxide/energy" are useful tools for evaluating this issue.

The reduction of the carbon dioxide footprint demands first of all the use of solar and wind energy, as applied in Princess Elisabeth Antarctica Station ↗ p.786 (PEA). This should be followed by a reduction of the embodied energy[2] in the construction and the reduction of the energy consumed annually over the building's lifetime.

The embodied energy in the shell and core is firstly a function of the choice of materials, which in turn is characterised by the ratio of its embodied energy per unit volume to its working stress: $Em = e_{mv}(J$ or $kWh/m^3) / \sigma (N/m^2)$. This ratio is known as the *material indicator* – and is a-dimensional.[3] This explains why PEA is built mainly of wood and steel. For the shell it is also a function of the compacity of the construction, $C_{(m)} = V/s$, i.e. the ratio of its enclosed volume $V_{(m^3)}$ to the surface $S_{(m^2)}$ of its envelope, which in turn is directly proportional to its size. For the core it is also a function of the forces $F(N)$ acting on the structure, the main span $L_{(m)}$ of its elements, and its geometry, the latter characterised by its volume indicator W – again a-dimensional.[4]

The forces F were reduced by giving PEA the most aerodynamic shape possible and also by raising it above the ground. The spans[5] L were reduced at PEA to the minimum acceptable dimension: 3.3 metres on a square grid. The geometry of the structure of PEA was optimised by ensuring that the supporting steel trusses are purely in compression, by bundling the forces coming from the columns to a limited number of foundation plots, by giving the maximum height to the wooden elements in bending, and by avoiding buckling issues.

The yearly energy demand of PEA, met mainly by solar and wind power, has also been limited to the minimum possible, which has been achieved largely by means of a highly insulated shell.[6]

The question of zero emissions relates directly to the population p served and thus to the net floor area P (m²) or to the internal net volume v_i (m³/p) allocated per person.
Dots are used here to indicate multiplications.

3. The **energy** needed **to clean** and **maintain** it.

Let us now further detail these three items, one by one:

1. The **embodied energy** *E1*

1.1 The embodied energy in the "n" elements of the structure – presenting, together, a volume of material Vm (in *m³***) – equals**

$Es = \sum_{i=1}^{i=n} (Fi \cdot Li \cdot Wi \cdot Emi)/(d \cdot p)$

where
• *Fi* is the maximum force acting on element *"i" (N)*;
• *Li* is the main dimension of element *"i" (m)*;
• *Wi = σ_i · Vmi/(Fi · Li)* is the Volume Indicator of the structural element *"i"*, with σ_i its working stress *(Pa)*, *Vmi* its volume of material *(m³)*, and *Vm = $\sum_{i=1}^{i=n}$ Vmi. Wi* is an a-dimensional number related only to the geometry of the construction;
• *Emi = emi / σ_i* is the Material Indicator: the ratio of *emi*, its embodied energy *(J/m³)*, to σ_i *(Pa)*;
• *d* is the total period of existence for the construction, in days *(day)*.

1.2 The embodied energy in the "m" elements of the envelope equals

$Ee = (\sum_{i=1}^{i=m} (emi \cdot ti)) \cdot V/(C \cdot d \cdot p)$

where
• *ti* is the thickness of the envelope elements *"i" (m)*;
• *V* is the enclosed volume of the construction *(m³)*;
• *C = V/S (m)* is the compacity of the construction, the ratio of *V* to the outside surface *S* of its envelope.

The methodology for a rigorous and precise evaluation of Es and Ee is now in development.

1.3 *Ef* and *Et*

The analysis of the embodied energy in the finishes *(Ef)* and the technical equipment *(Et)* is beyond our scope here.

1.4 Evaluation

In fact, the universal monetary unit is the *joule* or the *kWh* – money is always the expression of an energy expense.
Reciprocally the cost *(Ct)* of a construction or its elements might be a reasonable measure of its embodied energy: the less expensive, the lower *E1 · day · p*.
E1 · day · p is thus proportional to *Ct*.
At a constant *E1*, the cost may thus increase linearly with the life expectancy of the construction, which is, of course, related to

its robustness and ease of maintenance but is, first of all, dependant on its usefulness for humanity.

2. The daily energy consumption per person: *E*

The evaluation proposed here for E2 and E3 is based on rough figures which, although questionable, are useful for determining their orders of magnitude. E2 and E3 are here expressed per day and per person.

2.1 The daily thermal losses per person: *E2*

2.1.1 The energy to compensate the heat losses through the "r" elements of the construction envelope, neglecting the surface contributions, equals

Eh = (86400s) Δt·S / (p · R); with *S = V/C*, and *Vi = p·v_i* with *v_i = 60(m³/p)*
= 86400s·60m³ *Δt·(V/Vi)/(R·C)* =
(5.184·10⁶/s·m³) *Δt·(V/Vi)/(R·C)*

where
• *86400s = seconds in 24h;*
• *Δt = 45°K or 60°K,* the temperature difference between –25°C or –40°C outside[8] and 20°C inside;
• *S (m)* is the surface of the envelope composed of *i* elements of surface *Si*, such that
S = ($\sum_{i=1}^{i=r}$Si) m², each element presenting a thermal conduction resistance *Ri = ti/λi (m²°K/W)*, with
• *λi (W/(m°K))* its thermal conductivity;
• *R = 1/($\sum_{i=1}^{i=r}$Si/(S · Ri))(m²°K/W)* the average thermal conduction resistance of the envelope;
• *Vi (m³)* the net internal volume of the construction = *p·v_i* with
• *v_i (m³/p)* the volume per person, taken here as *60 m³/p*.

• for *Δt = 45°C*:
Eh₄₅ (kWh) = 64.8·(V/Vi)/(R·C) kWh

• and for *Δt = 60°C*:
Eh₆₀ (kWh) = 86.4·(V/Vi)/(R·C) kWh

2.1.2 The energy to heat the ventilation air

Assuming that *70%* of the energy from the exhaust air is recovered via a heat exchanger[9]

Ev = Δt·(0.3)·(24h)·(40m³/h)·t(1256J/(m³°K))

where
• *0.3* is the net energy needed, as 70% is recuperated;
• *24h* is the hours of the day;
• *40m³/h* is the volume of air exchange per hour and person;[10]
• *1256J/(m³°K)* is the calorific capacity of the air.

Ev (kWh) = Δt·0.1005 kWh

for *Δt = 45°C*:

Ev₄₅ (kWh) = 4.52 kWh

and for *Δt = 60°C*:

Ev₆₀ (kWh) = 6.03 kWh

2.1.3 The energy to humidify the air at 25% relative humidity (RH)

Ew = (0.3)·(24h)·(40m³/(h·p))·(0.005 kg/ m³)·(2.5·10⁶ J/kg) = p·3.6·10⁶

where
• *0.005 kg/m³* is the water content per *m³* of air required to humidify the air at *25% RH*,[11] and
• *2.5·10⁶ J/kg* is the energy required to vaporise the water.

EW (kWh) = 1 kWh

2.1.4 The energy released by one occupant

With a minimum power of *100 W/p*

Ep = 24h·(3600s/h)·(100W/p) = p·8.64·10⁶

Ep (kWh) = 2.4kWh

2.1.5 The total daily energy consumption per person: *E2*

E2= Eh + Ev + Ew – Ep

E2 (kWh) = [(1.44·(V/Vi)/(R·C) + 0.1005) Δt – 1.4] kWh

or for *Δt = 45°C*:

E2₄₅(kWh) = [64.8·(V/Vi)/(R·C)+3.125] kWh

and for *Δt = 60°C*:

E2₆₀(kWh) = [86.39·(V/Vi)/(R·C) + 4.63] kWh

2.2 The daily operational energy needs per person: *E3*

2.2.1 The energy for lighting

This area is currently evolving, as daylighting with high-performance vacuum glass *(U=0.4W/(m²°K))* is now available.[12] It is still reasonable to assume the provision of

El= (20m²)·(43200s)·(7.5 W/m²) = 6.48·10⁶

where
• *20 (m²)* is the net floor area per person;
• *43200 (s)* is the daily lighting time of *12 hours (in seconds)*;
• *7.5W/m²* is the power required for a performant lighting equipment providing a luminous flux per unit area of *500 Lux*.

El (kWh) = 1.8 kWh

2.2.2 The energy consumption for operating the station

It is reasonable to assume a power need, *10 hours* per day, of *250 W/p* (workstation and all other household equipment such as water supply and heating, treatment of disposal water, and sanitary and kitchen demands).

Eo = 10h·(3600 s/h)·(250 W/p) = 9·10⁶

Eo (kWh) = 2.5 kWh

2.2.3 The total daily operational energy needs per person: *E3*

E3= El + Eo

E3 (kWh) = 4.3 kWh

2.3 The global electrical energy needs per day and person: E

When $Eh \geq (E3 + Ep)$, then $E = E2 - E3$

• for $\Delta t = 45°C$:

E_{45} (kWh) = [64.8 (V/Vi)/(R·C) – 1.175] kWh

• and for $\Delta t = 60°C$:

E_{60} (kWh) = [86.39 (V/Vi)/(R·C) + 0.033] kWh

When $Eh < (E3 + Ep)$, then $E = E3 + Ev + Ew$

• for $\Delta t = 45°C$:

E_{45} (kWh) = 9.82 kWh

• and for $\Delta t = 60°C$:

E_{60} (kWh) = 11.33 kWh

The daily needs might in practice lead to the one or the other situation.
But while Δt is important for the energy calculation, it has little influence on the design strategy for energy saving, which still depends mainly on Eh, as I will conclude.

3. The energy needed for cleaning and maintenance

This energy depends on, and is closely related to, the investment in the embodied energy granted for the construction of the facility.
However, at present, we still lack a rigorous and precise evaluation methodology that allows for the careful selection of the building materials and elements and takes those aspects into account from the very early design stage.

PRINCESS ELISABETH STATION

1. Embodied energy

With respect to the design of the construction of the shell and core of the station, as well as the aspects related to prefabrication and logistics, we limited Es and Ee to a strict minimum:

first Es, by limiting
• the dimensions Li of the structural elements as well as the forces Fi to which they are subject
• the supporting foundation plots
• W, the volume indicator, by optimising the structural morphology
• Em, the material indicator using, for example, mainly wood with $300 \leq Em \leq 800$

and then Ee, by limiting

• Em, for all its materials, with the exception of the thin (ti =1.5 mm) stainless-steel sheeting for which $4,300 \leq Em \leq 5,000$, I would now use recycled aluminium sheets with $38 \leq Em \leq 300$

• ti, by using expanded polystyrene with graphite ($\lambda = 0.031W/(m°K)$) and by maximising C (the compacity V/S of the building) for the given volume V.

2. Theoretical daily energy consumption

Princess Elisabeth Station presents
• an outside volume $V = 2000\,m^3$ with a net inside volume $Vi = 1200\,m^3$, whereby $V/Vi = 1.667$
• a net floor area of $P = 400\,m^2$, intended for 20 persons
• an envelope of $S = 1400\,m^2$ with an average $R = 15\,m^2°K/W$
• a compacity $C = V/C = 1.429\,m$

for $\Delta t = 45°C$:

$E2_{45}$ (kWh) = [5.039 + 3.125] kWh/p = 8.164 kWh/(p·day)

• or for 20p: 163.28 kWh/day

and for $\Delta t = 60°C$:

$E2_{60}$ (kWh) = [6.719 + 4.63] kWh/p = 11.349 kWh/(p·day)

• or for 20p: 227 kWh/day

while the daily operational needs are $E3$ (kWh) = 4.3 kWh/(p·day)

• or for 20p: 86 kWh/day

The net theoretical daily energy needs are thus

for $\Delta t = 45°K$: 77.28 kWh/day, and

for $\Delta t = 60°K$: 141 kWh/day.

This very approximative theoretical and preliminary calculation method provides figures close to reality.
The daily demand of PEA (with its garages) indeed seems to fluctuate – as a function of the rate of occupation (p), the season (Δt) and the demand (E3) – between 70 and 233 kWh.

The solar panels and the windmills cover this normal energy need.

FURTHER LIMITING THE CARBON FOOTPRINT

The equations above show how the geometry and the material choices for the construction have a direct impact on the limitation of El and Eh, and hence its carbon footprint.

This calls for the minimisation of Fi, Li, Wi, Emi, and V/Vi and the maximisation[13] of the compacity C of the construction and its number p of inhabitants, and thus V (impacted by the necessary increase of Vi).

• F_h:[14] an **aerodynamic shape** for the envelope, with a low drag coefficient[15] (c_d) substantially reduces the horizontal forces (Fh) on the structure.
A **streamlined ellipsoidal volume**, detached from the ground, with its smallest section perpendicular to the catabic prevailing winds,[16] reduces Fh to a minimum.
• Li: the length of the structural elements, and in particular those in flexion, is as limited as possible;
• Wi is optimised by designing structural elements with a low W. Keeping fire resistance in mind, beams are as high as possible, preferably with trusses instead of rectangular sections;

• Emi is limited by using materials presenting a ratio emi/si that is as low as possible, and not only a low emi.
• V/Vi decreases with V. Indeed, as an example, for a cubical envelope of side D and thickness $t/2$: $V/Vi = 1/(1 – 3t/D + 3t^2/D^2 – t^3/D^3)$ or $V/Vi \cong 1/(1 – 3t/D)$.
• $C = V/S$, the **compacity**, increases linearly with the **volume** V (and hence the **population** p) and, for a given V, is at its maximum for a sphere.

Taking its diameter D as the reference dimension and with $C_s = D/6 = C$ for the sphere, the compacity for various basic shapes can be expressed as a proportion of this reference compacity C:
• for a cylinder of diameter and height D: $C_{c1}/C = (1/3)^{1/3} = 0.8736$
• for a cube of side D: $C_{p1}/C = (\pi/6)^{1/3} = 0.806$
• for a parallelepipedic volume of sides D_p, D_p, and xD_p: $C_{px}/C = 2.418·x^{2/3}/(1 + 2x)$; or

$C_{px}/C = 0.768$ for $x = 2$; 0.719 for $x = 3$; 0.677 for $x = 4$; and 0.642 for $x = 5$;
but its small surface exposed to the wind $D_{px}^2 = 0.65D^2/x^{2/3}$ is smaller than that of the sphere in the ratio: $D_{px}^2/(\pi.D^2/4) = 0.827/x^{2/3}$, or:
0.827 for $x = 1$; 0.521 for $x = 2$; 0.389 for $x = 3$; 0.329 for $x = 4$; and 0.283 for $x = 5$.

This reduces the projected area to the wind, F_h in proportion.
• for a cylindrical volume of diameter D and length xD:
$C_{cx}/C = 5.2415x^{2/3}/(2 + 4x)$, or:
$C_{cx}/C = 0.832$ for $x = 2$; 0.779 for $x = 3$; 0.734 for $x = 4$; and 0.697 for $x = 5$
(8.5% is better than C_{px}/C)

The ratio of its small surface exposed to the wind ($\pi D^2_{cx}/4$) to that of the sphere (=$0.763/x^{2/3}$) also reduces Fh compared to the parallelepipedic volume.
Again, a **large ellipsoid** might be the best answer!

• vi: the volume per person vi increases with V, but this parameter is much less important than V and C to allow for more circulation and service space per person.[17]

The daylighting issue also arises here as the increase of C means deeper floor plates. Regularly distributed light wells not only allow for efficient management of the daylighting and solar gain but also act as piazzas in the "village" while limiting to a minimum the increase of vi.

THE PERFORATED ELLIPSOID ON POSTS

All this leads us to consider inviting **a maximum of people** to gather and live in the **largest possible ellipsoidal volume** (its envelope equipped with **streamline spoilers**) and **detached from the ground** (to allow for the free flow of air and snow), to which it is smoothly connected by a small vertical elliptical cylinder.

This volume on braced posts is internally perforated by interconnected **light wells**.

Design for Removal.
The Life Cycle of Antarctic Stations

Thomas Schramm

THOMAS SCHRAMM is principal engineer at Ramboll Deutschland GmbH (Germany). Dip. Eng. Schramm is a civil engineer with over thirty years of experience in project management and construction supervision around the world, specialising in Antarctic stations, transport, marine services, logistics, and offshore installations.

Antarctica is a fragile ecosystem. Proof of its fragility can be exemplified by the establishment of the Protocol on Environmental Protection, ratified in 1998 as an essential addendum to the Antarctic Treaty. Prior to this protocol, limitations in terms of technology and budgets for the construction of Antarctic bases often led to large divergences in building quality control. Today, greater consciousness imposed by the protocol and recent "green-thinking" building practices, have led to a slow reverse-engineering of how stations in Antarctica are being built, operated, and maintained.

One of the basic requirements imposed by the Treaty System (see Annex III to the Protocol on Environmental Protection to the Antarctic Treaty, Article 1, clause 5) in relation to Antarctic architecture is a "legislation" which requires the removal of more or less everything that has been brought to Antarctica after it is no longer used. This does not apply just to objects, such as damaged vehicles or empty fuel drum canisters, but also includes entire decommissioned research stations.

The protocol has thus led to a substantial addition to the design brief of building in Antarctica. Alongside frequent requests to "provide accommodation for 40 people", "guarantee a building life-span of a minimum 25 years", and to ensure that "the station will operate using minimal energy consumption", the final client request today is to ensure "complete removal of the station after its intended life-span has been reached". Notwithstanding, this final condition is far easier said than done: it depends heavily on the location of the particular station. Local ground conditions (snow, shelf-ice, rocks) and anticipated (or even better, known) weather conditions must be taken into serious consideration during the design phase.

The knowledge of accurate snow accumulation levels and high wind speeds proves critical in anticipating the site's condition and the possibilities of successfully designing for removal. Snow accumulation can be a leading factor, as a station could have acceptable yearly levels of snow drift – which can be removed seasonally – but could slowly be irreparably engulfed by metres of snow, making its removal impossible.

"I think the whole point of constructing in the Antarctic is to minimize our footprint. […] The impact of buildings in the Antarctic is definitely improving. More thermally responsive envelopes, more efficient machinery, a greater reliance on renewable energy is all combining to reduce the carbon impact of these buildings on the Antarctic. Some of the buildings are designed to be removed so that there will be virtually no trace, whereas some of the other stations have concrete foundations, they are more permanent, the ground has been manipulated, and even after the building has been removed, there will be evidence of man's presence for decades."

Hugh Broughton (Architect, Hugh Broughton Architects), in conversation with UNLESS, Architectural Association School of Architecture, 2019

Such was the fate of Germany's Neumayer II Station [↗ p. 768], for example. Built on ice in 1992, within its lifespan the station was covered by over 20 metres of snow and ice that finally crushed it beyond repair. This lesson having been learned,[1] Neumayer III [↗ p. 796] was built in 2007 with a jacking system introduced that allows the station to be elevated every year to allow snow drift to run underneath the building and the structure to always remain above the surface.

Jacking up a station to elevate it above snow levels is a complex task. For Neumayer III this problem was solved by installing permanent hydraulically operated *pilotis,* which are regularly lifted to overcome the height variation. During the lifting process, for each *piloti,* the gap between the ground and the uplifted leg is filled with snow, which is compacted by the back-pushing feet until the station reaches the desired level.

The idea of elevated stations has subsequently been used in a variety of cases in Antarctica: for example, in the design of the New Russian Wintering Complex[2] at Vostok Station [↗ p. 862], which can be jacked if necessary by means of temporarily installed jack-up systems, which are then removed completely after it is deemed that the building lifespan has been reached. For coastal stations built on rock,[3] such as the Indian Bharati Station [↗ p. 804], the foundations consist of steel piles anchored into the rock to withstand wind speeds of more than 360 kilometres per hour. Naturally, these foundations are permanent.

The permanence of building foundations for coastal Antarctic stations built on rock, often accompanied by substantial cut-and-fill concrete surfaces that would have a grave impact on the environment when removed, seems to contradict the "design for removal" requirement enforced by the Madrid Protocol, as they will remain as a permanent human footprint on the Antarctic topography long after the buildings have been removed.

In accordance with the Antarctic Treaty, traces of human presence in Antarctica – unless preserved and listed as Historic Site and Monuments – are to be erased: simply removed. While this is now widely accepted, it still does not fully guarantee that "design for removal" is always accounted for when designing contemporary stations or that the principle is applied equally to the buildings and the topography that will anchor the structures to the stable or shifting "Antarctic ground". Although the Protocol on Environmental Protection obliges all national programmes to submit a detailed Environmental Impact Assessment (EIA) for any activity undertaken in the continent, including building, recent station designs offer evidence that even elevated structures which might appear very respectful to the environment and which can indeed be dismantled, rely on trench-and-fill foundations that will alter and contaminate, for good, the fragile polar landscape. The relationship with the "ground" is thus pivotal in assessing the temporality or permanence of Antarctic infrastructure. This is true, not only for the 74% of stations that are built on 3% of the exposed continental bedrock (represented on the opposite page by Casey Station and Great Wall Station), but also for the 16% of stations built on drifting ice (showcased below by Halley IV and Halley V). These, famously epitomised by Roald Amundsen's hut, which was rapidly buried under metres of snow, becoming the first hypo-glacial building to calve in the Southern Ocean, were historically left to their own fate to flow towards the ocean and morph into marine archaeology with all of their parts and contaminants. While this practice has been duly prohibited by the Madrid Protocol, and buildings on ice are mostly erected on jackable foundations, some of which even rest on skis to ensure their transportability and removal, the sustainability of structures built on bedrock (whether elevated or not) remains controversial and their impact on the territory – visible or concealed – is often irreversible.

Antarctic Foundations.
Towards Elevated Platforms

Hartwig Gernandt and Hans-Jürgen Meyer

HARTWIG GERNANDT is a scientist and polar researcher at the Alfred Wegener Institute for Polar and Marine Research (AWI) in Germany. As the head of AWI's polar infrastructure, Dr Gernandt oversaw the construction of Neumayer III Station.
HANS-JÜRGEN MEYER is a physicist who teaches building physics at the University of Applied Science in Bremen (Germany).

The Protocol on Environmental Protection to the Antarctic Treaty, signed in Madrid on the 4th of October 1991 and entered into force in 1998, designates Antarctica as a "natural reserve, devoted to peace and science" (Article 2). Article 3 of the protocol sets forth basic principles applicable to human activities in Antarctica while Article 7 prohibits all activities relating to Antarctic mineral resources, except for scientific research. Until 2048 the protocol can only be modified by unanimous agreement of all Consultative Parties to the Antarctic Treaty.

Complying with the strict environmental regulations that enforce the removal of every artefact within the Treaty area once its lifetime has expired poses an interesting challenge to all national Antarctic programmes. The protocol requires sound environmental operations for all activities and Antarctic constructions and the complete removal from the continent of everything at the end of usage.

The construction of partially permanent stations and temporary camps in the Antarctic region began slowly during the twentieth century. Most of these research bases were erected on bedrock or nunataks, a few more are founded on the inland ice sheet (currently Amundsen-Scott↗ p. 670, Concordia↗ p. 780, and Vostok↗ p. 862), and only two permanent facilities – the British Halley stations and the German Neumayer stations – have been erected, in 1956 and 1981 respectively, on coastal ice shelfs moving up to 100 times faster than the speed of inland glaciers (inland ice sheets). On these two sites, the snow accumulation rates differ between glacier outflows in the coastal region (Ekström Ice Shelf, for example, recorded an average 1.5 metres per year in 2009–2019) and inland ice sheet (Kohnen station recorded an average 0.3 metres per year between 2001 and 2018). Due to this extreme accumulation, the lifetime of any constructions is limited to approximately ten to fifteen years, maximum.

The complete removal of everything whose lifetime has expired, as set down in the protocol, induced the development of new design options to facilitate accessibility and allow for the removal of the facilities. This was true for Halley VI Station↗ p. 810 (which introduced pneumatic foundations resting on skis) and for Neumayer III↗ p. 796, which replaced its homonymous predecessor, buried more than 20 metres in the snow, after a lifespan of seventeen years. The design experimentation led to the definition of a system characterised by an extendable hydraulic support and elevation system (HSE system).

Situated on the Ekström Ice Shelf (70°40'S, 08°17'W) and commissioned on the 20th of February 2009 by the Alfred-Wegener-Institut (AWI) Helmholtz-Zentrum für Polar- und Meeresforschung to replace its predecessors Georg von Neumayer↗ p. 736 (1981–1992) and Neumayer II (1992–2009), Neumayer III marked a fundamental rethinking of the building's relationship with its foundations on the white continent. Founded on an ice shelf drifting slowly but continuously with an average speed of 145 metres per year, Neumayer III is not really a building but rather a moving entity like a research vessel. In fact, the station has been "sailing" northward over a distance of approx. 1,520 metres from its initial position on the 1st of March 2009 to its recent position on the 20th of April 2020. Designed and manufactured as a steel construction, Neumayer III consists of a platform and a trench structure. After a trial assembly, the elements of the framework were fitted together on-site, and, as required by the Environmental Protocol, it is therefore equally dismantlable without leaving any foundational remains in the ice ground.

The construction of the station consists of a platform with an insulated hall construction and a trench section with a highly specialised HSE system, with slabs shallowly founded approximately 8 metres deep in the firn. The

"During the first ten years of station operation, i.e. the years from 2009 until 2019, the specially designed HSE system worked reliably and fulfilled the technical requirements of keeping the station building at an adequate height position on the permanently growing accumulation of drifting and meteoric snow. A total of twenty-two jacking-up actions were carried out, gradually raising the station building by about 17.5 metres. At the same time the snow accumulation increased by about 10.8 metres, so that the station building is now located on a hill of accumulated snow, which overtops the surroundings by about 6.7 metres. The horizontal extension can be roughly estimated from Landsat images. The accumulated hill extends around the building within a radius of up to 115 metres on the windward side and more than several hundred metres on the leeward side."

Siegfried Rotthäuser (Engineer), during a technical on-site visit at Neumayer III in the 2019/2020 season

HSE system was designed in order to monitor and to control the challenges of a shallow sustainable foundation designed for a heavy, large-scale, multipurpose construction located on a moving ice shelf, in a site with an extremely high accumulation rate of drifting and meteoric snow, and moderately cold temperatures.

Weighing in total approximately 2,600 tons (for combined dead and service load), the entire structure can be lifted up and maintained in a position above the snow surface. The trench section is placed 8 metres deep in the firn and is 76 metres long and 26 metres wide. Access to the section is via a ramp covered by a 26-metre-long hydraulically hinged lid. With almost 4,900 square metres of protected space articulated in a total of six floors, the station houses the elements of the HSE system on the ground floor, alongside the fleet parking. Storage rooms and workshops, as well as the snowmelt facility and the central hydraulic unit of the HSE system, are accommodated one floor above. The hall – 68 metres long and 24 metres wide, built atop Neumayer III's 6-metre-high platform – consists of a three-dimensional steel structure with ribs and stringers supported by truss girders. This framework is completely encased in insulated sandwich panels designed as the protective outside hull. Inside the protected hall, 104 prefabricated isolated container modules are arranged over two floors. The modules are functionally equipped and arranged according to the required space for cogeneration plants, workshops and laboratories, social living and working areas, hospital, and supply units and other technical facilities.

The HSE system upon which the building rests has to fulfil several major tasks. Firstly, it has to elevate the entire built structure once a year in order to compensate for the yearly snow accumulation. Secondly, it is used to balance the individual settlement of the bipod foundation plates caused by the load impact on the firn and keep the station aligned for a defined period of time. Finally, the system has to continuously transfer the vertical loads (i.e. the dead and service loads), as well as the horizontal loads such as wind loads, into the firn ground.

Neumayer III was the first of the three Neumayer stations to be built 7 metres above ground. The platform of the station rests on hydraulic cylinders, which lead both the horizontal (wind) and vertical (dead and service) loads via the foundation plates into the ice. The station is built in an ice trench protected from ice drift, a space also used as a garage which is accessed via a ramp.

Altogether forty aprons, each with hydraulic cylinders, are installed in the steel framework alongside the trench walls. Likewise, sixteen bipods with two V-shaped hydraulic cylinders each bear the whole structure (platform and trench section) and allow the station to be lifted in accordance with the growing ice plain. Compacted snow is refilled under the elevated foundation pad to reset the full bearing load on the supporting bipod (HSE system).

The jacking-up sequence goes as follows: in the first step, the station is lifted by extending the hydraulic cylinders; subsequently each of the pairs of foundation steel plates is raised up, with compacted snow used to backfill the gap. While backfilling is in progress, the neighbour bipods take up the loads. For one lift-up operation the work involved takes approximately 16 man-days, with a mean net elevation per lift-up of 1.4 metres (initial snow height) and approximately 480 cubic metres of snow compacted by snow blower under sixteen foundation steel plates.

To ensure the proper operation of the station main building and to reset the internal stresses, the lifting procedures must be performed at least once a year. Over a period of ten years, the operational experiences have shown that a minimum of two lifting-up actions are required per year, regularly conducted at the beginning of the summer season (November and December). The most time-consuming work is the underfilling of the elevated foundation steel plates with compacted snow and the final backfilling of the garage ground with fresh snow. An additional three days are generally needed for lifting up the hydraulic jack-up device of the ramp lid. After backfilling, the fresh powdery snow solidifies within twenty-four hours and is capable of gradually taking up the original load within twelve hours. The two jacks of the bipods are finally used to compensate for the snow subsidence due to the actual load, which is about 25%. Using a special filling order which excludes filling neighbouring bipods the next day, it is possible to complete one lifting cycle in eight to ten days depending on the weather conditions.

While the levelling process is underway, the station's aprons have to be removed from the trench wall to enable the station structure to relax from the internal stresses that have accumulated in the steel structure. Following this, a levelling process for the renewal of the exact horizontal position has to be carried out. This procedure corrects and compensates for the different settlements of the individual bipods. During the station's ten years of operation the HSE system has worked reliably and fulfilled the technical requirements. The regular elevations and refilling procedures consistently keep the platform-trench structure well above the growing snow surface, as the station and its shifting foundations continue on their voyage heading 356° north on the Ekström Ice Shelf.

Ephemeralization.
"We left no footprint, even"

Giulia Foscari

GIULIA FOSCARI is an architect, researcher, and writer who has been practising in Europe, Asia, and the Americas. She is the founder of UNA, an architecture studio focused on cultural projects, and of its alter ego UNLESS, a not-for-profit agency for change devoted to interdisciplinary research on extreme environments. Foscari taught at Hong Kong University and at the Architectural Association, where she ran a Diploma Unit and founded the Polar Lab.

In the Heroic Age, the footprint – surprisingly durable (almost indelible) on the white snow surface of the continent, as Scott himself remarks in his journal on Camp 69 at –22°C – stood as a testament to the progress and achievements made by expedition parties in Antarctica. Metaphorically, human footprints, often detected as ski and sledge tracks,[1] traced the lines of an imperial geography on ice.

Manifested historically in the form of temporary wooden huts, which have been protected and restored as Historic Sites and Monuments (HSMs), human traces in Antarctica have since proliferated, in part due to the Antarctic Treaty System (ATS) prescription that renders a seat at the Antarctic governance meetings conditional upon the Contracting Parties demonstrating their "interest in Antarctica by conducting substantial scientific research activity, such as the establishment of a scientific station".[2] In this way the ATS *de facto* encouraged a substantial increment of built infrastructure in the continent, amplifying the notion of footprint well beyond that of centimetre-deep indents on the frozen topography.

In the name of science, to date seventy-six permanent structures belonging to thirty nation states punctuate the Antarctic landscape, enabling scientists to conduct groundbreaking research. Rarely international, the stations embody the geopolitical agendas of the nation they represent and their building design is often conceived to "project a sense of permanence".[3] Yet permanence is not only perceived; it is not simply an architectural feature which translates political aspiration harmlessly into matter; it is not a formal question. Topographic manipulations propaedeutic to the construction of sound foundations (through blasting and cut-and-fill methodologies) and resilient infrastructural systems that invariably pose contamination risks lead to consequences that are permanent. The operational side of the building industry unquestionably leaves lasting footprints on the Antarctic territory that should be given serious consideration.

By requiring all member states to the Treaty to present an Environmental Impact Assessment (EIA)[4] for all activities undertaken in Antarctica, Article VIII of the 1998 Protocol on Environmental Protection encourages Consultative Parties to invest in technologies and renewable energy to minimise anthropogenic environmental impacts on the continent. Yet the EIA, and the Comprehensive Environmental Evaluations (CEEs) required for activities "likely to have more than a minor or transitory impact",[5] offer simple guidelines rather than imposing any formal obligation, rendering Antarctica the only continent without formalised building design standards. Not surprisingly, as a consequence, the nations that take it upon themselves to assess the embodied energy of their stations, and that strive for performative buildings with zero emissions, are very few. Princess Elisabeth[↗ p.786], the Belgian summer-only station built on the Utsteinen Nunatak in Queen Maud Land, stands out in this context as a role model for sustainable design in the Antarctic.

While building optimisation and remediation are fundamental to sustaining the human presence in the continent "without ecological offence"[6] to its fragile ecosystem, in order, as Buckminster Fuller puts it, "to make the world work for 100% of humanity" – especially in a continent that could be argued to be a "global common" – we should strive to do "more and more with less and less until eventually [we] can do everything with nothing".

"Prognosticating" the future in 1938, Fuller's promotion of "ephemeralization"[7] is still contemporary, and his "dematerialization"[8] theory could reasonably find a literal application in the southernmost continent. A reduction in the number of permanent stations – for example – in favour of fewer international

Temporary structures have been deployed in Antarctica since the Heroic Age of Exploration. Perhaps the most iconic amongst them is the "pyramid tent" used by Robert Falcon Scott's party during their "Race to the Pole" and captured by Scott himself in the photograph shown at the side (top) taken near Beardmore Glacier and Mount Wild on the 20th of December 1911.[1] Originally weighing only 30 kilograms, the polar pyramid is still used by the British Antarctic Survey, which has since optimised the original model. Designed to accommodate two to three people, yet able to host up to four persons in case of emergency, the structure relies on its pyramidal form to resist winds forces of up to 100 kilometres per hour. Thermal insulation is guaranteed by a layer system that consist of an "outer tough, proofed, fire-resistant polyester fabric, a frost lining (which allows for separation and insulation between the outer tent, the inner and outer sleeve entrance), two ventilation cowls, double-sewn reinforced lap seams, a snow valance, a lower wall protection in heavy proofed Nylon, a separate heavy-duty Nylon fabric-butyl rubber sandwich groundsheet, and four poles normally supplied with a 30 cm overlength."[2] In addition to the standard structure, researchers often introduce a foam mat, a self-inflating Therm-a-Rest camping mat, a sheepskin rug, and an Arctic-grade sleeping bag to enhance insulation.[3] Easily transportable via Nansen sledge and Twin Otter airplanes, such tents have also been deployed, as shown at the side (bottom), in the recent Thwaites-Amundsen Regional Survey and Network (TARSAN) campaign in West Antarctica, within the context of the International Thwaites Glacier Collaboration (ITGC)[4] project which focuses on the infamous "doomsday glacier".[5]

"Each evening, radio operators in the station air tower check in with the scientists who camp out in the field, hundreds of miles from the base. Pinned to the wall is a map of Europe overlaid with dots showing how far away the scientists are working if one imagined Rothera as London. This year, one man was mapping the ice bed in the equivalent of Italy. Some were collecting equipment in what would have been Turkey, and another was drilling ice cores in a spot that would have been Saudi Arabia."

Pilita Clark (Business Columnist), "Inside Antarctica: The Continent Whose Fate Will Affect Millions", *The Financial Times*, 2018

bases (following the example of the International Space Station) would not only release state funding that could be redirected from logistic budgets to science, but would also reduce contamination of the white desert.

Such a strategy would not conflict with the development of temporary field-operation camps which allow scientist to reach remote sites in Antarctica, promoting crucial international research projects such as the International Thwaites Collaboration, which saw scientists camp for up to sixty days on the Thwaites Ice Shelf during the 2019–2020 season. On the contrary, temporary mobile structures, developed since Antarctica's discovery, would retain their pivotal role in the unwritten architectural history (and future) of the continent. Ranging from Amundsen's Polheim tent and Scott's "pyramidal tent" (still in use today by explorers such as Colin O'Brady and scientists alike), the 1980s Australian patent for Googie fibreglass structures and near-contemporary German Polybivouac huts, historic American single-container inhabitation units represented by stations such as Eights↗p.694 and Plateau↗p.698, as well as projective Japanese reinterpretations incarnated by the Antarctica Mobile Station Unit prototype, small-scale portable environments are key to the future of Antarctica.

Inhabitation within these micro-architectures brings to the fore reflections on the *Existenzminimum* first addressed at the 1929 Congrès Internationaux d'Architecture Moderne (CIAM), while embodying some of the innate qualities of Hannes Meyer's Co-op Zimmer which "evoke[s] a sense of ephemeral inhabitation driven not only by necessity but also by choice."[9] Even the presence of a gramophone in Meyer's project – in its surreal quality – recalls Herbert Ponting's photographs of Cape Evans↗p.632, one of the precursors of Antarctic architecture.

The scale of field-camp inhabitation units, mere speckles on the 14,200,000-square-kilometre territory, is compatible with Antarctica's fragile ecosystem. Their transportability, perhaps best represented by the historic French Charcot Station↗p.674 (which later inspired the structural units of Union Glacier↗p.828, the only permanent logistic hub that embraces an ephemeral architectural typology) and by the 1940 Antarctic Snow Cruiser (last sighted in 1963 by the crew of the *Edisto* icebreaker alongside the remains of Little America III), is key to their removal.

A leap forward in the process of ephemeralization is represented by the Halley VI Automation Project. Launched in March 2017 by the British Antarctic Survey (BAS), the unmanned unit, built alongside the only station which was truly "designed for removal" and powered by a microturbine, allowed scientists to receive consistent data streams on meteorology, ozone monitoring, tropospheric chemistry, and space weather. Developed out of necessity when the chasm within the Brunt Ice Shelf rendered life in Halley VI↗p.810 too dangerous during the winter (as the ice shelf could calve leaving the station and its inhabitants floating on an iceberg in the middle of the Southern Ocean), the automated unit represents a future in which we could "do everything with nothing".[10]

Built on the occasion of the International Geophysical Year (1957–1958), the 24-square-metre Charcot Station (top) was designed by French engineers Bertrand Imbert and Yves Vallette – seen conducting the final inspection of the preassembled station. With a semi-cylindrical section consisting of three Klégécel AG5 sandwich panels designed to withstand snow accumulation, the station rested on skis for easy transportation. Half a century later, the visionary principles that informed the French design – its mobility and temporality – are still valued in the continent, as proven by the recent collaboration between the Japan Aerospace Exploration Agency (JAXA), the National Institute of Polar Research (NIPR), Misawa Homes Co. Ltd. (Misawa Homes), and Misawa Homes Institute of Research and Development (MHIRD)[6] established for the development of "future-oriented, terrestrial housing [with] applied technologies developed for building manned bases on the moon and other planets". Tested at Syowa Station in February 2020, the two experimental units (bottom) "will be transported to Dome Fuji, located inland the Antarctic continent and 3,800 meters above sea level and used as the residential space (for maximum capacity of eighteen people) for the third Dome Fuji deep ice coring project".

The Googies

Tess Egan

TESS EGAN is library manager at the Australian Antarctic Division. With a background in information science and history, she is also an independent researcher and editor.

The Googie hut, designed and built by Australian Antarctic Division engineer Attila Vrana in the mid-1980s, has provided a robust and versatile field hut in the Australian Antarctic Territory for almost thirty years. Usually painted "international orange", a vivid colour chosen for high-visibility in white-out conditions, the name of the elliptical "Googies" comes from the Australian slang for egg.

Designed with a shape not unlike UFOs to prevent snowdrift build-up and withstand high winds, the Googies were made from moulded fibreglass cast in two halves and mounted on a steel-frame cradle, supported by three anchored legs. Lightweight and portable, at 600 kilograms without fittings, they were transported via ship and sling-loaded by helicopter to a field-work location. With an integrated fibreglass shell spanning 5 metres in diameter at bench height, the three-person hut maximised usable interior space with storage in its under-floor cabinets for food and other items. Under-floor heating and heat exchangers were installed around the walls, with a standard interior fit-out that included up to three couches doubling as beds, a work bench, showers, and toilet. The prefabricated huts could be configured to serve as accommodation, laboratory, mess area, or a shelter to conduct medical surgery. A metal stairway provided access. Large windows provided a light-filled experience but often made sleeping difficult in summer.

The Googies were more spacious than other huts designed at the time, and their insulation and weatherproofing provided comparatively better protection from extreme Antarctic elements than their predecessors.

Only five prototype Googies were made. They were deployed in February 1992 – the first at Béchervaise Island near Mawson Station to support the Adélie penguin monitoring programme, and the other four huts at Spit Bay, on Heard Island. Overwintering on Heard Island in 1992, Mr Vrana experimented with using a wind turbine to power the huts, the first Australian foray into using renewable energy. The wind turbine successfully provided most of the field camp's power needs on a regular basis. In February 1993 the four huts at Heard Island were dismantled and later sent to other locations. One was sited at Hop Island in the Rauer Group near Davis Station to support seabird research, while another was sent to Béchervaise Island to expand the field camp by providing a companion hut to the original Googie. The third Googie was sent to Waterfall Bay on Macquarie Island to replace an ageing hut at Lusitania Bay. The last Googie was re-sited at Brothers Point on Macquarie Island in 1996, as a replacement for the hut at Sandy Bay.

The Australian Antarctic Division's Macquarie Island Modernisation Project team is currently undertaking an assessment of existing field huts to determine whether each meets the structural, functional, and environmental criteria for future needs.[1] As part of a broader modernisation plan for Macquarie Island's infrastructure that includes building a new research station and decommissioning the existing seventy-year-old one, the assessment will recommend which huts should be refurbished or replaced, and where new huts might be sited. Despite some issues in the field, the Googies remain a sentimental favourite for many expeditioners. Though the Googie designs and moulds haven't survived, their iconic look has inspired inquiries from enthusiasts all over the world, cementing their unique place in Antarctic architectural history.

Designed by Helmut Ohnmacht, an Austrian engineer and alpinist, as an emergency accommodation for mountaineers, the Polybivouac Hut, a modular, expandable fibre-reinforced plastic system, was initially erected in 1970 in the Glockner area to be later deployed during the first German Antarctic North Victoria Land Expedition in 1979–1980. Known in the Antarctic as the Lillie Marleen Hut[1], the Polybivouac (shown under construction in the black and white photograph above) is still in use by the Federal Institute for Geosciences and Natural Resources (BGR) at the German Gondwana Station, as a temporary facility which acts as a logistic base to support BGR's flagship programme. Today the hut is complemented by "sixteen interconnected 20-foot containers [which] include repair, staff and working rooms, a kitchen, sanitary facilities, a generator station and a seawater treatment plant [with] scientists accommodated in tents near the station; [Lillie Marleen hut] is recognized as [one of the], 'Historic Sites and Monuments in Antarctic'."[2]

Union Glacier Field Camp

Marcelo Bernal, Paul Taylor, and Francisco Valdivia

MARCELO BERNAL and PAUL TAYLOR are architects and principals of ARQZE Ltd, a practice dedicated to designing, fabricating and installing mobile deployable infrastructure in Antarctica. Dr Bernal currently is Director of Design Process Lab at Perkins&Will. FRANCISCO VALDIVIA is trained as an architect and engineer, holding a Master of Engineering in Membrane Structures and Shell Technologies from Anhalt University of Applied Sciences (Germany).

In 1998 the Chilean Air Force commissioned Aquitectura de Zonas Extremas (ARQZE) to design, fabricate, and build the Estación Polar Teniente Arturo Parodi (EPTAP) in order to support the operation of the landing runway in blue ice at Patriot Hills, Antarctica. Owing to geopolitical issues the station was abandoned eight years later. In 2013, together with the Chilean Antarctic Institute (INACH), activities were resumed, and a new station dedicated to the research and development of polar technologies was consolidated near a second ice-landing strip at Union Glacier. Given the success of EPTAP's insertion, it was then decided to completely recycle the structure, testing the reversibility of its assembly process within a zero-impact logic. The components were transported on sledges and the Union Glacier Field Camp ↗ p. 828 was subsequently relocated 9 kilometres south, in a process that took fourteen days and with a mission of twenty people. In 2014 ARQZE developed a new housing unit for researchers (polar-helmet) and a proto-type of its enclosing skin (torsionoid) to receive the cold-dry sanitary system which freezes human residues and eliminates the use of water. The "polar-helmet" internally differentiates an area to sleep in half-light – counteracting the permanent daylight of the polar circle – and another area for work, naturally illuminated by two skylights. The "torsionoid" is a geodesic ellipsoidal structure composed of linear plywood elements and compressed by a PVC membrane with no insulation that keeps temperatures below zero. Each panel in the membrane has the form of a paraboloid of double curvature that coincides with the subdivision of the structural pattern, consolidating the geometry when it is tensed against the anchoring system. These points correspond to excavations in the snow which are filled with water that, once frozen, produces an efficient tension point with no impact on the environment. The structure has a ski on its base that distributes the loads to avoid the entire building sinking into the snow.

Halley and the Importance of Automation

Thomas Barningham

THOMAS BARNINGHAM is science coordinator for the British Antarctic Survey (United Kingdom) and project manager for the Halley Automation Project. He joined the British Antarctic Survey following the completion of his PhD in 2017 and began working as an atmospheric scientist at Halley. Dr Barningham oversees the delivery of science objectives and ensures the continued delivery of automated science throughout the unmanned winter.

The Antarctic transient population has been increasing each year since humankind first ventured onto the continent. This trend reflected an interest on the part of claimant nations, as well as a growing aspiration on the part of a greater number of countries, to participate in scientific research conducted in the southernmost continent. Predictions of continued growth remained unquestioned until automation was successfully applied in Halley VI ↗ p. 810. The new technology allowed scientists at the British Antarctic Survey (BAS) to receive consistent data sets throughout the year, including in wintertime when the station was evacuated owing to safety risks related to the potential calving of a portion of the Brunt Ice Shelf upon which Halley VI is built [shown at the side].

Initiated in March 2017, the Halley Automation Project aims to provide a duty and standby microturbine power supply and data link to a suite of autonomous scientific instruments that represent over 90% of the station's pre-relocation winter science output.

In 2018–2019, the installation and operation of the microturbine was successful, as was the set up and installation of almost 80% of the Halley winter science output with the associated real-time data link. Unmanned, the microturbine and the majority of science experiments ran uninterrupted throughout the nine-month winter period, with Halley Station preserving core science data streams such as Meteorology and Ozone Monitoring, Tropospheric Chemistry and Climate, and Space Weather and Upper Atmospheric Observations.

The microturbine power generation system (named Capstone C30) was chosen on account of its infrequent service schedule (approximately every 8,000 hours) allowing the system to run unmanned for the winter period. The system has a bespoke autonomous fuelling system that draws fuel from external tanks into a temperature-regulated container hosting the microturbine, a suite of internal monitoring systems, and control and data acquisition technologies. In addition to the automated science experiments, the microturbine also powers a real-time data link, allowing the science and system monitoring data to be viewed live from the British Antarctic Survey head office in Cambridge, United Kingdom.

The system proved its ability to withstand the Antarctic environment during the 2019 winter, unaffected by ambient temperatures as low as –55°C and winds gusting up to 70 knots. The system performed approximately 150 autonomous fuelling events throughout the 265-day period and provided an average of 9.5 kilowatts of power to the Halley Autonomous Long-Term Observational Science (HALOS) Platform.

The wintering science network consists of several external buildings powered by the microturbine. Some buildings are kept warm to protect and maintain the scientific instruments, whilst others are not. Those that are not contain smaller automated experiments contained within self-regulating boxes with smaller heaters, thereby conserving fuel on the microturbine where necessary.

The next phase of the project is to increase the resiliency of the electrical network and data capture systems, as well as ensure a duty and standby microturbine system to enable full power redundancy. Concurrently, the project will increase the winter science output, primarily focusing on radar instrumentation and optical instruments (Space Weather and Upper Atmospheric Observations) alongside bespoke equipment monitoring greenhouse gases (Tropospheric Chemistry and Climate).

The C30 microturbine is capable of providing up to 30 kilowatts of power and, given the current power consumption of the wintering science network, allows ample room for more experiments to be added to what will become known as the Halley Autonomous Long-Term Observational Science (HALOS) Platform.

"This is a truly innovative project and the fact that it has continued faultlessly […] is a major achievement for our engineers and scientists. I'm overjoyed that the crucial programme of long-term measurement of climate, ozone and space weather [is] continuing today because of our engineers' skill and ingenuity. The prospect of delivering such complex science from remote locations without the requirement to have people on the ground year-round opens so many opportunities."

David Vaughan (Science Director, British Antarctic Survey), "Engineers Automate Science from Remote Antarctic Station", 2019 (online)

The success of the Halley Automation Project is relevant for the Antarctic community at large as it offers proof that it is possible to control the key scientific functions of an Antarctic station remotely, paving the way for the potential optimisation, and thus reduction, of the future human footprint on the continent.

"Although it's on the ice, we have approached this in a similar way to designing a satellite in space, with multiple redundant components, with large amounts of data collection and control – it's been really interesting for everyone involved."

Mike Rose (Head of Engineering, British Antarctic Survey), "Engineers Automate Science from Remote Antarctic Station", 2019 (online)

A large rift, generally referred to as "Chasm 1" is extending along the Brunt Ice Shelf and is only 4.5 kilometres away from the so-called "Halloween Crack" and the "McDonald Ice Rumples". Their encounter will generate an iceberg twice as large as New York City upon which Halley VI, the British station designed by Hugh Broughton Architects (shown below in a quasi-frozen state), will float into the ocean. According to NASA,[1] the countdown has started, "putting at risk the life of researchers overwintering in the station".[2] While the prospect of a drastic reduction in the "building life-span due to the fragility of the environment"[3] and the loss of the visionary building would unquestionably be highly problematic, it gave impetus to the design process, leading to the development of an Automation Project, which guarantees scientific data collection during winter months from unmanned stations.

15
The Ideological Use of Relics

Questioning Antarctica's Museumification

Maria Ximena Senatore

MARIA XIMENA SENATORE is the vice-president of the International Polar Heritage Committee of the International Council on Monuments and Sites (ICOMOS), a professor of historical archaeology and heritage at the University of Buenos Aires and at the National University of Patagonia Austral (Argentina). As a senior researcher at the Argentine National Council of Research, Dr Senatore has served as principal investigator of multiple projects in Antarctica.

Humanity is certainly entering a new era dominated by change. In this context, Antarctic heritage and museums could be interestingly reimagined as part of a more equitable, inclusive, and sustainable future. But encouraging a new sense of equity and environmentalism means questioning not only the universalist model that prevails in Antarctic heritage but also the generalised ways in which humans have established relations with "things"[1] there.

Heritage is generally understood as a meaningful and influential cultural process[2] involving objects and places that contribute to the building of communities. However, in Antarctica the dominant vision seems to present heritage as objects that have inherent value, based on their links to certain relevant events or recognised persons. Most of the Antarctic Historic Sites and Monuments (HSMs) seem to be connected to a memorable past of science and exploration, in which national narratives and interests present fading boundaries.[3] What is more, recent policies have introduced "universal significance" as a desirable requirement that must be met if a site is to be designated an HSM. Should Antarctic narratives really include only inspiring stories of endurance and great scientific achievements, while excluding those that tell of capitalism, colonialism, and memories of unrestricted fishing, whaling, and sealing and of conflicts over territorial claims? In all likelihood it is time for new narratives to be brought forward, for a range of stories from multiple sources to be heard, for national interests to be transcended, for the multicultural character of Antarctica to be embraced, and for different communities and world visions to be embraced.

The paradigmatic approach to the restoration of the Heroic Age huts that began in the 1960s reflects the same conventional conception of heritage, i.e. saving objects from decay and attempting to stop the action of time.[4] In fact, the application of these heritage principles has worked to erase much of the non-human stories that had changed those same objects over the course of time. As a result, huts – for example Scott's Cape Evans hut ↗ p. 632 or Shackleton's Cape Royds hut ↗ p. 624 – have been staged as picturesque heritage sites, becoming time capsules that freeze the past.[5] The procedures devoted to protecting, restoring, and conserving these significant places were carried out over decades, and maintaining them will demand a considerable effort in the future. This approach fails to consider the fact that things are not inert, they change and establish relations, and that humans have often become wholly dependent on things or even subservient to them.[6] Sustainable environmentalist models need to simultaneously accept change and the effects of the passing of time. Conceptual frameworks should be introduced in order to interpret the stories released and disclosed by the instability, decay, and transformation of things.[7]

Antarctica is changing along with our ways of being there, our memories of it, and how we understand and designate the significance of places and things. Therefore, we need to be aware of the extent to which the application of heritage principles has, on the one hand, reinforced dominant historic narratives and neglected some other human stories while erasing the wild ones, and on the other, has led to human-thing entanglements that are currently difficult to reverse. We could learn from non-universalist perspectives and actively envision environmentalist models that include a diverse range of human and non-human stories, environments, and communities to build plural, multicultural, and more sustainable Antarctic futures through heritage.

"Today the cabin is once again full of snow. Behind her, we can still see some of the sledges made by previous expeditions. During my stay in McMurdo, I have often been drawn to this historic place; now surrounded by huge tanks of gasoline. Five of these tanks, with a capacity of over one million liters each, together with other surrounding tanks, constitute a reserve of over eight million liters. What a contrast to the heroic days of Scott!"

Emil Schulthess (Photographer), *Antarctica: A Photographic Survey*, translated by the editors, 1960

The Antarctic Heritage List

Francesco Bandarin

FRANCESCO BANDARIN is an architect and urban planner. He served as director of the UNESCO World Heritage Centre and as secretary of the World Heritage Convention (from 2000 to 2010) and as assistant director-general of UNESCO for culture (from 2010 to 2018). Bandarin is an associate member of the ICOMOS International Scientific Polar Committee. In 2019, he has been co-director of the Polar Lab United Kingdom.

The Antarctic Heritage List was established by the Antarctic Treaty System (ATS) to protect the traces of human exploration deemed of historical interest, to celebrate the conquest of the continent by humanity, and to commemorate important events and human losses. While the protection of Antarctic Heritage was not explicitly foreseen in the text of the Antarctic Treaty, already during the first Antarctic Treaty Consultative Meeting (ACTM) in 1961 the issue of the conservation of historical heritage was raised. The creation of the List of Historic Monuments was recommended by ATCM in 1968 and the list was established by ATCM in 1972, with the inscription of an initial group of forty-five sites. In the following years, new proposals were added to the list, reaching the current number of eighty-nine sites (ninety-four sites in total were inscribed but five have been removed because they have disappeared for natural reasons or have been incorporated in larger sites). The inscribed sites have been proposed by nineteen countries, and so far, no proposal has been rejected by the ATCM.

The management framework of the Antarctic Heritage List changed in 2002, with the entry into force of Annex V to the 1991 Protocol on Environmental Protection of the Antarctic Treaty (the so-called Madrid Protocol). Annex V of the protocol established the mechanisms for the designation of Antarctic Specially Protected Areas (ASPA) and the Antarctic Specially Managed Areas (ASMA). Both areas can include "sites and monuments of recognized historical value". As a result of this reform, Historic Sites and Monuments (HSMs) can now be located in an ASPA or ASMA, or outside these areas. The inclusion in a special area gives HSMs more protection than the HSM designation alone, as it requires management plans and the regulation of access and activities in the areas. The newly

The timeline and subsequent cartographies highlight the categories by which Historic Sites and Monuments (HSMs) were designated in Antarctica since 1972 when the formal list was drawn. The original 43 sites were predominantly addressing Monuments and Plaques, Heroic Age Huts and Shelters, Scientific Bases, Graves, omitting Archaeology as a category. Artefacts pertaining to the latter were introduced only in 1995 and are generally deemed to be still underrepresented.

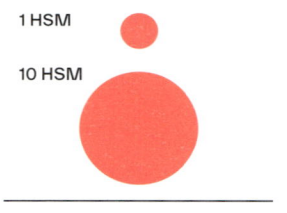

1 HSM

10 HSM

defined list incorporates all the sites and monuments inscribed in previous years by Consultative Meetings. The Madrid Protocol engages all the States Parties in removing from the continent any obsolete infrastructure, unless it is designated as a heritage site. This could prompt nations to propose heritage sites in order to reduce expenditure and to preserve traces of their activities in the continent.

The 2009 Consultative Meeting adopted a new set of "Guidelines for the designation and protection of Antarctic Historic Sites and Monuments", specifying the criteria for nominating sites in the list and for their proper management (Resolution 3, 2009, ATCM XXXII, CEP XII, Baltimore). These guidelines indicate seven criteria to be followed by states wishing to nominate a particular site or monument. The criteria are inspired by definitions followed internationally but have been adapted to the situation in Antarctica. The proposals to be considered for inclusion in the list of Historic Sites and Monuments need to meet certain requirements, whereby one or more of the following conditions should apply: 1) a particular event of importance in the history of science or the exploration of Antarctica occurred at the place; 2) there is a particular association with a person who played an important role in the history of science or the exploration in Antarctica; 3) there is a particular association with a notable feat of endurance or achievement; 4) the site is representative, or forms part, of some wide-ranging activity that has been important in the development and knowledge of Antarctica; 5) it has particular technical, historical, cultural or architectural value in its materials, design or method of construction; 6) it has the potential, through study, to reveal information or educate people about significant human activities in Antarctica; 7) it has symbolic or commemorative value for people of many different nations.

"The time elapsed between a past event and the attachment of a commemorative HSM designation [… is] approximately fifty-three years on average[…]. The longest gap has been 167 years, in the case of HSM 59 *San Telmo Cairn,* while the shortest time has been one year, in the case of HSM 44 *Dakshin Gangotri Plaque,* designated in 1983 to commemorate the First Indian Antarctic Expedition in 1982."

Ricardo Roura (Polar Regions Specialist and Consultant, ASOC), "Antarctic Cultural Heritage: Geopolitics and Management", in *Handbook on the Politics of Antarctica,* 2017

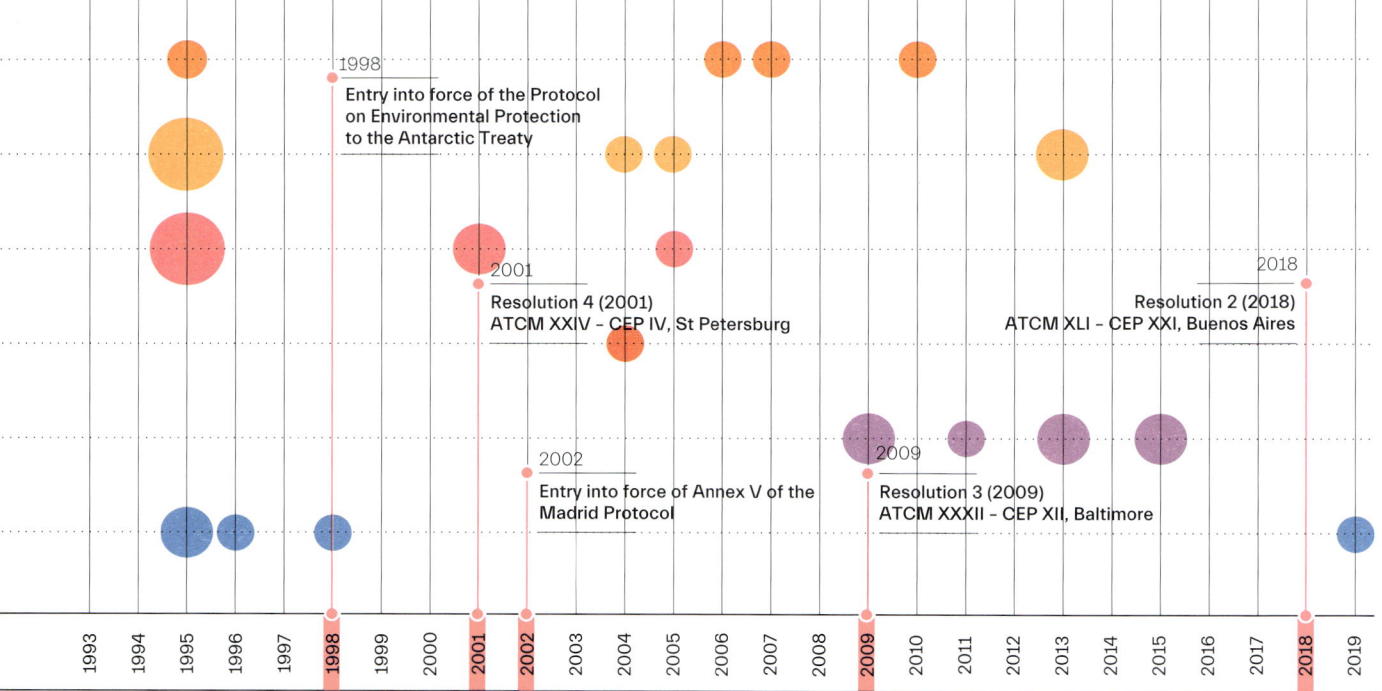

1998
Entry into force of the Protocol on Environmental Protection to the Antarctic Treaty

2001
Resolution 4 (2001)
ATCM XXIV – CEP IV, St Petersburg

2002
Entry into force of Annex V of the Madrid Protocol

2009
Resolution 3 (2009)
ATCM XXXII – CEP XII, Baltimore

2018
Resolution 2 (2018)
ATCM XLI – CEP XXI, Buenos Aires

1993 1994 1995 1996 1997 1998 1999 2000 2001 2002 2003 2004 2005 2006 2007 2008 2009 2010 2011 2012 2013 2014 2015 2016 2017 2018 2019

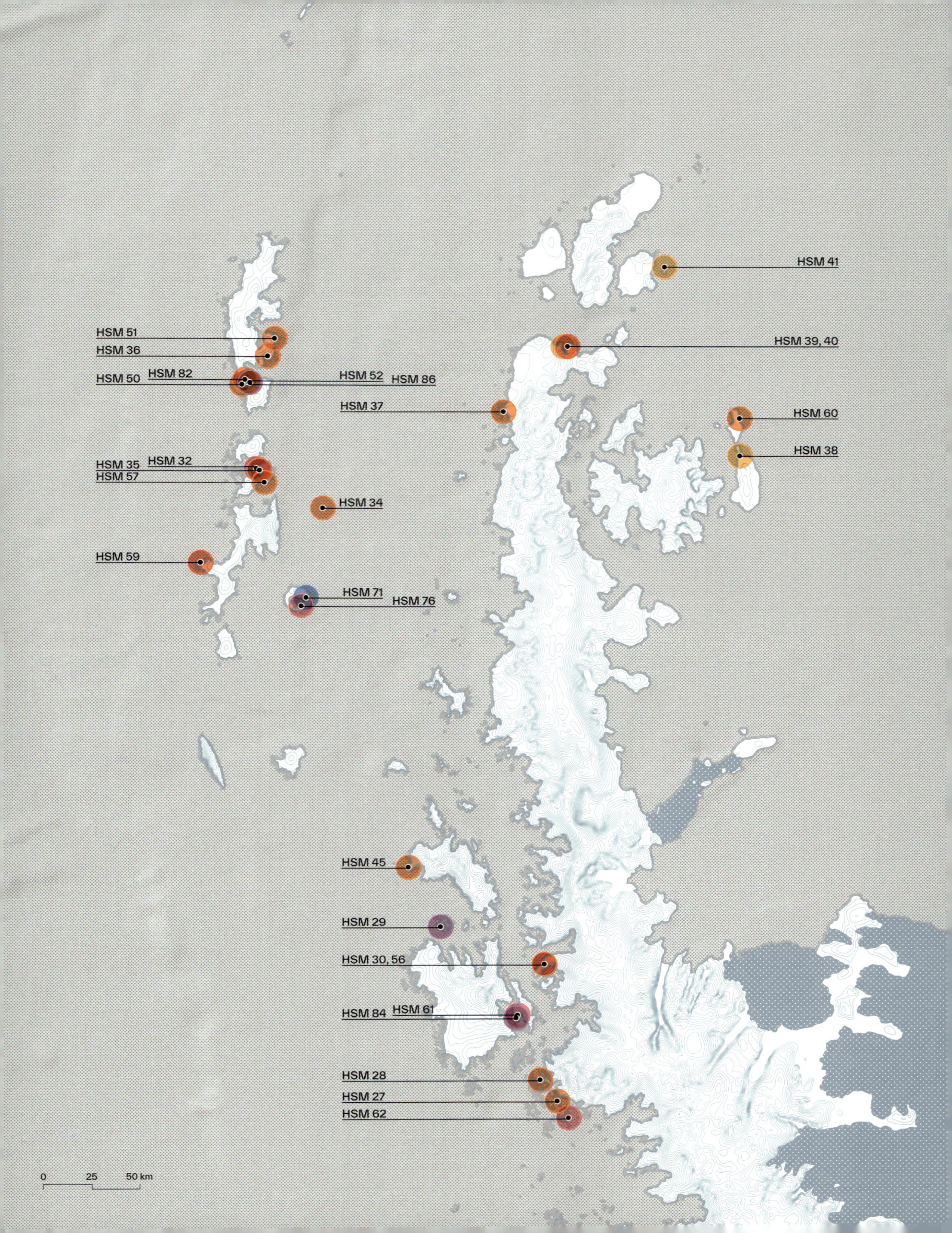

HSM 41

HSM 51
HSM 36

HSM 39, 40

HSM 50   HSM 82   HSM 52   HSM 86

HSM 37

HSM 60

HSM 38

HSM 35   HSM 32
HSM 57

HSM 34

HSM 59

HSM 71   HSM 76

HSM 45

HSM 29

HSM 30, 56

HSM 84   HSM 61

HSM 28

HSM 27

HSM 62

0   25   50 km

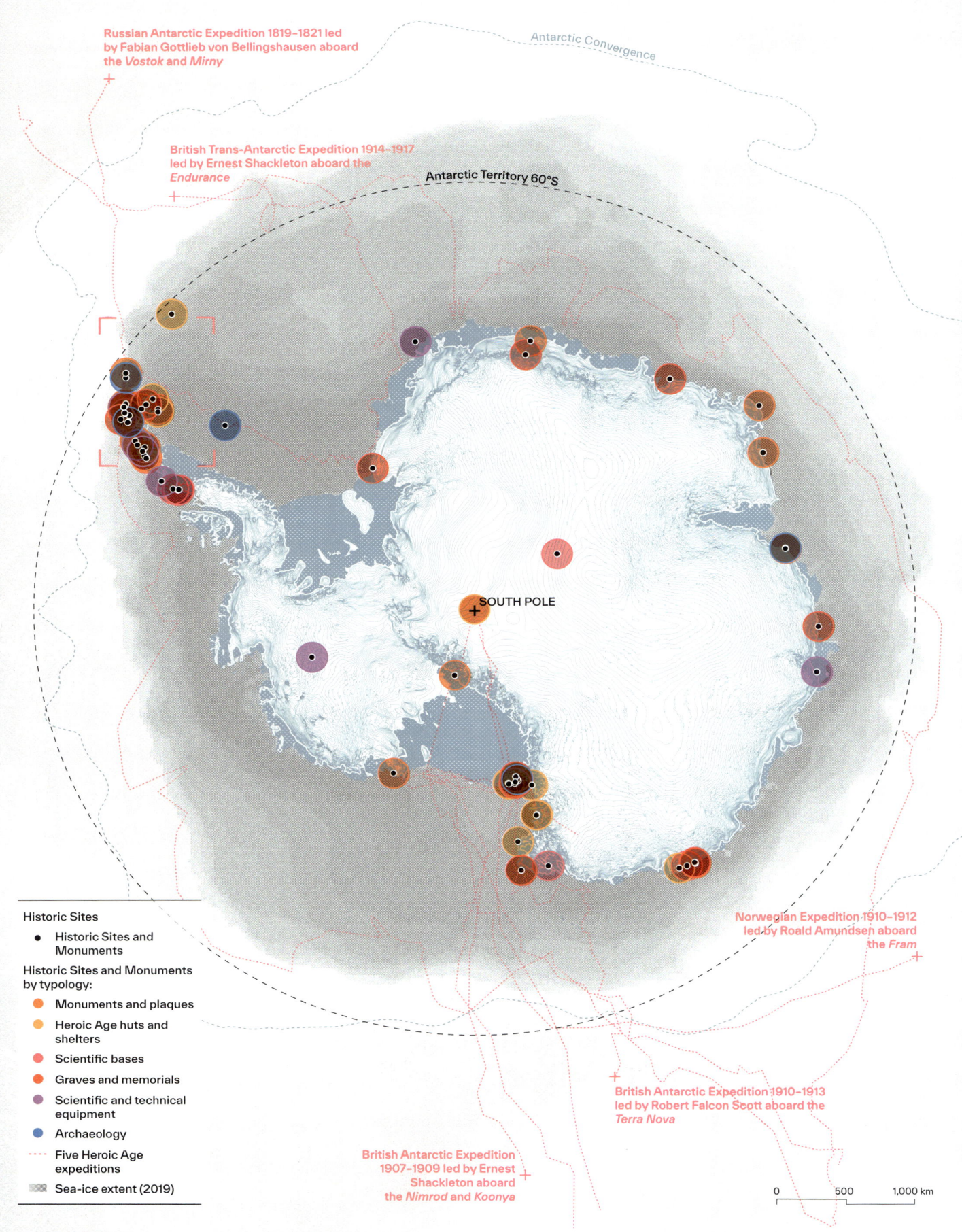

Russian Antarctic Expedition 1819–1821 led by Fabian Gottlieb von Bellingshausen aboard the *Vostok* and *Mirny*

Antarctic Convergence

British Trans-Antarctic Expedition 1914–1917 led by Ernest Shackleton aboard the *Endurance*

Antarctic Territory 60°S

SOUTH POLE

Norwegian Expedition 1910–1912 led by Roald Amundsen aboard the *Fram*

British Antarctic Expedition 1910–1913 led by Robert Falcon Scott aboard the *Terra Nova*

British Antarctic Expedition 1907–1909 led by Ernest Shackleton aboard the *Nimrod* and *Koonya*

Historic Sites

● Historic Sites and Monuments

Historic Sites and Monuments by typology:

● Monuments and plaques

● Heroic Age huts and shelters

● Scientific bases

● Graves and memorials

● Scientific and technical equipment

● Archaeology

--- Five Heroic Age expeditions

▨ Sea-ice extent (2019)

0   500   1,000 km

Prior to these guidelines, in 2001 a resolution was passed by ATCM regarding the review process of the HSM List (Resolution 4, 2001, ATCM XXIV, CEP IV, St Petersburg). Under these provisions, the state that has nominated and/or is undertaking management of an HSM should keep the site under review to assess whether the site still exists in whole or in part and continues to meet the guidelines. The state should also update the description of the site whenever necessary and provide accurate mapping of its location and boundaries. All these actions involve an overall management responsibility for the site, regardless of whether it is included in an ASPA or ASMA.

In 2018 the "Guidelines for the assessment and management of heritage in Antarctica" were adopted (Resolution 2, 2018, ATCM XLI, CEP XXI, Buenos Aires), providing a much more detailed set of principles and operational recommendations and fostering a more rigorous professional and scientific approach to heritage management. The guidelines address issues related to the assessment of heritage and historic values in Antarctica and consider the issue of *in situ* or *ex situ* conservation of artefacts and mobile elements belonging to heritage sites. A key focus of the guidelines, heritage conservation methodologies are discussed, as well as the nature and use of management plans. The guidelines also address the issue of visitor management, which has become very relevant with the increase of Antarctic tourism, as well as the promotion of education and outreach activities.

Interest in Antarctic heritage differs significantly among the countries taking part in the Treaty's activities. Indeed, the majority (75%) of the inscriptions have been proposed by only six countries (United Kingdom, New Zealand, Argentina, Chile, Norway, and Russia). The eighty-nine Antarctic HSMs currently in the list have been in large part proposed by a single country. However, twenty-seven HSMs were proposed jointly by two or more countries, indicating an interest in cooperating in the management process. Most of the binational sites involve the United Kingdom and New Zealand, and concern the huts built by the British expeditions in the Ross Sea region (sites HSM 15–22, for instance), while Norway and the United States have associations with these two countries in the listing of some other huts, shelters, and camps of the Heroic Age. Some wooden or stone huts in the Antarctic peninsula have been proposed jointly by Argentina and the United Kingdom (sites HSM 38 and 39) and managed in collaboration with Sweden. Often this collaboration stems from the proximity of the HSM to the base of one country (this is the case, for instance, for New Zealand, Argentina, and Italy) as this facilitates access, maintenance, and preservation activities.

Inscriptions of HSMs have characterised the life of the ATS in the past fifty years. With the exception of some years, inscriptions were added at a rate of one or two per year.

The alluvial diagram analyses data extracted from the Historic Sites and Monuments list to highlight which member states – in their capacities as proposing or managing countries – promote and undertake Antarctic conservation measures. As the graph clearly shows, the greatest advocates for the preservation of traces of human inhabitation on the continent are the United Kingdom and New Zealand.

"There is an interesting tension between the different people involved in the conservation of the huts. I was working for three masters: the Foreign Commonwealth Office, who very much felt that all of the huts should be kept but didn't want to spend any money on them and their interest was very territorial (although there is no territorial claim in Antarctica, they felt that by having the huts in the continent, should the territorial claims be reinstated, they would have a head start); I was also working for the British Antarctic Survey which is funded by the National Environmental Research Council and effectively saw the huts as an environmental problem and something which should be removed as cleanly and as swiftly as possible; and finally I was working for the Antarctic Heritage Trust, who saw the huts as historically significant and consequently sought to save the artefacts. Three very different points of view."

Michael Morrison (Conservation Architect), in conversation with UNLESS, Architectural Association School of Architecture, 2019

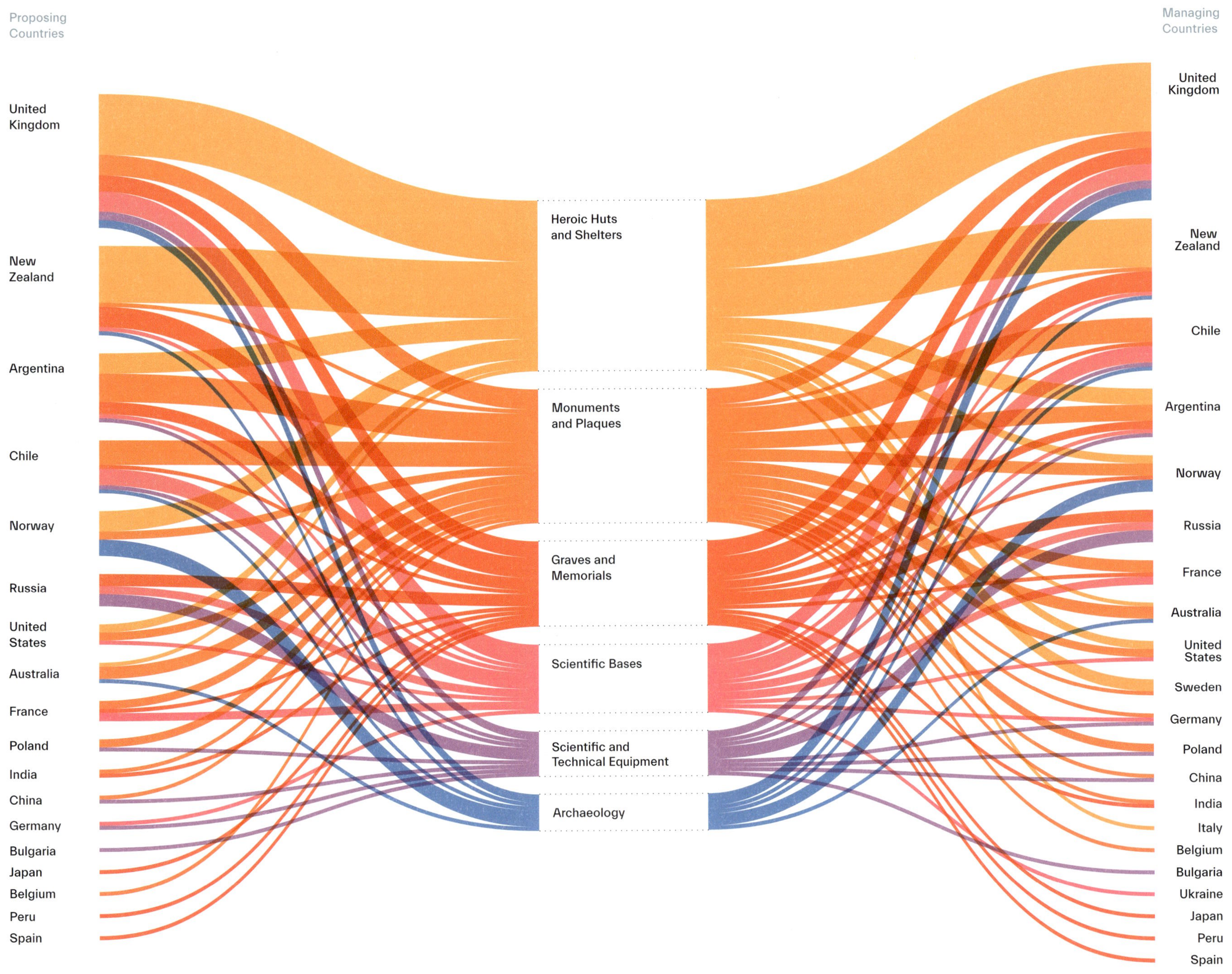
Proposing Countries
United Kingdom
New Zealand
Argentina
Chile
Norway
Russia
United States
Australia
France
Poland
India
China
Germany
Bulgaria
Japan
Belgium
Peru
Spain
Heroic Huts and Shelters
Monuments and Plaques
Graves and Memorials
Scientific Bases
Scientific and Technical Equipment
Archaeology
Managing Countries
United Kingdom
New Zealand
Chile
Argentina
Norway
Russia
France
Australia
United States
Sweden
Germany
Poland
China
India
Italy
Belgium
Bulgaria
Ukraine
Japan
Peru
Spain

Antarctic Archaeology, Heroic Age Huts, and Scientific Stations

Francesco Bandarin

FRANCESCO BANDARIN is co-director of the Polar Lab United Kingdom. He is an architect and urban planner. He served as director of the UNESCO World Heritage Centre and as secretary of the World Heritage Convention (from 2000 to 2010) and as assistant director-general of UNESCO for culture (from 2010 to 2018). Bandarin is an associate member of the ICOMOS International Scientific Polar Committee.

Antarctic archaeology is still largely under-represented in the Antarctic Heritage List. Many archaeological sites in the South Shetland Islands and the Antarctic Peninsula bear witness to seal hunting, the first human activity on the Antarctic continent, as practised in the first half of the nineteenth century. None of these sites is presently listed in the Antarctic Treat Heritage List, despite the fact that in the past decades they have been studied and documented by teams of archaeologists from Brazil, Argentina, Chile, the United Kingdom, Australia, New Zealand, the United States, South Africa, and other countries. While some of these sites are included in certain Antarctic Specially Protected Areas (like ASPA 125 in the Fieldes Peninsula, King George Island, or ASPA 126, Byers Peninsula in Livingston Island) their level of protection still remains insufficient.

Historic Sites or Monuments (also known as HSM sites) of the period from 1820 to 1897, that precede the Heroic Age of Antarctic Exploration, commemorate early explorers (examples are site HSM 39, commemorating the 1874 German Expedition; site HSM 37, which commemorates the landing of the Chilean vessel *Dragon* in 1820; site HSM 81, the "landing rock" in Terre Adélie of the French expedition of Dumont D'Urville in 1840). So far three sites linked to late nineteenth-century whaling activities have been registered in the list: an 1895 message post (site HSM 65), a boat wreck (site HSM 74), and Whalers Bay in Deception Island (site HSM 71 ↗ p. 622), which also includes remains from the early Chilean whaling period (1906–12) as well as of a Norwegian Whaling Station established in 1912 and operational until 1931.

Several huts and shelters from the Heroic Age of Antarctic Exploration, which conventionally starts with the 1898–1900 British Southern Cross Expedition led by Carsten Borchgrevink and ends with the death of Sir Ernest Shackleton during the *Quest* Expedition in 1922, have survived. Of these HSMs, eight are proper huts, while the other eleven are smaller structures built for temporary uses or in emergency situations. These huts are perhaps the best-known items in the Antarctic Heritage List and are among those that have been more carefully protected and restored by the Member States of the Antarctic Treaty or by specialised Antarctic Heritage NGOs. This group of huts includes the Cape Adare ↗ p. 614 huts built in 1899 by Borchgrevink (site HSM 22), Scott's Hut Point hut ↗ p. 618 built in 1902 (site HSM 18), Nordenkjöld's Hut at Snow Hill Island, built in 1902 (site HSM 38); Shackleton's Cape Royds hut ↗ p. 624 built in 1907 on Ross Island (site HSM 15), Scott's Cape Evans hut ↗ p. 632 1911 (site HSM 16) and Cape Denison ↗ p. 636, which includes Mawson's Hut built in 1912 at Boat Harbor, today included in ASPA 162.

During the Heroic Age of Exploration, parties sent to explore distant areas had to build temporary shelters to survive the winter in inaccessible regions. These shelters were built with locally available materials, like driftwood, bones and seal skins and ice, and therefore have only survived in part. Examples of these shelters are site HSM 14, "Ice cave at Inexpressible Island" at Terra Nova Bay, built in March 1912 by Victor Campbell's Northern Party during Scott's Terra Nova Expedition in 1910–1913; site HSM 41, the stone hut on Paulet Island built in February 1903 by survivors of the wrecked vessel *Antarctic* under Captain Carl A. Larsen, members of the Swedish South Polar Expedition led by Otto Nordenskjöld. Site HRM 56 is Waterboat Point, Danco Coast, Antarctic Peninsula, listing the remains and the immediate surroundings of the boat used as a shelter by the British two-man expedition of Thomas W. Bagshawe and Maxime C. Lester that overwintered there in 1921–1922. Perhaps the most unusual of all is site HSM 80, Polheim, Amundsen's tent at the South Pole, erected by Amundsen on reaching the South Pole on the 14th of December 1911. Only Amundsen's and Scott's parties

"There is an interesting discussion to be had around historic huts with regards to the [appropriate] degree of replication. […] Were they in a different location I don't think that we would be spending hundreds of millions of pounds on preserving them. […] The huts were abandoned from 1917 until the mid 1950s until an American expedition party in 1956 was the first to go back in. If you can call it conservation, it's conservation done with a shovel and pickaxe. Everything was […] frozen […] and so really what you see is not what it was: every single object has been […] repositioned using the historic photographs, which does not make it […] an authentic interior."

Michael Morrison (Conservation Architect), in conversation with UNLESS, Architectural Association School of Architecture, 2019

GHOSTLY CABIN FILLED WITH FROZEN HISTORY WAITS FOR EXPLORERS TO FINISH LAST MEAL
Rear Admiral Richard Byrd

Captain Robert Falcon Scott built the hut on Cape Evans in 1911. The last men to occupy it were seven survivors of Sir Ernest Shackleton's Ross Sea party. Shackleton came to their rescue in January 1917, with the ship *Aurora*. Sailing home, he wrote, "I had the hut put in order and locked up." Thirty-nine years later shore parties from Operation Deep Freeze were forbidden to enter, but photographer Fletcher aimed the camera and flash into the gloom through a broken windowpane and shot a series of pictures. This dramatic view developed, showing the table's wine, bread, and cheese perfectly preserved by Antarctica's natural icebox. A pot stands on the stove ready for cooking. King George V and Queen Mary appear on the wall in pre-coronation portraits. Snow, choking all but this room, drifts through a chink. This year's visitors sealed the cabin as a memorial to Scott and Shackleton.[1]

saw the tent in the location, and this object is presently in an unknown location and condition, buried under the ice.

For twenty years after Shackleton's death, and with the notable exception of the explorations of Richard Byrd (1928–1940), the first person to reach the South Pole by airplane, international interest in Antarctica declined, and there are very few listed from this period. By contrast, many stations, were built in Antarctica during and after World War II, initially for the purpose of military surveillance, but later with increasingly scientific aims, especially in the 1950s, culminating with the entry into force of the Antarctic Treaty in 1961. Many of the stations built in this period have been decommissioned as active bases and turned into heritage buildings, maintained and preserved in their original state, as far as possible. Examples of this type of heritage include several British bases: site HSM 61, Base A ↗ p. 644 at Port Lockroy, established in 1944 and today one of the most visited sites in Antarctica; site HSM 63, Base Y, built in 1955 at Horseshoe Island; site HSM 62, Base F, Wordie House, built in 1950 at Winter Island in the Antarctic Peninsula; site HSM 55, East Base ↗ p. 684 at Stonington Island, built by the US in 1939, near site HSM 64, the British E Base ↗ p. 684 of 1961.

Several bases of other countries are also represented in the list. Site HSM 26 is the abandoned installation of the 1951 Argentine Station General San Martin located on Barry Island, Debenham Islands, Marguerite Bay; site HSM 47, Base Marret, is a wooden building on the Île des Pétrels, Terre Adélie, where seven men overwintered in 1952 following the fire at Port Martin Base; site HSM 75 is the A Hut of Scott Base ↗ p. 678, located at Pram Point, Ross Island, built by the Commonwealth Trans-Antarctic Expedition of 1956–1957 led by Sir Vivian Fuchs and Sir Edmund Hillary. Site HSM 76 is the ruin of the Base Pedro Aguirre Cerda Station, a Chilean meteorological and vulcanological centre situated at Pendulum Cove, Deception Island, Antarctica, which was destroyed by volcanic eruptions in 1967 and 1969; site HSM 79 is Lillie Marleen Hut, Mt. Dockery, Everett Range, Northern Victoria Land, erected to support the work of the German Antarctic Northern Victoria Land Expedition (GANOVEX I) of 1979–1980. Site HSM 86 is the No. 1 Building at China's Great Wall Station, built in 1985 and situated in Fieldes Peninsula, King George Island, South Shetlands.

The Heroic Age huts are unique sites, as they are not only the first buildings to have been erected on the southernmost continent but are also time capsules, preserving the incredible endeavours and tales of "hardihood, endurance and courage" of the early explorers, who "would have stirred the heart of every Englishman".[1] Considered by many as the holy grail of Antarctic exploration, the hut at Cape Evans (above and opposite, bottom) stands as a memory and monument to Captain Scott's last expedition to the South Pole. With the aim of preserving this memory for the few who will have the possibility of visiting such a remote site, and more importantly for British culture at large, in 2004 the New Zealand Antarctic Heritage Trust embarked on detailing a conservation plan to preserve both the hut and the variety of more than 8,000 artefacts found within. Throughout the project the team engaged on philosophical issues surrounding the proposed work on the sites and the diverse approaches that could be adopted, ranging from no intervention whatsoever and pure documentation to substantial interventions on-site and the removal of all the structures so that they could be "reconstructed under museum conditions elsewhere in the world".[2] The conservation team settled on an intervention which saw "major repairs and the possible replication of some elements to preserve the overall ambience of the site".[3] A similar approach was deployed at Shackleton's Cape Royds hut (above and opposite, top) where conservators privileged maintenance and repair as well as the restoration of elements to reflect the "dominant period of occupation and in particular the time when the hut was abandoned in March 1909"[4] during the British Antarctic Expedition. A higher degree of replication compared to other sites was acknowledged with wooden crates and other objects being replicated to blend with authentic artefacts to pursue "historical authenticity".[5] The photographs by Shaun O'Boyle are testimony of the work done by the conservators in the past years. The "exceptionally difficult",[6] remote, and impractical site forces us to question the validity of investing millions in preserving such structures in the Antarctic and the cultural and political reasoning driving such investments.

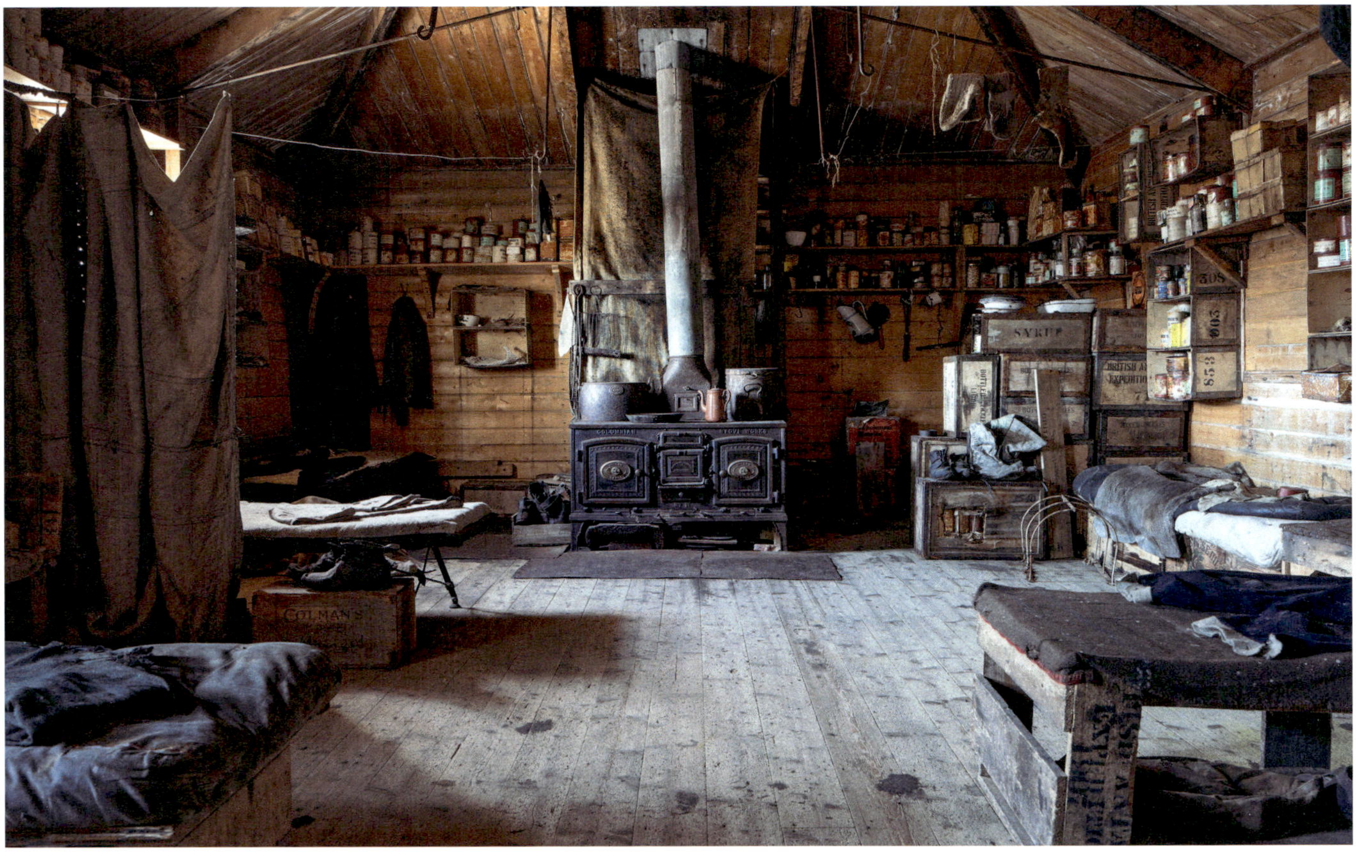
SYRUP
COLMAN'S

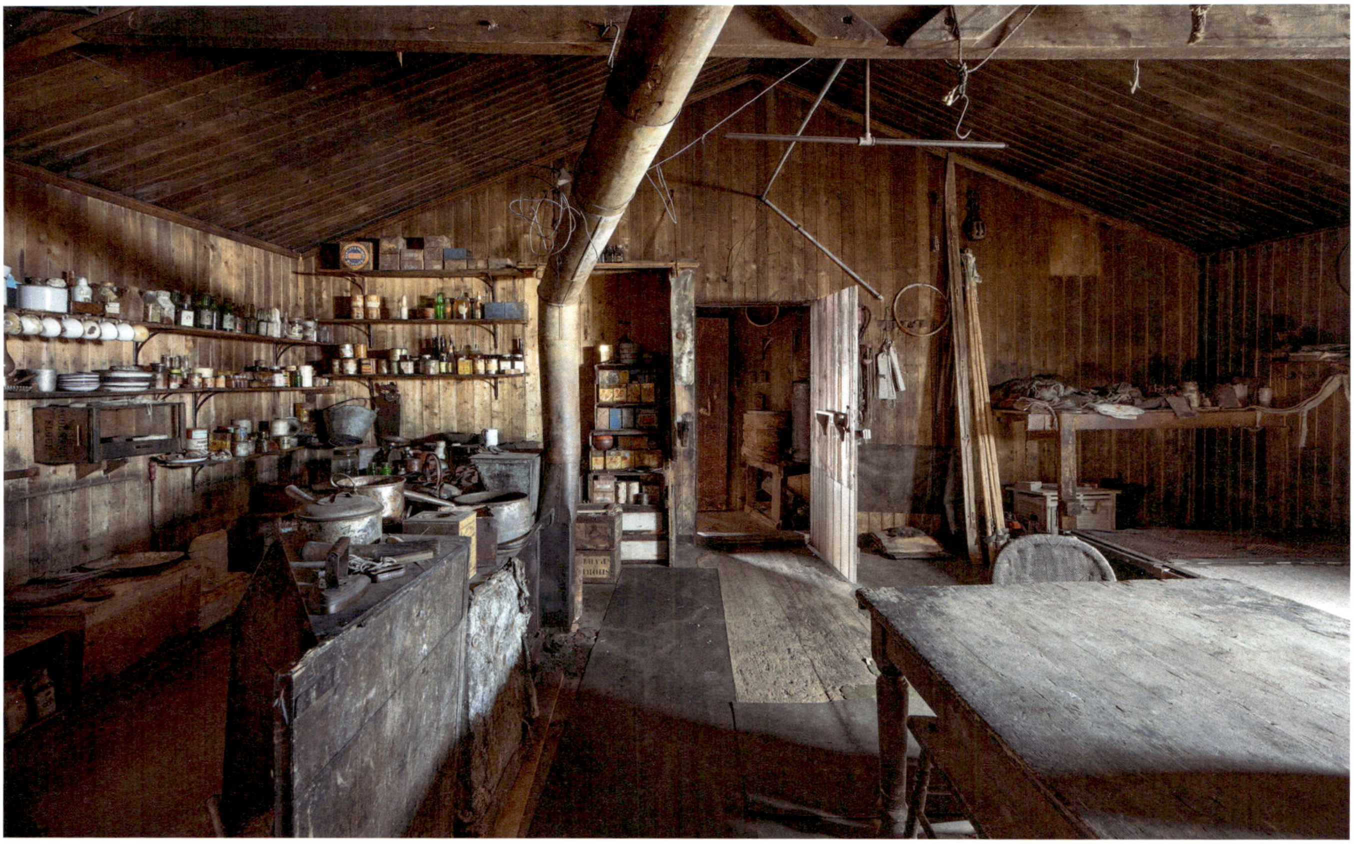

Antarctic Heritage Management

Camilla Nichol

CAMILLA NICHOL is the chief executive of the United Kingdom Antarctic Heritage Trust, a British charity in charge of the conservation of six heritage sites on the Antarctic Peninsula and promoting Antarctic heritage in the United Kingdom. Nichol holds an AMA from the Museums Association and is a fellow of the Royal Geographical Society, a trustee of the Burton Constable Foundation, and chair of the Cromwell Museum Trust (United Kingdom).

It is two hundred years since the first sightings of Antarctica – the Fimbul Ice Shelf by Fabian von Bellingshausen and the Antarctic Peninsula by Edward Bransfield. Since then, humankind has had a fascination with the Great White Continent – it has presented the greatest challenges in terms of endurance, survival, science, and exploration and today is the focus of our heightened concern for our planet.

In 1991 the Protocol on Environmental Protection to the Antarctic Treaty was agreed and in Annex III there is a stipulation that "past and present [...] abandoned work sites of Antarctic activities shall be cleaned up by the generator [...] and the user of such sites." Along with this was the opportunity to designate sites as Historic Sites and Monuments (HSMs), which protects them from damage, removal, or destruction. Today there are ninety-three international HSMs, six of which are looked after by the United Kingdom Antarctic Heritage Trust (UKAHT). Other nations take different approaches to managing heritage in Antarctica; some include it within their National Programmes, others have set up organisations devoted to this work, notably the New Zealand Antarctic Heritage Trust and the Mawson's Hut Foundation.

Not all heritage in Antarctica is looked after equally comprehensively. It is incumbent on the individual Treaty members to determine their approach to the care of their "national" Antarctic heritage – there are no detailed

guidelines, and the standards vary between nations. Beyond the six managed sites, other HSMs proposed by the United Kingdom are managed by other Treaty nations. The huts of Scott, Shackleton, and Borchgrevink in the Ross Sea region, for example, are managed by the New Zealand Antarctic Heritage Trust, while the HSM 71 at Whalers Bay ↗ p. 622 on Deception Island is co-managed by Argentina, Chile, Norway, Spain, and the United Kingdom. The latter is an interesting case in that the wider area, encompassing the geology (it is a live volcano), wildlife, and environment, is an Antarctic Specially Protected Area (ASPA), which means there is a detailed management plan for its ongoing protection. This does mean, conversely, that there is a policy of non-intervention in the heritage, since the site could be destroyed by a volcanic eruption, and as a result the heritage structures are "returning to nature" and may be lost in a few years. This demonstrates that the intricacies of caring for heritage in Antarctica are complex and multilayered, and hardly ever straightforward.

The UKAHT was established in 1993 with the aim of securing the legacy of past human endeavour in Antarctica for future generations. This work continues today through the care and conservation of six historic sites on the Antarctic Peninsula and public programming to share the extraordinary stories of extreme inhabitation in the continent.

The conservation effort requires considerable resources, expertise, and the dedication

of a range of people and organisations to ensure this heritage in the most hostile environment on Earth is preserved for future generations. Conservation of what were envisaged to be temporary structures supporting early exploration and science in Antarctica pushes to the extremes the understanding of materials resistance, and preservationists inevitably find themselves at the cutting edge of heritage management practice.

The sites managed by the UKAHT are former British science stations on the Antarctic Peninsula dating from World War II up to 1992 and together tell a compelling story of the twentieth-century scientific and exploration effort in Antarctica. They are Port Lockroy Base A ↗ p. 644 (1944), Wordie House Base F (1947), Horseshoe Island Base Y (1955), Detaille Base W(1956), Stonington Base E ↗ p. 684 (1961), and Damoy Refuge (1975).

The British-managed sites themselves are visited by tourists on cruise vessels travelling to the Antarctic Peninsula. Of the 50,000 visitors to Antarctica, some 16,000 will visit Port Lockroy, while far fewer will reach the other sites. The visitor experience is a highlight of most visitors' trips as they are immersed in these atmospheric spaces, sampling the reality of life in the continent before the advent of modern technologies and materials. The support of the visitors is vital to the ongoing conservation of the huts and the dissemination of their stories.

Port Lockroy, the most visited site, has been restored and is maintained and presented as a *living museum* offering displays in room settings, with original artefacts and ephemera which tell the story of the base in its heyday in the late 1950s. Alongside the museum, which is situated in a gentoo penguin colony, are the post office and gift shop, the proceeds of which support the work of the Trust.

The other sites are maintained and conserved *as found* and the focus lies on keeping them weathertight and preventing further dilapidation. The ambition is to stabilise the structures, weatherproof them to keep the interiors as dry as possible, and to conserve and monitor the artefacts in an ongoing maintenance and conservation regime.

Managing historic sites in Antarctica challenges everything we know about heritage management. These wooden huts, never designed to last beyond their operational life, are situated in the most unforgiving environment on Earth. They are at least 9,000 miles

The image on the opposite page reveals the state of Station E situated at Stonington Island in Marguerite Bay in 1994, prior to the renovation works undertaken by the British conservation team. The team is captured below during the 1996–1997 preservation campaign at Port Lockroy, the only Antarctic base which hosts a museum. The museum store, a major touristic attraction, is shown below, on the right.

GEOFF COOPER is Heritage Programme Manager for the United Kingdom Antarctic Heritage Trust, responsible for the conservation of six British Historic Sites on the Antarctic Peninsula. A conservation carpenter by trade, Cooper has spent five seasons working for the New Zealand and United Kingdom Antarctic Heritage Trusts in the Ross Sea and on the Antarctic Peninsula.

RESTORATION IN ANTARCTICA
Geoff Cooper

The United Kingdom Antarctic Heritage Trust (UKATH) has launched a four-phase conservation programme. Currently coming to the end of phase one (data capture, survey, and emergency repair), when completed, the overall programme will see the conservation of the six British historic sites and monuments which are under their care.

Prior to each field season the UKAHT conducts research into the history of each of the historic bases. Through access to archive materials, it is possible to track the history of each of the buildings and the numerous changes that have been made to them during their operational life. With this information the seasonal field team can ascertain if the features are still present at the bases and determine the measures that may be required during future conservation work.

away from the United Kingdom, plans to visit them are always subject to change or cancellation, and the operational window each season can be very short.

The Antarctic Heritage Trust can carry out this conservation work only with the assistance of stakeholders. The tourism industry provides most of the transport and delivery of cargo, as well as bringing visitors to the sites. The British Antarctic Survey provides specialist logistic and environmental expertise, as well as access to their archive, which contains valuable information and resources relating to the operation of the sites; and the Foreign and Commonwealth Office (FCO) scrutinises and supports the project by issuing permits to operate.

Entirely self-funded, the UKAHT secures its budget from a range of supporters as well as from trading activity at Port Lockroy and in the United Kingdom. Applications for short-term project funding target grant-giving bodies such as the FCO, but there is no regular or statutory funding source.

Complicated, intensive, and often unreliable, it would be reasonable to ask what justifies the conservation pursuit of a heritage which is so inaccessible and which most people will never experience in person. The underlying assumption is that without the physical heritage the people, events, and accomplishments of these places would disappear into the footnotes of Antarctic history along with a true understanding of the origins of the

advanced scientific programmes, which inform us of the critical changes in our world, the intricate geopolitics and, most powerfully, the human experience. The physicality of these places is thus of great value, and should be preserved, while simultaneously exploring the ability to create access to them in new ways, for example, through digital technology.

"We are looking at ways of using digital technologies such as point cloud not only for the preservation of these sites but also for public engagement. Although we do get 18,000 visitors at Port Lockroy yearly, there is a much larger population who would love to get down there but unfortunately it is extremely expensive and difficult to reach."

Geoff Cooper (Heritage Programme Manager, United Kingdom Antarctic Heritage Trust), in conversation with UNLESS, Architectural Association School of Architecture, 2019

Each austral summer season UKAHT sends a conservation team to at least one of the six Historic Sites that it manages in order to conduct building and artefact condition surveys and to schedule future conservation tasks. During these visits, the team conducts all emergency temporary repairs needed to ensure the buildings remain weathertight until the team returns.

Phase two of the programme will include the development of the UKAHT conservation policies, which are based upon national and international best conservation practices and principles developed specifically to suit the constraints of the Antarctic. Furthermore, phase two will result in the production of site-specific Conservation Management Plans, highlighting conservation objectives of the site based upon its history and significance. A fully costed site-specific Implementation Plan will be developed from these enabling the conservation repair programme to be implemented as well as the ongoing monitoring and maintenance work (fourth and final phase).

Preservation Contradictions on an Active Volcanic Crater. Deception Island

Arturo Lyon

ARTURO LYON is the director of the Polar Lab Chile. He is an assistant professor at the School of Architecture of the Pontifical Catholic University of Chile. Lyon is the founder of Lyon Bosch Martic, where he developed architecture projects including the design of Las Majadas Hotel, the Chena Public Park, and the winning proposal for Nueva Alameda Providencia.

Despite the high risks presented by constant and unpredictable volcanic activity, Deception Island still has a magnetic fascination for visitors. The island is the only place in the world where vessels can travel directly to the centre of a volcanic caldera and witness an active geological process as well as Antarctic wilderness, history, and science.[1]

Deception Island is an emerged active volcano, classified as a restless caldera with a significant volcanic risk, located amongst the South Shetland Islands. The flooded volcanic depression is horseshoe shaped, has barren volcanic slopes, steaming beaches, ash-layered glaciers, and decaying ruins from the whaling industry on its shores. Not surprisingly, this unique Antarctic landscape is one of the ten most visited touristic sites on the continent.[2] The island's name was first recorded by the American seal hunter Nathaniel Palmer in the year the continent was sighted (1820) and it refers to its deceptive appearance: what might have appeared as a conventional island was revealed to be a ring around a flooded caldera.[3] With a submerged basal diameter of 30 kilometres and an altitude of 1,500 metres from the sea bottom (and 540 metres above sea level), the island assumed its current physical form approximately 10,000 years ago, due to a violently explosive eruption that caused the collapse of the volcano summit and the consequent formation of the sea-flooded caldera depression known as Port Foster.[4] Due to its enclosed topography, which offers calm waters and protection from winds, Deception Island quickly became a point of interest for commercial, geopolitical, and scientific purposes.

During the first period of human presence on the island between 1820 and 1825, the anthropic activities were centred on the fur-sealing industry, almost leading to the extinction of the species. Accurately mapped in 1829 by the British Royal Navy, the island was used again for commercial purposes in 1906, when the Chilean-Norwegian company Sociedad Ballenera de Magallanes began its whaling activities, using Whalers Bay as an anchorage for factory ships. Shortly thereafter, in 1912, the British authorities granted permission to the Anglo-Norwegian Hektor Whaling Company to establish a shore-based whaling station ↗ p. 622. Employing approximately 150 people each austral summer and producing over 140,000 barrels of whale oil per season, the factory remained in operation until 1931. No other place in Antarctica has been so thoroughly identified with commercial activity.[5]

Having served as the launching platform for the very first Antarctic flight and aerial survey in the region in 1928 (with a plane piloted by Hubert Wilkins and Carl Ben Eilson taking off from Whalers Bay), the island was first documented though aerial photography in 1947 during the Chilean Air Force aerial survey of the South Shetland Islands, which was followed swiftly by the Falkland Islands and Dependencies Aerial Survey in 1955–1957, one of the first of the kind on the continent, conducted by the predecessor of the British Antarctic Survey. Due to the overlapping territorial claims on the Antarctic Peninsula made by Argentina, Chile, and the United Kingdom, Deception Island became a strategic territory in the international affairs of Antarctica. The secret wartime mission in 1944 (Operation Tabarin) led to the founding of the first British station in the continent (Base B ↗ p. 705) within the abandoned walls of the Hektor Whaling Company buildings.[6] The establishment of the British base was quickly followed in 1948 by the construction of the Decepción Station by Argentina at Fumarole Bay, and by a Chilean meteorological and volcanological centre – Pedro Aguirre Cerda Station – at Pendulum Cove, in 1955.

Both the Chilean and the British stations, together with the remains from the whaling period, were buried and partly swept away by the volcanic eruptions of 1967 and 1969; today their ruins are recognised as Historic Sites or Monuments and listed respectively as HSM 71 and 76. Based on the natural, scientific, historic, educational, and aesthetic values of the site, and in response to the pressures exerted by the combined imperatives of science, conservation, and tourism, Deception Island was declared an Antarctic Specially Managed Area (ASMA 4) in 2005. Yet, despite the HSMs and the ASMA designation, the site and the remains of the historic whaling settlements (the only remnants of this typology within the continent itself) are surprisingly in a state of disrepair due, perhaps, to the multinational origin and conflicting national interests of the proposing and managing parties of such sites. Joint initiatives for its conservation, such as the novel proposal for an open-air Antarctic Museum, located in Whalers Bay developed in 1996, sponsored by the Chilean Antarctic Institute (INACH) and the Norwegian Government,[7] have never been given concrete form.

As it offers unparalleled touristic activities within the continent such as hiking, kayaking, scuba diving, hot-springs bathing and wildlife watch, alongside exposure to architectural ruins, during the 2017–2018 season the island was visited by more than 20,000 tourists – most of whom landed in Whalers Bay.[8] As shown by recent studies, this unprecedented increase in the number of visitors poses a severe threat to the equilibrium of an already sensitive ecosystem, introducing worrisome disturbances to local bird nidification areas and risking the introduction of alien species of vegetation.[9] With all its contradictions, Deception Island, reveals clearly that the fierce independence of the current Antarctic governance system needs to respond with greater determination to successfully preserve the fragile ecosystem of 10% of Planet Earth and the remains of two hundred years of inhabitation in the extreme.

The sought-after touristic Mecca of Deception Island, a unique natural environment rich in biodiversity, is punctuated by derelict Historic Site and Monuments and Antarctic Specially Protected Areas (ASPA).

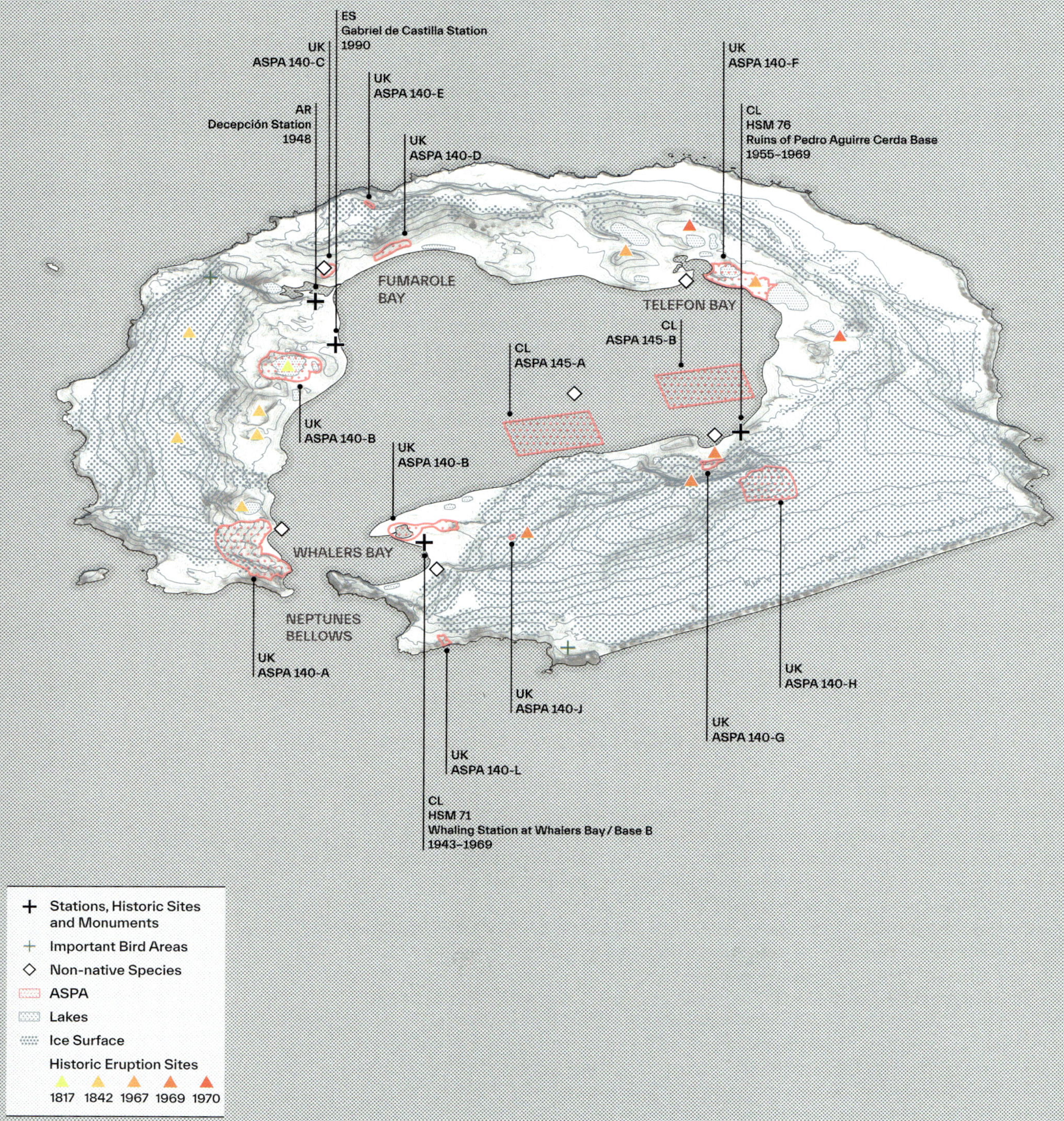

ES
Gabriel de Castilla Station
1990

UK
ASPA 140-C

UK
ASPA 140-E

AR
Decepción Station
1948

UK
ASPA 140-D

UK
ASPA 140-F

CL
HSM 76
Ruins of Pedro Aguirre Cerda Base
1955–1969

FUMAROLE
BAY

TELEFON BAY

CL
ASPA 145-A

CL
ASPA 145-B

UK
ASPA 140-B

UK
ASPA 140-B

WHALERS BAY

NEPTUNES
BELLOWS

UK
ASPA 140-A

UK
ASPA 140-J

UK
ASPA 140-L

UK
ASPA 140-H

UK
ASPA 140-G

CL
HSM 71
Whaling Station at Whalers Bay / Base B
1943–1969

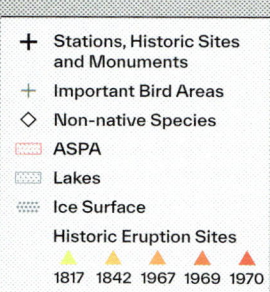

+ Stations, Historic Sites
  and Monuments
+ Important Bird Areas
◇ Non-native Species
▨ ASPA
▨ Lakes
▨ Ice Surface
Historic Eruption Sites

1817   1842   1967   1969   1970

The Fate of Whaling Stations in Antarctica

Michael Morrison

MICHAEL MORRISON is a conservation architect and the British representative at the International Polar Heritage Committee of the International Council on Monuments and Sites (ICOMOS). As the former chairman and senior partner of Purcell Architects, Prof. Morrison has worked, since 2002, on conservation projects devoted to Antarctic historic huts and whaling stations on the Peninsula, Ross Island, and South Georgia.

South Georgia is remote – some 1,300 kilometres south-east of the Falkland/Malvinas Islands, and a visit is not for the faint-hearted, taking some four or five days by ship across one of the world's stormiest seas. The island is approximately 160 kilometres long and 35 kilometres wide at its broadest point, rising sharply out of the sea with the highest point, Mount Paget at 2,935 metres above sea level. It is thought to have first been sighted in 1675 by a London merchant, Antoine de la Roché, and in 1775 Captain Cook sailed down the east coast. Cook made the first landing and "took possession" of the island in the name of His Majesty George III. Hunting for fur and elephant seals began on South Georgia in 1786 with peaks in the 1790s, late 1810s, and again in the 1870s; however, it is as a major centre of whaling that the island is best known.

In 1904 Norwegian Captain Carl Anton Larsen opened the first whaling station on South Georgia at Grytviken. This was with funding from Argentina, and the company was called Compania Argentina de Pesca (however throughout its long operating period from 1904 to 1959 it was manned mostly by Norwegians and Scotsmen). The great demand for whale oil made whaling a highly profitable enterprise – so much so that moored floating factories started to operate in sheltered anchorages, and between 1909 and 1916 further shore-based stations were set up along the east coast of the island at Leith and Ocean Harbour (in 1909), at Husvik (in 1910), at Stromness (in 1912), and at Prince Olav (in 1916).

Whale oil was much needed during World War I. Dominated by Norway, the whaling fleet operated mainly in British South Atlantic territory, allowing Britain to control most of the whale oil supply. While the main use of whale oil was to make soap, it was also used to produce high-quality lubricants of the sort very necessary for military equipment. A by-product of soap-making was nitroglycerine, a key component of cordite, the standard propellant used in British artillery shells and small arms ammunition.

Whaling continued at South Georgia until 1960, briefly interrupted in 1931 owing to a combination of oversupply and decreasing demand in the Great Depression, before stopping almost entirely during World War II. The shore-based stations gradually ceased to be effective in the 1950s as whale numbers fell, companies amalgamated, and sites were closed. Though the sites were shut, there was an expectation that they would eventually reopen and hence all the machinery, equipment, stores, and buildings were left behind.

South Georgia never had a permanent population; the residents who stayed longest were the managers and the senior staff of the whaling stations, some of whom were joined on the island by their wives and families. Larsen, for example, lived at Grytviken with his wife, three daughters, and two sons for several years, taking British citizenship in 1910. There were also British officials who controlled the whaling activities and acted as magistrates, providing policing and postal services. Most of the men who worked the stations were seasonal workers from Norway or the Scottish Islands.

Each of the stations was like a self-sufficient small town. The biggest buildings were those associated with rendering whale carcasses and storing the oil and "guano" (the confusingly named by-product of the rendering process – essentially bone meal for use as fertilizer). But there were many other buildings, stores, workshops, carpenters' shops, piggeries, poultry houses, kitchens and messes, libraries, cinemas, hospitals, and shops, and even a church at Grytviken. The larger buildings generally have steel frames and are clad in corrugated iron but most of the domestic ones are timber framed, clad with wooden boards and roofed with corrugated iron.

"Who discovered Antarctica we may never know. We remember the Shackletons and the Scotts, but it was the whalers and sealers who opened up the Antarctic, not the explorers. Because the whalers and sealers didn't talk much about their good hunting grounds, they have sifted between the seams of history."

Diane Ackerman (Poet, Essayist, and Naturalist), *The Moon by Whale Light*, 1991

To extract oil, whale carcasses were dragged up onto a timber working area, the "flensing plan" where the blubber was stripped, the meat cut away, and the bones sawn up. The three different elements were hauled up to platforms above large cast iron "cookers" where the pieces were fed down into boilers to be "rendered" to produce the oil. The process was initially quite crude and wasteful but gradually became more sophisticated, allowing every part of the carcass to be processed. The motive power for the stations was steam generated in large central boiler houses. In the early days coal was used as fuel, gradually a transition was made to using fuel oil – so one of the defining images of the stations is a series of large steel tanks – some to contain the fuel oil and others to store the extracted whale oil.

Steam was used for many purposes – the main one was to heat the cookers in which the whale carcasses were rendered – but steam was also used for powering winches and bone saws as well as for heating all the buildings. The steam was piped around the sites, generally in high-level pipes suspended from steel gantries, and everything was heavily insulated. The insulation is almost entirely asbestos as is the cladding on the boilers and much of the insulation inside the domestic buildings. This is a major part of the problem now facing the Government of South Georgia.

A research facility had been established in 1924 at King Edward Point, just across the bay from the Grytviken whaling station. This remained in operation until 1982 when Argentinian activity on South Georgia was the prelude to the Falklands/Malvinas War and all the research work ceased. The British retained a military presence on the island until 2000, when a new research station was opened. One of the researchers at a newly built station at King Edward Point detected that asbestos fibres from the Grytviken station were potentially contaminating the new research station. This led to a major asbestos removal programme in 2003, which entailed pulling down many of the buildings, burying the asbestos and all the contaminated material (pipes, corrugated iron, sections of boiler, etc.) in deep pits dug behind the station, and cleaning the remaining frames and elements of the machinery of any trace of insulation. Whilst ensuring safety for all station staff and for the considerable number of tourists who visit on Antarctic cruise ships, this campaign left a very "sterile" site compared to the other stations, which remain untouched. At present, the remaining whaling stations are all "off limits" for anyone not having a permit to enter and the right protective clothing. A 200-metre exclusion zone has been declared by the government for safety purposes, but this does not address questions about the long-term future of these places.

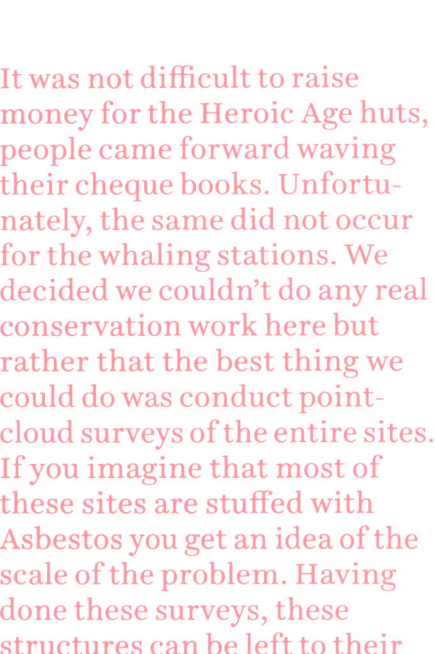

While today whaling may be something that most people regard with dismay, perhaps even horror, and wish to forget, the large-scale Antarctic whaling stations offer remarkable evidence of a major world industry which has (almost) completely disappeared. The scale of work needed to conserve even one of the stations is far beyond the means of the Government of South Georgia. With financial help from Norway, the government has commissioned a detailed laser survey of the major sites, which at least provides a comprehensive record of the present condition of the buildings and the major elements of the processing plant. At present this seems to be as much as is financially and practically possible, but the ongoing decay will inevitably lead to the complete loss of these structures, which are a testament to a key part of our industrial history.

"It was not difficult to raise money for the Heroic Age huts, people came forward waving their cheque books. Unfortunately, the same did not occur for the whaling stations. We decided we couldn't do any real conservation work here but rather that the best thing we could do was conduct point-cloud surveys of the entire sites. If you imagine that most of these sites are stuffed with Asbestos you get an idea of the scale of the problem. Having done these surveys, these structures can be left to their own devices to collapse."

Michael Morrison (Conservation Architect), in conversation with UNLESS, Architectural Association School of Architecture, 2019

The Neglected Sealing Heritage of Antarctica

Michael Pearson

MICHAEL PEARSON is the president of the International Council on Monuments and Sites (ICOMOS) International Polar Heritage Committee. He is a historical archaeologist and heritage planner who has worked in Australia and internationally. Dr Pearson has taken part in ten research expeditions to Antarctica and has published extensively on polar history and heritage, industrial heritage, and world heritage.

For anyone asked to name sites of Antarctic heritage value, the most common response is to refer to the expedition huts of Robert Falcon Scott, Roald Amundsen, or Ernest Shackleton, the best-known characters of what has been called the Heroic Age of Antarctic exploration in the first years of the last century. If the responder has more knowledge of the Heroic Age, they might also refer to other expedition huts, such as those of Douglas Mawson, Carsten Borchgrevink, or Otto Nordenskjöld. All of these huts, except Amundsen's, survive, are protected as Historic Sites and Monuments under the Antarctic Treaty System, are actively conserved by heritage trusts, and are tourist attractions.

But there is another aspect of Antarctic heritage that is little known and neglected: the makeshift huts of the first humans to live in Antarctica, the sealers of the 1820s, located in the South Shetland Islands, lying just off the Antarctic Peninsula. Their discovery in 1819 by British sealer William Smith led directly to the exploitation of the fur seals that bred there, bringing the Antarctic into the global fur-sealing trade. Between 1819 and 1827, some 144 sealing expeditions from Britain and the United States killed a minimum of 300,000 seals for their skins, effectively wiping out the local fur seal population.

Because of the scarcity of safe anchorages, sealing crews were dropped ashore by boat and left, with limited supplies, for periods ranging from days to months, and on two occasions for an entire winter. Once ashore, the sealers built shelters in caves, against cliff faces, or on open beaches, piling up dry-stone walls and roofing them with tarpaulins or seal skins over whale-rib beams. Over fifty sealing sites have been identified, a number have been excavated, and an archaeological research programme continues.

The sealing sites of the South Shetland Islands have been little disturbed and are a part of the world's shared heritage, though their location within the area controlled by the Antarctic Treaty, where no single nation has sovereignty, precludes them from being listed under the UNESCO World Heritage Convention. None of the sealing camp sites in the South Shetlands, however, are currently included in the List of Antarctic Historic Sites and Monuments protected under the Antarctic Treaty process. Similarly, none of the sealing sites have had any conservation attention. These sealing sites have been neglected for a number of reasons. Antarctic historiography has tended to emphasise exploration and science over exploitation. The sealing era is not often seen as being part of the "official" history of Antarctic endeavour, which is one of national aspirations, myths of individual heroism, and the romanticised struggle to wrest geographic and scientific knowledge from hostile nature: sealing, if mentioned at all, is only a footnote. Whereas the huts from the Heroic Age are the physical property of particular nations that accept responsibility for them, no nation claims ownership of the sealing sites, and the nationality of the builder is usually unknown. The individual sealers are simply names unassociated with any tales of heroism, and the struggle of the sealers to survive in Antarctic conditions is not regarded as "noble". Sealing was an industry involving the slaughter of seals, in total antipathy to the Antarctic Treaty's focus on the environmental stewardship of the continent. Even whaling has a more palatable history, because of the largely successful and highly publicised international conservation efforts to end it.

There are a number of threats to the survival of sealing sites in the South Shetlands. While tourism is often seen as a major threat, and the South Shetland Islands and the adjacent Antarctic Peninsula experience about 98% of the almost 52,000 annual tourist visits (from a 2018–2019 figure), the vast majority of beaches containing nineteenth-century sealing sites have never been visited by tourists, and on those that have there appears to have been little tourist impact on the sealing sites. This pattern may change if overall visitor numbers increase and more landing sites are needed to accommodate the demand: the 151 landing places visited in 2003–2004 grew to 219 in 2017–2018 (IAATO figures) and may increase further.

By far the most intensive and disturbing human presence in the South Shetland Islands is not that of tourists but of the scientific parties undertaking research work within the framework of the Antarctic Treaty system, and the related scientific programme support staff. The South Shetlands with the adjacent Antarctic Peninsula is the most intensely studied region of Antarctica, with ten permanent and nine summer national bases in the archipelago and widespread field camps on most islands each summer. While a number of important sealing sites were first relocated by scientific parties from the 1950s through to the 1980s, and their investigation was well intentioned, it resulted in extensive disturbance that reduces their archaeological research and overall heritage value. A large number of irreplaceable sealing artefacts removed from the sites in that period have since been disposed of or lost by the agencies involved.

While the Antarctic Treaty system has created the Historic Sites and Monuments List (1972), provided blanket protection of all human occupation sites created before 1958 until fully researched and assessed (2001), and established a Protected Areas system that can include cultural heritage sites (1998), there have been few guidelines or processes implemented to give effective protection to the less obvious heritage such as the sealing sites. The International Polar Heritage Committee (IPHC) of ICOMOS is working in collaboration with the Scientific Committee on Antarctic Research (SCAR) to develop further tools to assist the Antarctic Treaty System in heritage site identification, assessment, and protection.

Climate change is another risk factor. The region has experienced a rate of warming up to ten times the global average over the last fifty years. Coastal erosion can be accelerated by both sea level rise and by the reduction in the duration of sea-ice formation, which protects the coast from extremes of storm surge and wave action. A large number of sealing sites in the South Shetlands are at or near sea level, and typically sit on erodible sandy shorelines. The degradation of permafrost as a result of increasing soil temperature is a major problem in the Arctic, and it is being investigated in 2020 as one of the possible causes of ground instability at Nordenskjöld's 1901–1903 Snow Hill Island hut in the South Shetlands. It could also lead to instability at sealer shelter sites.

To unveil the censored history of Antarctica, the above-mentioned issues might suggest that a number of actions are needed. These include increased awareness training for national parties going to Antarctica; wider dissemination of guidelines on historic sites, their value and protection; revising management plans for Antarctic Special Protected Areas to provide greater protection of historic sites; widening the scope of the Historic Sites and Monuments listing process; and ensuring that the monitoring of climate change impacts includes its impact on cultural sites.

"It is estimated that 144 sealing expeditions worked in the area between 1819 and 1827, mostly from Britain and New England (United States). During the nineteenth century, the sealers working in the region of the Antarctic Peninsula left behind them the remains of their summer shelters, where small groups of men were set ashore from the sealing vessels to spend days or weeks killing and skinning fur seals. The simple shelters consisted of caves or hollows in the rock face, or simply spaces on a suitable beach, where rocks, timber from the ships, whalebones and sealskins were used for building materials. The modest remains today can be compared with early indigenous dwelling remains in the Arctic with their use of natural materials and the difficulty non-experts can have in recognising their significance."

Susan Barr, "Twenty Years of Protection of Historic Values in Antarctica under the Madrid Protocol", *The Polar Journal*, 2018

Māori Culture in Antarctica. Preserving Knowledge and Papatūānuku

Priscilla Wehi, Te Warihi Hetaraka, Fayne Robinson, Poutama Hetaraka, and James York

A collective worked over twelve months to conceive and complete the Māori carving (*whakairo*). Represented here are TE WARIHI HETARAKA, an elder of the northern people Ngātiwai and teacher of emerging carver POUTAMA HETARAKA (Ngātiwai, Ngāi Tahu); FAYNE ROBINSON and JAMES YORK from the southern people Ngāi Tahu have both been carving for more than thirty-five years; PRISCILLA WEHI is a conservation biologist.

Ko te unaunahi i whakapiripiri ki te ikanui a Maui[1]

The Indigenous Māori people are a tribal people descended from Polynesian ancestors who reached Aotearoa New Zealand in circa 1280 BCE, bringing with them philosophies, world views, and practices distinct from later settlers who arrived from the 1800s onwards. These expert voyagers traversed the Pacific Ocean from the tropics to Antarctic waters across thousands of years, with Hui-te-Rangiora one of those who voyaged in Antarctic waters around 700 years ago.[2] In recent times, recognition of Māori knowledge systems (*mātauranga Māori*) and what they offer has brought Māori expertise back into view, at a time when the world stands on a precipice of climate change and biodiversity loss. In Antarctica, this rich history and knowledge is reflected both in the magnificent carved *Te Kaiwhakatere o te Rangi*, or Navigator of the Heavens, that stands outside Scott Base, and in traditionally woven representations within the buildings. The recently completed *whakairo*, or Māori carving, that frames the entry into Scott Base ↗ p. 852 is the first built into the base itself, acting to welcome all who enter.

Toi whakairo, the unique Māori tradition of translating cultural knowledge in curvilinear symbols within carvings, is a form of recording and transferring knowledge through the generations, *before* the written word. The carving of symbols that evoke age-old stories and concepts is a task that requires the entire spirit and body to come together in an act of synchrony. As a repository of knowledge, *toi whakairo* draws on history and tradition but also leads to innovation and creativity. Each cut and symbol draws attention to stories and paradigms that act as a lens with which to interpret our world and challenge our behaviour. Thus, a *tohunga whakairo* is much more than a master carver; he is master of many strands of expertise.

Two *tohunga whakairo,* Te Warihi Hetaraka and Fayne Robinson, led the conception of the project and, together with expert carver James York and up-and-coming carver Poutama Hetaraka, collaborated with ecosystem scientists to understand the implications of the Ross Sea Marine Protected Area, which came into being in December 2017. Aiming to promote the protection of planet Earth for future generations, with *mātauranga* as the foundation, carvings were produced for Scott Base.

Formed by two panels, or *whakawae,* that frame the doorway and a lintel or *pare* above the door, the *whakairo* is overwhelmingly a story of love, of protection. On the left *whakawae, Papatūānuku,* Mother Earth, stands with her cloak only partially covering her body. The imagery recalls an ancient lesson from Ngātiwai that is incorporated into the "four scales of the fish". The scales represent the minerals that lie beneath the Earth as well as the different forms of life that move above the ground: flora, fauna, and humanity. Humans are part of the narrative, not just because of the damage they produce to *Papatūānuku* through their action, but because they are part of nature and kin to other living and non-living forms. As Poutama Hetaraka has said, "When we deforest and mine, we are scaling *Papatūānuku.*" On the right *whakawae,* the *taiaha,* a weapon of war, is transformed into a tool of peace, the *ko* or digging stick.

Referencing traditional Māori calendars, which take the moon as a guide, a warning about the effects of exploitation, that the world is out of kilter, was introduced on the lintel. The design symmetry reflects the partnership of Ngātiwai and Ngāi Tahu, as well as the essence of *mātauranga* and Antarctic science braiding together to protect the planet. It highlights how both Indigenous people and Antarctic scientists are working towards a common goal of protecting *Papatūānuku* and her ecosystems.

In *mātauranga,* knowledge is incomplete without experience. The carvers thus travelled to Antarctica and saw pressure ridges, formed by the interactions of the ice floes as they collide, as well as the intricate web of relationships between species, and re-created these in the *whakairo.* The logistical demands of creating a *whakairo* to carry these stories to Scott Base were immense. Because of the intense cold, early discussion ranged over issues from the hardness of different woods and the likelihood of cracking through to sealants. One of the skills of a carver is choosing wood that is appropriate to the task; in this case, *Robinia pseudoacacia* wood that had been cut and cured for many months was used for the planks on the sides of the doorway, and *tōtara* and *rimu* woods for the *pare.* Because of strict rules surrounding the use of organic materials in Antarctica, it was impossible to use traditional materials such as *harakeke* for the lashings.[3] Instead, the ties used were made from nylon cord. Since Scott Base itself will be rebuilt over the next ten years, the carvings were designed to ensure that the *whakairo* was of a size that would be well suited to the new buildings.

Ritual is an essential part of the journey for every *whakairo* of significance. The *whakairo* was transported to Antarctica as freight, but before leaving it was blessed with *karakia,* traditional chants, in a ceremony at Antarctica New Zealand, Christchurch. Once completed, the carvers led ritual *karakia* to open the *whakairo* to all those at Scott Base.

The work of all those involved has ensured a powerful visual message that conveys the urgency of protecting both Antarctica and our planet and acknowledges all those who work to do so. In this spirit, although the station of Scott Base will be demolished and rebuilt, the carvings will be preserved and reinstalled in the new building, connecting the past to the future in a way that is consistent with Māori philosophy and practice.

"We are using *whakairo* to have a conversation in and about the wellness of Antarctica. The well-being of Papatūānuku starts with Antarctica. It's an indicator, a litmus test for the rest of the world."

Te Warihi Hetaraka (Master Māori Carver), "Māori Whakairo to be Carved in Antarctica", 2019

Antarctic Memorials, Monuments, and Scientific Equipment

Francesco Bandarin

FRANCESCO BANDARIN is co-director of the Polar Lab United Kingdom. He is an architect and urban planner. He served as director of the UNESCO World Heritage Centre and as secretary of the World Heritage Convention (from 2000 to 2010) and as assistant director-general of UNESCO for culture (from 2010 to 2018). Bandarin is an associate member of the ICOMOS International Scientific Polar Committee.

Several sites in the Antarctic Heritage List commemorate people who lost their lives during the exploration of the continent or due to accidents. These memorials include graves or cemeteries as well as crosses or cairns. Examples of the first type are site HSM 23, the grave at Cape Adare ↗ p. 614 of Norwegian biologist Nicolai Hanson, a member of the British Southern Cross Expedition (1898–1900), led by Carsten E. Borchgrevink, and the site HSM 9, the Buromsky Island cemetery with tombs of citizens of the Soviet Union, Czechoslovakia, East Germany, and Switzerland who perished in Antarctica. Examples of the second type are site HSM-20, the cross on Observation Hill, Ross Island, erected in January 1913 by the British Antarctic Expedition in memory of the members of Captain Robert Falcon Scott's party, who perished on the return journey from the South Pole in March 1912; site HSM 7, a metal stele that commemorates the death of Ivan Khmara, a member of the first Soviet Antarctic Expedition in 1956; site HSM 59, the *San Telmo* Cairn on Half Moon Beach at Livingston Island, which commemorates the officers, soldiers and seamen aboard the Spanish vessel *San Telmo* that sank nearby in September 1819; site HSM 73, a memorial cross for the 1979 Mount Erebus air-crash victims at Lewis Bay, Ross Island, erected in memory of the 257 people of different nationalities who lost their lives in the tragedy.

The largest group of sites in the Antarctic Heritage List consists of monuments and plaques celebrating individuals or events that have marked Antarctic history, an element of great relevance in Antarctic heritage. However, several of these Historic Sites and Monuments (HSMs) have essentially a commemorative, and sometimes religious, meaning, and reflect more of an intention to celebrate a national achievement or mark a national claim rather than an interest in preserving a heritage site.

Many Heroic Age sites are represented in this category, such as for instance sites HSM 3 and HSM 5, the rock cairn erected at Proclamation Island, Enderby Land, and at Cape Bruce, Mac Robertson Land, in 1930 by Sir Douglas Mawson to commemorate the landings of the British, Australian, and New Zealand Antarctic Research Expedition of 1929–1931; site HSM 24, the Amundsen Cairn, on Mount Betty, Queen Maud Range, erected by the Norwegian explorer in January 1912 on the way back from the South Pole; site HSM 66, Prestrud's Cairn at Alexandra Mountains, Edward VII Peninsula, erected in December 1911 during the Norwegian Antarctic Expedition of 1910–1912; site HSM 28, a rock cairn at Port Charcot, Booth Island, erected by the French expedition led by Jean-Baptiste Charcot, which wintered here in 1904 aboard the ship *Le Français*. The last site (HSM 94) inscribed in the Antarctic Heritage List in 2019 is a rock cairn installed in 1892 by Norwegian Capt. Carl Anton Larsen during the first land exploration of the area around the current location of Argentina's Marambio base. All these elements have a high historic value and a high degree of authenticity, as they were erected directly by the early explorers.

Many other HSMs celebrate essentially national achievements since the end of the Heroic Age. Site HSM 34, Arturo Prat's bust on Greenwich Island commemorates a Chilean naval hero; site HSM 30, the shelter "Gabriel Gonzalez Videla" commemorates the visit of the Chilean President; site HSM 40 comprises the bust of General San Martín, hero of Argentinian independence, together with a statue of the Virgin of Lujan; site HSM 43 is a cross near the Argentinian Belgrano Station; site HSM 4, the Pole of Inaccessibility Station ↗ p. 682 Building, hosts a bust of Lenin commemorating the Soviet expedition of 1958; site HSM 50, the Polish Eagle Plaque, the national emblem of Poland, commemorates the landing of members of the first Polish Antarctic Marine Research Expedition on the vessels *Professor Siedlecki* and *Tazar* in February 1976; site

"Alternative methods must therefore be considered […]. Some monuments and sites can be digitally and archivally preserved for the future and thus [made] available to a far wider audience of scholars and the interested public through 3D animations and virtual reality tours. Others might be better conserved and certainly [made] more available by being removed to appropriate museums or other institutions outside Antarctica. Such a method could, for example, have been considered for the […] Russian over-snow heavy tractor 'Kharkovchanka' that was used in Antarctica from 1959 to 2010. The arguments for preserving such a mobile artefact at a polar museum in Russia seem to outweigh the challenges of preserving it in situ and making it available for the benefit of a wide audience, especially since it is described as 'a unique historical sample of engineering-technical developments made for exploration of Antarctica.'"

Susan Barr, "Twenty Years of Protection of Historic Values in Antarctica under the Madrid Protocol", *The Polar Journal*, 2018

Vladimir Ilyich Lenin, Captain Arturo Prat, and General San Martín are the only three individuals whose features will be perpetually preserved in Antarctica as monumental busts. While the Chilean and Argentine monuments (HSM 34, HSM 40) were erected respectively on the bedrock foundation shown on the left upon which Capitán Arturo Prat Base was constructed in 1947 and Base Esperanza was built in 1955, the bust of Lenin (HSM 4) appears to hover alone above the white plane surrounding the Pole of Inaccessibility (shown on the opposite page, as captured by Sebastian Copeland); the station on which it was erected has long been buried under the snow.

HSM 85 commemorates the PM-3A nuclear power plant at the US base of McMurdo Station ↗ p. 866 (removed according to the Treaty prescriptions); site HSM 52 is a monolith commemorating the establishment, in 1985, of the Great Wall Station, the first Chinese station in Antarctica; and finally, site HSM 82 is a monument, erected in 2011, to commemorate the 1959 signing of the Antarctic Treaty. Several other examples could be cited, also from more recent times: site HSM 1 commemorates the raising of a flagpole at the South Pole in 1965 by an Argentinian expedition, and site HSM 44, Dakshin Gangotri, commemorates the landing of the first Indian expedition in 1982.

A limited number of scientific equipment items have been added to the Antarctic Heritage List. Although these are not fully representative of the complex and innovative scientific and engineering infrastructure brought to and operated in the continent in the past century, they certainly reflect the main scope of the list: to help preserve the historic remains of human activity in science and exploration. Examples of these additions include site HSM 29, the lighthouse *Primero de Mayo* on Lambda Island, Melchior Islands, built in 1942, the first Argentine lighthouse in the Antarctic; site HSM 49, the concrete pillar erected by the first Polish Antarctic Expedition at Dobrolowski Station on the Bunger Hills to measure acceleration due to gravity; site HSM 88, Professor Kudryashov's Drilling Complex Building, constructed in the summer season of 1983–1984, to obtain ancient mainland ice samples; and site HSM 92, the Oversnow heavy tractor Kharkovchanka, used in Antarctica from 1959 to 2010 and specifically designed for the continent.

Under the Environmental Protocol of 1998 all obsolete structures and equipment are to be removed from the continent, unless declared as heritage. The declaration of an object as heritage exempts it from the obligation of removal, and some fear that in the future this could become a reason for proposing elements for inclusion in the list.

Even after the adoption of "Guidelines for the assessment and management of heritage in Antarctica" in 2018, the criteria for inclusion in the list are not subject to independent professional evaluations, as is the case with all other existing international Treaties related to cultural heritage. The Antarctic Treaty relies exclusively on its Members to assess whether items proposed for inclusion satisfy the official criteria, and the selection of new sites hinges on principles not always aligned with accepted international standards, as proven, for instance, by the inclusion of sites that cannot be seen or even located, such as Amundsen's Tent at the South Pole in 2005 (HSM 80) or the wreck of Shackleton's ship *Endurance* in 2019 (HSM 93).

"With 20 km to go our pains seemed to vanish – all that remained was the biting cold and an hour's kiting. At 6 km we spotted a dot on the horizon. Were our eyes deceiving us through a combination of exhaustion and so wanting something to be there? As we edged closer to the 'dot', it began to form into a noticeable pillar, an outline … a bust! With the realization that against all the odds Lenin was in fact still around to greet us we all burst into uncontrollable shouts and laughter. He is standing on a chimney of the old Soviet hut about 2 meters above the snow line. [He] is made of some plastic composite – he is totally frost free as if he was put there yesterday. It [is] so, so very surreal."

Rory Sweet, Henry Cookson, and Rupert Longden (Explorers), "Novo to Inaccessibility Antarctic Expedition", 2007

VLADIMIR LENIN AT THE POLE OF INACCESSIBILITY
Alexander Klepikov

Poles of Inaccessibility are sites that are difficult to reach due to their remoteness from the coastlines. Each continent has its own Pole of Inaccessibility, amounting to a geographical collection of some of the most inaccessible sites, rivalled only by "Point Nemo" – the Oceanic Pole of Inaccessibility renowned for being further away from humans than, at times, the International Space Station. On land, the Antarctic Pole – located at 82°06'S and 54°58'E – can be considered the most remote of them all. Conquered on the 14th of December 1958 during the third Soviet Antarctic expedition, under the leadership of the scientist Evgeny Tolstikov, the site was equipped with a temporary station to cement this achievement.

Determining the location of the relative Inaccessibility Pole point in Antarctica is a rather complex and ambiguous cartographic task. At the end of the twentieth century, specialists from the Scott Polar Research Institute determined as geographical coordinates of this point 85°50'S, 65°47'E, whilst specialists in poles believed that the pole was located at 82°53'S, 55°04'E, and scientists from the British Antarctic Survey thought it was located at 83°50'S, 65°43'E. On the 23rd of December 1957, a reconnaissance flight was carried out on dedicated aircraft to define, via barometric altitude measurements, its precise location. Three days later, on the 26th of December 1957, a sledge-caterpillar traverse consisting of ten tractors left Mirny Station ↗ p. 668 to reach the point.

In 1958, a year after the first traverse, the temporary Antarctic station Pole of Inaccessibility ↗ p. 682 was built by the second Antarctic Expedition of the Soviet Union Academy of Sciences, which was tasked with its construction. The station was operational until the 26th of December 1958, when the team – having completed the scheduled program of meteorological, geophysical and glaciological observations – returned to Mirny station. Subsequently, participants in the Soviet scientific expeditions visited the Pole of Inaccessibility station twice. The first visit was in 1964, when the participants of the scientific geological survey of the 9th Soviet Antarctic Expedition carried out a program of seismic, glaciological, meteorological, geomagnetic, and geodetic studies *en route* to Vostok; the second occurred between December 1966 and March 1967, when members of the twelfth Soviet Antarctic Expedition were sent to implement scientific geological surveys along the route between Molodezhnaya and Novolazarevskaya station.

To commemorate the Soviet discoverers of this unique geographical point, in 1964 a plaque was erected on the station premises of the seasonal station. On the roof of the building, a bronze bust of Vladimir Ilyich Lenin was mounted on a wooden pedestal to celebrate the achievement.

ALEXANDER KLEPIKOV is the head of the Russian Antarctic expedition. With a PhD in physics and mathematics, specialising in oceanology, he has participated in multiple Arctic and Antarctic expeditions, including the international expedition to the Weddell Sea. Dr Klepikov is a reviewer of European Science Foundation projects, a SCAR Team Expert on Antarctic Climate Change and Environment, and a member of the SCAR-IASC Group.

At the seventh Antarctic Treaty Consultative Meeting – held in Wellington, New Zealand, in 1972 – the monument to Lenin was assigned the status of a Historic Site and Monument (HSM) through Recommendation VII-9. Listed as HSM 4, the bust emerges in the white horizontal landscape, while the station that it crowns lies beneath metres of snow.

Roald Amundsen's Tent at the South Pole

Susan Barr

SUSAN BARR was the founding president of, and is now Arctic advisor to, the International Council on Monuments and Sites (ICOMOS) Polar Heritage Committee, which she founded in 2000. She has worked since 1979 solely with polar heritage and history, with extensive field work in both polar areas, authoring many publications. In 2019 Dr Barr was appointed a life member of ICOMOS Norway in recognition of her work for polar cultural heritage.

The tent left by Roald Amundsen and his expedition party at the South Pole in December 1911 is not only an international symbol of historical events of the greatest significance in terms of human endeavours but also a unique (intangible) heritage and architectural icon.

On the 14th of December 1911 Roald Amundsen and his four companions, Helmer Hanssen, Sverre Hassel, Oscar Wisting, and Olav Bjaaland, arrived at the South Pole. They spent three days there, and when they headed north again on the 18th of December a small tent was left on the spot, marking the first time that humans beings had reached this absolute extremity of the globe. The tent was later seen by Robert Falcon Scott and his companions (Edward Wilson, Henry Bowers, Edgar Evans, and Lawrence Oates) when they arrived at the Pole on the 17th of January 1912. It was photographed and sketched, and a comparison of the photographs taken by the two groups shows that by the time the British arrived, snow had already started to accumulate around the base of the tent. Since there is no natural physical marker at the Pole, the tent and the photographs were the material evidence that both the Norwegian and the British group had reached the geographic pole. Subsequently no one else has ever seen the tent standing at the Pole.

The tent itself was a small reserve tent sewn by sailmaker Martin Rønne on the *Fram* during the voyage south to Antarctica. The design resulted from Amundsen's experiences during his two previous polar expeditions. It was made of cotton cloth, pyramidal in shape and supported by a single tent pole and guy ropes.

American polar explorer Richard Evelyn Byrd and Norwegian pilot Bernt Balchen, who flew over the South Pole on the 29th of November 1929, were the first people in the area since Scott's party and saw no sign of the tent. In fact, as other standing structures in Antarctica have shown, it would have been completely covered by drifting snow within a short time. The ice around the Pole in Antarctica slides slowly towards the ocean and the tent is not only buried under an estimated eighteen metres of ice but has also been shifted away from its original position.

In the early 1990s Norwegian plans to find the tent by using ground-penetrating radar to locate any metal objects left by Amundsen did not come to fruition. It had been hoped to exhibit the tent at the 1994 Winter Olympics in Norway. The photographic images of the tiny tent standing bravely at the South Pole with flags flying and two groups of intrepid men standing beside it are powerful icons of man's attempts to reach the outermost limits of the globe. They can be compared with the first photographs of humans on the Moon. If the tent had been rediscovered, only pieces of canvas, ground to shreds by the ice, would have been found, detracting from the heroic image we have in our minds. As it is, we all know that the symbolic tent is still there, and the powerful intangible force of the historic object remains intact.

The plans to find and remove the tent led to public pressure and elicited the support of the Norwegian Directorate for Cultural Heritage to have the unseen tent in its unknown location protected within the Antarctic Treaty System of Historic Sites and Monuments (HSM). The public response showed how the intangible qualities were far more important than any material historic remains could ever be. The Treaty system works slowly, and it was only in 2005 that Amundsen's tent was declared a protected site and was entered in the HSM List as HSM 80, the first to be listed that cannot be seen and whose exact location is unknown. In 2019 Ernest Shackleton's sunken ship *Endurance*↗ p. 640 was added to the list, also without its exact position being known. It may perhaps be said that Amundsen's tent opened up this possibility.

> "Amundsen's Tent is a unique heritage object. All modern definitions of heritage refer to an existing relationship with humans, whether in a 'tangible' form, like monuments, archaeological areas, historic cities and cultural landscapes, or in an 'intangible' form, like cultural expressions and traditional crafts. Amundsen's Tent, an object carried away by ice drifting and probably dismembered, is a heritage category of its own, one that could inspire, in Antarctica as elsewhere, a new heritage brand: structures that are known to have existed, and possibly still exist in fragments, but are nowhere to be seen. An image, a memory."

Francesco Bandarin (Architect, United Nations Educational Scientific and Cultural Organization), in conversation with UNLESS, Architectural Association School of Architecture, 2019'

"So we arrived, and were able to raise our flag at the geographical South Pole – King Håkon VII's Vidda. Thanks be to God!"

Roald Amundsen (Explorer), 14 December 1911

"Great God! This is an awful place!"

Robert Falcon Scott (Explorer, Royal Navy Officer), in his diary, 14 December 1911

Photographed twice, firstly by the Norwegian party on the 14th of December 1911, and subsequently by the British on the 18th of January 1912, Roald Amundsen's tent exists as a unique visual record of the historic endeavour of the race to the South Pole. Even though the tent itself is no longer on the Antarctic plateau's surface – and lies under ice far away from the geographic South Pole where it once stood – the absent artefact (or else, its memory) was listed in 2005 as Historic Site and Monument number 80.

At the beginning of the twentieth century pioneering Heroic Age explorers marched over uncharted territories to reach the Pole fuelled purely by human willpower, dogs, and ponies. On the cusp of 1911, three explorers embarked on the expedition to the Geographic South Pole, a race that was ultimately won by the Norwegian party led by explorer Roald Amundsen on the 14th of December 1911, after having travelled for fifty-six days over an expanse of ice of approximately 1,700 km. To his dismay, Captain Robert Falcon Scott, leader of the British Expedition, arrived just a month later on the 18th of January 1912, whilst the Japanese explorer Nobu Shirase (whose attempt is often neglected) had to retreat to the top of the Ross Ice Shelf. The visualisations chart the three expeditions, emphasising the composition of the crew, equipment typologies, climatic and topographic conditions encountered en route, and the overall length of their voyage, as duly recorded in their journals.[1]

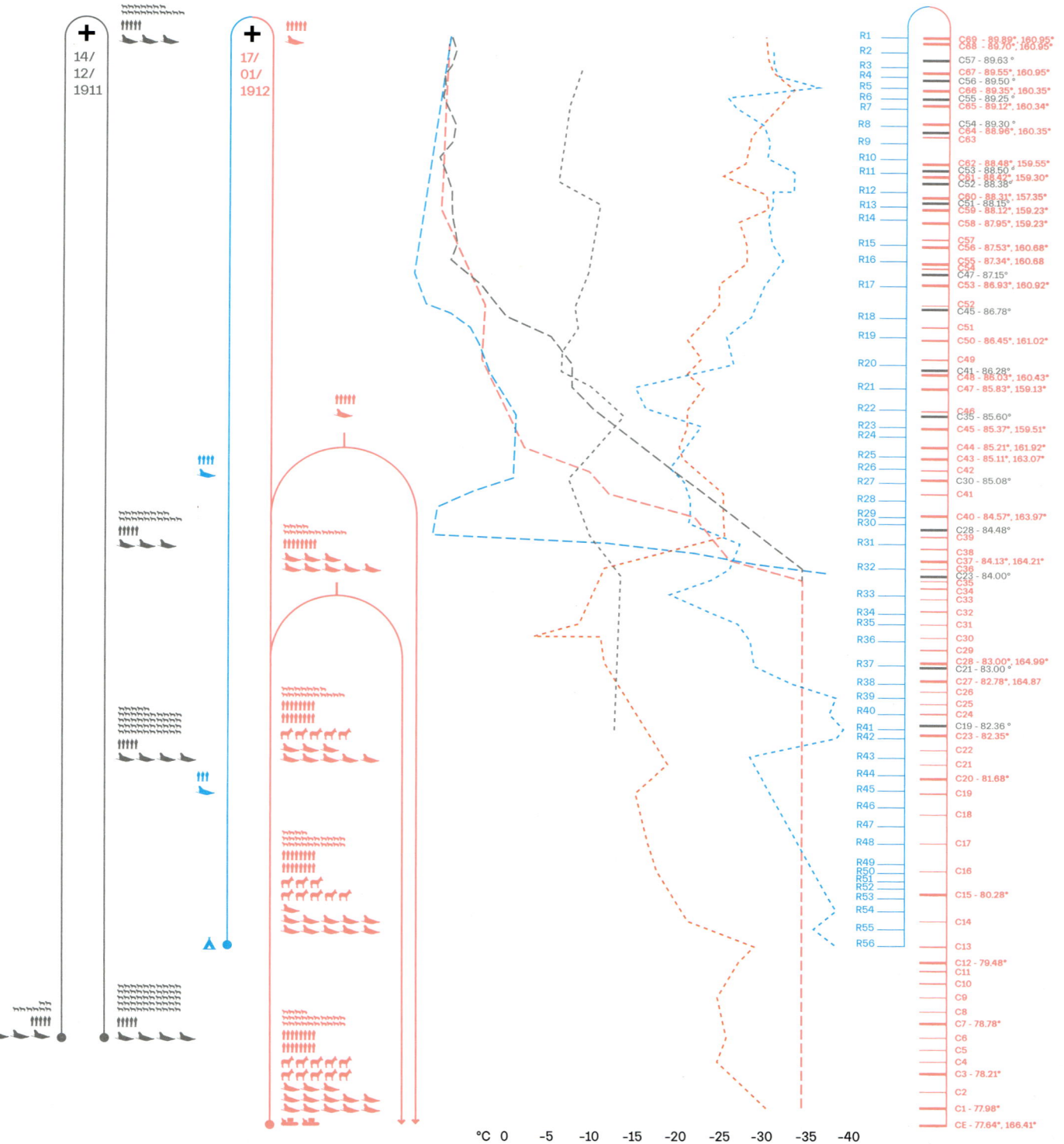

SOUTH POLE

Last Depot

Three Degree Depot

BEARDMORE
GLACIER

ROSS ICE SHELF

One Ton Depot

Kainan Maru
Framheim

Corner Camp

Cape Evans

ROSS ISLAND

The Wreck of Shackleton's *Endurance* in the Weddell Sea

Julian A. Dowdeswell

JULIAN A. DOWDESWELL is the director of the Scott Polar Research Institute and a professor of physical geography at the University of Cambridge (United Kingdom). He is the former chair of the UK National Committee on Antarctic Research and a past delegate to the Council of the Scientific Committee on Antarctic Research (SCAR). Having led more than 40 polar expeditions, Prof. Dowdeswell has received many awards, among them the Polar Medal and the Founder's Medal of the Royal Geographical Society.

Most heritage sites in Antarctica are found close to the coast and were established in the nineteenth and twentieth centuries to enable the exploration of the continent or the exploitation of its marine resources. More unusually, however, the submerged wrecks of vessels that sank during early sealing and whaling expeditions and the Heroic Age of Antarctic exploration may also be viewed as heritage sites; examples include Otto Nordenskjöld's *Antarctica* in 1903 and, arguably best known, Sir Ernest Shackleton's *Endurance*[↗ p. 640] in 1915. Both ships sank in the Weddell Sea after sustaining fatal damage from densely packed sea ice, which Shackleton famously described as "the evil ice that has scuppered our hopes". Photographs taken by Frank Hurley of the *Endurance* trapped in the ice just before it sank demonstrate that, while the masts and rigging had collapsed, the hull remained a coherent structure although compromised in terms of the vessel's ability to float and steer. Thus, once the ship slipped through the sea ice to the seafloor some 3,000 metres below, it is likely that the wreck remained largely intact and surrounded by a trail of debris.

Shackleton's *Endurance* expedition,[1] formally and rather grandly named the Imperial Trans-Antarctic Expedition (1914–1917), was conceived in the knowledge that both Roald Amundsen and Robert Falcon Scott had succeeded in reaching the South Pole one month apart in the austral summer of 1911–1912. Shackleton viewed a first crossing of Antarctica, from the Weddell to the Ross Sea, as the greatest remaining Antarctic challenge. The *Endurance* and her twenty-eight men sailed into the Weddell Sea from South Georgia, meeting severe sea-ice conditions and, just short of their planned landing site in Vahsel Bay, the ship became trapped. The vessel drifted helplessly with the sea ice for ten months, sustaining severe damage by ice pressure and finally sinking. After watching the *Endurance* slip beneath the ice, the expedition was left stranded with only tents, limited food and fuel, and three small lifeboats. Shackleton famously kept up the morale of the group while their camp drifted northwards with the ice. Finally, the three boats could be launched as the sea ice began to break up. There followed epic small-boat journeys, first by the whole party to the isolated Elephant Island and then, by six men led by Shackleton and navigator Frank Worsley in the 7-metre-long lifeboat *James Caird*. After an eighteen-day and 1,300-kilometre open-boat voyage across the stormbound seas of the Southern Ocean they reached South Georgia, exhausted. Shackleton, Worsley, and Tom Crean then had to climb across the unmapped alpine spine of the island to raise the alarm at Stromness whaling station. The remaining twenty-two men on Elephant Island, led by Frank Wild, were finally rescued by the Chilean ship *Yelcho* after more than three months marooned on what was little more than a storm beach backed by huge cliffs.

The limited amount of marine-geophysical data available from the floor of the western Weddell Sea in the area around the last known position of Shackleton's *Endurance* (68°39.5'S, 52°26.5'W according to Worsley's sextant and theodolite measurements[2]) suggests that the wreck has remained unburied and has sustained little or no further damage since it sank.[3] The submarine keels of icebergs drifting today in the Weddell Sea are no more than about 500 metres thick at most and sea-ice keels are only a few tens of metres deep – the wreck's great depth therefore protects it from any ice-keel damage.[4] The undisturbed stratigraphy of seafloor sediments indicates no recent down-slope sediment delivery from submarine landslides[5] and dated sediment cores from the Weddell Sea show that the rain-out of fine-grained sediment and dead microorganisms is very slow and limited to a few millimetres per year.[6] Thus, the wreck is unlikely to have been buried under subsequent sedimentation during the hundred or so

"The simple act of sailing had carried him beyond the world of reversals, frustrations, and inanities. And in the space of a few short hours, life had been reduced from a highly complex existence, with a thousand petty problems, to one of the barest simplicity in which only one real task remained – the achievement of the goal."

Alfred Lansing (Journalist and Writer), *Endurance: Shackleton's Incredible Voyage*, 1959

"There remained but one great object of Antarctic journeying – the crossing of the South Polar continent from sea to sea."

Sir Ernest Henry Shackleton (Explorer), *South: The Story of Shackleton's Last Expedition 1914–1917*, 1915

"Men go out into the void spaces of the world for various reasons. Some are actuated simply by the love of adventure, some have the keen thirst for scientific knowledge, and others again are drawn by the lure of little voices, the mysterious fascination of the unknown. I think in my own case it was a combination of these factors that determined me to try my fortune once again in the frozen south."

Sir Ernest Henry Shackleton (Explorer), extract from New Zealand Antarctic Heritage Trust, *Conservation Report: Shackleton's Hut, Cape Royds*, 2003

years since it sank to the seabed. In addition, the deep waters of the Southern Ocean are a cold, low-energy environment where the activity of currents is limited and there are low rates of microbial degradation; wood-boring "ship worms" (Xylophagainae bivalves), which often destroy deep-water wooden wrecks, have not been found in Antarctic waters.[7]

The site of the *Endurance* wreck is, therefore, both relatively well-known in terms of its latitude and longitude and the vessel is unlikely to have suffered from either physical reworking or biological degradation after the time of sinking. These factors, combined with Shackleton's extraordinary story of escape and survival,[8] mean that the *Endurance* has for many years represented a considerable challenge for those wishing to investigate Antarctica's marine heritage.

There have been a number of plans to search for the wreck of Shackleton's *Endurance,* which is viewed as one of the most difficult wreck sites on Earth to reach due to the almost continuous year-round sea-ice cover of the western Weddell Sea.[9] To the best of our knowledge, only three ships have been at or close to the site since 1915 – the German research icebreaker *Polarstern,* the Russian icebreaker *Kapitan Khlebnikov* on a tourist cruise and, most recently, the South African ice-strengthened *Agulhas II. Polarstern* acquired multibeam echo-sounding data of the seafloor in the vicinity of the last known position of the wreck, but, with a horizontal resolution of only about 100 metres, no evidence of the 44-metre-long and 7.6-metre-wide vessel was found.[10] The *Agulhas II* reached the wreck site on the 10th of February 2019 as part of the Weddell Sea Expedition 2019, which had two principal aims: scientific study of the Larsen Ice Shelf and western Weddell Sea and the search for the *Endurance.* Severe sea-ice and weather conditions meant that the ship stayed only thirty-seven hours at the wreck site, but an autonomous underwater vehicle (AUV) equipped with a high-resolution multibeam echo sounder, allowing mapping of the seafloor at sub-metre resolution, was launched. The AUV, navigating about 60 metres above the seafloor, completed a number of survey lines around the estimated position of the wreck with the aim of imaging the remains of the *Endurance.* The vehicle unfortunately ceased to transmit its location while it was completing the final survey lines and was lost along with the seafloor imagery it was acquiring. It should be noted that the aim of this exercise was to survey the wreck and surrounding seafloor in great detail in order to produce high-resolution visual imagery of the vessel as a historical site – the physical wreck was to remain undisturbed and nothing was to be taken from the site.

On the 9th of October 2019 the wreck of Sir Ernest Shackleton's *Endurance* was formally added to the list of Antarctic Historic Sites and Monuments.[11] The measure includes the wreck and "all artefacts contained within [...] or near the wreck within a 150 m radius. This includes all fixtures and fittings associated with the ship, including ship's wheel, bell, etc. The designation also includes all items of personal possessions left on the ship by the ship's company at the time of its sinking."[12] This protected status means that, under the Protocol on Environmental Protection to the Antarctic Treaty System, Annex V, the wreck "shall not be damaged, removed or destroyed".[13]

The wreck of the *Endurance,* lying 3,000 metres beneath the surface of the Weddell Sea, is thus protected by the national laws of each Antarctic Treaty nation in a similar way to, for example, the historic huts erected at Cape Royds ↗ p. 624 and Cape Evans ↗ p. 632 on Ross Island during the earlier *Nimrod* and *Terra Nova* expeditions of Shackleton and Captain Scott.

The Role of Photography in Antarctic Preservation

Shaun O'Boyle

SHAUN O'BOYLE is a photographer and architectural designer. His work uses photography to explore architecture and built environments in high-latitude regions. He is a three-time grantee of the National Science Foundation (NSF)'s Antarctic Artists and Writers Program and the recipient of a Guggenheim fellowship in photography. O'Boyle's work has been featured in numerous galleries, museums, books and magazines, including *Smithsonian* magazine.

One way of accessing Antarctica is through its literature, the narratives and journals written about the continent from the early expeditions to the present day, from circumnavigation to establishing bases for science and exploration. The subjects of my photography are primarily built environments and landscapes, but I found myself drawn to these early explorations, and I followed a few early paths, including Scott's discovery of the Dry Valleys and Cherry-Garrard's winter journey to Cape Crozier. The Antarctic landscape is an aggregate of these histories, where stories, buildings, settlements, infrastructure, and ruins shape zones of human-occupied landscapes among almost limitless regions of ice, snow, and mountains. Photography is an important medium that can preserve such histories in perpetuity, at times in ways more effective than canonical physical conservation practices.

My photographic work in Antarctica was made possible through the National Science Foundation's Antarctic Artists and Writers Program. The programme provides opportunities for scholars in the humanities to work in Antarctica and the Southern Ocean, and make observations at United States Antarctic Programme stations to enable serious writings and works of art. Through the programme I visited McMurdo Station ↗ p. 866 on Ross Island, Palmer Station on the Antarctic Peninsula, and Amundsen-Scott Station ↗ p. 790 at the geographic South Pole, spending a total of eleven weeks on the continent over a five-year period.

The project was driven by the poetic concept of establishing a presence on an unoccupied continent. A culture of science and exploration establishing itself in a place with no human history. I visited and photographed some of the first structures built on the continent and the tools used by the first inhabitants – the chemists, biologists, physicists, and meteorologists. These first inhabitants fashioned a pragmatic functional architecture that responded to the constraints and logistics of travelling to and living in this extreme environment. I could trace the transformation to our present-day science stations through orders of magnitude of scale, largely through advances in communication, technology, and logistics.

The image of Antarctica – the "most surveilled"[1] yet least visited continent on planet Earth – has been constructed largely through photography. Confronted with the challenges of shooting on an ever-reflecting landscape and developing plates in freezing darkrooms in which "the tank [had to] be kept warm or the plates [would have] become [enclosed] in an ice block", photographers have attempted, since the Heroic Age, to capture the essence of the continent – be it with a romantic idealism focused on the "pristine" environment or an objective critical perspective focused on the contaminating effects of Antarctic anthropic activities. In doing so, they have not only contributed to mapping the uncharted polar region – as was the case for Captain Ashley McKinley[2] (captured below with his large camera loaded with a 75-feet-long film roll during the 1928–1929 Byrd Antarctic Expedition[3]) – but have also contributed to recording historic events which would have seemed inconceivable. By documenting *Endurance*'s slow descent into the depths of the ocean with his Goerz-Anschütz plate camera, Frank Hurley has not only ensured an everlasting memory of Shackleton's astounding endeavour but has also enacted the most effective and sustainable form of preservation in the continent, one which does not entail disproportionate economic investments to conserve unattended matter but rather privileges an intangible, yet powerful, preservation of knowledge.

I was intensely aware that I was recording these places during a moment of historic change, with the threat of a warming climate creating an urgency to document these places before they change. This was brought into sharp focus while working alongside scientists studying the effects of climate change, including ocean acidification and melting glaciers. Some of the science projects I worked with are at the cutting edge of physics and cosmology, particularly the astronomy at the South Pole, yet I felt I was creating a record of a landscape and environment undergoing rapid transformation, rendered historic the moment it was recorded.

During this process, I reflected extensively on the fragility of the Antarctic Treaty. With long-established policies and treaties being bulldozed by current administrations around the world, could this treaty hold up over the long term? If the ice continues to melt, then opening tracts of land, prospecting, and exploitation of the mineral content seem bound to follow. This has happened in the Arctic and preventing it from happening in the Antarctic will mean a colossal battle. Thus, as a record amidst change, photography can act as an alternative form of preservation, a snapshot of a place in flux where change happening at a glacial pace is accelerating rapidly into a torrent.

"The United Kingdom recommends to the Committee for Environmental Protection (CEP) that it approves the listing of the 'Endurance, Wreck of the vessel owned and used by Sir Ernest Shackleton during his 1914–1915 Trans-Antarctic Expedition' as a new Historic Site and Monument. [...] The Endurance wreck meets several of the criteria set out in Resolution 3 (2009)"

Secretariat of the Antarctic Treaty, Measure 12 (2019) – ATCM XLII – CEP XXII, Prague, 2019

16
Antarctica and Outer Space. Laboratories for a Future as Global Commons

Exploring Antarctic and Extraterrestrial Connections in Fiction

Elizabeth Leane

ELIZABETH LEANE is Associate Dean Research in the College of Arts, Law and Education at the University of Tasmania (Australia), chief officer of the Standing Committee on Humanities and Social Sciences within the Scientific Committee on Antarctic Research (SCAR), and arts editor of *The Polar Journal*. With degrees in physics and literary studies, Prof. Leane has published seven books, including *Antarctica in Fiction* (Cambridge University Press, 2012).

The similarities between Antarctic and extraterrestrial environments are well known, with space scientists using the ice continent as a place to test how humans and machines might perform off-world. But connections between the Antarctic and Outer Space were formed in the mythological and literary imagination long before humans physically encountered the southern polar region.

The earliest of these connections relate to the Pole itself, which began as a celestial rather than a terrestrial concept. Although there is currently no bright star to mark the South Celestial Pole, people living in the southern hemisphere have seen its location in the night sky for many tens of thousands of years: the central Australian Aranda and Luritja tribes, for instance, were aware that stars relatively close to this point never moved below the horizon.[1] Looking at the northern sky, ancient Greek thinkers posited the existence of a *polos* – an axis around which the hemispherical cosmos revolved and also the end point of this axis. After the notions of a spherical cosmos and a spherical Earth were accepted, this idea was extended to the south.[2] The terrestrial poles became the places where the axis of the celestial sphere met the surface of the Earth in the north and south: conceptually, then, they were joined to the stars from the outset.

In literature, this connectedness was reimagined as a literal portal from Earth to other planets (a counterpart to the "hollow Earth" idea, in which the poles lead into the planet's interior). Thomas Erskine's utopian satire *Armata* (1817), for example, sees a ship sail to a twin Earth connected via a channel at or near the South Pole. The hero of Gustavus Pope's *Journey to Mars* (1894), stranded in the Antarctic region, travels with a group of Martians to their home world via "cosmo-magnetic currents" linking the poles of all the planets in the solar system.[3] Antarctica as a launching or landing point for astronauts became a familiar idea in popular fiction of the following century, such as in the children's novel *Mike Mars: South Pole Spaceman* (1962).

Aliens also figure prominently in twentieth-century Antarctic fiction (and film). The United States Antarctic expeditions of the 1930s led by Richard Byrd generated a number of science-fiction stories, including classic Antarctic tales by H. P. Lovecraft and John W. Campbell Jr, involving aliens frozen in or living under the ice. But Antarctic aliens also feature in literature produced in quite different cultural contexts. Aleksandr and Sergei Abramov's *Horsemen from Nowhere* (1969) – a translation of a Russian-language narrative published two years earlier in the magazine of the Youth Communist League – focuses on a Soviet expedition that discovers alien beings removing Antarctic ice, with the aim (it seems) of geo-engineering a planet elsewhere. In all cases, Antarctica functions in opposition to the mundane – as a remote, extreme, and unearthly place that might readily harbour alien life.

In the decades following the Antarctic Treaty, creative writers became interested in how human inhabitation of the southernmost continent might provide a model for the colonisation of other planets. Brian Aldiss's *White Mars* (1999) envisions Earth's neighbour becoming a "planet for science", much as the Antarctic has been preserved – at least to a great extent – as "unspoilt white wilderness", while the Martian colonisers of Kim Stanley Robinson's *Mars Trilogy* train in the Antarctic, debate the applicability of the Antarctic Treaty to the red planet, and consider the continent "a bit of Mars on Earth".[4] As space tourism and human travel to other planets come closer to realisation, Antarctica's physical and social similarities to extraterrestrial environments will no doubt continue to be the subject of literary as well as scientific experiments.

> "We came to probe the Antarctic's mystery, to reduce this land in terms of science, but there is always the indefinable which holds aloof yet which rivets our souls."
>
> Sir Douglas Mawson (Geologist, Explorer, Academic), *The Home of the Blizzard*, 1915

Narrating the story of the discovery, on behalf of American researchers stationed in Antarctica, of a frozen spacecraft as shown at the side, *The Thing* (as originally directed by John Carpenter in 1982 based on John Campbell Jr's 1938 novella *Who Goes There?*) exacerbates notions of isolation, claustrophobia, and overwintering paranoia (not infrequent in the remote continent) producing a disturbing sci-fi thriller. As the ice that contains the spaceship thaws, "the thing" – a parasitic extraterrestrial entity that assimilates the features of the humans it secretly inhabits, imitating them to perfection – takes over the scene, sowing utter distrust between the scientists.

The Antarctic as Analogue Site

William L. Fox

WILLIAM L. FOX is the founding director of the Center for Art + Environment at the Nevada Museum of Art in Reno, Nevada (United States). He is an art critic, science writer, and cultural geographer and has published sixteen books on cognition and landscape, including *Terra Antarctica: Looking into the Emptiest Continent* (Counterpoint, 2007). Fox spent two and a half months in Antarctica with the US Antarctic Artists and Writers Program.

In the early 1980s the American historian William H. Goetzmann proposed that there were two Great Ages of Discovery. The first coincided with the Renaissance and was an age of reconnaissance motivated by commerce and imperialism dating from the time of Christopher Columbus through the mid-1700s. The second was an era born of the Enlightenment, inaugurated by Captain James Cook's circumnavigations of the Pacific, and the concomitant global spread of optical technology and the scientific revolution. His student and noted historian Stephen Pyne proposed that during the 1957–1958 International Geophysical Year a Third Great Age of Discovery began. Fuelled by the Cold War and the launch of Sputnik, the era was characterised by remote sensing into places where humans are not found: the ocean depths, the Antarctic, and the solar system. Pyne saw the IGY as the moment of transition to the new era, and the Antarctic as the region on Earth with qualities closest to those of Outer Space. His assertion was confirmed by NASA's use of the frozen continent as an analogue environment for off-planet exploration protocols – that is, exploring in spacesuits, driving rovers, and living in environmentally closed habitats for extended periods of time.

Analogue sites, as characterised by international space programmes worldwide, offer fidelity to off-world conditions such as rugged terrain, extreme environments, working in pressurised suits, and long-term isolation in confined living spaces. Once humans ventured into space, three overlapping areas of analogue investigation were launched in the Antarctic: development of remote-sensing instruments; medical studies of humans living in confined quarters for extended periods, and the architecture and interior design of habitats in harsh environments.

The NASA Jet Propulsion Laboratory Extreme Environment Robotics Group has been testing robots in the Antarctic since 2017, but the Antarctic has also served as an analogue terrain for science experiments. A notable example was conducted by planetary scientist Michael Malin, who in 1983–1984 placed six thousand Antarctic rocks in racks above and below ground in the Dry Valleys to measure the effects of wind and cold on the samples during a ten-year period. The experiment was done in order to prepare visual guidelines for measuring the geomorphology in images made by photogeological satellite missions to Mars and other planets. The Antarctic studies allowed Malin to claim in 2000 that there was evidence of water having eroded the surface of Mars.

"Analogous environments are environments that can be compared with each other on one or more dimensions. But calling one of them, in this case Antarctica, an analogue of the other, Space, is not quite the same thing: it implies that Space is somehow the environment of actual importance, the real thing, of which Antarctica is a simulacrum. It is quite possible that the Antarctica-Space 'analogue' viewpoint implies that Antarctica, its stations, and its inhabitants, are of interest to the Extreme and Unusual Environments (EUE) / Isolated Confined Environments (ICE) psychologist only insofar as they duplicate, or come close to duplicating, their counterparts in Space. We should realize that Antarctic stations are not any more analogues of spacecraft (including space stations) than vice versa. Given that reciprocal relationship, we should start thinking of them as analogous situations – situations capable of being compared – rather than as either being an analogue, or imperfect imitation, of the other."

Peter Suedfeld (Professor Emeritus, University of British Colombia), *Antarctica and Space as Psychosocial Analogues*, 2018

Formal medical studies of personnel in the Antarctic were conducted by expeditions in the Heroic Age, but actual human analogue studies run by NASA started in 1997 in both hot and cold deserts: the arid and hot environment of Arizona, and the arid and cold one of Devon Island in the Canadian High Arctic. The Antarctic offers the harshest conditions on the planet, and NASA has been running staffed analogue studies at the American McMurdo Station ↗ p. 866 in the Ross Sea Region since the 2000s, as has the Franco-Italian Concordia Station ↗ p. 780 on the inland polar plateau. Concordia, which sits 600 kilometres away from any other humans, is more remote than the International Space Station and during the austral winter completely isolated, making it an ideal platform in which to measure the effects of both physical and psychological interiority.

In the Antarctic, as in Outer Space, the primary human experiences take place within the interiors of buildings and vehicles, which often must be endured for long durations of time. Space medicine, station infrastructure and interior design, and many other topics are studied year-round, human psychological and physiological responses being of great interest. The dangers of decreased cognitive function, which are measurable over time in Antarctica stations during winter, is of special concern to NASA and other space agencies.

The Antarctic has also served, at least metaphorically, as a location for science-fiction movies and novels (although frequently the filming is actually done in the logistically more approachable Arctic). The lack of vegetation and other objective correlatives found in inhabited terrestrial regions, such as diurnal sunlight cycles every twenty-four hours, make it difficult for humans to maintain emotional balance. The awareness of such phenomena facilitates the acceptance of fictional suspension of disbelief when the settings are in hot and cold deserts, the more extreme the better. Crashes of spacecraft bearing aliens in movies such as *The Thing* (1982, 2001) and *The X-Files* (1998) have become cult classics, as have novels such as Howard Phillips Lovecraft's *At the Mountains of Madness* (1931) or *The Ice Limit* (2000) by Lincoln Child and Douglas Preston. It is no accident that the author of the classic science-fiction trilogy about Mars, Kim Stanley Robinson, also spent time at McMurdo Station in 1995 as a National Science Foundation visiting writer.

The Antarctic serves as a launchpad into space for the imagination in both science and the arts. Whereas the continent was once believed to be the edge of the world, now it is a boundary between two realms, the Earth and Outer Space. The literal mapping of the Antarctic was, ironically, once far behind that done of Mars by Mike Malin from the mid-1990s onward. Now the Antarctic mapping is almost complete, and scientists practise Mars in the cold continent, where sometimes the surface temperatures of the two planets are on a par with one another.

SHAUN O'BOYLE is a photographer and architectural designer. His work uses photography to explore architecture and built environments in high-latitude regions. He is a three-time grantee of the National Science Foundation (NSF)'s Antarctic Artists and Writers Program and the recipient of a Guggenheim fellowship in photography. O'Boyle's work has been featured in numerous galleries, museums, books and magazines, including *Smithsonian* magazine.

"Going into the world's most extreme environments forces us to develop transformational innovations which invariably have tremendous value back here in the real world."

Scott Parazynski (Former NASA Astronaut), online

SIMPLE CAMP
Shaun O Boyle

Inside the Polar Haven tent structure Bill Stone is checking out Artemis, a NASA funded underwater robot being tested at SIMPLE camp. Britney Schmidt, primary investigator of the project, looks on. Artemis is a prototype of a robot that may be sent to Jupiter's moon, Europa, to explore the under-ice environment of that moon.

16   Antarctica and Outer Space. Laboratories for a Future as Global Commons

SURVIVING IN THE CRYOSPHERE   553

Dry Valleys, Permafrost, and Mars

Gary Wilson

GARY WILSON is the chief scientist at the Institute of Geological and Nuclear Sciences and a professor of geology and marine science at the University of Otago (New Zealand). He is the vice-president of the Scientific Committee on Antarctic Research (SCAR) and has undertaken more than thirty research expeditions to Antarctica. He holds a BMus and a PhD. Dr Wilson's research has helped to uncover the dynamic history of Antarctica's climate and ice sheets.

Less than 100 kilometres from New Zealand's Scott Base ⌐ p. 852 and the United States' McMurdo Station ⌐ p. 866 the largest ice-free oasis – of approximately 4,000 square kilometres – stretches from the coastline up into the Transantarctic Mountains. On a continent covered by thick ice sheets, the arid, desert-like landscape of the Dry Valleys has been considered an enigma by scientists. Their extreme dryness, cold, permanently frozen ground (permafrost), and apparent lack of life have led scientists to investigate the limits of life on Earth and, further, to see if the Dry Valleys region presents a potential analogue for where we might find life on other icy planets. The Antarctic permafrost includes ground that remains permanently frozen through successive summers. It can reach thicknesses greater than a kilometre and traps evidence of former and current life (as bacteria, fossil carbon, carbon dioxide, and methane). Much of the permafrost has a very low water content (< 5%) and, while still frozen, is considered dry.[1]

The main Dry Valleys are named after the men of Scott's Party (Taylor and Wright) who discovered them more than 100 years ago,[2] while the secondary valleys are named after parties from the International Geophysical Year (IGY) and subsequent expeditions[3] (including those of Balham, McKelvey, and Barwick, to name a few). The valleys exert a special fascination not only because they are mostly free of snow and ice, but also because in the summer months rivers run along their length to fill great lakes in depressions along the valley floors.[4] At some point in history, perhaps more than 15 million years ago,[5] the valleys were extensively glaciated, and at least one of those glaciations caused the Wright Valley to become over-deepened (with its mouth at a higher elevation than the main valley) and the Onyx River – Antarctica's longest – to flow inland from the Wilson-Piedmont Glacier at the Coast to Lake Vanda, some 20 kilometres further inland. The rivers only flow in the summer months, but that is enough to recharge the lakes from the winter ablation of ice from their surfaces caused by the famous katabatic winds that fall under their own cold weight from the ice sheet surface several kilometres above the valleys and blow relentlessly along the latter making their way to the coast. Surprisingly, at the peak of summer, the hyper-salinised bottom water of the lakes reaches temperatures as high as 22°C[6] and supports the growth of great cyanobacterial mats (multilayered sheets of microorganisms).[7] The ice cover that starts each spring, as clear as glass, becomes frosted with the oxygen produced through the summer and incorporated into the surface ice cover.[8]

The valleys have not always been as "dry" as they are today. Many studies have tried to decipher their long history and the glaciations that have carved them.[9] Some point to rivers carving the valleys before the continent became glaciated with more recent modification only by glaciers,[10] but the long-term glacial history still remains as much of an enigma as the present existence of these ice-free valleys on what is otherwise an ice-inundated continent. Perched deltas and lake shorelines high on the valley walls point to deep lakes filling the entire valleys, presumably dammed by a grounded ice sheet filling the Ross Sea at the most recent global glacial maximum only 18–20,000 years ago.[11] Further back in the Pliocene (~3 million years ago) the Wright Valley was inundated by a shallow marine inlet from the Ross Sea, with a scallop bed being deposited not far from Lake Vanda.[12]

Although early descriptions portray them as being barren of life compared to the coastal regions abundant with penguins and seals, the last few decades of study have shown that nothing could be further from the truth. The McMurdo Long-Term Ecological Research programme has been piecing together the link between the physical processes and the biodiversity of life in the lakes, streams, and soils of the valleys,[13] as well as the cyanobacterial mats and the light-limited phytoplankton communities[14] in the lakes, and the wide variety of biota including nematodes,[15] springtails,[16] and bacteria[17] as well as mosses and lichens,[18] supported by the soil where water becomes available in the Antarctic summer. A parallel set of studies looked at the extremes of life, trying to identify the extinction point of life in these cold, dry deserts. Regardless of the harsh conditions, the bacteria are highly adapted to the environment, surviving as colonies within the rocks themselves[19] and in suspended animation in the deep permafrost, with frozen soil samples older than 5 million years yielding viable bacteria.[20] Recently it has been thought that the permanently frozen ground at high elevations in the Dry Valleys could provide a potential analogue for the subsurface on Mars.[21] Frozen ground discovered by the Phoenix spacecraft at the Martian Arctic shares many features in common with the Antarctic Dry Valleys and may contain similar life forms.

The prolonged attention in the McMurdo Dry Valleys has led Antarctic Treaty members to develop a management system aimed at limiting the impact of the increased human activity so close to permanent stations.[22] Many of the specific locations have been designated as Antarctic Specially Protected Areas (ASPAs) with specific management plans and limits to the numbers of visitors and the types of activities that can be undertaken. The entire 4,000 square kilometres of the Dry Valleys is also included in an Antarctic Specially Managed Area (ASMA[23]), which provides guidance on limiting the number of campsites and number of samples collected for overlapping studies. It also provides advice on waste management and ensures the removal of all human waste, grey water, and fuels to prevent their release from disturbing the natural ecosystem. Long-term studies have developed facilities areas so that any impact is focused on discrete locations. Different nations share their plans, studies, and samples via the Dry Valleys Management Group that meets at the Antarctic Treaty Meeting each year to share knowledge and plans. The Dry Valleys ASMA also allows for managed tourist access with designated tourist landing sites, where larger groups can see a range of features representative of the region while the impact can still be managed. Integral to the success of the ASMA and the ASPAs contained within it is a plan for monitoring key locations and impacts of activities to ensure that the management plans remain fit for their purpose.

The McMurdo Dry Valleys have already been studied for more than one hundred years but will continue to hold the attention of scientists over the coming decades as they form an important barometer for changing Antarctic conditions due to the warming of the planet. The myriad of monitoring equipment and sites across the Valleys will help to refine models of heat distribution around the planet and to identify how much faster the poles are warming than the average temperature of the planet. The developing understanding of the Dry Valleys will also continue to help in the search for life on other planets, including Earth's nearest neighbours Venus and Mars.

The July 1997 images of Mars's Twin Peaks taken during "83 days of a planned seven-day"[1] Pathfinder mission exploring the Martian terrain (top), shows "bouldery ridges and swales or 'hummocks' of flood debris"[2] within the Red Planet landscape. "Places on Earth are often used by scientists as analogues for the kinds of environments that exist on Mars. Examples include Antarctica (which, like Mars is very cold and dry)"[3] as shown in an image of Mars's analogue *par excellence*, the Dry Valleys, photographed by Stuart Clipper (bottom).

"Many people talk about advancing humanity. Many people talk about spacefaring civilisation. I don't think we've earned Mars unless we can prove that we have taken care of this planet."

Neri Oxman (Designer, Professor, MIT), in conversation with Paola Antonelli, Design Emergency Instagram Live Interview, 29 January 2021

"Like the ISS [International Space Station] in low Earth orbit, [Antarctic Station] exists in a place where humans should not. Even bacterial life struggles to survive in the cold, dry climate of the polar desert. There's a third less oxygen than at sea level, impairing brain function. Smells are largely snuffed out and there is an overwhelming abundance of silence. For these and many other reasons it has been dubbed 'White Mars.'"

Thomas Page (Writer and Producer, CNN), "To Antarctica and Beyond", *CNN International*, 2019

16   Antarctica and Outer Space. Laboratories for a Future as Global Commons

SURVIVING IN THE CRYOSPHERE   555

Antarctica and the Exploration of the Solar System

Teasel Muir-Harmony

TEASEL MUIR-HARMONY is the curator of the Apollo Collection at the Smithsonian National Air and Space Museum (United States). Dr Muir-Harmony contributed to the volume *Globalizing Polar Science: Reconsidering the International Polar and Geophysical Years* (Palgrave, 2010) and is the author of *Apollo to the Moon: A History in 50 Objects* (National Geographic, 2018) and *Operation Moonglow: A Political History of Project Apollo* (Basic Books, 2020).

On a mild day in December 1984, at the farthest ice field of Allan Hills, Antarctica, a team of seven scientists searched for dark rocks against the blue ice. They had spent the morning riding Ski-Doo snowmobiles, spaced thirty metres apart, sweeping the flat terrain for meteorites. At midday, the meteorite hunters reached an area called the Pinnacles, where violent Antarctic winds and colliding ice had created five-metre-high frozen waves. Among these sculptural ice formations, scientist Roberta Score noticed a potato-sized rock lying exposed on the ice. Black glass covered part of its surface, making it look as if it had been partially dipped in tar. She carefully placed the rock inside a sterile nylon bag and sealed it. Like the other specimens collected that season, ALH84001, as the rock would come to be known, was kept frozen until it reached the Meteorite Curation Laboratory at Johnson Space Center in Houston, Texas. A dozen years later, ALH84001 would alter the trajectory of America's space programme and planetary exploration.[1]

Over 80% of meteorites found on Earth come from Antarctica. While meteorites fall on all areas of the planet, the rocks that land in Antarctica are concentrated, making it the ideal hunting ground for planetary geologists. Meteorites that land in Antarctica's high-altitude ice fields become embedded in an immense sheet of flowing ice. Over centuries they travel downhill within the ice sheet. When mountains block the sheet's path, part of the ice is pushed upward against the mountains. Fierce Antarctic winds carve away the uprisen ice, leaving a collection of meteorites behind. Over time, as this process is repeated, more meteorites collect alongside the mountains. Since the mid-1970s, the United States Antarctic Meteorite programme, a cooperative effort led by the National Aeronautics and Space Administration (NASA), the National Science Foundation (NSF), the Smithsonian Institution, and the Antarctic Search for Meteorites (ANSMET), has collected thousands of specimens for scientific study. Bountiful yearly expeditions recover between 30 and 1,200 meteorites from the Moon, asteroids, and Mars – like the one found in the Allan Hills region.

In 1996, after two years of analysing ALH84001, NASA investigator David McKay and a team of scientists published a paper in *Science*. They claimed that they had found potential traces of fossilised microbiological life. ALH84001, dated at over four billion years old, comes from the earliest crustal rocks on Mars. Roughly sixteen million years ago, when an asteroid or comet crashed into Mars, the meteorite was part of the debris flung out into the solar system. After orbiting around the Sun, it collided with Earth and landed in Antarctica. It is the oldest Mars specimen we have on Earth. Although the findings of organic material were not conclusive, the paper raised national and international interest in the search for signs of life on Mars.[2]

In the mid-1970s, NASA had sent two Viking spacecraft to Mars with the primary objective of finding biosignatures, the evidence of past or present life. Although the mission advanced knowledge of the Red Planet's atmosphere, geology, and evolution, it did not discover signs of life. These discouraging results had long-term consequences on space exploration. The scientific community turned their focus away from the search for life to the question of habitability. The exploration of Mars also slipped from the list of NASA's top priorities. But in 1996 the scientific study of the meteorite found on the Antarctic expedition catalysed new optimism and new political support for the search for life and the exploration of Mars.[3]

When the NASA administrator alerted President Clinton about ALH84001, the reaction was immediate – his "posture straightened, and his eyes opened wide".[4] In a press conference, Clinton promised NASA would put its "full intellectual power and technological prowess behind the search for further evidence of life on Mars."[5] Enthusiasm for the possible first evidence of life on another planet inspired extensive media coverage, with the *New York Times* postulating that "discovery of life on Mars, even primitive life in the very distant past, would have profound intellectual and philosophical implications."[6] Although the excitement was widespread, many within the scientific community remained sceptical. Scientist and science populariser Carl Sagan, among others, insisted on the need to verify the results through the future exploration of the Red Planet.

Enthusiasm for the meteorite found in Antarctica was enough to halt NASA's impending budget cuts. NASA's Jet Propulsion Lab (JPL)'s Mars programme received a new name, more funding, and a redirection of its research focus. The search for evidence that Mars once supported life became the priority, superseding the program's earlier emphasis on seismology and networked meteorology. NASA established a new Astrobiology Institute at its Ames Research Center in California, revitalising a field that had been moribund for decades. And consequently, NASA accelerated the goal of a Mars sample return mission.[7]

Eventually, the broader scientific community rejected the conclusions that ALH84001 contained traces of life. Nevertheless, the meteorite found in Antarctica had already fueled a major shift in planetary exploration, returning attention toward the question of life in the universe. It raised the status of astrobiology as a field, helping it move into mainstream scientific study. Since the discovery of the Mars meteorite in Antarctica, NASA has pursued a robust Mars program oriented around the search for life with a series of rovers and orbiters surveying the planet. Although a Mars sample return mission is costly and still in the distant future, the discovery of Mars meteorites may not. Meteorite hunters continue their yearly expeditions, advancing our exploration of our solar system among Antarctica's mountain ranges.

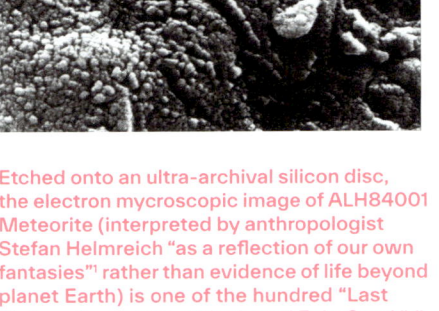

Etched onto an ultra-archival silicon disc, the electron mycroscopic image of ALH84001 Meteorite (interpreted by anthropologist Stefan Helmreich "as a reflection of our own fantasies"[1] rather than evidence of life beyond planet Earth) is one of the hundred "Last Pictures" sent into orbit aboard EchoStar XVI communication satellite by the artist Trevor Paglen. Seen as "warnings, [the picture's] composition enacts a paradox worthy of civilization's epitaph, one that desperately communicates [...] the dangers of a highly capable and creative society unchecked"[2] "long after the traces of human civilization have disappeared."[3]

Relational Trajectories of Antarctica and Outer Space

Juan Francisco Salazar

JUAN FRANCISCO SALAZAR is a professor of communications and media studies at Western Sydney University (Australia) and is currently an Australian Research Council Future Fellow. He is an environmental anthropologist and documentary film-maker with academic and artistic interests in social-ecological change, environmental futures and indigenous politics. Supported by the Chilean Antarctic Institute, Salazar has undertaken ethnographic and film work in Antarctica, where he directed *Nightfall on Gaia* (2015).

It is commonly agreed that the Space Age began sixty years ago when the Soviet Union launched Sputnik 1 on the 4th of October 1957, as the International Geophysical Year (IGY) was just getting underway. The IGY of 1957–1958 brought together Antarctica, the High Seas, the atmosphere, and Outer Space – the global commons subsequently defined in international law as new frontiers for scientific exploration. More than sixty-seven nations participated across more than four thousand research stations worldwide (with a focus on the polar regions) in a global cooperative endeavour without precedent. Among the pivotal political outcomes of the IGY were three groundbreaking legal frameworks, which today, six decades on, have withstood the test of time. The 1958 United Nations Convention on the High Seas provided, for the first time, an international governance regime for a space beyond sovereign jurisdictions.

In a period of heightened tensions in the early years of the Cold War, the Antarctic Treaty followed in 1959, laying down the principles of a legal regime for the governance and scientific exploration of Antarctica and the Southern Ocean. Importantly, in the very short period between the start of the IGY in July 1957 and the signing of the Antarctic Treaty in December 1959 it became clear how Antarctica was to provide both a model and a threshold for the geographical spaces beyond terrestrial inhabitation (through the launching of polar orbiting satellites, upper atmosphere testing, and seismic work). The Antarctic Treaty System has since provided lessons that are relevant to the governance of other extraterritorial spaces beyond sovereign jurisdictions, including Outer Space, as extraterritorial spaces that are "imaginatively, historically, and juridically interconnected."[1]

Both Antarctica and Outer Space have been key to modern understandings of Earth and to the visualisation of global environmental change. They are also intimately connected to each other in the search for biosignatures and microbial life forms in our solar system and in exoplanets and other celestial bodies. Since the 1980s Antarctica has been conceived by atmospheric scientists and physicists as a "window on Outer Space" and a unique laboratory for atmospheric studies. Additionally, since the 1980s Antarctica has become a primordial analogue and proxy for Outer Space environments, particularly for research in astrobiology and planetary sciences. The continent has thus always provided a "geopolitical test object for governance":[2] it not only acted as "a surrogate testing ground for technologies for potential use in Cold War Arctic contestations" in the 1940s and 1950s but also informed and shaped the Outer Space Treaty of 1967 and provided "a psychological test bed for habitation in Outer Space" in the 1980s.

These matters of the extraterrestrial, once mostly bound to the realm of literary studies, have increasingly become part of the remit of anthropologists, philosophers, historians, geographers, scholars in science and technology studies, and artistic researchers and practitioners. As several social studies of how astronomers and other natural scientists investigate the cosmos indicate, scientists never simply depict or describe the cosmos "just as it is". Their research is always characterised by a specific aesthetic style and by a "cosmic imagination". Scientific knowledge of the universe is also based on skilled judgements rather than on direct, unmediated perception. It is science, but it is also an art. Furthermore, and despite the fact that Antarctica and Outer Space have never been spaces "for humanity to attach to pre-existing flows of culture", contemporary social and cultural research is providing a rich body of work with accounts of how Antarctic places and cultures are emerging, with distinct modes of subjectivity and forms of sociality in extreme environments. These novel approaches

"Until this century, there was more known about the shape of ice on Mars than Antarctica. Satellites that could measure the Red Planet's ice caps in detail were launched in 1996 – seven years before an equivalent mission for Earth. It sounds crazy, but that's just the way it was back then."

Andrew Shepherd (Professor, University of Leeds), in "Inside Antarctica: The Continent Whose Fate Will Affect Millions", *The Financial Times*, 2018

"In the sixties, at the height of the Cold War, the Soviets used Vostok as a behavioural test bed for the Salyut space programme."

Sara Wheeler (Writer), *Terra Incognita: Travels in Antarctica*, 1999

in the human disciplines are providing new understandings of how planetary scientists and astrobiologists studying Outer Space environments rely on analogue environments on Earth (such as Antarctica) for their endeavours and are engaged in practices of place-making to make sense of other planets, such as Mars. Antarctica, as a place "outside the circuits of the known world that both precedes the Moon as a destination of otherworldly knowledge and is coterminous with 'outer space'",[3] has now been a sphere of human endeavour for well over a century; and Outer Space for just over fifty years.

Humans are now physically present in Antarctica year-round in the form of over a thousand transient and semi-permanent scientific and logistics personnel, a figure that expands – like Antarctic ice expands in winter – to five thousand people in summer, in addition to the more than thirty thousand tourists who visit the fringes of the Antarctic continent every year. Small numbers of individuals have "inhabited space" with relatively short hiatuses for the last two decades, and without interruption since 2000 by successive crews in the International Space Station, launched in 1998 and arguably the most expensive technological structure ever built.

In this regard, Antarctica and Outer Space come to represent an inherently future-oriented quandary as the most serious test to our collective and coordinated capacity to exercise foresight on a planet that for many is already damaged beyond repair. In the case of Antarctica, this foresight is necessary, not only to protect the unique and fragile Antarctic ecosystems but also to understand ongoing transformations in the framing of Antarctica. And on a more philosophical level to recast our species as part-of and in relation-with nature in ways that might provide novel experiments with living differently in the Anthropocene.

"Antarctica […] is the first place for the technological development of new representational practices, new modes of exploratory knowledge, and new ways of tying the unusable (or wasted) earth to the engines of capital accumulation that have propelled us into transnational global exchange."

Elena Glasberg (Professor at New York University, Essayist, Speaker), *Antarctica as Cultural Critique: The Gendered Politics of Science Exploration & Climate Change*, 2012

TERRAFORMING ANTARCTICA
Juan Francisco Salazar

To think of terraforming Antarctica involves examining coexistent processes of making Antarctica familiar. Not only as a mode of transforming Antarctica into a habitable world – a terraforming process of sorts – but also a way of opening up ways for thinking about human sociality beyond Earth. Since the mid-twentieth century, humans have been learning to live on the ice – inhabit the extreme. We tend to think of life on Earth at low, middle, and in some cases even high latitudes as the only model for a habitable world. Antarctica challenges this view, given its quasi extraterrestrial conditions for human inhabitability. The transformation of Antarctica into a habitable world invites reflection on how the Antarctic might be thought of as a stepping-stone to modes of projecting human habitability on other planets and celestial bodies. Turning to the notion of terraforming as a trope for examining the worlding processes of making Antarctica familiar, provides a tentative anthropological account of the ways in which worlds in extreme environments are made and unmade. Taking advantage of its conjectural elasticity, terraforming is a notion to rethink frontiers of life anthropologically, as it provides a novel platform from where to engage with notions of "extremes" and the expansion of "habitable environments".

Antarctica and Apollo. Heritage in the Extreme

Bryan Lintott

BRYAN LINTOTT serves as secretary-general of the ICOMOS International Polar Heritage Committee. He is a historian based at the Scott Polar Research Institute, University of Cambridge, and a member of the University of Tromsø in the Norwegian Arctic. Dr Lintott's research focuses on how heritage in extreme environments, beyond national boundaries, is governed, managed, conserved, and utilised.

Antarctica and Space are dual extreme environments where nations, groups, companies, and individuals have undertaken human and robotic exploration, while science and commercial activity have produced a legacy of physical and intangible heritage beyond national boundaries. Whilst the Antarctic nations have developed a system to protect Antarctic Historic Sites and Monuments (HSMs), there are no agreements or protocols to protect the historical and cultural values of non–terrestrial Historic Spacecraft, Sites, and Monuments.

In the aftermath of World War II, Argentina, Chile, and the United Kingdom all claimed the Antarctic Peninsula as territory. This disagreement manifested itself in the removal of competing claimants' huts, the painting of national flags on their buildings, and the displacement of objects. Prior to, and during, the International Geophysical Year (1957–1958), it was decided that such actions would cease in the higher interest of science. The avoidance of conflict was codified in the Antarctic Treaty (1959) and subsequent Antarctic Treaty System (ATS). Treaty signatories govern Antarctica by consensus, implementing their decisions through domestic laws. Over the decades, a mechanism for proposing, considering, and designating Historic Sites and Monuments has evolved. In 1991, the Protocol on Environmental Protection to the Antarctic Treaty was approved and the issue of HSMs incorporated in Article 8, Annex V; the latter reads that HSMs "shall not be damaged, removed or destroyed. In exceptional cases, items can be removed for human and wildlife safety, specialised conservation, or temporarily relocated in the event of a severe threat of damage or destruction."

President Eisenhower, speaking at the United Nations in 1960 stated that the Antarctic Treaty offered lessons in resolving or ameliorating international disputes and that these lessons be utilised to develop international cooperation in Space. During the Cold War, Space did become a region of technical and scientific competition. The United States became the first nation to land humans on the Moon, but at the cost of great sacrifice. In 1967, astronauts Virgil Grissom, Edward White, and Roger Chaffee perished in a fire while conducting ground tests in the command module of Apollo 1; and the financial cost for landing two humans on the Moon was estimated in 1969 at US\$ 21.25 billion – more than US\$ 100 billion in contemporary value. However, the rewards were significant: demonstrating technological superiority over the Soviet Union, pursuing significant advances in science, and achieving the greatest feat of human exploration in the twentieth century.

Space travel remains dangerous and expensive, and the lunar bases envisaged in the 1960s movie *2001: A Space Odyssey* remain in the realm of the imagination. At the turn of the millennium, advances in robotics, decreasing launch costs, and the need for historians to record first-hand accounts of the ageing cohort of Apollo resulted in several research projects. Beth O'Leary and her colleagues from the New Mexico Space Grant Lunar Legacy Project undertook "remote" archaeological research (using imagery, reports, and other material) on the Apollo 11 site, "Tranquillity Base". Scholars, including Peter Capelotti, Alice Gorman, Beth O'Leary, and others, examined Space heritage theoretically, conceptually, and legally. In April 2007, the topic of the Apollo sites and the use of Antarctica as an analogue in heritage management and conservation was presented at the Smithsonian's Mutual Concerns of Air and Space Museums Conference by the author. Later that year, Google announced a "Lunar XPRIZE" to encourage non-government robotic expeditions to the Moon, with a historic site bonus. Google, committed to preventing its contestants from damaging Apollo sites and announced that any visit to a historic site would have to be pre-approved and comply with "NASA's Recommendations to Space-Faring Entities: How to Protect and Preserve the Historic and Scientific Value of U.S. Government Lunar Artifacts". The criteria included avoidance of biological, chemical, or radiological contamination; strict limits on descent and landing profiles to minimise disruption or damage to the sites, by regolith (lunar rocks and dust) driven by the rocket blast, ensuring that a crash would impact outside the historic sites; and finally regulating that the Apollo 11 and Apollo 17 sites would remain in exclusion zones, and operational scientific equipment (i.e. Lunar Laser Ranging Retro-reflectors) would be protected. The criteria included NASA being able to approve scientific and/or engineering investigations in the context of the sites.

Whilst the ownership of the Apollo artefacts resides with the United States government under Article VIII of the Outer Space Treaty, site disruption (such as a robot driving across Neil Armstrong's footprints) is not illegal. The Lunar XPRIZE concluded without a winner, but one spacecraft did reach the Moon – where it crash-landed.

The Moon is profoundly embedded within human culture and history, but our visual experience of the Moon is at risk. Proposals to deploy thousands of small satellites around the Earth could cause intrusive reflections at sunset and sunrise – disrupting the majestic sight of the Moon close to the horizon. Technological advances may result in advertising being projected onto the Moon, in a similar way to floating LED "billboards" on rivers and at sea that mimic the visual onslaught of *Blade Runner*. In the future, surface mining on the Moon could obliterate features of deep cultural significance.

The Antarctic nations have developed a functioning system of Historic Site and Monument designation and management. The latter has proved to be a robust system, adaptable to change – albeit slowly – but there are still essential roles for ICOMOS, heritage academics, and practitioners to enhance Antarctic heritage conservation regulations. The Antarctic system was endorsed by all the major Space-faring nations, and hence there is the possibility that they could consider establishing a transferrable "Outer Space, Moon and Celestial Bodies Protocol on Historic Spacecraft, Sites and Monuments." With international vision and goodwill – as well as the expertise of heritage organisations and professionals – humanity can transfer the lessons learned in Antarctica to govern, protect, and utilise Space heritage.

"Apollo can best be understood as a vast machine for the creation of a single image; that of an American, and American Flag, on the surface of the moon. [...] The three-by-five-foot nylon flag so highlighted was obtained from a government supply catalog and altered only by the

addition of a hem across the top, to hold the crossbar allowing the flag to unfurl in the absence of all but the cosmic wind. Yet in describing the final flight of the Lunar Module's iconic, angular ascent stage, pilot Edwin E. 'Buzz' Aldrin remembered: 'There was no time to sight-see.' […] And even if it was still standing, subjected to decades of unfiltered radiation and solar wind, the mass-market pigments on the nylon surface have undoubtedly eroded to oblivion. Dennis Lacarrubba, whose New Jersey based flag-maker, Annin, likely sold the flag in question for $5.50, considered in 2008 the effect of thirty-nine years of UV light on even an earthbound flag: 'I gotta be honest with you,' he reported; 'It's gonna be ashes.'"

Nicholas de Monchaux, (Architect, Dean at Massachusetts Institute of Technology), *Kosmos* (online), 2020

Space and Suit

Nicholas de Monchaux

NICHOLAS DE MONCHAUX is a professor and head of the Department of Architecture at the Massachusetts Institute of Technology (MIT) and a partner in the Modem architecture practice. He is the author of *Spacesuit: Fashioning Apollo* (MIT Press, 2011). In 2012 he was named one of the "Public Interest Design 100" by *Good* magazine. Prof. de Monchaux's work has been widely exhibited, and he is a fellow of the American Academy in Rome.

Capital-S "Space" was first defined by Milton in 1667's *Paradise Lost;* "Space may produce new Worlds; whereof so rife / There went a fame in Heav'n."[1] The voice is Lucifer's, and "Space" the realm of the gods and angels, in which the "pendant Earth" hangs. From Space, Lucifer is exiled to another realm beyond living humanity, the "desert utmost Hell".[2] And the two are inextricably linked. Centuries later, standing on the surface of the Moon, Neil Armstrong observed of the extra-planetary landscape, "It has a stark beauty all its own. It's like much of the high desert of the United States."[3]

Antarctica is our planet's largest desert; for all its frozen water one of the driest places in the world. It occupies a space in the popular imagination that amplifies its emptiness with inaccessibility, approaching that of the heavens themselves. "Even now the Antarctic is to the rest of the earth as the Abode of the Gods was to the ancient Chaldees"; wrote Apsley Cherry-Garrard in his first-hand account of Scott's polar exhibition, *The Worst Journey in the World.*[4] And just like the heavens, because there is nothing there, we fill it with our dreams. "In a landscape where nothing officially exists", Reyner Banham wrote of the Mojave Desert where Neil Armstrong lived and worked as a test pilot, "absolutely anything becomes thinkable, and therefore possible."[5]

To dwell in uninhabitable places, we must be adapted to them. Despite early proposals to physically alter the bodies of astronauts, this has been done, historically, with clothing.[6] The word "suit" comes from the Anglo-French *siwte,* to follow ("pursue", as well as "sue"). Today, "suit" means a set of things that follow each other in style or purpose, as well as its second, essential meaning. Just as one wears a suit for a particular environment (boardroom or bathing party) one is better suited to it as well. The alchemy of this transformation is amplified the farther from the everyday limits of our bodies such suits take us – and with it the symbolic presence of the body that returns.

And the clothing as well. In 1969, the Italian skiwear company Tecnica introduced the "Moon Boot" insulated, après-ski-wear made to resemble the custom-designed footwear worn by Armstrong and Aldrin (and discarded by them on the lunar surface). More than twenty-five million pairs have been sold. In 2004, when the boots had one of their more recent returns to fashion, *The New York Times* recorded a shoe-seller observing, "People want to look like astronauts. They're the ugliest things I've ever seen in my life. And people love them."[7]

The members of Scott's *Terra Nova* expedition, by their own account, concluded that "our own clothing and equipment could not be bettered in any important respect".[8] In fact, the more "primitive" clothing of the Amundsen expedition, which drew extensively from native Inuit clothing, was more effective.[9] The physical demands that Cherry-Garrard's engineered oilcloth mediated were astonishing. Of the eponymous "Worst Journey in the World," actually not the summer journey to the pole, but a midwinter reconnaissance of the Ross Ice Shelf, Cherry-Garrard writes "The horrors of that return journey are blurred to my memory and I know they were blurred to my body at the time." A contemporaneous photograph of Cherry-Garrard shows him with his mouth shut. I don't know why our tongues never got frozen," he would write, "but all my teeth, the nerves of which had been killed, split to pieces."[10]

The *Terra Nova* expedition of 1911–1912 was no private matter, but nor was it strictly territorial. It was an exercise in asserting sovereignty over the public imagination by placing bodies where they were never meant to exist. So, too, was the vast architecture of Apollo.

When Neil Armstrong walked on the lunar surface in 1969, he wore, underneath the iconic brilliance of its outer white thermal layer, a twenty-one-layer

"[Cherry-Garrard] suffered from pronounced myopia or short-sightedness, photographs showing him habitually wearing metal-rimmed spectacles. In the temperatures he was to encounter, these would have iced up or frozen to his face and by necessity would have to have been used only selectively [...] As such, many of the spaces, hazards, and phenomena Cherry-Garrard experienced would have been to a great extent beyond his visual range. One thing struck me as inescapable when looking at [his] skis. Their tips would have been some two metres from his eyes and quite possibly blurred or indistinct. Without his spectacles, could he see their extent with any clarity, let alone what existed beyond?"

Tim O'Riley, *Endlessness*, 2020

assemblage which was given flexibility by a set of convoluted rubber joints developed from materials and techniques taken from women's underwear. Manufactured by what began as a small, industrial division of the Playtex Bra and Girdle company, the Apollo spacesuit affirmed the logic of clothing over equipment, as it was hand-sewn by former bra-and-girdle seamstresses on standard Singer machines – albeit to a near-superhuman 0.39-millimetre tolerance.[11] The victory of this epidermal, unglamorous assemblage over more engineered approaches mirrors, in fact, the victory of Amundsen's more literally epidermal furs over the *Terra Nova's* engineered oilcloth.

Yet there are close analogies with the exhaustion and extension of the *Terra Nova* expedition as well. A photograph of Neil Armstrong, taken on the 21st of June 1969, just minutes after his return from that first walk on the Moon to the safety of the Lunar Module, is witness to this. Standing at the outer limit of human exploration and inside his remarkable clothing, Armstrong stands at the end of his endurance as well. As the image reveals, the structured skin of contemporary expedition clothing is a testament not to power, but to the essential fragility of our bodies inside it. The "abodes of the gods" are not so much limits of distance, or geography, as limits of ourselves. We forget this at our peril.

"It was a wearable spacecraft. Cocooned within 21 layers of synthetics, neoprene rubber and metalized polyester films, Armstrong was protected from the airless Moon's extremes of heat and cold (plus 240 Fahrenheit degrees in sunlight to minus 280 in shadow), deadly solar ultra-violet radiation and even the potential hazard of micro-meteorites hurtling through the void at 10 miles per second. The Apollo suits were blends of cutting-edge technology and Old World craftsmanship. Each suit was hand-built by seamstresses who had to be extraordinarily precise; a stitching error as small as 1/32 inch could mean the difference between a space-worthy suit and a reject."

Andrew Chaikin (Writer), "Neil Armstrong's Spacesuit Was Made by a Bra Manufacturer", *Smithsonian Magazine*, 2013

Antarctic Attire

Conrad Anker

CONRAD ANKER is an accomplished alpinist who pushes the limits of mountaineering in the world's most inhospitable environments. With twelve trips to Antarctica since 1992, his first ascents include Mount Craddock in the Ellsworth Mountains and Rakekniven Peak in Queen Maud Land. In 2017 Anker established, with Jimmy Chin, a new route on Ulvetanna in Queen Maud Land.

Antarctica entered my awareness, curiosity, and imagination through an atlas and the occasional television documentary. As a child I would build forts with my siblings out of chairs, pillows, and blankets. We would crawl inside the improvised tent and dream of blizzards raking across a frozen plateau. We would stumble out, play-acting our resistance to the gale force wind in order to resupply cookies from the cupboard. Perhaps it was moments like these that ignited my dream of visiting the remote territory of Antarctica.

In 1992 my dreams slowly became reality. Guiding on Mount Vinson I had the opportunity to live in a real tent, be hammered by the wind, and bitten by the cold – with no cookies. While the ice stations of my childhood imagination had no consequence, reality was tough, yet fortunately I was equipped with the first line of defence any scientist, explorer, or climber would need on the frozen continent: clothing – the most essential equipment required for survival in Antarctica.

The first Antarctic explorers adapted the clothing of the indigenous people of the high northerly latitudes. Seals, foxes, and polar bears are a few of the animals that supported the Arctic communities; their fur and hides were sewn into clothing, allowing humans to exist in harsh winter climates. Since Antarctica has no indigenous peoples, to manage temperature, humidity, wind, and precipitation and guarantee their safe existence, pioneering Antarctic explorers used clothing (mostly) adapted from the northern hemisphere.

Temperature is an obvious factor, as bared skin is susceptible to frost injury after ten minutes of exposure in –40°C temperatures. While a frostbitten finger is not, of itself, lethal, it can be the start of a cascading series of events that limit one's ability to function; thus, having insulation and wind block is essential in preserving human life.

The low relative humidity in Antarctica – known to be the driest continent on planet Earth – implies that the moisture which humans need to manage is internal, i.e. generated by exertion and existence. Consequently, fabrics that can breathe and allow the escape of perspiration are essential to ensure an acceptable level of comfort, especially for one's feet. Moisture from saturated socks can also lead to frostbite, hence a liner boot in footwear that can be removed and dried out each night is key for journeys not punctuated by stays in heated buildings.

Wind is the invisible yet palpable adversary that increases the dangers presented by temperature. By reducing warmth, a 40-kilometre-per-hour wind, for example, can double –10°C to –20°C. The resultant danger to exposed skin is obvious and from a clothing standpoint the consequent challenge is to work with a fabric that blocks the wind while at the same time allowing perspiration to escape.

Finally, precipitation is and has always been a challenge for humans. Being wet is more than just miserable; it can cost you your life. Coastal Antarctica is a maritime environment with a greater likelihood of precipitation than the dry interior of the continent. When precipitation events do occur, they are mostly in the form of snow and while the likelihood of rain is increasing due to a warming climate, the demands on the fabric are not the same as those made by the environment of a temperate rainforest.

Ultimately, without a complex system of life support, from food to shelter to clothing, humans would not be able to visit this continent, and the first line of protection that humans need in Antarctica is, without doubt, insulated outerwear. Our ability to create tools that extend the physical environment in which we exist is our hallmark as a species, and clothing (from a functional standpoint) is the apex of this practice.

Antarctic explorers have struggled with frostbite, hypothermia, and snow blindness since the Heroic Age. Acting as the first architectural envelope, clothing and related accessories are essential to human survival in such an extreme environment. Snow goggles such as those worn by Borchgrevink during the Southern Cross Expedition to protect the eyesight from the intense radiation of reflected UV lights, and the fur mittens[1] and boots[2] worn by Edward A. Mackenzie during the British Antarctic Expedition (1910–1913), showcase early attempts to reinvent fashion as portable environments.

"You put on frozen mitts and frozen boots, stuffed with frozen grass and rime. There's a fascination about it all, but it can't be considered comfort."

Edward Wilson (Explorer, Orinthologist), extract from New Zealand Antarctic Heritage Trust, *Conservation Report: Discovery Hut*, 2004

1910–1913

1911–1913

1930

"One continues to wonder as to the possibilities of fur clothing as made by the Esquimaux, with a sneaking feeling that it may outclass our civilized garb. For us this can only be a matter of speculation, as it would have been quite impossible to have obtained such articles."

Robert Falcon Scott (Explorer, Royal Navy Officer), in his diary, 1912

"Our troubles were greatly increased by the state of our clothes. If we had been dressed in lead, we should have been able to move our arms and necks and heads more easily that we could now."

Apsley Cherry-Garrard (Explorer, Zoologist), *The Worst Journey in the World*, 1922

"The English have loudly and openly told the world that skis and dogs are unusable in these regions and that fur clothes are rubbish. We will see – we will see. I don't want to boast – it's not exactly in my line, but when people decide to attack the methods which have brought the Norwegians into the leader class as polar explorers – skis and dogs, well, then, one must be allowed to be irritated and try to show the world that it is not only luck that brought us through with the help of such means, but calculation and understanding of how to use them."

Roald Amundsen (Explorer), *The South Pole: An Account of the Norwegian Antarctic Expedition in the Fram 1910–1912*

THE CONQUEST OF ANTARCTICA BY AIR
Richard Evelyn Byrd

Clothing was one of numerous vital details that took days of thought, tests, and experiments. Clothes had to be warm, light, and roomy enough to permit perfect freedom, and windproof to protect us against the strongest and most piercing winds. For very cold weather, and for rest, fur clothes are far preferable to any others. Admiral Peary [...] copied the clothes of the Eskimos around Etah, Greenland, and so did we. I studied their methods when I went up there in 1925 for the National Geographic Society. The lightest and warmest, and therefore most practical kind of fur seems to be reindeer skin: so our mukluks (or boots), parkas (or coats), and many of our pants were made of reindeer skin. To give equal warmth, woollen clothes would have to be twice as heavy. [...] Clothed in a large, roomy, reindeer-skin parka, with armpits large enough to bring the arms inside, polar-bear pants and reindeer skin boots, one can sleep in the open without additional covering or tent in the coldest weather. [...] Contrary to general belief, the great problem in Antarctica arises, not from cold, but from the moisture. Moisture always forms and then freezes. That is why men lose their feet and fingers, and that is why they have such a difficult time with sleeping bags.[1]

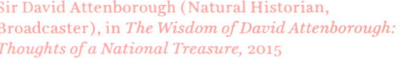

"At a time when it's possible for thirty people to stand on the top of Everest in one day, Antarctica still remains a remote, lonely and desolate continent. A place where it's possible to see the splendours and immensities of the natural world at its most dramatic and, what's more, witness them almost exactly as they were, long, long before human beings ever arrived on the surface of this planet. Long may it remain so."

Sir David Attenborough (Natural Historian, Broadcaster), in *The Wisdom of David Attenborough: Thoughts of a National Treasure*, 2015

Design and fabrication have made great improvements over time. With the advent of synthetic fabrics and closed-cell foam insulation, clothing has become lighter and more durable. Concurrently, the use of down as an insulation fortunately harks back to the use of animal products for insulation and comfort.

While I did not unlock the mysteries of the universe or find amber fields of grain, mountain climbing in Antarctica has humbled me. My childhood dream was realised and the cold, desperate environments tested my limits. Perhaps, understanding what one can accomplish with limited resources is exactly what can spark new designs, and – especially in the challenging times humanity faces today – Antarctica should be seen as a place in which we can tap into creativity borne of necessity. Let us learn from hardship.

Rethinking the Antarctic Suit

Lino Dainese

LINO DAINESE is the founder of Dainese and D-Air Lab (Italy), which are, respectively, a world-leading manufacturer of technical gear and protective equipment for dynamic sports founded in 1972 (a recipient of multiple design awards) and its research and development lab devoted to the prototyping of intelligent clothing for extreme environments and risk prevention in quotidian applications. In collaboration with UNLESS, D-Air Lab developed an "Antarctic Suit".

Rising to the challenge of rethinking Antarctic clothing requires thorough research undertaken with the determination to break with traditions, set new standards, and develop out-of-the-box solutions. With this mindset, the design of specialised equipment for extreme conditions produces "intelligent clothing" that embodies the potential offered by advanced wearable technologies and the intelligence of analogue handcrafting experience.

The first prototype of the Antarctic Suit developed by D-Air Lab and UNLESS (below) is being tested in Antarctica, at Concordia Station, during the XXXVI Italian Expedition in Antarctica (2020–2021).

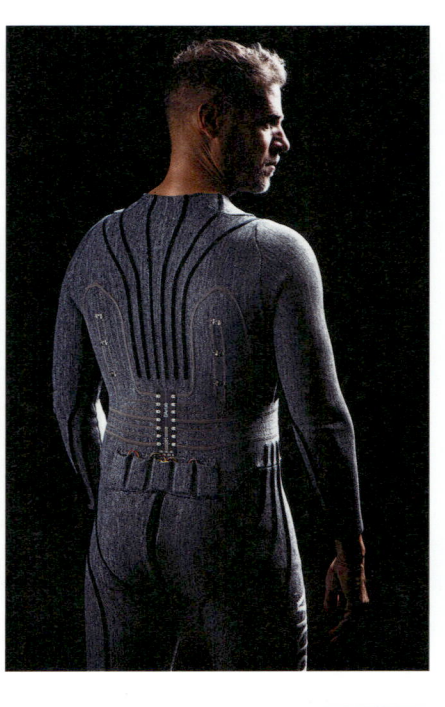

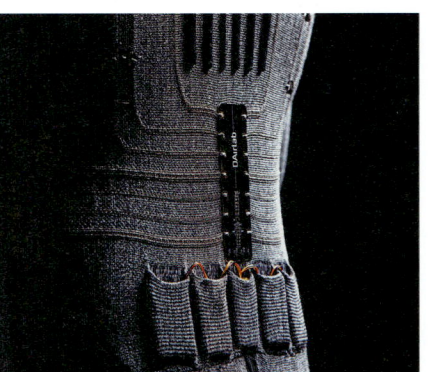

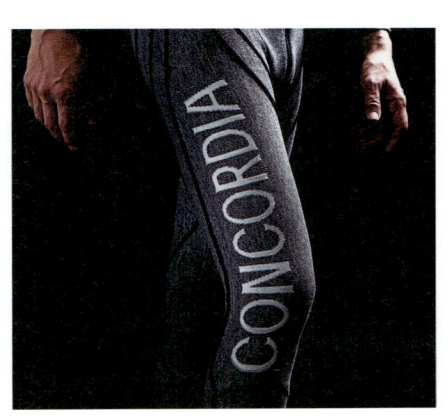

"A spacesuit is actually a miniature spaceship shaped like the human body."

Dava Newman (Engineer, former Deputy Administrator of NASA), "Building the Future Spacesuit", *ASK Magazine*, 2010

Dava Newman's famous statement that "a spacesuit is actually a miniature spaceship shaped like the human body" is true not only for Outer Space but also for another Global Commons: Antarctica. Taken by D-Air Lab and UNLESS as a starting point for conceiving a new technological armour for researchers working in the Antarctic – i.e. their first "home" on the white continent – the Antarctic Suit is intended as a first "architectural envelope".

Envisaged to enable scientists to work outside in conditions that can easily be as cold as 50°C below zero, the suit comprises two main parts: an under-suit – a very thin advanced layer which is in direct contact with the body; and an outer "cover", which remains in contact with the environment and insulates the wearer's body from the outside world.

The under-suit takes its inspiration from human skin's ability to continuously regulate temperature, ventilation, and transpiration in response to the conditions in which it finds itself. Such a layer makes use of an integrated series of different elements, combining silk yarn against the skin for comfort, a stiffer yarn to provide constant ventilation for the body, and silver thread to create conductive paths for generating heat. An electronic control system mounted on a flexible card receives and processes the information from a system of sensors distributed over the body, constantly monitoring temperature and vital functions (such as heartbeat and body temperature, for example) to adjust the energy released to heat the body and enable the organism to perform its vital functions normally. The ventilation needed to prevent moisture build-up is provided via "tunnels" with a "C"-shaped cross section that are in contact with the body and laid out in varying densities, proportional to the perspiration in different areas of the body. This network of channels allows air to be conveyed over the body, drawing excess moisture away and out at the shoulders. The suit's outer cover is a true "armour" against cold and inhospitable weather. This armour comprises two coats: the outer envelope's membranes and filling materials create two layers of air, which provide an initial barrier against the outside elements, and a more compact insulating layer, which traps the heat generated by our intelligent skin – or, in this case, the under-suit.

The under-suit, which features seamless technology, has been designed to be stretchy and body-hugging, with minimal thickness. The outer cover has likewise been designed to keep bulk to a minimum and facilitate easy movement for Antarctic scientists in the field. In addition, special care has been devoted to reducing the number of fabric panels, and therefore the amount of stitching and seams required, to optimise the garment's lightness, efficiency, and reliability over time.

The first prototype of the Antarctic Suit was tested at Concordia Station ↗ p.780 by members of the Italian National Antarctic Programme (PNRA) ENEA team on the occasion of the bicentenary of the first sighting of the continent. While this prototypical design stage was focused on body protection, the suit will further evolve by developing equally effective solutions for the protection of head, hands, and feet, with the aim of fully integrating all the parts in a true second skin.

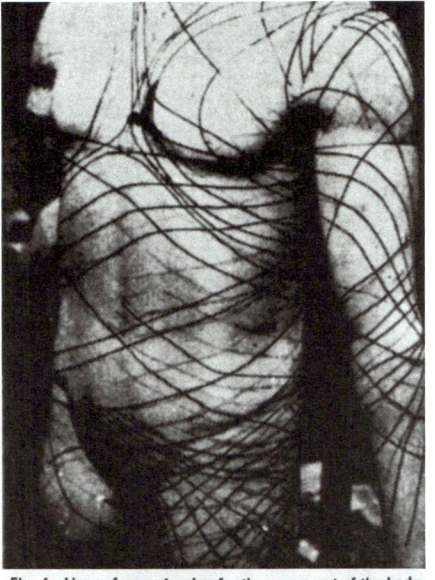

Fig. 4 Lines of nonextension for the upper part of the body

Building upon Dr Arthur Iberall's 1963–1964 theories of the "lines of nonextension" (above, top) – whereby "the human skin, [...] stretched during body motion, [has] virtually no stretch along certain lines"[1] – which can be used to develop "a full-pressure suit for a human being [, allowing] the wearer full mobility", in 2007 Dainese partnered with Professor Dava

Newman and Space Architect Guillermo Trotti to design the Biosuit®. Conceived to provide "the necessary one-third of an atmosphere [of pressure required] to keep you alive in the vacuum of space [...] through mechanical counterpressure",[2] the lightweight attire applies "pressure directly to the skin, thus avoiding gas pressure altogether".[3] By

combining "passive elastics with active materials" (i.e. a "stretchy garment, lined with tiny, muscle-like coils [...] made from a shape-memory alloy [...] that 'remembers' an engineered shape and [...] can spring back to this shape when heated"[4]), the suit does not limit the astronauts' freedom of movement like some of its predecessors.

UNDER-SUIT (932.6 G)

LAYERS (interwoven)
Silk, 100% polyester, and silver

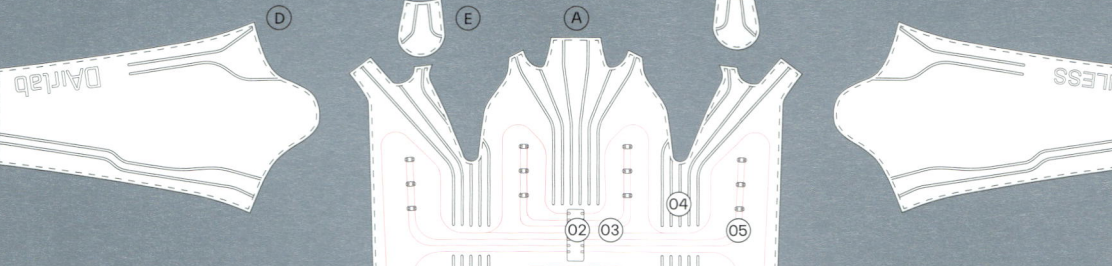

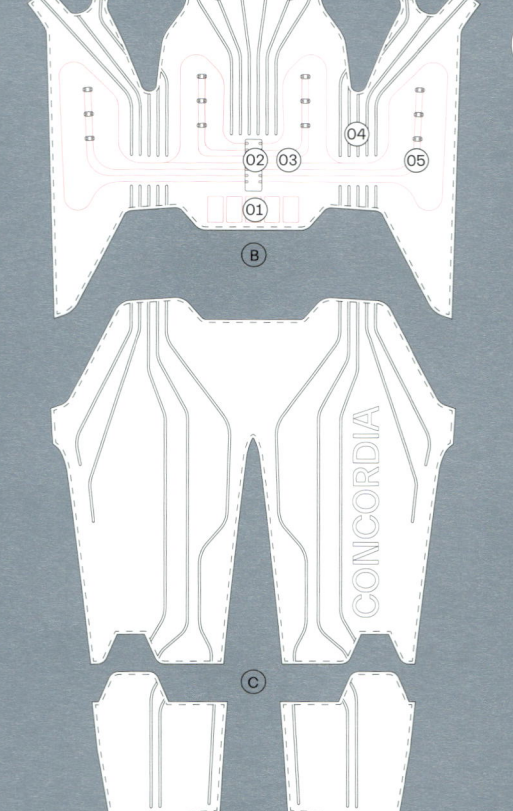

Under-Suit

A   Upper body element
B   Belt with batteries to guarantee 230-minute autonomy
01   Batteries (6 × 1000 mAh)
02   Printed circuit board
03   Heat resistance (3.8 m)
04   Hollow interspace with spiral coil for insulation
05   Biomedical sensor recording: heart rate; blood pressure; respiration; temperature, heat loss
C   Trouser element with station name
D   Sleeve element
E   Neck insert

Outer Envelope

F   Hood with embedded anti-shock technology and light
G   Face mask
H   Frontal element with double zip, station name, and Antarctic print
I   Upper sleeve with shoulder chimney effect
J   Rear element with Lee conformal projection print
K   Lower sleeve with embedded screen on wrist for biometric data and integrated gloves
L   Trousers with reinforced knee patch for high resistance and grip
M   Integrated boots

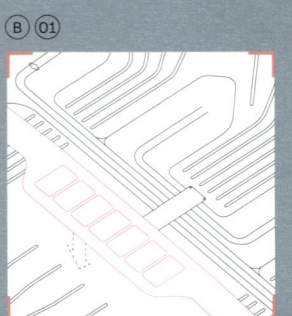

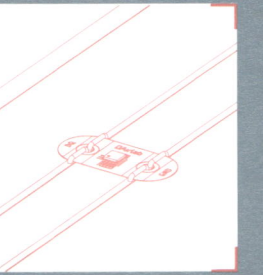

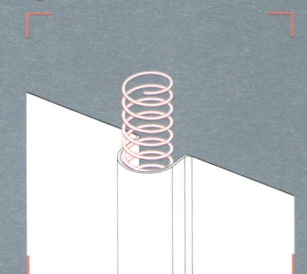

SOURCE
D-Air Lab

OUTER ENVELOPE (1,989.5 G)

LAYERS (from external to internal)
PA / PTFE 230 sqm (wind-stopper)
100% polyester 50 sqm (exterior envelope)
Camelux, 100 g (thermal exchange optimiser)
Twins graphene, 80 g (thermal distributor)
Mid-layer, 130 g (heat containment element)
Nativa Silkpad (human-body heat-radiation
reflective element)

0   15   30   60 cm

Do(o)med Interiors

Miranda Nieboer and Craig McCormack

MIRANDA NIEBOER is an architect, senior lecturer, and current PhD candidate at the University of Tasmania and the Institute for Marine and Antarctic Studies in Hobart (Australia). CRAIG MCCORMACK is a lecturer and PhD candidate at the University of Western Australia's School of Design (Australia).

Antarctica's environment is depicted as analogous to that of Outer Space. Both extreme milieus share the absence of indigenous human inhabitants and both have undergone human attempts to colonise their deadly environments. For a period in history, the South Pole and Outer Space shared a distinct typology for habitation: Buckminster Fuller's patented geodesic dome. The Amundsen-Scott South Pole Station ↗ p. 722, more affectionately known as the South Pole Dome (SPD), endured the harsh Antarctic environment from its dedication in 1975 to its dismantling in 2009. Even though extraterrestrial geodesic domes were never constructed, the dome has remained a speculative typology for NASA's ambitious colonisation plans throughout the solar system, including projected settlements on the Moon. The geodesic dome is a structure that galvanises the relationship between these two extreme environments at the limits of human existence.

The ultimate aim of the dome typology is to provide an autonomous enclosed world, or facsimile of the Earth's environment. Domes in the Antarctic and Outer Space both provide envelopes that temper extreme environmental characteristics that are otherwise fatal to humans. As such, the domes express a reciprocal relationship between the human body and architecture through both reduction and expansion. While the body is reduced to a series of inputs and outputs within a closed system, the geodesic interior becomes a prosthetic for extending human life into unlivable environments. The dome is thus a structure of control into which the human body has to integrate as a perfect and operational component. The geodesic dome expresses a utopian ambition of complete autonomy and control, both over the environment and humankind.

In this controlled relationship between the assumed ideal human and the domed artificial environment, paradoxically the weaknesses of the human body become emphasised. While the SPD was located within an environment that had the Earth's purest air, its inhabitants had to endure severe pollution from machinery occupying the same space.[1] Human life inside the SPD was further endangered when a sewage leak and the warmth generated by the associated human waste threatened to fracture the dome's foundational ring.[2]

Unlike their terrestrial counterparts, geodesic domes in Outer Space remain incredibly speculative projects, though there are examples of closed worlds that serve as inhabitation including the International Space Station (ISS). Aboard the ISS, unforeseen and increased levels of carbon dioxide off-gassed from materials that can cause headaches among astronauts cannot be recycled any quicker due to hardware constraints. This rise in carbon dioxide was an anomaly that the systems were not designed to accommodate and the inhabitants were obliged to simply endure the headaches.[3] A systematisation of the ideal human body thus has serious consequences for the real physical body in these closed interior worlds.

The SPD and the extraterrestrial geodesics illustrate that the production of such closed interiors results in an additional extreme environment when inhabited by the leaky and messy human bodies that they were designed to protect.

The drawing reveals a geodesic dome project featured in "Lunar Base Expansion", from Jesco von Puttkamer, "Developing Space Occupancy: Perspectives on NASA Future Space Program Planning", in *Space Manufacturing Facilities (Space Colonies)*, 1977.

Antarctica, Human Adaptation, and Space

Ron Roberts

RON ROBERTS is a chartered psychologist and an associate fellow of the British Psychological Society. Dr Roberts is an honorary lecturer in psychology at Kingston University (United Kingdom) and is the author of eleven books and numerous research articles, including *Foundations of Health Psychology* (Palgrave, 2001), *The New Politics of Experience* (PCCS Books, 2015), *Mental Health in Crisis* (Sage, 2019), and "Psychology at the End of the World: Mind and Behaviour in the Antarctic" (*The Psychologist*, 2011).

Antarctica's status as a prototypical extreme and unusual environment for human habitation has stimulated a good deal of scientific research predicated on its potential to serve as an analogue for off-world journeys and human settlement beyond the Earth,[1] given that the continent is "as close to Mars as we can get."[2] The first instance of Antarctica's potential to serve as a simulacrum for the behavioural and psychological environment of deep space can be traced back to the Soviet Union's use of its Vostok Station [↗ p. 862] in East Antarctica as a test bed for the Salyut space programme.

The anticipated psychological parallels between Antarctic habitation and space exploration are dependent upon the physical aspects of the Antarctic terrain, notably the length of the austral winter and the inhospitable environment. The human aspects of Antarctic life amongst those that winter on the continent are played out over the course of the season, which lasts approximately six months. The meaningfulness of comparisons with space missions flows from this lengthy period. In his paper "Polar Psychology: An Overview", published in *Environment and Behavior,* Peter Suedfeld outlined the key features of human inhabitation in Antarctica as follows: "Physical danger, a hostile climate, dependence on external supplies, isolation, enforced small-group togetherness, restricted mobility and social contact, and the disruption of normal recreational and professional activities."[3]

The enforced social isolation engenders further difficulties – how to deal with potential medical emergencies and equipment breakdowns, as well as isolation from routine family, social, and sexual relationships. To that end, in his account of Scott's doomed 1912 expedition the English polar explorer Apsley Cherry-Garrard wrote that "both sexually and socially the polar explorer must make up his mind to be starved."[4]

Since the Heroic Age, the physical as well as the psychological challenges for those who remain during the Antarctic winter have been known to be considerable, yet the problem of maladaptation was highlighted following the mental breakdown of an individual at one station and the interpersonal meltdown stemming from poor leadership at another. As a result, one of the more pragmatic research concerns has been to identify the people best suited to working in such extreme environments. It took some time, however, before psychological methods began to be properly deployed. Psychophysiological research initially grew following the International Geophysical Year (1957–1958) when the scale of Antarctic operations increased markedly – with more personnel working on the ice, appreciation of the human factors at work grew correspondingly.

To inform the decision-making process, in 1980–1981, as an adjunct to the psychological research they were conducting, the first International Biomedical Expedition to the Antarctic (IBEA)[5] undertook work in physiology, biochemistry, microbiology, immunology, and epidemiology. The expedition provided some of the first concrete scientific data that life on the frozen continent was stressful for humans, although there had, of course, been a wealth of personal and anecdotal testimony to this effect, dating back to the early explorers. Captain Scott, for example, had famously remarked "Great God! This is an awful place." The IBEA researchers used psychometric methods to assess intelligence and personality, as well as motivational questionnaires, interviews, and group discussions. They judged their subjects to be "not sufficiently skilled and experienced in human relationships to be reflective, expressive and supportive of each other."[6] From group discussions, morale was deemed to be "low and group cohesion poor".[7] One of the subjects departed early on psychological grounds because he felt homesick and depressed.

"Nine days later came the first summer visitors and replacements. Now again the atmosphere of the camp changed completely. It is hard to explain the feeling that comes over you when, after so many months of isolation, strangers suddenly appear. It is particularly difficult to fight down a feeling of annoyance when a 'foreigner' fails to show proper respect for some cherished trophy of the winter's toil. A new esprit de corps develops and forms a barrier between the veterans and the newcomers. To complicate things, the newcomers brought with them a variety of colds and allied ailments. For the first week most of them were laid low by their diseases and the effects of the altitude. When their germs struck the veterans. The colds we caught were worse by far than and I had ever seen before in my years in Antarctica."

Paul Allman Siple (Explorer, Geographer), "Man's First Winter at the South Pole", *National Geographic*, 1958

"You are sitting on a large white plate looking out to the edge and I would draw myself back up into the sky and into space and look down and think of myself as this atom on an ice cube in the middle of nowhere."

David Grann (Journalist), in "The White Darkness: A Solitary Journey across Antarctica", *The New Yorker*, 2018

Extensive attitude and personality tests have been employed by some nations to facilitate personnel selection. The United States has, in fact, used psychological evaluation since 1963. Other countries – France, Australia, and New Zealand – followed suit.

Overall psychological methods are considered to have enjoyed "reasonable success in avoiding misfits". Despite the reported success of these methods,[8] the British Antarctic Survey (BAS) continues to favour a formal application process and personal interview over both psychometric and psychiatric screening.

Psychological screening does not, however, eliminate all potential problems. The physical and psychological challenges in Antarctic life are by now well documented and the prevalence of psychiatric disorders at Antarctic research stations has been estimated to be around 5%[9] – mood and sleep disorders are the most oft-cited complaints. Psychological and physiological monitoring not only enables assessment of the nature of the people who are likely to be more resilient to the stresses and hardships endured but also helps us learn more about the processes of adaptation. A "winter-over syndrome" has been identified, which comprises psychosomatic symptoms, insomnia, depressed mood, irritability, reduced physical and cognitive tempo, social withdrawal, and fugue-like states – the latter sometimes referred to as the Antarctic stare (a 20-foot [6 m] stare in a 10-foot [3 m] room). Some of these have been linked to disrupted circadian rhythms, while others have been considered reasonable adaptations to an environment lacking in stimulation and demanding prolonged exposure to constant darkness.[10]

The existing research has concluded that three factors are critical to effective performance in the Antarctic – ability (competence), resistance to stress and emotional stability (mental health), and social compatibility (social skills).[11] Thorough psychological assessment, however, should not be limited to the use of quantitative methods. Beyond the measurement of personality, adaptation, social interaction, social support, leadership, and cognitive history, there remains the issue of identity – and this may be particularly pertinent to the comparisons made with deep space exploration. Just as the Antarctic landscape has appeared to some to offer freedom from "cultural moorings",[12] being viewed as "intact, complete and larger than my imagination could grasp", "sufficient unto itself, untainted by the inevitable tragedy of the human condition",[13] so, too, may long-duration space missions free prospective cosmic travellers from the shackles of national territorial bonds. Just as in our earthly life, in Space, culture and identity may prove to be of great importance.

"People who enter confinement simulations like the Human Exploration Research Analog at the Johnson Space Center in Houston can leave if the stress becomes too much. For Antarctic stations like Concordia that's not possible. 'If you want to leave, you're going to die […]. It adds a layer of complexity.'"

Thomas Page (Writer and Producer, CNN),
"To Antarctica and Beyond", *CNN International*, 2019

A researcher at Concordia Station (opposite page, below) is photographed during the six-month-long Antarctic night whilst an astronaut is captured Space-walking "outside" the International Space Station (ISS). Considering the extreme circumstances, both individuals are safely "connected" to their station, whether by means of a GPS tracking device linked to the Antarctic station or physically tethered to ISS. Due to the complexity of operating in both sites, researchers and astronauts are trained extensively prior to their departure. The fatigue – physical and psychological – that they sign up to can be traced in one of the final portraits of Henry Worsley, who attempted to cross Antarctica, unsupported, pushing himself "always a little further".[1]

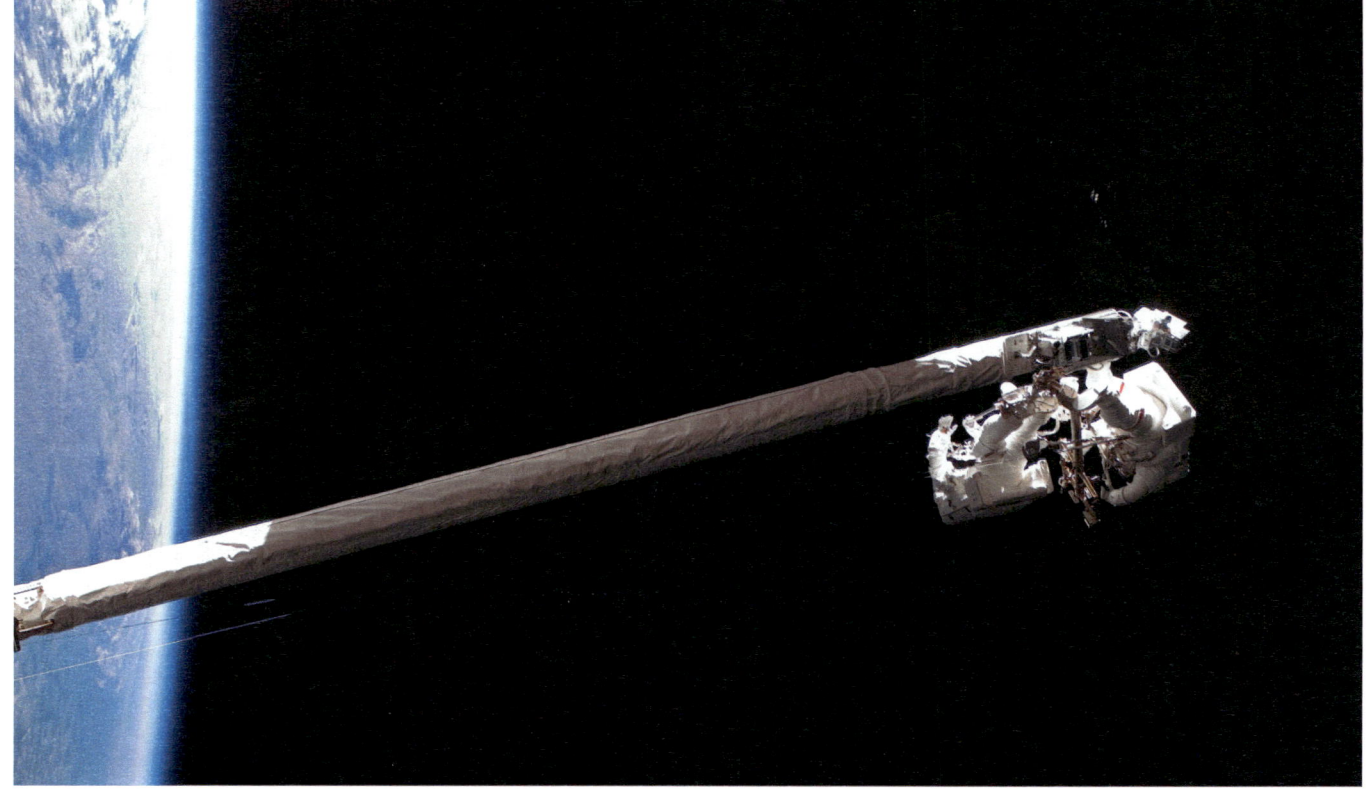

Humankind in the Extreme.
The Weakest Link

Didier Schmitt

DIDIER SCHMITT is the Strategy and Coordination group leader for Human and Robotic Exploration at the European Space Agency (ESA); he was a former member of the Space Task Force at the European External Action Service (EU diplomatic service) and foresight coordinator in the Bureau of European Policy Advisers. Dr Schmitt has authored several books, including *Terminus Antarctique* (2019) and a science fiction comic series called *Red Safari*, and is editor-in-chief of *Mook Mars* (Weyrich Editions).

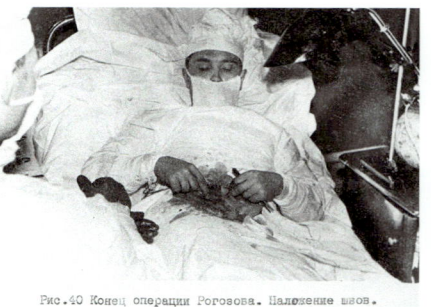

Рис. 40 Конец операции Рогозова. Наложение швов.

Antarctica is the largest stretch of inhospitable land on Earth. Among the thirty or so countries that have a *pied-à-terre* on the seventh continent, only about twenty have an uninterrupted presence and of these only three have been established at the very heart of the continent: the American Amundsen-Scott Station ↗ p. 670, 722, 790 at the South Pole, the Russian Vostok ↗ p. 862 and the Franco-Italian Concordia Station ↗ p. 780. Inaccessible by wide-body aircraft, this last is the most isolated and lies 1,100 kilometres inland, at an altitude of 3,200 metres above sea level.

The technical means and the know-how to survive in the most hostile environments are not available to everyone, whether it be Antarctica or its counterpart: Outer Space. To date, more than sixty countries have created a space agency, but only one in ten can send satellites independently. Only three of them were able to develop the ability to send men into space, and of these, only one reached the Moon. A permanent presence in Near Space – at the International Space Station – still remains a challenge that can be met only through a joint effort by the United States, Russia, Japan, Canada, and Europe.

Similar to astronauts in Outer Space, as soon as the members of crew step off the small Twin Otter plane on Dome C at Concordia Station, they notice the low pressure of the air and feel breathless in the oxygen-poor atmosphere – similar to altitudes of 3,800 metres, such as the Aiguille du Midi in the Alps. The high altitude, extreme cold, and dry air overwhelms the human brain as it attempts to adjust to the new environment. Day-night cycles are confusing, with 100 days without the sun between early May and early August, and 15 days of total darkness. Mobility outside the buildings is limited. The crew never goes further than three kilometres from the base in the winter months to carry out maintenance work. To venture outside, numerous layers of clothes must be donned, as at record temperatures below –80°C only a few seconds of exposure to the cold can cause frostbite and could jeopardise survival. The confinement is thus absolute: there is no way to push the "Mayday" distress button; for nine months of the year nobody will come to break the relentless isolation.

Despite all the hardships, up to sixteen European "hivernauts" volunteer to live in Concordia for around a year at a time in the name of science. A stay in the Franco-Italian station is the closest thing on Earth to living on the surface of another planet – it could be argued that the crew in Concordia is more isolated than the astronauts on the International Space Station. Without foreseeing it at its origin, this continental station has thus become a precious terrestrial "analogue" for preparing future manned Martian missions.

Contemporary ambitions of venturing up to 400 million kilometres away from Earth on a planet, Mars, whose CO_2-rich atmosphere is two hundred times thinner than ours and whose surface with an average temperature of –60°C is regularly battered by storms of ochre sand, would thus not be conceivable without applying the precious lessons learned at the Earth-based analogue par-excellence, Antarctica. Applicable at all scales, Antarctic prototypes – including inhabitable structures, clothing, scientific instrumentation, the ability to cope with remoteness, and logistic supply – are essential for extraterrestrial inhabitation.

Medical facilities are essential in Antarctica's hostile and remote environment. According to the Council of Managers of National Antarctic Programs (COMNAP) Antarctic Station Catalogue almost all stations have a dedicated medical area, albeit that in some cases this amounts to a surface as small as 6 square metres with only one member of staff with basic medical training. Learning from the infamous Antarctic health chronicle recording Russian doctor Leonid Rogozov's successful performance of an auto-appendectomy at Novolazarevskaya station in 1961 (shown above), and aware that the nearest hospital could be as far as 3,770 kilometres away and remain unreachable for entire months at a time, most National Programs running year-round stations ask their overwintering researchers to remove their appendix prior to deployment. Since then, telemedicine has revolutionised Antarctic healthcare, enabling physicians to provide treatment to scientists and support staff remotely. Doctors can now diagnose heart attacks, inspect a lesion, and provide psychiatric counselling via video-conference technology and specially designed medical instruments. Stethoscopes equipped with microphones enable doctors in remote environments to instantly hear a patient's heart or lungs, whilst special ophthalmoscopes allow them to see inside a patient's eye in real time. In addition to being able to monitor staff and scientists, these advances in technology also greatly assist the doctor on-site in performing complex procedures. In 2002, for example, doctors in Massachusetts aided a physician at Amundsen-Scott South Pole Station to surgically repair the damaged knee of an overwintering meteorologist. In preparation for Space deployment and missions to Mars, such remote systems are challenged even further in stations such as Belgrano II and Carlini, where the European Space Agency Astronaut Centre (ESA/EAC) is testing their latest Tempus Pro medical devices.

Behavioural Patterns in the Extreme. Concordia as Space Simulator

Beth Healey

BETH HEALEY was deployed by the European Space Agency (ESA) as a research medical doctor at space-flight analogue Concordia Station on the Antarctic plateau. Dr Healey's interests in extreme and remote environments have led her to be part of medical and logistical support teams for ski mountaineering expeditions and endurance races from Greenland to the North Pole.

It is an exciting period for the Space industry as the focus moves away from low Earth orbit, where the International Space Station (ISS) is currently orbiting, towards longer-duration, exploratory missions that travel deeper into space. When considering our capability for such missions, the greatest limiting factor is the sheer feat of engineering and cross-disciplinary collaboration needed to propel humans safely to and from Earth: the "rocket science". However, it's becoming increasingly apparent that our human capacity for these journeys could pose the greatest challenge of all. Mission success will depend on a crew's level of physiological and psychological resilience, allowing them to operate effectively in an isolated, extreme, and confined environments for a prolonged period of time. Crew habitat will play a vital role in determining crew resilience and providing the necessary regenerative life support systems to keep them alive.

In depth studies of behavioral patterns in extreme environments such as Antarctica have become increasingly important in the past decades. The trendline below reveals the number of papers published according to Web of Science since 1985 purely on psychological disorders in the southernmost continent.[1]

Antarctica is one of the most inaccessible environments on Earth and overwintering crews live at such extremes that they are often considered more isolated than astronauts on board the International Space Station. As such, Antarctic stations are regarded as high-fidelity analogue environments for long-duration space missions and some stations are used by space agencies as research platforms to help inform the likely physiological and psychological challenges future astronauts on long-duration missions may face, as well as to develop the medical models, tools, technologies, and countermeasures required to help facilitate successful adaptation to such an environment.

One example of such a Space analogue is Concordia Station ↗ p.780. Concordia is a European Space Agency (ESA) spaceflight analogue site, situated on the Dome Charlie (Dome C) plateau. It is one of only three inland Antarctic winter stations, with temperatures as low as –80°C and 105 days without sunlight. During the winter months, the crew members inhabiting Concordia are completely isolated even in case of emergency. Key stressors include the monotonous social and physical environment, confinement, physical and emotional deprivation, and limited privacy. In order to better understand the effects of these factors, a research

"Concordia station feels small in the summer when its over-crowded and very big in the winter when its almost empty. Designed to host more than eighty people throughout the summer months, in winter the station is quite alienating and it is frequent to get lost and have to message one's friends to set appointments in given areas. This is especially true for the canteen, which was also designed for maximum capacity without reflecting on the winter conditions when all you are left with is a tiny table within a hauntingly huge canteen. I found myself eating a piece of toast on the couch to feel somewhat normal."

Beth Healey (Doctor, European Space Agency), in conversation with UNLESS, Architectural Association School of Architecture, 2019

Number of research papers

1980-1981
During the International Biomedical Expedition to the Antarctic (IBEA) biographical variables, clinical ratings and psychological tests are used to predict performance of the eleven subjects involved in this first fully funded psychological analysis

1993
A collaborative Australian/United States research on immunology, microbiology, psychology and remote medicine is commenced throughout Australian National Antarctic Research Expeditions as outstanding analogues for space missions .

2000
The International Space Station is established creating the possibility of long duration manned missions to the outer reaches of Space.

2006
Medical, physiological and psychological research studies commence at the "Space Analogue" Concordia station.

Behavioural Patterns in the Extreme. Concordia as Space Simulator

576

programme has been running at the platform over a number of years to study the effects of what is a space simulator of sorts on the white continent.

Research protocols implemented at Concordia are often closely associated with the use of the habitat. For example, as part of one experiment the crew wore activity watches which, through the use of associated proximity data, were able to monitor crew activity levels, sleep-wake cycles, the location of crew members within the habitat (with particular regard to the proportional use of personal and social spaces), and crew cohesion. It is postulated that this data will provide an improved understanding of how crew cohesion and habitat use changed over time, facilitating our understanding of how to appropriately target countermeasures effectively aimed at time points when crew members are more likely to isolate themselves or where conflicts are more likely to occur. Research conducted at Concordia and similar platforms showed that there is variation in crew members' ability to grow or cope with the environmental and physiological stressors – their "psychological resilience" – on such missions. Some crew experienced a "winter-over syndrome", rendering them vulnerable to somatic and mental health illnesses, whilst others report personal growth and improvements in health, experiencing a feeling of accomplishment from successfully adapting to their new environment.

It has been hypothesised that some overwinterers achieve improved resilience

by entering an altered state of consciousness with pronounced absent-mindedness involving a "drifting" or wandering of attention. By reducing their interactions with other crew members and becoming more emotionally "flat", these crew members started to switch off mentally in a state of psychological hibernation.[1] While further research is required to confirm this theory, it is believed that there is an association between this mental state and reduced cognitive performance. This phenomenon has been noted as most pronounced during the third quarter of the mission, aligning with a period widely accepted as one of the more challenging for crew members. The "third quarter" is typically characterised by reduced psychological resilience, low mood and psychological change, increased crew tension and reduced motivation.

To enhance crew resilience, a number of architectural countermeasures have been developed, as exemplified by the British Halley VI Station ↗ p. 810, designed by Hugh Broughton Architects and opened in 2013. This base incorporated a number of environmental measures including large windows, specifically selected colour schemes, the incorporation of wood and natural materials and scents, and blue-enriched short-wavelength lights used to wake up the crew during the periods of prolonged darkness. Its design was based on research findings from previous studies on platforms like Concordia. It is also hypothesised that other stress-reducing techniques like yoga, meditation, and self-hypnosis could help reduce stress by aiding detachment and in turn

facilitating this "psychological hibernation" state; thus spaces encouraging these behaviours could be considered in the design of future habitats.

Recent experiments which see the implementation of sustainable life support systems in Antarctic stations – like Concordia's advanced grey-water recycling system or the German attempt with Eden ISS to use hydroponic facilities to grow fresh vegetables without the use of soil – make it clear that architecture has an integral role, both in these remote environments on Earth and on our journey deeper into Space, in facilitating the exploration and inhabitation of other planets.

"The first obvious psychological effect is that life in the outside world becomes totally insignificant, and with little to distract the mind other than cold and extreme discomfort, the mind becomes one's entire focus. […] All return from Antarctica changed – sometimes for the better but often for the worst, unable to take up their lives in normal society."

Catharine Hartley (Writer), *To the Poles Without a Beard,* 2004

16 Antarctica and Outer Space. Laboratories for a Future as Global Commons

SURVIVING IN THE CRYOSPHERE   577

Hydroponics. Greenhouses in the Extreme

Daniel Schubert

DANIEL SCHUBERT is deputy department leader of the System Analysis Space Segment at the German Aerospace Center, team leader of the EDEN Research Group, and responsible for the Space Habitation Plant Laboratory at the Institute of Space Systems. Dr Eng. Schubert led the EDEN-ISS project with 15 international partners and coordinated the deployment mission of the greenhouse system at the Neumayer III Antarctic research station.

Strong blizzards with wind speeds of up to 150 kilometres per hour, temperatures below –40°C, weeks spent in total darkness, and, most importantly, no escape capsule promising a quick return home as is the case on the International Space Station. These are just some of the many challenges faced by the EDEN ISS team during its space analogue mission at the German Neumayer III ↗ p. 796 in Antarctica, which is run by the Alfred Wegener Institute (AWI). The EU-funded research project led by the German Aerospace Centre (DLR) involves research into how plants could be cultivated in future habitats on the Moon and Mars, thus making a significant contribution towards the development of bio-regenerative life-support systems.

Acting as a prototyping laboratory for Outer Space, Antarctica became the site for experimentation with hydroponic systems that aimed not only at overcoming the challenges of growing plants during the dark winter months – in which the older practice of constructing traditional greenhouses in the continent was simply ineffective – but also represented a response to Article 3 of the Protocol for Environmental Protection, which states that "Activities in the Antarctic Treaty area shall be planned and conducted so as to avoid detrimental changes in the distribution, abundance or productivity of species or populations of species of fauna and flora; and further jeopardy to endangered or threatened species or populations of such species" (Section 2 IV and V). The prohibition of the importation of any living species to the continent (other than humans) made it necessary to research ways that would

allow plants to grow directly in Antarctica in the absence of soil and sun. EDEN ISS successfully achieved these two goals, contributing to both scientific development and the health of the teams stationed at Neumayer, who, like all other transient inhabitants of Antarctica, suffer from sensory deprivation and a reduced supply of fresh vitamins.

On the 3rd of January 2018, the South African research vessel S. A. Agulhas II rammed into the sea ice of the Atka Bay near Antarctica's Ekström Ice Shelf. On board were the two EDEN ISS greenhouse containers. The unloading of the ship was cause for celebration, as it marked the culmination of four years of intensive planning, hardware development, testing, and enhancements. This brought them to the cusp of the long-awaited Antarctic test mission. From the sea ice, a team of AWI engineers pulled the two greenhouse containers and a support container housing spare materials and consumable supplies over a distance of 24 kilometres to Neumayer III Station, which was to serve as the base for the upcoming isolation mission. The very next day, the AWI assembly team placed the two greenhouse containers on the preinstalled platform 400 metres south of the station. Together they form the EDEN ISS Mobile Test Facility (MTF). In the weeks that followed, the DLR EDEN team installed external stairways, cable ducts, thermal line, outside lighting, and the carbon dioxide storage and supply system. Power was supplied by a 7-centimetre-thick electrical cable running from Neumayer III Station to the greenhouse, which had been buried in the ice the previous summer.

Surprisingly, horticulture – something unimaginable on the sterile frozen territory – has been practised in Antarctica since the Heroic Age. The image above, taken from the 1911–1913 South Polar Times is proof of that. Banned in its traditional form since the entry into force of the Madrid Protocol in 1998 (which prohibits the introduction to the continent of non-native soils and species, as well as the use of penguin-guano manure) horticulture has moved from makeshift greenhouses to high-tech hydroponic structures, such as the "Future Exploration Greenhouse"[1] included in the high-cube container of EDEN-ISS, installed 400 metres south of Neumayer III Station, shown above.

The test facility went into operation on the 7th of February 2018, with the sowing of tomato, pepper, and cucumber plants. Radishes, kohlrabi, rocket salad, lettuce, leafy Asian vegetables, and herbs like parsley, basil, and chives followed over the next few days. The isolation phase began once the last summer crew left the station on the 18th of February 2018, leaving only the ten-strong overwintering team. From that point on, DLR researcher Paul Zabel ensured the smooth operation of the system on site.

On the 20th of May, the sun set for the last time and the polar night began for the German overwintering crew. The greenhouse continued to operate during this extreme phase. It became an endurance test, with the outside temperature regularly falling below -30°C. Both the greenhouse system and the overwintering crew had to contend with storms and darkness, but the occasional aurora borealis brought some light in the darkness.

The research greenhouse is packed with modern cultivation systems (controlled-environment agriculture, CEA). These allow independent and accelerated plant growth within the growing chamber of the Future Exploration Greenhouse (FEG).

The fully automated nutrient-supply system provides two precisely calibrated nutrient options – one for vegetative plants like lettuce, basil, and parsley, and the other for fruit-producing plants like tomatoes, cucumbers, and peppers. A mixing computer uses various micropumps to compile the exact amount of nutrients for the plants and adjusts the pH value of the nutrient solution. Eight high-pressure pumps then spray the fine nutrient mist into each plant vessel's lightproof root compartment. The roots hang freely in the air and are able to absorb the blend of nutrients and water directly. This soil-free irrigation method is referred to as aeroponics and is highly resource-efficient.

The internal atmosphere-management system regulates the temperature to 20°C and the humidity to 65% and actively sprays carbon dioxide (1,000 ppm) into the growing chamber. The plants need carbon dioxide for photosynthesis, supplemented by an artificial lighting rig consisting of specially developed LED systems.[1] The light spectrum of each lamp can be individually adjusted to the type of plant, with the wave-lengths blue, red, far-red, and white. The LEDs are water-cooled, creating thermal stability within the cultivation room. An innovative thermal control system ensures sufficient heat dissipation on the roof of the greenhouse. Air circulates around the greenhouse and is purified by various filter systems (preliminary, HEPA, and activated carbon filters). The water transpired from the plants is recovered by the system and fed back into the nutrient supply system, closing the water cycle. The only water that leaves the system is in the biomass of the harvested lettuce, herbs, and fruits.

The researchers regularly tested out different procedures for cultivating plants within closed systems, similar to habitats on the Moon and Mars. The working steps taking place in Antarctica were followed live from the mission control centre at the DLR Institute of Space Systems in Bremen, where the researchers were in constant contact with the greenhouse and had full remote control of its operations in the event of storms. Thirty-four cameras took pictures of the different plants every day, offering the project partners and specialists from all over the world the opportunity to observe the plants growing inside the Antarctic greenhouse and providing feedback to the mission team.

By mid-November 2018, around 270 kilograms of twenty kinds of vegetables had been harvested despite the harsh conditions. With almost 260 days of isolation, fifteen system and validation tests, over forty procedural tests, and more than twenty scientific experiments, the Antarctic analogue mission marked the highlight of the EDEN ISS greenhouse project. The experience acquired will be vital for a robust and reliably functioning greenhouse for plant cultivation in the Earth's desert regions, on the Moon, and on Mars.

16   Antarctica and Outer Space. Laboratories for a Future as Global Commons

SURVIVING IN THE CRYOSPHERE

579

Monitored closely from the Mission Control Room at the German Aerospace Centre (DLR) shown opposite, the vegetables growing at EDEN-ISS were planted carefully within the hydroponic system in which LED-lit crops were cultivated in the absence of soil by using mineral-rich nutrient solutions (top), testing a greenhouse concept developed for the Moon and Mars (centre). The new system developed for Space was conceived to replace existing facilities such as the Lada Validating Vegetable Production Unit[2] installed at the International Space Station (bottom). "Since 2002, the Lada greenhouse has been used to perform almost continuous plant growth experiments on the station. Fifteen modules containing root media, or root modules, have been launched to the station and 20 separate plant growth experiments have been performed. […] Root modules with seeds are launched to the space station on Russian Progress supply vehicles. Russian crew members water the plant seeds and perform maintenance. They also harvest the vegetables and place them in a station freezer before transferring them to a space shuttle freezer for return to Earth for analysis by U.S. investigators at the Space Dynamics Laboratory."[3]

"Future, long-term crewed space missions will require locally grown food. EDEN ISS has proven the feasibility of a space greenhouse in the Antarctic and thus demonstrated that this technology could also be used to produce food on the Moon and Mars […]. The space greenhouse concept now being presented is a valuable foundation on which we wish to develop further research work."

Hansjörg Dittus (Physicist, Space Research and Technology Executive), "EDEN ISS Project Presents Results of a New Greenhouse Concept for Future Space Missions", 2019 (online)

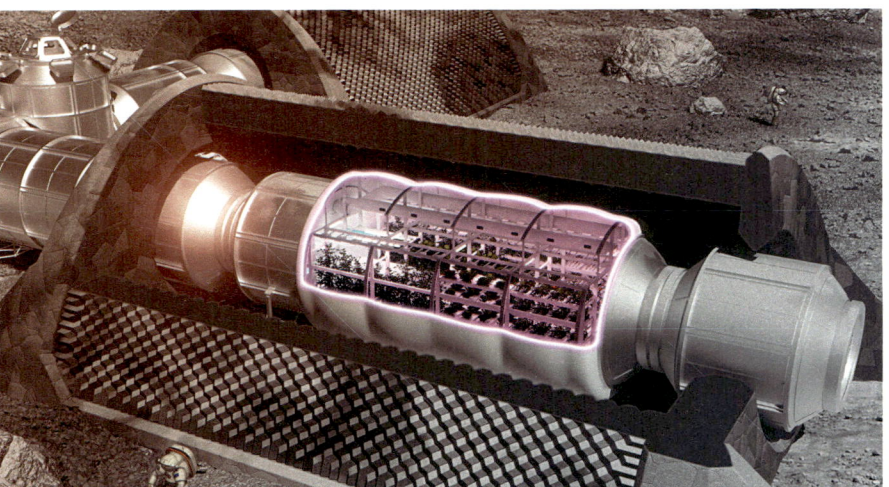

Inhabiting Space

Scott Parazynski

SCOTT PARAZYNSKI is a former NASA astronaut and a veteran of five space shuttle missions and seven spacewalks. Inducted into the US Astronaut Hall of Fame in 2016, he is the first to have both flown in space and summited Mount Everest. He was founding chief medical officer of the Center for Polar Medical Operations at UTMB-Galveston (United States), overseeing health and medical care for the US Antarctic Program. Dr Parazynski is currently the CEO of Fluidity Technologies (United States).

In an era of global pandemic, the concepts of quarantine and isolation have come into clear focus for every living soul on Earth. Yet for astronauts and polar explorers, they have been an integral part of the job for decades. Having had the privilege of being able to fly in Space on five occasions and travel broadly across the wilds of Antarctica, I note striking corollaries between these two extraordinary, extreme environments. I believe that the best training environment for future astronauts heading to the Moon or Mars would be a self-contained, remote outpost in Antarctica, where the explorers could learn to sustain their habitat and the cohesion of their crew. Conversely, our journeys into space teach us many lessons that can be applied to Antarctic exploration.

Any mission to space – and by extension to Antarctica – requires that a number of conditions be met before committing to launch, beginning with the health and fitness of the crew. Professional astronauts and cosmonauts are thoroughly screened for any preexisting condition that might jeopardise their ability to complete their mission assignments, from the rigors of launch to the physical stressors of extravehicular activity (EVA), also known as spacewalking. On Space Shuttle mission STS-95, for example, I had the good fortune to fly with then seventy-seven-year-old Mercury astronaut and later US Senator John Glenn; like all of his younger crewmates he, too, had to pass the flight physical. Moreover, since a crew occupies a small volume with a closed life support system for weeks to months, members are also required to quarantine several days before launch so as to not risk an entire crew succumbing to a flu, or even worse.

Robust spacecraft design, assembly, maintenance, and operation are next on the must-have list. Outside temperatures may vary by as much as 500°C in a single ninety-minute orbit of the Earth, as the International Space Station (ISS) alternates from orbital "day" (direct sunlight) to orbital "night" (in the shadow of Earth). Coupled with the vacuum of space and the presence of micrometeoroid orbital debris (MMOD, also known as "space junk"), it's remarkable that engineers can even design systems to reliably operate for years on end in such an inhospitable environment. That said, inevitably flight hardware and software can and will fail, so trying to design spacecraft systems with critical redundancies and the ability to repair them is paramount.

Training before a space mission takes on many forms, from the dramatic environment of the Neutral Buoyancy Laboratory (a giant swimming pool at the NASA Johnson Space Center, used to train spacewalking on full-scale mock-ups of the ISS) and virtual reality simulations, to more mundane computer-based lessons and spacecraft systems manuals. Astronauts and flight controllers have to absorb large amounts of information to understand not only how things are supposed to work but more importantly how they might fail. As I prepared for my first Space Shuttle flight on STS-66, we had an atmospheric science satellite called CRISTA-SPAS that we would be deploying for nine days of free flight, followed by capture with the robotic arm to return it and its vital science locked inside a small, mounted computer. It struck me that if, for some reason, we couldn't latch CRISTA-SPAS in our payload bay, priceless ozone-mapping data would be lost forever. I came up with a contingency plan to unbolt the computer by means of a spacewalk, but thankfully my forward thinking wasn't called into use. NASA always plans for success, but it also prepares for the inevitable failures that will certainly occur.

Crew composition and cross-training also requires a good deal of forethought. With a limited number of seats aboard any spacecraft and with astronauts from all over the world participating, each crew member has to serve in many different roles. Particularly for long-duration missions to the ISS, crew compatibility and cross-cultural understand-ing – not to mention some degree of fluency in a common language or two (with English and Russian the official languages aboard the ISS) – are vital to their success. Regarding cross-training, a physician like myself (for example) wouldn't be limited to the role of Crew Medical Officer based on my degree but would typically extend to such roles as flight engineer, spacewalker, roboticist, payload science operator, and in-flight maintenance lead. The mission Commander or Pilot, assigned essential piloting and leadership duties, might even be tasked with maintaining what is known euphemistically as the "waste containment system."

The most challenging days of my astronaut career, during Space Shuttle flight STS-120 to the ISS, evoke a number of important lessons for critical problem-solving under duress. Our crew and the team in Houston's Mission Control had just completed the third of five planned spacewalks, installing a solar array truss called P6 (the sixth truss element on the port side of the ISS) out at the very tip of the station. I recall elation and relief as Doug Wheelock and I floated back into the airlock at the end of that tiring EVA, thinking that the toughest part of our mission was complete and that we would be able to "coast" from that point forward. As the ISS crew and Mission Control Houston were commanding the solar panels we had just plugged in to unfurl, the nature of our mission changed character immediately. A snag in one of the arrays had developed, causing it

to rip apart in two places. Our two ISS space-craft Commanders Peggy Whitson and the Space Shuttle's Pam Melroy quickly stopped the process, and the full NASA team went into a fast-paced, problem-solving mode. Over the course of seventy-two hours, engineers and flight controllers across the country came up with a brilliant plan to get a space-walker outstretched to the site of the damage, to hopefully cut out the snag, and to install several repairs that would allow the solar panels to be fully extended.

Life in extreme conditions warrants a strong safety culture, but in exploration it's also important on occasion to be willing to accept higher risk in order to achieve greater rewards. For our STS-120 crew, this meant a complex interplay involving the spacewalkers (my colleague Doug Wheelock and I), a pair of robotic arm operators (Stephanie Wilson and Dan Tani), the Commander (Pam Melroy) guiding us through the procedures and moni-toring our every move, and our flight control team in Houston providing stellar oversight. Over the course of a seven-hour, nineteen-minute emergency repair spacewalk, the team safely removed the snag and installed repairs that would allow for us to safely undock our shuttle, and for ISS assembly to continue on subsequent missions. Rigorous preparation, engineering excellence, problem-solving creativity, teamwork, and resilience saved the day. Antarctica, past and future, has and will continue to benefit from these same attributes.

"Until recently, space programs were run by the government and focused on low-Earth orbits. […] After the official end of the NASA Space Shuttle Program in 2011, there has been a boom in the private sector to bring space exploration to civilians. Thanks to companies such as Virgin Galactic and Space X, civilians are able to go into orbit – for a small fortune. How-ever, […] the price will gradu-ally decrease, paving the way for affordable space travel in the future."

NASA EPDC, "The Future of Human Space Flight", 2019

International Antarctic Stations

Alan D. Hemmings

ALAN D. HEMMINGS is professor at the Gateway Antarctica Centre for Antarctic Studies and Research at the University of Canterbury (New Zealand). As a polar specialist, he focuses particularly on Antarctic geopolitics and governance. Dr Hemmings is the co-editor of *Handbook on the Politics of Antarctica* (Edward Elgar, 2017), *International Polar Law* (Edward Elgar, 2018), and *Philosophies of Polar Law* (Routledge, 2020).

The title of this essay is a misnomer – there are no "International Antarctic Stations". The actual focus of this contribution is on *why* this is so, given the supposed international nature of Antarctic science and the fact that we *do* have an International Space Station.[1]

Scientific research, and international cooperation around it, are core features of human presence in the Antarctic. This is seen in the grounding of the national presence in Antarctica;[2] the rationale for the 1959 Antarctic Treaty;[3] the repeated evocation of the value and role of science in that treaty,[4] in the subsequent instruments of the Antarctic Treaty System (ATS),[5] and in the wider "Continent for Science" framing.[6]

International collaboration in Antarctic research occurs across ships, aircraft, field camps, and stations. However, sixty years on from the adoption of the Antarctic Treaty, and almost thirty years after the commitment in the Madrid Protocol that "each Party shall endeavour to [...] undertake joint expeditions and share the use of stations and other facilities",[7] there is little evidence of internationalisation of Antarctic research stations.

Before the Antarctic Treaty only three joint stations had operated: Maudheim ↗ p. 648 (Norway, United Kingdom, and Sweden), 1949–1952; Hallett (United States and New Zealand), 1956–1964; Wilkes (United States and Australia), 1959–1961. None were established subsequently, until recently. Currently there are two: Concordia ↗ p. 780 (France and Italy) jointly developed, and jointly operated since 2005; and Law-Racovita-Negoita (Romania and Australia) – previously Australian only, but jointly operated from 2005 under a ten-year Memorandum of Understanding. To put the two joint stations in context, the Antarctic Station Catalogue of the Council of Managers of National Antarctic Programs currently itemises seventy-six operating Antarctic stations.[8]

Plainly, "sharing" can mean many things short of a *joint* facility, and there are various arrangements around visiting scientists and joint projects operating out of sole nation stations. However, given the seeming scientific, logistic, economic, and environmental benefits of reducing duplication, why have more joint stations not appeared, let alone any truly *international* stations? There are complexities around the integration of national research priorities, administrative processes, financial apportionment, and the determination of where legal responsibilities reside under various contingencies (and, in the case of the last, in a region already presenting legal challenges). But such complexities have been resolvable in the seemingly equally complex case of the International Space Station.

The particular difficulty in Antarctica appears to be the issue of territorial sovereignty. Whilst nobody claims Outer Space, most of the Antarctic is claimed by somebody, and the Peninsula by three states. Almost anywhere one wishes to place a station is considered by somebody to be "theirs". Non-recognition of claims, and Article IV of the Antarctic Treaty, enables states to locate stations where they will. Larger states do not need partners and, to the extent that their own Antarctic presence reflects longer-term geopolitical interests, may be disinclined to complicate things through joint facilities – which they may very well have to lead anyway. In addition, part of the rationale for the placement of stations by the United States, Russia, and China has been to countervail the territorial claimants, and, more generally, to exercise decisive influence by being everywhere in Antarctica. For them neither necessity nor strategic interests are advanced today by station sharing.

Claimants focus their attention (and their facilities) on "their" claimed sector precisely in order to reinforce their position. Internationalising an existing

> "Unlike the early explorers who devoted their efforts unilaterally and often competitively to geographic exploration of the continent, the modern scientist explorer in Antarctica understands the rewards that can be gained from cooperation and mutual assistance. It takes the financial resources, the trained manpower, the equipment, and the know-how of many nations working together to cover the Antarctic region with the network of observation points necessary to make certain data meaningful."
>
> R. A. Bennett (Lieutenant), in United States Navy *Operation Deep Freeze '65*, 1965

station, or creating a new joint station, may appear a zero-sum game. If one looks back at the functioning of the Antarctic Treaty System over the past sixty years, one sees a general tension amongst claimants between their support for that system and their concern that too great an internationalisation of Antarctic affairs risks a diminution of their autonomy and/or claimant status. In this view, a transition from an Antarctic world of National Antarctic Programs to a more integrated transnational science, with a corresponding multilateral infrastructure, might be the harbinger of a new dispensation less amenable to national territorial aspirations. Having visitors is an altogether different proposition to adding their names to the deeds of the property.

However, five of the seven claimants, Australia, France, New Zealand, Norway, United Kingdom, and the United States, *have* participated in joint facilities. Does this not undercut the foregoing argument? The detail suggests otherwise. The post-war Norwegian-British-Swedish Research Expedition involved two claimants who had coordinated their claim boundaries and a third non-claimant state (Sweden). Maudheim was located within the Norwegian claim. With the pre–Antarctic Treaty stations, the United States was in part looking to counter claimant pretensions. The claimants were between a rock and a hard place. The Americans were going to build these stations in "their" territories anyway. Better "joint" than United States alone. In the case of Law-Racovita-Negoita, Australia's ongoing engagement alongside Romania eases the problems. Australia retains ownership, control, and management; the station is only used during the summer; and its strategic importance as a balance to nearby Russian, Chinese, and Indian stations may now exceed its scientific value. Law-Racovita-Negoita is, on these terms, more like a national station hosting facilities and/or personnel from another state than a joint station. The Franco-Italian Concordia is certainly a joint facility. For claimant France the zero-sum problem does not arise because the station is not in the French claim – it is, in fact, in area claimed by Australia.

Are there scenarios where one might see the appearance of truly jointly operated Antarctic facilities? Over the next five to ten years, this might occur through three routes. The first is in relation to the establishment of a large astronomical facility, as these are often multinational projects everywhere in the world and Antarctica presents even greater costs. The second would be via a major European Union initiative in Antarctica which involved a number of member states; and, finally, in a post-COVID-19 Antarctic research reality, where economic rationalisation might be essential for the maintenance of Antarctic activity for at least some states.[9]

Beyond these scenarios, whilst other forms of scientific cooperation will continue, unless and until some resolution of the territorial sovereignty issues is found – and none is presently on the horizon – difficulties in establishing joint or international Antarctic stations will persist.

"However, the world of the 21st century is driven largely by commercial interests – whether directly by those interests or mediated through governments. National prestige is not manifesting itself in the same forms as it used to, and scientific endeavours are rarely undertaken without there being some good commercial (or military) reason behind them. Consequently, are commercial interests one of the factors driving some station proposals? If some of the motivation for new station construction stems primarily from reasons such as enhanced national prestige or the desire for increased influence within the [Antarctic Treaty System] ATS, it would be reasonable to question the necessity for such proposals. In addition, the notion that station proposals can be followed through without too much difficulty so long as science is used as a justification also warrants questioning."

Antarctic Southern Ocean Coalition in their report *Are More Antarctic Stations Justified?*, presented at the XXVII ACTM, March 2004

The Antarctic and Outer Space. From *Res Nullius* to *Res Communis*

Christy Collis

CHRISTY COLLIS is the associate director of the Office for Learning and Teaching at the University of Southern Queensland (Australia). She is a cultural geographer whose research focuses on the legal and cultural dynamics of territorial possession, specifically applied to Antarctica, outer space, and the deep seabed. Her writings include "Territories Beyond Possession? Antarctica and Outer Space" (*The Polar Journal*, 2017).

Three unique environments, together comprising 75% of the surface of the Earth and the planet's entire cosmic environment, have a unique, common significance for humanity. The first of these spaces, analysed in depth in this book, is Antarctica, a continent under ice. Accounting for approximately 10% of the Earth's surface, the Antarctic conceals minerals deep in its subglacial craton. Like Antarctica, the second space covers a vast portion of the Earth's surface and is rarely visited by humans. Accounting for 64% of the Earth's surface area, dark and pressurised and cold: the deep seabed is the ground subjacent to the international waters of the High Seas. The third – but no less important, comprising all known and unknown space beyond the ceiling of Earth's atmosphere, and including all celestial bodies as well as the vacuum between them – is Outer Space.

These three spaces share clear similarities with one another: they are all enormous, difficult for humans to access, and have no indigenous human populations. More interestingly, alongside differences in physical geographies, these three vast spaces also share legal and geopolitical geographies. Whether ice, compressed mud, or asteroidal rock, all these spaces belong to no one.

If, as Elden notes, "territory" refers to possessed space, Antarctica, Outer Space, and the deep seabed are spaces beyond territory.[1] So, if they are not possessions – not territory – what kind of spaces are they? They are *res nullius* (from Latin, nobody's thing). *Res nullius* refers to spaces that have not been transformed into a possession, into territories. Once a space has been defined as *res nullius,* there are three potential ways in which that space might be legally and geopolitically transformed. First, under the international law of territorial acquisition and possession, *res nullius* can be transformed into a state possession through a series of codified acts and documents. ("The discovery of *terra nullius* is the means through which European empires amassed substantial portions of the Earth). Second, *res nullius* can remain *res nullius,* i.e. it can remain un-owned, suspended in a balance between possession and non-possession. Finally, *res nullius* can be transformed into *res communis;* in this radical change, a space owned by no one becomes a space collectively owned by all of humanity. These legal terms have substantial histories, as well as operational specificities;[2] what is important here is that all three versions of *res nullius* posit spaces as un-possessed.

The story of these non-territorial spaces begins in Antarctica. Inspired by the international conviviality of the 1957–1958 International Geophysical Year (IGY) and its mitigation of Cold War tensions, the Antarctic Treaty of 1959 initiated *res nullius,* albeit a compromised version.[3] Under the Treaty, the seven states which had previously claimed slices of Antarctica as their own territory – as their sovereign state possessions – were allowed to maintain their claims, while other states opted to view and treat Antarctica as a *res nullius.* Antarctica thus embodies a hybrid version of *res nullius,* a palimpsest of legal geographies, and assumes a centrality within the geographies of non-territory.

The IGY was pivotal in triggering the redefinition not only of Antarctica but also of Outer Space. When Sputnik overflew Earth's state territories in 1957 and no state complained that its state airspace had been invaded, Outer Space began to be conceived as a new type of legal geography, setting a precedent of non-possession that was later ratified by the Outer Space Treaty (OST) of 1967. Deriving from the same non-territorial impetus as that which informed Antarctic space, the OST defines Outer Space as entirely *res nullius,* a space in which no state can claim any territory at all. Unlike the Antarctic Treaty, the OST does not compromise: it is not a hybrid *res nullius.* Although the more recent partitioning

"The question is often asked, 'What is the value of Antarctica?'. The Antarctic Continent today has no commercial value. The surrounding oceans offer a lucrative trade in the whaling industry. [...] Large deposits of coal, of a poor grade, have been found: also traces of gold, silver, tine, copper, manganese, lead, and other minerals – but of no economic value. Uranium bearing minerals have never been recorded ... It is conceivable that oil will be found in the Palmer Peninsula. Yet we must remember that before the IGY less than two tenths of one per cent of Antarctica had been surveyed geologically. [...] Antarctica is one of three areas that mankind must concentrate upon in the future. The other two are the ocean depths, and outer space. These three areas have one thing in common. None has any indigenous peoples. Man, in order to survive, must live peacefully in these areas. Perhaps in the Antarctic we can set the stage for international harmony."

George Dufek (Admiral), preface in Emil Schulthess's *Antarctica*, 1960

of the geostationary orbit into segments, where each segment is assigned to a state, represents a small diminution of Outer Space,[4] its *res nullius* remains almost entirely intact and it does not, and cannot, become a state possession.

The deep seabed represents a third, more radical, version of *res nullius.* Beyond the coastal Exclusive Economic Zones and the continental shelves off state territories – both of which are territorial possessions of the respective states – the deep seabeds lying under the High Seas are to remain beyond the domain of state territorial claims. Although the deep seabeds subtend the High Seas, they reflect a different type of legal geography than the water column above them. This geography has its genesis in the spatial visions of the Antarctic Treaty, and in the Treaty's creation of non-territorial spaces. In brief, under the terms of the 1994 United Nations' Convention on the Laws of the Sea, the International Seabed Authority (ISA) manages the deep seabeds on behalf of the United Nations and, by extension, on behalf of humanity. Any state or state-based entity wishing to extract commercial resources from the deep seabed must first submit a hugely detailed and costly application to the ISA, and in turn the ISA ensures that the profits of resource extraction are shared with developing states.[5] No one, neither a state nor a private company, can possess any portion of the deep seabeds because these are defined as the common property of humanity: they are *res humanitas,* global commons. Whether *res humanitas* will trigger the "tragedy of the commons" or a sustainable means of redefining the Earth's surface beyond state territorialism remains to be seen.

However, questioning what might happen to these geographies of non-, or compromised, state possession in the future is important in understanding the origins and implications of their common legal geographies and the underlying Roman concept of *res nullius.*

"The Antarctic is a world of suspended animation. Suspended between outer space and the fertile continents. Suspended in time – without a local civilization to make history. Civilization has been brought to it; it has never sustained any of its own. It sits suspended in a hanging nest of world politics. When things die in the Antarctic, they decay slowly. What has been is still there and will always be, unless we interfere. *Interfere* is such a simple world for what is happening."

Diane Ackerman (Poet, Essayist, and Naturalist), *The Moon by Whale Light*, 1991

The Antarctic through the Lens of Joan Myers

Photographs taken between 2001 and 2004, including as a grantee of the US National Science Foundation (NSF) Antarctic Artists and Writers Program in 2002

"After a five-and-a-half-hour flight, we land smoothly on an ice runway. The doors open, and light floods in. I put on my goggles and step out into more space and sky than I have ever seen."

Joan Myers (Photographer), *Wondrous Cold: An Antarctic Journey*, 2006

OF
C
TURE

It is hard to say with exactitude how many structures have been erected in Antarctica since our first landing on the continent in 1821, exactly 200 years ago, but it is a fact that there are too many. While advocating for the day in which nations will cease to overtly assert territorial claims through architecture, the Archive presents for the first time a selection of unedited drawings of experimental buildings constructed below the 60th parallel south. Conceived as an open-access platform, the Archive aspires to act as a tool that enables architects and engineers practising in the extreme southernmost territory to optimise their designs, develop solutions that reduce the exorbitant amount of energy required each year to construct and operate Antarctic buildings, and ultimately lessen our contaminating footprint on the continent.

ARCHIVE OF ANTARCTIC ARCHITECTURE INDEX

Year
STATION NAME
Country
Status (Active Station, Emergency
Shelter, Preserved/HSM, Closed,
Buried, Calved)

1899
CAPE ADARE HUT ↗ p. 614
United Kingdom
Preserved, HSM 22

1902
HUT POINT HUT ↗ p. 618
United Kingdom
Preserved, HSM 18

1902
SNOW HILL HUT
Sweden
Preserved, HSM 38

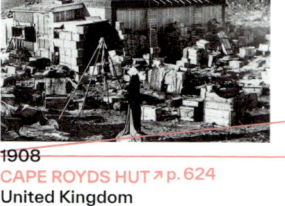

1908
CAPE ROYDS HUT ↗ p. 624
United Kingdom
Preserved, HSM 15

1911
FRAMHEIM HUT ↗ p. 628
Norway
Calved

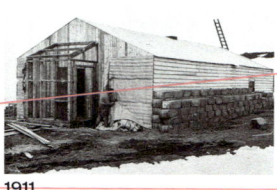

1911
CAPE EVANS HUT ↗ p. 632
United Kingdom
Preserved, HSM 16

1911
SOUTH POLE TENT
Norway
Buried / Preserved, HSM 80

1940
LITTLE AMERICA III (III/V)
United States of America
Calved

1941
EAST BASE
United States of America
Preserved, HSM 55

1944
BASE A ↗ p. 644
United Kingdom
Preserved, HSM 61

1944
BASE B ↗ p. 702
United Kingdom
Closed

1946
BASE E ↗ p. 684
United Kingdom
Replaced > Preserved, HSM 64

1947
CAPITAN ARTURO PRAT
Chile
Active Station

1947
MELCHIOR
Argentina
Active Station

1947
BASE H > SIGNY (1977)
United Kingdom
Active Station

1949
MAUDHEIM ↗ p. 648
Norway, Sweden &
United Kingdom
Calved

1950
PORT MARTIN
France
Preserved, HSM 46

1950
PARADISE HARBOUR SHELTER
Chile
Preserved, HSM 30

1951
ALMIRANTE BROWN
Argentina
Active Station

1903
HOPE BAY HUT
Sweden
Preserved, HSM 39

1903
OMOND HOUSE > BASE ORCADAS
Scotland > Argentina (1904)
Preserved, HSM 42 + Active Station

1903
PAULET ISLAND HUT
Sweden
Preserved, HSM 41

1906
AKTIESELSKABET HEKTOR
WHALING STATION ↗ p. 622
Norway
Preserved, HSM 71

1912
CAPE DENISON HUTS ↗ p. 636
Australia
Preserved, HSM 77

1914
THE *ENDURANCE* ↗ p. 640
United Kingdom
Sunk / Preserved, HSM 93

1929
LITTLE AMERICA I (I/V)
United States of America
Calved

1934
LITTLE AMERICA II (II/V)
United States of America
Calved

1945
BASE C1
United Kingdom
Closed

1945
BASE D > TENIENTE RUPERTO
ELICHIRIBEHETY
United Kingdom > Uruguay (1997)
Active Station

1946
LITTLE AMERICA IV (IV/V)
United States of America
Calved

1946
BASE C2
United Kingdom
Closed

1947
BASE F > VERNADSKY
United Kingdom > Ukraine (1996)
Preserved, HSM 62 +
Active Station

1947
BASE G
United Kingdom
Removed

1948
BASE DECEPCIÓN
Argentina
Active Station

1948
GENERAL BERNARDO
O'HIGGINS
Chile
Preserved, HSM 37 +
Active Station

1951
GABRIEL GONZÁLEZ VIDELA
Chile
Active Station

1951
SAN MARTIN
Argentina
Active Station

1953
DESTACAMENTO NAVAL BAHÍA
LUNA > TENIENTE CÁMARA
Argentina
Active Station

1953
ESPERANZA ↗ p. 734
Argentina
Active Station

1953
REFUGIO POTTER > CARLINI
Argentina
Active Station

1953
BASE V > GENERAL RAMON
CANAS MONTALVA
United Kingdom > Chile (1996)
Emergency Shelter

1954
MAWSON ↗ p. 652
Australia
Active Station

1954
LUIS RISOPATRÓN
Chile
Active Station

1956
DUMONT D'URVILLE
France
Active Station

1956
LITTLE AMERICA V (V/V)
↗ p. 654
United States of America
Calved

1956
HALLETT
United States of America
& New Zealand
Removed

1956
HALLEY I (I/VI) ↗ p. 658
United Kingdom
Calved > Replaced

1956
BASE O
United Kingdom
Removed

1956
BASE W
United Kingdom
Preserved, HSM 83

1957
AMUNDSEN-SCOTT
SOUTH POLE (I/III) ↗ p. 670
United States of America
Buried > Replaced

1957
BYRD (I/II)
United States of America
Buried > Replaced

1957
ROI BAUDOUIN
Belgium
Calved

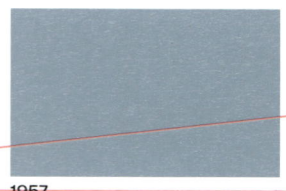

1957
KOMSOMOLSKAYA
Soviet Union > Russia
Active Support Structure

1957
SCOTT BASE ↗ p. 678
New Zealand
Preserved, HSM 75 + Active Station >
Project 2020+

1957
SYOWA ↗ p. 770
Japan
Active Station

1958
LITTLE ROCKFORD
United States of America
Closed

1958
POLE OF INACCESSIBILITY
↗ p. 682
Soviet Union > Russia
Preserved, HSM 4

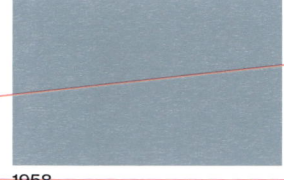

1958
SOVETSKAYA
Soviet Union
Closed

1959
ARRIVAL HEIGHTS LABORATORY
New Zealand
Active Support Structure

1955
GENERAL BELGRANO I (I/III)
Argentina
Calved > Replaced

1955
BASE N
United Kingdom
Removed

1955
BASE Y
United Kingdom
Preserved, HSM 63

1955
PRESIDENTE PEDRO AGUIRRE
CERDA
Chile
Preserved, HSM 76

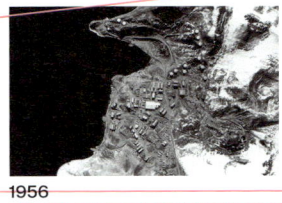

1956
NAVAL AIR FACILITY MCMURDO
> MCMURDO ↗ p. 666
United States of America
Active Station

1956
MIRNY ↗ p. 668
Soviet Union > Russia
Active Station

1956
PIONÉRSKAYA
Soviet Union
Closed

1956
OAZIS > A.B. DOBROWOLSKI
POLAR
Soviet Union > Poland (1959)
Preserved, HSM 10 + Closed

1957
CHARCOT ↗ p. 674
France
Closed

1957
DAVIS
Australia
Active Station

1957
ELLSWORTH
United States of America >
Argentina (1959)
Calved

1957
NORWAY
Norway > South Africa (1958)
Closed

1957
SOUTH ICE
United Kingdom
Closed

1957
BASE J
United Kingdom
Removed

1957
VOSTOK (I/II)
Soviet Union > Russia
Buried > Project 2020+

1957
WILKES
United States of America >
Australia (1961)
Buried

1959
LÁZAREV
Soviet Union
Closed

1960
DRUZHBA
Soviet Union
Closed

1960
POBEDA
Soviet Union
Closed

1960
SANAE I (I/IV)
South Africa
Closed > Replaced

ARCHIVE OF ANTARCTIC ARCHITECTURE INDEX

1961
BASE KG > FOSSIL BLUFF
United Kingdom
Active Support Structure

1961
BASE CONJUNTA TENIENTE
MATIENZO
Argentina
Active Station

1961
BASE T > TENIENTE CARVAJAL
United Kingdom > Argentina (1984)
Active Station

1961
NOVOLAZAREVSKAYA
Soviet Union > Russia
Active Station

1965
BROCKTON
United States of America
Closed

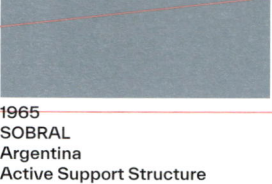

1965
PLATEAU ↗ p. 698
United States of America
Closed

1965
SOBRAL
Argentina
Active Support Structure

1965
OLD PALMER (I/II)
United States of America
Closed > Replaced

1964
CASEY REPSTAT (I/II) ↗ p. 712
Australia
Removed > Replaced

1969
CENTRO METEOROLÓGICO
EDUARDO FREI > BASE PRESIDENTE
EDUARDO FREI MONTALVA ↗ p. 746
Chile
Active Station

1969
MARAMBIO
Argentina
Active Station

1969
VANDA
New Zealand
Active Support Structure

1975
DAMOY POINT HUT
United Kingdom
Preserved, HSM 84

1973
SIPLE
United States of America
Closed

1973
HALLEY III (III/VI) ↗ p. 716
United Kingdom
Calved > Replaced

1975
AMUNDSEN-SCOTT
SOUTH POLE (II/III) ↗ p. 722
United States of America
Removed > Replaced

1977
HENRYK ARCTOWSKI (I/II)
Poland
Active Station > Project 2020+

1977
PRIMAVERA
Argentina
Active Station

1978
SALYUT
Soviet Union
Closed

1979
LILLIE MARLEEN HUT
Germany
Preserved, HSM 80

1962
MOLODYOZHNAYA
Soviet Union > Russia
Active Station

1962
YELCHO
Chile
Active Station

1962
NEW BYRD (II/II) ↗ p. 690
United States of America
Buried

1963
EIGHTS ↗ p. 694
United States of America
Closed

1967
HALLEY II (II/VI) ↗ p. 658
United Kingdom
Calved > Replaced

1968
BELLINGSHAUSEN
Soviet Union > Russia
Active Station

1968
PALMER (II/II)
United States of America
Active Station

1969
BORGA
South Africa
Closed

1970
MIZUHO
Japan
Closed

1971
LENINGRADSKAYA
Soviet Union
Closed

1971
SANAE II (II/IV)
South Africa
Closed > Replaced

1971
SODRÚZHESTVO
Soviet Union
Closed

1975
DRUZHNAYA I (I/IV)
Soviet Union
Calved

1975
BASE R > ROTHERA ↗ p. 726
United Kingdom
Active Station > Project 2020+

1976
GEORG FORSTER ↗ p. 730
East Germany > Germany
Removed

1976
GIACOMO BOVE
Italy
Closed

1979
GENERAL BELGRANO II (II/III)
Argentina
Active Station

1979
SANAE III (III/IV)
South Africa
Removed

1980
GENERAL BELGRANO III (III/III)
Argentina
Buried

1980
RUSSKAYA
Soviet Union > Russia
Active Station

ARCHIVE OF ANTARCTIC ARCHITECTURE INDEX

1981
GEORG VON NEUMAYER
↗ p. 736
Germany
Closed > Replaced

1982
DRUZHNAYA II (II/IV)
Soviet Union
Closed

1982
DRUZHNAYA III (III/IV)
Soviet Union
Closed

1982
FILCHNER
Germany
Calved

1983
HALLEY IV (IV/VI) ↗ p. 740
United Kingdom
Removed > Replaced

1984
ARTIGAS
Uruguay
Active Station

1984
COMANDANTE FERRAZ (I/II)
Brazil
Removed / Replaced

1984
DAKSHIN GANGOTRI
India
Active Support Structure

1986
LAW > LAW-RACOVIȚĂ
Australia > Romania (2006)
Active Station

1986
MARIO ZUCCHELLI ↗ p. 752
Italy
Active Station

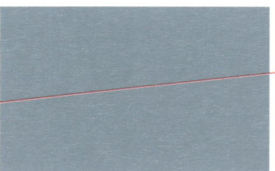

1987
DRUZHNAYA IV (IV/IV)
Soviet Union
Closed

1987
WORLD PARK
Greenpeace
Removed

1988
PROGRESS
Soviet Union > Russia
Active Station

1988
ST. KLIMENT OHRIDSKI (I/II)
Bulgaria
Active Station > Project 2020+

1988
SVEA
Sweden
Active Station

1989
ECO-NELSON
Czechoslovakia > Czech Republic
Active Station

1989
ZHONGSHAN ↗ p. 760
People's Republic of China
Active Station

1990
PEDRO VICENTE MALDONADO
Ecuador
Active Station

1990
TROLL
Norway
Active Station

1991
GARS LABORATORY AT
O'HIGGINS
Germany
Active Station

1982
JULIO RIPAMONTI
Chile
Active Station

1982
SARIE MARAIS
South Africa
Removed

1982
SOYUZ
Soviet Union
Closed

1983
GONDWANA
Germany
Active Support Structure

1985
ASUKA
Japan
Closed

1985
GREAT WALL
People's Republic of China
Active Station

1985
CAPTAIN PIETER J. LENIE BASE
United States of America
Active Support Structure

1986
DRESHER
Germany
Closed

1988
ABOA
Finland
Active Station

1988
JUAN CARLOS I (I/II)
Spain
Removed > Replaced

1988
CASEY (II/II) ↗ p. 756
Australia
Active Station

1988
KING SEJONG
South Korea
Active Station

1989
GABRIEL DE CASTILLA
Spain
Active Station

1989
MACHU PICCHU
Peru
Active Station

1989
MAITRI
India
Active Station

1989
WASA
Sweden
Active Station

1991
DR GUILLERMO MANN
Chile
Active Station

1991
JINNAH
Pakistan
Active Station

1991
SHIRREFF
United States of America
Active Support Structure

1992
HALLEY V (V/VI) ↗ p. 762
United Kingdom
Removed > Replaced

1992
NEUMAYER II (II/III) ↗ p. 768
Germany
Removed > Replaced

1992
ICE STATION WEDDELL N°1
United States of America & Russia
Removed

1993
TOR
Norway
Active Station

1994
DALLMANN LABORATORY
Germany
Active Station

1999
TENIENTE ARTURO PARODI
Chile
Removed

2001
KOHNEN
Germany
Active Station

2005
CONCORDIA ↗ p. 2005
France & Italy
Active Station

2006
COLLINS
Chile
Active Support Structure

2009
KUNLUN
People's Republic of China
Active Station

2009
NEUMAYER III (III/III) ↗ p. 796
Germany
Active Station

2012
BHARATI ↗ p. 804
India
Active Station

2013
DIRCK GERRITSZ LABORATORY
Netherlands
Active Station

2018
JUAN CARLOS I (II/II) ↗ p. 832
Spain
Active Station

2020
COMANDANTE FERRAZ (II/II)
↗ p. 838
Brazil
Active Station

2020+
ST. KLIMENT OHRIDSKI (II/II)
↗ p. 844
Bulgaria
Active Station > Project 2020+

2020+
HENRYK ARCTOWSKI (II/II)
↗ p. 848
Poland
Active Station > Project 2020+

2020+
MCMURDO (II/II) ↗ p. 866
United States of America
Active Station > Project 2020+

2020+
DOME FUJI ↗ p. 872
Japan
Active Support Structure >
Project 2020+

1994
PROFESSOR JULIO ESCUDERO
Chile
Active Station

1995
BEAVER LAKE
Australia
Active Support Structure

1995
DOME FUJI
Japan
Active Support Structure

1997
SANAE IV (IV/IV) ↗ p. 776
South Africa
Active Station

2007
JOHANN GREGOR MENDEL
Czech Republic
Active Station

2007
PRINCESS ELISABETH ↗ p. 786
Belgium
Active Station

2007
VECHERNYAYA
Republic of Belarus
Active Station

2008
AMUNDSEN-SCOTT
SOUTH POLE (III/III) ↗ p. 790
United States of America
Active Station

2013
HALLEY VI (VI/VI) ↗ p. 810
United Kingdom
Active Station

2014
JANG BOGO ↗ p. 820
South Korea
Active Station

2014
TAISHAN ↗ p. 824
People's Republic of China
Active Station

2014
UNION GLACIER FIELD CAMP
↗ p. 828
Chile
Active Field Camp

2020+
SCOTT BASE (II/II) ↗ p. 852
New Zealand
Active Station > Project 2020+

2020+
ROTHERA ↗ p. 856
United Kingdom
Active Station > Project 2020+

2020+
ROSS SEA STATION ↗ p. 860
People's Republic of China
Project 2020+

2020+
VOSTOK (II/II) ↗ p. 862
Russia
Active Station > Project 2020+

STATION NAME

↖ expanded upon on page XXX

Year of construction/YEAR of reconstruction

architecture image

"Antarctica is ground, not figure – it is nothingness, and nothingness cannot, by definition, be depicted. Any attempt to do so, to describe the continent as something, or even like something, is then interpreted as a sullying of its purity. […] If it is not the continent's blankness that prevents its depiction, it is its extremity."

Elizabeth Lean (Author), *Antarctica in Fiction: Imaginative Narratives of the Far South*, 2012

Designers
Engineers
Consultants
Contractors
Manufacturers
Site supervision

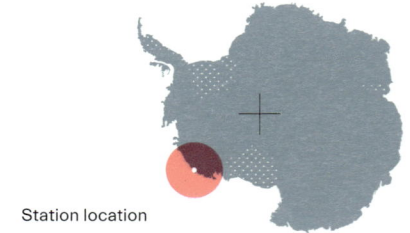

Station location

Country	Status	Geographic coordinates
Expedition / Mananging National Programme		Location
		Ground type, altitude, average yearly temperature

DRAWING TYPE | BUILDING NAME, YEAR

Ice-sheet plan view

Ice-shelf plan view (for above-surface architectures)

Ice-sheet and ice-shelf plan view (for below-surface architectures)

Ice-sheet and ice-shelf section view (for below-surface architectures)

Ice-free ground

Lake

Sea / Ocean

Topography lines (every 10 metres)

Master plan sectors and antenna fields

Infrastructure

Roads, vehicle tracks, and ropes

▲ Station access | at level

✕ Station access | from above

Please note that legend texts
reflect the wording of the
original archival documents.

01 Main station building
02 Meteorological
   observatory
03 Laboratory
04 Cold room
05 Aggregator,
   communication centre
06 Storage
07 Garage

08 New main station building

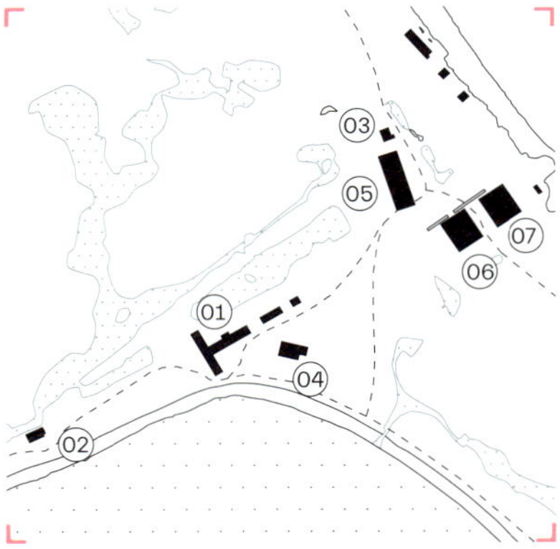

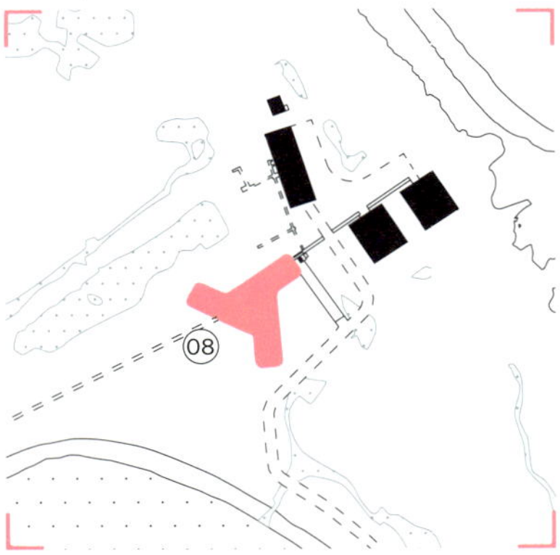

year

Compass pointing South
and prevailing wind direction

zoom, year

SOURCE
Drawing source, year

S

| 1 : 7,500 | 0 | 75 | 150 | 300 m |
| 1 : 5,000 | 0 | 50 | 100 | 200 m |

1899

CAPE ADARE HUT

↖ p. 34, 398, 404, 408, 426, 474, 524, 537

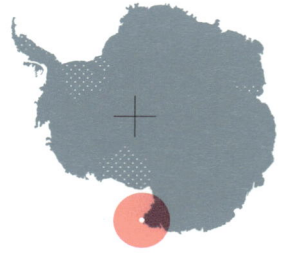

"We had two huts, constructed of pine logs in Norway, and put together at Cape Adare; they each measured about 15 ft × 15 ft [4.5 m × 4.5 m] and were well adapted to our purpose; one being used exclusively as a store house and the other as a living room."

Carsten Borchgrevink (Explorer) in his 1899 diary, reported in New Zealand Antarctic Heritage Trust, *Conservation Plan: The Historic Huts at Cape Adare,* 2004

Manufacturer: Strømmen Trævarefabrik, Norway

United Kingdom
British Antarctic Expedition, 1898–1900

Preserved, HSM 22
New Zealand Antarctic Heritage Trust (NZAHT)

71°189"S 170°09"E
Cape Adare, Adare Peninsula
Ice-free ground, 10 m.a.s.l.

SITE PLAN

01 Living hut
02 Stores hut
03 Latrines
04 Storage boxes
05 Bags of coal
06 Supply dump

SOURCE
New Zealand Antarctic Heritage Trust, 2004

S

1:1,250   0   12.5   25   50 m
1:5,000   0   50   100   200 m

PLAN | LIVING AND STORES HUTS

"The huts were constructed of interlocking boards that were made tight with steel tie rods whilst the living quarters had a double floor and walls insulated with *papier-mâché* with sliding panels and curtains giving the men some form of privacy. [...] When the huts had been built, some four yards [3.6 m] apart, I decided to use the north-western as the dwelling, and the other for provisions and outfit. The middle space between them I covered over with wood, seal-skin and canvas, and continued this cover towards the eastern side in the run with the slope of the roofs of both houses down to the ground, forming a continued sloping roof from the entire ridge of both houses, including that of the middle space and down to the ground. A strong framework of wood formed the rest for the canvas and seal-skin. By this means a great space was gained at the eastward side of the huts as well as a protected approach between the huts."

Carsten Borchgrevink (Explorer) in his 1899 diary, reported in New Zealand Antarctic Heritage Trust, *Conservation Plan: The Historic Huts at Cape Adare,* 2004

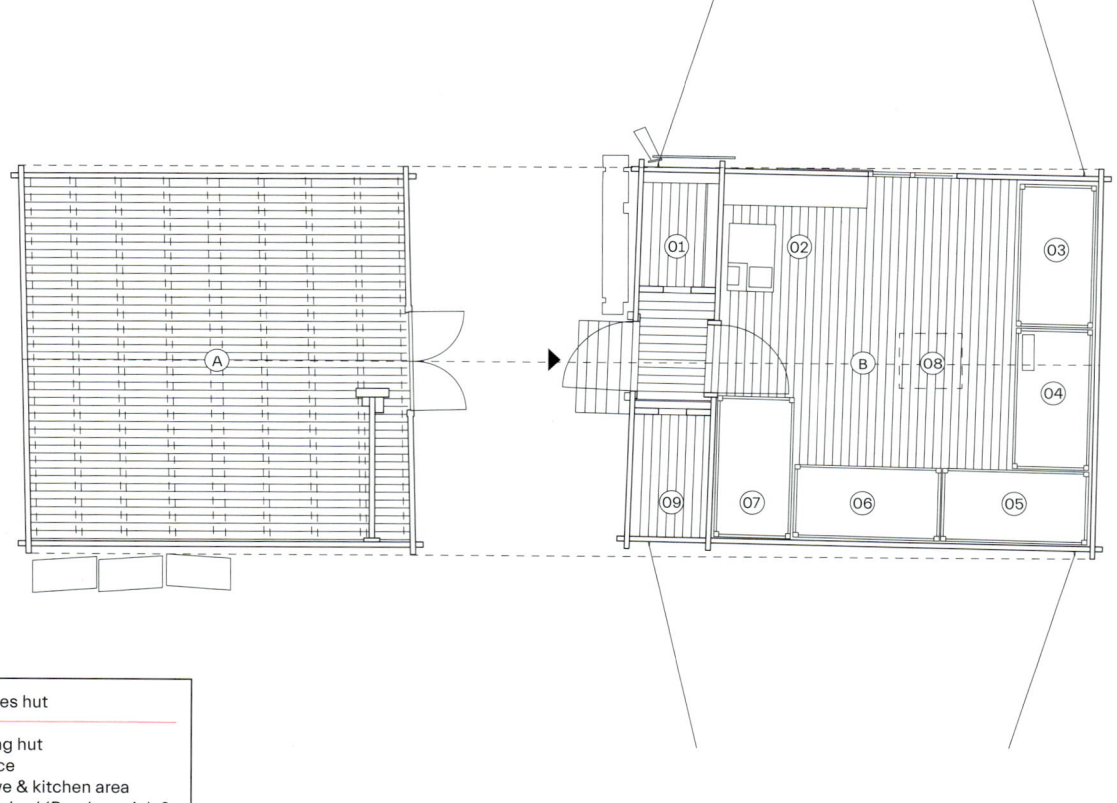

A Stores hut

B Living hut
01 Office
02 Stove & kitchen area
03 Bunk bed (Borchgrevink & Fougner)
04 Bunk bed (Colbeck & Hanson)
05 Bunk bed (Ellifsen & Evans)
06 Bunk bed (Klovstad & Bernacchi)
07 Bunk bed (Savi & Must)
08 Hatch to loft space above
09 Darkroom

Detail
01 Wall matchboard
02 Floor plank
03 Floor beam

1:100   0   1   2   4 m

ELEVATION, SECTION AND DETAIL | LIVING AND STORES HUTS

"The walls are made of pine planks 70 mm thick and 145 mm high which are interlocked at the corners and are fitted together in the same way as tongued-and-grooved boards. The walls are doubly strengthened by the addition of iron rods which are threaded down through pre-bored holes in the planks from wall plate level to the underside of the foundation beams. These rods were tightened by nuts at the base, effectively post-tensioning the construction."

New Zealand Antarctic Heritage Trust, Conservation Plan: The Historic Huts at Cape Adare, 2004

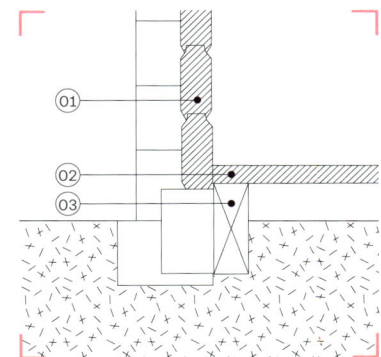

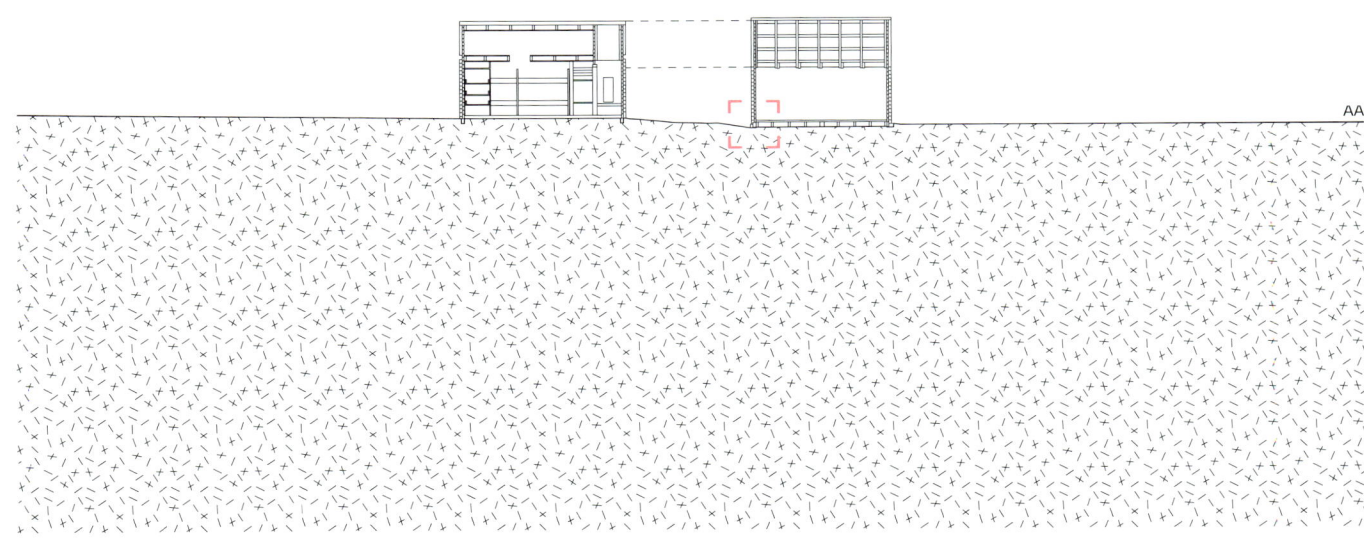

SOURCE
New Zealand Antarctic Heritage Trust, 2004

| 1 : 10 | 0 | 10 | 20 | 40 cm |
| 1 : 250 | 0 | 2.5 | 5 | 10 m |

HUT POINT HUT
1902

↖ p. 404, 410, 453, 524

![Hut Point Hut photograph]

"Scott's hut was definitely conceived as a very temporary building; we are not entirely sure of what its provenance is but we assume he picked up a bungalow in Australia because having a veranda around your hut isn't necessarily what you are after in the Antarctic. It was never quite satisfactory and most men from the *Discovery* expedition lived on board the ship rather than in the hut because of how uncomfortable it was. However, it was the furthest south hut so every Heroic Era expedition after that used it as a kicking off point."

Michael Morrison (Conservation Architect), in conversation with UNLESS, Architectural Association School of Architecture, 2019

Manufacturer: James Moore, Australia

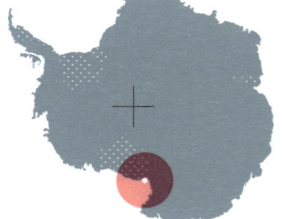

United Kingdom
British National Antarctic Expedition, 1901–1903

Preserved, HSM 18
New Zealand Antarctic Heritage Trust (NZAHT)

77°50'45"S 166°38'30"E
Hut Point, Ross Island
Ice-free ground, 3 m.b.s.l., –17°C

SITE PLAN

"The old hut has never been a cheerful place, even when we camped alongside it in the Discovery."

Sir Ernest Henry Shackleton (Explorer); *The Heart of the Antarctic: Being the Story of the British Antarctic Expedition 1907–1909*

01 Hut Point hut
02 Discovery ship

SOURCE
New Zealand Antarctic Heritage Trust, 2004

S

1 : 5,000   0   50   100   200 m

PLAN

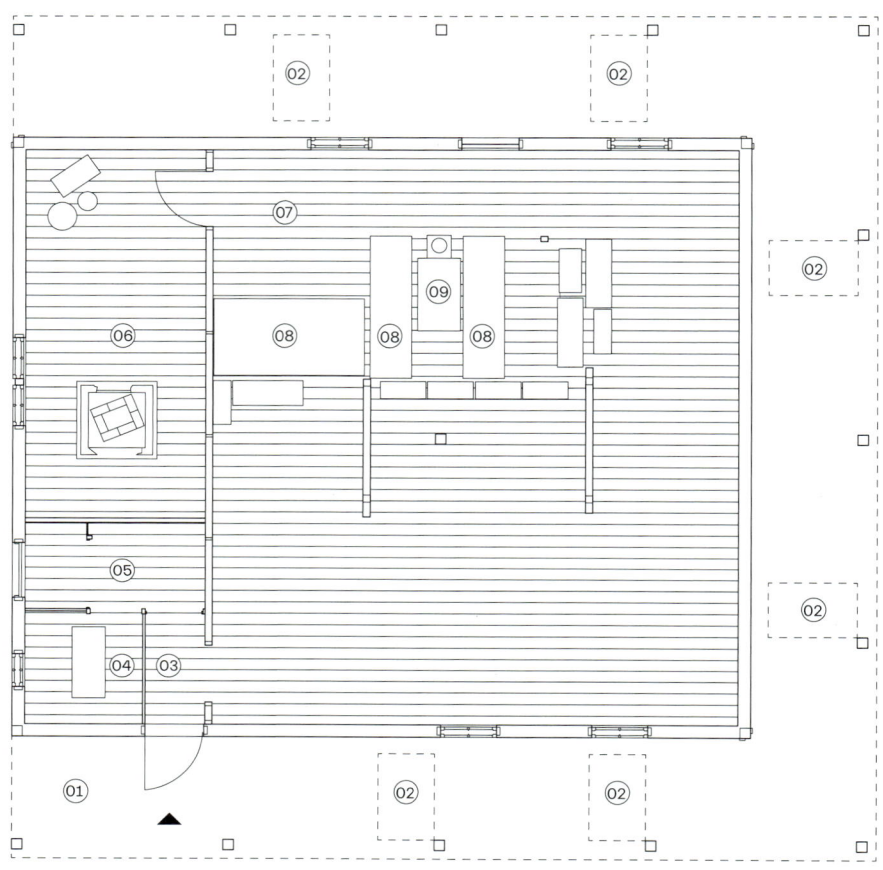

01 Verandah
02 Skylight
03 South vestibule
04 Store room
05 West vestibule
06 Physical laboratory
07 Main space
08 Sleeping platform
09 Blubber stove

"'Professor Gregory's Villa', as Scott's party called the hut, was described by Bernacchi as 'more adapted as a summer house than a polar hut', and by Armitage as a 'colonial shooting lodge'. It was […] very cold and, in the end, was used for scientific observations, for drying furs and tents after sledging, for skinning birds, as a repair shop and as a venue for entertainment, leading to it also being known as the 'Royal Terror Theatre.'"

New Zealand Antarctic Heritage Trust, "History of Scott's Expedition" (online)

SECTION AND ELEVATION

"The floor occupied a space of thirty-six feet square [3.3 sqm], but the overhanging eaves of the pyramidal roof rested on supports some four feet [1.2 m] beyond the sides, surrounding the hut with a covered verandah. […] We found that its erection was no light task, as all the main and verandah supports were designed to be sunk three or four feet [0.9–1.2 m] in the ground. We soon found a convenient site close to the ship on a small bare plateau of volcanic rubble, but an inch or two below the surface the soil was frozen hard, and many an hour was spent with pick, shovel, and crowbar before the solid supports were erected and our able carpenter could get to work on the frame."

Robert Falcon Scott (Explorer, Royal Navy Officer) in his diary, 1902

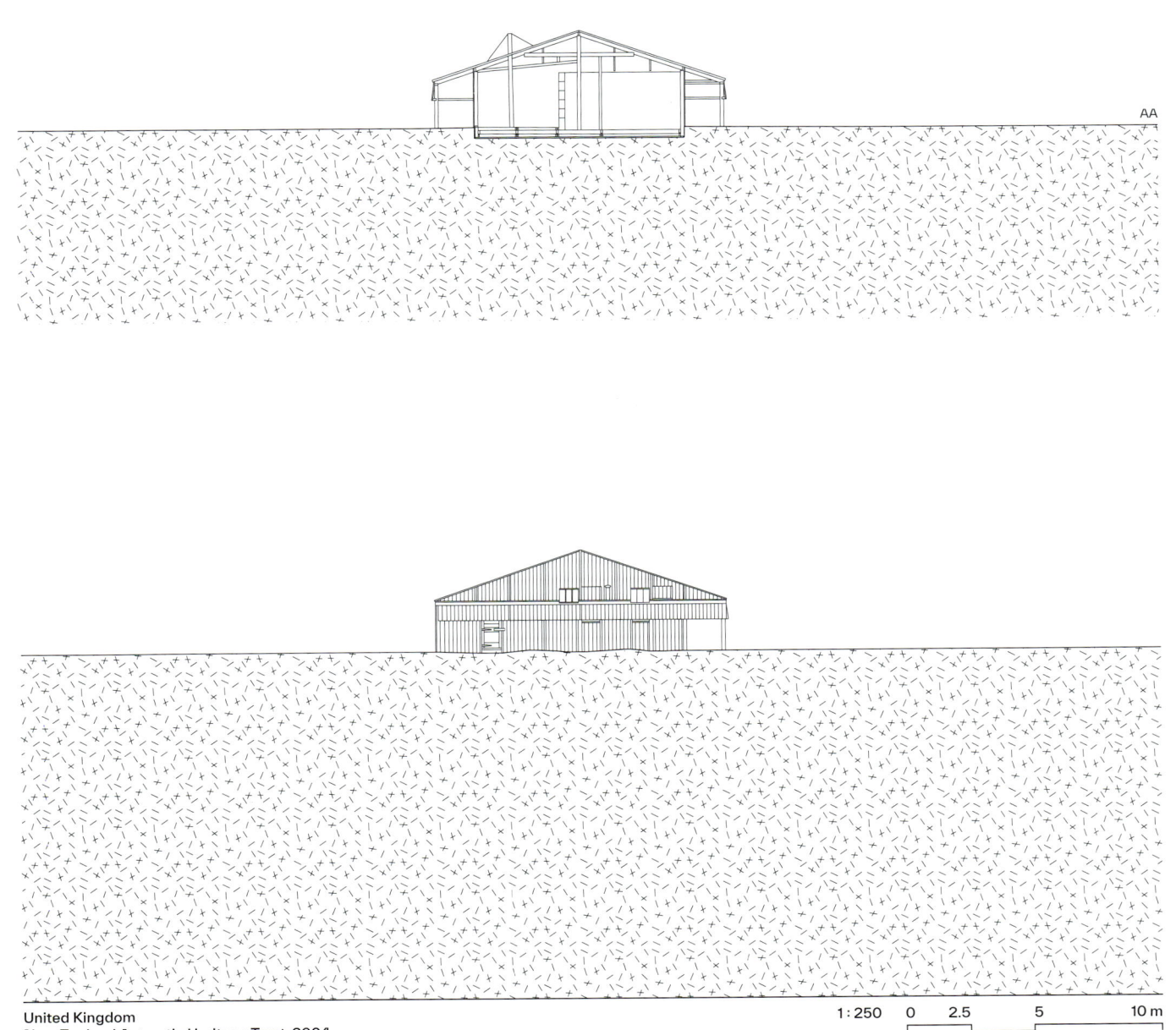

AA

United Kingdom
New Zealand Antarctic Heritage Trust, 2004

1 : 250   0   2.5   5   10 m

↖ p. 524, 528, 530

1906

"In 1906 [...] Adolf Andresen took a factory ship and two fast whale catchers to the island. By 1908 there were three whaling companies using the bay, two Norwegian and one Chilean, which required 200 men [...]. These early ships could process only the whale blubber by boiling out the oil in their open tanks. The carcasses [...] were then cast adrift to float in the bay. [...] Such waste was prohibited from 1912 [...], and the Hektor Whaling company of Tønsberg in Norway established a land station at Whalers Bay to deal specifically with [the carcasses]. Whales were harpooned at sea and floated [...] into Whalers Bay [...] to be flensed and dismembered there. The bone, meat and entrails were loaded into the thirty-six pressure cookers to extract as much oil as possible and the waste bones were crushed down for fertilizer. With the increasing number of operators at Deception Island, a small Norwegian prefabricated house and post office was erected for the British magistrate and Deception Island became the first port of entry to Antarctica for all ships working in the area. The Hektor whaling station operated until 1931, when a combination of recession and the invention of the factory ship slipway put the land factory out of business."

United Kingdom Antarctic Heritage Trust, "Whalers Bay Deception Island: A Brief History", video, 2014

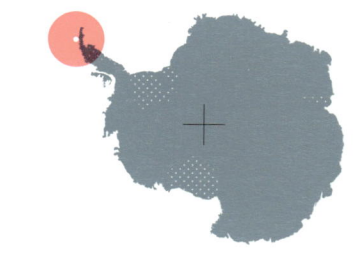

Norway
Aktieselskabet Hektor

Preserved, HSM 71
Chile, Norway, United Kingdom

62°59'00"S 60°34'00"W
Whalers Bay, Deception Island
Ice-free ground, 7 m.a.s.l., –3°C

SITE PLAN

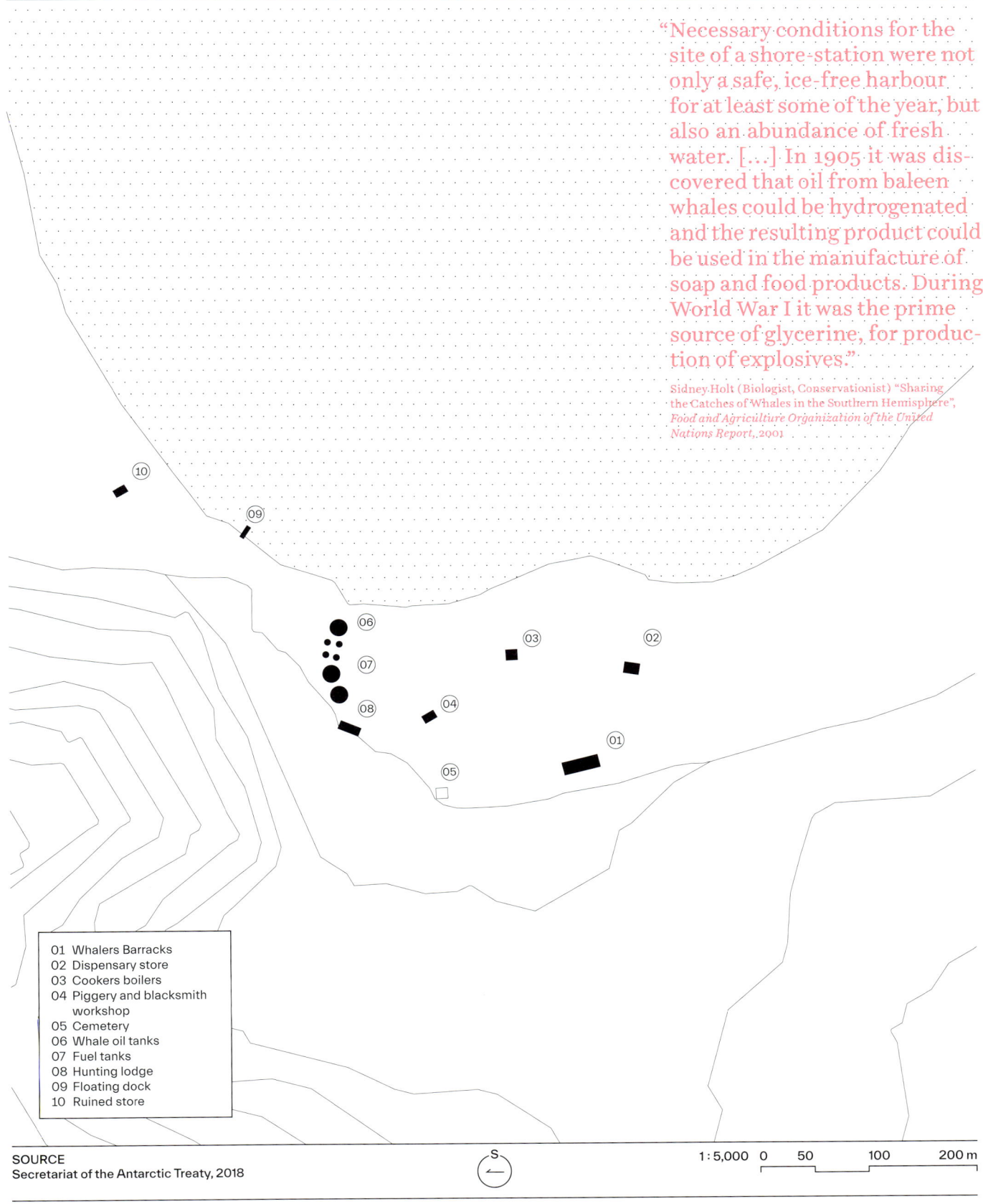

"Necessary conditions for the site of a shore-station were not only a safe, ice-free harbour for at least some of the year, but also an abundance of fresh water. [...] In 1905 it was discovered that oil from baleen whales could be hydrogenated and the resulting product could be used in the manufacture of soap and food products. During World War I it was the prime source of glycerine, for production of explosives."

Sidney Holt (Biologist, Conservationist) "Sharing the Catches of Whales in the Southern Hemisphere", *Food and Agriculture Organization of the United Nations Report*, 2001

01 Whalers Barracks
02 Dispensary store
03 Cookers boilers
04 Piggery and blacksmith
     workshop
05 Cemetery
06 Whale oil tanks
07 Fuel tanks
08 Hunting lodge
09 Floating dock
10 Ruined store

SOURCE
Secretariat of the Antarctic Treaty, 2018

S

1 : 5,000   0   50   100   200 m

CAPE ROYDS HUT

↖ p. 398, 404, 408, 426, 516, 524, 545

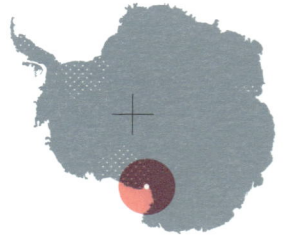

"Erection of the hut began with the digging of 22 holes for the foundations […] This work was made difficult by permafrost beneath a veneer of scoria and it was necessary to use a hammer and chisel to prepare the foundations. Erection of the hut was […] completed in a little over two weeks – nearly all the walls were up and windows installed by 13 February, and quarters in the hut were allocated on 24 February. Although it was intended to paint the exterior, this was never done."

New Zealand Antarctic Heritage Trust, *Shackleton's Hut Conservation Report*, 2003

Manufacturer: Humphreys Limited of Knightsbridge, United Kingdom

United Kingdom
British Antarctic Expedition, 1907–1909

Preserved, HSM 15
New Zealand Antarctic Heritage Trust (NZAHT)

77°33'00"S 166°10'00"E
Cape Royds, Ross Island
Ice-free ground, 10 m.a.s.l., −17°C

SITE PLAN

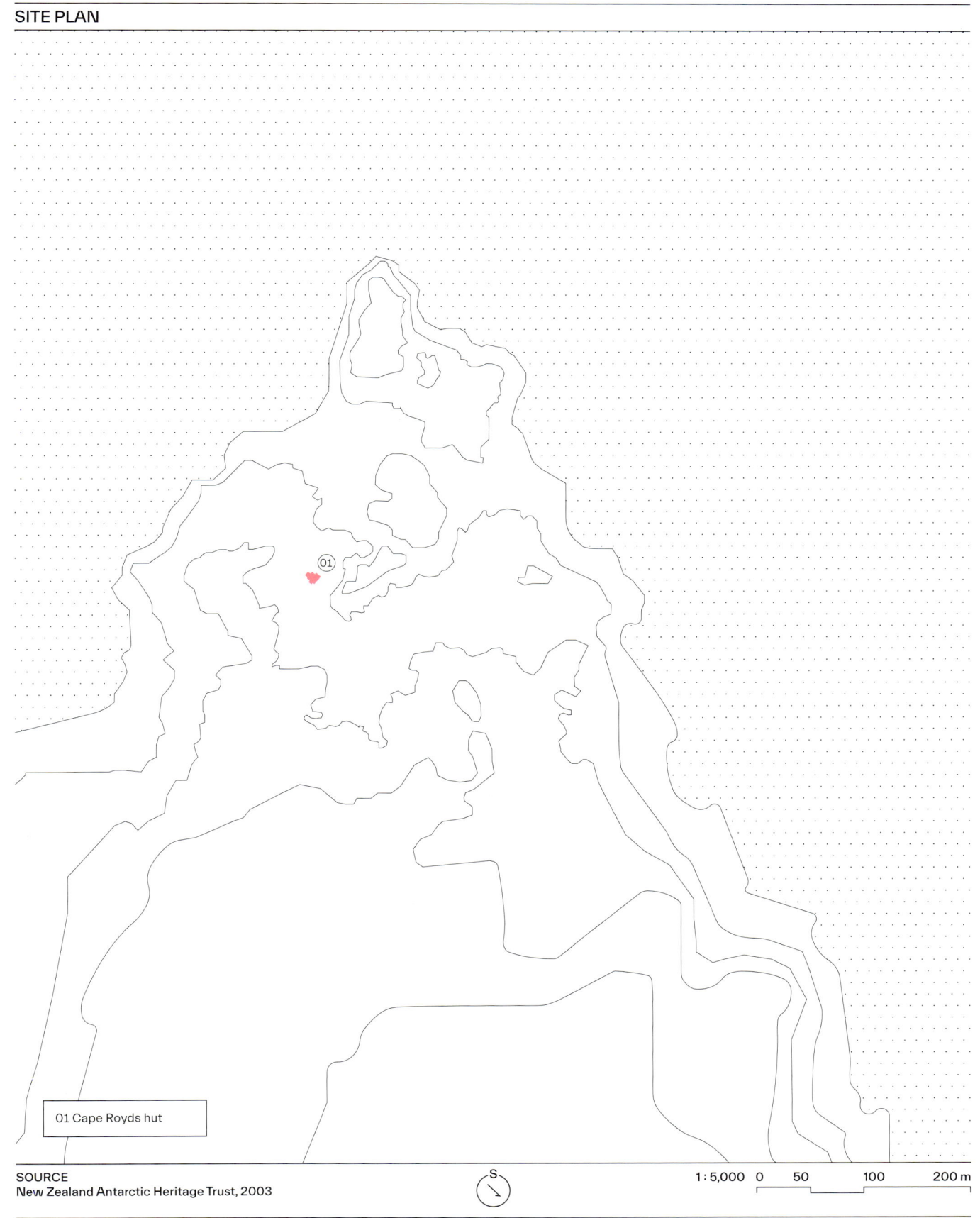

01 Cape Royds hut

SOURCE
New Zealand Antarctic Heritage Trust, 2003

1 : 5,000   0   50   100   200 m

PLAN

"The lee side of the hut ultimately became the wall of the stables [...]. A double row of cases of maize, built at one end to a height of five feet eight inches [1.73 m], made one end, and then the longer side of the shelter was composed of bales of fodder. Over all this was stretched the canvas tarpaulin which we had previously used in the fodder hut, and with planks and battens on both sides to make it windproof the stable was complete."

Sir Ernest Shackleton (Explorer), *The Heart of the Antarctic: Being the Story of the British Antarctic Expedition 1907–1909*

A   Living quarters
01 Vestibule
02 Laboratory
03 Darkroom
04 Bed
05 Coal Box
06 Stove
07 Pantry
08 Pantry table
09 Cook's table
10 Dining table
11 Lithograph machine
12 Printing press
13 Shackleton's cubicle

B   Stables

C   Garage
14 Venesta boxes

SOURCE
New Zealand Antarctic Heritage Trust, 2003

1:100   0   1   2   4 m

SECTIONS AND ELEVATION

"[The hut] was made of stout
fir timbering of the best quality
in walls, roofs and floors, and
the parts were all morticed and
tenoned to facilitate erection
in the Antarctic. The walls were
strengthened with iron cleats
bolted to main posts and horizontal timbering, and the roof
principals were provided with
strong iron tie rods."

*Sir Ernest Shackleton (Explorer), The Heart of
the Antarctic: Being the Story of the British Antarctic
Expedition 1907–1909*

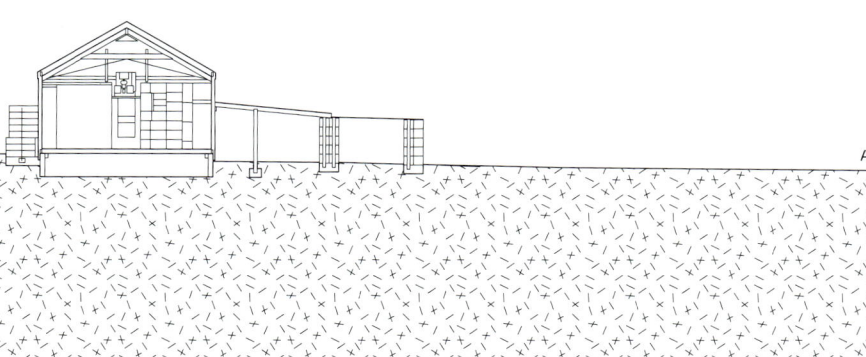

AA

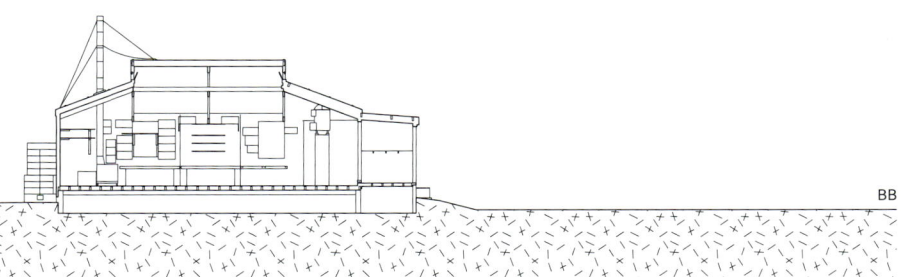

BB

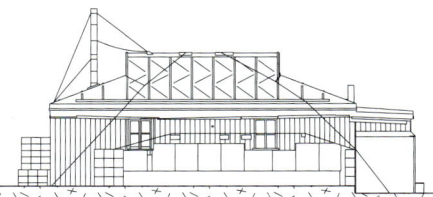

SOURCE
New Zealand Antarctic Heritage Trust, 2003

1 : 250   0   2.5   5   10 m

↖ p. 350, 404, 458

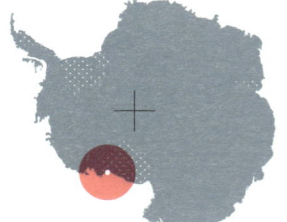

"After thorough consideration, I fixed upon the Bay of Whales as a winter station, for several reasons. In the first place, because we could there go farther south in the ship than at any other point – a whole degree farther south than Scott could hope to get in McMurdo Sound, where he was to have his station. And this would be of great importance in the subsequent sledge journey toward the Pole. Another great advantage was that we came right on to our field of work, and could see from our hut door the conditions and surface we should have to deal with. […] In addition, animal life in the Bay of Whales was, according to the descriptions, extraordinarily rich, and offered all the fresh meat we required in the form of seals, penguins, etc.
Besides these purely technical and material advantages […], it offered a specially favourable site for an investigation of the meteorological conditions, since here one would be unobstructed by land on all sides. […] Such interesting phenomena as the movement, feeding, and calving of this immense mass of ice could, of course, be studied very fully at this spot."

Roald Amundsen (Explorer), *The South Pole: An Account of the Norwegian Antarctic Expedition in the Fram 1910–1912*

Manufacturer: Hans and Jørgen Stubberud, Norway

Norway
Norwegian Antarctic Expedition, 1911–1912

Calved

78°30"S 164°00"W
Bay of Whales, Ross Ice Shelf
Ice shelf, –22.2°C

SITE PLAN

(01)

(02)

01 Framheim hut
02 Tent camp

SOURCE
Fram Museum, 1913 (reconstruction)

S

1 : 5,000   0   50   100   200 m

PLAN, ABOVE SNOW LEVEL | FRAMHEIM HUT, 1911

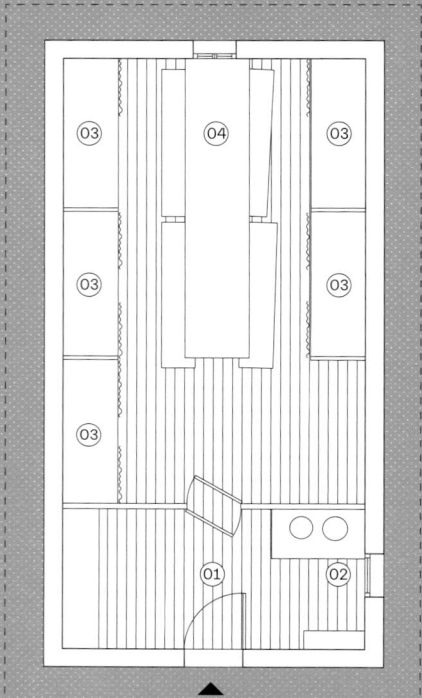

01 Pantry
02 Stove, kitchen
03 Bunk bed with curtain
04 Dining table

---

01 Vestibule
02 Bathroom
03 Workshop
04 Steam bath
05 Metal workshop
06 Laundry
07 Food storage
08 Pendel observatory
09 Coal storage
10 Timber and oil storage
11 Ice dome
12 Leader's room
13 Storage
14 Crystal palace
15 Dog tent
16 Meat tent
17 Fish tent
18 Puppies tent
19 Thermometer hut

SOURCE
Fram Museum, 1913

1 : 100   0   1   2   4 m

PLAN, BELOW SNOW LEVEL | FRAMHEIM HUT AND TENT CAMP, 1912

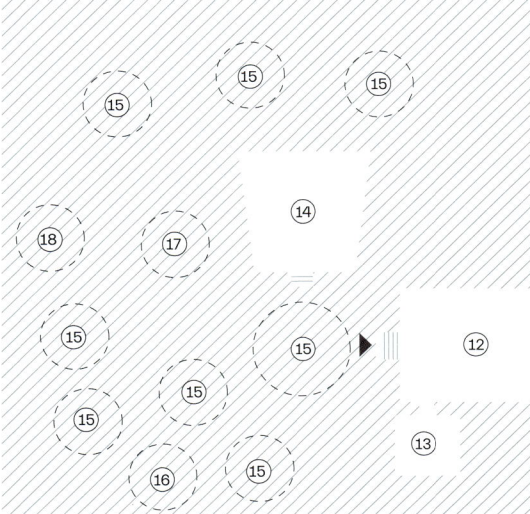
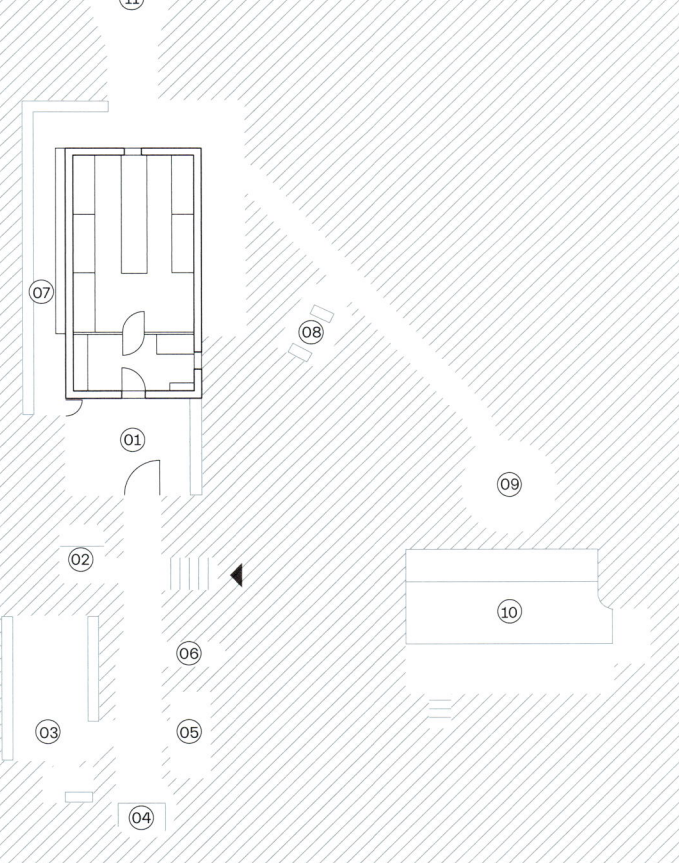

"The planing-bench was cut out in the wall […]. The workshop terminated at its western end in a little room, where the carpenters kept their smaller tools. A broad stairway, cut in the snow and covered with boards, led from the shop into the passage. […] Opposite the carpenters came the smithy, dug to the same depth as the other […]. On the other side […], a deep hole was dug to receive all of the waste water from the kitchen. Between the Carpenters' Union and the entrance to the pent-house opposite the ascent to the Barrier, we built a little room […]. By fixing the door-frame into the wall in an oblique position […] we were secured against an invasion of dogs. Four snow steps covered with boards led from the door down into the passage. In addition to all these new rooms, we had thus gained an extra protection for our house."

Roald Amundsen (Explorer), *The South Pole: An Account of the Norwegian Antarctic Expedition in the Fram 1910–1912*

SOURCE
Fram Museum, 1913

1 : 250   0   2.5   5   10 m

1911

↖ p. 350, 366, 404, 408, 426, 509, 516, 524, 545

"In a bank of ice behind the magnetic hut, two caves were excavated – one for storage of meat and the other for Simpson's and Wright's magnetic observations. Meteorological stations were set up on top of Wind Vane Hill, on The Ramp and between the main hut and magnetic hut. A local telephone system was established with lines to the ice cave where Wright made his observations, and from the sea-ice where Nelson obtained water temperatures, and later in June, where Lieutenant Evans observed the 'occulation' of Jupiter. Electric wires came from an anemometer on Wind Vane Hill down to the hut and a large cache of stores was placed on a low ridge about 100 metres south of the hut."

New Zealand Antarctic Heritage Trust, "History of Scott's Expedition" (online)

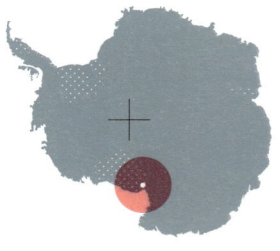

United Kingdom
British Antarctic Expedition, 1910–1913

Preserved, HSM 16
New Zealand Antarctic Heritage Trust (NZAHT)

77°38'S 166°24'E
Cape Evans, Ross Island
Ice-free ground, 6 m.a.s.l., –1.6°C

SITE PLAN

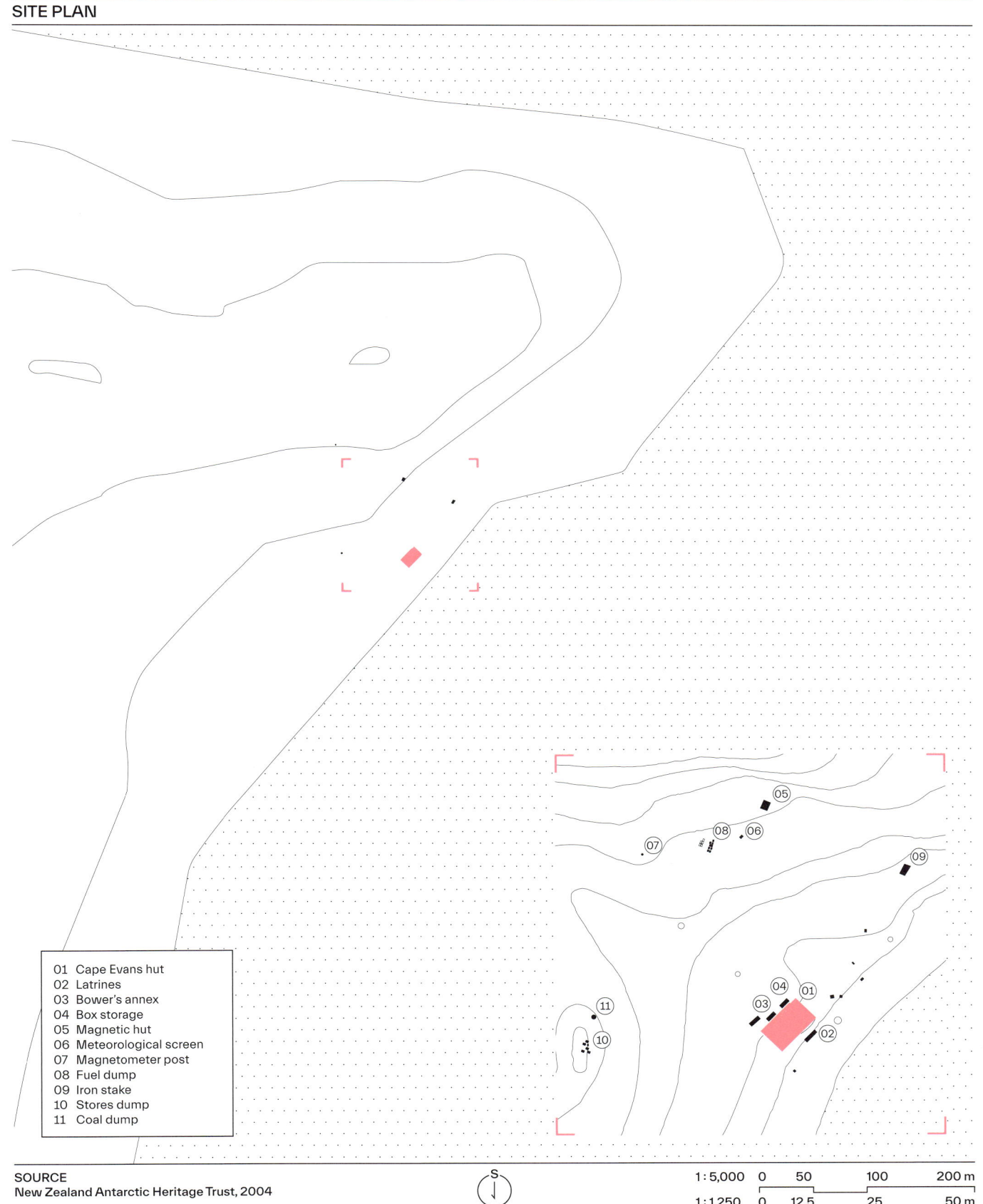

01 Cape Evans hut
02 Latrines
03 Bower's annex
04 Box storage
05 Magnetic hut
06 Meteorological screen
07 Magnetometer post
08 Fuel dump
09 Iron stake
10 Stores dump
11 Coal dump

SOURCE
New Zealand Antarctic Heritage Trust, 2004

1 : 5,000   0   50   100   200 m

1 : 1,250   0   12.5   25   50 m

PLAN

"The hut is becoming the most comfortable dwelling-place imaginable. We have made unto ourselves a truly seductive home, within the walls of which peace, quiet and comfort reign supreme. Such a noble dwelling transcends the word 'hut', and we pause to give it a more fitting title only from lack of the appropriate suggestion. What shall we call it? The word 'hut' is misleading. Our residency is really a house of considerable size, in every respect the finest that has ever been erected in the Polar regions; 50 ft [15 m] long by 25 [7.6 m] wide and 9 ft [2.7 m] to the eaves."

Robert Falcon Scott (Explorer, Royal Navy Officer)
in his diary, 1902

"The interior of the Hut was divided by a wall, shelved for stores, into two large apartments, the second being about twice the area of the first. The first room, in which were the men's quarters and the galley, was known as the Mess-deck. Nine men lived and messed in this room; which was the warmest part of the building, as it was comfortably heated by the cook's range."

Herbert Ponting (Photographer, Explorer),
The Great White South, 1921

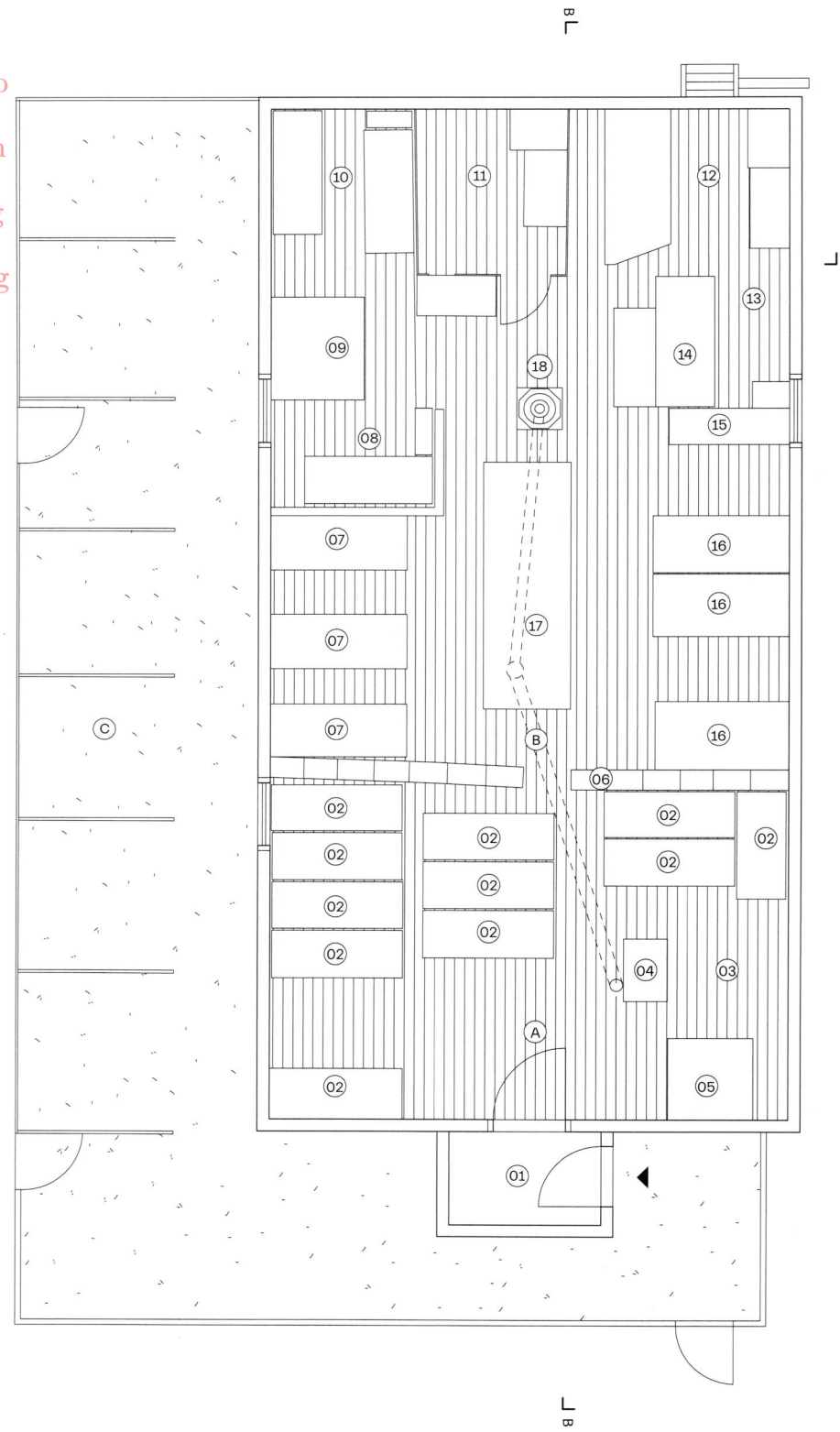

SOURCE
New Zealand Antarctic Heritage Trust, 2004

1:100 0   1   2   4 m

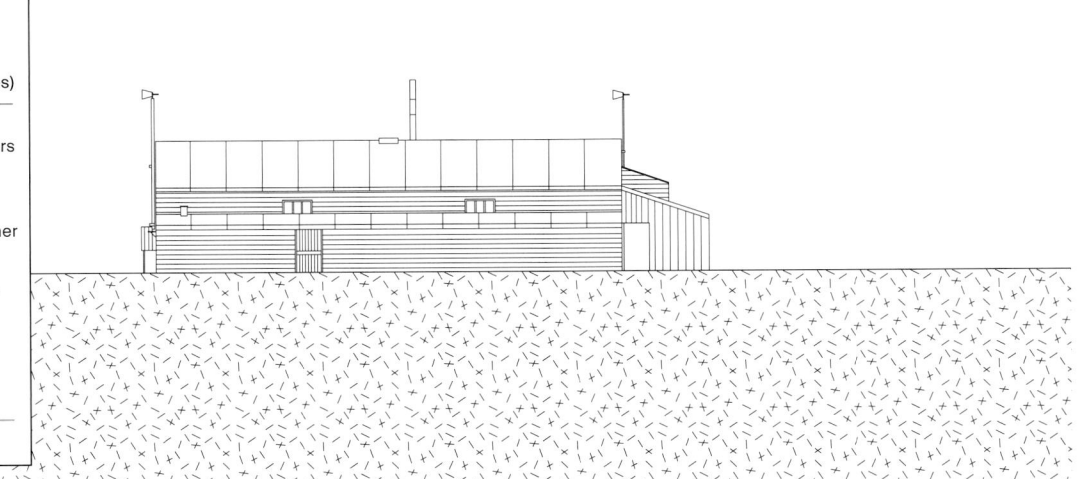

SECTIONS AND ELEVATION

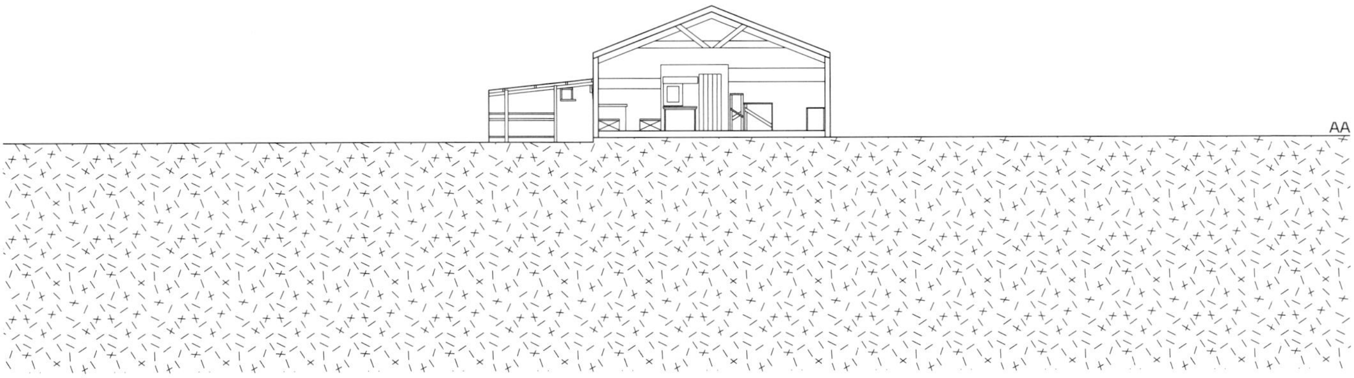

AA

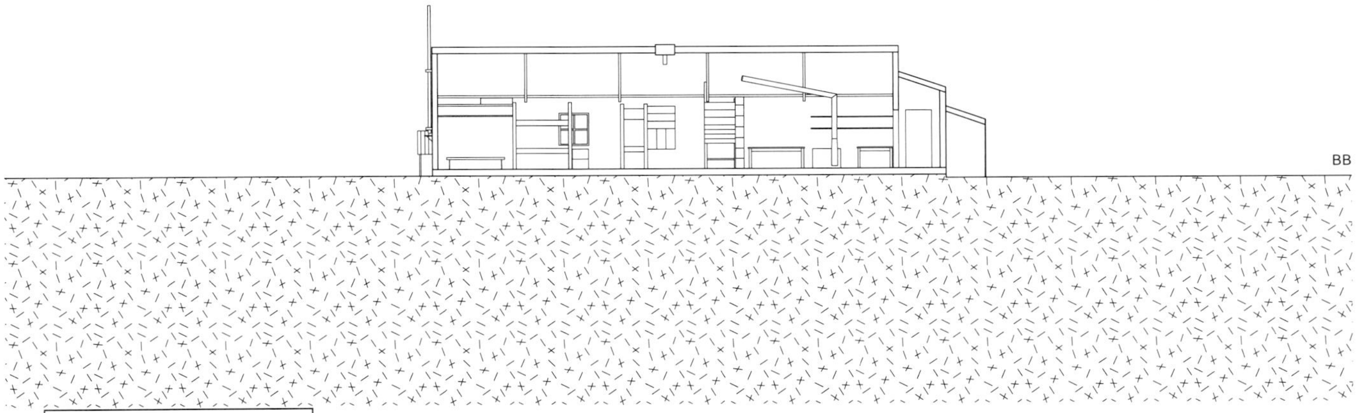

BB

A   Mess deck
01   Vestibule
02   Folding bed
03   Galley
04   Stove
05   Table
06   Bulkhead (Venesta boxes)

B   Wardroom
07   "The Tenements" quarters
       (bunk beds)
08   Scott's cubicle
09   Chart table
10   Wilson and Evans's corner
11   Ponting's darkroom
12   Physical laboratory
13   Meteorology laboratory
14   Biology bench
15   Geology laboratory
16   Bunk bed
17   Dining/work table
18   Stove

C   Stables

SOURCE
New Zealand Antarctic Heritage Trust, 2004

1 : 250   0   2.5   5   10 m

↖ p. 356, 404, 524

1912

"Wind is the dominant feature that has shaped the occupation of this place, and continues to define the landscape. The wind makes it different from most other Antarctic landscapes, and sets it apart from the sites of other Heroic Era huts. Sun, cloud and seasonal changes in daylight and darkness are largely irrelevant compared to the cycle of katabatic winds that creates an annual average daily maximum wind speed of 71 km per hour. Frequent blizzards and gusts exceed 100 km per hour: in 1913 the wind was recorded at 143 km per hour for twelve continuous hours."

Godden Mackay Logan (Heritage Consultant), *Mawson's Huts Historic Site Conservation Management Plan*, 2001

Architect: (main living quarters) Alfred Hodgeman, Australia
Manufacturers: (main living quarters) George Hudson & Son; (Western Base) Messrs Anthony; Walter and Morris Limited; (Workroom) Sarnia Timber Yards; (Magnetograph hut) Risby Brothers, Australia

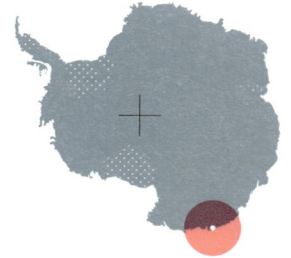

Australia
Australasian Antarctic Expedition, 1911–1914

Preserved, HSM 77
Australian Antarctic Program (AAD)

67°00'31.6"S 142°39'39.7"E
Cape Denison, Commonwealth Bay
Ice-free ground, 10 m.a.s.l., –12.8°C

SITE PLAN

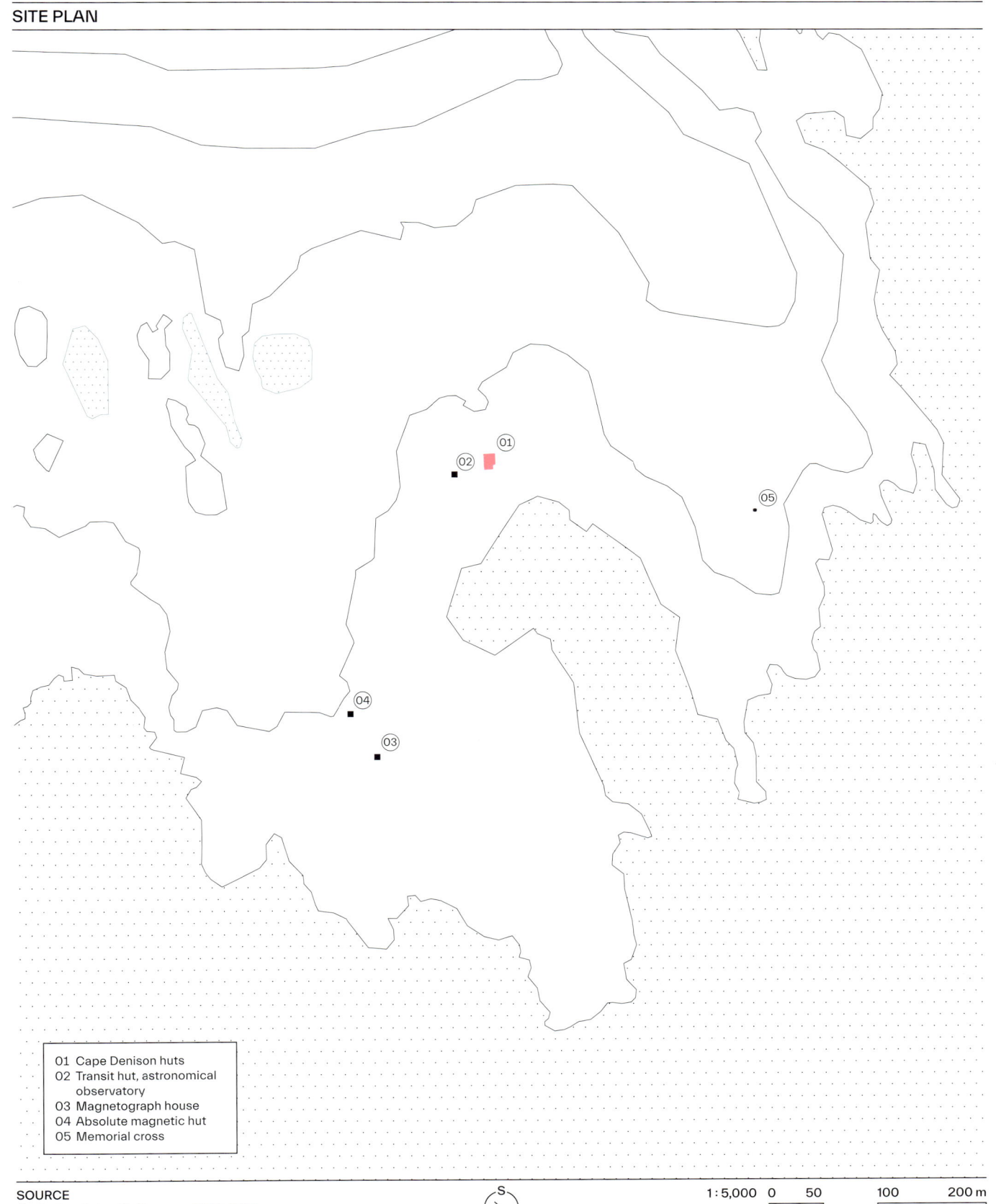

01 Cape Denison huts
02 Transit hut, astronomical
     observatory
03 Magnetograph house
04 Absolute magnetic hut
05 Memorial cross

S

1 : 5,000   0   50   100   200 m

PLAN, ABOVE AND BELOW SNOW LEVEL | MAIN LIVING QUARTERS AND WORKROOM

"The living section, prefabricated by George Hudson & Son (Sydney), is 7.3 × 7.3 m with a pyramid shaped roof supported in the centre by four 100 mm by 100 mm posts. There are four skylights, all with glass and timber covers which can be opened. Bunks are arranged in two tiers around the walls, leaving a central living space for the stove and dining table. There are two separate rooms: Mawson's cubicle and Hurley's small dark room."

Australian Antarctic Division, Mawson's Hut Historic Site Management Plan 2013–2018, 2013

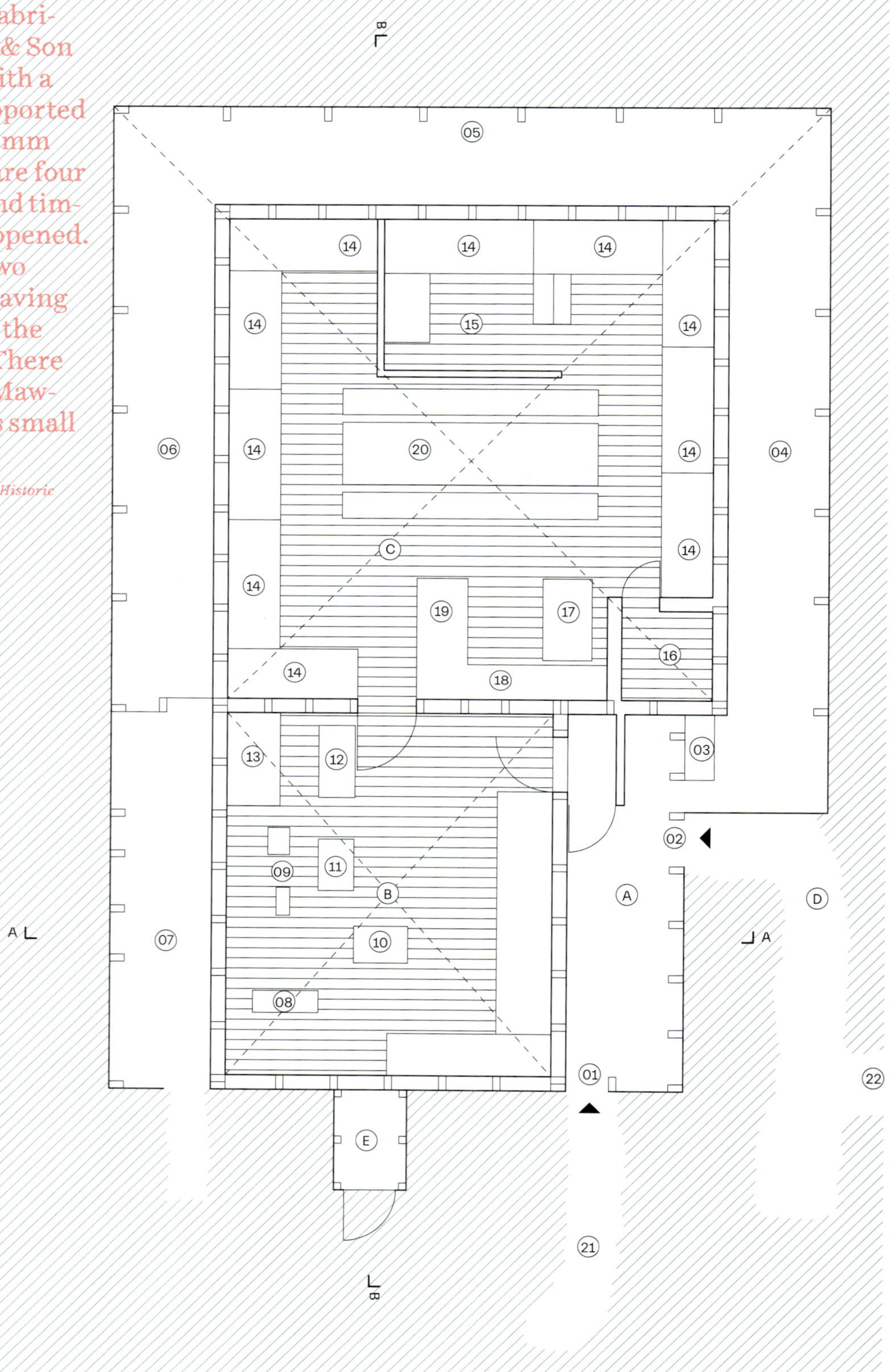

A   Verandah
01   Summer entry
02   Winter entry
03   Latrine
04   Biological store
05   Food store
06   General store
07   Dogs

B   Workroom
08   Engine
09   Generators
10   Stove
11   Large sawing machine
12   Lathe
13   Wireless operating bench

C   Main living quarters
14   Bunk bed
15   Mawson's cubicle
16   Darkroom
17   Stove
18   Dresser
19   Cook's table
20   Dining table

D   Tunnels
21   Stores tunnel
22   Dog tunnel

E   Aurora observatory

SOURCE
Australian Antarctic Program, 2013–2018

1 : 100   0   1   2   4 m

SECTIONS | MAIN LIVING QUARTERS AND WORKROOM

"By that time the Hut had been further protected by a crescent of cases, erected behind the first break-wind. In height the erection stood above the hangar, and, when the snow became piled in a solid ramp on the leeward side, it was more compact than ever. Inside the hut extra struts were introduced, stiffening the principal rafters on the southern side. […] On May 15, the wind blew at an average velocity of ninety miles per hour [145 km/h] throughout the whole twenty-four hours."

Sir Douglas Mawson (Geologist, Explorer, Academic), *The Home of the Blizzard*, 1915

AA

BB

SOURCE
Australian Antarctic Program, 2013–2018

1 : 250   0   2.5   5   10 m

THE *ENDURANCE*
1914

"Huge blocks of ice, weighing
many tons, were lifted into
the air and tossed aside as other
masses rose beneath them.
We were helpless intruders in
a strange world, our lives dependent upon the play of grim
elementary forces that made
a mock of our puny efforts."

Sir Ernest Henry Shackleton (Explorer), *South: The
Story of Shackleton's Last Expedition 1914–1917*, 1915

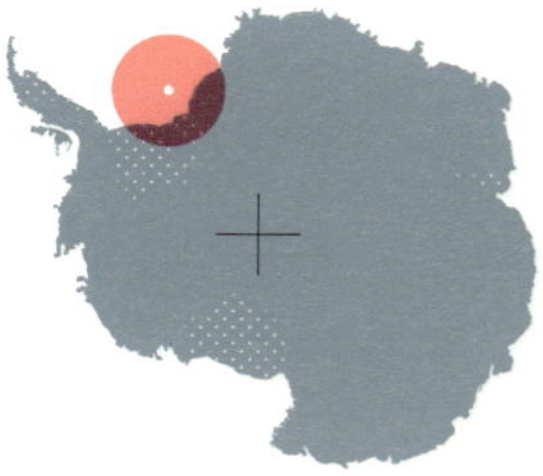

Designer: Ole Aanderud Larsen, Norway
Manufacturer: Christian Jacobsen, Norway

United Kingdom
The Imperial Trans-Antarctic Expedition, 1914–1917

Marine Archaeology/Preserved HSM 93
United Kingdom Antarctic Heritage Trust (UKAHT)

64°14'13"S 56°37'7"W
Weddell Sea

SITE PLAN

01 (01)

01 The *Endurance*

SOURCE
Vestfold Musenee (reconstruction)

1 : 5,000   0   50   100   200 m

PLAN | UPPER, MAIN, LOWER DECK

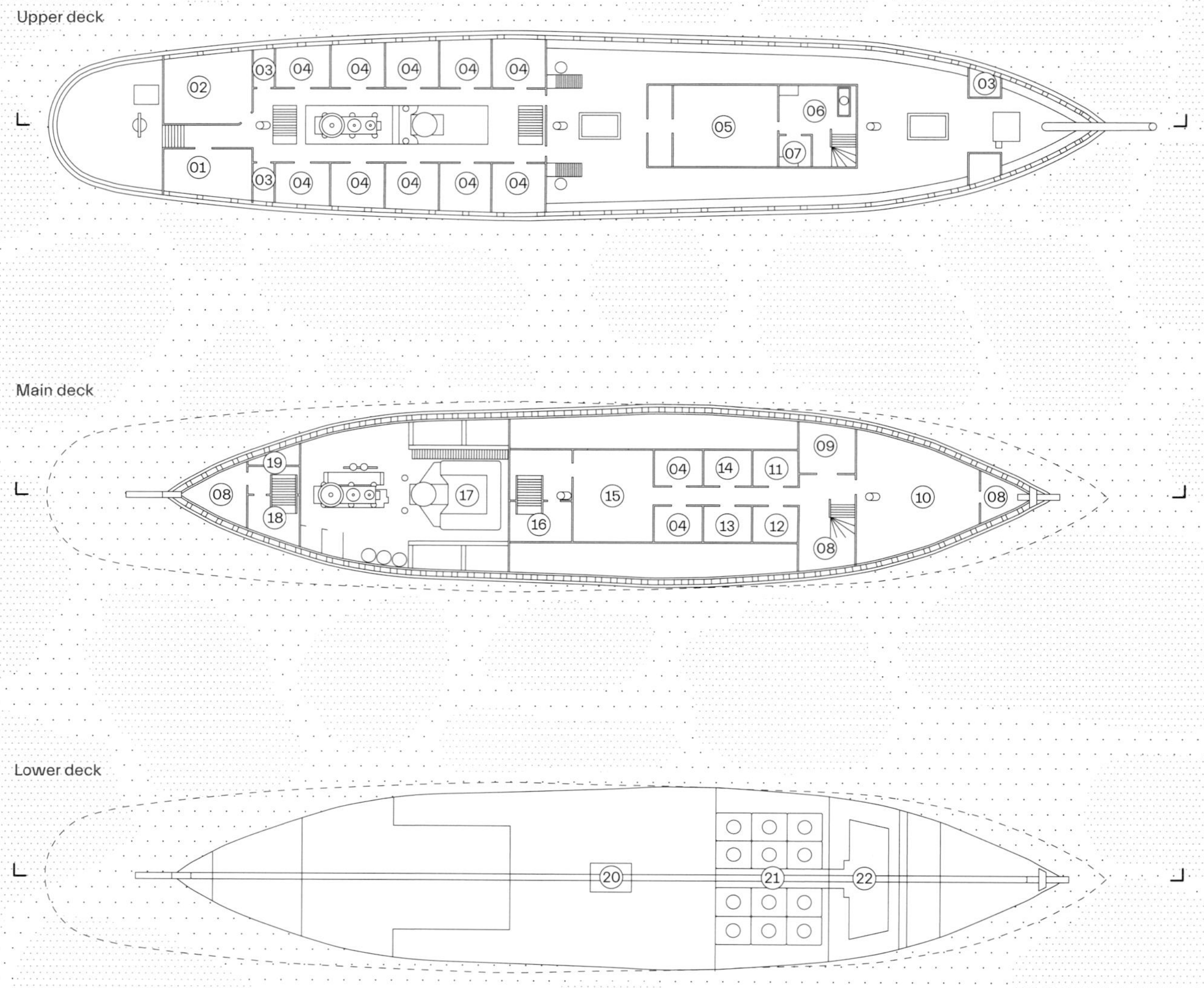

Upper deck

Main deck

Lower deck

Upper deck	Main deck	16 Doctor's cabin
01 Captain's cabin	08 Storage	17 Boiler
02 Smoking room	09 Mess room	18 Baths
03 Toilet	10 Crew's quarter	19 Darkroom
04 Berth / bed	11 Officer's cabin	
05 Dining room	12 Cook's cabin	Lower deck
06 Galley	13 Servant's cabin	20 Coal storage
07 Pantry	14 Engineer's cabin	21 Fresh water storage
	15 Laboratory	22 Provisions storage

SOURCE
Vestfold Musenee, 1911

1 : 250   0   2.5   5   10 m

SECTION

"Just at daybreak I went over to the *Endurance* with Wild and Hurley, in order to retrieve some tins of petrol that could be used to boil up milk for the rest of the men. The ship presented a painful spectacle of chaos and wreck. The jib-boom and bowsprit had snapped off during the night and now lay at right angles to the ship, with the chains, martingale, and bob-stay dragging them as the vessel quivered and moved in the grinding pack. The ice had driven over the forecastle and she was well down by the head."

Sir Ernest Henry Shackleton (Explorer), *South: The Story of Shackleton's Last Expedition 1914–1917*, 1915

"At 5 pm she went down by the head: the stern the cause of all the trouble was the last to go under water. I cannot write about it."

Sir Ernest Henry Shackleton (Explorer), *South: The Story of Shackleton's Last Expedition 1914–1917*, 1915

BASE A

1944

↖ p. 37, 88, 92, 409, 427, 526, 528

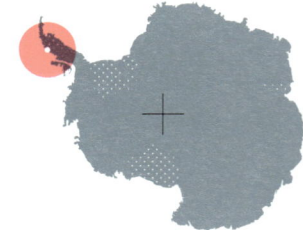

"Base A, Port Lockroy, Wiencke Island, established on the 11th of February 1944. Occupied by nine men, winter 1944, and by four men, winter 1945. The main base was to have been at Hope Bay on the mainland but the second support vessel, SS *Fitzroy*, was not ice-strengthened and could not risk the sea ice in the bay."

British Antarctic Survey, "History of Port Lockroy (Station A)" (online)

Manufacturer: (Bransfield house) Boulton and Paul, United Kingdom

United Kingdom
Operation Tabarin, 1943–1944

Preserved, HSM 61
United Kingdom Antarctic Heritage Trust (UKAHT)

64°49'31"S 63°29'37"W
Port Lockroy, Wiencke Island
Ice-free ground, 33 m.a.s.l., 1.8 °C

SITE PLAN

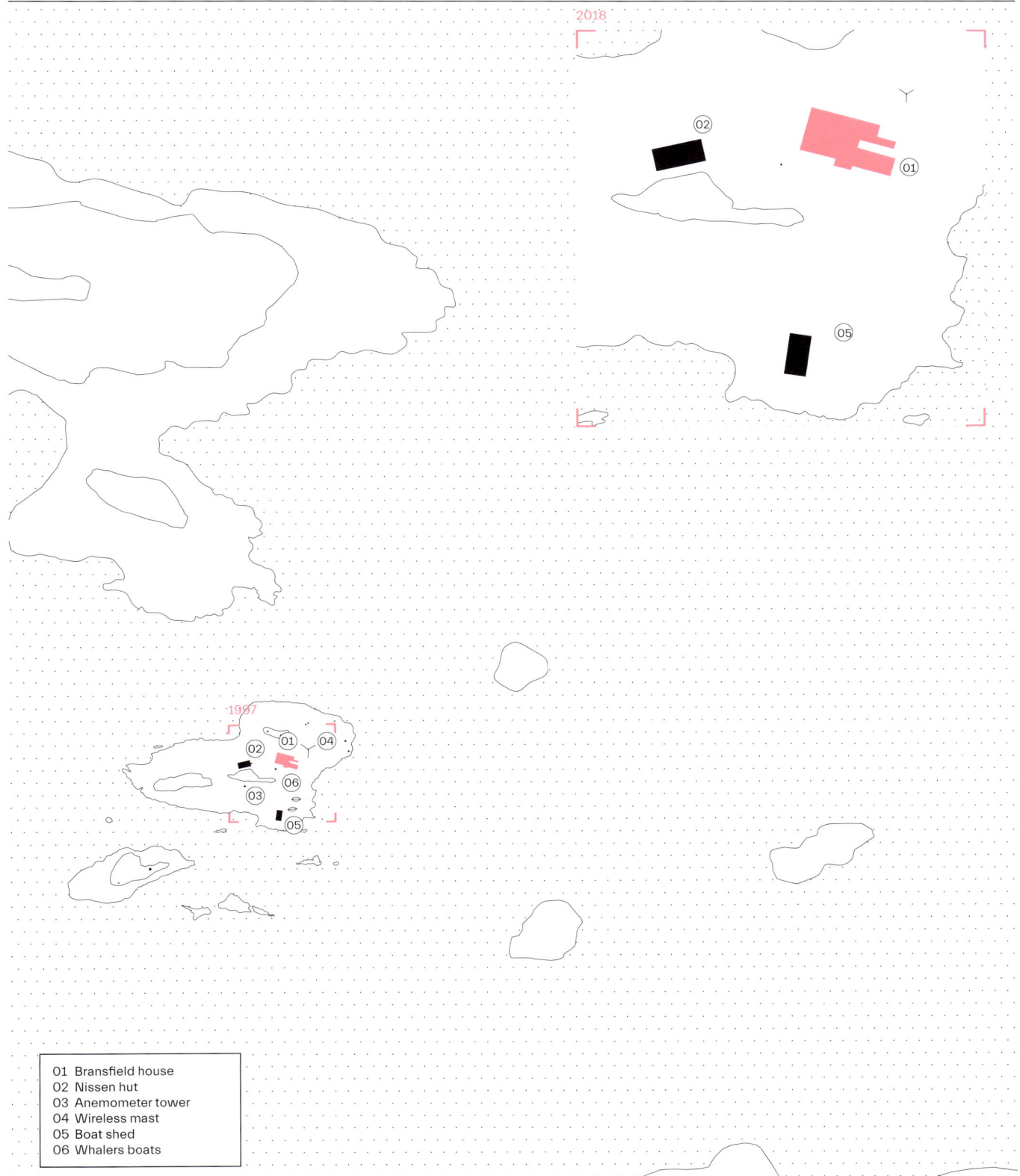

01 Bransfield house
02 Nissen hut
03 Anemometer tower
04 Wireless mast
05 Boat shed
06 Whalers boats

SOURCE
British Antarctic Survey, 1997/2018

1 : 1,250   0   25   50   100 m

1 : 5,000   0   50   100   200 m

PLAN AND SECTION | BRANSFIELD HOUSE, 1943

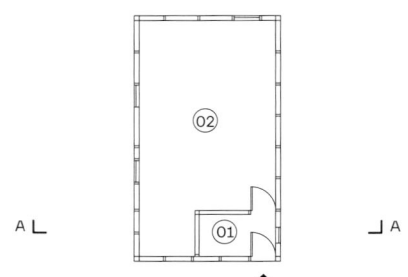

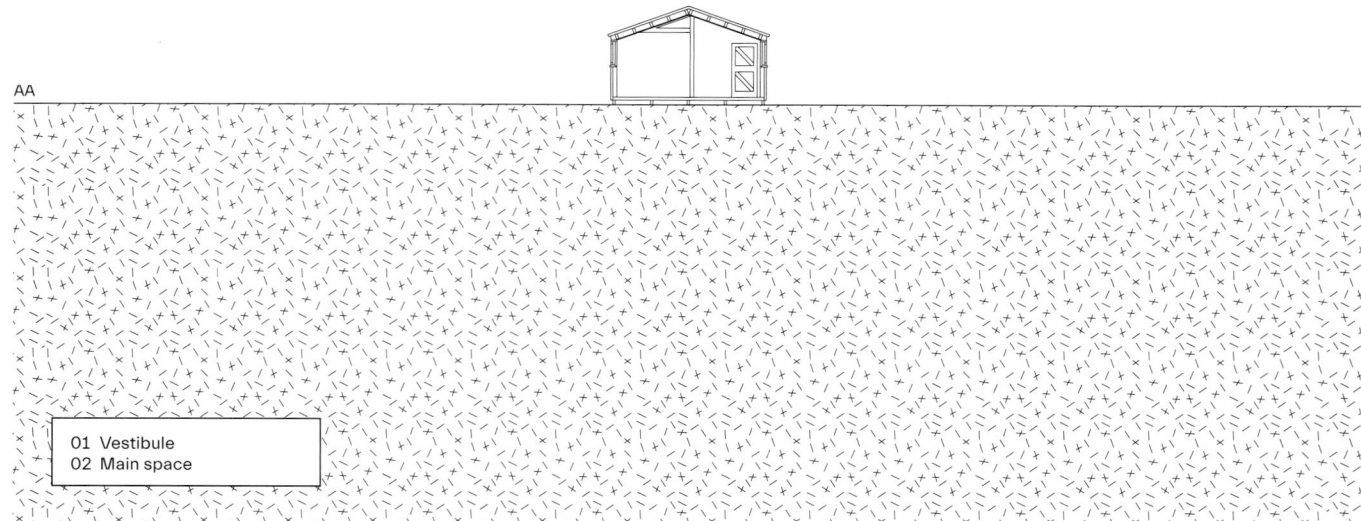

01 Vestibule
02 Main space

SOURCE
Falkland Islands Dependencies Survey, 1943

1 : 250   0   2.5   5   10 m

PLAN AND ELEVATION | BRANSFIELD HOUSE, 1959

"This building, for it can hardly be described as a hut […] is strong and commodious and is possibly a more elaborate building than has hitherto been erected in Antarctica […]. Of the whole structure only the […] workroom, together with the kitchen, is the original Boulton and Paul hut we brought with us from England. The remainder, comprising mess room, cabins, store room, scullery and the rest, was put together from a motley collection of materials – corrugated iron, timber and lining paper from Deception, heavy beams dating from the whaling days dug up out of the ice on Wiencke Island, the woodwork and beaver board of the second Nissen hut which was not erected, packing cases and junk of every description, the whole eked out with a quantity of timber, sisalkraft and aluminium foil supplied as good measure by Boulton and Paul along with the original hut."

William Flett (Geologist, Base Leader at Deception Island), *Base "B" Deception Island Report* for the Falkland Islands Dependencies Survey, 1944

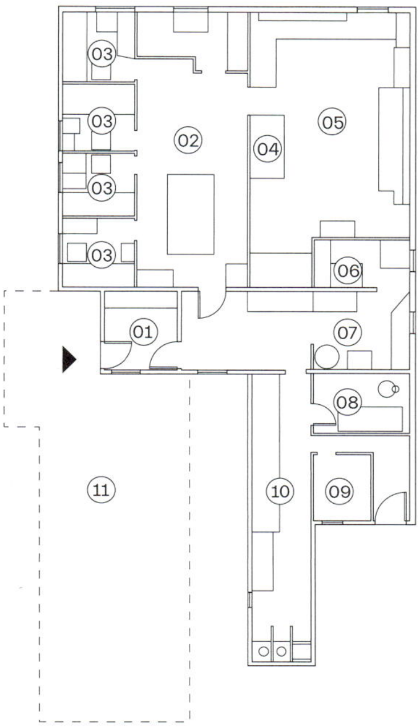

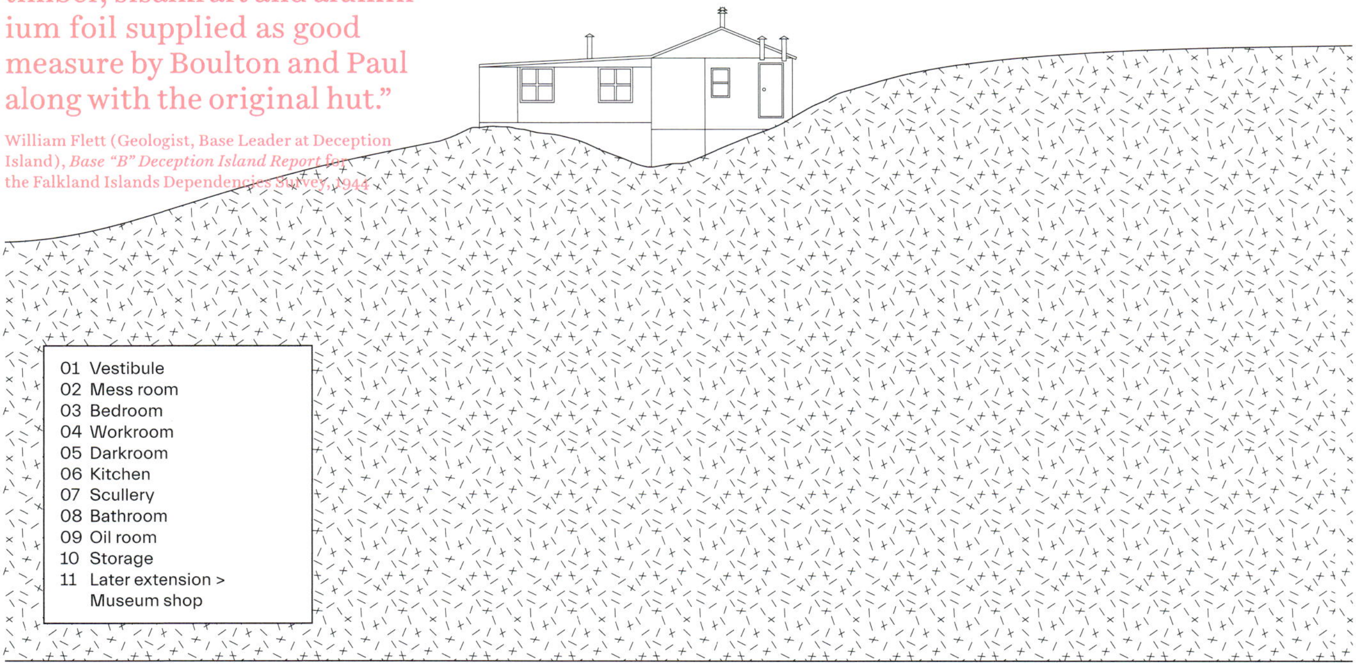

01 Vestibule
02 Mess room
03 Bedroom
04 Workroom
05 Darkroom
06 Kitchen
07 Scullery
08 Bathroom
09 Oil room
10 Storage
11 Later extension >
   Museum shop

SOURCE
Falkland Islands Dependencies Survey, 1959

1 : 250   0   2.5   5   10 m

MAUDHEIM
1949

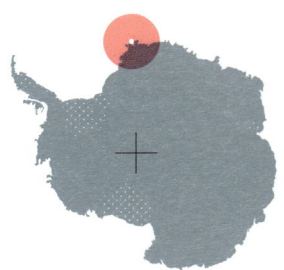

"The Norwegian Polar Institute announced here tonight that, with the permission of King Haakon, the winter base of the joint British-Scandinavian Expedition in the Antarctic has been named Maudheim, in memory of Norway's late Queen Maud. The name localizes the station as situated in Queen Maud Land. [...] It is expected that eight or ten days will elapse before wireless communication is established with Maudheim. [...] When the Norsel left for Capetown, fifteen men remained with the Maudheim party. [...] The composition of the party now is six Norwegians, five Britons and four Swedes."

North American Newspaper Alliance, "Maudheim Is Base of Antarctic Party", *The New York Times*, 1950

Architect: John Engh, Norway

Norway, United Kingdom, Sweden
Norwegian-British-Swedish Antarctic Expedition,
1949–1952

Calved

71°01'S 10°53'W
Quar Ice Shelf, Queen Maud Land
Ice shelf, –25 °C

SITE PLAN

01 Maudheim base

SOURCE
Norwegian Polar Institute (reconstruction)

S

1 : 5,000   0   50   100   200 m

PLAN, BELOW SNOW SURFACE

"Designed by Norwegian architect John Engh, the [Norwegian-Swedish-British hut] was purpose-built for Antarctic conditions and state of the art for its day. The pre-cut timber for the hut was easily transported, with sub-floor framing designed specifically to be built on an ice platform. The hut was able to withstand full burial in snow."

Australian Antarctic Division, "Biscoe Hut" (online), 2018

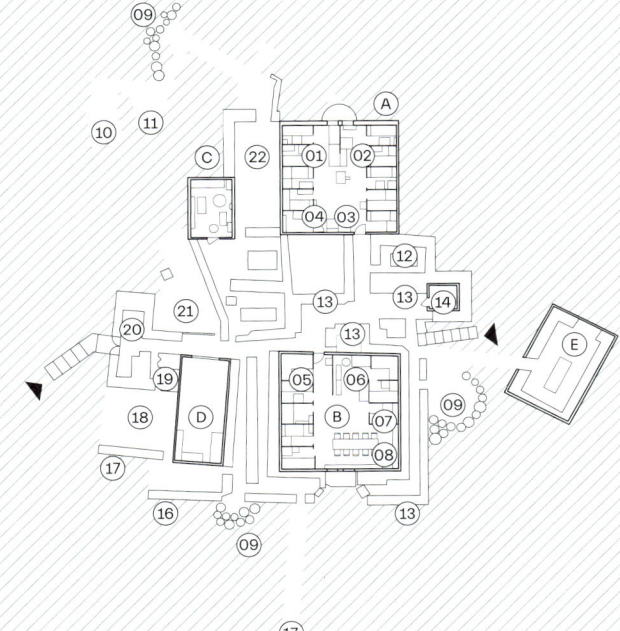

A	Command hut	09	Fuel drums
01	Meteorological department	10	Balloon filling room
02	Telecommunications department	11	Hydrogen generation room
03	Medical department	12	Meteorological store
04	Darkroom	13	Provisions
		14	Glaciological lab
B	Kitchen/Dining hut	15	Snow tunnel for dogs
05	Pantry	16	Miscellaneous
06	Kitchen	17	Clothing
07	Dining room	18	Garage
08	Library	19	Bathroom
		20	Radio and electrical store
C	Generator hut	21	Puppy pen
D	Workshop	22	Sledge, ski & dog supplies
E	Drilling machine hut		

SOURCE
Norwegian Polar Institute, 1960

1:500   0   5   10   20 m

↖ p. 411, 427

MAWSON
1954

"Impressed by the Norwegian designed pre-fabricated huts [at Maudheim], [the first Australian National Antarctic Research Expeditions (ANARE) Director, Dr Phillip Law] persuaded the Australian government to purchase these for Mawson station. While the ANARE-designed huts took only a few days to erect, the NSB (Norwegian-Swedish-British) took over a month. Over 30 tons of rock and gravel had to be gathered and moved by the Ferguson tractor just for the foundations. A level surface was critical because the hut consisted of squared prefabricated panels so everything within the structure had to be square to make them fit. Reassembly of the prefabricated hut proved to be a complex task. [...] By the end of March 1954, construction was sufficiently advanced for the 10 wintering expeditioners to occupy the building. Once completed, [Biscoe] hut was one of five buildings first erected at Mawson in 1954. With subsequent expeditions, the number of buildings at Mawson rapidly multiplied and the role of Biscoe Hut changed many times."

Australian Antarctic Program, "Biscoe Hut" (online), 2018

Architect: (Biscoe hut), John Engh, Norway
Manufacturer: (aluminium-clad post-tensioned boxes) Australian National Antarctic Research Expeditions, Commonwealth Department of Works & Olympic General Products, Australia

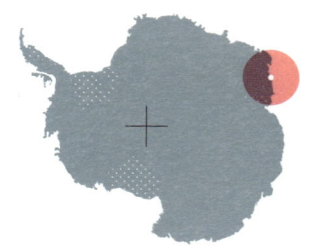

Australia
Australian National Antarctic Research Expeditions
> Australian Antarctic Program (AAD)

Active Station

67°36'12"S 62°52'27"E
Horseshoe Harbour, Mac. Robertson Land
Ice-free ground, 42 m.a.s.l., –8.3 °C

SITE PLAN

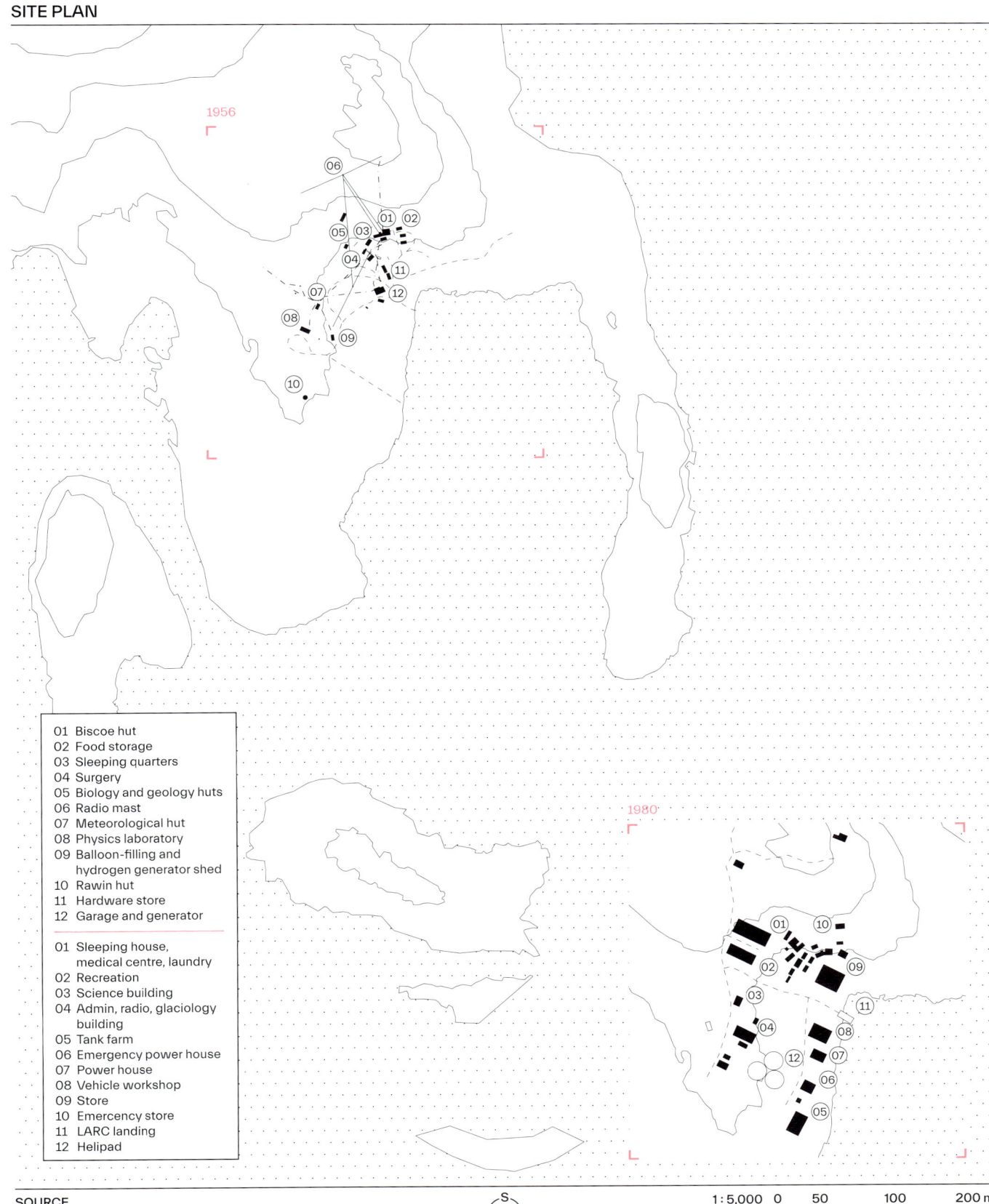

1956

1980

01 Biscoe hut
02 Food storage
03 Sleeping quarters
04 Surgery
05 Biology and geology huts
06 Radio mast
07 Meteorological hut
08 Physics laboratory
09 Balloon-filling and
   hydrogen generator shed
10 Rawin hut
11 Hardware store
12 Garage and generator

01 Sleeping house,
   medical centre, laundry
02 Recreation
03 Science building
04 Admin, radio, glaciology
   building
05 Tank farm
06 Emergency power house
07 Power house
08 Vehicle workshop
09 Store
10 Emercency store
11 LARC landing
12 Helipad

SOURCE
Australian Antarctic Program, 1956/1980

S

1 : 5,000   0   50   100   200 m

LITTLE AMERICA V
1929/1956

↖ p. 435, 458

"The base consisted of 15 buildings, 2 towers, miscellaneous shops, and covered storage areas. Some distance away from the main group of buildings lay a 6000-ft [1828 m] runway and two additional buildings. When [it] was evacuated in 1959 there was as much as 8 ft [2.4 m] of snow lying over the rooftops, which themselves had been about 10 ft [3 m] above the snow surface at the time of construction in 1956."

Malcolm Mellor (Glaciologist, Polar Explorer), *Methods of Building on Permanent Snowfield*, 1968

Engineer and Architect: Thomas B. Bourne Associates, Inc., United States of America
Construction: United States Navy Seabees, United States of America

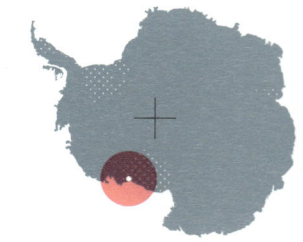

United States of America
Operation Deep Freeze, 1955–1956

Calved > Replaced

78°12'S 162°12'W
Bay of Whales, Ross Ice Shelf
Ice shelf, 14 m.a.s.l., –22.2°C

SITE PLAN

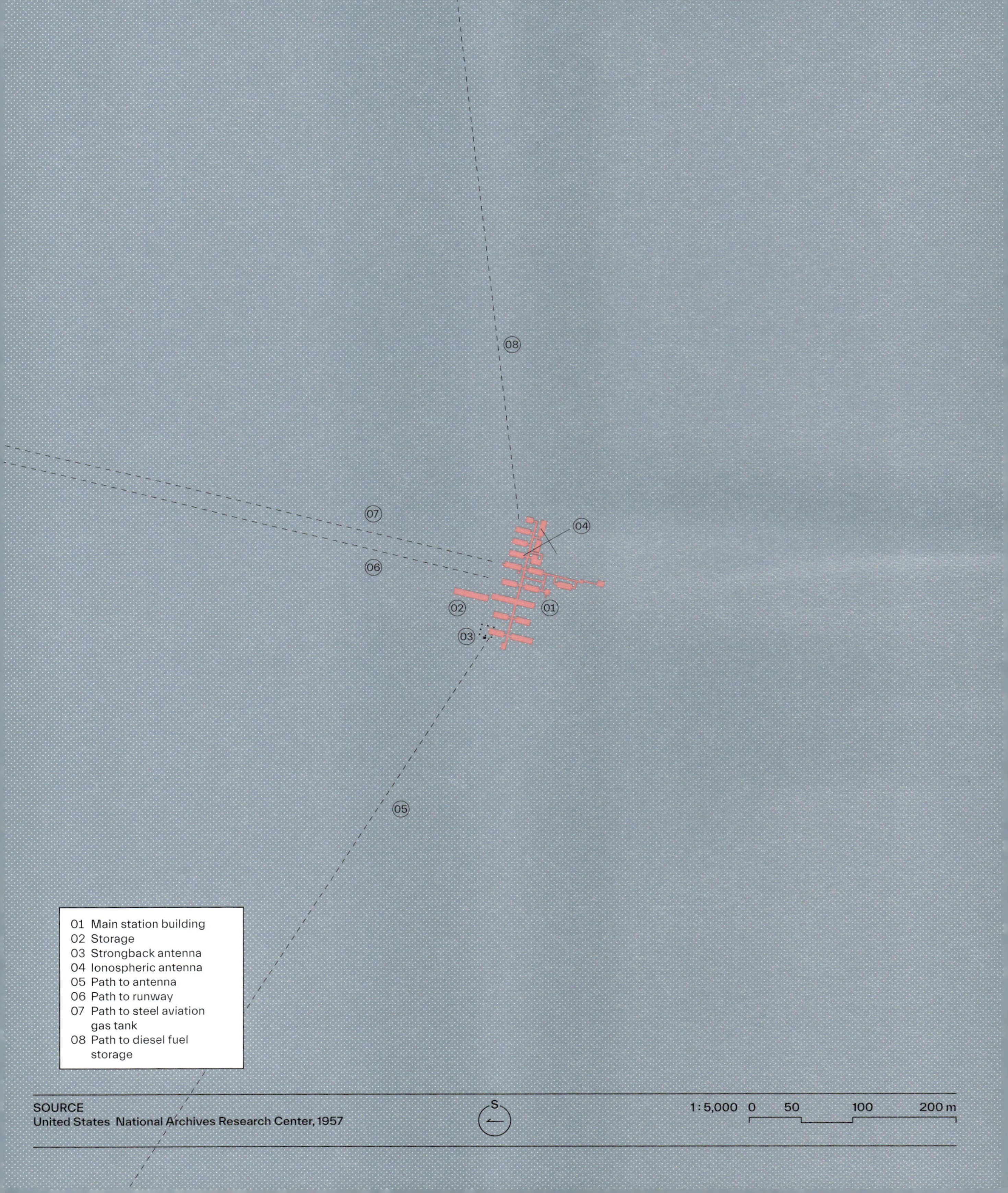
08
07
04
06
02
01
03
05
01 Main station building
02 Storage
03 Strongback antenna
04 Ionospheric antenna
05 Path to antenna
06 Path to runway
07 Path to steel aviation gas tank
08 Path to diesel fuel storage
S
1:5,000 0 50 100 200 m

SOURCE
United States National Archives Research Center, 1957

PLAN, BELOW SNOW LEVEL | MAIN STATION BUILDINGS

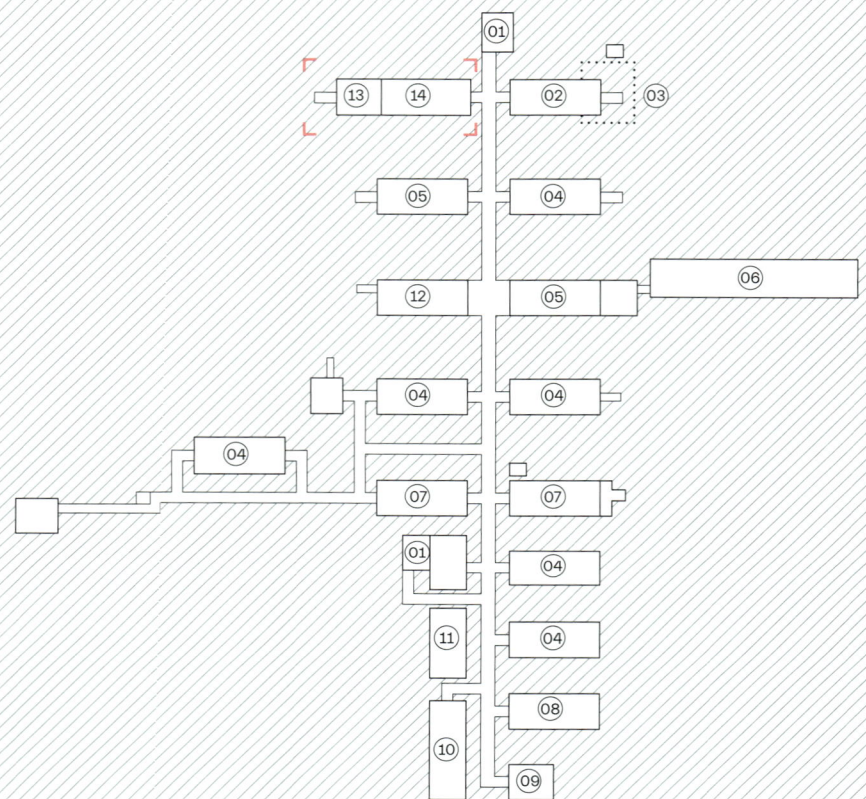

01 Heads, vertical shaft
   entrance
02 Communication building
03 Strongback antenna
04 Living quarters
05 Mess room
06 Storage
07 Laboratory
08 Recreation hall
09 Auxiliary power
10 Garage
11 Power house
12 Living quarters and
   Sick bay
13 Rawin dome
14 Meteorological laboratoy

SOURCE
United States National Archives Research Center, 1957

1 : 1,250   0   12.5   25   50 m

PLAN | METEOROLOGICAL LABORATORY

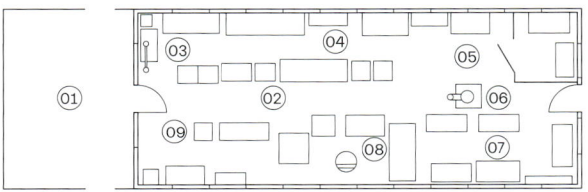

01 Rawin tower
02 Meteorological observations area
03 Jet heater, balloon conditioner, radio activity recorder
04 CO_2 analyser
05 Weather area
06 Space heater
07 Map desk
08 Ladder to plastic dome
09 Rawin sonde

SOURCE
United States National Archives Research Center, 1957

1 : 250   0   2.5   5   10 m

HALLEY I
1957

↖ p. 305, 399, 458, 466

"A new main hut and dog kennels were built close to the original IGY [International Geophysical Year] buildings in February 1961, by which time the latter were completely covered by snow. It was closed early 1968."

British Antarctic Survey, "History of Halley (Station Z)" (online)

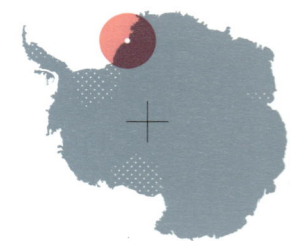

Architect and Engineer: The Crown Agents for Oversea Governments and Administrations Civil Engineering Department, United Kingdom
Manufacturer: William Harbrow Bermondsey, United Kingdom

United Kingdom
Royal Geographic Society (RGS) > Falkland Islands Dependencies Survey (FIDS) > British Antarctic Surey (BAS)

Calved > Replaced

75°31'S 26°36'W
Brunt Ice Shelf, Halley Bay
Ice shelf, 37 m.a.s.l., –20°C

SITE PLAN

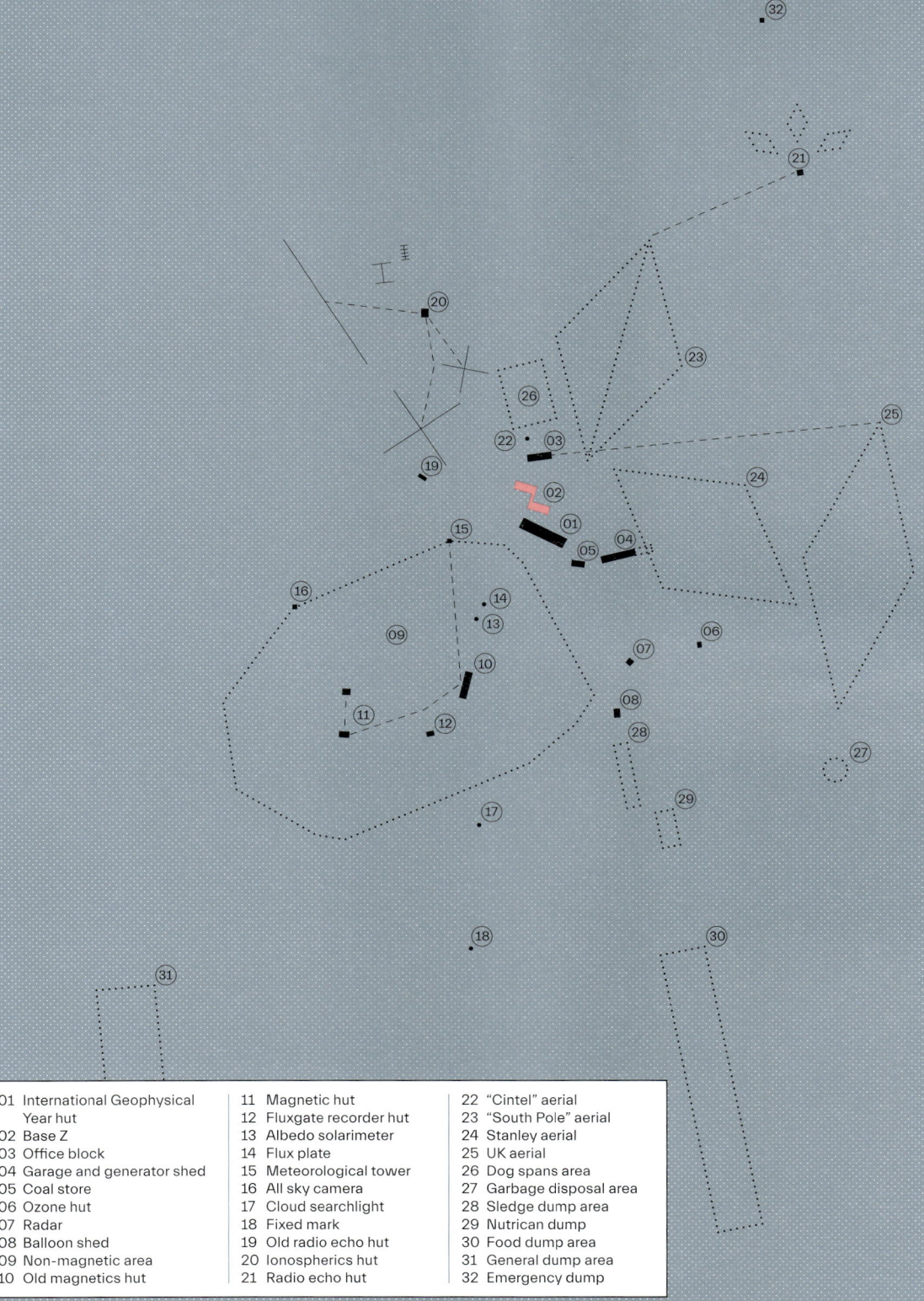

01 International Geophysical
    Year hut
02 Base Z
03 Office block
04 Garage and generator shed
05 Coal store
06 Ozone hut
07 Radar
08 Balloon shed
09 Non-magnetic area
10 Old magnetics hut

11 Magnetic hut
12 Fluxgate recorder hut
13 Albedo solarimeter
14 Flux plate
15 Meteorological tower
16 All sky camera
17 Cloud searchlight
18 Fixed mark
19 Old radio echo hut
20 Ionospherics hut
21 Radio echo hut

22 "Cintel" aerial
23 "South Pole" aerial
24 Stanley aerial
25 UK aerial
26 Dog spans area
27 Garbage disposal area
28 Sledge dump area
29 Nutrican dump
30 Food dump area
31 General dump area
32 Emergency dump

SOURCE
Falkland Islands Dependencies Survey, 1965

S

1 : 5,000   0   50   100   200 m

PLAN, BELOW SNOW LEVEL | BASE Z, 1960

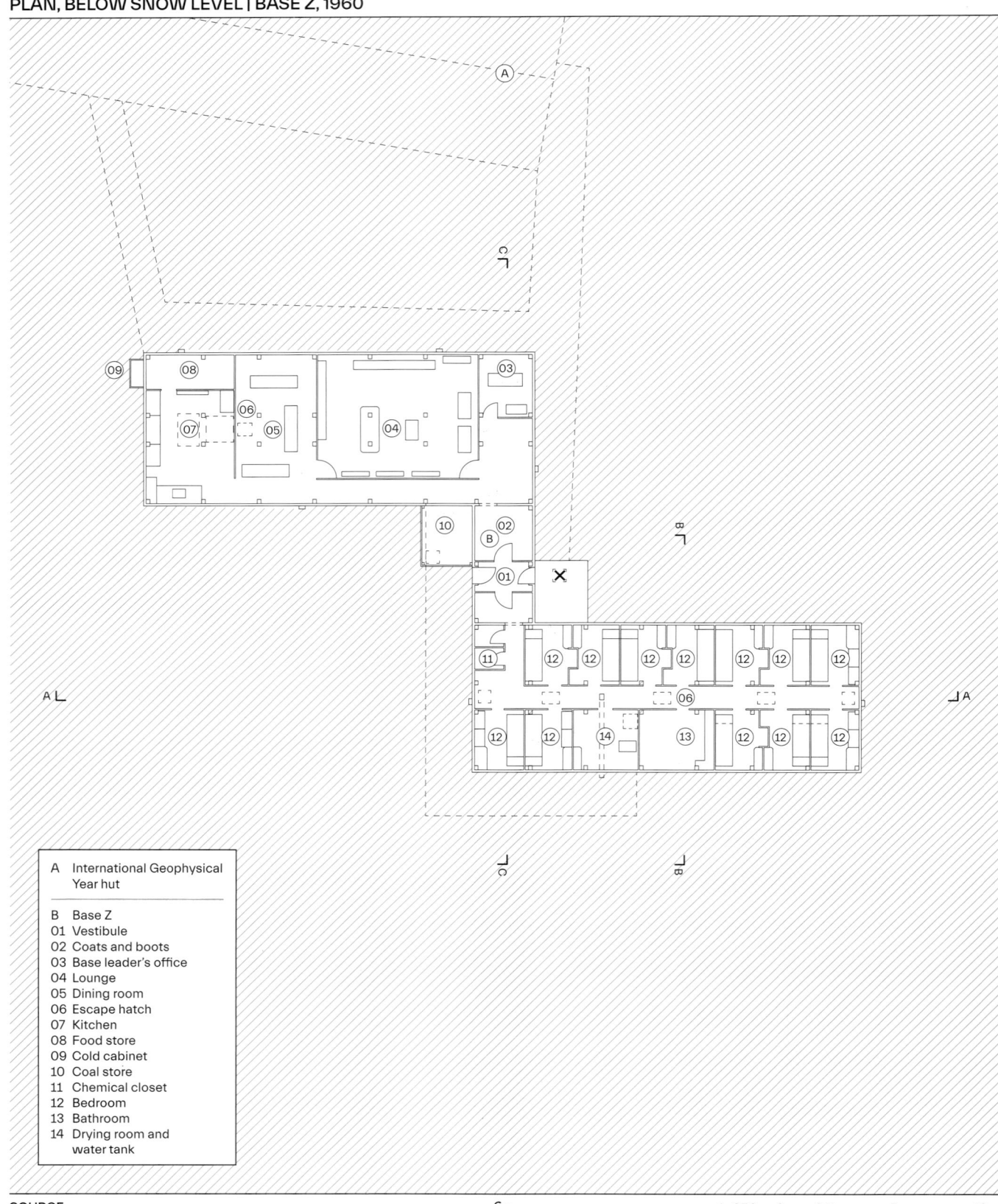

A   International Geophysical
    Year hut

B   Base Z
01 Vestibule
02 Coats and boots
03 Base leader's office
04 Lounge
05 Dining room
06 Escape hatch
07 Kitchen
08 Food store
09 Cold cabinet
10 Coal store
11 Chemical closet
12 Bedroom
13 Bathroom
14 Drying room and
    water tank

SOURCE
Falkland Islands Dependencies Survey, 1960

1 : 250   0   2.5   5   10 m

SECTION | BASE Z, 1960

"Ideally all hut entrances should be at least 20 yards [18 m] away from the buildings, connected to them by tunnels. These tunnels could be supplied with the hut and need not be excessively strong as the materials could be stripped from the walls and roofs once the snow cover was great enough. […] The normal type of door soon becomes unusable because of drift accumulation and even before this happens the hatch will prove invaluable in high winds, as it has been found that little or no drift will get in through the hatch."

Walter H. Townsend (Meteorologist), *Base Z Building Report for Falkland Islands Dependencies Survey*, 1960

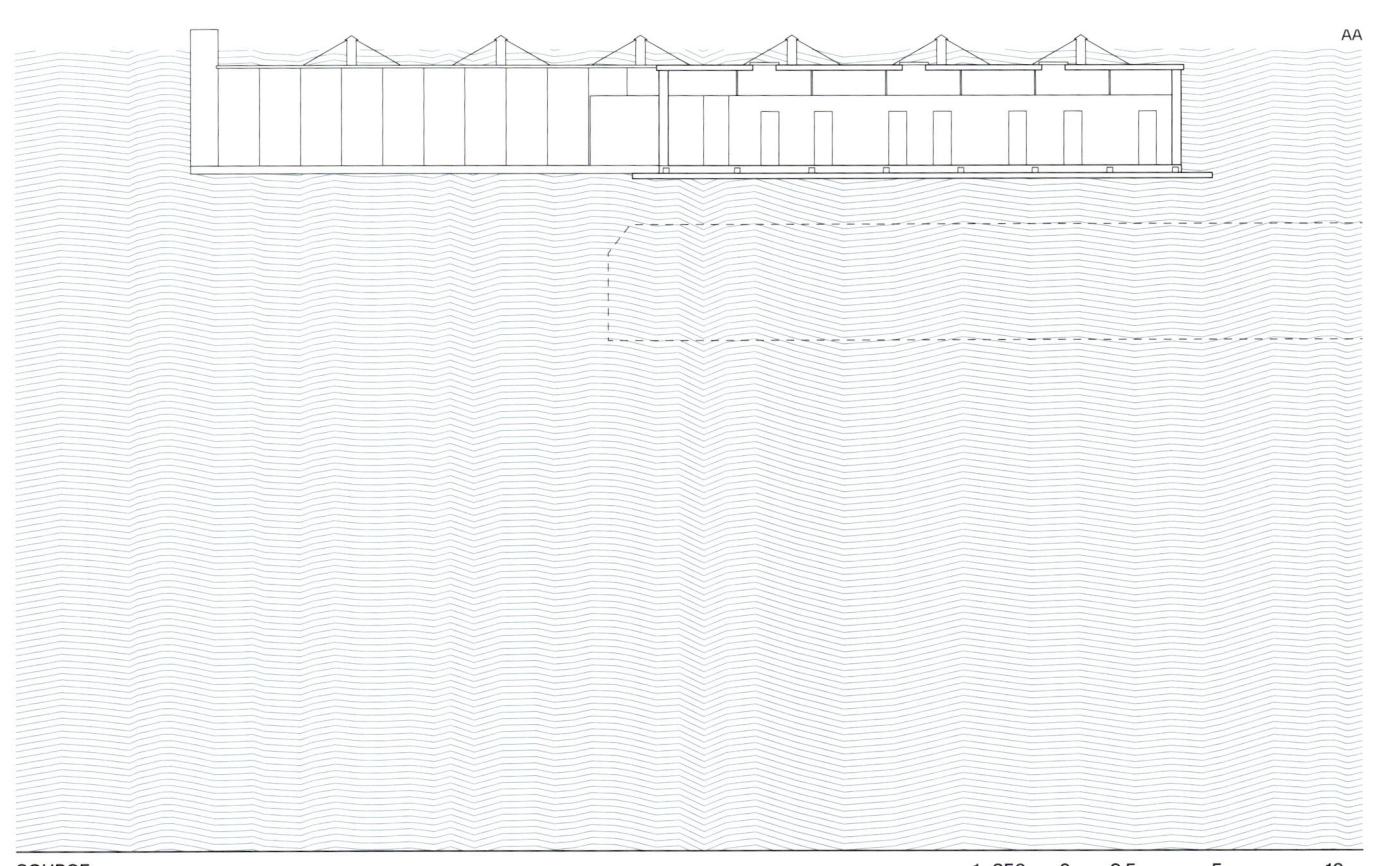

AA

SECTION DETAIL | BASE Z, 1960

BB

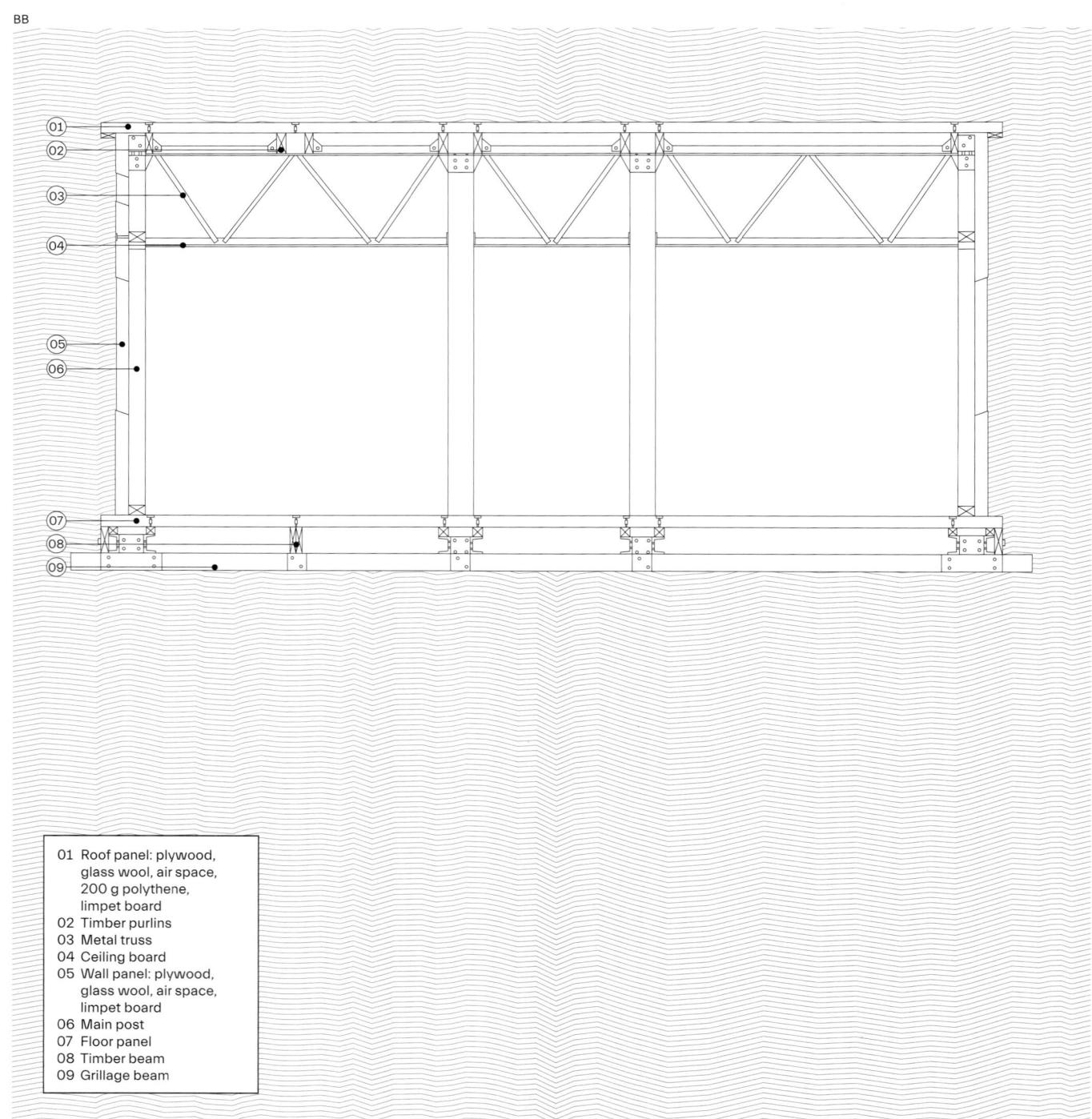

01 Roof panel: plywood,
   glass wool, air space,
   200 g polythene,
   limpet board
02 Timber purlins
03 Metal truss
04 Ceiling board
05 Wall panel: plywood,
   glass wool, air space,
   limpet board
06 Main post
07 Floor panel
08 Timber beam
09 Grillage beam

SOURCE
Falkland Islands Dependencies Survey, 1960

1:50   0   0.5   1   2 m

SECTION | BASE Z, 1960

"The special flat roofed, prefabricated living accommodation was built in two units joined by a linkpiece to form a 'z' shape. […] The structure consists of insulated plywood panels on wood and steel framing which is erected upon an expanded metal carpet laid on snow. [By 1965] the structure is sinking unevenly and is buried with 10 to 12 feet [3.6 m of snow] over the entire building. It is in a very poor condition. Lack of bolts in foundations and roof trusses has caused accelerated deformation of the structure. [The] foundations are settling unevenly opening up floor and roof joists [allowing] water in by the gallon. Roof panels are saturated with water inside as it flows through nail holes in the intermediate framing. Roof trusses bearing down presumably due to the pressure of ice on the roof [has led to] unstable foundations and [the] buckling [of] interior partitions. [The] pressure of ice on the roof is breaking the interior framing of the panels (5 [have been] damaged). [The] ingress of water through [the] roof [is] shorting out electrical fittings. [The] hut is extremely cold in winter due to draughts underneath the floor coming through open panel joints. In conclusion it is deemed that the foundations and flat roof are unsuitable, the roof panels are not strong enough and that, should the foundations not move, double tongue-and-groove flooring is deemed better than the floor panels used."

A. Baker & D. R. Jean, *British Antarctic Survey Accommodation Buildings Report,* 1964

CC

SECTION | NON-MAGNETIC HUT

"The great problem at Halley Bay is drifting snow. The object of all hut designs should be to produce buildings with as few projections from the roof as possible, chimneys and ventilators being the absolute maximum. The flat roofed buildings now being supplied are a great improvement on the ridge-roofed types, but a building of semi-circular cross section as the present Non-Magnetic Hut, would appear to be ideal. This type of building is very strong and the shape is such that it will 'scour' for many months before it becomes covered."

G. R. Lush (Explorer, Handyman), *Base Z Building Report for Falkland Islands Dependencies Survey*, 1959

01 Roof panels: curved resin bonded plywood with adesive strips, felt outer final covering
02 Isolation panels: plywood top-bottom, timber framing, three layers of glasswool
03 Supporting framework extension with entry hatch
04 Wall panel with hardwood covering strip equipped with bituminous felt: internal and external plywood covering, insulation panels of resin-bonded fibreglass

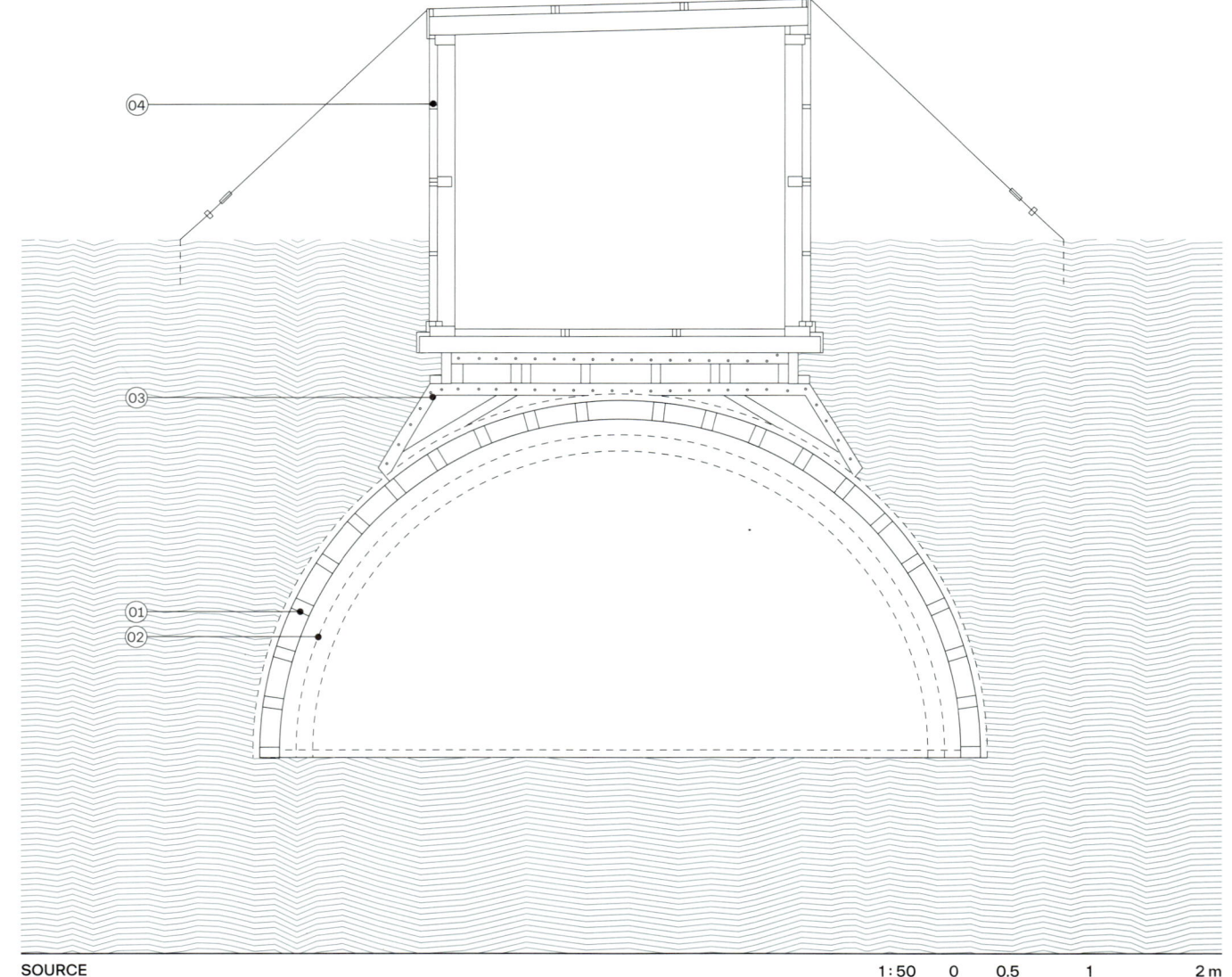

SOURCE
Falkland Islands Dependencies Survey, 1958

1:50   0   0.5   1   2 m

PLAN AND SECTION | SPECTROPHOTOMETER HUT

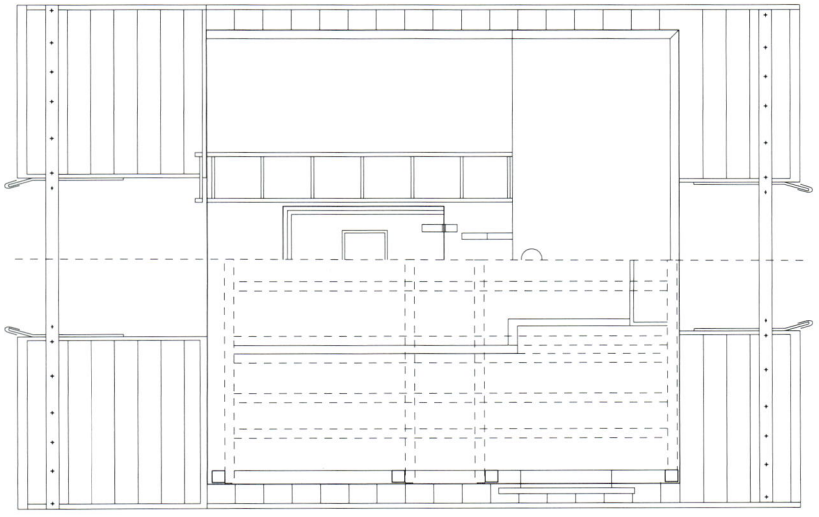

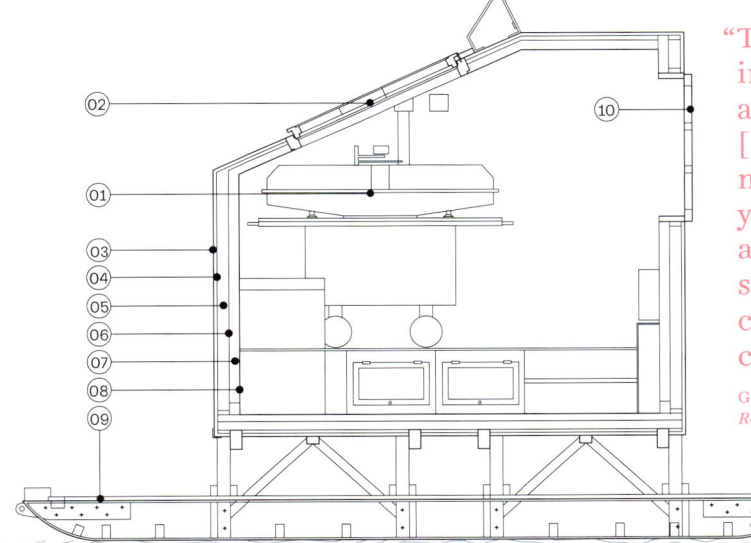

"This mobile unit was built early in the year. The design is good and it has been a great success. […] It scours well and has been moved five times during the year, the whole operation taking about 10–12 minutes. The sledge runners were given two coats of ski lacquer during construction."

G. R. Lush (Explorer, Handyman), *Base Z Building Report for Falkland Islands Dependencies Survey*, 1959

01 Spectrophotometer

02 Trapdoor
03 Resin-bonded plywood
05 Insulation board
05 Framing
06 Insulation
07 Polythene film
08 Resin-bonded plywood
09 Sledge
10 Window (single- and
      double-glazed glass)

SOURCE
Falkland Islands Dependencies Survey, 1958

1 : 50   0   0.5   1   2 m

NAVAL AIR FACILITY MCMURDO

↖ p. 399, 453

1956

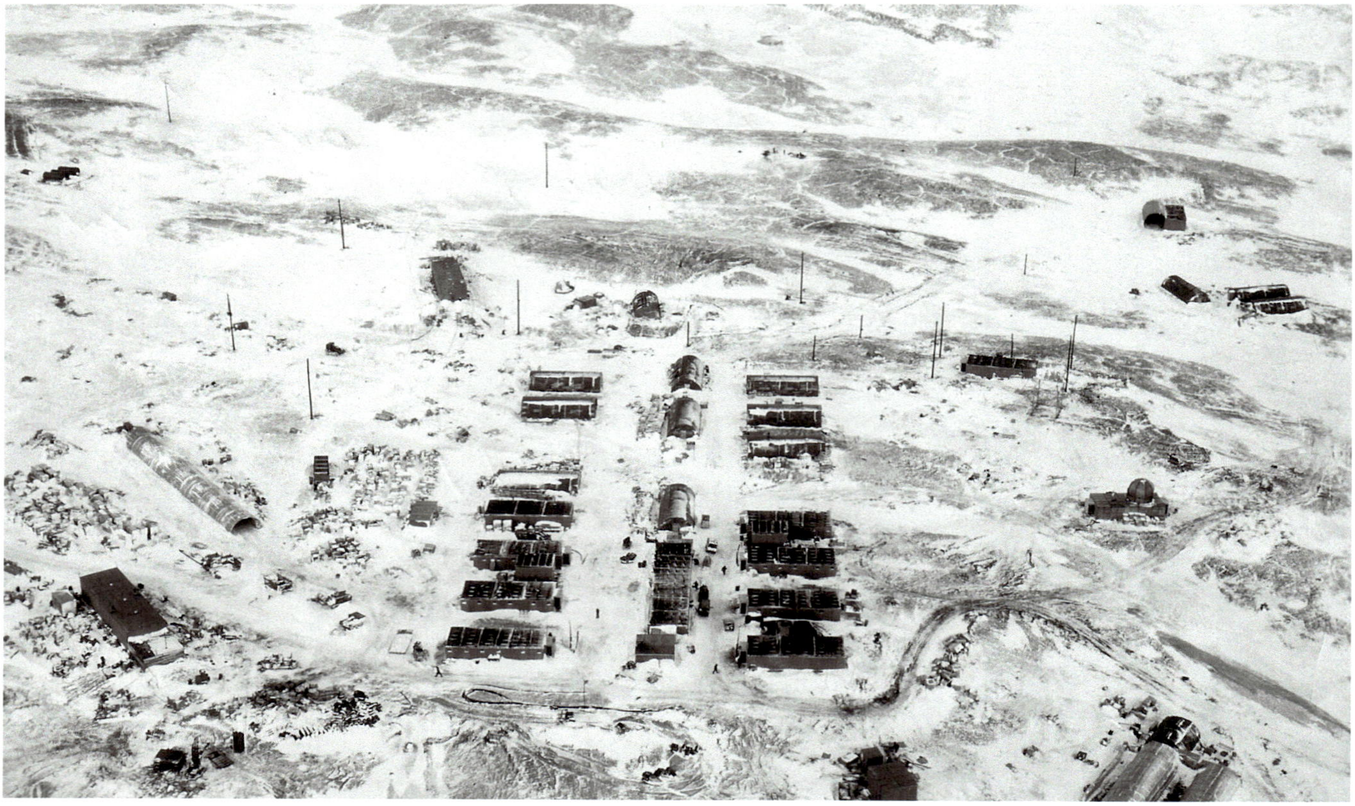

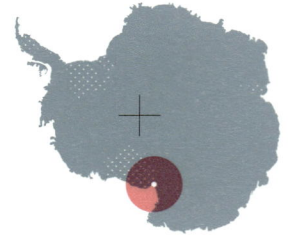

"Principal port of entry, main air staging base, nerve centre … no wonder that Naval Air Facility McMurdo has been called 'The New York of Antarctica'. Strictly speaking, McMurdo is not an IGY [International Geophysical Year] station as it had no resident scientist aboard. But without it, there would be no IGY station at any point in Antarctica save possibly the granddaddy of them all, Little America. For McMurdo is an air base and support center extraordinary, communications center, and meteorological center with four or five support aerographers on full-time duty. To sum it up, 'McMudhole' is the air hub of the Antarctic universe."

United States Navy, *Operation Deep Freeze Task Force*, 1958

Engineer: Richard Bowers, United States Navy Mobile Construction Battalion Special, United States of America
Construction: United States Navy Seabees, United States of America

United States of America

Active Station

77°50'53.5"S 166°40'06"E
Hut Point, Ross Island
Ice-free ground, 0 m.a.s.l., -17 °C

SITE PLAN

01 McMurdo
02 Hut Point hut and
    tent camp

SOURCE
United States National Archives Research Center, 1960

S

1 : 5,000   0   50   100   200 m

↖ p. 398, 539

MIRNY
1956

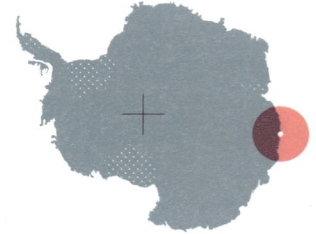

"The general plan for the construction of the base at Mirny Station was based on the following principles: maximum concentration of all the vital centres of the village within the limits defined by fire safety and protection from excessive snow drifts; placing the maximum number of structures, and especially the most vital ones, on a solid rock base; construction of all other buildings on the glacier sections, to exclude the possibility of them sliding into the sea in the future."

Grigory Luchansky, "Experience of Soviet Antarctic Stations" (online; translated from Russian by the editor)

Designer: Arctic project team, Soviet Union
Engineer: V. M. Kunin, Soviet Union
Manufacturer: Zharkovsky factory, Soviet Union

Soviet Union > Russia
Arctic and Antarctic Research Institute (AARI)

Active Station

66°33'11"S 93°01'00"E
Mirny Peninsula, Davis Sea shore
Ice-free ground, 35 m.a.s.l., −11.4°C

SITE PLAN

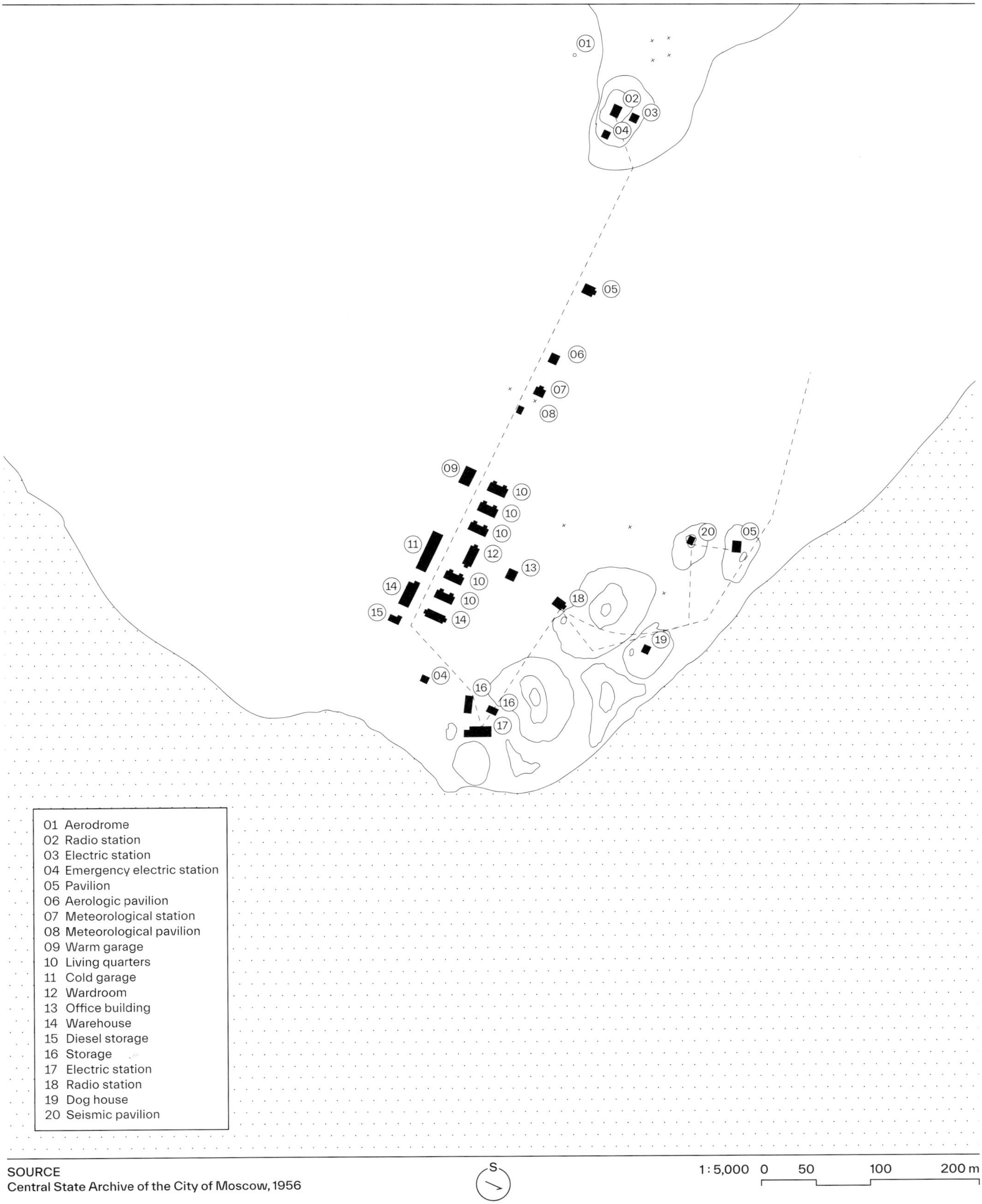

01 Aerodrome
02 Radio station
03 Electric station
04 Emergency electric station
05 Pavilion
06 Aerologic pavilion
07 Meteorological station
08 Meteorological pavilion
09 Warm garage
10 Living quarters
11 Cold garage
12 Wardroom
13 Office building
14 Warehouse
15 Diesel storage
16 Storage
17 Electric station
18 Radio station
19 Dog house
20 Seismic pavilion

SOURCE
Central State Archive of the City of Moscow, 1956

1 : 5,000   0   50   100   200 m

AMUNDSEN-SCOTT SOUTH POLE

⤹ p. 102, 574

1957

"Even in the bitterest storms
inhabitants of the base
can follow passageways from
building to building. These
icy aisles also serve as storage
houses for goods and equipment not affected by the cold.
Burlap and chicken wire,
later reinforced by parachutes,
cover the two by four framework. Snow drifts across the
roof."

Paul Allman Siple (Explorer, Geographer), "We Are
Living at the South Pole", *National Geographic,* 1958

Engineer: Richard Bowers, United States Navy
Mobile Construction Battalion Special, United
States of America
Construction: United States Navy Seabees, United
States of America

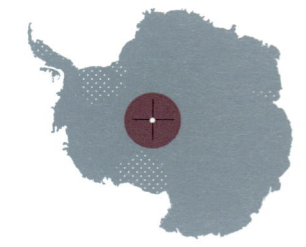

United States of America
United States Antarctic Research Program (USAP)

Buried > Replaced

90°00'00"S
Geographic South Pole, Antarctic Plateau
Ice sheet, 2,835 m.a.s.l., –49°C

SITE PLAN

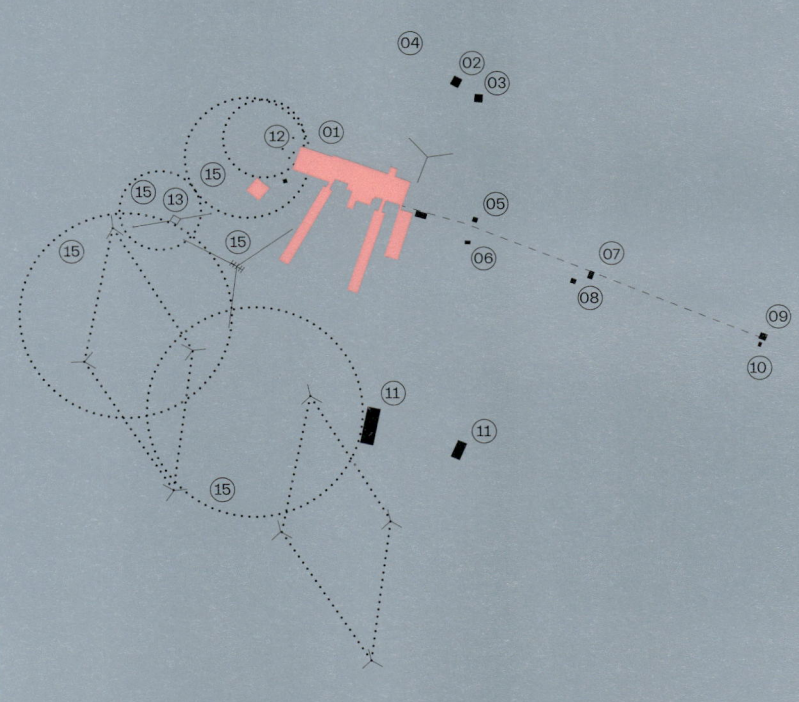

01 Main station buildings
02 Harrington hut
03 Cosmic ray building
04 Survival gear wannigan
05 NRL building
06 Dobson building
07 Absolute building
08 Variations building
09 Seismographic building
10 Seismography recording
     building
11 Garage
12 Ultra and very high
     frequency antenna field
13 Low frequency homer
14 Forward scatter antenna
15 Restricted area
16 South Pole

SOURCE
United States National Archives Research Center, 1965

S

1 : 5,000   0   50   100   200 m

PLAN, BELOW SNOW LEVEL | MAIN STATION AND WEATHER BALLOON INFLATION TOWER, 1956

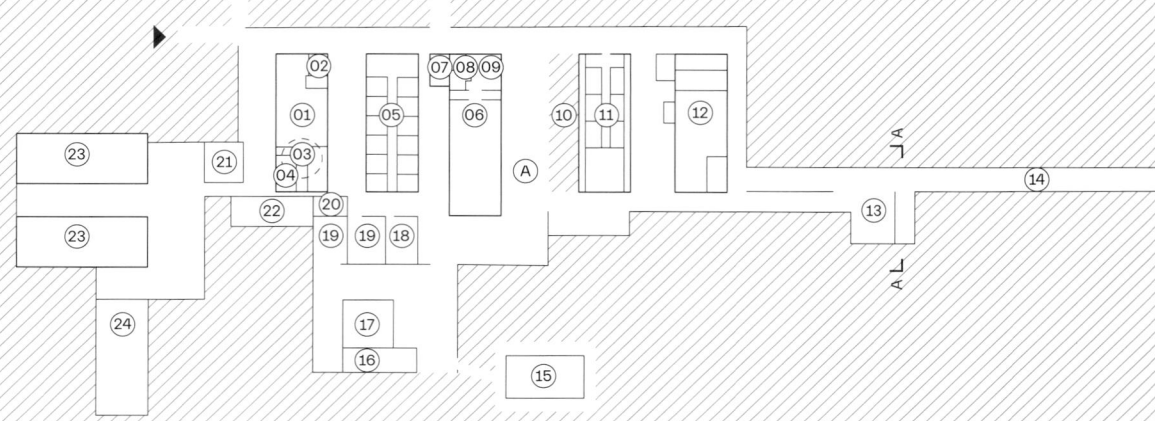

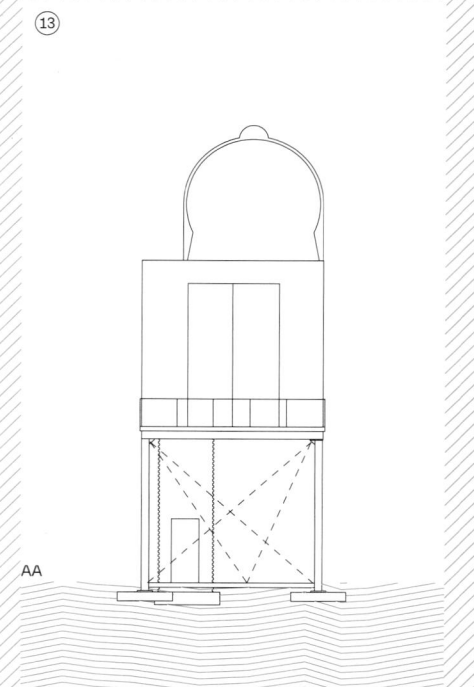

A   Original buildings
01 Galley, mess deck
02 Post office
03 Rawin dome
04 Radio shack
05 Bedrooms and sick bay
06 Utility shop, garage
07 Snow mine
08 Scientific photo lab
09 Bathroom
10 Snow wall
11 Temporary barracks
12 Science building
13 Weather balloon inflation
    tower
14 Tunnel to geomagnetism
    building
15 Emergency shelter with
    supplies for six months
16 Snow melter
17 Generator
18 Building material cache
19 Electrical cache
20 Carpenter shop
21 Chapel
22 Store
23 Food cache
24 Garage

25 Communication room
26 Snow mine
27 Snow melter
28 Generator
29 Head

30 Incenerator vault
31 Fuel cache
32 Food cache

SOURCE
US Cold Regions Research and Engineering Laboratory, 1968
United States National Archives Research Center, 1969

| 1 : 750 | 0 | 7.5 | 15 | 30 m |
| 1 : 250 | 0 | 2.5 | 5 | 10 m |

PLAN, BELOW SNOW LEVEL | MAIN STATION BUILDINGS, 1964

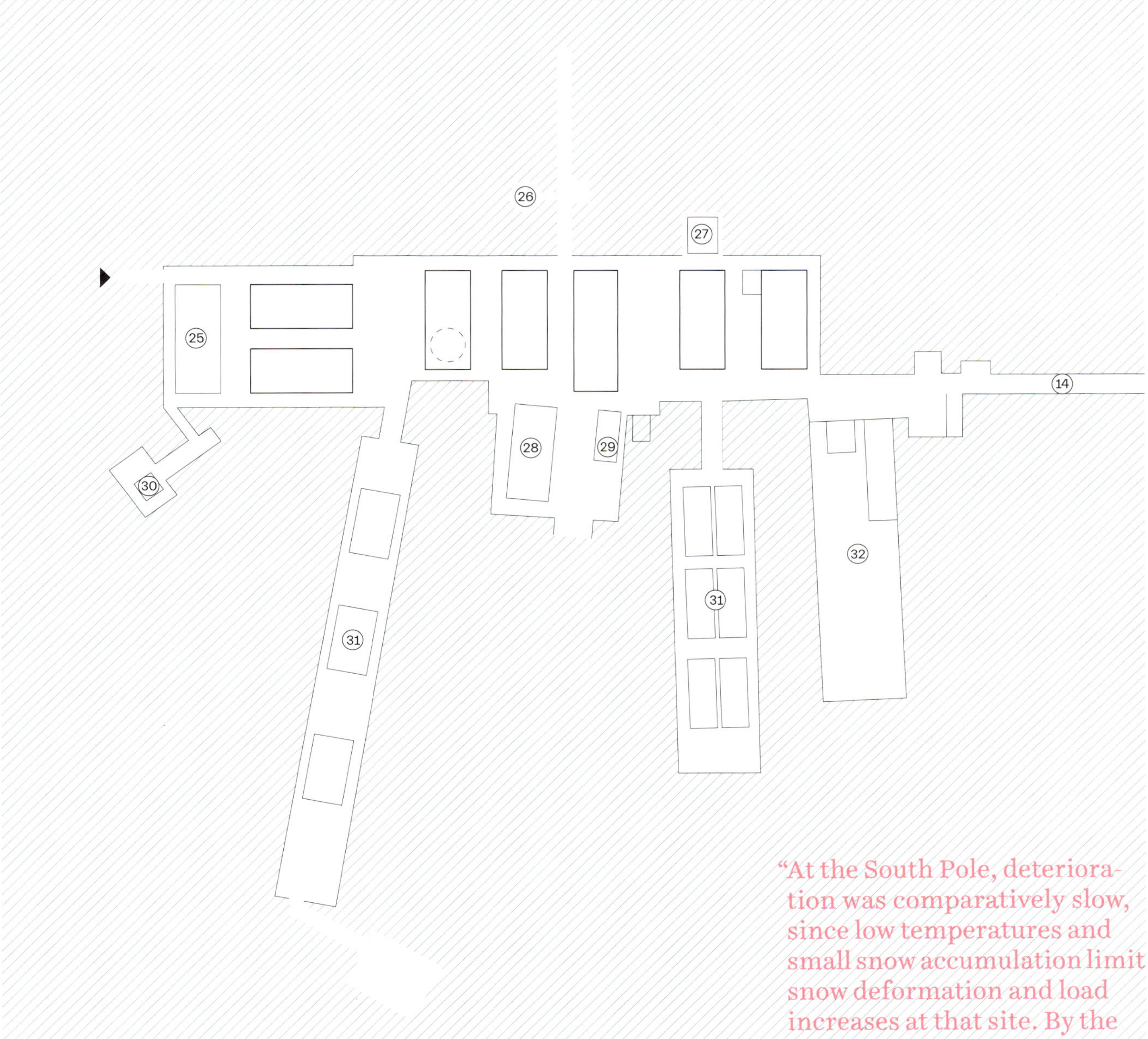

"At the South Pole, deterioration was comparatively slow, since low temperatures and small snow accumulation limit snow deformation and load increases at that site. By the end of 1959 some replacements for damaged buildings had to be provided, but it was decided eventually that the station could be maintained by a program of annual renovation."

Malcolm Mellor (Glaciologist, Polar Explorer), *Methods of Building on Permanent Snowfield*, 1968

SOURCE
United States National Archives Research Center, 1969

1:750   0   7.5   15   30 m

1957

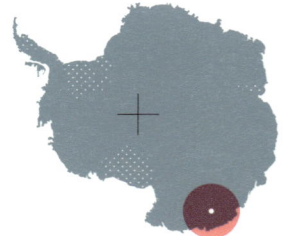

"Roland [Schlich] and Claude [Lorius] dug every day, moving ice as hard as rock out of the way. It was –40°C in the gallery and the cold burned their fingers. But with the fast-approaching winter, these corridors became a priority. […] Every day we had to spend some hours doing this work that was vital for our scientific research and for the life of the station. […] The dug-up ice was loaded onto a sledge and dragged out to be emptied far off from the station to avoid it freezing up around the station again. This took three months, during which Claude and Roland dug out 30 m of tunnel and got rid of 120 tonnes of snow to prepare Charcot's underground storage system for the winter. […] The barrack was buried under the snow to protect it from blizzards. Here, on the glacier the snow was as hard as concrete, and the team needed several days to dig the hole to bury the barrack. The storm increased again […] and the morning after, there was nothing. Everything was buried, it was completely blocked. The snow was at the roof's height."

Enterrés Volontaires Au Coeur De L'Antartique (Documentary), 2008

Engineer: Bertrand Imbert and Yves Vallette, France
Manufacturer: ACNAM, Château-du-Loir, France

France
French Antarctic Expedition (EPF)

Closed

69°22'30"S 139°01'00"E
Adélie Land
Ice sheet, –37.1°C

SITE PLAN

01 Charcot

1 : 5,000   0   50   100   200 m

SITE PLAN

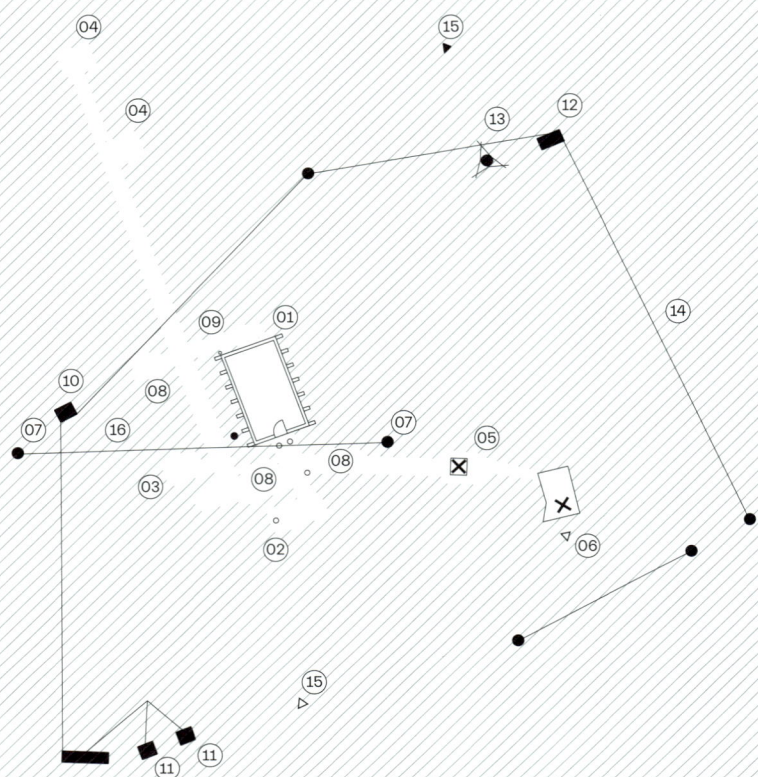

01 Living house
02 Electric station
03 Glaciological laboratory
04 Magnetic laboratory
05 Gas generator
06 Theodolite
07 Radio mast
08 Store
09 Pole with wind vane
10 Aerological tower
11 Actinometric instruments
12 Meteorological booth
13 Windmill
14 Rope along the camp
   boundary
15 Heliograph
16 Antenna

"All the material containers were grouped together in different categories for usage around the station. Continuously exposed to blizzards, they were covered with snow more and more each day. […] Jacques made a call that all of the essential materials needed for our survival were left outside. We could not possibly cram our little 24 sqm space with those boxes, where there was a pile of scientific equipment, food, energy sources […]. The only solution was to dig corridors between the barracks to access the materials, thus creating underground tunnels to move around in."

Roland Schlich (Scientist) in *Enterrés volontaires au cœur de l'Antarctique* (documentary), 2008

SOURCE
French Antarctic Expedition

1 : 500   0   5   10   20 m

SECTION AND ELEVATION | LIVING HOUSE

"With the heater on, the living temperatures inside were 0°C at the floor, and 8°C degrees at head level."

Enterrés volontaires au coeur de l'Antarctique (documentary), 2008

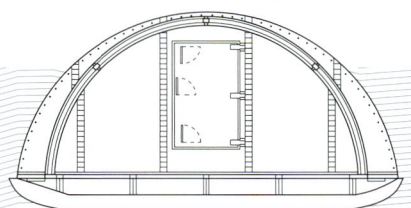

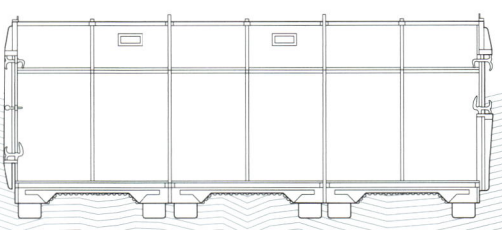

SOURCE
French Antarctic Expedition

1 : 100   0   1   2   4 m

SCOTT BASE
1957

↖ p. 398, 411, 423, 427, 452, 456, 526

![Scott Base 1957 photograph showing the prefabricated huts and connecting tunnel]

"The combination of both the TAE [Commonwealth Trans-Antarctic Expedition] and IGY [International Geophysical Year] parties from New Zealand into the New Zealand Antarctic Expedition and the construction of Scott Bas to house both has two most important results. In the first place New Zealand put into Antarctic an expedition whose capabilities on all phases of field and observatory work was most comprehensive. The laboratory aspects of the IGY were complementary to the field aspects of the TAE. Even more important was the result that when at the end of 1958 both undertakings had been success-fully completed, New Zealand was left with a base in Antarctic which was able to support observatory and field studies of the wildest demands, and this nation was able to concur with other larger nations in a resolution to continue research in Antarctic, indefinitely."

A. S. Helm and J. H. Miller, *Antarctica*, 1964

Architect: Frank Ponder (Head Architect), Ministry of Works, New Zealand
Manufacturers: (insulation) Explastics Insulations of Melbourne, Australia; (base frames, interior partitions, beams, built-in furniture) Hitchens, Fletcher Construction Co.Ltd., Australia; (structural steel) Dorman Long and Co.Ltd, United Kingdom; (main door) N O Piersen Ltd (refrigeration and air conditioning engineers), New Zealand
Suppliers: (shacklock 501) H E Shacklock Ltd, New Zealand; (No. 2 president cooker) Smith and Wellstood, United Kingdom

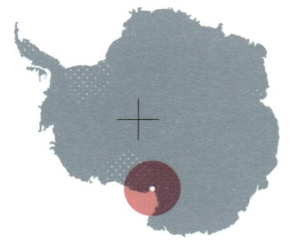

New Zealand
Commonwealth Trans-Antarctic Expedition, 1955–1958

Preserved, HSM 75 + Active Station > Project 2020+
New Zealand Antarctic Heritage Trust +
Antarctica New Zealand

77°50'53.5"S 166°40'06.3"E
Pram Point, Ross Island
Ice-free ground, 26 m.a.s.l., –19.8°C

SITE PLAN

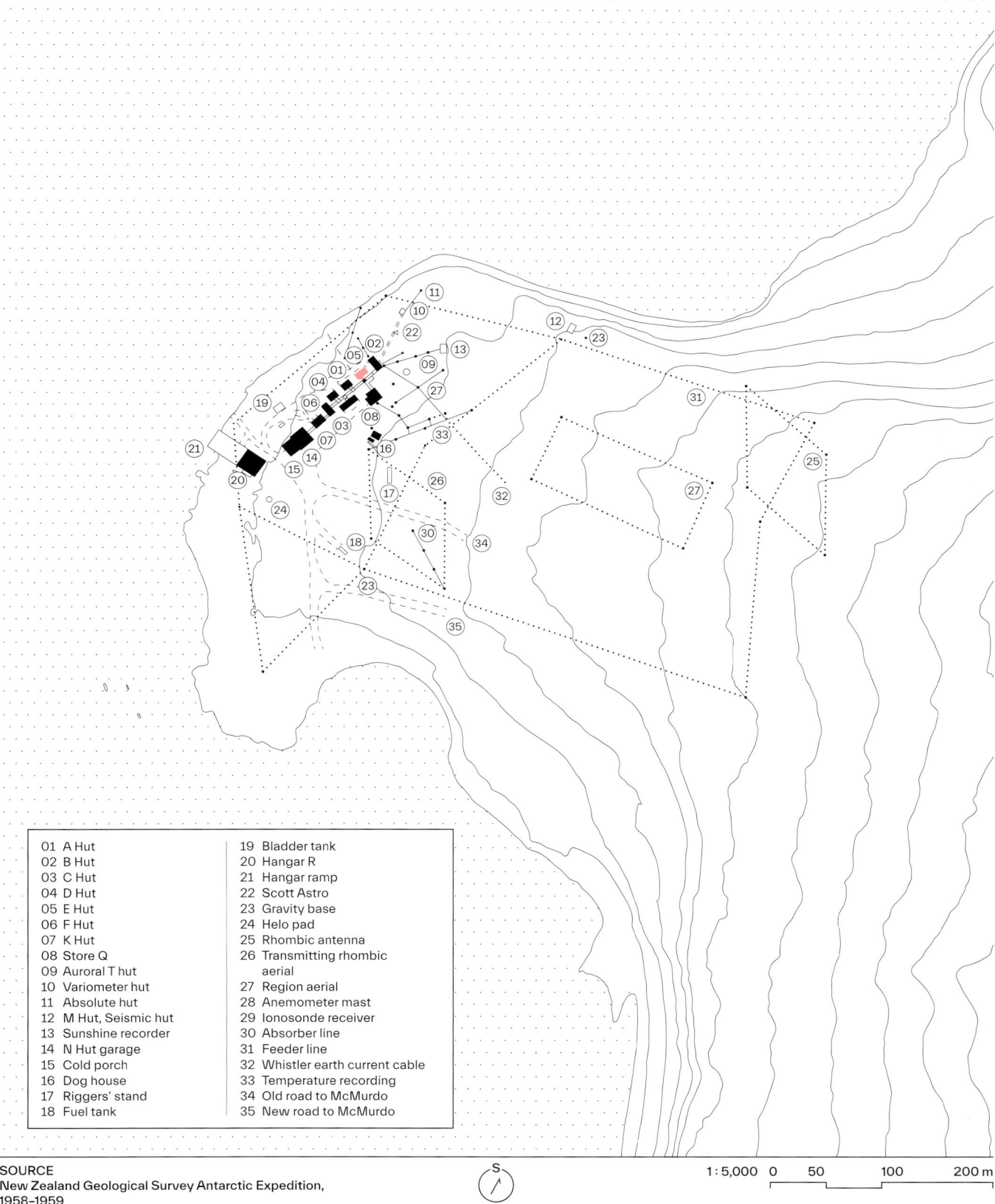

01 A Hut	19 Bladder tank
02 B Hut	20 Hangar R
03 C Hut	21 Hangar ramp
04 D Hut	22 Scott Astro
05 E Hut	23 Gravity base
06 F Hut	24 Helo pad
07 K Hut	25 Rhombic antenna
08 Store Q	26 Transmitting rhombic
09 Auroral T hut	aerial
10 Variometer hut	27 Region aerial
11 Absolute hut	28 Anemometer mast
12 M Hut, Seismic hut	29 Ionosonde receiver
13 Sunshine recorder	30 Absorber line
14 N Hut garage	31 Feeder line
15 Cold porch	32 Whistler earth current cable
16 Dog house	33 Temperature recording
17 Riggers' stand	34 Old road to McMurdo
18 Fuel tank	35 New road to McMurdo

SOURCE
New Zealand Geological Survey Antarctic Expedition,
1958–1959

S

1 : 5,000   0   50   100   200 m

PLAN AND ELEVATION | A HUT

"The designing and building of the base huts and their fittings was probably the biggest problem that faced the expedition before its departure. Two basic types of hut were considered. The first, which Fuchs was using, was the traditional type of British hut that has been used for many years by the Falkland Islands Dependencies Survey on the Grahamland Peninsula. It is a prefabricated wooden building with every precut piece of timber carefully numbered. It is a strong and utilitarian type of construction and is relatively cheap. […] In contrast was the second type which is widely used now by the Americans, the Russians and the Australians. This uses large insulated panels which fit one into the other and are then bolted or wedged together […]. This method has the major advantage of being very quick to erect."

Sir Edmund Hillary (Explorer, Mountaineer), *No Latitude for Error*, 1961

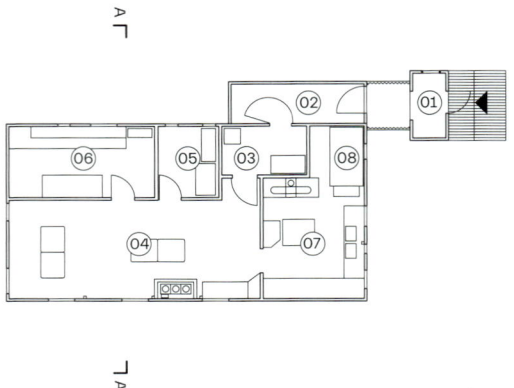

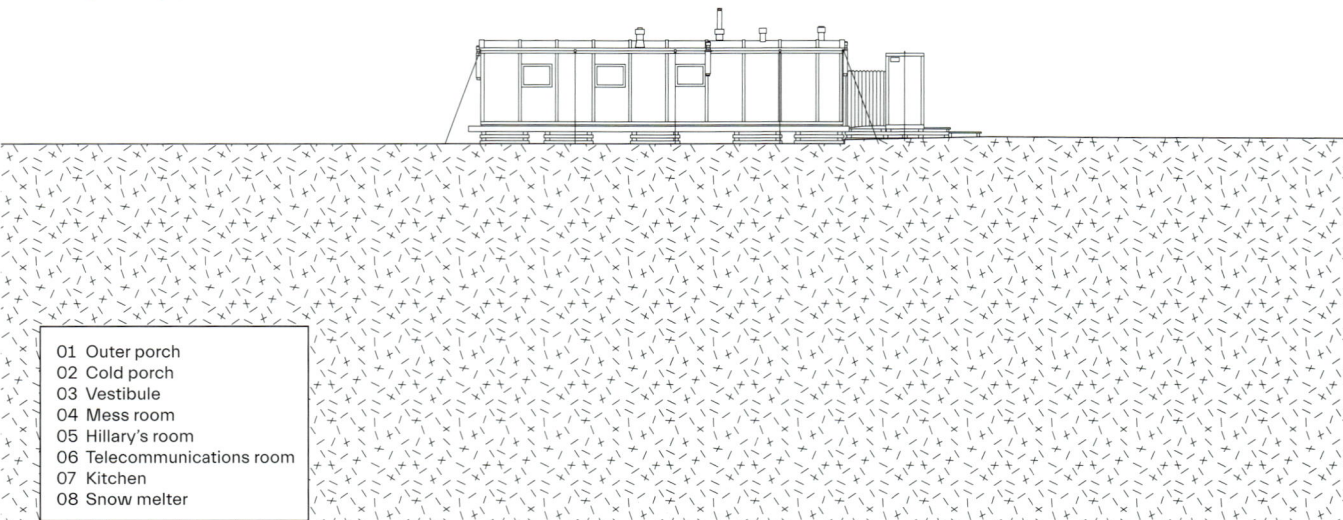

01 Outer porch
02 Cold porch
03 Vestibule
04 Mess room
05 Hillary's room
06 Telecommunications room
07 Kitchen
08 Snow melter

SOURCE
New Zealand Antarctic Heritage Trust, 2015

1 : 250   0   2.5   5   10 m

SECTION AND ELEVATION | A HUT

"Scott Base was itself a remarkable achievement, made of six interconnected buildings and three other detached science huts. Hut A, which comprised the original kitchen and mess, as well as Hillary's room, survives today as part of the modern and much expanded Scott Base. It has been listed as an Historic Monument under the Antarctic Treaty, in recognition of its important in the history of exploration and science in Antarctica. It still serves a useful purpose in the management of the base, principally as a refuge."

New Zealand Antarctic Heritage Trust, *Conservation Plan: Hilary's Hut, Scott Base,* 2015

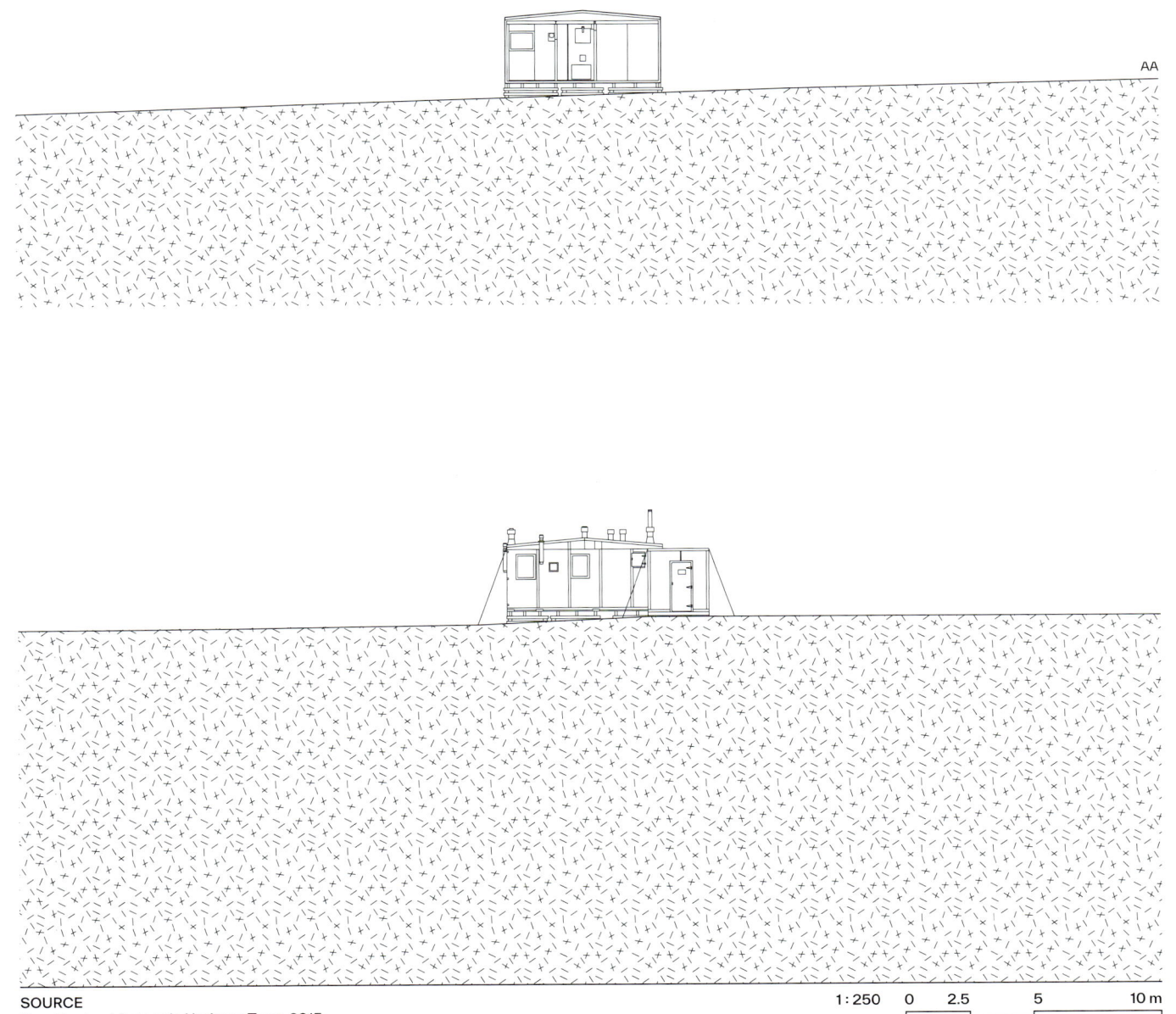

1 : 250   0   2.5   5   10 m

POLE OF INACCESSIBILITY

1958

↖ p. 537, 539

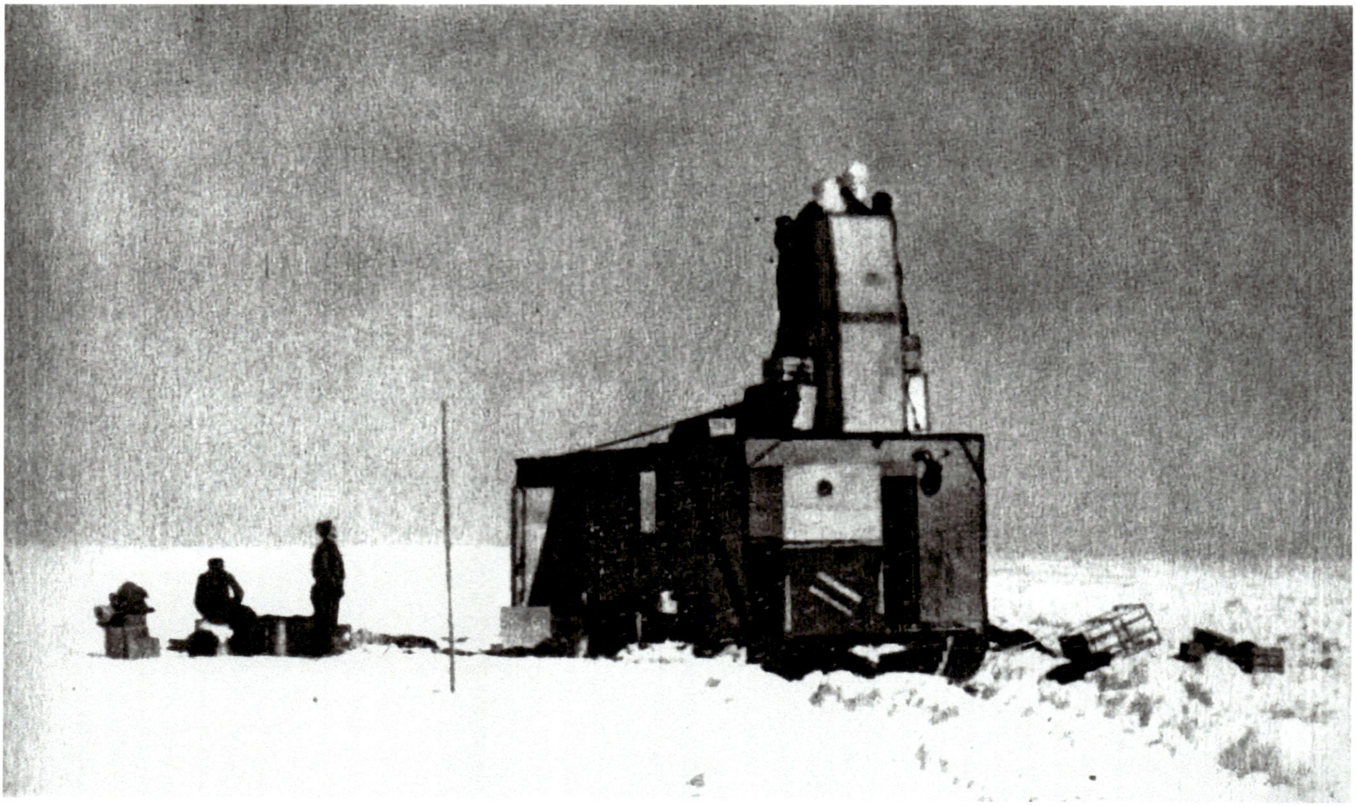

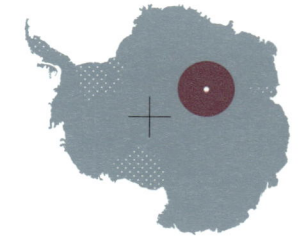

"The station consists of a small house with 24 m² floor space set up on steel sledges containing an electric station of 13 kw power, electric kitchen, radio station and living quarters for four persons. The quarters are heated by electric stove, and it has a stand-by stove working on solar oil. A stock of fuel to last for 6 months is kept at the station. A storeroom has been improvised for stocking provisions. It holds enough stock of provisions for six months for four persons."

L. I. Dubrovin and V. N. Petrov, *Scientific Stations in Antarctica 1882–1963*, 1967

Soviet Union > Russia
Arctic and Antarctic Research Institute (AARI)

Soviet Union
Buried

82°06'S 54°58'E
Kemp Land, Kemp Coast
Ice sheet, 3,718 m.a.s.l., −56.8 °C

SITE PLAN

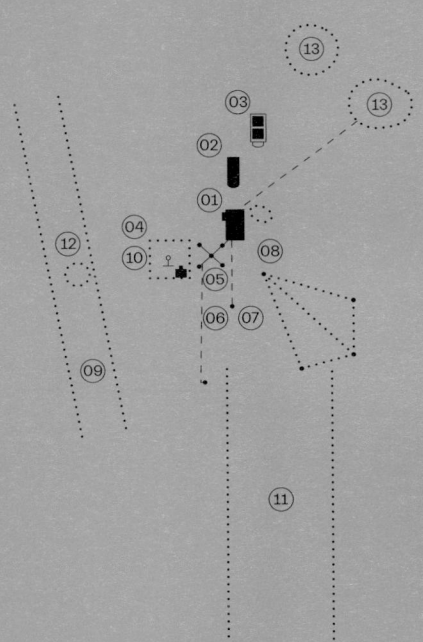

01 Sleeping quarters sledge
02 Preserved tractor HSM
   No. 15
03 Two-container sledge
04 Weather station
05 Radio mast
06 Thermometer thermowells
07 Blast hole
08 Astro-gravimetric point
09 Soviet Union runway
10 Mast with weather vane
11 United States of America
   runway
12 United States of America
   empty barrel warehouse
13 United States of America
   food warehouses

SOURCE
Arctic and Antarctic Research Institute, 1967

1 : 5,000   0   50   100   200 m

↖ p.528

BASE E
1946/1961

"The original building was known as Trepassey House after the ship MV *Trepassey* in which it was transported. The station was re-sited when a new main hut was erected on 4 Mar 1961. The new hut was the first two-storey building to be erected by FIDS [Falkland Islands Dependencies Survey]. It was unnamed. Two single-storey extensions were added, one in 1965 and another begun on 27 Jan 1972. Buildings from East Base were also used as workshops and stores. These were known as Passion Flower Hotel, Jenny's Roost and Finn Ronne. The derelict Trepassey House was burnt down in stages between Jan 1973 and Jan 1974."

British Antarctic Survey, "History of Stonington Island (Station E)" (online)

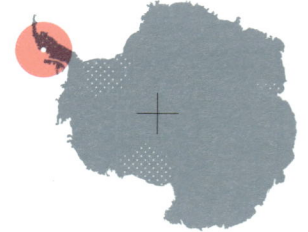

Manufacturer: Boulton and Paul, United Kingdom

United Kingdom
Falkland Islands Dependencies Survey (FIDS)

Preserved, HSM 64
United Kingdom Antarctic Heritage Trust (UKAHT)

68°11'S 67°00'W
Stonington Island, Marguerite Bay
Ice-free ground, 5 m.a.s.l., −11°C

SITE PLAN

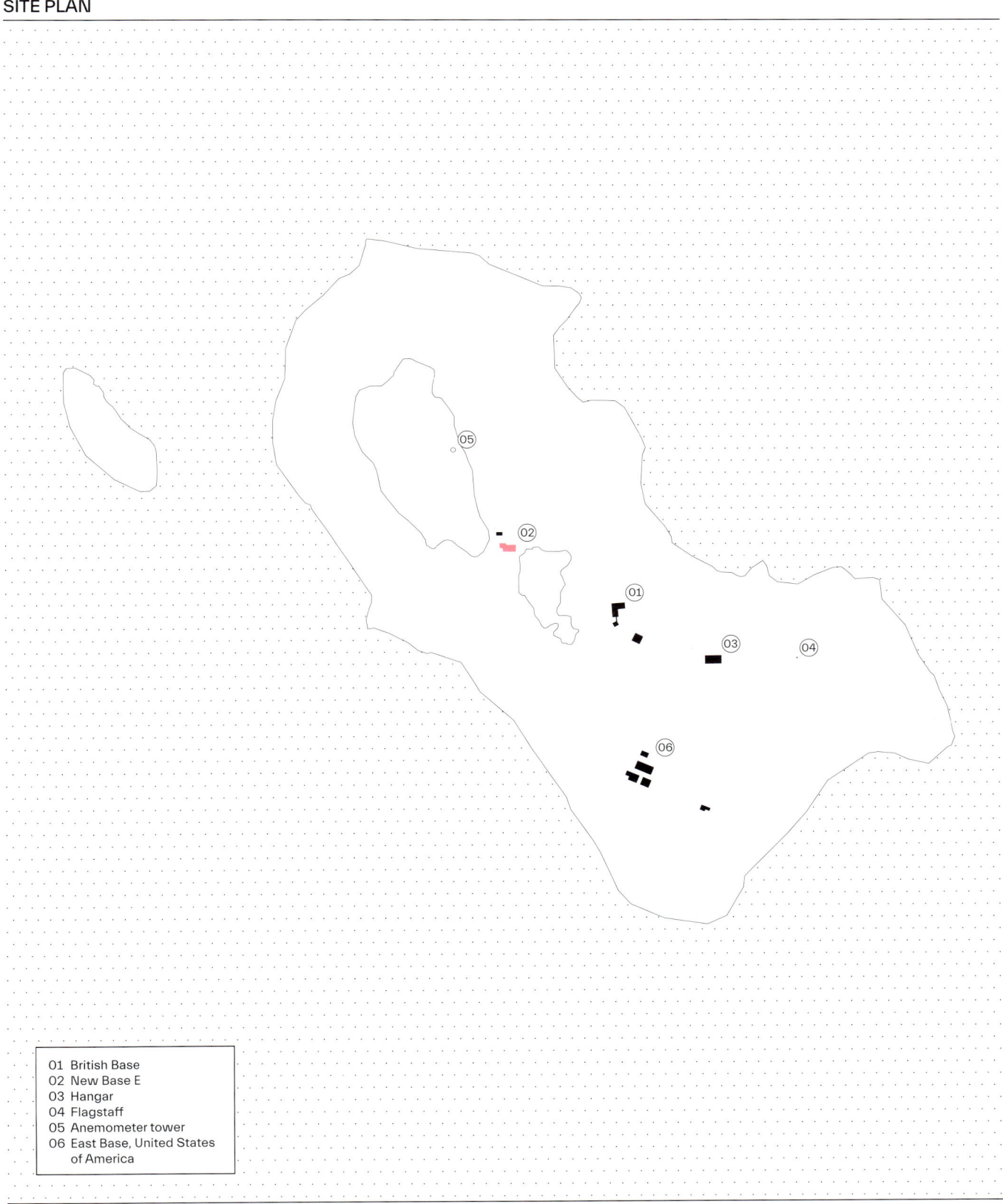

01 British Base
02 New Base E
03 Hangar
04 Flagstaff
05 Anemometer tower
06 East Base, United States
     of America

SOURCE
Falkland Islands Dependencies Survey, 1961

S

1 : 5,000   0   50   100   200 m

PLAN, 1ST AND ROOF LEVEL | TREPASSEY HOUSE, 1946

"Established 250 yards [228 m] from the US East Base. The station was closed in 1950 as sea ice conditions had prevented access to relieve the station in 1949. It was reopened in 1960 as the centre for field work in the south Antarctic Peninsula area when Horseshoe Island (Station Y) was closed. The original intention had been to build Station E on the east coast of Graham Land."

British Antarctic Survey, "History of Stonington Island (Station E)" (online)

roof level

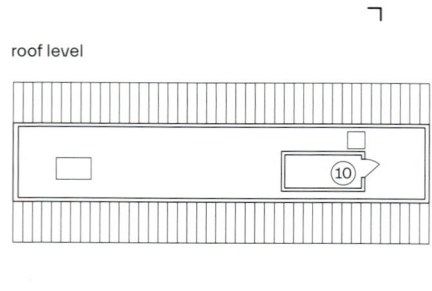

1st level

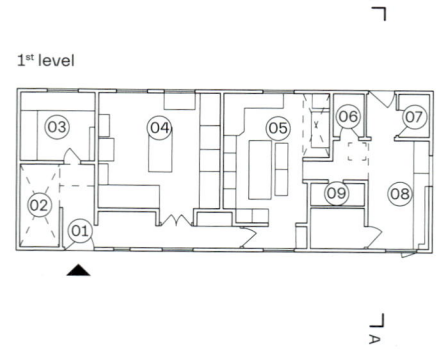

1st level
01 Vestibule
02 Oil store
03 Geological laboratory and survey room
04 Sleeping quarters and living room
05 Kitchen and dining room
06 Bathroom
07 Chemical lavatory
08 Workshop
09 Coal store

Roof level
10 Darkroom

SOURCE
Falkland Islands Dependencies Survey, 1943

1 : 250   0   2.5   5   10 m

SECTION AND ELEVATION | TREPASSEY HOUSE, 1946

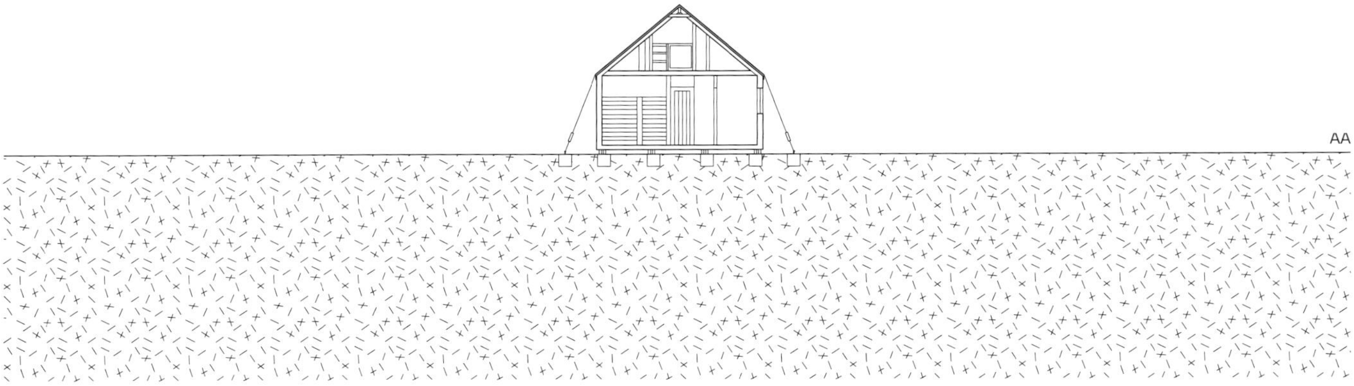

AA

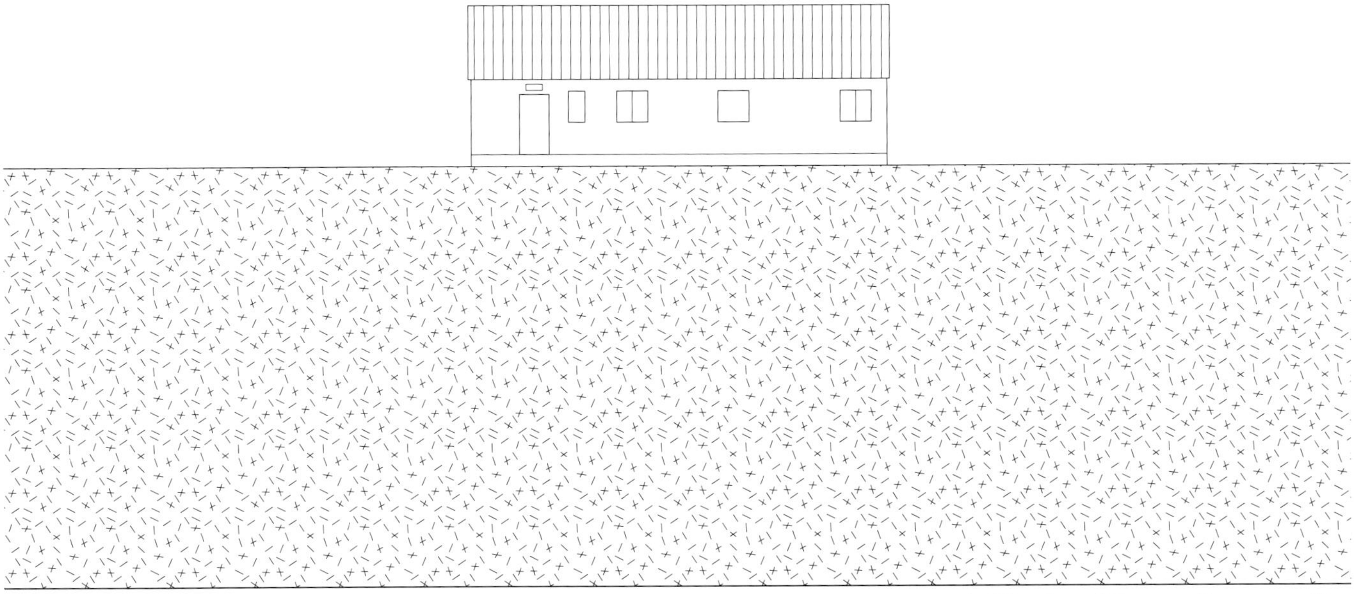

SOURCE
Falkland Islands Dependencies Survey, 1943

1 : 250   0   2.5   5   10 m

PLAN, 1ST AND 2ND LEVEL | NEW BASE E, 1961

"Framing = Of an extremely strong arrangement with all plated half-lapped and dwangs and studs connected by mortise and tenon joints. The heavy timbers requiring a certain amount of determination and strength to erect, the structure as a whole was assembled in the manner of a 'Jig Toy.' Only one criticism can be levelled at the manufacturers – the framework had not been completely assembled in the United Kingdom, and a large number of joints had to be cleaned before assembly. Once assembled the framework was completely solid without a sign of looseness anywhere."

John Crabbe Cunningham (Climber), *Falkland Islands Dependencies Survey Report on the Erection of the New Base Hut at Stonington Island,* 1961

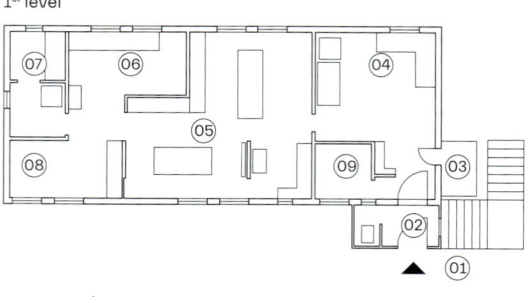

2nd level

1st level

1st level
01 Storage under stairs
02 Vestibule
03 Cold food storage
04 Kitchen
05 Dining and sitting room
06 Survey room
07 Bathroom
08 Metereology room
09 Telecommunications room

2nd level
10 Darkroom
11 Sleeping quarters
12 Snow hatch

SOURCE
Falkland Islands Dependencies Survey, 1960

1 : 250   0   2.5   5   10 m

SECTION AND ELEVATION | NEW BASE E, 1961

"The upper hut is built up of insulated panels with roofing felt and wooden strips over the joints, the corners, eaves and gable ends were all covered with galvanized sheet and metal flashing. The door, as with the door on the lower hut was a heavy arrangement with a refrigeration type door handle with rubberzote around the door opening (to make it drift proof); the rubberzote perished very quickly as the base pups found it very tasty and I think an essential part of their diet if one judges by the rate of disappearance."

John Crabbe Cunningham (Climber), *Falkland Islands Dependencies Survey Report on the Erection of the New Base Hut at Stonington Island,* 1961

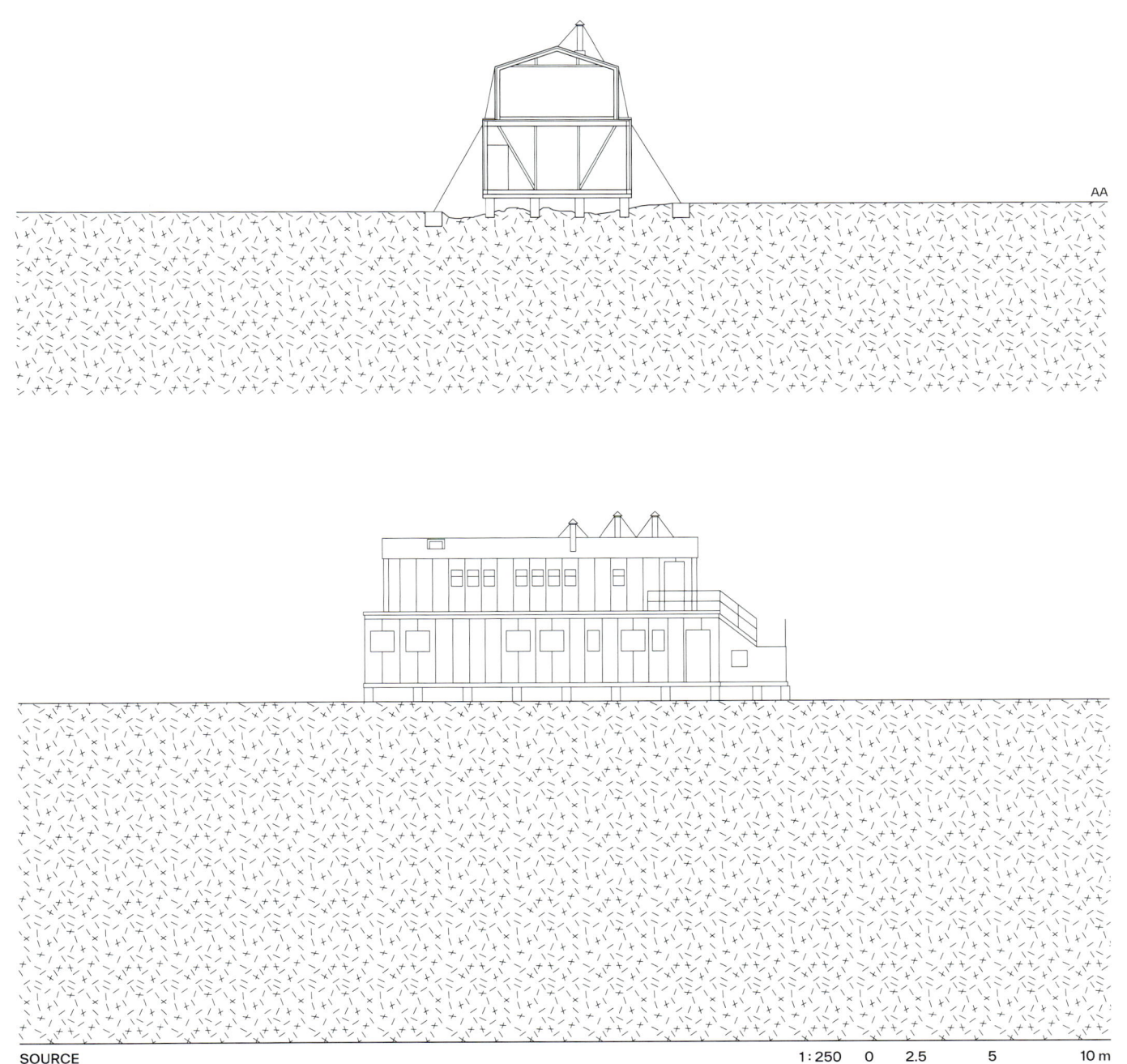

AA

SOURCE
Falkland Islands Dependencies Survey, 1960

1 : 250   0   2.5   5   10 m

NEW BYRD
1957/1962

↖ p. 424, 428, 460

*"The new Byrd station, commis-
sioned in 1961, was basically an
under-snow facility constructed
inside a network of tunnels,
following the principles of Camp
Century, Greenland."*

Malcolm Mellor (Glaciologist, Polar Explorer), *Methods
of Building on Permanent Snowfield,* 1968

Architect and Engineer: Thomas B. Bourne
Associates, United States of America
Construction: United States Navy Seabees, United
States of America

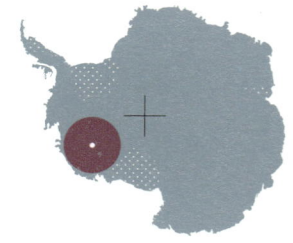

United States of America
Federal Government of the United States, United
States Navy

Buried

80°00'53"S 119°33'56"W
Marie Byrd Land, West Antarctica
Ice sheet, 1,514 m.a.s.l., –25 °C

SITE PLAN

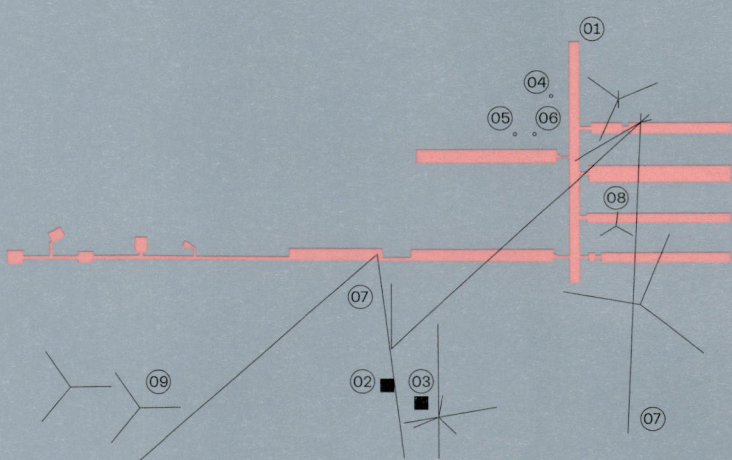

01 Main station buildings
02 Radio noise laboratory
03 Ionosphere laboratory
04 Whip antenna
05 Ultra high frequency
   antenna
06 Very high frequency
   antenna
07 V-beam antenna
08 Homer antenna
09 Meteorological antenna

SOURCE
United States National Archives Research Center, 1965

S

1 : 5,000　0　　50　　　100　　　200 m

PLAN, BELOW SNOW LEVEL

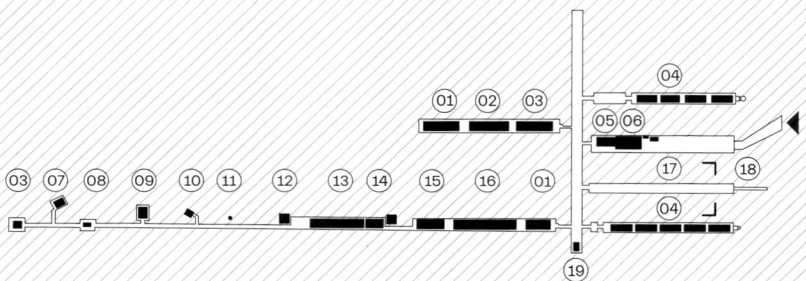

01 Administration and
   hospital
02 Dining room and kitchen
03 Communications
04 Fuel trench with inflatable
   fuel bladders
05 Public maintenance
06 Vehicle maintenance shop
07 Magnetic variation
08 Geomagnetic calibration
09 Seismology laboratory
10 Hydrogen generator, radio
   noise vaults, balloon
   inflation shelter
11 Cargo hatch
12 Rawin building and tower
13 Science, meteorology
14 Recreation
15 Science, darkroom
16 Sleeping quarters
17 Deep-drill facility
18 Cargo chute
19 Emergency diesel

SOURCE
United States National Archives Research Center, 1965

S

1 : 1,250   0   25   50   100 m

SECTION

"While most facilities (e.g. quarters, mess, galley, workshops, offices, power plant, storage rooms) are housed inside the tunnels, some small buildings (aurora tower, balloon pavilion, rawin dome, radio noise laboratory) are elevated above the ever-rising snow surface on extensible columns. Access from the tunnels to elevate buildings is provided by vertical steel-lined shafts. […] The tunnels were constructed by the 'cut and cover' method. Trenches were excavated rapidly and precisely to a depth of about 25 ft [7.6 m] by a Swiss Peter rotary snowplow. There were two principal types of trench cross section: wide trenches with vertical walls for roofing by 30- and 40-ft- [9–12 m] span Wonder arch, and narrower undercut trenches for roofing by 14-ft- [4.2 m] span corrugated steel arch. Timber stills were laid on prepared snow abutments for the support of roof arches. Snow from the excavations was blown over the arches as a backfill to give a fixed snow cover of 2 to 3 ft [0.6–0.9 m] on the arch crown."

Malcolm Mellor (Glaciologist, Polar Explorer), *Methods of Building on Permanent Snowfield,* 1968

SOURCE
United States National Archives Research Center, 1965

1 : 250   0   2.5   5   10 m

1963

↖ p. 413, 509

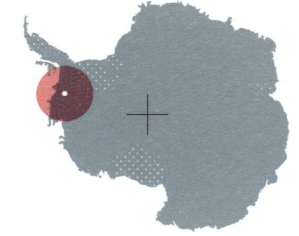

"The basic station units were 8- × 8- × 27-ft [2.4 × 2.4 × 8.2 m] skid-mounted 'vans'. These factory-built units, which were essentially similar to house trailers, were shipped completely fitted with electrical wiring, stoves, cabinets, and basic furnishings. The vans were flown to the site in LC-130 Hercules aircraft from McMurdo Sound [...]. Eight of the basic van units were arranged at the site into two parallel ranks of four vans each, and the 8-ft- [2.4 m] wide space between was fitted with roof and floor. [...] The station was designed so that it could be disassembled and repositioned on the surface if snow loads became excessive."

Malcolm Mellor (Glaciologist, Polar Explorer), *Methods of Building on Permanent Snowfield*, 1968

Architect and Engineer: Bureau of Yards & Docks, Department of the Navy, United States of America
Manufacturer: Alberta Trailer Company, United States of America
Construction: United States Navy Seabees, United States of America

United States of America
United States Antarctic Program (USAP)

Closed

75°14'00"S 77°10'00"W
Queen Maud Land
Ice sheet, 420 m.a.s.l., −25.6 °C

SITE PLAN

01 (01)

01 Main station building

SOURCE
United States National Archives Research Center
(reconstruction)

S

1 : 5,000   0   50   100   200 m

PLAN, 1ST LEVEL AND ROOF | AIR PORTABLE SCIENTIFIC FACILITY

Roof level

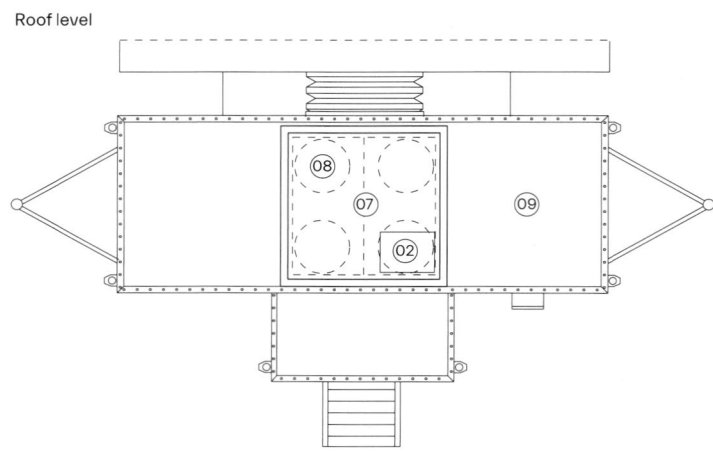

1st level

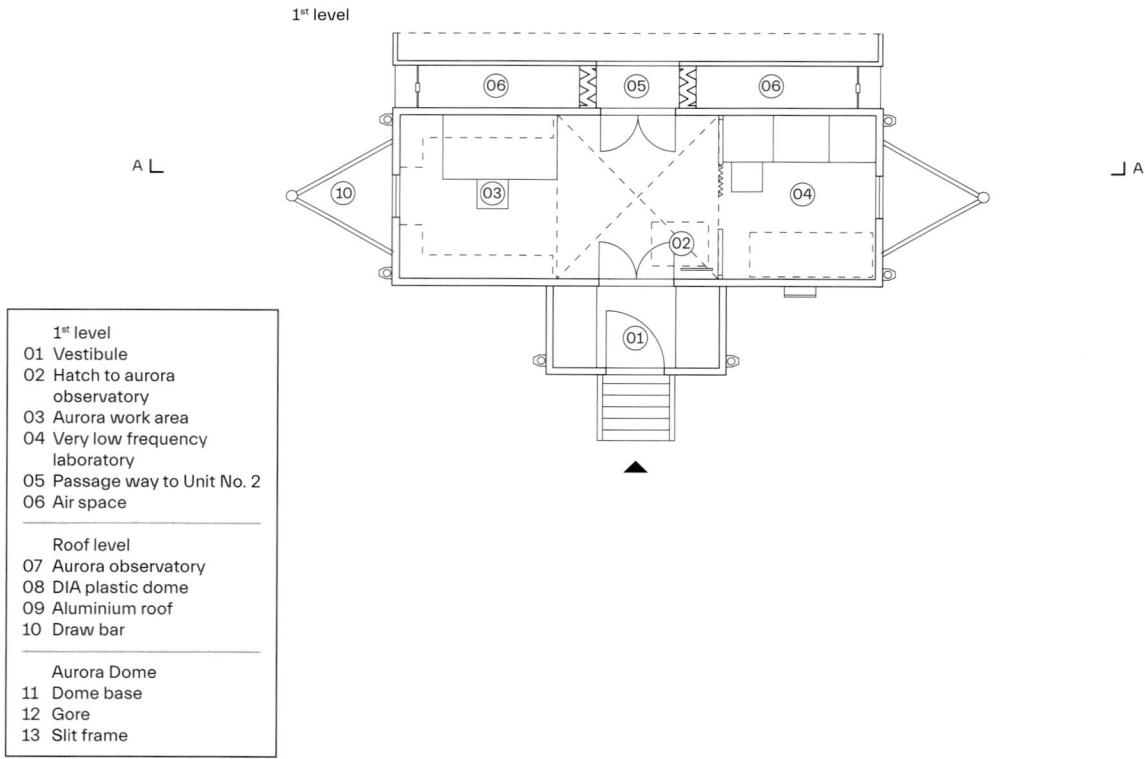

1st level
01 Vestibule
02 Hatch to aurora
   observatory
03 Aurora work area
04 Very low frequency
   laboratory
05 Passage way to Unit No. 2
06 Air space

Roof level
07 Aurora observatory
08 DIA plastic dome
09 Aluminium roof
10 Draw bar

Aurora Dome
11 Dome base
12 Gore
13 Slit frame

SOURCE
United States National Archives Research Center

1 : 100   0   1   2   4 m

SECTION AND DETAIL | AIR PORTABLE SCIENTIFIC FACILITY

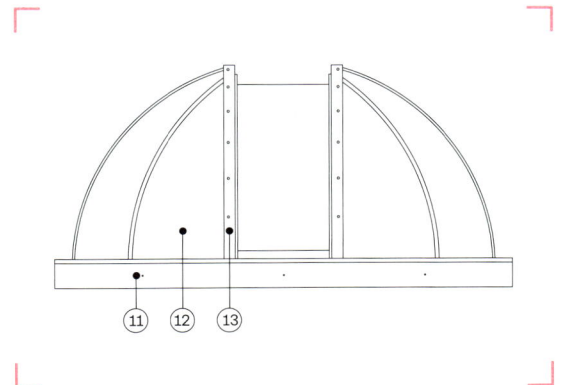

"The scientific programme for the new station entails purely research in upper atmosphere physics, and its position will allow for it alone to observe visually the northern extension of the aurora. It is to play an active part in the International Year of the Quiet Sun – a period of intensive study comparable to that of I.G.Y. [International Geophysical Year] but occurring during the period of minimum solar activity."

New Zealand Antarctic Society, *Antarctic: A News Bulletin,* June 1963

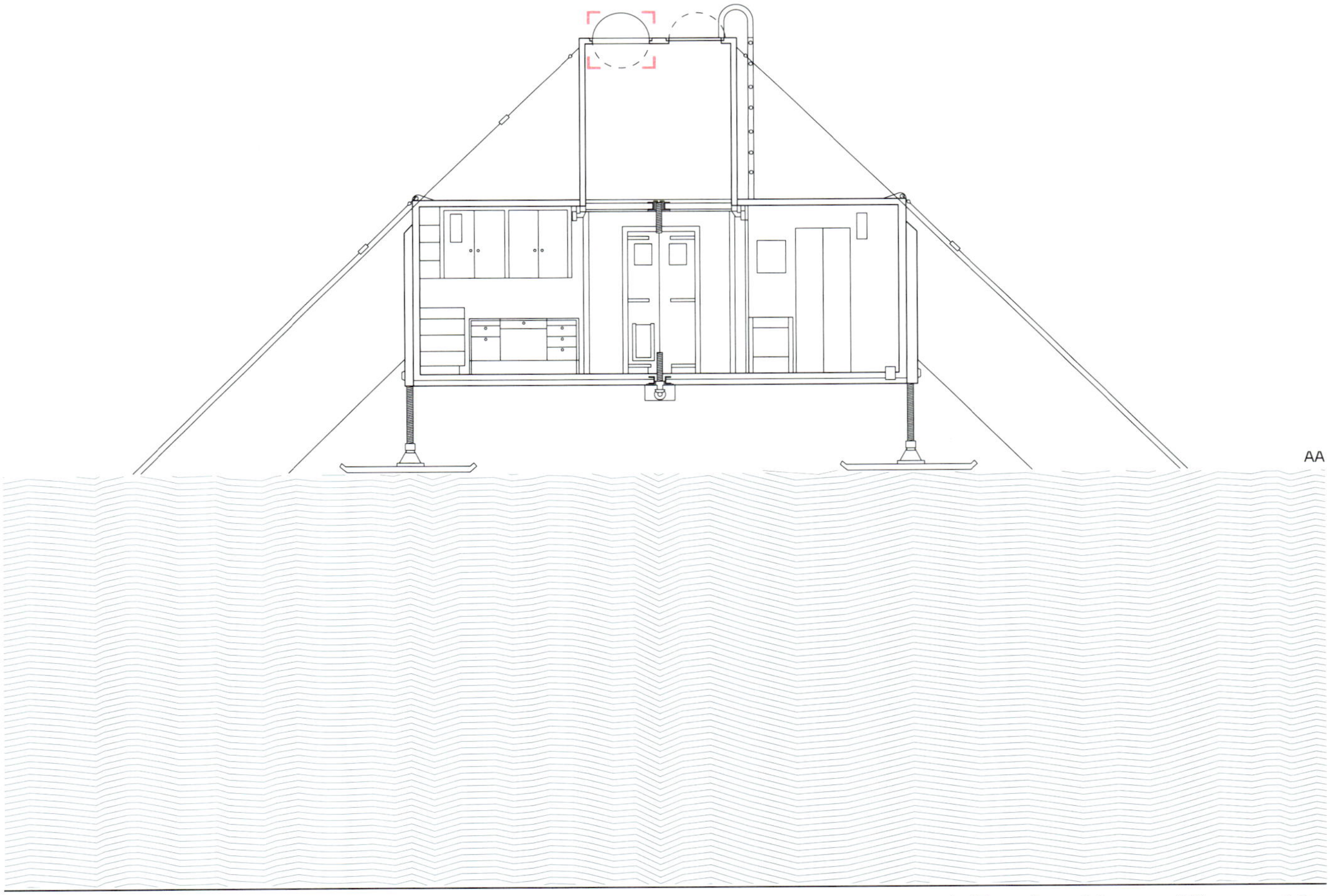

AA

SOURCE
United States National Archives Research Center

| 1:25 | 0 | 0.4 | 0.8 | 1.6 m |
| 1:100 | 0 | 1 | 2 | 4 m |

↖ p. 413, 509

PLATEAU
1965

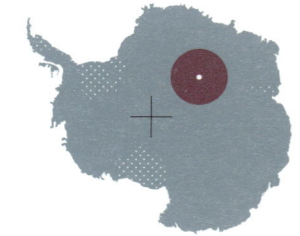

"The design concept – compact, prefabricated vans – emphasized ease of construction. Plumbing and electrical connections between vans were minimized. Assembly of the vans into a station had to be simple. Shipping schedules and austral summer temperatures dictated the beginning and end of the construction season: the job had to be done in January, Because of support problems, the station construction crew was limited to 10 men and the period to assemble the vans to 14 days."

United States National Science Foundation, "Design for Survival: The Story of Plateau Station", *Antarctic Journal of the United States,* 1966

Architect and Engineer: Bureau of Yards & Docks, Department of the Navy, United States of America
Manufacturer: Alberta Trailer Company, United States of America
Construction: United States Navy Seabees, United States of America

United States of America
United States Antarctic Program (USAP)

Closed

79°15'03"S 40°33'38"E
Queen Maud Land, Dome A
Ice sheet, 3,624 m.a.s.l., −56.7 °C

SITE PLAN

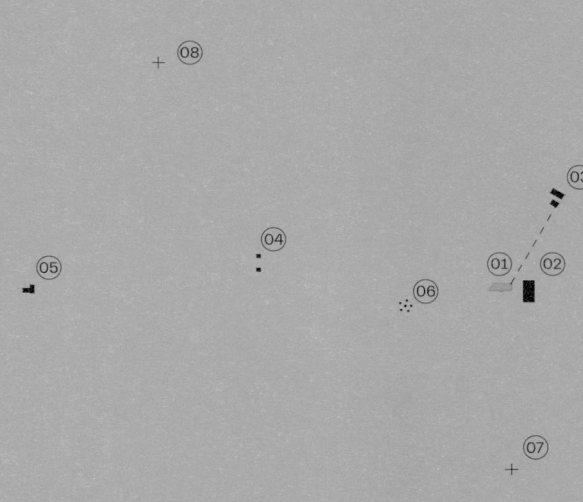

01 Main station building
02 Snow dike, fuel bladder
03 Storage jamesway
04 Geomagnetic buildings
05 Summer camp
06 Communication antenna
07 Micrometeorology tower
08 Very low frequency
    antenna
09 Runway

SOURCE
United States National Archives Research Center
(reconstruction)

S

1 : 5,000   0   50   100   200 m

PLAN AND SECTION

"The vans were to be prefabricated of a wood framework and plywood wall construction. The outside was to be covered with sheet aluminum. Rigid polyurethane insulation, three inches [7.6 cm] thick, would be used between the wood studs to insure very low heat transmission. Careful attention was paid to the nailing so that there would be no through connection of any kind between the outside and inside walls. A strip of ¼-inch [0.6 cm] cork was to be placed between each stud and the plywood panel to help eliminate the possibility of thermal conductivity."

United States National Science Foundation, "Design for Survival: The Story of Plateau Station", *Antarctic Journal of the United States*, 1966

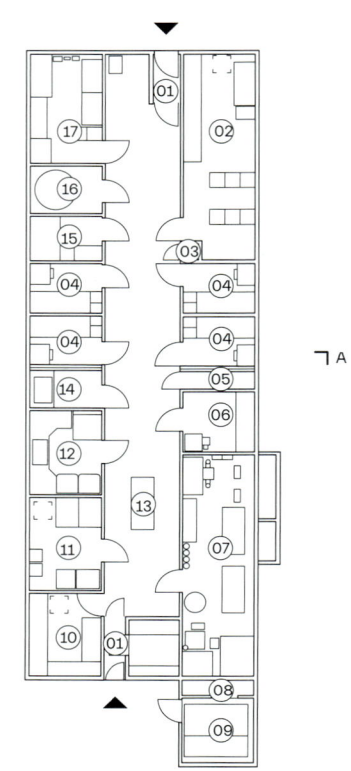

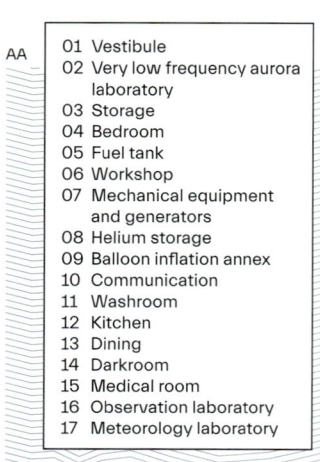

01 Vestibule
02 Very low frequency aurora laboratory
03 Storage
04 Bedroom
05 Fuel tank
06 Workshop
07 Mechanical equipment and generators
08 Helium storage
09 Balloon inflation annex
10 Communication
11 Washroom
12 Kitchen
13 Dining
14 Darkroom
15 Medical room
16 Observation laboratory
17 Meteorology laboratory

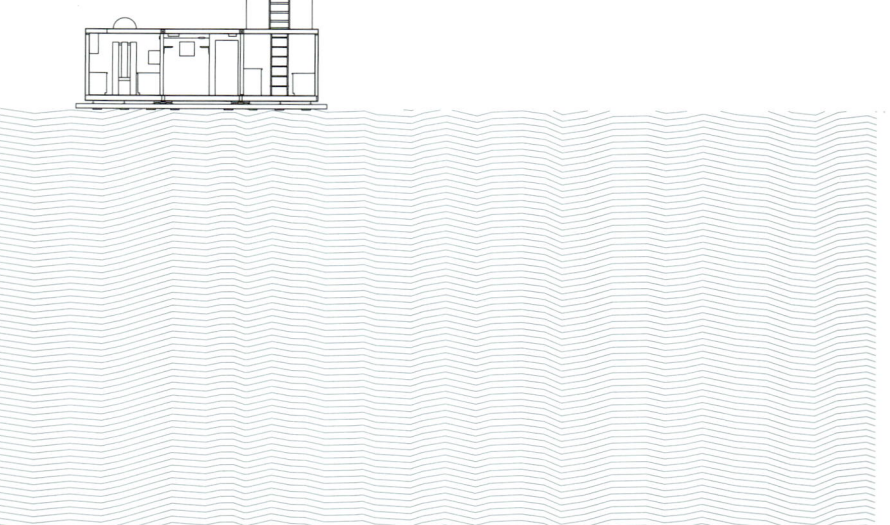

AA

SOURCE
United States National Archives Research Center, 1965

1 : 250   0   2.5   5   10 m

ELEVATIONS

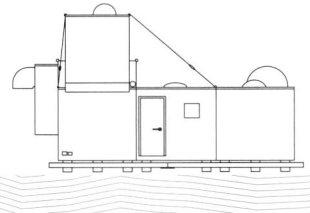

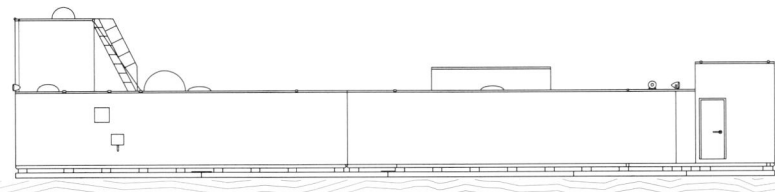

SOURCE
United States National Archives Research Center, 1965

1 : 250   0   2.5   5   10 m

BASE B
1944/1966

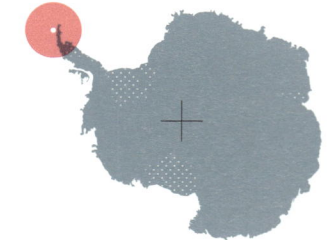

"Originally buildings from the former Norwegian Aktieselskabet Hektor whaling station were used. These included Bleak House, a former dormitory which was converted into the main accommodation and office building. It was destroyed by fire on 8 Sep 1946. Another former dormitory from the whaling station was then used as the main accommodation building and became known as Biscoe House after John Biscoe […]. The Magistrate's Villa, the name reflecting its previous use, was used as a store. A new hut known as the FIDASE hut or Hunting Lodge was erected on 13 Dec 1955. It was used by members of the Falkland Islands Dependencies Aerial Survey Expedition, 1955–57, employed by Hunting Aerosurveys Ltd. On completion of the survey this hut became the property of FIDS. An aircraft hangar was completed in Mar 1962. A plastic accommodation building known as Priestley House after Sir Raymond Priestley, Acting Director of FIDS 1955–59 and geologist on Scott's expedition 1910–13, was erected in Jan 1966 and found to be missing (sic) on 22 Mar 1985 when RRS *John Biscoe* visited."

British Antarctic Survey, "History of Deception Island (Station B)" (online)

Manufacturers: Mickleover Transport Ltd.,
United Kingdom

United Kingdom
Falkland Islands Dependencies Aerial Survey
Expedition (FIDS) > British Antarctic Survey (BAS)

Closed

62°59'00"S 60°34'00"W
Whalers Bay, Deception Island
Ice-free ground, 7 m.a.s.l., −3°C

SITE PLAN

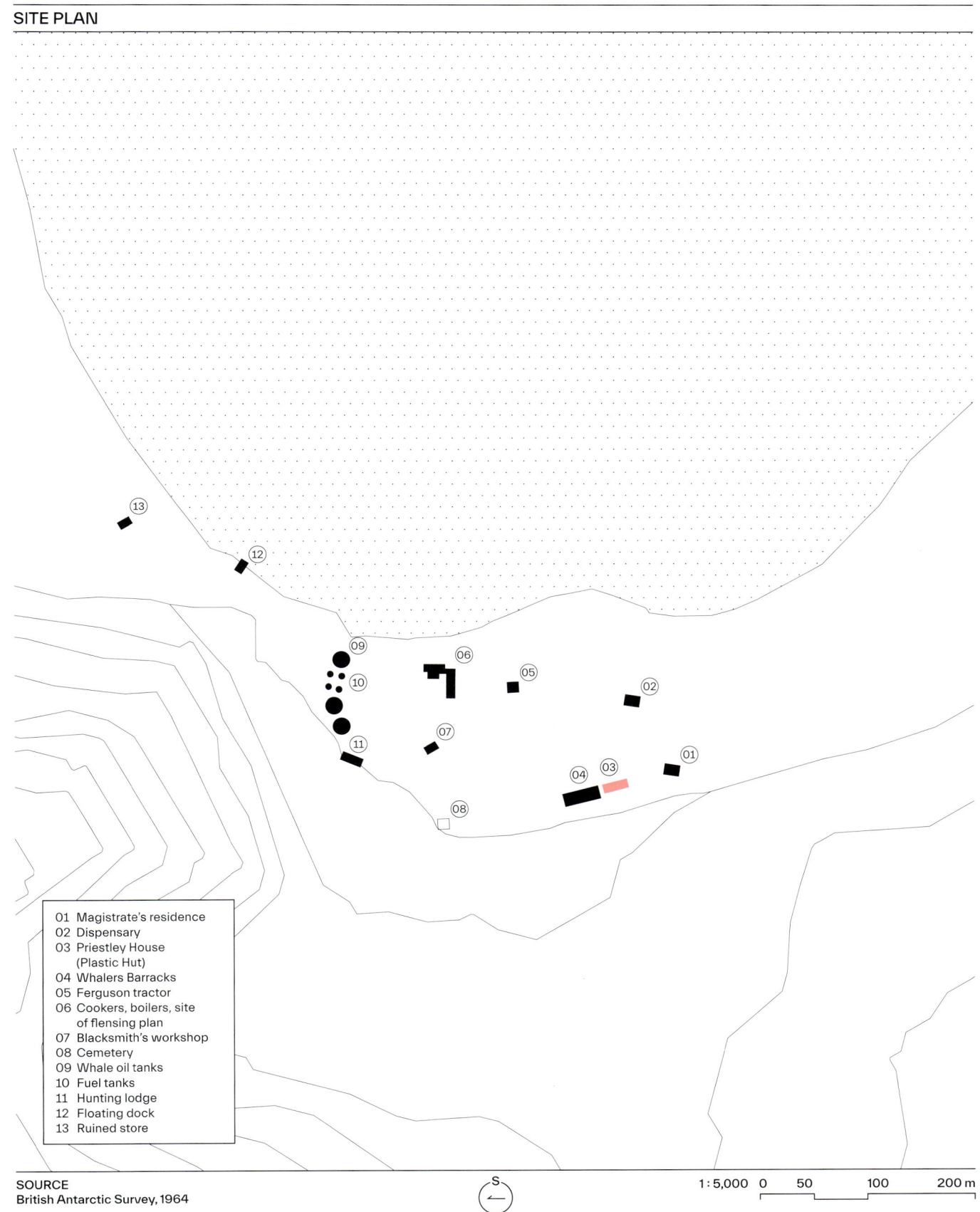

01 Magistrate's residence
02 Dispensary
03 Priestley House
     (Plastic Hut)
04 Whalers Barracks
05 Ferguson tractor
06 Cookers, boilers, site
     of flensing plan
07 Blacksmith's workshop
08 Cemetery
09 Whale oil tanks
10 Fuel tanks
11 Hunting lodge
12 Floating dock
13 Ruined store

SOURCE
British Antarctic Survey, 1964

S

1 : 5,000   0   50   100   200 m

PLAN AND SECTION | PRIESTLEY HOUSE (PLASTIC HUT)

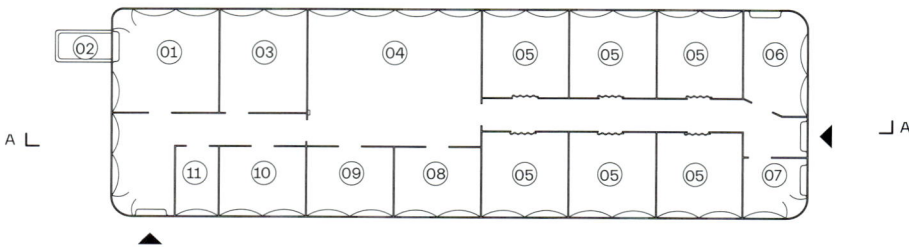

Main building
01 Kitchen
02 Cold storage
03 Dining room
04 Living room
05 Bedroom
06 Washroom
07 Bathroom
08 Base leader's office
09 Communication room
10 Meteorological lab
11 Store

Detail
01 Fibreglass panel
02 Rag bolts
03 1.9 cm thick tongue-and-groove boards
04 5 × 5 cm floor battens
05 0.3 cm thick oil tempered hardboard panels
06 Polythene sheet
07 Angle bracket screwed into building and floor battens
08 Concrete base

AA

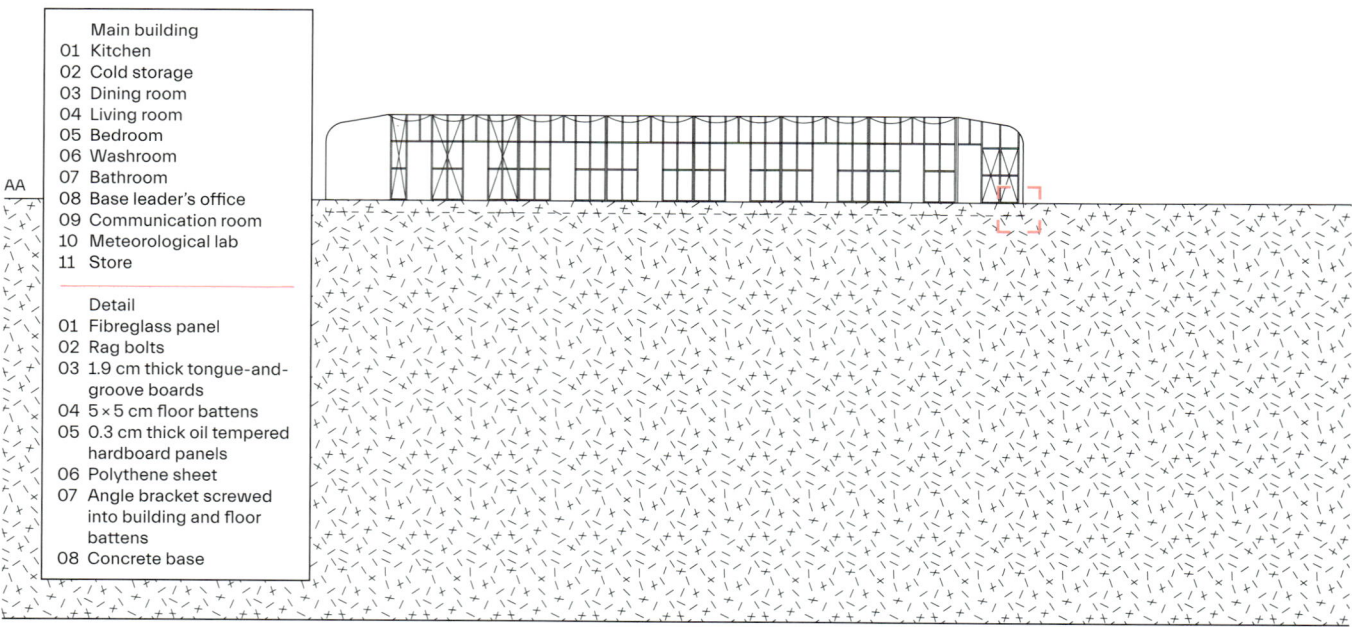

SOURCE
British Antarctic Survey, 1964

1 : 250   0   2.5   5   10 m

ELEVATIONS AND DETAIL | PRIESTLEY HOUSE (PLASTIC HUT)

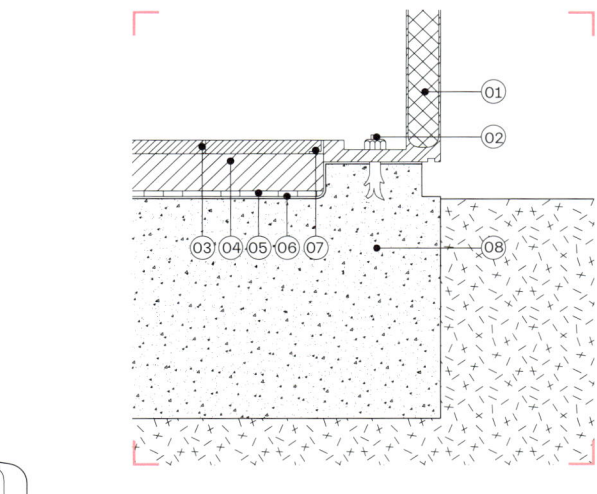

SOURCE
British Antarctic Survey, 1964

| 1:10 | 0 | 0.1 | 0.2 | 0.4 m |
| 1:250 | 0 | 2.5 | 5 | 10 m |

HALLEY II
1957/1967

↖ p. 425, 458, 466

"With a total of 79 days with drift between the month of March and June the base complex was completely covered with snow and is now at a depth of approx. 12 feet [3.65 m] from eaves to surface. Caverns have occurred along the side of the building."

L. V. James, *British Antarctic Survey Accommodation Building Report*, 1967

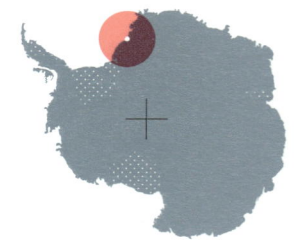

Suppliers: (timber) William Brown and Co. Ltd.;
(roof sheeting) Butyl Products Ltd.; (generators)
Rolls Royce, United Kingdom

United Kingdom
British Antarctic Survey (BAS)

Buried > Replaced

75°31'S 26°39'W
Brunt Ice Shelf, Halley Bay
Ice shelf, 37 m.a.s.l., −20°C

SITE PLAN

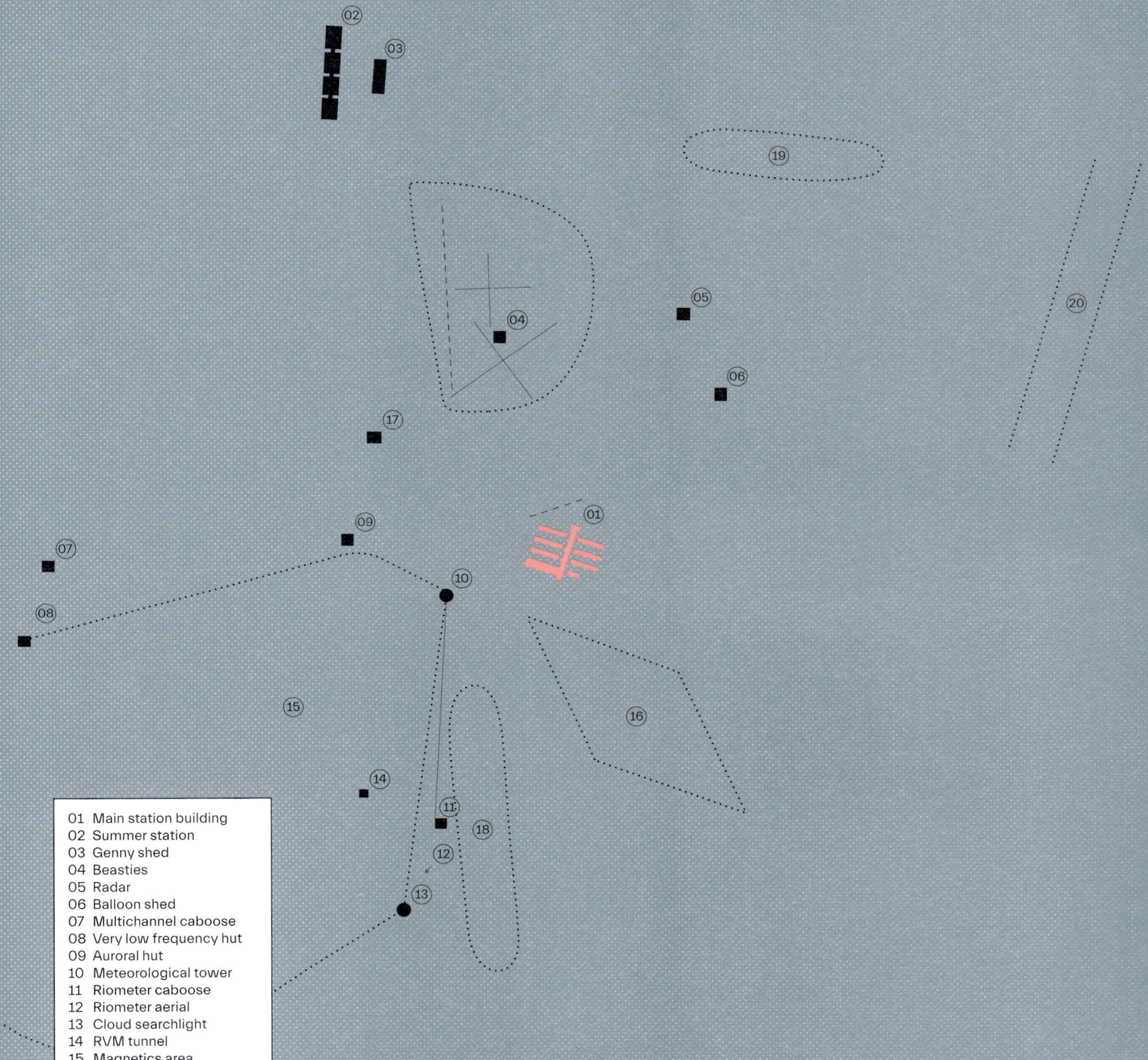

01 — Main station building
02 — Summer station
03 — Genny shed

19 — Gash dump

20 — Fuel dump

05 — Radar

04 — Beasties

06 — Balloon shed

17 — Dog tunnel

09 — Auroral hut

01 — Main station building

07 — Multichannel caboose

10 — Meteorological tower

08 — Very low frequency hut

15 — Magnetics area

16 — Communication rhombic

14 — RVM tunnel

11 — Riometer caboose

18 — General dump

12 — Riometer aerial

13 — Cloud searchlight

01	Main station building
02	Summer station
03	Genny shed
04	Beasties
05	Radar
06	Balloon shed
07	Multichannel caboose
08	Very low frequency hut
09	Auroral hut
10	Meteorological tower
11	Riometer caboose
12	Riometer aerial
13	Cloud searchlight
14	RVM tunnel
15	Magnetics area
16	Communication rhombic
17	Dog tunnel
18	General dump
19	Gash dump
20	Fuel dump

SOURCE
British Antarctic Survey, 1968

S

1 : 5,000   0   50   100   200 m

PLAN, BELOW SNOW LEVEL

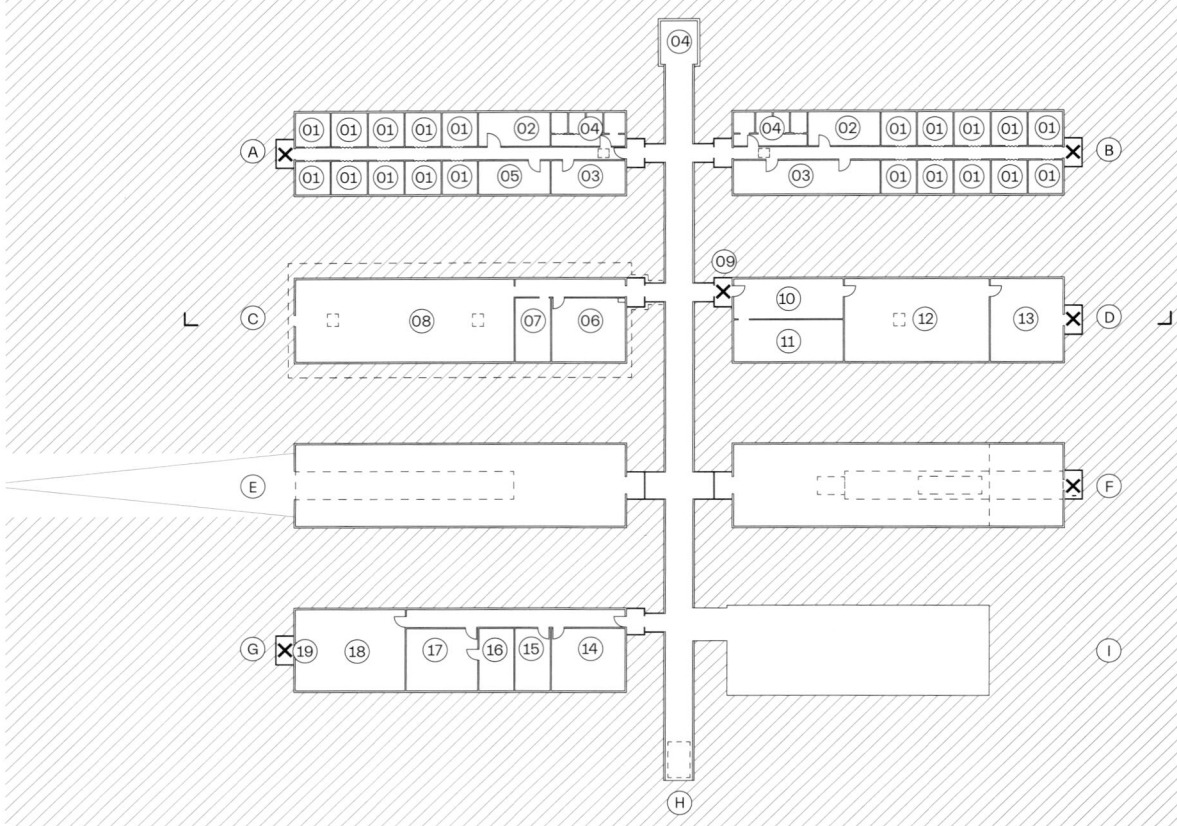

A/B	Sleeping quarters	12	Dining room
01	Bedroom	13	Kitchen
02	Drying room		
03	Bathroom	E/F	Garage
04	Toilet		
05	Hospital and psychology room	G	Science
		14	Survey and geology office
C/D	Living quarters	15	Darkroom
06	Telecommunications cabin	16	Geophysics darkroom
07	Base leader's office	17	Geophysics office
08	Lounge	18	Meteorological office
09	Main access shaft	19	Observation point
10	Lobby and cloakroom		
11	Glaciology and ice sounding	H	Gash shaft
		I	Generator building

SOURCE
British Antarctic Survey, 1966

1:500   0   5   10   20 m

SECTION

"Structurally the design of the building for the new Halley Bay Base Complex is not so different from previous recent base huts erected in the Antarctic. They retain the steel portals linked with timber rails and purlins, and are clad overall with factory-built insulated panels. The main differences lie in the great strength of steel and timber members used, compound floor beams crossing the buildings, and the use of corrugated iron sheeting to support and spread the load over the levelled snow sites. The seven buildings supplied were identical in design detail.

L.J. Shirtcliffe (Explorer, Meteorologist, Builder), *British Antarctic Survey Building New Base Complex, Halley Bay,* 1967

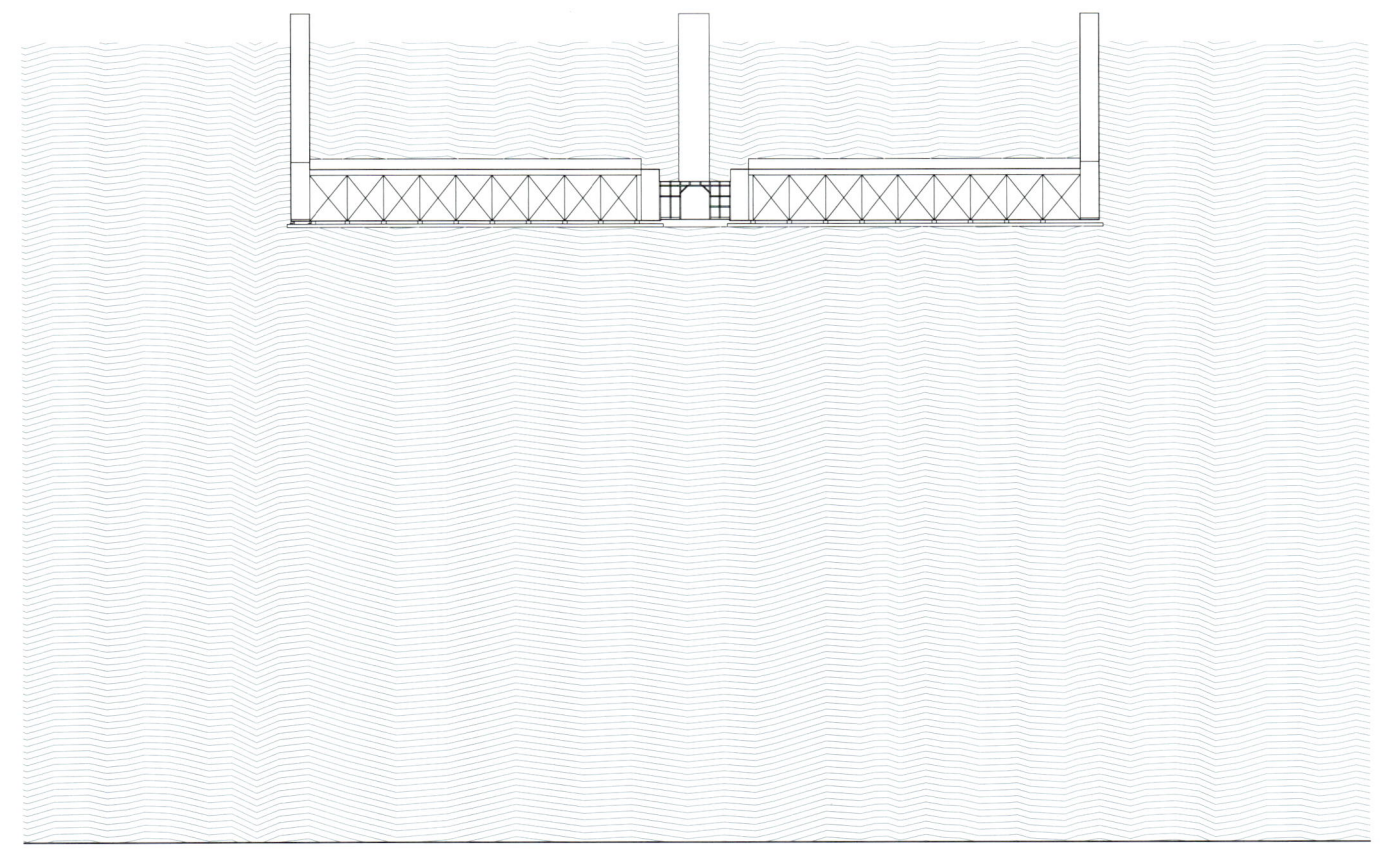

ELEVATION | MAIN STATION BUILDINGS

"The first two portal frames were assembled flat and raised one at a time by man power […] and positioned 8 ft. [2.4 mm] centres, and temporarily connected with 6" × 3" [15 × 8 cm] cross rails. The winch cable was passed over the ridge of the standing portals to pull up the next frame. Once three portals were standing, the cross rails had to be removed to permit the fixing of the inner and outer rails and the eaves beams, because the working tolerances were too critical. […] With three portal frames standing, the purlins could be loosely positioned to tie in the rafters. When the ten portal frames of the building were standing inner and outer rails and eaves beams fixed, the cross rails could be dropped into place and bolted, and tightening up overall finally accomplished. The nine intermediate floor bearers were roughly positioned as the building progressed, but, as the gable frames were being assembled, they were finally fixed with the blocking piece. […] The bottom plates of the gable frames were spiked in place to mark the boundaries of the floor panels."

L.J. Shirtcliffe (Explorer, Meteorologist, Builder), *British Antarctic Survey Building New Base Complex, Halley Bay,* 1967

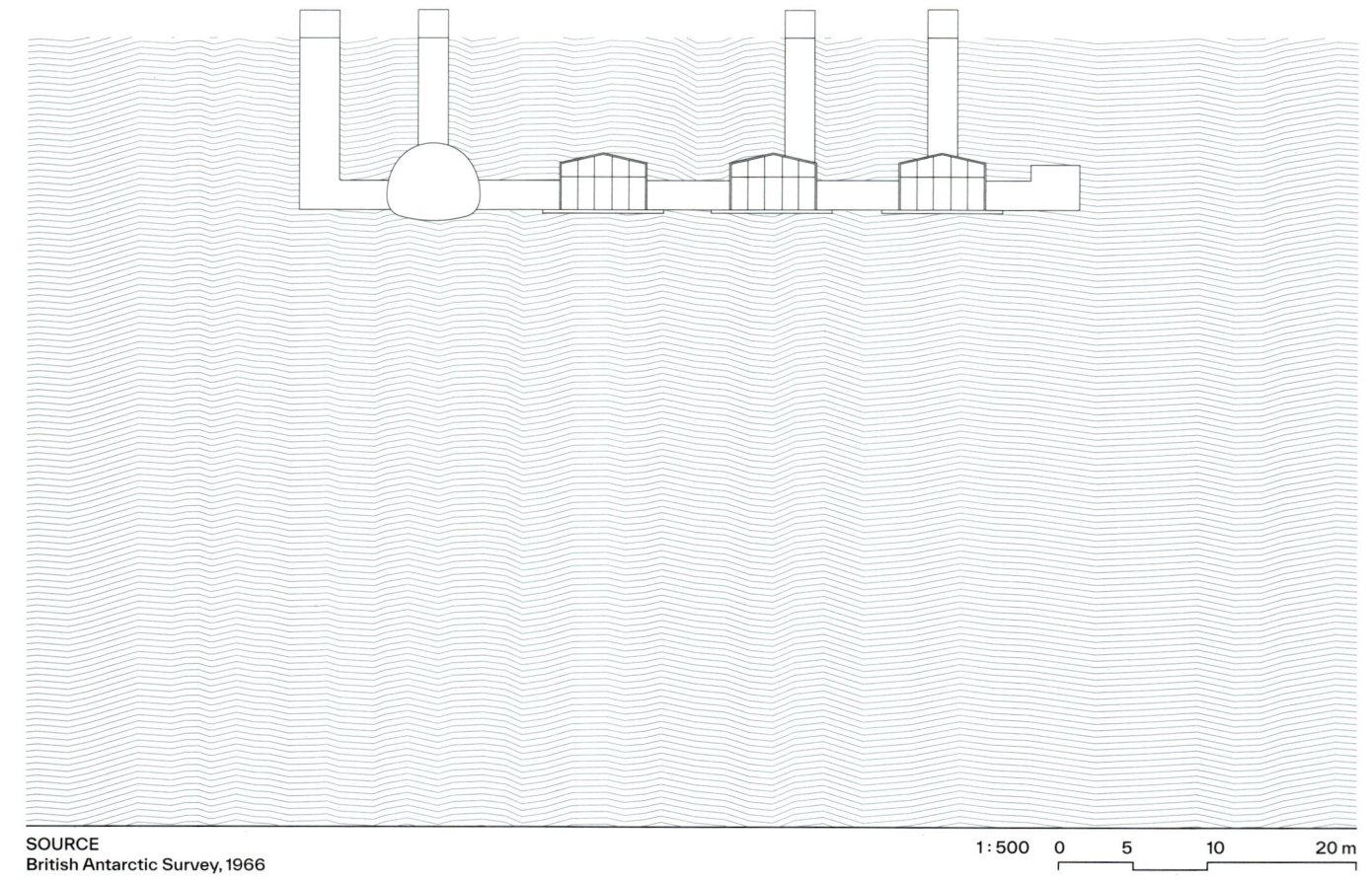

SECTION | MAIN STATION BUILDING AND ACCESS SHAFT

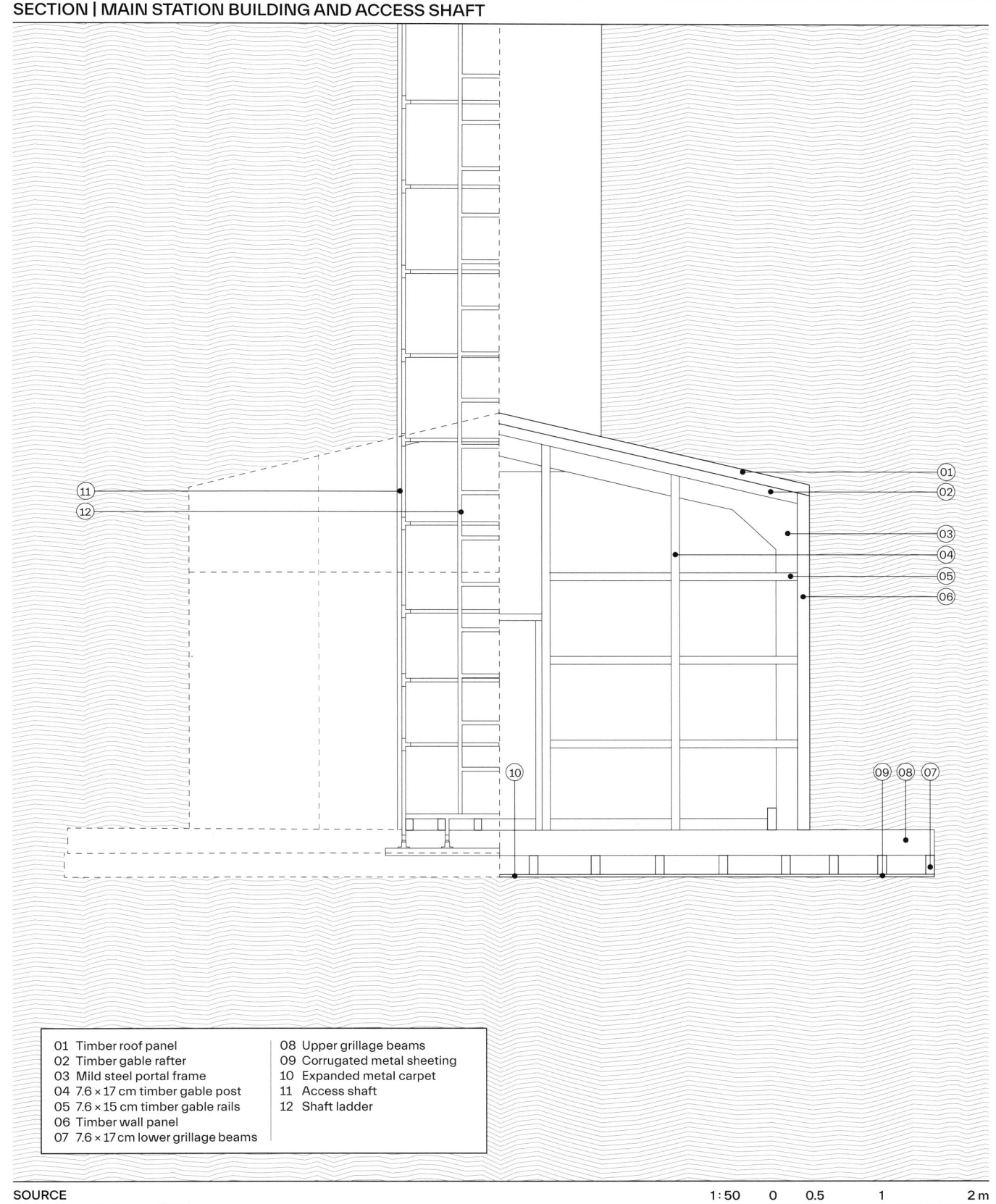

01 Timber roof panel
02 Timber gable rafter
03 Mild steel portal frame
04 7.6 × 17 cm timber gable post
05 7.6 × 15 cm timber gable rails
06 Timber wall panel
07 7.6 × 17 cm lower grillage beams
08 Upper grillage beams
09 Corrugated metal sheeting
10 Expanded metal carpet
11 Access shaft
12 Shaft ladder

SOURCE
British Antarctic Survey, 1966

1:50   0   0.5   1   2 m

CASEY REPSTAT
1969

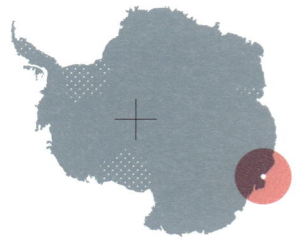

"Casey station was designed following wind tunnel tests to determine the drift-free nature of this design, as a line of buildings to be elevated some 3 metres above ground. Orientation was North-South with its long dimension across the direction of the drift-bearing winds from the East. Individual buildings were connected by a sheltered walkway along the western edge of the structure formed by corrugated galvanised steel sheets (as used in Australia to make rain-water tanks)."

Philip Incoll (Architect), *The Influence of Architectural Theory on the Design of Australian Antarctic Stations,* 1990

Australia
Australian Antarctic Program (AAD)

Removed > Replaced

66°17'S 110°32'E
Bailey Peninsula, East Antarctica
Ice-free ground, −9.3°C

SITE PLAN

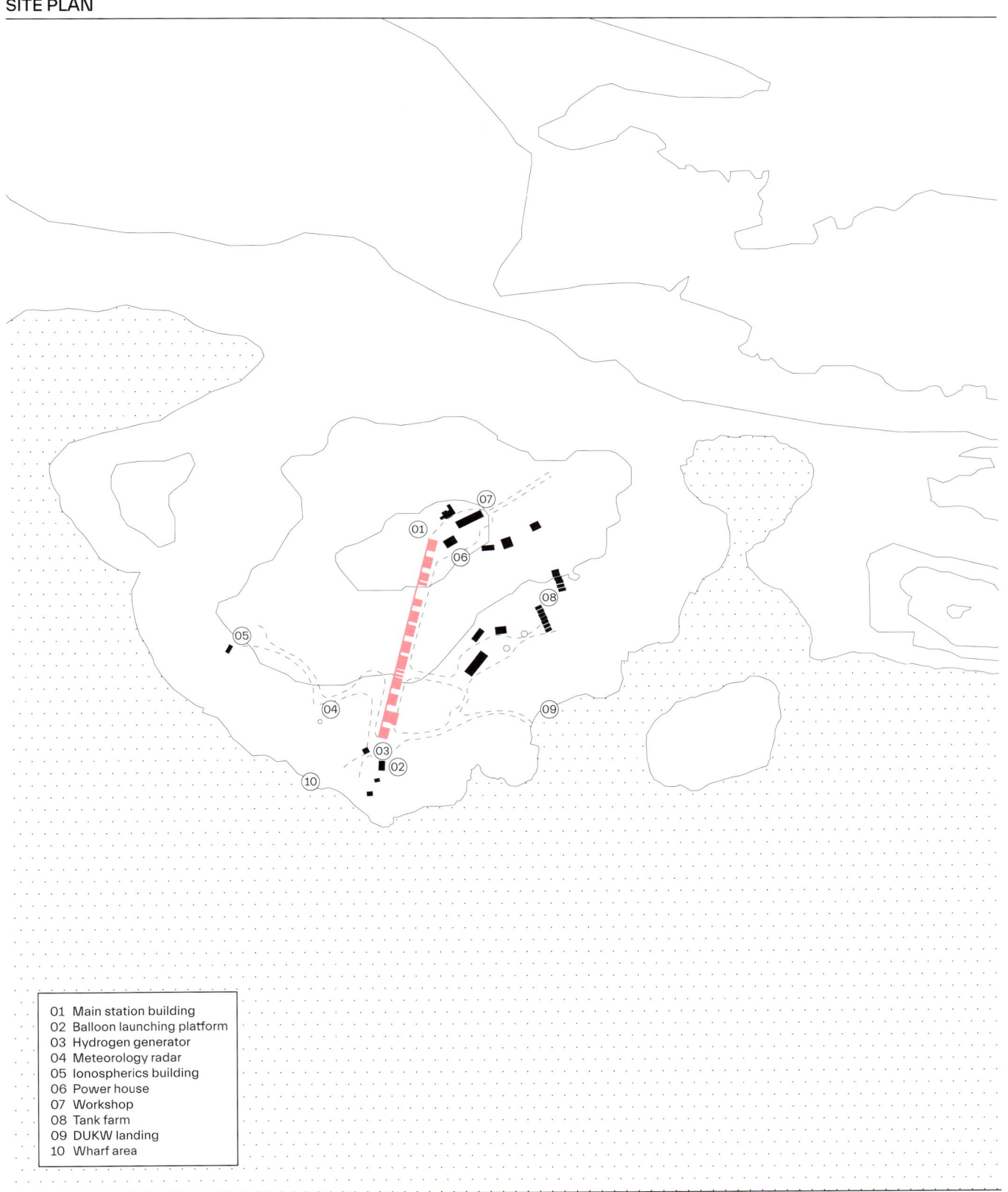

01 Main station building
02 Balloon launching platform
03 Hydrogen generator
04 Meteorology radar
05 Ionospherics building
06 Power house
07 Workshop
08 Tank farm
09 DUKW landing
10 Wharf area

SOURCE
Australian Antarctic Program, 1991

1 : 5,000 0   50   100   200 m

SITE PLAN

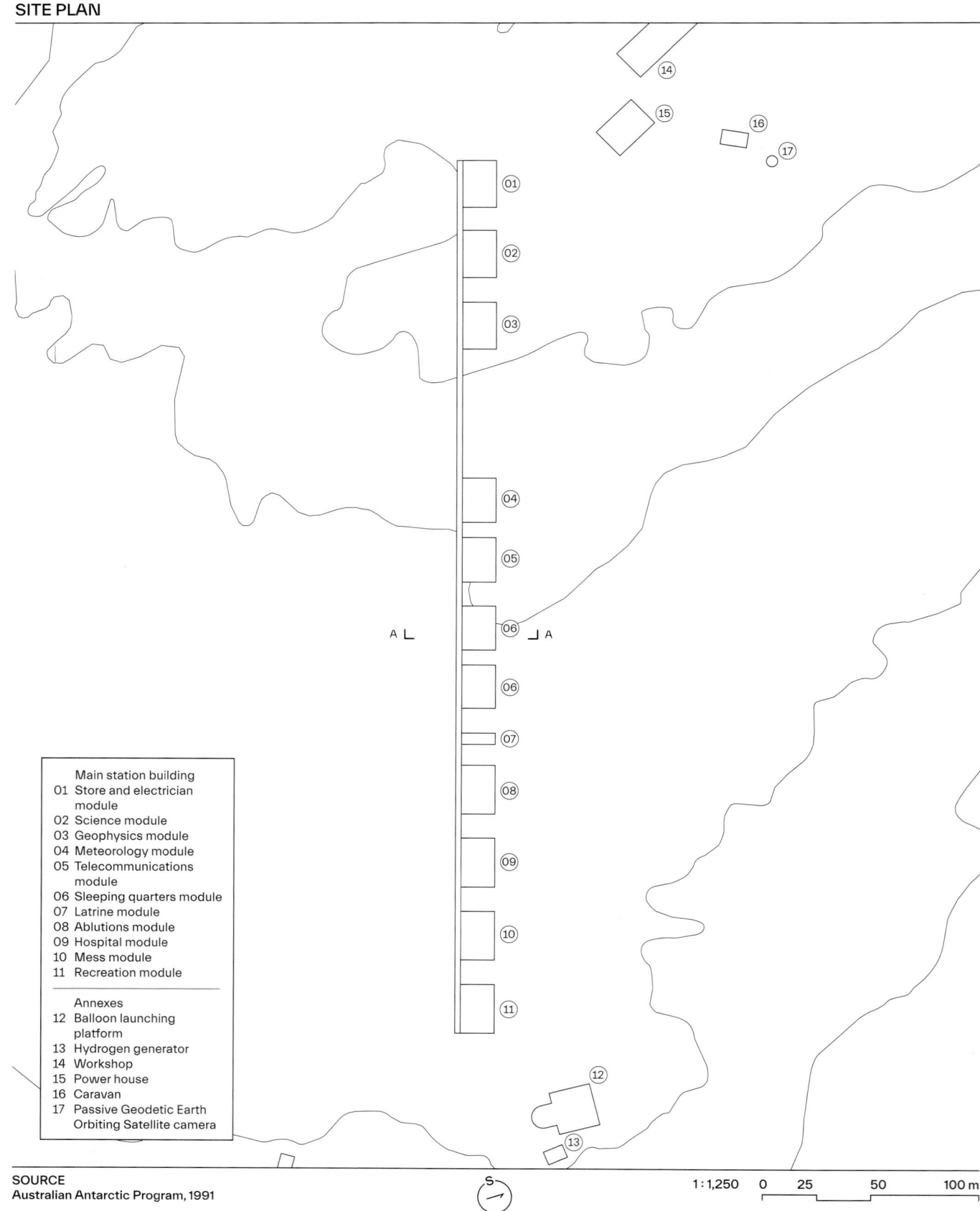

Main station building
01 Store and electrician module
02 Science module
03 Geophysics module
04 Meteorology module
05 Telecommunications module
06 Sleeping quarters module
07 Latrine module
08 Ablutions module
09 Hospital module
10 Mess module
11 Recreation module

Annexes
12 Balloon launching platform
13 Hydrogen generator
14 Workshop
15 Power house
16 Caravan
17 Passive Geodetic Earth Orbiting Satellite camera

SOURCE
Australian Antarctic Program, 1991

1 : 1,250   0   25   50   100 m

SECTION

"Some years after Casey was occupied the Antarctic Division believed that the occupants of work areas included in the main elevated and walkway connected structure were less productive than similar expeditioners at the other stations. Tasks which required to be performed outside the structure by these people were often not done and the conclusion was reached that a form of 'cabin fever' was being encountered. So the requirement was laid down that an expeditioner's normal day would include a walk to work across the snow – providing him with daily evidence that he could cope with the outdoor environment, and that the Abominable Snowman did not lurk behind the next snowdrift."

Philip Incoll (Architect), *The Influence of Architectural Theory on the Design of Australian Antarctic Stations,* 1990

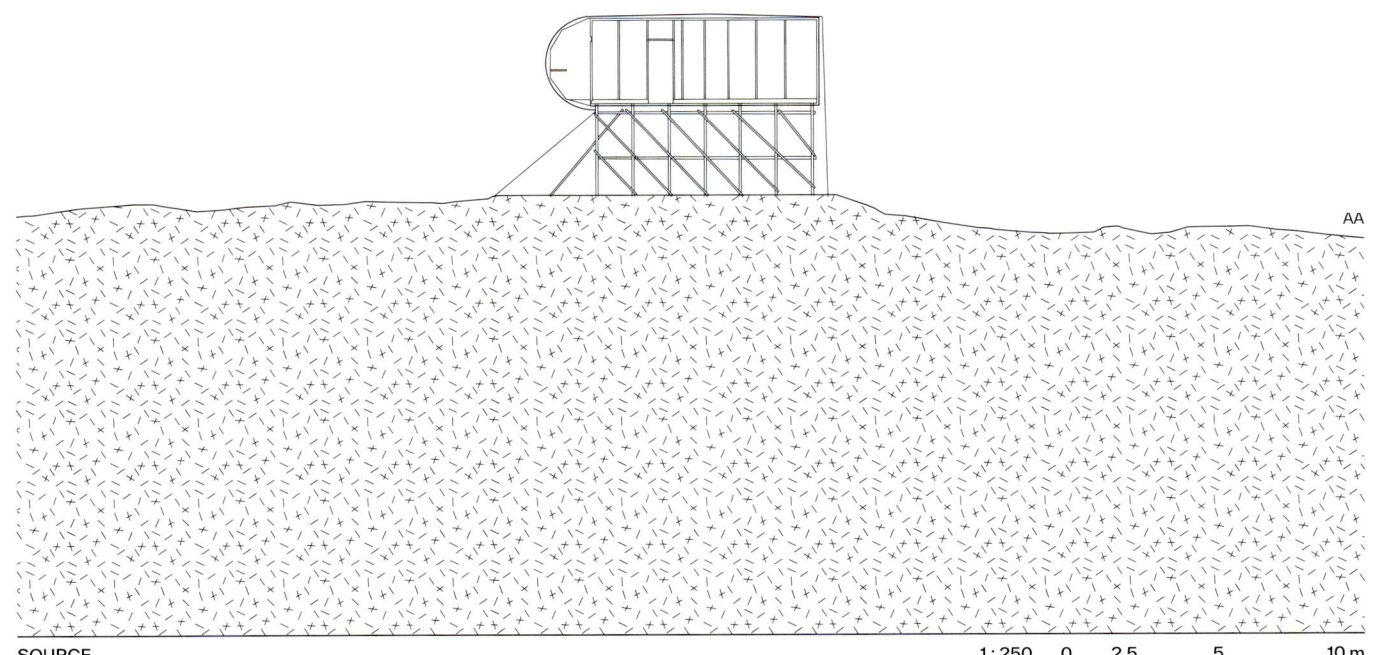

AA

SOURCE
Australian Antarctic Program, 1991

1 : 250   0   2.5   5   10 m

HALLEY III
1956/1973

"For two years I liaised with the South Africans giving them full design specifications and faults, from Halley [II]. They have used the data to design and build their new Antarctic station. Last year we liaised with a German company of consultants who have also based their design for a new Antarctic Station on Design Technical Data received from us and the South Africans. Both these organisations have published their specifications and using these I have designed a new station complex for building on a floating ice shelf. I have borne in mind our limited finances ... Each project is based on a complement of 24 men. The station consists of a number of prefabricated timber buildings erected inside Armco steel tubes."

Alan Smith (Builder, Head of British Antarctic Survey Building Services), *Proposed Building Replacement for Halley*, 1979

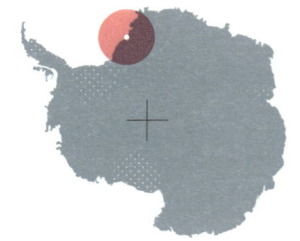

Manufacturers: (Armco steel tubes) Armco Ltd.; (prefabricated timber buildings) Boulton & Paul; (sureBeam system) Lyndon SGB, United Kingdom

United Kingdom
British Antarctic Survey (BAS)

Calved > Replaced

75°31'S 26°43'W
Brunt Ice Shelf, Caird Coast
Ice shelf, 37 m.a.s.l., −20°C

SITE PLAN

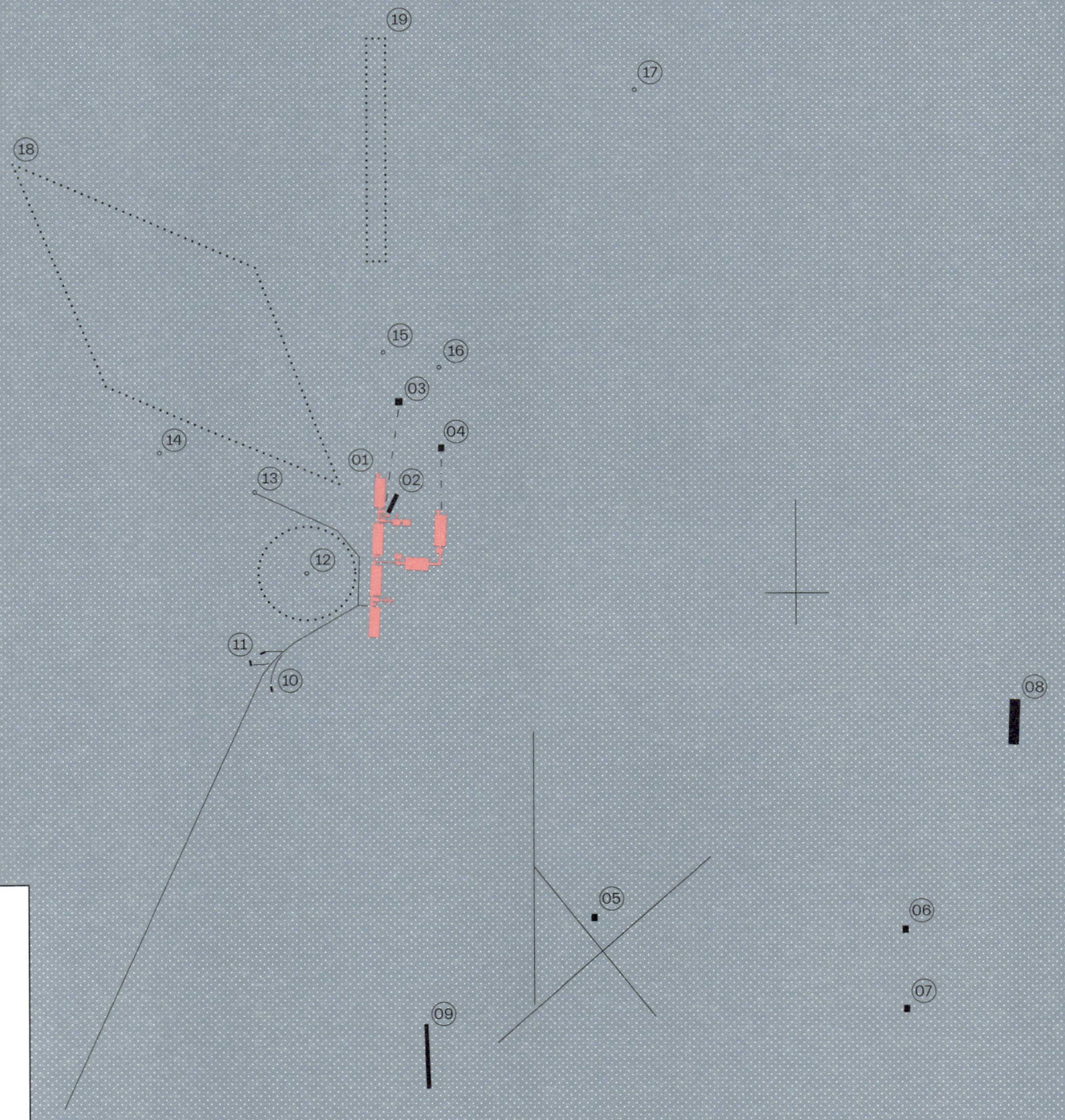

01 Main station building
02 Food tunnel
03 Aurora hut
04 Ozone hut
05 Beastie hut
06 Radar
07 Balloon shed
08 Food storage
09 Dog tunnel
10 Textran
11 Aerials
12 Conical monopole
13 Meteorological tower
14 Net flux
15 Albedo
16 Sunshine stand
17 Whistler cables brought
      to surface
18 Rhombic antenna
19 General dump

SOURCE
British Antarctic Survey, 1974

S

1 : 5,000   0   50   100   200 m

PLAN, BELOW SNOW LEVEL

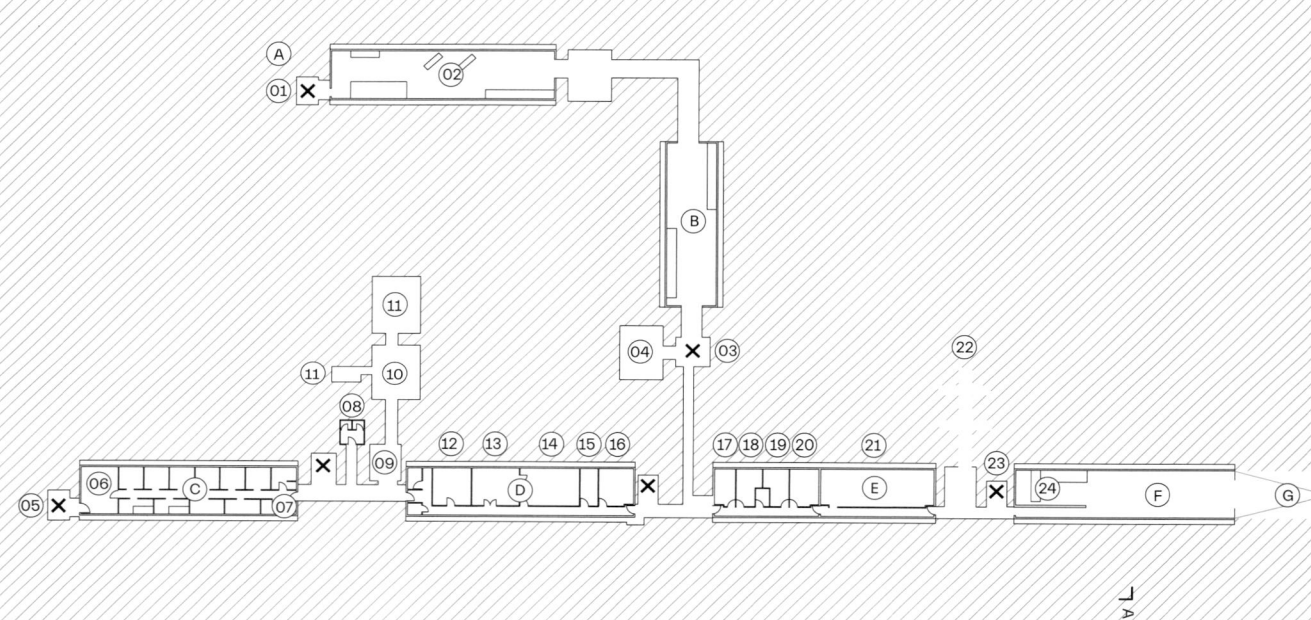

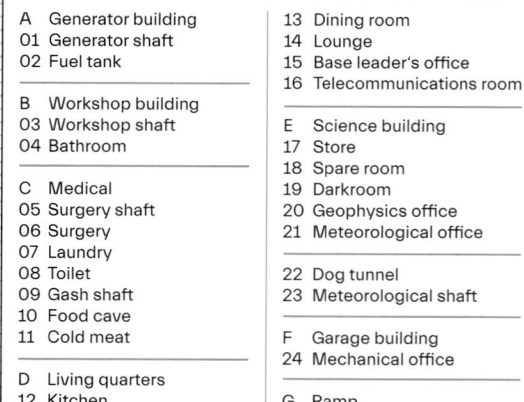

A	Generator building		13	Dining room
01	Generator shaft		14	Lounge
02	Fuel tank		15	Base leader's office
			16	Telecommunications room
B	Workshop building			
03	Workshop shaft		E	Science building
04	Bathroom		17	Store
			18	Spare room
C	Medical		19	Darkroom
05	Surgery shaft		20	Geophysics office
06	Surgery		21	Meteorological office
07	Laundry			
08	Toilet		22	Dog tunnel
09	Gash shaft		23	Meteorological shaft
10	Food cave			
11	Cold meat		F	Garage building
			24	Mechanical office
D	Living quarters			
12	Kitchen		G	Ramp

SOURCE
British Antarctic Survey, 1972

1 : 750   0   7.5   15   30 m

ELEVATION

"At the design stage we were aware that there would be a number of deficiencies due to the amount of finance available to date. The following inadequacies have occurred. The buildings are too flimsy to take the imposed loads and have inadequate thermal insulation. There is inadequate working, living and storage space; there is difficulty in maintaining the building's level; there is inadequate ventilation in the buildings; there is inadequate ventilation in the void which causes meltwater to drip through Armco bolt holes and freezes on the bottom of the Armco void. Manhours are wasted in removing this ice; the generator building is too small to accommodate present layout of the generators; there is insufficient power as the generators are too small; during the summer months approximately 500 gallons of water per day are pumped out from the powerhouse building."

Alan Smith (Builder, Head of British Antarctic Survey Building Services), *Proposed Building Replacement for Halley*, 1979

SOURCE
British Antarctic Survey, 1972

1:750 0 7.5 15 30 m

SECTION

"The main outer shell consists of galvanised corrugated steel plates which are bolted together to form a pipe profile. The main function of the pipe profile is to accommodate all pressures exerted by the surrounding snow and ice. This profile is called the Armco Multiplate Vehicular Underpass type with a height of 4535 mm and a span of 6197 mm. The circular profile selected helps the snow to form a natural arching effect without causing any excessive pressure points on the steel shell. […] All the passages are of the Armco Multiplate Pedestrian Underpass type with a width of 1,788 mm and 2,336 mm high. The floors of the passage are fitted with timber walkways that provide a flat surface for easy movement. The floors, walls and roofs of these buildings consist of interlocking self-supporting panels to the following specifications: softwood framing, covered on outer face with 9 mm external quality resin bonded plywood, filled in with 50 mm resin bonded fibreglass with a polythene vapour barrier, covered on the inner face with 3 mm flame guard hardboard and all glued and nailed together."

Alan Smith (Builder, Head of British Antarctic Survey Building Services), *Proposed Building Replacement for Halley*, 1979

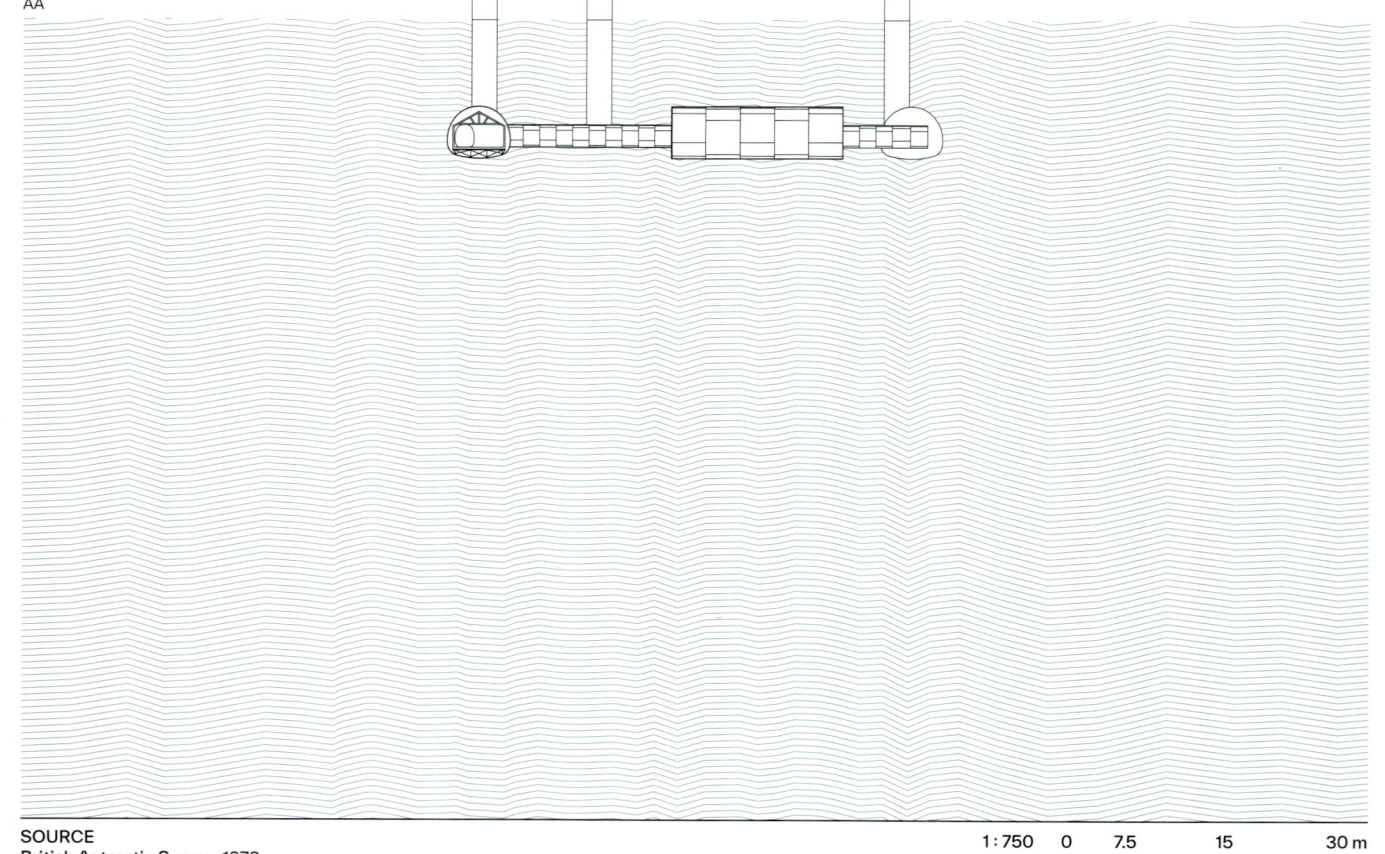

AA

SOURCE
British Antarctic Survey, 1972

1:750   0   7.5   15   30 m

DETAILS | ARMCO TUNNELS

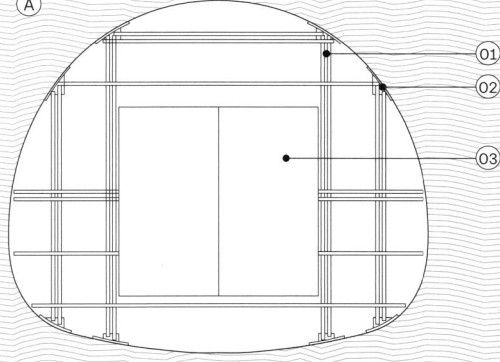

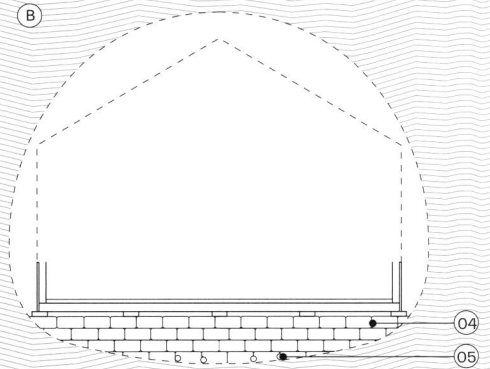

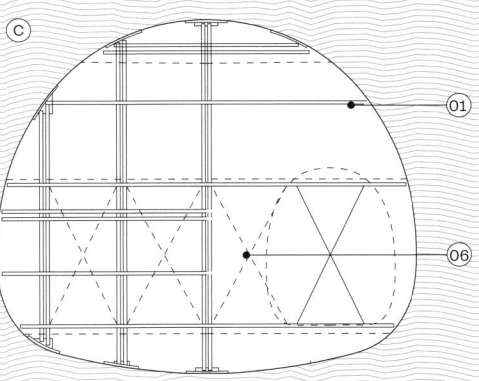

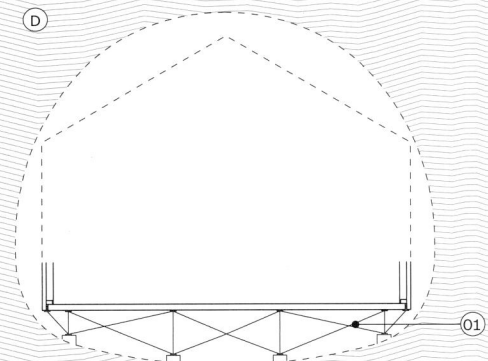

A Scaffold and plywood layout to end of ARMCO tunnel with workshop, garage, and generator block 01 Lyndon SGB sure beam 02 Welded "T" plate cleat fixed to ARMCO 03 Spacing of SGB sure beams to suit width of garage and generator doors	B Aglite aggregate sub-floor to workshop, garage, and generator blocks 04 Coarse grade Aglite aggregate provided in polythene sacks 05 Drainage ducts ――― C Scaffold and plywood layout to end of ARMCO tunnel with sleeping quarters, kitchen, and meteoreology block	01 Lyndon SGB sure beam 06 Possible opening positions for ARMCO tunnel ――― D Steel grillage subfloor to sleeping quarters and kitchen/living and MET office blocks 01 Lyndon SGB sure beam

1:100   0   1   2   4 m

AMUNDSEN-SCOTT SOUTH POLE
1957/1975

"The dome, 50 metres in diameter at its base, would serve as a protective covering for three two-story buildings made from prefabricated modules sized to fit in the LC-130s. These buildings would contain a communications centre, a store, a library and a recreation room, science spaces, single-room quarters for 16 (later 23) persons, a galley, a post office, a photographic darkroom and laboratory, and a meeting hall."

United States National Science Foundation,
Antarctic Journal of the United States, 1966

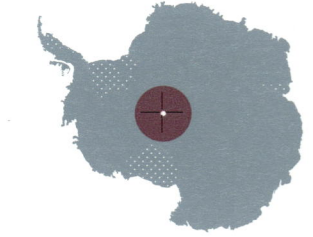

Engineers: Donal Richter (Chief Engineer) TEMCOR;
United States Navy Facilities Engineering Command,
United States of America
Contractor: Holmes and Nervier Inc., United States
of America

United States of America
United States Antarctic Program (USAP)

Removed > Replaced

90°00'00"S
Geographic South Pole, Antarctic Plateau
Ice sheet, 2,835 m.a.s.l., –49°C

SITE PLAN

01 Geodesic dome
02 Archways
03 Balloon inflation tower
04 Clean air facility
05 Emergency camp
06 Ice core drilling site
07 Ceremonial South Pole
08 True Geographic Pole,
   1985
09 Fuel bladder
10 Emergency camp
11 Ground control building
12 Tractors with drag chain
13 Airstrip
14 Antenna field

SOURCE
United States National Archives Research Center, 1983

S

1 : 5,000   0   50   100   200 m

PLAN, BELOW SNOW LEVEL | GEODESIC DOME AND ANNEXES

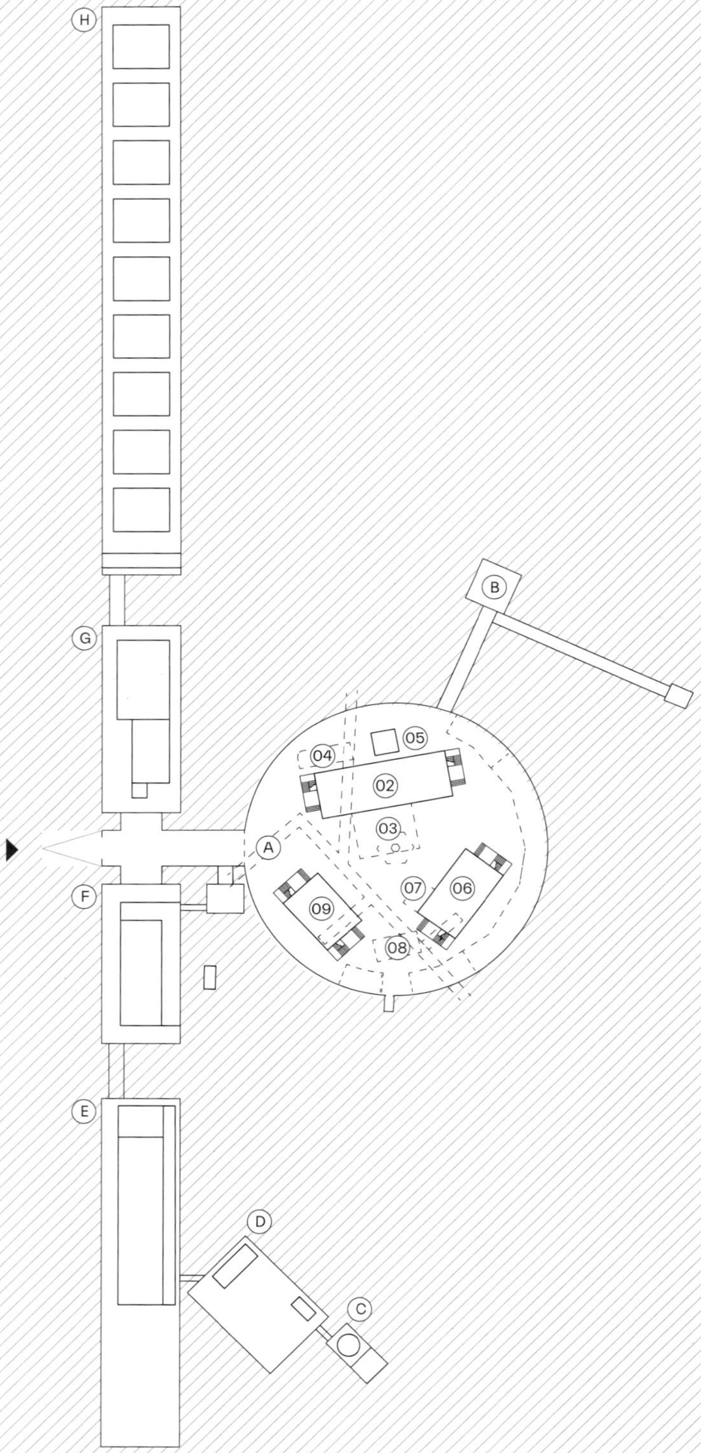

A Geodesic dome
01 Sleeping quarters and
science building
02 Annex building
03 Weight room
05 Gravity vault
05 Communications and
library building
06 Freshie shack
07 Greenhouse
08 Dining and bar building

B Skylab
C Balloon inflation tower
D Cargo arch
E Garage arch
F Power plant arch
G Biomedical arch
H Fuel arch

SOURCE
United States National Archives Research Center, 1970

1 : 1,250   0   12.5   25   50 m

ELEVATION, SECTION AND DETAIL | GEODESIC DOME

"The structures were planned for a maximum wind-load of 35 meters per second. […] Roof design was for a maximum of 1.5 meters of snow cover, with the top of the dome designed for no snow cover at all. Snow would be processed and compacted to a depth of 2.5 meters to provide a uniform foundation with a shear strength of over 250 kilograms per square meter."

United States National Science Foundation, *Antarctic Journal of the United States*, 1966

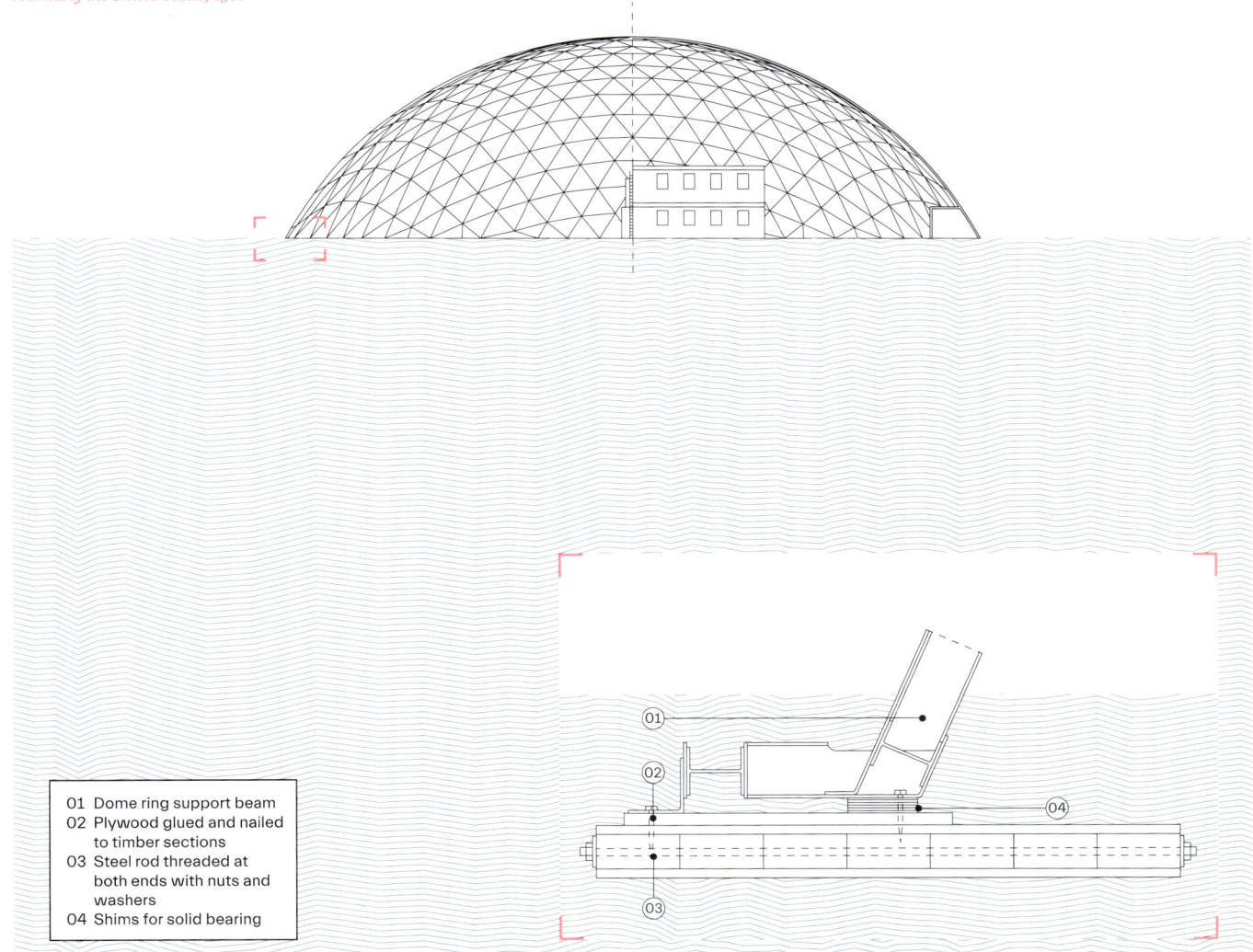

01 Dome ring support beam
02 Plywood glued and nailed to timber sections
03 Steel rod threaded at both ends with nuts and washers
04 Shims for solid bearing

SOURCE
United States National Archives Research Center, 1970

1 : 500   0   5   10   20 m
1 : 25   0   0.25   0.5   1 m

BASE R

1975

"Rothera Station, known as Rothera Point until 15 August 1977, was established in 1975 to replace Adelaide, where the glacier ski-way had deteriorated rendering the operation of ski-equipped aircraft hazardous. A party camped at Rothera Point in the 1975/76 austral summer to open up the air facility. There was a phased construction programme so that by 1980 the station provided accommodation, electrical power generation, vehicle workshops, scientific offices and a store for travel equipment."

British Antarctic Survey, "History of Rothera (Station R)" (online)

Architects: (Fuchs House) Mosse and Kesteven; (Accommodation Building, Bonner Laboratory) Top Housing; (New Bransfield House) Hall Grey Architects, United Kingdom
Suppliers: (fuel tanks) British Hovercraft Corporation; (generators) Kolfor Power; (Fuchs House timber) Structaply of Gloucester, United Kingdom
Contractors: (Accommodation Building, Bonner Laboratory) Morrison Construction; (air operations tower) Rothwell Robinson, United Kingdom

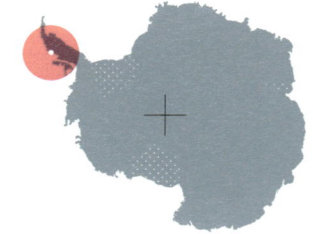

United Kingdom
British Antarctic Survey (BAS)

Active Station > Project 2020+

67°35'8"S 68°7'59"W
Rothera Point, Adelaide Island
Ice-free ground, 10 m.a.s.l., –3.7°C

SITE PLAN

01 Phase I: First hut
02 Phase II: Main living quarters block, power house
03 Phase III: Fuchs house

1 : 5,000   0   50   100   200 m

PLAN AND ELEVATION | FUCHS HOUSE, 1979

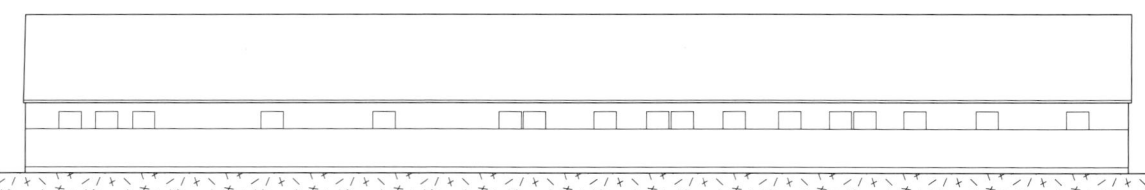

01 Sledge workshop
02 Travelling store
03 Clothing store
04 Tent store
05 Glaciology laboratory
06 Geology laboratory
07 Geophysics laboratory
08 Earth sciences laboratory
09 Cold store

SOURCE
British Antarctic Survey, 1977

1 : 250   0   2.5   5   10 m

SECTION AND ELEVATION | FUCHS HOUSE 1979

"First hut erected on 1 Feb 1976. Phase II was built in 1976/77. This included the main accommodation block, power house and tractor shed. An old storage shed from Adelaide (Station T) was erected close to Phase I and known as the Bingham building after Surgeon Commander E W Bingham, leader of FIDS 1945–47. Phase III was erected 1978/79. This building included scientific offices, travel store and a coldroom."

British Antarctic Survey, "History of Rothera (Station R)" (online)

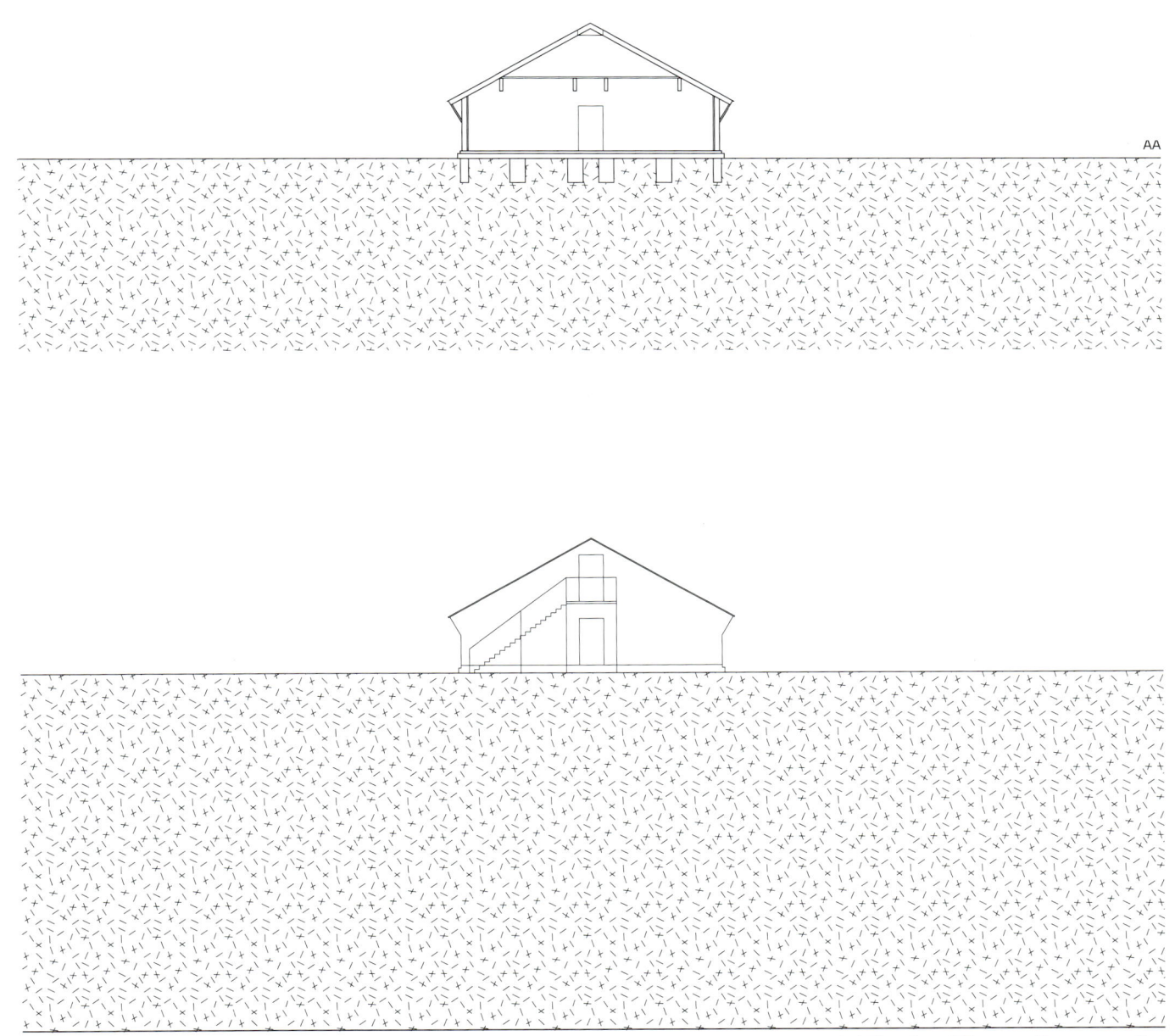

AA

SOURCE
British Antarctic Survey, 1977

1 : 250   0   2.5   5   10 m

↖ p. 412, 413

GEORG FORSTER

1976

"Georg Forster Station [was] originally established as an annex to the Soviet station Novolazarevskaya, which was commissioned on 20 April 1976. At that time the construction concept was pioneering. Prefabricated container modules for laboratories, small power plant and accommodation were carried on sledges from the unloading site at the ice edge of the Lazarev Ice Shelf over a distance of 120 km into the Oasis. There the base was assembled on rocks within six weeks."

Hartwig Gernandt, Saad El Naggar, Jürgen Janneck, Thomas Matz, and Cord Drücker, *From Georg Forster Station to Neumayer Station III – A Sustainable Replacement at Atka Bay for Future*, 2007

Designers: Hartwig Gernandt, Aerological Observatory Lindenberg (MD); Armin Grafe; Geomagnetic Observatory Niemegk (ZISTP) AdW, Germany; Johannes Weiss; Ionospheric Observatory Juliusruh (ZISTP) AdW, Germany
Engineers: Bodo Tripphahn, Operations & Engineering Division, Potsdam of Academy of Sciences of the GDR, East Germany

East Germany
Academy of Sciences of the German Democratic Republic

Removed

70°46'39"S 11°51'03"E
Schirmacher Oasis, Queen Maud Land
Ice-free ground

SITE PLAN

01 Main station building
02 Radio mast

SOURCE
Alfred Wegener Institute, Helmholtz Centre
for Polar and Marine Research (reconstruction)

1 : 5,000   0   50   100   200 m

PLAN

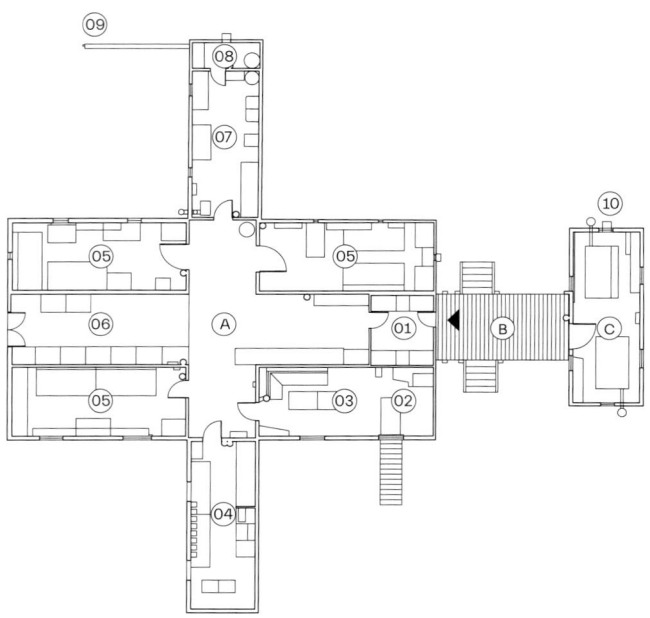

A Main quarters
01 Vestibule
02 Communication room
03 Living room
04 Laundry, science, photo
   laboratory
05 Bedroom
06 Workshop
07 Science laboratory
08 Bathroom
09 Waste water pipe

B Connecting platform

C Diesel generator unit
10 Pipeline to diesel tank

SOURCE
Alfred Wegener Institute, Helmholtz Centre
for Polar and Marine Research, 1976

1 : 250   0   2.5   5   10 m

ELEVATION

"Long-term studies of magneto-spheric-ionospheric processes, geophysical investigations, biological studies and sea-ice observations using satellite imaging were performed [at Georg Forster Station]. In 1985 this station became known to the international scientific community when the vertical extension of the strong ozone depletion (ozone hole) in the southern polar stratosphere was firstly recorded by regular balloon-borne ozone observations. These ozone measurements were performed at Georg Forster Station until 1992 and continued at Neumayer Station afterwards."

Hartwig Gernandt, Saad El Naggar, Jürgen Janneck, Thomas Matz, and Cord Drücker, *From Georg Forster Station to Neumayer Station III – A Sustainable Replacement at Atka Bay for Future*, 2007

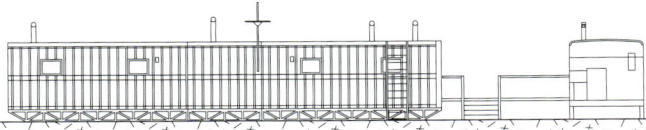

SOURCE
Alfred Wegener Institute, Helmholtz Centre
for Polar and Marine Research, 1976

1 : 250   0   2.5   5   10 m

ESPERANZA
1953/1978

[image: photograph of Base Esperanza station with red buildings]

"Nineteen seventy-eight sig-
nalled a new period for the
presence of Argentina in
Antarctica with the founding
of a small town called Fortín
'Sargento Cabral', which included single-family houses
and the Provincial School
No. 38 'Presidente Raúl Ricardo
Alfonsín'. At Esperanza the
first Antarctic child was conceived."

Gustavo Lezcano (Undersecretary of Antarctic
Management, Provincial Argentine Government),
"Buscan a nacidos en la base Esperanza para reconstruir un registro quemado hace 40 años" (online;
translated by the editor), 2019

"Each of these events – births,
baptisms, marriages, the installation of institutions such as
the Civil Registry office itself or
the national radio station – are
milestones that seek to reaffirm
Argentina's sovereignty over
Antarctica."

Agencia Nacional de Noticias, "Buscan a nacidos en
la base Esperanza para reconstruir un registro quemado
hace 40 años" (online; translated by the editor), 2019

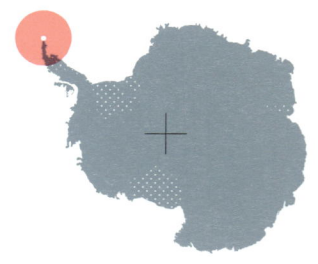

Argentina
Argentine Antarctic Institute (IAA)

Active Station

63°23'51"S 56°59'52"W
Hope Bay, Trinity Peninsula
Ice-free ground, 10 m.a.s.l., −4.5°C

SITE PLAN

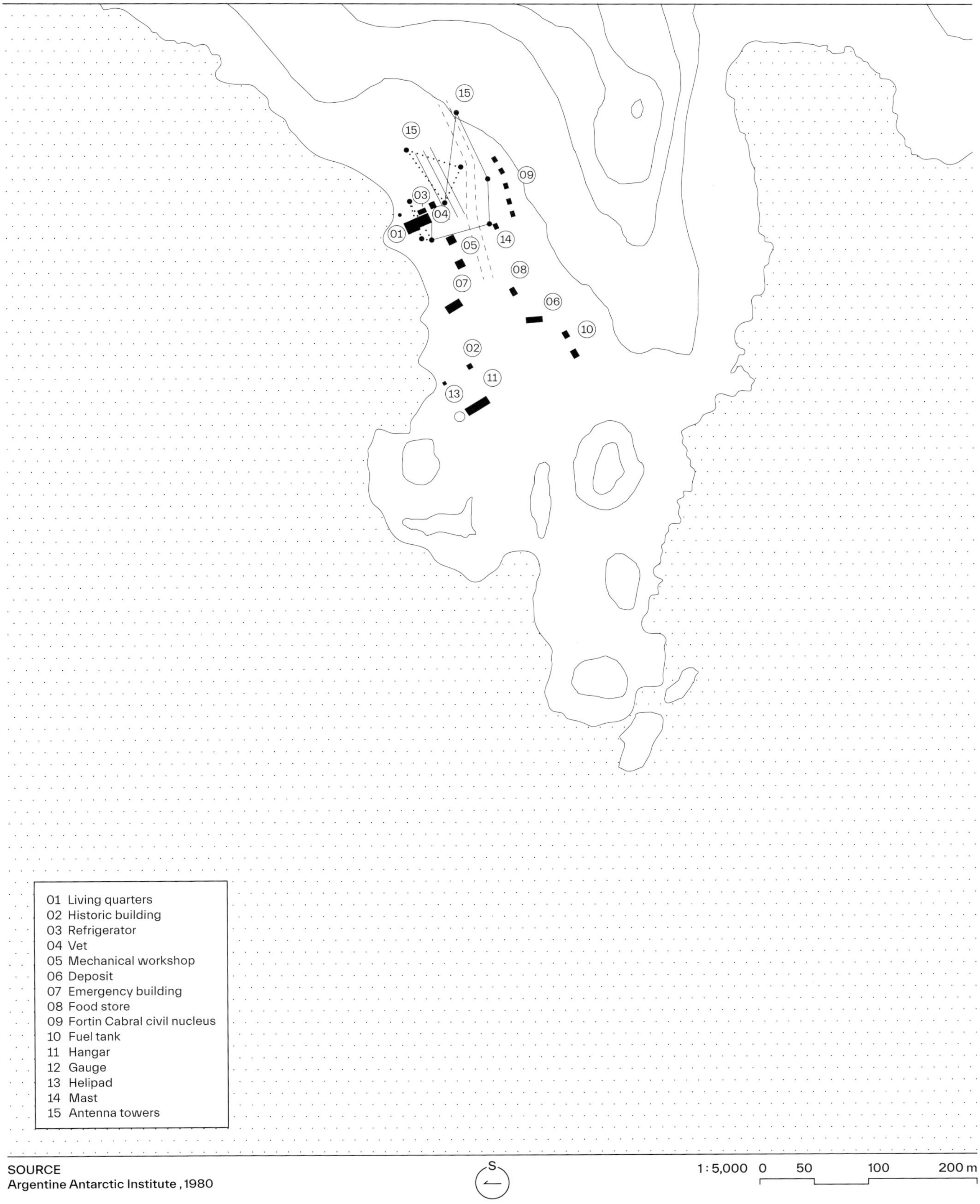

01 Living quarters
02 Historic building
03 Refrigerator
04 Vet
05 Mechanical workshop
06 Deposit
07 Emergency building
08 Food store
09 Fortin Cabral civil nucleus
10 Fuel tank
11 Hangar
12 Gauge
13 Helipad
14 Mast
15 Antenna towers

SOURCE
Argentine Antarctic Institute , 1980

1 : 5,000   0   50   100   200 m

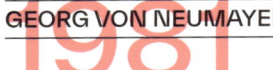

GEORG VON NEUMAYER

↖ p. 412, 413, 424, 460

1981

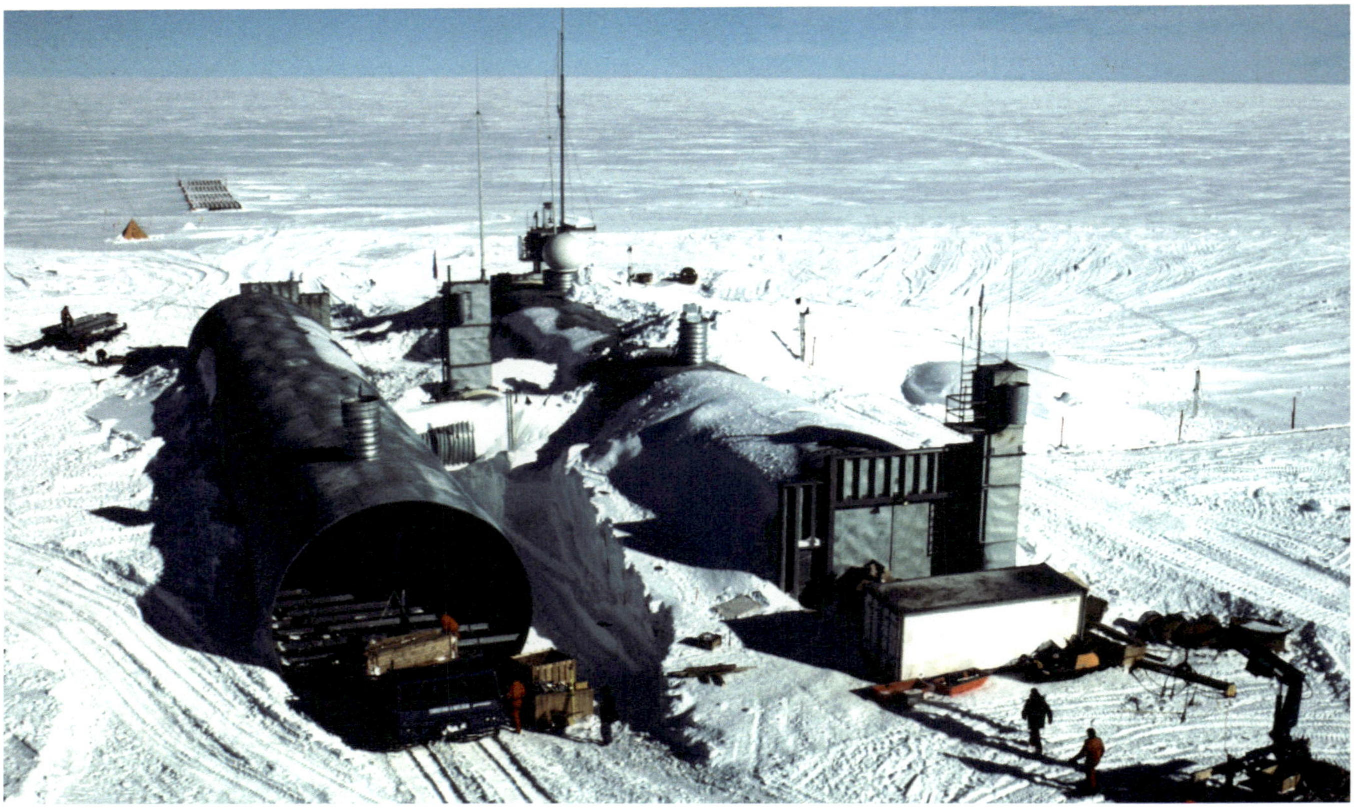

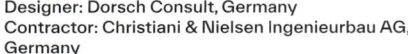

Designer: Dorsch Consult, Germany
Contractor: Christiani & Nielsen Ingenieurbau AG, Germany

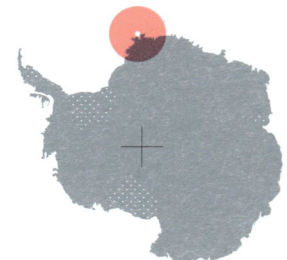

Germany
Alfred Wegener Institute for Polar and Marine Research (AWI)

Closed > Replaced

70°37'00"S 8°22'00"W
Ekström Ice Shelf, Queen Maud Land
Ice shelf, 43 m.a.s.l., –16°C

SITE PLAN

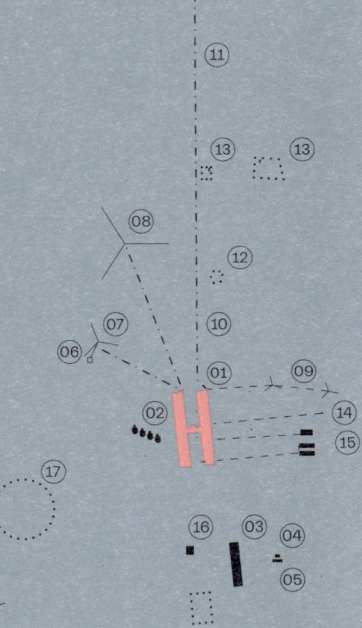

01 Main station building
02 Living quarters
03 Deep drilling site
04 Balloon caboose
05 Helium store container
06 Radiation caboose
07 Meteorological mast, 15 m
08 Meteorological mast, 45 m
09 Dipole antennas
10 English weather hut
11 To geophysics and
   chemistry observatory
12 Oil storage
13 Diesel storage
14 Drainage
15 Storage
16 Skidoos container
17 Helipad
18 To Atka Bay airstrip

SOURCE
Alfred Wegener Institute, Helmholtz Centre
for Polar and Marine Research, 1983

S

1 : 5,000   0   50   100   200 m

PLAN, BELOW SNOW LEVEL

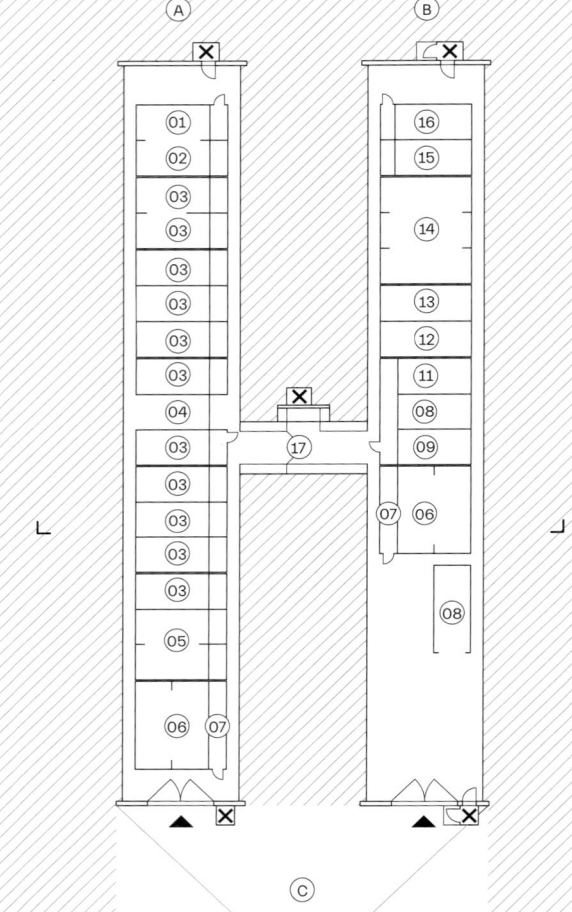

A   Tunnel 01
01   Meteorological laboratory
02   Geophysical laboratory
03   Bedroom
04   Medical room
05   Storage
06   Generator room
07   Technical room

B   Tunnel 02
08   Workshop
09   Snow melt
10   Laundry
11   Bathroom
12   Food storage
13   Kitchen
14   Dining room
15   Geology laboratory
16   Telecommunications
     room

17   Connecting tunnel

C   Ramp

SOURCE
Alfred Wegener Institute, Helmholtz Centre
for Polar and Marine Research, 1983

1 : 500   0   5   10   20 m

SECTION

"For the German station the buildings were designed in container form. They were placed in large tubes made of wavy steel plates and therefore protected against the influence of the atmospheric conditions. The steel tubes compensated the resulting forces from the snow pressure. The air gap between the buildings and the tube walls resulted in a certain insulation of the buildings. The real insulation was situated on the outer faces of the buildings. […]

The steel tubes, having a length of 50 m for the protection of the building modules consisted of cambered and zinc coated wavy steel plates. In each case 14 plates were mounted to a circle about 5 m in diameter; each plate had a weight up to 370 kg."

Eberhard Kohlberg and Jürgen Janneck, *Georg von Neumayer Station (GvN) and Neumayer Station II (NM-II) German Research Stations on Ekström Ice Shelf, Antarctica,* Report, 2007

SOURCE
Alfred Wegener Institute, Helmholtz Centre
for Polar and Marine Research, 1983

1:500   0   5   10   20 m

HALLEY IV
1956/1983

↖ p. 428, 460, 467

"During the trial erection period at Ross on Wye many fundamental lessons were learned with respect to procedures and working methods. The most important of these was the degree of lateral restraint required in order to maintain the true tube shape and how easy it was for the tube to distort if this restraint was ineffective. It was intended that the Base be constructed with its longitudinal axis into the prevailing wind in order that uneven drift on the tubes was kept to a minimum. Records show that the Base was unfortunately installed 20 degrees off the prevailing wind direction."

Alan Smith (Builder, Head of British Antarctic Survey Building Services), *Halley No. 4: Brief Report on the Establishment of the Station Complex at Halley*, 1983

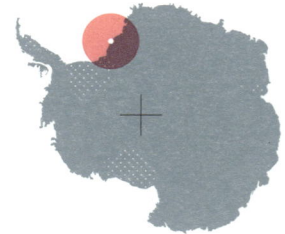

Architect: Jamieson Associates Designers,
United Kingdom

United Kingdom
British Antarctic Survey (BAS)

Removed > Replaced

77°50'53.5"S 166°40'06.3"E
Brunt Ice Shelf, Caird Coast
Ice shelf, 37 m.a.s.l., −20°C

SITE PLAN

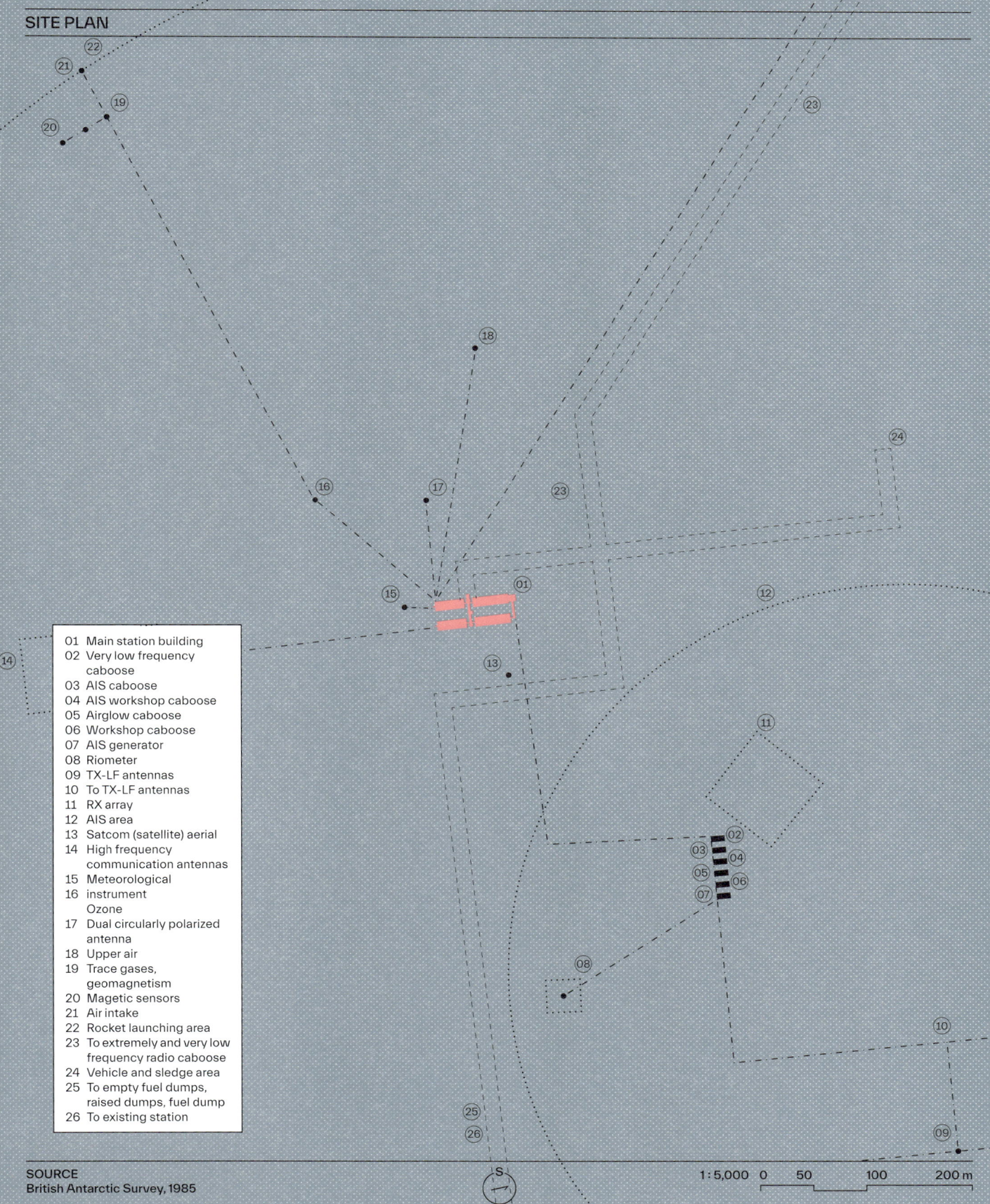

01 Main station building
02 Very low frequency
     caboose
03 AIS caboose
04 AIS workshop caboose
05 Airglow caboose
06 Workshop caboose
07 AIS generator
08 Riometer
09 TX-LF antennas
10 To TX-LF antennas
11 RX array
12 AIS area
13 Satcom (satellite) aerial
14 High frequency
     communication antennas
15 Meteorological
16 instrument
     Ozone
17 Dual circularly polarized
     antenna
18 Upper air
19 Trace gases,
     geomagnetism
20 Magetic sensors
21 Air intake
22 Rocket launching area
23 To extremely and very low
     frequency radio caboose
24 Vehicle and sledge area
25 To empty fuel dumps,
     raised dumps, fuel dump
26 To existing station

SOURCE
British Antarctic Survey, 1985

S

1 : 5,000   0   50   100   200 m

PLAN, 1ST AND 2ND LEVEL BELOW SNOW LEVEL | MAIN STATION BUILDINGS

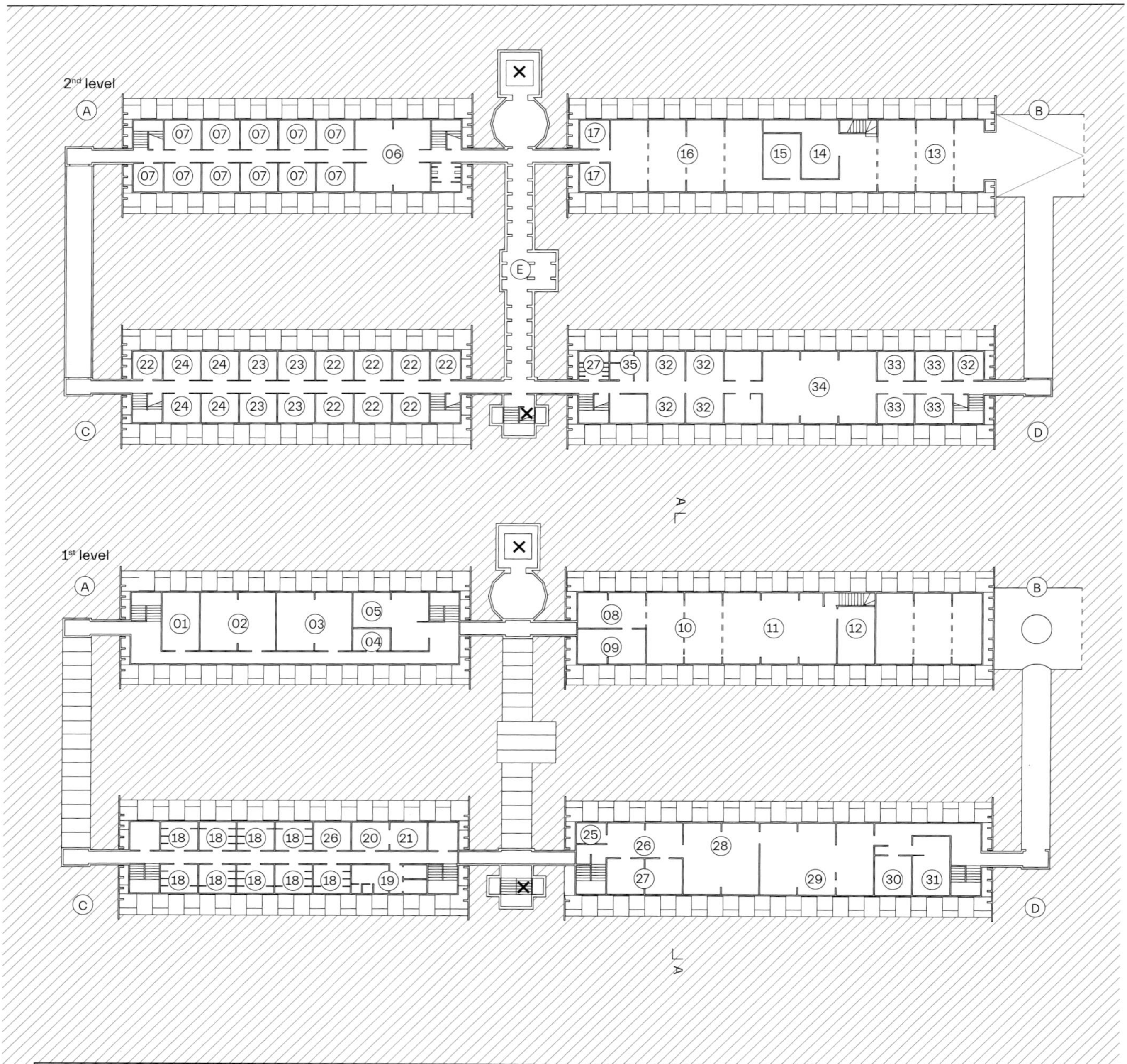

2nd level

1st level

A	Science building	10	Storage	20	Laundry	30	Base leader's office
01	Scientific library	11	Water tanks	21	Drying room	31	Telecommunications
02	Meteorological office	12	Mechanical officer	22	Storage	32	Storage
03	Geographic office	13	Garage, workshop	23	Travelling store	33	Food store
04	Darkroom	14	Machine room	24	Clothes store	34	Recreation store
05	Surgery	15	Electrical workshop			35	Bathroom
06	Workshop store	16	Generator	D	Living quarters		
07	Scientific store	17	Storage	25	Pantry	E	Connecting volume
				26	Wash up		
B	Miscellaneous	C	Sleeping quarters	27	Kitchen		
08	Magnetism zone	18	Bedroom	28	Dining area		
09	Technical room	19	Bathroom	29	Lounge		

SOURCE
British Antarctic Survey, 1983

1 : 500   0   5   10   20 m

ELEVATION | MAIN STATION BUILDINGS

"The information received to date, in our opinion, indicates that the tube has failed to maintain its shape primarily due to consolidation of the foundation medium and a lack of lateral restraint being provided to the tube sides. [...] The tube in its deflected form is unlikely to achieve the same predicted life span as it would had if remained circular. This is primarily due to a change in loading pattern on that the tube was designed to act in compression and now finds itself having to accommodate a greater degree of bending earlier in its life."

Alan Smith (Builder, Head of British Antarctic Survey Building Services), *Halley No. 4: Brief Report on the Establishment of the Station Complex at Halley,* 1983

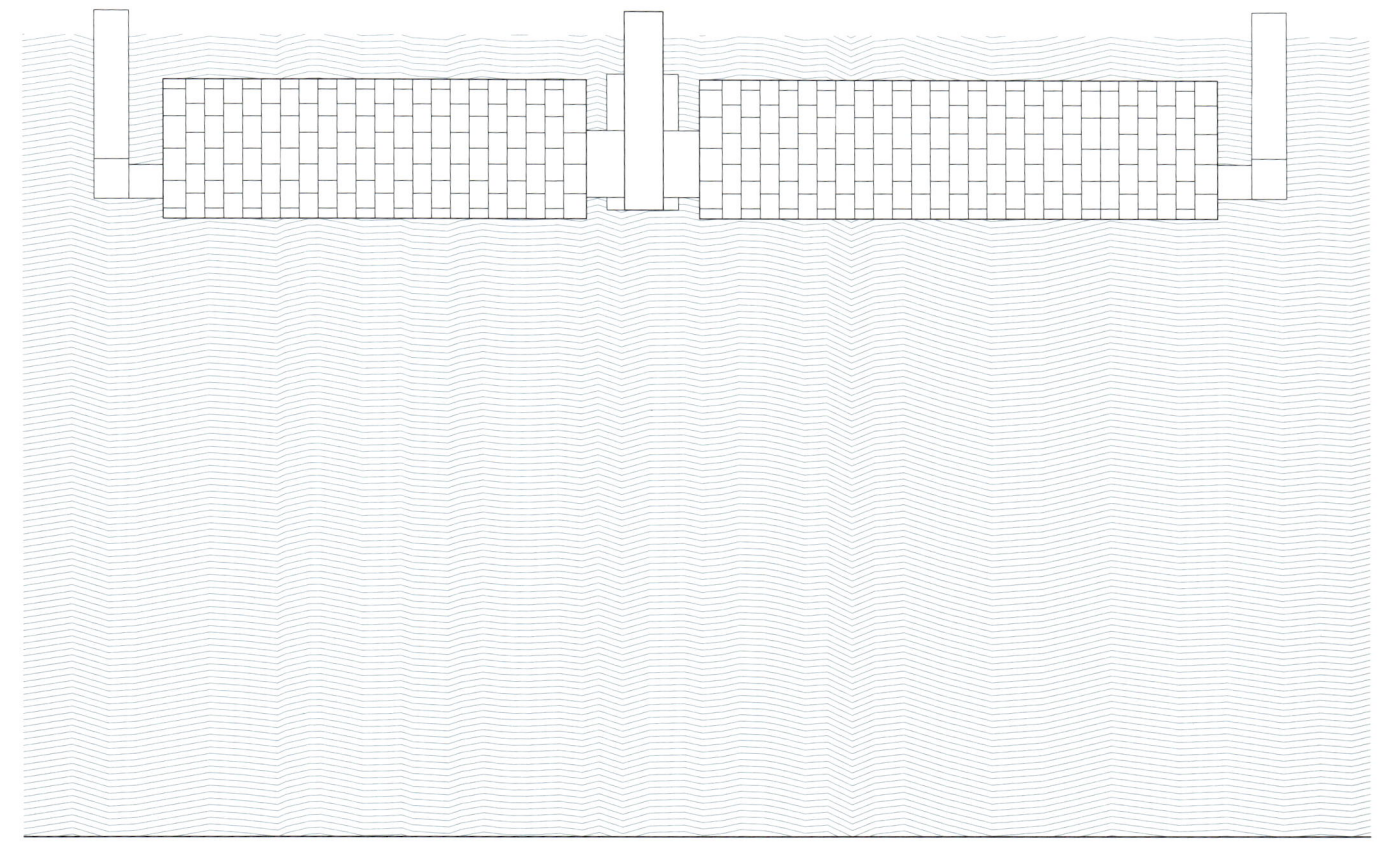

SOURCE
British Antarctic Survey (reconstruction)

1 : 500   0   5   10   20 m

SECTION

"It was the intention during the design period to provide a consistent foundation level upon which the base was constructed which provided a density of 0.5 g/cm³. From the reports received to date it would appear that a density of this order was available at 5 m depth but not necessarily at surface level or at 2 m depth."

Alan Smith (Builder, Head of British Antarctic Survey Building Services), *Halley No. 4: Brief Report on the Establishment of the Station Complex at Halley,* 1983

AA

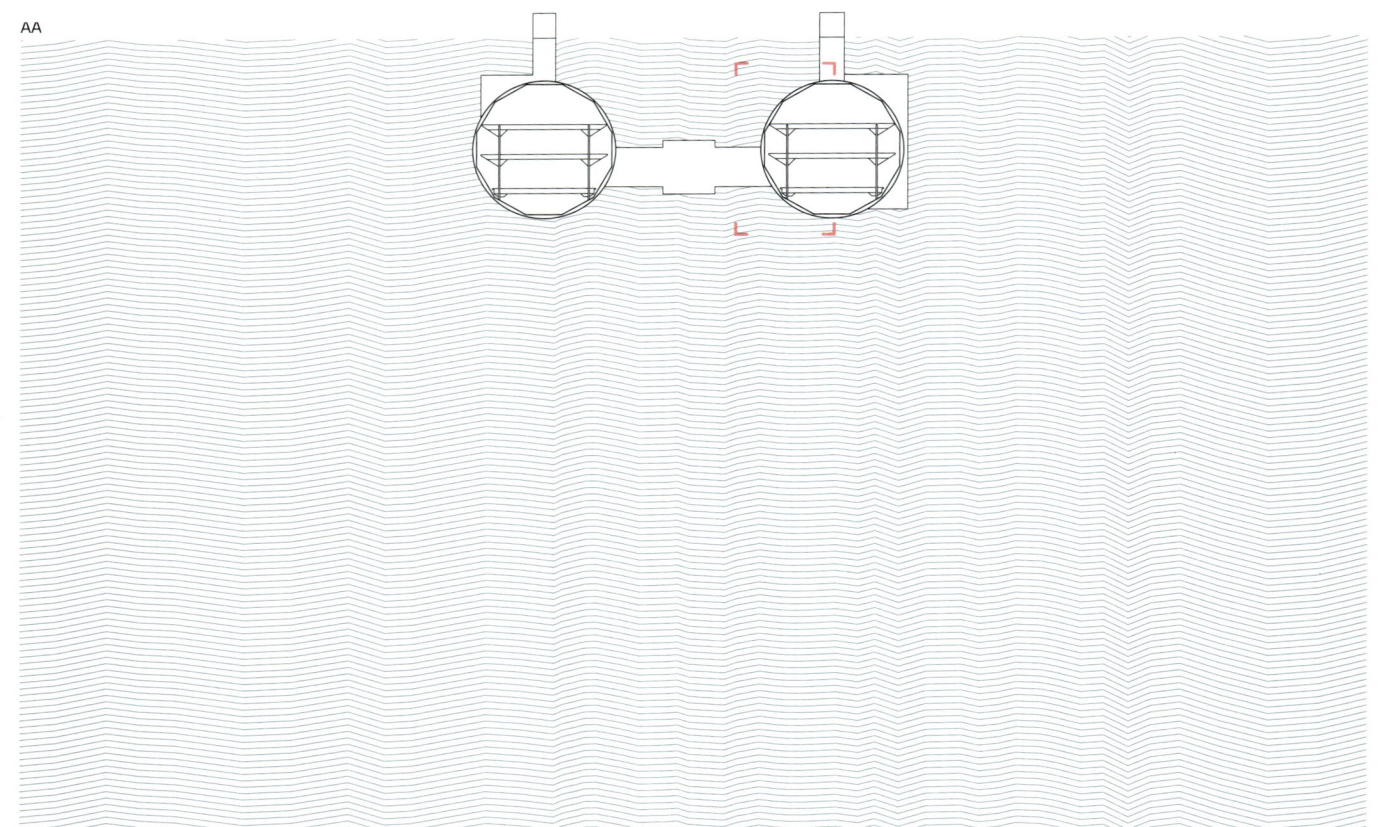

SOURCE
British Antarctic Survey, 1983

1:500   0   5   10   20 m

DETAIL

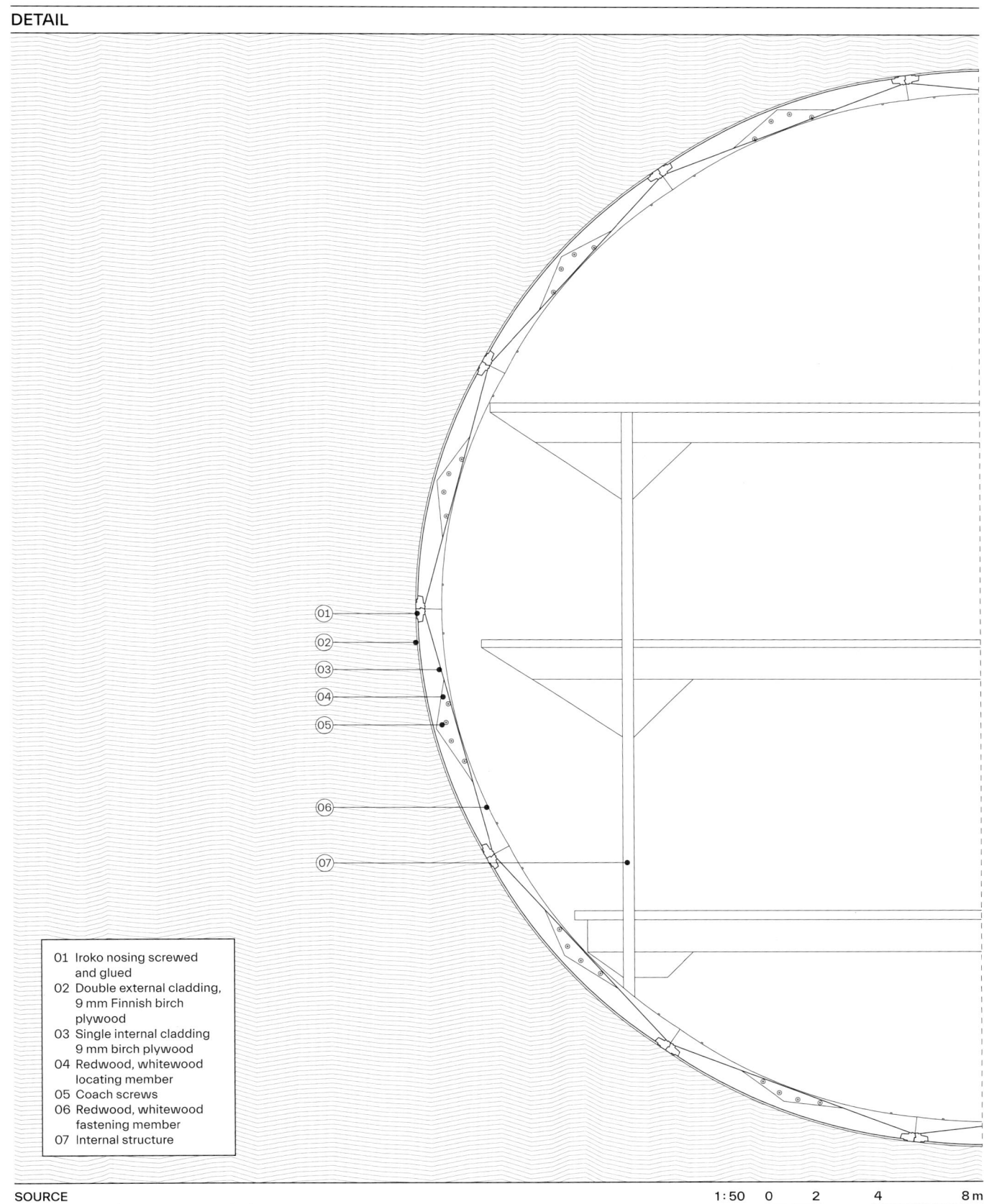

01 Iroko nosing screwed
     and glued
02 Double external cladding,
     9 mm Finnish birch
     plywood
03 Single internal cladding
     9 mm birch plywood
04 Redwood, whitewood
     locating member
05 Coach screws
06 Redwood, whitewood
     fastening member
07 Internal structure

SOURCE
British Antarctic Survey (reconstruction)

1 : 50   0   2   4   8 m

BASE PRESIDENTE EDUARDO FREI MONTALVA

↖ p. 444

1969/1984

"Frei Base is a key hub for the supply of Antarctic bases on the Peninsula and, since the 1980s, significant efforts have been made to improve and increase the infrastructure of the station. Since 1984, the civilian settlement of Villa Las Estrellas, has provided housing and services to aviation personnel, scientists, technicians, and other professionals who, together with their families, live and work in the 'Antarctic town.'"

Casa Museo Frei, "50 Years of Science, Collaboration and Peace on the White Continent" (online; translated)

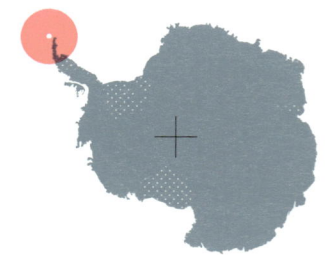

Chile
Chilean Antarctic Institute (INACH)

Active Station

62°12'03"S 58°57'37"W
Fildes Bay, King George Island, South Shetland Islands
Ice-free ground, 10 m.a.s.l., −2.3 °C

SITE PLAN

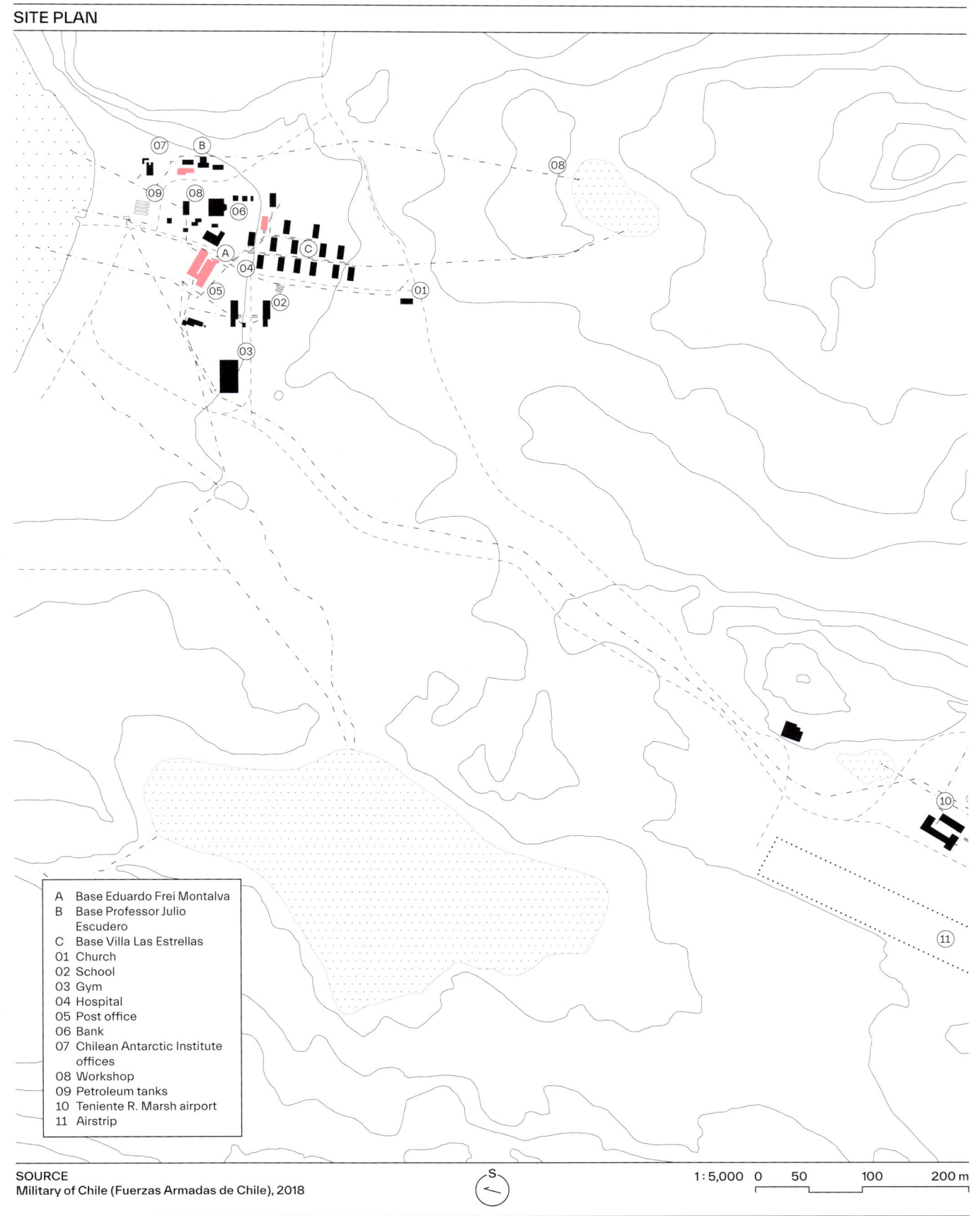

A   Base Eduardo Frei Montalva
B   Base Professor Julio
    Escudero
C   Base Villa Las Estrellas
01 Church
02 School
03 Gym
04 Hospital
05 Post office
06 Bank
07 Chilean Antarctic Institute
    offices
08 Workshop
09 Petroleum tanks
10 Teniente R. Marsh airport
11 Airstrip

PLAN AND ELEVATION | VILLA LAS ESTRELLAS LIVING UNIT, 1984

"Children at the schoolhouse here study under a portrait of Bernardo O'Higgins, Chile's independence leader. The bank manager welcomes deposits in Chilean pesos. The cellphone service from the Chilean phone company Entel is so robust that downloading iPhone apps works like a charm. […] Fewer than 200 people live in this outpost founded in 1984 during the dictatorship of Gen. Augusto Pinochet, when Chile was seeking to bolster its territorial claims in Antarctica. Since then, the tiny hamlet has been at the center of one of Antarctica's most remarkable experiments: exposing entire families to isolation and extreme conditions in an attempt to arrive at a semblance of normal life at the bottom of the planet."

Simon Romero (Journalist), "Antarctic Life: No Dogs, Few Vegetables and 'a Little Intense' in the Winter", *The New York Times*, 2016

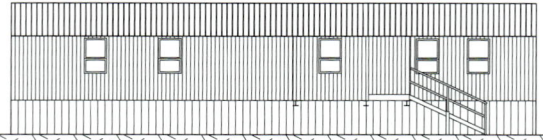

01 Vestibule
02 Storage
03 Double bedroom
04 Bedroom
05 Family room
06 Bathroom
07 Kitchen
08 Dining
09 Living room

PLAN AND ELEVATION | PROFESSOR JULIO ESCUDERO MULTIPURPOSE BUILDING , 1995

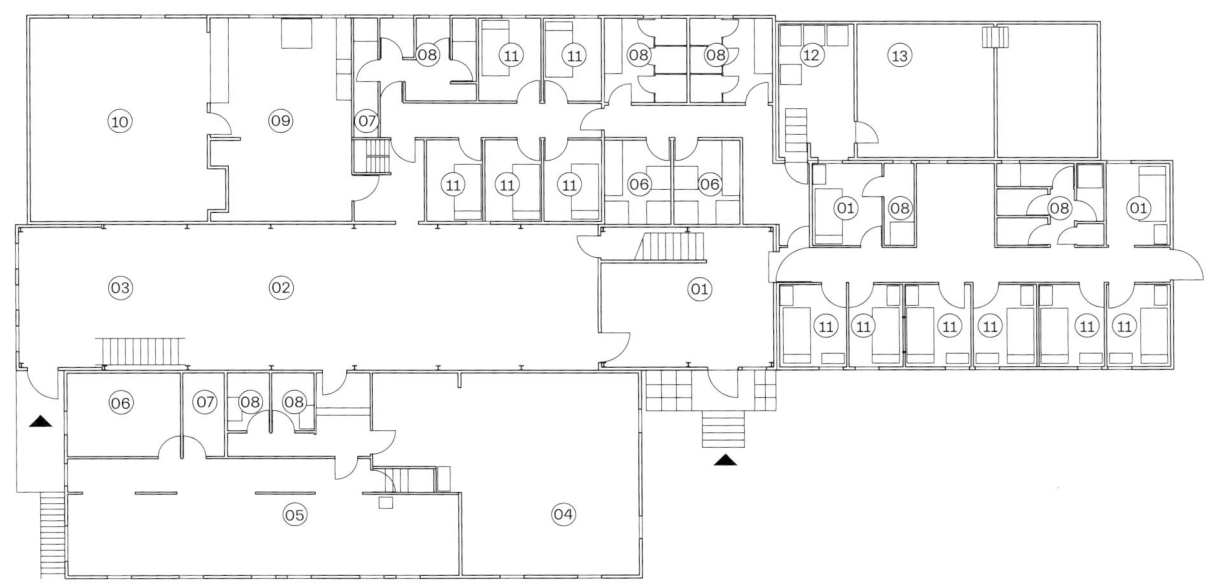

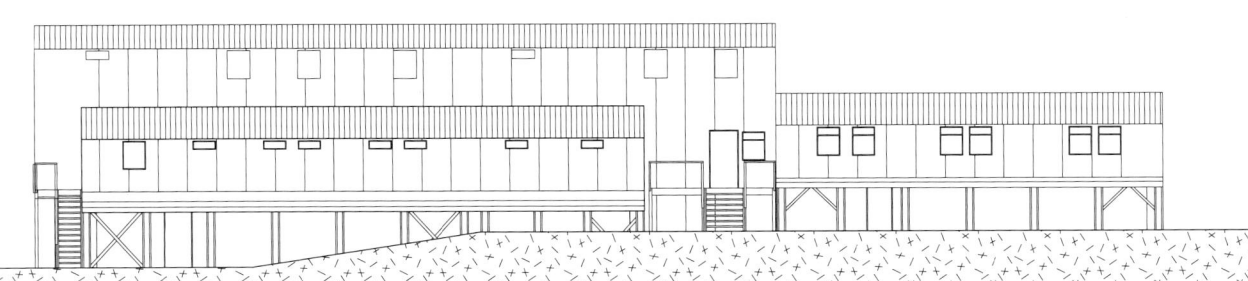

01 Vestibule
02 Recreation area
03 Living room
04 Molecular biology
     laborarory
05 Microbiology laboratory
06 Multipurpose laboratory
07 Storage
08 Bathroom
09 Kitchen
10 Dining
11 Bedroom
12 Laundry
13 Water tanks

SOURCE
Military of Chile (Fuerzas Armadas de Chile), 2011

1 : 250   0   2.5   5   10 m

PLAN | CENTRO METEOROLÓGICO EDUARDO FREI, 2011

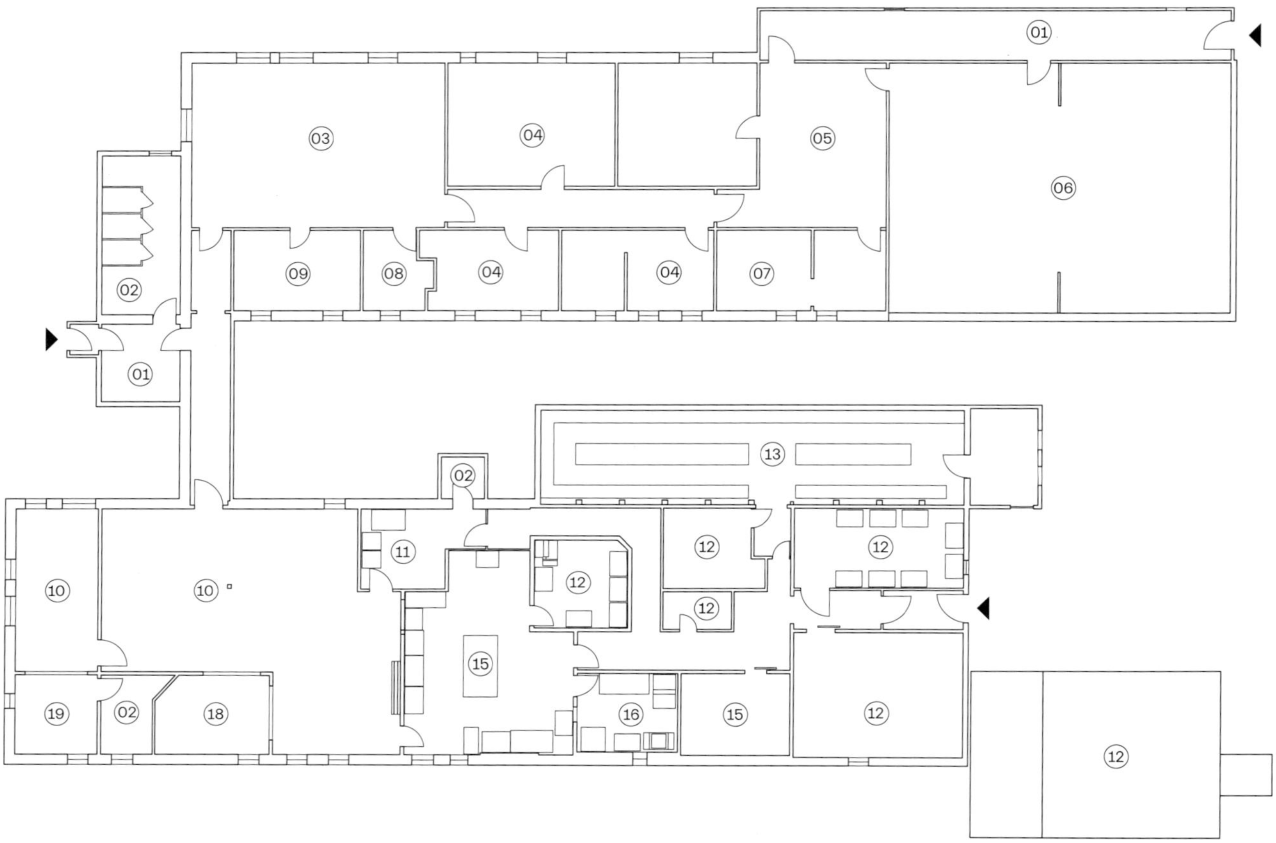

01 Vestibule
02 Bathroom
03 Games room
04 Office
05 Tool depot
06 Water storage room
07 Electric room
08 Hairdresser
09 Library
10 Dining room
11 Laundry
12 Storage
13 Supermarket
14 Souvenir store
15 Fruit shop
16 Bakery
17 Kitchen
18 Bar
19 Living room

SOURCE
Military of Chile (Fuerzas Armadas de Chile), 2019

1 : 250   0   2.5   5   10 m

ELEVATION | CENTRO METEOROLÓGICO EDUARDO FREI, 2011

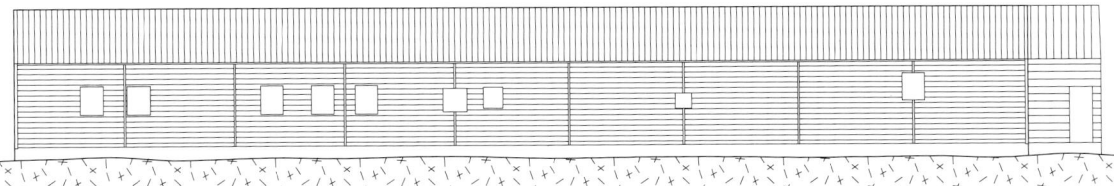

SOURCE
Military of Chile (Fuerzas Armadas de Chile), 2019

1 : 250   0   2.5   5   10 m

MARIO ZUCCHELLI
1986

↖ p. 375, 413, 415

"The station is made by joining ISO 20' containers 1CC type (2.45 × 6.05 × 2.60 metres), a design which allows the ensemble to be erected cheaply and quickly. Each container had a removable wooden wall to cover the joint between the individual modules, which was achieved by bolting the panels together and sealing them with silicone."

Francesco Pellegrino (Engineer, ENEA), in conversation with UNLESS, Architectural Association School of Architecture, 2019

Consultant: (preliminary studies) Aquater and ENEA, Italy
Construction: (production of modules) SNAM Progetti, Franzisella Spa, Italy

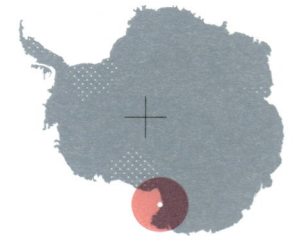

Italy
Italian National Antarctic Research Program (PNRA)

Active Station

74°41'39"S 164°06'50"E
Terra Nova Bay, Ross Sea
Ice-free ground, 8 m.a.s.l., −14°C

SITE PLAN

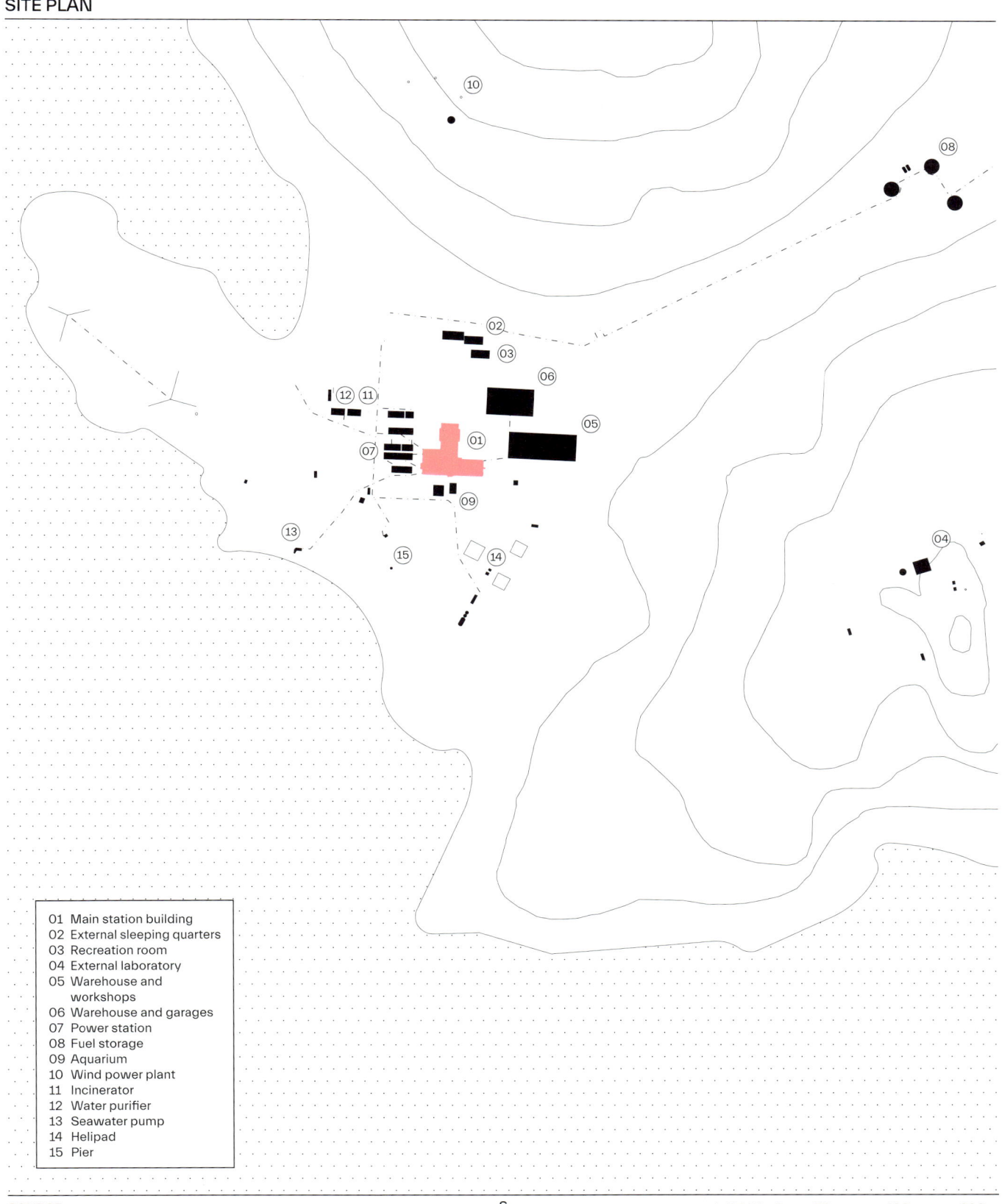

01 Main station building
02 External sleeping quarters
03 Recreation room
04 External laboratory
05 Warehouse and
   workshops
06 Warehouse and garages
07 Power station
08 Fuel storage
09 Aquarium
10 Wind power plant
11 Incinerator
12 Water purifier
13 Seawater pump
14 Helipad
15 Pier

SOURCE
Italian National Antarctic Research Program, 2004

S

1 : 5,000   0   50   100   200 m

PLAN, 1ST AND 2ND LEVEL | MAIN STATION BUILDING AND CONTROL TOWER

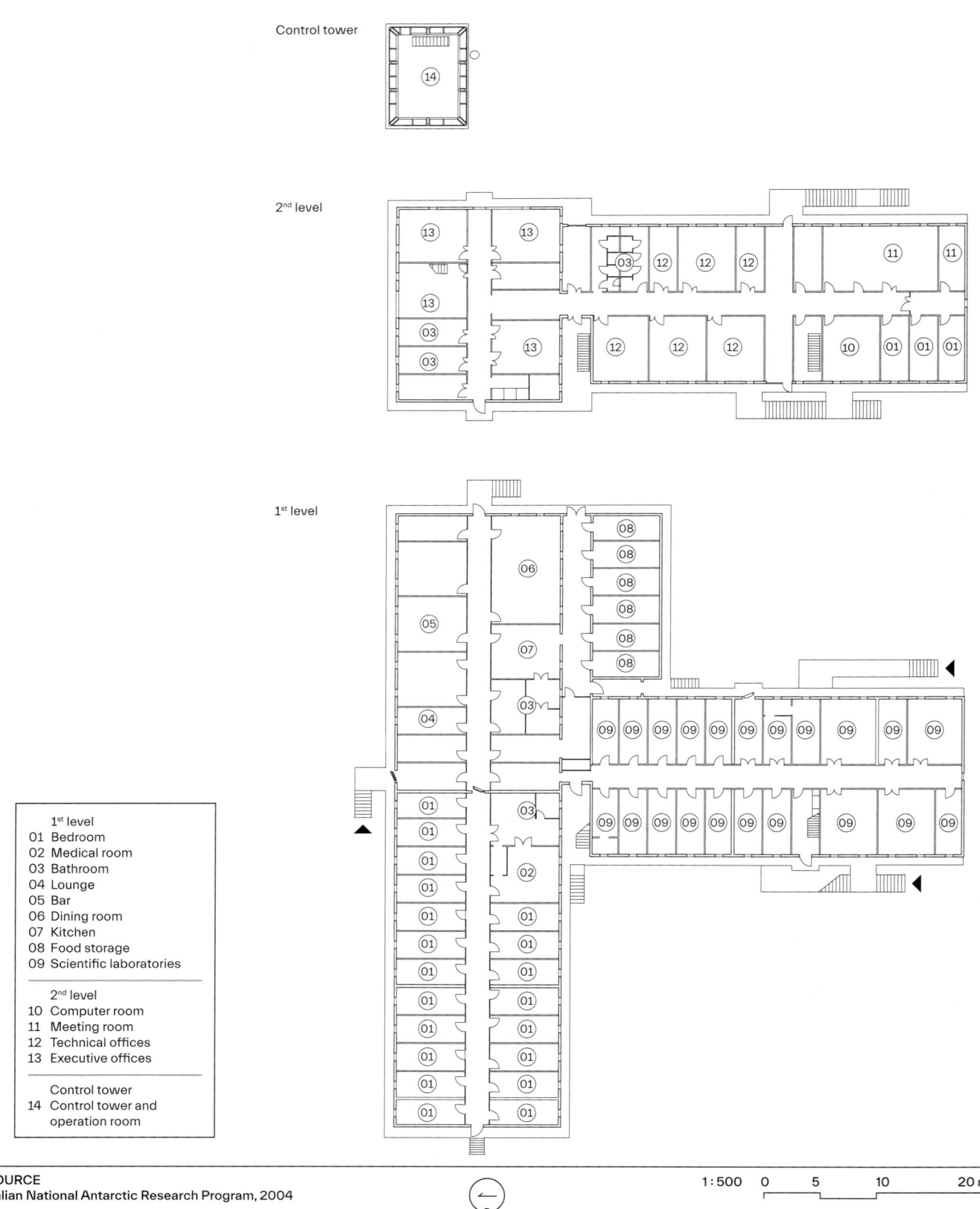

Control tower

2nd level

1st level

1st level
01 Bedroom
02 Medical room
03 Bathroom
04 Lounge
05 Bar
06 Dining room
07 Kitchen
08 Food storage
09 Scientific laboratories

2nd level
10 Computer room
11 Meeting room
12 Technical offices
13 Executive offices

Control tower
14 Control tower and
operation room

SOURCE
Italian National Antarctic Research Program, 2004

1:500   0   5   10   20 m

ELEVATIONS | MAIN STATION BUILDING AND CONTROL TOWER

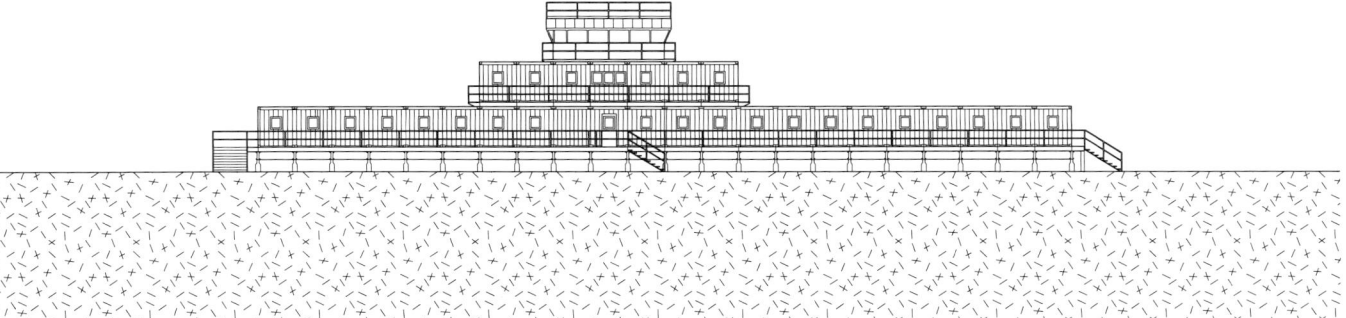

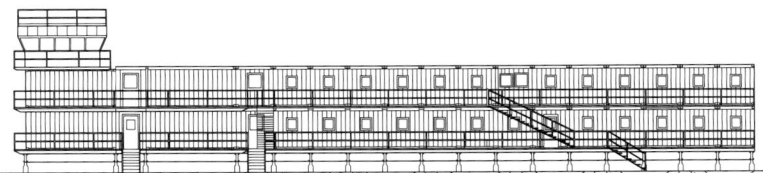

SOURCE
Italian National Antarctic Research Program, 2004

1:500   0   5   10   20 m

↖ p. 428

CASEY
1969/1988

"The Antarctic Division also considered that interconnection of buildings at Casey was a moral factor and that as a general rule expeditioners should have to go to work through the snow each day – thereby assuring themselves that conditions outdoors were not bad as one would imagine when inside listening to the wind whistling around the building."

Philip Incoll (Architect), *The Influence of Architectural Theory on the Design of Australian Antarctic Stations,* 1990

Architect and Engineer: Australian Construction Services
Supplier: (prefabricated components and internal partitions) Burton Turner Coolrooms; (interior fittings) Inggall and Tribe P/L; (heating and ventilation) J. H. Hill

"Australian Construction Services proposed that Antarctic buildings could be built faster and better with a modular system where large pieces (3.6 metres × 6 metres × 4 metres high) were completely finished and fitted out in Australia before being shipped to Antarctica. [...] The new Casey station, consisting of AANBUS [Australian Antarctic Building System] modular buildings, was opened in December 1988."

Australian Antarctic Program, "Australia's Antarctic Buildings: AANBUS" (online)

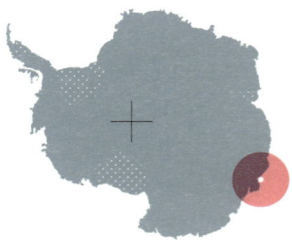

Australia
Australian Antarctic Program (AAD)

Active Station

66°16'57"S 110°31'36"E
Bailey Peninsula
Ice-free ground, 32 m.a.s.l., –5.9°C

SITE PLAN

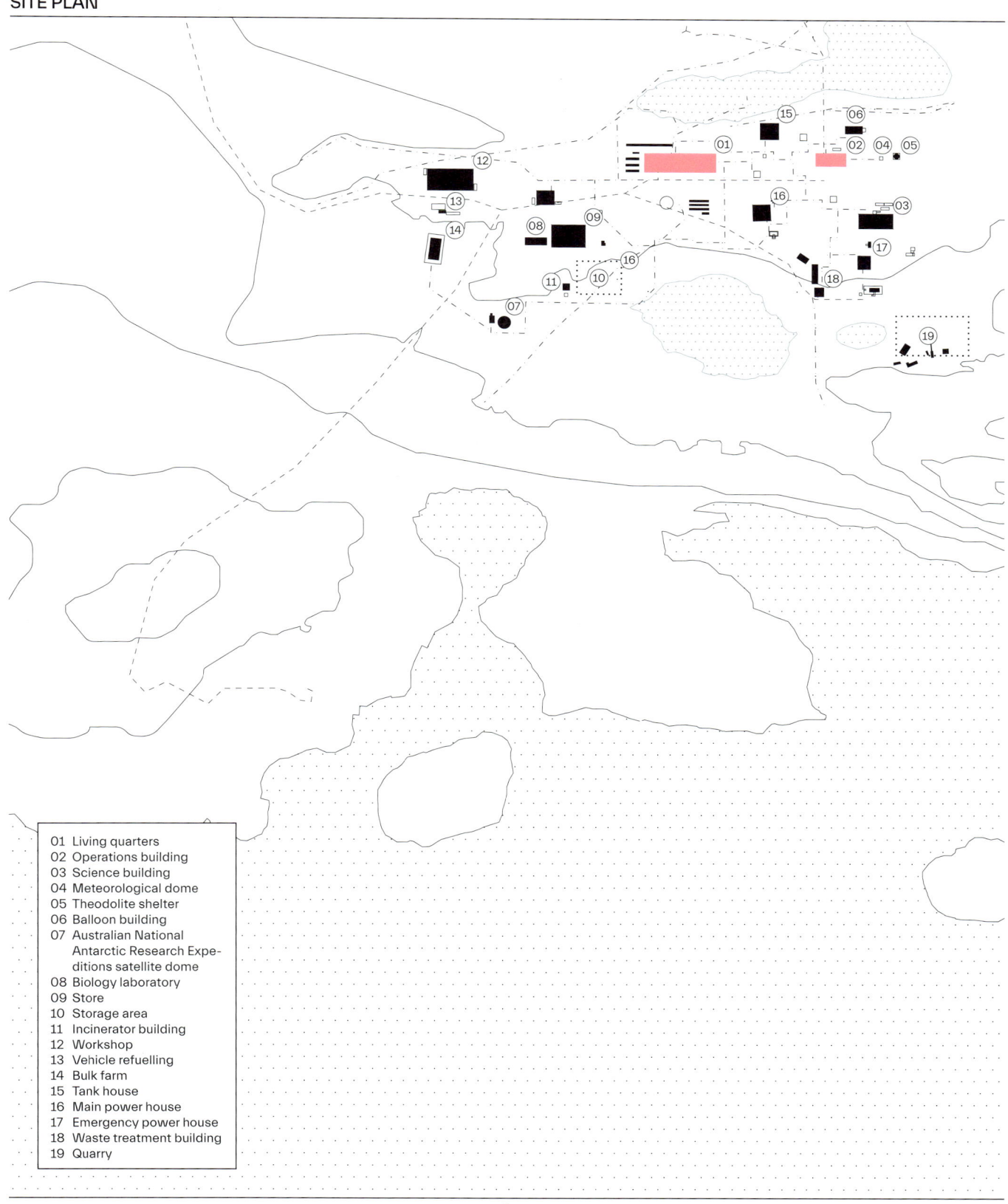

01 Living quarters
02 Operations building
03 Science building
04 Meteorological dome
05 Theodolite shelter
06 Balloon building
07 Australian National
   Antarctic Research Expeditions satellite dome
08 Biology laboratory
09 Store
10 Storage area
11 Incinerator building
12 Workshop
13 Vehicle refuelling
14 Bulk farm
15 Tank house
16 Main power house
17 Emergency power house
18 Waste treatment building
19 Quarry

SOURCE
Australian Construction Services, 1980

S

1 : 5,000   0   50   100   200 m

PLAN AND ELEVATION | OFFICE BUILDING

"The graded hierarchy of privacy allows individuals a choice of the extent of socialization that they desire. […] The range of privacy extends from fully private (bedrooms) through partly secluded (small lounge areas and rooms) to fully public (the centre of the main lounge area). […] At Casey all who wish to eat must pass through the door of the Mess which opens to the ground floor of the Lounge. All expeditioners must pass through the lounge on the ground floor or down the central stairs into the lounge to reach this point. […] This concentration of traffic and facilities creates the strong focal point and the compulsory socialization. The design of the Casey lounge provides a central 2-storey-high space lit by full height windows. […] Around this completely open and unprivate space are the smaller scale spaces."

Philip Incoll (Architect), *The Influence of Architectural Theory on the Design of Australian Antarctic Stations,* 1990

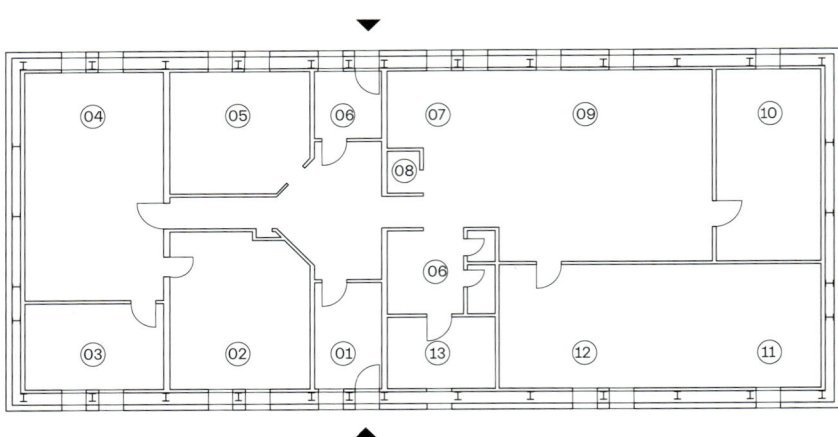

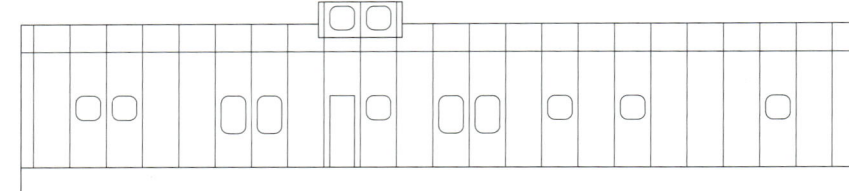

01 Vestibule
02 Base leader's office
03 Meteorology workshop
04 Meteorology office
05 Meteorology store
06 Bathroom
07 Phone booth
08 Waiting area
09 Radio operation
10 Radio workshop
11 Teletype
12 Radio store
13 Plant room

SOURCE
Australian Construction Services, 1980

1 : 250   0   2.5   5   10 m

PLAN, 1ST AND 2ND LEVEL | DOMESTIC BUILDING

2nd level

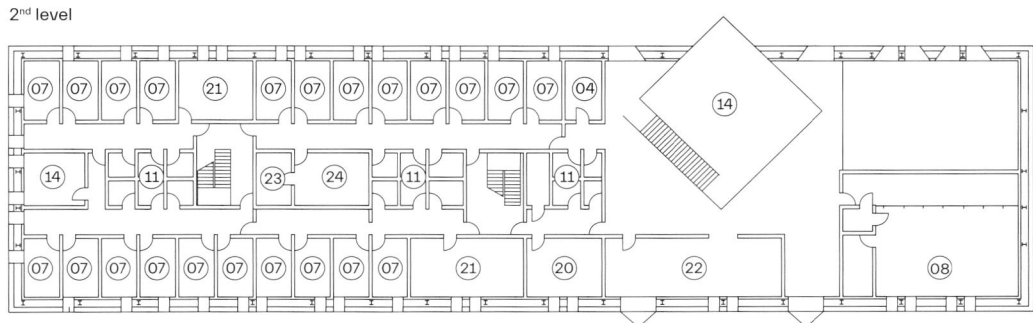

1st level

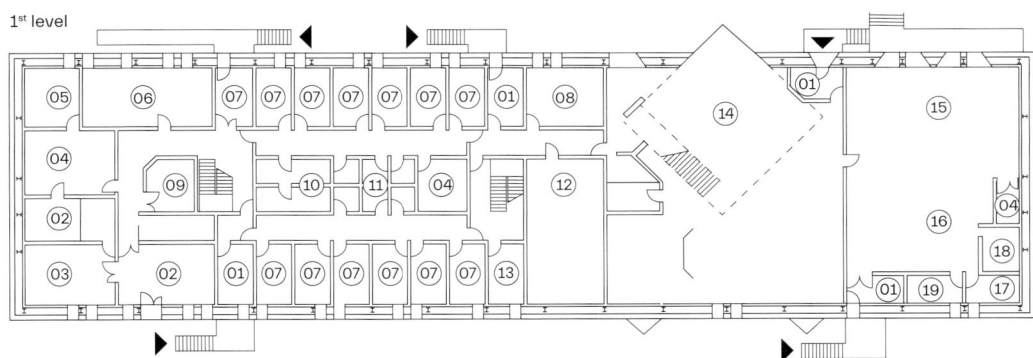

1st level	14 Lounge
01 Vestibule	15 Dining room
02 Medical room	16 Kitchen
03 Surgery room	17 Dry store
04 Storage	18 Cool store
05 Laboratory	19 Office
06 Dental examination room	
07 Bedroom	2nd level
08 Recreation	20 Sewing and hobbies
09 Consultation room	21 Dormitory & lounge
10 Darkroom	22 Library
11 Bathroom	23 Drying room
12 Plant room	24 Laundry
13 Hydroponics room	

SOURCE
Australian Construction Services, 1980

1:500   0   5   10   20 m

ZHONGSHAN
1989

↖ p. 99, 366, 418

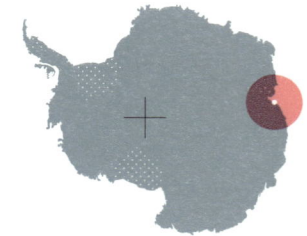

Architect: China National Building Material Group
Corporation (1989); Architectural Design and
Research Institute of Tsinghua University (since
2002), People's Republic of China
Manufacturer: China Railway Construction
Engineering Group, People's Republic of China

People's Republic of China   Active Station   69°22'25"S 76°22'18"E
Polar Research Institute of China (PRIC)   Larsemann Hills, Prydz Bay
                                                                               Ice-free ground, 60 m.b.s.l., –11.2°C

SITE PLAN

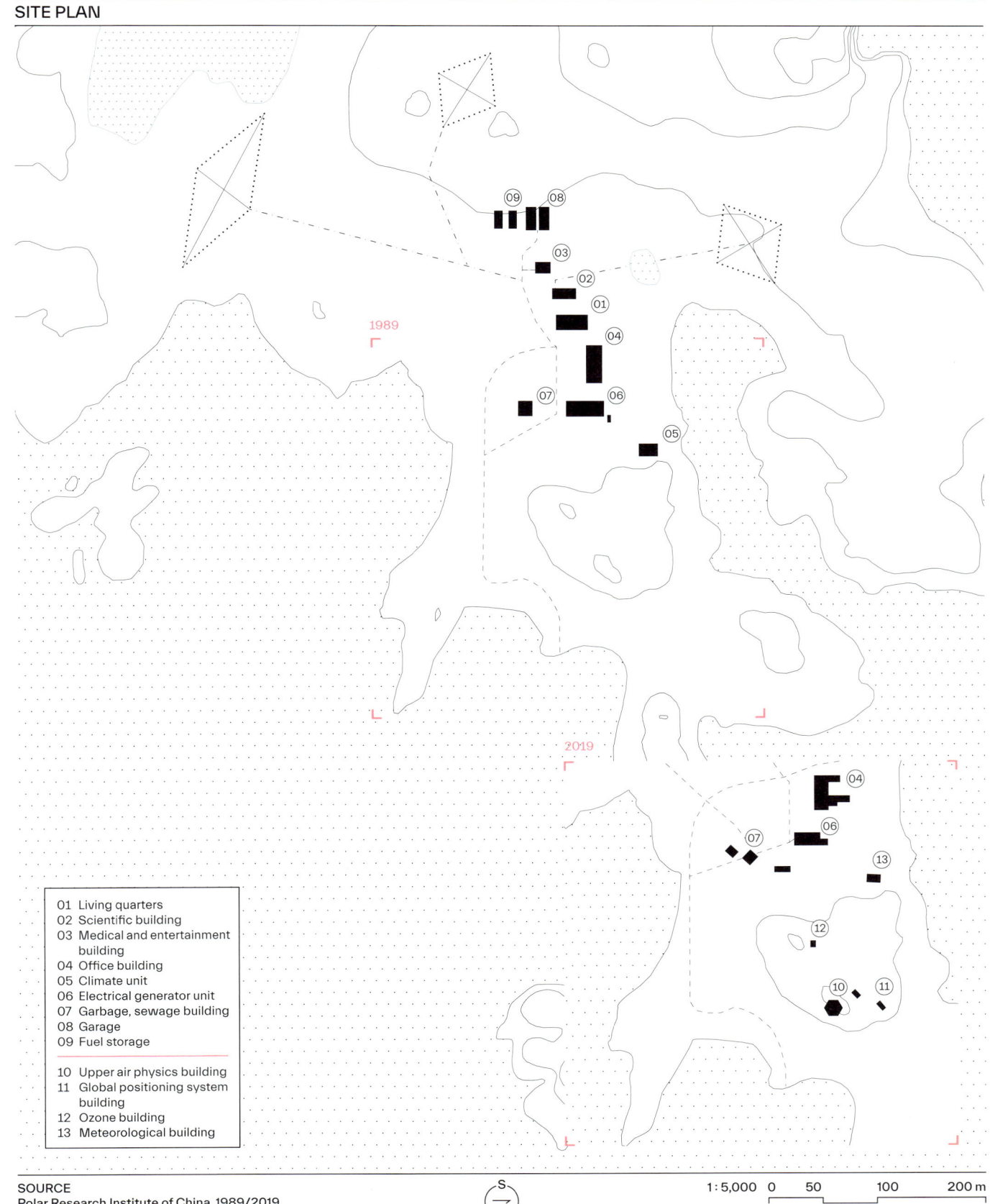

01 Living quarters
02 Scientific building
03 Medical and entertainment
    building
04 Office building
05 Climate unit
06 Electrical generator unit
07 Garbage, sewage building
08 Garage
09 Fuel storage

10 Upper air physics building
11 Global positioning system
    building
12 Ozone building
13 Meteorological building

SOURCE
Polar Research Institute of China, 1989/2019

1 : 5,000   0   50   100   200 m

HALLEY V
1956/1992

↖ p. 468, 469

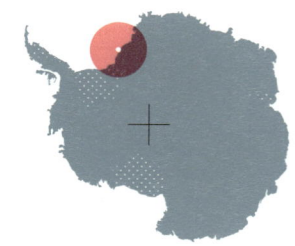

"Construction of the three jackable platforms in Phase I was completed in the 1988–89 season. Phase II (1989–90) saw the completion of the main Accommodation Platform (AP), with final mechanical and electrical installation, flooring, furniture and furnishings and fittings to be completed over the 1990 Winter.

[…] The Ice and Climate Science Platform (ICP) was complete to the same extent. The AP/ICP tunnel and services shaft had been completed, and the tunnel from the Upper Atmospheric Science Platform (UAP) was largely completed."

Ian Lovegrove (Station Manager, British Antarctic Survey), *Halley 5 – Phase III 1990/9 Report*, 1991

Designer and Engineer: Christiani & Nielsen Ingenieurbau AG, Germany
Consultants: (MEP) Günther Mattiessen GmbH, Germany; (wind tunnel testing) Cold Regions Research and Engineering Laboratory, United States
Manufacturer: (panels) OKAL GmbH, Germany

United Kingdom
British Antarctic Survey (BAS)

Removed > Replaced

75°35'S 26°14'W
Brunt Ice Shelf, Caird Coast
Ice shelf, 37 m, –20°C

SITE PLAN

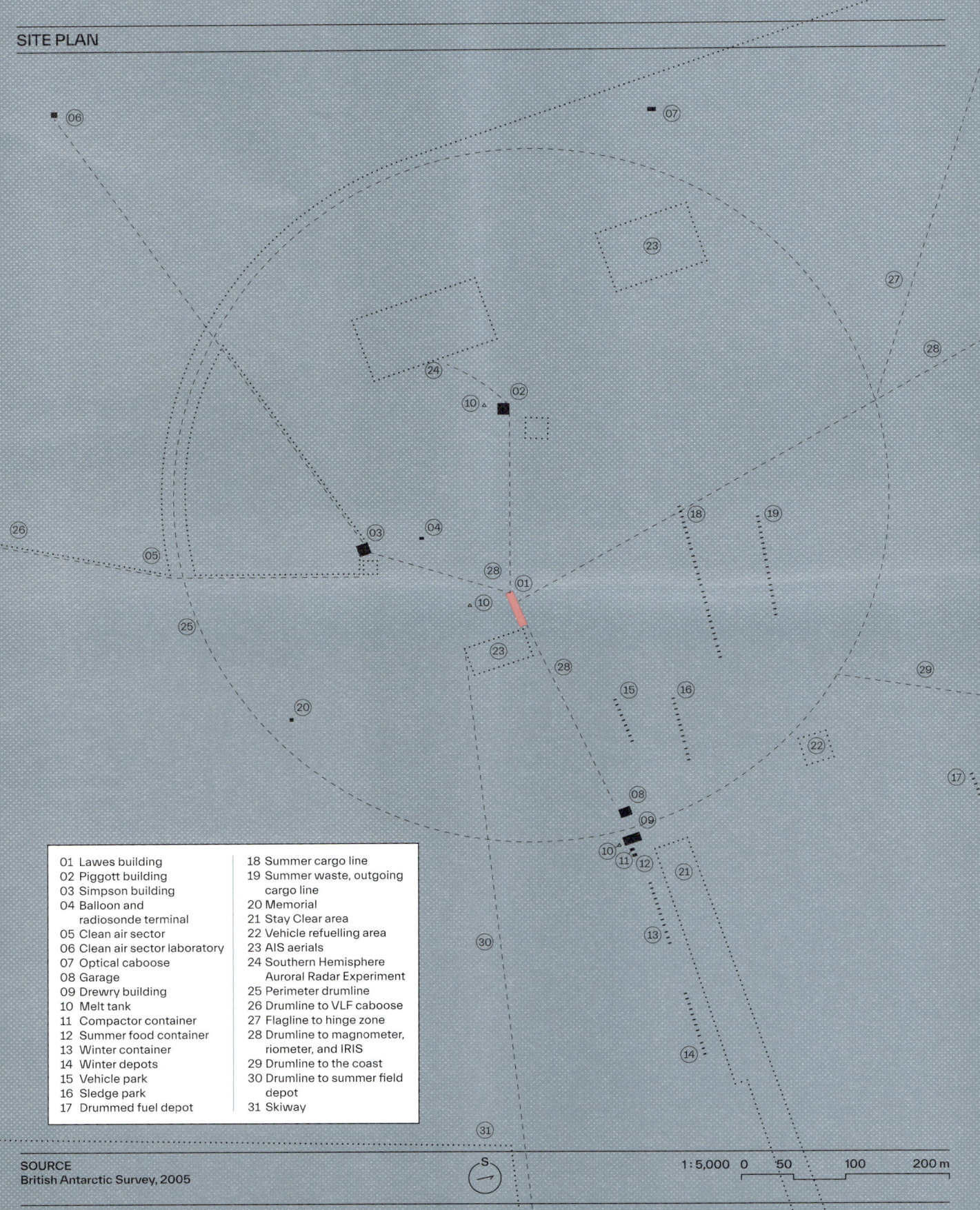

01 Lawes building
02 Piggott building
03 Simpson building
04 Balloon and
   radiosonde terminal
05 Clean air sector
06 Clean air sector laboratory
07 Optical caboose
08 Garage
09 Drewry building
10 Melt tank
11 Compactor container
12 Summer food container
13 Winter container
14 Winter depots
15 Vehicle park
16 Sledge park
17 Drummed fuel depot

18 Summer cargo line
19 Summer waste, outgoing
   cargo line
20 Memorial
21 Stay Clear area
22 Vehicle refuelling area
23 AIS aerials
24 Southern Hemisphere
   Auroral Radar Experiment
25 Perimeter drumline
26 Drumline to VLF caboose
27 Flagline to hinge zone
28 Drumline to magnometer,
   riometer, and IRIS
29 Drumline to the coast
30 Drumline to summer field
   depot
31 Skiway

SOURCE
British Antarctic Survey, 2005

1 : 5,000   0   50   100   200 m

PLAN | LAWES BUILDING

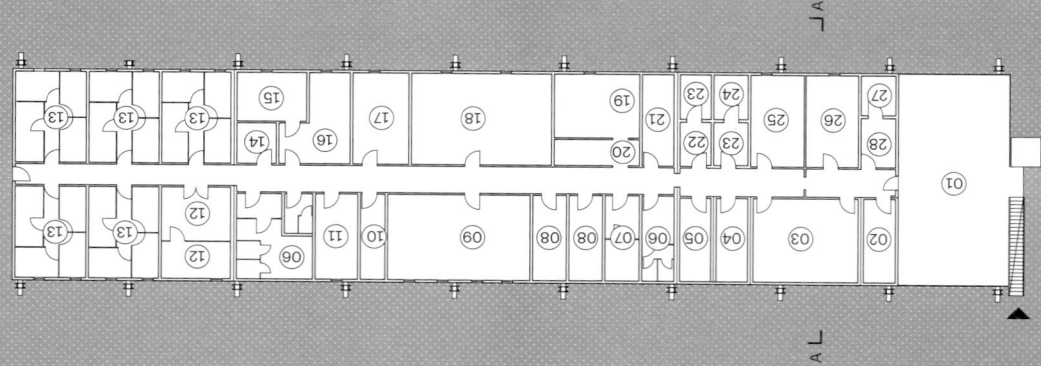

SOURCE
British Antarctic Survey, 1987

1:500   0   5   10   20 m

01	Vestibule	15	Communication room
02	Fuel storage	16	Computer room
03	Generator	17	Recreation room
04	Service plant	18	Dining room
05	Laundry and dry room	19	Kitchen
06	Bathroom	20	Pantry and dry stores
07	Darkroom	21	Food stores
08	Food stores	22	Storage
09	Lounge	23	Frozen provisions, −20°C
10	Bar	24	Provisions, +4°C
11	Library	25	Electrical workshop
12	Medical room	26	Mechanical workshop
13	Bedroom	27	Refuse collection
14	Base leader's room	28	Cloakroom

ELEVATION | LAWES BUILDING

"The station contains a mix of building technologies. Four buildings sit on platforms 4 m above the snow surface on legs that are jacked up annually to keep them clear of the accumulating snowfall. A further two buildings are mounted on skis and winched by tracked vehicles each year to a new position on the snow surface and then winched back to their original position the following year."

British Antarctic Survey and Natural Environment Research Council, *Proposed Construction and Operation of Halley VI Research Station, and Demolition and Removal of Halley V Research Station, Brunt Ice Shelf Antarctica: Final Comprehensive Environmental Evaluation*, 2007

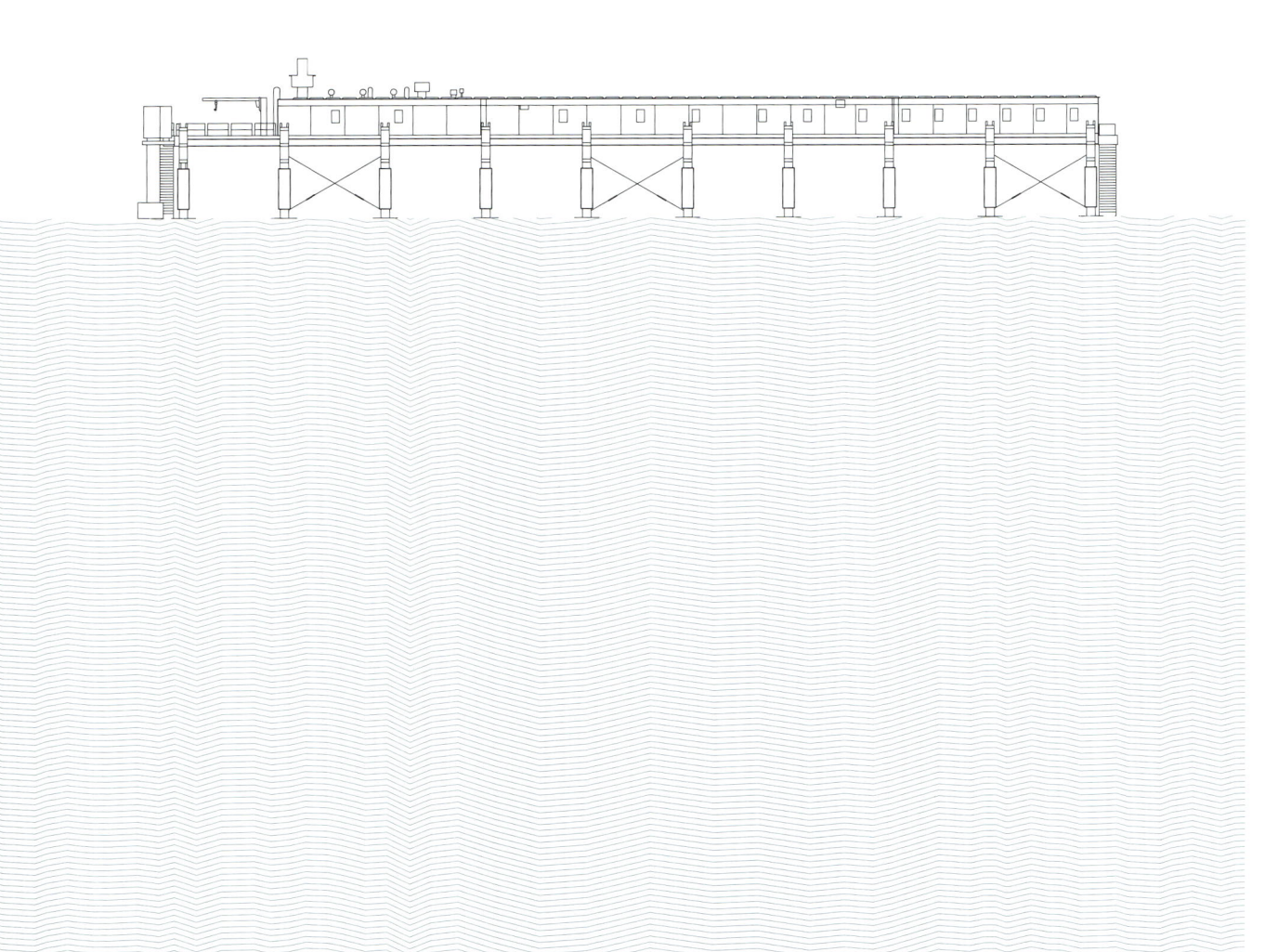

PLAN, FOUNDATIONS | LAWES BUILDING

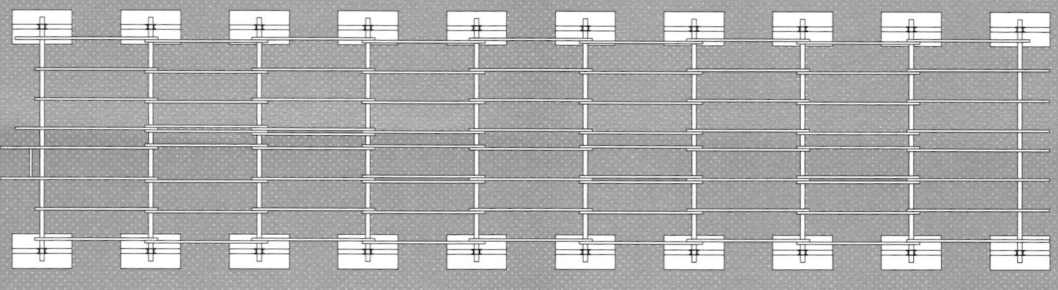

SECTION AND DETAIL | LAWES BUILDING

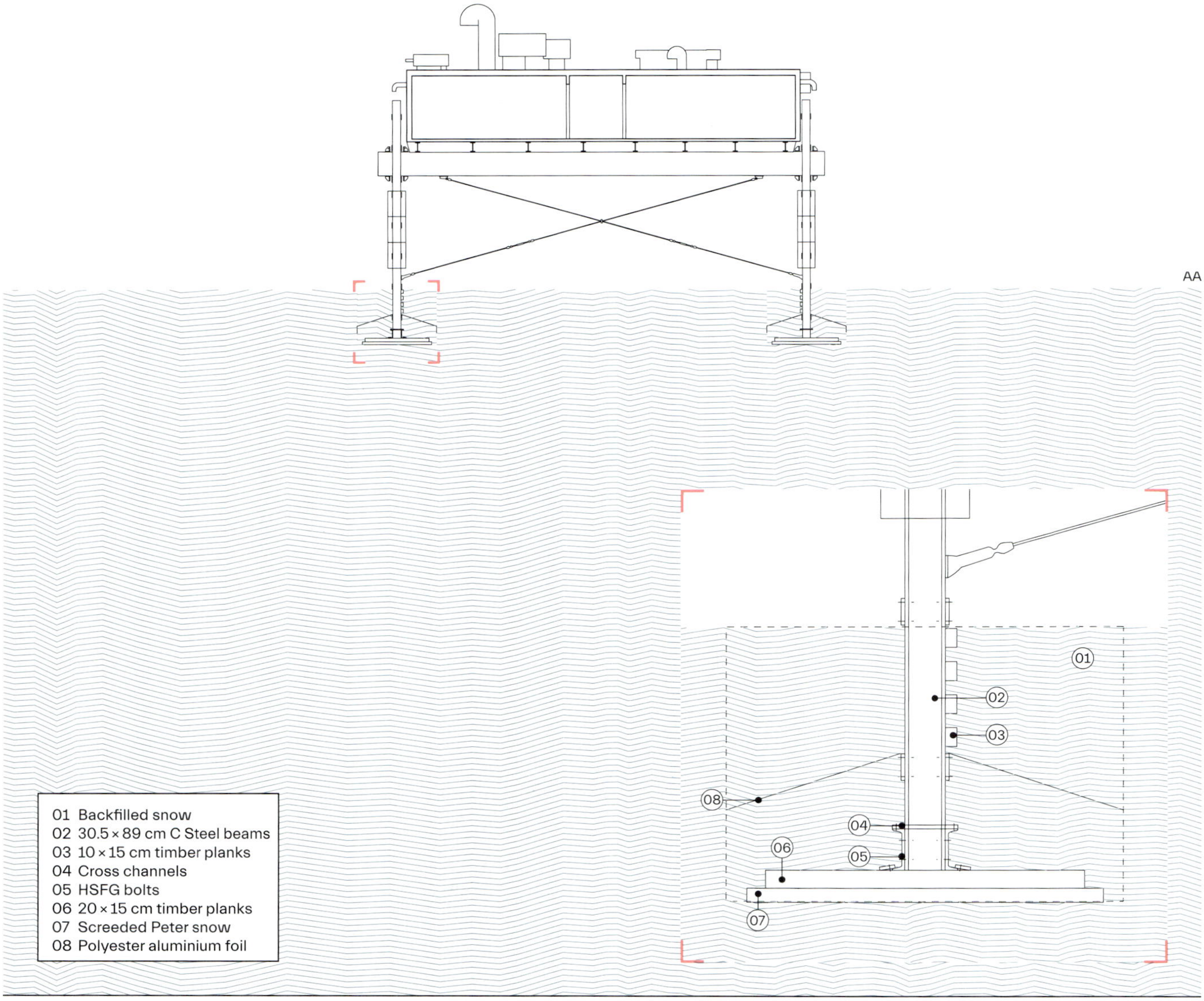

AA

01 Backfilled snow
02 30.5 × 89 cm C Steel beams
03 10 × 15 cm timber planks
04 Cross channels
05 HSFG bolts
06 20 × 15 cm timber planks
07 Screeded Peter snow
08 Polyester aluminium foil

SOURCE
British Antarctic Survey, 1987

| 1 : 250 | 0 | 2.5 | 5 | 10 m |
| 1 : 50 | 0 | 2 | 4 | 8 m |

NEUMAYER II
1981/1992

↖ p. 414, 460, 503, 504

"The central building is a steel tube system consisting of two main tubes (eastern tube 82 m, western tube 92 m in length), a 92 m long cross tube with stores and fuel tanks and a garage in a snow trench."

Hartwig Gernandt, Saad El Naggar, Jürgen Janneck, Thomas Matz, and Cord Drücker, *From Georg Forster Station to Neumayer Station III – A Sustainable Replacement at Atka Bay for Future*, 2007

"It was an underground station made of corrugated steel tubes with containerised and heated accommodation insides. This construction has certain advantages in the rough environment of an Antarctic ice shelf, but it is also exposed to the ever increasing loads of accumulating snow, which eventually will lead to the unavoidable destruction of the building."

Hartwig Gernandt, Saad El Naggar, Jürgen Janneck, Thomas Matz, and Cord Drücker, *From Georg Forster Station to Neumayer Station III – A Sustainable Replacement at Atka Bay for Future*, 2007

Design: Polarmar GmbH, Germany
Contractor: ARGE; Christiani & Nielsen Ingenieur-bau AG, J. H. K. GmbH & Co. KG, Germany
On site supervision: Polarmar GmbH, Germany
Certification: Germanischer Lloyd, Germany

Germany
Alfred Wegener Institute for Polar and
Marine Research (AWI)

Removed > Replaced

70°40'S 8°16'W
Ekström Ice Shelf, Queen Maud Land
Ice shelf, 43 m.a.s.l., −16°C

SITE PLAN

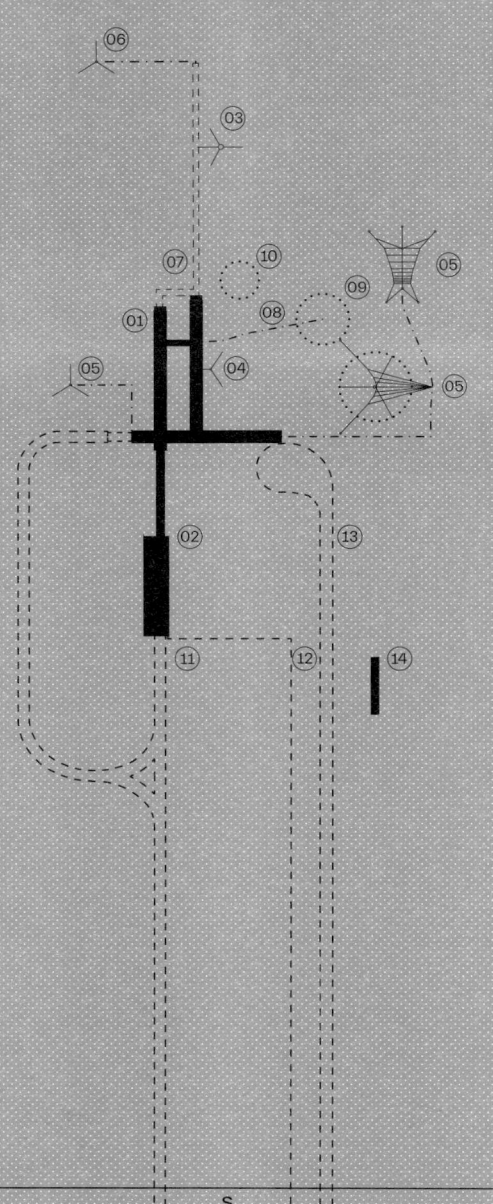

01 Main station building
02 Vehicle hall
03 Wind generator
04 Ultra-short-wave antenna
05 Short-wave receiving
   antenna
06 Meteorological mast
07 Climate tunnel
08 Waste water pipe
09 Septic tank
10 Helipad
11 To ship berth
12 To airstrip
13 Trace to tank storage
14 Tank storage

SOURCE
Alfred Wegener Institute, Helmholtz Centre for
Polar and Marine Research, 1991

S

1 : 5,000   0   50   100   200 m

SYOWA
1957/1993

↖ p. 423, 498

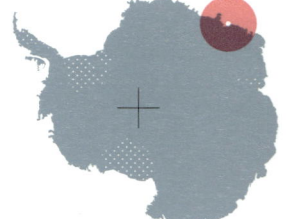

Architect: Takashi Asada, Japan
Contractor: Misawa Homes, Japan

Japan
National Institute of Polar Research (NIPR)

Active Station

69°00'16"S 39°34'54"E
East Ongul Island, Queen Maud Land
Ice-free ground, 29 m.a.s.l., –10.4°C

SITE PLAN

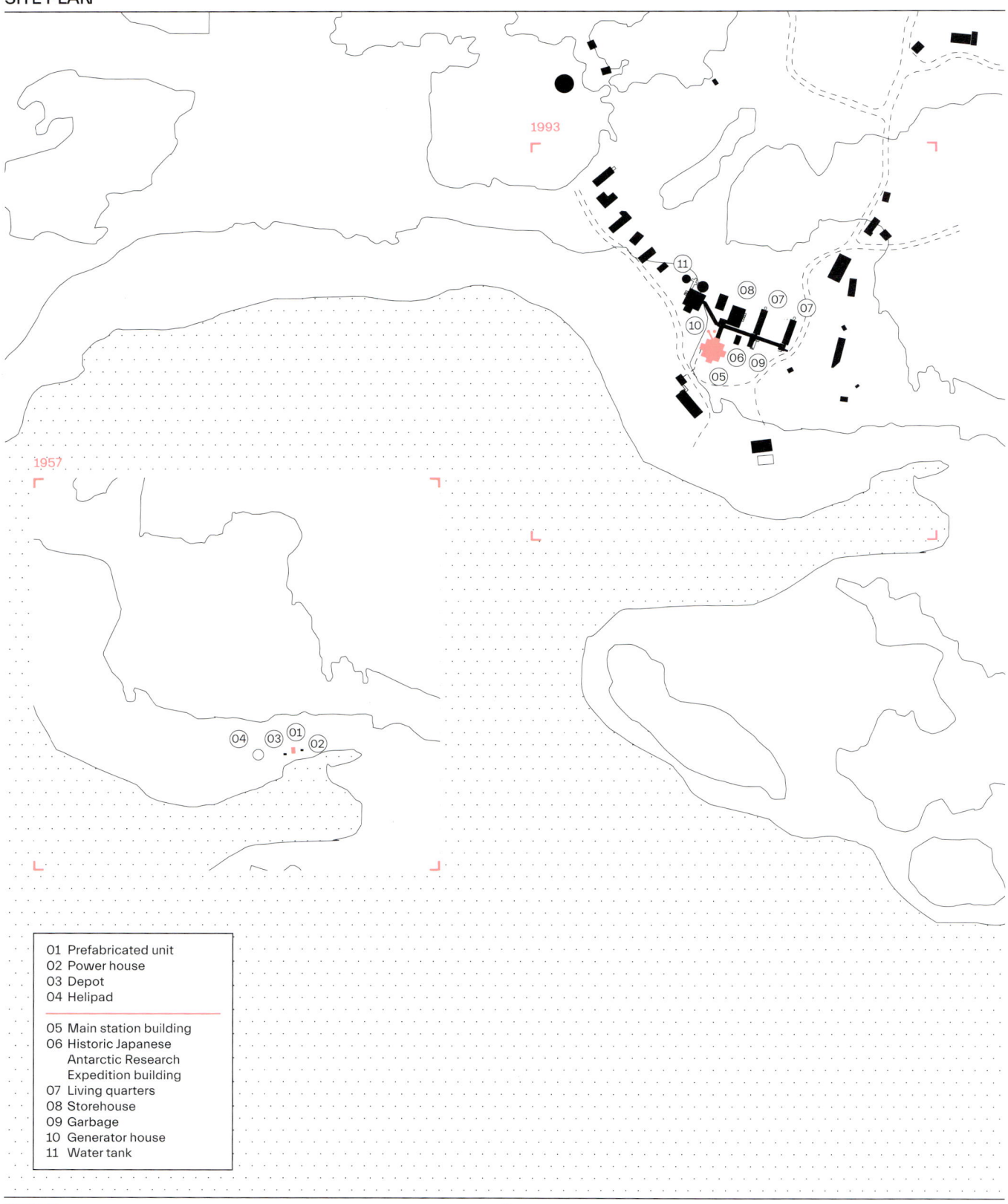

1993

1957

01 Prefabricated unit
02 Power house
03 Depot
04 Helipad

05 Main station building
06 Historic Japanese
     Antarctic Research
     Expedition building
07 Living quarters
08 Storehouse
09 Garbage
10 Generator house
11 Water tank

SOURCE
National Institute of Polar Research, 1993

S

1 : 5,000 0   50   100   200 m

PLAN AND ELEVATION | PREFABRICATED UNIT, 1957

"This building, which survived
the icy wind and snow and
returned safely to Japan, paved
the way for industrialised housing in Japan as the country's
first prefabricated housing and
can be said to symbolise the
challenge of the unknown."

Takenaka Corporation, "Building for the Japanese
Antarctic Research Expedition: From the South
Pole to a New Kind of Housing in Japan" (online;
translated by the editor)

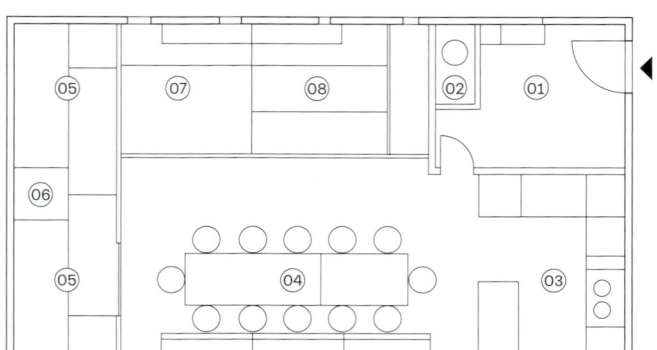

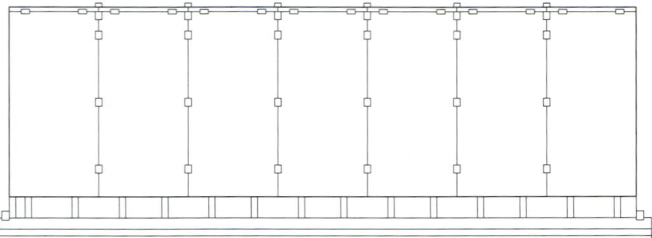

01 Vestibule
02 Bathroom
03 Kitchen
04 Dining room
05 Bedroom
06 Wardrobe and desk
07 Tatami
08 Heating unit

SOURCE
National Academy of Sciences, 1957

1:100   0   1   2   4 m

SECTION AND DETAIL | PREFABRICATED UNIT, 1957

"Japanese scholars suggested that the feasibility of using industrialized capsules was demonstrated at first with the Syowa station for the Soya expedition to Antarctica designed by Asada Takashi in 1956. […] This temporary shelter was 'Japan's first serious attempt to create industrialized housing.' Because the Syowa station had to be shipped to site and erected in extreme weather conditions, Asada developed together with Misawa Homes an innovative prefabricated construction system applied for housing purposes."

Michael Gibert, Naziaty Mohd Yaacob, and Sr Zuraini Md Ali, "The Capsule Living Unit Reconsidered a Utopia Transformed Reality", 2018

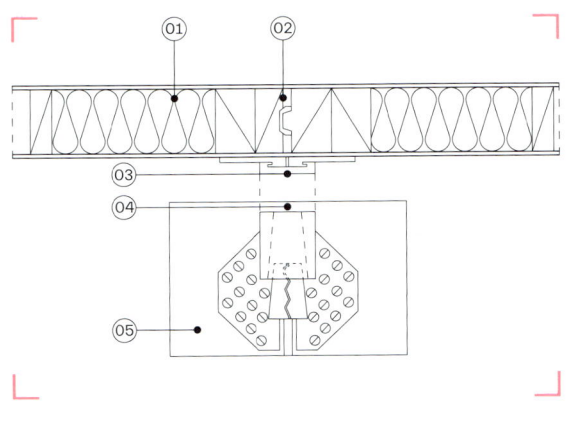

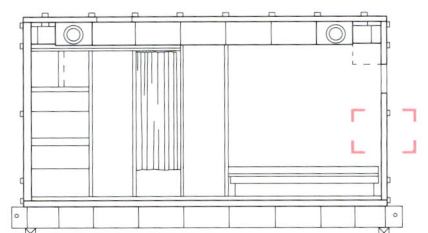

AA

01	Insulation
02	Rubber seal
03	Panel joint
04	Connector
05	Wall panel

SOURCE
Takenaka, 1957

1 : 10   0   2.5   5   10 m
1 : 100   0   1   2   4 m

PLAN, 1ST, 2ND, AND 3RD LEVEL | ADMINISTRATIVE BUILDING, 1993

3rd level

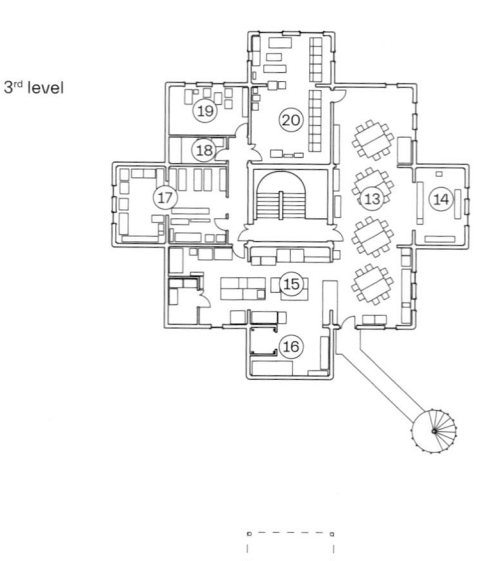

2nd level

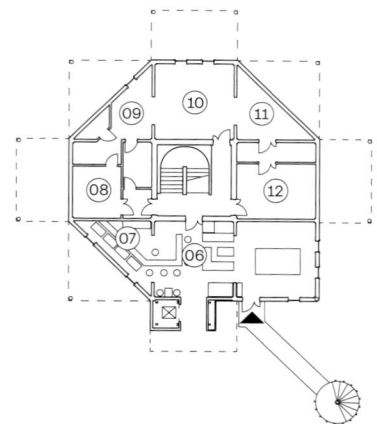

1st level

01 Vestibule
02 Food storage
03 Air-conditioning control
04 Water supply unit
05 Storage

2nd level
06 Recreation room
07 Bar
08 X-ray room
09 Dental room
10 Medical room
11 Medical storage room
12 Surgery room

3rd level
13 Dining room
14 Lounge
15 Kitchen
16 Food stores
17 Library
18 Telephone room
19 Base leader's office
20 Telecommunications room

1st level

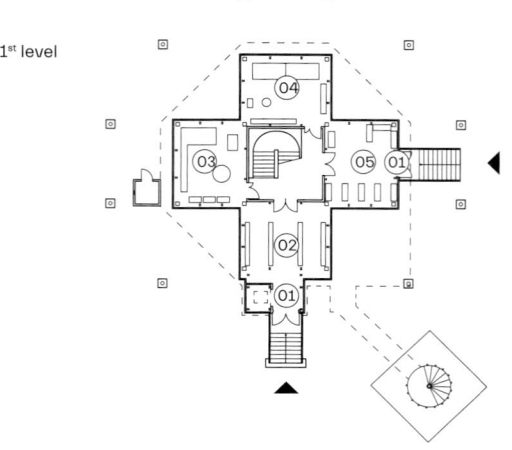

SOURCE
National Institute of Polar Research, 1993

1 : 250   0   2.5   5   10 m

ELEVATION | ADMINISTRATIVE BUILDING, 1993

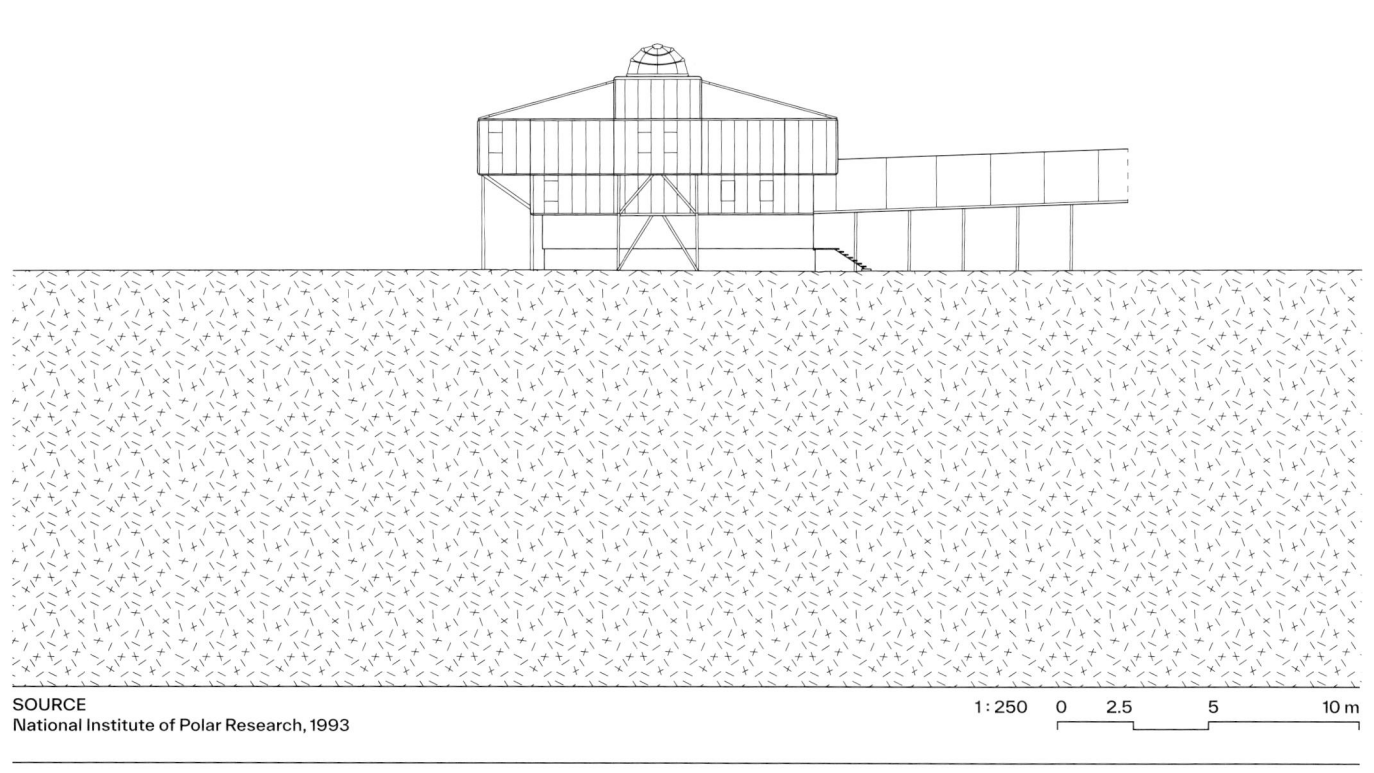

SOURCE
National Institute of Polar Research, 1993

1 : 250   0   2.5   5   10 m

SANAE IV
1960/1997

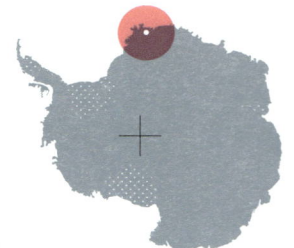

"The present South-African base [SANAE III] is built on the Fimbulisen ice-shelf that covers the continental terrace. [...] The search for an alternative site proposed moving away from this area in search of a natural bay with a solid rock coastline.

[...] This is in all respects the most economic and suitable option if an appropriate site could be identified."

South African Department of Environmental Affairs, *SANAE IV Environmental, Health and Safety Impact Assessment: Initial Environmental Evaluation Report,* 1991

Engineers: (structural design) Endecon; (mechanical systems) Royal Haskoning DHV, South Africa
Contractors: Nolitha; CAT ; Martin Membrane Systems; Jets; TFD, South Africa
Suppliers: TFD; Wilo; CAS; Johnson Controls, South Africa

South Africa
South African National Antarctic Program (SANAP)

Active Station

71°40'22"S 2°50'26"W
Vesleskarvet, Queen Maud Land
Ice-free ground, 41 m.a.s.l., –16.5°C

SITE PLAN

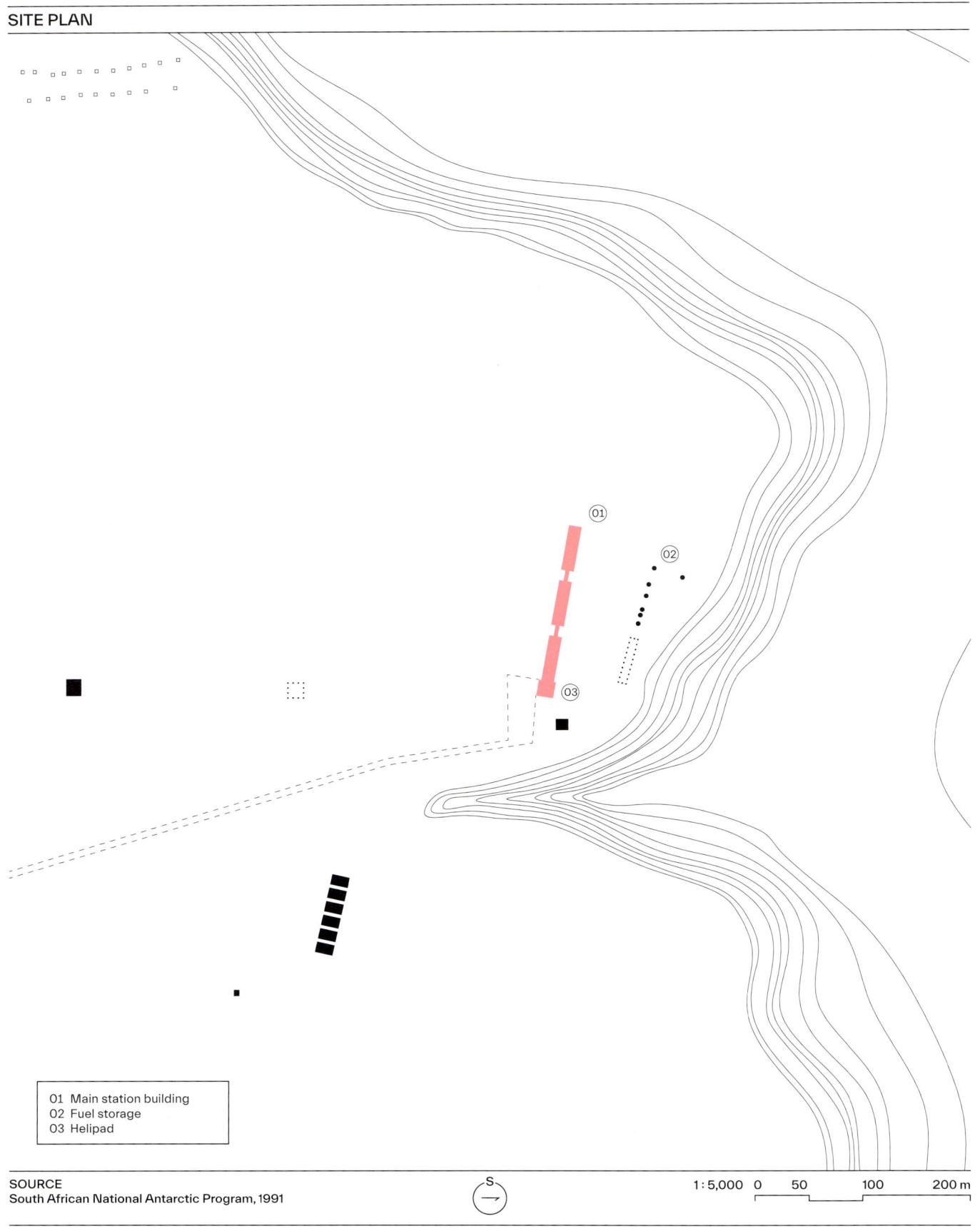

01 Main station building
02 Fuel storage
03 Helipad

SOURCE
South African National Antarctic Program, 1991

S

1 : 5,000   0   50   100   200 m

PLAN AND ELEVATION

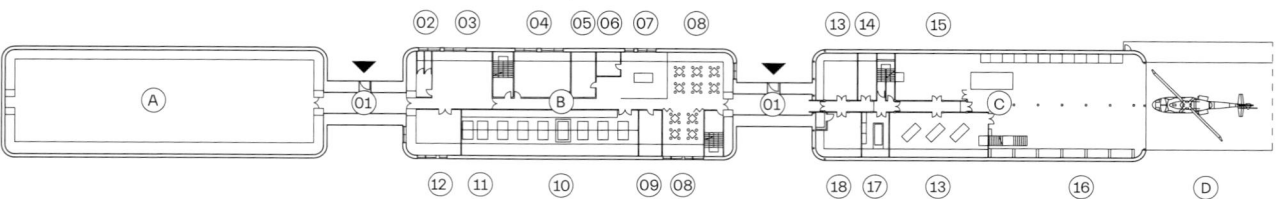

A L

02 03   04 05 06 07   08   13 14   15

01   B   01   C

12 11   10   09 08   18 17   13   16   D

L
A

A	Block A	05	Cold room	C	Block C
		06	Pantry	13	Plant room
B	Block B	07	Kitchen	14	Woodwork store
01	Vestibule and	08	Dining room	15	Workshop
	interconnecting	09	Frozen store	16	Hangar
	link	10	Dry store	17	Diesel day tank
02	Bathroom	11	Bar lounge	18	Storage
03	Lounge	12	Dining room		
04	TV lounge			D	Helipad

1:1,000   0   10   20   40 m

SECTION AND ELEVATION

"The structure was optimised for strength, layout, fabrication and erection. […] The final structure consists of braced structural steel frames clad with fibre-glass skinned insulated panels. Various countries were visited i.e. England, France, Germany, Italy and the USA to discuss and analyse various existing alternative design concepts. The final design for the new SANAE base was developed for the particular set of conditions prevailing at Vesleskarvet and may be regarded as a product moulded by concepts used by the other Antarctic nations in the design of their bases but employs several unique designs and techniques."

South African Department of Environmental Affairs, *SANAE IV Environmental, Health and Safety Impact Assessment: Initial Environmental Evaluation Report,* 1991

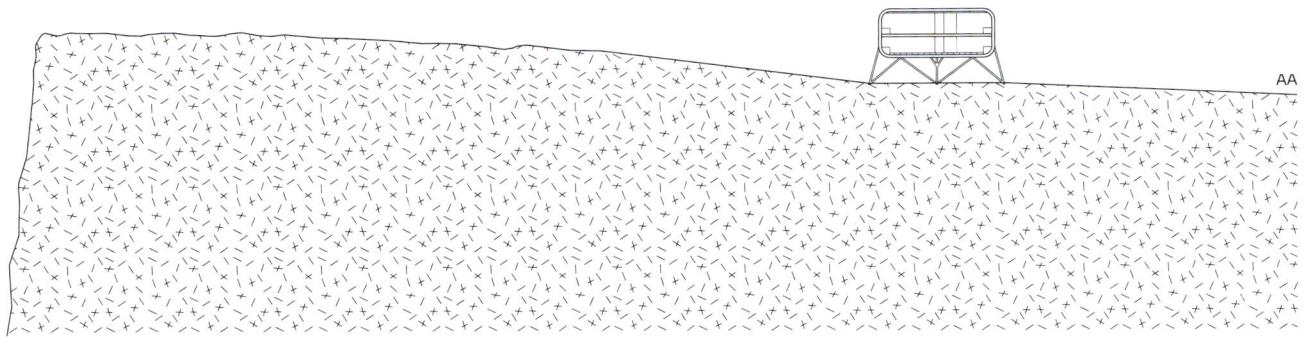

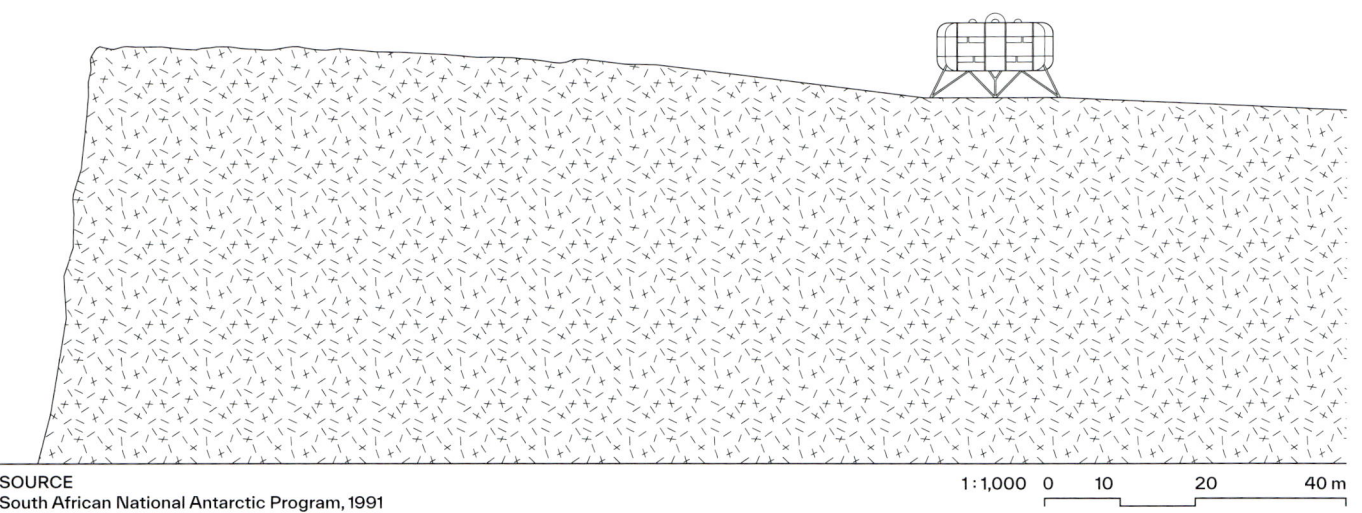

CONCORDIA 2005

↖ p. 242, 250, 366, 394, 504, 552, 557, 567, 574, 575, 582

"Concordia Station, jointly constructed and administered by IPEV [Institut polaire français Paul-Émile-Victor] and the Italian Antarctic program ENEA-UTA, consists of one core of three 'winter' buildings flanked by a summer camp that also acts as an emergency station. All structures are on or above ground. Two of the three winter buildings are unique integral self-elevating buildings forming the station's main living and working areas. The third winter-building houses the main power plant and technical services. The summer/emergency camp and all peripheral structures are modular units set low to the ground and skid-mounted. These units can be towed away to avoid progressive burial by snow accumulation."

Patrice Godon (Polar Engineer), "Concordia / The Station" (online)

Concept: Patrice Godon, Institut Polaire Français
Paul-Émile Victor, France
Design: Jean Paul Fave, France
Engineer: French Polar Expeditions (EPF), France
Interiors: Gianluca Pompili, Italy
Manufacturer: (structure) Bureau d'études Structure
Métallique, France

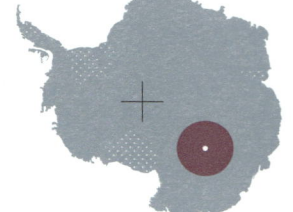

France and Italy
French Polar Institute Paul-Émile Victor (IPEV) and
National Antarctic Research Program (PNRA)

Active Station

75°05'59"S 123°19'56"E
Dome C, Antarctic Plateau
Ice sheet, 3,233 m.a.s.l., −52.1°C

SITE PLAN

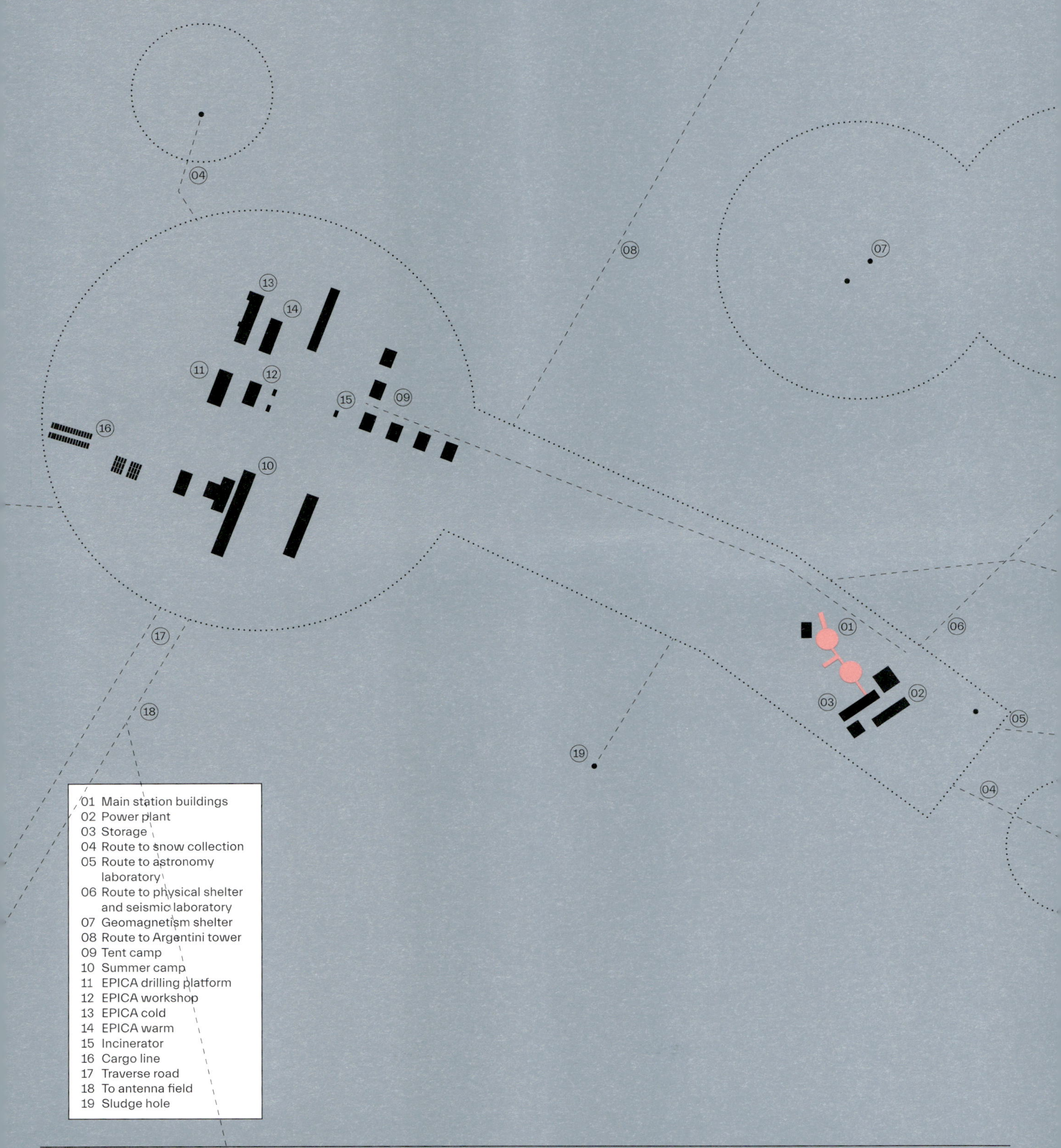

01 Main station buildings
02 Power plant
03 Storage
04 Route to snow collection
05 Route to astronomy
   laboratory
06 Route to physical shelter
   and seismic laboratory
07 Geomagnetism shelter
08 Route to Argentini tower
09 Tent camp
10 Summer camp
11 EPICA drilling platform
12 EPICA workshop
13 EPICA cold
14 EPICA warm
15 Incinerator
16 Cargo line
17 Traverse road
18 To antenna field
19 Sludge hole

SOURCE
National Antarctic Research Program, 2005

1 : 5,000   0   50   100   200 m

PLAN, FOUNDATIONS

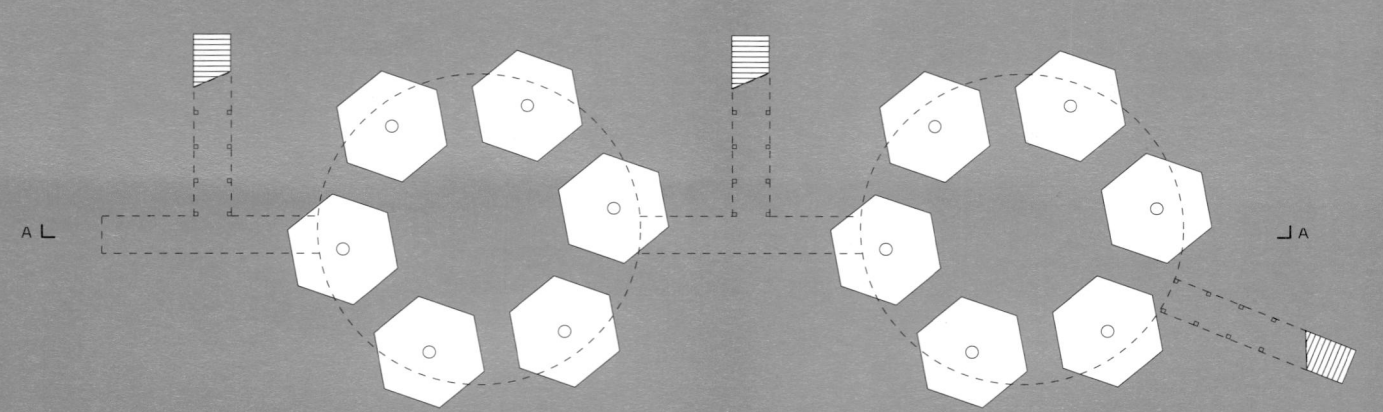

A Noisy building	12 Gym	21 Pre-operatory room	32 Physical atmosphere
01 Generators and servers	13 Bathrooms	22 Surgery room	laboratory
02 Workshop	14 Dining room	23 Base leader's room	33 Astrophysics laboratory
03 Battery room	15 Lounge	24 Converter room	34 Glaciology laboratory
04 Technical room	16 Dishwashing	25 Mail room	35 Antenna laboratory
05 Waste storage	17 Food storage	26 Technical room	36 Radio and information
06 Waste compactor	18 Kitchen	27 Toilet	laboratory
07 Toilet		28 Bathroom	37 Human biology laboratory
08 Dry storage	B Quiet building	29 Bedroom	
09 Cold room	19 First aid, consulting	30 Storage	
10 Storage	room, pharmacy	31 Magnetology and	
11 TV room	20 Medical room	seismology laboratory	

SOURCE
National Antarctic Research Program, 2005

1 : 500   0   5   10   20 m

PLAN, 1ST, 2ND, AND 3RD LEVEL

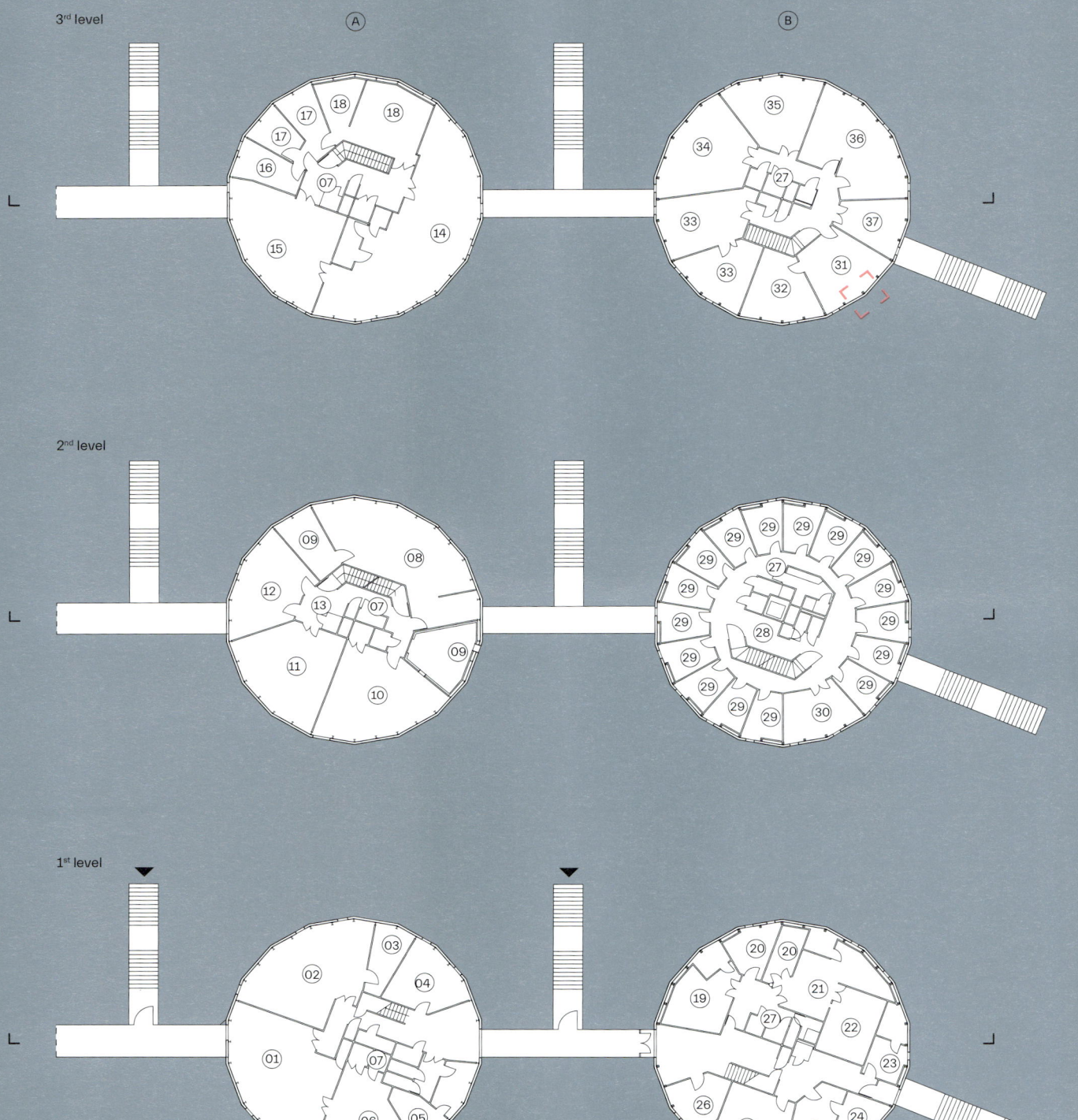

SECTION AND ENVELOPE DETAIL

"Each of the two self-elevating buildings is based on a roughly cylindrical body (18 sides) supported by six legs. Each leg sits on a large 'footing' pad spreading the load over the snow. Each leg can move up and down relative to the body of the building via hydraulic jacks. This allows the horizontality of the building to be adjusted but also to have the entire structure leap frog its way up over the ice as the ground level rises with snow accumulation."

Patrice Godon (Polar Engineer), "Concordia / The Station" (online)

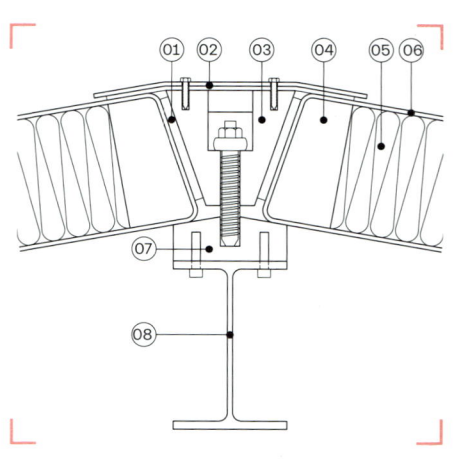

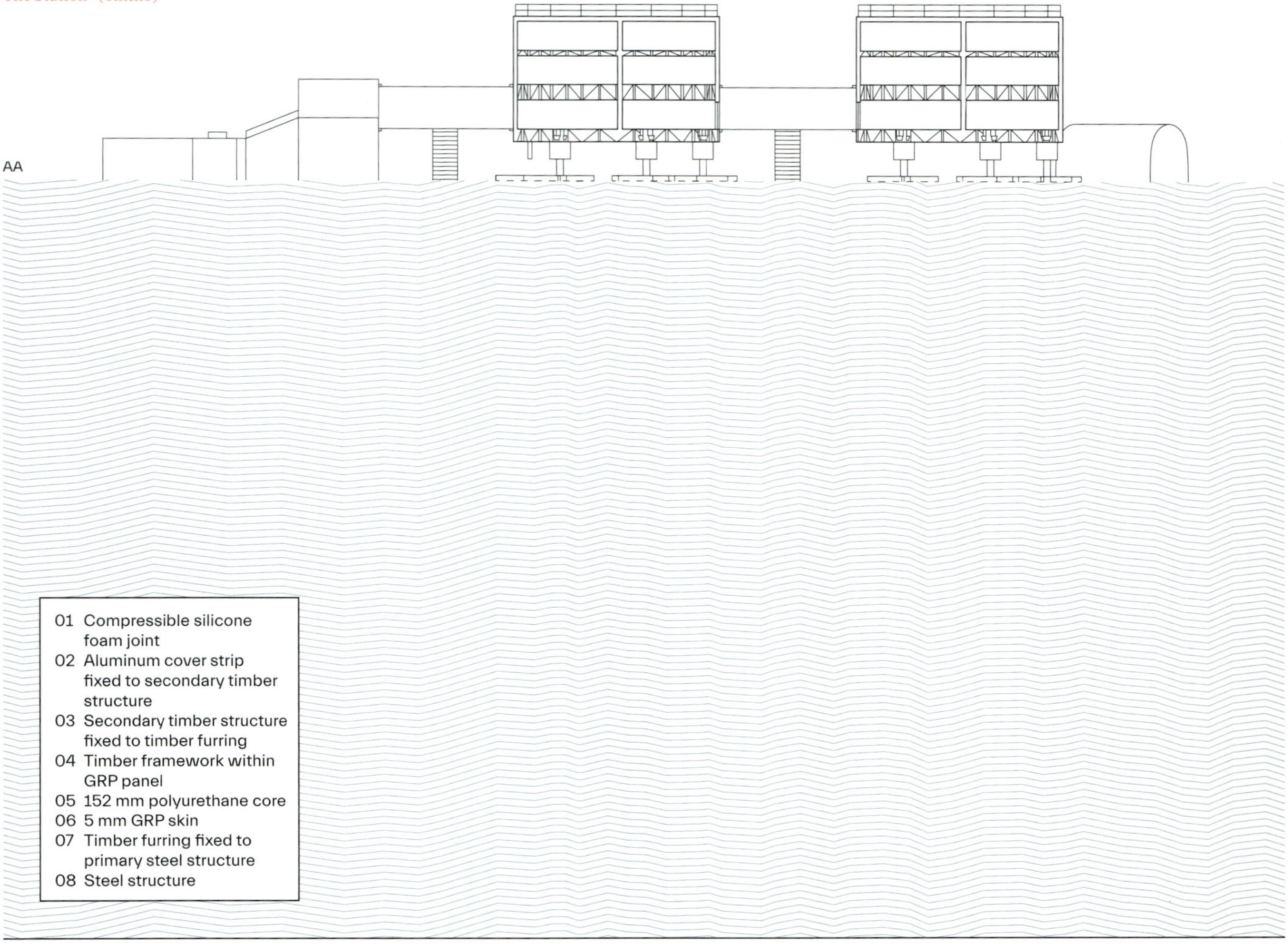

01 Compressible silicone foam joint
02 Aluminum cover strip fixed to secondary timber structure
03 Secondary timber structure fixed to timber furring
04 Timber framework within GRP panel
05 152 mm polyurethane core
06 5 mm GRP skin
07 Timber furring fixed to primary steel structure
08 Steel structure

AA

SOURCE
National Antarctic Research Program, 2005

1 : 10	0	10	20	40 cm
1 : 750	0	7.5	15	30 m

ELEVATION

"The two buildings will be connected at Level 1 by an aerial
intercommunication tunnel
about 10 m long. This solution
provides a clear separation
between areas where noise is
produced and areas where
peace and quiet are wanted by
locating them in two well separated buildings, the 'noisy'
and the 'quiet' buildings. It also
allows to separate the activities with high fire risk (kitchen,
workshops) from the living
quarters."

French Polar Institute, *Concordia: A New Permanent,*
International Research Support Facility High on the
Antarctic Ice Cap: Technical Overview Report, 2000

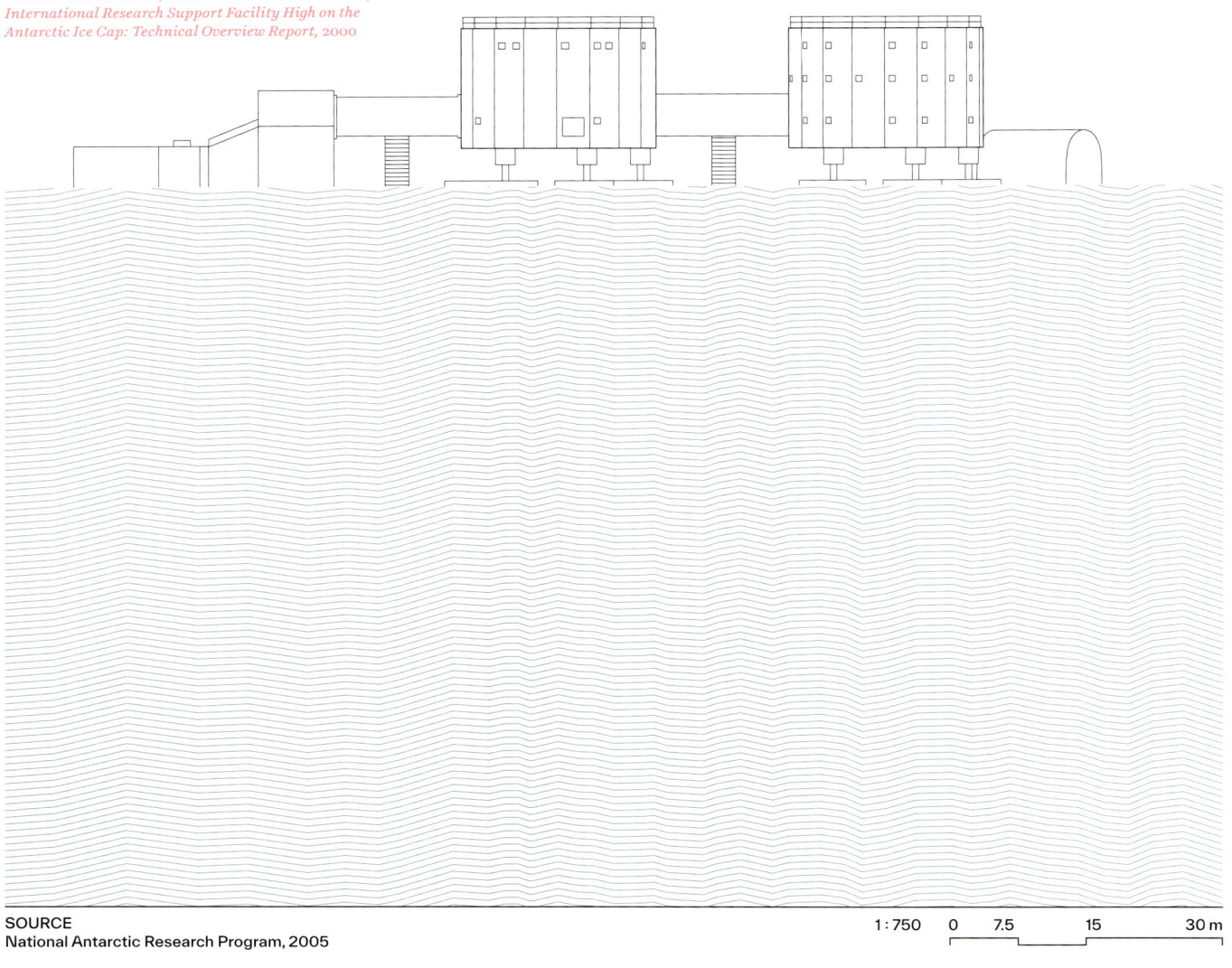

SOURCE
National Antarctic Research Program, 2005

1 : 750   0   7.5   15   30 m

PRINCESS ELISABETH 2007

↖ p. 366, 425, 429, 498, 499, 507

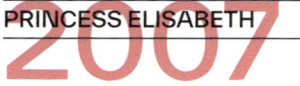

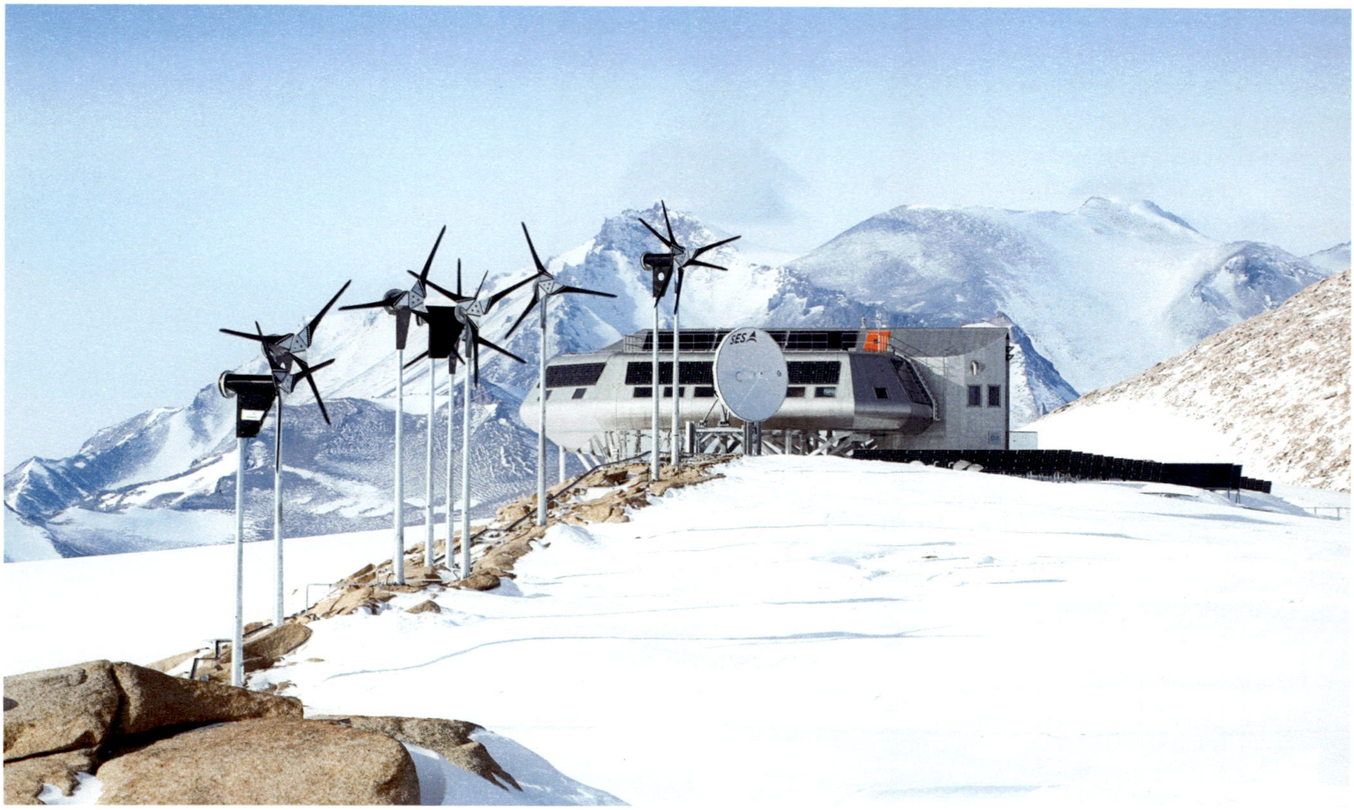

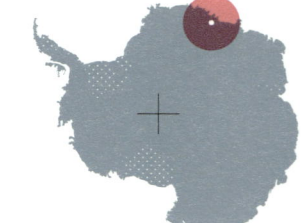

Design and Construction: International Polar Foundation; Alain Hubert (General Supervisor); Johan Berte (Project Manager); Nighat Amin (Programme Administrator), Belgium
Engineers: (building core and skin) Philippe Samyn and Partners; (building physics and active systems) 3E; (aerodynamics) von Karman Institute; (control systems and power network) Schneider Electric; (electrical protection and power storage) Laborelec; (hydraulics systems) EPAS; (soil mechanics) Smet-Boring; (fire safety) SECO, Belgium; (fluid mechanics) Dr D. Olivari, Italy
Technical control: SECO, Belgium
Contractors: Besix (lead contractor), Belgium; (wood structure and building enclosure) Prefalux, Luxembourg; (steel structures) Lemants; (ground anchorage) Smet-Boring; (electric systems) Schneider Electric; (water distribution) Aquasanit & Polet; (interior designer) Cherbai, Belgium

Belgium
International Polar Foundation (IPF)

Active Station

71°57'00"S 23°20'49"E
Sør Rondane Mountains, Queen Maud Land
Ice-free ground, 820 m.a.s.l., –18°C

SITE PLAN

"The station's skin, insulation, shape, orientation and window disposition allow a comfortable ambient temperature to be maintained inside the building with little energy input. Sophisticated ventilation and air circulation systems are an integral part of temperature management. Princess Elisabeth Antarctica was conceived to take full advantage of currently available passive building techniques. […] Princess Elisabeth Antarctica's Micro Smart Grid, the key feature that makes it a zero emission station, is a unique system based on a Demand Power Management System."

International Polar Foundation, *Press Release for Princess Elizabeth Station*, 2008

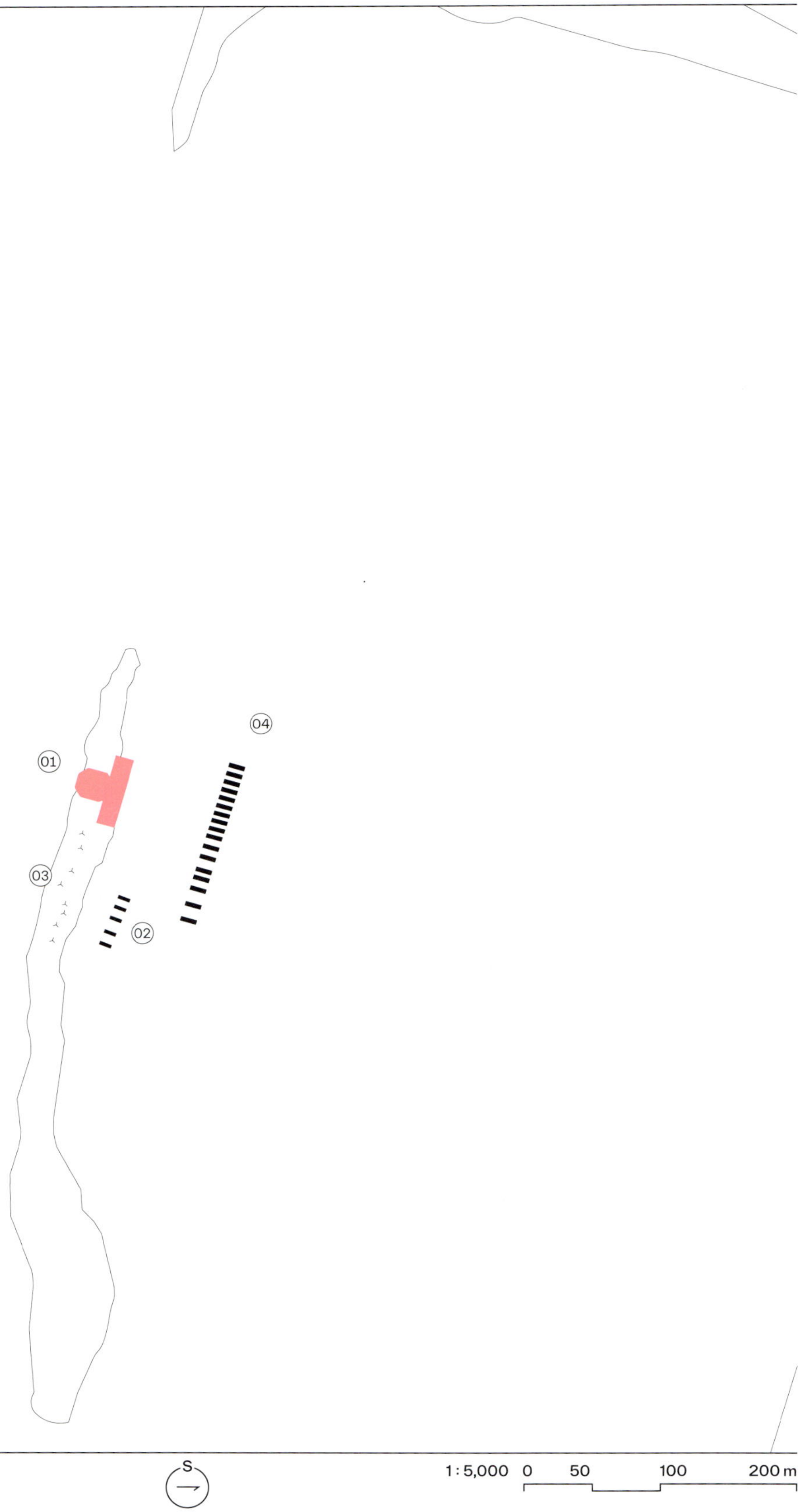

01 Main station building
02 Solar panels
03 Wind turbines
04 Storage containers

SOURCE
Philippe Samyn and Partners (reconstruction)

S

1 : 5,000   0   50   100   200 m

PLAN

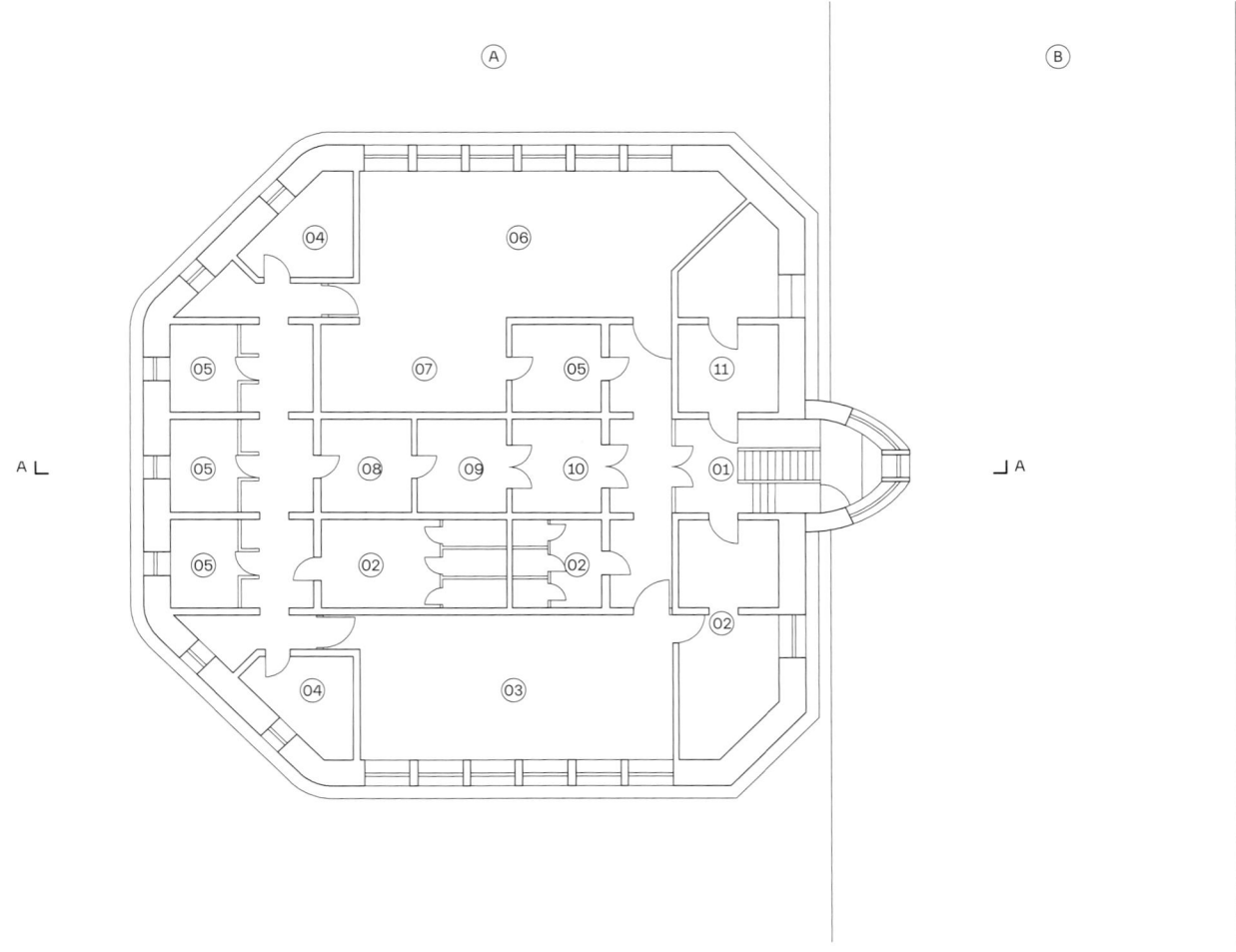

A Living module
01 Vestibule
02 Laboratory
03 Office
04 Polyvalent room
05 Bedroom
06 Lounge
07 Kitchen
08 Eneregy storage unit
09 Water treatment unit
10 Station control unit
11 Polar gear area

B Garage and services

SOURCE
Philippe Samyn and Partners

1 : 250   0   2.5   5   10 m

SECTION AND ELEVATION

"The sub-structure […] is composed of four steel trestles, which may expand and contract independently of each other, and which support a large wooden superstructure. The trestles are anchored in the non-uniform, surface weathered, granite bedrock and are shored by 6 m deep tie-rods such as to provide a reaction to the important wind uplift experienced by the building. […] The envelope of the superstructure covers an orthogonal grid of trusses expanding from floor to ceiling in laminated wood elements assembled with Blumer type connectors. […] 25 tons of stainless steel 304 B [compose] the final layer of the envelope."

Philippe Samyn Architects, "Solaripedia" (online)

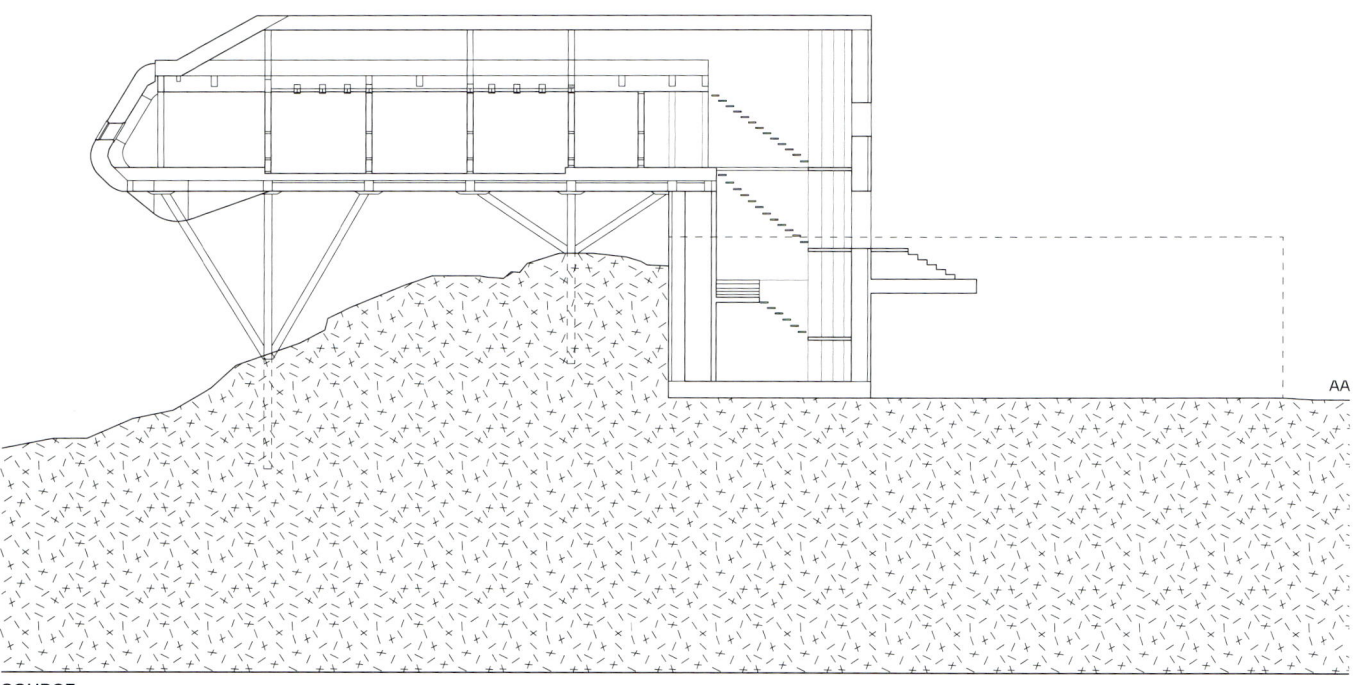

SOURCE
Philippe Samyn and Partners

| 1 : 500 | 0 | 5 | 10 | 20 m |
| 1 : 250 | 0 | 2.5 | 5 | 10m |

AA

AMUNDSEN-SCOTT SOUTH POLE
1957/2008

↖ p. 140, 318, 366, 377, 390, 438, 442, 478, 482, 504, 546, 574

"The Amundsen-Scott South Pole Station (hereafter referred to as South Pole Station), operated by the United States, is located on the polar plateau at an elevation of 2835 m near the geographic South Pole at 90°S. An area of ~26,344 km² around the South Pole Station is designated as an Antarctic Specially Managed Area (hereafter referred to as 'the Area'). The Area has been designated in order to maximize the valuable scientific opportunities at the Pole, protect the near-pristine environment and ensure that all activities, including those to experience the extraordinary qualities of the South Pole, can be conducted safely, environmentally responsibly and without disruption to scientific programs."

South Pole ASMA V, Measure 8: Management Plan for Antarctic Specially Managed Area No. 5 Amundsen-Scott South Pole Station, South Pole, 2017

Architect: Ferraro Choi and Associates Ltd, Hawaii
Engineers: (structural) BBFM Engineers Inc.; (mechanical and electrical) PDC Consultants, United States
Consultants: (communications) AlliedSignal Technical Service Corp.; (snow studies and acoustics) RWDI Consulting Engineers and Scientists; (materials handling) Semco Sweet & Mayers Inc., United States
Suppliers: (kitchens) George Matsumoto Inc., United States

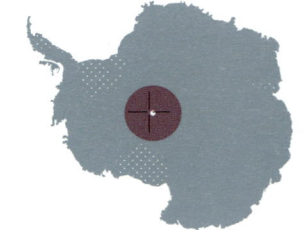

United States of America
United States Antarctic Program (USAP)

Active Station

89°59'51.19"S 139°16'22.41"E
Geographic South Pole, Antarctic Plateau
Ice sheet, 2,835 m.a.s.l., –49°C

SITE PLAN

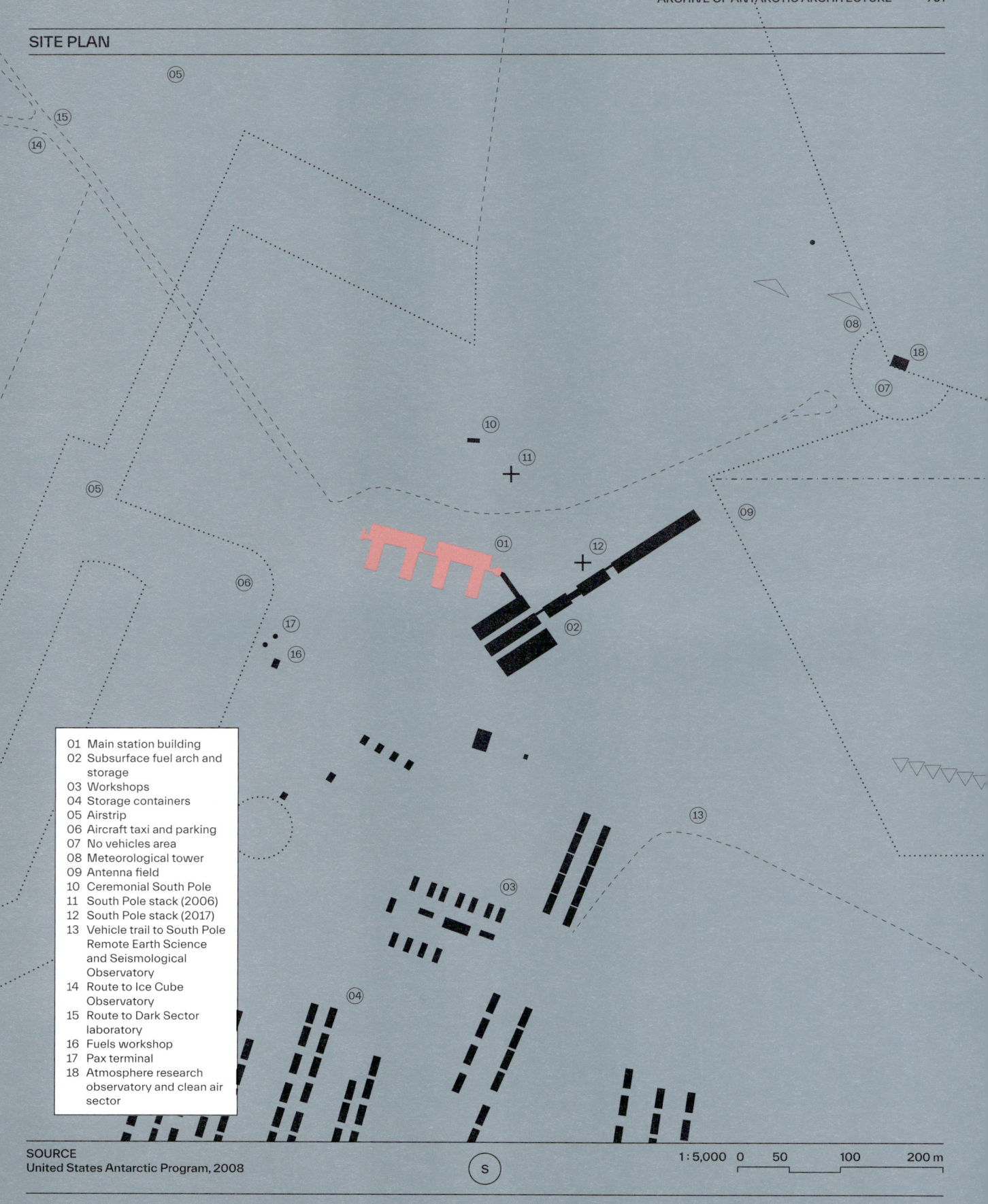

01 Main station building
02 Subsurface fuel arch and
    storage
03 Workshops
04 Storage containers
05 Airstrip
06 Aircraft taxi and parking
07 No vehicles area
08 Meteorological tower
09 Antenna field
10 Ceremonial South Pole
11 South Pole stack (2006)
12 South Pole stack (2017)
13 Vehicle trail to South Pole
    Remote Earth Science
    and Seismological
    Observatory
14 Route to Ice Cube
    Observatory
15 Route to Dark Sector
    laboratory
16 Fuels workshop
17 Pax terminal
18 Atmosphere research
    observatory and clean air
    sector

SOURCE
United States Antarctic Program, 2008

S

1 : 5,000   0   50   100   200 m

PLAN, 1ST AND 2ND LEVEL | MAIN STATION BUILDING

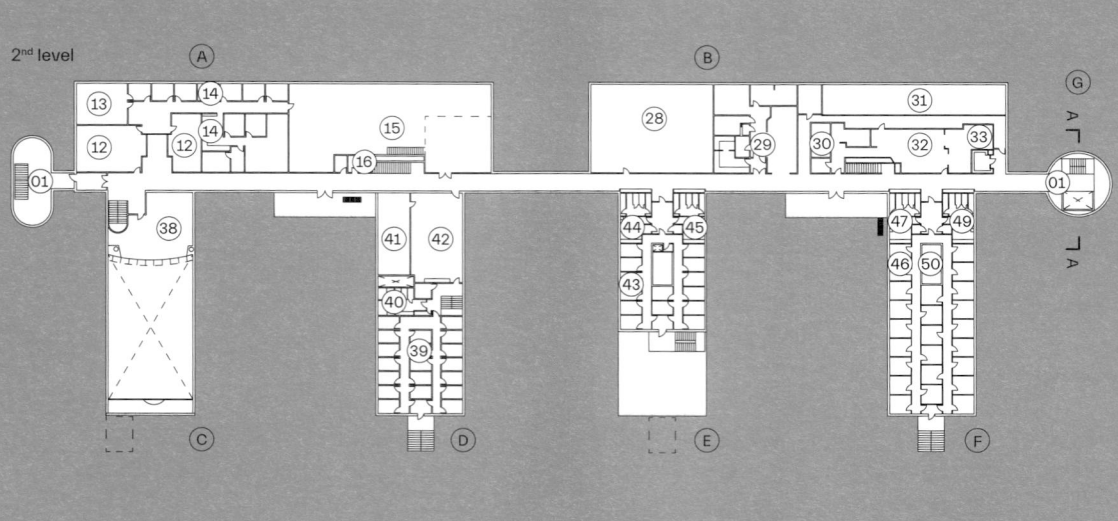

2nd level

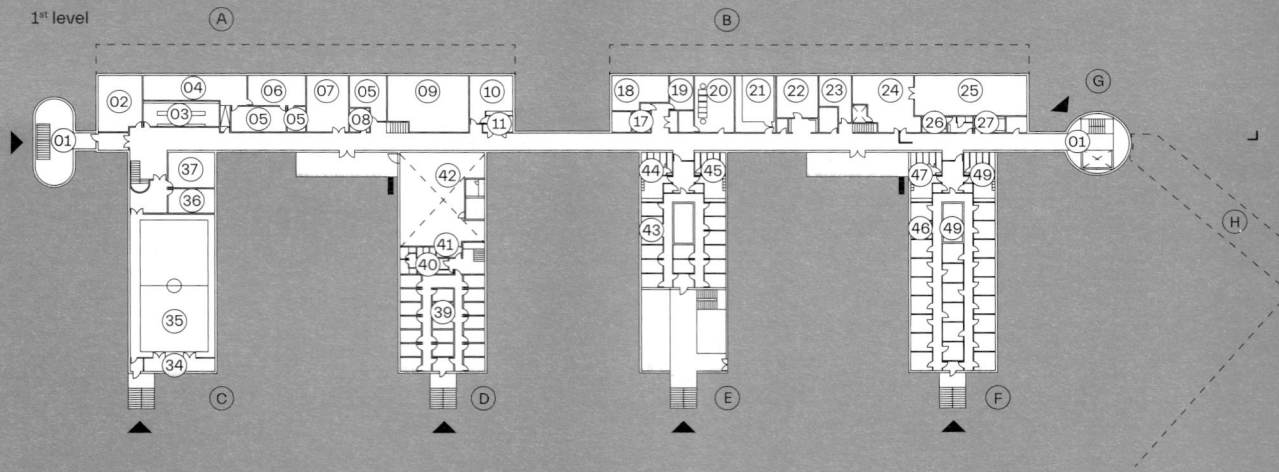

1st level

01	Access core	13	Communication room	23	Electrical room	36	Fan room	F	
		14	Offices (resident manager, operations manager, NSF operations manager)	24	Fan room	37	Activity room	46	Bedroom
A				25	Mechanical room	38	Exercise room	47	Men's bathroom
02	Lounge			26	Sauna			48	Women's bathroom
03	Cloakroom	15	Science Laboratories	27	Cloakroom	D		49	Fan room
04	Electronics workshop	16	Electricity closet	28	Computer laboratory	39	Bedroom	50	Doctor's room
05	Electrical room			29	Medical room	40	Bathroom		
06	Network centre			30	Laundry	41	Laundry	G	Access core to sub-surface fuel arch and storage
07	Science storage	B		31	Dining room	42	Emergency power generation fan room		
08	Recyling room	17	Darkroom	32	Kitchen				
09	Fan room	18	Quiet room	33	Cold store				
10	Arts and crafts room	19	Medical storage			E		H	Subsurface fuel arch and storage
11	Communications closet	20	Prefabricated plant growth chamber	C		43	Sleeping quarters		
12	Conference room	21	Storage	34	Storage	44	Kitchen		
		22	Storage	35	Multipurpose gymnasium	45	Cold room		

SOURCE
United States Antarctic Program, 1998

1 : 1,000   0   10   20   40 m

SECTION | ACCESS CORE

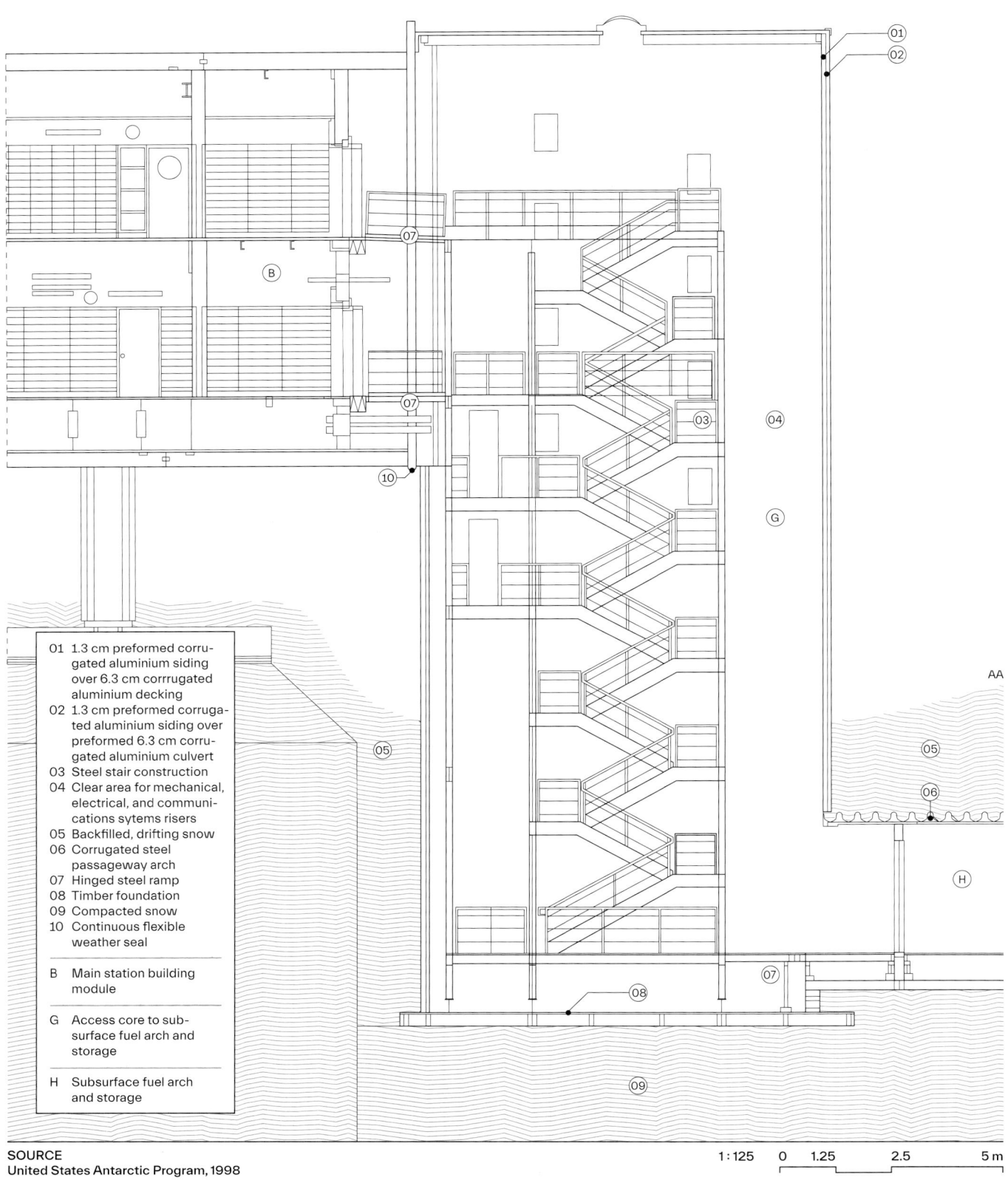

01 1.3 cm preformed corru-
     gated aluminium siding
     over 6.3 cm corrrugated
     aluminium decking
02 1.3 cm preformed corruga-
     ted aluminium siding over
     preformed 6.3 cm corru-
     gated aluminium culvert
03 Steel stair construction
04 Clear area for mechanical,
     electrical, and communi-
     cations sytems risers
05 Backfilled, drifting snow
06 Corrugated steel
     passageway arch
07 Hinged steel ramp
08 Timber foundation
09 Compacted snow
10 Continuous flexible
     weather seal

B   Main station building
     module

G   Access core to sub-
     surface fuel arch and
     storage

H   Subsurface fuel arch
     and storage

SOURCE
United States Antarctic Program, 1998

1 : 125   0   1.25   2.5   5 m

ELEVATIONS | MAIN STATION BUILDING

“Ultimately, the […] objective […] was to limit the number of times the facility would need to be raised during its design life of 25 years to a maximum of twice. […] The final structural/ jacking design involves an integral main building floor ‘platform’, supported by double trusses that straddle primary 914 mm diameter steel pipe columns located outboard of the building envelope. The primary columns transfer building loads to welded steel box beams on timber raft footings. Jacking involves adding a 4 meter column extension to the top of each column, placing hydraulic jacks under spreader beams at the top of each extended column, connecting the spreader beam to the trusses with steel rods, disengaging the trusses from the columns, hoisting the station up a full floor’s height (3 m), and then securing the trusses again at the top of the column extension.”

Ferraro Choi and Associates, “Elevated Station Design for the South Pole Redevelopment Project at Amundsen-Scott South Pole Station” (online)

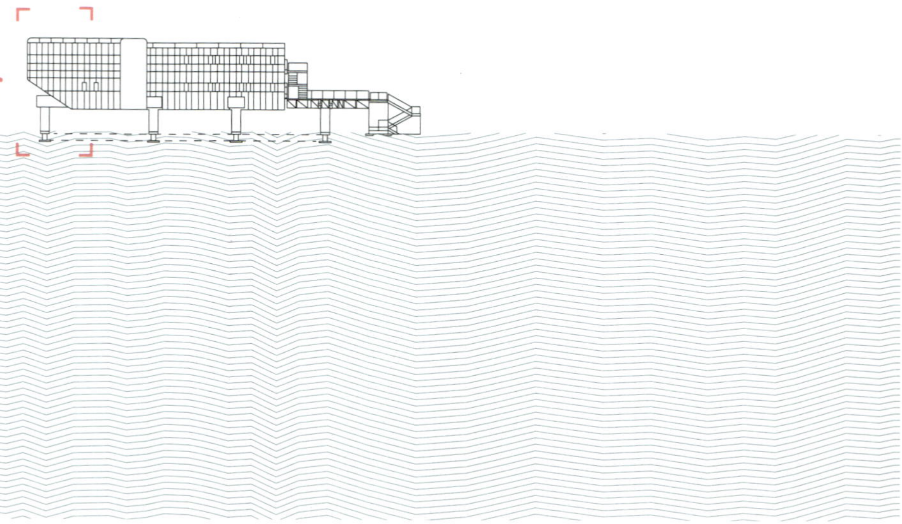

“Outboard columns conceal jacking systems which allow the whole station to be raised periodically, keeping it above the plateau that continues to gain elevation at a rate of 0.2 meter per year.”

Ferraro Choi and Associates, “Amundsen-Scott South Pole Station” (online)

SOURCE
United States Antarctic Program, 1998

1 : 1,000 0 10 20 40 m

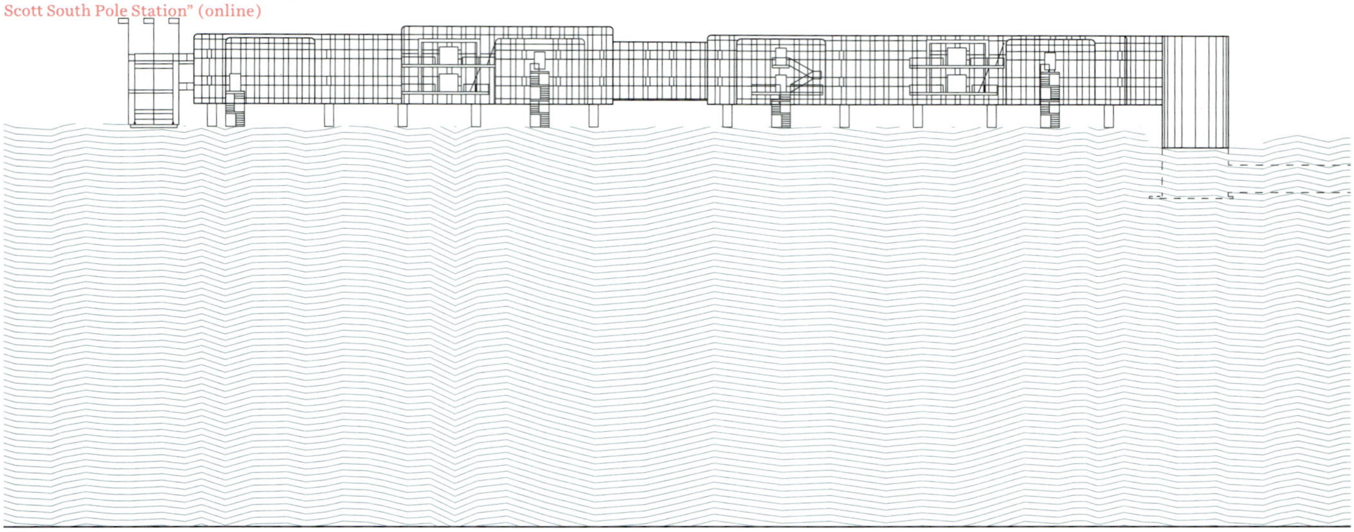

SECTION DETAIL | BUILDING ENVELOPE

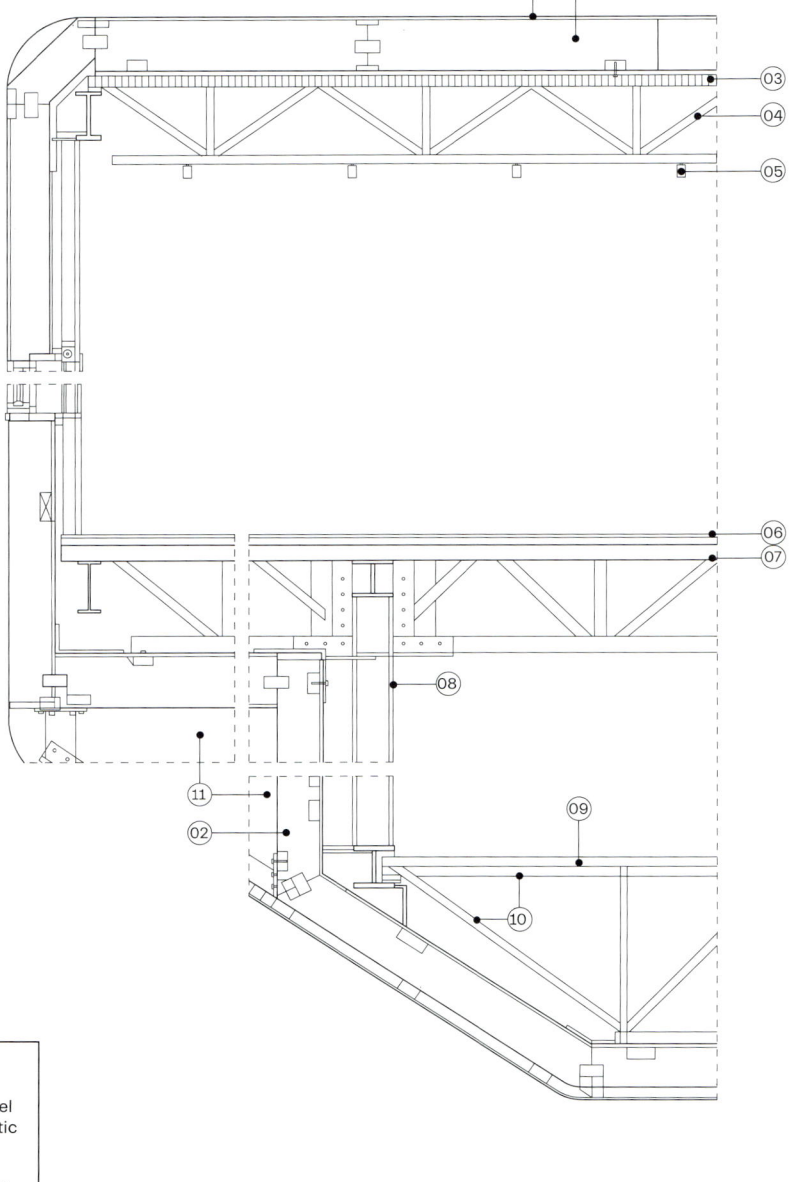

01 Field-installed
    prefinished cladding
02 Insulation building panel
03 Duct liner board acoustic
    treatment
04 Steel roof joist
05 Structural steel channels
    for lighting and miscella-
    neous system support
06 Decking over steel
    joist/truss
07 Spray-on insulation
08 Truss
09 Floor finish over
    bottom chord of truss
10 Steel floor joist
11 Void space

"To prevent burial by annual snow drift, [...] its windward face is chamfered similar to an airplane wing. The unique design works by increasing wind speed above and below the station, effectively scouring drifting snow by preventing it from settling until it is well beyond the station on the downwind side."

Ferraro Choi and Associates, "Amundsen-Scott South Pole Station" (online)

SOURCE
United States Antarctic Program, 1998

1 : 50   0   1   2.5   4 m

NEUMAYER III
1981/2009

↖ p. 414, 418, 503, 504, 577

"The detailed engineering planning addressed construction details such as the statics of the whole steel structure and the stability of the aerodynamic shaped hull with respect to the anticipated wind loads as well as the performance of the service systems: power supply, heating, air conditioning, water supply, waste water treatment, etc. […] Mainly because of economical constraints the whole compound became shorter in length [than] previously planned, and only 16 legs will bear the platform."

Hartwig Gernandt, Saad El Naggar, Jürgen Janneck, Thomas Matz, and Cord Drücker, *From Georg Forster Station to Neumayer Station III – A Sustainable Replacement at Atka Bay for Future,* 2007

Designers: Dietrich Enss (Polar Engineering Consultant); Hartwig Gernandt; Saad El Naggar; Ralf Siegmund; Alfred Wegener Institute – Helmholtz Center for Polar and Marine Research, Germany
Engineers: IMS Ingenieurgesellschaft mbH; GH Consult GmbH; M+P consulting Nord GmbH & Co. KG; KSF Ingenieurbüro für Bauwesen, Germany
Contractors: ARGE Neumayer III; J.H.K. Engineering GmbH & Co. KG; KAEFER Construction GmbH, Germany
Sub-contractors: (hydraulics) Lingk & Sturzebecher GmbH; (cladding) TELEDOOR MELLE Isoliertechnik GmbH; (power/air-conditioning) Imtech Deutschland GmbH & Co. KG, Germany
Consultants: (wind tunnel testing) Meteorologisches Institut, Universität Hamburg, Zentrum für Meeres- und Klimaforschung, Germany

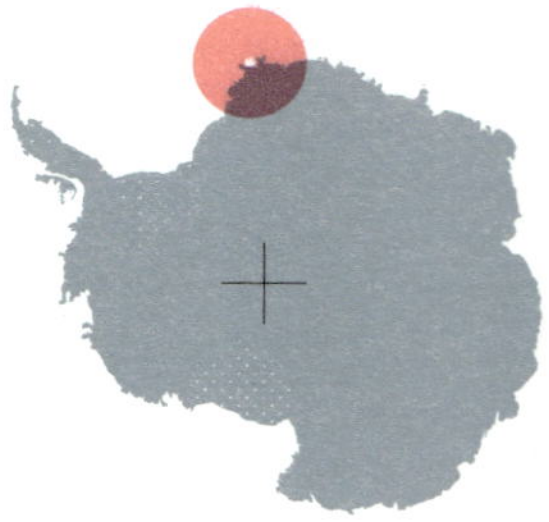

Germany
Alfred Wegener Institute for Polar and Marine Research (AWI)

Active Station

70°39'54"S 8°16'52"W
Ekström Ice Shelf, Queen Maud Land
Ice shelf, 43 m.a.s.l., –16°C

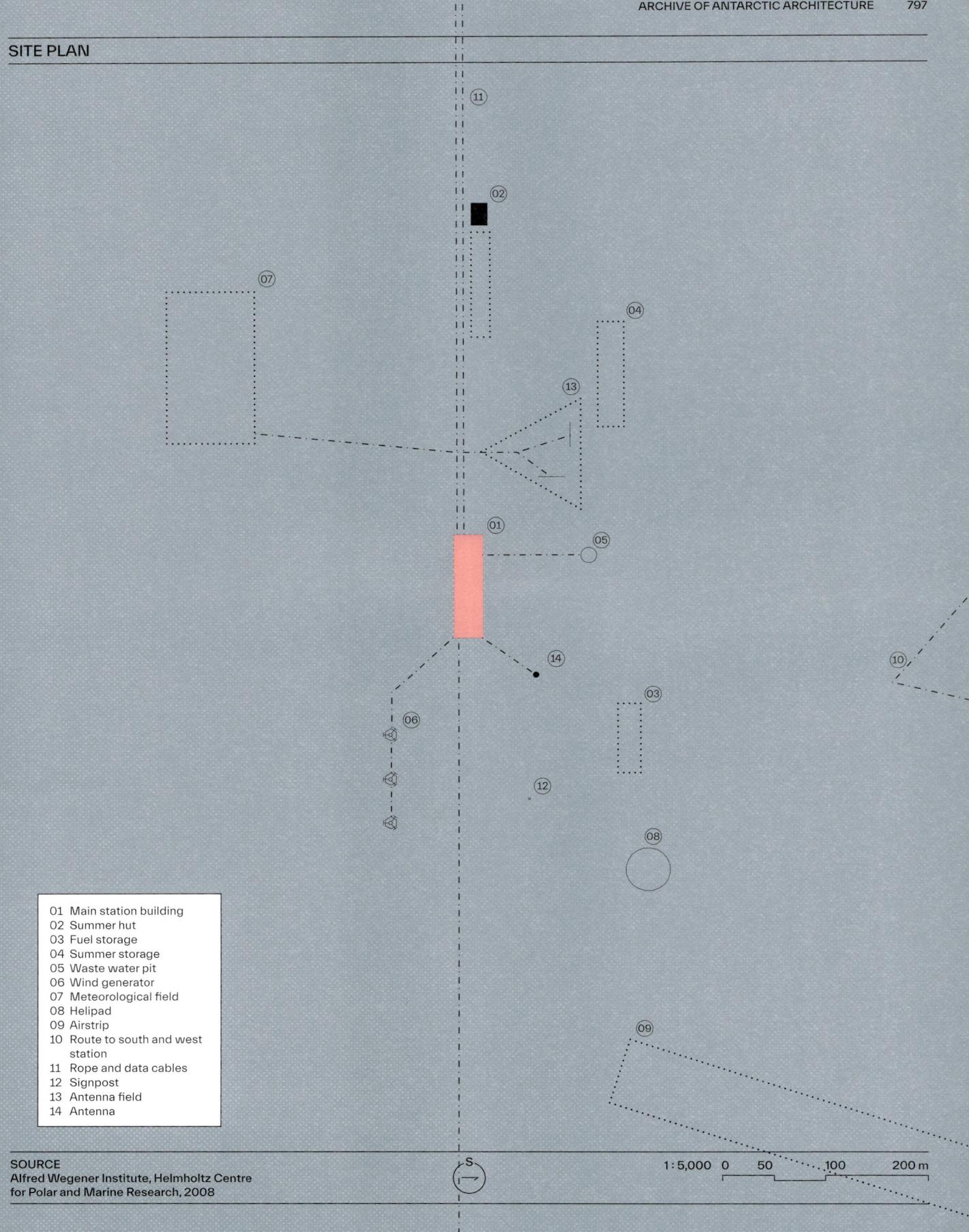

SITE PLAN

01 Main station building
02 Summer hut
03 Fuel storage
04 Summer storage
05 Waste water pit
06 Wind generator
07 Meteorological field
08 Helipad
09 Airstrip
10 Route to south and west
     station
11 Rope and data cables
12 Signpost
13 Antenna field
14 Antenna

SOURCE
Alfred Wegener Institute, Helmholtz Centre
for Polar and Marine Research, 2008

S

1 : 5,000   0   50   100   200 m

PLAN, FOUNDATIONS AND GARAGE

A	Ramp	12	Office	27	Technical office
		13	Workshop	28	Dining room
	1st level	14	Power station	29	Kitchen
01	Stair and elevator core	15	Fresh water		
02	Cold storage −25°C	16	Grey water		3rd level
03	Pump station	17	Power supply system	30	Storage
04	Battery charging station	18	Server	31	Bedroom
05	Gym	19	Bathroom	32	Sauna
06	Technical storage	20	Changing room	33	Laundry
07	Cold storage +5°C	21	Food storage	34	Multipupose laboratory
08	Workshop	22	Radio	35	Clean room
09	Warm storage	23	Operating room	36	Air chemistry laboratory
10	Cold Storage	24	Medical room	37	Meteorological laboratory
11	Hydraulic station	25	Lounge	38	Geophysical laboratory
		26	Commander's and	39	Forecast laboratory
	2nd level		doctor's room	40	Ozone laboratory

SOURCE
Alfred Wegener Institute, Helmholtz Centre
for Polar and Marine Research, 2008

1 : 750   0   7.5   15   30 m

PLAN , 1ST, 2ND, AND 3RD LEVEL

3rd level

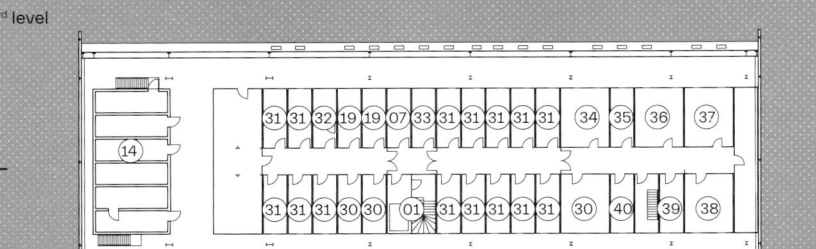

2nd level

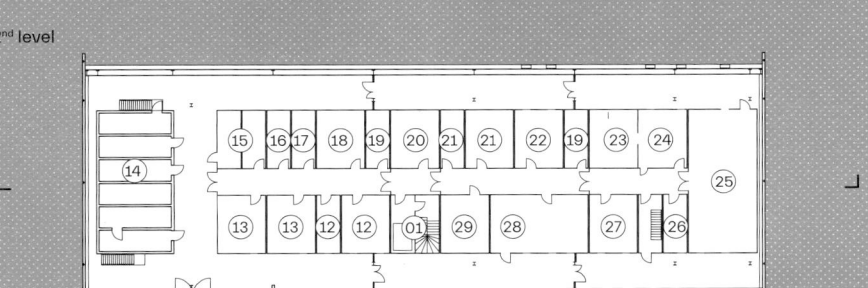

1st level

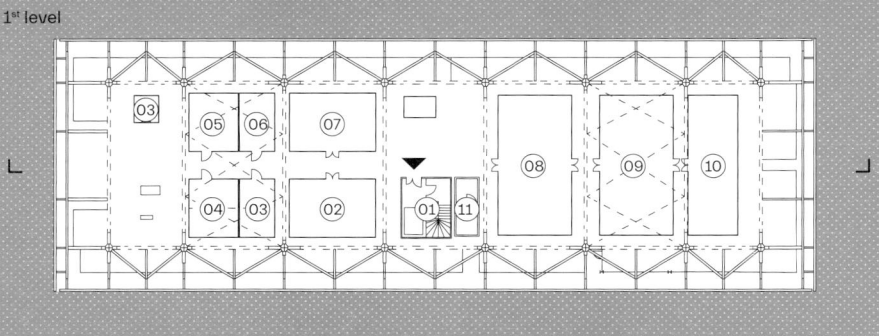

SOURCE
Alfred Wegener Institute, Helmholtz Centre
for Polar and Marine Research, 2008

1 : 750   0   7.5   15   30 m

ELEVATION

"Deck U1 stretches over the full length of the garage trench, where mainly workshops and provision storage rooms as well as some technical installations (snow-melter, fuel pumping station, hydraulic control room) will be accommodated. Deck U2 is not only the garage for the tracked vehicle but also for additional storage."

Hartwig Gernandt, Saad El Naggar, Jürgen Janneck, Thomas Matz, and Cord Drücker, *From Georg Forster Station to Neumayer Station III – A Sustainable Replacement at Atka Bay for Future,* 2007

"Neumayer Station III will feature above-ground and below-ground facilities combined in one large building, which can be raised hydraulically to compensate [for] snow accumulation. […] A shell is to protect this building from wind and, because of its aerodynamic shape, to reduce snow accumulation or erosion around the base. A trench in the snow under the platform, accessible via a ramp, will serve as garage and

cold storage room and provide room for an intermediate deck with various facilities and technical installations hanging to the flat, rigid roof that covers the trench […] level with the snow surface."

Hartwig Gernandt, Saad El Naggar, Jürgen Janneck, Thomas Matz, and Cord Drücker, *From Georg Forster Station to Neumayer Station III – A Sustainable Replacement at Atka Bay for Future,* 2007

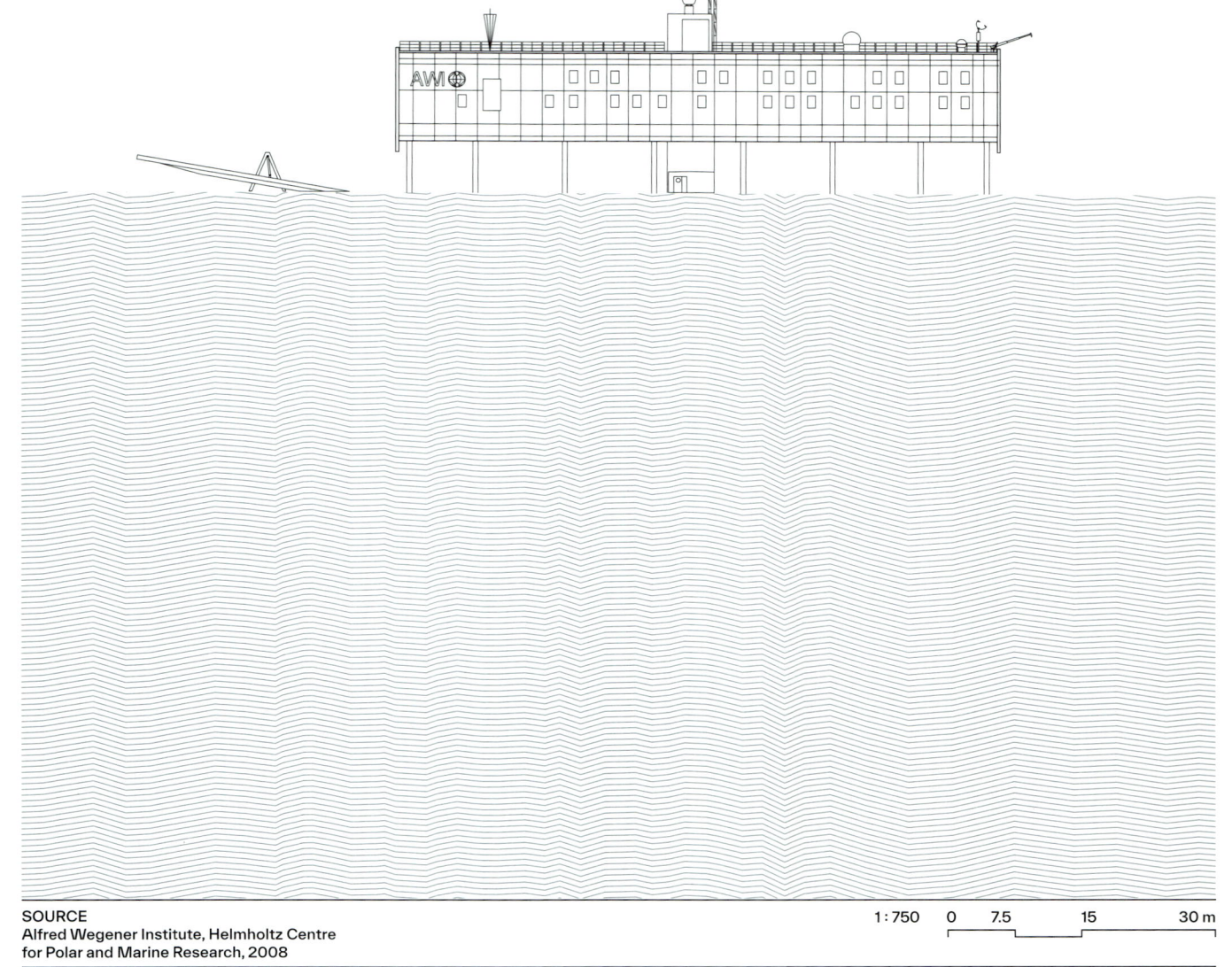

SOURCE
Alfred Wegener Institute, Helmholtz Centre
for Polar and Marine Research, 2008

1 : 750   0   7.5   15   30 m

SECTION

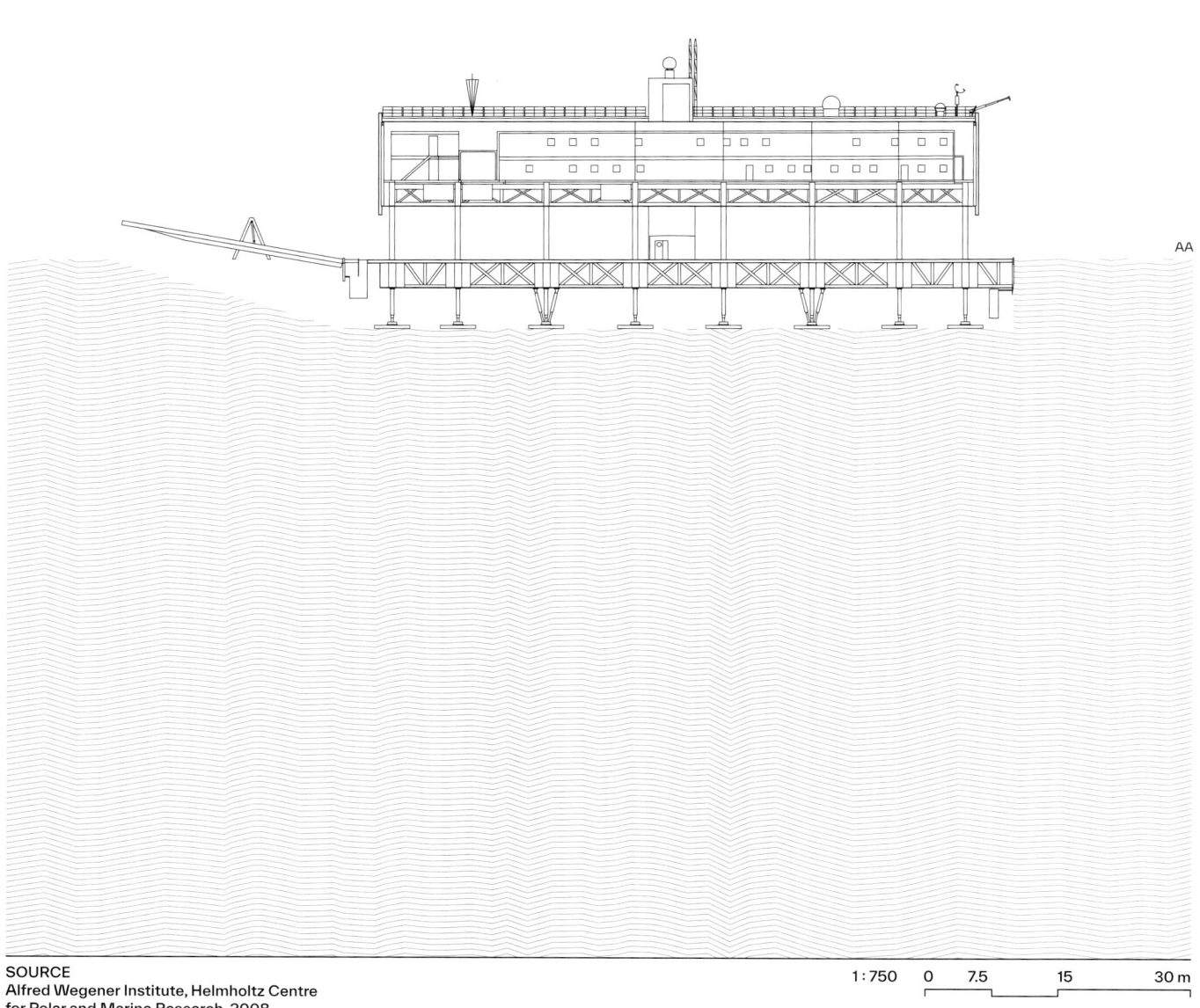

AA

SOURCE
Alfred Wegener Institute, Helmholtz Centre
for Polar and Marine Research, 2008

1 : 750   0   7.5   15   30 m

SECTION

"Exceptional care had to be given to the design of the foundation of the new station. The accumulation rate, the bearing capacity for constant loading and the settling behaviour of snow determine the design of the foundations."

Hartwig Gernandt, Saad El Naggar, Jürgen Janneck, Thomas Matz, and Cord Drücker, *From Georg Forster Station to Neumayer Station III – A Sustainable Replacement at Atka Bay for Future,* 2007

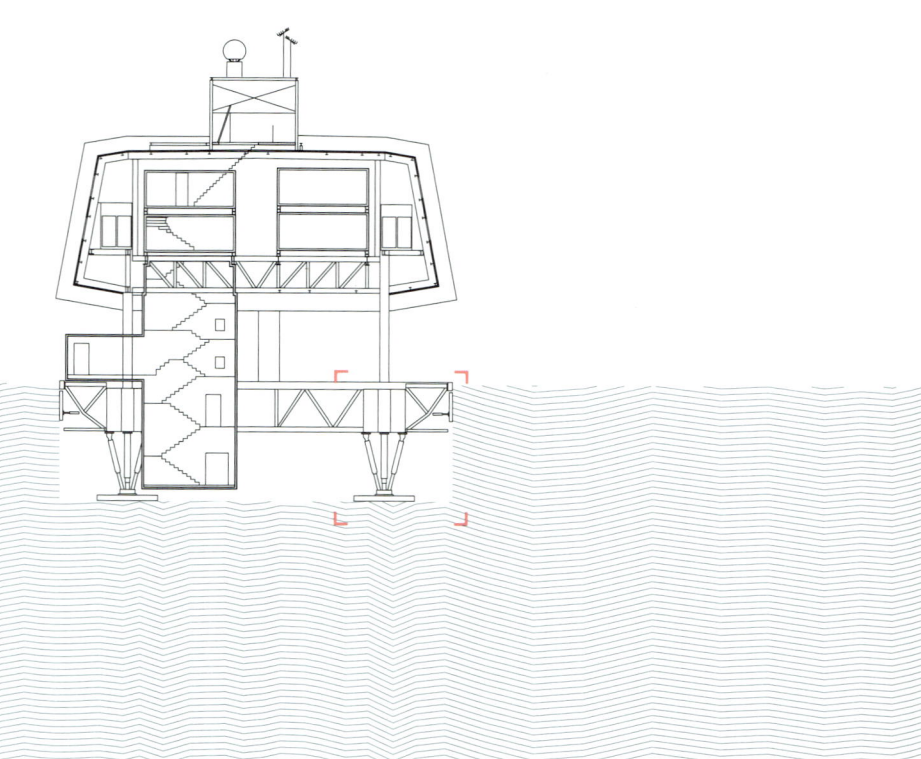

BB

SOURCE
Alfred Wegener Institute, Helmholtz Centre
for Polar and Marine Research, 2008

1:500   0   5   10   20 m

DETAIL | FOUNDATIONS JACKING SYSTEM

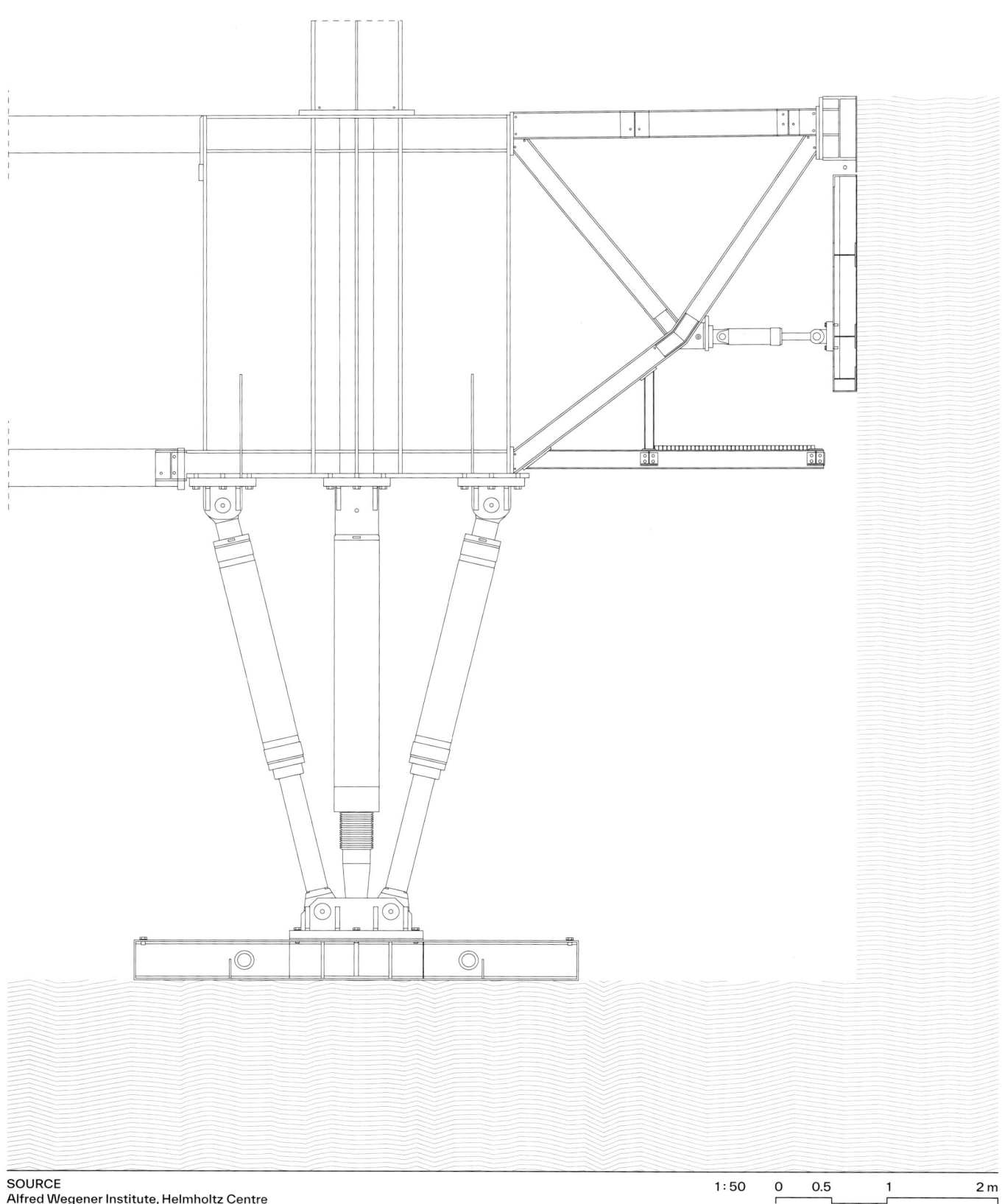

SOURCE
Alfred Wegener Institute, Helmholtz Centre
for Polar and Marine Research, 2008

1:50   0   0.5   1   2 m

BHARATI **2012**

"[Bharati is] a modular, three story structure with total floor area of 2900 m² over a small footprint of 1650 m² […]. The station consists of one main building, fuel farm, fuel station, sea water pump house, a summer camp and a number of smaller containerized modules."

Council of Managers of National Antarctic Programs (COMNAP), *Antarctic Station Catalogue*, 2017

Architect: bof Architekten, Germany
Engineer: Ramboll IMS Ingenieurgesellschaft, Germany
Consultants: M+P Consulting; (building physics) Ingenieurbüro Axel C. Rahn GmbH; (wind tunnel testing) Meteorologisches Institut, Universität Hamburg, Zentrum für Meeres- und Klimaforschung, Germany
Manufacturers: KAEFER Construction, Germany
Suppliers: (windows) Wicona; (facade) Lenderoth; (flooring) Nora; (flooring) Roma, Germany

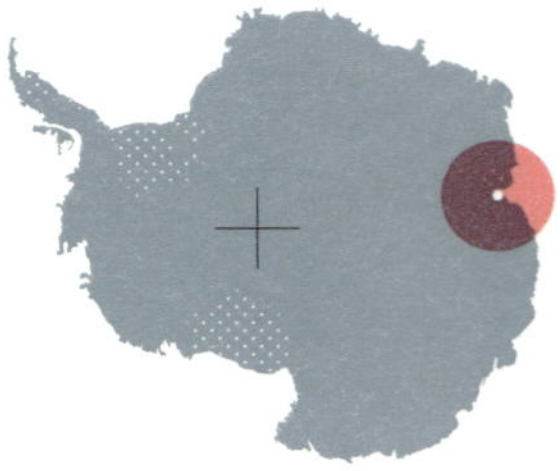

India
Indian Antarctic Program

Active Station

69°24'29"S 76°11'14"E
Larsemann Hills, Prydz Bay
Ice-free ground, 35 m.a.s.l., –10.2°C

SITE PLAN

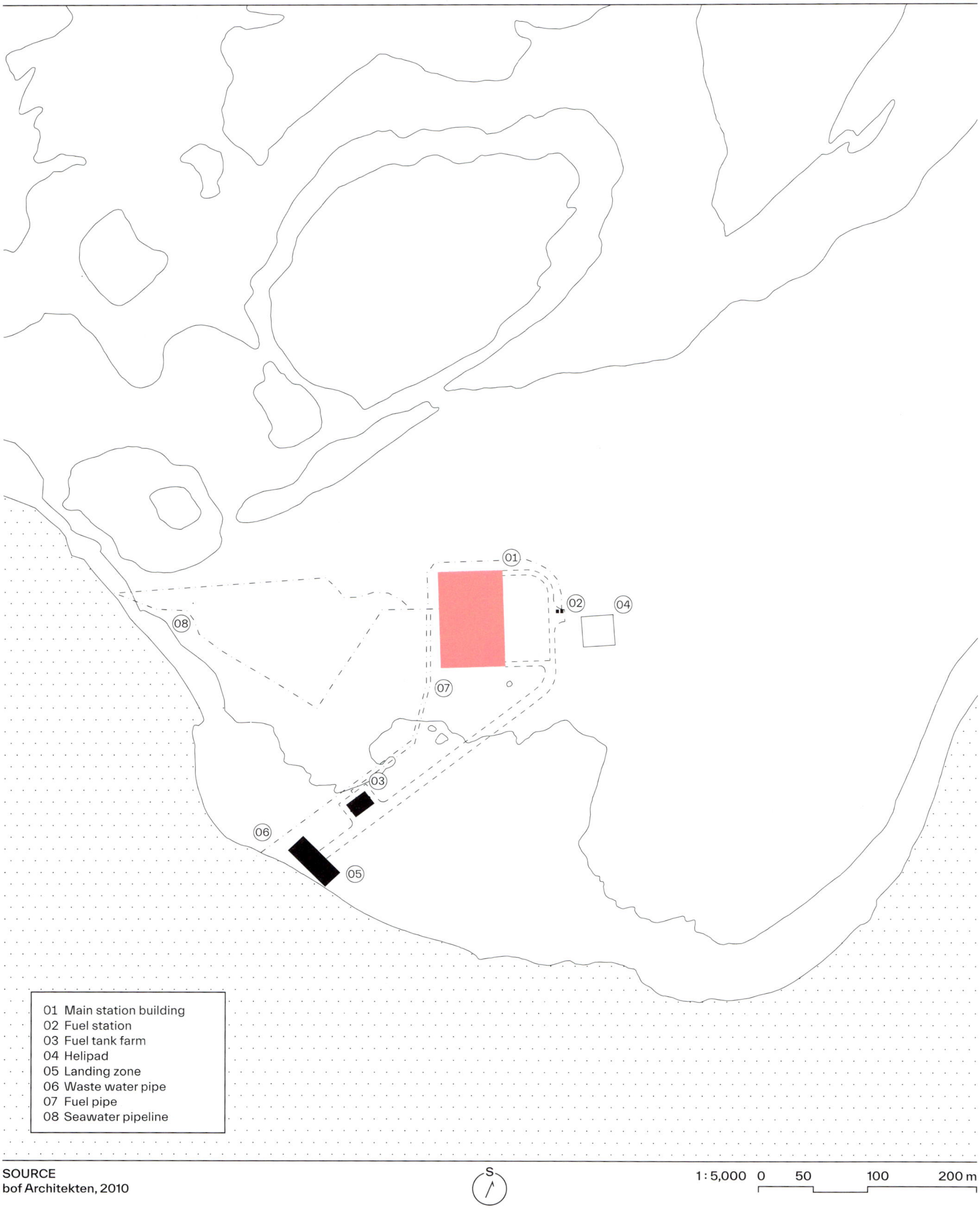

01 Main station building
02 Fuel station
03 Fuel tank farm
04 Helipad
05 Landing zone
06 Waste water pipe
07 Fuel pipe
08 Seawater pipeline

SOURCE
bof Architekten, 2010

S

1 : 5,000   0   50   100   200 m

PLAN, 1ST AND 2ND LEVEL

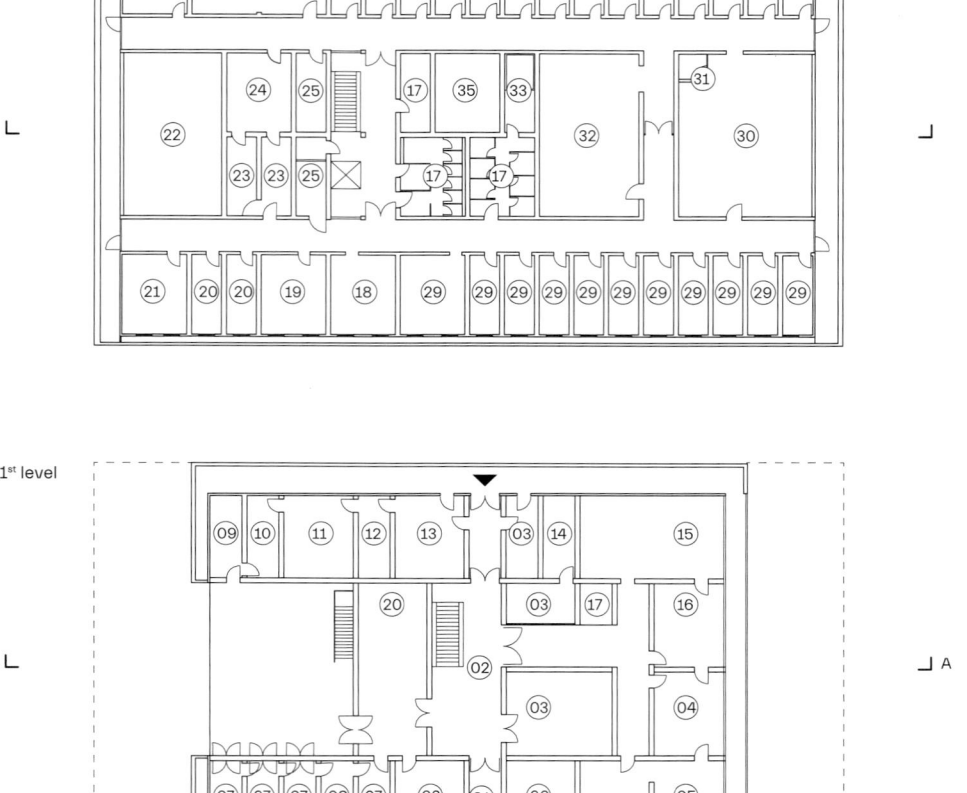

2nd level

1st level

A

A

1st level		07 Generator	15 Meteorology and	20 Communication	29 Bedroom
01	Vestibule	08 Fuel	geomagnetism	21 Library	30 Lounge and bar
02	Stair core	09 Waste water	laboratory	22 Dining room	31 Smoking area
03	Storage	10 Grey water	16 Seismic laboratory	23 Cold storage	32 Entertainment
04	Biology and environment	11 Water storage	17 Bathroom	24 Kitchen	33 Sauna
	laboratory	12 Distribution		25 Electrical appliance	34 Laundry
05	Human physiology	13 Waste	2nd level	26 Gym	
	laboratory	14 Changing room	18 Office	27 Medical room	
06	Earth science laboratory		19 Computer	28 Prayer room	

SOURCE
bof Architekten, 2010

S

1 : 500   0   5   10   20 m

ELEVATION

"The building consists of 134 standard shipping containers that not only define the individual spaces, but which also account for the structural system. The high degree of mobility and flexibility associated with such containers provides for an optimal means of transport and extremely short assembly periods. [...]

Due to the extreme conditions in the Antarctic, the extremely low temperatures and the powerful winds, the containers have been clad with an insulated, aerodynamic skin consisting of metal panels. The form of the facade was tested in a wind canal and was optimized for this purpose."

boF Architekten, "Bharati" (online)

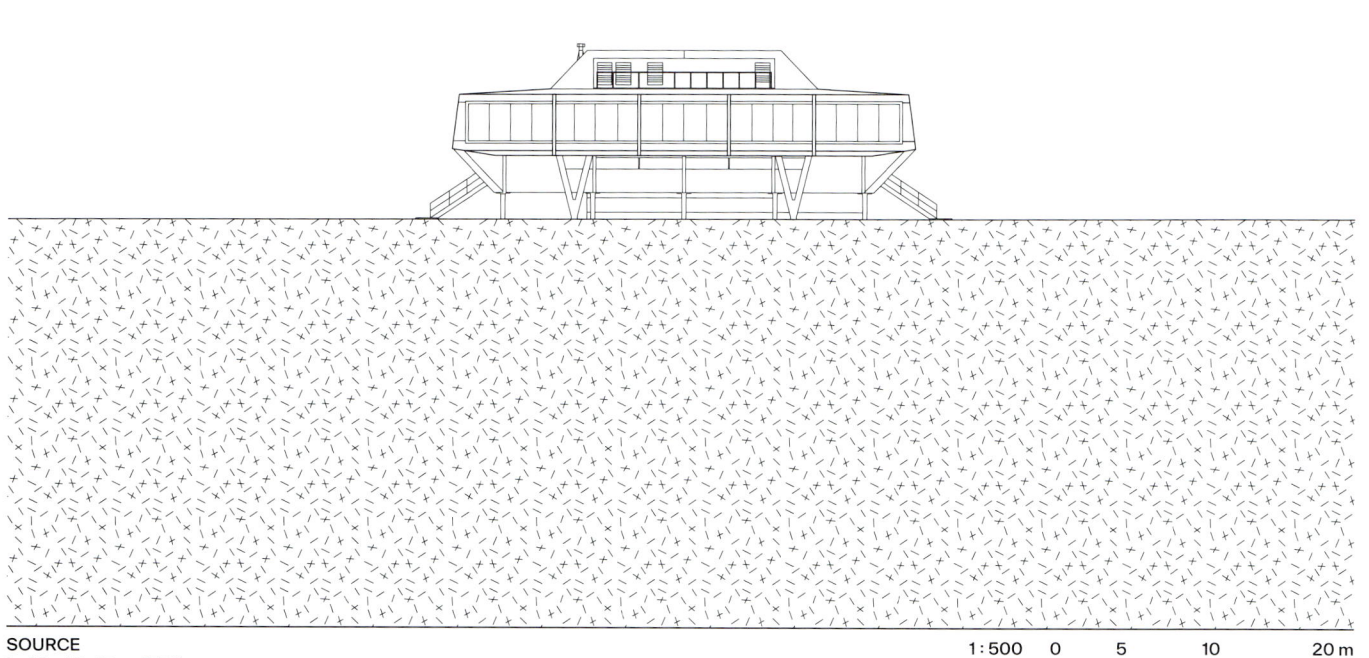

SOURCE
bof Architekten, 2010

1 : 500   0   5   10   20 m

SECTION

"To avoid water entering via the
gap in between the metal panels
which constitute the flat roof
of the structure the company in
charge of the construction came
up with a product used for the
infrastructure of pipelines.
A very special material which
helped us to seal every gap."

Bert Bücking (Architect, bof Architekten), in conversation with UNLESS, Architectural Association School of Architecture, 2019

01 Bent panel element
02 Prefabricated QuickBuild
    insulated panel elements,
    rigid polyurethane foam
03 Aluminium cover strip
04 Neoprene strips
05 Specially shaped sealing
    profile
06 Aluminium cover strip
07 Prefabricated triple-glazed
    panel elements, aluminium
    frame, powder coated,
    with concealed electrical
    frame heating
08 Frame heating
09 Window frame
10 Fixation and alignment of
    glazing panel
11 Steel sheet, galvanised

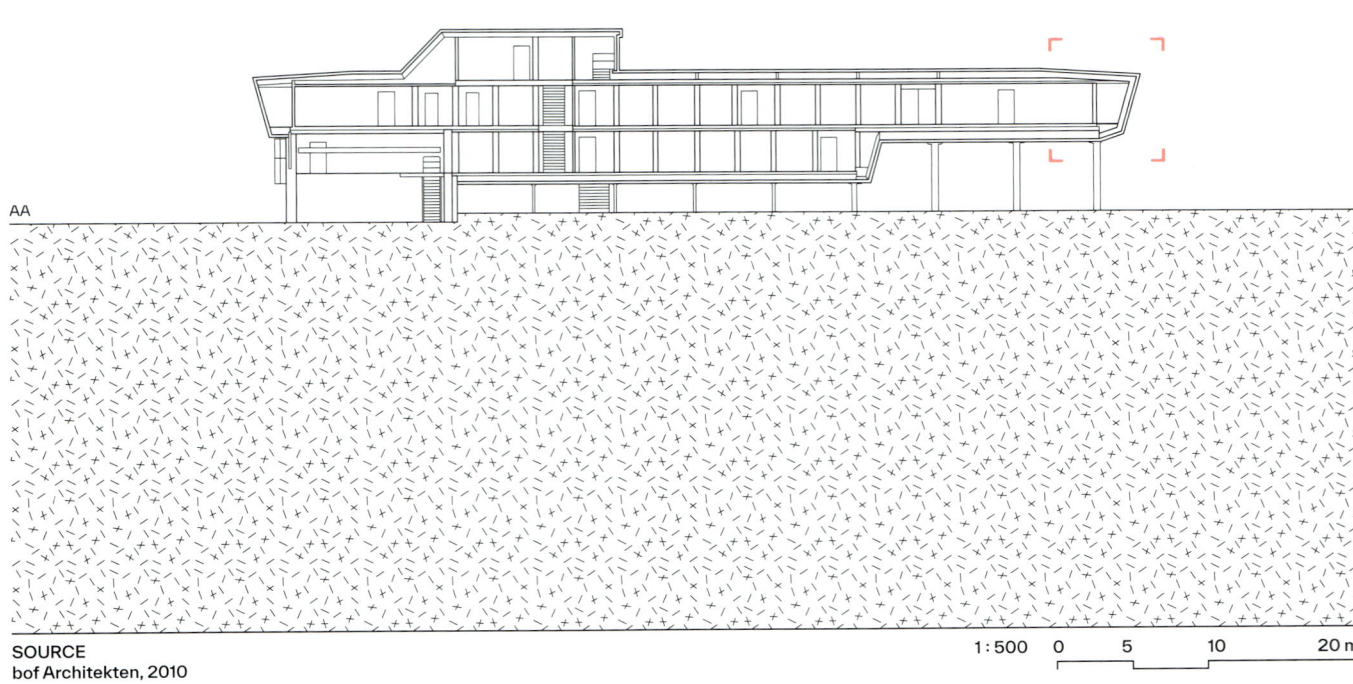

AA

1 : 500   0   5   10   20 m

DETAIL | BUILDING ENVELOPE

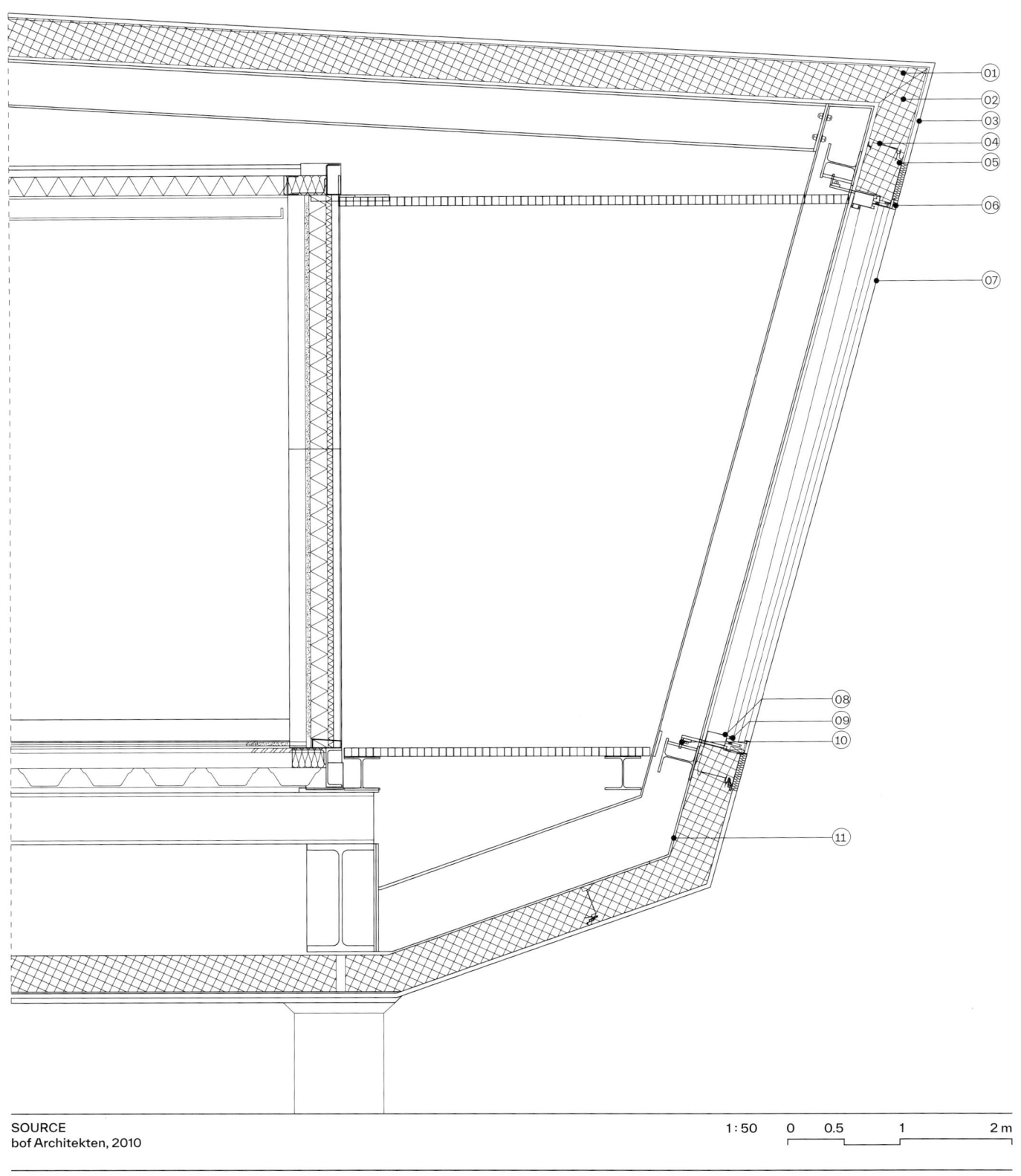

SOURCE
bof Architekten, 2010

1 : 50   0   0.5   1   2 m

HALLEY VI
1956/2013

↖ p. 366, 372, 418, 422, 425, 461, 504, 509, 512, 576

"The design for Halley VI Research Station is based on a number of light-weight, semi-autonomous modular buildings that can be plugged together in a variety of ways. [...] Science modules are placed to the southern end of the station, close to the clean air sector, optical dark zone and geospace radar arrays."

British Antarctic Survey and Natural Environment Research Council, *Proposed Construction and Operation of Halley VI Research Station, and Demolition and Removal of Halley V Research Station, Brunt Ice Shelf Antarctica: Final Comprehensive Environmental Evaluation*, 2007

Architect: Hugh Broughton Architects, United Kingdom
Engineer: AECOM, formerly Faber Maunsell, United Kingdom
Contractor: Galliford Try International, United Kingdom
Consultants: Billings Design Associates, United Kingdom; (colour) Colour Effects, United Kingdom

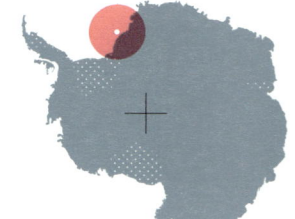

United Kingdom
British Antarctic Survey (BAS)

Active Station

75°35'S 26°14' W
Brunt Ice Shelf, Caird Coast
Ice shelf, 37 m.a.s.l., −20°C

SITE PLAN

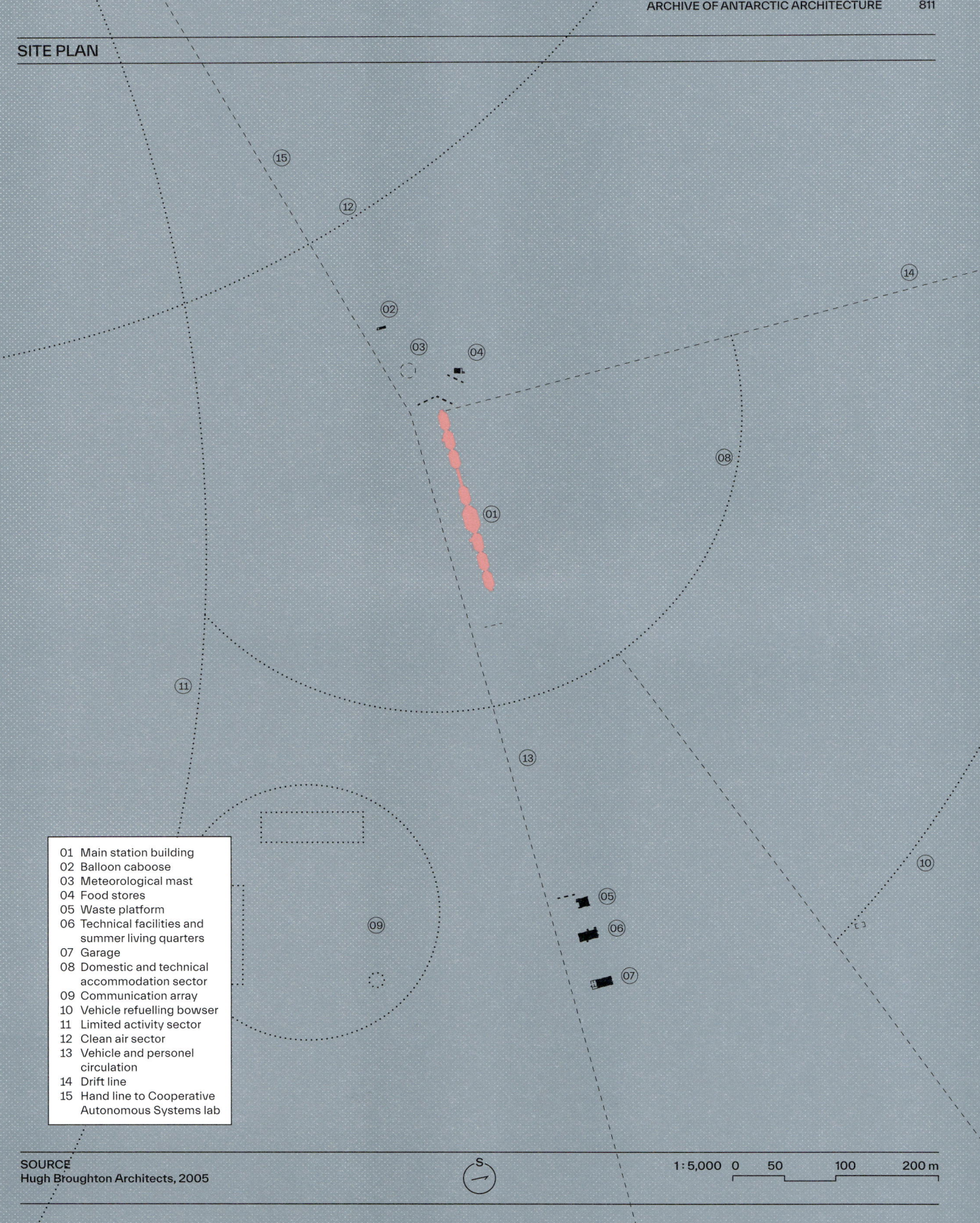

01 Main station building
02 Balloon caboose
03 Meteorological mast
04 Food stores
05 Waste platform
06 Technical facilities and
    summer living quarters
07 Garage
08 Domestic and technical
    accommodation sector
09 Communication array
10 Vehicle refuelling bowser
11 Limited activity sector
12 Clean air sector
13 Vehicle and personel
    circulation
14 Drift line
15 Hand line to Cooperative
    Autonomous Systems lab

SOURCE
Hugh Broughton Architects, 2005

S

1 : 5,000   0   50   100   200 m

PLAN AND ELEVATION

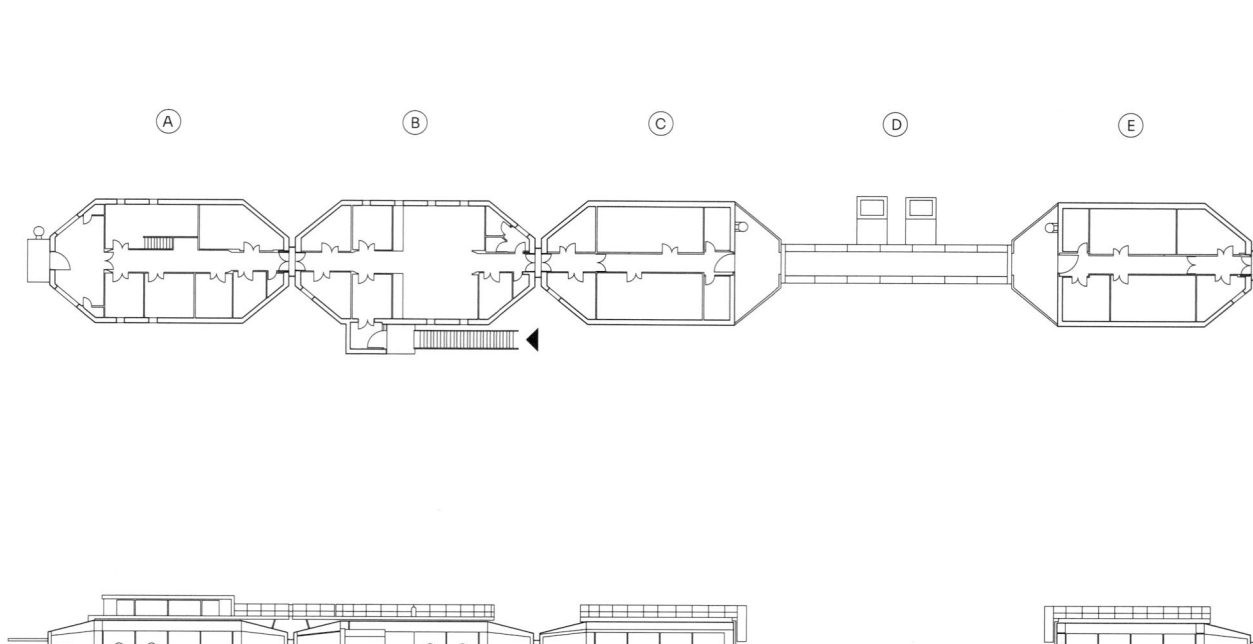

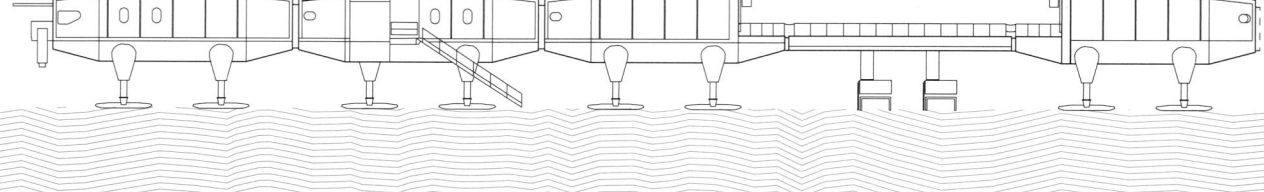

A   Science module
B   Science module
C   Generator and plant
D   Service link
E   Generator and plant
F   Social module
G   Command module
H   Sleeping module
I   Sleeping module

SOURCE
Hugh Broughton Architects, 2005

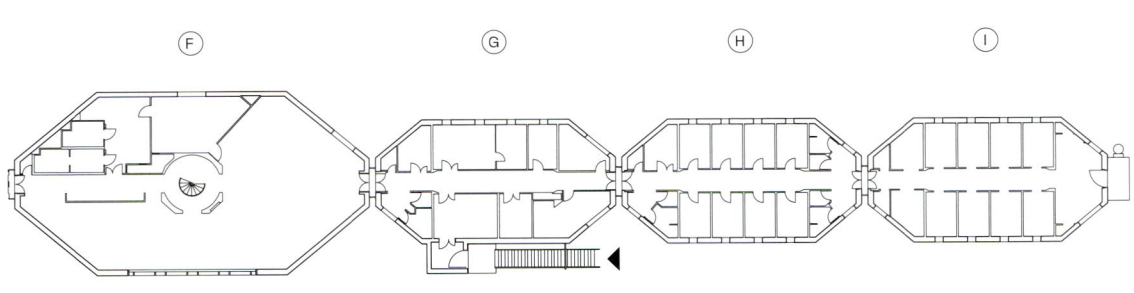

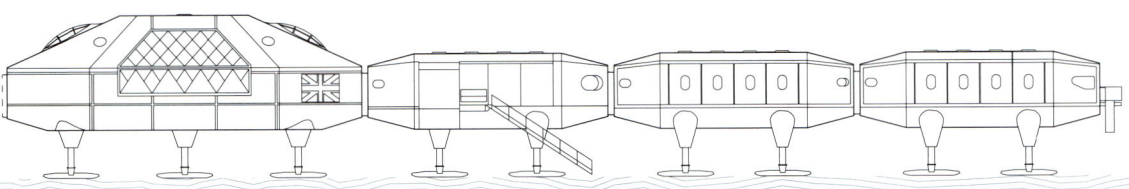

"The station modules are connected to form two main platforms. The northern platform provides the principal living accommodation. The southern platform contains the science modules. Separation into two platforms creates a refuge in case of catastrophic failure of one platform or energy module. Stand alone structures provide garage, technical, waste management and summer accommodation facilities."

British Antarctic Survey and Natural Environment Research Council, *Proposed Construction and Operation of Halley VI Research Station, and Demolition and Removal of Halley V Research Station, Brunt Ice Shelf Antarctica: Final Comprehensive Environmental Evaluation*, 2007

S

1 : 600   0   6   12   24 m

PLAN, 1ST LEVEL AND SECTION | SOCIAL MODULE F

"A large two-storey light-filled red module provides the social heart of the station and is used for living, dining and recreation."

Hugh Broughton (Architect, Hugh Broughton Architects), in conversation with UNLESS, Architectural Association School of Architecture, 2019

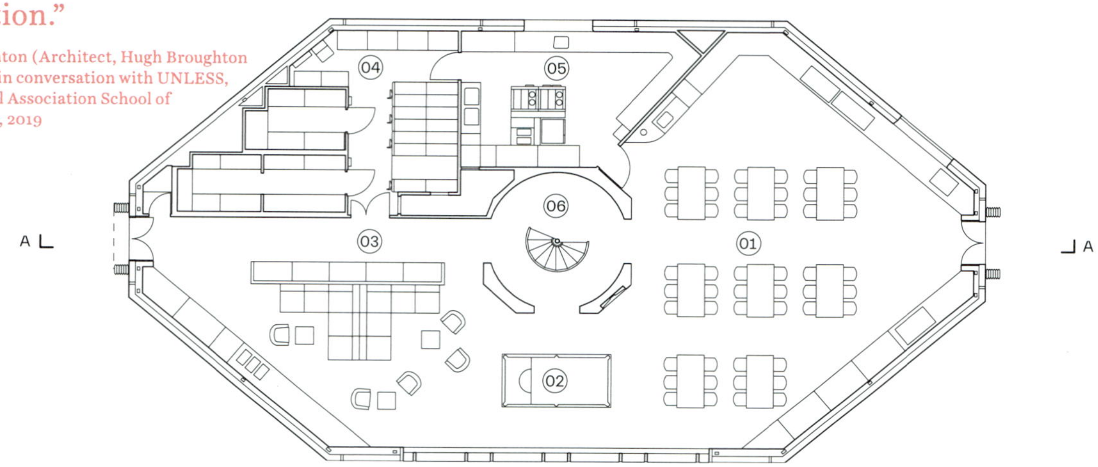

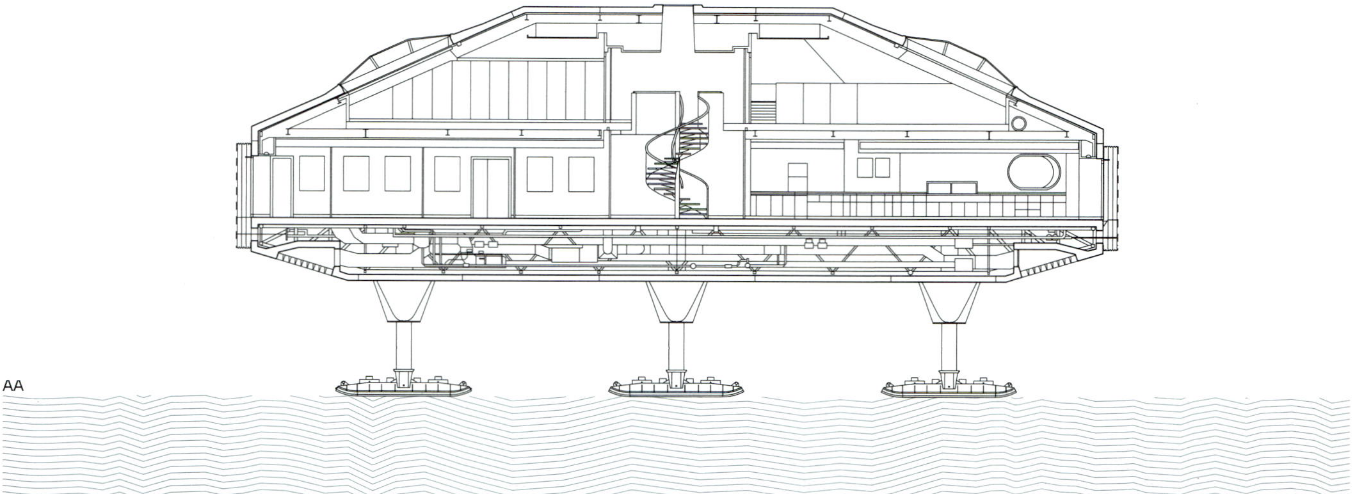

01 Dining room
02 Recreation room in double-height space
03 Bar
04 Food stores
05 Kitchen
06 Stairs

07 TV lounge, meeting space
08 Cockpit rooflight
09 Gym
10 Local plant room

SOURCE
Hugh Broughton Architects, 2005

1 : 250   0   2.5   5   10 m

PLAN, 2ND LEVEL AND ELEVATION | SOCIAL MODULE F

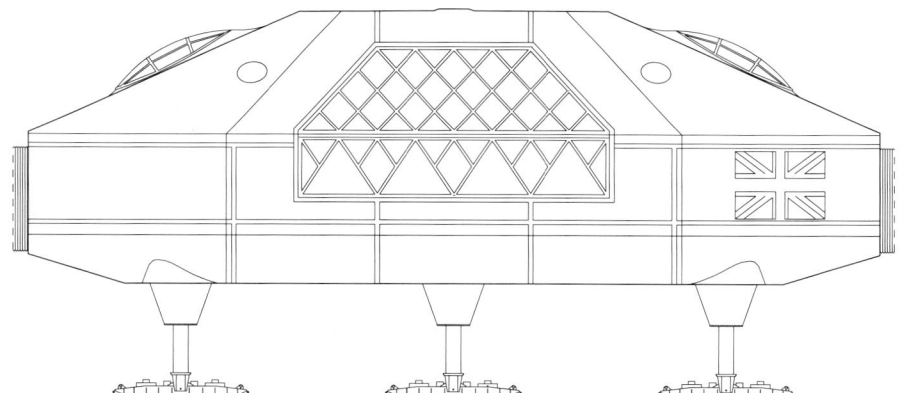

SOURCE
Hugh Broughton Architects, 2005

1:250   0   2.5   5   10 m

PLAN, 1ST LEVEL AND SECTION | SCIENCE MODULE A

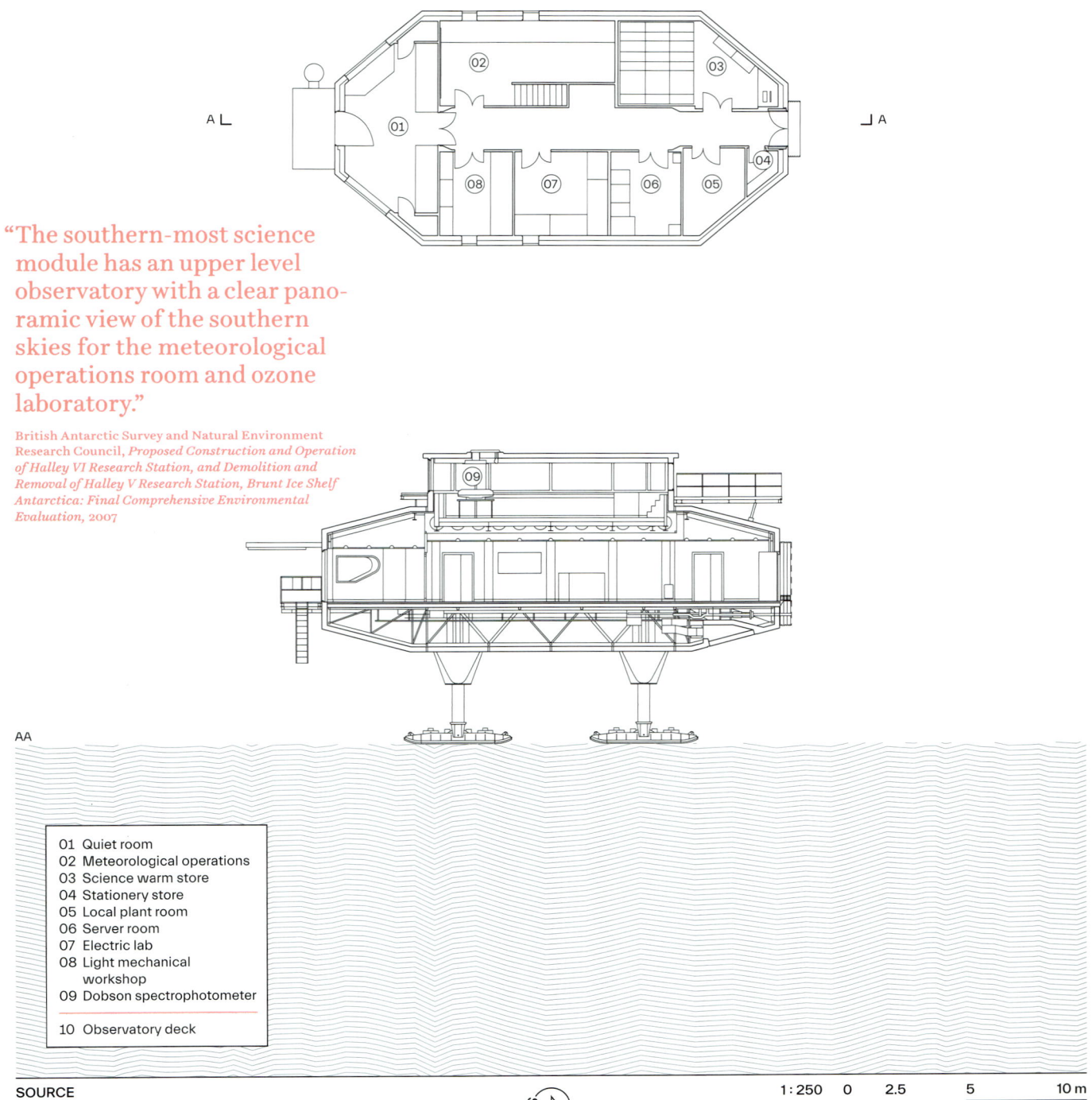

"The southern-most science module has an upper level observatory with a clear panoramic view of the southern skies for the meteorological operations room and ozone laboratory."

British Antarctic Survey and Natural Environment Research Council, *Proposed Construction and Operation of Halley VI Research Station, and Demolition and Removal of Halley V Research Station, Brunt Ice Shelf Antarctica: Final Comprehensive Environmental Evaluation,* 2007

01 Quiet room
02 Meteorological operations
03 Science warm store
04 Stationery store
05 Local plant room
06 Server room
07 Electric lab
08 Light mechanical workshop
09 Dobson spectrophotometer

10 Observatory deck

SOURCE
Hugh Broughton Architects, 2005

1 : 250   0   2.5   5   10 m

PLAN, ROOF LEVEL AND ELEVATION | SCIENCE MODULE A

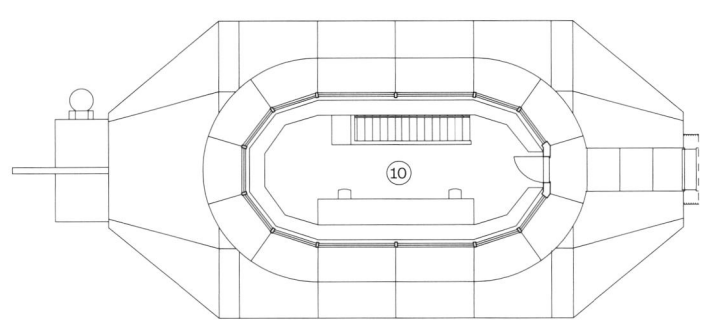

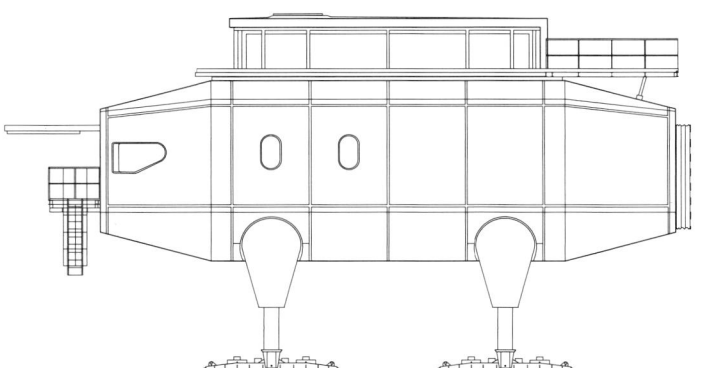

SOURCE
Hugh Broughton Architects, 2005

1:250   0   2.5   5   10 m

ELEVATION | SCIENCE MODULE A

"The modules will be raised on hydraulically operated steel legs to be clear of the snow surface and are designed to be relocated to deal with snow accumulation and movement of the ice shelf. Each leg will sit on a specially developed ski so that the modules can be towed by tracked vehicles. The skis will be designed to be manually handled and inter-changeable to allow for future flexibility and mobility. Each ski will be secured with a dagger board driven through the ski into the ice to prevent sliding of the modules once in position."

British Antarctic Survey and Natural Environment Research Council, *Proposed Construction and Operation of Halley VI Research Station, and Demolition and Removal of Halley V Research Station, Brunt Ice Shelf Antarctica: Final Comprehensive Environmental Evaluation*, 2007

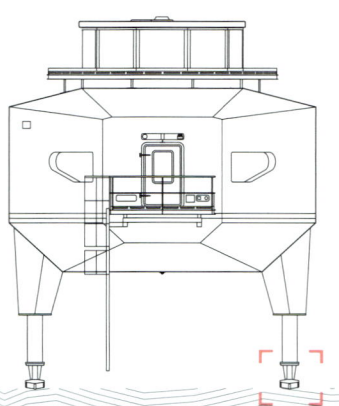

SOURCE
Hugh Broughton Architects, 2005

1 : 250   0   2.5   5   10 m

DETAIL | FOUNDATIONS / SKI

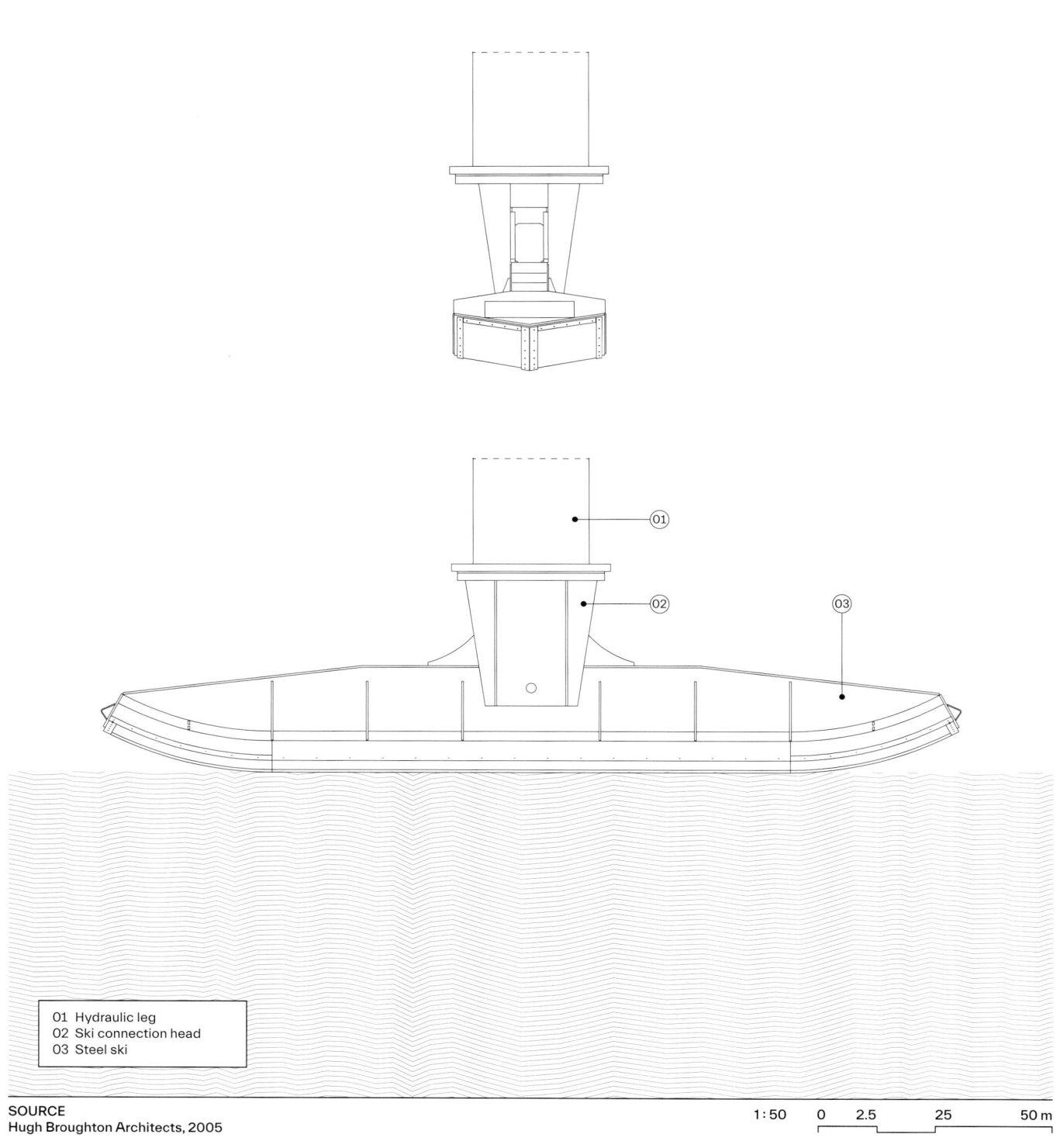

01 Hydraulic leg
02 Ski connection head
03 Steel ski

SOURCE
Hugh Broughton Architects, 2005

1 : 50   0   2.5   25   50 m

↖ p. 366, 377, 425, 429, 430

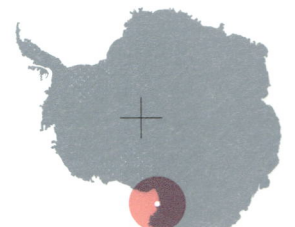

Architect: Space Group; ADD, South Korea
Engineers: Hyunday ENG.; Arup; (structural) Dawon;
(mechanical) Hanil EMC; (electrical) Nara ENG;
Duksung ALPHA (civil engineering), South Korea
Consultants: (snow drift simulation) RWDI;
(environmental) EAN Technology; (landscape) Group
HAN, South Korea
Manufacturers: Hyundai Engineering & Construction;
Kyeryong Construction; Kolon Engineering &
Construction, South Korea
Construction Planning: Team Focus Korea, South
Korea

South Korea
Korea Polar Research Institute (KOPRI)

Active Station

74°37'26"S 164°13'44"E
Terra Nova Bay, Ross Sea
Ice-free ground, 36.6 m.a.s.l., –15.1°C

SITE PLAN

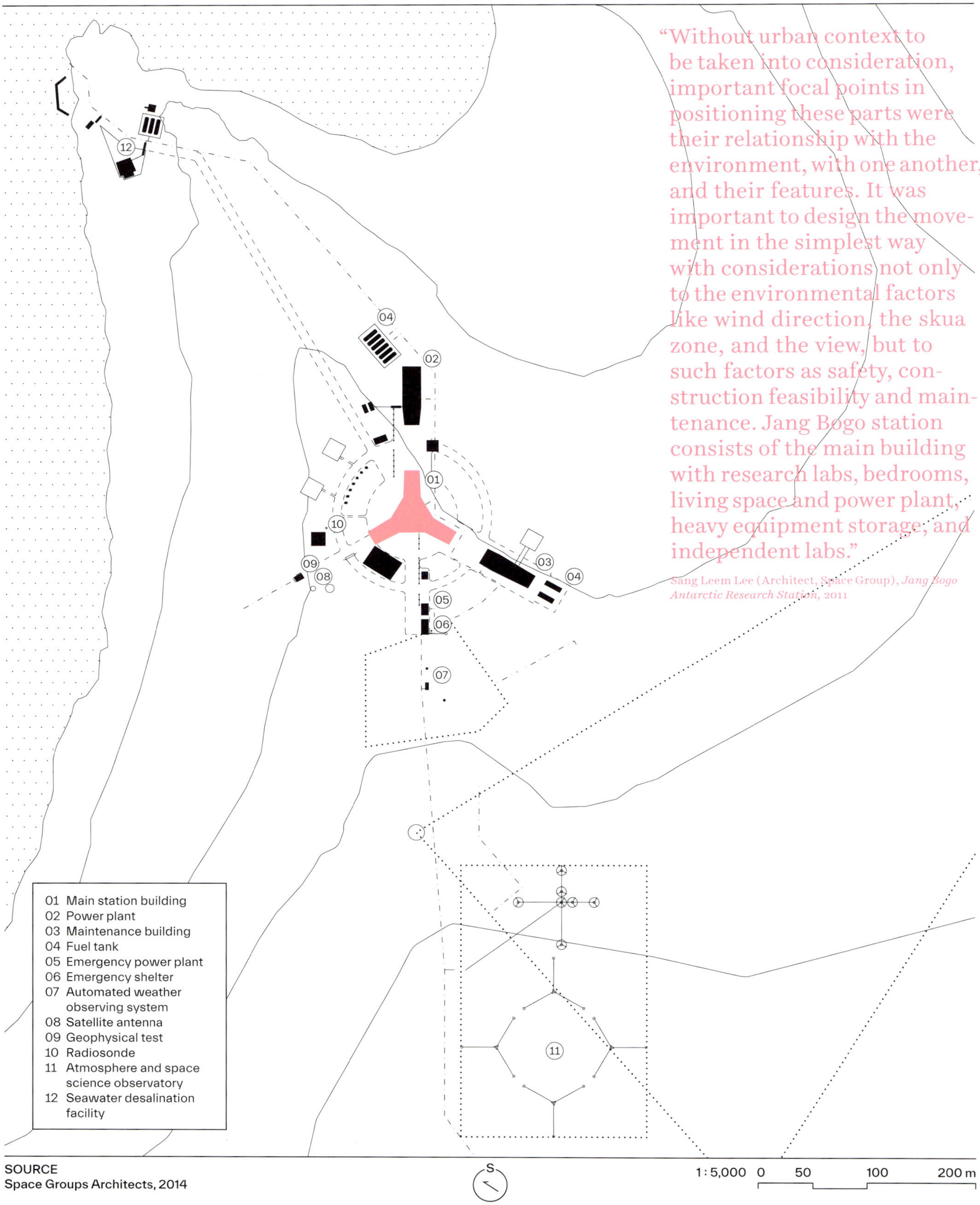
01
02
03
04
05
06
07
08
09
10
11
12
01 Main station building
02 Power plant
03 Maintenance building
04 Fuel tank
05 Emergency power plant
06 Emergency shelter
07 Automated weather observing system
08 Satellite antenna
09 Geophysical test
10 Radiosonde
11 Atmosphere and space science observatory
12 Seawater desalination facility
S
1:5,000 0 50 100 200 m

"Without urban context to be taken into consideration, important focal points in positioning these parts were their relationship with the environment, with one another, and their features. It was important to design the movement in the simplest way with considerations not only to the environmental factors like wind direction, the skua zone, and the view, but to such factors as safety, construction feasibility and maintenance. Jang Bogo station consists of the main building with research labs, bedrooms, living space and power plant, heavy equipment storage, and independent labs."

Sang Leem Lee (Architect, Space Group), *Jang Bogo Antarctic Research Station*, 2011

SOURCE
Space Groups Architects, 2014

PLAN, 1ST LEVEL

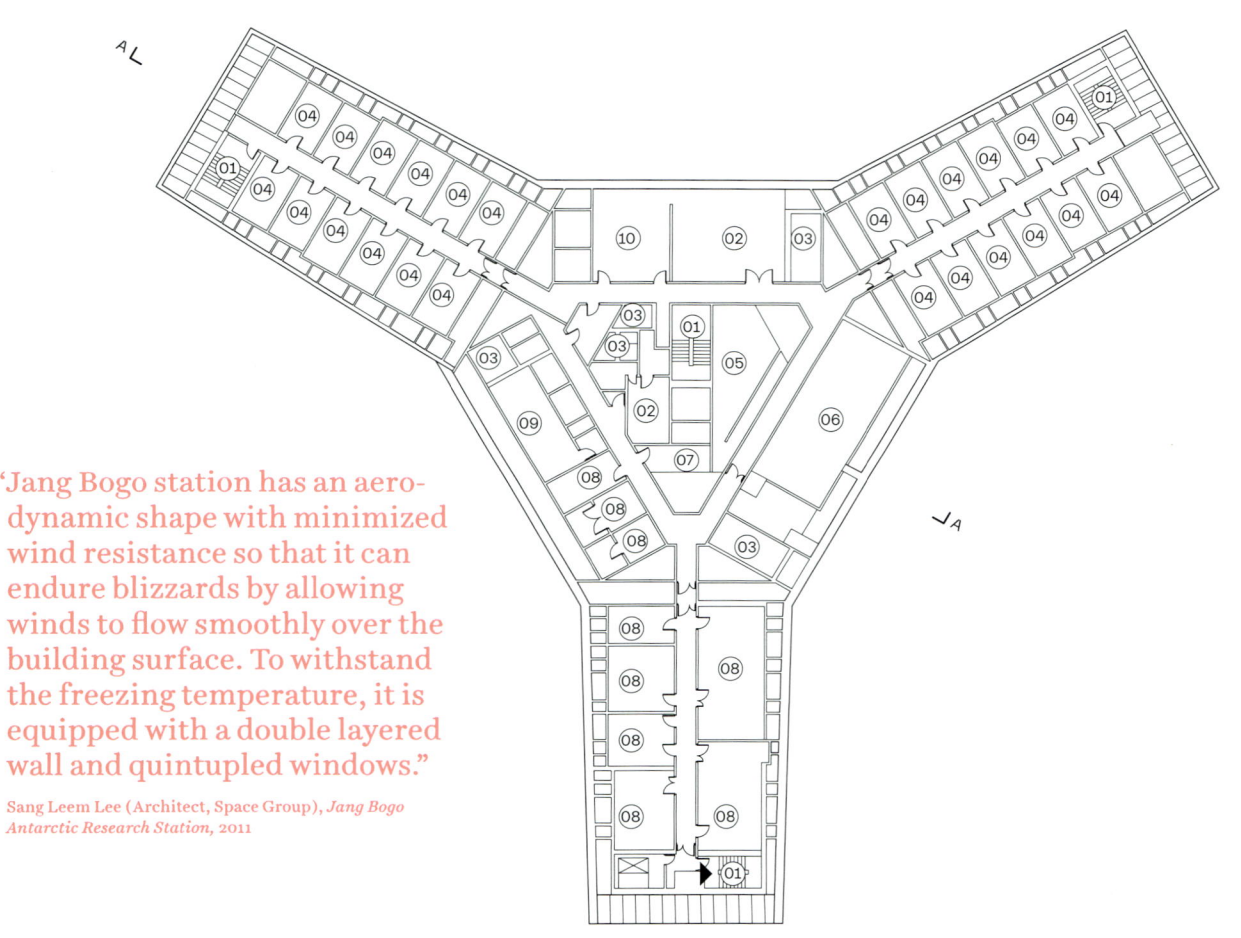

"Jang Bogo station has an aerodynamic shape with minimized
wind resistance so that it can
endure blizzards by allowing
winds to flow smoothly over the
building surface. To withstand
the freezing temperature, it is
equipped with a double layered
wall and quintupled windows."

Sang Leem Lee (Architect, Space Group), *Jang Bogo
Antarctic Research Station,* 2011

01 Stair core
02 Gym
03 Bathroom
04 Bedroom
05 Kitchen
06 Dining room
07 Laundry
08 Research laboratories
09 Medical room
10 Meeting room

SOURCE
Space Groups Architects, 2014

1 : 500   0   5   10   20 m

SECTION AND ELEVATION

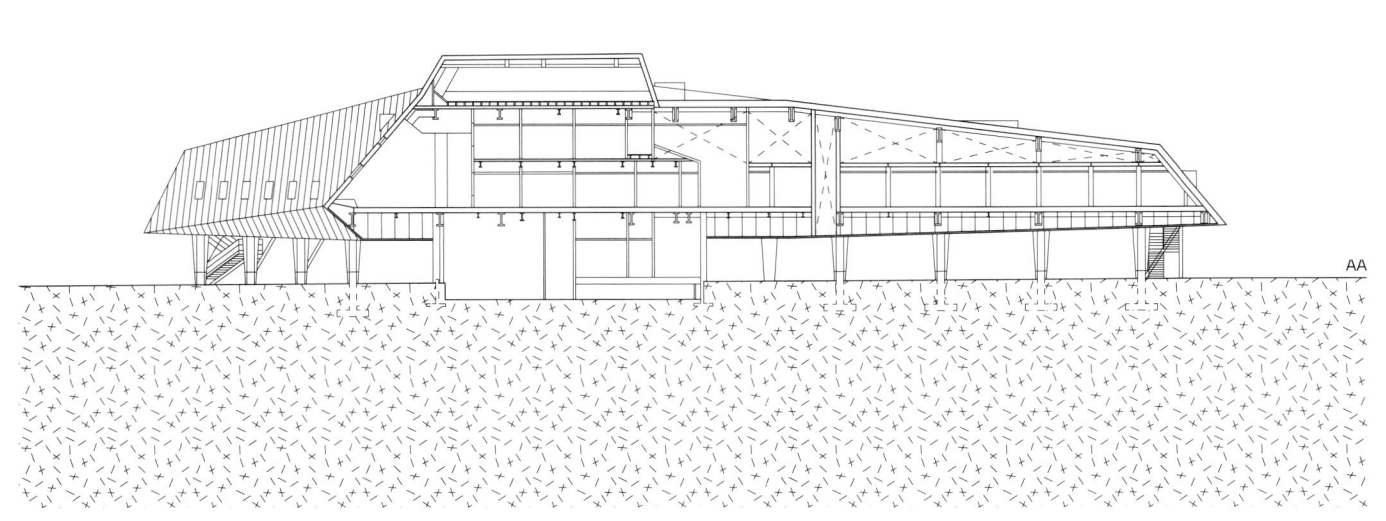

AA

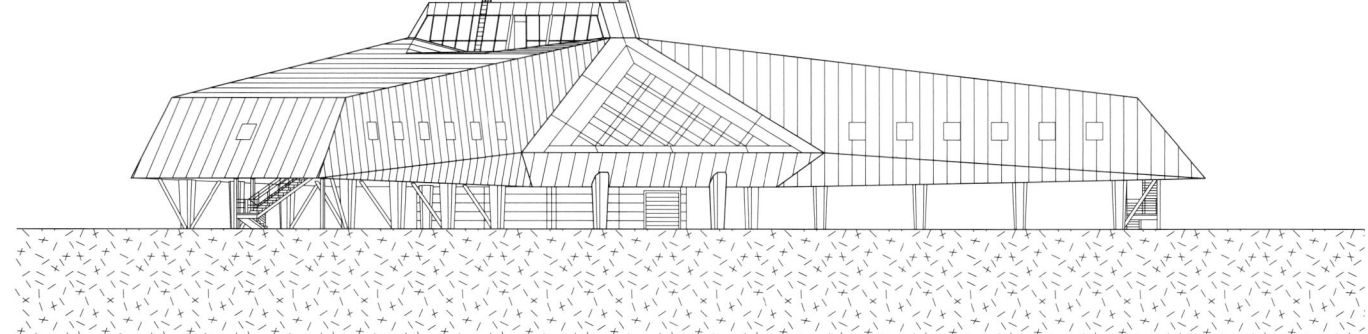

SOURCE
Space Groups Architects, 2014

1 : 500   0   5   10   20 m

2014

"What makes life more convenient at the station is built under
the ice sheet. The supporting
facilities – at first sight a long
pipeline corridor with doors that
open on both sides to equipment
rooms – were just completed on
Friday to light the 'red lantern'
in a sustainable manner."

Li Xia (Editor, *Xinhua News*), "Feature: China's
Taishan Station Offers Conveniences of Modern Life
Deep in Antarctic Ice Sheet", *Xinhua News,* 2019

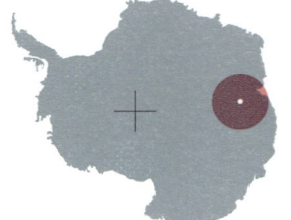

Architect, Engineer, Manufacturer: Wang Zhongjun;
Baosteel Group Corporation People's Republic of
China

People's Republic of China   Active Station   73°51'00"S 76°58'27"E
Polar Research Institute of China (PRIC)   Princess Elizabeth Land
                                                                          Ice sheet, 2,621 m.a.s.l., −30.3°C

SOURCE
Polar Research Institute of China

1:5,000   0   50   100   200 m

S

01 Main station building
02 Technical addition
03 Windmill
04 Storage area

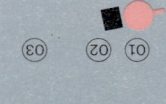 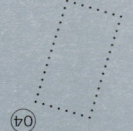

03   02   01

04

SITE PLAN

SOURCE
Polar Research Institute of China (reconstruction)

01 Vestibule
02 Bedroom
03 Bathroom
04 Lounge

1:250   0   2.5   5   10 m

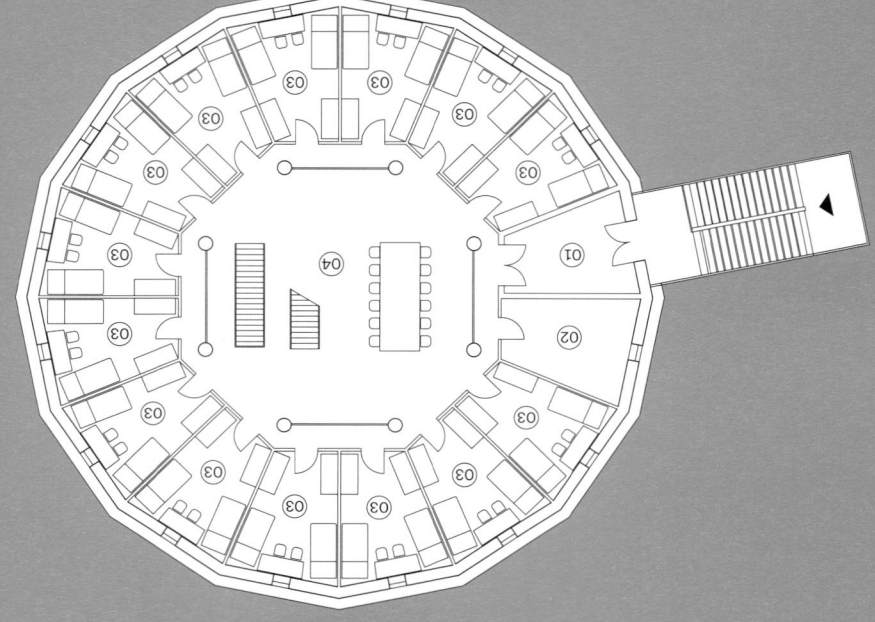

ELEVATION

1 : 250   0   2.5   5   10 m

UNION GLACIER
1998/2014

↖ p. 509, 511

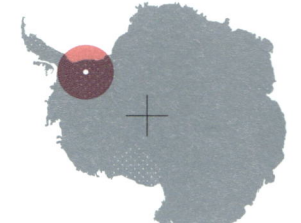

"In 2014 we developed a new housing unit for researchers 'cascopolar' [polar-helmet] and a prototype of its enclosing skin 'torsionoid' to receive the cold-dry sanitary system which freezes human residues and eliminates the use of water. The 'cascopolar' internally differentiates an area to sleep in half-light – counteracting the permanent daylight of the polar circle – and another area for work, naturally illuminated by two skylights.

The 'torsionoid' is a geodesic ellipsoidal structure composed of linear elements made of plywood and compressed by a PVC membrane with no insulation which resists low temperatures. Each panel in the membrane is a paraboloid of double curvature that coincides with the subdivision of the structural pattern, consolidating the geometry when it's tensed against the anchoring system. These points correspond to excavations in the snow which are filled with water that, once frozen, produces an efficient tension point with no impact on the environment. The structure has a ski on its base that distributes the loads to avoid sinking."

Marcelo Bernal, Paul Taylor, and Francisco Valdivia (Architects), *Ilaia: Estación Polar Científica Conjunta Glaciar Unión*, 2014

Architects: Marcelo Bernal, Chile; Paul Taylor, United Kingdom; Francisco Valdivia, Chile
Manufacturer: Aquitectura de Zonas Extremas, Chile
Consultants: (construction and plumbing) ARQZE, Chile; (structural design) Francisco Valdivia, Chile

---

Chile
Chilean Antarctic Institute (INACH)

Active Field Camp*

79°46'00"S 82°52'00"W
Ellsworth Mountains, Union Glacier
Ice sheet, 700 m.a.s.l., –19.2°C

SITE PLAN

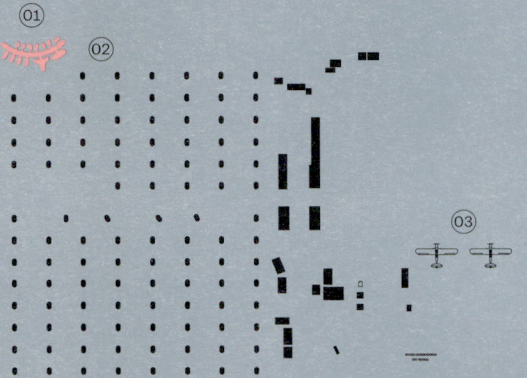

01 Union Glacier joint
   scientific polar station
02 Clam tents
03 Twin airplanes

SOURCE
Antarctic Logistics & Expeditions, 2014 (reconstruction)

1 : 5,000   0   50   100   200 m

PLAN | UNION GLACIER JOINT SCIENTIFIC POLAR STATION

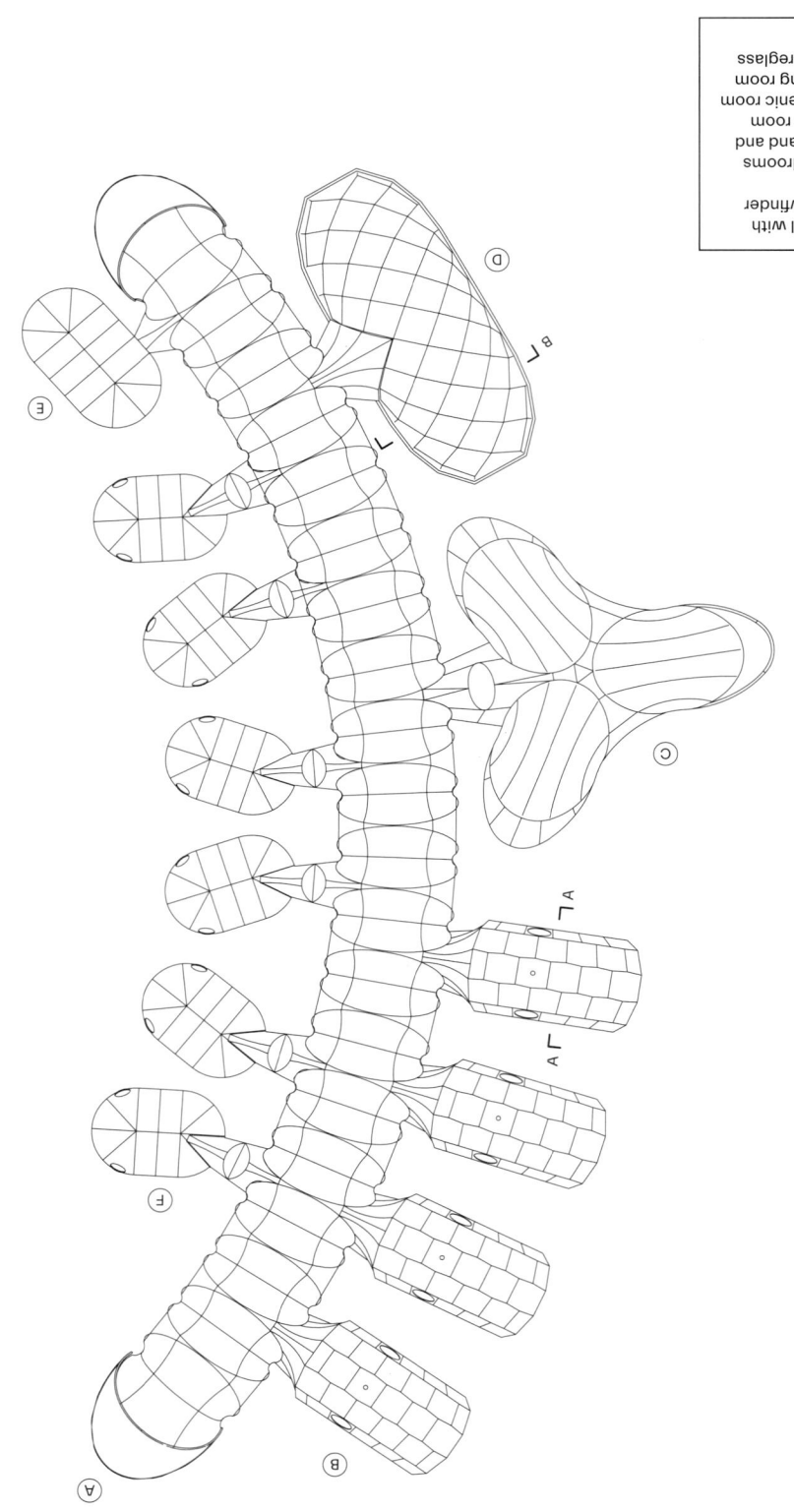

SOURCE
Ilaia: Estación Polar Científica Conjunta Glaciar Unión, 2014

1 : 250   0   2.5   5   10 m

A   Technical tunnel with
    retractable viewfinder
    access
B   Cascopolar bedrooms
C   Sastrugi command and
    communication room
D   Torsionoid hygienic room
E   Tricolor operating room
F   Bedrooms in fibreglass
    capsules

SECTION | CASCOPOLAR BEDROOM (B) AND TORSIONOID HYGIENIC ROOM (D)

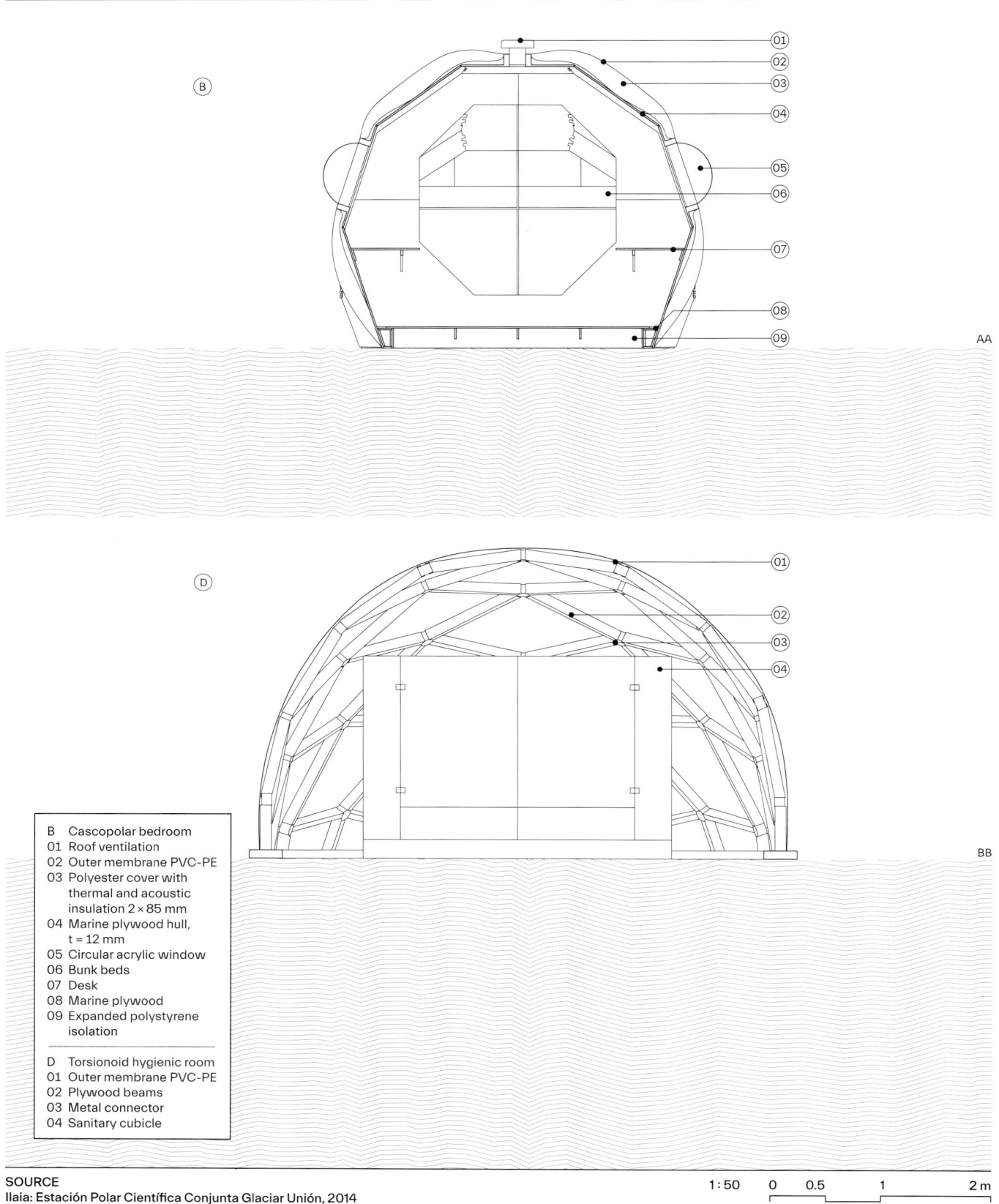

B   Cascopolar bedroom
01   Roof ventilation
02   Outer membrane PVC-PE
03   Polyester cover with thermal and acoustic insulation 2 × 85 mm
04   Marine plywood hull, t = 12 mm
05   Circular acrylic window
06   Bunk beds
07   Desk
08   Marine plywood
09   Expanded polystyrene isolation

D   Torsionoid hygienic room
01   Outer membrane PVC-PE
02   Plywood beams
03   Metal connector
04   Sanitary cubicle

SOURCE
Ilaia: Estación Polar Científica Conjunta Glaciar Unión, 2014

1:50   0   0.5   1   2 m

JUAN CARLOS I
1988/2018

"The new base comprises a habitat module, separate science module and a series of support modules for services and storage. The habitat building has three wings of accommodation arranged around a central core while the science building is a separate structure far enough away to provide a refuge in case of a major fire within the habitat. […] The orientation of the buildings makes best use of the site topography, with windows framing wonderful views of the surrounding land and seascapes."

Hugh Broughton (Architect, Hugh Broughton Architects), "Juan Carlos 1 Spanish Antarctic Base" (online)

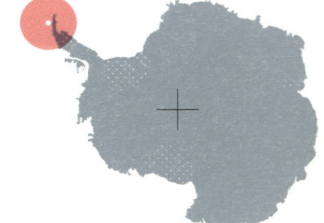

Architect: Hugh Broughton Architects, United Kingdom
Engineer: AECOM, United Kingdom

Spain
Spanish Polar Committee (CPE)

Active Station

62°39'48"S 60°23'17"
Hurd Peninsula, Livingston Island
Ice-free ground, 12 m.a.s.l., −1.2°C

SITE PLAN

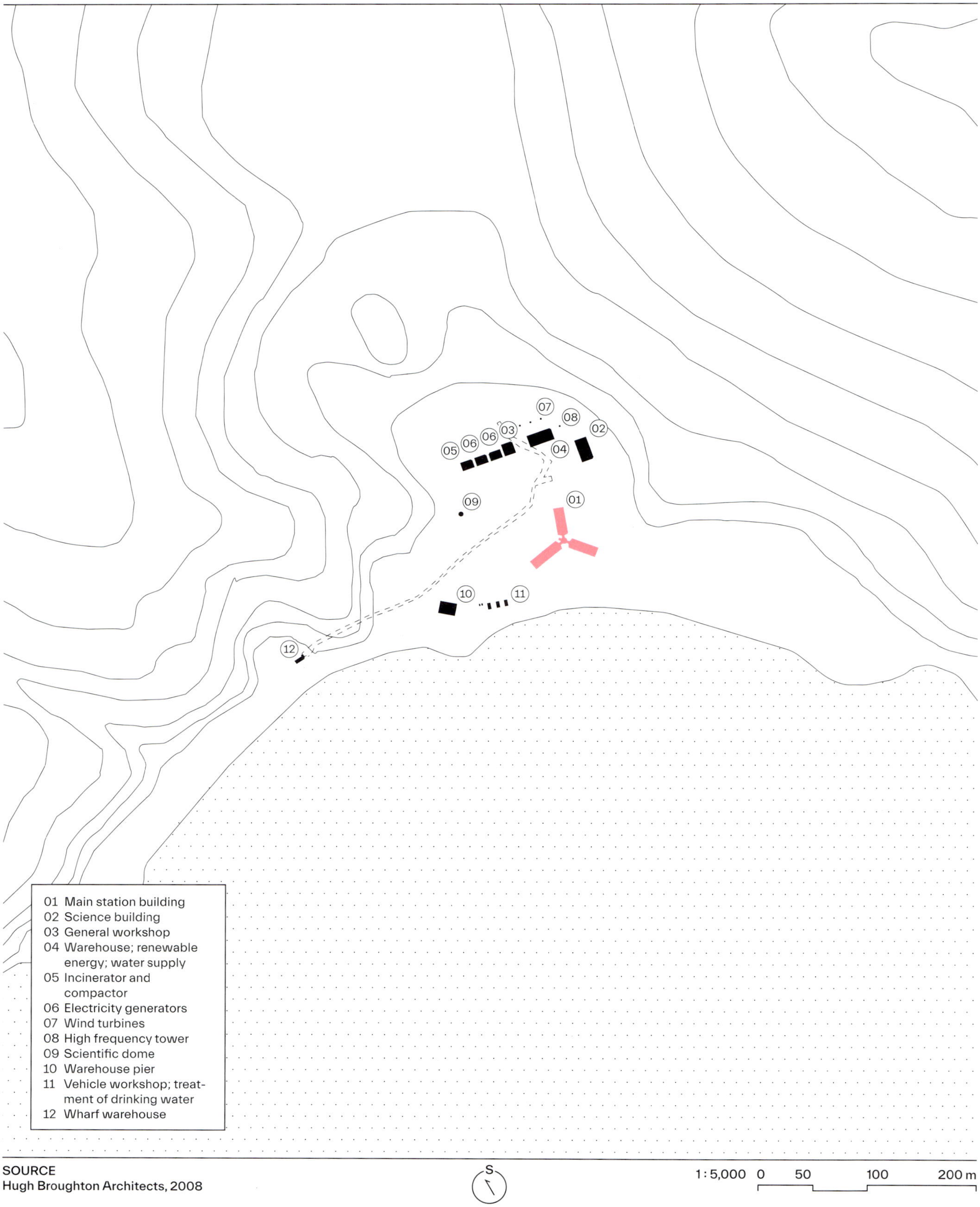

01 Main station building
02 Science building
03 General workshop
04 Warehouse; renewable
   energy; water supply
05 Incinerator and
   compactor
06 Electricity generators
07 Wind turbines
08 High frequency tower
09 Scientific dome
10 Warehouse pier
11 Vehicle workshop; treatment of drinking water
12 Wharf warehouse

SOURCE
Hugh Broughton Architects, 2008

S

1 : 5,000   0   50   100   200 m

PLAN, 1ST LEVEL

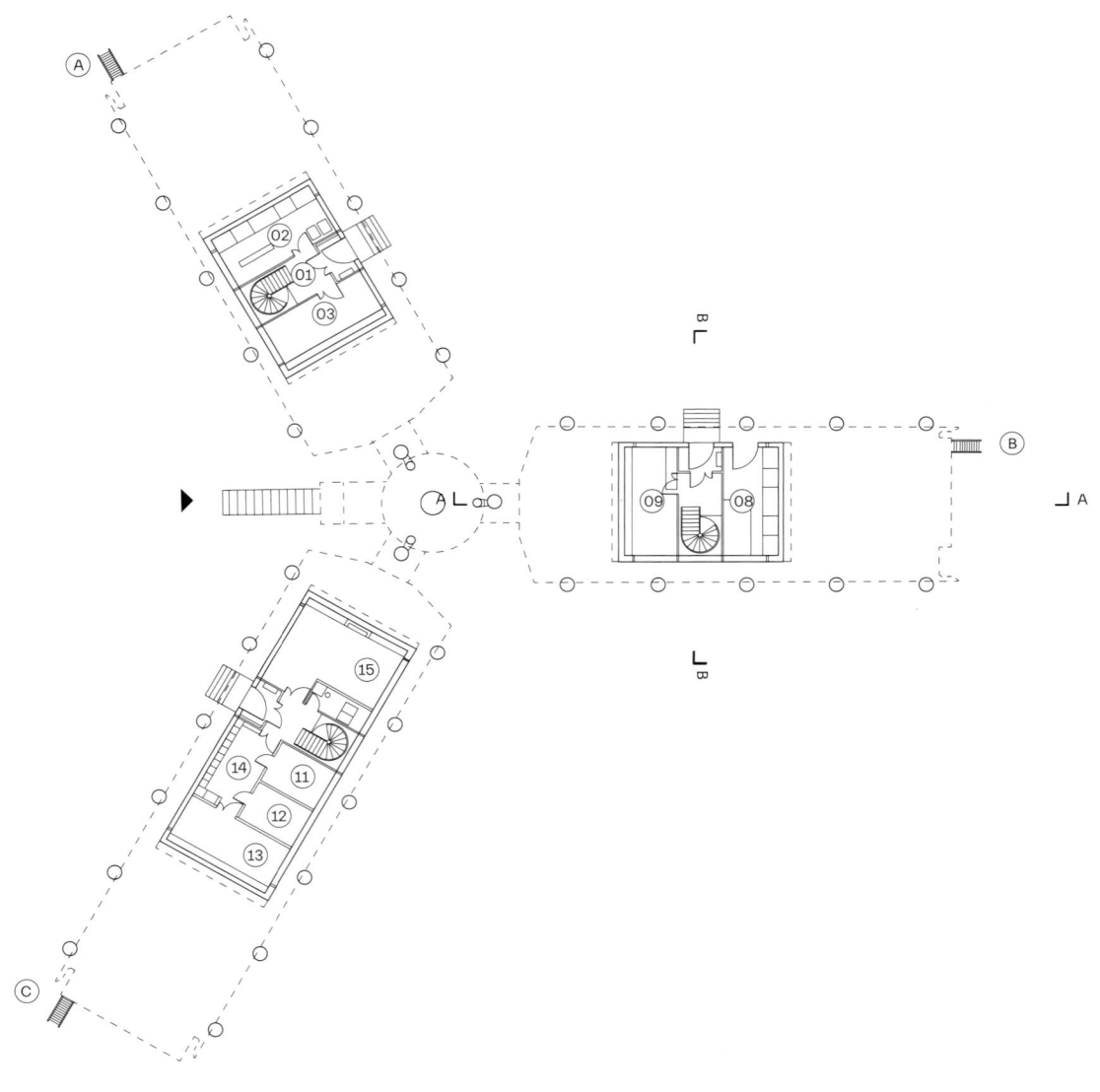

A Sleeping quarters	08 Plant room	15 Gym
01 Stair core	09 Water treatment	16 Kitchen
02 Laundry	10 Lounge	17 Lounge
03 Plant room		18 Washroom
04 Bedroom	C Social quarters	19 Pantry
05 Bathroom	11 Fridges	20 Smoking room
06 Balcony	12 Dry food	21 Medical room
	13 Plant room	22 Communications
B Sleeping quarters	14 Boot room	23 Base leader's office

SOURCE
Hugh Broughton Architects, 2008

1 : 500   0   5   10   20 m

PLAN, 2ND LEVEL

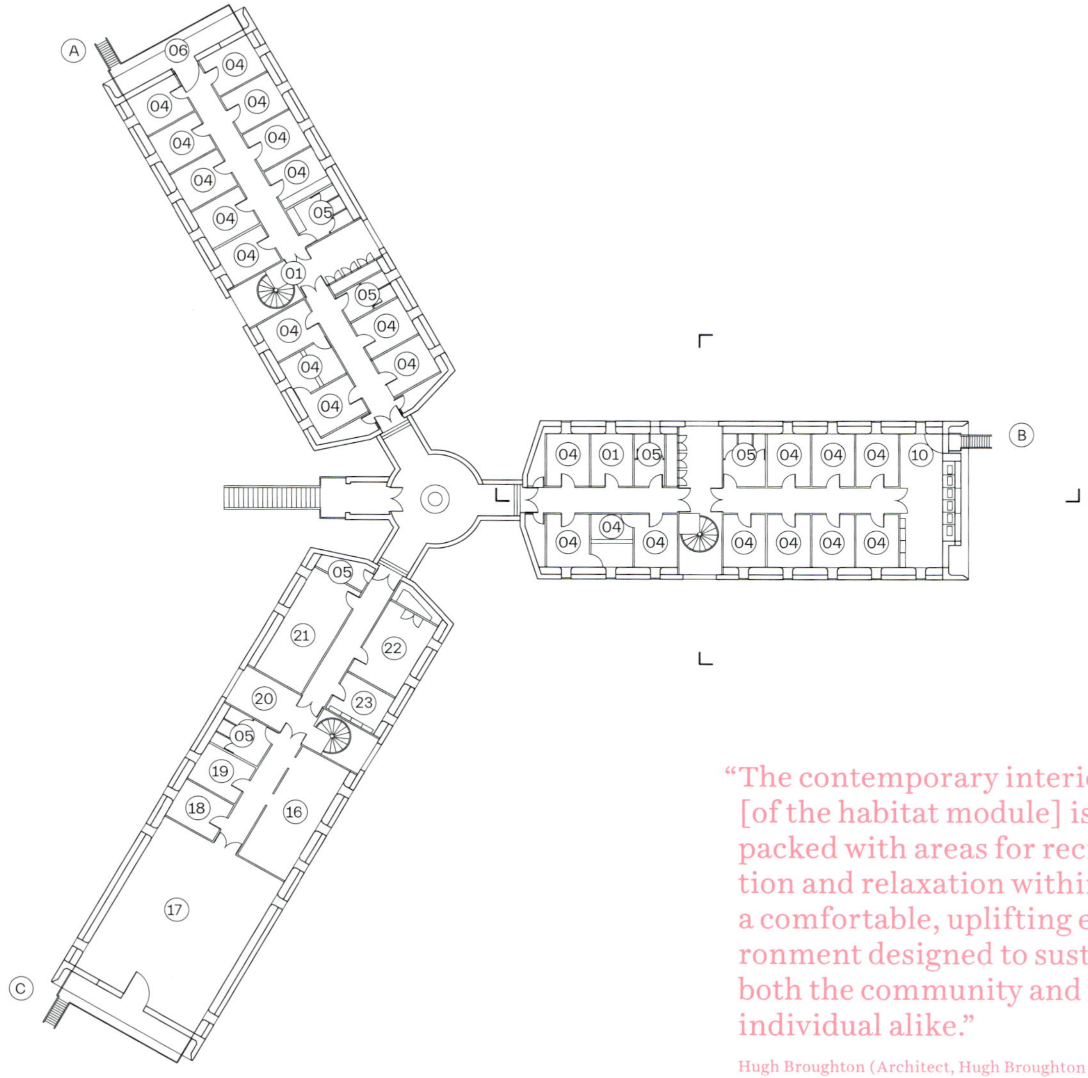

"The contemporary interior [of the habitat module] is packed with areas for recreation and relaxation within a comfortable, uplifting environment designed to sustain both the community and the individual alike."

Hugh Broughton (Architect, Hugh Broughton Architects), "Juan Carlos 1 Spanish Antarctic Base" (online)

SOURCE
Hugh Broughton Architects, 2008

1:500   0   5   10   20 m

SECTION AND ELEVATION | SLEEPING QUARTERS

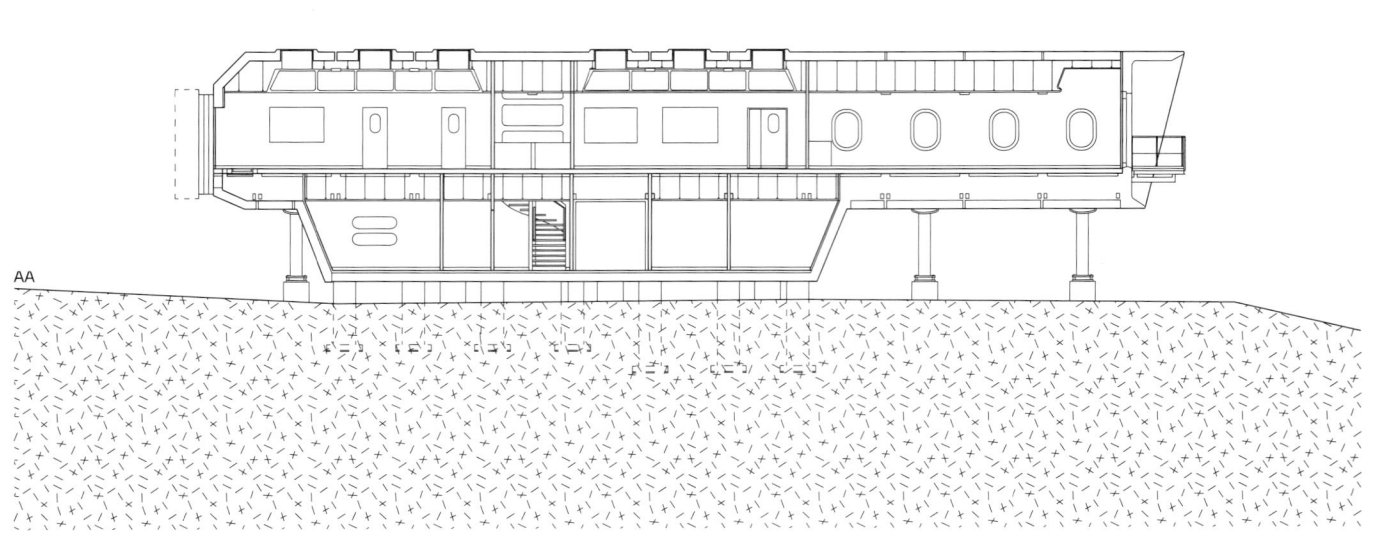

AA

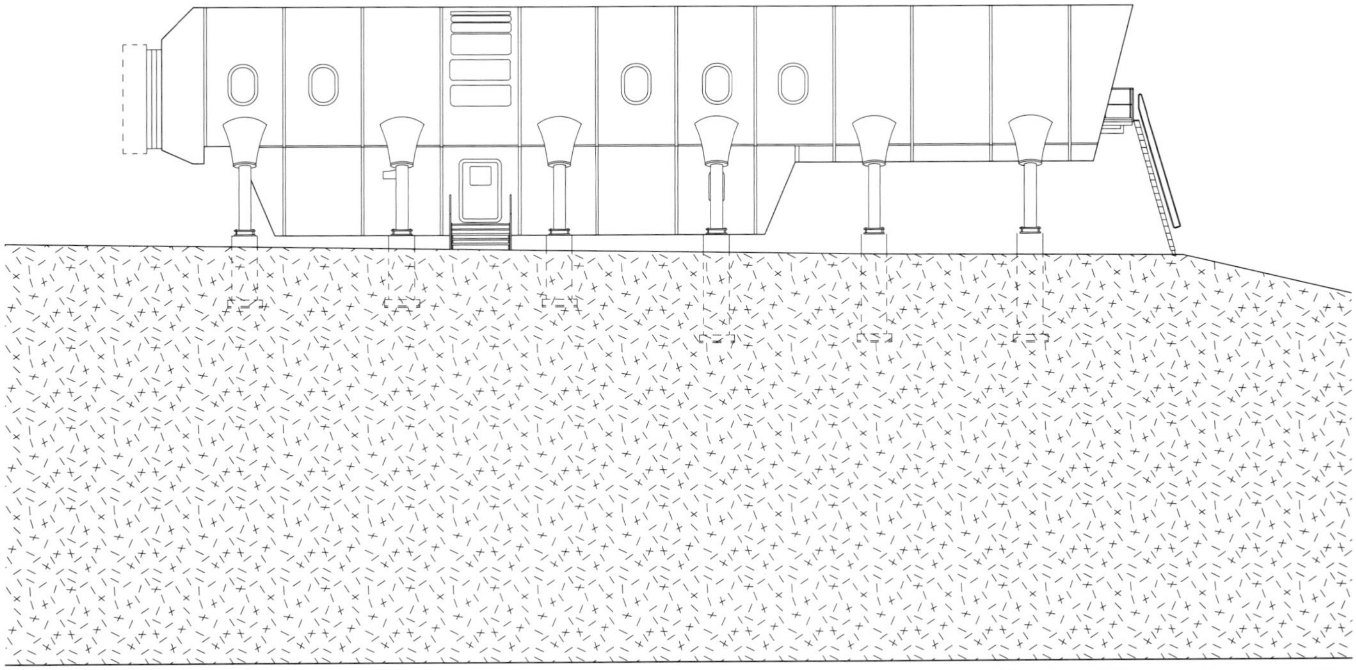

SECTION AND ELEVATION | SLEEPING QUARTERS

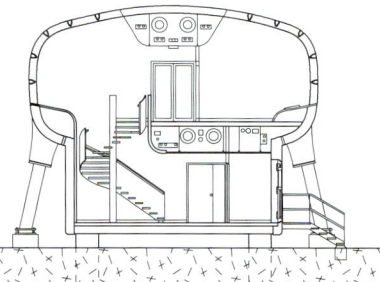

"The habitat and science buildings are clad with fibre reinforced plastic panels fixed to a steel frame, supported on legs, with ancillary space suspended below. [...] Walls are fabricated in cassette form to ease construction. [...] Roof-lights and glazed entrance areas maximise daylight, reducing energy consumption and allowing the crew to continually engage with their surroundings."

Hugh Broughton (Architect, Hugh Broughton Architects), "Juan Carlos 1 Spanish Antarctic Base" (online)

BB

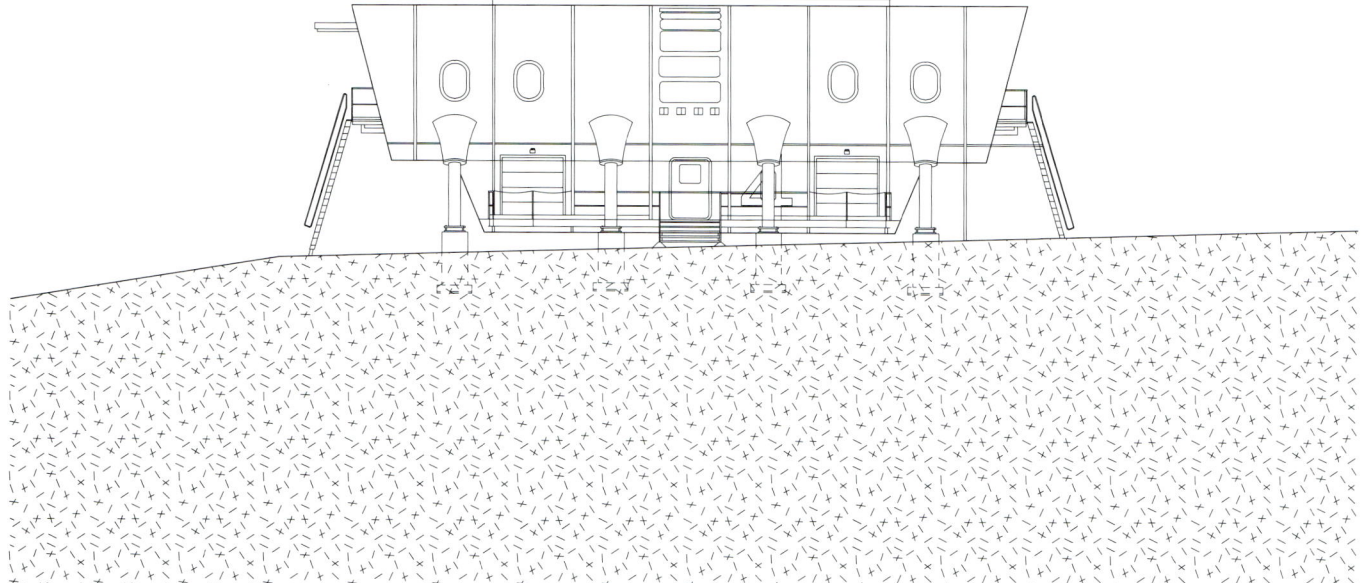

SOURCE
Hugh Broughton Architects, 2008

1:250   0   2.5   5   10 m

COMANDANTE FERRAZ
1984/2020

↖ p. 416

> "Located near the tip of the Antarctic Peninsula, the teal-hued 4,500-square-metre facility houses the naval service's Brazilian Antarctic Program, replacing its 1980s base that was ravaged by fire in 2012."

Lizzie Crook (Reporter), "Estúdio 41 Completes Prefabricated Antarctic Research Station for Brazil", *Dezeen*, 2020

Architect: Estúdio 41, Brazil
Engineers: (structural, geotechnical, hydraulic, mechanical, electrical, telecommunications, fire safety, solid waste) AFA CONSULT, Portugal
Consultants: (building envelope) Stephan Heinlein; (geotechnics) Pedro Huergo and Josiele Patias; (comfort and energy) Guido Petinelli (Pettinelli); (mechanical engineer) Eduardo Brofman, (facilities) Eduardo Ribeiro; (security and fire prevention) Carlos Garmatter; (structures) Ricardo Dias

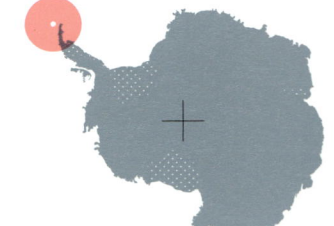

Brazil
Brazilian Antarctic Program (PROANTAR)

Active Station

62°05'07"S 58°23'29"W
King George Island, South Shetland Islands
Ice-free ground, 8 m.a.s.l., –2°C

SITE PLAN

"The proposal for the Ferraz Station is based on the interpretation of the land and local geographical conditions. The proposed site plan contemplates the Keller Peninsula's topography and local animal and plant life preservation requirements, among other factors. In order to minimize impacts on nature, several environmental zoning requirements have been met. [...] The site plan is complemented by photovoltaic panels to the north and [...] wind turbines to the southwest."

Estúdio 41, "Comandante Ferraz Antarctic Station", *ArchDaily Online*

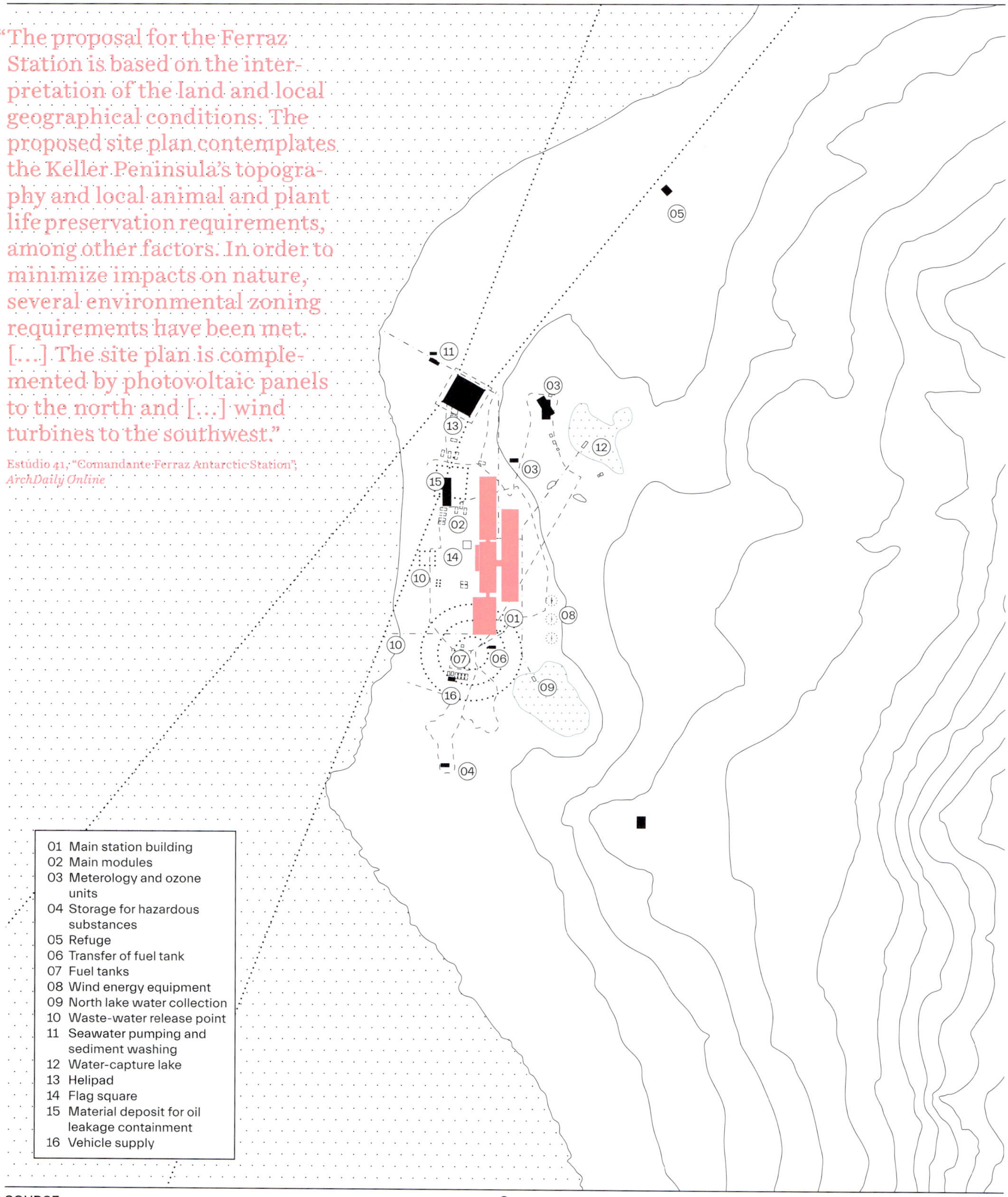

05
11
03
13
12
03
15
02
14
10
01
08
10
07
06
09
16
04
01 Main station building
02 Main modules
03 Meterology and ozone units
04 Storage for hazardous substances
05 Refuge
06 Transfer of fuel tank
07 Fuel tanks
08 Wind energy equipment
09 North lake water collection
10 Waste-water release point
11 Seawater pumping and sediment washing
12 Water-capture lake
13 Helipad
14 Flag square
15 Material deposit for oil leakage containment
16 Vehicle supply
S
1:5,000 0 50 100 200 m

SOURCE
Estúdio 41, 2014

PLAN, 1ST LEVEL

A
01 Central compartment
02 Pumping station
03 Technical and mechanical room
04 Firefighting tank
05 Drinking and saltwater reservoir
06 Bedroom (en suite)
07 Library
08 Conference room
09 TV Lounge
10 Documentation centre
11 Gym
12 COPA
13 Communications
14 Computer room
15 Breakdown control

B
16 Solid waste storage
17 Workshop
18 Batteries room
19 Technical and mechanical room
20 Cold room
21 Electric boiler
22 Mezzanine
23 Fire control room
24 Mechanical room

C
25 Bathroom
26 Laundry
27 Maintenance room
28 Medical room
29 Cloakroom
30 Storage
31 Cold room
32 Kitchen
33 Coffee room
34 Technical and mechanical room
35 Dining room
36 Deck
37 Access room
38 Ski and mountaineering equipment
39 Operations control centre
40 Electrical room
41 Breakdown control

D
42 Molecular biology laboratory
43 Biosciences laboratory
44 Technical room
45 Communal space
46 Research deposit
47 Sample deposit
48 Super freezers
49 Ultra freezers
50 Access room
51 Screening room
52 Biological assays
53 Chemistry laboratory
54 Plankton studies
55 Microbiology laboratory

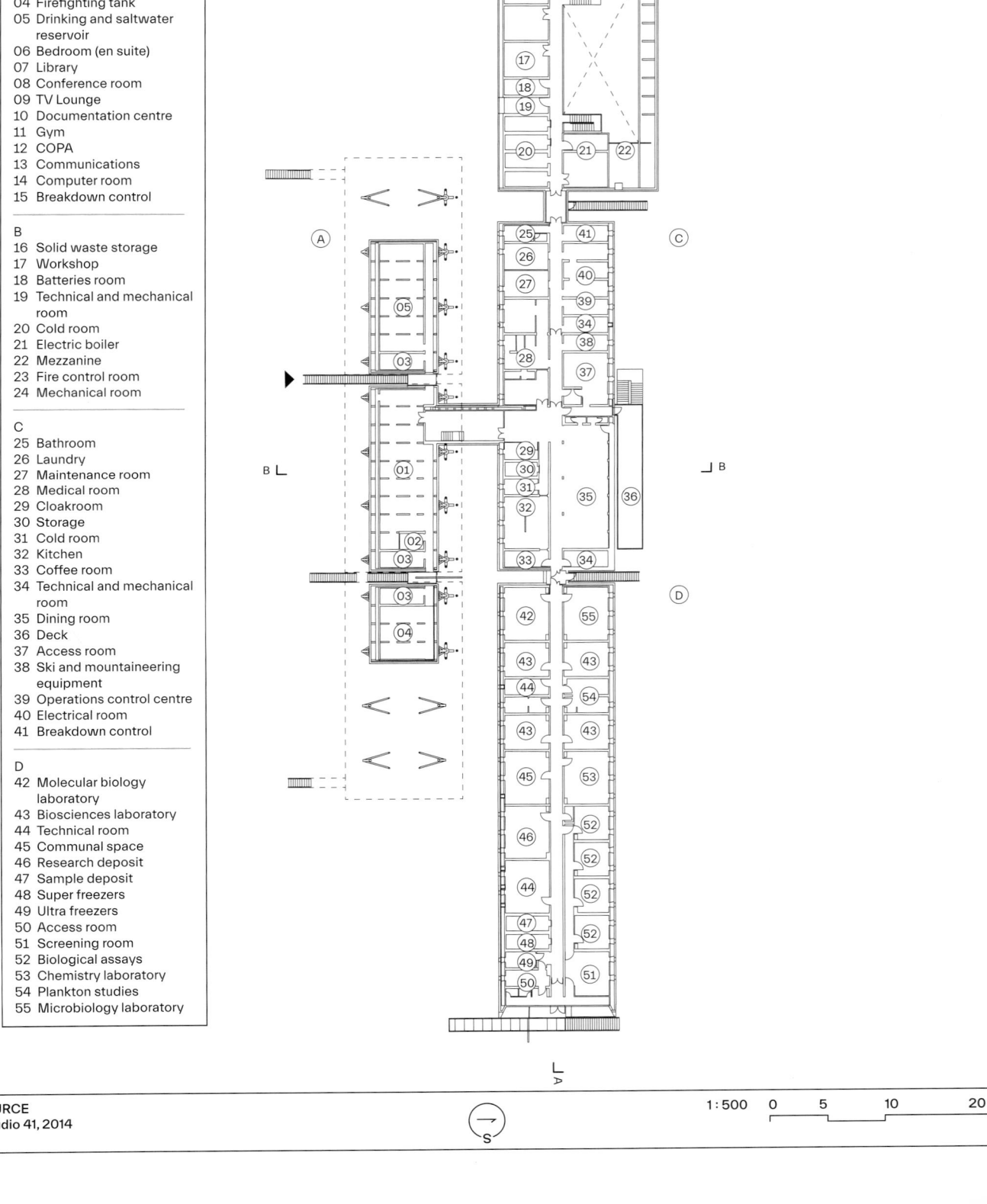

SOURCE
Estúdio 41, 2014

1 : 500   0   5   10   20 m

PLAN, 2ND LEVEL

"The operational sectors are organized in blocks in which different purposes are distributed. The
cabins, laundry areas, and dining/
living rooms are located in the
upper block, at level +9.10. The
lower block, at level +5.95, consists of the laboratories and the
operational and maintenance
areas. This same block houses
the garages and the central storeroom, located on level +2.50.
A transversal block, also at level
+5.95, combines the living and
meeting areas. The video-room/
auditorium, internet café, meet-
ing/conference room, library,
and living room are located in
this section."

Estúdio 41, "Commandante Ferraz Antarctic Station",
ArchDaily Online

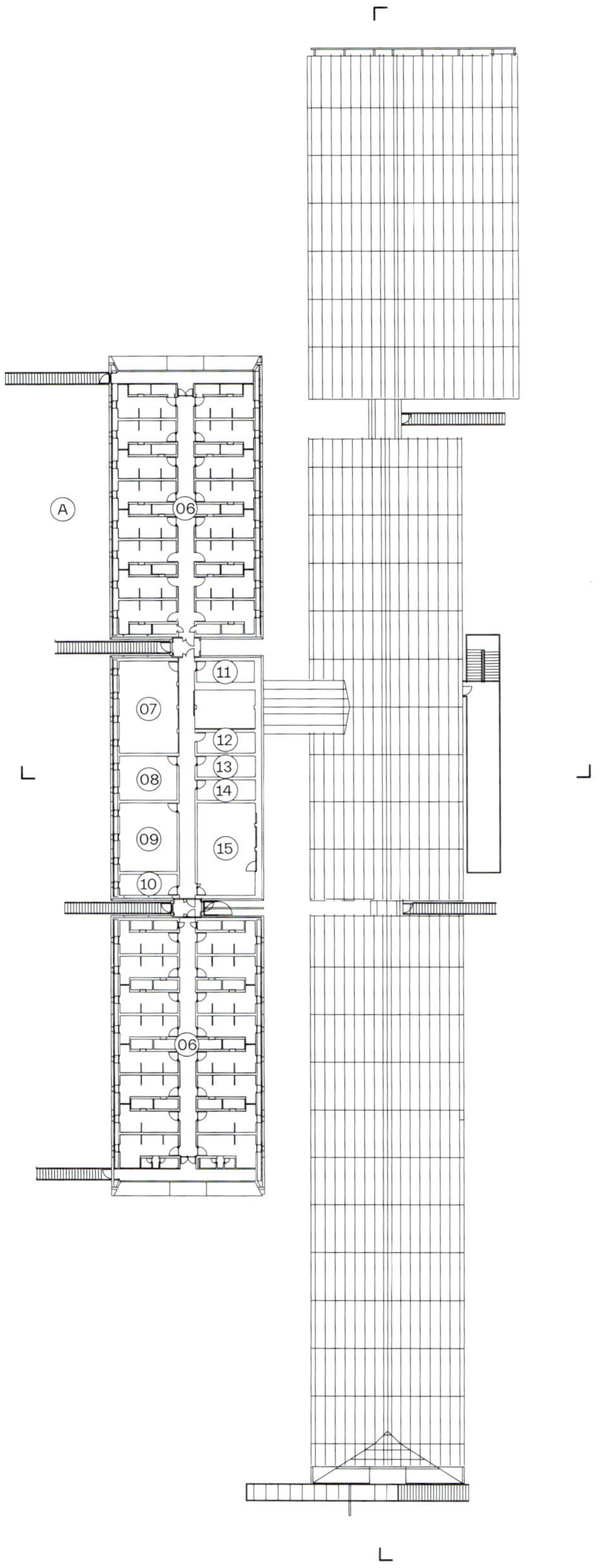

1 : 500   0   5   10   20 m

SECTION AND ELEVATION

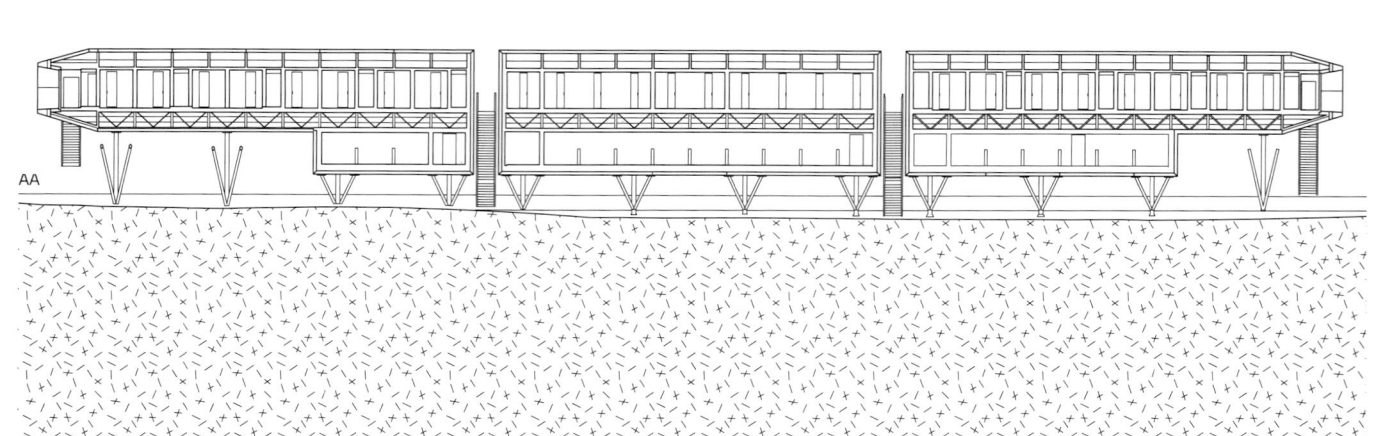

AA

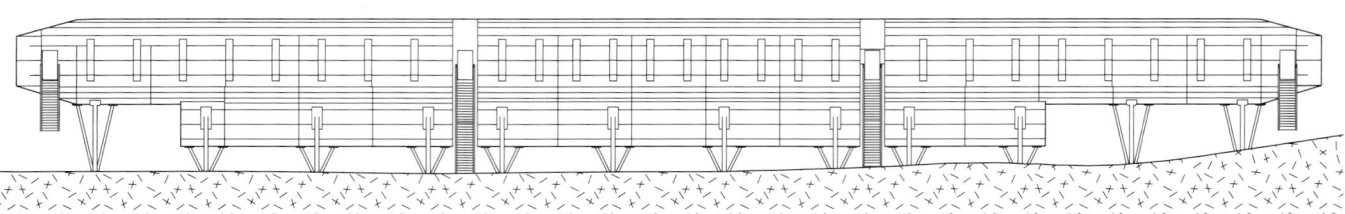

1 : 500   0   5   10   20 m

SECTION AND ELEVATION

"The challenges of designing a building for this landscape was creating a shelter, a safe place [...]. In certain parts of the planet, nature sometimes creates harsh conditions for the human body [...]. To design a building in these places is almost like building a garment, an artefact that protects and comforts. It is an issue of technological performance, but it must go hand-in-hand with aesthetics."

Eron Costin (Studio Architect) featured in "Estúdio 41 Completes Prefabricated Antarctic Research Station for Brazil", *Dezeen*, 2020

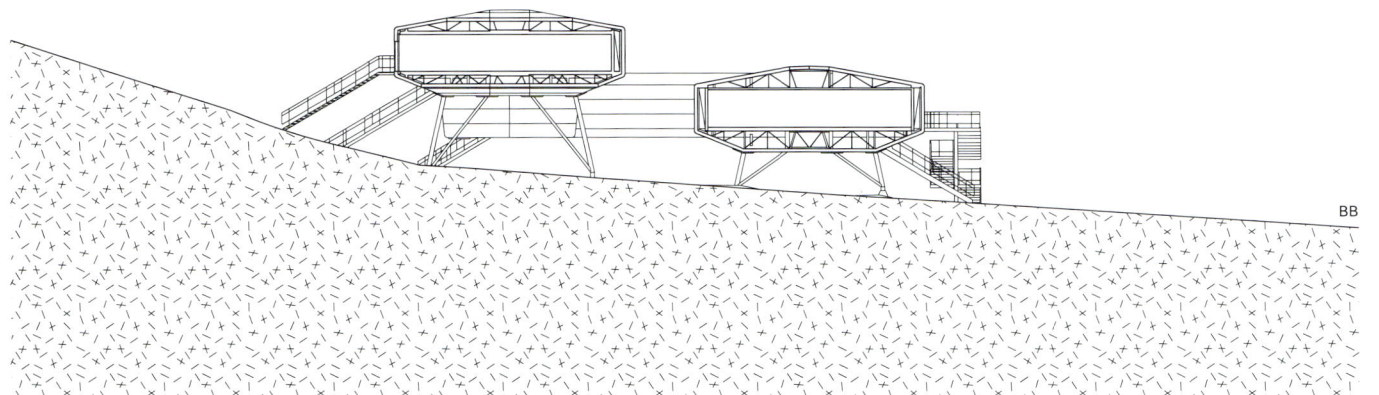

BB

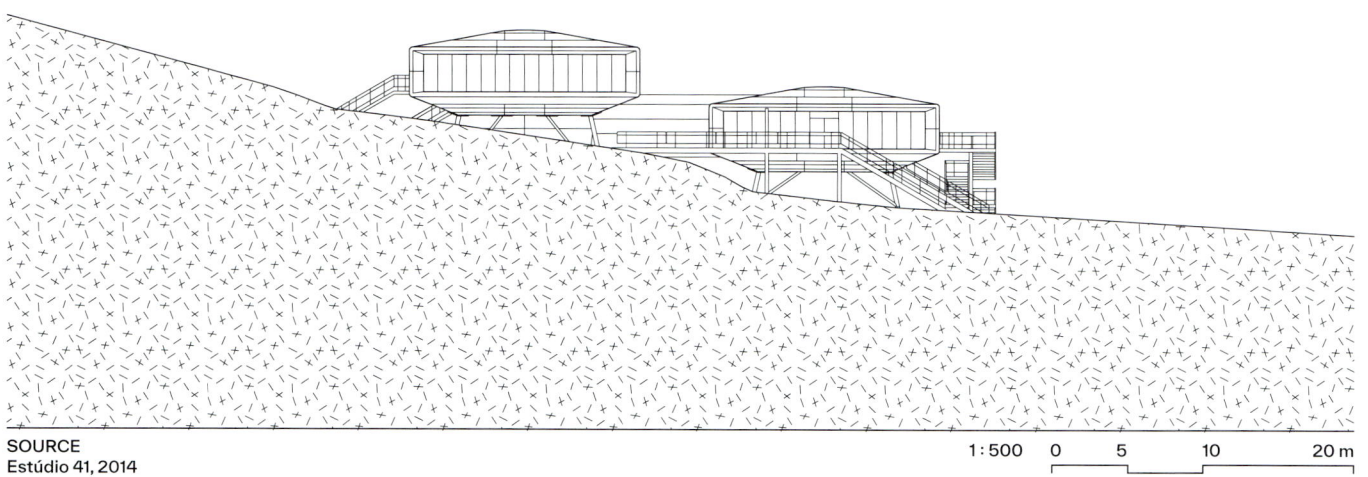

SOURCE
Estúdio 41, 2014

1 : 500   0   5   10   20 m

ST. KLIMENT OHRIDSKI
1988/2020+

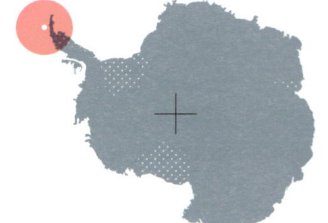

Architect: PS Architects, Bulgaria
Engineer: Jivko Ivanov, Bulgaria

Bulgaria
Bulgarian Antarctic Institute

Active Station > Project 2020+

62°38'27"S 60°21'55"W
Livingston Island, South Shetland Islands
Ice-free ground, 15 m.a.s.l., –2.8°C

SITE PLAN

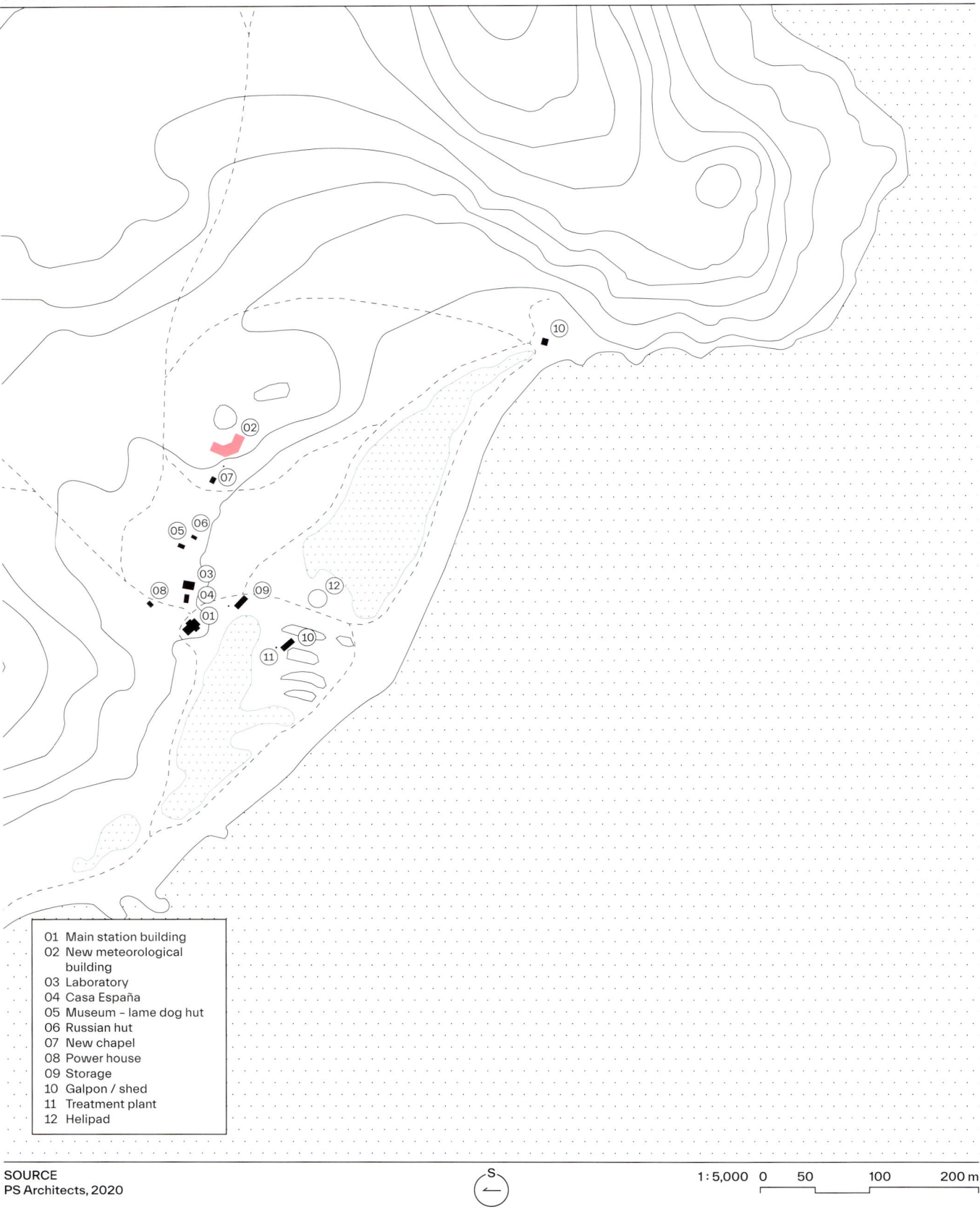

01 Main station building
02 New meteorological
    building
03 Laboratory
04 Casa España
05 Museum – lame dog hut
06 Russian hut
07 New chapel
08 Power house
09 Storage
10 Galpon / shed
11 Treatment plant
12 Helipad

SOURCE
PS Architects, 2020

S

1 : 5,000   0   50   100   200 m

PLAN | NEW MAIN STATION BUILDING

"The new base will be a single-storey building, reminiscent of the pitched-roof houses whose architecture is typical of more remote locations and colder climates. [The ambition was] to replicate the silhouette of the existing station building, but also to extend and modernize it. The materials used are a metal structure for the load-bearing part of the building and thermal panels in order to insulate the interior and reduce the loss of heat to the maximum."

Penka Stancheva (Architect, PS Architects), "Architect Penka Stancheva Designs New Laboratory for Researchers at Bulgarian Antarctic Base", *Radio Bulgaria Online*, 2019

Main building
01 Vestibule
02 Drying room
03 Bathroom
04 Bedroom
05 Storage
06 Lounge and dining room
07 Kitchen
08 Laboratory
09 Technical room
10 Electrical room

Annexes
11 Water tank
12 Diesel generator

SOURCE
PS Architects, 2020

1 : 250   0   2.5   5   10 m

SECTION AND ELEVATION | NEW MAIN STATION BUILDING

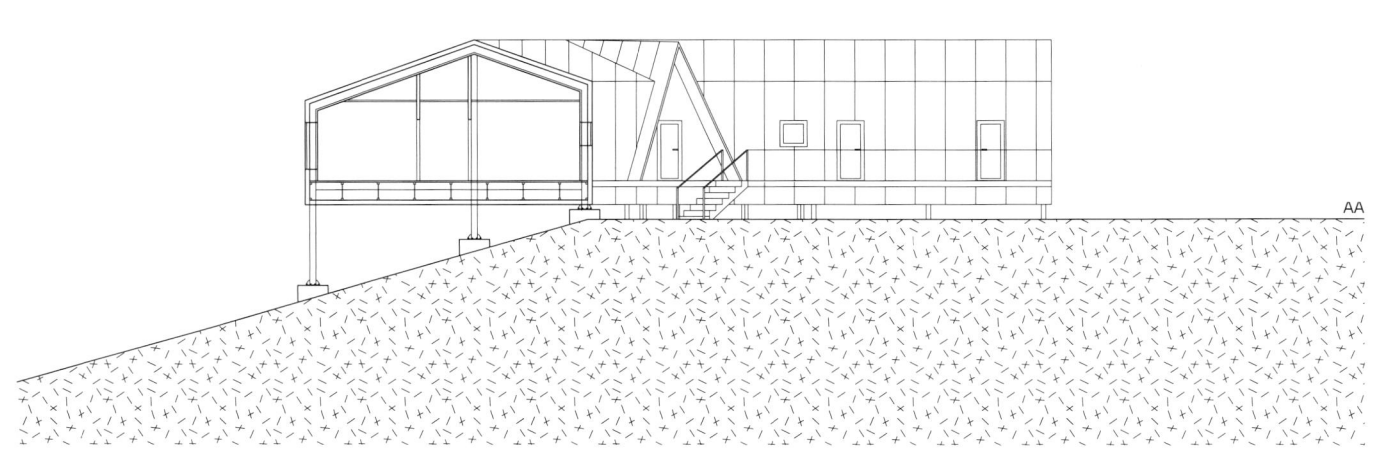

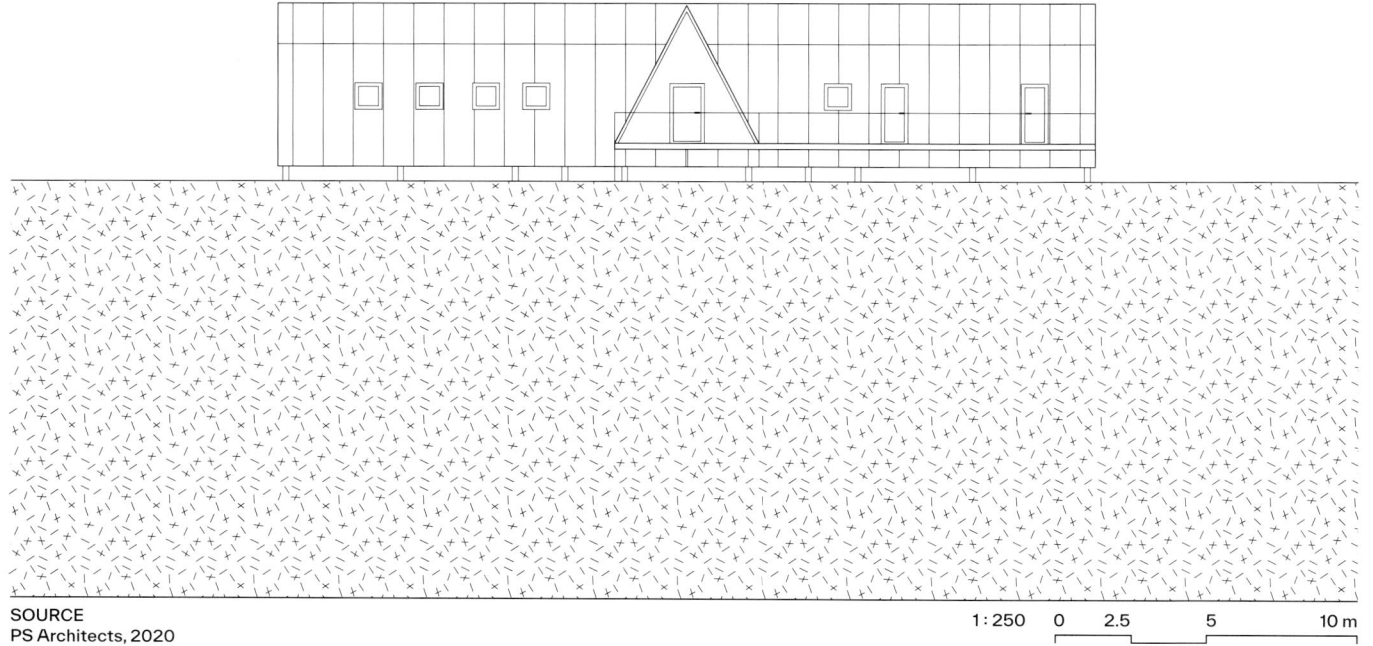

1 : 250   0   2.5   5   10 m

Poland
Polish Academy of Sciences

Active Station > Project 2020+

62°09'35"S 58°28'24"W
King George Island, South Shetland Islands
Ice-free ground, 2 m.a.s.l., −1.6°C

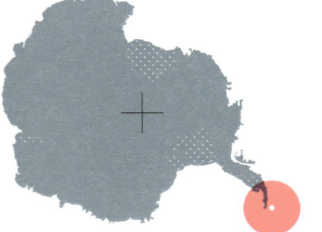

Architect: Kuryłowicz & Associates, Poland
Engineer: (structural and building services) Buro Happold, United Kingdom; (acoustics) ewKAkustika, Poland

"The design of the station is shaped by a detailed functional program, the extreme environmental conditions of the site and by a modular construction strategy which has been imposed by site access restrictions [...]. The result is a highly efficient tripartite ensemble that provides scientists with research space and a 'home away from home' atmosphere,

while the sinuous exterior seeks to capture the mystery of arctic life."

Kuryłowicz & Associates in "Kuryłowicz & Associates Reveals Design for Golden Antarctic Research Centre," *Dezeen*, 2019

HENRYK ARCTOWSKI
1977/2020+

SITE PLAN

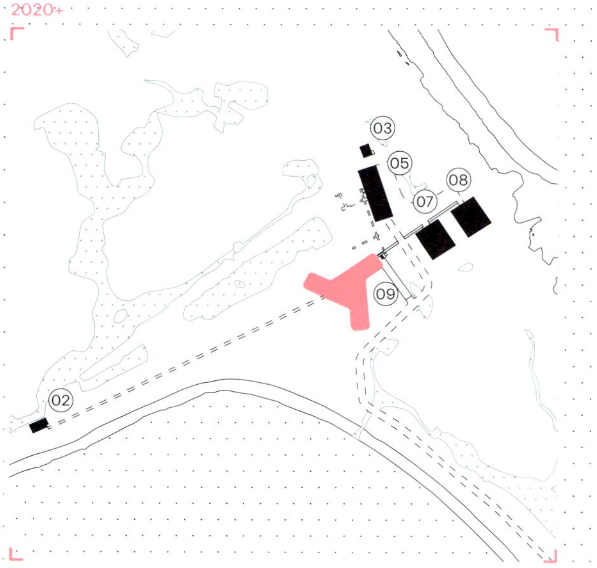

1977

2020+

01 Main station building
02 Meteorological
   observatory
03 Laboratory
04 Cold room
05 Aggregator,
   communication centre
06 Lighthouse, tourist
   information centre
07 Storage
08 Garage

09 New main station building

SOURCE
Kuryłowicz & Associates, 1977/2020

1 : 5,000   0   50   100   200 m

PLAN, 1ST LEVEL | NEW MAIN STATION BUILDING

"Making a minimal impact on the environment, the station is elevated 3 m over the landscape. […] This is done using a steel lattice sub structure which allows for the free flow of water, wind and snow underneath the station. The entrance has been situated in a central location, where it is protected from wind and high levels of snowfall. The building itself is orientated to reduce the impact of wind blowing from three main directions [and it's] sectional profile acts as an upside-down 'aircraft wing.'"

Kurylowicz & Associates, press release, 2019

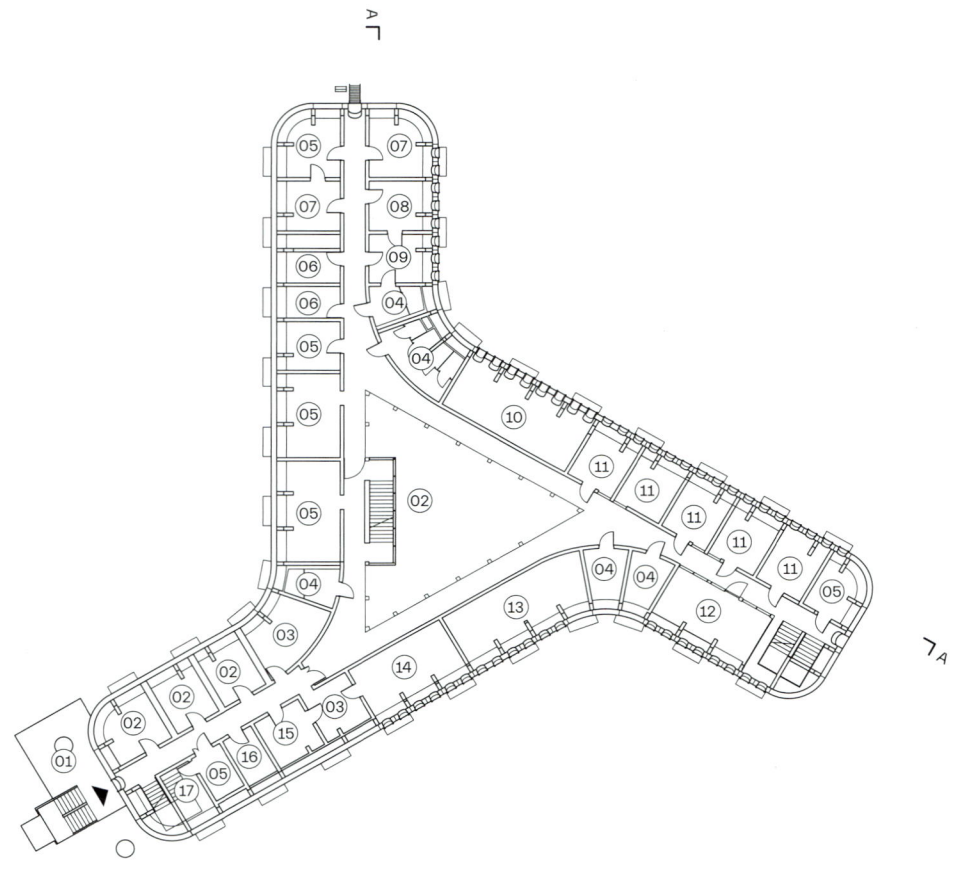

01	Entrance	10	Mess
02	Cold room	11	Laboratories
03	Refrigerator	12	Library
04	Toilets	13	Dining room
05	Storage	14	Kitchen
06	Technical room	15	Dumpster
07	Office	16	Goods lift
08	Surgery	17	Delivery
09	Isolation ward		

SOURCE
Kuryłowicz & Associates, 2020

1 : 500   0   5   10   20 m

SECTION AND ELEVATION | NEW MAIN STATION BUILDING

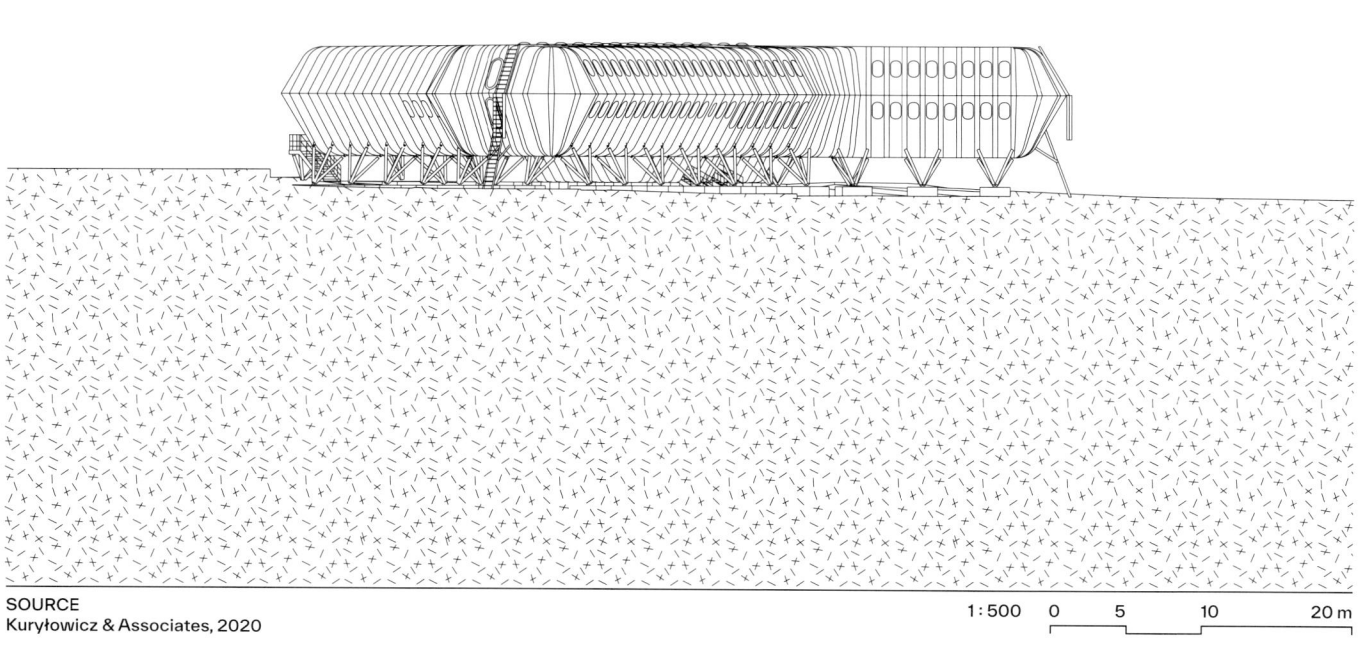

AA

SOURCE
Kuryłowicz & Associates, 2020

1 : 500   0   5   10   20 m

SCOTT BASE
1957/**2020+**

↖ p. 154, 366, 422, 435, 456, 476, 536, 553

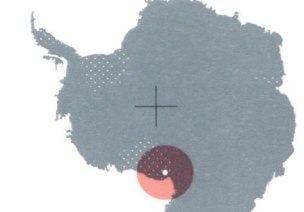

"The design proposes that the existing 12-building ice station, built in the early 1980s, is replaced by three large interconnected buildings (one for accommodation, dining and welfare, one for science and management, and one for engineering and storage) and a separate helicopter hangar. The new base could house 100 people."

Kate Youde, "Antarctic Return: Hugh Broughton Unveils Designs for New Zealand's Ice Base", *Architects Journal*, 2019

Architect: Hugh Broughton Architects (principal architect), United Kingdom; Jasmax (partner architect), New Zealand
Engineer: (structural and civil engineering) WSP, United Kingdom; (building services) Steensen Varming, Denmark
Consultants: (quantity surveyors) Rawlinsons, Australia; The Building Intelligence Group (project managers), New Zealand

New Zealand
Antarctica New Zealand

Preserved, HSM 75 + Active Station >
Project 2020+

77°50'53.5"S 166°40'06.3"E
Pram Point, Ross Island
Ice-free ground, 10 m.a.s.l., −19.8°C

SITE PLAN

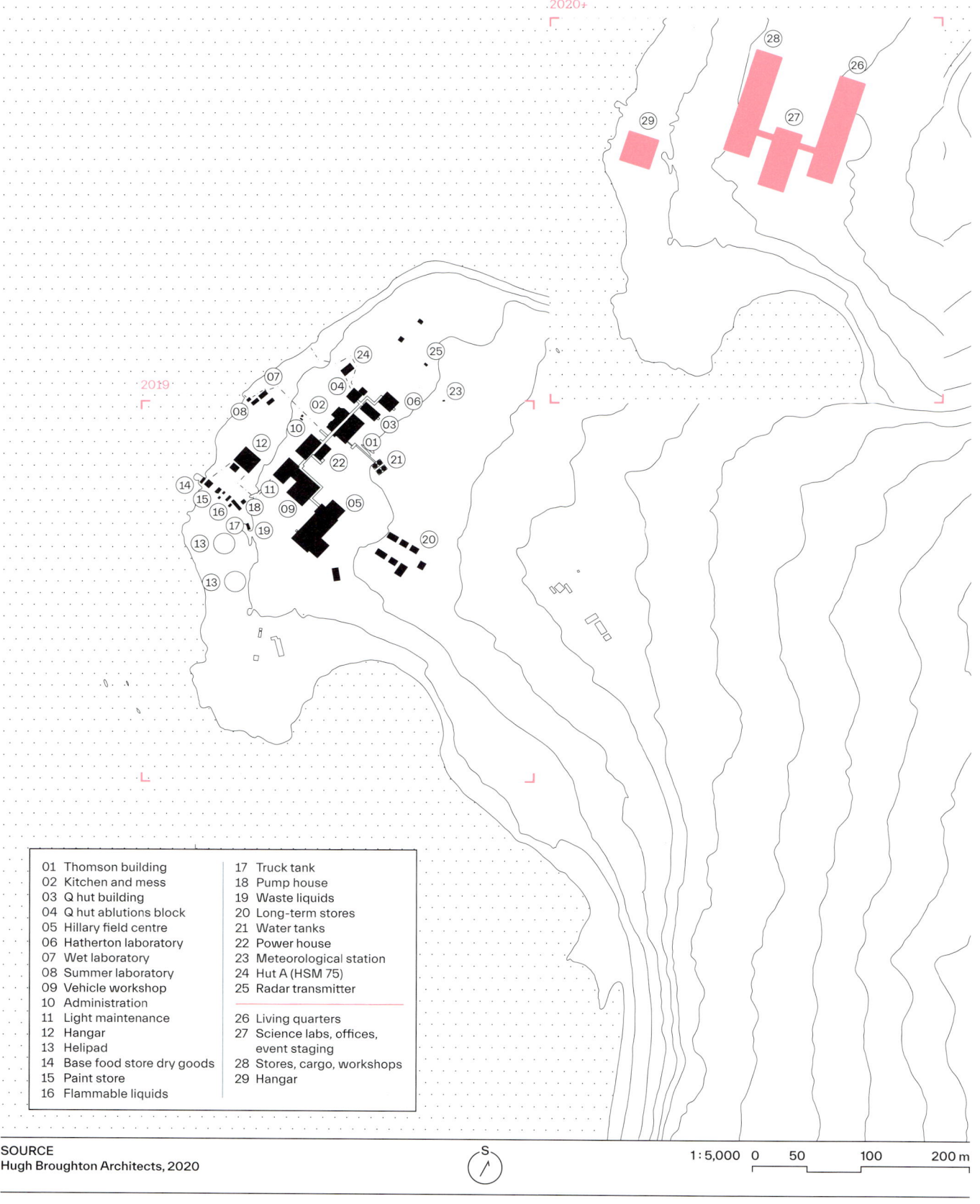

01	Thomson building	17	Truck tank
02	Kitchen and mess	18	Pump house
03	Q hut building	19	Waste liquids
04	Q hut ablutions block	20	Long-term stores
05	Hillary field centre	21	Water tanks
06	Hatherton laboratory	22	Power house
07	Wet laboratory	23	Meteorological station
08	Summer laboratory	24	Hut A (HSM 75)
09	Vehicle workshop	25	Radar transmitter
10	Administration		
11	Light maintenance	26	Living quarters
12	Hangar	27	Science labs, offices,
13	Helipad		event staging
14	Base food store dry goods	28	Stores, cargo, workshops
15	Paint store	29	Hangar
16	Flammable liquids		

SOURCE
Hugh Broughton Architects, 2020

1 : 5,000   0   50   100   200 m

SECTION | NEW MAIN STATION BUILDING

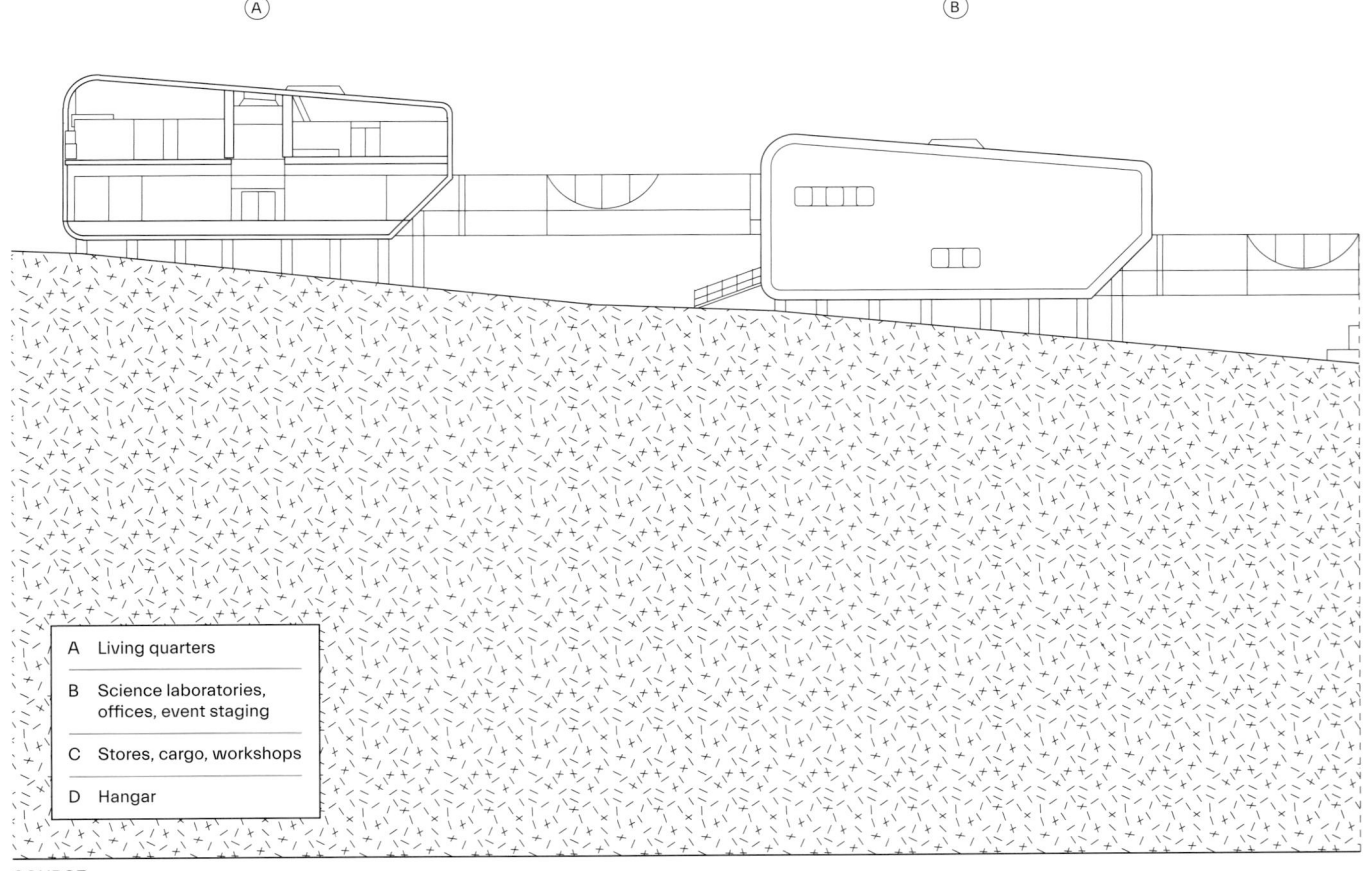

A Living quarters

B Science laboratories, offices, event staging

C Stores, cargo, workshops

D Hangar

SOURCE
Hugh Broughton Architects, 2020

"The design solution comprises three inter-connected aerodynamically shaped two-storey buildings, which step down the hillside of Pram Point. The three buildings are offset from each other to minimise risk of snowdrift between, and are connected with enclosed links. All the buildings are elevated above the ground to encourage wind to flow under, thereby minimising snow accumulation."

Hugh Broughton (Architect, Hugh Broughton Architects), "Scott Base Redevelopment" (online)

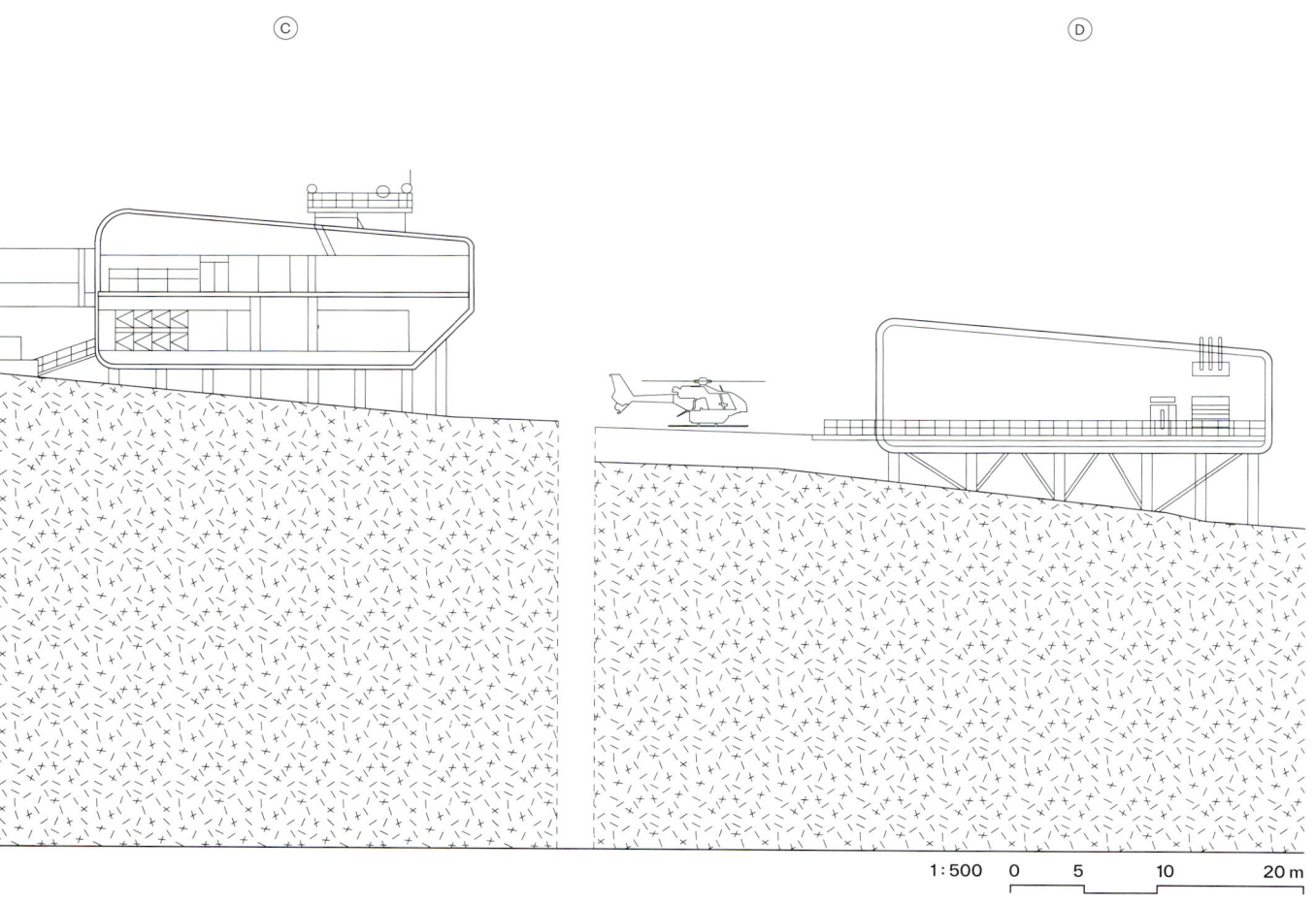

1 : 500   0   5   10   20 m

ROTHERA

1975/2020+

"The two-storey 4,500 sqm building will contain preparation areas for field expeditions, a central store, medical facility, offices, recreational spaces, workshops and areas for plant."

British Antarctic Survey, *Rothera Modernisation – Phase 1 Initial Environmental Evaluation: Report,* 2019

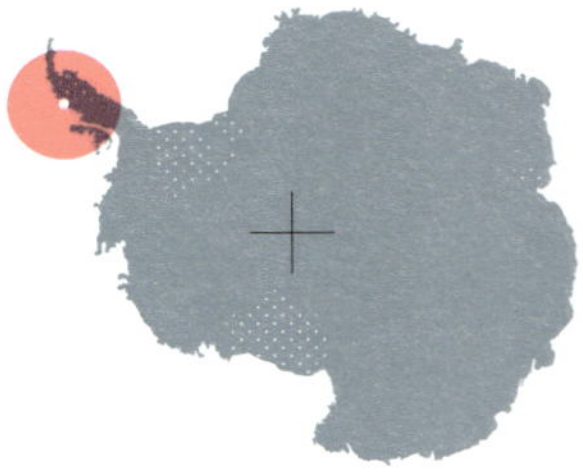

Architect: (Discovery building) Hugh Broughton Architects, United Kingdom
Engineer: Sweco, Sweden
Consultants: (cladding) Billings Design Associates, Ireland; NORR architects; Ramboll; (technical) Turner & Townsend, United Kingdom
Manufacturers: BAM Construction, United Kingdom

United Kingdom
British Antarctic Survey (BAS)

Active Station > Project 2020+

67°34'00"S 68°07'59"W
Rothera Point, Adelaide Island
Ice-free ground, 16 m.a.s.l., –3.7°C

SITE PLAN

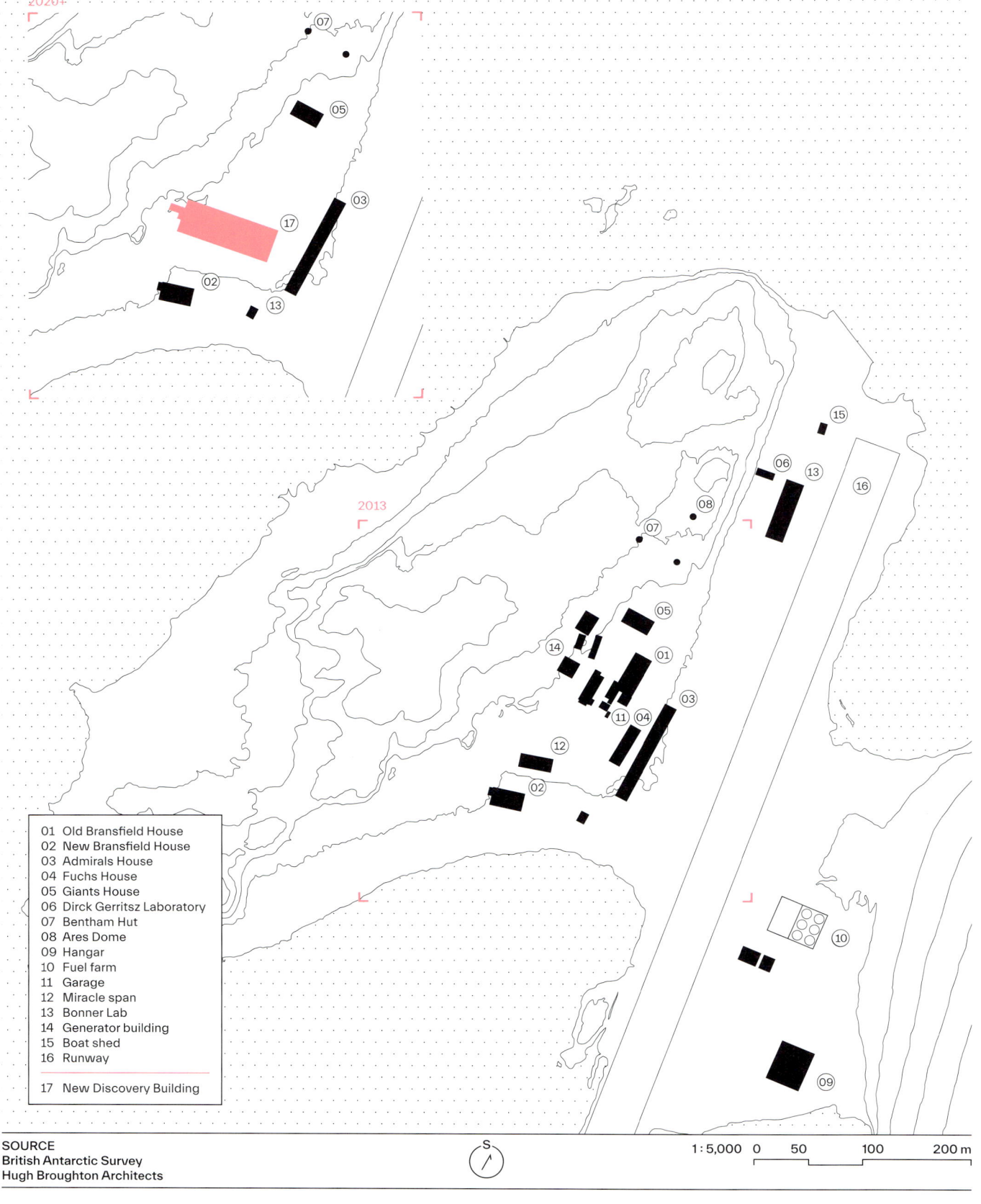

2020+

2013

01 Old Bransfield House
02 New Bransfield House
03 Admirals House
04 Fuchs House
05 Giants House
06 Dirck Gerritsz Laboratory
07 Bentham Hut
08 Ares Dome
09 Hangar
10 Fuel farm
11 Garage
12 Miracle span
13 Bonner Lab
14 Generator building
15 Boat shed
16 Runway

17 New Discovery Building

SOURCE
British Antarctic Survey
Hugh Broughton Architects

S

1 : 5,000   0   50   100   200 m

PLAN, 1ST AND 2ND LEVEL | NEW DISCOVERY BUILDING

2nd level

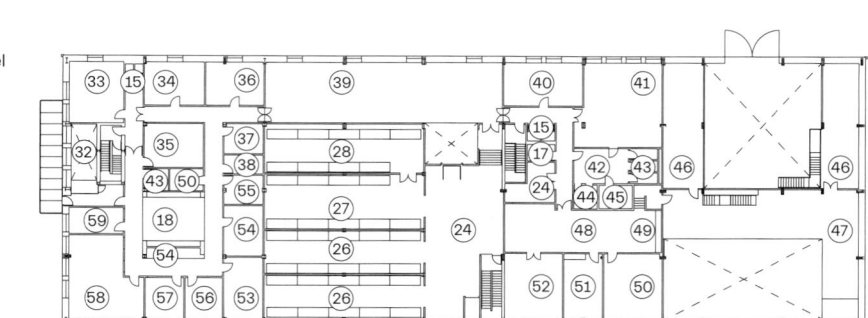

1st level

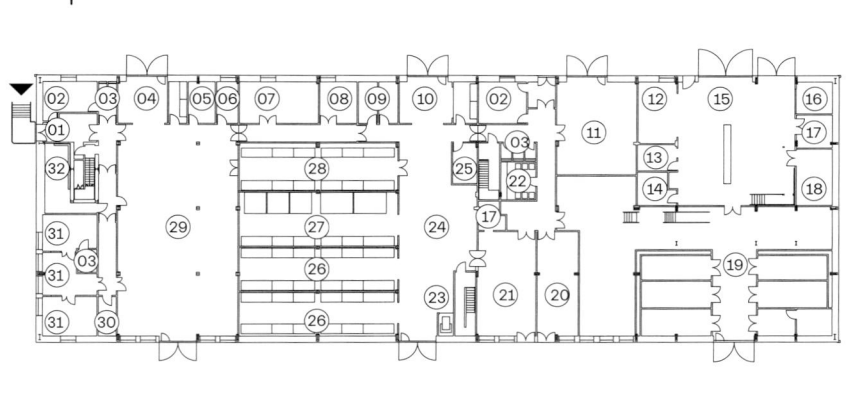

1st level		
01 Vestibule	12 Fabrication workshop	26 Pallet store
02 Boot room	13 Office	27 Freezer and pallet store
03 Bathroom	14 Consumables store	28 Heated store
04 Incoming and outgoing science field cargo	15 Workshop	29 Preparation area
	16 Power wash	30 Pharmacy
05 Ski store	17 Hazardous store	31 Medical room
06 Hot works	18 Clean room	32 Climbing wall
07 Field preparations workshop	19 Plant room	
	20 Estates workshop	2nd level
08 Electronics workshop	21 Wood workshop	33 Winter MO & engineers
09 Paint and battery stores	22 Laundry	34 Field guide office
10 Unpacking & biosecurity	23 Elevator shaft	35 Field guide locker room
11 Waste management	24 Central store	36 Visitors office
	25 Manager's office	37 Field Comms office

38 ICT Store	52 Education and training centre
39 Open plan office	53 Briefing and meeting room
40 Estates office	54 Server
41 Gym	55 UPS
42 Locker area	56 ICT Office
43 Bathrooms	57 WSLO and Summer administration
44 Sports store	58 Operations room
45 Sauna	59 Base leader's office
46 Vehicle mezzanine	
47 Plant mezzanine	
48 Multipurpose store	
49 Tea point	
50 Music room	
51 Arts and crafts store	

SOURCE
Hugh Broughton Architects

1:1,000   0   10   20   40 m

SECTION AND ELEVATION | NEW DISCOVERY BUILDING

"The energy-efficient, aero-dynamic design is oriented into the prevailing wind and utilises a deflector to channel air at higher speeds down the leeward face, minimising snow accumulation around the entire perimeter of the building. It is the first time a wind deflector has been used at this scale in Antarctica. […] The pale blue colour of the building […] minimises the impacts of degradation from high levels of UV, which is a feature of Antarctica."

British Antarctic Survey, *Rothera Modernisation – Phase 1 Initial Environmental Evaluation: Report,* 2019

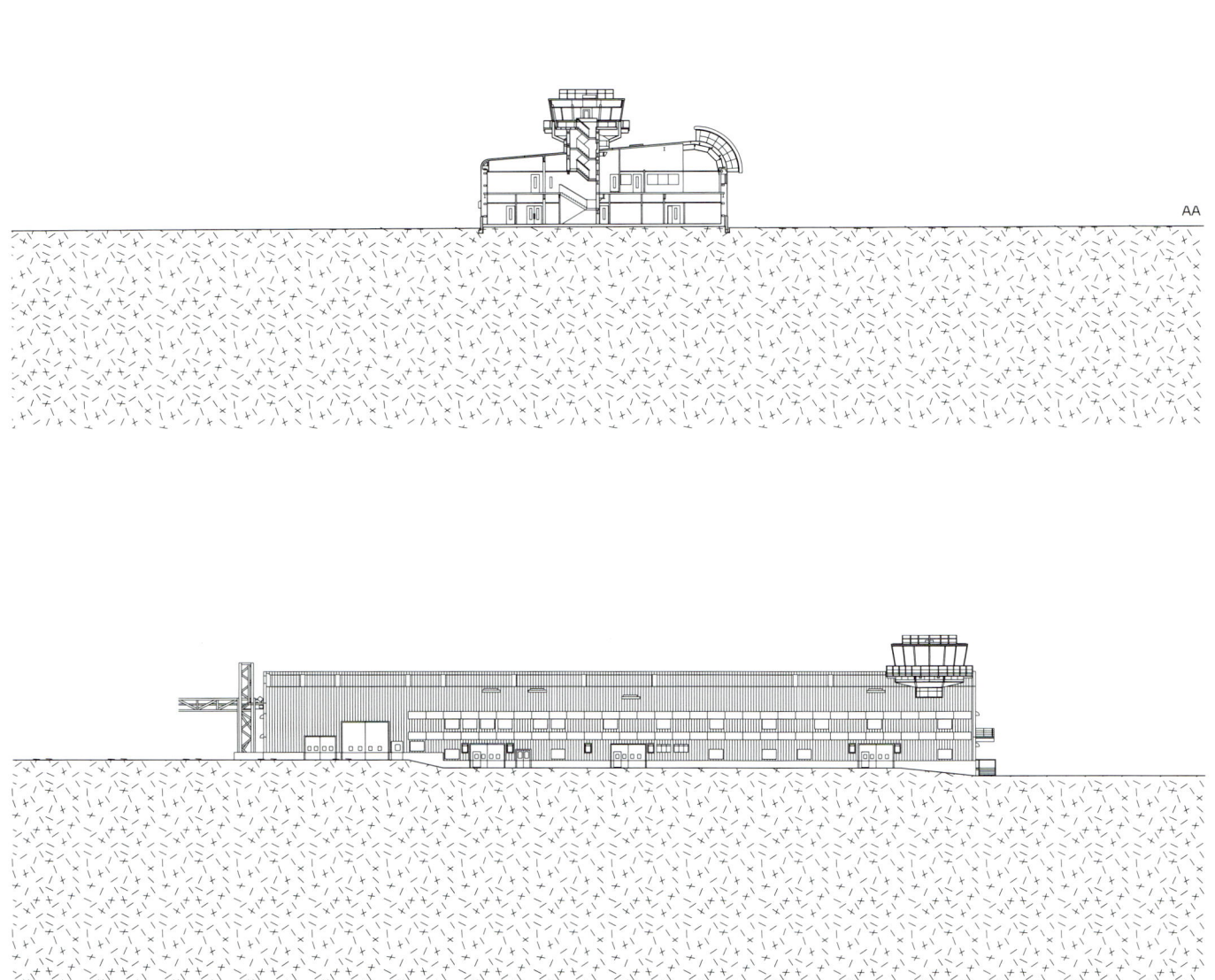

AA

SOURCE
Hugh Broughton Architects

1 : 1,000   0   10   20   40 m

ROSS SEA STATION
2020+

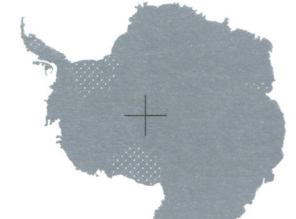

Architects: China Architecture Design & Research
Group, People's Republic of China
Contractors: China Railway Construction
Engineering Group, People's Republic of China

People's Republic of China
Chinese Arctic and Antarctic Administration (PRIC)

Project 2020+

location unknown

SITE PLAN

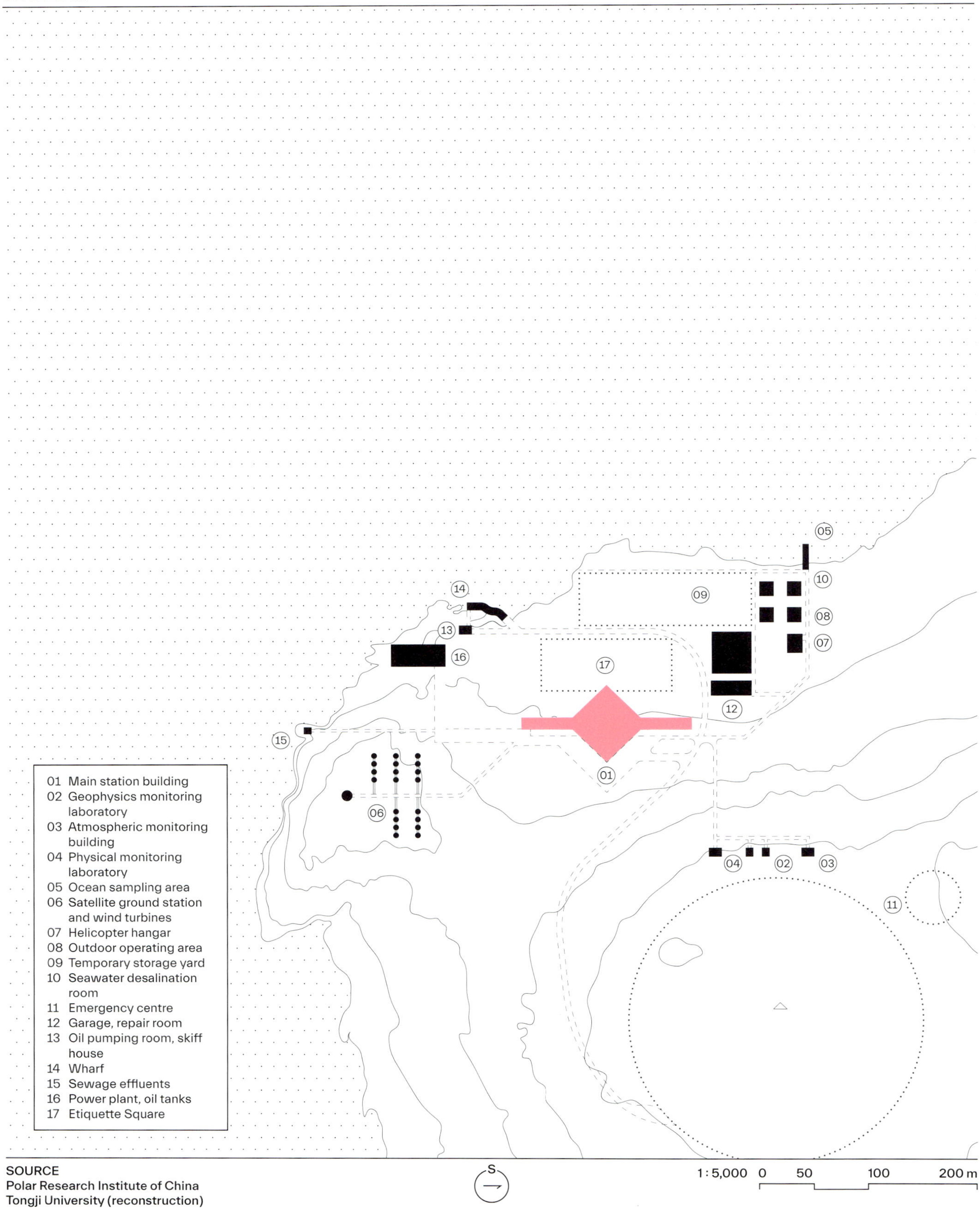

01 Main station building
02 Geophysics monitoring
    laboratory
03 Atmospheric monitoring
    building
04 Physical monitoring
    laboratory
05 Ocean sampling area
06 Satellite ground station
    and wind turbines
07 Helicopter hangar
08 Outdoor operating area
09 Temporary storage yard
10 Seawater desalination
    room
11 Emergency centre
12 Garage, repair room
13 Oil pumping room, skiff
    house
14 Wharf
15 Sewage effluents
16 Power plant, oil tanks
17 Etiquette Square

SOURCE
Polar Research Institute of China
Tongji University (reconstruction)

S

1 : 5,000   0   50   100   200 m

VOSTOK 2020+

↖ p. 237, 395, 491, 503, 504, 571, 574

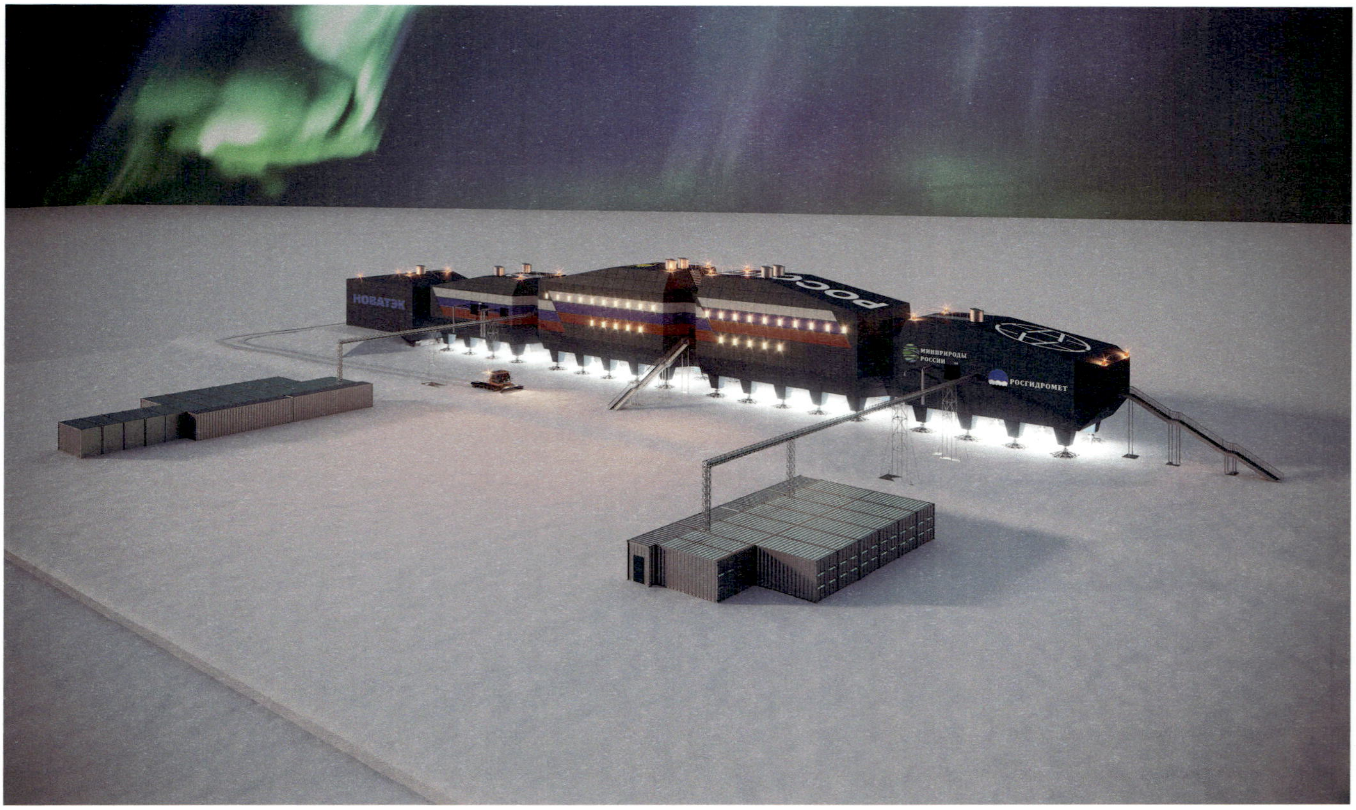

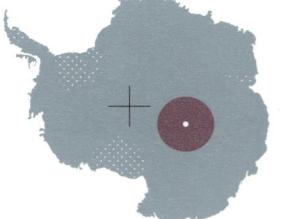

"The complex will consist of residential modules, a medical building with an equipped operating room, a laboratory, a relaxation area with a greenhouse and a gym, a power station and a garage. The new complex will occupy an area of two and a half thousand square metres and take 15 people for the winter."

Travel Reports, "Antarctica: Walking beyond the Three Poles; Part 3" (online)

Russia
Arctic and Antarctic Research Centre (AARI)

Active Station > Project 2020+

78°28'00"S 106°48'00"E
Pole of Cold, South Geophysical Pole
Ice sheet, 3,700 m.a.s.l., -55.4°C

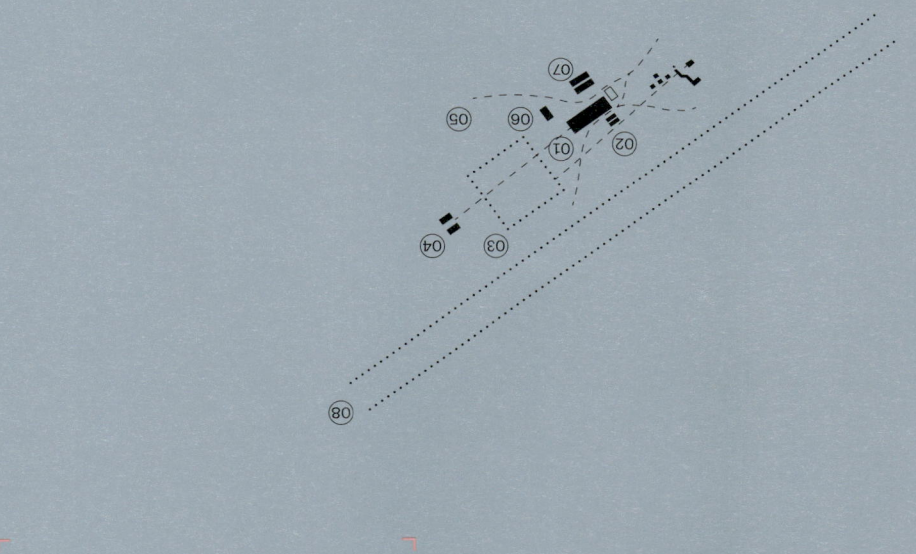

01 Living quarters
02 Aerology
03 Meteorological site
04 Pavilion
05 Access ways
06 Tractor shed
07 Fuel tank
08 Runway

09 New Vostok Station
10 Fuel tank

1:5,000

0   50   100   200 m

2020+

PLAN, 1ST LEVEL AND ELEVATION

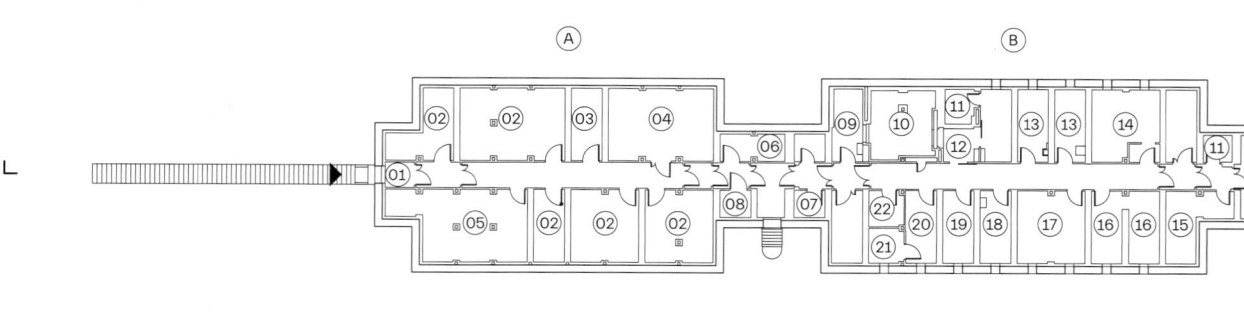

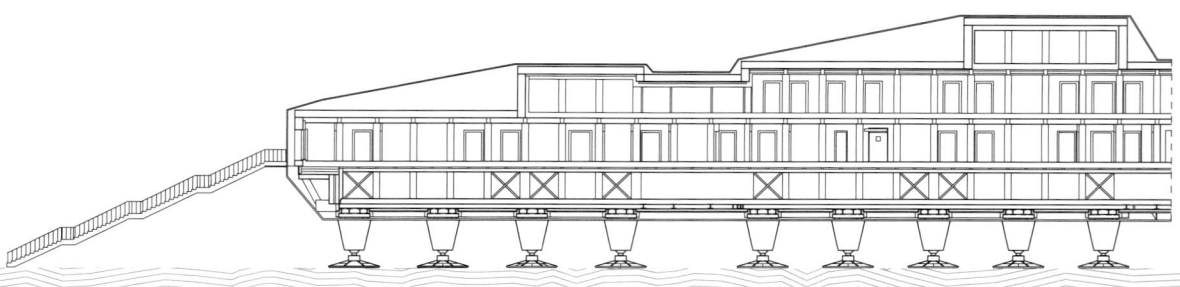

A	Block A	11	Bathroom	C	Block C	37	Server
01	Vestibule	12	Sluice	25	Bathroom and showers	38	Hot water storage
02	Cold storage equipment	13	Doctor's cabinet	26	Gym		
	(–40°C)	14	Bio-laboratory	27	Lounge	D	Block D
03	Fuel storage room	15	Wardrobe	28	Grocery warehouse	39	Water treatment room
04	Diesel generator	16	Commander's room		(+5°C)	40	Cleaning of black water
05	Equipment storage room	17	Laboratory	29	Bulk products warehouse		room
06	Switchboard	18	Office		(+5°C)	41	Water treatment room
07	Uninterruptible power	19	Dentist's cabinet	30	Vegetable warehouse	42	Boiler room
	supply for medical room	20	Inspection room		(+4°C)	43	Machine-cleaning room
08	Storage room for	21	Treatment room	31	Hot shop	44	Cold storage (-18°C)
	household supplies	22	Sanitary room	32	Kitchen		
		23	Storage of personal	33	Hall and dining room	E	Block E
B	Block B		belongings	34	Laundry	45	Waste sorting
09	Preoperative room	24	Smoking room	35	Beverage warehouse	46	Equipment storage
10	Small operating room			36	Warm storage	47	Ventilation chamber

SOURCE
Arctic and Antarctic Research Centre

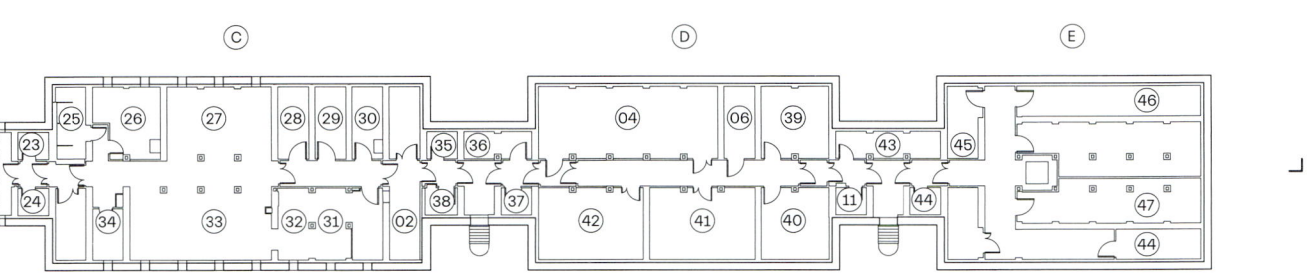

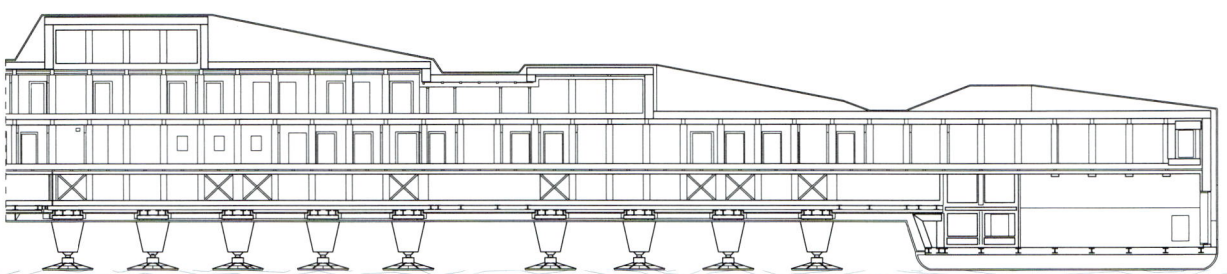

"Hydraulics are vulnerable to temperatures noticeably below minus 60 degrees Celsius: fluids that power the hydraulic jacks would thicken too much. Mechanical lifting devices are rather weak for the legs of the modules and in extreme conditions can also be unreliable. Therefore, […] it was decided to make the supports normal, stationary, but with a sufficiently large gap between their steel 'heels' and the bottom of the residential modules, so that snow removal equipment could pass there, reducing the height of the sediment."

Konstantin Alexandrov, "Stay in Antarctica:
What Will Be the Updated Russian Station 'Vostok'"
(online; translated by the editor), 2019

1:500   0   5   10   20 m

MCMURDO
1956/2020+

↖ p. 154, 219, 266, 268, 308, 344, 362, 366, 377, 390, 422, 435, 446, 450, 453, 456, 538, 546, 552, 553

"McMurdo comprises 105 buildings and sits on a site which spans approx. 650,000 sqms. Our role [as design architects] is to increase efficiency and decrease the footprint, [by] removing approximately 17 structures with a 40% reduction in square footage."

Rick Petersen (Architect, OZ Architecture), in conversation with UNLESS, Architectural Association School of Architecture, 2019

Architect: OZ Architecture, United States
Engineer: Merrick and Company, United States
Consultants: (sustainability) Rocky Mountain Institute and YRG/WSP; (wind testing and engineering) CPP; (building envelope) Cold Weather Regions and Engineering Laboratory, United States

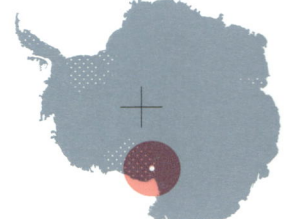

United States of America
United States Antarctic Program (USAP)

Active Station > Project 2020+

77°50'54"S 166°40'06"E
Hut Point Peninsula, Ross Island
Ice-free ground, 10 m.a.s.l., –17°C

SITE PLAN

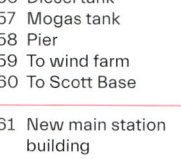

01	Crary Science and Engineering Center
02	CSEC emergency generator set
03	Core storage
04	Science support centre
05	Chapel of the snow
06	Hut X
07	Equipment operations
08	Warm-up shack
09	Cargo warehouse/food
10	Gym
11	Coffee house
12	Aerobics facility
13	Recompression chamber
14	Southern exposure
15	Gallagher's club
16	Food warehouse
17	Electrical wharehouse
18	Skua central
19	National science foundation lodging
20	Helicopter hangar
21	Supply warehouse
22	Facilities engineering, maintenance, and construction warehouse
23	Modular office
24	Post office
25	Fuel department office
26	Medical dispensary
27	Heavy vehicle maintenance facility
28	Dive shack
29	Fire pump house
30	Galley, store, laundry, and library
31	Air guard supply
32	Cold storage warehouse
33	Comms warehouse
34	Berg Field Center
35	Frozen food storage
36	Operations, air traffic, and weather
37	National Science Foundation
38	Auto unheated storage warehouse
39	Flammable storage warehouse
40	National Science Foundation office
41	Cold food storage
42	Paint barn
43	Fire house
44	Waste management
45	Joint spacecraft operation centre
46	Logistics
47	Carpentry workshop
48	Workshop
49	Safety
50	Power plant
51	Water distillation plant
52	Waste water treatment plant
53	Lodging
54	Unheated storage
55	Construction storage warehouse
56	Diesel tank
57	Mogas tank
58	Pier
59	To wind farm
60	To Scott Base
61	New main station building

SOURCE
OZ Architecture, 2015/2020

S

1 : 5,000   0   50   100   200 m

PLAN, 1ST LEVEL | NEW MAIN STATION BUILDING

A	Fire/Medical Building	C	Central Services	11	Cool storage	16	Warehouse	23	Ice cores
01	Fire prevention area	05	Post office	12	Frozen storage	17	Science cargo	24	Concourse
02	Medical area	06	Gear issue	13	Waste preperation	18	Storage pods		
		07	Computer area	14	Central warehousing	19	Tent wash	E	Trade Shop
B	Leisure	08	Dining	15	Retail store	20	Conference room	25	Warehouse
03	Gym	09	Lecture hall			21	Administration	26	Workshop
04	Lounge/recreation area	10	Kitchen	D	Field Science Support	22	Field gear		

SOURCE
OZ Architecture

1:1,000   0   10   20   40 m

PLAN, 2ND LEVEL | NEW MAIN STATION BUILDING

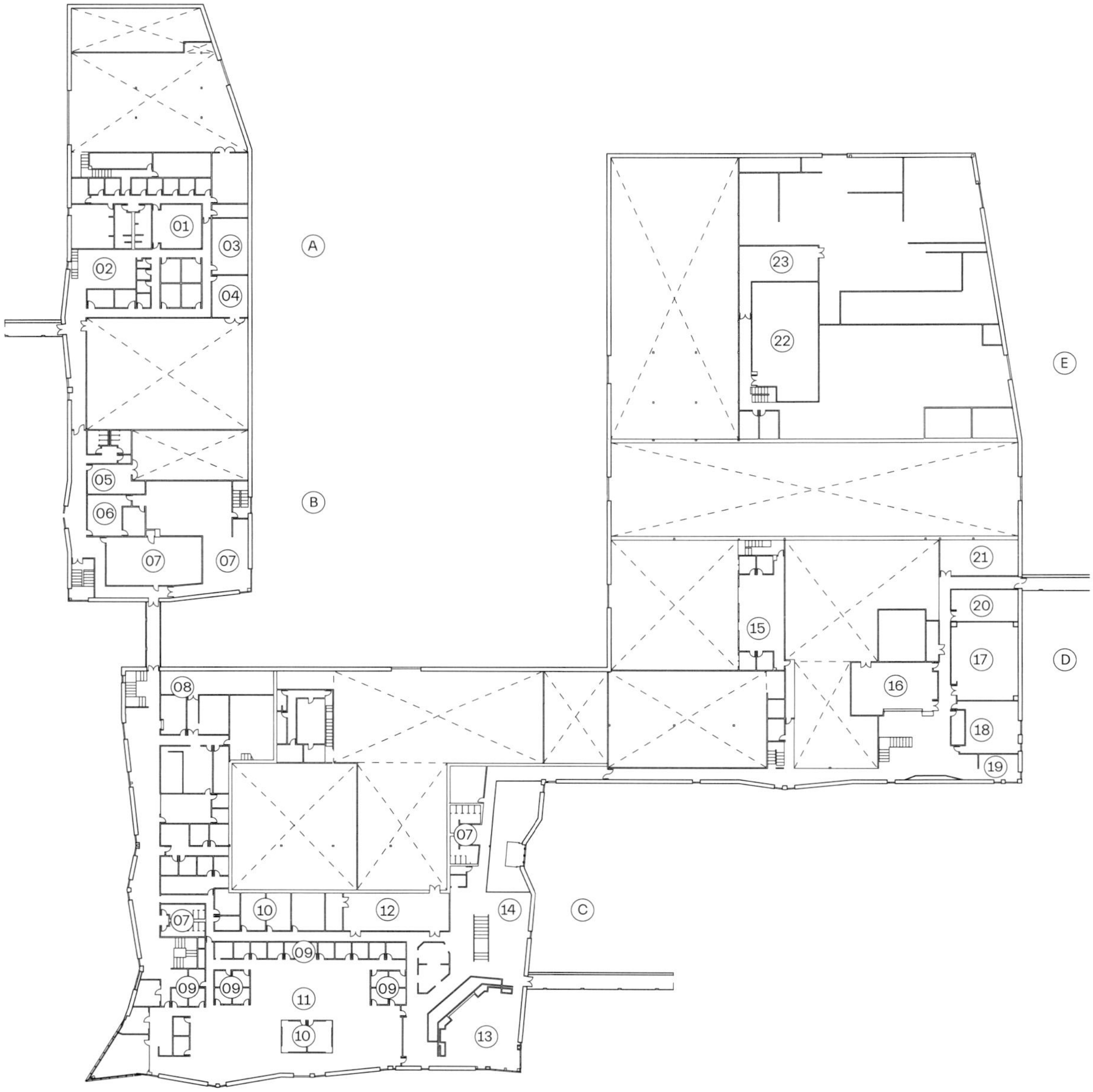

A Fire/Medical Building
01 Conference room
02 Lounge
03 Craft room
04 Mechanical room

B Leisure

05 Mechanical
06 Physical therapy
07 Gym

C Central services
08 Data centre
09 Offices
10 Conference room
11 Open office
12 Air handling units
13 Cafe
14 Library
15 Open office

D
16 Field training
17 Classroom
18 Break room
19 Storage
20 Electrical
21 Air handling unit

E Trade Shop
22 Air handling unit
23 Storage

SOURCE
OZ Architecture

S

1 : 1,000 0 10 20 40 m

SECTION AND ELEVATION | NEW MAIN STATION BUILDING

"We are floating the building over the ground so as to prevent any heat transfer into the ice crystals and rock surface and allow cold air to flow beneath it.

The interesting aspect of this is that the footprint of our station is such that there is quite an expanded area between the building and the ground

[which] does not allow for the usual scouring. Walls are thus designed to come down to the ground and prevent snow and ice from building up

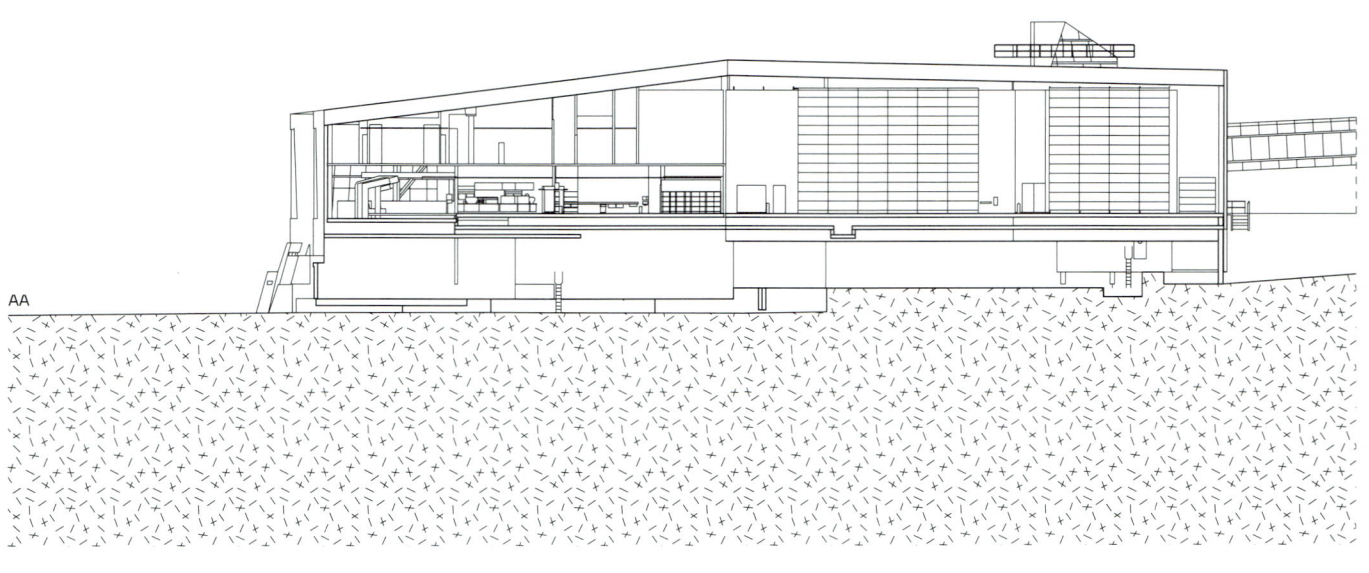

AA

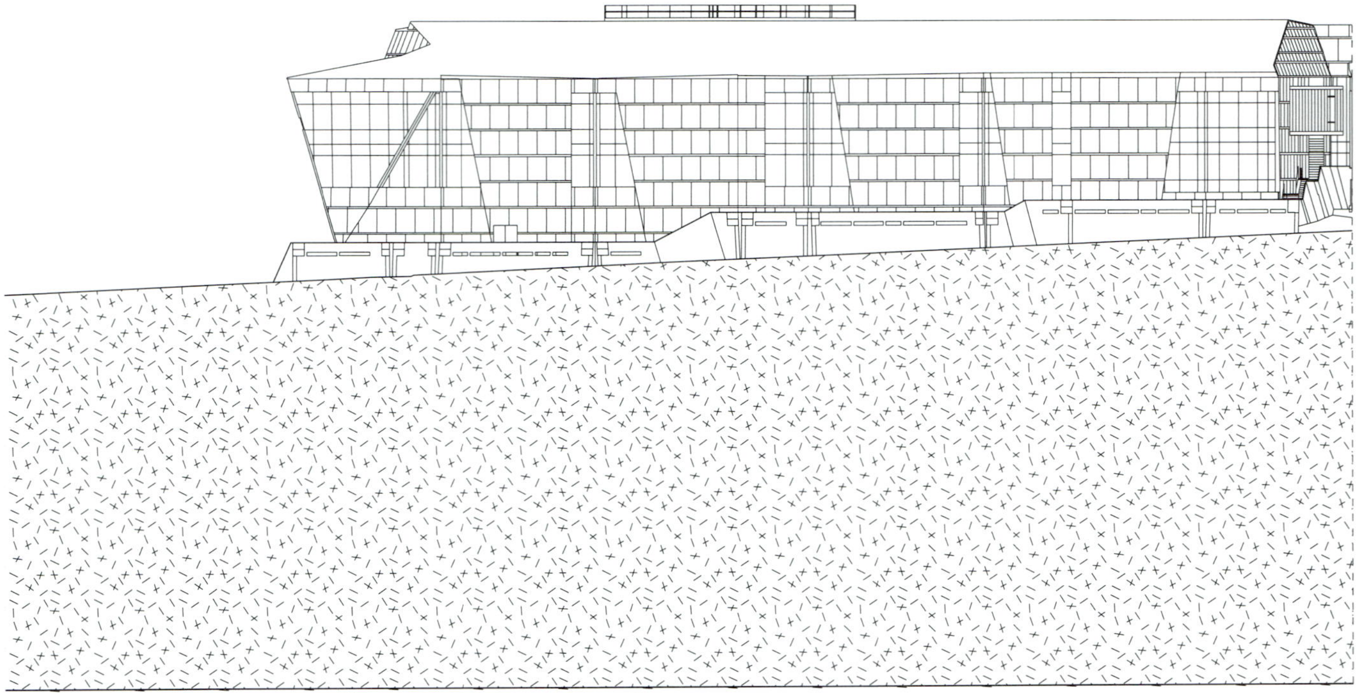

SOURCE
OZ Architecture

beneath the station. The station hence appears to be sitting directly on the ground; whilst, on the one hand, this helps functionality, on the other it does create a sense of permanence to the station that the National Science Foundation wants to reflect."

Rick Petersen (Architect, OZ Architecture), in conversation with UNLESS, Architectural Association School of Architecture, 2019

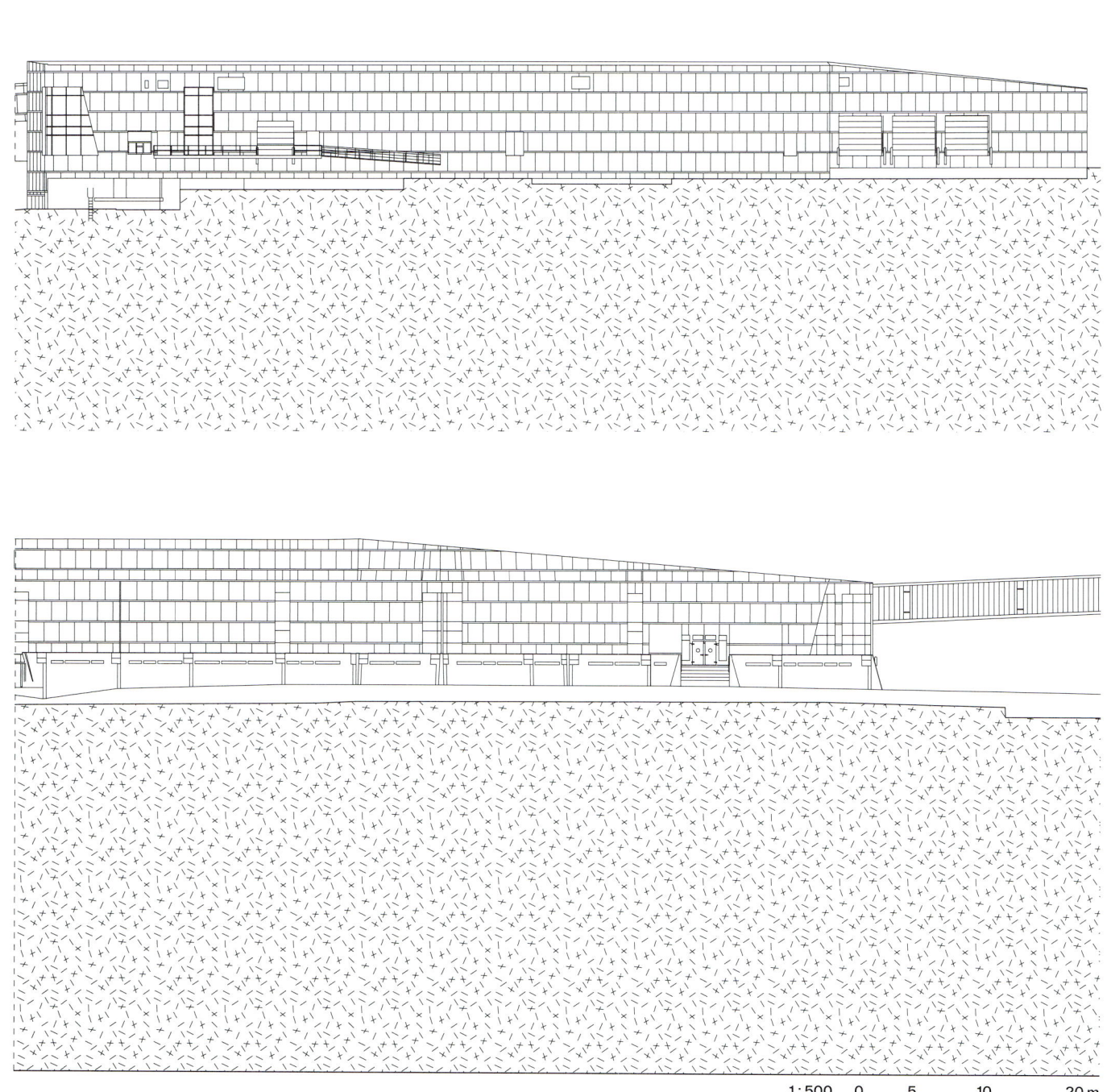

1 : 500   0   5   10   20 m

DOME FUJI
2020+

↖ p.509

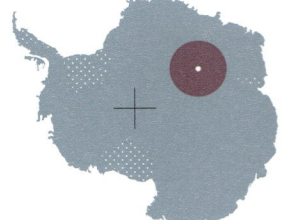

"After the demonstration test at the Syowa Station, the Antarctica mobile station unit will be transported to Dome Fuji, located inland the Antarctic continent and 3,800 meters above sea level and used as the residential space (max. capacity of 18 people) for the third Dome Fuji deep ice coring project. The sensors will continue to be used for monitoring the conditions during the project and the data will be utilized for [the] future construction of stations in Antarctica."

National Research & Development Agency et al., "A Joint Project of JAXA, NIPR, Misawa Homes, and MHIRD: Demonstration Test of Antarctica Mobile Station Unit" (online), 2019

Manufacturer: Misawa Homes Co., Japan
Research: National Research & Development Agency; National Institute of Polar Research; Inter-university Research Institute Corporation Research Organization of Information and Systems; Misawa Homes Institute of Research and Development Co., Ltd.; Space Exploration Innovation Hub Center: Japan Aerospace Exploration Agency, Japan

Japan
National Institute of Polar Research (NOPRI)

Active Station > Project 2020+

77°18'59"S 39°42'04"E
Dome F, Queen Maud Land,
Ice sheet, 3,810 m.a.s.l., −54.3°C

SITE PLAN

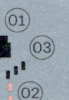

01 Main station building
02 Mobile station unit
03 Transporting vehicle

SOURCE
Misawa Homes (reconstruction)

S

1 : 5,000   0   50   100   200 m

ELEVATION | MOBILE STATION UNIT

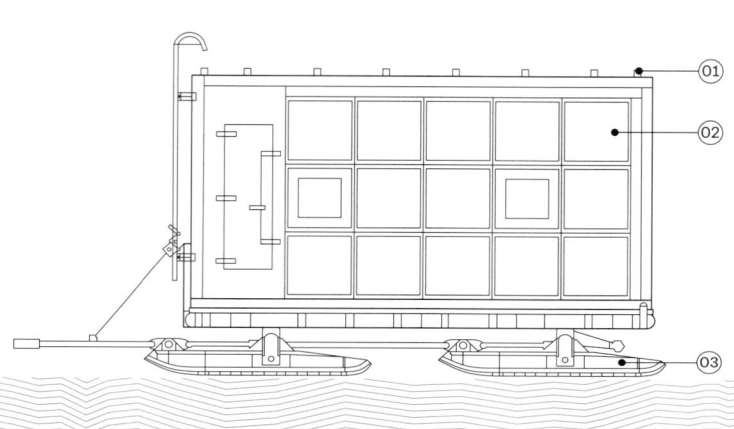

01 Safety monitoring sensors
   for research (temperature,
   humidity, CO_2 detection,
   fire detection)
02 Photovoltaic panels (PV)
03 Transport sledge

04 Gear for coupling with
   snow vehicle

ELEVATION | MOBILE STATION UNIT

"A demonstration test will be conducted with the aim of developing a sustainable housing system that withstands the extreme environment of Antarctica. The test will verify simplified construction, optimized energy based on a natural energy system, monitoring sensors, and other technologies developed for the unit. The goals of the joint research of the four organizations [National Research & Development Agency, Japan Aerospace Exploration Agency (JAXA), Inter-university Research Institute Corporation Research Organization of Information and Systems, and National Institute of Polar Research] is to develop future-oriented housing, a station in Antarctica, and a manned base on the moon."

National Research & Development Agency et al., "A Joint Project of JAXA, NIPR, Misawa Homes, and MHIRD: Demonstration Test of Antarctica Mobile Station Unit" (online), 2019

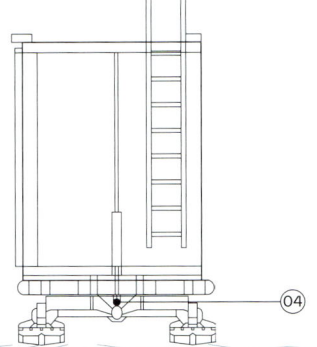

04

SOURCE
Misawa Homes

1:100   0   1   2   4 m

The Antarctic through the Lens of Jean de Pomereu

Photographs taken whilst reporting on the 25th Chinese Antarctic Research Expedition (CHINARE) during the 4th International Polar Year, 2007-2008

"Traveling through this ice-scape felt like entering a lost city, resembling Atlantis, where the icebergs replaced monumental ruins. The silence and desolation were profound. It was as if not just matter, but time itself had frozen."

Jean de Pomereu (Photographer), "Letter from Antarctica", *Some Magazine*, 2014

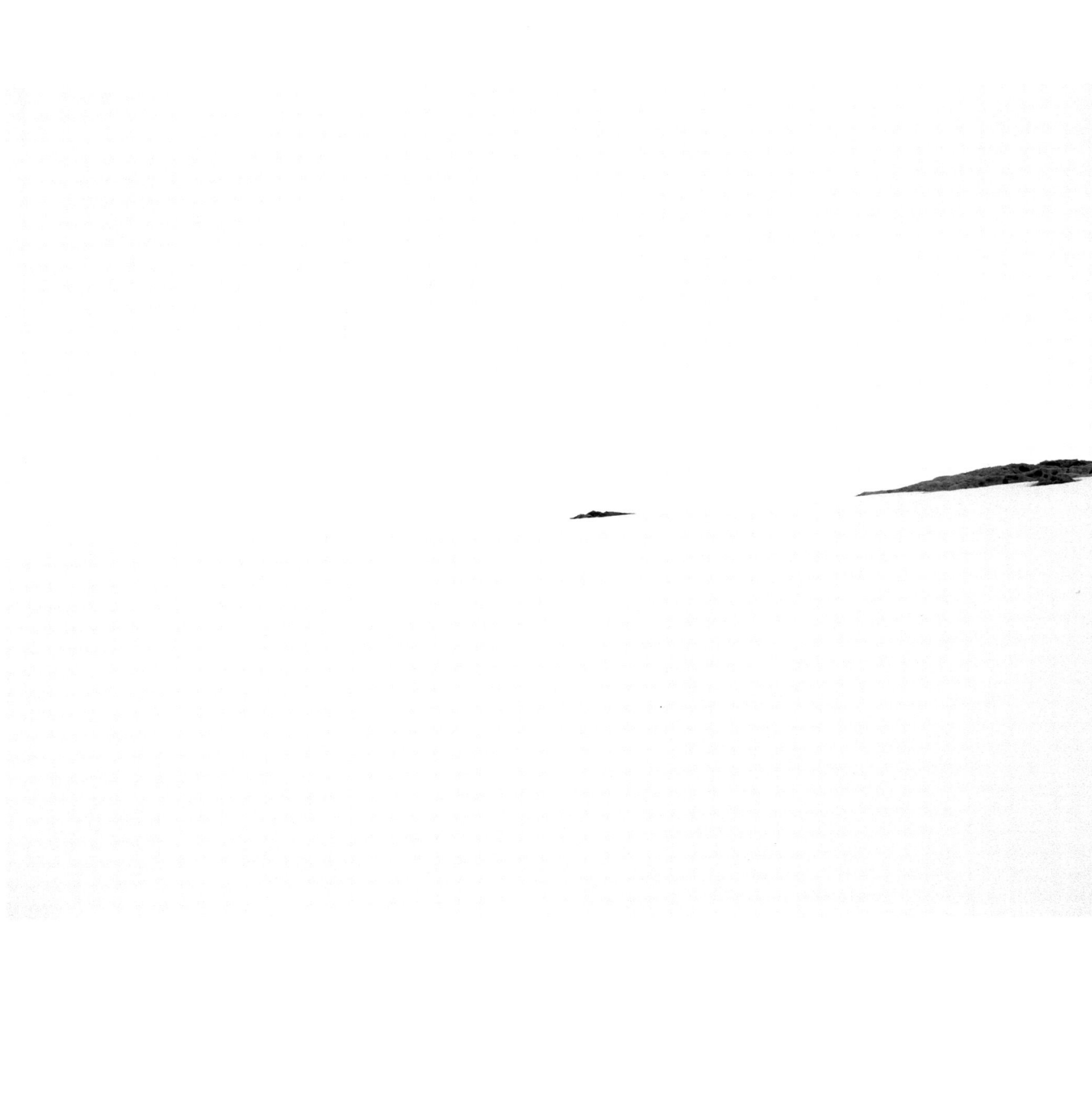

APPENDIX

THANK YOU TO OUR SPONSORS

D-Air Lab

Fondazione Giuseppe
e Pericle Lavazza

Furthermore: a program
of the J. M. Kaplan Fund

Graham Foundation
for Advanced Studies
in the Fine Arts

Thyssen-Bornemisza Art
Contemporary Foundation
(TBA21)

Acknowledgments

Antarctic Resolution is a transnational effort.
The book would not have been possible without the determination and generous collaboration of each and every one of the individuals and organisations listed below. On behalf of the next generation, we say thank you to all of you.

Our deepest gratitude goes to …

FRANCESCO BANDARIN, who ignited our curiosity about the Antarctic and offered at all times his most welcomed experienced advice.

THE AUTHORS of *Antarctic Resolution*, who believed in the project and entrusted us with their invaluable knowledge even when our unwavering level of commitment to the Antarctic cause was still inversely proportional to any evidence we could offer of expertise on the subject. The selfless generosity with which have all contributed is a lesson in itself of what is possible when people come together to act for the greater good. We will treasure such a lesson and work tirelessly to ensure that their voices will be heard.

Doaa Abdel-Motaal (World Trade Organisation), Conrad Anker (Mountaineer), Francesco Bandarin (formerly UNESCO), Carlo Barbante (University Ca' Foscari), James N. Barnes (Antarctic and Southern Ocean Coalition), Thomas Barningham (British Antarctic Survey), Carlo Baroni (University of Pisa), Susan Barr (International Council on Monuments and Sites), Elisa Bergami (University of Siena), Marcelo Bernal (ARQZE Ltd), Anne-Marie Brady (University of Canterbury), Ralf Brauner (Jade University of Applied Sciences), Cassandra M. Brooks (University of Colorado Boulder), Shaun T. Brooks (University of Tasmania), Hugh Broughton (Hugh Broughton Architects), Bert Bücking (bof architekten), David Burrows (Environmental Systems Research Institute), Sol Camacho (RADDAR), Sanjay Chaturverdi (South Asian University), Swhadheet Chaturvedi (Architectural Association School of Architecture), Christy Collis (Queensland University of Technology), Peter Convey (British Antarctic Survey), Geoff Cooper (United Kingdom Antarctic Heritage Trust), Gabriele Coppi (University of Pennsylvania), Ilaria Corsi (University of Siena), Lino Dainese (Dainese and D-Air Lab), Klaus Dodds (Royal Holloway, University of London), Christel Misund Domaas (University of Tromsø), Julian Dowdeswell (Scott Polar Research Institute, Cambridge University), Chris Drury (Artist), Juan Du (University of Hong Kong), Graeme Eagles (Alfred Wegener Institute for Polar and Marine Research), Tess Egan (Australian Antarctic Division), Alexey A. Ekaykin (Arctic and Antarctic Research Institute), Fausto Ferraccioli (British Antarctic Survey), Joe Ferraro (Ferraro Choi), James Rodger Fleming (Boston University), Giulia Foscari (UNLESS), Adrian Fox (British Antarctic Survey), William L. Fox (Nevada Museum of Art), Bob Frame (University of Canterbury), Peter T. Fretwell (British Antarctic Survey), Lutz Frisch (Artist), Jacopo Gabrielli (National Research Council), Hartwig Gernandt (Alfred Wegener Institute for Polar and Marine Research), Andrew J. Gerrard (New Jersey Institute of Technology), Neil Gilbert (Constantia Consulting), El Glasberg (New York University), Karsten Gohl (Alfred Wegener Institute for Polar and Marine Research), Francis Halzen (University of Wisconsin–Madison), Kael Hanson (Wisconsin, IceCube Particle Astrophysics Center), Ursula Harris (Australian Antarctic Data Centre), Judith Hauck (Alfred, Wegener Institute for Polar and Marine Research), Robert Headland (Scott Polar Research Institute, Cambridge University), Beth Healey (European Space Agency) , Alan D. Hemmings (University of Canterbury), Poutama Hetaraka (Hihiaua Cultural Centre), Te Warihi Hetaraka (*Tohunga whakairo*), Alfred Hiatt (Queen Mary University of London),

Gretchen Hoffman (University of California, Santa Barbara), Adrian Howkins (University of Bristol), Kevin A. Hughes (British Antarctic Survey), Andrew T. Hynous (National Aeronautics and Space Administration), Julia Jabour (University of Tasmania), Stéphanie Jenouvrier (Woods Hole Oceanographic Institution), Solan Jensen (Quark Expeditions), Kevin Johnson (University of California, Santa Barbara), Andrea Kavanaugh (The Pew Charitable Trusts), Daniel Kiss (Architectural Association School of Architecture), Georg Kleinschmidt (Alfred Wegener Institute for Polar and Marine Research), Alexander Klepikov (Arctic and Antarctic Research Institute), Peter Landschützer (Max Planck Institute for Meteorology), Louis John Lanzerotti (New Jersey Institute of Technology), Elizabeth Leane (University of Tasmania), Sang Leem Lee (Space Group), Inti Ligabue (Giancarlo Ligabue Foundation), Daniela Liggett (University of Canterbury), Bryan Lintott (Scott Polar Research Institute, Cambridge University), Vladimir Lipenkov (Arctic and Antarctic Research Institute), Cornelia Lüdecke (University of Hamburg), Arturo Lyon (Pontifical Catholic University of Chile), James Madsen (Wisconsin IceCube Particle Astrophysics Center), Craig McCormack (University of Western Australia), Tony McGlory (Ramboll), Hans-Jürgen Meyer (Alfred Wegener Institute for Polar and Marine Research), Joseph Micallef (Military.com), Nicholas de Monchaux (Massachusetts Institute of Technology), Chiara Montanari (Antarctic Expedition Leader), Michael Morrison (Purcell), Teasel Muir-Harmony (National Air and Space Museum, Smithsonian Institution), John Nelson (Environmental Systems Research Institute), Camilla Nichol (United Kingdom Antarctic Heritage Trust), Miranda Nieboer (University of Tasmania), Anne Noble (Artist), Dirk Notz (Max Planck Institute for Meteorology), Shaun O'Boyle (Photographer), Madeleine O'Keefe (Wisconsin IceCube Particle Astrophysics Center), Lucy + Jorge Orta (Artists), Lawrence A. Palinkas (University of Southern California), Scott Parazynski (National Aeronautics and Space Administration), Carolina Passos (RADDAR), Michael Pearson (International Council on Monuments and Sites, International Polar Heritage Committee), Francesco Pellegrino (National Agency for New Technologies, Energy and Sustainable Economic Development), Rick Petersen (OZ Architecture, Inc.), Katherina Petrou (University of Technology Sydney), Andrea Piñones (Austral University of Chile), Jean-Yves Pirlot (Council of European Geodetic Surveyors), Ceisha Poirot (Antarctica New Zealand), Jean de Pomereu (Photographer), Alexander Ponomarev (Artist, Antarctic Biennale), Brian Rauch (Washington University in St. Louis), Ron Roberts (Kingston University), Fayne Robinson (Artist), Donald R. Rothwell (Australian National University College of Law), Juan Francisco Salazar (Western Sydney University), Jean-Baptiste Sallée (French National Centre for Scientific Research), Philippe Samyn (Philippe Samyn and Partners), Bojan Šavrič (Environmental Systems Research Institute), Mirko Scheinert (Dresden University of Technology), Didier Schmitt (European Space Agency), Thomas Schramm (Ramboll), Daniel Schubert (Alfred Wegener Institute for Polar and Marine Research), Karen Nadine Scott (University of Canterbury), Cara Seitchek (American University), Maria Ximena Senatore (International Council on Monuments and Sites), Jonathan Shanklin (British Antarctic Survey), Yuri Shibaev (Arctic and Antarctic Research Institute), Santiago Sierra (Artist), Tim Stephens (University of Sydney), Pavel Grigorievich Talalay (Jilin University), Paul Taylor (ARQZE Ltd.), Steve Theno (United States Antarctic Program), Paul Thur (United States Antarctic Program), Philip Trathan (British Antarctic Survey), Francisco Valdivia (Federico Santa María Technical University), David Vaughan (British Antarctic Survey), Emerson Vidigal (Estudio 41), Priscilla Wehi (University of Otago), Claudio Williams (Amancio Williams Archive), Gary Wilson (Department of Marine Science,

University of Otago), Juliet Wong (University of California, Santa Barbara), Gillian Wratt (Independent Consultant), Angela Wright (Colour Affects), James York (Artist), and Federica Sofia Zambeletti (UNLESS).

We would like to express special thanks to Alan Hemmings, whose contributions and advice, not least in reviewing the introduction, have been invaluable, and to David Vaughan whose authoritative factual illustration of the scientific phenomena underlying the Antarctic ice sheet's accelerated rate of thinning has fuelled our determination to launch a call for action. We are deeply grateful for this and for his close collaboration with UNLESS and Arcangelo Sassolino on the *Antarctic Resolution* exhibition at the 2021 Venice Architecture Biennale.

THE UNLESS TEAM, who have worked on the project with relentless determination, endurance and patience. To put into words the gratitude I feel towards Federica Sofia Zambeletti, who has masterfully managed the colossal project of *Antarctic Resolution* from its conception, is hard. Suffice it to say, that the book *really* would not have seen the light of day without her optimism, her critical thinking, and her uncompromising commitment, which even led her to part with her healthy appendix to join our Antarctic expedition. Alongside Federica Zambeletti, I would like to thank Sabrina Syed for the accuracy of her editorial work and attentive collaboration with all authors; Giulio Marchetti and Olympia Simopoulou for their incredible archival research on the Archive of Antarctic Architecture; Olimpia Presutti and Nicolyn Moffitt for their tactful persistence in obtaining image copyright clearance for all the photographs; to Lloyd Suk-Gyo Lee, Sonja Draskovic, and Ines Molinari for their contribution to the research and for the production of compelling visualisations for the publication; and to Eleonora Cappuccio for her data-mining and infographic efforts.

THE POLAR LAB – especially its directors Francesco Bandarin, Sol Camacho, Juan Du, Giulia Foscari, Arturo Lyon, and Florencia Rodriguez – for curating transnational lectures, workshop, seminars, symposia, and master programmes focusing on the Antarctic, and for producing the Antarctic Atlas.

The Architectural Association School of Architecture and its former director Eva Franch i Gilabert for their invaluable support in launching UNLESS's experimental academic platform: the Polar Lab. Directed by Francesco Bandarin and Giulia Foscari, the English Polar Lab research cluster was also generously supported by the British Antarctic Survey, the Scott Polar Research Institute, the United Kingdom Antarctic Heritage Trust, the AA Landscape Urbanism MArch, the AA Public Programme, the AA Visiting School, and the London Design Festival, which invited the Polar Lab to be partner in their 2019 edition. The body of work produced with the support of these organisations would not have translated into publication content without the hard work of the following students: Aleksandar Aksentijevic, Callum Campbell, Swadheet Chaturvedi, Ryan Darius, Noah Gottlieb, Othmane Kandri, Daniel Kiss, Jane Ling, Giulio Marchetti, Patricia Roig, Sebastian Serzysko, Olympia Simopoulou, Younseo Song, Sabrina Syed, and Sze May Wong. Our gratitude goes to all the tutors: Marco Bernardi (Accurat), Maria Shéhérazade Giudici (Architectural Association School of Architecture), Lloyd Suk-Gyo Lee (UNLESS), Jean de Pomereu (Scott Polar Research Institute), Alfredo Ramirez (Architectural Association School of Architecture), Eduardo Rico (Architectural Association School of Architecture), Federica Sofia Zambeletti (UNLESS), Alessandro Zotta (Accurat); to all the lecturers: Hugh Broughton (Hugh Broughton Architects), Bert Bücking (bof architekten), Geoff Cooper (United Kingdom Antarctic Heritage Trust), Oliver

Darke (British Antarctic Survey), Klaus Dodds (Royal Holloway, University of London), Beth Healey (European Space Agency), Bryan Lintott (Scott Polar Research Institute), Tony McGlory (Ramboll), Michael Morrison (Purcell), Camilla Nichols (United Kingdom Antarctic Heritage Trust), Francesco Pellegrino (National Agency for New Technologies, Energy and Sustainable Economic Development, ENEA), Rick Petersen (OZ Architecture), Philippe Samyn (Philippe Samyn and Partners), David Winston Harris Walton (British Antarctic Survey); and to all the guest critics: David Burns (Royal College of Art), Anna Somers Cocks (*The Art Newspaper*), Hanif Kara (AKT II), Barbara Campbell Lange (Bartlett, University College London), Valerio Massaro (Architectural Association School of Architecture), John Palmesino (Architectural Association School of Architecture), and Christopher Pierce (Architectural Association School of Architecture) – all of whom have patiently guided, inspired, and challenged the Polar Lab UK. To David Winston Harris Walton, who mentored us during our early Antarctic steps and honoured us with his last lecture, and to Hugh Broughton, who has contagiously transferred knowledge and enthusiasm during the two years of research, goes our deepest gratitude.

The Department of Architecture of the University of Hong Kong for hosting the Polar Lab HK directed by Juan Du and incorporated into the Master of Architecture design studio (autumn 2019). Supported by the Chinese Arctic and Antarctic Administration, the HKU Department of Architecture Publications Committee, and the People's Republic of China State Oceanic Administration, the body of research produced by the Hong Kong Polar Lab team is the outcome of the hard work of the following researchers and students: Nesia Cheung, Chihing Hu, Peng Hu, Karina Kan, Joyce Cheuk Yan Lai, Sardonna Leung, Julian Long, Kelly Mok, Subin Park, Huiqing Song, Rachel Wong, Mei Yang, and Connie Yeung. They were aided in their work by University of Hong Kong research assistants Wei Chen and Eric Yuen and received advice and archival material support from: Gangying Chen, Zhijian Tang (China Central Television), Limin Xia (Chinese Arctic and Antarctic Administration), Jiansong Zhang (Xinhua News Agency), and Yi Zhang (Tsinghua University), to whom we are very grateful.

The Pontifical Catholic University of Chile, for incorporating the Polar Lab CL within the Master in Landscape Architecture programme (MAPA-PUC 2019–2020). Directed by Arturo Lyon and taught with Osvaldo Moreno and Tomas Tironi, the Chilean edition of the Polar Lab counted on the research support of Francesca Dalla Mora and Francesco Famá and on the collaboration with Rafaela Behrens, Marcelo Bernal, Leonardo Gabrielli, and the Escobar Meza Family. The thorough research work was developed by the following students: Antonella Bernucci, Francesca Dalla Mora, Domenica Debarbieri, Gregoire Dorthe, Carlos Escamilla, Francesco Famá, Alejandro Matamala, Pascal Mondion, Micaela Muchnick, Consuelo Nuñez, Juan Oyarzún, Alberto Pérez, Mario Toledo, and Sofía Valenzuela. Their efforts were supported by the following institutional collaborators, to whom we are very grateful: Roberto Avendaño, Eduardo Celedón, Fernando Machuca, Santiago Madrid, Jeannette Möller, and Charles Torrejón from the Chilean Air Force; Humberto Julio and Eduardo Villalón from the Chilean Army; Cesar Alveal, Lars Christiansen, and Alejandro Cornejo from the Chilean Navy; Gustavo Durán and Raúl Irarrázabal from the Ministry of Public Works; Felipe de La Lastra and Edgardo Vega from the Chilean Antarctic Institute (INACH).

RADDAR for hosting the Polar Lab BR. Directed by Sol Camacho, the Brazilian Polar Lab team counted on the skilful mapping arts of Carolina Passos, the specialist expertise and enthusiasm of Anál04 Amorim, and the organisational role of Ana Cecília Tourinho. The Antarctic Atlas would not have seen the light of day without their guidance and the hard work of Julia Lima Albuquerque, Caio Aquinaga, Vitor Fernandes, Caio França, and Karina Yuki Kagohara; a special thought is devoted to Julia Lima Albuquerque, who will be remembered dearly. The work research produced in Brazil would not have been possible without the collaboration with the Norwegian Polar Institute and with Quantarctica, and the support of Jonathan Franklin and Jan Melchior van Wessem.

   -NESS for hosting the Polar Lab AR. Directed by Florencia Rodriguez, the Argentine Polar Lab counted on the collaboration of Martina de Barba, Santiago Bogani, Pablo Gerson, Francisca Martina Gil Sosa, Rodrigo Kommers Wender, Natalia La Porta, Isabella Moretti, Victoria Nuviala, Violeta Nuviala, Mariam Samur, and, to whom we are grateful. The Polar Lab AR lecture series, held at Monte and as part of the programme of the La Bienal Arquitectura, included enlightening lectures by Ignacio Fleurquin (Bulla), Sabrina Juárez (Servicio Meteorológico Nacional Argentino), Mauricio Nicolás Laurizi (Servicio Meteorológico Nacional Argentino), Cristian Lorenzo (Conicet), Fabio Ayerra Muzikantas (Fábrica de Paisaje), Diego Perez (Fábrica de Paisaje), and Silvina Righetti (Servicio Meteorológico Nacional Argentino).

THE FOLLOWING INSTITUTIONS AND INDIVIDUALS, who opened their archives for us and generously shared their historically unique documentation: Alexander Turnbull Library National Library of New Zealand; Alfred Wegener Institute for Polar and Marine Research (AWI), Germany; American Geographical Society, United States of America; Amon Carter Museum of American Art, United States of America; Antarctica New Zealand; Antarctic Biennale; Amancio Williams Archive, Chile; Arctic and Antarctic Research Institute (AARI), Russia; Argentine Antarctic Institute (IAA), Argentina; ARZQE Ltd., Chile; Australian Antarctic Program (AAD), Australia; bof Architekten, Germany; Boulton & Paul Archive, United Kingdom; British Antarctic Survey (BAS), United Kingdom; British Library, United Kingdom; British Telecom Heritage & Archives, United Kingdom; Canterbury Museum, New Zealand; Chilean Air Force, Chile; Chilean Armed Forces, Chile; Chinese Arctic and Antarctic Administration (CAA), China; Clean 2 Antarctica, Netherlands; Contemporary China Press, China; Cool Antarctica, United Kingdom; Cornell University Library, United States of America; Elmer M. Cranton; D-Air Lab, Italy; Dietrich Enss; Ellis Family Archive, United Kingdom; European Space Agency (ESA); Estúdio 41, Brazil; Federal Institute for Geosciences and Natural Resources (BGR), Germany; Ferraro Choi, United States of America; Fram Museum, Norway; French Polar Institute (IPEV), France; Galerie l'Atelier d'Artistes, France; German Aerospace Centre (DLR), Germany; Government of the British Antarctic Territory (BAT), United Kingdom; Hugh Broughton Architects, United Kingdom; International Polar Foundation (IPF), Belgium; Italian National Antarctic Program (PNRA), Italy; Jean-Pierre Jacquin, French Polar Expeditions (EPF), France; Larry Johnson; Korea Polar Research Institute (KOPRI), South Korea; Lia Rumma Gallery, Italy; Misawa Homes, Japan; Mori Art Museum, Japan; Museums Victoria, Australia; National Academies Press, Unites States of America; National Aeronautics and Space Administration (NASA), United States of America; National Archives of Chile, Ministry of Foreign Affairs, Chile; National Centre for Scientific Research (CNRS), France; National Geospatial Intelligence Agency, United States of America; National Library of Norway, Norway; National Maritime Museum, United Kingdom; National Science Foundation (NSF), United States of America; *New York Times*, United States of America; Norwegian Museum of Cultural History, Norway; Norwegian Polar Institute (NPI), Norway;

Ohio State University (OSU), United States of America; OZ Architecture, Inc., United States of America; People's Republic of China State Oceanic Administration, China; Philippe Samyn and Partners, Belgium; Pontifical Catholic University of Chile (PUC), Chile; Prussian Heritage Image Archive, Germany; Ramboll, Germany; Barbara Ronte; Royal Geographical Society (RGS), United Kingdom; Scott Polar Research Institute (SPRI), University of Cambridge, United Kingdom; Sotheby's, United Kingdom; William Sumrall; Space Group, South Korea; State Library, United Kingdom; Swiss Photo Foundation, Switzerland; Taylor and Francis, United Kingdom; United Kingdom Antarctic Heritage Trust (UKAHT), United Kingdom; United States Antarctic Program (USAP), United States of America; United States Navy (USN), United States of America; University of Chicago, United States of America; University of Nebraska, United States of America; University of Oxford, United Kingdom; Richard Wolack.

Within the institutions mentioned above a special thanks goes to: Aleksandr Makarov and Aleksandra Mušta (Arctic and Antarctic Research Institute, AARI, Russia), who invited us to access their archive in St Petersburg and uncover historical documentation; Samantha Chapman (Antarctica New Zealand, New Zealand) for helping us connect to numerous scientists in New Zealand who subsequently contributed essays; Tess Egan (Australian Antarctic Program, AAD, Australia) who patiently and resourcefully connected us with valuable archival documents from the Australian Antarctic Program; Antje Boetius (Alfred Wegener Institute for Polar and Marine Research, AWI, Germany) who opened the doors at AWI for us, establishing precious contacts with its scientists, and establishing a contact for us with Christian Salewski at the Alfred Wegener Institute Helmholtz Centrum; Linda Capper and Athena Dinar (British Antarctic Survey, BAS, United Kingdom), who were pivotal in welcoming us in at BAS and introducing us to their incredible research community; Ieuan Hopkins (British Antarctic Survey, BAS, United Kingdom), who led the BAS archives with the support of Alysa Hulbert and Kevin Roberts and who overcame Herculean challenges to patiently and generously help us source, select, and receive unedited archival images and drawings of historic Antarctic bases; Roxane Bailet and Luca Rendina (Hugh Broughton Architects, United Kingdom), who shared detailed architectural drawings and images of contemporary Antarctic stations; Paola Potena (Lia Rumma Gallery, Italy), who helped us establish a contact with William Kentridge; Andreas Nitschke (Ramboll, Germany), who introduced us to the engineering complexity of building on ice; Joy Wheeler (Royal Geographical Society, RGS, United Kingdom), who helped us navigate the rich archives of RGS; Lucy Martin (Scott Polar Research Institute, SPRI, United Kingdom), who assisted us in navigating SPRI's Archives. A special thanks goes to Larry Johnson, who granted us the right to use his images while also gently reminding us of the beauty of analogue communication by religiously writing only all-caps letters as he "learned typing in the Navy on a communication typewriter which had no lower-case letters".

JEAN DE POMEREU, who introduced us to Antarctic photography and has curated the portfolios by Joan Myers, Paolo Pellegrin, Herbert George Ponting, and Eliot Porter, along with his own.

THE FOLLOWING ARTISTS AND PHOTOGRAPHERS, who generously shared their work with us in support of the collective project: Lita Albuquerque, Todd Anderson, Laurent Ballesta, Jimmy Chin, Sebastian Copeland, Chris Drury, Lutz Frisch, Andrea Izzotti, William Kentridge, Stuart Klipper, An-My Lê, Spencer Lowell,

James Morris, Joan Myers, Anne Noble, Shaun O'Boyle, Lucy + Jorge Orta, Paolo Pellegrin, Ricardo Roura, Sebastião Salgado, Connie Samaras, Santiago Sierra, Douglas Sonders, David Stephenson, Dave Walsh, and John Weller.

THE PLATFORMS that host public domain images, from satellite to small objects, championing the free dissemination of Antarctic knowledge and allowing us to share them across our book.

POMO for collaborating closely with UNLESS and the Polar Lab on the production of the publication's info-graphic visualisations. A special thanks goes to director Marco Cendron and to Gabriele Gastaldin, who led the following team: Teresa Cremonesi, Stefano Falconi, Nayla Giannakopoulos, Nils Oh, Carlo Ottaviani, Arianna Smaron, and Simone Verduci.

NOLAN PAPARELLI for having designed the typeface Everett and having allowed us to launch it officially with *Antarctic Resolution*.

LARS MÜLLER and his incredible team, especially Martina Mullis and Maya Rüegg, for accepting the challenge of designing an extremely dense thousand-page book with us, and for doing so, as ever, with passion and precision.

OUR GENEROUS SPONSORS, the D-Air Lab, the Fondazione Giancarlo Ligabue, the Fondazione Giuseppe e Pericle Lavazza, Furthermore: a program of the J. M. Kaplan Fund, the Graham Foundation for Advanced Studies in the Fine Arts, Arend and Brigitte Oetker, and TBA21, for their support in the production of the book.

Amongst them, we would like to extend a special thanks to D-Air Lab for embracing UNLESS's challenge to rethink Antarctic clothing and shift the design agenda from fashion to the realm of architecture, conceiving suits as portable environments. Our deepest gratitude goes to Lino Dainese, founder of Dainese and D-Air Lab, for pushing the idea to another level and producing prototypes of a new generation "Antarctic Suit". The development of the suits would not have been possible without the expertise of Luigi Ronco and Alberto Piovesan and the support of Vittorio Cafaggi. We are sincerely grateful to the PNRA/ENEA team comprising Rocco Ascione, Gianluca Bianchi Fasani, and Francesco Pellegrino for allowing D-Air Lab and UNLESS to test the prototype of the "Antarctic Suit" in Antarctica at Concordia Station in the 2020–2021 season.

THE FOLLOWING INDIVIDUALS who believed in the project and supported us in different ways: Hashim Sarkis (Dean of the School of Architecture and Planning at the Massachusetts Institute of Technology, MIT, and Director of the 17th International Architecture Biennale of Venice), and his curatorial assistants Gabriel Kozlowski and Roi Salgueiro, for inviting UNLESS to exhibit the research developed for *Antarctic Resolution* in the Central Pavilion of the Venice Biennale; Arcangelo Sassolino (artist) for abstracting the message of *Antarctic Resolution* and transforming it, with UNLESS and David Vaughan, into a highly powerful large-scale installation, which will be presented alongside the publication at the Biennale; Boris Hermann, who – prior to stepping on board his sailing boat to complete the 2020 Vendée Globe, while at the same time gathering rare ocean climate data with a SubCtech ocean laboratory device – established contacts for us with scientists from the Max Planck Society, amongst them Peter Landschützer, to whom we are very grateful for our long conversations about the Southern Ocean; Claire Scoville, who has established contacts for us with environmental activists and Outer Space experts, amongst them NASA Flight Director Zebulon Scoville, and has assisted our research in the National Archives in Washington, DC; Jacob Scherr, who reached out to Antarctic Southern Ocean Coalition (ASOC) founder James N. Barnes, inviting him on our behalf to participate in our publication; Sebastiano Arlotta Tarino, who patiently reviewed his National Geographic collection to unearth important Antarctic-related articles; Teresa de Anchorena, who kindly tried to help source information on Argentine Antarctic stations; Lisa Immordino Vreeland for her precious advice; Philip Oetker and Ferigo Foscari who have supported the project in countless ways; Bonnie Burnham, Mario Cerutti, Barbara Del Vicario, Richard Dunn, Tonci Foscari, Thomas Grunzke, David Hrankovic, David Landau, Salvatore Nastasi, Michael Neumann, Georgia Oetker, Gianluca Passi, Marilyn Perry, Felicitas von Peter, Markus Reymann, Herbert Scheidt, and Guido Venturini for helping us secure partners who could support the project.

Finally, I would like to extend a special thank you to Olympia and Ato. Looking at the world through their eyes – through the perspective of the next generations – one fully appreciates the gravity of the planetary crisis and is compelled to act. *Antarctic Resolution* is the first milestone of a cause that UNLESS will pursue relentlessly, driven by the conviction that not only the Antarctic is a Global Commons, but so is our Future.

Giulia Foscari and the UNLESS team

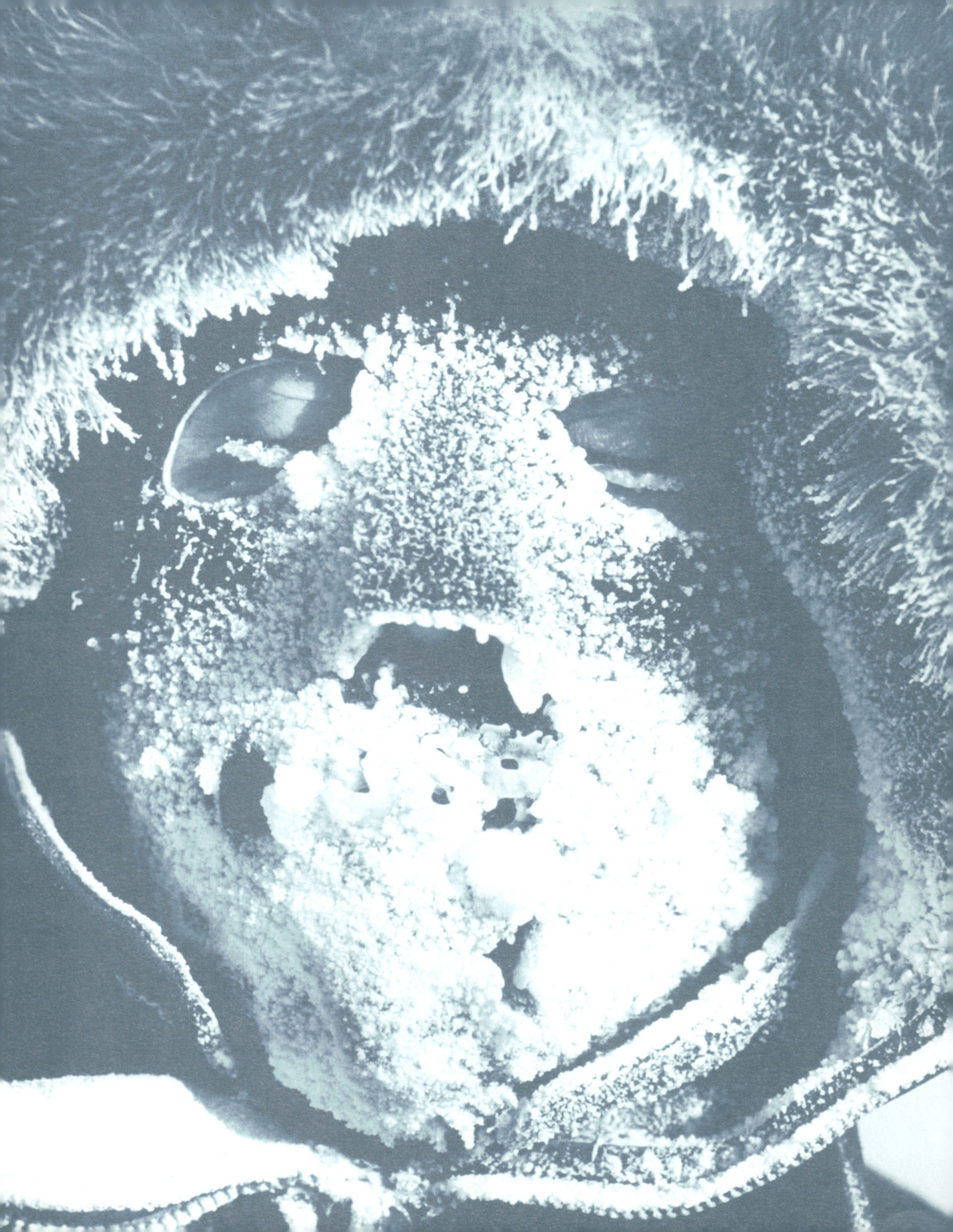

About the Authors

DOAA ABDEL-MOTAAL is the former executive director of the Rockefeller Foundation Economic Council on Planetary Health at the University of Oxford (United Kingdom) and has served in various high-level positions in the United Nations and other multilateral organisations. Dr Abdel-Motaal is the author of *Antarctica: The Battle for the Seventh Continent* (Praeger, 2016), nominated for the 2018 Mountbatten Award for Best Book. ↖ pp. 136, 182

CONRAD ANKER is an accomplished alpinist who pushes the limits of mountaineering in the world's most inhospitable environments. With twelve trips to Antarctica since 1992, his first ascents include Mount Craddock in the Ellsworth Mountains and Rakekniven Peak in Queen Maud Land. In 2017 Anker established, with Jimmy Chin, a new route on Ulvetanna in Queen Maud Land. ↖ p. 562

FRANCESCO BANDARIN is an architect and urban planner. He served as director of the UNESCO World Heritage Centre and as secretary of the World Heritage Convention (from 2000 to 2010) and as assistant director-general of UNESCO for culture (from 2010 to 2018). Bandarin is an associate member of the ICOMOS International Scientific Polar Committee. In 2019, he has been co-director of the Polar Lab United Kingdom. ↖ pp. 518, 524, 537

CARLO BARBANTE is the director of the Institute of Polar Sciences of the National Research Council (Italy), a professor at the University Ca' Foscari of Venice and a professor of Earth's climate at Ca' Foscari Harvard Summer School (United States). Barbante has participated in several expeditions to polar regions and in the Alps and has authored over 280 publications in scientific journals. ↖ pp. 242, 250

JAMES N. BARNES is chair of the board of the Antarctic and Southern Ocean Coalition (ASOC), which he co-founded in 1978. He served on the board of Greenpeace USA from 1982 to 1985 and became Greenpeace International's UN representative in 1983, serving in that position for five years. Barnes has devoted more than forty years of his work to protecting Antarctica. ↖ pp. 186, 187

THOMAS BARNINGHAM is science coordinator for the British Antarctic Survey (United Kingdom) and project manager for the Halley Automation Project. He joined the British Antarctic Survey following the completion of his PhD in 2017 and began working as an atmospheric scientist at Halley. Dr Barningham oversees the delivery of science objectives and ensures the continued delivery of automated science throughout the unmanned winter. ↖ p. 512

CARLO BARONI is a professor of geomorphology at the University of Pisa (Italy). He is a member of the Italian Scientific Commission on Antarctic Research and a delegate to the Standing Committee on Antarctic Geographic Information of the Scientific Committee on Antarctic Research (SCAR). Prof. Baroni has participated in fifteen national and international Antarctic expeditions, acting as scientific coordinator at Italy's Mario Zucchelli Station. ↖ p. 82

SUSAN BARR was the founding president of, and is now Arctic advisor to, the International Council on Monuments and Sites (ICOMOS) Polar Heritage Committee, which she founded in 2000. She has worked since 1979 solely with polar heritage and history, with extensive field work in both polar areas, authoring many publications. In 2019 Dr Barr was appointed a life member of ICOMOS Norway in recognition of her work for polar cultural heritage. ↖ pp. 106, 540

ELISA BERGAMI is a postdoctoral fellow at the University of Siena (Italy). She has been researching the impact of nanoplastics on marine organisms for the last six years, and has joined four international Antarctic expeditions to study plastic pollution. Dr Bergami is a member of the Steering Committee of the Plastic in Polar Environments Action Group of the Scientific Committee on Antarctic Research (SCAR). ↖ p. 159

MARCELO BERNAL is an architect and principal of ARQZE Ltd, a practice dedicated to designing, fabricating and installing mobile deployable infrastructure in Antarctica. Before obtaining his PhD in design computing at the Georgia Institute of Technology (United States), he was the director of the undergraduate program at the Federico Santa María Technical University (Chile). Dr Bernal currently is the director of the Design Process Lab at Perkins&Will, an interdisciplinary research-based international architecture firm (United States). ↖ p. 511

ANNE-MARIE BRADY is a professor of political science and international relations and a research associate at Gateway Antarctica at the University of Canterbury, Christchurch (New Zealand). Prof. Brady is a global fellow at the Wilson Center in Washington, DC (United States), and executive editor of *The Polar Journal.* ↖ pp. 98, 99

RALF BRAUNER is a professor of marine meteorology at the Jade University of Applied Sciences (Germany) in the Department of Maritime and Logistic Studies. Prof. Brauner has participated in several expeditions to Antarctica over the past twenty years, forecasting weather on the continent for several months at a time at research stations and on scientific vessels. ↖ p. 237

CASSANDRA M. BROOKS is an assistant professor of environmental studies at the University of Colorado Boulder (United States). Her expertise lies at the intersection of marine science, environmental policy and science communication. With a PhD from Stanford University on international ocean policy (*Policies for Managing the Global Commons: The Case of Marine Protected Areas in Antarctica*), Brooks was awarded a Switzer Fellowship in Environmental Leadership. In the past fifteen years she has published over 200 articles in scientific journals, books and popular outlets. ↖ p. 149

SHAUN T. BROOKS is an associate at the University of Tasmania (Australia). He is an environmental scientist specialising in conservation with field experience across East Antarctica and the subantarctic. Dr Brooks' research focuses on the effects of the anthropic footprint on Antarctica as a means of promoting informed environmental protection policies. ↖ p. 487

HUGH BROUGHTON is an architect and the founder of Hugh Broughton Architects (United Kingdom). He is the designer of the Halley VI Antarctic Research Station, the world's first relocatable polar base, and the Juan Carlos I Spanish Antarctic Base. Broughton is currently developing the designs for the Scott Base for Antarctica New Zealand, the Discovery Building at Rothera for the British Antarctic Survey (United Kingdom), and a master plan for Davis Station for the Australian Antarctic Division. ↖ pp. 422, 456, 466, 469

BERT BÜCKING is an architect and a co-founder of bof Architekten (Germany). His experience focuses on the interface between civil engineering and architecture. Bücking's interest in modular and sustainable building techniques led him to design and build, in collaboration with Ramboll Engineering, Bharati Station for the Indian Antarctic Program. ↖ p. 418

DAVID BURROWS is a geographer and senior principal software developer at the Environmental Systems Research Institute (United States), where he leads the development team focused on map projections and coordinate transformations. ↖ p. 54

SOL CAMACHO is the director of the Polar Lab Brazil. She is an architect, urban designer, researcher and curator based in São Paulo, where she directs RADDAR, a studio working on multi-scalar architecture, landscape heritage conservation and adaptive reuse projects. Arch. Camacho is co-curator and editor of the "Walls of Air" project developed for the Brazilian Pavilion at the 16th Venice Architecture Biennale. ↖ p. 49

SANJAY CHATURVEDI is a professor of international relations and dean of the Faculty of Social Sciences at South Asian University, New Delhi (India). He is the chairman of Indian Ocean Research Group, Inc. (IORG), chief editor of the *Journal of the Indian Ocean Region* (Routledge), and regional editor of *The Polar Journal.* Chaturvedi's publications include *The Polar Regions: A Political Geography* (John Wiley & Sons, 1996) and, co-authored, *Climate Terror: A Critical Geopolitics of Climate Change* (Palgrave Macmillan, 2015). ↖ p. 174

CHRISTY COLLIS is the associate director of the Office for Learning and Teaching at the University of Southern Queensland (Australia). She is a cultural geographer whose research focuses on the legal and cultural dynamics of territorial possession, specifically applied to Antarctica, outer space, and the deep seabed. Her writings include "Territories Beyond Possession? Antarctica and Outer Space" (*The Polar Journal,* 2017). ↖ pp. 172, 584

PETER CONVEY is a polar ecologist based at the British Antarctic Survey (United Kingdom), where he is currently the deputy leader of the Survey's Biodiversity, Evolution and Adaptation Team. He has over thirty-one years' experience in Antarctic and Arctic environments. Dr Convey's research focuses on the evolution of polar biodiversity and biogeography, analysing the polar regions to model past and future global consequences of climate change. ↖ p. 144

GEOFF COOPER is Heritage Programme Manager for the United Kingdom Antarctic Heritage Trust, responsible for the conservation of six British Historic Sites on the Antarctic Peninsula. A conservation carpenter by trade, Cooper has spent five seasons working for the New Zealand and United Kingdom Antarctic Heritage Trusts in the Ross Sea and on the Antarctic Peninsula. ↖ p. 409

GABRIELE COPPI is a researcher at the University of Pennsylvania (United States), having been awarded a PhD from the University of Manchester (United Kingdom). He has managed the power system and the in-flight mapping software of the Antarctic Balloon-Borne Large-Aperture Submillimeter Telescope since 2018. Dr Coppi's writings include "The Balloon-Borne Large Aperture Submillimeter Telescope Observatory" and "In-Flight Performance of the BLAST-TNG Telescope Platform" (SPIE, 2020). ↖ p. 315

ILARIA CORSI is the coordinator of the Plastics in Antarctic Environment project (PLANET) funded by the Italian National Antarctic Research Programme. Prof. Corsi has been involved in Antarctic research for more than fourteen years and is a member of the Scientific Committee on Antarctic Research (SCAR)'s Steering Committee of the Plastic in Polar Environments Action Group. ↖ p. 159

LINO DAINESE is the founder of Dainese and D-Air Lab (Italy), which are, respectively, a world-leading manufacturer of technical gear and protective equipment for dynamic sports founded in 1972 (a recipient of multiple design awards) and its research and development lab devoted to the prototyping of intelligent clothing for extreme environments and risk prevention in quotidian applications. In collaboration with UNLESS, D-Air Lab developed an "Antarctic suit". ↖ p. 565

KLAUS DODDS is a professor of geopolitics and the director of research and knowledge exchange at Royal Holloway, University of London (United Kingdom). He is co-editor of *Handbook on the Politics of Antarctica* (Edward Elgar, 2017), co-author of *The Scramble for the Poles* (Polity, 2016) and the editor-in-chief of *Territory, Politics, Governance* (Routledge). Prof. Dodds is an honorary fellow of the British Antarctic Survey and a trustee of the Royal Geographical Society (United Kingdom) and Regional Studies Association (United Kingdom). ↖ p. 114

CHRISTEL MISUND DOMAAS is an advisor at the University Library in Tromsø (Norway). Misund Domaas holds a master's degree in history from the University of Tromsø and a postgraduate certificate in Antarctic studies from the University of Canterbury (New Zealand). ↖ p. 108

JULIAN A. DOWDESWELL is the director of the Scott Polar Research Institute and a professor of physical geography at the University of Cambridge (United Kingdom). He is the former chair of the UK National Committee on Antarctic Research and a past delegate to the Council of the Scientific Committee on Antarctic Research (SCAR). Having led more than forty polar expeditions, Prof. Dowdeswell has received many awards, among them the Polar Medal from HM The Queen and the Founder's Medal of the Royal Geographical Society. ↖ p. 544

CHRIS DRURY is an eco land artist who has worked on every continent on Earth, including Antarctica, where he spent three months with the British Antarctic Survey on the Artists' and Writers' Fellowship in 2006–2007. In 2018 Drury was awarded the Lee Krasner Lifetime Achievement Award. ↖ p. 222

JUAN DU is the director of the Polar Lab Hong Kong. She is an associate professor and associate dean of architecture at the University of Hong Kong, where she directs the Urban Ecologies Design Laboratory. The projects within her practice, IDU, range from community centres to informal settlements upgrades. Prof. Du's extensive writings include *The Shenzhen Experiment* (Harvard University Press, 2020). ↖ pp. 154, 366, 377, 450

GRAEME EAGLES is a senior scientist at the Alfred Wegener Institute for Polar and Marine Research in Bremerhaven (Germany). He is the head of Airborne Geophysics at the Alfred Wegener Institute and has participated in numerous Antarctic marine and air-borne expeditions. Dr Eagles completed a PhD on Antarctic plate tectonics at the British Antarctic Survey and the University of Leeds (United Kingdom). ↖ p. 288, 293

TESS EGAN is library manager at the Australian Antarctic Division. With a background in information science and history, she is also an independent researcher and editor. ↖ p. 510

ALEXEY A. EKAYKIN is a researcher at the Consortium of European Research Libraries at the Arctic and Antarctic Research Institute (Russia) with over twenty years of experience in the fields of glaciology, paleogeography, and isotope geochemistry. Dr Ekaykin has participated in 16 expeditions to the Antarctic plateau to conduct glaciological works, mass-balance studies, ice drilling and ice core studies, and to explore subglacial lakes. ↖ p. 256

FAUSTO FERRACCIOLI is the science leader of the Geology and Geophysics Team and the head of Airborne Geophysics at the British Antarctic Survey (United Kingdom). He has led aerogeophysical studies of Antarctic subglacial geology over the last twenty-five years and has published 130 papers, including in *Nature* and *Science*. In 2010, Ferraccioli was awarded the Polar Medal by HRH Prince Philip for outstanding dedication and achievements in polar science. ↖ p. 295

JOE FERRARO is an architect and a founding partner of Ferraro Choi and Associates Ltd (United States). He has been a lead designer of buildings and infrastructure in Antarctica since 1984. In addition to the Amundsen-Scott South Pole Station, Ferraro was the architect for the Albert P. Crary Science and Engineering Center at McMurdo Station and is currently the principal in charge of the design of the stations' Information Technology and Communications Primary Facility. ↖ p. 442

JAMES FLEMING is the Charles A. Dana Professor of Science, Technology, and Society at Colby College and a research associate at the Smithsonian Institution (United States). Having earned degrees in astronomy at Penn State University and atmospheric science at Colorado State University as well as a PhD in history at Princeton University (United States), Prof. Fleming has written extensively on the social, cultural and intellectual history of the International Geophysical Year and its predecessors. ↖ p. 110

GIULIA FOSCARI WIDMANN REZZONICO is an architect, researcher, and writer who has been practising in Europe, Asia, and the Americas. She is the founder of UNA, an architecture studio focused on cultural projects, and of its alter ego UNLESS, a not-for-profit agency for change devoted to interdisciplinary research on extreme environments. Foscari taught at Hong Kong University and at the Architectural Association, where she ran a Diploma Unit and founded the Polar Lab. She authored *Elements of Venice* and is the editor of *Antarctic Resolution* (Lars Müller Publishers). ↖ pp. 11, 350, 435, 448, 458, 507

ADRIAN FOX is the head of the Mapping and Geographic Information Centre at the British Antarctic Survey (United Kingdom) and secretary of the United Kingdom Antarctic Place-Names Committee. He was co-chief officer of the Standing Committee on Antarctic Geographic Information of the Scientific Committee on Antarctic Research (SCAR) from 2012 to 2020. Dr Fox has led thirteen aerial photography and survey fieldwork campaigns in Antarctica since 1990. ↖ p. 82

WILLIAM L. FOX is the founding director of the Center for Art + Environment at the Nevada Museum of Art in Reno, Nevada (United States). He is an art critic, science writer, and cultural geographer and has published sixteen books on cognition and landscape, including *Terra Antarctica: Looking into the Emptiest Continent* (Counterpoint, 2007). Fox spent two and a half months in Antarctica with the US Antarctic Artists and Writers Program. ↖ pp. 39, 303, 551

BOB FRAME is an adjunct associate professor at the Gateway Antarctica Centre for Antarctic Studies and Research at the University of Canterbury (New Zealand). A futures specialist, he has a particular interest in Antarctic scenarios, climate change and cultural heritage. Prof. Frame has written numerous journal articles on various aspects of long-term futures as they relate to climate change, sustainability and complexity. ↖ p. 130

PETER T. FRETWELL is a geographic information officer with the British Antarctic Survey (United Kingdom). He chairs several international Antarctic groups, including the Bedmap3 project and the Censusing Animal Populations from Space group. Dr Fretwell's interests include Antarctic satellite remote sensing of wildlife, mapping the terrestrial, ocean, and subglacial environment, polar geographic analysis, and GIS. He has recently published the *Antarctic Atlas* (Penguin, 2020). ↖ p. 290

LUTZ FRITSCH is an artist renowned for his large-scale sculptures, among them his landmark *Library in Ice* (2005), which is the only permanent piece of art in Antarctica. ↖ p. 412

JACOPO GABRIELI is a researcher at the Institute of Polar Sciences of the National Research Council (Italy) involved in the study of climatic archives in ice cores with the implementation of innovative analytical techniques. Gabrieli has participated in various key research projects on Alpine glaciers, in Greenland and Antarctica and on the Svalbard islands. ↖ pp. 242, 250

HARTWIG GERNANDT has been a scientist and polar researcher at the Alfred Wegener Institute for Polar and Marine Research (Germany) since 1992. Active in polar science for over fifty years, he led an East German expedition to Antarctica from 1975 until 1977, during which his containerised concept of a small permanent research station was realised. Later, as the head of AWI's polar infrastructure, Dr Gernandt oversaw the construction of Neumayer III Station. ↖ pp. 412, 504

ANDREW J. GERRARD is a professor and Physics Department chair at the New Jersey Institute of Technology and the director of the Center for Solar-Terrestrial Research (United States). Prof. Gerrard's research interests include upper atmospheric and magnetospheric processes at high latitudes and involve numerous geospace instrumentation suites distributed across the Antarctic, with an emphasis on the Eastern High Plateau. ↖ p. 300

NEIL GILBERT has worked on polar issues in various roles for over thirty years. A former environmental manager at Antarctica New Zealand, he attended numerous meetings of the Antarctic Treaty and its Committee for Environmental Protection (CEP), representing both the United Kingdom and New Zealand. As the founder and director of Constantia Consulting, Dr Gilbert focuses on a range of Antarctic environmental and policy projects. ↖ pp. 474, 477

EL GLASBERG is a senior lecturer in expository writing for the Tisch School of the Arts at New York University (United States). Glasberg first discovered Antarctica through Edgar Allan Poe's 1838 sea journey *The Narrative of Arthur Gordon Pym*, and authored *Antarctica as Cultural Critique: The Gendered Politics of Scientific Exploration and Climate Change* (Palgrave, 2012). ↖ p. 63

KARSTEN GOHL is a geophysicist and senior scientist at the Alfred Wegener Institute for Polar and Marine Research (Germany). Dr Gohl conducts and supervises research on geodynamic, tectonic and glacial sedimentation processes of polar and subpolar regions, leading numerous ship expeditions to Antarctica and other parts of the world. ↖ p.282

FRANCIS HALZEN is principal investigator at the IceCube Neutrino Observatory at the South Pole, and Hilldale and Gregory Breit Distinguished Professor of Physics and the director of the Institute for Elementary Particle Physics at the University of Wisconsin–Madison (United States). Prof. Halzen has been a fellow of the American Physical Society since 1994 and is the recipient of numerous awards and honorary doctorates from several universities. ↖ p.319

KAEL HANSON is a professor of physics and the director of the Wisconsin IceCube Particle Astrophysics Center (United States). Prof. Hanson has been a researcher in the field of neutrino astrophysics, with a specialisation in detector development, for over twenty-five years and has been an IceCube collaborator for twenty years. ↖ p.319

URSULA HARRIS is a mapping and spatial data manager at the Australian Antarctic Data Centre (Australia). Harris is a cartographer who has visited Antarctica in different roles. She is interested in mapping and place names, in particular the important role they play in search and rescue operations, management and science. ↖ p.82

JUDITH HAUCK is the head of the Helmholtz Young Investigator Group for Marine Carbon and Ecosystem Feedbacks in the Earth System (MarESys) and deputy head of Marine Biogeosciences at the Alfred Wegener Institute for Polar and Marine Research (Germany). Dr Hauck is responsible for the ocean carbon sink estimate in the annual Global Carbon Budget report (GCB). ↖ p.206

ROBERT HEADLAND is a fellow of the Royal Geographical Society who has served as an archivist and curator for the Scott Polar Research Institute, University of Cambridge (United Kingdom). Headland became involved with the Antarctic Heritage Trust, authored *A Chronology of Antarctic Exploration* (Cambridge University Press, 2009) and is the recipient of the Polar Medal (United Kingdom, 1984). ↖ pp.30, 72

BETH HEALEY was deployed by the European Space Agency (ESA) as a research medical doctor at space-flight analogue Concordia Station on the Antarctic plateau. Dr Healey's interests in extreme and remote environments have led her to be part of medical and logistical support teams for ski mountaineering expeditions and endurance races from Greenland to the North Pole. ↖ p.575

ALAN D. HEMMINGS is professor at the Gateway Antarctica Centre for Antarctic Studies and Research at the University of Canterbury (New Zealand). As a polar specialist, he focuses particularly on Antarctic geopolitics and governance. Dr Hemmings is the co-editor of *Handbook on the Politics of Antarctica* (Edward Elgar, 2017), *International Polar Law* (Edward Elgar, 2018), and *Philosophies of Polar Law* (Routledge, 2020). ↖ pp.130, 179, 582

TE WARIHI HETARAKA is an elder and tōhunga whakairo from Ngātiwai, Ngāpuhi and Tainui in the north of New Zealand. He was a graduate of the first intake at the NZ Māori Arts and Crafts Institute in the 1960s. Hetaraka has been recognised by Creative New Zealand for his lifetime contribution to Māori arts and culture. ↖ p.536

POUTAMA HETARAKA is Ngātiwai, Ngāpuhi and Tainui on his father's side and Ngāi Tahu on his mother's side. He is an emerging carver and artist being trained by his father, Te Warihi, at the Hihiaua Cultural Centre in Whangārei. Hetaraka's involvement in the Antarctica whakairo project has expanded his interest in issues of climate change and global warming. ↖ p.536

ALFRED HIATT is a professor of medieval studies at Queen Mary University of London (United Kingdom). His publications on maps and spatial representation include *Terra Incognita: Mapping the Antipodes before 1600* (University of Chicago Press, 2008). ↖ pp.41, 42, 43

ADRIAN HOWKINS is co-principal investigator at the United States' National Science Foundation (NSF)-funded McMurdo Dry Valleys Long-Term Ecological Research project in Antarctica, and a reader of environmental history at the University of Bristol (United Kingdom). Focusing on the environmental history of the polar regions, Dr Howkins' publications include *Frozen Empires: An Environmental History of the Antarctic Peninsula Region* (Oxford University Press, 2017). ↖ p.88

KEVIN A. HUGHES is vice-chair of the Committee for Environmental Protection (CEP), a member of the United Kingdom delegation to the Antarctic Treaty Consultative Meeting (ATCM), and an environmental researcher at the British Antarctic Survey (United Kingdom). His Antarctic interests include conservation, area protection, geological heritage, non-native species and environmental monitoring. Dr Hughes has visited the polar regions eleven times and overwintered once. ↖ p.139

ANDREW T. HYNOUS is mission operations manager for the United States' National Aeronautics and Space Administration (NASA) Balloon Program Office, which conducts Long-Duration Balloon campaigns from McMurdo Station. Hynous directly supervises coordination efforts with the Columbia Scientific Balloon Facility, the National Science Foundation (NSF), the Antarctic Support Contract, and the United States Antarctic Program for campaign planning. ↖ p.308

JULIA JABOUR is a senior lecturer in law and policy at the University of Tasmania (Australia). She has served as an observer with the Australian delegation at Antarctic Treaty Consultative Meetings (ATCM) on a number of occasions. She has been researching, writing and lecturing on polar governance for more than 20 years and hosted the annual Polar Law Symposium in Hobart (Australia) in 2014 and 2019. Dr Jabour has visited Antarctica six times. ↖ p.160

STÉPHANIE JENOUVRIER is an associate scientist at the Woods Hole Oceanographic Institution (United States). Her work as an ecologist focuses on predicting the effect of climate change on population dynamics, specifically on seabirds in the Southern Ocean. Dr Jenouvrier contributes to characterising species' responses to climate change, which informs the Intergovernmental Panel on Climate Change (IPCC) assessment. ↖ p.225

SOLAN JENSEN is a professional expedition leader on the Peninsula of Western Antarctica and in the Arctic region. He received a degree in philosophy, leading him to a career in the polar territories. Jensen first visited Antarctica in 2002 as part of the construction team of the new Amundsen-Scott South Pole Station and has been working on the continent since. ↖ p.167

ANDREA KAVANAGH is the director of The Pew Charitable Trusts. Her work promotes efforts to protect Antarctica's Southern Ocean, one of the world's last great wilderness areas. Kavanagh has worked for more than two decades to establish solid marine protections and precautionary toothfish and krill fishing regulations in the Southern Ocean. ↖ p.152

WILLIAM KENTRIDGE is an internationally acclaimed artist based in Johannesburg, South Africa. His work spans a variety of media which include drawing, writing, film, performance, music, theatre and collaborative practices to create works of art that are grounded in politics, science, literature and history, while maintaining a space for contradiction and uncertainty. His

aesthetics are drawn from the medium of film's own history, from stop-motion animation to early special effects. ↖ p.48

DANIEL KISS and SWADHEET CHATURVEDI hold M.Arch. degrees in landscape urbanism from the Architectural Association School of Architecture (United Kingdom), where they took an active part in the Polar Lab. Their thesis, "Dynamic Domains of Antarctica", envisions the dynamic management of commercial krill fishing activities through the models of regulatory systems of global common resources. ↖ p.156

GEORG KLEINSCHMIDT is a professor emeritus of geology at the Goethe University of Frankfurt (Germany), where he was a professor from 1985 to 2003. His research focuses on the geological evolution of Gondwana, the Antarctic orogens and ancient cratons. Prof. Kleinschmidt has published extensively on Antarctica and has participated in ten Antarctic expeditions, including multiple editions of GANOVEX (the German Antarctic North Victoria Land Expedition). ↖ p.282

ALEXANDER KLEPIKOV is the head of the Russian Antarctic expedition. With a PhD in physics and mathematics, specialising in oceanology, he has participated in multiple Arctic and Antarctic expeditions, including the international expedition to the Weddell Sea. Dr Klepikov is a reviewer of European Science Foundation projects, a SCAR Team Expert on Antarctic Climate Change and Environment, and a member of the SCAR-IASC Group. ↖ p.539

PETER LANDSCHÜTZER is a marine biogeochemist at the Max Planck Institute for Meteorology (Germany), specialising in the exchange of carbon dioxide between the ocean and the atmosphere. He works with ocean observations and novel data interpolation techniques such as neural networks. Dr Landschützer is particularly interested in the Southern Ocean and the processes driving the air-sea carbon dioxide flux in high-latitude waters. ↖ p.212

LOUIS J. LANZEROTTI is a distinguished research professor of physics at the New Jersey Institute of Technology (United States). He served as principal investigator on several NASA interplanetary and planetary missions and has a long career of space- and ground-based research, including in the Antarctic at Siple Station, South Pole Station, Arrival Heights, and the deep field at Automatic Geophysical Observatories. Prof. Lanzerotti has chaired numerous committees and is the recipient of several awards, including two NASA Distinguished Public Service Medals. ↖ p.300

ELIZABETH LEANE is Associate Dean Research in the College of Arts, Law and Education at the University of Tasmania (Australia), chief officer of the Standing Committee on Humanities and Social Sciences within the Scientific Committee on Antarctic Research (SCAR), and arts editor of *The Polar Journal*. With degrees in physics and literary studies, Prof. Leane has published seven books, including *Antarctica in Fiction* (Cambridge University Press, 2012). ↖ pp.57, 356, 550

SANG LEEM LEE is an architect and the chairman of the Space Group architectural practice (South Korea). He is the honorary president of the Korean Institute of Architects and was a UNESCO chair professor (Social Sustainability of Historic Districts). He has participated in numerous exhibitions, including at the Venice Biennale. Lee's architectural projects include the Jang Bogo Antarctic Research Station in Antarctica. ↖ p.430

INTI LIGABUE is the president of the Giancarlo Ligabue Foundation (Italy), an organisation committed to pursuing and promoting cultural activities in various fields from anthropology to the figurative arts. Ligabue has been the chairman and chief executive officer of the Ligabue Group since 2012, a leading company in the food service and life support sector of the maritime and industrial market that has been supplying Antarctic stations since the late 1980s. ↖ p.374

DANIELA LIGGETT is the chief officer of the Standing Committee on the Humanities and Social Sciences at the Scientific Committee on Antarctic Research (SCAR). She is a social scientist with a background in environmental management, Antarctic politics and tourism. Dr Liggett is a senior lecturer at the University of Canterbury (New Zealand) and is on the editorial boards of *Polar Geography*, *The Polar Journal* and *Advances in Polar Science*. ↖ p.162

BRYAN LINTOTT serves as secretary-general of the ICOMOS International Polar Heritage Committee. He is a historian based at the Scott Polar Research Institute, University of Cambridge, and a member of the University of Tromsø in the Norwegian Arctic. Dr Lintott's research focuses on how heritage in extreme environments, beyond national boundaries, is governed, managed, conserved, and utilised. ↖ p.558

VLADIMIR LIPENKOV is the head of the Consortium of European Research Libraries at Russia's Arctic and Antarctic Research Institute. He supervised deep drilling of the Antarctic glacier at Vostok Station, is a scientific advisor to the Russian-French Glacial Archives of Climate and Environment Data Lab, and represents Russia on the International Partnership in Ice Core Sciences committee. Dr Lipenkov was knighted in France for his service to Antarctic research. ↖ p.256

CORNELIA LÜDECKE is the chief officer and founder of the Expert Group on the History of Antarctic Research within the Scientific Committee on Antarctic Research (SCAR). She is a professor of the history of natural sciences at the University of Hamburg (Germany), focusing on the history of polar research and meteorology. Prof. Lüdecke is a member of the editorial boards of scientific journals, including *Polarforschung*, *Polar Record* and *The Polar Journal*. ↖ p.76

ARTURO LYON is the director of the Polar Lab Chile. He is an assistant professor at the School of Architecture of the Pontifical Catholic University of Chile. Lyon is the founder of Lyon Bosch Martic, where he developed architecture projects including the design of Las Majadas Hotel, the Chena Public Park, and the winning proposal for Nueva Alameda Providencia. ↖ pp.286, 361, 530

JAMES MADSEN is the executive director of the Wisconsin IceCube Particle Astrophysics Center and the associate director of the IceCube education and outreach programme (United States). He has been involved in efforts to detect neutrinos at the South Pole for over twenty years. In the past decade, Dr Madsen has worked on a smaller project to study low-energy cosmic rays using neutron monitors. He has deployed to Antarctica five times. ↖ p.319

CRAIG MCCORMACK is a lecturer and PhD candidate at the University of Western Australia's School of Design (Australia). A recent Fulbright postgraduate scholar and co-creative director of the 2014 Venice Architecture Biennale Australian Pavilion, McCormack focuses on the possibility of architecture in outer space. ↖ p.570

TONY MCGLORY is the director of Ramboll United Kingdom's Building Services Engineering team. Eng. McGlory has over thirty years' experience designing complex engineering projects. His responsibilities at Ramboll include leading Building Services Engineering design for projects in Antarctica and the subantarctic islands. ↖ pp.478, 484, 490, 494

HANS-JÜRGEN MEYER is a physicist who teaches building physics at the University of Applied Science in Bremen (Germany). Prof. Dr Meyer participated in the planning and construction of Neumayer III Station and has been a technical consultant to its operation until now. ↖ p.412, 504

JOSEPH MICALLEF is a historian, author, keynote speaker, syndicated columnist and war correspondent. He has been an advisor to the American government and is currently the honorary consul to Oregon from the Republic of Malta. Micallef has also appeared as a commentator on a variety of broadcast venues including CNN, Fox News and Fox News Radio. He is an opinion columnist for Military.com and a contributor to *Forbes*. ↖ p.94

NICHOLAS DE MONCHAUX is a professor and head of the Department of Architecture at the Massachusetts Institute of Technology (MIT) and a partner in the Modem architecture practice. He is the author of *Spacesuit: Fashioning Apollo* (MIT Press, 2011). In 2012 he was named one of the "Public Interest Design 100" by *Good* magazine. Prof. de Monchaux's work has been widely exhibited, and he is a fellow of the American Academy in Rome. ↖ p.560

CHIARA MONTANARI has seventeen years of work experience in the polar environment, which saw her leading expeditions to the Italo-French Concordia Station and the Belgian Princess Elisabeth Station. An engineer and explorer by training, Montanari founded Complexity Aware, a consultancy for enterprises interested in developing an "Antarctic mindset", and authored *Chronicles from the Ice: 90 Days in Antarctica* (Mondadori, 2015). ↖ p.394

MICHAEL MORRISON is a conservation architect and the British representative at the International Polar Heritage Committee of the International Council on Monuments and Sites (ICOMOS). As the former chairman and senior partner of Purcell Architects, Prof. Morrison has worked, since 2002, on conservation projects devoted to Antarctic historic huts and whaling stations on the Peninsula, Ross Island, and South Georgia. ↖ pp.408, 532

TEASEL MUIR-HARMONY is the curator of the Apollo Collection at the Smithsonian National Air and Space Museum (United States). Dr Muir-Harmony contributed to the volume *Globalizing Polar Science: Reconsidering the International Polar and Geophysical Years* (Palgrave, 2010) and is the author of *Apollo to the Moon: A History in 50 Objects* (National Geographic, 2018) and *Operation Moonglow: A Political History of Project Apollo* (Basic Books, 2020). ↖ p.555

JOHN NELSON is a cartographer and user experience designer at the Environmental Systems Research Institute (United States). Nelson creates maps, experimental cartographic techniques and related Web experiences. ↖ p.54

CAMILLA NICHOL is the chief executive of the United Kingdom Antarctic Heritage Trust, a British charity in charge of the conservation of six heritage sites on the Antarctic Peninsula and promoting Antarctic heritage in the United Kingdom. Nichol holds an AMA from the Museums Association and is a fellow of the Royal Geographical Society, a trustee of the Burton Constable Foundation, and chair of the Cromwell Museum Trust (United Kingdom). ↖ pp.92, 528

MIRANDA NIEBOER is an architect, a lecturer at the University of Tasmania and an affiliated researcher at the Institute for Marine and Antarctic Studies in Hobart (Australia). Dr Nieboer's interdisciplinary research on "Antarctic interiors" is informed by her participation in the logistical traverse of French Polar Institute Paul-Émile-Victor to Concordia Station in the continental interior of Antarctica. ↖ p.570

ANNE NOBLE is an artist, a professor of fine arts at Massey University (New Zealand), and an NZ Arts laureate. She has visited Antarctica as an NZ Arts fellow and a US National Science Foundation (NSF) fellow. Her work examines the imagination and representation of Antarctica. She has exhibited and published widely on the photographic construction of the Antarctic imaginary. Her books include *Ice Blink: An Antarctic Imaginary* (Clouds, 2011), *These Rough Notes* (VUW, 2012), and *The Last Road* (Clouds, 2013). ↖ pp.66, 400

DIRK NOTZ is a professor of cryosphere research at the University of Hamburg, group leader in the Ocean in the Earth System Department at the Max Planck Institute for Meteorology (Germany), and a lead author of the sixth report of the Intergovernmental Panel on Climate Change (IPCC). Dr Notz investigates the past and future evolution of Antarctic and Arctic sea ice, both in the field in polar regions and in large-scale model simulations. ↖ p.228

SHAUN O'BOYLE is a photographer and architectural designer. His work uses photography to explore architecture and built environments in high-latitude regions. He is a three-time grantee of the National Science Foundation (NSF)'s Antarctic Artists and Writers Program and the recipient of a Guggenheim fellowship in photography. O'Boyle's work has been featured in numerous galleries, museums, books and magazines, including *Smithsonian* magazine. ↖ p.546

MADELEINE O'KEEFE is a communications specialist at the IceCube Neutrino Observatory (United States). She has been communicating science, especially particle physics and astrophysics, for over three years. O'Keefe holds a BA in astronomy and an MS in science journalism and has previously written for the CMS Experiment at CERN (Switzerland) and Argonne National Laboratory (United States). ↖ p.319

LUCY + JORGE ORTA are artists. They operate a collaborative visual arts practice and employ a diversity of media including painting, sculpture, film and performance to realise bodies of work that address key social and ecological challenges. Aided by the logistical crew and scientists stationed at Marambio Base in the Antarctic, the pair realised the ephemeral art piece *Antarctic Village No Borders* and raised the *Antarctic Flag*. ↖ p.79

LAWRENCE A. PALINKAS is the Albert G. and Frances Lomas Feldman Professor of Social Policy and Health at the USC Suzanne Dworak-Peck School of Social Work at the University of Southern California (United States). He also serves as a professor of anthropology and preventive medicine at the University of Southern California (United States). A polar researcher since 1985, Dr Palinkas was awarded the Antarctic Service Medal by the National Science Foundation (NSF) and the US Navy. ↖ p.344

SCOTT PARAZYNSKI is a former NASA astronaut and a veteran of five space shuttle missions and seven spacewalks. Inducted into the US Astronaut Hall of Fame in 2016, he is the first to have both flown in space and summited Mount Everest. He was founding chief medical officer of the Center for Polar Medical Operations at UTMB-Galveston (United States), overseeing health and medical care for the US Antarctic Program. Dr Parazynski is currently the CEO of Fluidity Technologies (United States). ↖ p.580

CAROLINA PASSOS is an architect, urban designer and data scientist. With combined expertise in planning, design and GIS research, she is the founder of, and an urban data expert at, Mapping Lab, which has participated at the 2018 Venice Architecture Biennale, the 2019 Biennale of São Paulo and the 2019 Seoul Biennale. ↖ p.49

MICHAEL PEARSON is the president of the International Council on Monuments and Sites (ICOMOS) International Polar Heritage Committee. He is a historical archaeologist and heritage planner who has worked in Australia and internationally. Dr Pearson has taken part in ten research expeditions to Antarctica and has published extensively on polar history and heritage, industrial heritage, and world heritage. ↖ pp. 404, 534

FRANCESCO PELLEGRINO is the technical manager of Mario Zucchelli Station in Antarctica. A mechanical engineer by training, he is a member of ENEA's Antarctic Unit (Italy), a construction manager of renewable energy plants, and a site supervisor of structural maintenance works. Having visited the continent six times, Eng. Pellegrino is an expert on renewable energy and energy-saving plants for application at extreme sites. ↖ p. 415

RICK PETERSEN is an architect who is a partner at OZ Architecture (United States). Petersen led the master planning of McMurdo and Palmer Stations, as well as the building design of McMurdo Station under the Antarctic Infrastructure and Modernization for Science project, an initiative of the National Science Foundation (NSF)'s United States Antarctic Program. ↖ p. 453

KATHERINA PETROU is a phytoplankton eco-physiologist. She is a senior lecturer at the School of Life Sciences, programme director at Marine Biology and leader of the Marine Microphycology Lab at the University of Technology Sydney (Australia). Specialising in Southern Ocean and Antarctic phytoplankton responses to climate change, Dr Petrou's interests include phytoplankton phenotypic plasticity and phytoplankton-bacteria interactions. ↖ p. 215

ANDREA PIÑONES is an oceanographer at the Research Center Dynamics of High-Latitude Marine Ecosystems (IDEAL) at the Universidad Austral de Chile and at the Center for Oceanographic Research (COPAS Sur-Austral) at the Universidad de Concepción (Chile). Piñones specialises in biological-physical interactions in high-latitude systems, physical oceanography, and climate change. ↖ p. 226

JEAN-YVES PIRLOT is a former deputy director general of the Belgian Mapping Agency for the National Geographic Institute (Belgium). While at NGI he became co-chief officer of the Standing Committee on Antarctic Geographic Information of the Scientific Committee on Antarctic Research (SCAR) in 2012. Pirlot is the current chair of the working group in charge of finalising the International Principles and Procedures for Antarctic Place Names. ↖ p. 82

CEISHA POIROT is New Zealand's representative to the Committee for Environmental Protection (CEP) and the expert group leader of the Council of Managers of National Antarctic Programs (COMNAP) Environmental Protection Group. Poirot has been involved in Antarctic policy and environmental management for twelve years, working for Antarctica New Zealand. ↖ pp. 474, 477

JEAN DE POMEREU is a photographer, an associate of the Scott Polar Research Institute at Cambridge University and a fellow of the Royal Geographical Society (United Kingdom). His interest in Antarctica includes historical research, photography, curating, publishing and journalism, spanning the history of Antarctic science and visual culture. Dr de Pomereu has participated in several expeditions to Antarctica and his work has been widely exhibited. ↖ p. 60

ALEXANDER PONOMAREV is an artist and professional sailor. He has staged several art projects mostly in the oceans, the Arctic and Antarctica. He represented Russia in the National Pavilion at the Venice Biennale and conceived the first Contemporary Art Biennale in the Antarctic, held on board the research vessel *Akademik Sergey Vavilov*. Ponomarev is a member of the Russian Academy of Arts and an Officier des Arts et des Lettres awarded by France. ↖ p. 270

PREFALUX is a general construction company specialising in wood construction, roofing and finishing, based in Luxembourg. Prefalux played an important role in the creation of Belgium's Princess Elisabeth Antarctica station, in particular in the development, production, and construction of the structure's wooden frame, its exterior walls, and its roof. ↖ p. 498

BRIAN RAUCH is principal investigator for the National Aeronautics and Space Administration (NASA) of the SuperTIGER stratospheric balloon-borne instrument flown from Antarctica and a co-investigator on the ANITA and X-Calibur balloons. Since he obtained his PhD for his work on the first two flights of the original TIGER instrument, Dr Rauch has been deployed to Antarctica four times. ↖ p. 314

RON ROBERTS is a chartered psychologist and an associate fellow of the British Psychological Society. Dr Roberts is an honorary lecturer in psychology at Kingston University (United Kingdom) and is the author of eleven books and numerous research articles, including *Foundations of Health Psychology* (Palgrave, 2001), *The New Politics of Experience* (PCCS Books, 2015), *Mental Health in Crisis* (Sage, 2019), and "Psychology at the End of the World: Mind and Behaviour in the Antarctic" (*The Psychologist*, 2011). ↖ p. 571

FAYNE ROBINSON graduated from the NZ Māori Arts and Crafts Institute in 1984. He combines both contemporary and traditional forms in his work. Robinson is currently working to incorporate Māori ethos and philosophy into the new design and rebuild of Scott Base. ↖ p. 536

DONALD R. ROTHWELL is a professor of international law at the Australian National University College of Law. He is a leading expert on the law of the sea and of the polar regions. Prof. Rothwell is the author of twenty-four books and over 200 book chapters and articles including, with Tim Stephens, *The International Law of the Sea* (Hart Publishing, 2016) and *International Polar Law* (Edward Elgar, 2018), co-edited with Alan Hemmings. ↖ pp. 102, 132

JUAN FRANCISCO SALAZAR is a professor of communications and media studies at Western Sydney University (Australia) and is currently an Australian Research Council Future Fellow. He is an environmental anthropologist and documentary film-maker with academic and artistic interests in social-ecological change, environmental futures and indigenous politics. Supported by the Chilean Antarctic Institute, Salazar has undertaken ethnographic and film work in Antarctica, where he directed *Nightfall on Gaia* (2015). ↖ pp. 53, 382, 556

JEAN-BAPTISTE SALLÉE is an oceanographer at the French National Centre for Scientific Research (CNRS) and researcher at the Pierre-Simon Laplace Institute (France). His work tackles questions of the dynamics of the ocean and climate with active research efforts on the study of the Southern Ocean. Dr Sallée is a lead author of the sixth report of the Intergovernmental Panel on Climate Change (IPCC). ↖ p. 209

PHILIPPE SAMYN is a civil engineer, architect, and urban planner. He is the founder of the Samyn and Partners architectural and engineering practice (Belgium) and is involved in a range of international projects. Dr Samyn was responsible for the shell and core of the Belgian station, Princess Elisabeth, in Antarctica. For his technical research on sustainability, he was awarded the Global Award for Sustainable Architecture (2008). ↖ p. 499

BOJAN ŠAVRIČ is a geodesist and software developer at the Environmental Systems Research Institute (United States). Dr Šavrič is the co-author of several map projections for world maps and an enthusiastic lover of the maths behind maps. ↖ p. 54

MIRKO SCHEINERT is a co-chair of the Geodetic Infrastructure of Antarctica (GIANT) group at the Scientific Committee on Antarctic Research (SCAR), and a senior scientist in the Group of Geodetic Earth System Research at the Department of Geosciences, Dresden University of Technology (Germany). With a PhD in orbital dynamics of low-flying satellites from the University of Stuttgart (Germany), Dr Scheinert has participated in more than fifteen polar expeditions. ↖ p. 288

DIDIER SCHMITT is the Strategy and Coordination group leader for Human and Robotic Exploration at the European Space Agency (ESA); he was a former member of the Space Task Force at the European External Action Service (EU diplomatic service) and foresight coordinator in the Bureau of European Policy Advisers. Dr Schmitt has authored several books, including *Terminus Antarctique* (2019) and a science fiction comic series called *Red Safari*, and is editor-in-chief of *Mook Mars* (Weyrich Editions). ↖ p. 574

THOMAS SCHRAMM is principal engineer at Ramboll Deutschland GmbH (Germany). Dipl. Eng. Schramm is a civil engineer with over thirty years of experience in project management and construction supervision around the world, specialising in Antarctic stations, transport, marine services, logistics, and offshore installations. ↖ pp. 384, 502

DANIEL SCHUBERT is deputy department leader of the System Analysis Space Segment at the German Aerospace Center, team leader of the EDEN Research Group, and responsible for the Space Habitation Plant Laboratory at the Institute of Space Systems. Dr Eng. Schubert led the EDEN-ISS project with 15 international partners and coordinated the deployment mission of the greenhouse system at the Neumayer III Antarctic research station. ↖ p. 577

KAREN N. SCOTT is a professor of law at the University of Canterbury (New Zealand) and the president of the Australian and New Zealand Society of International Law. Prof. Scott researches and teaches in the areas of Antarctic law, law of the sea and international environmental law, and is editor-in-chief of *Ocean Development & International Law* (Taylor and Francis). ↖ pp. 102, 132

CARA SEITCHEK works for the Smithsonian Institution and is an adjunct instructor at the University of California, Los Angeles and American University (United States). Seitchek holds an MA in history, an MPA in arts administration, and an MA in writing. ↖ p. 110

MARIA XIMENA SENATORE is the vice-president of the International Polar Heritage Committee of the International Council on Monuments and Sites (ICOMOS), a professor of historical archaeology and heritage at the University of Buenos Aires and at the National University of Patagonia Austral (Argentina). As a senior researcher at the Argentine National Council of Research, Dr Senatore has served as principal investigator of multiple projects in Antarctica. ↖ p. 516

JONATHAN SHANKLIN is one of the scientists who discovered the ozone hole, leading to a global shift in climate policy. He is an emeritus fellow at the British Antarctic Survey (United Kingdom). During his career, he has made twenty visits to the Antarctic, most recently visiting Halley, where he calibrated a new automated Dobson spectrophotometer. Shanklin's awards include the Polar Medal (United Kingdom) and the EPA Montreal Protocol Award (United States). ↖ p. 305

YURI A. SHIBAEV is the head of construction of the new wintering complex at Vostok Station on behalf of the Russian Arctic and Antarctic Research Institute (Russia). He acted as scientific director of the first Progress-East trip in the 53rd Russian Antarctic Expedition. Shibaev has participated in the implementation of the federal World Ocean Program, the creation of the Consortium of European Research Libraries, and various projects on palaeoclimate. ↖ pp. 395, 491

SANTIAGO SIERRA is an artist currently based in Madrid (Spain). After graduating in fine arts at the Complutense University (Spain), he completed his artistic training in Hamburg (Germany). Sierra's career unfolded primarily in Mexico and Italy. ↖ p. 79

TIM STEPHENS is a professor of international law at the University of Sydney Law School and deputy director at the University of Sydney Marine Studies Institute (Australia). He teaches public international law, with a particular focus on the law of the sea, the law of the polar regions and international environmental law. He is a Fellow of the Australian Academy of Law. Dr Stephens is the co-editor, with Ben Saul, of *Antarctica in International Law* (Hart Publishing, 2015). ↖ p. 175

PAVEL GRIGORIEVICH TALALAY is a professor and the director of the Polar Research Center at Jilin University (China). Prof. Talalay has taken part in six polar expeditions and was involved in drilling the deepest hole in ice, at Vostok Station, Antarctica. ↖ p. 252

PAUL TAYLOR is an entrepreneur, architect and principal of ARQZE Ltd. Having read philosophy and psychology at Oxford University, he then studied architecture in Edinburgh, Barcelona, London and Glasgow, where he received the title of architect from the Royal Institute of British Architects in 1994. ↖ p. 511

STEVE THENO is a professional engineer based in Alaska with over forty years' experience in the planning and design of facilities, energy systems, and infrastructure, predominantly in remote, cold regions. He is a former president of PDC Engineers (United States) who led the engineering design for the mechanical, electrical, and utilities systems for the third-generation Amundsen-Scott South Pole Station. Theno continues to be involved in Antarctic projects. ↖ p. 482

PAUL THUR served as Traverse Operations Manager for the United States Antarctic Program for six years. He is a Certified Government Financial Manager and a Society for Human Resource Management Certified Professional. Thur directed the development and refinement of the United States Antarctic Program traverse technologies in conjunction with the National Science Foundation (NSF), the Cold Regions Research and Engineering Laboratories, the United States Antarctic Program, Antarctic-support contract personnel, and other national programmes. ↖ p. 389

PHILIP TRATHAN is the head of conservation biology at the British Antarctic Survey (United Kingdom) and a UK delegate to the Commission for the Conservation of Antarctic Marine Living Resources (CCAMLR). With over 250 peer-reviewed publications and twenty trips to the Antarctic, PhD DSc Dr Trathan has been made an Officer of the Order of the British Empire, and a portion of Antarctica has been named after him: the Trathan Coast. ↖ p. 221

FRANCISCO VALDIVIA studied architecture at the Federico Santa María Technical University (Chile) and holds a Master of Engineering in Membrane Structures and Shell Technologies from Anhalt University of Applied Sciences (Germany). ↖ p. 511

DAVID VAUGHAN is the director of science for the British Antarctic Survey (United Kingdom). He was a coordinating lead author for the 4th and 5th assessment of the Intergovernmental Panel on Climate Change (IPCC) and sits on the Science Board of the UK Natural Environment Research Council. Vaughan has participated in twelve Antarctic scientific field campaigns. He was awarded the Polar Medal and

was made an Officer of the Order of the British Empire (United Kingdom). ↖ pp. 259–70, 292

EMERSON VIDIGAL is an architect and a co-founder, with Fabio Faria, Eron Costin, João Gabriel Rosa, and Martin Goic, of Estudio 41 (Brazil). Estudio 41 designed and built the Brazilian Antarctic station, Comandante Ferraz. Vidigal is also a professor at the Federal University of Paraná (Brazil). ↖ p. 416

PRISCILLA WEHI is a conservation scientist who leads the Vision Mātauranga programme of the Ross Sea Research and Monitoring Programme. She is the incoming director of Te Pūnaha Matatini, the Centre of Research Excellence for Complex Systems. Dr Wehi has visited the Antarctic Peninsula as part of the Homeward Bound women in science leadership programme. ↖ p. 536

CLAUDIO WILLIAMS worked from a very young age with his father, Argentine architect Amancio Williams, until the end of his days. He faithfully directed, preserved and shared the work of the Williams Archive until 2020, when it was donated to the Canadian Centre for Architecture (CCA). He is the author of *Amancio Williams* (Gaglianone, 1990). Williams is also the owner of the company Lece Construcciones S.A. (Argentina). ↖ p. 449

GARY WILSON is the chief scientist at the Institute of Geological and Nuclear Sciences and a professor of geology and marine science at the University of Otago (New Zealand). He is the vice-president of the Scientific Committee on Antarctic Research (SCAR) and has undertaken more than thirty research expeditions to Antarctica. He holds a BMus and a PhD. Dr Wilson's research has helped to uncover the dynamic history of Antarctica's climate and ice sheets. ↖ pp. 124, 284, 553

GILLIAN WRATT is a past chief executive of Antarctic New Zealand, the chair of the Council of Managers of National Antarctic Programs (COMNAP), and the vice-chair of the Committee for Environmental Protection (CEP). She is the author of *A Story of Antarctic Co-operation: 25 Years of the Council of Managers of National Antarctic Programs* (COMNAP, 2013). Wratt now chairs the Steering Group for the NZ Antarctic Science Platform. ↖ p. 126

ANGELA WRIGHT is the founder of Colour Affects (United Kingdom). She is an established expert on the unconscious effects of colour. In 1985, after studying both Freudian psychology and colour theory, she founded Colour Affects in order to test her colour theory empirically. Today, Wright advises blue chip companies across the world on colour psychology and she is an established spokesperson on the subject. ↖ p. 372

JAMES YORK has been a Māori carver for more than thirty-five years. He lives on New Zealand's southern coast, where he sees erosion and changing weather patterns from climate change every day. York's work spans both isolated and urban settings, including the entranceway to Toitū Otago Settlers Museum in Dunedin. ↖ p. 536

FEDERICA SOFIA ZAMBELETTI is an architect at UNA, an international architectural office focusing on cultural projects based in Hamburg (Germany), and a researcher at UNLESS, a not-for-profit research agency devoted to interdisciplinary research on extreme environments, where she oversaw *Antarctic Resolution* as project manager. She is the founder of KooZA/rch, a digital platform researching the architectural imaginary. ↖ p. 423

"Most importantly we have learned that from here on it is success for all or for none, for it is experimentally proven by physics that 'unity is plural.'"

Richard Buckminster Fuller, *Operating Manual for Spaceship Earth*, 1969

Bibliography

The desire to describe a profoundly unfamiliar territory in which time is dilated to an unrecognisable degree (with a single day expanding to one year), collective certainties waver (as poles that are presumed to be fixed waltz on the scale-less landscape), and matter changes state (the ever-present flowing water running through every other continent is found motionless here, having lain frozen for millions of years) leads inevitably to the invention of new words and the definition of an Antarctic vocabulary. The terms dispersed within the bibliography thus challenge the acritical use of inherited words and offer a sometimes humorous insight into the Antarctic language.

The Antarctic legal regime is referred to as the "Antarctic Treaty System" (ATS), and this is defined in Article 1 of the Madrid Protocol as comprising 1959 Antarctic Treaty, the measures in effect under that Treaty, its associated separate international instruments in force [so, the 1972 Convention for the Conservation of Antarctic Seals (CCAS); the 1980 Convention on the Conservation of Antarctic Marine Living Resources (CCAMLR); and the 1991 Protocol on Environmental Protection to the Antarctic Treaty (Madrid Protocol) – the 1988 Convention on the Regulation of Antarctic Mineral Resource Activities (CRAMRA) did not enter into force and was abandoned in favour of the Madrid Protocol, so it is not formally art of the ATS] and the measures in effect under these instruments. All of these instruments may most conveniently be accessed via the Secretariat of the Antarctic Treaty at https://www.ats.aq/e/key-documents.html.

A Antarcticitis

"The yearning for Antarctica."[1] A sentiment perhaps best expressed by Admiral Richard Evelyn Byrd who wrote, "Part of me remained forever at latitude 80°08' South" – one that is presumably shared by all who have been lucky to venture below the 60th parallel South.

Aartsen, Mark, Markus Ackermann, James Adams, Juan Aguilar, Markus Ahlers, M. Ahrens, C. Alispach et al. "Neutrino Astronomy with the Next Generation IceCube Neutrino Observatory." November 2019. arXiv.org. [online]

Abdel-Motaal, Doaa. *Antarctica: The Battle for the Seventh Continent.* Denver: Praeger, 2016.

Abram, Nerilie, Jean-Pierre Gattuso, Anjal Prakash et al. "Framing and Context of the Report." Chap. 1 in Pörtner, Roberts, Masson-Delmotte et al., *IPCC Special Report on the Ocean and Cryosphere in a Changing Climate.*

Abrams, Peter A. "How Precautionary Is the Policy Governing the Ross Sea Antarctic Toothfish (*Dissostichus mawsoni*) Fishery?" *Antarctic Science* 26, no. 1 (2014): 3–14.

Abrams, Peter A. "Necessary Elements of Precautionary Management: Implications for the Antarctic Toothfish." *Fish and Fisheries* (2016): 1–23.

Adam, David. "The Rush to Sock Away Glacier Ice Before It All Melts." *The Atlantic,* 31 May 2020.

Adams, Edward, John Priscu, Christian Fritsen, Scott Smith, and Steven Brackman. "Permanent Ice Covers of the McMurdo Dry Valley Lakes: Bubble Formation and Metamorphism." In *Ecosystem Dynamics in a Polar Desert: The McMurdo Dry Valleys, Antarctica.* Edited by John Priscu, 281–95. Antarctic Research Series 72. Washington, DC: American Geophysical Union, 1998.

Aguirre de Cárcer, Daniel, Alberto López-Bueno, David A. Pearce, and Antonio Alcamí. "Biodiversity and Distribution of Polar Freshwater DNA Viruses." *Science Advances* 1, no. 5 (2015).

Ainley, David, Joellen Russell, Stéphanie Jenouvrier, Eric Woehler, Philip Lyver, William Fraser, and Gerald Kooyman. "Antarctic Penguin Response to Habitat Change as Earth's Troposphere Reaches 2°C above Pre-industrial Levels." *Ecological Monographs* 80 (2010): 49–66.

Air Liquide Fondation. "ICE MEMORY, Preserving Ice Memory for Future Generations." Accessed 7 January 2021. fondationairliquide.com. [online]

Aker BioMarine. "Our Exclusive Eco-Harvesting Method." Aker BioMarine, 2020. akerbiomarine.com. [online]

Akers, Pete. "Life and Work on the Frozen Continent: Antarctic Research Stations." 17 January 2020. wunderground.com. [online]

Albuquerque, Lita. *Stellar Axis and Other Projects.* New York: Skira Rizzoli, 2014.

Aldiss, Brian, and Roger Penrose. *White Mars, or The Mind Set Free: A 21st-Century Utopia.* London: Little, Brown, 1999.

Alexandrov, Konstantin. "Остаться В Антарктиде" [Ostatsia v Antarktide | Stay in Antarctica]. Translated by UNLESS. 28 May 2019. nplus1.ru. [online]

Alighieri, Dante. *The Divine Comedy.* Translated by Charles Sisson. Oxford: Oxford University Press, 1980.

Allison, Ian, Michel Béland, David Carlson, Dahe Qin, Eduard Sarukhanian, and C. Smith. "International Polar Year 2007–2008." World Meteorological Organization, *Bulletin* 56, no. 4 (2007). public.wmo.int. [online]

Amin, Nighat. *Princess Elisabeth Antarctica: And the Zero Emission Quest.* Brussels: Racine, 2013.

Amos, Jonathan. "Climate Change: 'Stunning' Seafloor Ridges Record Antarctic Retreat." *BBC News,* 29 May 2020. bbc.com. [online]

Amos, Jonathan. "Dinosaurs Walked through Antarctic Rainforests." *BBC News,* 1 April 2020.

Amos, Jonathan. "Is the World's Biggest Iceberg about to Break Up?" *BBC News,* 23 April 2020.

Amos, Jonathan. "Scientists Count Whales from Space." BBC News, 1 November 2018. bbc.com. [online]

Amos, Jonathan. "Thwaites: 'Doomsday Glacier' Vulnerability Seen in New Maps" *BBC News,* 8 September 2020.

Amundsen, Roald. *The South Pole: An Account of the Norwegian Antarctic Expedition in the Fram 1910–1912.* New York: Cooper Square Press, 2001.

Andrady, Anthony L. "The Plastic in Microplastics: A Review." *Marine Pollution Bulletin* 119, no. 1 (2017): 12–22.

Andrews, Robin George. "Antarctica vs. Science." *The New York Times,* 2 May 2020.

Antarctic and Southern Ocean Coalition (ASOC). "Are More Antarctic Stations Justified?" XXVII ATCM: ATCM Information Paper, March 2004. asoc.org. [online]

Antarctic and Southern Ocean Coalition (ASOC). "The Convention Proceeds," *ECO* 48, no. 12, June 1988.

Antarctic and Southern Ocean Coalition (ASOC). "Ocean Acidification." Accessed 9 January 2021. asoc.org. [online]

Antarctic Heritage Trust (NZ). *Conservation Plan: The Historic Huts at Cape Adare; Borchgrevink's British Antarctic Expedition, 1898–1900, Northern Party of Scott's British Antarctic Expedition, 1910–1913, Cape Adare, Antarctica.* Christchurch, NZ: Antarctic Heritage Trust, 2004.

Antarctic Heritage Trust (NZ). "History of Scott's Expedition: British Antarctic (Terra Nova) Expedition 1910–1913." Accessed 17 January 2021. nzaht.org. [online]

Antarctic Heritage Trust (NZ). "History of Scott's Expedition: The National Antarctic Expedition 1901–04." Accessed 17 January 2021. nzaht.org. [online]

Antarctic Heritage Trust (NZ). *Shackleton's Hut Conservation Report.* Christchurch, NZ: Antarctic Heritage Trust, 2003.

Antarctica New Zealand, United States Antarctic Program. *McMurdo Dry Valleys ASMA Manual (Fourth Edition).* Christchurch: Antarctica New Zealand, 2015. era.gs. [online]

Antonioni, Michelangelo. *That Bowling Alley on the Tiber: Tales of a Director.* Translated by William Arrowsmith. London: Oxford University Press, 1986.

Antony, Jason. *Hoosh: Roast Penguin, Scurvy Day and Other Stories of Antarctic Cuisine.* Lincoln: University of Nebraska Press, 2012.

Archer, David, Michael Eby, Victor Brovkin, Andy Ridgwell, Long Cao, Uwe Mikolajewicz, Ken Caldeira et al. "Atmospheric Lifetime of Fossil Fuel Carbon Dioxide." *Annual Review of Earth and Planetary Sciences* 37 (2009): 117–34.

Arends, Hans Juergen, Elkin Andres, Xiaoqiong Bai, G. Barouch, S. Barwick, Ryan Bay, T. Becka et al.

"Observation of High Energy Atmospheric Neutrinos with the Antarctic Muon and Neutrino Detector Array." *Physical Review D: Particle and Fields* 66, no. 1 (July 2002).

Argentinian Government. "Base Esperanza." Accessed 13 February 21. argentina.gob.ar. [online]

Armstrong, Sue, Robert Lockhart, R. B. Woodward, Sonia Ritter, Catherine W., Colleen Y. Wallis, and Paul David Tuff. Comments on the article *Licence to Krill.* Greenpeace International, 2018. Accessed 1 February 2020. greenpeace.org. [online]

Arndt, Jan, Hans Werner Schenke, Martin Jokobsson, Frank Nitsche, Gwen Buys, Bruce Goleby, Michele Rebesco et al. "The International Bathymetric Chart of the Southern Ocean (IBCSO) Version 1.0 – A New Bathymetric Compilation Covering Circum-Antarctic Waters." *Geophysical Research Letters* 40, no. 12 (June 2013): 3111–17.

Aronson, Richard B., Sven Thatje, James B. McClintock, and Kevin A. Hughes. "Anthro-pogenic Impacts on Marine Ecosystems in Antarctica." *Annals of the New York Academy of Sciences* 1223 (2011): 82–107.

Ashton, Peter, Peter Ade, Francesco Angilè, Steven Benton, Mark Devlin, Bradley Dober, Laura Fissel et al. "First Observation of the Submilli-meter Polarization Spectrum in a Translucent Molecular Cloud." *The Astrophysical Journal* 857, no. 1 (April 2018).

Aston, Felicity. "Women on the White Continent." *Geographical* 77, no. 9 (2005): 26.

Atkinson, Angus, Volker Siegel, Evgeny Pakhomov, and Peter Rothery. "Long-Term Decline in Krill Stock and Increase in Salps within the Southern Ocean." *Nature* 432 (November 2004): 100–103.

Atkinson, Angus, Simeon Hill, Evgeny Pakhomov, Volker Siegel, Christian Reiss, Valerie Loeb, Deborah Steinberg et al. "Krill (*Euphausia superba*) Distribution Contracts Southward During Rapid Regional Warming." *Nature Climate Change* 9, no. 2 (2019): 142–47.

Atlas Obscura. "South Pole Ice Tunnels." Accessed 8 June 2020. atlasobscura.com. [online]

Atlas of Marine Protection. "Explore the World's Marine Protected Areas." Accessed 1 March 2020. mpatlas.org. [online]

Augustin, Laurent, Carlo Barbante, Piers Barnes, Jean Marc Barnola, Matthias Bigler, Emiliano Castellano, Olivier Cattani et al. "Eight Glacial Cycles from an Antarctic Ice Core." *Nature* 429 (June 2004): 623–28.

Aureli, Pier Vittorio. "The Theology of Tabula Rasa: Walter Benjamin and Architecture in the Age of Precarity." 9 May 2015. thecityasaproject.org. [online]

Australia, Czech Republic, SCAR, United States. *Co-Conveners' Report of the Joint SCAR/CEP Workshop on Further Developing the Antarctic Protected Area System.* Prague, 2019. scar.org. [online]

Australian Antarctic Data Centre, *The SCAR Composite Gazetteer.* Accessed 3 February 2020. aad.gov.au. [online]

Australian Antarctic Program (AAD). "ANARE Communications 1947–1985." Last modified 16 February 2017. antarctica.gov.au. [online]

Australian Antarctic Program (AAD). "Australia's Antarctic Buildings: AANBUS." Last modified 31 July 2018. antarctica.gov.au. [online]

Australian Antarctic Program (AAD). "Australia's Antarctic Buildings: The DIY Era." Last modified 31 July 2018. antarctica.gov.au. [online]

Australian Antarctic Program (AAD). "Biscoe Hut." Last modified 31 July 2018. antarctica.gov.au. [online]

Australian Antarctic Program (AAD). "Do You Need Your Appendix Removed Before You Go?" Last modified 27 November 2015. antarctica.gov.au. [online]

Australian Antarctic Program (AAD). "Shelters from the Stormy Blast." Last modified 15 September 2011. mawsonshuts.antarctica.gov.au. [online]

Australian Antarctic Program (AAD). "Sørsdal Glacier Dynamics." Last modified 22 November 2016. antarctica.gov.au. [online]

Australian Antarctic Program (AAD). "Waste Management." Last modified 24 August 2012. antarctica.gov.au. [online]

Australian Government. Scott Base and McMurdo Station: Report of an Inspection under Article VII of the Antarctic Treaty and Article 14 of the Protocol on Environmental Protection. Kingston: Australian Antarctic Division, 2005.

Automobilist Magazine. "More Than Snow-Play: The Story Behind the Trans Antarctic Expedition." 1 November 2019. automobilist.com. [online]

B Bioindicator

"Bioindicators include biological processes, species, or communities and are used to assess the quality of the environment and how it changes over time. Changes in the environment are often attributed to anthropogenic disturbances (e.g., pollution, land use changes) or natural stressors (e.g., drought, late spring freeze). The widespread development and application of bioindicators has occurred primarily since the 1960s. Over the years, [...] the repertoire of bioindicators [has been expanded to assist the study of] all types of environments (i.e., aquatic and terrestrial), using all major taxonomic groups."[2]

Baker, A., and D. R. Jean. *British Antarctic Survey Accommodation Buildings Report*. Cambridge: BAS Archives, 1964.

Bakker, Dorothee, Benjamin Pfeil, Camilla Landa, Nicolas Metzl, Kevin O'Brien, Are Olsen, Karl Smith et al. "A Multi-Decade Record of High-Quality *f*CO$_2$ Data in Version 3 of the Surface Ocean CO$_2$ Atlas (SOCAT)." *Earth System Science Data* 8 (2016): 383–413.

Balch, Edwin Swift. "Antarctic Names." *Bulletin of The American Geographical Society* 44 (8) (1912): 561.

Ballard, J. G. *The Crystal World*. New York: Berkley, 1967.

Ballard, Susan. "Gathered Scratchings." DATE. suballard.net.nz. [online]

Bamsey, Matthew, Paul Zabel, Conrad Zeidler, Lucie Poulet, Daniel Schubert, Eberhard Kohlberg, and Thomas Graham. "Design of a Containerized Greenhouse Module for Deployment to the Neumayer III Antarctic Station." Paper presented at the 44th International Conference on Environmental Systems, July 2014.

Bamsey, Matthew, Paul Zabel, Conrad Zeidler, Dávid Gyimesi, Daniel Schubert, Eberhard Kohlberg, Dirk Mengedoht et al. "Review of Antarctic Greenhouses and Plant Production Facilities: A Historical Account of Food Plants on the Ice." Paper presented at the 45th International Conference on Environmental Systems, Bellevue, July 2015.

Bamsey, Matthew, Paul Zabel, Conrad Zeidler, Vincent Vrakking, Daniel Schubert, Eberhard Kohlberg, Michael Stasiak et al. "Early Trade-Offs and Top-Level Design Drivers for Antarctic Greenhouses and Plant Production Facilities." Paper presented at the 46th International Conference on Environmental Systems, July 2016.

Ban Ki-moon. "United Nations Secretary-General Visit to Antarctica." Speech, 2007.

Banham, Reyner. *Scenes in America Deserta*. Cambridge, MA: The MIT Press, 1989.

Bannova, Olga. *Designing for Extremes: A Methodological Approach to Planning in Arctic Regions.* PhD diss. Gothenburg: Chalmers University of Technology, 2016.

Baraniuk, Chris. "What the *Diamond Princess* Taught the World about COVID-19." *British Medical Journal* (April 2020): 369.

Barbosa, Andrés, Arvind Varsani, Virginia Morandini, Wray Grimaldi, Ralph Vanstreels, Julia Diaz, Thierry Boulinier et al. "Risk Assessment of SARS-CoV-2 in Antarctic Wildlife." *Science of the Total Environment* (2020): 143352.

Barbraud, Christophe, and Henri Weimerskirch. "Emperor Penguins and Climate Change." *Nature* 411 (May 2001): 183–86.

Barkaszi, Irén, Endre Takács, István Czigler, and László Balázs. "Extreme Environment Effects on Cognitive Functions: A Longitudinal Study in High Altitude in Antarctica." *Frontiers in Human Neuroscience* 10 (2016): 331.

Barker, Stephen and Andy Ridgwell, A. "Ocean Acidification." *Nature Education Knowledge* 3 (10) (2012): 21.

Barlow, William Henry. "On the Spontaneous Electrical Currents Observed in the Wires of the Electric Telegraph." *Philosophical Transactions of the Royal Society of London* 139 (1849): 61–72.

Barnes, James N. "The Emerging Convention on the Conservation of Antarctic Marine Living Resources: An Attempt to Meet the New Realities of Resource Exploitation in the Southern Ocean." In *The New Nationalism and the Use of Common Spaces,* edited by Jonathan I. Charney. Totowa, NJ: Allanheld, Osmun, 1982.

Barnes, James N. *Let's Save Antarctica.* Appendix Q, p. 81. Richmond, Australia: Greenhouse Publications, 1982.

Barnes, James N. "A Reminiscence on Antarctic Governance and Transparency: The NGO Role." *Antarctic Affairs* 5 (2018): 1–3.

Barnett, Lewis A. K. and Marissa L. Baskett, "Marine Reserves Can Enhance Ecological Resilience." *Ecology Letters* 18, no. 12 (2015): 1301–10.

Barr, Susan. "Twenty Years of Protection of Historic Values in Antarctica under the Madrid Protocol." *The Polar Journal* 8, no. 2 (2018): 241–64.

Barr, Susan, and Paul Chaplin, eds. *Cultural Heritage in the Arctic and Antarctic Regions.* Lørenskog: ICOMOS, 2004.

Barr, Susan, and Paul Chaplin, International Polar Heritage Committee, eds. *Historical Polar Bases: Preservation and Management.* International Polar Heritage Committee, 2008.

Barr, Susan, and Cornelia Lüdecke, eds. *The History of the International Polar Years (IPYs).* Berlin and Heidelberg: Springer-Verlag, 2010.

Barrett, Peter. "The Devonian to Jurassic Beacon Supergroup of the Transantarctic Mountains and Correlatives in Other Parts of Antarctica." In *The Geology of Antarctica,* edited by Robert Tingey, 120–52. Oxford: Clarendon Press, 1991.

Barrett, Peter. Introduction to *Antarctic Cenozoic History from the CIROS-1 Drillhole, McMurdo Sound.* DSIR Bulletin 245, edited by Peter Barrett, 5–6. Wellington: DSIR Publishing, 1989.

Barrett, Peter, Donald Elston, David Harwood, Barrie McKelvey, and Peter-Noel Webb. "Cenozoic Climate and Sea Level History from Glacimarine Strata off the Victoria Land Coast, Cape Roberts Project, Antarctica." In *Glacial Sedimentary Processes and Products,* edited by Michael Hambrey, Poul Christoffersen, Neil Glasser, and Bryn Hubbard. International Association of Sedimentologists Special Publication 39, 259–87. New York: Wiley, 2009.

Barrett, Peter, Donald Elston, David Harwood, Barrie McKelvey, and Peter-Noel Webb. "Mid-Cenozoic Record of Glaciation and Sea-Level Change on the Margin of the Victoria Land Basin, Antarctica." *Geology* 15, no. 7 (July 1987): 634–37.

Bartlett, Jesamine, Peter Convey, Arlie McCarthy, and Scott Hayward. "Aliens in Antarctica." *The Biologist* 66 (2019): 22–25.

BAS Environment Office. "Rothera Modernisation: Phase 1 Initial Environmental Evaluation." Cambridge: British Antarctic Survey, Natural Environment Research Council, 2019.

Basset, Sue. "Shackleton's Car." *Antarctic Blog.* 3 February 2014. nzaht.org. [online]

Bastmeijer, Kees. *The Antarctic Environmental Protocol and its Domestic Legal Implementation.*

The Hague, London, and New York: Kluwer Law International, 2003.

Bastmeijer, Kees. "Introduction: The Madrid Protocol 1998–2018; Need to Address 'the Success Syndrome.'" *The Polar Journal* 8, no. 2 (2018): 230–40.

Bastmeijer, Kees. "Regulation of Tourism in Antarctica (Regulering van Toerisme in Antarctica)." *Circumpolar Journal* 8, no. 3–4 (1993): 13–33.

Bastmeijer, Kees, and Ricardo Roura. "Environmental Impact Assessment in Antarctica." In *Theory and Practice of Transboundary Environmental Impact Assessment*, edited by Kees Bastmeijer and Timo Koivurova, 175–219. Leiden and Boston: Brill and Martinus Nijhoff Publishers, 2008.

Bateman, Sam. "Strategic Competition and Emerging Security Risks: Will Antarctica Remain De-militarised?" In *Antarctic Security in the Twenty-First Century: Legal and Policy Perspectives*, edited by Alan D. Hemmings, Donald R. Rothwell, and Karen N. Scott, 116–134. London: Routledge, 2012.

Bechtel, Robert, and Amy Berning. "The Third-Quarter Phenomenon: Do People Experience Discomfort after Stress has Passed?" In *From Antarctica to Outer Space: Life in Isolation and Confinement*. Edited by Albert Harrison, Yvonne Clearwater, and Christopher McKay, 261–66. New York: Springer Verlag, 1991.

Beck, Peter J. "Antarctica and the United Nations." In *Handbook on the Politics of Antarctica*, 255–264. Cheltenham: Edger Elgar Publishing Inc., 2017.

Beck, Peter J. *The International Politics of Antarctica*. London and Sydney: Croom Helm, 1986.

Beck, Peter J. "Twenty Years On: The UN and the 'Question of Antarctica' 1983–2003." *Polar Record* 40 (2004): 205–12.

Beebe, Morton, ed. *Operation Deep Freeze III: The Story of Task Force 43; Third Phase, 1957–58*. Paoli, PA: Dorville Corporation, 1957.

Belanger, Dian. *Deep Freeze: The United States, the International Geophysical Year, and the Origins of Antarctica's Age of Science*. Boulder: University Press of Colorado, 2010.

Bender, Nicole A., Kim Crosbie, and Heather J. Lynch. "Patterns of Tourism in the Antarctic Peninsula Region: A 20-Year Analysis." *Antarctic Science* 28, no. 3 (2016): 194–203.

Beniston, Martin. "Impacts of Climatic Change on Water and Associated Economic Activities in the Swiss Alps." *Journal of Hydrology* 412–13 (January 2012): 291–96.

Benjamin, Walter. "The Destructive Character." In *Walter Benjamin: Selected Writings*. Vol. 2, part 2, *1931–1934*, edited by Michael Jennings, Howard Eiland, and Gary Smith. Translated by Rodney Livingstone. Cambridge, MA.: Belknap Press, 1999.

Bennett, R. A., ed. *Operation Deep Freeze '65: Ten Years of Progress*. Boston, MA: Burdette & Company, 1965.

Bergami, Elisa, Andrews Krupinski Emerenciano, Marcelo González-Aravena, César A. Cárdenas, Pablo Hernández-Almaraz, José Roberto Machado Cunha da Silva, and Ilaria Corsi. "Polystyrene Nanoparticles Affect the Innate Immune System of the Antarctic Sea Urchin *Sterechinus neumayeri*." *Polar Biology* 42, no. 4 (2019): 743–57.

Bergami, Elisa, Clara Manno, Simone Cappello, Maria Luisa Vannuccini, and Ilaria Corsi. "Nanoplastics Affect Moulting and Faecal Pellet Sinking in Antarctic Krill (*Euphausia superba*) Juveniles," *Environment International* 143 (October 2020).

Bergami, Elisa, Emilia Rota, Tancredi Caruso, Giovanni Birarda, Lisa Vaccari, and Ilaria Corsi. "Plastics Everywhere: First Evidence of Polystyrene Fragments Inside the Common Antarctic Collembolan *Cryptopygus antarcticus*." *Biology Letters* 16, no. 6 (June 2020).

Bergman, Lars, and Robin Stuart. "On the Location of Shackleton's Vessel Endurance." *Journal of Navigation* 72, no. 2 (2018): 257–68.

Bergstrom, Dana, Dominic Hodgson, and Peter Convey. "The Physical Setting of the Antarctic." In *Trends in Antarctic Terrestrial and Limnetic Ecosystems: Antarctica as a Global Indicator*, edited by Dana Bergstrom, Peter Convey, and Ad Huiskes, 15–33. Dordrecht: Springer Netherlands, 2006.

Berkman, Paul Arthur, Michael A. Lang, David W. H. Walton, and Oran R. Young, eds. Science Diplomacy: Science, Antarctica, and the Governance of International Spaces. Washington, DC: Smithsonian Institution Scholarly Press, 2011.

Bernacchi, Louis. *To the South Polar Regions: Expedition of 1898-1900*. 1901. Reprint, Denton: Bluntisham, 1991.

Bernal, Marcelo, Pol Taylor, and Francisco Valdivia. "Ilaia; Joint Scientific Polar Station Unión Glacier; Antártica, Chile, 2013–2014." *ARQ* 90 (August 2015): 76–79. scielo.conicyt.cl. [online]

Bessa, Filipa, Norman Ratcliffe, Vanessa Otero, Paula Sobral, João C. Marques, Claire M. Waluda, Philip N. Trathan, and José C. Xavier. "Micro-plastics in Gentoo Penguins from the Antarctic Region." *Scientific Reports* 9 (2019): 1–7.

Beutler, Gerhard, Markus Rothacher, Stefan C. Schaer, Timothy A. Springer, Jan Kouba, and Ruth E. Neilan, "The International GPS Service (IGS): An Interdisciplinary Service in Support of Earth Sciences," *Advances in Space Research* 23, no. 4 (1999): 631–53.

Beyond EPICA Oldest Ice. "Home page." Accessed 10 March 2021. beyondepica.eu. [online]

BGR. "Development of the BGR's Polar Research Activities." Accessed 17 February 2021. bgr.bund. de. [online]

Bin Mohamad, Mahathir. "The Question of Antarctica." Presentation, New York United Nations General Assembly. 29 September 1982.

Black Mountain College Project. "Buckminster Fuller." Accessed 17 February 2021. blackmountaincollegeproject.org. [online]

Blay, Samuel, and B. Martin Tsamenyi, "Australia and the Convention for the Regulation of Antarctic Mineral Resource Activities (CRAMRA)." *Polar Record* 26, no. 158 (1990): 195–202.

Blazeski, Goran. "In 1977 Argentina Sent a Pregnant Woman to Antarctica in an Attempt to Claim Partial Possession of the Continent." *The Vintage News*, 12 October 2016.

Blum, Hester. *News at the Ends of the Earth: The Print Culture of Polar Exploration*. Durham, NC: Duke University Press, 2019.

Bockheim, James, and Charles Tarnocai, "Nature, Occurrence and Origin of Dry Permafrost," In *Permafrost: Seventh International Conference (Proceedings), Yellowknife (Canada)* Quebec: Centre d'études nordiques, Université Laval, 1998.

BoF Architekten. "Bharati – New Indian Research Station on Larsemann Hills." Accessed 13 February 2021. archello.com. [online]

Bokhorst, Stef, Peter Convey, and Rien Aerts. "Nitrogen Inputs by Marine Vertebrates Drive Abundance and Richness in Antarctic Terrestrial Ecosystems." *Current Biology* 29 (2019).

Boothe, Joan. *The Storied Ice: Exploration, Discovery and Adventure in Antarctica's Peninsula Region*. Berkeley: Regent Press, 2011.

Borchgrevink Carsten Egeberg. *First on the Antarctic Continent: Being an Account of the British Antarctic Expedition, 1898-1900*. London: G. Newnes, 1901.

Boscheri, Giorgio, Marco Volponi, Matteo Lamantea, Cesare Lobascio, Daniel Schubert, and Paul Zabel. "Main Performance Results of the EDEN ISS Rack-like Plant Growth Facility." Paper presented at the 47th International Conference on Environmental Systems, July 2017.

Bourke, Deborah. *Mawson's Huts Historic Site Management Plan 2013–2018*. Commonwealth of Australia, 2013. legislation.gov.au. [online]

Bourrel, Marie, Torsten Thiele, and Duncan Currie. "The Common Heritage of Mankind as a Means to Assess and Advance Equity in Deep Sea Mining." *Marine Policy* 95 (2016).

Bourzac, Katherine. "The Race to Preserve Earth's Historical Climate Record – Its Ice." *Chemical & Engineering News* 98, no. 23 (June 2020).

Boy, Jens, Roberto Godoy, Olga Shibistova, Diana Boy, Robert McCulloch, Alberto Andrino de la Fuente, Mauricio Aguirre Morales et al. "Successional Patterns along Soil Development Gradients Formed by Glacier Retreat in the Maritime Antarctic, King George Island." *Revista Chilena de Historia Natural* 89 (2016): 1–18.

Brady, Anne-Marie. *China as a Great Power*. Cambridge: Cambridge University Press, 2017.

Brady, Anne-Marie. "China, Russia Push GPS Rivals into Antarctica." *The Australian*, 6 September 2018.

Brady, Anne-Marie. *China's Expanding Antarctic Interests: Implications for Australia*. Special report. Barton, ACT: ASPI, 2017. aspi.org.au. [online]

Brady, Anne-Marie "Cooperation or Conflict? China's Position on Points of Contention in the Polar Regions." In *China as a Polar Great Power*, 187–219.

Brady, Anne-Marie, ed. *The Emerging Politics of Antarctica*. Abingdon: Routledge, 2012.

Brandt, Anthony. *The South Pole: A Historical Reader*. National Geographic Adventure Classics. Washington DC: National Geographic, 2004.

Braun, Christina, et al. "Environmental Monitoring and Management Proposals for the Fildes Region, King George Island, Antarctica." *Polar Research* 31 (2012): 18206.

BREEAM. "Home page." Accessed 18 February 2021. breeam.com. [online]

Brier, Frank, Robert Cunningham, Joyce Jatko, Art Jung, Peter Karasik, John Maier et al. *Modernization of the Amundsen-Scott South Pole Station Antarctica: Final Environmental Impact Statement*. Arlington, VA: Office of Polar Programs National Science Foundation, 1998.

British Antarctic Survey. "Antarctic Peninsula Glaciers in Widespread Retreat." 21 April 2005. bas.ac.uk. [online]

British Antarctic Survey. "The Antarctic Peninsula's Retreating Ice Shelves." 1 January 2012. bas. ac.uk. [online]

British Antarctic Survey. "Bedmap2 Gives Scientists a More Detailed View of Antarctica's Landmass." 8 March 2013. bas.ac.uk. [online]

British Antarctic Survey. "Engineers Automate Science from Remote Antarctic Station." 25 June 2019. phys.org. [online]

British Antarctic Survey. "First Steps." Accessed 17 February 2021. bas.ac.uk. [online]

British Antarctic Survey. "History of Deception Island (Station B)." Accessed 13 February 2021. bas. ac.uk. [online]

British Antarctic Survey. "History of Halley (Station Z)." Accessed 27 January 2021. bas.ac.uk. [online]

British Antarctic Survey. "History of Port Lockroy (Station A)." Accessed 17 January 2021. bas.ac.uk. [online]

British Antarctic Survey. "History of Rothera (Station R)." Accessed 13 February 2021. bas.ac.uk. [online]

British Antarctic Survey. "History of Signy (Station H)." Accessed 17 February 2021. bas.ac.uk. [online]

British Antarctic Survey. "History of Stonington Island (Station E)." Accessed 27 January 2021. bas. ac.uk. [online]

British Antarctic Survey "The Larsen C Ice Shelf." 10 February 2018. bas.ac.uk. [online]

British Antarctic Survey. "New High-Precision Map of Antarctica's Bed Topography." 12 December 2019. bas.ac.uk. [online]

British Antarctic Survey. "Operation Tabarin Overview." Accessed 17 February 2021. bas. ac.uk. [online]

British Antarctic Survey. *The Ozone Hole*. Science Briefing (April 2017). bas.ac.uk. [online]

British Antarctic Survey. *Rothera Wharf Reconstruction & Coastal Stablisation: Final Comprehensive Environmental Evaluation.* Cambridge, 2018. ats.aq. [online]

British Antarctic Survey. "Satellite Spies on Doomed Antarctic Ice Shelf." 19 March 2002. bas.ac.uk. [online]

British Antarctic Survey. "Scientists Drill for First Time on Remote Antarctic Glacier." 28 January 2020. bas.ac.uk. [online]

British Antarctic Survey. "Sea-Level Rise." 21 May 2019. bas.ac.uk. [online]

British Antarctic Survey. "Space Weather 'Super Storms' Occurred Every 25 Years." 29 January 2020. bas.ac.uk. [online]

British Antarctic Survey. "Warm Ocean Currents Cause Majority of Ice Loss from Antarctica." 25 April 2012. bas.ac.uk. [online]

British Antarctic Survey, National Environment Research Council. "The Ozone Hole." *Public Information Leaflet.* Cambridge: National Environment Research Council, 2012.

British Antarctic Survey, Natural Environment Research Council. *Proposed Construction and Operation of Halley VI Research Station, and Demolition and Removal of Halley V Research Station, Brunt Ice Shelf Antarctica: Final Comprehensive Environmental Evaluation.* Cambridge: British Antarctic Survey, Natural Environment Research Council, 2007.

British Geological Survey (BGS). "Geological Timechart: Cenozoic Era | Palaeogene to Quaternary." Accessed 9 January 2021. bgs.ac.uk. [online]

Brooks, Cassandra. "Competing Values on the Antarctic High Seas: CCAMLR and the Challenge of Marine-Protected Areas." *The Polar Journal* 3, no. 2 (2013): 277–300.

Brooks, Cassandra. "Cracks in the Future of the Antarctic." *National Geographic Blogs,* 2018. nationalgeographic.org. [online]

Brooks, Cassandra, and David Ainley, "Fishing the Bottom of the Earth: The Political Challenges of Ecosystem-Based Management," in *Handbook on Antarctic Politics,* edited by Klaus J. Dodds, Alan D. Hemmings, and Peder Roberts, 422–38. Cheltenham: Edward Elgar Publishing, 2017.

Brooks, Cassandra, David Ainley, Peter Abrams, Paul Dayton, Robert Hofman, Jennifer Jacquet, and Donald Siniff. "Watch over Antarctic Waters." *Nature* 558 (2018): 177–180.

Brooks, Cassandra, Steven Chown, Lucinda Douglass, Ben Raymond, Justine Shaw, Zephyr Sylvester, and Christa Torrens. "Progress Towards a Representative Network of Southern Ocean Protected Areas." *PLOS ONE* 15, no. 4 (2020).

Brooks, Cassandra, Larry Crowder, Lisa Curran, Robert Dunbar, David Ainley, Klaus Dodds, Kristina Gjerde, and Rashid Sumaila. "Science-Based Management in Decline in the Southern Ocean." *Science* 354, no. 6309 (2016): 185–87.

Brooks, Cassandra, Larry Crowder, Henrik Österblom, and Aaron Strong. "Reaching Consensus for Conserving the Global Commons: The Case of the Ross Sea, Antarctica." *Conservation Letters* (2019).

Brooks, Shaun. "Our Footprint on Antarctica – Buildings, Disturbance: Version 2." Australian Antarctic Data Centre, 2019.

Brooks, Shaun, and Julia Jabour. "Australia Wants to Build a Huge Concrete Runway in Antarctica: Here's Why That's a Bad Idea." 17 July 2020. theconversation.com. [online]

Brooks, Shaun, Julia Jabour, and Dana M. Bergstrom. "What Is 'Footprint' in Antarctica: Proposing a Set of Definitions." *Antarctic Science* 30, no. 4 (2018): 227–35.

Brooks, Shaun, Julia Jabour, John van den Hoff, and Dana Bergstrom. "Our Footprint on Antarctica Competes with Nature for Rare Ice-Free Land." *Nature Sustainability* 2, no. 3 (2019): 185.

Brooks, Shaun, Pablo Tejedo, and Tanya O'Neill. "Insights on the Environmental Impacts Associated with Visible Disturbance of Ice-Free Ground in Antarctica." *Antarctic Science* 31, no. 6 (2019).

Broughton, Hugh. Lecture in conversation with UNLESS. "The Extreme Challenge of Building in Antarctica," Architectural Association School of Architecture Polar Lab Workshop, London, February 2019.

Broughton, Hugh. Lecture in conversation with UNLESS. "Architecture in the Extreme," Architectural Association School of Architecture Polar Lab Workshop, London, September 2019.

Bücking, Bert. Lecture in conversation with UNLESS. "Architecture in the Extreme," Architectural Association School of Architecture Polar Lab Workshop, London, September 2019.

Buckminster Fuller Institute. "World Game." Accessed 20 February 2021. bfi.org. [online]

Burton-Johnson, Alex, Martin Black, Peter T. Fretwell, and Joseph Kaluza-Gilbert. "An Automated Methodology for Differentiating Rock from Snow, Clouds and Sea in Antarctica from Landsat 8 Imagery: A New Rock Outcrop Map and Area Estimation for the Entire Antarctic Continent." *The Cryosphere* 10, no. 4 (2016): 1665–77.

Bushinsky, Seth, Peter Landschützer, Christian Rödenbeck, Alison Gray, David Baker, Matthew Mazloff, Laure Resplandy, Kenneth Johnson, and Jorge Sarmiento. "Reassessing Southern Ocean Air-Sea CO_2 Flux Estimates with the Addition of Biogeochemical Float Observations." *Global Biogeochemical Cycles* 33, no. 11 (November 2019): 1370–88.

Bussert, James C. and Bruce A. Elleman, *People's Liberation Army Navy Combat Systems Technology, 1949–2010.* Annapolis, MD: Naval Institute Press, 2011.

Butcher, Ginger. "Fra Mauro's *Mappamundi* – Landsat Science." NASA, 10 January 2014. Accessed 15 June 2020. landsat.gsfc.nasa.gov. [online]

Byrd, Richard. "All-Out Assault on Antarctica." *National Geographic* (August 1956).

Byrd, Richard. *Alone: The Classic Polar Adventure.* 1938. Reprint, Washington DC: Island Press, 2003.

Byrd, Richard. "The Conquest of Antarctica by Air." *National Geographic* (January 1930).

Byrd, Richard. "Exploring the Ice Age in Antarctica." *The National Geographic Magazine* 68, no. 4 (October 1935).

Byrd, Richard. *Exploring with Byrd: Episodes from an Adventurous Life.* New York: G. P. Putnam's Sons, 1938.

Byrd, Richard. "To the Men at South Pole Station." *National Geographic* (July 1957).

Byrd, Richard, and Harold Eugene Saunders. "The Flight to Marie Byrd Land: With a Description of the Map." *Geographical Review* 23 no. 2 (April 1933): 177–209.

C Convergence

The circumpolar boundary known as Antarctic Convergence is regarded by many as the true border of the Antarctic. Also referred to as the "Polar Front", the approximately 40- to 50-kilometre-wide area of water marks the zone of encounter between cold and dense Antarctic surface waters and warmer sub-Antarctic surface waters. Unlike the rigidity of the geographic and geopolitical frontiers imposed by humans, the biological boundary, which has existed for over twenty million years, fluctuates relentlessly between the 41st and 61st latitude South.

Calkin, Parker, Robert Behling, and Colin Bull. "Glacial History of Wright Valley, Southern Victoria Land, Antarctica." *Antarctic Journal of the United States* 5 no. 1 (1970): 22–27.

Calvo-López, Antonio, Eva Arasa-Puig, Mar Puyol, Joan Manuel Casalta, and Alonso Chamarro. "Biparametric Potentiometric Analytical Microsystem for Nitrate and Potassium Monitoring in Water Recycling Processes for Manned Space Missions." *Analytica Chimica Acta* 804 (2013): 190–96.

Cameron, Henry Alan D. *Falkland Islands Dependencies Survey: Base Building Report 1959 – Base A.* Cambridge: BAS Archives, 1959.

Cameron, R., J. King, and C. N. David, "Microbiology, Ecology and Microclimatology of Soil Sites in Dry Valleys of Southern Victoria Land, Antarctica," in *Antarctic Ecology,* edited by M. W. Holdgate, 2:702–16. London: Academic Press, 1970.

Campbell, Nancy. *The Library of Ice: Readings from A Cold Climate.* New York: Simon & Schuster, 2018.

Cantrill, David, and Imogen Poole. *The Vegetation of Antarctica through Geological Time.* Cambridge: Cambridge University Press, 2012.

Carey, Peter. "Is It Time for a Paradigm Shift in How Antarctic Tourism Is Controlled?" *Polar Perspectives* 1 (July 2020).

Carmoodie, Gillian. "The Incredible Story of America's Lost 1939 Antarctic Snow Cruiser." 28 April 2019. buy.motorious.com. [online]

Casa Museo Eduardo Frei Montalva. "50 años de ciencia, colaboración y paz en el continente blanco." 4 March 2019. casamuseoeduardofrei.cl. [online]

Caspers, H. "Marr, James: The Natural History and Geography of the Antarctic Krill (*Euphausia superba* Dana)." *Hydrobiology* 48, no. 4 (January 1963): 637.

Cavicchioli, Ricardo. "Microbial Ecology of Antarctic Aquatic Systems." *Nature Reviews in Microbiology* 13 (2015): 691–706.

Chaikin, Andrew. "Neil Armstrong's Spacesuit Was Made by a Bra Manufacturer." *Smithsonian Magazine* (November 2013).

CEEQUAL. "Home page." Accessed 18 February 2021. ceequal.com. [online]

Chaloner, Bill, and Paul Kenrick. "Did Captain Scott's *Terra Nova* Expedition Discover Fossil *Nothofagus* in Antarctica?" *The Linnean* 31, no. 2 (October 2015): 11–17.

Chaplin, Paul. "Polar Heritage Sites at Risk: Politics, Principles and Practical Problems." In *Cultural Heritage in the Arctic and Antarctic Regions,* edited by Susan Barrand and Paul Chaplin, 24–28. Lørenskog: Icomos, 2004.

Chapman, Sydney. *IGY: Year of Discovery.* Ann Arbor: University of Michigan Press, 1959.

Chaturvedi, Sanjay. "Antarctic as 'Global Knowledge Commons': Geopolitics, Science and Trusteeship." In *Scientific and Geopolitical Interests in Arctic and Antarctic,* edited by Rengaswamy Ramesh, Maruthadu Sudhakar, and Sulagna Chattopadhyay. New Delhi: Lights Research Foundation, 2013.

Chaturvedi, Sanjay. "India and Antarctica: Towards Post-Colonial Engagement." In *The Emerging Politics of Antarctica,* edited by Anne-Marie Brady, 50–74. London: Routledge, 2013.

Cherry-Garrard, Apsley. *The Worst Journey in the World.* London: Constable & Co., 1922.

Cheung, Wilson, Thomas Bauer, and Jinyang Deng. "The Growth of Chinese Tourism to Antarctica: A Profile of Their Connectedness to Nature, Motivations, and Perceptions." *The Polar Journal* 9, no. 1 (2019): 197–213.

Child, Jack. *Miniature Messages: The Semiotics and Politics of Latin American Postage Stamps.* Durham, NC: Duke University Press, 2008.

Chinn, Trevor. "Hydrology and Climate in the Ross Sea Area." *Journal of the Royal Society of New Zealand* 11, no. 4 (1981): 373–86.

Choi, Charles. "Antarctica Shines as Icy Bastion of Space Science." 9 December 2011. space.com. [online]

Chown, Steven, Cassandra Brooks, Aleks Terauds, Céline Le Bohec, Céline van Klaveren--Impagliazzo, Jason Whittington, Stuart Butchart, et al. "Antarctica and the Strategic Plan for Biodiversity." *PLOS Biology* 15 (2017).

Chown, Steven, Andrew Clarke, Ceridwen Fraser, Craig Cary, Katherine Moon, and Melodie

McGeoch. "The Changing Form of Antarctic Biodiversity." *Nature* 522 (2015): 431–38.

Chown, Steven, and Peter Convey. "Antarctic Entomology." *Annual Reviews of Entomology* 61 (2016): 119–37.

Chown, Steven, Jennifer Lee, Kevin Hughes, James Barnes, Peter Barrett, Dana Bergstrom, Peter Convey et al. "Challenges to the Future Conservation of the Antarctic." *Science* 337 (2012): 158–59.

Chu, Jennifer. "Shrink-Wrapping Spacesuits." *MIT News*. 18 September 2014. news.mit.edu. [online]

Clancy, Robert, John Manning, and Henk Brolsma. *Mapping Antarctica: A Five-Hundred-Year Record of Discovery*. Chichester: Praxis Publishing, 2014.

Clark, Elmer. *Technical Report 174: Camp Century Evolution of Concept and History of Design Construction and Performance*. Hanover, NH: U.S. Army Materiel Command, Cold Regions Research & Engineering Laboratory, 1965.

Clark, Pilita. "Inside Antarctica: The Continent Whose Fate Will Affect Millions." *Financial Times*, 28 March 2018. ft.com. [online]

Clarke, Andrew, David K. A. Barnes, and Dominic A. Hodgson. "How Isolated Is Antarctica?" *Trends in Ecology and Evolution* 20 (2005): 1–3.

Clarke, Andrew, and J. Alistair Crame. "Evolutionary Dynamics at High Latitudes: Speciation and Extinction in Polar Marine Faunas." *Philosophical Transactions of the Royal Society Series B* 365 (2010): 3655–66.

Clarke, Andrew, and Nadine M. Johnston. "Antarctic Marine Benthic Diversity." *Oceanography and Marine Biology Annual Review* 41 (2003): 47–114.

Cochran, Christopher. *Conservation Plan: Hillary's Hut, Scott Base*. Christchurch: New Zealand Antarctic Heritage Trust, 2015.

Coetzee, Bernard, Peter Convey, and Steven Chown. "Expanding the Protected Area Network in Antarctica Is Urgent and Readily Achievable." *Conservation Letters* 10 (2017): 670–80.

Cohen, Harlan. "Public Participation in Antarctica: The Role of Nongovernmental and Intergovernmental Organizations." In *Science Diplomacy: Antarctica, Science and the Governance of International Spaces*, edited by Paul Arthur Berkman, Michael Lang, David Walton, and Oran Young. Washington, DC: Smithsonian Press, 2010.

Coleridge, Samuel Taylor. *Rime of The Ancient Mariner and Select Poems*. Reprint, Waiheke Island: Floating Press, 2009.

Collis, Christy. "Critical Legal Geographies of Possession: Antarctica and the International Geophysical Year 1957–1958." *GeoJournal* 75, no. 4 (2010): 387–95.

Collis, Christy. "The Geostationary Orbit: A Critical Legal Geography of Space's Most Valuable Real Estate." In *Down to Earth: Satellite Technologies, Industries, and Cultures*, edited by Lisa Parks and James Schwoch, 61–81. New Brunswick, NJ: Rutgers University Press, 2012.

Collis, Christy, and Klaus Dodds. "Assault on the Unknown: The Historical and Political Geographies of the International Geophysical Year (1957–58)." *Journal of Historical Geography* 34 (2008): 555–73.

Colten, Harvey, and Bruce Altevogt. *Sleep Disorders and Sleep Deprivation: An Unmet Public Health Problem*. Washington, DC: National Academies Press, 2006.

Commission for the Conservation of Antarctic Marine Living Resources. ccamlr.org. [online]

Commission for the Conservation of Antarctic Marine Living Resources. "Conservation Measure 91-03 (2009): Protection of the South Orkney Islands Southern Shelf." *CCAMLR-XXVIII*. 2009. Accessed 1 March 2020. ccamlr.org. [online]

Commission for the Conservation of Antarctic Marine Living Resources. "Conservation Measure 91-05 (2016): Ross Sea Region Marine Protected Area." *CCAMLR-XXXV*. 2016. Accessed 1 March 2020. ccamlr.org. [online]

Commission for the Conservation of Antarctic Marine Living Resources. "Krill Fishery Report 2018." Accessed 1 March 2020. ccamlr.org. [online]

Commission for the Conservation of Antarctic Marine Living Resources. "Marine Protected Areas (MPAs)." Last modified 3 July 2020. ccamlr.org. [online]

Commission for the Conservation of Antarctic Marine Living Resources. "Report of the Thirty-Seventh Meeting of the Commission." Hobart, Tas, 2018.

Commission for the Conservation of Antarctic Marine Living Resources. "Report of the Thirty-Eighth Meeting of the Commission." Hobart, Tas, 2019.

COMNAP: Council of Managers of National Antarctic Programs. "Antarctic Facility List." n.d. Accessed 3 February 2020. comnap.aq. [online]

COMNAP: Council of Managers of National Antarctic Programs. "Antarctic Roadmap Challenges Project." comnap.aq. [online]

COMNAP: Council of Managers of National Antarctic Programs. *Antarctic Station Catalogue*. Christchurch: COMNAP Secretariat, 2017. comnap.aq. [online]

COMNAP: Council of Managers of National Antarctic Programs. *COMNAP Constitution*. 4 July 2008. comnap.aq [online].

COMNAP: Council of Managers of National Antarctic Programs. *International Scientific and Logistic Collaboration in Antarctica*. Antarctic Treaty Consultative Meeting 37, XXXVII ATCM/IP047, 2014.

Connor, Mike. "Wastewater Treatment in Antarctica." *Polar Record* 44, no. 2 (2008): 165–71.

Conrad, Joseph. *Lord Jim*. Oxford: Oxford University Press, 2002.

Constable, Andrew, Daniel Costa, Oscar Schofield, Louise Newman, Edward Urban Jr., Elizabeth Fulton, Jessica Melbourne-Thomas et al. "Developing Priority Variables ('ecosystem Essential Ocean Variables' – eEOVs) for Observing Dynamics and Change in Southern Ocean Ecosystems." *Journal of Marine Systems* 161 (September 2016): 26–41.

Convention on the Conservation of Antarctic Marine Living Resources, 20 May 1980, 1329 UNTS 47.

Convention on the Law of the Sea, Dec. 10, 1982, 1833 UNTS 397.

Convey, Peter. "Antarctic Ecosystems." *Encyclopedia of Biodiversity* 1 (2017): 179–87.

Convey, Peter. "Current Changes in Antarctic Ecosystems." In *Encyclopedia of the World's Biomes*, edited by Michael Goldstein and Dominick Della-Sala. Amsterdam: Elsevier, 2020.

Convey, Peter, David K. A. Barnes, Huw James Griffiths, Susie M. Grant, Katrin Linse, and David N. Thomas. "Biogeography and Regional Classifications of Antarctica." In *Antarctica: An Extreme Environment in a Changing World*, edited by Alex David Rogers, Nadine M. Johnston, Eugene J. Murphy, and Andrew Clarke. Oxford: Blackwell, 2012.

Convey, Peter, Vanessa C. Bowman, Steven L. Chown, Jane E. Francis, Ceridwen I. Fraser, John L. Smellie, Bryan Storey, and Aleks Terauds. "Ice-Bound Antarctica: Biotic Consequences of the Shift from a Temperate to a Polar Climate." In *Mountains, Climate, and Biodiversity*, edited by Carina Hoorn, Allison Perrigo, and Alexandre Antonelli. Hoboken, NJ: Wiley, 2018.

Convey, Peter, Angelika Brandt, and Steve Nicol. "Life in a Cold Environment." In *Antarctica: Global Science from a Frozen Continent*, edited by David W. H. Walton. Cambridge: Cambridge University Press, 2013.

Convey, Peter, Steven L. Chown, Andrew Clarke, David K. A. Barnes, Stef Bokhorst, Vonda Cummings, Hugh W. Ducklow, Francesco Frati, et al. "The Spatial Structure of Antarctic Biodiversity." *Ecological Monographs* 84 (2014): 203–44.

Convey, Peter, John A. E. Gibson, Claus-Dieter Hillenbrand, Dominic A. Hodgson, Philip J. A. Pugh, John L. Smellie, and Mark I. Stevens. "Antarctic Terrestrial Life: Challenging the History of the Frozen Continent?" *Biological Reviews* 83 (2008): 103–17.

Convey, Peter, Kevin A. Hughes, and Tina Tin. "Continental Governance and Environmental Management Mechanisms under the Antarctic Treaty System: Sufficient for the Biodiversity Challenges of This Century?" *Biodiversity* 5 (2012).

Convey, Peter, and Lloyd S. Peck. "Antarctic Environmental Change and Biological Responses." *Science Advances* 5, no. 11 (2019).

Conway, Erik. *Exploration and Engineering: The Jet Propulsion Library and the Quest for Mars*. Baltimore: Johns Hopkins University Press, 2016.

Cook, Alison, Paul Holland, Michael Meredith, Tavi Murray, Adrian Luckman, and David Vaughan. "Ocean Forcing of Glacier Retreat in the Western Antarctic Peninsula." *Science* 353, no. 6296 (15 July 2016): 283–86.

Cook, Frederick Albert. *Through the First Antarctic Night, 1896–1899*. Reprint, New York: Doubleday & McClure, 1900.

Cook, James. *Logbook of Lieut. James Cook, Add Ms 27885, F. 55*. Reprint, The British Library, 1770.

Cool Antarctica. "Antarctic Travel: Practicalities and Modern Vehicles." Accessed 20 February 2021. coolantarctica.com. [online]

Cool Antarctica. "The First Motor Car in Antarctica – 1908: Shackleton's *Nimrod* Expedition – 1907–1909." Accessed 20 February 2021. coolantarctica.com. [online]

Cool Antarctica. "Hoosh, an Antarctic Recipe Food from the Deep South." Accessed 8 March 2021. coolantarctica.com. [online]

Cool Antarctica. "Human Impacts on Antarctica and Threats to the Environment: Mining and Oil." coolantarctica.com. 2020. [online]

Cool Antarctica. "The Threat to Antarctica from Mining and Exploiting Oil and Gas." -coolantarctica.com. 2020. [online]

Cooper, Geoff. Lecture in conversation with UNLESS. "Architecture in the Extreme," Architectural Association School of Architecture Polar Lab Workshop, London, September 2019.

Copeland, Sebastian. *Antarctica: A Call to Action*. San Rafael: Earth Aware Editions, 2008.

Crawford, James. *Brownlie's Principles of Public International Law*, 9th ed. Oxford: Oxford University Press, 2019.

Crook, Lizzie. "Estúdio 41 Completes Prefabricated Antarctic Research Station for Brazil." 11 November 2020. dezeen.com. [online]

Crook, Lizzie. "Kuryłowicz & Associates Reveals Design for Golden Antarctic Research Centre." 22 January 2019. dezeen.com. [online]

Cunningham, John Crabbe. *Falkland Islands Dependencies Survey Report on the Erection of the New Base Hut at Stonington Island*. Cambridge: BAS Archives, 1961.

Curtis, Gary Noble. "South Pole Dome: Let's Keep It!" *The Polar Times* 2, no. 7 (1996): 16.

D Dark Sector

Situated at the Geographic South Pole, the "Dark Sector" encompasses an area of 383 square kilometres that is kept free of any sources of interference that might disrupt electromagnetic signalling and hamper the extraordinary scientific instruments deployed within the specially managed site. This includes the IceCube Neutrino Observatory with its 5,160 "electronic pearls" and other large-scale devices used by scientists residing at the nearby Amundsen-Scott Station to observe, record, and study cosmic rays whose origin and acceleration is one of the biggest mysteries in astroparticle physics today. The "Dark Sector" is adjacent to the "Clean Air Sector", the "Quiet Sector", and the "Downwind Sector".[3]

Dalziell, Janet, and Lyn Goldsworthy, "World Park Antarctica: Does It Have a Future?" *Forum for Applied Research and Public Policy* 9, no. 1 (1994): 71–75.

D'Arcy Wood, Gillen. "Arctic Voyages (1838–1842): Sir Edward Sabine, James Ross, and the Magnetic Sublime." *The Wordsworth Circle* 50, no. 2 (Spring 2019): 237–46.

Darke, Oliver. Lecture in conversation with UNLESS. "Architecture in the Extreme," Architectural Association School of Architecture Polar Lab Workshop, London, September 2019.

Darrieussecq, Marie. *White.* London: Faber and Faber, 2005.

David, Bruno, and Thomas Saucède. *Biodiversity of the Southern Ocean.* Amsterdam: Elsevier, 2015.

Davies, Bethan. "Grounding Lines." Last modified 28 October 2020. antarcticglaciers.org. [online]

Davis, Georgina A. "A History of McMurdo Station through Its Architecture." *Polar Record* 53, no. 269 (2017): 167–85.

Davis, Nicola. "Iceberg Twice the Size of Luxembourg Breaks Off the Antarctic Ice Shelf." *The Guardian,* 12 July 2017.

Davis, Ruth. "The Whaling Dispute in the South Pacific: An Australian Perspective." *Journal of East Asia and International Law* 4, no. 2 (2011): 419–47.

Dawson, Amanda, So Kawaguchi, Catherine K. King, Kathy A. Townsend, Robert King, Wilhelmina M. Huston, and Susan M. Bengtson Nash. "Turning Microplastics into Nanoplastics through Digestive Fragmentation by Antarctic Krill." *Nature Communications* 9 (2018).

Day, David. *Antarctica: A Biography.* Oxford: Oxford University Press, 2013.

Day, David. *Antarctica: What Everyone Needs to Know.* Oxford: Oxford University Press. 2019.

De Broyer, Claude, Philippe Koubbi, Huw James Griffiths, Ben P. Raymond, Cédric d'Udekem d'Acoz, Anton Pieter Van de Putte, Bruno Danis, Bruno David, et al., eds. *Biogeographic Atlas of the Southern Ocean.* Cambridge: Scientific Committee on Antarctic Research, 2014, xii, 498.

Deception Island Antarctic Specially Managed Area. "Volcanic Activity." Accessed 19 February 2020. deceptionisland.aq. [online]

DeLoughrey, Elizabeth. "Satellite Planetarity and the Ends of the Earth." *Public Culture* 26, no. 2 (73) (2014): 257–80.

Delaqua, Victor. "Estação Antártica Comandante Ferraz International Competition 1st Place / Estúdio 41." Translated by Sebastian Jordana. 30 April 2013. archdaily.com. [online]

Denyer, Simon, and Akiko Kashiwagi. "Japan to Leave International Whaling Commission, Resume Commercial Hunting." *The Washington Post,* 26 December 2018. washingtonpost.com. [online]

Department of Environmental Affairs. *SANAE IV Environmental Health and Safety Impact Assessment: Initial Environmental Evaluation Report.* Pretoria: South African Antarctic Programmes, 1991.

Deretsky, Zina. "Map of the South Pole Station and Attached Science Instruments." Accessed 10 March 2021. nsf.gov. [online]

Desbruyères, Damien, Sarah Purkey, Elaine McDonagh, Gregory Johnson, and Brian King. "Deep and Abyssal Ocean Warming from 35 Years of Repeat Hydrography." *Geophysical Research Letters* 43, no. 19 (October 2016).

Design Emergency (@design.emergency). Paola Antonelli interviews Neri Oxman on the future of making. Instagram Live Interview. 29 January 2021.

DeSilvey, Caitlin. "Observed Decay: Telling Stories with Mutable Things." *Journal of Material Culture* 11, no. 3 (2006): 318–38.

Devlin, Mark, Peter Ade, Itziar Aretxaga, James Bock, Edward Chapin, Matthew Griffin, Joshua Gundersen et al. "Over Half of the Far-Infrared Background Light Comes from Galaxies at z ≥ 1.2." *Nature* 458 (April 2009): 737–39.

DeVries, Tim, Mark Holzer, and Francois Primeau. "Recent Increase in Oceanic Carbon Uptake Driven by Weaker Upper-Ocean Overturning." *Nature* 542 (February 2017): 215–18.

De Wit, Maarten. *Minerals and Mining in Antarctica: Science and Technology, Economics and Politics.* Oxford: Clarendon Press, 1985.

Dibbern, Stephen. "Fur Seals, Whales and Tourists: A Commercial History of Deception Island, Antarctica." *Polar Record* 46, no. 3 (2010): 210–21.

Dilke, Oswald. "The Culmination of Greek Cartography in Ptolemy." *The History of Cartography.* Vol. 1. Chicago: University of Chicago Press, 1992.

Ding Huang, ed. *Jidi guojia zhengce yanjiu baogao 2012–2013* [Annual Report on National Polar Policy Research, 2012–2013]. Beijing: Kexue chubanshe, 2013.

Ding Huang, ed. *Jidi guojia zhengce yanjiu baogao 2013–2014* [Annual Report on National Polar Policy Research, 2013–2014]. Beijing: Kexue chubanshe, 2014.

DLR: Institute of Space Systems. "EDEN ISS." Accessed 20 February 2021. dlr.de. [online]

DLR: German Aerospace Center. "EDEN ISS Project Presents Results of a New Greenhouse Concept for Future Space Missions." 23 August 2019. dlr.de. [online]

DMJM. *McMurdo Station Long-Range Development Plan: June 2003 Update.* Washington: National Science Foundation Office of Polar Programs, 2003.

Doake, Christopher, Hugh Corr, Helmut Rott, Pedro Skvarca, and Neal Young. "Breakup and Conditions for Stability of the Northern Larsen Ice Shelf, Antarctica." *Nature* 391 (February 1998): 778–80.

Dober, Bradley Jerald. "The Next Generation Balloon-Borne Large Aperture Submillimeter Telescope (BLAST-TNG)." PhD diss., University of Pennsylvania, 2016.

Dodds, Klaus. "'Awkward Antarctic Nationalism': Bodies, Ice Cores and Gateways in and beyond Australian Antarctic Territory/East Antarctica." *Polar Record* 53, no. 1 (2017): 16–30.

Dodds, Klaus. "Governing Antarctica: Contemporary Challenges and the Enduring Legacy of the 1959 Antarctic Treaty." *Global Policy* 1: 2010.

Dodds, Klaus. "The Great Game in Antarctica: Britain and the 1959 Antarctic Treaty." *Contemporary British History* 21 (2007): 1–24.

Dodds, Klaus. *Pink Ice: Britain and the South Atlantic Empire.* London: I. B. Tauris, 2002.

Dodds, Klaus. "Post-colonial Antarctica: an emerging engagement." *Polar Record* 42 (2006): 59–70.

Dodds, Klaus. "Reflecting on the 60th Anniversary of the Antarctic Treaty." *Polar Record* 55, no. 5 (2019): 311–16.

Dodds, Klaus, and Christy Collis. "Post-Colonial Antarctica." In *Handbook on the Politics of Antarctica,* edited by Klaus J. Dodds, Peder Roberts, and Alan D. Hemmings, 50–68. Cheltenham: Edward Elgar Publishing, 2017.

Dodds, Klaus, and Alan Hemmings. "Antarctic Diplomacy in a Time of Pandemic." *The Hague Journal of Diplomacy* 15 (2020): 530–41.

Dodds, Klaus, and Alan Hemmings, "Arctic and Antarctic Regionalism." In *Handbook on the Geographies of Regions and Territories,* edited by Anssi Paasi, John Harrison, and Martin Jones, 499–500. Cheltenham: Edward Elgar, 2018.

Dodds, Klaus, Alan Hemmings, and Peder Roberts, eds. *Handbook on the Politics of Antarctica.* Cheltenham: Edward Elgar, 2017.

Dodds, Klaus, and Mark Nuttall. *The Scramble for the Poles.* Cambridge: Polity Press, 2016.

Dodds, Klaus, and Juan Francisco Salazar. "Gateway to Antarctica? Geopolitics, Infrastructure and Tourism in Hobart and Australian Antarctic Territory/East Antarctica." In *Tourism Geopolitics: Assemblages of Infrastructure, Affect, and Imagination,* edited by Mary Mostafanezhad, Matilde Córdoba Azcárate, and Roger Norum. Tucson: University of Arizona Press, 2021.

Dömel, Jana S., Roland R. Melzer, Avril M. Harder, Andrew R. Mahon, and Florian Leese. "Nuclear and Mitochondrial Gene Data Support Recent Radiation within the Sea Spider Species Complex *Pallenopsis patagonica.*" *Frontiers in Ecology and Evolution* 4 (2017).

Donohue, Kathleen, Karen Tracey, Dave Watts, Maria Chidichimo, and Teresa Chereskin. "Mean Antarctic Circumpolar Current Transport Measured in Drake Passage." *Geophysical Research Letters* 43, no. 22 (November 2016).

Dowdeswell, Julian, and Jonathan Bamber. "Keel Depths of Modern Antarctic Icebergs and Implications for Sea-Floor Scouring in the Geological Record." *Marine Geology* 243, nos. 1–4 (2007): 120–31.

Dowdeswell, Julian, Christine Batchelor, Boris Dorschel, Toby Benham, Frazer Christie, Evelyn Dowdeswell, Aleksandr Montelli, Jan Erik Arndt, and Catalina Gebhardt. "Sea-Floor and Sea-Ice Conditions in the Western Weddell Sea, Antarctica, around the Wreck of Sir Ernest Shackleton's *Endurance.*" *Antarctic Science* 32, no. 4 (2020): 301–13.

Dudeney, John, and David Walton. "Leadership in Politics and Science within the Antarctic Treaty," *Polar Research* 31 (2012): 11075.

Dufek, George. "What We've Accomplished in Antarctica." The National Geographic Magazine 116, no. 4 (October 1959).

Drewry, David, S. Jordan, and E. Jankowski. "Measured Properties of the Antarctic Ice Sheet: Surface Configuration, Ice Thickness, Volume and Bedrock Characteristics." *Annals of Glaciology* 3 (1982): 83–91.

Dryden, Hugh Latimer. "The International Geophysical Year." *The National Geographic Magazine* 109, no. 2 (February 1956).

Dubrovin, Leonid Ivanovich, and Vladimir Nikolayevich Petrov. *Scientific Stations in Antarctica, 1882–1963.* Translated from Russian. New Delhi: Indian National Scientific Documentation Centre, 1971.

Dudeney, John. "Operation Tabarin: Britain's Secret Wartime Expedition to Antarctica." *Polar Record* 51, no. 1 (January 2015).

Duggan, Jennifer. "Inside the 'Doomsday' Vault." Accessed 9 January 2021. time.com. [online]

Dupont, T. K., and Richard Alley. "Assessment of the Importance of Ice-Shelf Buttressing to Ice-Sheet Flow." *Geophysical Research Letters* 32, no. 4 (February 2005).

Durack, Paul, and Susan Wijffels. "Fifty-Year Trends in Global Ocean Salinities and Their Relationship to Broad-Scale Warming." *Journal of Climate* 23, no. 16 (August 2010).

E Environmental Protocol

"The Protocol on Environmental Protection to the Antarctic Treaty was signed in Madrid on October 4, 1991 and entered into force in 1998. It designates Antarctica as a 'natural reserve, devoted to peace and science' (Art. 2). Article 3 of the Environment Protocol sets forth basic principles applicable to human activities in Antarctica and Article 7 prohibits all activities relating to Antarctic mineral resources, except for scientific research. Until 2048 the Protocol can only be modified by unanimous agreement of all Consultative Parties to the Antarctic Treaty. In addition, the prohibition on mineral resource activities cannot be removed unless a binding legal regime on Antarctic mineral resource activities is in force (Art. 25.5).[4]

Eastman, Joseph T., and Amy R. McCune. "Fishes on the Antarctic Continental Shelf: Evolution of a Marine Species Flock?" *Journal of Fish Biology* 57 (2000): 84–102.

Economist. "Banking against Doomsday." 10 March 2012.

Economist. "Carbon Dioxide Emissions Are Rising. Reducing Them Is a Monumental Challenge." The Climate Issue, 21 September 2019.

Economist. "An Expedition Reveals the Perils of Reading Dostoyevsky in Antarctica." 10 October 2019. economist.com. [online]

Economist. "A Quest to Drill the Oldest Ice Core in Antarctica Is Beginning." 26 September 2019.

Edensor, Tim. *Industrial Ruins: Space, Aesthetics and Materiality.* Oxford: Berg, 2005.

Eddington, Arthur. *The Nature of the Physical World.* Cambridge: Cambridge University Press, 1928.

Egan, Tess. "Going Gaga over Googies." *Australian Antarctic Magazine* 38 (June 2020): 24–25.

Egorov, Boris. "How Soviet Off-Road Vehicles Conquered the South Pole" *Russia Beyond,* 26 February 2018. rbth.com. [online]

Eijgelaar, Eke, Carla Thaper, and Paul Peeters. "Antarctic Cruise Tourism: The Paradoxes of Ambassadorship, 'Last Chance Tourism' and Greenhouse Gas Emissions." *Journal of Sustainable Tourism* 18, no. 3 (2010): 337–54.

Ejército de Chile. *Base Militar "General O'Higgins."* [Santiago]: Instituto Geográfico Militar [de] Chile, 1948.

Ekstein, Nikki. "Travel Industry Sees Glimmers of Recovery in Africa, Antarctica." *Bloomberg*, 22 February 2021. bloomberg.com. [online]

Ekuan, Kenji. *Research into Existing Forms of Living Space and the Development of New Forms More Suitable to Contemporary Society.* Unpublished raw data, 1965. Cited in Michael, Yaacob, and Ali, "The Capsule Living Unit."

Elden, Stuart. "Missing the Point: Globalization, Deterritorialization and the Space of the World." *Transactions of the Institute of British Geographers* 30, no. 1 (2005): 8–19.

Elliot, David, Edwin Colbert, William Breed, James Jensen, and Jon Powell. "Triassic Tetrapods from Antarctica: Evidence for Continental Drift." *Science* 169 (1970): 1197–201.

Elzinga, Aant. "Antarctica: The Construction of a Continent by and for Science." In *Denationalizing Science: The Contexts of International Scientific Practice,* edited by Elisabeth Crawford, Terry Shinn, and Sverker Sörlin, 73–106. Dordrecht, Boston, and London: Kluwer Academic Publishers, 1992.

Elzinga, Aant. "The Continent for Science." In *Handbook on the Politics of Antarctica,* edited by Klaus J. Dodds, Alan D. Hemmings and Peder Roberts, 103–124. Cheltenham: Edward Elgar, 2017.

Elzinga, Aant. "Punta Arenas and Ushuaia: Early Explorers and the Politics of Memory in Constructing Antarctic Gateway Cities." *The Polar Journal* 3, no. 1 (2013): 227–56.

Elzinga, Aant, and Ingemar Bohlin. "The Politics of Science in Polar Regions." In *Changing Trends in Antarctic Research,* edited by Aant Elzinga, 7–27. Dordrecht, Boston, and London: Kluwer Academic Publishers, 1993.

EPICA Community Members. "Eight Glacial Cycles from an Antarctic Ice Core." *Nature* 429 (2004): 623–28.

European Space Agency. "From Antarctica to Space: Telemedicine at the Limit." 30 January 2020. esa.int. [online]

Everson, Inigo. "Antarctic Fisheries." *Polar Record* 19 (1978): 233–51.

ExplorersWeb. "'It [Is] So, So Very Surreal,' Team N2i Meets Lenin at the POI! Marines Arrive [in] Patriot Hills." 22 January 2007. explorersweb. com. [online]

F Fuel Bladder

As the name suggests, fuel bladders owe their name to the membranous hollow organ found in all animals as a liquid receptacle. Made of rubber and incorporated in lightweight, high-efficiency sledges, fuel bladders are used to supply fuel to Antarctic stations.

Having largely replaced airborne fuel deliveries, Antarctic National Programmes make extensive use of fuel bladders during the Antarctic traverses, i.e. to undertake seasonal journeys on ice consisting of "a group of vehicles and their loads moving in convoy across the Antarctic continent in complete autonomy".[5]

Falkland Islands Dependencies Survey. *Horseshoe Island Building Report.* Cambridge: British Antarctic Survey Archives, 1955.

Falkowski, Paul, Tom Fenchel, and Edward Delong. "The Microbial Engines That Drive Earth's Biogeochemical Cycles." Review. *Science* 320, no. 5879 (23 May 2008): 1034–39.

Farman, Joseph C., Brian Gardiner, and Jonathan D. Shanklin. "Large Losses of Total Ozone in Antarctica Reveal Seasonal ClO_x/NO_x Interaction." *Nature* 315 (1985): 207–210.

Feiger, Leah, and Mara Wilson. "The Countries Taking Advantage of Antarctica during the Pandemic." *The Atlantic,* 15 May 2020. theatlantic.com. [online]

Ferraro Choi. "Elevated Station Design for the South Pole Redevelopment Project at Amundsen-Scott South Pole Station." Accessed 13 February 2021. ferrarochoi.com [online]

Ferraro Choi. *Conceptual Design and Programming Study South Pole Replacement Facilities for Peer Review.* March 1994.

Field, Christopher, Michael Behrenfeld, James Randerson, and Paul Falkowski. "Primary Production of the Biosphere: Integrating Terrestrial and Oceanic Components." *Science* 281, no. 5374 (10 July 1998): 237–40.

Fisher, Richard "The Icy Village Where You Must Remove Your Appendix: Welcome to Villas Las Estrellas." *BBC Future*, 3 September 2018. bbc.com. [online]

Fissel, Laura, Peter Ade, Francesco Angilè, Peter Ashton, Steven Benton, Che-Yu Chen, Maria Cunningham et al. "Relative Alignment between the Magnetic Field and Molecular Gas Structure in the Vela C Giant Molecular Cloud Using Low- and High-Density Tracers." *The Astrophysical Journal* 878, no. 2 (June 2019).

Flecker, James Elroy. *The Story of Hasan of Baghdad and How He Came to Make the Golden Journey to Samarkand.* London: Goschen, 1913

Fleming, Thomas, Ariel Heimann, Kenneth Foland, and David Elliot. "$^{40}Ar/^{39}Ar$ Geochronology of Ferrar Dolerite Sills from the Transantarctic Mountains, Antarctica: Implications for the Age and Origin of the Ferrar Magmatic Province." *Geological Society of America Bulletin* 109, no. 5 (May 1997): 533–46.

Flett, William. *Base B: Deception Island; Report, 1944.* BAS Archive (AD6/1B/1944/A), Cambridge, 1944.

Florindo, Fabio, and Martin Siegert, eds., *Antarctic Climate Evolution.* Developments in Earth and Environmental Sciences, Volume 8. Amsterdam: Elsevier Science & Technology, 2008.

Fogg, Gordon Elliott. *A History of Antarctic Science.* Studies in Polar Research. Cambridge: Cambridge University Press, 1992.

Fontana, Pablo. *La Pugna Antártica, El Conflicto Por El Sexto Continente: 1939-1959.* Buenos Aires: Guazuvira Ediciones, 2014.

Food and Agriculture Organization of the United Nations. "FAO Fisheries & Aquaculture – Aquatic Species." 2020. fao.org. [online]

Forcadam, Jaum, and Philip Trathan. "Penguin Responses to Climate Change in the Southern Ocean." *Global Change Biology* 15, no. 7 (June 2009): 1618–30.

Foscari, Giulia. "Research." In *AA Files 76,* edited by Maria Shéhérazade Giudici. London: Architectural Association, 2019.

Fox, William. *Terra Antarctica: Looking into the Emptiest Continent.* San Antonio, TX: Trinity University Press, 2005.

Frame, Bob, and Alan D. Hemmings. "Coronavirus at the End of the World: Antarctica Matters." *Social Sciences & Humanities Open* 2 (2020): 100054.

Fraser, Ceridwen I., Adele K. Morrison, Andrew McC. Hogg, Erasmo C. Macaya, Erik van Sebille, Peter G. Ryan, et al. "Antarctica's Ecological Isolation Will Be Broken by Storm-Driven Dispersal and Warming." *Nature Climate Change* 8 (2018): 704–08.

Fretwell, Peter. *Antarctic Atlas.* London: Penguin Books, 2020.

Fretwell, Peter, Hamish Pritchard, David Vaughan, Jonathan Bamber, Nicholas Barrand, Robin Bell, Cesidio Bianchi et al. "Bedmap2: Improved Ice Bed, Surface and Thickness Datasets for Antarctica." *The Cryosphere* 7, no. 1 (February 2013): 375–93.

Friedlingstein, Pierre, Matthew Jones, Michael O'Sullivan, Robbie Andrew, Judith Hauck, Glen Peters, Wouter Peters et al. "Global Carbon Budget 2019." *Earth System Science Data* 11, no. 4 (December 2019): 1783–838.

Friedmann, Imre. "Endolithic Microorganisms in the Antarctic Cold Desert," *Science* 215, no. 4356 (26 February 1982): 1045–53.

Friedman, Robert Marc. "Å spise kirsebær med de store." In *Norsk Polarhistorie: 2: Vitenskapene,* edited by Einar-Arne Drivenes, Harald Dag Jølle, Ketil Zachariassen, and Norsk Polarhistorie. Oslo: Gyldendal, 2004.

Frölicher, Thomas, Jorge Sarmiento, David Paynter, John Dunne, John Krasting, and Michael Winton. "Dominance of the Southern Ocean in Anthropogenic Carbon and Heat Uptake in CMIP5 Models." *Journal of Climate* 28, no. 2 (January 2015): 862–86.

Fu, Caihong, et al. "Risky Business: The Combined Effects of Fishing and Changes in Primary Productivity on Fish Communities." *Ecological Modelling* 368 (2017): 265–76.

Fuller, Richard Buckminster. *Nine Chains to the Moon.* 1938. Reprint, New York: Anchor Books Edition, 1971.

Fuller, Buckminster. *Operating Manual for Spaceship Earth.* Baden: Lars Müller Publishers, 2008. First published 1969.

Fuller, Richard Buckminster. *Utopia or Oblivion: The Prospects for Humanity,* edited by Jaime Snyder. Baden: Lars Müller Publishers, 1969.

G Global Commons

"The generic term 'Global Commons', when viewed through the lens of state jurisdiction, invokes the idea of open-access spaces (i.e. Antarctica, the high seas, the deep seabed, the atmosphere, the radio-frequency spectrum, the internet, and Outer Space) and resources (i.e. biodiversity, carbon, genes, water or forests) beyond the sovereignty of any state. Normatively speaking, Global Commons are perceived as being held in common, or in trust, for both the present and future generations, and endowed with the capacity to provide public goods."[6]

Galitzki, Nicholas. "Magnetic Fields in Molecular Clouds: The Blastpol and Blast-Tng Experiments." PhD diss., University of Pennsylvania, 2016.

Gandilo, Natalie, Peter Ade, Francesco Elio Angilè, Peter Ashton, Steven Benton, Mark Devlin, Bradley Dober et al. "Submillimeter Polarization Spectrum in the Vela C Molecular Cloud." *The Astrophysical Journal* 824, no. 2 (June 2016)

Gascón, Virginia, and Rodolfo Werner. "Antarctic Krill: A Case Study on the Ecosystem Implications of Fishing." An article prepared for the Lighthouse Foundation. Puerto Madryn, 2005.

Gaul, Kenneth, and British Antarctic Survey. *FIDS 1955 Building Report Base Y, Horseshoe Island.* BAS Archive (AD6/2Y/1955/C), Cambridge, 1955.

Gell, Fiona R., and Callum M. Roberts. "Benefits Beyond Boundaries: The Fishery Effects of Marine Reserves." *Trends in Ecology & Evolution* 18, no. 9 (2003): 448–55.

Gendall, John. "The Coolest Architecture on Earth Is in Antarctica." *The New York Times*. 6 January 2020. nytimes.com. [online]

Gerber, Margaret, B. N. Ramamurti, Frances Calcraft, and Taylor Hattori. Comments on "30×30: A Blueprint for Ocean Protection." Greenpeace International, 2020. greenpeace.org. [online]

German Embassy Wellington. "GONDWANA Station in Terra Nova Bay and Lillie Marleen Hut at Everett Range, Northern Victoria Land." 23 May 2019. wellington.diplo.de. [online]

Gernandt, Hartwig, Saad El Naggar, Jürgen Janneck, Thomas Matz, and Cord Drücker. "From Georg Forster Station to Neumayer Station III: A Sustainable Replacement at Atka Bay for Future." *Polarforschung* 76, nos. 1–2 (2007).

Giæver, John Schjelderup. *The White Desert: The Official Account of the Norwegian-British-Swedish Antarctic Expedition.* New York: E. P. Dutton & Co., 1955.

Gilichinsky, David, Gary Wilson, Imre Friedmann, Christopher McKay, Ronald Sletten, Elizaveta Rivkina, Tatiana Vishnivetskaya et al. "Microbial Populations in Antarctic Permafrost: Biodiversity, State, Age, and Implication for Astrobiology," *Astrobiology* 7, no. 2 (2007): 275–311.

Glasberg, Elena. *Antarctica as Cultural Critique: The Gendered Politics of Scientific Exploration and Climate Change.* New York: Palgrave Macmillan, 2012.

Glasson, John, Riki Therivel, and Andrew Chadwick. *Introduction to Environmental Impact Assessment.* 1st ed. London: Routledge, 2005.

Glenday, James, and Kathryn Diss. "Australians Stuck on Cruise Ship off Uruguay Worry They Will Be Forgotten amid Coronavirus Pandemic." *ABC News* [Australia], 31 March 2020. abcnews.net.au. [online]

Glover, Adrian G., Helena Wiklund, Sergio Taboada, Conxita Avila, Javier Cristobo, Craig R. Smith, Kristy M. Kemp, Alan J. Jamieson, and Thomas G. Dahlgren. "Bone-Eating Worms from the Antarctic: The Contrasting Fate of Whale and Wood Remains on the Southern Ocean Seafloor." *Proceedings of the Royal Society B* (2013).

Gobiet, Andreas, Sven Kotlarski, Martin Beniston, Georg Heinrich, Jan Rajczak, and Markus Stoffel. "21st Century Climate Change in the European Alps: A Review." *Science of the Total Environment* 493 (15 September 2014): 1138–51.

Godden Mackay Logan. *Mawson's Huts Historic Site, Cape Denison, Commonwealth Bay, Australian Antarctic Territory: Conservation Management Plan.* Sydney: Godden Mackay Logan, 2001.

Godfrey, Mark. "Shen Lan Launches Antarctic Krill Shipping Vessel." Seafood Source, 20 May 2020. seafoodsource.com. [online]

Goetzmann, William. *New Lands, New Men: America and the Second Great Age of Discovery.* New York: Viking, 1986.

Goldstein, Joseph. "Bronx Zoo Tiger Is Sick with the Coronavirus." *The New York Times,* 6 April 2020. nytimes.com. [online]

Goldsworthy, Lyn. "Achieving a Ban on Mining in Antarctica." Paper presented at the Symposium on Commemoration of the 20th anniversary of the Hawke Government Initiative to Prevent Mining in Antarctica. Australian National Maritime Museum, Darling Harbour, Sydney, 14 December 2009.

Goldsworthy, Lyn. "The Madrid Protocol: An NGO Perspective," paper presented at Symposium to Commemorate the 20th Anniversary of the Adoption of the Protocol on Environmental Protection to the Antarctic Treaty. Australian Antarctic Division, Kingston, 4 October 2011.

Golynsky, Alexander, Fausto Ferraccioli, Jong Kuk Hong, Dmitry Golynsky, Ralph von Frese, Duncan Young, Don Blankenship et al. "New Magnetic Anomaly Map of the Antarctic." *Geophysical Research Letters* 45, no. 13 (16 July 2018): 6437–49.

Golynsky, Alexander, Dmitry Golynsky, Fausto Ferraccioli, Tom Jordan, Detlef Damaske, Don

Blankenship, Jack Holt et al. "ADMAP-2: The Next-Generation Antarctic Magnetic Anomaly Map." EGU General Assembly, Vienna, 2017. *Geophysical Research Abstracts* 19 (April 2017): 2444.

González Monte, Lucas. "La vida en la Base Esperanza, el pueblo más austral argentino." 11 February 2016. telam.com.ar. [online]

González Videla, Gabriel. *Memorias.* Santiago: Gabriela Mistral, 1975.

Gordon, Kenny. *The Wisdom of David Attenborough: Thoughts of a National Treasure.* CreateSpace Independent Publishing Platform, 2014.

Gordon, Samuel. "Highly Multiplexed Superconducting Detectors and Readout Electronics for Balloon-Borne and Ground-Based FarInfrared Imaging and Polarimetry." PhD diss., Arizona State University, 2019.

Gössling, Stefan, Daniel Scott, and C. Michael Hall. "Pandemics, Tourism and Global Change: A Rapid Assessment of COVID-19." *Journal of Sustainable Tourism* 29, no. 5 (April 2020).

Grann, David. "The White Darkness: A Solitary Journey across Antarctica." *The New York Times,* 12 and 19 February 2018.

Graven, Heather, Nicolas Gruber, Robert Key, Samar Khatiwala, and Xavier Giraud. "Changing Controls on Oceanic Radiocarbon: New Insights on Shallow-to-Deep Ocean Exchange and Anthropogenic CO_2 Uptake." *Journal of Geophysical Research: Oceans* 117, no. 10 (October 2012).

Green, Bill. *Water, Ice and Stone.* Michigan: Harmony Books, 1995.

Greenpeace Australia Pacific. "Greenpeace History in the Antarctic." Accessed 15 June 2020. documents.ats.aq. [online]

Greenpeace International. *Licence to Krill: The Little-Known World of Antarctic Fishing.* March 2018. greenpeace.org. [online]

Griffiths, Huw James, David K. A. Barnes, and Katrin Linse. "Towards a Generalised Biogeography of the Southern Ocean Benthos." *Journal of Biogeography* 36 (2009): 162–77.

Griggs, Kim. "Where U.S., Kiwis Are Neighbors." *Wired,* 12 June 2001. wired.com. [online]

Grilly, Emily, et al. "The Price of Fish: A Global Trade Analysis of Patagonian (*Dissostichus eleginoides*) and Antarctic Toothfish (*Dissostichus mawsoni*)." *Marine Policy* 60 (2015): 186–96.

Gruber, Nicolas, Dominic Clement, Brendan Carter, Richard Feely, Steven van Heuven, Mario Hoppema, Masao Ishii et al. "The Oceanic Sink for Anthropogenic CO_2 from 1994 to 2007." *Science* 363, no. 6432 (15 March 2019): 1193–99.

Gruber, Nicolas, Manuel Gloor, Sara MikaloffFletcher, Scott Doney, Stephanie Dutkiewicz, Michael Follows, Markus Gerber et al. "Oceanic Sources, Sinks, and Transport of Atmospheric CO_2." *Global Biogeochemical Cycles* 23, no. 1 (March 2009).

Gruber, Nicolas, Peter Landschützer, and Nicole Lovenduski. "The Variable Southern Ocean Carbon Sink." *Annual Review of Marine Science* 11 (2019): 159–86.

Guenter, Clarence, Albert Joern, Jay Shurley, and Chester Pierce. "Cardiorespiratory and Metabolic Effects in Men on the South Polar Plateau." *Archives of Internal Medicine* 125, no. 4 (April 1970): 630–7.

Gulland, J. A. *Report of a Meeting of the SCAR Group of Specialists on Marine Living Resources of the Southern Ocean.* Scientific Committee on Antarctic Research, March 2016.

Guojia ziyuan xinxi zhongxin, *Quanchuan ziyuan yu guojia anquan* [Mineral Resources and National Security], 149. Beijing: Dizhi chubanshe, 2000.

Guthridge, Guy. "A New Research Station at the South Pole." *Antarctic Journal of the United States 10, no. 2* (March/April 1975): 37–44.

Guyomard, Ann-Isabelle. "Ethics and Bioprospecting in Antarctica." *Ethics in Science and Environmental Politics* 10 (2010): 31–41.

Hoosh

Known as one of the most traditional "recipes" of the Antarctic specifically during the Heroic Age, hoosh was commonly eaten throughout the expeditions led by Amundsen, Mawson, Scott and Shackleton. The warm stew was generally cooked with the limited ingredients available, which included a dry concentrated mix of fat and meat known as pemmican (replaced in extreme situations by horse or dog meat), a thickener (mostly biscuits or oatmeal), and water.[7]

Haazen, Henk. *Greenpeace World Park Base: Antarctica 1987–1992. Treading Lightly: A Minimal Impact Antarctic Station.* Greenpeace International, March 1994.

Haddelsey, Stephen, and Alan Carroll. *Operation Tabarin: Britain's Secret Wartime Expedition to Antarctica 1944–46.* Stroud: The History Press, 2014.

Haene, Eduardo, and Pablo Reggio. *Visitors' Reception Plan at the Antarctic Base Esperanza: A Proposal for Discussion.* Buenos Aires: Aves Argentinas, 2003.

Haeuplik-Meusburger, Sandra, Carrie Paterson, Daniel Schubert, and Paul Zabel. "Greenhouses and Their Humanizing Synergies." *Acta Astronautica* 96 (March–April 2014): 138–150.

Häkkinen, Sirpa, Peter Rhines, and Denise Worthen. "Warming of the Global Ocean: Spatial Structure and Water-Mass Trends." *Journal of Climate* 29, no. 13 (July 2016): 4949–63.

Halpern, Benjamin, Shaun Walbridge, Kimberly Selkoe, Carrie Kappel, Fiorenza Micheli, Caterina D'Agrosa, John Bruno et al. "A Global Map of Human Impact on Marine Ecosystems." *Science* 319, no. 5865 (February 2008): 948–52.

Halpern, Benjamin, Melanie Frazier, John Potapenko, Kenneth Casey, Kellee Koenig, Catherine Longo, Julia Stewart Lowndes et al. "Spatial and Temporal Changes in Cumulative Human Impacts on the World's Ocean," *Nature Communications* 6, no. 7615 (July 2015).

Halzack, Sarah. "Telemedicine Makes New Advances, All the Way to Antarctica." *The Washington Post*, 7 April 2013.

Halzen, Francis, John Learned, and Todar Stanev. "Neutrino Astronomy." *AIP Conference Proceedings* 198 (January 1990): 39–51.

Hamre, Ivar. "The Japanese South Polar Expedition of 1911–1912: A Little-Known Episode in Antarctic Exploration." *The Geographical Journal* 82, no. 5 (November 1933).

Hancock, Alyce, Andrew Davidson, John McKinlay, Andrew McMinn, Kai Schulz, and Rick van den Enden. "Ocean Acidification Changes the Structure of an Antarctic Coastal Protistan Community." *Biogeosciences* 15 (2018): 2393–410.

Hanessian, John. "The Antarctic Treaty 1959." *The International and Comparative Law Quarterly* 9, no. 3 (1960): 436–80.

Hao, Luoxi, Yi Lin, Junli Xu, Kun Zeng, and Zhe Cui. "Antarctic and Lighting Technology." *China Illuminating Engineering Journal* 25, no. 1 (2014): 1–7.

Harkell, Louis. "Aker Biomarine's New $118m Vessel Set for Next Antarctic Krill Fishing Season." *Undercurrent News,* 5 November 2018. undercurrentnews.com. [online]

Harris, Colin. "Science and Environmental Management in the McMurdo Dry Valleys." In *Ecosystem Dynamics in a Polar Desert: The McMurdo Dry Valleys, Antarctica,* edited by John Priscu, 337–50. Antarctic Research Series 72. Washington, DC: American Geophysical Union,1998.

Harris, Richard. "Modes of Informal Urban Development: A Global Phenomenon." *Journal of Planning Literature* 33, no. 3 (2018): 267–86.

Harrison, Albert, Yvonne Clearwater, and Christopher McKay, eds. *From Antarctica to Outer Space:*

Life in Isolation and Confinement. New York: Springer, 1991.

Harrowfield, David. *Icy Heritage: The Historic Sites of the Ross Sea Region.* Christchurch: Antarctic Heritage Trust, 1995.

Hartley, Catharine. *To the Poles Without a Beard.* London: Simon & Schuster, 2002.

Harvard CMB Group. "Keck Array Overview." Accessed 9 January 2021. cfa.harvard.edu. [online]

Hauck, Judith, Andrew Lenton, Clothilde Langlais, and Richard Matear. "The Fate of Carbon and Nutrients Exported out of the Southern Ocean." *Global Biogeochemical Cycles* 32, no. 10 (October 2018): 1556–73.

Hauteurs UGA. *Claude Lorius and Paleoclimatology.* YouTube video, 4 November 2015. youtube.com. [online]

Hayashi, Moritaka. "The Antarctica Question to the United Nations." *Cornell International Law Journal* 19 (1986): 275–90.

Hayes, Dennis, Lawrence Frakes, Peter Barrett, Derek Burns, Pei-Hsin Chen, Arthur Ford, Ansis Kaneps et al. Introduction to *Initial Reports of the Deep Sea Drilling Project,* vol. 28, 5–18. Washington DC: US Government Printing Office, 1975.

Haynes, Roslynn. "Astronomy and the Dreaming: The Astronomy of the Aboriginal Australians." In *Astronomy Across Cultures: The History of Non-Western Astronomy,* edited by Helaine Selin. Dordrecht: Springer, 2000.

Headland, Robert. "Chronological List of Antarctic Expeditions and Related Historical Events." *Choice Reviews Online* (2005), 16.

Headland, Robert. *A Chronology of Antarctic Exploration: A Synopsis of Events and Activities from the Earliest Times until the International Polar Years, 2007–09.* London: Bernard Quaritch, 2009.

Headland, Robert, ed. *Historical Antarctic Sealing Industry: Proceedings of an International Conference in Cambridge, 16–21 September 2016.* Organised by Bryan Lintott. Cambridge: Scott Polar Research Institute, 2018.

Headland, Robert. "Historical Development of Antarctic Tourism." *Annals of Tourism Research* 21, no. 2 (1994): 269–80.

Headland, Robert. "Sealers Wintering in the South Shetland Islands." In Headland, *Historical Antarctic Sealing Industry,* 61–71.

Headland, Robert. "Territory and Claims in the Antarctic Treaty Region: A Disquisition on Historical and Recent Developments." *The Cartographic Journal* 35 (2020): 1–16.

Healey, Beth. Lecture in conversation with UNLESS. "Architecture in the Extreme," Architectural Association School of Architecture Polar Lab Workshop, London, September 2019.

Heidegger, Martin. *Hölderlin's Hymn "The Ister."* Translated by William McNeill and Julia Davis. Bloomington: Indiana University Press, 1996. Originally published as *Hölderlins Hymne »Der Ister«* (Frankfurt am Main: Vittorio Klostermann, 1984).

Heldmann, Jennifer, Wayne Pollard, Christopher McKay, Margarita Marinova, Alfonso Davila, Kaj Williams, Denis Lacelle, and Dale Andersen. "The High Elevation Dry Valleys in Antarctica as Analog Sites for Subsurface Ice on Mars," *Planetary and Space Science* 85 (September 2013): 53–58.

Hellmer, Hartmut, Frank Kauker, Ralph Timmermann, Jürgen Determann, and Jamie Rae. "Twenty-First-Century Warming of a Large Antarctic Ice-Shelf Cavity by a Redirected Coastal Current." *Nature* 485, no. 7397 (May 2012): 225–28.

Helm, A. S., and J. H. Miller, *Antarctica.* Wellington: Government Printer, 1964.

Hemmings, Alan. "After the Party: The Hollowing of the Antarctic Treaty System and the Governance of Antarctica," Paper presented at the Symposium on Antarctic Politics at the University of Canterbury, Christchurch, New Zealand, 8–9 July 2010.

Hemmings, Alan. "Antarctic Governance in a Time of Coronavirus." *ANZSIL Perspective* 13 (15 May 2020): 3–5.

Hemmings, Alan. "The Antarctic Treaty System." *New Zealand Yearbook of International Law* 16 (2018): 362–70.

Hemmings, Alan. "Does Bioprospecting Risk Moral Hazard for Science in the Antarctic Treaty System?" *Ethics in Science and Environmental Politics* 10 (2010): 5–12.

Hemmings, Alan. "Evolution of A Polar Law." In *Research Handbook on Polar Law,* edited by Karen Scott and David VanderZwaag. Cheltenham: Edward Elgar, 2020.

Hemmings, Alan. "The Philosophy of Law in the Antarctic." In *Philosophies of Polar Law,* edited by Dawid Bunikowski and Alan Hemmings. London: Routledge, 2020.

Hemmings, Alan. "Re-justifying the Antarctic Treaty System for the 21st Century: Rights, Expectations and Global Equity." In *Polar Geopolitics? Knowledges, Resources and Legal Regimes,* edited by Richard Powell and Klaus Dodds. Cheltenham: Edward Elgar, 2014.

Hemmings, Alan. "Southern Horizons: South Asia in the South Indian Ocean." *Panjab University Research Journal Social Sciences* 24 (2016): 129–153.

Hemmings, Alan. "Why Did We Get an International Space Station Before an International Antarctic Station?" *The Polar Journal* 1, no. 1 (2011): 5–16.

Hemmings, Alan, and Neil Gilbert, "Antarctica's Unclaimed Sector," *Antarctic* 33, no. 4 (2015).

Hemmings, Alan, and Lorne Kriwoken. "High Level Antarctic EIA under the Madrid Protocol: State Practice and the Effectiveness of the Comprehensive Environmental Evaluation Process." *International Environmental Agreements: Politics, Law and Economics* 10, no. 3 (September 2010): 187–208.

Hemmings, Alan., Donald Rothwell, and Karen Scott, eds. *Antarctic Security in the Twenty-First Century: Legal and Policy Perspectives.* New York: Routledge, 2012.

Hemmings, Alan, and Ricardo Roura. "A Square Peg in a Round Hole: Fitting Impact Assessment under the Antarctic Environmental Protocol to Antarctic Tourism." *Impact Assessment and Project Appraisal* 21, no. 1 (2003): 13–24.

Herber, Bernard P. "The Economic Case for an Antarctic World Park in Light of Recent Policy Developments." *Polar Record* 28, no. 167 (1992): 293–300.

Herr, Richard, and Robert Hall. "Science and Currency and the Currency of Science." In *Antarctica: Policies and Policy Development,* edited by John Handmer, 13–24. Canberra: Centre for Resource and Environmental Studies, Australia National University, 1989.

Herzog, Werner. *Encounters at the End of the World.* Documentary, Discovery Films, 2007.

Hill, Simeon, Kathryn Keeble, Angus Atkinson, and Eugene Murphy. "A Foodweb Model to Explore Uncertainties in the South Georgia Shelf Pelagic Ecosystem." *Deep Sea Research Part II: Topical Studies in Oceanography* 59–60 (2012): 237–52.

Hillary, Sir Edmund. *No Latitude for Error.* New York: E. P. Dutton & Co. Inc., 1961.

Hince, Bernadette. *The Antarctic Dictionary: A Complete Guide to Antarctic English.* Reprint, Collingwood: CSIRO Publishing, 2000.

Hodder, Ian. *Studies in Human-Thing Entanglement.* eBook, 2016. ian-hodder.com. [online]

Hofman, Robert. "Sealing, Whaling and Krill Fishing in the Southern Ocean." *Polar Record* 53, no. 1 (2017): 88–99.

Holderith, Peter. "There's a Massive Antarctic Exploration Vehicle Lost Somewhere at the Bottom of the World." 12 May 2020. thedrive.com. [online]

Holmes, Tao Tao. "How A Baby Staked Argentina's Claim on Antarctica." 25 February 2016. atlasobscura.com. [online]

Holt, Emily, and Scott Miller. "Bioindicators: Using Organisms to Measure Environmental Impacts". *Nature Education Knowledge* 3(10):8 (2010).

Holt, Sidney. "Sharing the Catches of Whales in the Southern Hemisphere." In *Case Studies on the Allocation of Transferable Quota Rights in Fisheries,* edited by Ross Shotton. FAO Fisheries Technical Paper 411. Rome: FAO, 2001. fao.org. [online]

Hook, Leslie. "Why Penguins May Help Us Predict the Impact of Climate Change." *Financial Times,* 26 February 2020. ft.com. [online]

Hook, Leslie, and Benedict Mander. "The Fight to Own Antarctica." *Financial Times,* 23 May 2018. ft.com. [online]

Hooper, Meredith. *The Longest Winter: Scott's Other Heroes.* Berkeley, CA: Counterpoint, 2010.

Howat, Ian, Claire Porter, Benjamin Smith, Myoung-Jong Noh, and Paul Morin. "The Reference Elevation Model of Antarctica." *The Cryosphere* 13 (2019): 665–74.

Howell, Elizabeth. "Sputnik: The Space Race's Opening Shot." 29 September 2020. space.com. [online]

Howkins, Adrian. *Frozen Empires: An Environmental History of the Antarctic Peninsula.* New York: Oxford University Press, 2017.

Howkins, Adrian. "Melting Empires? Climate Change and Politics in Antarctica Since the International Geophysical Year." *Osiris* 26 (2011): 180–97.

Howkins, Adrian. *The Polar Regions: An Environmental History.* Cambridge: Polity Press, 2016.

Howkins, Adrian. "Politics and Environmental Regulation in Antarctica: A Historical Perspective." In *Handbook on the Politics of Antarctica.* Cheltenham: Edward Elgar, 2017.

Howkins, Adrian. "Reluctant Collaborators: Argentina and Chile in Antarctica during the IGY." *Journal of Historical Geography* 34 (2008): 596–617.

Hugh Broughton Architects. "Halley VI British Antarctic Research Station." Accessed 21 February 2021. hbarchitects.co.uk. [online]

Hugh Broughton Architects. "Juan Carlos 1 Spanish Antarctic Base." Accessed 13 February 2021. hbarchitects.co.uk. [online]

Hugh Broughton Architects. "Scott Base Redevelopment." Accessed 21 February 2021. hbarchitects.co.uk. [online]

Hughes, Kevin, Peter Convey, Luis Pertierra, Greta Vega, Pedro Aragón, and Miguel Olalla-Tárraga. "Human-Mediated Dispersal of Terrestrial Species between Antarctic Biogeographic Regions: A Preliminary Risk Assessment." *Journal of Environmental Management* 232 (2019): 73–89.

Hughes, Kevin, Don Cowan, and Annick Wilmotte. "Protection of Antarctic Microbial Communities: 'Out of Sight, Out of Mind.'" *Frontiers in Microbiology* 6 (2015).

Hughes, Kevin, and Susie Grant. "Current Logistical Capacity Is Sufficient to Deliver the Implementation and Management of a Representative Antarctic Protected Area System." *Polar Research* 37, no. 1 (2018).

Hughes, Kevin, and Susie Grant. "The Spatial Distribution of Antarctica's Protected Areas: A Product of Pragmatism, Geopolitics, or Conservation Need?" *Environmental Science & Policy* 72 (2017): 41–51.

Hughes, Kevin, Louise Ireland, Peter Convey, and Andrew Fleming. "Assessing the Effectiveness of Specially Protected Areas for Conservation of Antarctica's Botanical Diversity." *Conservation Biology* 30 (2015): 113–20.

Hughes, Kevin, Jerónimo López-Martínez, Jane Francis, Alistair Crame, Luis Carcavilla, Kazuyuki Shiraishi, Tomokazu Hokada, and Akira Yamaguchi. "Antarctic Geoconservation: A Review of Current Systems and Practices." *Environmental Conservation* 43 (2016): 97–108.

Hughes, Kevin, Luis Pertierra, Marco Molina-Montenegro, and Peter Convey. "Biological Invasions in Terrestrial Antarctica." *Biodiversity and Conservation* 24 (2015): 1031–55.

Hughes, Kevin, Luis Pertierra, and David Walton, "Area Protection in Antarctica: How Can Conservation and Scientific Research Goals Be Managed Compatibly?" *Environmental Science and Policy* 31 (2013): 120–32.

Hughes, Kevin, Oliver Pescott, Jodey Peyton, Tim Adriaens, Elizabeth Cottier-Cook, Gillian Key, Wolfgang Rabitsch et al. "Invasive Non-Native Species Likely to Threaten Biodiversity and Ecosystems in the Antarctic Peninsula Region." *Global Change Biology* 26 (2020).

Hughes, Peter. "Antarctica: The World's Coldest Cruise." *Financial Times,* 24 March 2016. ft.com. [online]

Hunter, John. Rise & Shine: Diary of John George Hunter, Australasian Antarctic Expedition 1911–1913. Edited by Jenny Hunter. Hinton, NSW: Hunter House: 2011.

Huntford, Roland. *The Last Place on Earth: Scott and Amundsen's Race to the South Pole.* Rev. ed. New York: Modern Library, 1999.

Hurley, Frank. Shackleton's Argonauts: A Saga of the Antarctic Ice-Packs. Sydney: Angus and Robertson, 1948.

Hyde, Timothy. "Architecture at the End of the World: The Pasts and Futures of Heritage Preservation in Antarctica." *Future Anterior: Journal of Historic Preservation, History, Theory, and Criticism* 14, no. 2 (Winter 2017): 73–86.

Hynous, Andrew. "More Fun Than a BARREL of Electrons." NASA: Earth Observatory, 31 December 2019. earthobservatory.nasa.gov. [online]

Ice Core

"Antarctica is the largest planetary archive. Ice cores drilled on the Antarctic Plateau allow palaeo-climatologists to extract from the ice sheet scientific values on historic temperature fluctuations and reconstruct trends in atmospheric greenhouse-gas concentrations from past ice ages and 'warm' interglacial eras. These frozen time capsules of our planet's climate history are the foundation upon which future sustainable environmental policies must, as a matter of urgency, be constructed."

Iberall, Arthur. "The Use of Lines of Nonextension to Improve Mobility in Full-Pressure Suits." *Aerospace Medical Research Laboratories* AMRL-TR-64-118 (November 1964). apps.dtic.mil. [online]

IceCube Collaboration. "Evidence for High-Energy Extraterrestrial Neutrinos at the IceCube Detector." *Science* 342, no. 6161 (22 November 2013).

IceCube Collaboration. "Multi-messenger Observations of a Flaring Blazar Coincident with High-Energy Neutrino IceCube-170922A." *Science* 361, no. 6398 (13 July 2018).

IceCube Collaboration. "Neutrino Emission from the Direction of the Blazar TXS 0506+056 prior to the IceCube-170922A Alert." *Science* 361, no. 6398 (13 July 2018): 147–51.

IceCube Collaboration. "Time-Integrated Neutrino Source Searches with 10 Years of IceCube Data." *Physical Review Letters* 124 (February 2020).

IceCube: South Pole Neutrino Observatory. "Antarctic Weather." Accessed 21 February 2021. icecube.wisc.edu. [online]

IceCube: South Pole Neutrino Observatory. "Detector." Accessed 9 January 2021. icecube. wisc.edu. [online]

IceCube: South Pole Neutrino Observatory. "Digital Optical Module (DOM) Development." Accessed 9 January 2021. icecube.wisc.edu. [online]

IceCube: South Pole Neutrino Observatory. "Meet the Collaboration." Accessed 9 January 2021. icecube.wisc.edu. [online]

Incoll, Philip. *AANBUS: The Creation of a Building System for Antarctica.* Canberra: Commonwealth of Australia, Department of Housing & Construction, 1980.

Incoll, Philip. *The Influence of Architectural Theory on the Design of Australian Antarctic Stations.* Melbourne: Australian Construction Services, 1990.

Ing, Alvin, Christine Cocks, and Jeffery Peter Green. "COVID-19: In the Footsteps of Ernest Shackleton." *Thorax* (27 May 2020). thorax.bmj.com. [online]

Institut Polaire. *Concordia: A New Permanent, International Research Support Facility High on the Antarctic Ice Cap; Technical Overview.* Plouzane: French Polar Institute, 2020. latitude.aq. [online]

Intergovernmental Panel on Climate Change (IPCC). "Choices Made Now Are Critical for The Future of Our Ocean and Cryosphere." 25 September 2019. ipcc.ch. [online]

Intergovernmental Panel on Climate Change (IPCC). *Climate Change 2014: Synthesis Report; Contribution of Working Groups I, II and III to the Fifth Assessment Report of the Intergovernmental Panel on Climate Change.* Geneva: Intergovernmental Panel on Climate Change, 2015.

Intergovernmental Panel on Climate Change (IPCC). "Special Report on the Ocean and Cryosphere in a Changing Climate." Geneva: IPCC, 2019. ipcc.ch. [online]

Intergovernmental Panel on Climate Change (IPCC). "Summary for Policymakers." In Masson-Delmotte, Zhai, Pörtner et al., *Global Warming of 1.5°C.*

International Association of Antarctica Tour Operators (IAATO). iaato.org. [online]

International Association of Antarctica Tour Operators (IAATO). *IAATO Overview of Antarctic Tourism: 2017–18 Season and Preliminary Estimates for 2018–19 Season.* Information Paper 71. Antarctic Treaty Consultative Meeting XLI, Buenos Aires, 2018.

International Association of Antarctica Tour Operators (IAATO). *IAATO Overview of Antarctic Tourism: 2018–19 Season and Preliminary Estimates for 2019–20 Season,* Information Paper 140. Antarctic Treaty Consultative Meeting XLII, Prague, 2019.

International Court of Justice. *Whaling in the Antarctic (Australia v Japan; New Zealand Intervening).* Judgement of 31 March 2014.

International Polar Foundation. "BELARE 2007–2008." Press release, 10 March 2008. antarcticstation.org. [online]

International Polar Foundation. "Princess Elisabeth Antarctica: The First Zero Emission Polar Research Station." Brussels, 2013. antarcticstation.org. [online]

International Polar Foundation. "Technical Sheet 2: Passive Building." 1 October 2008. antarcticstation.org. [online]

International Science Council. "Scientific Committee on Antarctic Research (SCAR)." Accessed 16 March 2021. council.science. [online]

International Thwaites Glacier Collaboration. "Home page." Accessed 9 January 2021. thwaitesglacier.org. [online]

International Union for Conservation of Nature. *Second World Conference on National Parks.* Morges: International Union for Conservation of Nature and Natural Resources, 1974.

Ishihara, Aya. "The IceCube Upgrade: Design and Science Goals." *36th International Cosmic Ray Conference (ICRC2019), July 24th – August 1st, 2019, Madison, WI, U.S.A.* Trieste: SISSA, 2019.

Itten, Johannes. The Art of Color: The Subjective Experience and Objective Rationale of Color. New York: Van Nostrand Reinhold, 1993.

Ivar do Sul, Juliana, David Barnes, Monica Costa, Peter Convey, Erli Schneider Costa, and Lúcia Campos. "Plastics in the Antarctic Environment: Are We Looking Only at the Tip of the Iceberg?" *Oecologia Australis* 15, no. 1 (March 2011): 150–70.

Jamesway Huts

First designed during World War I by Peter Norman Nissen, the Jamesway huts were deployed in the Antarctic during the US Navy military operation known as Deep Freeze, launched during the 1955–1956 season. Featuring wooden ribs covered with insulated blankets, the structures were fastened on the floor to fibreglass insulated plywood panels obtained by recycling shipping crates. Also known as Nissen huts, such 5-metre-long inhabitation modules weighed 540 kilograms.

Jabour, Julia. "Biological Prospecting: The Ethics of Exclusive Reward from Antarctic Activities." *Ethics in Science and Environmental Politics* 10 (2010): 19–29.

Jabour, Julia. "Biological Prospecting in Antarctica: Fair Game?" In *The Emerging Politics of Antarctica,* edited by Anne-Marie Brady, 242–57. Abingdon: Routledge, 2013.

Jabour, Julia. "Case Study Antarctica: Up Against the Ice Barrier; Antarctic Tourism Operators Prepare for the Polar Shipping Code." *Global Climate Change and Coastal Tourism: Recognizing Problems, Managing Solutions and Future Expectations* (2017): 273.

Jabour, Julia. "The Potential to Regulate Bioprospecting for Marine Genetic Resources: Two Case Studies." In *Handbook of Maritime Regulation and Enforcement,* edited by Robin Warner and Stuart Kaye, 324–41. Abingdon: Routledge, 2016.

Jabour-Green, Julia, and Diane Nicol. "Bioprospecting in Areas Outside National Jurisdiction: Antarctica and the Southern Ocean." *Melbourne Journal of International Law* 4, no. 1 (2003): 76–111.

Jacobs, Stanley, Adrian Jenkins, Claudia Giulivi, and Pierre Dutrieux. "Stronger Ocean Circulation and Increased Melting under Pine Island Glacier Ice Shelf." *Nature Geoscience* 4 (June 2011): 519–23.

Jacquet, Jennifer, et al. "'Rational Use' in Antarctic Waters." *Marine Policy* 63 (2016): 28–34.

James, L. V. "Accommodation Building Report." British Antarctic Survey. 1967.

Japan Aerospace Exploration Agency. "A Joint Project of JAXA, NIPR, Misawa Homes, and MHIRD: Demonstration Test of Antarctica Mobile Station Unit." Press release. 26 August 2019. jaxa.jp. [online]

JAXA: Japan Aerospace Exploration Agency. "A Joint Project of JAXA, NIPR, Misawa Homes, and MHIRD Demonstration Test of Antarctica Mobile Station Unit." Press release, 26 August 2019. global.jaxa.jp. [online]

Jenouvrier, Stéphanie, Hal Caswell, Christophe Barbraud, Marika Holland, Julienne Stroeve, and Henri Weimerskirch. "Demographic Models and IPCC Climate Projections Predict the Decline of an Emperor Penguin Population." *PNAS* 106, no. 6 (February 2009): 1844–47.

Jenouvrier, Stéphanie, Jimmy Garnier, Florian Patout, and Laurent Desvillettes. "Influence of Dispersal Processes on the Global Dynamics of Emperor Penguin, a Species Threatened by Climate Change." *Biological Conservation* 212 (August 2017): 63–73.

Jenouvrier, Stéphanie, Marika Holland, David Iles, Sara Labrousse, Laura Landrum, Jimmy Garnier, Hal Caswell et al. "The Paris Agreement Objectives Will Likely Halt Future Declines of Emperor Penguins." *Global Change Biology* 26, no. 3 (March 2020): 1170–84.

Jenouvrier, Stéphanie, Marika Holland, Julienne Stroeve, Christophe Barbraud, Henri Weimerskirch, Mark Serreze, and Hal Caswell. "Effects of Climate Change on an Emperor Penguin Population: Analysis of Coupled Demographic and Climate Models." *Global Change Biology* 18, no. 9 (September 2012): 2756–70.

Jenouvrier, Stéphanie, Marika Holland, Julienne Stroeve, Mark Serreze, Chistophe Barbraud,

Henri Weimerskirch, and Hal Caswell. "Projected Continent-Wide Declines of the Emperor Penguin under Climate Change." *Nature Climate Change* 4, no. 8 (August 2014): 715–18.

Jidi zhanlue yanjiu dongtai [Polar Strategy Research] 3 (2013), 16. Internal Report, Chinese Arctic and Antarctic Administration. Cited in Brady, "Cooperation or Conflict?"

Johnson, Dan. "Antarctic Treasure: The Underwater Images of Norbert Wu." National Science Foundation, 9 July 2004. nsf.gov. [online]

Johnson, Hamish. "Cosmic Neutrinos Named Physics World 2013 Breakthrough of the Year." 13 December 2013. physicsworld.com. [online]

Johnson, Nicholas. *Big Dead Place: Inside the Strange and Menacing World of Antarctica.* Los Angeles: Feral House, 2005.

Johnston, Paul, David Santillo, Richard Pages, and Cat Dorey. "Gambling with Krill Fisheries in the Antarctic: Large Uncertainties Equate with High Risks." *Greenpeace Research Laboratories Technical* Note. Reprint, Exeter: School of Biological Sciences, University of Exeter, 2009.

Jones-Williams, Kirstie, Tamara Galloway, Matthew Cole, Gabriele Stowasser, Claire M. Waluda, and Clara Manno. "Close Encounters: Microplastic Availability to Pelagic Amphipods in Sub-Antarctic and Antarctic Surface Waters." *Environment International* 140 (2020).

Jordan, Tom, David Porter, Kirsty Tinto, Romain Millan, Atsuhiro Muto, K. Hogan, Robert Larter, Alastair Graham, and John Paden. "New Gravity-Derived Bathymetry for the Thwaites, Crosson and Dotson Ice Shelves Revealing Two Ice Shelf Populations." *The Cryosphere* 14 (2020): 2869–82.

Jourdan, Michael. "That's One Stale Biscuit." *National Geographic Society Newsroom,* 21 September 2011. blog.nationalgeographic.org. [online]

Jouzel, Jean. "A Brief History of Ice Core Science over the Last 50 Yr." *Climate of the Past* 9 (2013): 2525–47.

Joyner, Christopher. "Antarctica and the Law of the Sea: An Introduction and Overview." *Ocean Development and International Law* 13, no. 3 (1983): 277–89.

Joyner, Christopher. "The Antarctic Minerals Negotiating Process." *American Journal of International Law* 81, no. 4 (October 1987): 888–904.

Joyner, Christopher. "The Antarctic Treaty and the Law of the Sea: Fifty Years On." *Polar Record* 46 (2010): 14–17.

Joyner, Christopher. *Eagle over the Ice: The U.S. in the Antarctic.* Hanover, NH: University Press of New England, 1997.

K Krill

Antarctic krill, also referred to as *Euphausia Superba,* is a shrimp-like planktonic crustacean found in abundance in the Southern Ocean. Known for being the dominant animal species on Earth and for playing a vital role in the marine ecosystem (representing the predominant food source for many species including whales), krill are dangerously overexploited by national fishing enterprises, which commodify them as fish food for aquacultures and Omega-3 supplements for humans.

Kagge, Erling. *Philosophy for Polar Explorers.* Translated by Kenneth Steven. New York: Pantheon Books, 2020.

Kaplan, Joseph. "The Scientific Program of the International Geophysical Year." *Proceedings of the National Academy of Sciences* 40, no. 10 (1954): 926–31.

Kaufman, Marc. "The South Pole Is a Great Place to View Space." 20 March 2014. nationalgeographic.com. [online]

Kauffman, Steven, and Arthur Weber. "Design for Survival: The Story of Plateau Station."

Antarctic Journal of the United States 1, no. 4 (July–August 1966).

Keeling, Charles, Stephen Piper, Robert Bacastow, Martin Wahlen, Timothy Whorf, Martin Heimann, and Harro Meijer. "Exchanges of Atmospheric CO_2 and $^{13}CO_2$ with the Terrestrial Biosphere and Oceans from 1978 to 2000: I. Global Aspects." UC San Diego: Library – Scripps Digital Collection, 2001. escholarship.org. [online]

Kelly, Anna, Delphine Lannuzel, Thomas Rodemann, Klaus M. Meiners, and Heidi J. Auman. "Microplastic Contamination in East Antarctic Sea Ice." *Marine Pollution Bulletin* 154 (2020).

Kennett, James, Robert Houtz, Peter Andrews, Anthony Edwards, Victor Gostin, Marta Hajós, Monty Hampton et al. "Cenozoic Paleoceanography in the Southwest Pacific Ocean, Antarctic Glaciation, and the Development of the Circum-Antarctic Current." In *Initial Reports of the Deep Sea Drilling Project,* vol. 29, 1155–69. Washington DC: US Government Printing Office, 1974.

Keppler, Lydia, and Peter Landschützer. "Regional Wind Variability Modulates the Southern Ocean Carbon Sink." *Scientific Reports* 9 (2019).

Kerr, Ian. *Campbell Island: A History.* Wellington: A. H. Reed, 1976.

Kerry, Knowles, and Martin Riddle, eds. *Health of Antarctic Wildlife: A Challenge for Science and Policy.* Dordrecht: Springer, 2009.

Key, Robert, Alexander Kozyr, Christopher Sabine, Kitack Lee, Rik Wanninkhof, John Bullister, Richard Feely, Frank Millero, C. Mordy, and Tsung-Hung Peng. "A Global Ocean Carbon Climatology: Results from Global Data Analysis Project (GLODAP)." *Global Biogeochemical Cycles* 18, no. 4 (December 2004).

Kezina, Darya. "New GLONASS Stations to Appear in Antarctica," *Russia Beyond,* 24 April 2015. rbth.com. [online]

Khatiwala, Samar, François Primeau, and T. Hall. "Reconstruction of the History of Anthropogenic CO_2 Concentrations in the Ocean." *Nature* 462, no. 7271 (19 November 2009): 346–49.

Klages, Johann, Ulrich Salzmann, Torsten Bickert, Claus-Dieter Hllenbrand, Kartsen Gohl, Gerhard Kuhn, Steven Bohaty et al., "Temperate Rainforests near the South Pole during Peak Cretaceous Warmth." *Nature* 580, no. 7801 (2 April 2020): 81–86.

Klein, Emily, Simeon Hill, Jefferson Hinke, Tony Phillips, and George Watters. "Impacts of Rising Sea Temperature on Krill Increase Risks for Predators in the Scotia Sea." *PLOS ONE* 13, no. 1 (2018).

Klein, Natalie. "International Law Perspectives on Cruise Ships and COVID-19." *Journal of International Humanitarian Legal Studies* 11, no. 1 (2020): 1–13.

Knowles, Kerry, and Martin Riddle, eds. *Health of Antarctic Wildlife: A Challenge for Science and Policy.* Dordrecht: Springer, 2009.

Knowles, Kerry, and Martin Riddle, eds. "Infectious Diseases." In *Health of Antarctic Wildlife: A Challenge for Science and Policy.* Dordrecht: Springer, 2009, 15–21.

Koelmans, Albert , Ellen Besseling, and Won Shim. "Nanoplastics in the Aquatic Environment: Critical Review." In *Marine Anthropogenic Litter,* edited by Melanie Bergmann, Lars Gutow, and Michael Klages, 325–40. Cham: Springer, 2015.

Kögel, Tanja, Ørjan Bjorøy, Benuarda Toto, André M. Bienfait, and Monica Sanden. "Micro- and Nanoplastic Toxicity on Aquatic Life: Determining Factors." *Science of the Total Environment* 709 (2020).

Koh, Tommy. "A Constitution for the Oceans." United Nations. Accessed 10 February 2020. un.org. [online]

Kohlberg, Eberhard, and Jürgen Janneck. "Georg Von Neumayer Station (Gvn) and Neumayer Station II (NM-II) German Research Stations on Ekström Ice Shelf, Antarctica." *Polarforschung* 76, nos. 1–2 (2007).

Koolhaas, Rem. *Elements of Architecture.* Cologne: Taschen, 2018.

Krupnik, Igor, Michael A. Lang, and Scott E. Miller, eds. *Smithsonian at the Poles: Contributions to International Polar Year Science.* Washington, DC: Smithsonian Institution, 2009.

Kumar, Alexander. "The Journey to White Mars." *The New York Times,* 2 October 2012.

Kuznetsov, Nikita. *Russian Antarctica: History in Illustrations.* Moscow: Paulsen Publishers, 2020.

L Little American

On a continent that has no indigenous peoples and thus is – or should be – alien to the very notion of nation state, the transient Antarctic citizen builds a sense of belonging into the scientific station in which they are deployed. Residents of Little America, which was erected in its first iteration in 1929 at the Bay of Whales and occupied up until 1958 in its later incarnation at Kainan Bay, thus called themselves Little Americans. Similarly, scientists deployed at McMurdo Station were referred to as MacTownites or McMurdoites, while inhabitants of Australia's Casey Station were known as Caseyites. Applied in similar ways to most stations, the peculiar naming system was transferred to larger territories, as exemplified by the names attributed to the scientists deployed in Adélie Land, who were nicknamed the Adelians.

La Ferla, Ruth. "Moon Boots Back on Earth," *The New York Times,* 17 October 2004.

La Nación Argentina: Justa. Libre. Soberana. Buenos Aires: Talleres Gráficos Peuser, 1949.

Lacerda, Ana, Lucas Rodrigues, Erik van Sebille, Fábio Lameiro Rodrigues, Lourenço Ribeiro, Eduardo Secchi, Felipe Kessler, and Maíra Proietti. "Plastics in Sea Surface Waters around the Antarctic Peninsula." *Scientific Reports* 9 (2019).

Laganà, Pasqualina, Gabriella Caruso, Ilaria Corsi, Elisa Bergami, Valentina Venuti, Domenico Majolino, Rosabruna La Ferla, Maurizio Azzaro, and Simone Cappello. "Do Plastics Serve as a Possible Vector for the Spread of Antibiotic Resistance? First Insights from Bacteria Associated to a Polystyrene Piece from King George Island (Antarctica)," *International Journal of Hygiene and Environmental Health* 222, no. 1 (January 2019): 89–100.

Lagorio, Juan Jose. "U.N.'s Ban Says Global Warming Is 'An Emergency.'" Reuters, 10 November 2007. reuters.com. [online]

Lambright, Henry. *Why Mars: NASA and the Politics of Space Exploration.* Baltimore: Johns Hopkins University Press, 2014.

Landschützer, Peter, Nicolas Gruber, and Dorothee Bakker. "Decadal Variations and Trends of the Global Ocean Carbon Sink." *Global Biogeochemical Cycles* 30, no. 10 (October 2016): 1396–417.

Langway, Chester. *The History of Early Polar Ice Cores,* US Army Corps of Engineers, Cold Regions Research and Engineering Laboratory, Report TR-08-1 (2008), 47.

Lansing, Alfred. *Endurance: Shackleton's Incredible Voyage.* New York: Carroll & Graf, 1959.

Launius, Roger, James Roger Fleming, and David DeVorkin. *Globalizing Polar Science: Reconsidering the Social and Intellectual Implications of the International Polar and Geophysical Years.* New York: Palgrave, 2010.

Law, Jennifer, Mary Van Baalen, Millennia Foy, Sara Mason, Claudia Mendez, Mary Wear, Valerie Meyers, and David Alexander. "Relationship between Carbon Dioxide Levels and Reported Headaches on the International Space Station." *Journal of Occupational & Environmental Medicine* 56, no. 5 (May 2014): 477–83.

Law, Phillip. "Techniques of Living, Transport and Communication." In *Antarctica: A New Zealand*

Antarctic Society Survey, edited by Trevor Hatherton. Wellington: Reed, 1965.

Lazzara, Matthew, and Charles Stearns. *The Future of the Next Generation Satellite Fleet and the McMurdo Ground Station.* Madison: University of Wisconsin-Madison Space Science and Engineering Center, 2004.

Leane, Elizabeth. *Antarctica in Fiction: Imaginative Narratives of the Far South.* Cambridge: Cambridge University Press, 2012.

Leane, Elizabeth. *South Pole: Nature and Culture.* London: Reaktion, 2016.

Leane, Elizabeth, and Jeffrey McGee, eds. *Anthropocene Antarctica: Perspectives from the Humanities, Law and Social Sciences.* London: Routledge, 2019.

Leane, Elizabeth, and Graeme Miles. "The Poles as Planetary Places." *The Polar Journal* 7, no. 2 (2017): 270–86.

Leary, David. "Agreeing to Disagree on What We Have Not Agreed on: The Current State of Play of the BBNJ Negotiations on the Status of Marine Genetic Resources in Areas beyond National Jurisdiction." *Marine Policy* 99 (2019): 21–29.

Lefeber, René. "Marine Scientific Research in the Antarctic Treaty System." In *Law of the Sea and the Polar Regions: Interactions between Global and Regional Regimes,* edited by Donald R. Rothwell, Alex G. Oude Elferink, and Erik Jaap Molenaar, 321–41. Leiden: Martinus Nijhoff, 2013.

Le Guen, Camille, Giuseppe Suaria, Richard B. Sherley, Peter G. Ryan, Stefano Aliani, Lars Boehme, and Andrew S. Brierley. "Microplastic Study Reveals the Presence of Natural and Synthetic Fibres in the Diet of King Penguins (*Aptenodytes patagonicus*) Foraging from South Georgia." *Environment International* 134 (2020).

Le Guin, Ursula K. "Sur: A Summary Report of the Yelcho Expedition to the Antarctic 1909–1910." *The New Yorker,* 1 February 1982.

Le Quéré, Corinne, Christian Rödenbeck, Erik Buitenhuis, Thomas Conway, Ray Langenfelds, Antony Gomez, Casper Labuschagne et al. "Saturation of the Southern Ocean CO_2 Sink Due to Recent Climate Change." *Science* 316, no. 5832 (22 June 2007): 1735–38.

Lee, Sang Leem. *Jangbogo Antarctic Research Station.* Seoul: Space Group, 2011.

Lentati, Sarah. "The Man Who Cut Out His Own Appendix." *BBC News Magazine,* 4 May 2015. bbc.com. [online]

Lever, James, Jason Weale, Thomas Kaempfer, and Monica Preston. *Advances in Antarctic Sled Technology.* U.S. Army Corps of Engineers. February 2016. usace.contentdm.oclc.org. [online]

Levick, Murray. Journals kept on the Northern Party of the British Antarctic Expedition, 1910–1913. 9 January –14 November 1912. Vol. 2, Rough Journal, 29 March–28 September 1912. MS 1423/2. Cambridge: Scott Polar Research Institute.

Levy, Richard, et al. "Antarctic Ice Sheet Sensitivity to Atmospheric CO_2 Variations in the Early to Mid-Miocene," *Proceedings of the National Academy of Sciences* 113 (2016), 3453–58.

Lewis, Richard. *A Continent for Science: The Antarctic Adventure.* London: Secker & Warburg, 1965.

Li, Xia. "Feature: China's Taishan Station Offers Conveniences of Modern Life Deep in Antarctic Ice Sheet." 2 February 2019. xinhuanet.com. [online]

Liggett, Daniela. "Tourism in the Antarctic: Modi Operandi and Regulatory Effectiveness." In *Gateway Antarctica.* Saarbrücken: VDM, 2009.

Liggett, Daniela, and Alan Hemmings, eds., *Exploring Antarctic Values: Proceedings of the Workshop "Exploring Linkages between Environmental Management and Value Systems: The Case of Antarctica"* held at the University of Canterbury, Christchurch, New Zealand on 5 December 2011. Christchurch: University of Canterbury, 2013.

Liggett, Daniela, Alison McIntosh, Anna Thompson, Neil Gilbert, and Bryan Storey. "From Frozen Continent to Tourism Hotspot? Five Decades of Antarctic Tourism Development and Management, and a Glimpse into the Future." *Tourism Management* 32, no. 2 (2011): 357–66.

Liggett, Daniela, and Emma Stewart. "The Changing Face of Political Engagement in Antarctic Tourism." In *Handbook on the Politics of Antarctica.* Cheltenham: Edward Elgar Publishing, 2017.

Liggett, Daniela, and Emma Stewart. "Sailing in Icy Waters: Antarctic Cruise Tourism Development, Regulation and Management." In *Cruise Ship Tourism,* edited by Ross Dowling and Clare Weeden, 484–504. 2nd ed. Wallingford, Oxfordshire: CABI, 2017.

Lindsey, Rebecca, and Michon Scott. "What Are Phytoplankton?" 13 July 2010. earthobservatory.nasa.gov. [online]

Liu, Nengye, and Cassandra Brooks. "China's Changing Position towards Marine Protected Areas in the Southern Ocean: Implications for Future Antarctic Governance." *Marine Policy* 94 (August 2018): 189–95.

Liu, Nengye, Cassandra Brooks, and Tianbao Qin, eds. *Governing Marine Living Resources in the Polar Regions.* Cheltenham: Edward Elgar, 2019.

Liu Yijian, *Zhi haiquan yu haijun zhanlue* [The Command of the Sea and the Strategic Employment of Naval Forces]. Beijing: Jiefangjun guofang daxue chubanshe, 2004.

Lohan, Dagmar, and Sam Johnston. *Bioprospecting in Antarctica.* Yokohama: UNU-IAS, 2005. collections.unu.edu. [online]

Lord, Thomas. "The Antarctic Treaty System and the Peaceful Governance of Antarctica: The Role of the ATS in Promoting Peace at the Margins of the World," *The Polar Journal* (May 2020).

Lorenzo, Cristian, Daniela Liggett, Bob Frame, Andrea Herbert, Ilan Kelman, Jennifer Pickett, and Yelena Yermakova. "Antarctica and the COVID-19 Pandemic: Taking a Social Sciences and Humanities Perspective." *ECO: Environment Coastal & Offshore* (2020): 116–19.

Lorius, Claude. "Home page." Accessed 9 January 2021. claude-lorius.com. [online]

Lourie, Nathan. "Building the Next-Generation BLAST Experiment." PhD diss., University of Pennsylvania, 2018.

Lourie, Nathan. "Thermal Modeling of the BLAST-TNG Balloon Telescope." Accessed 9 January 2021. crtech.com. [online]

Lourie, Nathan, Peter Ade, Francisco Angilè, Peter Ashton, Jason Austermann, Mark Devlin, Bradley Dober et al. "Preflight Characterization of the BLAST-TNG Receiver and Detector Arrays." *SPIE Proceedings Volume 10708, Millimeter, Submillimeter, and Far-Infrared Detectors and Instrumentation for Astronomy IX* (2018).

Lourie, Nathan, Francisco Angilè, Peter Ashton, Brian Catanzaro, Mark Devlin, Simon Dicker, Joy Didier et al. "Design and Characterization of a Balloon-Borne Diffraction-Limited Submillimeter Telescope Platform for BLAST-TNG." *SPIE Proceedings Volume 10700, Ground-Based and Airborne Telescopes VII* (2018).

Lovecraft, Howard. *At the Mountains of Madness.* Serialised in *Astounding Stories Magazine* (February–April 1936).

Lovegrove, Ian. *Halley 5 – Phase III 1990/91 Report.* BAS Archives (AD6/22/1990/C). Cambridge: British Antarctic Survey, 1991.

Luchansky, Grigory. "Opyt Antarkticheskikh Sovetskikh Stantsii" | Опыт Антарктических Советских Станций. Accessed 27 January 2021. geolmarshrut.ru. [online]

Lucibella, Michael. "Podcast: Fuels; Powering the South Pole." *The Antarctic Sun,* 18 September 2019. antarcticsun.usap.gov. [online]

Lüdecke, Cornelia, and Colin Summerhayes. *The Third Reich in Antarctica: The Story of the German Antarctic Expedition of 1938/39.* Eccles: Erskine Press, 2012.

Lugg, Desmond. "Behavioral Health in Antarctica: Implications for Long-Duration Space Missions." *Aviation, Space, and Environmental Medicine* 76, suppl. 1 (June 2005): B74–77.

Lush, G. R. *Base Z, Building Report.* Falkland Islands Dependencies Survey. Cambridge: BAS Archives, 1959.

Lüthi, Dieter, Martine Le Floch, Bernhard Bereiter, Thomas Blunier, Jean-Marc Barnola, Urs Siegenthaler, Dominique Raynaud et al. "High-Resolution Carbon Dioxide Concentration Record 650,000–800,000 Years before Present." *Nature* 453, no. 7193 (15 May 2008): 379–82.

Lutz, Diana. "Super-TIGER Shatters Scientific Balloon Record in Antarctica." *The Source – Washington University in St. Louis,* 22 January 2013. source.wustl.edu. [online]

Lythe, Matthew, and David Vaughan. "BEDMAP: A New Ice Thickness and Subglacial Topographic Model of Antarctica." *Journal of Geophysical Research: Solid Earth* 106, no. B6 (June 2001): 11335–52.

M Man-Hauling

Man-hauling refers to the action of travelling by foot while dragging a load, i.e. wearing a harness that is connected to the object being towed (typically a sledge) without relying on any mechanical or animal assistance. Encouraged vehemently by Sir Clements Markham,[8] president of the British Royal Geographic Society towards the end of the nineteenth century, this technique was used predominantly by British explorers throughout the Heroic Age and was often seen as a key factor in determining the demise and death of Captain Robert Falcon Scott's party during the Race to the South Pole.

Maas, Elizabeth, Cliff Law, Julie Hall, Stu Pickmere, Kim Currie, F. H. Chang, Matt Voyles, and Dianna Caird. "Effect of Ocean Acidification on Bacterial Abundance, Activity and Diversity in the Ross Sea, Antarctica." *Aquatic Microbial Ecology* 70, no. 1 (2013): 1–15.

Mabie, Justin (jmabie). "Arrival at Jang Bogo Antarctic Research Station." *Building a Space Weather Radar in Antarctica* (blog), 27 January 2015. ciresblogs.colorado.edu. [online]

Madigan, Cecil. *Madigan's Account: The Mawson Expedition; The Antarctic Diaries of C. T. Madigan 1911–1914.* Transcribed by J. W. Madigan. North Hobart, Tas: Wellington Bridge Press: 2012.

Madison, D. W., ed. *Operation Deep Freeze '64: Task Force 43.* Boston, MA: Burdette & Company, 1964.

Madrigal, Alexis. "Designing a Better Antarctic Base for Science." *The Atlantic,* 21 September 2017. amp.theatlantic.com. [online]

Malewar, Amit. "Warm Water Found Beneath Antarctica's Doomsday Glacier." 31 January 2020. techexplorist.com. [online]

Malin, Michael, and Kenneth Edgett. "Evidence for Recent Groundwater Seepage and Surface Runoff on Mars." *Science* 288, no. 5475 (2000): 2330–35.

Manhire, Bill, ed. *The Wide White Page: Writers Imagine Antarctica.* Wellington: Victoria University Press, 2004.

Marit ASA, "History / About Aker." 2020. eng.akerasa.com. [online]

Marsching, Jane, and Andrea Polli, *Far Field: Digital Culture, Climate Change, and the Poles.* Bristol: Intellect, 2012.

Market Watch. "Global Krill Oil Market Size, Share 2019: Industry Analysis by Key Competitors, Production Overview, Supply Demand and Shortage, Trends, Growth, Regional Outlook and Forecast 2023." 25 January 2019. marketwatch.com. [online]

Markov, Moisey. "On High Energy Neutrino Physics." In *Proceedings, 10th International Conference* on High-Energy Physics (ICHEP 60), Rochester, NY, USA, 25 Aug – 1 Sep 1960, edited by George Sudarshan, John Tinlot, and Adrian Melissinos, 578–81. Rochester, NY: University of Rochester, 1960.

Marr, James. First Report on the Work of Operation Tabarin: Part 1 The Work of Base A 1943–44. AD6/1A/1944/A. Cambridge: British Antarctic Survey Archives. 1944. bas.ac.uk. [online]

Marr, James. The Natural History and Geography of the Antarctic Krill (Euphausia superba Dana), Discovery Reports 32, 33–464. Cambridge: Cambridge University Press, 1962.

Martos, Yasmina, Manuel Catalán, Tom Jordan, Alexander Golynsky, Dmitry Golynsky, Graeme Eagles, and David Vaughan. "Heat Flux Distribution of Antarctica Unveiled." Geophysical Research Letters 44, no. 22 (28 November 2017): 11417–26.

Masson-Delmotte, Valérie, Panmao Zhai, Hans-Otto Pörtner, Debra Roberts, Jim Skea, Priyadarshi Shukla, Anna Pirani et al. Global Warming of 1.5°C: An IPCC Special Report on the Impacts of Global Warming of 1.5°C above Pre-industrial Levels and Related Global Greenhouse Gas Emission Pathways, in the Context of Strengthening the Global Response to the Threat of Climate Change, Sustainable Development, and Efforts to Eradicate Poverty. Geneva: Intergovernmental Panel on Climate Change, 2018.

Matignon, Louis de Gouyon. "The Legal Status of Antarctica." 25 January 2019. spacelegalissues.com. [online]

Matthews, Samuel. "Antarctica's Nearer Side." National Geographic 140, no. 5 (November 1971): 622–54.

Mauerer, Mareike, Daniel Schubert, Paul Zabel, Matthew Bamsey, Eberhard Kohlberg, and Dirk Mengedoht. "Initial Survey on Fresh Fruit and Vegetable Preferences of Neumayer Station Crew Members: Input to Crop Selection and Psychological Benefits of Space-Based Plant Production Systems." Open Agriculture 1 (2016): 179–88.

Mawson, Douglas. The Home of the Blizzard: Being the Story of the Australasian Antarctic Expedition 1911–1914. London: Heinemann, 1915.

Mawson, Douglas. Mawson's Antarctic Diaries. Edited by Fred and Eleanor Jacka. 1988. Reprint, Crows Nest, NSW: Allen & Unwin, 2008.

May, John. The Greenpeace Book of Antarctica: A New View of the Seventh Continent. London: Dorling Kindersley, 1988.

Mayer, Jim. Shackleton: A Life in Poetry. Oxford: Signal, 2014.

McCarthy, Arlie, Lloyd S. Peck, Kevin A. Hughes, and David C. Aldridge. "Antarctica: The Final Frontier for Marine Biological Invasions." Global Change Biology 25 (2019): 1–21.

McDonald, Adrian, and Luke Cairns. "A New Method to Evaluate Reanalyses Using Synoptic Patterns: An Example Application in the Ross Sea/Ross Ice Shelf Region." Earth and Space Science 7, no. 1 (November 2019).

McGaughran, Angela, Mark Stevens, Ian Hogg, and Antonio Carapelli. "Extreme Glacial Legacies: A Synthesis of the Antarctic Springtail Phylogeographic Record." Insects 2 (April 2011): 62–82.

McGlory, Tony. Lecture in conversation with UNLESS. "Architecture in the Extreme," Architectural Association School of Architecture Polar Lab Workshop, London, September 2019.

McKay, David, Everett Gibson, Kathie Thomas-Keprta, Hojatollah Vali, Christopher Romanek, Simon Clemett, Xavier Schillier, Claude Maechling, and Richard Zare. "Search for Past Life on Mars: Possible Relic Biogenic Activity in Martian Meteorite ALH84001." Science 273 no. 5277 (August 1996): 924–30.

McKelvey, Barrie, and Peter Webb. "Geological Investigations in Southern Victoria Land, Antarctica: Part 3 - Geology of Wright Valley." New Zealand Journal of Geology and Geophysics 5, no. 1 (1962): 143–62.

McMurdo Dry Valleys LTER. "McMurdo Dry Valleys LTER Overview." Accessed 22 February 2021. mcm.lternet.edu. [online]

McNeil, Ben, and Richard Matear. "Southern Ocean Acidification: A Tipping Point at 450-ppm Atmospheric CO_2." PNAS: Proceedings of the National Academy of Sciences of the United States of America 105, no. 48 (2 December 2008): 18860–64.

Medeiros, Julia. "New Sights in the Second Field Season of the International Thwaites Glacier Collaboration" 15 April 2020. thwaitesglacier.org. [online]

Meggs, Lori. "Growing Plants and Vegetables in a Space Garden." 15 June 2010. nasa.gov. [online]

Meijde, Mark van der, Roland Pail, Rory Bingham, and Rune Floberghagen. "GOCE Data, Models, and Applications: A Review." International Journal of Applied Earth Observation and Geoinformation 35, Part A (March 2015): 4–15.

Mellor, Malcolm. Methods of Building on Permanent Snowfields. Hanover, NH: U.S. Army Materiel Command Terrestrial Sciences Center, Cold Regions Research and Engineering Laboratory, 1968.

Melville, Stephen. Robert Smithson: Time Crystals. Edited by Amelia Barikin and Chris McAuliffe. Melbourne: Monash University Publishing, 2018

Meredith, Michael, Martin Sommerkorn, Sandra Cassotta, Chris Derksen, Alexey Ekaykin, Anne Hollowed, Gary Kofinas et al. "Polar Regions." Chap. 3 in Pörtner, Roberts, Masson-Delmotte et al., IPCC Special Report on the Ocean and Cryosphere in a Changing Climate.

Metcalfe, Tom. "Hidden Beneath a Half Mile of Ice, Antarctic Lake Teems with Life." 15 January 2019. livescience.com. [online]

Meunier, Tony. U.S. Geological Survey Scientific Activities in the Exploration of Antarctica: Introduction to Antarctica (Including USGS Field Personnel: 1946–59). Reston, VA: United States Geologic Survey, 2007. Accessed 18 February 2020. pubs.usgs.gov. [online]

Meyer, Bettina, Angus Atkinson, D. Stübing, B. Oetti, Wilhelm Hagen, and Ulrich Bathmann. "Feeding and Energy Budgets of Antarctic Krill Euphausia superba at the Onset of Winter: I. Furcilia III Larvae." Limnology and Ocean-ography 47, no. 4 (July 2002): 943–52.

Meyer, Judith, and Ulf Riebesell. "Reviews and Syntheses: Responses of Coccolithophores to Ocean Acidification: A Meta-Analysis." Biogeosciences 12 (2015): 1671–82.

Meyer, Robinson. "The New Video of One of the Scariest Places on Earth." 30 January 2020. theatlantic.com. [online]

Michael, Gibert, Naziaty Mohd Yaacob, and Zuraini Md Ali. "The Capsule Living Unit Reconsidered a Utopia Transformed Reality." Pertanika Journal of Social Sciences and Humanities 26, no. 3 (2018): 1405–17.

Mikaloff, Fletcher, Nicolas Gruber, Andrew Jacobson, Manuel Gloor, Scott Doney, Stephanie Dutkiewicz, Markus Gerber et al. "Inverse Estimates of the Oceanic Sources and Sinks of Natural CO_2 and the Implied Oceanic Carbon Transport." Global Biogeochemical Cycle 21, no. 1 (March 2007).

Milken Institute School of Public Health. "Infographic: CO_2 Emissions vs. Vulnerability to Climate Change." Infographic from George Washington University. 17 May 2017. planetforward.org. [online]

Milton, John. Paradise Lost. London: Samuel Simmons, 1667.

Mingasson, Nicolas, Agnès Voltz, and Vincent Gaullier. L'aventure des pôles: Charcot, explorateur visionnaire. Paris: Larousse, 2017.

Misund Domaas, Christel. "The Norwegian-British-Swedish Antarctic Expedition: Science and Politics." Master's thesis, UiT Norges Arktiske Universitet, 2019.

Mitchell, Barbara. Frozen Stakes: The Future of Antarctic Minerals. London: IIED, 1983.

Molenaar, Erik Jaap. "Sea-Borne Tourism in Antarctica: Avenues for Further Intergovernmental Regulation." International Journal for Marine and Coastal Law 20, no. 2 (2005): 247–95.

Monchaux, Nicholas de. "Kosmos." e-flux architecture. Accessed 20 February 2021. e-flux.com. [online]

Monchaux, Nicholas de. Spacesuit: Fashioning Apollo. Cambridge, MA: MIT Press, 2011.

Montalva, Ramon Cañas. "El valor geopolitico de la posición Antártica de Chile." Revista Geográfica de Chile 9 (1953).

Moore, Keith, Weiwei Fu, Francois Primeau, Gregory Britten, Keith Lindsay, Matthew Long, Scott Doney, Natalie Mahowald, Forrest Hoffman, and James Randerson. "Sustained Climate Warming Drives Declining Marine Biological Productivity." Science 359, no. 6380 (9 March 2018): 1139–43.

Morlighem, Mathieu, Eric Rignot, Tobias Binder, Donald Blankenship, Reinhard Drews, Graeme Eagles, Olaf Eisen et al. "Deep Glacial Troughs and Stabilizing Ridges Unveiled beneath the Margins of the Antarctic Ice Sheet." Nature Geoscience 13, no. 2 (February 2020): 132–37.

Morrison, Michael. Lecture in conversation with UNLESS. "Architecture in the Extreme," Architectural Association School of Architecture Polar Lab Workshop, London, September 2019.

Mueller, Robert, Stephen Hoffman, and Paul Thur. "A Study of Parallels between Antarctica South Pole Traverse Equipment and Lunar/Mars Sur-face Systems." In Earth and Space 2010: Engineering, Science, Construction, and Operations in Challenging Environments. Edited by Gangbing Song and Ramesh Malla, 908–40. Reston, VA: American Society of Civil Engineers, 2010.

Muñoz, Guillermo. "Museo en la Antártica." Boletín Antártico Chileno 15, no. 1 (May 1996).

Murray, James, and George Marston. Antarctic Days: Sketches of the Homely Side of Polar Life by Two of Shackleton's Men. London: Andrew Melrose, 1913.

Murray, John. Address to the Royal Geographic Society, London, 1893.

Myers, Joan. Wondrous Cold: An Antarctic Journey. Washington: Smithsonian Institution Press, 2006.

Myles, Robert. "What Antarctica Looks Like Without Ice - The Bedmap 2 Project." 30 March 2013. digitaljournal.com. [online]

N Night

"Because the Earth rotates on a tilted axis as it revolves around the sun, sunlight is experienced in extremes at the poles. [...] From the South Pole, the sun is always above the horizon in the summer and below the horizon in the winter. This means the region experiences up to 24 hours of sunlight in the summer and 24 hours of darkness in the winter."[9] Thus, in a context in which one experiences only one sunrise and sunset per year (or one "sunsight" and one "sunclipse",[10] to use Buckminster Fuller's definition), familiar terms such as "day" and "night" assume a radically difference significance.

Naish, Timothy, Ross Powell, Richard Levy, and Gary Wilson. "Obliquity-Paced Pliocene West Antarctic Ice Sheet Oscillations." Nature 458, no. 7236 (19 March 2009): 322–28.

Naish, Timothy, Ken Woolfe, Peter Barrett, Gary Wilson, Cliff Atkins, Steven Bohaty, Christian Bücker et al. "Orbitally Induced Oscillations in the East Antarctic Ice Sheet at the Oligocene/Miocene Boundary." Nature 413, no. 6857 (18 October 2001): 719–23.

"Nanji zhoubian haiyu yu dalu zonghe pinggu" [A Comprehensive Evaluation of Antarctic Waters]. Jidi zhuanying jianbao 10 (January 2016): 3.

NASA. "Oxygen Factories in the Southern Ocean." 13 January 2016. earthobservatory.nasa.gov. [online]

NASA. "Superconducting Detectors for Study of Infant Universe." 17 March 2014. nasa.gov. [online]

NASA. "2019 Ozone Hole Is the Smallest on Record Since Its Discovery." 21 October 2019. nasa.gov. [online]

NASA Earth Observatory. "Countdown to Calving at Brunt Ice Shelf." Accessed 22 February 2021. earthobservatory.nasa.gov. [online]

NASA Earth Observatory. "Rules of Engagement." 30 August 2005. earthobservatory.nasa.gov. [online]

NASA Earth Observatory. "World of Change: Collapse of the Larsen-B Ice Shelf." Accessed 22 February 2021. earthobservatory.nasa.gov. [online]

NASA EPDC: STEM. "The Future of Human Spaceflight." 1 November 2019. [online]

NASA/Goddard Space Flight Center Scientific Visualization Studio. "NASA Research Leads to First Complete Map of Antarctic Ice Flow." 18 August 2011. svs.gsfc.nasa.gov. [online]

NASA Science: Mars Exploration Program. "Mars's Exploration Program." Accessed 22 February 2021. mars.nasa.gov. [online]

NASA Science: Mars Exploration Program. "Twin Peaks in Super Resolution – Right Eye." 8 September 1999. mars.nasa.gov. [online]

Natani, Kirmach, and Jay Shurley. "Sociopsychological Aspects of a Winter Vigil at South Polar Station." In *Human Adaptability to Antarctic Conditions,* edited by Eric Gunderson, 89–114. Washington, DC: American Geophysical Union, 1974.

National Aeronautics and Space Administration. *Apollo 11: Technical Air-to-Ground Voice Transcription (GOSS NET 1).* Houston, TX: Manned Spacecraft Center, 1969.

National Geographic. "Ice Coring." Accessed 9 January 2021. nationalgeographic.org. [online]

National Geographic. "Society's New Map Updates Antarctica." *The National Geographic Magazine* 112, no. 3 (September 1957).

National Geographic, "South Pole" Accessed 8 March 2021. nationalgeographic.org. [online]

National Marine Environmental Forecasting Center, "Guojia Nanji 'shiwu' nengli jianshe zhongdian xiangmu Nanji Changcheng zhan, Zhongshan zhan, dimian weixing jieshou chuli xitong" [National 15th Five-Year Plan Capacity Key Building Project to Install Satellite Ground Receiving and Processing Stations Successfully Completed]. 18 March 2010.

National Science Foundation. "Fire Destroys Garage at McMurdo Station." *Antarctic Journal of the United States* vol. 16 no. 4 (December 1981).

National Science Foundation. *More and Better Science in Antarctica through Increased Logistical Effectiveness.* Report of the U.S. Antarctic Program Blue Ribbon Panel. Washington, DC, July 2012.

National Science Foundation. "NSF, University of Wisconsin-Madison Complete Construction of the World's Largest Neutrino Observatory." 17 December 2010. nsf.gov. [online]

National Science Foundation. "Telemedicine Link with South Pole Allows Remote Knee Surgery." 17 July 2002. nsf.gov. [online]

National Science Foundation et al. *McMurdo Station Master Plan 2.1.* 16 December 2015. future.usap.gov. [online]

Nayler, Simon, Katrina Dean, and Martin Siegert. "The IGY and the Ice Sheet: Surveying Antarctica." *Journal of Historical Geography* 34 (2008): 574–95.

Neruda, Pablo. *Piedras Antárticas.* 1938. Accessed 12 March 2021. neruda.uchile.cl. [online]

Newman, Dava. "Building the Future Spacesuit." *ASK* 45 (Winter 2012): 37–40.

Newman, Louise, Petra Heil, Rowan Trebilco, Katsuro Katsumata, Andrew Constable, Esmee van Wijk,

Karen Assmann et al. "Delivering Sustained, Coordinated, and Integrated Observations of the Southern Ocean for Global Impact." 8 August 2019. frontiersin.org. [online]

New York Times. "Life on Mars?" 8 August 1996.

New Zealand Antarctic Heritage Trust. *Conservation Plan: Cape Evans.* Christchurch: NZAHT, 2004.

New Zealand Antarctic Heritage Trust. *Conservation Plan: Hillary's Hut, Scott Base.* Christchurch: NZAHT, 2015.

New Zealand Antarctic Heritage Trust. *Conservation Plan: The Historic Huts at Cape Adare.* Christchurch: NZAHT, 2004.

New Zealand Antarctic Heritage Trust. *Conservation Report: Discovery Hut, Hut Point.* Christchurch: NZAHT, 2004.

New Zealand Antarctic Heritage Trust. *Conservation Report: Shackleton's Hut, Cape Royds.* Christchurch: NZAHT, 2003.

New Zealand Antarctic Heritage Trust. "History of Scott's Expedition." Accessed 23 February 2021. nzaht.org. [online]

New Zealand Antarctic Society. "Deep Freeze '63 Has Ended." *Antarctic: A News Bulletin* 3, no. 6 (June 1963).

New Zealand Green Building Council. "Green Star." Accessed 23 February 2021. nzgbc.org.nz. [online]

Nicol, Stephen. "Krill, Currents, and Sea Ice: *Euphausia superba* and Its Changing Environment." *Bioscience* 56, no. 2 (2006): 117.

Nieboer, Miranda. "Antarctic Interiors: Practices of Inhabitation through Embodied Interactions with the Ice." PhD diss., University of Tasmania, 2020.

Nieboer, Miranda, and Craig McCormack. "Under Geodesic Skies: A Cultural Perspective on the Former South Pole Dome and Geodesic Domes in Outer Space." *The Polar Journal* 7, no. 2 (2017): 351–73.

Nielsen, Jerri. *Icebound.* London: Ebury Press, 2001.

Noble, Anne. *Ice Blink.* Auckland: Clouds, 2011.

Noble, Anne. *The Last Road.* Auckland: Clouds, 2014.

Nordenskjöld, Otto. *Antarctica, or Two Years amongst the Ice of the South Pole.* 1905. Reprint, Brisbane: University of Queensland Press, 1977.

North American Newspaper Alliance. "Maudheim Is Base of Antarctic Party." *The New York Times,* 23 February 1950.

Novak, Matt. "Sunday Funnies Blast Off into the Space Age." *Smithsonian,* 2012.

Noyes, Jenny. "More Than 80 Passengers on Board Greg Mortimer Cruise Ship Test Positive for COVID-19." *The Sydney Morning Herald,* 6 April 2020. smh.com.au [online]

Nuclear Power. "What Is Péclet Number." Accessed 23 February 2021. nuclear-power.net. [online]

O Ocean Acidification

Since the Industrial Revolution, collectively the oceans have absorbed from the atmosphere approximately 40% of all carbon dioxide emissions produced through the burning of fossil fuels and land-use changes such as deforestation and cement production. Becoming the largest carbon reservoir on Earth, i.e. the greatest "carbon sink", the oceans are altering the chemistry of their waters, resulting in a lowered pH value known as ocean acidification. Such chemical imbalance severely impacts the oceanic biology imperilling food webs, nutrient cycling, and long-term global climate regulations capabilities.[11]

Oceanwide Expeditions. "The Dirty Details of Antarctica's Dry Valleys." Accessed 23 February 2021. oceanwide-expeditions.com. [online]

Office of the President of the United States. "Memorandum on Safeguarding U.S. National Interests in the Arctic and Antarctic Regions." White House Presidential Memorandum, 9 June 2020.

Ohneiser, Christian, and Gary Wilson, "Revised Magnetostratigraphic Chronologies from New Harbour Drill Cores, Southern Victoria Land, Antarctica." *Global and Planetary Change* 82–83 (February 2012): 12–24.

Olds, Andrew, Kylie Pitt, Paul Maxwell, Russ Babcock, David Rissik, and Rod Connolly. "Marine Reserves Help Coastal Ecosystems Cope with Extreme Weather." *Global Change Biology* 20, no. 10 (2014): 3050–58.

Oliver, Amy. "Lochmuir Salmon? It Doesn't Exist: How Supermarkets Invent Places and Farms to Trick Shoppers into Buying Premium Food." *Mail Online,* 15 February 2012. dailymail.co.uk. [online]

Olsen, Bjørnar. *In Defense of Things: An Archaeology and the Ontology of Objects.* Lanham, MD: AltaMira Press, 2010.

Oman, Rok, and Spela Videcnik. *Habitation in Extreme Environments.* Cambridge, MA: Harvard University Graduate School of Design, 2015.

O'Neill, Natalie. "Antarctic Scientist Stabbed Colleague for Spoiling Book Endings." *New York Post,* 30 October 2018. nypost.com. [online]

O'Reilly, Jessica. *The Technocratic Antarctic: An Ethnography of Scientific Expertise and Environmental Governance.* Cornell: Cornell University Press, 2017.

O'Reilly, Jessica, and Juan Francisco Salazar. "Inhabiting the Antarctic." *The Polar Journal* 7, no. 1 (2017): 9–25.

O'Riley, Tim. *Endlessness.* Eindhoven: Peter Foolen Editions, 2020.

Orrego Vicuña, Francisco, ed. *Antarctic Resources Policy: Scientific, Legal, and Political Issues.* Cambridge: Cambridge University Press, 1983.

Orta, Jorge. "Antarctica 2000 Manifesto." *46th Venice Art Biennale Argentine Pavilion.* 46th Venice Art Biennale Catalogue, 1995.

Ortelius, Abraham. *Thesaurus Ortelius.* Antwerp: Plantin Press, 1587.

P Papatūānuku

Oral traditions suggest that the South Pacific Polynesian ancestors of Māori indigenous peoples might have sailed in Antarctic waters as early as seven hundred years ago. Their reverence and spiritual bond to Mother Earth, Papatūānuku, induces a sense of indebtedness and promotes the protection of our land.[12]

Pacheco, Luis. "ATIC-1 (Advanced Thin Ionization Calorimeter)." 27 March 2020. stratocat.com.ar. [online]

Pacheco, Luis. "BOOMERANG (Balloon Observations of Millimetric Extragalactic Radiation and Geophysics)." 27 March 2020. stratocat.com.ar. [online]

Pacheco, Luis. "CREAM I (Cosmic Ray Energetics and Mass)." 27 March 2020. stratocat.com.ar. [online]

Pacheco, Luis. "GRAD (Gamma Ray Advanced Detector)." 27 March 2020. stratocat.com.ar. [online]

Pacheco, Luis. "Hiregs I (High Resolution Gamma-Ray and Hard X-Ray Spectrometer)." Accessed 13 March 2021. stratocat.com.ar. [online]

Pacheco, Luis. "JACEE 10 + HIREGS + PAX." 27 March 2020. stratocat.com.ar. [online]

Page, Thomas. "To Antarctica and Beyond." Accessed 23 February 2021. edition.cnn.com. [online]

Paglen, Trevor. *Blank Spots on the Map: The Dark Geography of the Pentagon's Secret World.* New York: Penguin, 2009.

Paglen, Trevor. *The Last Pictures.* Berkeley and Los Angeles, CA: University of California Press, 2012.

Palinkas, Lawrence. "Going to Extremes: The Cultural Context of Stress, Illness and Coping in Antarctica." *Social Science and Medicine* 35, no. 5 (September 1992): 651–64.

Palinkas, Lawrence, Mark Cravalho, and Dierdre Browner. "Seasonal Variation of Depressive Symptoms in Antarctica." *Acta Psychiatrica Scandinavia* 91, no. 6 (June 1995): 423–29.

Palinkas, Lawrence, Frederic Glogower, Mark Dembert, Kendall Hansen, and Robert Smullen. "Incidence of Psychiatric Disorders after Extended Residence in Antarctica." *International Journal of Circumpolar Health* 63, no. 2 (2004): 157–68.

Palmer, Ada. *The Will to Battle: Terra Ignota, Book III.* New York: Tor Books, 2017.

Parks, Lisa. "Between Orbit and the Ground: Conflict Monitoring, Google Earth and the 'Crisis in Darfur'." In *Documentary Testimonies: Global Archives of Suffering,* edited by Bhaskar Sarkar and Janet Walker, 245–67. London and New York: Routledge, 2010.

Patrice Godon Polar Engineering. "Concordia / The Station." Accessed 13 February 2021. patricegodonpolarengineering.eu. [online]

Pattyn, Nathalie, Martine Van Puyvelde, Helio Fernandez-Tellez, Bart Roelands, and Olivier Mairesse. "From the Midnight Sun to the Longest Night: Sleep in Antarctica." *Sleep Medicine Reviews* 37 (February 2018): 159–72.

Pearson, Michael. "Artefact or Rubbish? A Dilemma for Antarctic Managers." In *Cultural Heritage in the Arctic and Antarctic Regions,* edited by Susan Barr and Paul Chaplin, 39–43. Lørenskog: ICOMOS, 2004.

Pearson, Michael. "Expedition Huts in Antarctica: 1899–1917." *Polar Record* 28, no. 167 (October 1992): 261–76.

Pearson, Michael, and Ruben Stehberg. "Nineteenth Century Sealing Sites on Rugged Island, South Shetland Island." *Polar Record* 42, no. 4 (October 2006): 335–47.

Pearson, Michael, Ruben Stehberg, Andrés Zarankín, Maria Ximena Senatore, and Carolina Gatica. "Conserving the Oldest Historic Sites in the Antarctic: The Challenges in Managing the Sealing Sites in the South Shetland Islands." *Polar Record* 46, no. 1 (January 2010): 57–64.

Peck, Lloyd. "Antarctic Marine Biodiversity: Adaptations, Environments and Responses to Change." *Oceanography and Marine Biology Annual Review* 56 (2018): 105–236.

Peck, Lloyd, Peter Convey, and David Barnes. "Environmental Constraints on Life Histories in Antarctic Ecosystems: Tempos, Timings and Predictability." *Biological Reviews* 81 (2006): 75–109.

Peeken, Ilka, Sebastian Primpke, Birte Beyer, Julia Gütermann, Christian Katlein, Thomas Krumpen, Melanie Bergmann, Laura Hehemann, and Gunnar Gerdts. "Arctic Sea Ice Is an Important Temporal Sink and Means of Transport for Microplastic." *Nature Communications* 9 (2018).

Pellegrino, Francesco. Lecture in conversation with UNLESS. "Architecture in the Extreme," Architectural Association School of Architecture Polar Lab Workshop, London, September 2019.

Perlez, Jane. "China, Pursuing Strategic Interests, Builds Presence in Antarctica." *The New York Times,* 3 May 2015. nytimes.com. [online]

Pertierra, Luis, Kevin Hughes, Greta Vega, and Miguel Olalla-Tárraga. "High-Resolution Spatial Mapping of Human Footprint across Antarctica and Its Implications for the Strategic Conservation of Avifauna." *PLOS ONE* 12, no. 3 (2017).

Peter, Hans-Ulrich, Christina Braun, Susann Janowski, Anja Nordt, Anke Nordt, and Michael Stelter. *The Current Environmental Situation and Proposals for the Management of the Fildes Peninsula Region.* Dessau-Roßlau: Federal Environment Agency (Umweltbundesamt), March 2013. umweltbundesamt.de. [online]

Petersen, Rick. Lecture in conversation with UNLESS. "Architecture in the Extreme," Architectural Association School of Architecture Polar Lab Workshop, London, September 2019.

Petrou, Katherina, Kirralee Baker, Daniel Nielsen, Alyce Hancock, Kai Schulz, and Andrew

Davidson. "Acidification Diminishes Diatom Silica Production in the Southern Ocean." *Nature Climate Change* 9, no. 10 (October 2019): 781–96.

Phillips, Richard, and Claire Waluda. "Albatrosses and Petrels at South Georgia as Sentinels of Marine Debris Input from Vessels in the South-west Atlantic Ocean." *Environment International* 136 (2020).

Phyne, John. "A Comparative Political Economy of Rural Capitalism." *Acta Sociologica* 53, no. 2 (2010): 160–80.

Piñones, Andrea, and Alexey Fedorov. "Projected Changes of Antarctic Krill Habitat by the End of the 21st Century." *Geophysical Research Letters* 43, no. 16 (28 August 2016): 8580–89.

Pinsky, Malin, and Nathan Mantua, "Emerging Adaptation Approaches for Climate-Ready Fisheries Management," *Oceanography* 27, no. 4 (2014): 146–59.

Poe, Edgar Allan. *The Narrative of Arthur Gordon Pym.* New York: Harper and Brothers, 1838.

Polar Research Board. *Antarctic Treaty System – An Assessment: Proceedings of a Workshop Held at Beardmore South Field Camp, Antarctica, January 7–13, 1985.* Washington, DC: National Academy Press, 1986.

Pollard, David, and Robert DeConto, "Modelling West Antarctic Ice Sheet Growth and Collapse through the Past Five Million Years." *Nature* 458 (2009), 329–32.

Ponting, Herbert George. *The Great White South: Being an Account of Experiences with Captain Scott's South Pole Expedition and of the Nature Life of the Antarctic – With 164 Photographic Illustrations,* xxvi, 305. London: Duckworth & Co., 1921.

Ponting, Herbert George. *With Scott to the Pole: The* Terra Nova *Expedition 1910–1913; The Photographs of Herbert Ponting.* Crows Nest, NSW: Allen & Unwin, 2004.

Poore, Richard, Richard Williams Jr., and Christopher Tracy. "Sea Level and Climate." US Geological Survey Fact Sheet 002-00 (2000).

Pope, Gustavus. *Journey to Mars: The Wonderful World; Its Beauty and Splendor; Its Mighty Races and Kingdoms; Its Final Doom.* Westport, CT: Hyperion, 1974.

Porcelli, Simone, Mauro Marzorati, Beth Healey, Laura Terraneo, Alessandra Vezzoli, Silvia Della Bella, Roberto Dicasillati, and Michele Samaja. "Lack of Acclimatization to Chronic Hypoxia in Humans in the Antarctica." *Scientific Reports* 7, no. 1 (2017).

Porter, Stephen, and James Beget. "Provenance and Depositional Environments of Late Cenozoic Sediments in Permafrost Cores from Lower Taylor Valley." In *Dry Valley Drilling Project,* edited by Lyle McGinnis, Antarctic Research Series 33, 351–63. Washington, DC: American Geophysical Union, 1981.

Pörtner, Hans-Otto, Debra Roberts, Valérie Masson-Delmotte, Panmao Zhai, Melinda Tignor, Elvira Poloczanska, Katja Mintenbeck et al., eds. *IPCC Special Report on the Ocean and Cryosphere in a Changing Climate.* Geneva: Intergovernmental Panel on Climate Change, 2019.

Poulin, Elie, and Lea Cabrol. *Covid-19 en la Antártica: Un continente hasta ahora libre de virus, pero vulnerable.* Santiago: Documento del Instituto de Ecología y Biodiversidad (IEB), 2020.

Powell, Robert, Stephen Kellert, and Sam Ham. "Antarctic Tourists: Ambassadors or Consumers?." *Polar Record* 44, no. 3 (2008): 233–41.

Powell, Robert, Matthew Brownlee, Stephen Kellert, and Sam Ham. "From Awe to Satisfaction: Immediate Affective Responses to the Antarctic Tourism Experience." *Polar Record* 48, no. 2 (2012): 145–56.

Presidencia de la Nación. *La nación Argentina: Justa, libre, soberana.* Buenos Aires: Talleres Gráficos Peuser, 1949.

Priestley, Raymond. "Copy of Part of His General Diary of the British Antarctic Expedition, 1910–13:

1 January – 3 October 1912. Typescript (unbound), MS 298/6/3. Scott Polar Research Institute, Cambridge.

Priscu, John, Linda Priscu, Warwick Vincent, and Clive Howard-Williams. "Photosynthate Distribution by Microplankton in Permanently Ice-Covered Antarctic Desert Lakes." *Limnology and Oceanography* 32, no. 1 (January 1987): 260–70.

Pritchard, Hamish, Stefan Ligtenberg, Helen Fricker, David Vaughan, Michiel Van den Broeke, and Laurie Padman. "Antarctic Ice-Sheet Loss Driven by Basal Melting of Ice Shelves." *Nature* 484, no. 7395 (26 April 2012): 502–5.

Protocol on Environmental Protection to the Antarctic Treaty, 4 October 1991, 30 ILM 1461.

Pugh, Philip J. A., and Peter Convey. "Surviving Out in the Cold: Antarctic Endemic Invertebrates and Their Refugia." *Journal of Biogeography* 35 (2008): 2176–86.

Purkey, Sarah G., and Gregory C. Johnson. "Antarctic Bottom Water Warming and Freshening Contributions to Sea Level Rise, Ocean Freshwater Budgets, and Global Heat Gain." *Journal of Climate* 26, no. 16 (2013): 6105–22.

Pynchon, Thomas. *V.* Philadelphia: J. B. Lippincott & Co., 1963.

Pyne, Stephen. *The Ice: A Journey to Antarctica.* Washington: University of Washington Press, 1986.

Pyne, Stephen. *Voyager: Seeking Newer Worlds in the Third Great Age of Discovery.* New York: Viking, 2010.

Q Question of Antarctica

Placed on the agenda of the United Nations for over two decades (between 1983 and 2005), the "Question of Antarctica" addressed the international community's concern that the Antarctic Treaty encourage a limited number of (developed) countries to manage the Antarctic on behalf of mankind. While the discussion became less heated with the entry into force of the Environment Protocol in 1991, the UN General Assembly confirmed that the United Nations will "remain seized of the matter."[13]

Quigg, Philip W. *A Pole Apart: The Emerging Issue of Antarctica.* New York: McGraw-Hill, 1983.

R Radio-Echo Sounding

"Mapping the topography and features of land beneath the ice in Antarctica relies on radio-echo sounding (RES). This technique normally requires an aircraft to fly over a region, using wing-mounted radar antennae to emit a radio signal that penetrates the ice, bouncing back from the point at which the ice meets the rock, sediment, or water beneath. Since its development in the 1960s, this technique has led to the discovery of subglacial lakes under the ice sheet, as well as the mapping of the land beneath the ice sheet."[14]

Ramboll. "Modernising Infrastructure in Antarctica." Accessed 24 February 2021. uk.ramboll.com. [online]

Ramboll. *Sustainable Buildings Market Study 2019.* Copenhagen: Ramboll, 2019. uk.ramboll.com. [online]

Ramboll Deutschland GmbH (formerly IMS Ingenieurgesellschaft mbH). "Design of German Neumayer III Station." Internal report, 2005–8.

Ramboll Deutschland GmbH (formerly IMS Ingenieurgesellschaft mbH). "Design of Indian Bharati Station." Internal report, 2008–13.

Ramboll Deutschland GmbH. "Design of New Wintering Complex at Vostok Station." Internal report, 2018–20.

Rayfuse, Rosemary. "Climate Change and Antarctic Fisheries: Ecosystem Management in CCAMLR." *Ecology Law Quarterly* 45, no. 53 (2018).

Raymond, Bruce / United States Navy. *Photograph Caption, Antarctic Photo Library.* National Science Foundation Archives, 1 January 1961.

Reed, Lester, Kenneth Burman, Mohamed Shakir, and John O'Brian. "Alterations in the Hypothalamic-Pituitary-Thyroid Axis after Prolonged Residence in Antarctica." *Clinical Endocrinology* 25, no. 1 (July 1986): 55–65.

Regional State Archives in Tromsø. Archive materials on Antarctic Expeditions from Norges Svalbard- og Ishavsundersøkelser (NSIU) 1928–1948 and Norsk Polarinstitutt (NP) 1948–1985. Norsk Polarinstitutt, Tromsø, 2004.

Reid, Tyler. "Nuclear Power at McMurdo Station, Antarctica." 21 March 2014. large.stanford.edu. [online]

Rejcek, Peter. "Stamp on History." *The Antarctic Sun,* 15 October 2020. antarcticsun.usap.gov. [online]

Republic of Liberia. "Report of Investigation in the Matter of the Sinking of Passenger Vessel *Explorer* (O.N. 8485) 23 November 2007 in the Bransfield Strait near the South Shetland Islands." Monrovia: Bureau of Maritime Affairs, 2009.

Research and Markets. "Omega-3 Market (Fish & Krill Oil): Insights, Trends and Forecast (2020–2024)." 23 April 2020. globenewswire.com. [online]

Rester, Carl, Robert Coldwell, Jack Trombka, Richard Starr, Guenther Eichhorn, and George Lasche. "Performance of Bismuth Germanate Active Shielding on a Balloon Flight over Antarctica." *IEEE Transactions on Nuclear Science* 37, no. 2 (April 1990): 559–65.

Riches, Caroline. "Life on Ice: What Esperanza Base in Antarctica Can Teach Us about Isolation." *The Guardian*, 11 April 2020. theguardian.com. [online]

Richter, Andreas, Sergey Popov, Mathias Fritsche, and Valery Lukin. "Height Changes over Subglacial Lake Vostok, East Antarctica: Insights from GNSS Observations." *Journal of Geophysical Research: Earth Surface* 119, no. 11 (November 2014): 2460–80.

Rigg, Kelly. "Environmentalists' Perspectives on the Protection of Antarctica." In *The Future of Antarctica: Exploitation Versus Preservation,* edited by Grahame Cook. Manchester, England: Manchester University Press, 1990.

Rintoul, Stephen. "The Global Influence of Localized Dynamics in the Southern Ocean." *Nature* 558, no. 7709 (14 June 2018): 209–18.

Rintoul, Stephen, Steven Chown, R. Deconto, M. England, H. Fricker, V. Masson-Delmotte, T. Naish, Martin Siegert, and José Xavier. "Choosing the Future of Antarctica." *Nature* 558, no. 7709 (14 June 2018): 233–41.

Risopatrón, Luis. *La Antartida Americana.* Santiago: Imp. Cervantes, 1908.

Rivolier, Jean. *Man in the Antarctic: The Scientific Work of the International Biomedical Expedition to the Antarctic (IBEA).* London: Taylor & Francis, 1988.

Roberts, Brian. "International Cooperation for Antarctic Development: The Test for The Antarctic Treaty." *Polar Record* 19 (1978): 107–20.

Roberts, Callum, and Julie Hawkins. "Effects of Marine Reserves on Adjacent Fisheries." *Science* 294 (December 2001).

Roberts, Callum, Bethan O'Leary, Douglas McCauley, Philippe Maurice Cury, Carlos Duarte, Jane Lubchenco, Daniel Pauly et al. "Marine Reserves Can Mitigate and Promote Adaptation to Climate Change." *Proceedings of the National Academy of Sciences* 114, no. 24 (2017): 6167–75.

Roberts, Peder. *A Frozen Field of Dreams: Science, Strategy, and the Antarctic in Norway, Sweden, and the British Empire, 1912–1952.* PhD diss., Stanford University, 2010.

Roberts, Peder, Lize-Marié van der Watt, and Adrian Howkins, eds. *Antarctica and the Humanities.* London: Palgrave McMillan, 2016.

Robinson, Kim Stanley. *Antarctica.* London: Harper Collins, 1997.

Robinson, Kim Stanley. *Blue Mars.* London: Harper Collins, 1996.

Rödenbeck, Christian, Dorothee Bakker, Nicolas Metzl, Are Olsen, Chris Sabine, Nicolas Cassar, Friedemann Reum, R. Keeling, and Martin Heimann. "Interannual Sea-Air CO_2 Flux Variability from an Observation-Driven Ocean MixedLayer Scheme." *Biogeosciences* 11 (2014): 4599–613.

Rogers, Alex David. "Evolution and Biodiversity of Antarctic Organisms: A Molecular Perspective." *Philosophical Transactions of the Royal Society B* 362 (2007): 2191–214.

Rogers, Alex David, Betina A. V. Frinault, David K. A. Barnes, Nathaniel Lee Bindoff, Rod Downie, Hugh W. Ducklow, Ari Seth Friedlaender et al. "Antarctic Futures: An Assessment of Climate-Driven Changes in Ecosystem Structure, Function, and Service Provisioning in the Southern Ocean." *Annual Review of Marine Science* 12, no. 1 (2020): 87–120.

Rogers, Alex David, Nadine Johnston, Eugene Murphy and Andew Clarke, eds. *Antarctic Ecosystems: An Extreme Environment in a Changing World.* Hoboken: Wiley-Blackwell, 2012.

Roman, Benjamin. "Serious About Sports at End of the Earth." 3 November 2010. espn.com. [online]

Romero, Simon. "Antarctic Life: No Dogs, Few Vegetables and 'a Little Intense' in the Winter." *The New York Times,* 6 January 2016.

Roscoe, John H. "Contributions to the Study of Antarctic Surface Features by Photographical Methods." PhD diss., University of Maryland, 1952.

Rose, Lisle. *Assault on Eternity: Richard E. Byrd and the Exploration of Antarctica, 1946–47.* Washington, DC: Naval Institute Press, 1980.

Rosenberg, Nathan. *Inside the Black Box: Technology and Economics.* Cambridge: Cambridge University Press, 1982.

Ross, Robin, and Langdon Quetin. "Energetic Cost to Develop to the First Feeding Stage of *Euphausia superba* Dana and the Effect of Delays in Food Availability." *Journal of Experimental Marine Biology and Ecology* 133, nos. 1–2 (14 December 1989): 103–27.

Ross, Sandra, ed. *Ice Lab: New Architecture and Science in Antarctica.* Manchester: The British Council, 2013.

Rossiter, Ned, and Soenke Zehle. "The Aesthetics of Algorithmic Experience." In *The Routledge Companion to Art and Politics,* edited by Randy Martin, 214–21. New York: Routledge, 2015.

Rothwell, Donald. "The Boundaries of the Polar Regions." In *Research Handbook on Polar Law,* edited by Karen N. Scott and David L. Vander-Zwaag. Cheltenham: Edward Elgar, 2020.

Rothwell, Donald. "Law Enforcement in Antarctica." In *Antarctic Security in the Twenty-First Century: Legal and Policy Perspectives,* edited by Alan D. Hemmings, Donald R. Rothwell, and Karen N. Scott, 135–53. London: Routledge, 2012.

Rothwell, Donald, and Alan Hemmings. "Evolution of a Polar Law." In *Research Handbook on Polar Law,* 454–73. Cheltenham: Edward Elgar, 2020.

Rothwell, Donald, and Alan Hemmings. "Introduction: The Context of International Polar Law." In *International Polar Law,* edited by Donald Rothwell and Alan Hemmings, xviii–xxiv. Cheltenham: Edward Elgar Publishing, 2018.

Rothwell, Donald, and Tim Stephens. "Illegal Southern Ocean Fishing and Prompt Release: Balancing Coastal and Flag State Rights and Interests." *International and Comparative Law Quarterly* 53 (2004): 171–87.

Rothwell, Donald, and Tim Stephens. *The International Law of the Sea,* 2nd ed. Oxford: Hart, 2016.

Rotthäuser, Siegfried. In conversation with Hartwig Gernandt, in the context of an on-site technical visit to Neumayer III during the 2019/20 season.

Roura, Ricardo. "Antarctic Cultural Heritage: Geopolitics and Management." In *Handbook on the Politics of Antarctica,* edited by Klaus Dodds,

Alan Hemmings, and Peder Roberts. Cheltenham: Edward Elgar, 2017.

Roura, Ricardo. "Shared Research Stations in Antarctica: The Holy Grail of International Cooperation, or Just a Nice Idea?" (2015). Accessed 3 March 2020. lecerclepolaire.com. [online]

Roura, Ricardo, and Tina Tin. "Strategic Thinking and the Antarctic Wilderness: Contrasting Alternative Futures." In *Antarctic Futures: Human Engagement with the Antarctic Continent,* edited by Tina Tin, Daniela Liggett, Patrick Maher, and Machiel Lamers. Dordrecht: Springer, 2014.

Rousseau, Bryant. "Cold Cases: Crime and Punishment in Antarctica." *The New York Times,* 29 September 2016. Accessed 15 June 2020. nytimes.com. [online]

Rowlatt, Justin. "Journey to the 'Doomsday Glacier'." *BBC News,* 28 January 2020.

Royal Society, "Clements Markham." Accessed 14 March 2021. makingscience.royalsociety.com. [online]

S   South Pole

The South Pole, also known as Geographic or Terrestrial South Pole, is the southernmost point at which lines of longitude converge at the latitude of 90° South. Corresponding to the point at which the Earth's axis of rotation intersects with its surface, the South Pole is distinct from its neighbouring South Magnetic Pole and South Geomagnetic Pole. The former is the rapidly shifting point in the southern hemisphere where the earth's magnetic field lines are vertically aligned, whilst the latter is the theoretical point at which the magnetic field lines converge, defining the ideal magnetic dipole axis of the Earth.

Sabine, Christophe, Richard Feely, Nicolas Gruber, Robert Key, Kitack Lee, John Bullister, Rik Wanninkhof et al. "The Oceanic Sink for Anthropogenic CO_2." *Science* 305, no. 5682 (16 July 2004): 367–71.

Sabine, Edward. *Report on the Variation of the Magnetic Intensity at Different Points on the Earth's Surface,* 83–84. London: R. and J. E. Taylor, 1838.

Salazar, Juan Francisco. "Geographies of Place-Making in Antarctica: An Ethnographic Approach." *The Polar Journal* 3, no. 1 (2013): 53–71.

Salazar, Juan Francisco. "Polar Infrastructures." In *The Routledge Companion to Digital Ethnography*, edited by Larissa Hjorth, Heather Horst, Anne Galloway, and Genevieve Bell, chap. 34. New York: Routledge, 2017.

Sallée, Jean-Baptiste. "Southern Ocean Warming." *Oceanography* 31, no. 2 (June 2018): 52–62.

Samyn, Philippe. "Structural Engineering and Embodied Energy." *Steel Construction Journal* 12, no. 3. (August 2019): 174–75.

Samyn, Philippe. Lecture in conversation with UNLESS. "Architecture in the Extreme," Architectural Association School of Architecture Polar Lab Workshop, London, September 2019.

Samyn, Philippe. Étude de la morphologie des structures à l'aide des indicateurs de volume et de déplacement. Brussels: Académie Royale de Belgique Classe des Sciences, 2004.

Sandal, Gro Mjeldheim, Fons van deVijver, and Nathan Smith. "Psychological Hibernation in Antarctica." *Frontiers in Psychology.* 20 November 2018. frontiersin.org. [online]

Santos, António, Matthew Bamsey, Virgínia Infante, and Daniel Schubert. *A Case Study in the Application of Failure Analysis Techniques to Antarctic Systems: EDEN ISS.* IEEE International Symposium on Systems Engineering. Edinburgh, 2016.

Sarmiento, Jorge, and Nicolas Gruber. *Ocean Biogeochemical Dynamics.* Princeton, NJ: Princeton University Press, 2006.

Satchell, Michael. "Women Who Conquer the South Pole." *San Bernardino County Sun,* 5 June 1983, 95.

Saucède, Thomas, and Bruno David. *Biodiversity of The Southern Ocean.* Amsterdam: Elsevier, 2016.

Saul, Ben, and Tim Stephens, *Antarctica in International Law.* Oxford: Hart Publishing, 2015.

Šavrič, Bojan, David Burrows, and Melita Kennedy. "The Spilhaus World Ocean Map in a Square." Last modified 7 February 2020. storymaps.arcgis.com. [online]

Scambos, Ted. "Plateau Station." Norwegian-U.S. Scientific Traverse of East Antarctica. Accessed 25 February 2021. traverse.npolar.no (web.archive.org). [online]

Scambos, Ted, and Clarence Novak. "On the Current Location of the Byrd 'Snow Cruiser' and Other Artifacts from Little America I, II, III and Framheim.*" Polar Geography* 29, no. 4 (2005): 237–52.

Scheinert, Mirko, Fausto Ferraccioli, Joachim Schwabe, Robin Bell, Michael Studinger, Detlef Damaske, Wilfried Jokat et al. "New Antarctic Gravity Anomaly Grid for Enhanced Geodetic and Geophysical Studies in Antarctica." *Geophysical Research Letters* 43, no. 2 (28 January 2016): 600–610.

Schillat, Monika, Marie Jensen, Marisol Vereda, Rodolfo A. Sánchez, and Ricardo Roura. *Tourism in Antarctica: A Multidisciplinary View of New Activities Carried Out on the White Continent.* Springer, 2016.

Schlacht, Irene Lia, Bernard Foing, Olga Bannova, Frans Blok, Alexandre Mangeot, Kent Nebergall, Ayako Ono, and Daniel Schubert. "Existing and New Proposals of Space Analog, Off-Grid and Sustainable Habitats with Space Applications." Paper presented at the 46th International Conference on Environmental Systems, Vienna, 2016.

Schmidtko, Sunke, Karen Heywood, Andrew Thompson, and Shigeru Aoki. "Multidecadal Warming of Antarctic Waters." *Science* 346, no. 6214 (5 December 2014): 1227–31.

Schneider, Darryn. "WYSSA – All my Love Darling." Antarctica. Accessed 14 March 2021. antarctica.kulgun.net. [online]

Schoof, Christian. "Ice Sheet Grounding Line Dynamics: Steady States, Stability, and Hysteresis." *Journal of Geophysical Research* 112, no. F3 (September 2007).

Schrijver, Nico. "Managing the Global Commons: Common Good or Common Sink?" *Third World Quarterly* 37, no. 7 (2016).

Schubert, Daniel. "Greenhouse Production Analysis of Early Mission Scenarios for Moon and Mars Habitats." *Open Agriculture* 2 (2017): 91–115.

Schubert, Daniel, Matthew Bamsey, Paul Zabel, Conrad Zeidler, and Vincent Vrakking. "Status of the EDEN ISS Greenhouse after On-Site Installation in Antarctica." Paper presented at the 48th International Conference on Environmental Systems, Albuquerque, NM, 2018.

Schulthess, Emil. *Antarctica: A Photographic Survey.* New York: Simon and Schuster, 1960.

Schwarz, Anne-Maree, Thomas Green, and Rod Seppelt. "Terrestrial Vegetation at Canada Glacier, Southern Victoria Land, Antarctica." *Polar Biology* 12, nos. 3–4 (September 1992): 397–404.

Schytt, Anna. *Med känsla för is: Om polarforskaren Valter Schytt och gåtorna hans Antarktisexpedition bidrog till att lösa.* Stockholm: Fri Tanke, 2018.

Scientific Committee on Antarctic Research (SCAR). "Antarctic Place Names." Accessed 26 February 2021. scar.org. [online]

Scientific Committee on Antarctic Research (SCAR). "Horizon Scan." Accessed 9 January 2021. scar.org. [online]

Scientific Committee on Antarctic Research (SCAR). "Plastic in Polar Environments (Plastic AG)." Accessed 9 October 2020. scar.org. [online]

Scientific Committee on Antarctic Research (SCAR). "What Is SCAR?" Accessed 9 January 2021. scar.org. [online]

Score, Roberta. "The Thrill of the Search: Finding ALH84001." *The Planetary Report* 17, no. 1 (January/February 1997): 5–7.

Scotcher, Jack. *British Antarctic Survey Building Report* (Halley III). Report (C/1979/Z). Cambridge: BAS Archives, 1979.

Scott, Karen. "Regulating Subglacial Aquatic Research under the Antarctic Treaty System." *New Zealand Universities Law Review* 23 (2008): 134–54.

Scott, Karen. "Scientific Rhetoric and Antarctic Security." In *Antarctic Security in the Twenty-First Century: Legal and Policy Perspectives,* edited by Alan D. Hemmings, Donald R. Rothwell, and Karen N. Scott, 384–306. London: Routledge, 2012.

Scott Polar Research Institute. "Antarctic Daily Sledge Ration, 1910s." Science Photo Library. Accessed 25 February 2021. sciencephoto.com. [online]

Scott Polar Research Institute. "Boot." Museum Catalogue: Antarctic Collection. Accessed 25 February 2021. spri.cam.ac.uk. [online]

Scott Polar Research Institute. "Mitten." Museum Catalogue: Antarctic Collection. Accessed 25 February 2021. spri.cam.ac.uk. [online]

Scott, Robert Falcon. "Camp under the Wild Range." 20 December 1911. In *Scott's Last Expedition.* London: Smith, Elder & Co., 1913. alamy.com. [online]

Scott, Robert Falcon. "Message to the Public." 29 March 1912. spri.cam.ac.uk. [online]

Scott, Robert Falcon. *Scott's Last Expedition.* 2 vols. London: Smith, Elder & Co., 1913.

Scott, Robert Falcon. *The Voyage of the "Discovery."* London: John Murray, 1913.

Scott, Shirley. "The Evolving Antarctic Treaty System: Implications of Accommodating Developments in the Law of the Sea." In *The Law of the Sea and the Polar Regions: Interactions between Global and Regional Regimes,* edited by Erik Jaap Molenaar, Alex G. Oude Elferink, and Donald R. Rothwell, 17–34. Leiden: Martinus Nijhoff, 2013.

Scottish Sea Farms. "Our Locations." Accessed 14 March 2021. scottishseafarms.com. [online]

"Sea Level Rise World Map." Accessed 18 December 2020. vnitourist.com. [online]

Searle, David. "Horseshoe Diaries." UK Antarctic Heritage Trust, 2017. Last modified 15 June 2020. ukaht.org. [online]

Sears, Mary, and Daniel Merriman. *Oceanography: The Past.* New York, NY: Springer, 1980.

Secretariat of the Antarctic Treaty. "Antarctic Specially Managed Area No 5 (Amundsen-Scott South Pole Station, South Pole): Revised Management Plan." *Measure 8 – ATCM XL – CEP XX.* Beijing, 2017. ats.aq. [online]

Secretariat of the Antarctic Treaty. "Antarctic Specially Protected Area No 120 (Pointe-Géologie Archipelago, Terre Adélie): Revised Management Plan." *Measure 2 – ATCM XXXIX – CEP XIX.* Santiago, 2016. ats.aq. [online]

Secretariat of the Antarctic Treaty. "Area Protection and Management / Monuments." Accessed 10 March 2021. ats.aq. [online]

Secretariat of the Antarctic Treaty. "ATCM Recommendation VI-4: Man's Impact on the Antarctic Environment." Tokyo, 1970.

Secretariat of the Antarctic Treaty. "ATCM Recommendation VIII-1: Man's Impact on the Antarctic Environment." Oslo, 1975.

Secretariat of the Antarctic Treaty. *ATCM XL – CEP XX.* Beijing, 2017. ats.aq. [online]

Secretariat of the Antarctic Treaty. *ATCM XXXVII – CEP XVII.* Brasilia, 2014. ats.aq. [online]

Secretariat of the Antarctic Treaty. "CEP Handbook 2019." Accessed 14 March 2021. ats.aq. [online]

Secretariat of the Antarctic Treaty. *Compilation of Key Documents of the Antarctic Treaty System.* 2nd ed. Buenos Aires, 2004.

Secretariat of the Antarctic Treaty. "Designation of ASMA 4 (Deception Island) and Revision of Management Plans for ASPA 140 (Parts of Deception Island) and 145 (Port Foster)." *Measure 3 – ATCM XXVIII – CEP VIII.* Stockholm, 2005. ats.aq. [online]

Secretariat of the Antarctic Treaty. "Designation of ASMA 5 (Amundsen Scott South Pole Station) and 6 (Larsemann Hills): ASMA 5 – Management Plan." *Measure 2 – ATCM XXX – CEP X.* New Delhi, 2007. ats.aq. [online]

Secretariat of the Antarctic Treaty. "Measure 19: Annex – Revised List of Historic Sites and Monuments." In *Final Report of the Thirty-Eighth Antarctic Treaty Consultative Meeting.* Buenos Aires: Secretariat of the Antarctic Treaty, 2015. ats.aq. [online]

Secretariat of the Antarctic Treaty. "Prague Declaration on the Occasion of the Sixtieth Anniversary of the Antarctic Treaty." Appendix 1 in *Final Report of the Forty-Second Antarctic Treaty Consultative Meeting, Prague, Czech Republic, 1–11 July 2019.* Vol. 1, 209–11. Buenos Aires: Secretariat of the Antarctic Treaty, 2019.

Secretariat of the Antarctic Treaty. "Resolution 5: Reducing Plastic Pollution in Antarctica and the Southern Ocean." Forty-Second Antarctic Treaty Consultative Meeting – Twenty-Second Committee for Environmental Protection Meeting. Prague, 2019. ats.aq. [online].

Secretariat of the Antarctic Treaty. "Revised Guidelines for Environmental Impact Assessment in Antarctica." *Resolution 1 – ATCM XXXIX – CEP XIX.* Santiago, 2016. ats.aq. [online]

Secretariat of the Antarctic Treaty. "Revised List of Antarctic Historic Sites and Monuments: The Wreck of Sir Ernest Shackleton's Vessel *Endurance* and C. A. Larsen Multiexpedition Cairn." *Measure 12 – ATCM XLII – CEP XXII.* Prague, 2019. ats.aq. [online]

Seed, Patricia. *Ceremonies of Possession in Europe's Conquest of the New World, 1492–1640.* Cambridge: Cambridge University Press, 1995.

Senatore, Maria Ximena. "Archaeologies in Antarctica from Nostalgia to Capitalism: A Review." *International Journal of Historical Archaeology* 23, no. 3 (September 2019): 755–71.

Senatore, Maria Ximena, and Andrés Zarankín. "Widening the Scope of the Antarctic Heritage: Archaeology and the 'The Ugly, The Dirty and The Evil' in Antarctic History." In *Polar Settlements: Location, Techniques and Conservation,* edited by Susan Barr and Paul Chaplin, 51–59. Oslo: ICOMOS, 2011.

Senthilingam, Meera. "The Closest Thing on Earth to a Mission to Mars." *CNN,* 9 December 2015. edition.cnn.com. [online]

Sfriso, Andrea A., Yari Tomio, Beatrice Rosso, Andrea Gambaro, Adriano Sfriso, Fabiana Corami, Eugenio Rastelli, Cinzia Corinaldesi, Michele Mistri, and Cristina Munari. "Microplastic Accumulation in Benthic Invertebrates in Terra Nova Bay (Ross Sea, Antarctica)." *Environment International* 137 (2020).

Shabecoff, Philip. "U.S. Seeks Moratorium on Antarctic Minerals." *The New York Times,* 14 November 1990.

Shackleton, Edward. "The New Continent." *United Nations World* 1, no. 10 (1949): 380–82.

Shackleton, Ernest Henry. *The Heart of the Antarctic: Being the Story of the British Antarctic Expedition 1907–1909.* London: Heinemann, 1909.

Shackleton, Ernest Henry. *South: The Story of Shackleton's Last Expedition 1914–17.* London: Heinemann, 1919.

Shah, Rohani Mohd. "Public Perceptions of Antarctic Values: Shaping Future Environmental Protection Policy." *Procedia: Social and Behavioral Science* 168 (January 2015): 211–18.

Shanklin, Jonathan. "The Antarctic Ozone Hole – 25 Years On." 16 March 2012. nerc.ukri.org. [online]

Shanklin, Jonathan. "The Discovery of The Antarctic Ozone Hole: Scientific Data Reveal Startling Atmospheric Changes." Accessed 9 January 2021. digital.nls.uk. [online]

Shariff, Jamil, Peter Ade, Francesco Angilè, Peter Ashton, Steven Benton, Mark Devlin, Bradley Dober et al. "Submillimeter Polarization Spectrum of the Carina Nebula." *The Astrophysical Journal* 872, no. 2 (20 February 2019): 197

Shaw, Justine D., Aleks Terauds, Martin J. Riddle, Hugh P. Possingham, and Steven L. Chown. "Antarctica's Protected Areas Are Inadequate, Unrepresentative, and at Risk." *PLOS Biology* 12, no. 6 (17 June 2014).

Shirase, Nobu. "Lt. Shirase's Calling Card – from *Nankyokuki* (1913)." In *The Ends of the Earth: An Anthology of the Finest Writing on the Arctic and the Antarctic*, edited by Elizabeth Kolbert, 55–62. New York: Bloomsbury, 2007.

Shirtcliffe, L. J. "Building New Base Complex, Halley Bay." British Antarctic Survey. 1967.

Shortis, Emma. "'Who Can Resist This Guy?' Jacques Cousteau, Celebrity Diplomacy, and the Environmental Protection of the Antarctic." *Australian Journal of Politics and History* 61, no. 3 (2015): 366–80.

Sidney, Philip. "Scott's Last Expedition and the Literature of Cold." PhD diss., University of Cambridge, 2013.

Siegel, Volker, ed. *Biology and Ecology of Antarctic Krill*, Advances in Polar Ecology. Basel: Springer International Publishing, 2016.

Simpson-Housley, Paul. *Antarctica: Exploration, Perception and Metaphor*. London: Routledge, 1992.

"Sinan Beidou daohang jieshouji biaoxian youyi" [Sinan BeiDou Navigation Receiver's Outstanding Performance], *Xian Beidou xun*. Online document (now expired). Cited in Brady, *China's Expanding Antarctic Interests*.

Siple, Paul Allman. "We Are Living at the South Pole." *National Geographic Magazine* (July 1957).

Siple, Paul Allman. "We Are Living at The South Pole." *National Geographic* (July 1957).

Siple, Paul Allman. "Man's First Winter at the South Pole." *National Geographic* (January 1958).

Slavid, Ruth. Ice Station: The Creation of Halley VI; Britain's Pioneering Antarctic Research Station. Zurich: Park Books, 2015.

Slezak, Michael. "World's Largest Marine Park Created in Ross Sea in Antarctica in Landmark Deal." *The Guardian*, 2020. theguardian.com. [online]

Smith, Alan. *Antarctic Station Halley Report No. 1 Structural Defects*. Cambridge: British Antarctic Survey, 1979

Smith, Alan. *Halley no. 4: Brief Report on the Establishment of the Station Complex at Halley*. Cambridge: British Antarctic Survey, 1983.

Smith, Alan. *Proposed Replacement Station for Halley*. Report, 31 October 1979. BAS Archives (AD6/2Z/1979/C2). Cambridge: British Antarctic Survey, 1979.

Smith, Laurajane. *The Uses of Heritage*. London: Routledge, 2006.

Smith, Roff. "How Polar Explorers Survived Months of Isolation without Cracking." *National Geographic*. 26 May 2020. nationalgeographic.com. [online]

Smith, Ronald. "Terrestrial Biology of the Antarctic and Sub-Antarctic." in *Antarctic Ecology*, edited by Richard Laws, 61–162. London: Academic Press, 1984.

Smith, Ronald, and Hugh Simpson. "Early Nineteenth Century Sealers' Refuges on Livingston Island, South Shetland Islands." *British Antarctic Survey Bulletin* 74 (1987): 48–72.

Smith, Walker, David Ainley, Kevin Arrigo, and Michael Dinniman. "The Oceanography and Ecology of the Ross Sea." *Annual Review of Marine Science* 6 (2014): 469–87.

Smith, Yvette. "Mars's Twin Peaks." *NASA* 2 September 2020. nasa.gov. [online]

Smithson, Robert. "Entropy and the New Monuments." In *Robert Smithson: The Collected Writings*, edited by Jack Flam. Berkeley: University of California Press, 1966.

Solaripedia. "Project: Princess Elisabeth Antarctica Polar Station." 2009. Accessed 14 March 2021. solaripedia.com. [online]

Sörlin, Sverker. "Polare jubileer: Kunnskap og politikk i nord." In *Ottar,* edited by Marit Anne Hauan. Tromsø: Tromsø Museum – Universitetsmuseet, 2011.

South Africa Department of Environmental Affairs. *SANAE IV Environmental, Health and Safety Impact Assessment: Initial Environmental Evaluation Report*. 1991.

South Polar Times. London: The Folio Society, 2018.

South-Pole.com, "Operation Highjump." Accessed 26 February 2021. south-pole.com. [online]

Spigel, Robert, and John Priscu, "Physical Limnology of the McMurdo Dry Valleys Lakes." In *Ecosystem Dynamics in a Polar Desert: The McMurdo Dry Valleys, Antarctica*, edited by John Priscu, 153–87. Washington, DC: American Geophysical Union, 1998).

Spilhaus, Athelstan. "To See the Oceans, Slice Up the Land." *Smithsonian,* November 1979, 54–63.

Spilhaus, Athelstan. "World Ocean Maps: The Proper Places to Interrupt." *Proceedings of the American Philosophical Society,* vol. 127, no.1 (1983): 50–60. Accessed 3 March 2020.

Spufford, Francis. *I May Be Some Time: Ice and the English Imagination*. London: Faber, 1996.

Stafford, Ed. *Expeditions Unpacked: What the Great Explorers Took into the Unknown*. London: White Lion Publishing: 2019.

Stam, David. *Adventures in Polar Reading*. New York: Grolier Club, 2019.

Stammerjohn, Sharon, Robert Massom, David Rind, and Douglas Martinson. "Regions of Rapid Sea Ice Change: An Inter-hemispheric Seasonal Comparison." *Geophysical Research Letters* 39, no. 6 (28 March 2012).

Stehberg, Ruben. *Arqueología histórica antártica: Aborígenes sudamericanos en los mares subantárticos en el siglo XIX*. Santiago: Centro de Investigaciones Diego Barros Arana, 2003.

Stephens, Tim. "The Antarctic Treaty System and the Anthropocene." *The Polar Journal* 8, no. 1 (January 2018): 29–43.

Stephens, Tim. "An Icy Reception or a Warm Embrace? The Antarctic Treaty System and the International Law of the Sea." In *Handbook on the Politics of Antarctica,* edited by Klaus Dodds, Alan Hemmings, and Peder Roberts, 439–52. Cheltenham: Edward Elgar Publishing, 2017.

Stephens, Tim. "Polar Continental Shelves: Australian and Canadian Challenges and Opportunities." In *Polar Oceans Governance in an Era of Environmental Change,* edited by Tim Stephens and David VanderZwaag, 146–65. Cheltenham: Edward Elgar, 2014.

Stevens, Alex. "Governments Cannot Just 'Follow the Science' on COVID-19." *Nature Human Behaviour* 4 (2020): 560.

Stewart, John. *Antarctica: An Encyclopedia*. Jefferson, NC: McFarland & Company, 2011.

Stilwell, Jeffrey, and John Long. *Frozen in Time: Prehistoric Life in Antarctica*. Collingwood: CSIRO Publishing, 2011.

Stone, Madeleine. "A Huge Iceberg Just Broke Off West Antarctica's Most Endangered Glacier." 11 February 2020. nationalgeographic.com. [online]

Stonehouse, Bernard, and John Snyder. *Polar Tourism: An Environmental Perspective. Aspects of Tourism.* Bristol, UK: Channel View Publications, 2010.

Strange, Carolyn, and Alison Bashford. *Griffith Taylor: Visionary, Environmentalist, Explorer*. Toronto: University of Toronto Press, 2008.

Strange, Robert, and S. A. Youngman. "Emotional Aspects of Wintering Over." *Antarctic Journal of the United States* 6, no. 6 (November–December 1971): 255–57.

Student, Jillian, Bas Amelung, and Machiel Lamers. "Towards a Tipping Point? Exploring the Capacity to Self-Regulate Antarctic Tourism Using Agent-Based Modelling." *Journal of Sustainable Tourism* 24, no. 3 (2016): 412–29.

Stuiver, Minze, George Denton, Terence Hughes, and James Fastook. "History of the Marine Ice Sheet in West Antarctica during the Last Glaciation: A Working Hypothesis". In *The Last Great Ice Sheets*, edited by George Denton and Terence Hughes. New York: Wiley, 1981.

Stuster, Jack, Claude Bachelard, and Peter Suedfeld. "The Relative Importance of Behavioral Issues during Long-Duration ICE Missions." *Aviation, Space, and Environmental Medicine* 71, no. 9 (September 2000): A17–A25.

Suaria, Giuseppe, Vonica Perold, Jasmine R. Lee, Fabrice Lebouard, Stefano Aliani, and Peter G. Ryan. "Floating Macro- and Microplastics around the Southern Ocean: Results from the Antarctic Circumnavigation Expedition." *Environment International* 136 (2020).

Suedfeld, Peter. "Antarctica and Space as Psychosocial Analogues." *REACH* 9–12 (December 2018): 1–4.

Suedfeld, Peter. "Polar Psychology: An Overview." *Environment and Behavior* 23, no. 6 (November 1991): 653–65.

Suedfeld, Peter, and Karine Weiss. "Antarctica: Natural Laboratory and Space Analogue for Psychological Research." *Environment and Behavior* 32, no. 1 (2000): 7–17.

Sugden, David, George Denton, and David Marchant, "Landscape evolution of the Dry Valleys, Transantarctic Mountains: Tectonic Implications." *Journal of Geophysical Research* 100, no. B6 (June 1995): 9949–67.

Sullivan, Walter. *Assault on the Unknown: The International Geophysical Year*. 1st ed. New York: McGraw-Hill, 1961.

Summerhayes, Colin. "International Collaboration in Antarctica: The International Polar Years, the International Geophysical Year, and the Scientific Committee on Antarctic Research," *Polar Record* 44, no. 4 (2008): 321–34.

Svalbard Global Seed Vault. "Home page." Accessed 9 January 2021. seedvault.no. [online]

Swart, Neil, Sarah Gille, John Fyfe, and Nathan Gillett. "Recent Southern Ocean Warming and Freshening Driven by Greenhouse Gas Emissions and Ozone Depletion." *Nature Geoscience* 11, no. 11 (November 2018): 836–41.

Sykora-Bodie, Seth, and Tiffany Morrison. "Drivers of Consensus-Based Decision-Taking in International Environmental Regimes: Lessons from the Southern Ocean." *Aquatic Conservation: Marine and Freshwater Ecosystems* 29, no. 12 (2019): 2147–61.

T Terra Australis Incognita

In the classical Greco-Roman and medieval European tradition, Terra Incognita (unknown land) could refer either to land within or at the edges of the known world of Europe, Asia, and Africa, or to hypothesised "antipodal" land masses located in the southern hemisphere. Opinions differed as to whether antipodal land was inhabited or not. With the emergence of European colonialism, terra incognita became a standard feature on maps, typically found in the New World, but also in parts of the "old world", such as sub-Saharan Africa. Terra Australis Incognita referred to an enormous Antarctic land mass which in the sixteenth century was conjectured to extend from the South Pole to latitudes as high as 15 degrees South. It incorporated Tierra del Fuego as well as a number of essentially fictional locations, such as the "Region of Parrots".[15]

Tahi, Djamel, dir. *Enterrés volontaires au cœur de l'Antartique*. Paris: Centre National de la Recherche Scientifique (CNRS), 2008.

Takenaka Group. "Building for the Japanese Antarctic Research Expedition: From the South Pole to a New Kind of Housing in Japan." Accessed 13 February 2021. takenaka.co.jp. [online]

Talalay, Pavel. *Mechanical Ice Drilling Technology.* Singapore: Springer Geophysics, 2016.

Talalay, Pavel. *Thermal Ice Drilling Technology.* Singapore: Springer Geophysics, 2020.

Taylor, Adam, and Stefano Pitrelli. "One Continent Remains Untouched by the Coronavirus: Antarctica." *The Washington Post,* 24 March 2020. washingtonpost.com. [online]

Taylor, Alan. "The Antarctic Snow Cruiser: Updated", *The Atlantic*, 20 January 2016. theatlantic.com. [online]

Taylor, Anthony. *Antarctic Psychology.* Wellington: Department of Scientific and Industrial Research, 1987.

Taylor, Matthew. "Antarctic's Future in Doubt After Plan for World's Biggest Marine Reserve Is Blocked." *The Guardian,* 2 November 2018. theguardian.com. [online]

Taylor, Matthew. "Decline in Krill Threatens Antarctic Wildlife, from Whales to Penguins." *The Guardian,* 14 February 2018. theguardian.com. [online]

Taylor, Thomas Griffith. *The Physiography of the McMurdo Sound and Granite Harbour Region.* London: Harrison and Sons, 1922.

Te Ahukaramū Charles Royal. "Story: Papatūānuku – The Land." *The Encyclopedia of New Zealand.* 24 September 2007. Accessed 9 March 2021. teara.govt.nz. [online]

Télam: Agencia nacional de noticias. "Buscan a nacidos en la base Esperanza para reconstruir un registro quemado hace 40 años." Accessed 18 February 2021. telam.com.ar. [online]

Telegraph. "Early Mobile Phones Help Robert Falcon Scott in Successful South Pole Bid." 18 January 2017. telegraph.co.uk. [online]

Terauds, Aleks, Steven L. Chown, Fraser Morgan, Helen J. Peat, David J. Watts, Harry Keys, Peter Convey, and Dana M. Bergstrom. "Conservation Biogeography of the Antarctic." *Diversity and Distributions* 18 (2012): 726–41.

Terauds, Aleks, and Jasmine R. Lee, "Antarctic Biogeography Revisited: Updating the Antarctic Conservation Biogeographic Regions." *Diversity and Distributions* 22 (2016): 836–40.

Thompson, Chris. "Inspired by Thunderbirds: Halley VI Antarctic Research Station." 28 January 2017. gerryanderson.co.uk. [online]

Thompson, Tabitha, and Ed Compton, "Joint NASA Study Reveals Leaks in Antarctic 'Plumbing System.'" 15 February 2007. nasa.gov. [online]

Thomson, Janet, and Alexander Cooper. "The SCAR Antarctic Digital Topographic Database." *Antarctic Science* 5, no. 3 (September 1993): 239–44.

Thomson, John. *Elephant Island and Beyond: The Life and Diaries of Thomas Orde Lees.* Norwich: Erskine Press, 2003.

Thoreau, Henry David. "Walking." *The Atlantic,* May 1862, 657–74.

Times. "Sir Ernest Shackleton on Poetry." 28 October 1911.

Tin, Tina, Zoe Louise Fleming, Kevin Hughes, David Ainley, Peter Convey, Carlos Moreno, Simone Pfeiffer, J. Scott, and Ian Snape. "Impacts of Local Human Activities on the Antarctic Environment: A Review." *Antarctic Science* 21 (2009): 3–33.

Tin, Tina, Daniela Liggett, Patrick Mahers, and Machiel Lamers, eds. *Antarctic Futures: Human Engagement with the Antarctic Environment.* Dordrecht: Springer, 2014.

Torii, Tetsuya. "A Review of the Dry Valley Drilling Project, 1971–76." *Polar Record* 20, no. 129 (September 1981): 533–41.

Torres, Ella. "What Life Is Like on Antarctica, the Only Continent without a Case of Coronavirus." *ABC News* [USA], 20 March 2020. abcnews.go.com. [online]

Toyokawa, Saikaku. "Cold Climate Housing Research and the Syowa Station Building in Antarctica: Asada Takashi and the Beginnings of Capsule Architecture." In *Metabolism: The City of the Future,* translated by Nathan Elchert, 234–41. Tokyo: Mori Art Museum, 2011. Exhibition catalogue.

Trathan, Philip, Pablo García-Borboroglu, Dee Boersma, Charles-André Bost, Robert Crawford, Glenn Crossin, Richard Cuthbert et al. "Pollution, Habitat Loss, Fishing, and Climate Change as Critical Threats to Penguins." *Conservation Biology* 29, no. 1 (2015): 31–41.

Trathan, Philip, and Susie Grant. "The South Orkney Islands Southern Shelf Marine Protected Area: Towards the Establishment of Marine Spatial Protection within International Waters in the Southern Ocean." In *Marine Protected Areas: Science, Policy and Management,* edited by John Humphreys and Robert Clark, 67–98. Amsterdam: Elsevier, 2020.

Trathan, Philip, Barbara Wienecke, Christophe Barbraud, Stéphanie Jenouvrier, Gerald Kooyman, Céline Le Bohec, David Ainley et al. "The Emperor Penguin: Vulnerable to Projected Rates of Warming and Sea Ice Loss." *Biological Conservation* 241 (January 2020).

Travel Reports. "Antarctica: Walking Beyond the Three Poles; Part 3." 29 February 2020. travelrept.ru. [online]

Treves, Tullio. "Historical Development of the Law of the Sea." In *The Oxford Handbook of the Law of the Sea,* edited by Donald R. Rothwell, Alex G. Oude Elferink, Karen N. Scott, and Tim Stephens, 1–24. Oxford: Oxford University Press, 2015.

Tsankova, Diana. "Architect Penka Stancheva Designs New Laboratory for Researchers at Bulgarian Antarctic Base." 1 September 2019. bnr.bg. [online]

Turner, John, et al. *Antarctic Climate Change and the Environment.* Cambridge: Scientific Committee on Antarctic Research, 2009.

Tyree, David M. "New Era in the Loneliest Continent." *National Geographic* 123, no. 2 (February 1963).

Tyszczuk, Renata. *Provisional Cities: Cautionary Tales for the Anthropocene.* London: Routledge, 2017.

U Ultraviolet

Ultraviolet (UV) radiation is a form of short-wave electromagnetic radiation emitted by the Sun which passes through the magnetosphere and meets resistance in the ozone layer within the stratosphere. This layer shields the earth and its inhabitants from the harmful effects of UV radiation which are known to damage human DNA, increasing the risks of skin cancer. The discovery of the "ozone hole" detected at Halley Station in Antarctica in 1985, and the subsequent understanding that ozone depletion was caused by a chemical reaction induced by man-made chlorofluorocarbons, led to the signing in 1987 of one of the greatest environmental policies, the Montreal Protocol on Substances That Deplete the Ozone Layer.

Undercurrent News. "Aker Biomarine: New Study Shows Krill Is the Most Effective Growth Enhancer for Shrimp." 20 September 2019. undercurrentnews.com. [online]

UN-Habitat. "Informal Urbanism." Accessed 3 March 2020. uni.unhabitat.org. [online]

United Kingdom Antarctic Heritage Trust. *Britain's Antarctic Heritage.* Cambridge: UKAHT, 2019.

United Kingdom Antarctic Heritage Trust. "UKAHT: Port Lockroy." Accessed 15 March 2020. ukaht.org. [online]

United Kingdom Government. "Environmental Impact Assessments: Update on Broader Policy Discussions." Working Paper number 41 submitted to the 40th Antarctic Treaty Consultative Meeting. Prague, Czech Republic, 2017.

United Nations. Index to Proceedings of the General Assembly: Sixtieth Session 2005/2006, New York: United Nations, 2007.

United Nations. *Question of Antarctica: Report of the Secretary-General.* 39 U.N. GAOR (Agenda Item 66), U.N. Doc. A/39/583 (1984).

United Nations General Assembly. "The Question of Antarctica." Resolutions, 1983–2006.

United Nations General Assembly, "Development of an International Legally Binding Instrument under the United Nations Convention on the Law of the Sea on the Conservation and Sustainable Use of Marine Biological Diversity of Areas beyond National Jurisdiction." Resolution A/RES/69/292. Accessed 9 February 2020. un.org. [online]

United States Army. *The Story of Camp Century; The City under Ice.* Research and Development Progress Report No. 6, documentary film, 31:48. 1963. usace.contentdm.oclc.org. [online]

United States Army Corps of Engineers, "Polar Traverse Technology." 12 February 2016. erdc.usace.army.mil. [online]

United States Congress, Office of Technology Assessment. *Polar Prospects: A Minerals Treaty for Antarctica.* Washington, DC: US Government Printing Office, September 1989.

United States Navy. *Antarctica: A New Look.* Vol 2, *Winter-Over Program.* Seabee Museum Archives, Port Hueneme, CA, 1972. history.navy.mil. [online]

United States Navy. *Operation Deepfreeze | The Story of Task Force 43 – Third Phase: 1957–58.* Paoli, PA: The Dorville Cooperation, 1958.

United States Navy. *A Season South: The Inside and Outside of Duty in Antarctica.* Antarctica: ASA and CBU 201 in Operation Deep Freeze '70, 1970. history.navy.mil. [online]

United States State Department. "Remarks at the Joint Session of the Antarctic Treaty Consultative Meeting and the Arctic Council, 50th Anniversary of the Antarctic Treaty." Washington, DC: US Department of State, 2020.

Uptime Institute. "Data Center Certification." Accessed 27 February 2021. uptimeinstitute.com. [online]

Urwin, Rosamund. "Meeting the Doctor Who Braved Antarctica in the Name of Space Exploration." *Evening Standard,* 8 September 2016.

U.S. Green Building Council. "Home page." Accessed 27 February 2021. usgbc.org. [online]

V Veranda

The inexperience of early explorers venturing into the Antarctic is perhaps epitomised by the Australian-design hut erected at Hut Point by Captain Robert Falcon Scott's party during the 1902–1903 British Expedition. In good tropical tradition, the prefabricated structure featured a veranda on three sides of the building. Impossible to inhabit within the harsh polar environment, the poorly insulated hut and its veranda were soon abandoned and the party relocated to the *Discovery* vessel.

Vairo, Carlos, Ricardo Capdevila, Veronica Aldazabal, and Pablo Pereyra. *Antarctica and Argentina Cultural Heritage: Museums, Sites and Shelters of Argentina.* Buenos Aires: Zagier & Urruty, 2007.

Van Allen, James A. "Genesis of the International Geophysical Year." *Eos Trans. AGU* 64, no. 50 (1983): 977.

Van der Watt, Lizé-Marie. "Antarctica." *Encyclopædia Britannica.* 2020.

Van Ombergen, Angelique, Andrea Rossiter, and Thu Jennifer Ngo-Anh "'White Mars' – Nearly Two Decades of Biomedical Research at the Antarctic Concordia Station." *Experimental Physiology* 106, no. 1 (January 2021): 6–17.

Van Rattinghe, Kristoff. "Princess Elisabeth Research Station at Antarctica: Renewable Energy Systems Design, Simulation and Optimization." Master's thesis, TU Delft, 2008.

Van Sant Hall, Marshall. "Argentine Policy Motivations in the Falklands War and the Aftermath." *Naval War College Review* 36, no. 6 (1983).

Vanstappen, Nils. "Legitimacy in Antarctic Governance: The Stewardship Model," *Polar Record* 55 (2019): 358–60.

Vasiliev, Nikolai, Pavel Talalay, N. E. Bobin, V. K. Christyakov, V. M. Zubkov, A. V. Krasilev, A. N. Dmitriev, Svetlana Yankilevich, and V. Y. Lipenkov. "Deep Drilling at Vostok Station, Antarctica: History and Recent Events," *Annals of Glaciology* 47 (2007): 10–23.

Vaughan, David. "Ice Sheets: Indicators and Instruments of Climate Change." In *Climate Change: Observed Impacts on Planet Earth,* edited by Trevor Letcher, 391–400. Amsterdam: Elsevier, 2009.

Vaughan, David, Karen Heywood, Doug Benn, and Ted Scambos. "Thwaites: Antarctica's Doomsday Glacier." Lecture, Palace of Westminster, 11 June 2019.

Vaughan, David, Gareth Marshall, William Connolley, Claire Parkinson, Robert Mulvaney, Dominic Hodgson, John King, Carol Pudsey and John Turner. "Recent Rapid Regional Climate Warming on the Antarctic Peninsula." *Climatic Change* 60, no. 3 (October 2003): 243–74.

Verne, Jules. *Twenty Thousand Leagues Under the Sea,* translated by David Coward. Reprint, Penguin Classics: 2018.

Verne, Jules. *An Antarctic Mystery (Also Called the Sphinx of the Ice Fields).* Translated by Mrs Cashel HOey. Lector House, 2019.

Vidas, Davor. *Implementing the Environmental Protection Regime for the Antarctic.* Dordrecht: Springer, 2000.

Vigni, Patrizia. "Antarctic Maritime Claims: 'Frozen Sovereignty' and the Law of the Sea." In *The Law of the Sea and Polar Maritime Delimitation and Jurisdiction,* edited by Alex G. Oude Elferink and Donald R. Rothwell, 85–101. Leiden: Martinus Nijhoff, 2001.

Vila, Mar, Gerard Costa, Carlos Angulo-Preckler, Rafael Sarda, and Conxita Avila. "Contrasting Views on Antarctic Tourism: 'Last Chance Tourism' or 'Ambassadorship' in the Last of the Wild." *Journal of Cleaner Production* 111 (2016): 451–60.

Villemain, Aude, and Patrice Godon. "Logistic Transport in Extreme Environments: The Evolution of Risk and Safety Management over 27 Years of the Polar Traverse." *Ergonomics* 63, no. 10 (2020): 1257–70.

Vrakking, Vincent, Matthew Bamsey, Conrad Zeidler, Paul Zabel, Daniel Schubert, and Oliver Romberg. "Service Section Design of the EDEN ISS Project." Paper presented at the 47th International Conference on Environmental Systems, Charleston, SC, July 2017.

W White Mars

Antarctica is often referred to as "White Mars" and as "Space Analogue", i.e. as sites, "as characterised by international space programmes worldwide, which offer fidelity to off-world conditions such as rugged terrain, extreme environments, working in pressurised suits, and long-term isolation in confined living spaces."[16] In anticipation of prototyping in-habitation models for the red planet, the Antarctic desert has been used extensively as a research site to understand the psychophysical effects induced by living in isolated, confined, and extreme (ICCE) environments in which individuals are faced with prolonged periods of complete isolation, altered day-night cycles, chronic hypobaric hypoxia, and exposure to extreme cold temperatures. The principal site for such studies in the Antarctic is the Franco-Italian Concordia Station.[17]

Walker, Gabrielle. *Antarctica: An Intimate Portrait of The World's Most Mysterious Continent.* London: Bloomsbury, 2013.

Walton, David. Lecture in conversation with UNLESS. "Constructing an Antarctic Atlas Part I," Architectural Association School of Architecture Polar Lab Workshop, London, February 2019.

Walton, David. *Antarctica: Global Science from a Frozen Continent.* Cambridge: Cambridge University Press, 2013.

Walton, David, and Peter Clarkson. *Science in the Snow: Fifty Years of International Collaboration through the Scientific Committee on Antarctic Research.* Cambridge: Scientific Committee on Antarctic Research, 2011.

Wathern, Peter, ed. *Environmental Impact Assessment: Theory and Practice.* London: Routledge, 1988.

Watson, Kirk F. *Whalers Bay Deception Island: A Brief History.* UK Antarctic Heritage Trust, 2014. YouTube video, 3 November 2014. youtube.com. [online]

Watters, George, Jefferson Hinke, and Christian Reiss. "Long-Term Observations from Antarctica Demonstrate That Mismatched Scales of Fisheries Management and Predator-Prey Interaction Lead to Erroneous Conclusions about Precaution." *Scientific Reports* 10, no. 1 (2020).

Wauchope, Hannah, Justine Shaw, and Aleks Terauds, "A Snapshot of Biodiversity Protection in Antarctica," *Nature Communications* 10 (2019).

Webb, Peter. "Wright Fjord, Pliocene Marine Invasion of an Antarctic Dry Valley." *Antarctic Journal of the United States* 7, no. 6 (November–December 1972): 227–34.

Webber, Alex. "Poland's New Polar Ice Station Set to Take Shape in Antarctica." 19 July 2020. thefirstnews.com. [online]

Wenger, Michael. "Dark COVID-19 Clouds over Antarctic Season." *The Polar Journal* (2020).

Westwood, Karen, Paul Thomson, Rick van den Enden, Linsey Mahler, Simon Wright, and Andrew Davidson. "Ocean Acidification Impacts Primary and Bacterial Production in Antarctic Coastal Waters During Austral Summer." *Journal of Experimental Marine Biology and Ecology* 498 (January 2018): 46–60.

Wexler, Harry. "Antarctic Research during the International Geophysical Year," *Antarctica in the International Geophysical Year: Based on a Symposium on the Antarctic, Volume 1,* Geophysical Monograph series, vol. 1, edited by Albert P. Crary et al., 7–12. Washington, DC: American Geophysical Union, 1956.

Wharton, David, and Ian Brown, "A Survey of Terrestrial Nematodes from the McMurdo Sound Region." *New Zealand Journal of Zoology* 16, no. 3 (1989): 467–70.

Wharton, R. A. "Stromatolitic Mats in Antarctic Lakes." In *Phanerozoic Stromatolites II,* edited by Janine Bertrand-Sarfati and Claude Monty, 53–70. Dordrecht: Springer, 1994.

Wheeler, Sara. *Terra Incognita: Travels in Antarctica.* New York: Modern Library, 1999.

White House Office of the Press Secretary. "Barack Obama: Statement by the President on the Montreal Protocol." 15 October 2016. obamawhitehouse.archives.gov. [online]

Wigley, Mark. *Buckminster Fuller Inc: Architecture in the Age of Radio.* Zurich: Lars Müller Publishers, 2015.

Wikipedia contributors. "Jamesway Hut." Wikipedia. Accessed 14 March 2021. en.wikipedia.org. [online]

Wikipedia contributors. "Pleistocene." Wikipedia. Accessed 9 January 2021. en.wikipedia.org. [online]

Williams, Nancy, Lauren Juranek, Richard Feely, Kenneth Johnson, Jorge Sarmiento, Lynne Talley, Andrew Dickson et al. "Calculating Surface Ocean pCO_2 from Biogeochemical Argo Floats Equipped with pH: An Uncertainty Analysis." *Global Biogeochemical Cycles* 31, no. 3 (March 2017): 591–604.

Wilson, David. *The Lost Photographs of Captain Scott: Unseen Images from the Legendary Antarctic Expedition.* New York: Little Brown and Company, 2011.

Wilson, Edward. *With Scott in the Antarctic.* UK: The History Press, 2009.

Wilson, Gary, Richard Levy, Tim Naish, Ross Powell, Fabio Florindo, Christian Ohneiser, Leonardo Sagnotti et al. "Neogene Tectonic and Climatic Evolution of the Western Ross Sea, Antarctica: Chronology of Events from the AND-1B Drill Hole." *Global and Planetary Change* 96–97 (October–November 2012): 189–203.

Wilson, Gary, Andrew Roberts, Kenneth Verostub, Florindo Fabio, and Leonardo Sagnotti. "Magnetobiostratigraphic Chronology of the Eocene-Oligocene Transition in the CIROS-1 Core, Victoria Land Margin, Antarctica: Implications for Antarctic Glacial History." *Geological Society of America Bulletin* 110, no. 1 (January 1998): 35–47.

Wilson, John T. *IGY: The Year of the New Moons.* London: Michael Joseph, 1961.

World Geography. "Antarctic Religion?" Accessed 27 February 2021. theworldgeography.com. [online]

World Green Building Council. "Bringing Embodied Carbon Upfront." Accessed 27 February 2021. worldgbc.com. [online]

World Health Organization. "Coronavirus Disease (COVID-19) Pandemic." Accessed 27 February 2021. who.int. [online]

Worsley, Frank Arthur. *Endurance: An Epic of Polar Adventure.* London: Allan, 1931.

Wratt, Gillian. *A Story of Antarctic Co-operation: 25 Years of the Council of Managers of National Antarctic Programs.* Christchurch: COMNAP, 2013.

Wright, Asia. "Southern Exposure: Managing Sustainable Cruise Ship Tourism in Antarctica." *California Western International Law Journal* 39 (2008): 43–86.

Wright, Christopher. "The Rise and Rise and Rise of Polar Cruising." *The Maritime Executive,* 2018. maritime-executive.com. [online]

Wu Yilin. "Guanyu fazhan Nanji luyouye de sikao" [Thoughts on Expanding Antarctic Tourism]. *Zhongguo haiyang daxue xuebao, shehuixuebao* (2010): 5–8.

X X-ray Spectrometer

X-ray spectrometers are deployed in long-duration ballooning missions, mostly launched in proximity to McMurdo station. The spectrometers are attached to balloons the size of football stadiums which in turn are released into the stratosphere "looking to see if there are any other Earth-like planets in our galaxy, measuring and counting the atomic particles flung from supernovas in far-off galaxies, or exposing bacteria to the harsh environment of the upper atmosphere to understand what would happen to a microorganism hitchhiker that caught a ride on the next rover mission to Mars".[18] When deployed on missions such as HIREGS I (High Resolution Gamma-Ray and Hard X-Ray Spectrometer) X-ray spectrometers are used to study solar flare activities and understand their relation to solar activities, the solar cycle, and the field particle environments of the earth.[19]

Xie Yu, "China Steps Up Efforts in Antarctica to Benefit from Krill Bonanza," *China Daily,* 5 March 2015. usa.chinadaily.com.cn. [online]

Y YIKLA

Before satellite communications were introduced in the mid 1980s, ensuring greater "communications intimacy"[20] between researchers deployed in Antarctica and their families back home, telegraphs were used on a regular basis. Prior to this, the Australian National Antarctic Research Expeditions (ANARE) had developed a special ANARE five-letter code to reduce the word count – using this code, YIKLA, for instance, stood for "This is life".[21]

Yan Qide and Zhu Jiangang, eds., *Nanjizhou ziyuan lingtu zhuquan yu ziyuan quan shu wenti yanjiu* [Research on Antarctic Sovereignty and Resources Rights]. Shanghai: Kexue jishu chubanshe, 2009.

"Yinggai peizhi zhenzheng de Zhongguo Nanji luyou shichang" [We Should Foster the Development of a Truly Chinese Antarctic Tourism Market]. *Zhongguo luyou bao,* 3 July 2014. henanci.com. [online]

Youde, Kate. "Antarctic Return: Hugh Broughton Unveils Designs for New Zealand's Ice Base." 2 July 2019. architectsjournal.co.uk. [online]

Yusoff, Kathryn. "Test Landscapes and the 'Geographical Gift' of a Continent to Science." Paper presented at the workshop Polar Field Stations and International Polar Year (IPY) History: Culture, Heritage, Governance (1882–Present), Cambridge, May 2007.

Z
Zucchini

In the mid-1980s, the Australian Antarctic Programme (AAD) developed a series of moulded fibreglass lightweight inhabitation modules for the Antarctic that ironically were named after vegetables, fruits, and animal products that are notably absent in the southernmost desert. Departing from the elevated "Googies" (which is an Australian slang term for eggs), the portable rounded structures were at times installed on the ground, transforming into "Apples", or extended by 0.75 metres, morphing into "Zucchini".

Zabel, Paul, Matthew Bamsey, Conrad Zeidler, Vincent Vrakking, Bernd-Wolfgang Johannes, Petra Rettberg, Daniel Schubert et al. "Introducing EDEN ISS: A European Project on Advancing Plant Cultivation Technologies and Operations." Paper presented at the 45th International Conference on Environmental Systems. Bellevue, WA, July 2015.

Zabel, Paul, Matthew Bamsey, Conrad Zeidler, Vincent Vrakking, Daniel Schubert, and Oliver Romberg. "Future Exploration Greenhouse Design of the EDEN ISS Project." Paper presented at the 47th International Conference on Environmental Systems. Charleston, SC, July 2017.

Zabel, Paul, Matthew Bamsey, Conrad Zeidler, Vincent Vrakking, Daniel Schubert, Oliver Romberg, Giorgio Boscheri, and Tom Dueck. (2016) "The Preliminary Design of the EDEN ISS Mobile Test Facility: An Antarctic Greenhouse." Paper presented at the 46th International Conference on Environmental Systems. Vienna, July 2016.

Zarankín, Andrés, Melisa Salerno, and Adrian Howkins. "From the Antarctic to New England: Remembrance of Sealing and Sealers." In Headland, *Historical Antarctic Sealing Industry,* 107–19.

Zarankín, Andrés, and Maria Ximena Senatore. "Archaeology in Antarctica: Nineteenth-Century Capitalism Expansion strategies." *International Journal of Historical Archaeology* 9, no. 1 (March 2005): 43–56.

Zarankín, Andrés, and Maria Ximena Senatore. *Historias de un pasado en blanco: arqueología histórica antártica.* Belo Horizonte: Argumentum, 2007.

Zeidler, Conrad, Vincent Vrakking, Matthew Bamsey, Lucie Poulet, Paul Zabel, Daniel Schubert, Christel Paille, Erik Mazzoleni, and Niko Domurath. "Greenhouse Module for Space Systems: A Lunar Greenhouse Design", *Open Agriculture* 2 (2017): 116–32.

Zhu Jiangang, Yan Qide, and Ling Xiaoliang, "Nanji ziyuan ji qi kaifa liyong qianjing fenxi" [An Analysis of Antarctic Resources, Their Exploitation and Potential Utilisation]. *Zhongguo ruan kexue* 8 (2005): 18, 20.

Ziegelmayer, Eric. "Capitalist Impact on Krill in Area 48 (Antarctica)." *Capitalism Nature Socialism* 25, no. 4 (2014): 36–53.

Zimmer, Marilene, João Carlos Centurion Rodrigues Cabral, Fernanda Czarneski Borges, Karen Gonçalves Côco, and Bianca da Rocha Hameister. "Psychological Changes Arising from an Antarctic Stay: Systematic Overview." *Estudos de Psicologia (Campinas)* 30, no. 3 (July–September 2013): 415–23.

Zorn, Stephen A. "Antarctic Minerals: A Common Heritage Approach," *Resources Policy* 10, no. 1 (1984): 2–18.

1   Bernadette Hince, *The Antarctic Dictionary: A Complete Guide to Antarctic English* (Clayton, Vic: CSIRO, 2020).
2   Emily Holt and Scott Miller, "Bioindicators: Using Organisms to Measure Environmental Impacts," *Nature Education Knowledge* 3, no. 10 (2010): 8.
3   Zina Deretsky, *Map of the South Pole Station and Attached Science Instruments,* National Science Foundation, accessed 9 March 2021, nsf.gov [online].
4   Secretariat of the Antarctic Treaty, "The Protocol on Environmental Protection to the Antarctic Treaty," accessed 9 March 2021, ats.aq [online].
5   Aude Villemain and Patrice Godon, "Logistic Transport in
6   Sanjay Chaturvedi, see p. 174.
7   Cool Antarctica, "Hoosh, an Antarctic Recipe Food from the Deep South," accessed 9 March 2021, coolantarctica.com [online].
8   The Royal Society, "Clements Markham," accessed 9 March 2021, makingscience.royalsociety.com [online].
9   National Geographic, "South Pole," accessed 8 March 2021, nationalgeographic.org [online].
10   Joachim Krausse and Claude Lichtenstein, eds., *Your Private Sky: R. Buckminster Fuller; The Art of Design Science* (Zurich: Lars Müller Publishers, 2017)
11   Katherina Petrou and Peter Landschutzer, see pp. 212, 215, and Stephen Barker and Andy Ridgwell, "Ocean Acidification," *Nature Education Knowledge* 3, no. 10 (2012): 21.
12   Te Ahukaramū Charles Royal, "Story: Papatūānuku – The Land," *The Encyclopedia of New Zealand*, 24 September 2007, accessed 9 March 2021, teara.govt.nz [online].
13   Doaa-Abdel Motaal, "The 'Question of Antarctica' at the United Nations," see p. 182.
14   David Vaughan, see p. 292.
15   Alfred Hiatt, in conversation with the editor, 2020..
16   William L. Fox, see p. 551.
17   Angelique Van Ombergen, Andrea Rossiter, and Thu Jennifer Ngo-Anh "'White Mars': Nearly Two Decades of Biomedical Research at the Antarctic Concordia Station" *Experimental Physiology* 106, no. 1 (January 2021): 6–17; Meera Senthilingam "The Closest Thing on Earth to a Mission to Mars," *CNN,* 9 December 2015, edition.cnn.com [online]; Alexander Kumar, "The Journey to White Mars," *The New York Times,* 2 October 2012, scientistatwork.blogs.nytimes.com [online].
18   Andrew Hynous, see p. 308.
19   StratoCat, "Hiregs I (High Resolution Gamma-Ray and Hard X-Ray Spectrometer)," 27 March 2020, stratocat.com.ar [online].
20   Richard Buckminster Fuller, *Operating Manual for Spaceship Earth* (1969; repr. Baden: Lars Müller Publishers, 2008), 31.
21   Daryn Schneider, "WYSSA – All My Love Darling," accessed 9 March 2021, antarctica.kulgun.net [online].

Notes

DOMINANCE OR RESEARCH

01
Ant-Arctica In-Cognita

The Discovery of Antarctica

1   Nobu Shirase, *Nankyokuki* (1913), quoted in translation in Elizabeth Kolbert and Francis Spufford, *The Ends of the Earth: An Anthology of the Finest Writing on the Arctic and the Antarctic* (New York: Bloomsbury, 2007), 55, 62.
2   Ibid.

Asymptotic Cartography

1   Derek Searle, "Horseshoe Diaries," UK Antarctic Heritage Trust, 2017.
2   *The South Polar Times,* "Ballooning in the Antarctic," vol. 3 (June 1902): 2.

Timeline
1   Ginger Butcher, "Fra Mauro's *Mappamundi* – Landsat Science," NASA, 10 January 2014.
2   Abraham Ortelius, *Thesaurus Geographicus* (Antwerp: Plantin Press, 1587).
3   Ivar Hamre, "The Japanese South Polar Expedition of 1911–1912: A Little-Known Episode in Antarctic Exploration," *The Geographical Journal* 82, no. 5 (November 1933): 411.
4   August Petermann, *Geographische Mitteilungen* (March 1855).
5   David Boyer, "Year of Discovery Opens in Antarctica," *The National Geographic Magazine* 112, no. 3 (September 1957): 339–82.
6   Richard Evelyn Byrd, "Exploring the Ice Age in Antarctica," *The National Geographic Magazine* 68, no. 4 (October 1935): 399 ff. The quote is from National Geographic, "Society's New Map Updates Antarctica," *The National Geographic Magazine* 112, no. 3 (September 1957): 380–81.

Climate Models. Antarctica and the Politics of Digital Representations

1   Ned Rossiter and Soenke Zehle, "The Aesthetics of Algorithmic Experience," in *The Routledge Companion to Art and Politics,* ed. R. Martin (New York: Routledge, 2015), 214.
2   Rossiter and Zehle, "The Aesthetics of Algorithmic Experience," 215.
3   Lisa Parks, "Between Orbit and the Ground: Conflict Monitoring, Google Earth and the 'Crisis in Darfur,'" in *Documentary Testimonies Global Archives of Suffering* (London, New York: Routledge, 2010), 260.

1   Jonathan Amos, "Climate Change: 'Stunning' Seafloor Ridges Record Antarctic Retreat," *BBC News,* 29 May 2020.

One Map, One Ocean. The Athelstan Spilhaus Projection

1   Matt Novak, "Sunday Funnies Blast Off into the Space Age," *Smithsonian,* 2012.
2   Athelstan Spilhaus, "To See the Oceans, Slice Up the Land," *Smithsonian,* November 1979, 54–63.
3   Spilhaus, "World Ocean Maps: The Proper Places to Interrupt," *Proceedings of the American Philosophical Society* 127, no. 1 (1983): 50–60.
4   Bojan Šavrič, David Burrows, and Melita Kennedy, "The Spilhaus World Ocean Map in a Square," n.d.

Constructing a Place through Literature. The Image of Antarctica

1   Joseph Conrad, *Lord Jim* (Oxford: Oxford University Press, 2002), 128.

Constructing a Place through Photography. In Wilderness

1   Thoreau, "Walking," *The Atlantic,* May 1862, 657–74.
2   Joan Myers, *Wondrous Cold: An Antarctic Journey* (Washington, DC: Smithsonian Books, 2006), 11.

Photography's Trace on Ice

1   Stephen J. Pyne, *The Ice: A Journey to Antarctica* (Iowa City: University of Iowa Press, 1986), 20.
2   Trevor Paglen pushes photography's documentary function to its limits to reveal covert architectures and what he calls "invisible images". See, for example, his *Blank Spots on the Map: The Dark Geography of the Pentagon's Secret World* (New York: Penguin, 2009).
3   See Werner Herzog's 2007 documentary *Encounters at the End of the World.*

Imagined, Recreated, and Rephotographed. Ice Blink

1   Anne Noble, *Ice Blink* (Auckland: Clouds, 2011).

02
Antarctic Pie. The Menacing Geometry of Power

Flags
1   Elena Glasberg, *Antarctica as Cultural Critique: The Gendered Politics of Scientific Exploration and Climate Change* (New York: Palgrave Macmillan, 2012).
2   Jorge Orta, "Antarctica 2000 Manifesto," 46th Venice Art Biennale Catalogue, 1995.

Piss Poles
1   Anne Noble, "The Expedition Photographer," in *The Last Road* (Auckland: Clouds, 2014): 25–28.

Political Conflict in the Antarctica Peninsula

1   Klaus Dodds, *Pink Ice: Britain and the South Atlantic Empire* (London: I. B. Tauris, 2002).
2   Adrian Howkins, *Frozen Empires: An Environmental History of the Antarctic Peninsula* (New York: Oxford University Press, 2017).
3   Howkins, *Frozen Empires.*
4   Ibid.
5   Pablo Fontana, *La Pugna Antártica, El Conflicto Por El Sexto Continente: 1939–1959* (Buenos Aires: Guazuvira Ediciones, 2014).
6   Adrian Howkins, "Reluctant Collaborators: Argentina and Chile in Antarctica during the IGY," *Journal of Historical Geography* 34 (2008): 596–617.
7   Christopher Joyner, *Eagle over the Ice: The U.S. in the Antarctic* (Hanover, NH: University Press of New England, 1997).
8   Howkins, *Frozen Empires.*
9   Walter Sullivan, *Assault on the Unknown: The International Geophysical Year* (New York: McGraw-Hill, 1961); Roger Launius, James Fleming, and David DeVorkin, *Globalizing Polar Science: Reconsidering the Social and Intellectual Implications of the International Polar and Geophysical Years* (New York: Palgrave, 2010).
10   Dodds, *Pink Ice.*
11   Howkins, "Melting Empires? Climate Change and Politics in Antarctica Since the International Geophysical Year," *Osiris* 26 (2011): 180–97.
12   Howkins, *Frozen Empires.*

1   Gabriel González Videla, *Memorias* (Santiago: Gabriela Mistral, 1975), 809.
2   Ibid., 826.
3   Luis Risopatrón, *La antártida americana* (Santiago: Imp. Cervantes, 1908).
4   Howkins, *Frozen Empires: An Environmental History of the Antarctic Peninsula* (New York: Oxford University Press, 2017), 11.
5   Ramón Cañas Montalva, "El valor geopolítico de la posición Antártica de Chile," *Revista Geográfica de Chile* 9 (June 1953).
6   Argentina, Presidencia de la Nación, *La Nación Argentina: Justa. Libre. Soberana* (Buenos Aires: Talleres Gráficos Peuser, 1949).

How to Claim a Continent. Operation Tabarin and the Post Office

1   Peter Rejcek, "Stamp on History," *The Antarctic Sun,* 15 October 2010

A Concise Military History of Antarctica

1   David Day, *Antarctica: A Biography* (Oxford: Oxford University Press, 2013), 383.
2   "Memorandum on Safeguarding U.S. National Interests in the Arctic and Antarctic Regions," White House Presidential Memorandum, 9 June 2020.
3   Ibid.

Strategic Interests in Antarctica. The Example of China

1   Guojia ziyuan xinxi zhongxin, *Quanchuan ziyuan yu guojia anquan* [Mineral Resources and National Security], 149.

2   Anne-Marie Brady, *The Emerging Politics of Antarctica* (Abingdon: Routledge, 2012).
3   Zhu Jiangang, Yan Qide, and Ling Xiaoliang, "Nanji ziyuan ji qi kaifa liyong qianjing fenxi" [An Analysis of Antarctic Resources, Their Exploitation and Potential Utilisation], *Zhongguo ruan kexue* 8 (2005): 18 and 20.
4   Yan Qide and Zhu Jiangang, *Nanjizhou ziyuan lingtu zhuquan yu ziyuan quan shu wenti yanjiu* [Research on Antarctic Sovereignty and Resources Rights] (Shanghai: Kexue jishu chubanshe, 2009), 183.
5   Ding Huang, *Jidi guojia zhengce yanjiu baogao 2012–2013* [Annual Report on National Polar Policy Research, 2012–2013] (Beijing: Kexue chubanshe, 2013), 79; and Ding Huang, *Jidi guojia zhengce yanjiu baogao 2013–2014* [Annual Report on National Polar Policy Research, 2013–2014] (Beijing: Kexue chubanshe, 2014), 138.
6   *Jidi zhanlue yanjiu dongtai* 3 (2013): 16.
7   "Nanji zhoubian haiyu yu dalu zonghe pinggu" [A Comprehensive Evaluation of Antarctic Waters] *Jidi zhuanying jianbao* 10 (January 2016): 3.
8   Commission for the Conservation of Antarctic Marine Living Resources, "Krill Fishery Report, 2018," 2018.
9   Xie Yu, "China Steps Up Efforts in Antarctica to Benefit from Krill Bonanza," *China Daily,* March 5, 2015.
10   Mark Godfrey, "Shen Lan Launches Antarctic Krill Shipping Vessel," *Seafood Source,* 20 May 2020.
11   Wu Yilin, "Guanyu fazhan Nanji luyouye de sikao" [Thoughts on Expanding Antarctic Tourism] *Zhongguo haiyang daxue xuebao, shehuixuebian* (2010): 5–8.
12   "Yinggai peizhi zhenzheng de Zhongguo Nanji luyou shichang" [We Should Foster the Development of a Truly Chinese Antarctic Tourism Market], *Zhongguo luyou bao,* 3 July 2014.

Antarctica as a Hybrid Warfare Environment?

1   Liu Yijian, *Zhi haiquan yu haijun zhanlüe* [The Command of the Sea and the Strategic Employment of Naval Forces] (Beijing: Jiefangjun guofang daxue chubanshe, 2004), 233.
2   Gerhard Beutler et al., "The International GPS Service (IGS): An Interdisciplinary Service in Support of Earth Sciences," *Advances in Space Research* 23, no. 4 (1999): 631–653.
3   National Marine Environmental Forecasting Center, "Guojia Nanji 'shiwu' nengli jianshe zhongdian xiangmu Nanji Changcheng zhan, Zhongshan zhan, dimian weixing jieshou chuli xitong" [National 15th Five-Year Plan Capacity Key Building Project to Install Satellite Ground Receiving and Processing Stations Successfully Completed], news release, 18 March 2010.
4   "Sinan Beidou daohang jieshouji biaoxian youyi" [Sinan BeiDou Navigation Receiver's Outstanding Performance], *Xi'an Beidou xun,* 17 May 2013.
5   Darya Kezina, "New GLONASS Stations to Appear in Antarctica," *Russia Beyond,* 24 April 2015.
6   James Bussert and Bruce Elleman, *People's Liberation Army Navy Combat Systems Technology, 1949–2010* (Annapolis, MD: Naval Institute Press, 2011), 161.
7   Prague Declaration on the Occasion of the Sixtieth Anniversary of the Antarctic Treaty, "recognising that it is in the interest of all humankind that Antarctica continue to be used exclusively for peaceful purposes."

03
The Governance System
of a Citizenless Continent

Science, Politics, and Governance in the Antarctic Treaty System. A Symbiotic Relationship

1   Aant Elzinga, "Antarctica: The Construction of a Continent by and for Science," in *Denationalizing Science: The Contexts of International Scientific Practice,* eds. Elisabeth Crawford, Terry Shinn, and Sverker Sörlin, 77 (Dordrecht/Boston/London: Kluwer Academic Publishers, 1992) (emphasis added).
2   This essay draws heavily from Karen Scott, "Scientific Rhetoric and Antarctic Security," in *Antarctic Security in the Twenty-First Century: Legal and Policy Perspectives,* eds. Alan Hemmings, Donald Rothwell, Karen Scott, 284–306 (London: Routledge, 2012).
3   See generally Elzinga, "The Continent for Science," in *Handbook on the Politics of Antarctica,* eds. Klaus Dodds, Alan Hemmings, Peder Roberts, 103–124 (Cheltenham: Edward Elgar, 2017).
4   See Klaus Dodds, "Post-colonial Antarctica: An Emerging Engagement," *Polar Record* 42 (2006): 59–70.
5   Scott, "Scientific Rhetoric," 284, quoting Elzinga, "Antarctica," 86.
6   Antarctic Treaty, 1 December 1959, United Nations Treaty Series 402, 71.
7   Klaus Dodds, "The Great Game in Antarctica: Britain and the 1959 Antarctic Treaty," *Contemporary British History* 21 (2007): 2.
8   Simon Nayler, Katrina Dean, and Martin Siegert, "The IGY and the Ice Sheet: Surveying Antarctica," *Journal of Historical Geography* 34 (2008): 591.
9   See Christy Collis and Klaus Dodds, "Assault on the Unknown: The Historical and Political Geographies of the International Geophysical Year (1957–58)," *Journal of Historical Geography* 34 (2008): 555–73.
10   See Peter Beck, *The International Politics of Antarctica* (London and Sydney: Croom Helm, 1986), 53.
11   Protocol to the Antarctic Treaty on Environmental Protection, 4 October 1991, *ILM* 30 (1991): 802.
12   1991 Environmental Protocol, Annex II, Articles 3(2)(a) and (c) and 3(5); Annex V, Article 7.
13   1991 Environmental Protocol, Annex V, Articles 3(2)(a) and (e).
14   1959 Antarctic Treaty, Article XI(2).
15   Richard Herr and Robert Hall, "Science and Currency and the Currency of Science," in *Antarctica: Policies and Policy Development,* ed. J. Handmer, 13–24 (Canberra: Centre for Resource and Environmental Studies, Australia National University, 1989).
16   Aant Elzinga and Ingemar Bohlin, "The Politics of Science in Polar Regions," in *Changing Trends in Antarctic Research,* ed. Aant Elzinga, 22 (Dordrecht/Boston/London: Kluwer Academic Publishers, 1993).
17   Scott, "Scientific Rhetoric," 287.
18   The "Question of Antarctica" was a regular United Nations General Assembly agenda item for over 20 years. See Moritaka Hayashi, "The Antarctica Question to the United Nations," *Cornell International Law Journal* 19 (1986): 275–290; Peter Beck, "Twenty Years On: The UN and the 'Question of Antarctica,'" *Polar Record* 40 (2004): 217–227.
19   Scott, "Scientific Rhetoric," 300–305. See also Scott, "Regulating Subglacial Aquatic Research under the Antarctic Treaty System," *New Zealand Universities Law Review* 23 (2008): 134–154.
20   See Sam Bateman, "Strategic Competition and Emerging Security Risks: Will Antarctica Remain Demilitarised?" in *Antarctic Security in the Twenty-First Century: Legal and Policy Perspectives,* eds. Alan Hemmings, Donald Rothwell, and Karen Scott, 116–134 (London: Routledge, 2012).
21   See Brady, "China, Russia Push GPS Rivals into Antarctica," *The Australian,* 6 September 2018.

The Precursor of the International Geophysical Year. The Norwegian-British-Swedish Antarctic Expedition

1   This essay has been adapted from the author's master's thesis: Christel Misund Domaas, *The Norwegian-British-Swedish Antarctic Expedition: Science and Politics* (Tromsø: UiT Norges arktiske universitet, 2019).
2   Robert Friedman, "Å spise kirsebær med de store," in *Norsk Polarhistorie: 2: Vitenskapene,* eds. Einar-Arne Drivenes, Harald Dag Jølle, Ketil Zachariassen and Norsk Polarhistorie: 334 (Oslo: Gyldendal, 2004).
3   Peder Roberts, *A Frozen Field of Dreams: Science, Strategy, and the Antarctic in Norway, Sweden, and the British Empire, 1912–1952,* PhD diss. Stanford, 2010, 316.
4   Friedman, "Å spise kirsebær med de store," 334; Domaas, *The Norwegian-British-Swedish Antarctic Expedition,* 19.
5   The Regional State Archives in Tromsø (SATø), Archive from Norges Svalbard- og Ishavsundersøkelser (NSIU) 1928–1948 and Norsk Polarinstitutt (NP) 1948–1985 (Tromsø: Norsk Polarinstitutt, 2004), Db 0194/12 10B/1 (Antarktisekspedisjonene – Industridept: planlegging, budsjett mv.).
6   SATø, Archive from NSIU and NP, Db 0199 10B/16 Antarktis (Antarktiskomiteen, Norsk geografisk selskap, forberedelser mv.): "For å avgjøre om den nåtidige klimaforandringer av regional eller universal karakter, er det av største betydning å få undersøkt breene i Antarktis."
7   John Schjelderup Giæver, *The White Desert: The Official Account of the Norwegian-British-Swedish Antarctic Expedition* (New York: E. P. Dutton & Co., 1955), 11.
8   Friedman, "Å spise kirsebær med de store," 336.
9   Ibid. 344; SATø, Archive from NSIU and NP, Db 0197 10B/13 Antarktis (Svenska Antarktiska Kommitten).
10   Roberts, *A Frozen Field of Dreams,* 303–05; SATø, Archive from NSIU and NP, Db 0199 10B/16 Antarktis (Antarktiskomiteen, Norsk geografisk selskap, forberedelser mv.), 1946–48.
11   Friedman, "Å spise kirsebær med de store," 405.
12   Sverker Sörlin, "Polare jubileer: kunnskap og politikk i nord," in *Ottar,* ed. Marit Anne Hauan, 12–18 (Tromsø: Tromsø Museum – Universitetsmuseet, 2011).
13   Roberts, *A Frozen Field of Dreams,* 348.
14   Anna Schytt, *Med känsla för is: om polarforskaren Valter Schytt och gåtorna hans Antarktisexpedition bidrog till att lösa* (Stockholm: Fri Tanke, 2018), 10; Roberts, *A Frozen Field of Dreams,* 332–49; Gordon Elliott Fogg, *A History of Antarctic Science* (Cambridge: Cambridge University Press, 1992), 168.

The Pivotal Role of the International Geophysical Year (IGY) 1957–1958

1   Elizabeth Howell, "Sputnik: The Space Race's Opening Shot," space.com, 22 August 2018.
2   *The Illustrated London News,* 27 July 1957.
3   James Van Allen, "Genesis of the International Geophysical Year," *Eos Trans. AGU* 64, no. 50 (1983): 977.

The Scientific Committee on Antarctic Research

1   "Research (SCAR)," International Science Council, https://council.science/what-we-do/research-programmes/thematic-organizations/scientific-committee-on-antarctic-research-scar/.
2   Harry Wexler, "Antarctic Research During the International Geophysical Year", *Antarctica in the International Geophysical Year: Based on a Symposium on the Antarctic,* Volume 1, Geophysical

Monograph series, vol. 1, ed. Albert P. Crary et al. (Washington, DC: American Geophysical Union, 1956), 7–12.
3   "Antarctic Treaty," Secretariat of the Antarctic Treaty, https://www.ats.aq/index_e.html.
4   Colin Summerhayes, "International Collaboration in Antarctica: The International Polar Years, the International Geophysical Year, and the Scientific Committee on Antarctic Research," *Polar Record* 44, no. 4 (2008): 321–334.
5   "Welcome to the Council of Managers of National Antarctic Programs (COMNAP) Website," Council
of Managers of National Antarctic Programs, https://www.comnap.aq/.
6   Chester Langway, *The History of Early Polar Ice Cores,* US Army Corps of Engineers, Cold Regions Research and Engineering Laboratory, Report TR-08-1, 2008, 47.
7   Nikolai Vasiliev et al., "Deep Drilling at Vostok Station, Antarctica: History and Recent Events," *Annals of Glaciology* 47 (2007): 10–23.
8   David Elliot et al., "Triassic Tetrapods from Antarctica: Evidence for Continental Drift," *Science* 169 (1970): 1197–1201.
9   Inigo Everson, "Antarctic Fisheries," *Polar Record* 19 (1978): 233–51.
10   "Specially Protected and Manged Areas in Antarctica," Antarctic Environments Portal, https://environments.aq/publications/specially-protected-and-managed-areas-in-antarctica/.
11   Joseph Farman, Brian Gardiner, and Jonathan Shanklin, "Large Losses of Total Ozone in Antarctica Reveal Seasonal ClO_x/NO_x Interaction," *Nature* 315 (1985): 207–210.
12   David Walton and Peter Clarkson, *Science in the Snow: Fifty Years of International Collaboration through the Scientific Committee on Antarctic Research* (Cambridge: Scientific Committee on Antarctic Research, 2011), 258.
13   Fabio Florindo and Martin Siegert, *Antarctic Climate Evolution: Developments in Earth and Environmental Sciences,* Volume 8 (Amsterdam: Elsevier Science & Technology, 2008), 606.
14   https://www.environments.aq/information-summaries/antarctic-subglacial-lakes/.
15   "Antarctic Climate Change and the Environment," Scientific Committee on Antarctic Research, https://www.scar.org/library/scar-publications/occasional-publications/3508-antarctic-climate-change-and-the-environment-1/.
16   Ian Allison et al., "International Polar Year 2007–2008," *World Meteorological Organization Bulletin* 56, no. 4 (2007).
17   Association of Polar Early Career Scientists, https://www.apecs.is/.
18   EPICA Community Members, "Eight Glacial Cycles from an Antarctic Ice Core," 623–28.
19   Timothy Naish et al., "Obliquity-Paced Pliocene West Antarctic Ice Sheet Oscillations," *Nature* 458 (2009): 322–28; David Pollard and Robert DeConto, "Modelling West Antarctic Ice Sheet Growth and Collapse through the Past Five Million Years," *Nature* 458 (2009): 329–32; Richard Levy et al., "Antarctic Ice Sheet Sensitivity to Atmospheric CO_2 Variations in the Early to Mid-Miocene," *Proceedings of the National Academy of Sciences* 113 (2016): 3453–58.

Facilitating International Collaboration in Antarctic Science. The Role of COMNAP

1   Council of Managers of National Antarctic Programs, "COMNAP Constitution," 4 July 2008. [online].
2   Dudeney, John R., and David W. H. Walton, "Leadership in Politics and Science within the Antarctic Treaty," *Polar Research* 31 (2012): 11075.
3   Council of Managers of National Antarctic Programs, *Antarctic Station Catalogue* (Christchurch: COMNAP, 2017). comnap.aq. [online].
4   Scientific Committee on Antarctic Research, "Horizon Scan," scar.org. [online].

5   Council of Managers of National Antarctic Programs, "Antarctic Roadmap Challenges Project," comnap.aq. [online].
1   Alan Hemmings, Donald Rothwell, and Karen N. Scott, eds., *Antarctic Security in the Twenty-First Century* (London: Routledge, 2013).
2   Council of Managers of National Antarctic Programs, "Welcome to the Council of Managers of National Antarctic Programs (COMNAP) Website," 2020.

Antarctic Law Enforcement

1   This essay draws heavily from Donald Rothwell, "Law Enforcement in Antarctica," in *Antarctic Security in the Twenty-First Century: Legal and Policy Perspectives,* ed. by Alan Hemmings, Donald Rothwell, and Karen Scott (London: Routledge, 2012), 135–53.
2   CGTN, "Antarctica: An Increasingly Popular Destination for Chinese Tourists," 1 November 2019.
3   See Antarctic Treaty (Environmental Protection) Act 1980 (Australia), ss. 19A, 19B.
4   See Australian Antarctic Territory Act 1954 (Australia), s. 6.
5   See generally James Crawford, *Brownlie's Principles of Public International Law* (Oxford: Oxford University Press, 2019), chapter 5.
6   See e.g. Secretariat of the Antarctic Treaty, "Visitor Site Guidelines," undated, available online at https://www.ats.aq/devAS/Ats/VisitorSiteGuidelines?lang=e.
7   Antarctic Treaty, 1 December 1959, 402 UNTS 71.
8   Antarctic Treaty, Article VIII(1).
9   Convention on the Conservation of Antarctic Marine Living Resources, 20 May 1980, 1329 UNTS 47 (hereinafter CCAMLR).
10   Protocol on Environmental Protection to the Antarctic Treaty, 4 October 1991, 30 ILM 1461 (hereinafter Madrid Protocol).
11   See Kees Bastmeijer, *The Antarctic Environmental Protocol and Its Domestic Legal Implementation* (The Hague/London/New York: Kluwer Law International, 2003).
12   Donald Rothwell and Tim Stephens, "Illegal Southern Ocean Fishing and Prompt Release: Balancing Coastal and Flag State Rights and Interests," *International and Comparative Law Quarterly* 53 (2004): 171–87.
13   Davis, "The Whaling Dispute in the South Pacific: An Australian Perspective," *Journal of East Asia and International Law* 4, no. 2 (2011): 419–47.

1   Bryant Rousseau, "Cold Cases: Crime and Punishment in Antarctica," *The New York Times,* 29 September 2016.

04

Antarctic Resources. Temptations and Accountability

Antarctic Resources and the Protocol for Environmental Protection

1   Doaa Abdel-Motaal, *Antarctica: The Battle for the Seventh Continent* (Denver: Praeger, 2016), 8.
2   Robert Clancy, John Manning, and Henk Brolsma, *Mapping Antarctica: A Five-Hundred-Year Record of Discovery* (Chichester: Praxis Publishing, 2014).
3   Chapter 4, "Potential Mineral Resources in Antarctica," in United States Congress, Office of Technology Assessment, *Polar Prospects: A Minerals Treaty for Antarctica,* 96.
4   Article 7 of the Protocol on Environmental Protection to the Antarctic Treaty, in *Compilation of Key Documents of the Antarctic Treaty System,* 39.

5   US Congress, Office of Technology Assessment, *Polar Prospects,* 8–9.
6   Maarten de Wit, *Minerals and Mining in Antarctica: Science and Technology, Economics and Politics* (Oxford: Clarendon Press, 1985), 17.
7   US Congress, Office of Technology Assessment, *Polar Prospects,* 122.
8   Tore Gjelsvik, "The Mineral Resources of Antarctica: Progress in Their Identification," in *Antarctic Resources Policy: Scientific, Legal, and Political Issues,* ed. Francisco Orrego Vicuña, 62–63 (Cambridge: Cambridge University Press, 1983).
9   Egil Bergsager, "Basic Conditions for the Exploration and Exploitation of Mineral Resources in Antarctica: Options and Precedents," *Polar Record* 22, no. 136 (January 1984): 25–49; and de Wit, *Minerals and Mining in Antarctica.*
10   Abdel-Motaal, *Antarctica,* 20–25.
11   Between 1904 and 1940, these claims were laid by Australia, which claimed the lion's share of the continent, France, New Zealand, Chile, the United Kingdom, Argentina and Norway.
12   Articles I and IV of the Antarctic Treaty and Article 7 of the Environmental Protocol, Secretariat of the Antarctic Treaty, 21, 22 and 39.
13   Council of Managers of National Antarctic Programs' Antarctic Facility List, accessed February 3, 2020, https://www.arcgis.com/home/item.html?id=77e95a444681402f94011ad8756c13ac.
14   Abdel-Motaal, *Antarctica,* 89–94.
15   On 7 February 2020, a temperature of 18.3° C was recorded at Esperanza Station on the northern tip of the Peninsula – the highest temperature ever recorded in Antarctica; Ibid., 2.
16   Alan Hemmings, "Why Did We Get an International Space Station Before an International Antarctic Station?" *The Polar Journal* 1, no. 1 (2011): 5–7.
17   Australian Antarctic Data Centre, *The SCAR Composite Gazetteer,* accessed 3 February 2020, https://data.aad.gov.au/aadc/gaz/scar/.
18   Chaturvedi, "India and Antarctica: Towards Post-Colonial Engagement," in *The Emerging Politics of Antarctica,* ed. Anne-Marie Brady, 50–74 (London: Routledge, 2013).
19   Abdel-Motaal, *Antarctica,* 201–61.
20   Abdel-Motaal, *Antarctica,* 201–61. Under Article VIII of the Antarctic Treaty, a person in Antarctica is only subject to the penal jurisdiction of the country of which he is a national; see Secretariat of the Antarctic Treaty, 24. This creates a jurisdictional vacuum that prevents countries from dealing with the transgressions of third parties.
21   Justine Shaw et al., "Antarctica's Protected Areas Are Inadequate, Unrepresentative, and at Risk," *PLOS Biology* 12, no. 6 (17 June 2014).
22   Abdel-Motaal, *Antarctica,* 3.

1   Cool Antarctica, "Human Impacts on Antarctica and Threats to the Environment: Mining and Oil," 2020; David Boyer, "Year of Discovery Opens in Antarctica," *The National Geographic Magazine* 112, no. 3 (September 1957): 354; David Tyree, "New Era in the Loneliest Continent," *National Geographic* 123, no. 2 (February 1963): 288.
2   Leslie Hook and Benedict Mander, "The Fight to Own Antarctica," *Financial Times,* 23 May 2018; Tyree, "New Era," 288.

The Role and Impact of Antarctic Specially Protected Areas and Antarctic Specially Managed Areas

1   Ronald Smith, "Terrestrial Biology of the Antarctic and Sub-Antarctic," in *Antarctic Ecology,* ed. Richard Laws, 61–162 (London: Academic Press, 1984); Peter Convey, "Antarctic Ecosystems," *Encyclopedia of Biodiversity* 1 (2017): 179–87.
2   De Broyer et al., *Biogeographic Atlas of the Southern Ocean* (Cambridge: Scientific Committee on Antarctic Research, 2014), xii, 498.
3   Nicole Bender, Kim Crosbie and Heather Lynch, "Patterns of Tourism in the Antarctic Peninsula

Region: A 20-Year Analysis," *Antarctic Science* 28, no. 3 (2016): 194–203; International Association of Antarctica Tour Operators, *IAATO Overview of Antarctic Tourism: 2018–19 Season and Preliminary Estimates for 2019–20 Season,* Information Paper 140, Antarctic Treaty Consultative Meeting XLII, Prague, 27–28 June 2019.

4   Kevin Hughes, Luis Pertierra, and David Walton, "Area Protection in Antarctica: How Can Conservation and Scientific Research Goals Be Managed Compatibly?" *Environmental Science and Policy* 31 (2013): 120–32.

5   Antarctic Treaty Secretariat, *Area Protection and Management* (2020).

6   Justine Shaw et al., "Antarctica's Protected Areas Are Inadequate, Unrepresentative and at Risk," *PLOS Biology* 12, no. 6 (17 June 2014): e1001888; Kevin Hughes et al., "Assessing the Effectiveness of Specially Protected Areas for Conservation of Antarctica's Botanical Diversity," *Conservation Biology* 30 (2016): 113–20.

7   Aleks Terauds and Jasmine Lee, "Antarctic Biogeography Revisited: Updating the Antarctic Conservation Biogeographic Regions," *Diversity and Distributions* 22 (2016): 836–40.

8   Hannah Wauchope, Justine Shaw, and Aleks Terauds, "A Snapshot of Biodiversity Protection in Antarctica," *Nature Communications* 10 (2019): article no. 946.

9   Antarctic Treaty Secretariat, *Area Protection and Management* (2020).

10   Kevin Hughes and Susie Grant, "Current Logistical Capacity Is Sufficient to Deliver the Implementation and Management of a Representative Antarctic Protected Area System," *Polar Research* 37, no. 1 (2018): 1521686.

11   Co-Conveners' Report of the Joint SCAR/CEP Workshop on Further Developing the Antarctic Protected Area System, Prague, 27–28 June 2019.

12   Christina Braun et al., "Environmental Monitoring and Management Proposals for the Fildes Region, King George Island, Antarctica," *Polar Research* 31 (2012): 18206; Anne-Marie Brady, *China as a Great Polar Power* (Cambridge: Cambridge University Press, 2017).

13   Klaus Dodds, Alan Hemmings, and Peder Roberts, *Handbook on the Politics of Antarctica* (Cheltenham: Edward Elgar, 2017).

Antarctic Biodiversity

1   Andrew Clarke, David Barnes, and Dominic Hodgson, "How Isolated Is Antarctica?," *Trends in Ecology and Evolution* 20 (2005): 1–3; Convey et al., "Ice-Bound Antarctica: Biotic Consequences of the Shift from a Temperate to a Polar Climate," in *Mountains, Climate, and Biodiversity,* ed. Carina Hoorn, A. Perrigo, and A. Antonelli (Hoboken, NJ: Wiley, 2018); Lloyd Peck, "Antarctic Marine Biodiversity: Adaptations, Environments and Responses to Change," *Oceanography and Marine Biology Annual Review* 56 (2018): 105–236.

2   Lloyd Peck, Peter Convey, and D. K. A. Barnes, "Environmental Constraints on Life Histories in Antarctic Ecosystems: Tempos, Timings and Predictability," *Biological Reviews* 81 (2006): 75–109; Peter Convey and Lloyd Peck, "Antarctic Environmental Change and Biological Responses," *Science Advances* 5, no. 11 (2019).

3   Peter Convey et al., "The Spatial Structure of Antarctic Biodiversity," *Ecological Monographs* 84 (2014): 203–44.

4   Steven Chown et al., "The Changing Form of Antarctic Biodiversity," *Nature* 522 (2015): 431–38; Peter Convey, "Antarctic Ecosystems," *Encyclopedia of Biodiversity* 1 (2017): 179–87.

5   Daniel Aguirre de Cárcer et al., "Biodiversity and Distribution of Polar Freshwater DNA Viruses," *Science Advances* 1 (2015): e1400127; Ricardo Cavicchioli, "Microbial Ecology of Antarctic Aquatic Systems," *Nature Reviews in Microbiology* 13 (2015): 691–706.

6   Andrew Clarke and Nadine M. Johnston, "Antarctic Marine Benthic Diversity," *Oceanography and Marine Biology Annual Review* 41 (2003): 47–114; Broyer et al., *Biogeographic Atlas of the Southern Ocean* (Cambridge: Scientific Committee on Antarctic Research, 2014), xii, 498.

7   Jana Dömel et al., "Nuclear and Mitochondrial Gene Data Support Recent Radiation within the Sea Spider Species Complex *Pallenopsis patagonica,*" *Frontiers in Ecology and Evolution* 4 (2017): 9139.

8   Philip J. A. Pugh and Peter Convey, "Surviving Out in the Cold: Antarctic Endemic Invertebrates and Their Refugia," *Journal of Biogeography* 35 (2008): 2176–86; Peter Convey, "Biogeography and Regional Classifications of Antarctica," in *Antarctica: An Extreme Environment in a Changing World,* ed. Alex D. Rogers et al., Chapter 15, 471–91 (Oxford: Blackwell, 2012).

9   Bernard Coetzee et al., "Expanding the Protected Area Network in Antarctica Is Urgent and Readily Achievable," *Conservation Letters* 10 (2017): 670–80.

10   Peter Convey, "Current Changes in Antarctic Ecosystems," in *Encyclopedia of the World's Biomes,* ed. Michael Goldstein and Dominick DellaSala (Elsevier, 2020; in press); Peter Convey, A. Brandt, and S. Nicols, "Life in a Cold Environment," in *Antarctica: Global Science from a Frozen Continent,* ed. David W. H. Walton, 161–210 (Cambridge: Cambridge University Press, 2013); Convey et al., "Ice-Bound Antarctica."

11   Stef Bokhorst, Peter Convey, and Rien Aerts, "Nitrogen Inputs by Marine Vertebrates Drive Abundance and Richness in Antarctic Terrestrial Ecosystems," *Current Biology* 29 (2019).

12   Convey, "Antarctic Ecosystems."

13   Convey et al., "Antarctic Terrestrial Life: Challenging the History of the Frozen Continent?," *Biological Reviews* 83 (2008): 103–17; Convey et al., "Ice-Bound Antarctica."

14   Convey and Peck, "Antarctic Environmental Change and Biological Responses"; Huw Griffiths and Catherine Waller, "The First Comprehensive Description of the Biodiversity and Biogeography of Antarctic and Sub-Antarctic Intertidal Communities," *Journal of Biogeography* 43, no. 6 (10 February 2016): 1143–55.

15   Joseph Eastman and A. R. McCune, "Fishes on the Antarctic Continental Shelf: Evolution of a Marine Species Flock?," *Journal of Fish Biology* 57 (2000): 84–102.

16   Peck, "Antarctic Marine Biodiversity"; Convey and Peck, "Antarctic Environmental Change."

17   Andrew Clarke and J. Alistair Crame, "Evolutionary Dynamics at High Latitudes: Speciation and Extinction in Polar Marine Faunas," *Philosophical Transactions of the Royal Society Series B* 365 (2010): 3655–66.

18   Convey, "Antarctic Ecosystems"; Aleks Terauds et al., "Conservation Biogeography of the Antarctic," *Diversity and Distributions* 18 (2012): 726–41; Aleks Terauds and Jasmine Lee, "Antarctic Biogeography Revisited: Updating the Antarctic Conservation Biogeographic Regions," *Diversity and Distributions* 22 (2016): 836–40.

19   See also Steven Chown et al., "The Changing Form of Antarctic Biodiversity" *Nature* 522 (2015): 431–38; Steven Chown and Peter Convey, "Antarctic Entomology," *Annual Reviews of Entomology* 61 (2016): 119–37.

20   Steven Chown et al., "Antarctica and the Strategic Plan for Biodiversity," *PLOS Biology* 15 (2017): e2001656; Convey and Peck, "Antarctic Environmental Change"; Jes Bartlett et al., "Aliens in Antarctica," *The Biologist* 66 (2019): 22–25; Kevin Hughes et al., "Human-Mediated Dispersal of Terrestrial Species between Antarctic Biogeographic Regions: A Preliminary Risk Assessment," *Journal of Environmental Management* 232 (2019): 73–89; Kevin Hughes et al., "Invasive Non-Native Species Likely to Threaten Biodiversity and Ecosystems in the Antarctic Peninsula Region," *Global Change Biology* 26 (2020); A. McCarthy et al., "Antarctica:

The Final Frontier for Marine Biological Invasions," *Global Change Biology* 25 (2019): 1–21.

21   For example, Shaw et al., "Antarctica's Protected Areas Are Inadequate, Unrepresentative, and at Risk," *PLOS Biology* 12, no. 6 (17 June 2014); Hughes et al., "Assessing the Effectiveness of Specially Protected Areas for Conservation of Antarctica's Botanical Diversity," *Conservation Biology* 30 (2015): 113–120.

1   Bruno David and Thomas Saucède, *Biodiversity of the Southern Ocean* (Amsterdam: Elsevier, 2015), 61.

2   David and Saucède, *Biodiversity of the Southern Ocean,* 61.

3   Herbert George Ponting and Robert Falcon Scott, *The Great White South: Being an Account of Experiences with Captain Scott's South Pole Expedition and of the Nature Life of the Antarctic* (London: G. Duckworth & Co., 1921), xxvi, 305.

4   Dan Johnson, "Antarctic Treasure: The Underwater Images of Norbert Wu," National Science Foundation, 9 July 2004.

Fishing in the Southern Ocean and the Role of CCAMLR

1   See, for example, Stephen Rintoul et al., "Choosing the Future of Antarctica," *Nature* 558 (2018): 233–41.

2   Claude De Broyer et al., *Biogeographic Atlas of the Southern Ocean* (Cambridge: Scientific Committee on Antarctic Research, 2014), xii, 498.

3   B. S. Halpern et al., "A Global Map of Human Impact on Marine Ecosystems," *Science* 319, no. 5865 (2008): 948–52.

4   Cassandra Brooks et al., "Watch over Antarctic Waters," *Nature* 558 (2018): 177–80.

5   R. Hofman, "Historical Overview of Sealing, Whaling and the Krill Fishery in the Southern Ocean," *Polar Record* 53, no.1 (2017): 88–99.

6   Intergovernmental Panel on Climate Change (IPCC), "Special Report on the Ocean and Cryosphere in a Changing Climate" (Geneva: IPCC, 2019), https://www.ipcc.ch/report/srocc/.

7   See, for example, Rintoul et al., "Choosing the Future of Antarctica," 233–41; Angus Atkinson et al., "Krill (*Euphausia superba*) Distribution Contracts Southward during Rapid Regional Warming," *Nature Climate Change* 9, no. 2 (2019): 142–47.

8   CCAMLR, "Report of the Thirty-Eighth Meeting of the Commission," para. 5.7–5.8.

9   Emily Grilly et al., "The Price of Fish: A Global Trade Analysis of Patagonian (*Dissostichus eleginoides*) and Antarctic Toothfish (*Dissostichus mawsoni*)," *Marine Policy* 60 (2015): 186–96.

10   Peter Abrams, "How Precautionary Is the Policy Governing the Ross Sea Antarctic Toothfish (*Dissostichus mawsoni*) Fishery?" *Antarctic Science* 26, no. 1 (2014): 3–14; P. A. Abrams, "Necessary Elements of Precautionary Management: Implications for the Antarctic Toothfish," *Fish and Fisheries* (2016): 1–23.

11   See, for example, Caihong Fu et al., "Risky Business: The Combined Effects of Fishing and Changes in Primary Productivity on Fish Communities," *Ecological Modelling* 368 (2017): 265–76; Benjamin Halpern et al., "Spatial and Temporal Changes in Cumulative Human Impacts on the World's Ocean," *Nature Communications* 6, no. 7615 (2015); Malin Pinsky and N. J. Mantua, "Emerging Adaptation Approaches for Climate-Ready Fisheries Management," *Oceanography* 27, no. 4 (2014): 146–59.

12   See, for example, Callum Roberts et al., "Marine Reserves Can Mitigate and Promote Adaptation to Climate Change," *Proceedings of the National Academy of Sciences* 114, no. 24 (2017): 6167–75; Andrew Olds et al., "Marine Reserves Help Coastal Ecosystems Cope with Extreme Weather," *Global Change Biology* 20, no. 10 (2014): 3050–58; Lewis Barnett and M. L. Baskett, "Marine Reserves Can Enhance Ecological Resilience," *Ecology Letters* 18, no. 12 (2015): 1301–10.

13   Cassandra Brooks et al., "Reaching Consensus for Conserving the Global Commons: The Case of the Ross Sea, Antarctica," *Conservation Letters* (2019).
14   Brooks, "Reaching Consensus."
15   Cassandra Brooks et al., "Progress Towards a Representative Network of Southern Ocean Protected Areas," *PLOS ONE* 15, no. 4 (2020): e0231361.
16   See, for example, Rosemary Rayfuse, "Climate Change and Antarctic Fisheries: Ecosystem Management in CCAMLR," *Ecology Law Quarterly* 45, no. 53 (2018); CCAMLR, "Report of the Thirty-Eighth Meeting of the Commission"; CCAMLR, "Report of the Thirty-Seventh Meeting of the Commission"; Cassandra Brooks, "Cracks in the Future of the Antarctic," *National Geographic Blogs,* 2018.
17   Cassandra Brooks and D. Ainley, "Fishing the Bottom of the Earth: The Political Challenges of Ecosystem-Based Management," in *Handbook on Antarctic Politics,* ed. Klaus Dodds, Alan Hemmings, and Peder Roberts, 422–38 (Cheltenham: Edward Elgar Publishing, 2017); Jennifer Jacquet et al., "'Rational Use' in Antarctic Waters," *Marine Policy* 63 (2016): 28–34.
18   Klaus Dodds and M. Nuttall, *The Scramble for the Poles* (Cambridge: Polity Press, 2016).
19   Cassandra Brooks, "Competing Values on the Antarctic High Seas: CCAMLR and the Challenge of Marine-Protected Areas," *The Polar Journal* 3, no. 2 (2013): 277–300.
20   See, for example, Brooks et al., "Reaching Consensus"; Brooks et al., "Science-Based Management in Decline in the Southern Ocean," *Science* 354, no. 6309 (2016): 185–87.
1   Sidney Holt, "Sharing the Catches of Whales in the Southern Hemisphere," in *Case Studies on the Allocation of Transferable Quota Rights in Fisheries,* FAO Fisheries Technical Paper 411, ed. Ross Shotton (Rome: FAO, 2001).
2   Simon Denyer and Akiko Kashiwagi, "Japan to Leave International Whaling Commission, Resume Commercial Hunting," *The Washington Post,* 26 December 2018.

Marine Protected Areas

1   Rogers et al., "Antarctic Futures: An Assessment of Climate-Driven Changes in Ecosystem Structure, Function, and Service Provisioning in the Southern Ocean," 87–120.
2   Watters et al., "Long-Term Observations from Antarctica Demonstrate That Mismatched Scales of Fisheries Management and Predator-Prey Interaction Lead to Erroneous Conclusions about Precaution."

Spatial Exceptions at Sea

1   "Explore the World's Marine Protected Areas," in *Atlas of Marine Protection,* accessed 1 March 2020, http://mpatlas.org/.
2   CCAMLR, Conservation Measure 91-03 (2009): Protection of the South Orkney Islands Southern Shelf, 2009, accessed 1 March 2020, https://www.ccamlr.org/en/measure-91-03-2009.
3   Philip Trathan and S. M. Grant, "Chapter 4 – The South Orkney Islands Southern Shelf Marine Protected Area: Towards the Establishment of Marine Spatial Protection within International Waters in the Southern Ocean," in *Marine Protected Areas: Science, Policy and Management,* ed. by John Humphreys and Robert Clark, 67–98 (Elsevier, 2020).
4   Walker Smith et al., "The Oceanography and Ecology of the Ross Sea," *Annual Review of Marine Science* 6 (2014): 469–87.
5   Smith et al., "The Oceanography."
6   Sarah Purkey and Gregory Johnson, "Antarctic Bottom Water Warming and Freshening Contributions to Sea Level Rise, Ocean Freshwater Budgets,

and Global Heat Gain," *Journal of Climate* 26, no. 16 (2013): 6105–22.
7   Nengye Liu and Cassandra Brooks, "China's Changing Position towards Marine Protected Areas in the Southern Ocean: Implications for Future Antarctic Governance," *Marine Policy* 94 (August 2018): 189–95.
8   Liu and Brooks, "China's Changing Position."
9   CCAMLR, Conservation Measure 91-05 (2016): Ross Sea Region Marine Protected Area, 2016, accessed 1 March 2020, https://www.ccamlr.org/en/measure-91-05-2016.
1   Commission for the Conservation of Antarctic Marine Living Resources, "Marine Protected Areas (MPAs)," n.d.

The Krill Market

1   "What Are Phytoplankton?", earthobservatory.nasa.gov, 2020, https://earthobservatory.nasa.gov/features/Phytoplankton/page2.php.
2   Sue Armstrong et al., comments on the article "Licence to Krill – Greenpeace International", 2018; Simeon Hill et al., "A Foodweb Model to Explore Uncertainties in the South Georgia Shelf Pelagic Ecosystem," *Deep Sea Research Part II: Topical Studies in Oceanography* 59–60 (2012): 237–52.
3   Virginia Gascón and Rodolfo Werner, "Antarctic Krill: A Case Study on the Ecosystem Implications of Fishing. An Article Prepared for the Lighthouse Foundation," 2006.
4   "Our Exclusive Eco-Harvesting Method," Aker BioMarine, 2020, https://video.akerbiomarine.com/our-exclusive-eco-harvesting-method.
5   Eric Ziegelmayer, "Capitalist Impact on Krill in Area 48 (Antarctica)," *Capitalism Nature Socialism* 25, no. 4 (2014): 36–53.
6   Ziegelmayer, "Capitalist Impact on Krill."
7   Gascón and Werner, "Antarctic Krill."
8   John Phyne, "A Comparative Political Economy of Rural Capitalism," *Acta Sociologica* 53, no. 2 (2010): 160–180.
9   "Statistical Bulletin | CCAMLR," ccamlr.org, 2020, https://www.ccamlr.org/en/publications/statistical-bulletin.
10   "Environmental Protocol | Antarctic Treaty," ats.aq, 2020, https://www.ats.aq/e/protocol.html.
11   Margaret Gerber et al., comments on "30×30: A Blueprint for Ocean Protection," *Greenpeace International,* 2020.
12   Callum Roberts and Julie Hawkins, "Effects of Marine Reserves on Adjacent Fisheries," *Science* 294 (December 2001); Fiona Gell and Callum Roberts, "Benefits Beyond Boundaries: The Fishery Effects of Marine Reserves," *Trends in Ecology & Evolution* 18, no. 9 (2003): 448–55.
1   Louis Harkell, "Aker BioMarine's New $118M Vessel Set for Next Antarctic Krill Fishing Season," *Undercurrent News,* 5 November 2018.
2   Aker BioMarine, "Aker BioMarine – Home," n.d.

The Emerging Issue of Plastic Pollution in Antarctica

1   Ceridwen Fraser et al., "Antarctica's Ecological Isolation Will Be Broken by Storm-Driven Dispersal and Warming," *Nature Climate Change* 8 (2018): 704–8.
2   Anna Kelly et al., "Microplastic Contamination in East Antarctic Sea Ice," *Marine Pollution Bulletin* 154 (2020).
3   Ilka Peeken et al., "Arctic Sea Ice Is an Important Temporal Sink and Means of Transport for Microplastic," *Nature Communications* 9 (2018).
4   SCAR: Scientific Committee on Antarctic Research, "Plastic in Polar Environments (Plastic AG)," scar.org [online].
5   Elisa Bergami et al., "Polystyrene Nanoparticles Affect the Innate Immune System of the Antarctic Sea Urchin *Sterechinus neumayeri*," *Polar Biology* 42, no. 4 (2019): 743–57.

6   Amanda Dawson et al., "Turning Microplastics into Nanoplastics through Digestive Fragmentation by Antarctic Krill," *Nature Communications* 9 (2018).
7   Elisa Bergami et al., "Nanoplastics Affect Moulting and Faecal Pellet Sinking in Antarctic Krill (*Euphausia superba*) Juveniles," *Environment International* 143 (October 2020).
8   Elisa Bergami et al., "Plastics Everywhere: First Evidence of Polystyrene Fragments inside the Common Antarctic Collembolan *Cryptopygus antarcticus*," *Biology Letters* 16, no. 6 (June 2020).
9   Pasqualina Laganà et al., "Do Plastics Serve as a Possible Vector for the Spread of Antibiotic Resistance? First Insights from Bacteria Associated to a Polystyrene Piece from King George Island (Antarctica)," *International Journal of Hygiene and Environmental Health* 222, no. 1 (January 2019): 89–100.

Bioprospecting in Antarctica

1   Alex Rogers, "Evolution and Biodiversity of Antarctic Organisms: A Molecular Perspective," *Philosophical Transactions of the Royal Society B* 362 (2007): 2191.
2   Protocol on Environmental Protection to the Antarctic Treaty, adopted 1991, in force 1998, Article 1(e). Antarctic Treaty Secretariat, *Key Documents,* accessed 8 February 2020, https://ats.aq/e/key-documents.html.
3   United Nations Convention on the Law of the Sea (LOSC), adopted 1982, in force 1994, available online from https://www.un.org/depts/los/convention_agreements/texts/unclos/unclos_e.pdf (accessed 8 February 2020); Convention on Biological Diversity (CBD), adopted 1992, in force 1983, available online at https://www.cbd.int/doc/legal/cbd-un-en.pdf (accessed 8 February 2020).
4   Julia Jabour, "Biological Prospecting: The Ethics of Exclusive Reward from Antarctic Activities," *Ethics in Science and Environmental Politics* 10 (2010): 21.
5   Four distinct phases of bioprospecting, of which sample collection is the first, are outlined in Julia Jabour-Green and Diane Nicol, "Bioprospecting in Areas outside National Jurisdiction: Antarctica and the Southern Ocean," *Melbourne Journal of International Law* 4, no. 1 (2003): 85–87.
6   Antarctic Treaty Secretariat, *Final Report of the Fortieth Antarctic Treaty Consultative Meeting* (Beijing, 2017), paragraph 173, accessed 9 February 2020, https://www.ats.aq/devAS/Info/FinalReports?lang=e.
7   René Lefeber, "Marine Scientific Research in the Antarctic Treaty System," in *Law of the Sea and the Polar Regions: Interactions between Global and Regional Regimes,* ed. Donald Rothwell, Alex Oude Elferink and Erik Molenaar, 341 (Leiden: Martinus Nijhoff, 2013).
1   Dagmar Lohan and Sam Johnston, *Bioprospecting in Antarctica* (Yokohama: United Nations University Institute of Advanced Studies, 2005).

Antarctic Tourism

1   Robert Headland, "Historical Development of Antarctic Tourism," *Annals of Tourism Research* 21, no. 2 (1994): 271; Robert Headland, *Chronological List of Antarctic Expeditions and Related Historical Events, Choice Reviews Online* (2005):16.
2   Daniela Liggett et al., "From Frozen Continent to Tourism Hotspot? Five Decades of Antarctic Tourism Development and Management, and a Glimpse into the Future," *Tourism Management* 32, no. 2 (2011): 357–66.
3   Liggett et al., "From Frozen Continent to Tourism Hotspot?"

4 Republic of Liberia, Bureau of Maritime Affairs, *Report of Investigation in the Matter of the Sinking of Passenger Vessel* Explorer *(O.N. 8485) 23 November 2007 in the Bransfield Strait near the South Shetland Islands* (2009), Monrovia, Liberia.
5 Daniela Liggett and Emma Stewart, "The Changing Face of Political Engagement in Antarctic Tourism," in *Handbook on the Politics of Antarctica* (Cheltenham: Edward Elgar Publishing, 2017), 368–91.
6 Liggett and Stewart, "The Changing Face."
7 Daniela Liggett, "Tourism in the Antarctic: Modi Operandi and Regulatory Effectiveness" in *Gateway Antarctica* (Saarbrücken: VDM, 2009).
8 Wilson Cheung, Thomas Bauer and Jinyang Deng, "The Growth of Chinese Tourism to Antarctica: A Profile of Their Connectedness to Nature, Motivations, and Perceptions," *The Polar Journal* 9, no. 1 (2019): 197–213; Nicole Bender, Kim Crosbie and Heather Lynch, "Patterns of Tourism in the Antarctic Peninsula Region: A 20-Year Analysis," *Antarctic Science* 28, no. 3 (2016): 194–203.
9 Measure 15 (2009) focuses on restricting landings in Antarctica to vessels carrying 500 or fewer passengers and allowing a maximum of 100 passengers to be on shore at the same time. Measure 15 (2009) is not yet in effect, as currently only 14 of the 29 ATCPs required to implement this measure for it to enter into effect have done so (status: 3 March 2019).
10 Section C of IAATO's binding bylaws specifies operational procedures including limiting the number of passengers on shore at any one time to not more than 100 and requiring vessels that wish to schedule landings in the Antarctic to carry no more than 500 passengers.
11 Julia Jabour, "Case Study Antarctica: Up Against the Ice Barrier; Antarctic Tourism Operators Prepare for the Polar Shipping Code," *Global Climate Change and Coastal Tourism: Recognizing Problems, Managing Solutions and Future Expectations* (2017): 273–87.
12 Liggett and Stewart, "The Changing Face," 368–91.
13 Liggett and Stewart, "The Changing Face," 368–91.
14 Robert Powell, Stephen Kellert and Sam Ham, "Antarctic Tourists: Ambassadors or Consumers?," *Polar Record* 44, no. 3 (2008): 233–41.
15 The International Code for Ships Operating in Polar Waters (Polar Code), adopted on 24 November 2014, entered into force 1 January 2017, IMO Doc. MEPC 68/21/Add.1, Annex 10.
16 The International Convention for the Safety of Life at Sea (SOLAS), adopted 1 November 1974, entered into force 25 May 1980, 1184 UNTS 2.
17 The International Convention for the Prevention of Pollution from Ships and the Protocol of Amendment (MARPOL), 1973/1978, 1340 UNTS 61.
18 Protocol to the Convention on the Prevention of Marine Pollution by Dumping of Wastes and Other Matter (London Protocol), adopted 17 November 1996, entered into force 24 March 2006, and amended 2 November 2006 (amendments entered into force 10 February 2007).
19 Christopher Wright, "The Rise and Rise and Rise of Polar Cruising," *The Maritime Executive*, 2018.
20 Eke Eijgelaar, Carla Thaper, and Paul Peeters, "Antarctic Cruise Tourism: The Paradoxes of Ambassadorship, 'Last Chance Tourism' and Greenhouse Gas Emissions," *Journal of Sustainable Tourism* 18, no. 3 (2010): 337–54.

1 Peter Hughes, "Antarctica: The World's Coldest Cruise," *Financial Times*, 24 March 2016.

Ship-Based Antarctic Tourist Landing Procedures

1 Data from the International Association of Antarctica Tour Operators, iaato.org, 2020.

05

"The Question of Antarctica"

Antarctica. Territory Beyond Possession?

1 Klaus Dodds, "'Awkward Antarctic Nationalism': Bodies, Ice Cores and Gateways in and beyond Australian Antarctic Territory/East Antarctica," *Polar Record* 53, no. 1 (2017): 16–30.
2 Jane Perlez, "China, Pursuing Strategic Interests, Builds Presence in Antarctica," *The New York Times,* 3 May 2015.
3 Alan Hemmings and Neil Gilbert, "Antarctica's Unclaimed Sector," *Antarctic* 33, no. 4 (2015).

Antarctica as a Global-Commons. Promises and Pitfalls

1 The generic term "global commons" "denotes areas and natural resources that are not subject to the national jurisdiction of a particular state": Nico Schrijver, "Managing the Global Commons: Common Good or Common Sink?" *Third World Quarterly* 37, no. 7 (2016).
2 Nils Vanstappen, "Legitimacy in Antarctic Governance: The Stewardship Model," *Polar Record* 55 (2019): 358–60.
3 Gordon Fogg, *A History of Antarctic Science* (Cambridge: Cambridge University Press, 1992).
4 Stephen Zorn, "Antarctic Minerals: A Common Heritage Approach," *Resources Policy* 10, no. 1 (1984): 2–18.
5 Thomas Lord, "The Antarctic Treaty System and the Peaceful Governance of Antarctica: The Role of the ATS in Promoting Peace at the Margins of the World," *The Polar Journal* (May 2020).
6 The Antarctic Treaty (1959), accessed June 25, 2020, https://www.bas.ac.uk/about/antarctica/the-antarctic-treaty/the-antarctic-treaty-1959/.
7 Sanjay Chaturvedi, "Antarctic as 'Global Knowledge Commons': Geopolitics, Science and Trusteeship," in *Scientific and Geopolitical Interests in Arctic and Antarctic,* ed. Rengaswamy Ramesh, M. Sudhakar, and S. Chattopadhyay (New Delhi: Lights Research Foundation, 2013).

Where the Ice Meets the Waves. Antarctica and the Law of the Sea

1 Ben Saul and Tim Stephens, *Antarctica in International Law* (Oxford: Hart Publishing, 2015).
2 The seven claimant states are Argentina, Australia, Chile, France, New Zealand, Norway and the United Kingdom. Both the United States and Russia have reserved the right to assert a territorial claim.
3 Tullio Treves, "Historical Development of the Law of the Sea," in *The Oxford Handbook of the Law of the Sea,* ed. by Donald R. Rothwell, Alex G. Oude Elferink, Karen N. Scott, and Tim Stephens, 1–24 (Oxford: Oxford University Press, 2015).
4 Donald Rothwell and Tim Stephens, *The International Law of the Sea* (Oxford: Hart, 2016), 10–14.
5 Christy Collis and Klaus Dodds, "Assault on the Unknown: The Historical and Political Geographies of the International Geophysical Year (1957–58)," *Journal of Historical Geography* 34 (2008): 559.
6 Tommy Koh, "A Constitution for the Oceans," United Nations, n.d.
7 See especially Christopher Joyner, "Antarctica and the Law of the Sea: An Introduction and Overview," *Ocean Development and International Law* 13, no. 3 (1983): 277–89.
8 Patrizia Vigni, "Antarctic Maritime Claims: 'Frozen Sovereignty' and the Law of the Sea," in *The Law of the Sea and Polar Maritime Delimitation and Jurisdiction,* ed. by Alex Oude Elferink and Donald Rothwell, 85–101 (Leiden: Martinus Nijhoff, 2001).
9 Christopher Joyner, "The Antarctic Treaty and the Law of the Sea: Fifty Years On," *Polar Record* 46 (2010): 15.
10 Christopher Joyner, "Antarctic Treaty," 15.
11 Vigni, "Antarctic Maritime Claims."
12 Stephens, "Polar Continental Shelves: Australian and Canadian Challenges and Opportunities," in *Polar Oceans Governance in an Era of Environmental Change,* ed. Tim Stephens and David VanderZwaag, 146–165 (Cheltenham: Edward Elgar, 2014).
13 Tim Stephens, "An Icy Reception or a Warm Embrace? The Antarctic Treaty System and the International Law of the Sea," in *Handbook on the Politics of Antarctica,* ed. Klaus Dodds, Alan Hemmings, and Peder Roberts, 439–452 (Cheltenham: Edward Elgar Publishing, 2017).
14 *Whaling in the Antarctic (Australia v Japan; New Zealand Intervening),* International Court of Justice, Judgement of 31 March 2014.
15 Shirley Scott, "The Evolving Antarctic Treaty System: Implications of Accommodating Developments in the Law of the Sea," in *The Law of the Sea and the Polar Regions: Interactions between Global and Regional Regimes,* ed. Erik Molenaar, Alex Oude Elferink, and Donald Rothwell, 28 (Leiden: Martinus Nijhoff, 2013).

1 Rothwell and Stephens, *The International Law of the Sea* (Oxford: Hart, 2016).

Bounding Antarctica. The Greater Southern Ocean

1 Alan Hemmings, "Southern Horizons: South Asia in the South Indian Ocean," *Panjab University Research Journal Social Sciences* 24 (2016): 136.
2 The eastern part of the claim extended further north to include South Georgia and the South Sandwich Islands. A 60° S boundary avoided including any part of South America. This claim south of 60° S is now termed the "British Antarctic Territory".
3 Donald Rothwell, "The Boundaries of the Polar Regions," in *Research Handbook on Polar Law,* ed. Karen N. Scott and David VanderZwaag (Cheltenham: Edward Elgar, 2020); Donald Rothwell and Alan Hemmings, "Evolution of a Polar Law," in *Research Handbook on Polar Law,* 454–73 (Cheltenham: Edward Elgar, 2020).
4 Klaus Dodds and Alan Hemmings, "Arctic and Antarctic Regionalism," in *Handbook on the Geographies of Regions and Territories,* ed. Anssi Paasi, John Harrison, and Martine Jones, 499–500 (Cheltenham: Edward Elgar, 2018).
5 Hemmings, "Southern Horizons" 143–44.
6 Dodds and Hemmings, "Arctic and Antarctic Regionalism," 499.

The "Question of Antarctica" at the United Nations

1 UN General Assembly, *Question of Antarctica, Report of the Secretary-General (Part I),* 35.
2 UN General Assembly, *Question of Antarctica;* and Peter Beck, "Twenty Years On: The UN and the 'Question of Antarctica' 1983–2003," *Polar Record* 40 (2004): 205–12.
3 UN General Assembly, *Question of Antarctica,* 3–21.
4 Article IX(2) of the Antarctic Treaty. Secretariat of the Antarctic Treaty, *Compilation of Key Documents of the Antarctic Treaty System,* 25.
5 UN General Assembly, *Question of Antarctica,* 5.
6 Doaa Abdel-Motaal, *Antarctica: The Battle for the Seventh Continent* (Denver: Praeger, 2016), 97.
7 UN General Assembly Resolution, A/RES/41/88, "The Question of Antarctica," 4 December 1986. The General Assembly "calls upon the Antarctic Treaty

Consultative Parties to impose a moratorium on the negotiations to establish a minerals regime until such time as all members of the international community can participate fully in such negotiations."
8   Abdel-Motaal, *Antarctica,* 83–119.
9   Abdel-Motaal, *Antarctica,* 83–111 and 154.

Timeline
1   Lisle Rose, *Assault on Eternity: Richard E. Byrd and the Exploration of Antarctica, 1946–47* (Washington, DC: Naval Institute Press, 1980), 219.
2   Peter Beck, *The International Politics of Antarctica* (London/Sydney: Croom Helm, 1986), 271.
3   John Hanessian, "The Antarctic Treaty 1959," The *International and Comparative Law Quarterly* 9, no. 3 (1960): 448–49; Edward Shackleton, "The New Continent," *United Nations World* 1, no. 10 (1949): 380–82.
4   Hemmings, *Re-justifying the Antarctic Treaty System for the 21st Century: Rights, Expectations and Global Equity,* in *Polar Geopolitics? Knowledges, Resources and Legal Regimes,* ed. Richard C. Powell and Klaus Dodds, 62 (Cheltenham: Edward Elgar, 2014).
5   Hanessian, "The Antarctic Treaty 1959," 450.
6   UN 1956n1; UN 1956b2; UN 1958, 5.
7   UN 1956c; Larus, "India Claims a Role in Antarctica," 45–56; Sanjay Chaturvedi, "Antarctic as 'Global Knowledge Commons': Geopolitics, Science and Trusteeship," in *Scientific and Geopolitical Interests in Arctic and Antarctic,* ed. Rengaswamy Ramesh, M. Sudhakar, and S. Chattopadhyay, 304–13 (New Delhi: Lights Research Foundation, 2013).
8   Brian Roberts, "International Cooperation for Antarctic Development: The Test for the Antarctic Treaty," *Polar Record* 19 (1978): 107–20.
9   Philip Quigg, *A Pole Apart: The Emerging Issue of Antarctica* (New York: McGraw-Hill, 1983); Alan Hemmings, Donald Rothwell, and Karen Scott, *Antarctic Security in the Twenty-First Century: Legal and Policy Perspectives* (New York: Routledge, 2012), 87–91.
10   Antarctic Treaty Consultative Meeting 25, 1977.
11   Tim Stephens and Ben Saul, *Antarctica in International Law* (Oxford: Hart Publishing, 2015).
12   Antarctic Treaty Consultative Meeting 25, 1977.
13   Foreign and Commonwealth Office, 5, 1977.
14   UN 1982a, 17–20.
15   *Index to Proceedings of the General Assembly: Sixtieth Session 2005/2006.*
16   Antarctic Treaty Secretariat 2007, 4: UN News Centre, 2008.
17   United States State Department, "Remarks at the Joint Session of the Antarctic Treaty Consultative Meeting and the Arctic Council, 50th Anniversary of the Antarctic Treaty."

The Role of NGOs in the Antarctic Treaty System

1   James Barnes, "A Reminiscence on Antarctic Governance and Transparency: The NGO Role," *Antarctic Affairs* 5 (2018): 1–3. On the UN relationship generally, see Beck, "Antarctica and the United Nations," in *Handbook on the Politics of Antarctica* (Cheltenham: Edger Elgar Publishing Inc., 2017), 255–64.
2   The report of this important meeting is at https://www.ccamlr.org/en/document/organisation/report-meeting-scar-group-specialists-marine-living-resources-southern-ocean.
3   That programme and its successor, FIBEX, provided unrivalled scientific information about krill and other Southern Ocean species. See Gordon Fogg, *A History of Antarctic Science;* Siegel, *Biology and Ecology of Antarctic Krill* (Cambridge: Cambridge University Press, 1992).
4   I served as a member of US delegations during those years, and other ASOC colleagues were on the Australian and New Zealand delegations.
5   James Barnes, *Let's Save Antarctica* (Richmond, Australia: Greenhouse Publications, 1982), appendix Q, 81.

6   James Barnes, "The Emerging Convention on the Conservation of Antarctic Marine Living Resources: An Attempt to Meet the New Realities of Resource Exploitation in the Southern Ocean," in *The New Nationalism and the Use of Common Spaces,* ed. by Jonathan I. Charney (Totowa, NJ: Allanheld, Osmun, 1982).
7   ASOC maintains an archive of *ECOs* at https://www.asoc.org/news-and-publications/archives/58.
8   See Barbara Mitchell, *Frozen Stakes: The Future of Antarctic Minerals* (London: IIED, 1983); Kimball's articles appeared in journals and conference proceedings, including the 8th Annual Conference of the University of Rhode Island's Center for Ocean Management (1984); American Geophysical Union's Antarctic Research Series (vol. 51, 1990); and the *Yearbook of International Environmental Law* (1990, 1991).
9   See Polar Research Board, *Antarctic Treaty System – An Assessment: Proceedings of a Workshop Held at Beardmore South Field Camp, Antarctica, January 7–13, 1985.*
10   ATCPs sometimes argued that because IUCN is an umbrella NGO, no others needed to be admitted, but eventually they gave up that false argument.
11   CCAMLR, *Report of the Seventh Meeting of the Commission* (1988), paragraphs 153–156. See generally Harlan Cohen, "Public Participation in Antarctica: The Role of Nongovernmental and Intergovernmental Organizations," in *Science Diplomacy: Antarctica, Science and the Governance of International Spaces,* ed. by Paul Arthur Berkman, Michael A. Lang, David W. H. Walton, and Oran R. Young (Washington, DC: Smithsonian Press, 2010).
12   ASOC's formal papers presented to ATCM and CCAMLR meetings, negotiations and all special meetings can be read at https://www.asoc.org/news-and-publications/59.
13   Many people have written about the role of NGOs in Antarctic decision-making, including: Ricardo Roura and Tina Tin, "Strategic Thinking and the Antarctic Wilderness: Contrasting Alternative Futures," in *Antarctic Futures: Human Engagement with the Antarctic Continent,* eds. Tina Tin, Daniela Liggett, Patrick Maher, and Machiel Lamers (Dordrecht: Springer, 2014), 253–71.

Antarctica as World Park

1   James Barnes, *Let's Save Antarctica!* (Richmond, Australia: Greenhouse Publications, 1982), appendix Q, 76.
2   Barnes, *Let's Save Antarctica!,* 79.
3   James Barnes, "The Emerging Convention on the Conservation of Antarctic Marine Living Resources: An Attempt to Meet the New Realities of Resource Exploitation in the Southern Ocean," in *The New Nationalism and the Use of Common Spaces,* ed. Jonathan I. Charney (Totowa, NJ: Allanheld, Osmun, 1982).
4   See Cassandra Brooks et al., "Science-Based Management in Decline in the Southern Ocean," *Science* 354, no. 6309 (2016): 185–87.
5   James Barnes, "A Reminiscence on Antarctic Governance and Transparency: The NGO Role," *Antarctic Affairs* 5 (2018): 19–22. ASOC published a global NGO paper called *ECO* at each meeting. This is the first one, published in Wellington: https://www.asoc.org/storage/documents/ECOs/1983/xxii.1_amr.pdf.
6   See John May, *The Greenpeace Book of Antarctica: A New View of the Seventh Continent* (London: Dorling Kindersley, 1988); Kelly Rigg, "Environmentalists' Perspectives on the Protection of Antarctica," in *The Future of Antarctica: Exploitation Versus Preservation,* ed. Grahame Cook (Manchester, England: Manchester University Press, 1990).
7   Cited in Christopher Joyner, "The Antarctic Minerals Negotiating Process," *American Journal of International Law* 81, no. 4 (October 1987): 888–904. The UN Secretary-General compiled a detailed report with substantial assistance from ASOC. See

Question of Antarctica: Report of the Secretary-General, 39 U.N. GAOR (Agenda Item 66), U.N. Doc. A/39/583 (1984).
8   Henk Haazen, *Greenpeace World Park Base: Antarctica 1987-1992; Treading Lightly: A Minimal Impact Antarctic Station* (Greenpeace International, March 1994).
9   Convention on the Regulation of Antarctic Minerals Resource Activities (CRAMRA). ASOC's analysis of the Convention was published in "The Convention Proceeds," *ECO* 48, no. 12 (June 1988).
10   Bernard Herber, "The Economic Case for an Antarctic World Park in Light of Recent Policy Developments," *Polar Record* 28, no. 167 (1992): 293–300.
11   See Lyn Goldsworthy, "Achieving a Ban on Mining in Antarctica," 14 December 2009. See also Samuel Blay and B. Martin Tsamenyi, "Australia and the Convention for the Regulation of Antarctic Mineral Resource Activities (CRAMRA)," *Polar Record* 26, no. 158 (1990): 195–202.
12   Emma Shortis, "'Who Can Resist This Guy?' Jacques Cousteau, Celebrity Diplomacy, and the Environmental Protection of the Antarctic," *Australian Journal of Politics and History* 61, no. 3 (2015): 366–80.
13   https://www.ats.aq/e/protocol.html. Although the minerals ban could be reviewed in 2048 if one or more ATCPs desire, the rules for negotiating a new minerals convention and getting it ratified are very difficult.
14   See Lyn Goldsworthy, "The Madrid Protocol: An NGO Perspective," 4 October 2011; Dalziell and Goldsworthy, "World Park Antarctica: Does It Have a Future?" *Forum for Applied Research and Public Policy* 9, no. 1 (1994): 71–75; and Hemmings, "After the Party: The Hollowing of the Antarctic Treaty System and the Governance of Antarctica," Symposium on Antarctic Politics at the University of Canterbury, Christchurch, New Zealand, 8–9 July 2010.
15   Klaus Dodds, "Governing Antarctica: Contemporary Challenges and the Enduring Legacy of the 1959 Antarctic Treaty," *Global Policy* 1 (2010); Daniela Liggett and Alan Hemmings, *Exploring Antarctic Values: Proceedings of the Workshop "Exploring Linkages between Environmental Management and Value Systems: The Case of Antarctica"* held at the University of Canterbury, Christchurch, New Zealand, 5 December 2011; Tina Tin et al., "Impacts of Local Human Activities on the Antarctic Environment," *Antarctic Science* 21 (2009): 3–33; Brooks et al., "Science-Based Management"; Jennifer Jacquet et al., "'Rational use' in Antarctic Waters," *Marine Policy* 63 (2016): 28–34; Tina Tin et al., *Antarctic Futures: Human Engagement with the Antarctic Environment* (Dordrecht: Springer, 2014).

FOUR ELEMENTS

06

The Revolutionary Ocean

The Global Carbon Budget

1   "Carbon Dioxide Emissions Are Rising. Reducing Them Is a Monumental Challenge," The Climate Issue, *The Economist,* September 2019.
2   Carbon Dioxide Information Analysis Center; Research Institute for Environment, Energy and Economics, "Infographic: CO₂ Emissions vs. Vulnerability to Climate Change," infographic from George Washington University, 2010. planetforward.org. [online]
3   Pierre Friedlingstein et al., "Global Carbon Budget 2019," *Earth System Science Data* 11, no. 4 (2019): 1783–838.

The Variable Southern Ocean Carbon Sink

1   Peter Landschützer, Nicolas Gruber, and Dorothee Bakker, "Decadal Variations and Trends of the Global Ocean Carbon Sink," *Global Biogeochemical Cycles* 30, no. 10 (2016): 1396–417.
2   Christian Rödenbeck et al., "Interannual Sea–Air CO$_2$ Flux Variability from an Observation-Driven Ocean Mixed-Layer Scheme," *Biogeosciences* 11, no. 17 (2014): 4599–613.
3   Corinne Le Quéré et al., "Global Carbon Budget 2018," *Earth System Science Data* 10, no. 4 (December 2018): 2141–94.
4   Nicolas Gruber et al., "Oceanic Sources, Sinks, and Transport of Atmospheric CO$_2$," *Global Biogeochemical Cycles* 23, no. 1 (2009).
5   Robert Key et al., "A Global Ocean Carbon Climatology: Results from Global Data Analysis Project (GLODAP)," *Global Biogeochemical Cycles* 18, no. 4 (December 2004).
6   Heather Graven et al., "Changing Controls on Oceanic Radiocarbon: New Insights on Shallow-to-Deep Ocean Exchange and Anthropogenic CO$_2$ Uptake," *Journal of Geophysical Research: Oceans* 117, no. 10 (2012).
7   Estimated by Gruber et al., "Oceanic Sources."
8   Figure adapted from Landschützer, Gruber, and Bakker, "Decadal Variations."

Antarctica's Bio-indicators

1   Samuel Matthews, "Antarctica's Nearer Side," *National Geographic* 140, no. 5 (1971): 622–54.

The Impact of Climate Change on the Antarctic Krill

1   Greenpeace International, *Licence to Krill: The Little-Known World of Antarctic Fishing*, 5.

Antarctic Sea Ice and Its Impact on the Global Climate System

1   David Wilson, *The Lost Photographs of Captain Scott* (New York: Little, Brown, and Co., 2011).
2   David Walton, *Antarctica: Global Science from a Frozen Continent* (Cambridge: Cambridge University Press, 2013).

07

Twenty-Six Quadrillion Tons of Ice. The Urgencies of Contemporaneity vs. the Truths of "Deep Time"

Antarctica. The World's Coldest Desert

1   Carolyn Strange and Alison Bashford, *Griffith Taylor: Visionary, Environmentalist, Explorer* (Toronto: University of Toronto Press, 2008), 60.

Antarctica as Environmental Archive

1   Laurent Augustin et al., "Eight Glacial Cycles from an Antarctic Ice Core," *Nature* 429 (2004): 623–28.
2   See the Beyond EPICA website, beyondepica.eu.

1   Jean Jouzel, "A Brief History of Ice Core Science over the Last 50 Yr," *Climate of the Past* 9, no. 6 (2013): 2525–47.
2   *National Geographic*, "Ice Coring," 2020.

Ice Memory. An International Salvage Program

1   Valérie Masson-Delmotte et al., *Global Warming of 1.5°C* (IPCC, 2018 – in press).
2   Martin Beniston, "Impacts of Climatic Change on Water and Associated Economic Activities in the Swiss Alps," *Journal of Hydrology* 412 (2012): 291–96.
3   Andreas Gobiet et al., "21st Century Climate Change in the European Alps: A Review," *Science of the Total Environment* 493 (2014).

1   *The Economist*, "Banking against Doomsday," 10 March 2012.
2   Svalbard Global Seed Vault; Air Liquide Foundation, "ICE MEMORY, Preserving Ice Memory for Future Generations."

Ice Shelf Retreat in Antarctica

1   Australian Antarctic Program, "Sørsdal Glacier Dynamics."
2   David Walton, *Antarctica: Global Science from a Frozen Continent* (Cambridge: Cambridge University Press, 2013); Bethan Davies, "Grounding Lines."

The Larsen Ice Shelf

1   Walton, *Antarctica;* NASA/Goddard Space Flight Center Scientific Visualization Studio, "NASA Research Leads to First Complete Map of Antarctic Ice Flow."
2   NASA Earth Observatory, "World of Change: Collapse of the Larsen-B Ice Shelf." earthobservatory.nasa.gov. [online]

Antarctica and Global Sea Level Rise

1   Richard Z. Poore, Richard. S. Williams Jr., and Christopher Tracy, "Sea Level and Climate," US Geological Survey Fact Sheet 002-00 (2000).
2   Lea Rekow, "Sea Level Rise," BifrostOnline. bifrostonline.org. [online]

08

A Continental Archipelago under Ice

Gondwana, Antarctica, and Continental Drift

1   David Walton, *Antarctica: Global Science from a Frozen Continent* (Cambridge: Cambridge University Press, 2013), 54–55.

Scientific Drilling on the Antarctic Margin

1   Tetsuya Torii, "A Review of the Dry Valley Drilling Project, 1971–76," *Polar Record* 20, no. 129 (September 1981): 533–41.
2   Peter Barrett, "The Devonian to Jurassic Beacon Supergroup of the Transantarctic Mountains and Correlatives in Other Parts of Antarctica," in *The Geology of Antarctica*, ed. Robert Tingey (Oxford: Clarendon Press, 1991), 120–52.
3   Thomas Fleming et al., "^{40}Ar/^{39}Ar Geochronology of Ferrar Dolerite Sills from the Transantarctic Mountains, Antarctica: Implications for the Age and Origin of the Ferrar Magmatic Province," *Geological Society of America Bulletin* 109, no. 5 (May 1997): 533–46.
4   Stephen Porter and James Beget, "Provenance and Depositional Environments of Late Cenozoic Sediments in Permafrost Cores from Lower Taylor Valley," in *Dry Valley Drilling Project,* ed. Lyle McGinnis,

Antarctic Research Series 33 (Washington, DC: American Geophysical Union, 1981), 351–63.
5   Christian Ohneiser and Gary Wilson, "Revised Magnetostratigraphic Chronologies from New Harbour Drill Cores, Southern Victoria Land, Antarctica," *Global and Planetary Change* 82–83 (February 2012): 12–24.
6   Dennis Hayes et al., introduction to *Initial Reports of the Deep Sea Drilling Project,* vol. 28 (Washington DC: US Government Printing Office, 1975), 5–18.
7   James Kennett et al., "Cenozoic Paleoceanography in the Southwest Pacific Ocean, Antarctic Glaciation, and the Development of the Circum-Antarctic Current," in *Initial Reports of the Deep Sea Drilling Project,* vol. 29 (Washington DC: US Government Printing Office, 1974), 1155–69.
8   Peter Barrett et al., "Mid-Cenozoic Record of Glaciation and Sea-Level Change on the Margin of the Victoria Land Basin, Antarctica," *Geology* 15, no. 7 (July 1987): 634–37.
9   Peter Barrett, introduction to *Antarctic Cenozoic History from the CIROS-1 Drillhole, McMurdo Sound,* ed. Peter Barrett, DSIR Bulletin 245 (Wellington: DSIR Publishing, 1989), 5–6.
10   Gary Wilson et al., "Magnetobiostratigraphic Chronology of the Eocene-Oligocene Transition in the CIROS-1 Core, Victoria Land Margin, Antarctica: Implications for Antarctic Glacial History," *Geological Society of America Bulletin* 110, no. 1 (January 1998): 35–47.
11   Peter Barrett et al., "Cenozoic Climate and Sea Level History from Glacimarine Strata off the Victoria Land Coast, Cape Roberts Project, Antarctica," in *Glacial Sedimentary Processes and Products,* ed. Michael Hambrey et al., International Association of Sedimentologists Special Publication 39 (New York: Wiley, 2009), 259–87.
12   Timothy Naish et al., "Orbitally Induced Oscillations in the East Antarctic Ice Sheet at the Oligocene/ Miocene Boundary," *Nature* 413, no. 6857 (18 October 2001): 719–23.
13   Gary Wilson et al., "Neogene Tectonic and Climatic Evolution of the Western Ross Sea, Antarctica: Chronology of Events from the AND-1B Drill Hole," *Global and Planetary Change* 96–97 (October 2012): 189–203.
14   Timothy Naish et al., "Obliquity-Paced Pliocene West Antarctic Ice Sheet Oscillations," *Nature* 458, no. 7236 (19 March 2009): 322–28; David Pollard and Robert DeConto, "Modelling West Antarctic Ice Sheet Growth and Collapse through the Past Five Million Years," in ibid., 329–32.
15   Johann Klages et al., "Temperate Rainforests near the South Pole during Peak Cretaceous Warmth," *Nature* 580, no. 7801 (2 April 2020): 81–86.

1   Jonathan Amos, "Dinosaurs Walked through Antarctic Rainforests," *BBC News,* April 2020.

Plant Migration

1   Jeffrey Stilwell and John Long, *Frozen in Time: Prehistoric Life in Antarctica* (Collingwood: CSIRO Publishing, 2011).
2   David Cantrill and Imogen Poole, *The Vegetation of Antarctica through Geological Time* (Cambridge: Cambridge University Press, 2012).
3   Bill Chaloner and Paul Kenrick, "Did Captain Scott's *Terra Nova* Expedition Discover Fossil *Nothofagus* in Antarctica?" *The Linnean* 31, no. 2 (2015): 11–17.

Big Geo-Scientific Data in the Antarctic

1   Peter Fretwell et al., "Bedmap2: Improved Ice Bed, Surface and Thickness Datasets for Antarctica," *The Cryosphere* 7 (2013): 375–93.
2   Alexander Golynsky et al., "New Magnetic Anomaly Map of the Antarctic," *Geophysical Research Letters* 45, no. 13 (2018): 6437–49.
3   Mirko Scheinert et al., "New Antarctic Gravity Anomaly Grid for Enhanced Geodetic and

Geophysical Studies in Antarctica," *Geophysical Research Letters* 43, no. 2 (2016): 600–610.
4   Mark van der Meijde et al., "GOCE Data, Models, and Applications: A Review," *International Journal of Applied Earth Observation and Geoinformation* 35 (November 2013).
5   See, for example, Andreas Richter et al., "Height Changes over Subglacial Lake Vostok, East Antarctica: Insights from GNSS Observations," *Journal of Geophysical Research: Earth Surface* 119, no. 11 (November 2014): 2460–80.
6   Yasmina Martos et al., "Heat Flux Distribution of Antarctica Unveiled," *Geophysical Research Letters* 44, no. 22 (2017): 11417–26.
7   Tom Jordan et al., "New Gravity-Derived Bathymetry for the Thwaites, Crosson and Dotson Ice Shelves Revealing Two Ice Shelf Populations," *The Cryosphere Discussions* (2020).

The Magnetic Crusades
1   Edward Sabine, *Report on the Variation of the Magnetic Intensity at Different Points on the Earth's Surface* (London: R. and J. E. Taylor, 1838), 83–84.
2   Gillen D'Arcy Wood, "Arctic Voyages (1838–1842): Sir Edward Sabine, James Ross, and the Magnetic Sublime," *The Wordsworth Circle* 50, no. 2 (Spring 2019): 237–46.

1   Matthew Lythe and David Vaughan, "BEDMAP: A New Ice Thickness and Subglacial Topographic Model of Antarctica," *Journal of Geophysical Research: Solid Earth* 106, no. B6 (June 2001): 11335–52; Peter Fretwell et al., "Bedmap2: Improved Ice Bed, Surface and Thickness Datasets for Antarctica," *The Cryosphere* 7 (2013): 375–93.

09
The White Desert. A Unique Viewing Platform into the Cosmos

The Magnetosphere

1   William Henry Barlow, "On the Spontaneous Electrical Currents Observed in the Wires of the Electric Telegraph," *Philosophical Transactions of the Royal Society of London* 139 (1849): 66.

Atmospheric Chemistry. The Discovery of the Ozone Hole

1   British Antarctic Survey, *The Ozone Hole* (Cambridge: National Environment Research Council, 2012).

Long-Duration Ballooning

1   Herbert George Ponting, *The Great White South: Being an Account of Experiences with Captain Scott's South Pole Expedition and of the Nature Life of the Antarctic* (London: Duckworth & Co., 1921), 115–16.
2   Diana Lutz, "Super-TIGER Shatters Scientific Balloon Record in Antarctica," *The Source – Washington University in St. Louis,* 22 January 2013.

Timeline
1   Luis Pacheco, "GRAD (Gamma Ray Advanced Detector)"; A. Rester et al., "Performance of Bismuth Germanate Active Shielding on a Balloon Flight over Antarctica," *IEEE Transactions on Nuclear Science* 37, no. 2 (April 1990): 559–65.

The Balloon-Borne Large Aperture Submillimeter Telescope

1   Mark Devlin et al., "Over Half of the Far-Infrared Background Light Comes from Galaxies at z ≥1.2," *Nature* 458, no. 7239 (April 2009): 737–39.
2   Nathan Lourie et al., "Design and Characterization of a Balloon-Borne Diffraction-Limited Submillimeter Telescope Platform for BLAST-TNG," Proceedings Volume 10700, Ground-Based and Airborne Telescopes VII (2018).
3   Nathan Lourie et al., "Preflight Characterization of the BLAST-TNG Receiver and Detector Arrays," *Proceedings Volume 10708, Millimeter, Submillimeter, and Far-Infrared Detectors and Instrumentation for Astronomy IX* (2018).

1   Nathan Lourie, "Thermal Modeling of the BLAST-TNG Balloon Telescope," n.d.
2   Harvard CMB Group, "Keck Array Overview," n.d.

IceCube Neutrino Observatory. Transforming Antarctic Ice into a Science Experiment

1   M. Markov, "On High Energy Neutrino Physics," *Proceedings, 10th International Conference on High-Energy Physics (ICHEP 60), Rochester, NY* (1960): 578–81.
2   Francis Halzen, John Learned, and Todar Stanev, "Neutrino Astronomy," *AIP Conference Proceedings* 198 (1990): 39–51.
3   Hans-Juergen Arends et al. "Observation of High Energy Atmospheric Neutrinos with the Antarctic Muon and Neutrino Detector Array," *Physical Review D: Particle and Fields* 66, no. 1 (2002).
4   National Science Fondation, "University of Wisconsin–Madison Complete Construction of the World's Largest Neutrino Observatory," 17 December 2010.
5   IceCube Collaboration, "Evidence for High-Energy Extraterrestrial Neutrinos at the IceCube Detector," *Science* 342, no. 6161 (2013).
6   Hamish Johnson, "Cosmic Neutrinos Named Physics World 2013 Breakthrough of the Year," *Astroparticle Physics, Physics World,* 13 December 2013.
7   Aya Ishihara, "The IceCube Upgrade – Design and Science Goals," 36th International Cosmic Ray Conference (ICRC2019) (July 2019).
8   Markus Aartsen et al. "Neutrino Astronomy with the Next Generation IceCube Neutrino Observatory," *The IceCube-Gen 2 Collaboration* (July 2019).

Ice Cube Laboratory
1   IceCube, "Digital Optical Module (DOM) Development," n.d.

Cosmic Rays and Neutrino Astronomy

1   IceCube Collaboration, "Multi-messenger Observations of a Flaring Blazar Coincident with High-Energy Neutrino IceCube-170922A," *Science* 361, no. 6409 (2018).
2   IceCube Collaboration, "Neutrino Emission from the Direction of the Blazar TXS 0506+056 prior to the IceCube-170922A Alert," *Science* 361, no. 6398 (2018): 147–51.
3   IceCube Collaboration, "Time-Integrated Neutrino Source Searches with 10 years of IceCube Data," *Physical Review Letters* 124 (2020).

1   IceCube: South Pole Neutrino Observatory, "Detector," n.d.
2   IceCube: South Pole Neutrino Observatory, "Meet the Collaboration," n.d.

SURVIVING IN THE CRYOSPHERE

10
Extreme Habitation. Compressed Spaces and Dilated Times

Inhabitation in the Extreme. The Winter-Over Syndrome

1   Frederick Cook, *Through the First Antarctic Night 1898–1899* (New York: Doubleday, Page & Co., 1909).
2   Lawrence Palinkas, "Going to Extremes: The Cultural Context of Stress, Illness and Coping in Antarctica," *Social Science and Medicine* 35, no. 5 (September 1992): 653.
3   Lawrence Palinkas, Mark Cravalho, and Dierdre Browner, "Seasonal Variation of Depressive Symp-
toms in Antarctica," *Acta Psychiatrica Scandinavia* 91, no. 6 (June 1995): 426; Robert Bechtel and Amy Berning, "The Third-Quarter Phenomenon: Do People Experience Discomfort after Stress Has Passed?," in *From Antarctica to Outer Space: Life in Isolation and Confinement,* ed. Albert Harrison, Yvonne Clearwater, and Christopher McKay (New York: Springer Verlag, 1991), 264.
4   Robert Strange and S. Youngman, "Emotional Aspects of Wintering Over," *Antarctic Journal of the United States* 6 (November–December 1971): 255.
5   Kirmach Natani and Jay Shurley, "Sociopsy-chological Aspects of a Winter Vigil at South Polar Station," in *Human Adaptability to Antarctic Conditions,* ed. Eric Gunderson (Washington: American Geophysical Union, 1974), 110; Jack Stuster, Claude Bachelard, and Peter Suedfeld, "The Relative Importance of Behavioral Issues during Long-Duration ICE Missions," *Aviation, Space, and Environmental Medicine* 71, no. 9. (September 2000): A22; Anthony Taylor, *Antarctic Psychology* (Wellington: Department of Scientific and Industrial Research, 1987).
6   Clarence Guenter, Albert Joern, Jay Shurley, and Chester Pierce. "Cardiorespiratory and Metabolic Effects in Men on the South Polar Plateau," *Archives of Internal Medicine* 125, no. 4 (April 1970): 630; Lester Reed, Kenneth Burman, K. Shakir, and John O'Brian, "Alterations in the Hypothalamic-Pituitary-Thyroid Axis," *Clinical Endocrinology* 25, no. 1 (July 1986): 55.

1   Herbert George Ponting, *The Great White South: Being an Account of Experiences with Captain Scott's South Pole Expedition and of the Nature Life of the Antarctic* (London: Duckworth & Co., 1921).
2   United States Navy, Operation Deep Freeze III Task Force 43, 1958 report.
3   Diary entry quoted in Morton Beebe, ed., "The Perplexing Poles (All of Them)," in *Operation Deep Freeze III: The Story of Task Force 43; Third Phase, 1957–58* (Paoli: Dorville Corporation, 1957), 28.
4   See n. 2.
5   Ibid.

Interior Urbanism at the Pole. Framheim vs. Cape Evans

1   Roald Amundsen, *The South Pole: An Account of the Norwegian Antarctic Expedition in the Fram 1910–1912.* (1912; repr., New York: Cooper Square Press, 2001).
2   Amundsen, *The South Pole.*
3   Edward Wilson, *With Scott in the Antarctic* (UK: The History Press, 2009).
4   Scott, *Journals: Captain Scott's Last Expedition,* reissue from original 1913 edition (Oxford: Oxford University Press, 2009).
5   Herbert George Ponting, *The Great White South: Being an Account of Experiences with Captain Scott's South Pole Expedition and of the Nature Life of the Antarctic* (London: Duckworth & Co., 1921).

6   Amundsen, *The South Pole.*
7   Ibid.
8   Ibid.
9   Ibid.
10   Ibid.
11   Ibid.
12   Scott, *Journals.*
13   Amundsen, *The South Pole.*
14   Theodore Scambos and Clarence Novak, "On the Current Location of the Byrd 'Snow Cruiser' and Other Artifacts from Little America I, II, III and Framheim," *Polar Geography* 29, no. 4 (2005): 237–52.

1   Herbert George Ponting, *The Great White South: Being an Account of Experiences with Captain Scott's South Pole Expedition and of the Nature Life of the Antarctic* (London: Duckworth & Co., 1921).
2   Martin Heidegger, *Hölderlin's Hymn "The Ister"* (Bloomington: Indiana University Press, 1996).
3   Ponting, *The Great White South.*

The Medicinal Role of Reading in the Heroic Age

1   Natalie O'Neill, "Antarctic Scientist Stabbed Colleague," *New York Post,* 30 October 2018.
2   "Sir Ernest Shackleton on Poetry," *The Times,* 28 October 1911, 11.
3   Shackleton, *South: The Story of Shackleton's Last Expedition 1914–17* (London: Heinemann, 1919), 250.
4   Quoted in John Thomson, *Elephant Island and Beyond* (Norwich: Erskine Press, 2003), 248.
5   Raymoned Priestley, *Copy of Part of His General Diary of the British Antarctic Expedition, 1910–13: 1 January – 3 October 1912,* typescript (unbound), Ms 298/6/3. Scott Polar Research Institute, Cambridge.
6   George Murray Levick, *Journals Kept on the Northern Party of the British Antarctic Expedition, 1910–1913: 9 January –14 November 1912,* vol. 2, *Rough Journal, 29 March – 28 September 1912,* Ms 1423/2. Scott Polar Research Institute, Cambridge.
7   Cecil Madigan, *Madigan's Account: The Mawson Expedition; The Diaries of C.T. Madigan 1911–1914,* ed. Julia Butler (Wellington Bridge Press, 2012), 5 June 1912; John Hunter, *Rise & Shine: Diary of John George Hunter, Australasian Antarctic Expedition 1911–1913,* ed. Jenny Hunter (Hunter House, 2011), 27 May 1912.
8   Douglas Mawson, *Mawson's Antarctic Diaries,* ed. Fred and Eleanor Jacka (Crows Nest: Allen & Unwin, 1988), 7 July 1913.
9   Frank Hurley, *Shackleton's Argonauts,* in *Antarctic Eyewitness: Charles Laseron's South with Mawson and Frank Hurley's Shackleton's Argonauts* (Sydney: Angus & Robertson-HarperCollins, 1999), 285.
10   Apsley Cherry-Garrard, *The Worst Journey in the World* (London: Pimlico, 1922), 203; Douglas Mawson, *The Home of the Blizzard, Being the Story of the Australasian Antarctic Expedition 1911–1914,* vol. 1 (London: Heinemann, 1915), 260.

1   Apsley Cherry-Garrard, *The Worst Journey in the World* (London: Pimlico, 1922).
2   Robert Falcon Scott, *Journals: Captain Scott's Last Expedition* (Oxford: Oxford University Press, 2009).
3   Ernest Shackleton, *The Heart of the Atlantic* (London: Heinemann, 1909).
4   "Early Mobile Phones Help Robert Falcon Scott in Successful South Pole Bid," *The Telegraph,* 18 January 2017.
5   Godden Mackay Logan, *Mawson's Huts Historic Site, Cape Denison, Commonwealth Bay, Australian Antarctic Territory: Conservation Management Plan* (Sydney: Godden Mackay Logan, 2001).
6   Shackleton, *The Heart of the Atlantic.*

Communication Abstinence

1   Ejército de Chile, *Base Militar "General O'Higgins"* ([Santiago]: Instituto Geográfico Militar [de] Chile, 1948).
2   Phillip Law, "Techniques of Living, Transport and Communication," in *Antarctica: A New Zealand Antarctic Society Survey,* ed. Trevor Hatherton (Wellington: Reed, 1965).

1   British Antarctic Survey, "Telecommunications in Antarctica," accessed 4 March 2021, bas.ac.uk [online].
2   United States Antarctic Program, "South Pole Satellite Communications and Pass Schedules," accessed 4 March 2021, usap.gov [online].
3   United States Antarctic Program, "SPTR Satellite," accessed 4 March 2021, usap.gov [online].
4   Cool Antarctica, "Telecommunications in Antarctica," accessed 4 March 2021, coolantarctica.com [online].

Extreme Habitation. Antarctica's "Architecture of Sleep"

1   Timothy Hyde, "Architecture at the End of the World: The Pasts and Futures of Heritage Preservation in Antarctica," *Future Anterior: Journal of Historic Preservation, History, Theory, and Criticism* 14, no. 2 (2017): 73–86.
2   Nathalie Pattyn et al., "From the Midnight Sun to the Longest Night: Sleep in Antarctica," *Sleep Medicine Reviews* 37, (2018): 159–172.
3   Ibid.
4   Ibid.
5   Harvey Colten and Bruce Altevogt, *Sleep Disorders and Sleep Deprivation: An Unmet Public Health Problem* (Washington: National Academies Press, 2006).
6   Luoxi Hao et al., "Antarctic and Lighting Technology," *China Illuminating Engineering Journal* 25, no.1 (2014): 1–7.

1   Felicity Aston, "Women of the White Continent," *Geographical* 77, no. 9 (September 2005): 6.
2   Michael Satchell, "Women Who Conquer the South Pole," *Parade Magazine* (5 June 1983): 16–17.

Sensory Deprivation and the Role of Colour in Antarctica

1   Johannes Itten, *The Art of Color: The Subjective Experience and Objective Rationale of Color,* (New York: Van Nostrand Reinhold, 1993), 16.

1   Emil Schulthess, *Antarctica: A Photographic Survey* (New York: Simon and Schuster, 1960).

Antarctic Nourishment

1   Robert Falcon Scott, *Journals: Captain Scott's Last Expedition* (Oxford: Oxford University Press, 2009).
2   "Antarctic Daily Sledge Ration, 1910s," Science Photo Library, accessed 14 February 2021, sciencephoto.com [online].
3   Scott, *Journals.*

Confined Interiors and the Importance of Collective Spaces

1   Marilene Zimmer et al., "Psychological Changes Arising from an Antarctic Stay," *Estudos de Psicologia* 30, no. 3 (July 2013): 415–23."
2   Pete Akers, "Life and Work on the Frozen Continent: Antarctic Research Stations," *Weather Underground,* 17 January 2020, wunderground.com [online].

3   "Working in Antarctica," *Victor's Travels* (blog), accessed 14 February 2021, victorstravels.com [online].
4   "Arrival at Jang Bogo Antarctic Research Station," *Building a Space Weather Radar in Antarctica* (blog), CIRES, 27 January 2015, colorado.edu [online].
5   Atlas Obscura, "South Pole Ice Tunnels," accessed 14 February 2021, atlasobscura.com [online].

1   United States Navy, "Antarctica: A New Look," Report (California: Seabee Museum Archives, 1972).
2   Photograph taken by Bruce Raymond, United States Navy, 7 January 1961, Antarctic Photo Library, National Science Foundation Archives, photolibrary.usap.gov [online].
3   Richard Donovan, quoted in Benjamin Roman, "Serious about Sports at End of the Earth," *ESPN,* 3 November 2010.
4   Ibid.
5   "Antarctic Religion?," accessed 14 February 2021, theworldgeography.com [online]. 2020.

11
Importing and Transporting the Possibility of Life

From Polar Gateways to Antarctic Cities

1   Klaus Dodds and Juan Francisco Salazar, "Gateway to Antarctica? Geopolitics, Infrastructure and Tourism in Hobart and Australian Antarctic Territory/East Antarctica," in *The Geopolitics of Tourism: Assemblages of Power, Mobility and the State,* ed. M. Cordoba, M. Mostafanezhad, and R. Norum. (Arizona: University of Arizona Press, 2020).
2   Aant Elzinga, "Punta Arenas and Ushuaia: Early Explorers and the Politics of Memory in Constructing Antarctic Gateway Cities," *The Polar Journal* 3, no. 1 (2013): 227–56.

Construction Seasonality and Building Phasing

1   Paul Siple, "Man's First Winter at the South Pole," *National Geographic* (1 January 1958).
2   Emil Schulthess, *Antarctica: A Photographic Survey* (New York: Simon and Schuster, 1960).

Traverses and Intermodal Transportation

1   James Lever et al., "Advances in Antarctic Sled Technology," U.S. Army Engineer Research and Development Center, February 2016.
2   Robert Mueller, Stephen Hoffman, and Pual Thur, "A Study of Parallels between Antarctica South Pole Traverse Equipment," *Earth and Space: Engineering, Science, Construction, and Operations in Challenging Environments,* ed. Gangbing Song and Ramesh Malla (Reston: American Society of Civil Engineers, 2010).

1   US Army Corps of Engineers, "Polar Traverse Technology," 12 February 2016, erdc.usace.army.mil [online].
2   Ibid.

White Noise
1   Text excerpt from a series of vignettes titled "The Expedition Photographer," in Anne Noble, *The Last Road* (Auckland: Clouds, 2014), 25–28.

A Day in the Life of an Antarctic Traverser

1   *Automobilist,* "More Than Snow-Play," n.d.

Antarctic Mindset
1   Alfred Lansing, *Endurance: Shackleton's Incredible Voyage* (London: Weidenfeld & Nicolson, 2002); Robert Falcon Scott, *Scott's Last Expedition* (London: Smith, Elder & Co., 1913); Robert Huntford, *The Race for the South Pole* (London: Bloomsbury Publishing PLC, 2014).
2   Warren Bennis and Burt Nanus, *Leaders: Strategies for Taking Charge* (New York: Harper and Row, 1985).

The Antarctic Snow Cruiser
1   Gillian Carmoodie, "The Incredible Story of America's Lost 1939 Antarctic Snow Cruiser," 28 April 2019.
2   Gillian Carmoodie, "The Incredible Story of America's Lost 1939 Antarctic Snow Cruiser," 28 April 2019, motorious.com [online].
3   Peter Holderith, "There's a Massive Antarctic Exploration Vehicle Lost Somewhere at the Bottom of the World," *The Drive* (12 May 2020).
4   Alan Taylor, "The Antarctic Snow Cruiser," *The Atlantic* (23 June 2015).

Timeline
1   Sue Bassett, "Shackleton's Car," Antarctic Heritage Trust, 3 February 2014, nzaht.org [online].
2   "The First Motor Car in Antarctica – 1908: Shackleton's *Nimrod* Expedition, 1907–1909," accessed February 2021, coolantarctica.com [online].
3   "Operation Highjump 1946–47," accessed 16 February 2021, south-pole.com [online].
4   Boris Egorov, "How Soviet Off-Road Vehicles Conquered the South Pole," *Russia Beyond,* 26 February 2018.
5   "Antarctic Travel Practicalities and Modern Vehicles," accessed 16 February 2021, coolantarctica.com [online].

Bitch in Slippers
1   Text excerpt from a series of vignettes titled "The Expedition Photographer," in Anne Noble, The Last Road (Auckland: Clouds, 2014), 25–28.

12
Towards a Temperate "Existenzminimum". From Seaweed to Polystyrene

Prefabrication. From Timber-Frame Structures to Panelised Construction

1   Stephen Haddelsey and Alan Carroll, *Operation Tabarin: Britain's Secret Wartime Expedition to Antarctica 1944–46* (Stroud: The History Press, 2014).
2   Kenneth Gaul, *FIDS 1955 Building Report Base Y, Horseshoe Island. AD6/2Y/1955/C* (Cambridge: BAS Archive, 1955).

1   New Zealand Antarctic Heritage Trust, "Conservation Plan: Hillary's Hut, Scott Base."
2   Ibid.
3   Ibid.

Containerisation in Polar Architecture

1   Tony McGlory, "AA Polar Lab," in conversation with UNLESS, lecture at the Architectural Association, 2019.
2   Ibid.

Timeline
1   Ted Scambos, "Plateau Station – Norwegian-U.S. Scientific Traverse of East Antarctica," 2010, traverse.npolar.no [online].

The Evolution of the Architectural Envelope

1   "What Is Péclet Number," accessed 14 February 2021, nuclear-power.net [online].
2   Rem Koolhaas, *Elements of Architecture* (Cologne: Taschen, 2018).
3   New Zealand Antarctic Heritage Trust, *Conservation Plan: Hillary's Hut, Scott Base* (Christchurch: New Zealand Antarctic Heritage Trust, 2004).
4   Ibid.
5   Ibid.
6   Gibert Michael, Naziaty Mohd Yaacob, and Zuraini Md Ali, "The Capsule Living Unit Reconsidered a Utopia Transformed Reality," *Pertanika Journals* 26, no. 3 (2018): 1405–17.
7   Takenaka Group, "Building for the Japanese Antarctic Research Expedition: From the South Pole to a New Kind of Housing in Japan," accessed 16 February 2021, takenaka.co.jp [online].
8   Japan Aerospace Exploration Agency, "A Joint Project of JAXA, NIPR, Misawa Homes, and MHIRD: Demonstration Test of Antarctica Mobile Station Unit," 26 August 2019.
9   Malcolm Mellor, *Methods of Building on Permanent Snow Fields* (Hanover, NH: U.S. Army Materiel Command Terrestrial Sciences Center, Cold Regions Research and Engineering Laboratory, 1968).
10   British Antarctic Survey, "History of Halley (Station Z)," 2020.
11   Sang Leem Lee, *Jangbogo Antarctic Research Station* (Seoul: Space Group, 2011).
12   Hugh Broughton Architects, "Halley VI British Antarctic Research Station," 2020.
13   Koolhaas, *Elements of Architecture.*

1   British Antarctic Survey, "History of Signy (Station H)," n.d.
2   British Antarctic Survey, "History of Deception Island (Station B)," n.d.

Timeline
1   Courtesy of Hugh Broughton Architects, 2005.

13
Geography of Science, "Domed Cities", and Mobile Architectures

Abstract Master Plan

1   Stephen Melville, "Thing of the Past," in *Robert Smithson: Time Crystals,* ed. Amelia Barikin and Chris McAuliffe (Melbourne: Monash University Publishing, 2018); J. G. Ballard, *The Crystal World* (New York: Berkley Books, 1967).
2   Elena Glasberg, *Antarctica as Cultural Critique: The Gendered Politics of Scientific Exploration & Climate Change* (New York: Palgrave Macmillan, 2012).
3   Richard Buckminster Fuller, *Utopia or Oblivion: The Prospects for Humanity* (1969; repr. Baden Lars Müller Publishers, 2008).
4   Black Mountain College Project, "Buckminster Fuller," accessed 17 February 2021, blackmountaincollegeproject.org [online].
5   Buckminster Fuller, *Utopia or Oblivion,* 431.
6   Ibid.
7   Secretariat of the Antarctic Treaty, "Antarctic Specially Managed Area No 5 (Amundsen-Scott South Pole Station, South Pole): Revised Management Plan," *Measure 8 – ATCM XL – CEP XX,* Beijing, 2017, ats.aq [online].
8   Ibid.

9   Robert Smithson, "Entropy and the New Monuments," in *Robert Smithson: The Collected Writings,* edited by Jack Flam (Berkeley: University of California Press, 1966).
10   Arthur Eddington, *The Nature of the Physical World* (Cambridge: Cambridge University Press, 1928), 79.

1   Werner Herzog, *Encounters at the End of the World,* film, 2007.

Drawing
1   Antarctic Treaty Secretariat, *Management Plan for Antarctic Specially Managed Area No.5,* 2017.

A Supra-regional Dispersed City. The Fildes Peninsula

1   Jens Boy et al., "Successional Patterns along Soil Development Gradients Formed by Glacier Retreat in the Maritime Antarctic, King George Island," *Revista Chilena de Historia Natural* 89 (2016): 1-18.
2   Hans-Ulrich Peter et al., *The Current Environmental Situation and Proposals for the Management of the Fildes Peninsula Region* (Dessau-Roßlau: Federal Environment Agency, 2012).
3   Council of Managers of National Antarctic Programs, *Antarctic Station Catalogue* (Christchurch: COMNAP, 2017).
4   Peter et al., *The Current Environmental Situation.*

1   BBC, "Welcome to Villas Las Estrellas," 3 September 2018.

Domesticity in the Extreme

1   Jack Child, *Miniature Messages: The Semiotics and Politics of Latin American Postage Stamps* (Durham, NC: Duke University Press, 2008).
2   Ibid.

The "Informal Settlements" of Antarctica

1   Georgina Davis, "A History of McMurdo Station," *Polar Record* 53, no. 269 (2017): 167–85.; DMJM, *McMurdo Station Long-Range Development Plan: June 2003 Update* (Washington: National Science Foundation Office of Polar Programs, 2003).
2   Harris, "Modes of Informal Urban Development: A Global Phenomenon," *Journal of Planning Literature* 33, no. 3 (2018): 267–86.
3   UN-Habitat, "Informal Urbanism," n.d.
4   Davis, "A History of McMurdo Station."
5   Alan Hemmings and Lorne Kriwoken, "High Level Antarctic EIA under the Madrid Protocol: State Practice and the Effectiveness of the Comprehensive Environmental Evaluation Process." *International Environmental Agreements: Politics, Law and Economics* 10 (2010): 187–208.
6   Australian Government, *SCOTT BASE and McMURDO STATION – Report of an Inspection under Article VII of the Antarctic Treaty and Article 14 of the Protocol on Environmental Protection* (Kingston: Australian Antarctic Division, 2005).
7   Seth Sykora-Bodie and Tiffany Morrison, "Drivers of Consensus – Based Decision – Making in International Environmental Regimes," *Aquatic Conservation: Marine and Freshwater Ecosystems* 29, no. 12 (2019): 2147–61; Rohani Mohd Shah, "Public Perceptions of Antarctic Values," *Procedia: Social and Behavioral Sciences* 168 (2015): 211–18.

1   Elena Glasberg, *Antarctica as Cultural Critique* (New York: Palgrave MacMillan, 2012).

Lessons in Temporal Permanence. McMurdo Station

1   U.S. Antarctic Program, *Report of the U.S. Antarctic Program Blue Ribbon Panel: More and Better Science in Antarctica through Increased Logistical Effectiveness* (Washington: White House Office of Science and National Science Foundation, 2012).

1   Hugh Broughton, in conversation with UNLESS, lecture at Architectural Association, 2019.

Calving Architecture. Archaeology of the Debris

1   British Antarctic Survey, "History of Halley (Station Z)," accessed 17 February 2021, bas.ac.uk [online].
2   Ursula Le Guin, "Sur: A Summary Report of the Yelcho Expedition to the Antarctic 1909–1910," *The New Yorker,* 1 February 1982.
3   Malcolm Mellor, *Methods of Building on Permanent Snow Fields* (Hanover, NH: U.S. Army Materiel Command Terrestrial Sciences Center, Cold Regions Research and Engineering Laboratory, 1968).
4   Mellor, *Methods of Building.*
5   Elmer Clark, *Technical Report 174: Camp Century Evolution of Concept and History of Design Construction and Performance* (Hanover, NH: U.S. Army Materiel Command, Cold Regions Research & Engineering Laboratory, 1965); U.S. Army, "The U.S. Army's Top Secret Arctic City Under the Ice! 'Camp Century'," restored classified film, 1964, youtube.com [online].
6   Clark, *Technical Report 174.*
7   Alan Smith, *Proposed Replacement Station for Halley* (Cambridge: British Antarctic Survey, 1979).
8   Dante Alighieri, *The Divine Comedy: Inferno,* trans. Charles H. Sisson (Oxford: Oxford University Press, 1980), canto 3, st. 1–7.
9   David Walton, "Constructing an Antarctic Atlas," lecture, Architectural Association School of Architecture, London, February 2019.
10   London Transport Museum, *Poster by Man Ray,* 1938, ltmuseum.co.uk [online].
11   Dante Alighieri: *The Divine Comedy: Paradiso,* trans. Charles H. Sisson (Oxford: Oxford University Press, 1980), canto 28, st. 16–21.
12   Annex II to the Protocol on Environmental Protection to the Antarctic Treaty: *Conservation of Antarctic Fauna and Flora,* 4 October 1991.
13   Annex III to the Protocol on Environmental Protection to the Antarctic Treaty: *Waste Disposal and Waste Management,* 4 October 1991.

14

From Anthropic Footprints to Disembodied Technologies

Antarctic Building Design Standards

1   David Harrowfield, *Icy Heritage: The Historic Sites of the Ross Sea Region* (Christchurch: Antarctic Heritage Trust, 1995).

2   Council of Managers of National Antarctic Programs, *Antarctic Station Catalogue* (Christchurch: COMNAP, 2019).
3   Mike Connor, "Wastewater Treatment in Antarctica," *Polar Record* 44, no. 229 (2008): 165–71.
4   Antarctic Treaty Secretariat, *Guidelines for Environmental Impact Assessment in Antarctica Adopted through ATCM Resolution 1, 2016.*

5   Gillian Wratt, *A Story of Antarctic Co-operation: 25 Years of the Council of Managers of National Antarctic Programs,* e-book, 2013.
6   Kristoff Van Rattinghe, "Princess Elisabeth Research Station at Antarctica: Renewable Energy Systems design, Simulation and Optimization," master's thesis, TU Delft, 2008.
7   Antonio Calvo-López et al., "Biparametric Potentiometric Analytical Microsystem for Nitrate and Potassium Monitoring in Water Recycling Processes for Manned Space Missions," *Analytica Chimica Acta* 804 (2013): 190–96.
8   CEEQUAL, ceequal.com.
9   British Antarctic Survey, *Rothera Wharf Reconstruction and Coastal Stablisation: Final Comprehensive Environmental Evaluation* (Cambridge: British Antarctic Survey Environment Office and National Environment Research Council, 2018)
10   New Zealand Green Building Council, "Green Star," accessed 9 June 2020, nzbc.org.nz [online].
11   U.S. Green Building Council, accessed 9 June 2020, usgbc.org [online].
12   Building Research Establishment Environmental Assessment Method (BREEAM), accessed 9 June 2020, www.breeam.com [online].

Waste Management
1   Australian Antarctic Program, "Waste Management," 2020.

Antarctic Environmental Impact Assessment

1   ATCM Recommendation VI-4 (1970).
2   ATCM Recommendation VIII-11 (1975).
3   Protocol on Environmental Protection to the Antarctic Treaty (1991), Article 3.
4   Protocol on Environmental Protection to the Antarctic Treaty (1991), Article 2.
5   Secretariat of the Antarctic Treaty, *CEP Handbook,* 2019, ats.aq [online].
6   United Kingdom, *Environmental Impact Assessments: Update on Broader Policy Discussions,* Working Paper number 41, submitted to the 40th Antarctic Treaty Consultative Meeting (Prague, 2017).
7   Kees Bastmeijer and Ricardo Roura, "Environmental Impact Assessment in Antarctica," in *Theory and Practice of Transboundary Environmental Impact Assessment,* ed. Kees Bastmeijer and Timo Koivurova, 175–219 (Leiden: Brill, 2008).
8   Alan Hemmings and Ricardo Roura, "A Square Peg in a Round Hole: Fitting Impact Assessment under the Antarctic Environmental Protocol to Antarctic Tourism," *Impact Assessment and Project Appraisal* 21, no. 1 (2003): 13–24.
9   Peter Wathern, *Environmental Impact Assessment: Theory and Practice* (New York: Routledge, 1998), e-book; John Glasson, Riki Therivel, and Anrew Chadwick, *Introduction to Environmental Impact Assessment* (New York: Routledge, 2005).
10   See, for example, the reports of CEP IV (para. 31); CEP VI (paras. 50 to 54); CEP VIII (para. 112); and CEP XVIII (para. 118).
11   Bastmeijer and Roura, "Environmental Impact Assessment in Antarctica."
12   Antarctic Treaty Secretariat, *Guidelines for Environmental Impact Assessment in Antarctica adopted through ATCM Resolution 1,* 2016.
13   Bastmeijer and Roura, "Environmental Impact Assessment in Antarctica."
14   Steven Chown et al., "Challenges to the Future Conservation of the Antarctic," *Science* 337 (2012): 158–59.

The Footprint of Antarctic Stations

1   Antarctic Treaty Secretariat, *Protocol on Environmental Protection to the Antarctic Treaty,* 1991.
2   Current estimates of the ice-free area range from just 0.18% to 0.44% of the continent's overall land area.

1   New Zealand Antarctic Heritage Trust, "Conservation Plan: Cape Evans" (Christchurch: NZATH, 2004).

Built-in Redundancies or Infrastructure Resilience

1   Uptime Institute, *Data Center Site Infrastructure Tier Standard: Topology* (2018).

1   Richard Byrd, "To the Men at South Pole Station," *National Geographic* (July 1957).

The Last Road
1   Text excerpt from a series of vignettes titled "The Expedition Photographer," in Anne Noble, *The Last Road* (Auckland: Clouds, 2014), 25–28.

Building Performance and Carbon Emissions

1   World Green Building Council, "Bringing Embodied Carbon Upfront," 2020.
2   Ramboll, "Sustainable Buildings Market Study 2019," accessed 16 February 2021, uk.ramboll.com [online].
3   Ramboll, "Modernising Infrastructure in Antarctica," accessed 16 February 2021, uk.ramboll.com [online].

1   British Antarctic Survey, Natural Environment Research Council, *Proposed Construction and Operation of Halley VI Research Station.*
2   Antarctic Treaty Secretariat, *Management Plan for Antarctic Specially Managed Area No.5,* 2017.
3   Tyler Reid, "Nuclear Power at McMurdo Station," 2014.

Striving for Zero Carbon Emission in Princess Elisabeth Antarctica Station and the Ellipsoid

1   The joule is a derived unit of the 6 base units (m, kg, s, A, K, Cd):
$1 N = 1 kg.m/ s^2$; $1 J = 1 N.m = 1 kg\ m^2/ s^2$; $1 W = 1 J/s = 1 N\ m/s = 1 kg\ m^2/ s^3$;$1 kWh = 3.6\ 10^6\ J.$
2   As the energy (the money) needed to bring anything to Antarctica (including for logistics) is an important part of the embodied energy, the design of PEA, and, in particular, its shell and core, focused on reducing this cost to a minimum. The related costs also depend on the density (d in kg / m^3) and on the volume (V in m^3) of the transported item.
3   Philippe Samyn, "Structural Engineering and Embodied Energy," *Steel Construction Journal* 12, no. 3 (2019): 174–75.
4   Wikipedia: "The volume and displacement indicator of an architectural structure." Full theory in Philippe Samyn, *Étude de la morphologie des structures à l'aide des indicateurs de volume et de déplacement* (Brussels: Académie Royale de Belgique Classe des Sciences, 2004).
5   All other parameters unchanged, the optimum for L depends on the lifetime of the structure, which is directly proportional to the cost. Indeed, for a given Power P [the energy or the money per unit of time: $E/t = (kWh/s)$ or cost/time ($€/s$)] the longer the lifetime expectation, the longer L might be.
6   It is still difficult to imagine increasing the thermal resistance of a shell in Antarctica above the $R = 15\ m^2\ °K/W$ achieved at PEA, but I am working on it.
7   Solar gains are related to the amount of daylight enjoyed during the Antarctic summer. Further studies that go into this interesting issue in depth are not taken into account here.
8   This annual average value is, of course, dependant on the location in the Antarctic and may vary from −15°C to −50°C.
9   This includes the electrical energy of the fans and auxiliary equipment.
10   It might also be 30 or 50 $m^3/(h \cdot p)$.

11 At 20°C and atmospheric pressure, it might be lowered to 15% RH (as at PEA).
12 One must, however, be careful with regard to the possible importance of the solar gains resulting from the low path of the sun.
13 It also calls for R to be maximised, which does not seem easy above 15 m²K/W but perhaps I lack imagination here.
14 The design has no influence on the given vertical exploitation loads F_v: hence, only F_h matters.
15 c_d equals 1.05 for a cube, 0.82 for a long cylinder, 0.47 for a sphere, 0.09 for a streamlined half-body on the ground, and 0.04 for a streamlined body detached from the ground.
16 It is therefore of the greatest importance to precisely examine the wind forces arising for F_h.
17 In just the same way as the gross build areas and volumes per person in a town increase in accordance with its size to allow for additional infrastructure and equipment spaces.

Design for Removal. The Life Cycle of Antarctic Stations

1 Ramboll Deutschland GmbH, *Design of German Neumayer III Station,* report, 2005–2008.
2 Ramboll Deutschland GmbH, *Design of New Wintering Complex at Vostok Station,* report, 2018–2020.
3 Ramboll Deutschland GmbH, *Design of Indian Bharati Station,* report, 2008–2013.

Ephemeralization. "We left no footprint, even"

1 Robert Falcon Scott, *Journals: Captain Scott's Last Expedition,* (1913; repr., Oxford: Oxford University Press, 2009).
2 Antarctic Treaty, Article IX (2).
3 Rick Petersen, "AA Polar Lab: Inhabitation in the Extreme Workshop," lecture, London, 2019.
4 Guidelines for Environmental Impact Assessment in Antarctica adopted through ATCM Resolution 1.
5 Protocol on Environmental Protection to the Antarctic Treaty, 4 October 1991.
6 Richard Buckminster Fuller, "The World Game," 1961.
7 Richard Buckminster Fuller, *Nine Chains to the Moon* (1938; repr., New York: Anchor Books Edition, 1971).
8 Nathan Rosenberg, *Inside the Black Box* (Cambridge: Cambridge University Press, 1982).
9 Pier Vittorio Aureli, "The Theology of Tabula Rasa: Walter Benjamin and Architecture in the Age of Precarity," in *The City as a Project (A Research Collective),* 9 May 2015.
10 Jaime Snyder, introduction to Richard Buckminster Fuller, *Operating Manual for Spaceship Earth* (1969; repr., Baden: Lars Müller Publishers, 2015).

1 Alamy, image description: "Camp under the Wild Range", 20 December 1911, (1913). Artist: Robert Falcon Scott.
2 Ratcliffe Ellis, "Scott Pyramid Tents – Ratcliffe Ellis," 2020.
3 British Antarctic Survey, "Field Training Camp – British Antarctic Survey," 2020.
4 International Thwaites Glacier Collaboration, "New Sights in the Second Field Season of the International Thwaites Glacier Collaboration," press release, 15 April 2020.
5 Jonathan Amos, "Thwaites: 'Doomsday Glacier' Vulnerability Seen in New Maps," *BBC News,* 8 September 2020.
6 JAXA, *A Joint Project of JAXA, NIPR, Misawa Homes, and MHIRD Demonstration Test of Antarctica Mobile Station Unit,* 26 August 2019.

The Googies

1 *Australian Antarctic Magazine.*

1 BGR, "Development of the BGR's Polar Research Activities," n.d.
2 German Embassy Wellington, "GONDWANA Station in Terra Nova Bay and Lillie Marleen Hut at Everett Range, Northern Victoria Land," 23 May 2019.

Halley and the Importance of Automation

1 NASA Earth Observatory, "Countdown to Calving at Brunt Ice Shelf," 23 January 2019.
2 Hugh Broughton, in conversation with UNLESS, lecture at Architectural Association, London 2019.
3 Ibid.

15

The Ideological Use of Relics

Questioning Antarctica's Museumification

1 Bjørnar Olsen, *In Defense of Things: An Archaeology and the Ontology of Objects* (Plymouth: Altamira Press, 2010).
2 Laurajane Smith, *The Uses of Heritage* (London: Routledge, 2006).
3 Maria Ximena Senatore and Andrés Zarankin, "Widening the Scope of the Antarctic Heritage Archaeology and 'The Ugly, the Dirty and The Evil' in Antarctic History," in *Polar Settlements – Location, Techniques and Conservation,* ed. Susan Barr and Paul Chaplin (Oslo: IPHC-ICOMOS, 2011), 51–59.
4 Caitlin De Silvey, "Observed Decay: Telling Stories with Mutable Things," *Journal of Material Culture* 11, no. 3 (2006): 318–38.
5 Maria Ximena Senatore, "Archaeologies in Antarctica from Nostalgia to Capitalism: A Review," *International Journal of Historical Archaeology* 23, no. 3 (September 2019): 755–71.
6 Ian Hodder, *Studies in Human-Thing Entanglement,* e-book, 2016.
7 Tim Edensor, *Industrial Ruins: Space, Aesthetics and Materiality* (Oxford: Berg, 2005).

Antarctic Archaeology, Heroic Age Huts, and Scientific Stations

Ghostly Cabin Filled with Frozen History Waits for Explorers to Finish Last Meal
1 Richard Byrd, "All-Out Assault on Antarctica," National Geographic (August 1956): 156.

1 Scott Polar Research Institute, "Scott's Last Expedition," archival press release, 29 March 1912.
2 New Zealand Antarctic Heritage Trust, "Conservation Report: Shackleton's Hut" (Christchurch: NZATH, 2003).
3 Ibid.
4 Ibid.
5 Ibid.
6 Ibid.

Preservation Contradictions on an Active Volcanic Crater. Deception Island

1 Deception Island, "Deception Island Management Package," 2005, deceptionisland.aq [online].
2 International Association of Antarctica Tour Operators (IAATO), "IAATO Overview of Antarctic Tourism: 2017–18 Season and Preliminary Estimates for 2018–19 Season" (2018).
3 SCAR Composite Gazetteer of Antarctica, "Antarctic Place Names," 2020.
4 Deception Island, "Volcanic Activity," accessed 16 February 2021, deceptionisland.aq [online].
5 Stephen Dibbern, "Fur Seals, Whales and Tourists: A Commercial History of Deception Island, Antarctica," *Polar Record* 3 (July 2010): 210–21.
6 John Dudeney, "Operation Tabarin: Britain's Secret Wartime Expedition to Antarctica," *Polar Record* 51, no. 1 (January 2015).
7 Guillermo Muñoz, "Museo en la Antártica," *Boletín Antártico Chileno* 15, no. 1 (May 1996).
8 IAATO, "IAATO Overview of Antarctic Tourism."
9 Kevin Hughes et al., "Biological Invasions in Terrestrial Antarctica: What Is the Current Status and Can We Respond?" *Biodiversity and Conservation* 24, no. 5 (March 2015).

Māori Culture in Antarctica. Preserving Knowledge and Papatūānuku

1 In Ngātiwai creation genealogies, *ko te unaunahi i whakapiripiri ki te ikanui a Maui* refers to the "four fish scales of Maui" symbolised in Ngātiwai carvings, which are tied together with *te aho tapu,* the sacred binding thread.
2 Priscilla Wehl et al., "Transforming Antarctic Management and Policy with an Indigenous Māori Lens," *Nature Ecology and Evolution* (forthcoming).
3 *Tōtara* (*Podocarpus totara*), rimu (*Dacrydium cupressinum*), and harakeke (*Phormium tenax*).

Roald Amundsen's Tent at the South Pole

1 Roald Amundsen, *The South Pole: An Account of the Norwegian Antarctic Expedition in the Fram 1910–1912* (New York: Cooper Square Press, 2001); Robert Falcon Scott, *Scott's Last Expedition* (London: Smith, Elder & Co., 1913); Nobu Shirase, *Nankyokuki* (1913), trans. Elizabeth Kolbert and Francis Spufford, in *The Ends of the Earth: An Anthology of the Finest Writing on the Arctic and the Antarctic,* ed. Elizabeth Kolbert (New York: Bloomsbury, 2007), 55–62.

The Wreck of Shackleton's *Endurance* in the Weddell Sea

1 Ernest Shackleton, *South: The Story of Shackleton's Last Expedition 1914–17* (London: Heinemann, 1919), 376; Frank Arthur Worsley, *Endurance: An Epic of Polar Adventure* (London: Allan, 1931), 316.
2 Lars Bergman and Robin G. Stuart, "On the Location of Shackleton's Vessel Endurance," *Journal of Navigation* 72, no. 2 (2018): 257–68.
3 Julian Dowdeswell et al., "Sea-Floor and Sea-Ice Conditions in the Western Weddell Sea, Antarctica, around the Wreck of Sir Ernest Shackleton's Endurance," *Antarctic Science* (2020): 32.
4 Ibid.; Julian Dowdeswell and Jonathan Bamber, "Keel Depths of Modern Antarctic Icebergs and Implications for Sea-Floor Scouring in the Geological Record," *Marine Geology* 243 (2007): 120–31.
5 Dowdeswell et al., "Sea-Floor and Sea-Ice Conditions," 32.
6 Ibid.
7 Adrian Glover et al., "Bone-Eating Worms from the Antarctic: The Contrasting Fate of Whale and Wood Remains on the Southern Ocean Seafloor," *Proceedings of Biological Sciences* (2013): 280.
8 Shackleton, *South,* 376; Worsley, *Endurance,* 316.
9 Dowdeswell et al., "Sea-Floor and Sea-Ice Conditions," 32.
10 Ibid.
11 Antarctic Treaty Secretariat, *Antarctic Treaty Forty-Second Consultative Meeting, Measure 12: Revised List of Antarctic Historic Sites and Monuments: The Wreck of Sir Ernest Shackleton's Vessel Endurance and C. A. Larsen Multiexpedition Cairn,* 2019.

12   Ibid.
13   The Protocol on Environmental Protection to the Antarctic Treaty, Annex V (1991), Article 8.4.

The Role of Photography in Antarctic Preservation

1   Elena Glasberg, *Antarctica as Cultural Critique* (New York: Palgrave MacMillan, 2012).
2   John Roscoe, *Contributions to the Study of Antarctic Surface Features by Photographical Methods,* master's thesis, University of Maryland, 1952.
3   Richard Byrd and Harold Saunders, "The Flight to Marie Byrd Land: With a Description of the Map", *Geographical Review* 23, no. 2 (April 1933): 177–209.

16

Antarctica and Outer Space. Laboratories for a Future as Global Commons

Exploring Antarctic and Extraterrestrial Connections in Fiction

1 Roslynn Haynes, "Astronomy and the Dreaming: The Astronomy of the Aboriginal Australians," in *Astronomy Across Cultures: The History of Non-Western Science,* ed. Helain Selin (London: Springer, 2000), 59.
2 For more detail, see Elizabeth Leane, *South Pole: Nature and Culture* (London: Reaktion, 2016), ch. 1.
3 Gustavus Pope, *Journey to Mars. The Wonderful World,* (Westport: Hyperion, 1974), 90.
4 Brian Aldiss, and Roger Penrose, *White Mars, or the Mind Set Free: A 21st-Century Utopia* (London: Little, Brown, 1999), 323; Kim Stanley Robinson, *Blue Mars* (London: HarperCollins, 1996), 778.

Dry Valleys, Permafrost, and Mars

1   James Bockheim and C. Tarnocai, "Nature, Occurrence and Origin of Dry Permafrost," *Collection Nordicana* 55 (1998): 57–63.
2   Griffith Taylor, *The Physiography of the McMurdo Sound and Granite Harbour Region* (London: Harrison and Sons, 1922).
3   B. McKelvey and P. Webb, "Geological Investigations in Southern Victoria Land," *New Zealand Journal of Geology and Geophysics* (1959), 143–62, 361–87.
4   Trevor Chinn, "Hydrology and Climate in the Ross Sea Area," *Journal of the Royal Society of New Zealand* 11, no. 4 (January 1981): 373–86.
5   David Sugden, George Denton, and David Marchant, "Landscape Evolution of the Dry Valleys," *Journal of Geophysical Research Atmospheres* 100 (June 1995): 9949–67.
6   Robert Spigel and John Priscu, "Physical Limnology of the McMurdo Dry Valley Lakes," in *Ecosystem Dynamics in a Polar Desert: The McMurdo Dry Valleys, Antarctica,* ed. John Priscu (Washington: American Geophysical Union, 1998), 153–87.
7   Robert Wharton, "Stromatolitic Mats in Antarctic Lakes," in *Phanerozoic Stromatolites II,* ed. Janine Bertrand-Sarfati and C. Monty (Dordrecht: Springer, 1994), 53–70.
8   Edward Adams et al. "Permanent Ice Covers of the McMurdo Dry Valley Lakes," in *Ecosystem Dynamics in a Polar Desert: The McMurdo Dry Valleys, Antarctica,* ed. John Priscu (Washington, DC: American Geophysical Union, 1998), 281–95.
9   P. Calkin, R. Behling, and C. Bull, "Glacial History of Wright Valley," *Antarctic Journal of the United States* 5, no. 1 (1970): 22–27.
10   Sugden, Denton, and Marchant, "Landscape Evolution of the Dry Valleys."
11   M. Stuiver et al., "History of the Marine Ice Sheet in West Antarctica," in *The Last Great Ice Sheets* (New York: Wiley, 1981), 319–436.
12   P. Webb, "Wright Fiord, Pliocene Marine Invasion of an Antarctic Dry Valley," *Antarctic Journal of the United States* 7, no. 6 (1972): 227–34.
13   McMurdo Dry Valleys LTER, "Home page," accessed 16 February 2021, mcm.lternet.edu [online].
14   John Priscu et al., "Photosynthate Distribution by Microplankton in Permanently Ice-Covered Antarctic Desert Lakes," *Limnology and Oceanography* 32, no. 1 (January 1987): 260–70.
15   David Wharton and Ian Brown, "A Survey of Terrestrial Nematodes from the McMurdo Sound Region," *New Zealand Journal of Zoology* 16, no. 3 (July 1989): 467–70.
16   Angela McGaughran et al., "Extreme Glacial Legacies: A Synthesis of the Antarctic Springtail Phylogeographic Record," *Insects* 2 (April 2011): 62–82.
17   R. Cameron, J. King, and Charles David, "Microbiology, Ecology and Microclimatology of Soil Sites in Dry Valleys of Southern Victoria Land, Antarctica," in *Antarctic Ecology,* ed. Martin Holdgate (New York: Academic Press, 1970), 702–716.
18   A. Schwartz, T. Green, and Rod Seppe, "Terrestrial Vegetation at Canada Glacier, Southern Victoria Land, Antarctica," *Polar Biology* 12, no. 3 (September 1992): 397–404.
19   E. Imre Friedmann, "Endolithic Microorganisms in the Antarctic Cold Desert," *Science* 215, no. 4536 (March 1982): 1045–53.
20   D. Gilichinsky, Gary Wilson, E. Friedmann, and C. McKay, "Microbial Populations in Antarctic Permafrost: Biodiversity, State, Age, and Implication for Astrobiology," *Astrobiology* 7, no. 2 (May 2007): 275–311.
21   J. Heldmann, et al., "The High-Elevation Dry Valleys in Antarctica as Analog Sites for Subsurface Ice on Mars," *Planetary and Space Science* 85 (1 September 2013): 53–58.
22   Colin Harris, "Science and Environmental Management in the McMurdo Dry Valleys, Southern Victoria Land, Antarctica," in *Ecosystem Dynamics in a Polar Desert: The McMurdo Dry Valleys, Antarctica,* ed. John Priscu (Washington: American Geophysical Union, 1998), 337–50.
23   Antarctica New Zealand, United States Antarctic Program, *McMurdo Dry Valleys ASMA Manual* (Christchurch: Office of Polar Programs, 2015).

1   NASA, "Twin Peaks in Super Resolution – Right Eye," 8 September 1999.
2   NASA, "Mars's Twin Peaks," 2 September 2020.
3   NASA, "Mars Exploration Program," n.d.

Antarctica and the Exploration of the Solar System

1   Roberta Score, "The Thrill of the Search: Finding ALH84001," *The Planetary Report* 17, no. 1 (January/February 1997): 5–7.
2   David McKay et al., "Search for Past Life on Mars: Possible Relic Biogenic Activity in Martian Meteorite ALH84001," *Science* 273, no. 5277 (16 August 1996): 924–30.
3   W. Henry Lambright, *Why Mars: NASA and the Politics of Space Exploration* (Baltimore: Johns Hopkins University Press, 2014), 67–70.
4   Journalist Kathy Sawyer quoted in Lambright, *Why Mars,* 132.
5   President Bill Clinton quoted in "Life on Mars?" *New York Times,* 8 August 1996, A26.
6   "Life on Mars?" *New York Times,* 8 August 1 996, A26.
7   Erik Conway, *Exploration and Engineering: The Jet Propulsion Laboratory and the Quest for Mars* (Baltimore: Johns Hopkins University Press, 2015), 115–20.

1   Trevor Paglen, *The Last Pictures* (New York: Creative Time Books, 2012).
2   Noble and Thompson, foreword in Paglen, *The Last Pictures.*
3   Paglen, *The Last Pictures.*

Relational Trajectories of Antarctica and Outer Space

1   Elizabeth DeLoughrey, "Satellite Planetarity and the Ends of the Earth," *Public Culture* 26, no. 273 (2014): 260.
2   Kathryn Yusoff, "Test Landscapes and the 'Geographical Gift' of a Continent to Science," paper presented at the workshop "Polar Field Stations and International Polar Year (IPY) History, Culture, Heritage, Governance (1882–Present)," University of Cambridge, May 3–4, 2007, 2.
3   Elena Glasberg, *Antarctica as Cultural Critique: The Gendered Politics of Scientific Exploration & Climate Change* (New York: Palgrave Macmillan, 2012), 4.

Space and Suit

1   John Milton, *Paradise Lost* (London: Jacob Tonson, 1730), bk. 1, line 650.
2   Ibid., bk. 10, line 437.
3   United States et al., *Apollo 11 Technical Air-to-Ground Voice Transcription: (GOSS NET 1)* (Houston: Manned Spacecraft Center, 1969), 249, timecode 01 13 34 56.
4   Apsley Cherry-Garrard, *The Worst Journey in the World: Antarctic, 1910–1913* (New York: John H. Doran, 1922), 40.
5   Reyner Banham, *Scenes in America Deserta* (London : Thames and Hudson, 1982), 44.
6   See Nicholas de Monchaux, *Spacesuit : Fashioning Apollo* (Cambridge: MIT Press, 2011), chap. 6, "Cyborgs and Space."
7   Ruth La Ferla, "Moon Boots Back on Earth," *The New York Times: Style,* 17 October 2004.
8   Cherry-Garrard, *The Worst Journey,* 334.
9   Ed Stafford, *Expeditions Unpacked: What the Great Explorers Took into the Unknown* (London: White Lion Publishing, 2019), 40.
10   Cherry-Garrard, *The Worst Journey,* 448.
11   See Monchaux, *Spacesuit* chapter.

Antarctic Attire

1   Scott Polar Research Institute Museum Catalogue, "Mitten," archival image.
2   Scott Polar Research Institute Museum Catalogue, "Boot," archival image.

Timeline
1   Richard Evelyn Byrd, "The Conquest of Antarctica by Air," *National Geographic* (August 1930): 129–33.

Rethinking the Antarctic Suit

1   Arthur Iberall, "The Use of Lines of Nonextension to Improve Mobility in Full-Pressure Suits," *Aerospace Medical Research Laboratories* AMRL-TR-64-118 (November 1964).
2   Jennifer Chu, "Shrink-Wrapping Spacesuits" *MIT News* (18 September 2014).
3   Ibid.
4   Ibid.

Do(o)med Interiors

1   Jerri Nielsen, *Ice Bound* (London: Ebury Press, 2001), 49. The quality of the air inside the SPD was frequently "worse than an L.A. freeway during rush hour".
2   Gary Curtis, "South Pole Dome: Let's Keep It!" *The Polar Times* 2, no. 7 (1996): 16.

3   Jennifer Law et al. "Relationship between Carbon Dioxide Levels and Reported Headaches on the International Space Station," *Journal of Occupational & Environmental Medicine* 56, no. 5 (2014): 477–83.

Antarctica, Human Adaptation, and Space

1   See Peter Suedfield and Karine Weiss, "Antarctica: Natural Laboratory and Space Analogue for Psychological Research," *Environment and Behavior* 32, no. 1 (2000): 7–17; Gro Mjeldheim Sandal, Fons van deVijver, and Nathan Smith, "Psychological Hibernation in Antarctica," *Frontiers in Psychology* (20 November 2018).
2   Sara Wheeler, *Terra Incognita* (New York: The Modern Library, 1999), 61.
3   Peter Suedfield, "Polar Psychology: An Overview," *Environment and Behavior* 23, no. 6 (1991): 653–65.
4   Apsley Cherry-Garrard, *The Worst Journey in the World* (London: Pimlico, 1922).
5   Jean Rivolier et al., *Man in the Antarctic: The Scientific Work of the International Biomedical Expedition to the Antarctic (IBEA)* (London: Taylor & Francis, 1988).
6   Rivolier et al., *Man in the Antarctic,* 82–83.
7   Ibid.
8   See Gordon Elliott Fogg, *A History of Antarctic Science* (Cambridge: Cambridge University Press, 2005), 382.
9   Though an incidence of one in twenty might not seem high, the people manifesting problems belong to an already highly screened population. See Lawrence Palinkas et al., "Incidence of Psychiatric Disorders after Extended Residence in Antarctica," *International Journal of Circumpolar Health* 63, no. 2 (2004): 157–68; Desmond Lugg, "Behavioral Health in Antarctica: Implications for Long-Duration Missions in Space," *Aviation Space and Environmental Medicine* 76, suppl. 6 (2005): B89–93.
10   Suedfield and Weiss, "Antarctica."
11   Suedfield, "Polar Psychology."
12   Wheeler, *Terra Incognita,* 68.
13   Ibid.

1   James Flecker, *The Golden Journey to Samarkand,* poem, 1913.

Behavioural Patterns in the Extreme. Concordia as Space Simulator

1   Gro Mjeldheim Sandal, Fons van deVijver, and Nathan Smith, "Psychological Hibernation in Antarctica," *Frontiers in Psychology* (20 November 2018).

1   Web of Science, accessed 18 December 2020, webofknowledge.com [online].

Hydroponics. Greenhouses in the Extreme

1   A total of forty-two LED lights are installed in the greenhouse.

1   DLR Institute of Space Systems, "EDEN ISS," n.d.
2   Lori Meggs, "Growing Plants and Vegetables in a Space Garden," 15 June 2010, nasa.gov [online].
3   Ibid.

International Antarctic Stations

1   See Alan Hemmings, "Why Did We Get an International Space Station before an International Antarctic Station?" *The Polar Journal 1* (2011): 5–16.
2   Alan Hemmings, "The Philosophy of Law in the Antarctic," in *Philosophies of Polar Law,* ed. Dawid Bunikowski and Alan Hemmings (London: Routledge, 2020).
3   Antarctic Treaty, Preamble: "CONVINCED that the establishment of a firm foundation for the continuation and development of such cooperation on the basis of freedom of scientific investigation in Antarctica as applied during the International Geophysical Year accords with the interests of science and the progress of all mankind."
4   Antarctic Treaty, Articles II, III, IX.
5   See, for example, Convention on the Conservation of Antarctic Marine Living Resources, Preamble: "Considering that it is essential to increase knowledge of the Antarctic marine ecosystem and its components so as to be able to base decisions on harvesting on sound scientific information."
6   See, for example, Richard Lewis, *A Continent for Science* (London: Secker & Warburg, 1965); Paul Berkman et al., eds., *Science Diplomacy: Science, Antarctica, and the Governance of International Spaces* (Washington, DC: Smithsonian Institution Scholarly Press, 2011).
7   Article 6 1(e), Protocol on Environmental Protection to the Antarctic Treaty.
8   Council of Managers of National Antarctic Programs, *Antarctic Station Catalogue* (Christchurch: COMNAP, 2019).
9   See Alan Hemmings and Bob Frame, "Antarctica's Latest Governance Challenge: Coronavirus," in this volume, p. 130.

The Antarctic and Outer Space. From *Res Nullius* to *Res Communis*

1   Stuart Elden, "Missing the Point: Globalization, Deterritorialization and the Space of the World," *Transactions of the Institute of British Geographers* 30, no. 1 (2005): 8–19.
2   Patricia Seed, *Ceremonies of Possession in Europe's Conquest of the New World, 1492–1640* (Cambridge: Cambridge University Press, 1995).
3   Klaus Dodds and Christy Collis, "Post-colonial Antarctica," in *Handbook on the Politics of Antarctica,* ed. Klaus Dodds, P. Roberts, and Alan Hemmings (Cheltenham: Edward Elgar Publishing, 2017), 50–68.
4   Christy Collis, "The Geostationary Orbit: A Critical Legal Geography of Space's Most Valuable Real Estate," in *Down to Earth: Satellite Technologies, Industries, and Cultures,* ed. L. Parks and J. Schwoch (New Brunswick: Rutgers University Press, 2012), 61–81.
5   Marie Bourrel, Torsten Thiele, and Duncan Currie, "The Common Heritage of Mankind as a Means to Assess and Advance Equity in Deep Sea Mining," *Marine Policy* 95 (2016).

Visualisation and Drawing Sources

The copyright of all visualisations and drawings published in this book is held by Giulia Foscari / UNLESS.

INTRODUCTION

p. 15
Australian Antarctic Data Centre. "Southern Ocean Zones." 1995. Accessed November 2019. [online]
British Antarctic Survey. "Antarctic Coastline." 2019. Accessed November 2019. bas.ac.uk. [online]
British Antarctic Survey. "Antarctic Contours." 2019. Accessed November 2019. bas.ac.uk. [online]
British Antarctic Survey. "Antarctic Surface DEM." 2014. Accessed November 2019. bas.ac.uk. [online]
British Antarctic Survey. "Data Limit 60S." 2012. Accessed November 2019. bas.ac.uk. [online]
Council of Managers of National Antarctic Programs. Antarctic Station Catalogue. Christchurch: COMNAP, 2017. comnap.aq. [online]
Fretwell, Peter. *Antarctic Atlas: New Maps and Graphics That Tell the Story of a Continent.* London: Particular Books, 2021.
National Oceanic and Atmospheric Administration. "Geomagnetism." Accessed: November 2019. noaa.gov. [online]
National Oceanic and Atmospheric Administration. *Sea Ice and Snow Cover Extent, 1980–2019, 1981–2019.* Accessed: November 2019. [online]
Natural Earth Data. "World Shaded Relief." 2012. Accessed December 2019. naturalearthdata.com. [online]
Norwegian Polar Institute, 2017. "Polar Place Names." Accessed December 2019. npolar.no. [online]
QGIS. "Quantarctica: An Antarctic GIS Package." 2020. Accessed 18 December 2020. qgis.org. [online]
United States Geological Survey. "Tectonic Plates." 2019. Accessed December 2019. [online]

p. 16
Global Carbon Project. "Global Carbon Budget 2020." Accessed November 2019. globalcarbonproject.org. [online]
Le Quéré, Corinne, Robbie M. Andrew, Pierre Friedlingstein, Stephen Sitch, Judith Hauck, Julia Pongratz, Penelope A. Pickers et al. "Global Carbon Budget 2018." *Earth System Science Data* 10, no. 4 (December 2018): 2141–94.
Ahn, Jinho, and Edward J. Brook. "Siple Dome Ice Reveals Two Modes of Millennial CO_2 Change During the Last Ice Age." *Nature Communications* 5, no. 1 (2014).
Bereiter, Bernhard, Dieter Lüthi, Michael Siegrist, Simon Schüpbach, Thomas F. Stocker, and Hubertus Fischer. "Mode Change of Millennial CO_2 Variability During the Last Glacial Cycle Associated with a Bipolar Marine Carbon Seesaw." *PNAS* 109, no. 25 (2012): 9755–60.
Bereiter, Bernhard, H. Fischer, J. Schwander, and Thomas Stocker. "Diffusive Equilibration of N2, O2 and CO_2 Mixing Ratios in A 1.5-Million-Years-Old Ice Core." *The Cryosphere* 8, no. 1 (2014): 245–56.
Buizert, C., Benjamin Keisling, J. E. Box, F. He, Anders E. Carlson, Gaylen Sinclair, R. M. DeConto. "Greenland-Wide Seasonal Temperatures During the Last Deglaciation." *Geophysical Research Letters* (2018).

Gaspari, Vania, Carlo Barbante, Giulio Cozzi, Paolo Cescon, Claude Boutron, Paolo Gabrielli, Gabriele Capodaglio et al. "Atmospheric Iron Fluxes over the Last Deglaciation: Climatic Implications." *Geophysical Research Letters* 33, no. 3 (2006): L03704.
Jouzel, Jean, Valerie Maison-Delmotte, O. Cattani and Gabrielle Dreyfus. "Orbital and Millennial Antarctic Climate Variability Over the Past 800,000 Years." *Science* 317, no. 5839 (2007): 793–96.
MacFarling, Meure C., David Etheridge, Cathy Trudinger, P. Steele, R. Langenfelds, T. van Ommen et al. "Law Dome CO2, Ch4 and N2O Ice Core Records Extended to 2000 Years BP." *Geophysical Research Letters* 33, no. 14 (2006).
Petit, Jean-Robert, Jean Jouzel, Doninique Raynaud, N. Barkov, I. Basile, M. Bender, J. Chappellaz et al. "Climate and Atmospheric History of the Past 420,000 Years from the Vostok Ice Core, Antarctica." *Nature* 399 (1999): 429–36.
Ritchie, Hannah, and Max Roser. "CO_2 Emissions." Accessed 17 March 2021. ourworldindata.org. [online]
Rubino, Mauro, David Etheridge, Cathy Trudinger, Colin Allison, Mark Battle, R. L. Langenfelds, L. P. Steele et al. "A Revised 1000 Year Atmospheric $\delta_{13}C$-CO_2 Record from Law Dome and South Pole, Antarctica." *Journal of Geophysical Research: Atmospheres* 118, no. 15 (2013): 8482–99.
Schneider, Robert, Jochen Schmitt, Peter Koehler, Fortunat Joos, and Hubertus Fischer. "A Reconstruction of Atmospheric Carbon Dioxide and Its Stable Carbon Isotopic Composition from the Penultimate Glacial Maximum to the Glacial Inception." *Climate of the Past* 9, no. 6 (2013): 2507–23.
Siegenthaler, Urs, Thomas Stocker, Eric Monnin, Jakob Schwander, Bernhard Stauffer, Dominique Raynaud, Jean-Mac Barnola et al. "Stable Carbon Cycle-Climate Relationship During the Late Pleistocene." *Science* 310, no. 5752 (2005): 1313–17.

p. 22
Flanders Marine Institute. "Exclusive Economic Zones (EEZ)." Accessed January 2020. marineregions.org. [online]
Database of Global Administrative Areas (GADM), 2018. Accessed December 2019. gadm.org. [online]

DOMINANCE OR RESEARCH

01
Ant-Arctica In-Cognita

The Discovery of Antarctica

pp. 32–33
Australian Antarctic Division. "Macquarie Island Station: A Brief History". Accessed March 2020. antarctica.gov.au. [online]

British Antarctic Survey. "Antarctic Coastline." 2019. Accessed November 2019. bas.ac.uk. [online]
British Antarctic Survey. *Two Hundred Years Since the Discovery of the South Shetland Islands.* 19 February 2019. bas.ac.uk. [online]
Cool Antarctica. *Ross Island and Mount Erebus Antarctic Regions – Maps and Pictures.* Accessed 1 March 2020. coolantarctica.com. [online]
Database of Global Administrative Areas (GADM), 2018. Accessed December 2019. gadm.org. [online]
Dater, Henry M. "History of Antarctic Exploration and Scientific Investigation." In *Antarctic Map Folio Series,* edited by Vivian C. Bushnell. New York: American Geographical Society, 1975. [online]
Falkland Islands Government. "Historical Dates." Accessed 1 March 2020. falklands.gov.fk. [online]
Headland, Robert K. *A Chronology of Antarctic Exploration: A Synopsis of Events and Activities from the Earliest Times until the International Polar Years, 2007–09.* London: Bernard Quaritch, 2009.
Scribd. "Historical Timeline of Antarctic Exploration | Antarctica | Ernest Shackleton." Accessed 1 March 2020. scribd.com. [online]
Antarctic Logistics & Expeditions. "Sir James Clark Ross (1800–1862)." 28 August 2010. antarctic-logistics.com. [online]
Government of South Georgia. "South Georgia & the South Sandwich Islands." Accessed 1 March 2020. gov.gs. [online]
Tristan de Cunha Government. "Gough Island." Accessed April 2020. tristandc.com. [online]

Drawing the Antarctic Resolution Atlas

pp. 50–51
Antarctic Treaty Secretariat. *Antarctic Treaty Forty-second Consultative Meeting, Measure 12: Revised List of Antarctic Historic Sites and Monuments: The Wreck of Sir Ernest Shackleton's* Vessel Endurance *and C.A. Larsen Multiexpedition Cairn.* 2019. Accessed 15 May 2020. ats.aq. [online]
Antarctic Treaty System. *ATCM XXXVIII Final Report, Annex: Revised List of Historic Sites and Monuments.* ATCM Sofia, 1–10 June 2015. ats.aq. [online]
Australian Antarctic Data Centre. "Southern Ocean Zones." 1995. Accessed November 2019. data.aad.gov.au. [online]
British Antarctic Survey. "Antarctic Coastline." 2019. Accessed November 2019. bas.ac.uk. [online]
British Antarctic Survey. "Antarctic Contours." 2019. Accessed November 2019. bas.ac.uk. [online]
British Antarctic Survey. "Antarctic Surface DEM." 2014. Accessed November 2019. bas.ac.uk. [online]
British Antarctic Survey. "Data Limit 60S." 2012. Accessed November 2019. bas.ac.uk. [online]
Council of Managers of National Antarctic Programs. *Antarctic Station Catalogue.* Christchurch: COMNAP, 2017. comnap.aq. [online]

National Oceanic and Atmospheric Administration. *Sea Ice Extent 1980–2019.* Accessed January 2020. ftp://sidads.colorado.edu. [online]
Natural Earth Data. "World Shaded Relief." 2012. Accessed December 2019. naturalearthdata. com. [online]

One Map, One Ocean. The Athelstan Spilhaus Projection

p. 55
Drawing courtesy of Bojan Šavrič, John Nelson and David Burrows for UNLESS

02
Antarctic Pie.
The Menacing Geometry of Power

National Claims in Antarctica

pp. 74–75
Australian Antarctic Data Centre. "Southern Ocean Zones." 1995. Accessed November 2019. data.aad.gov.au. [online]
British Antarctic Survey. "Antarctic Coastline." 2019. Accessed November 2019. bas.ac.uk. [online]
British Antarctic Survey. "Antarctic Contours." 2019. Accessed November 2019. bas.ac.uk. [online]
British Antarctic Survey. "Antarctic Surface DEM." 2014. Accessed November 2019. bas.ac.uk. [online]
British Antarctic Survey. "Data Limit 60S." 2012. Accessed November 2019. bas.ac.uk. [online]
Council of Managers of National Antarctic Programs. "Antarctic Facilities." 2019. Accessed November 2019. comnap.aq. [online]
Council of Managers of National Antarctic Programs. *Antarctic Station Catalogue.* Christchurch: COMNAP, 2017. comnap.aq. [online]
Headland, R. K. *Antarctic Winter Stations Operating During the International Geophysical Year (1957/1958 winter).* University of Cambridge, 2012.
Headland, R. K. *Territorial Claims in the Antarctic Treaty Region.* Scott Polar Research Institute. 12 April 2019. [online]
National Oceanic and Atmospheric Administration. *Sea Ice Extent 1980–2019.* Accessed January 2020. ftp://sidads.colorado.edu. [online]
Natural Earth Data. "World Shaded Relief." 2012. Accessed December 2019. naturalearthdata.com. [online]

Neuschwabenland

pp. 76–77
Headland, R. K. *Territorial Claims in the Antarctic Treaty Region.* Scott Polar Research Institute. 12 April 2019. [online]
Saul, Ben, and Tim Stephens. *Antarctica in International Law.* Reprint, London: Bloomsbury Publishing, 2015.
Feedback provided by Robert Keith Headland to UNLESS, 2020

Naming the Antarctic

pp. 83, 85
Scientific Committee on Antarctic Research. "Antarctic Place Names." Accessed 18 December 2020. scar.org. [online]

Scientific Committee on Antarctic Research. "Antarctic Place Names." Accessed 18 December 2020. scar.org. [online]

pp. 86–87
Scientific Committee on Antarctic Research. "Antarctic Place Names." Accessed 18 December 2020. scar.org. [online]

A Concise Military History of Antarctica

pp. 94–95 bottom
Council of Managers of National Antarctic Programs. *Antarctic Station Catalogue.* Christchurch: COMNAP, 2017. comnap.aq. [online]
Encyclopædia Britannica Online. "Operation High Jump." Accessed 18 December 2020. britannica.com. [online]
Fogg, Gordon Elliott. *A History of Antarctic Science.* Studies in Polar Research. Cambridge: Cambridge University Press, 1992.
Kearns, David. *Where Hell Freezes Over: A Story of Amazing Bravery and Survival.* New York: St. Martin's Press, 2005.

03
The Governance System
of a Citizenless Continent

Science, Politics, and Governance in the Antarctic Treaty System. A Symbiotic Relationship

pp. 104–5
Council of Managers of National Antarctic Programs. 2020. comnap.aq.[online]

The Pivotal Role of the International Geophysical Year (IGY) 1957–1958

p. 113
Datasets obtained from: "International Polar Year Publications Database." 2020. nes.biblioline.com. [online]

The Antarctic Treaty

pp. 116–17
Australian Antarctic Data Centre. "Southern Ocean Zones." 1995. Accessed November 2019. data.aad.gov.au. [online]
British Antarctic Survey. "Antarctic Coastline." 2019. Accessed November 2019. bas.ac.uk. [online]
British Antarctic Survey. "Antarctic Contours." 2019. Accessed November 2019. bas.ac.uk. [online]
British Antarctic Survey. "Antarctic Surface DEM." 2014. Accessed November 2019. bas.ac.uk. [online]
British Antarctic Survey. "Data Limit 60S." 2012. Accessed November 2019. bas.ac.uk. [online]
Council of Managers of National Antarctic Programs. "Antarctic Facilities." 2019. Accessed November 2019. comnap.aq. [online]
Council of Managers of National Antarctic Programs. *Antarctic Station Catalogue.* Christchurch: COMNAP, 2017. comnap.aq. [online]
Headland, R. K. *Antarctic Winter Stations Operating During the International Geophysical Year (1957/1958 winter).* University of Cambridge, 2012.
Headland, R. K. *Territorial Claims in the Antarctic Treaty Region.* Scott Polar Research Institute. 12 April 2019. [online]
National Oceanic and Atmospheric Administration. *Sea Ice Extent 1980–2019.* Accessed January 2020. ftp://sidads.colorado.edu. [online]

Natural Earth Data. "World Shaded Relief." 2012. Accessed December 2019. naturalearthdata.com. [online]

p. 117
Secretariat of the Antarctic Treaty. Accessed 18 December 2020. ats.aq. [online]
Worldometer. "GDP by Country." Accessed 18 December 2020. worldometers.info. [online]

pp. 118–19
Secretariat of the Antarctic Treaty. Accessed 18 December 2020. ats.aq. [online]

pp. 120–21
Flowchart adapted from an image shared by Klaus Dodds. "The Antarctic Treaty System Map." Era 2011.

The Scientific Committee on Antarctic Research

p. 125
Database of Global Administrative Areas (GADM), 2018. Accessed December 2019. gadm.org. [online]
Natural Earth Data. "World Shaded Relief." 2012. Accessed December 2019. naturalearthdata.com. [online]
Scientific Committee on Antarctic Research. "Antarctic Place Names." Accessed 18 December 2020. scar.org. [online]

Facilitating International Collaboration in Antarctic Science. The Role of COMNAP

pp. 127–29
Council of Managers of National Antarctic Programs. Accessed 18 December 2020. comnap.aq. [online]
Database of Global Administrative Areas (GADM), 2018. Accessed December 2019. gadm.org. [online]
Natural Earth Data. "World Shaded Relief." 2012. Accessed December 2019. naturalearthdata.com. [online]

04
Antarctic Resources. Temptations
and Accountability

The Role and Impact of Antarctic Specially Protected Areas and Antarctic Specially Managed Areas

p. 141
Secretariat of the Antarctic Treaty. "Status of Antarctic Specially Protected Area and Antarctic Specially Managed Area Management Plans." Accessed 18 December 2020. ats.aq. [online]

pp. 142–43
Australian Antarctic Data Centre. "Southern Ocean Zones." 1995. Accessed November 2019. data.aad.gov.au. [online]
British Antarctic Survey. "Antarctic Coastline." 2019. Accessed November 2019. bas.ac.uk. [online]
British Antarctic Survey. "Antarctic Contours." 2019. Accessed November 2019. bas.ac.uk. [online]
British Antarctic Survey. "Antarctic Surface DEM." 2014. Accessed November 2019. bas.ac.uk. [online]
British Antarctic Survey. "Data Limit 60S." 2012. Accessed November 2019. bas.ac.uk. [online]

Cutler, Paul, and Igor Krupnik, eds. "Planning and Implementing IPY 2007–2008." International Science Council. Accessed March 2020. [online]

National Oceanic and Atmospheric Administration. *Sea Ice Extent 1980–2019.* Accessed January 2020. ftp://sidads.colorado.edu. [online]

Natural Earth Data. "World Shaded Relief." 2012. Accessed December 2019. naturalearthdata.com. [online]

Secretariat of the Antarctic Treaty. "ASMAs Polygons." 2017. Accessed December 2019. [online]

Antarctic Biodiversity

pp. 146–47
Atkinson, Angus, Simeon L. Hill, Evgeny A. Pakhomov, Volker Siegel, Ricardo Anadon, Sanae Chiba, Kendra L. Daly, et al. "KRILLBASE: A Circumpolar Database of Antarctic Krill and Salp Numerical Densities, 1926–2016." *Earth System Science Data* 9, no. 1 (2017): 193–210.

Australian Antarctic Data Centre. "Southern Ocean Zones." 1995. Accessed November 2019. data.aad.gov.au. [online]

British Antarctic Survey. "Antarctic Coastline." 2019. Accessed November 2019. bas.ac.uk. [online]

British Antarctic Survey. "Antarctic Contours." 2019. Accessed November 2019. bas.ac.uk. [online]

British Antarctic Survey. "Antarctic Surface DEM." 2014. Accessed November 2019. bas.ac.uk. [online]

British Antarctic Survey. "Data Limit 60S." 2012. Accessed November 2019. bas.ac.uk. [online]

Fretwell, Peter T., Michelle A. LaRue, Paul Morin, Gerald L. Kooyman, Barbara Wienecke, Norman Ratcliffe, Adrian J. Fox, et al. "An Emperor Penguin Population Estimate: The First Global, Synoptic Survey of a Species from Space." *PLOS ONE* 7, no. 4 (2012): e33751.

Harris, Colin, Lincoln Fishpool, Ben Lascelles, Katherina Lorenz. "Important Bird Areas in Antarctica." *BirdLife International*, 2016. Accessed December 2019. [online]

Johnson, R., Sumner, M., Raymond, B. "Southern Ocean Summer Chlorophyll – A Climatology." 2017. Accessed December 2019. data.gov.au. [online]

Roquet, F., C. Wunsch, G. Forget, P. Heimbach, C. Guinet, G. Reverdin, J.-B. Charrassin, et al. "Seal tracks." MEOP Consortium, 2013. Accessed December 2019. meop.net. [online]

National Oceanic and Atmospheric Administration. *Sea Ice Extent 1980–2019.* Accessed January 2020. ftp://sidads.colorado.edu. [online]

Natural Earth Data. "World Shaded Relief." 2012. Accessed December 2019. naturalearthdata.com. [online]

Fishing in the Southern Ocean and the Role of CCAMLR

p. 151
Antarctic and Southern Ocean Coalition. "Protecting the Southern Ocean." Accessed 18 December 2020. asoc.org. [online]

British Antarctic Survey. "Antarctic Coastline." 2019. Accessed November 2019. bas.ac.uk. [online]

Commission for the Conservation of Antarctic Marine Living Resources. "Statistical Areas." 2017. Accessed January 2020. data.ccamlr.org. [online]

Commission for the Conservation of Antarctic Marine Living Resources. "Small-scale Research Units (SSRUs)." 2018. Accessed January 2020. data.ccamlr.org. [online]

Commission for the Conservation of Antarctic Marine Living Resources. "Small-scale Management Units (SSMUs)." 2017. Accessed January 2020. data.ccamlr.org. [online]

Commission for the Conservation of Antarctic Marine Living Resources. "Research Blocks."

2018. Accessed January 2020. data.ccamlr.org. [online]

Commission for the Conservation of Antarctic Marine Living Resources. "Marine Protected Areas (MPAs)." 2017. Accessed January 2020. data.ccamlr.org. [online]

Commission for the Conservation of Antarctic Marine Living Resources. "Krill Fisheries | CCAMLR." 2020. ccamlr.org. [online]

Database of Global Administrative Areas (GADM), 2018. Accessed December 2019. gadm.org. [online]

Greenpeace International. *Licence to Krill: The Little-Known World of Antarctic Fishing.* March 2018. greenpeace.org [online].

World Bank Group. "World Surface Area and GDP." 2014. Accessed January 2020. data.worldbank.org. [online]

Fishing route data from FleetMon. "Live AIS Vessel Tracker with Ship and Port Database." Accessed 18 December 2020. fleetmon.com. [online]

Additional research on MPAs shared to UNLESS by Juan Du

Marine Protected Areas

pp. 152–53
Protected Planet. "Explore the World's Marine Protected Areas." Accessed 18 December 2020. protectedplanet.net. [online]

Spatial Exceptions at Sea

pp. 154–55
British Antarctic Survey. "Antarctic Coastline." 2019. Accessed November 2019. bas.ac.uk. [online]

Database of Global Administrative Areas (GADM), 2018. Accessed December 2019. gadm.org. [online]

Commission for the Conservation of Antarctic Marine Living Resources. "Statistical Areas." 2017. Accessed January 2020. data.ccamlr.org. [online]

Commission for the Conservation of Antarctic Marine Living Resources. "Small-scale Research Units (SSRUs)." 2018. Accessed January 2020. data.ccamlr.org. [online]

Commission for the Conservation of Antarctic Marine Living Resources. "Small-scale Management Units (SSMUs)." 2017. Accessed January 2020. data.ccamlr.org. [online]

Commission for the Conservation of Antarctic Marine Living Resources. "Research Blocks." 2018. Accessed January 2020. data.ccamlr.org. [online]

Commission for the Conservation of Antarctic Marine Living Resources. "Marine Protected Areas (MPAs)." 2017. Accessed January 2020. data.ccamlr.org. [online]

Greenpeace International. *Licence to Krill: The Little-Known World of Antarctic Fishing.* March 2018. greenpeace.org [online].

World Bank Group. "World Surface Area and GDP." 2014. Accessed January 2020. data.worldbank.org. [online]

Fishing route data from FleetMon. "Live AIS Vessel Tracker with Ship and Port Database." Accessed 18 December 2020. fleetmon.com. [online]

Ross Sea MPA information shared courtesy of Juan Du for UNLESS

The Krill Market

pp. 156, 158
Commission for the Conservation of Antarctic Marine Living Resources. "Krill Fisheries | CCAMLR." 2020. ccamlr.org. [online]

p. 157
Natural Earth Data. "World Shaded Relief." 2012. Accessed December 2019. naturalearthdata.com. [online]

Bioprospecting in Antarctica

p. 161
Diagram reproduced by UNLESS for Julia Jabour

Antarctic Tourism

pp. 162–63
International Association of Antarctica Tour Operators. 2020. iaato.org. [online]

Ship-Based Antarctic Tourist Landing Procedures

pp. 168–69
Council of Managers of National Antarctic Programs. *Antarctic Station Catalogue.* Christchurch: COMNAP, 2017. comnap.aq. [online]

British Antarctic Survey. "Antarctic Coastline." 2019. Accessed November 2019. bas.ac.uk. [online]

British Antarctic Survey. "Antarctic Contours." 2019. Accessed November 2019. bas.ac.uk. [online]

British Antarctic Survey. "Antarctic Surface DEM." 2014. Accessed November 2019. bas.ac.uk. [online]

British Antarctic Survey. "Data Limit 60S." 2012. Accessed November 2019. bas.ac.uk. [online]

National Oceanic and Atmospheric Administration. *Sea Ice Extent 1980–2019.* Accessed January 2020. ftp://sidads.colorado.edu. [online]

Natural Earth Data. "World Shaded Relief." 2012. Accessed December 2019. naturalearthdata.com. [online]

International Association of Antarctica Tour Operators. "Vessel Tourist Site Visits by Activity." 2019. iaato.org. [online]

Antarctic Treaty System. *Revised List of Historic Sites and Monuments.* Report, 2015. Accessed February 2020. ats.aq. [online]

05

"The Question of Antarctica"

Where the Ice Meets the Waves. Antarctica and the Law of the Sea

pp. 176–77
Australian Antarctic Data Centre. "Southern Ocean Zones." 1995. Accessed November 2019. data.aad.gov.au. [online]

British Antarctic Survey. "Antarctic Coastline." 2019. Accessed November 2019. bas.ac.uk. [online]

British Antarctic Survey. "Antarctic Contours." 2019. Accessed November 2019. bas.ac.uk. [online]

British Antarctic Survey. "Antarctic Surface DEM." 2014. Accessed November 2019. bas.ac.uk. [online]

British Antarctic Survey. "Data Limit 60S." 2012. Accessed November 2019. bas.ac.uk. [online]

Council of Managers of National Antarctic Programs. 2020. comnap.aq. [online]

Headland, R. K. *Peri-Antarctic Islands.* Scott Polar Research Institute, 2019. Accessed March 2020. [online]

Headland, R. K. *Territorial Claims in the Antarctic Treaty Region.* Scott Polar Research Institute, 12 April 2019. [online]

Marine Regions. *World EEZ.* 2019. Accessed January 2020. marineregions.org. [online]

National Oceanic and Atmospheric Administration. *Sea Ice Extent 1980–2019.* Accessed January 2020. ftp://sidads.colorado.edu. [online]

Natural Earth Data. "World Shaded Relief." 2012. Accessed December 2019. naturalearthdata.com. [online]

Rothwell, Donald, and Tim Stephens. *International Law of The Sea.* Reprint, Hart Publishing, 2016.
Tuyet Mai, Le Thi. "UNCLOS Sets Legal Regimes for Maritime Rights and Exploitation within EEZ." *Hanoi Times,* 10 August 2019. [online]

Bounding Antarctica. The Greater Southern Ocean

pp. 180–81
Commission for the Conservation of Antarctic Marine Living Resources. "Home Page | CCAMLR." Accessed 18 December 2020. ccamlr.org. [online]
Boundaries sourced from:
The Antarctic Treaty, opened for signature on 1 December 1959.
Convention for the Conservation of Antarctic Seals, signed 1 June 1972. Entered into force 1978.
Convention on the Conservation of Antarctic Marine Living Resources, 20 May 1980, 1329 UNTS 47.
Conservation of the Living Marine Resources of the High Seas of the South Pacific (Galapagos Agreement), enacted 14 August 2000.
Convention on the Conservation and Management of Fishery Resources in the Southeast Atlantic, signed 20 April 2001.
Convention on the Conservation and Management of High Seas Fishery Resources in the South Pacific Ocean, closed for signature 31 January 2011. Entered into force 24 August 2012.
Protocol on Environmental Protection to the Antarctic Treaty, 4 October 1991, 30 ILM 1461.
Southern Indian Ocean Fisheries Agreement, signed 7 July 2006.

FOUR ELEMENTS

06

The Revolutionary Ocean

The Global Carbon Budget

p. 207 top
Milken Institute School of Public Health. "[Graphic] CO$_2$ Emissions v. Vulnerability to Climate Change, by Nation." 8 September 2015. [online]

p. 207 bottom
Graphic from *The Economist,* "The Past, Present and Future of Climate Change," 21 September 2019. [online]

p. 208
Figure adapted by UNLESS for Judith Hauck, 2020
Data provided by Judith Hauck, 2020

The Southern Ocean Overturning Circulation

p. 210
Figure adapted by UNLESS for Jean-Baptiste Sallée, 2020

p. 211
Hook, Leslie. "Why Penguins May Help Us Predict the Impact of Climate Change." *Financial Times,* 26 February 2020. ft.com. [online]
National Oceanic and Atmospheric Administration. "Ocean Temperatures." 2013. Accessed December 2019. noaa.gov. [online]
Natural Earth Data. "World Shaded Relief." 2012. Accessed December 2019. naturalearthdata. com. [online]

The Variable Southern Ocean Carbon Sink

p. 213 top
Figure adapted by UNLESS for Peter Landschützer, 2020
Gruber, Nicolas, Peter Landschützer, and Nicole Lovenduski. "The Variable Southern Ocean Carbon Sink." *Annual Review of Marine Science* 11 (2019): 159–86.

p. 213 bottom
Gruber, Nicolas, Peter Landschützer, and Nicole Lovenduski. "The Variable Southern Ocean Carbon Sink." *Annual Review of Marine Science* 11 (2019): 159–86.
Figure adapted by UNLESS for Peter Landschützer, 2020

p. 214
Figure adapted by UNLESS for Peter Landschützer, 2020
Gruber, Nicolas, Peter Landschützer, and Nicole Lovenduski. "The Variable Southern Ocean Carbon Sink." *Annual Review of Marine Science* 11 (2019): 159–86.

The Impact of Climate Change on the Antarctic Krill

p. 226
Figure and data adapted by UNLESS for Andrea Piñones, 2020

Antarctic Sea Ice and Its Impact on the Global Climate System

p. 230
Australian Antarctic Division. "Sea Ice." Accessed March 2020. antarctica.gov.au. [online]
United States Coast Guard Navigation Center. "Iceberg Formation." Report. Undated. [online]

pp. 232–33
Australian Antarctic Data Centre. "Southern Ocean Zones." 1995. Accessed November 2019. data. aad.gov.au. [online]
British Antarctic Survey. "Antarctic Coastline." 2019. Accessed November 2019. bas.ac.uk. [online]
British Antarctic Survey. "Antarctic Contours." 2019. Accessed November 2019. bas.ac.uk. [online]
British Antarctic Survey. "Antarctic Surface DEM." 2014. Accessed November 2019. bas.ac.uk. [online]
British Antarctic Survey. "Data Limit 60S." 2012. Accessed November 2019. bas.ac.uk. [online]
National Oceanic and Atmospheric Administration. *Sea Ice Extent 1980–2019.* Accessed January 2020. ftp://sidads.colorado.edu. [online]

07

Twenty-Six Quadrillion Tons of Ice. The Urgencies of Contemporaneity vs. the Truths of "Deep Time"

Antarctica. The World's Coldest Desert

pp. 238–39
High Latitude Southern Hemisphere Meteorology, Ralf Brauner (all figures)
All figures adapted by UNLESS for Ralf Brauner, 2020

Antarctica as Environmental Archive

p. 243
British Antarctic Survey. "Antarctic Coastline." 2019. Accessed November 2019. bas.ac.uk. [online]
Database of Global Administrative Areas (GADM), 2018. Accessed December 2019. gadm.org. [online]
Das, Indrani, Robin Bell, Theodore Scambos, and M. Wolovick. "Influence of Persistent Wind Scour on the Surface Mass Balance of Antarctica." *Nature Geoscience* 6, no. 5 (2013): 367–71.
National Oceanic and Atmospheric Administration. *Ice Cores.* Accessed April 2020. noaa.gov. [online]
Van Wessem, J. M., C. H. Reijmer, J. T. M. Lenaerts, W. J. van de Berg, M. R. van den Broeke, and E. van Meijgaard. "Updated Cloud Physics in a Regional Atmospheric Climate Model Improves the Modelled Surface Energy Balance of Antarctica." *The Cryosphere* 8 (2014): 125–35.
Van Wessem, J. M., C. H. Reijmer, M. Morlighem, J. Mouginot, E. Rignot, B. Medley, I. Joughin et al. "Improved Representation of East Antarctic Surface Mass Balance in a Regional Atmospheric Climate Model." *Journal of Glaciology* 60, no. 222 (2014): 761–70.

pp. 244–45
Beyond EPICA. "Home Page." 2016. beyondepica.eu. [online]
Figure adapted by UNLESS for Carlo Barbante

pp. 246–47
Figure adapted by UNLESS for Carlo Barbante
Ahn, Jinho, and Edward J. Brook. "Siple Dome Ice Reveals Two Modes of Millennial CO$_2$ Change During the Last Ice Age." *Nature Communications* 5, no. 1 (2014).
Bereiter, Bernhard, Dieter Lüthi, Michael Siegrist, Simon Schüpbach, Thomas F. Stocker, and Hubertus Fischer. "Mode Change of Millennial CO$_2$ Variability During the Last Glacial Cycle Associated with a Bipolar Marine Carbon Seesaw." *PNAS* 109, no. 25 (2012): 9755–60.
Bereiter, Bernhard, H. Fischer, J. Schwander, and Thomas Stocker. "Diffusive Equilibration of N$_2$, O$_2$ and CO$_2$ Mixing Ratios in A 1.5-Million-Years-Old Ice Core." *The Cryosphere* 8, no. 1 (2014): 245–56.
Buizert, C., Benjamin Keisling, J. E. Box, F. He, Anders E. Carlson, Gaylen Sinclair, R. M. DeConto. "Greenland-Wide Seasonal Temperatures During the Last Deglaciation." *Geophysical Research Letters* (2018).
Gaspari, Vania, Carlo Barbante, Giulio Cozzi, Paolo Cescon, Claude Boutron, Paolo Gabrielli, Gabriele Capodaglio et al. "Atmospheric Iron Fluxes over the Last Deglaciation: Climatic Implications." *Geophysical Research Letters* 33, no. 3 (2006): L03704.
Jouzel, Jean, Valerie Maison-Delmotte, O. Cattani and Gabrielle Dreyfus. "Orbital and Millennial Antarctic Climate Variability Over the Past 800,000 Years." *Science* 317, no. 5839 (2007): 793–96.
MacFarling, Meure C., David Etheridge, Cathy Trudinger, P. Steele, R. Langenfelds, T. van Ommen et al. "Law Dome CO$_2$, CH$_4$ and N$_2$O Ice Core Records Extended to 2000 Years BP." *Geophysical Research Letters* 33, no. 14 (2006).
Petit, Jean-Robert, Jean Jouzel, Dominique Raynaud, N. Barkov, I. Basile, M. Bender, J. Chappellaz et al. "Climate and Atmospheric History of the Past 420,000 Years from the Vostok Ice Core, Antarctica." *Nature* 399 (1999): 429–36.
Rubino, Mauro, David Etheridge, Cathy Trudinger, Colin Allison, Mark Battle, R. L. Langenfelds, L. P. Steele et al. "A Revised 1000 Year Atmospheric δ^{13}C-CO$_2$ Record from Law Dome and South Pole, Antarctica." *Journal of Geophysical Research: Atmospheres* 118, no. 15 (2013): 8482–99.

Schneider, Robert, Jochen Schmitt, Peter Koehler, Fortunat Joos, and Hubertus Fischer. "A Reconstruction of Atmospheric Carbon Dioxide and Its Stable Carbon Isotopic Composition from the Penultimate Glacial Maximum to the Glacial Inception." *Climate of the Past* 9, no. 6 (2013): 2507–23.

Siegenthaler, Urs, Thomas Stocker, Eric Monnin, Jakob Schwander, Bernhard Stauffer, Dominique Raynaud, Jean-Mac Barnola et al. "Stable Carbon Cycle-Climate Relationship During the Late Pleistocene." *Science* 310, no. 5752 (2005): 1313–17.

Antarctic Subglacial Lakes. The Case of Vostok

pp. 256–57
Database of Global Administrative Areas (GADM), 2018. Accessed December 2019. gadm.org. [online]

Smith, B. E., H. A. Fricker, I. R. Joughin, and S. Tulaczyk. "An Inventory of Active Subglacial Lakes in Antarctica Detected by ICESat (2003–2008)." *Journal of Glaciology* 55, no. 192 (September 2009): 573–95. [online]

Ice Shelf Retreat in Antarctica

p. 260
Hook, Leslie, Stephen Bernard, and Ian Bott. "Climate Change: What Antarctica's 'Doomsday Glacier' Means for the Planet." *Financial Times,* 13 July 2020. ft.com. [online]

Hulbe, Christina. "Is Ice Sheet Collapse in West Antarctica Unstoppable?" *Science* 356, no. 6341 (2017): 910–11.

pp. 263–64
Amos, J. "World's Biggest Iceberg Makes a Run for It." *British Broadcasting Corporation – BBC,* 2020. Accessed February 2020. [online]

Australian Antarctic Division. "Sea Ice." Accessed March 2020. antarctica.gov.au. [online]

British Antarctic Survey. "Antarctic Coastline." 2019. Accessed November 2019. bas.ac.uk. [online]

British Antarctic Survey. "Antarctic Contours." 2019. Accessed November 2019. bas.ac.uk. [online]

British Antarctic Survey. "Antarctic Surface DEM." 2014. Accessed November 2019. bas.ac.uk. [online]

British Antarctic Survey. "Data Limit 60S." 2012. Accessed November 2019. bas.ac.uk. [online]

British Antarctic Survey. "The Antarctic Peninsula's Retreating Ice Shelves." 2012. Accessed January 2020. bas.ac.uk. [online]

Fretwell, Peter. *Antarctic Atlas: New Maps and Graphics That Tell the Story of a Continent.* London: Particular Books, 2020.

National Aeronautics and Space Administration. "Drainage System." 2012. Accessed February 2020. nasa.gov. [online]

Natural Earth Data. "World Shaded Relief." 2012. Accessed December 2019. naturalearthdata.com. [online]

National Oceanic and Atmospheric Administration. *Sea Ice Extent 1980–2019, 1981–2019.* Accessed November 2019. ftp://sidads.colorado.edu. [online]

Mouginot, Jérémie, Bernd Scheuchl, and Eric Rignot. "MEaSUREs Antarctic Boundaries for IPY 2007–2009 from Satellite Radar, Version 2" [Ice Velocity Map]. NASA National Snow and Ice Data Center Distributed Active Archive Center. nsidc.org. Accessed November 2019. [online]

Rignot, Eric, Jérémie Mouginot, Bernd Scheuchl, Michiel van den Broeke, Melchior van Wessem, Mathieu Morlighem. "Four Decades of Antarctic Ice Sheet Mass Balance from 1979-2017."

Proc Natl Acad Sci USA 116, no. 4 (2019): 1095–1103.

Rowlatt, J. "Antarctica Melting: Climate Change and the Journey to the 'Doomsday Glacier.'" *British Broadcasting Corporation – BBC,* 2020.

United States Coast Guard Navigation Center. "Iceberg Formation". Report. Undated. [online]

West Antarctica Ice Shelf. The Thwaites Glacier

p. 267
British Antarctic Survey. "Antarctic Coastline." 2019. Accessed November 2019. bas.ac.uk. [online]

ESRI / Antarctic REMA Explorer Dataset, Accessed November 2019.

Rowlatt, Justin. "Antarctica Melting: Climate Change and the Journey to the 'Doomsday Glacier.'" *BBC News,* 28 January 2020.

Slater, Thomas, Andrew Shepherd, Malcolm Mcmillan, Alan Muir, Lin Gilbert, Anna E. Hogg, Hannes Konrad et al. "A New Digital Elevation Model of Antarctica Derived from CryoSat-2 Altimetry." *The Cryosphere* 12 (2018): 1551–62.

Antarctica and Global Sea Level Rise

pp. 272–73
Weller, Richard J., Claire Hoch, and Chieh Huang. "Sea Level Rise." 2017. atlas-for-the-end-of-the-world.com. [online]

pp. 274–75
Asmelash, Leah. "This Giant Glacier in Antarctica Is Melting, and It Could Raise Sea Levels by 5 Feet, Scientists Say." CNN, 25 March 2020. cnn.com. [online]

Bennett, Sukee. "Scientists Find Warm Water Beneath Antarctica's Most At-risk Glacier." *Nova,* 22 April 2020. pbs.org. [online]

Cockburn, Harry. "'Teetering at the Edge': Scientists Warn of Rapid Melting of Antarctica's 'Doomsday Glacier.'" *The Independent,* 13 July 2020. independent.co.uk. [online]

Ferreira, Fernanda. "Melting Glaciers Cool the Southern Ocean." *MIT News,* 17 May 2020. news.mit.edu. [online]

Fogwill, Chris, Chris Turney, and Zoë Thomas. "Ancient Antarctic Ice Melt Caused Extreme Sea Level Rise 129,000 Years Ago – and It Could Happen Again." *The Conversation*, 12 February 2020. theconversation.com. [online]

Fountain, Henry. "Even the South Pole Is Warming, and Quickly, Scientists Say." *The New York Times*, 29 June 2020. nytimes.com. [online]

Garrison, Cassandra. "Antarctica's Giant Ice Shelves Are Vulnerable to Collapse, Study Warns." Reuters and World Economic Forum, 28 August 2020. weforum.org. [online]

"Global Sea Levels to Rise Drastically by 2100 Due to Greenland, Antarctica's Melting Ice Sheets." 22 September 2020. firstpost.com. [online]

Gohd, Chelsea. "Melting Ice Sheets Will Add Over 15 inches to Global Sea Level Rise by 2100." 20 September 2020. space.com. [online]

Harvey, Fiona. "Greenland's Ice Melting Faster Than at Any Time in Past 12,000 Years." *The Guardian*, 30 September 2020. theguardian.com. [online]

Hook, Leslie, Steven Bernard, and Ian Bott. "Climate Change: What Antarctica's 'Doomsday Glacier' Means for the Planet." *Financial Times*, 12 July 2020. ft.com. [online]

NASA Earth Observatory. "Extensive Early Melting Detected Along the Antarctic Peninsula." 3 December 2020. scitechdaily.com. [online]

Rowlatt, Justin. "Antarctica Melting: Climate Change and the Journey to the 'Doomsday Glacier.'" *BBC News,* 28 January 2020. bbc.com. [online]

Saplakoglu, Yasemin. "Giant Crack Frees a Massive Iceberg in Antarctica." 26 February 2021. liveScience.com. [online]

Tenenbaum, Laura. "Greenland and Antarctica Are Melting Six Times Faster Than in the 1990s." *Forbes,* 30 March 2020. forbes.com. [online]

Vaughan, Adam. "Antarctic Ice Melt Could Push Sea Levels to Rise 1.5 Metres by 2100." *New Scientist,* 13 February 2020. newscientist.com. [online]

Watts, Jonathan. "Climate Change Is Turning Parts of Antarctica Green, Say Scientists." *The Guardian*, 20 May 2020. theguardian.com. [online]

Woodyatt, Amy. "Antarctic Ice Sheets Capable of Much Faster Melting Than We Thought." *CNN,* 28 May 2020. cnn.com. [online]

08
A Continental Archipelago under Ice

pp. 278–79
Australian Antarctic Data Centre. "Southern Ocean Zones." 1995. Accessed November 2019. data.aad.gov.au. [online]

British Antarctic Survey. "Data Limit 60S." 2012. Accessed November 2019. bas.ac.uk. [online]

British Antarctic Survey. "Antarctic Bedmap DEM." 2014. Accessed November 2019. bas.ac.uk. [online]

British Antarctic Survey. "Antarctic Coastline." 2019. Accessed November 2019. bas.ac.uk. [online]

British Antarctic Survey. "Antarctic Contours." 2019. Accessed November 2019. bas.ac.uk. [online]

National Oceanic and Atmospheric Administration. *Sea Ice Extent 1980–2019, 1981–2019.* Accessed November 2019. ftp://sidads.colorado.edu. [online]

Natural Earth Data. "World Shaded Relief." 2012. Accessed December 2019. naturalearthdata.com. [online]

Norwegian Polar Institute, 2017. "Polar Place Names." Accessed December 2019. npolar.no. [online]

United States Geological Survey. "Tectonic Plates." 2019. Accessed December 2019. [online]

United States Geological Survey. "Earthquakes 1900–2017." 2017. Accessed November 2019. [online]

Gondwana, Antarctica, and Continental Drift

p. 282
Diagram adapted from original supplied by Karsten Gohl and Georg Kleinschmidt

p. 283
Walton, David W. H. Antarctica: *Global Science from A Frozen Continent.* Cambridge: Cambridge University Press, 2013.

Scientific Drilling on the Antarctic Margin

p. 285
Original diagrams sourced from Fretwell, Peter. *Antarctic Atlas: New Maps and Graphics That Tell the Story of a Continent.* London: Particular Books, 2020.

p. 286 bottom
Original diagrams sourced from Fretwell, Peter. *Antarctic Atlas: New Maps and Graphics That Tell the Story of a Continent.* London: Particular Books, 2020.

Wikipedia. "Pleistocene." wikipedia.org. [online]

p. 287
Natural Earth Data. "World Shaded Relief." 2012. Accessed December 2019. naturalearthdata.com. [online]
Walton, David W. H. *Antarctica: Global Science from A Frozen Continent*. Cambridge: Cambridge University Press, 2013.

Two Diagrams adapted from:
Storey, Bryan. "A Keystone in a Changing World." In *Antarctica: Global Science from a Frozen Continent*, edited by David W. H. Walton, 35–66. Cambridge: Cambridge University Press, 2013.
Convey, Peter, Angelika Brandt, and Steve Nicol in Walton, David W. H. *Antarctica: Global Science from a Frozen Continent*. Cambridge: Cambridge University Press, 2013.

Exploring Antarctic Subglacial Geology and Lithosphere

p. 295
Chegg. "Question: Questions Related to Material from Lab #20 (Plate Tectonics)." 2020. chegg.com. [online]

09

The White Desert. A Unique Viewing Platform into the Cosmos

The Magnetosphere

p. 300–1
Lanzerotti, Louis J. "Space Weather: Historical and Contemporary Perspectives." *Space Science Revision* 212, nos. 3–4 (2017): 1253–70.
Earthlabs Carleton College. "2B: Following the Energy Flow." *Climate and the Biosphere,* 2020. serc.carleton.edu. [online]
European Space Agency. "Schematic of Earth's Night-Side Magnetosphere." 2007. esa.int. [online]
Figure adapted by UNLESS for Louis J. Lanzarotti and Andrew J Gerrard.

p. 304
Figure adapted by UNLESS for Louis J. Lanzarotti and Andrew J Gerrard.
British Antarctic Survey. "Halley Weather Balloons." 2020. bas.ac.uk. [online]

Atmospheric Chemistry. The Discovery of the Ozone Hole

pp. 306–7
Ritchie, Hannah, and Max Roser. "Ozone Layer." 2018. ourworldindata.org. [online]
UNEP. "The Montreal Protocol on Substances that Deplete the Ozone Layer." 2020. ozone.unep.org. [online]
Datasets obtained from: "Web of Science." 2020. [online]

Long-Duration Ballooning

p. 313
Australian Antarctic Data Centre. "Southern Ocean Zones." 1995. Accessed November 2019. data.aad.gov.au. [online]
Binns, W. R., Ranita Bose, D. Braun, T. Brandt, W. Daniels, P. F. Dowkontt et al. "The SUPERTIGER Instrument: Measurement of Elemental Abundances of Ultra-Heavy Galactic Cosmic Rays." *The Astrophysical Journal* 788, no. 1 (2014): 18.
British Antarctic Survey. "Antarctic Coastline." 2019. Accessed November 2019. bas.ac.uk. [online]
British Antarctic Survey. "Antarctic Contours." 2019. Accessed November 2019. bas.ac.uk. [online]
British Antarctic Survey. "Antarctic Surface DEM." 2014. Accessed November 2019. bas.ac.uk. [online]
British Antarctic Survey. "Data Limit 60S." 2012. Accessed November 2019. bas.ac.uk. [online]
National Aeronautics and Space Administration. "NASA's SuperTIGER Balloon Flies Again to Study Heavy Cosmic Particles." 6 December 2017. nasa.gov. [online]
National Aeronautics and Space Administration. "Antarctic Anticyclone Sends Balloons Flying in Circles." Press Release. 14 December 2015. spaceref.com. [online]
Natural Earth Data. "World Shaded Relief." 2012. Accessed December 2019. naturalearthdata.com. [online]
National Oceanic and Atmospheric Administration. *Sea Ice Extent 1980–2019, 1981–2019.* Accessed November 2019. ftp://sidads.colorado.edu. [online]

IceCube Neutrino Observatory. Transforming Antarctic Ice into a Science Experiment

p. 320
IceCube South Pole Neutrino Observatory. "Home Page." Accessed March 2020. icecube.wisc.edu. [online]
Figure shared by Madeleine O'Keefe and adapted by UNLESS for IceCube Neutrino Observatory, 2020

p. 321
IceCube South Pole Neutrino Observatory. "Home Page." Accessed March 2020. icecube.wisc.edu. [online]
Figure shared by Madeleine O'Keefe and adapted by UNLESS for IceCube Neutrino Observatory, 2020

p. 323
Natural Earth Data. "World Shaded Relief." 2012. Accessed December 2019. naturalearthdata.com. [online]
Database of Global Administrative Areas (GADM), 2018. Accessed December 2019. gadm.org. [online]
IceCube South Pole Neutrino Observatory. "Home Page." Accessed March 2020. icecube.wisc.edu. [online]

SURVIVING IN THE CRYOSPHERE

10

Extreme Habitation. Compressed Spaces and Dilated Times

p. 342
Council of Managers of National Antarctic Programs. *Antarctic Station Catalogue.* Christchurch: COMNAP, 2017. comnap.aq. [online]

Inhabitation in the Extreme. The Winter-Over Syndrome

pp. 348–49
Timeanddate.com. "Sund & Moon: Sunrise & Sunset." Accessed 18 December 2020. timeanddate.com. [online]

Interior Urbanism at the Pole. Framheim vs. Cape Evans

p. 351
Ponting, Herbert George, and Robert Falcon Scott. *The Great White South: Being an Account of Experiences with Captain Scott's South Pole Expedition and of the Nature Life of the Antarctic – With 164 Photographic Illustrations,* xxvi, 305. London: Duckworth & Co., 1921.
Scott, Robert Falcon. *Journals: Captain Scott's Last Expedition.* 1913. Reprint, Oxford: Oxford University Press, 2008.
Timeanddate.com. "Home Page." Accessed 18 December 2020. timeanddate.com. [online]
Drawing prototypes developed during AA PolarLab Workshop: *Pioneering Modes of Inhabitation in the Extreme,* London, June 2019.

pp. 354–55
Framheim, Fram Museum, 1911
Amundsen, Roald. *The South Pole: An Account of the Norwegian Antarctic Expedition in the Fram 1910–1912.* 1912. Reprint, New York: Cooper Square Press, 2001.
Ponting, Herbert George, and Robert Falcon Scott. *The Great White South: Being an Account of Experiences with Captain Scott's South Pole Expedition and of the Nature Life of the Antarctic – With 164 Photographic Illustrations,* xxvi, 305. London: Duckworth & Co., 1921.
Ponting, Herbert. *With Scott to the Pole: The* Terra Nova *Expedition 1910–1913; The Photographs of Herbert Ponting.* 1913. Reprint, Crows Nest, NSW: Allen & Unwin, 2004.
Scott, Robert Falcon. *Journals: Captain Scott's Last Expedition.* 1913. Reprint, Oxford: Oxford University Press, 2008.
Timeanddate.com. "Home Page." Accessed 18 December 2020. timeanddate.com. [online]
For further information, please see "Archive of Antarctic Architecture," p. 628.
Drawing prototypes developed during AA PolarLab Workshop: *Pioneering Modes of Inhabitation in the Extreme,* London, June 2019.

Extreme Habitation. Antarctica's "Architecture of Sleep"

pp. 368–69
Axonometric studies produced by Polar Lab HK at the Urban Ecology Design Lab, University of Hong Kong.

11

Importing and Transporting the Possibility of Life

From Polar Gateways to Antarctic Cities

p. 383
Brooks, Shaun. "Our Footprint on Antarctica – Buildings, Disturbance." Australian Antarctic Data Centre, 2019. Accessed 1 November 2019. data.aad.gov.au. [online]
British Antarctic Survey. "Antarctic Coastline." 2019. Accessed November 2019. bas.ac.uk. [online]

Cool Antarctica. "Antarctica Travel." Accessed 1 March 2020. coolantarctica.com. [online]
Database of Global Administrative Areas (GADM), 2018. Accessed 18 December 2019. gadm.org. [online]
Fretwell, Peter. *Antarctic Atlas: New Maps and Graphics That Tell the Story of a Continent.* London: Particular Books, 2021.
Swoop Antarctica. "Routes to Antarctica." Accessed 1 March 2020. swoop-antarctica.com. [online]

Traverses and Intermodal Transportation

pp. 390–91
Original figure adapted from Patrice Godon Polar Engineering. "Home Page." Accessed 18 December 2020. patricegodonpolarengineering.eu. [online]
Figure adapted by UNLESS for Paul Thur

p. 397
Freitag, Dean, and Stephen Dibbern. "Dr. Poulter's Antarctic Snow Cruiser." Last updated 18 November 2009. joeld.net. [online]

12

The Three Typologies of the Heroic Age Huts

pp. 406–7
Figure adapted by UNLESS for Mike Pearson
Information courtesy of Mike Pearson

Comandante Ferraz Station

p. 417
Figure adapted by UNLESS for Estudio41

Bharati Station

pp. 418–19
Figure adapted by UNLESS courtesy of BoF Architekten

13

Abstract Master Plan

p. 434
Australian Antarctic Data Centre. "Southern Ocean Zones." 1995. Accessed 1 November 2019. data.aad.gov.au. [online]
British Antarctic Survey. "Antarctic Coastline." 2019. Accessed 1 November 2019. bas.ac.uk. [online]
British Antarctic Survey. "Antarctic Contours." 2019. Accessed 1 November 2019. bas.ac.uk. [online]
British Antarctic Survey. "Antarctic Surface DEM." 2014. Accessed 1 November 2019. bas.ac.uk. [online]
British Antarctic Survey. "Data Limit 60S." 2012. Accessed 1 November 2019. bas.ac.uk. [online]
Council of Managers of National Antarctic Programs. *Antarctic Station Catalogue.* Christchurch: COMNAP, 2017. comnap.aq. [online]

National Oceanic and Atmospheric Administration. *Sea Ice Extent 1980–2019, 1981–2019.* Accessed 1 November 2019. ftp://sidads.colorado.edu. [online]
Natural Earth Data. "World Shaded Relief." 2012. Accessed 18 December 2019. naturalearthdata.com. [online]

pp. 440–41
South Pole ASMA. "Maps: Location & Topography." 2018. Accessed 18 December 2020. southpole.aq. [online]

A Supra-regional Dispersed City. The Fildes Peninsula

pp. 444–45
British Antarctic Survey. "Antarctic Coastline." 2019. Accessed 17 February 2020. bas.ac.uk. [online]
British Antarctic Survey. "Antarctic Contours." 2019. Accessed 17 February 2020. bas.ac.uk. [online]
Further information courtesy of Chilean Air Force, Infrastructure Divison (February 2019)

Lessons in Temporal Permanence. McMurdo Station

p. 455
Caryl, Ed. "'Urban Heat Island' Effect Appears to Be Far More Pronounced in Polar Regions Than You Might Think!" 14 August 2014. notrickszone.com. [online]
Cheung, Nesia. Diagram of McMurdo Station drawn at the Urban Ecology Design Lab, University of Hong Kong, 2020.
Google Maps. Accessed 18 December 2020. google.com/maps. [online]
Sinclair, M. R. "Local Topographic Influence on Low-Level Wind at Scott Base, Antarctica." *New Zealand Journal of Geology and Geophysics* 31, no. 2 (1988): 237–45.
OZ Architects. *McMurdo Station Modernization Study: Building Shell & Fenestration Study.* Report, 29 April 2016. [online]
OZ Architects and National Science Foundation. *McMurdo Station Master Plan 2.1.* Report, National Science Foundation, 2015.
QGIS. "Quantarctica: An Antarctic GIS Package." 2020. Accessed 18 December 2020. qgis.org. [online]
Further information provided courtesy of Hugh Broughton Architects

Calving Architecture. Archaeology of the Debris

p. 462
Broughton, Hugh. "The Extreme Challenge of Building in Antarctica." Hugh Broughton in conversation with Giulia Foscari and Francesco Bandarin / AA Polar Lab. Lecture, London, Architectural Association School of Architecture, 7 March 2019.
Gough, Evan. "Antarctica Is About to Unleash an Iceberg Twice the Size of New York City." *Universe Today,* 21 February 2019. universetoday.com. [online]
Hugh Broughton Architects. "Halley VI British Antarctic Research Station." 2020. Accessed 18 December 2020. hbarchitects.co.uk. [online]
King, J. C. "Control of Near-Surface Wind over an Antarctic Ice Shelf." *Journal of Geophysical Research* 98, no. D7 (1993): 12949–53.

p. 463
British Antarctic Survey. "Graphic Showing Changes to the Brunt Ice Shelf and the location of Halley VI Research Station." 31 October 2017. bas.ac.uk. [online]
Further information provided courtesy of the British Antarctic Survey

pp. 464–65
British Antarctic Survey. Archives, 1960–2005. Cambridge.
Council of Managers of National Antarctic Programs. *Antarctic Station Catalogue.* Christchurch: COMNAP, 2017. comnap.aq. [online]
King, J. C. "Control of Near-Surface Winds over an Antarctic Ice Shelf." *Journal of Geophysical Research* 98, no. D7 (1993): 12949–53.
Hugh Broughton Architects, 2005
For further information, please see "Archive of Antarctic Architecture," pp. 658, 716, 740, 762, 810.

14

Water Management in the Extreme

p. 481
British Antarctic Survey, Natural Environment Research Council. *Proposed Construction and Operation of Halley VI Research Station, and Demolition and Removal of Halley V Research Station, Brunt Ice Shelf Antarctica: Final Comprehensive Environmental Evaluation.* Cambridge: British Antarctic Survey, 2007.
Ritchie, Hannah, and Max Roser. "Water Use and Stress." Last modified July 2018. ourworldindata.org. [online]

Rodriguez Wells and Water Management

p. 483
Hoffman, Stephen. "Subsurface Water Well Development: Rodwell Approach." 2020. Accessed 18 December 2020. erdc.usace.army.mil. [online]
Marquis, David. "ERDC Supports NASA'S Mission to Mars." 5 August 2020. erdc.usace.army.mil. [online]
Rice, Michael. "South Pole Water Source and More Station Pictures." 1 February 2013. southpolemike.blogspot.com. [online]

The Footprint of Antarctic Stations

pp. 486–87
Council of Managers of National Antarctic Programs. *Antarctic Station Catalogue.* Christchurch: COMNAP, 2017. comnap.aq. [online]

Built-in Redundancies or Infrastructure Resilience

p. 490
Figure adapted by UNLESS for Tony McGlory

Building Performance and Carbon Emissions

p. 494
Council of Managers of National Antarctic Programs. *Antarctic Station Catalogue.* Christchurch: COMNAP, 2017. comnap.aq. [online]

p. 495
Australian Antarctic Data Centre. "Southern Ocean Zones." 1995. Accessed 1 November 2019. data.aad.gov.au. [online]
Beccario, Cameron. National Oceanic and Atmospheric Administration. *Earth Wind Map.* Accessed 18 January 2020. earth.nullschool.net. [online]

Das, Indrani, Robin Bell, Ted Scambos, and M. Wolov-
    ick. "Influence of Persistent Wind Scour on the
    Surface Mass Balance of Antarctica." *Nature
    Geoscience* 6, no. 5 (2013): 367–71.
British Antarctic Survey. "Antarctic Coastline." 2019.
    Accessed 1 November 2019. bas.ac.uk. [online]
British Antarctic Survey. "Antarctic Contours." 2019.
    Accessed 1 November 2019. bas.ac.uk. [online]
British Antarctic Survey. "Antarctic Surface DEM."
    2014. Accessed 1 November 2019. bas.ac.uk.
    [online]
British Antarctic Survey. "Data Limit 60S." 2012.
    Accessed 1 November 2019. bas.ac.uk. [online]
National Oceanic and Atmospheric Administration.
    Sea Ice Extent 1980–2019, 1981–2019.
    Accessed 1 November 2019. ftp://sidads.
    colorado.edu. [online]
National Oceanic and Atmospheric Administration.
    Wind Data. 2018. Accessed 18 December 2019.
    esrl.noaa.gov. [online]
Natural Earth Data. "World Shaded Relief." 2012.
    Accessed 18 December 2019. naturalearthdata.
    com. [online]
Van Wessem, Melchoir, Carleen Reijmer, Jan
    Lenaerts, Willem Jan van de Berg, Michiel van
    den Broeke, and E. van Meijgaard. "Updated
    Cloud Physics in a Regional Atmospheric
    Climate Model Improves the Modelled Surface
    Energy Balance of Antarctica." *The Cryosphere*
    8 (2014): 125–35.
Van Wessem, Melchoir, Carleen Reijmer, Mathieu
    Morlighem, Jeremie Mouginot, E. Rignot, Brooke
    Medley, I. Joughin et al. "Total Precipitation
    Rate." 2014. Accessed November 2019.
    cambridge.org. [online]
Van Wessem, Melchoir, Carleen Reijmer, Mathieu
    Morlighem, Jeremie Mouginot, E. Rignot, Brooke
    Medley, I. Joughin et al. "Improved Representa-
    tion of East Antarctic Surface Mass Balance in
    a Regional Atmospheric Climate Model." *Journal
    of Glaciology* 60, no. 222 (2014): 761–70.

15
The Ideological Use of Relics

The Antarctic Heritage List

pp. 518–19
Secretariat of the Antarctic Treaty. Accessed 18
    December 2020. ats.aq. [online]

pp. 520–21
Secretariat of the Antarctic Treaty. "Measure 19:
    Annex – Revised List of Historic Sites and
    Monuments." In *Final Report of the Thirty-
    Eighth Antarctic Treaty Consultative Meeting.*
    Buenos Aires: Secretariat of the Antarctic
    Treaty, 2015. Accessed 18 December 2020.
    ats.aq. [online]
Australian Antarctic Data Centre. "Southern
    Ocean Zones." 1995. Accessed November 2019.
    data.aad.gov.au. [online]
British Antarctic Survey. "Antarctic Coastline."
    2019. Accessed 1 November 2019. bas.ac.uk.
    [online]
British Antarctic Survey. "Antarctic Contours."
    2019. Accessed 1 November 2019. bas.ac.uk.
    [online]
British Antarctic Survey. "Antarctic Surface DEM."
    2014. Accessed 1 November 2019. bas.ac.uk.
    [online]
British Antarctic Survey. "Data Limit 60S." 2012.
    Accessed 1 November 2019. bas.ac.uk. [online]
National Oceanic and Atmospheric Administration.
    Sea Ice Extent 1980–2019, 1981–2019.
    Accessed 1 November 2019. ftp://sidads.
    colorado.edu. [online]
Natural Earth Data. "World Shaded Relief." 2012.
    Accessed 18 December 2019.
    naturalearthdata.com. [online]

p. 523
Secretariat of the Antarctic Treaty. Accessed
    18 December 2020. ats.aq. [online]

Preservation Contradictions on an Active Volcanic Crater. Deception Island

p. 531
Antarctic Treaty System. *Revised List of Historic
    Sites and Monuments.* Report, 2015. Accessed
    17 February 2020. ats.aq. [online]
Berrocos, Manuel, Cristina Torrecillas, Bismarck
    Jigena, and Alberto Fernandez-Ros. "Determi-
    nation of Geomorphological and Volumetric
    Variations in the 1970 Land Volcanic Craters
    Area (Deception Island, Antarctica) from 1968
    Using Historical and Current Maps, Remote
    Sensing and GNSS." *Antarctic Science* 24,
    no. 4 (2012): 24.
British Antarctic Survey. "Antarctic Coastline."
    2019. Accessed 1 November 2019. bas.ac.uk.
    [online]
British Antarctic Survey. "Antarctic Contours."
    2019. Accessed 17 February 2020. bas.ac.uk.
    [online]
British Antarctic Survey. "Antarctic Lakes." 2019.
    Accessed 17 February 2020. bas.ac.uk. [online]
Harris, Colin, Lincoln Fishpool, Ben Lascelles,
    and Katherina Lorenz. "Important Bird Areas in
    Antarctica." 26 May 2016. environments.aq.
    [online]
Council of Managers of National Antarctic Programs.
    Antarctic Station Catalogue. Christchurch:
    COMNAP, 2017. comnap.aq. [online]
Hughes, Kevin, Luis Pertierra, Marco Molina-
    Montenegro, Peter Convey. "Biological
    Invasions in Terrestrial Antarctica: What Is the
    Current Status and Can We Respond?"
    Biodiversity and Conservation 24, no. 5 (2015):
    1031–55.
Secretariat of the Antarctic Treaty. "ASMAs
    Polygons." 2017. Accessed 17 February 2020.
    antarctictreaty.maps.arcgis.com [online]
Secretariat of the Antarctic Treaty. "Measure 10:
    Annex – Deception Island Management
    Package." In *Final Report of the Thirty-Fifth
    Antarctic Treaty Consultative Meeting.* Hobart:
    Secretariat of the Antarctic Treaty, 2012.
    ats.aq. [online]

Roald Amundsen's Tent at the South Pole

p. 542
Original traverse route and data plotted by Callum
    Campbell during AA PolarLab Workshop:
    *Pioneering Modes of Inhabitation in the
    Extreme,* London, June 2019.

p. 543
American Museum of Natural History. "Timeline:
    Amundsen Expedition to the South Pole." Part
    of the exhibition *Race to the End of the* Earth.
    American Museum of Natural History, 29 May
    2010–2 January 2011. Accessed 18 December
    2020. amnh.org. [online]
Artem Gorbunov Design Bureau. "Diagram of
    Amundsen and Scott's Polar Expeditions."
    February 2015. bureau.ru. [online]
Australian Antarctic Data Centre. "Southern
    Ocean Zones." 1995. Accessed 1 November
    2019. data.aad.gov.au. [online]
British Antarctic Survey. "Antarctic Coastline." 2019.
    Accessed 1 November 2019. bas.ac.uk. [online]
British Antarctic Survey. "Antarctic Contours." 2019.
    Accessed 1 November 2019. bas.ac.uk. [online]
British Antarctic Survey. "Antarctic Surface DEM."
    2014. Accessed 1 November 2019. bas.ac.uk.
    [online]

British Antarctic Survey. "Data Limit 60S." 2012.
    Accessed 1 November 2019. bas.ac.uk.
    [online]
National Oceanic and Atmospheric Administration.
    Sea Ice Extent 1980–2019, 1981–2019.
    Accessed 1 November 2019. ftp://sidads.
    colorado.edu. [online]
Natural Earth Data. "World Shaded Relief." 2012.
    Accessed 18 December 2019.
    naturalearthdata.com. [online]

16
Antarctica and Outer Space. Laboratories for a Future as Global Commons

Rethinking the Antarctic Suit

pp. 568–69
Drawings courtesy of D-Air Lab

Behavioural Patterns in the Extreme. Concordia as Space Simulator

p. 575
Web of Science. "Home Page." 2020. Accessed
18 December 2020.webofknowledge.com. [online]

ARCHIVE OF ANTARCTIC ARCHITECTURE

1899, Cape Adare huts
pp. 614–17

SITE PLAN
Primary Sources:
Antarctic Heritage Trust (NZ). *Map A – Cape Adare,
Antarctic Specially Protected Area: regional map.*
2004. 21 × 29.7 cm. In Antarctic Heritage Trust (NZ),
Conservation Plan: The Historic Huts at Cape Adare.
Christchurch, NZ: Antarctic Heritage Trust, 2004.

DIAGRAM
Primary Sources:
Antarctic Heritage Trust (NZ). *Map B – Cape Adare,
Antarctic Specially Protected Area: site map.* 2004.
21 × 29.7 cm. In Antarctic Heritage Trust (NZ),
Conservation Plan: The Historic Huts at Cape Adare.
Christchurch, NZ: Antarctic Heritage Trust, 2004.

PLAN
Primary Sources:
Purcell Miller Tritton & Antarctic Heritage Trust
(NZ). *PLAN OF STORES HUT & LIVING HUT AS
EXISTING 1:20. Drawing No. 002.* 14 July 2003.
59.4 × 84.1. In Antarctic Heritage Trust (NZ), *Con-
servation Plan: The Historic Huts at Cape Adare.*
Christchurch, NZ: Antarctic Heritage Trust, 2004.

Purcell Miller Tritton & Antarctic Heritage Trust
(NZ). *PLAN OF STORES HUT & LIVING HUT
– PROPOSED REPAIRS 1:20. Drawing No. 012.* 26
November 2003. 59.4 × 84.1. In Antarctic Heritage
Trust (NZ), *Conservation Plan: The Historic Huts
at Cape Adare.* Christchurch, NZ: Antarctic
Heritage Trust, 2004.

ELEVATION
Primary Sources:
Purcell Miller Tritton & Antarctic Heritage Trust
(NZ). *LIVING HUT EXTERNAL ELEVATIONS –
PROPOSED REPAIRS 1:20. Drawing No. 015.* 14 July
2003. 59.4 × 84.1. In Antarctic Heritage Trust (NZ),

Conservation Plan: The Historic Huts at Cape Adare. Christchurch, NZ: Antarctic Heritage Trust, 2004.

SECTIONS
Primary Sources:
Purcell Miller Tritton & Antarctic Heritage Trust (NZ). *LIVING HUT SECTIONAL INTERNAL ELEVATIONS A, B, C & D AS EXISTING 1:20. Drawing No. 006.* 20 August 2003. 59.4 × 84.1. In Antarctic Heritage Trust (NZ), *Conservation Plan: The Historic Huts at Cape Adare.* Christchurch, NZ: Antarctic Heritage Trust, 2004.

Purcell Miller Tritton & Antarctic Heritage Trust (NZ). *STORES HUT SECTIONAL INTERNAL ELEVATIONS – PROPOSED REPAIRS 1:20. Drawing No. 014.* 27 November 2003. 59.4 × 84.1. In Antarctic Heritage Trust (NZ), *Conservation Plan: The Historic Huts at Cape Adare.* Christchurch, NZ: Antarctic Heritage Trust, 2004.

DETAIL
Primary Sources:
Purcell Miller Tritton & Antarctic Heritage Trust (NZ). *STORES HUT SECTIONAL INTERNAL ELEVATIONS – PROPOSED REPAIRS 1:20. Drawing No. 014.* 27 November 2003. 59.4 × 84.1. In Antarctic Heritage Trust (NZ), *Conservation Plan: The Historic Huts at Cape Adare.* Christchurch, NZ: Antarctic Heritage Trust, 2004.

Secondary Sources:
Borchgrevink, Carsten E. *First on the Antarctic Continent: Being an Account of the British Antarctic Expedition 1898–1900.* Cambridge: Cambridge University Press, 1901.

1902, Hut Point hut
pp. 618–21

SITE PLAN
Primary Sources:
QGIS. "Quantarctica: An Antarctic GIS Package." August 2013. qgis.org. [online]

Cochran, Christopher. *Conservation Plan: Discovery Hut, Hut Point: National Antarctic Expedition, 1901–1904, Hut Point, Ross Island, Antarctica.* Christchurch, NZ: Antarctic Heritage Trust, 2004.

PLAN
Primary Sources:
Cochran, Christopher. *Conservation Plan: Discovery Hut, Hut Point: National Antarctic Expedition, 1901–1904, Hut Point, Ross Island, Antarctica.* Christchurch, NZ: Antarctic Heritage Trust, 2004.

SECTION & ELEVATION
Primary Sources:
Cochran, Christopher. *Conservation Plan: Discovery Hut, Hut Point: National Antarctic Expedition, 1901–1904, Hut Point, Ross Island, Antarctica.* Christchurch, NZ: Antarctic Heritage Trust, 2004.

Secondary Sources:
Antarctic Heritage Trust (NZ). *Conservation Plan: The Historic Huts at Cape Adare; Borchgrevink's British Antarctic Expedition, 1898–1900, Northern Party of Scott's British Antarctic Expedition, 1910–1913, Cape Adare, Antarctica.* Christchurch, NZ: Antarctic Heritage Trust, 2004.

National Science Foundation. *McMurdo Station: Master Plan 2.1.* 16 December 2015. future.usap.gov. [online]

1906, Aktieselskabet Hektor Whaling Station
pp. 622–23

SITE PLAN
Primary Sources:
Google Maps. Accessed 17 January 2021. google.com/maps. [online]

Secretariat of the Antarctic Treaty. WHALERS BAY – DECEPTION ISLAND (9mage) from "17. Whalers Bay." Accessed 21 January 2021. ats.aq. [online]

Secondary Sources:
Antarctic Treaty Visitors Guide, 2018.

Deception Island: Antarctic Specially Managed Area. "Climate." Accessed 21 January 2021. deceptionisland.aq. [online]

1908, Cape Royds hut
pp. 624–27

SITE PLAN
Primary Sources:
Jasmax & Antarctic Heritage Trust (NZ). *SHACKLETON'S HUT CAPE ROYDS ROSS ISLAND: LOCATION PLAN, SITE PLAN.* 28 May 2002. 21×29 cm. In Antarctic Heritage Trust (NZ), *Shackleton's Hut, Cape Royds.* Christchurch, NZ: Antarctic Heritage Trust, 2003.

PLAN
Primary Sources:
Jasmax & Antarctic Heritage Trust (NZ). *SHACKLETON'S HUT CAPE ROYDS ROSS ISLAND: LOCATION PLAN, SITE PLAN.* 28 May 2002. 21×29 cm. In Antarctic Heritage Trust (NZ), *Shackleton's Hut, Cape Royds.* Christchurch, NZ: Antarctic Heritage Trust, 2003.

SECTIONS
Primary Sources:
Jasmax & Antarctic Heritage Trust (NZ). *SHACKLETON'S HUT CAPE ROYDS ROSS ISLAND: CROSS SECTION INTERNAL ELEVATIONS EAST AND WEST (1909 Configuration) 1:25.* 1 August 2002. 59.4 × 84.1 cm. In Antarctic Heritage Trust (NZ), *Shackleton's Hut, Cape Royds.* Christchurch, NZ: Antarctic Heritage Trust, 2003.

Jasmax & Antarctic Heritage Trust (NZ). *SHACKLETON'S HUT CAPE ROYDS ROSS ISLAND: LONGITUDINAL INTERNAL ELEVATION SOUTH (1909 Configuration) 1:25.* 1 August 2002. 59.4 × 84.1 cm. In Antarctic Heritage Trust (NZ), *Shackleton's Hut, Cape Royds.* Christchurch, NZ: Antarctic Heritage Trust, 2003.

ELEVATION
Primary Sources:
Jasmax & Antarctic Heritage Trust (NZ). *SHACKLETON'S HUT CAPE ROYDS ROSS ISLAND: NORTH ELEVATION (1909 Configuration) 1:25.* 1 August 2002. 59.4 × 84.1 cm. In Antarctic Heritage Trust (NZ). *Shackleton's Hut, Cape Royds.* Christchurch, NZ: Antarctic Heritage Trust, 2003.

Secondary Sources:
Antarctic Heritage Trust (NZ). *Conservation Plan: The Historic Huts at Cape Adare; Borchgrevink's British Antarctic Expedition, 1898–1900, Northern Party of Scott's British Antarctic Expedition, 1910–1913, Cape Adare, Antarctica.* Christchurch, NZ: Antarctic Heritage Trust, 2004.

1911, Framheim hut
pp. 628–31

SITE PLAN & PLAN
Primary Sources:
Fram Museum. *Fig 4. Plan of Amundsen's Framheim, Bay of Whales, 1911. Conjectural reconstructions based on descriptions and photographs (Amundsen 1912).* Oslo: Fram Museum.

PLAN TUNNELS
Primary Sources:
Amundsen, Roald. *Plan of the Winter Quarters of the Expedition.* Hand drawing. in Gran, Tryggve, *Kampen om Sydpolen.* Oslo: Ernst G. Mortensen, 1961.

Secondary Sources:
Fletcher, Daina. "Midwinter in Antarctica with Roald Amundsen." Australian National Maritime Museum. 23 June 2017. sea.museum. [online]

1911, Cape Evans hut
pp. 632–35

SITE PLAN
Primary Sources:
Jasmax & Antarctic Heritage Trust (NZ). *SCOTT'S HUT, CAPE EVANS, ROSS ISLAND: LOCATION PLAN, SITE PLAN.* 14 May 2003, 21 × 29 cm. In Antarctic Heritage Trust (NZ), *Scott's Hut, Cape Evans.* Christchurch, NZ: Antarctic Heritage Trust, May 2004.

Google Maps. Accessed 17 January 2021. google.com/maps. [online]

QGIS. "Quantarctica: An Antarctic GIS Package." August 2013. qgis.org. [online]

PLAN
Primary Sources:
Jasmax & Antarctic Heritage Trust (NZ). *SCOTT'S HUT, CAPE EVANS, ROSS ISLAND: OVERALL BUILDING PLAN 1:50.* 7 August 2003. 59.4 × 84.1 cm. In Antarctic Heritage Trust (NZ), *Scott's Hut, Cape Evans.* Christchurch, NZ: Antarctic Heritage Trust, May 2004.

SECTIONS
Primary Sources:
Jasmax & Antarctic Heritage Trust (NZ). *SCOTT'S HUT, CAPE EVANS, ROSS ISLAND: CROSS SECTIONS, INTERNAL ELEVATION EAST AND WEST 1:25.* 7 August 2003. 59.4 × 84.1 cm. In Antarctic Heritage Trust (NZ), *Scott's Hut, Cape Evans.* Christchurch, NZ: Antarctic Heritage Trust, May 2004.

Jasmax & Antarctic Heritage Trust (NZ). *SCOTT'S HUT, CAPE EVANS, ROSS ISLAND: LONGITUDINAL SECTION, INTERNAL ELEVATION SOUTH (PROPOSED) 1:50.* 7 August 2003. 59.4 × 84.1 cm. In Antarctic Heritage Trust (NZ), *Scott's Hut, Cape Evans.* Christchurch, NZ: Antarctic Heritage Trust, May 2004.

ELEVATION
Primary Sources:
Jasmax & Antarctic Heritage Trust (NZ). *SCOTT'S HUT, CAPE EVANS, ROSS ISLAND: ELEVATION NORTH (PROPOSED) 1:50.* 7 August 2003. 59.4 × 84.1 cm. In Antarctic Heritage Trust (NZ), *Scott's Hut, Cape Evans.* Christchurch, NZ: Antarctic Heritage Trust, May 2004.

1912, Cape Denison huts
pp. 636–39

SITE PLAN
Primary Sources:
Google Maps. Accessed 17 January 2021. google.com/maps. [online]

Hodgeman, Alfred. *Plans of the Main Hut (from the Home of the Blizzard: Being the Story of the Australasian Antarctic Expedition 1911–1914).* In Bourke, Deborah, *Mawson's Huts Historic Site: Management Plan 2013–2018.* Kingston: Australian Antarctic Program (AAD), 2013.

PLAN
Primary Sources:
Hodgeman, Alfred. *Plans of the Main Hut (from the Home of the Blizzard: Being the Story of the Australasian Antarctic Expedition 1911–1914).* In Bourke, Deborah, *Mawson's Huts Historic Site: Management Plan 2013–2018.* Kingston: Australian Antarctic Program (AAD), 2013.

SECTIONS
Primary Sources:
Hodgeman, Alfred. *Plans of the Main Hut.* In Mawson, Douglas, *The Home of the Blizzard: Being the Story of the Australasian Antarctic Expedition 1911–1914.* London: Heinemann, 1915.

1914, The *Endurance*
pp. 640–43

ALL DRAWINGS
Primary Sources:
Framnæs Mekaniske Værksted, Sandefjord, Polar Wooden Screw Yacht: "Polaris." Circa 1912. 97.5 × 71.2 cm. Vestfold Musenee. Last modified 20 April 2018. vestfoldmuseene.no. [online]

1944, Base A
pp. 644–47

SITE PLAN
Primary Sources:
Burkitt, D. M. *Plane Table Survey of Port Lockroy. Coastline surveyed by prismatic compass and tape measure.* January 1997. Scale 1:500. Drawing. Falkland Islands Dependencies Survey, British Antarctic Survey Archives, Cambridge.

To implement site plan:
Google Maps. Accessed 17 January 2021. google.com/maps. [online].

PLAN & SECTION (small)
Primary Sources:
Boulton & Paul Ltd. *Timber Building Required by Lieutenant J.W.S. Marr., Port Lockroy.* Amended 24 October 1943. Drawings. Falkland Islands Dependencies Survey, British Antarctic Survey Archives, Cambridge, July 1944.

PLAN & SECTION (large)
Primary Sources:
L. Ashton & Boulton & Paul Ltd. *Plan of Building as Constructed at Goudier Islet, Port Lockroy Graham Land.* Drawing. Falkland Islands Dependencies Survey, British Antarctic Survey Archives, Cambridge, July 1944.

1949, Maudheim
pp. 648–51

SITE PLAN & PLAN 1
Recreation by UNLESS from Photograph:
Swithinbank, Charles. "Maudheim 3 Days Old" (presentation slide). In *Parallels Between Antarctic Travel in 1950 and Planetary Travel in 2050 (to Accompany Notes on "The Norwegian British-Swedish Antarctic Expedition 1949–52")*, conference paper presented at the Cambridge University, UK, 2013.

PLAN 2
Primary Sources:
Swithinbank, Charles. "Plan of Maudheim, our habitat for two years" (presentation slide. In *Parallels Between Antarctic Travel in 1950 and Planetary Travel in 2050 (to Accompany Notes on "The Norwegian British-Swedish Antarctic Expedition 1949–52")*, conference paper presented at the Cambridge University, UK, 2013.

Secondary Sources:
Norsk Polar Institut, 1960 (archives).

1954, Mawson
pp. 652–53

SITE PLAN
Primary Sources:
Australian Antarctic Program. *First Agreed Mawson Layout*, 1956.

Australian Antarctic Data Centre. "Mawson Station, Station Area Map." 1997.

Secondary Sources:
Mawson, Sir Douglas. *The Home of the Blizzard: Being the Story of the Australasian Antarctic Expedition, 1911–1914.* Oxford: Benediction Classics, 2008.

Australian Antarctic Program (AAD). "Shelters from the Stormy Blast." Last modified 15 September 2011. mawsonshuts.antarctica.gov.au. [online]

1929/1956, Little America V (V/V)
pp. 654–57

SITE PLAN & PLAN
Primary Sources:
Mellor, Malcolm. *Figure 22. Layout of Little America IGY Station on Ross Ice Shelf, Antarctica.* In *Methods of Building on Permanent Snow Fields.* Hanover, NH: US Army Materiel Command, Cold Regions Research & Engineering Laboratory, 1968.

United States Navy. "EXPEDITION DEEPFREEZE USNC IGY MAIN BASE PLOT PLAN, ARCHITECTURAL." 22 September 1955. Thomas B. Bourne Associates Inc., Engineers and Architects. Department of the Navy Bureau of Yards and Docks. United States National Archives Research Center, Washington, 1955. [online]

PLAN 2
Drawing shared from United States National Archives Research Center, Washington, 1957.

Mellor, Malcolm. *Figure 22. Layout of Little America IGY Station on Ross Ice Shelf, Antarctica.* In *Methods of Building on Permanent Snow Fields.* Hanover, NH: US Army Materiel Command, Cold Regions Research & Engineering Laboratory, 1968.

Secondary Sources:
UNITED STATESDepartment of Commerce/ Environmental Science Services Administration. Collected Reprints, Essa Institute for Oceanography. Atlantic-Pacific Oceanographic Laboratories, Miami, 1967.

1957, Halley I (I/VI)
pp. 658–65

SITE PLAN
Primary Sources:
Cotton, J.P.D. and D.P. Wild. *Base Z Halley Bay 1:1250.* Hand-drawn plan. April 1964. Falkland Islands Dependencies Survey, British Antarctic Survey Archives, Cambridge, 1965.

PLAN
Primary Sources:
F.I.D.S. *New Base Hut, Halley Bay General Arrangement.* Hand-drawn plan. Falkland Islands Dependencies Survey, British Antarctic Survey Archives, Cambridge, 1965.

SECTIONS
Reconstruction Based on documents from:
F.I.D.S. *New Base Hut, Halley Bay General Arrangement.* Hand-drawn plan. Falkland Islands Dependencies Survey, British Antarctic Survey Archives, Cambridge, 1965.

SECTION DETAIL, BASE Z
F.I.D.S. *New Base Hut – Halley Bay Typical Details.* Hand-drawn sections. Falkland Islands Dependencies Survey, British Antarctic Survey Archives, Cambridge [year unknown].

SECTION, Non-magnetic Hut
Original Drawing from
F.I.D.S. *Details of Extension to Non-Magnetic Hut for Base Z Halley Bay – Declinometer Room.* Hand-drawn sections. Falkland Islands Dependencies Survey, British Antarctic Survey Archives, Cambridge, 1958.

PLAN & SECTION, Spectrophotometer Hut
Primary Sources:
F.I.D.S. *Spectrophotometer Hut: General Arrangement & Typical Details.* Hand-drawn sections. Falkland Islands Dependencies Survey, British Antarctic Survey Archives, Cambridge, 1958.

Secondary Sources:
King, John. "Control of Near-Surface Winds over an Antarctic Ice Shelf." *Journal of Geophysical Research* 98 no. D7 (1993): 12949–53.

1956, McMurdo
pp. 666–67

SITE PLAN
Primary Sources:
OZ Architecture and National Science Foundation. "McMurdo Station Master Plan 2.1." Report. National Science Foundation, 2015.

QGIS. "Quantarctica: An Antarctic GIS Package." Accessed 27 January 2021. qgis.org. [online]

Beebe, Morton, ed. *Operation Deep Freeze III: The Story of Task Force 43; Third Phase, 1957–58.* Paoli, PA: Dorville Corporation, 1957.

Secondary Sources:
Davis, Georgina A. "A History of McMurdo Station through Its Architecture." *Polar Record* 53, no. 2 (2017): 167–85.

Hood, Elaine. "Appreciation: Dick Bowers, 1928–2019; The Seabee Who Built South Pole Station." 1 April 2019. antarcticsun.usap.gov. [online]

1956, Mirny
pp. 668–69

SITE PLAN
Primary Sources:
"Mirny Station." Hand drawing. 1956. Stored in Arctic and Antarctic Research Institute (AARI) & Central State Archive of the City of Moscow. Posted on *Polar Forum* (original in Russian). Accessed 17 February 2021. polarpost.ru. [online]

Secondary Sources:
Arctic and Antarctic Research Institute (AARI). "Mirny Observatory." Accessed 27 January 2021. aari.aq. [online]

Luchansky, Grigory. "Opyt Antarkticheskikh Sovetskikh Stantsii." | Опыт Антарктических Советских Станций. Accessed 27 January 2021. geolmarshrut.ru. [online]

1957, Amundsen-Scott South Pole (I/III)
pp. 670–73

SITE PLAN
Primary Sources:
Department of the Navy, Washington D.C. Bureau of Yards and Docks. *Pole Station Antarctica, PLOT PLAN.* Hand-drawn plan. 1964.

PLAN MAIN STATION & BALLOON INFLATION BUILDING
Primary Sources:
Mellor, Malcolm. *Figure 25. Original layout of South Pole Statio.* In *Methods of Building on Permanent Snow Fields.* Hanover, NH: U.S. Army Materiel Command, Cold Regions Research & Engineering Laboratory, 1968.

Department of the Navy, Washington. D.C. Naval Facilities Engineering Command. *New South Pole Station, Antarctica: GMD, BALLOON INFLATION BLDG. & TOWER SCHEMATIC,* 1969.

PLAN MAIN STATION BUILDINGS
Primary Sources:
Drawing shared from United States National Archives Research Center, Washington.

Secondary Sources:
Hood, Elaine. "Appreciation: Dick Bowers, 1928–2019; The Seabee Who Built South Pole Station." 1 April 2019. antarcticsun.usap.gov. [online]

National Science Foundation. *Draft Environmental Impact Statement: Modernization of the Amundsen-Scott South Pole Station Antarctica.* Arlington, VA: Office of Polar Programs, 1998.

1957, Charcot
pp. 674–77

SITE PLAN
Primary Sources:
Dubrovin, Leonid Ivanovich, and Vladimir Nikolaye-vich Petrov. *Figure 90. Plan of Charcot Station.* In *Scientific Stations in Antarctica, 1882–1963,* p. 354. Translated from Russian. New Delhi: Indian National Scientific Documentation Centre, 1971.

SECTION & ELEVATION, LIVING HOUSE
Primary Sources:
Jacquin, Jean-Pierre. *Conception et construction de la baraque de la station Charcot.* Expéditions polaires françaises. Document shared with UNLESS courtesy of Jean-Pierre Jacquin, 2020.

1957, Scott Base
pp. 678–81

SITE PLAN
Primary Sources:
Henderson, G. & New Zealand Geological Survey Antarctic Expedition. *Scott Base.* Hand drawing. 1959–60.

QGIS. "Quantarctica: An Antarctic GIS Package." Accessed 27 January 2021. qgis.org. [online]

PLAN
Primary Sources:
Cochran, Chris, and Antarctic Heritage Trust (NZ), *Restored Floor Plan 1:50.* 2013. 29.7 × 42 cm. In Antarctic Heritage Trust (NZ), *Conservation Plan: Hillary's Hut, Scott Base.* Christchurch: Antarctic Heritage Trust, January 2015.

ELEVATION & SECTIONS
Primary Sources:
Cochran, Chris, and Antarctic Heritage Trust (NZ) *South & East Elevations 1:50.* July 2001. 29.7 × 42 cm. In Antarctic Heritage Trust (NZ), *Conservation Plan: Hillary's Hut, Scott Base.* Christchurch: Antarctic Heritage Trust, January 2015.

Cochran, Chri, and Antarctic Heritage Trust (NZ) *North & West Elevations 1:50.* July 2001. 9.7 × 42 cm. In Antarctic Heritage Trust (NZ), *Conservation Plan: Hillary's Hut, Scott Base.* Christchurch: Antarctic Heritage Trust, January 2015.

Secondary Sources:
McKay, Bill. "Extreme Conservation: Antarctic Huts of the Heroic Age." *Architecture New Zealand* 6 (November 2018).

Sinclair, M. R. "Local Topographic Influence on Low-Level Wind at Scott Base, Antarctica." *New Zealand Journal of Geology and Geophysics* 31, no. 2 (1988): 237–45.

1957, Pole of Inaccessibility
pp. 682–83

SITE PLAN
Primary Sources:
L. M. Savatiygin, M. A. Preobrajendkaya, *Russian Research in Antarctica, Tom I (1st–20th Soviet Antarctic Expedition), Image 66, Plan of Pole of Inaccessibility 1967.* Arctic and Antarctic Research Centre (AARI), Central State Archive of the City of Moscow.

1946/1961, Base E
pp. 684–89

SITE PLAN
Primary Sources:
Mason, D. P. *Base E and Vicinity 1:2000.* Hand-drawn plan. 1946. Approx. 42 × 59.4 cm. Falkland Islands Dependencies Survey, British Antarctic Survey Archives, Cambridge.

ROOF PLAN
Primary Sources:
Boulton & Paul LTD. *Base Hut Base E.* Hand-drawn plan. November 1957. Falkland Islands Dependencies Survey, British Antarctic Survey Archives, Cambridge.

ROOF PLAN
Primary Sources:
Boulton & Paul LTD. *Base E: PLAN.* Hand-drawn plan. November 1954. Falkland Islands Dependencies Survey, British Antarctic Survey Archives, Cambridge.

SECTION & ELEVATION
Primary Sources:
Base Huts, Bases N & E. Undated. Falkland Islands Dependencies Survey, British Antarctic Survey Archives, Cambridge.

PLANS 1st and 2nd LEVEL, SECTION & ELEVATION NEW BUILDING
Primary Sources:
Base E Stonington Is. BASE HUT. Hand-drawn plans. 1986. Falkland Islands Dependencies Survey, British Antarctic Survey Archives, Cambridge.

1957/1962, New Byrd (II/II)
pp. 690–93

SITE PLAN & PLAN
Primary Sources:
United States Navy. *Byrd Station Antarctica PLOT PLAN: As built conditions proposed constructions.* Hand-drawn plan. 25 August 1965. Department of the Navy Bureau of Yards and Docks. United States National Archives Research Center, Washington, DC.

SECTION
Primary Sources:
United States Navy. Facilities Engineering Command. *Byrd Station Antarctica Deep Core Facility Workshop Sections and Detail. Section Tunnel L-2* (Sheet 2 of 4). Hand-drawn. 1 July 1966. Department of the Navy Bureau of Yards and Docks. United States National Archives Research Center, Washington, DC.

Secondary Sources:
Mock, S. J., and W. F. Weeks. "The Distribution of Ten-Meter Snow Temperatures on the Greenland Ice Sheet." Research Report 170. U.S. Army Materiel Command, Cold Regions Research & Engineering Laboratory, Hanover, NH, 1965.

1963, Eights
pp. 694–97

SITE PLAN & PLANS
Primary Sources:
United States Navy. *Eights Station: Air Portable Scientific Facility, VLF – Aurora Laboratory … Unit No.1 – Plans, Elevations and Sections.* Hand-drawn. 1 March 1962. Department of the Navy Bureau of Yards and Docks. United States National Archives Research Center, Washington.

SECTION, DETAIL
Primary Sources:
United States Navy. *Eights Station: Air Portable Scientific Facility, VLF – Aurora Laboratory … Unit No.1 – Plans, Elevations and Sections.* Hand-drawn. 1 March 1962. Department of the Navy Bureau of Yards and Docks. United States National Archives Research Center, Washington.

United States Navy. *Eights Station: Air Portable Scientific Facilty. Pibal Dome … Details, Plans, Elevations & Sections.* Hand-drawn. 1 March 1962. Department of the Navy Bureau of Yards and Docks. United States National Archives Research Center, Washington.

Secondary Sources:
Mellor, Malcolm, ed. *Antarctic Snow and Ice Studies.* Washington: American Geophysical Union, 1964.

National Science Foundation. *Antarctic Journal of the United States* 1, no. 4 (July–August 1966).

1965, Plateau
pp. 698–701

SITE PLAN
Primary Sources:
United States National Archives (unauthored). *Figure 2: Plateau Station Plot Plan.* Hand drawing. United States National Archives Research Center, Washington, 1965.

PLAN
Primary Sources:
Marconi, J. *POLAR PLATEAU BASE CAMP FLOOR PLAN. ARCHITECTURAL.* Hand-drawn. 12 July 1965. ATCO Industries / Northland Camps Inc. United States National Archives Research Center, Washington, 1965.

SECTION & ELEVATIONS
Primary Sources:
Weibe, J. R. *POLAR PLATEAU STATION BASE CAMP – EXTERIOR ELEVATIONS.* Hand-drawn. 21 June 1965. ATCO Industries / Northland Camps Inc. United States National Archives Research Center, Washington, 1965.

Secondary Sources:
Cunde, Xiao, Li Yuansheng, Ian Allison, Hou Shugui, Gabrielle Dreyfus, Jean-Marc Barnola, Ren Jiawen et al. "Surface Characteristics at Dome A, Antarctica: First Measurements and a Guide to Future Ice-Coring Sites." *Annals of Glaciology* 48 (2008): 82–87.

1944/1966, Base B
pp. 702–5

SITE PLAN
Google Maps. Accessed 17 January 2021. google.com/maps. [online]

PLAN & ELEVATIONS
Primary Sources:
British Antarctic Survey. *General Arrangement of G.R. Plastic Building for British Antarctic Survey.* Hand-drawn. 1964. British Antarctic Survey Archives, Cambridge.

SECTION
Primary Sources:
British Antarctic Survey. *Partitioning Layout of G.R. Plastic Building. British Antarctic Survey.* Hand-drawn. 1964. British Antarctic Survey Archives, Cambridge.

DETAIL
Primary Sources:
British Antarctic Survey. *Title Unknown. British Antarctic Survey.* Hand-drawn. 1964. British Antarctic Survey Archives, Cambridge.

Secondary Sources:
Antarctic Treaty Visitors Guide. "Site Plan Reconstruction." 2018.

"Climate." Deception Island – Antarctic Specially Managed Area. Accessed 13 February 2021. deceptionisland.aq. [online]

1956/1967, Halley II (II/VI)
pp. 706–11

SITE PLAN
Primary Sources:
British Antarctic Survey. *(Map of) Halley Bay – Iono Aerials.* Scaled approx. 1 cm : 21 m. Hand drawing.

Undated. British Antarctic Survey Archives, Cambridge.

PLAN BELOW SNOW LEVEL
Original Drawing from :
British Antarctic Survey. Preliminary Copy: *New Base at Halley Bay.* Hand drawing. Undated. British Antarctic Survey Archives, Cambridge.

SECTION
Primary Sources:
British Antarctic Survey. Preliminary Copy: *New Base at Halley Bay – Typical Details in Buildings.* Hand drawing. Undated. British Antarctic Survey Archives, Cambridge.

ELEVATION
Recreated from:
British Antarctic Survey. *Base Z 1967–73, Halley Bay.* Hand drawing. Undated. British Antarctic Survey Archives, Cambridge.

SECTION ACCESS SHAFT
Primary Sources:
British Antarctic Survey. *New Base at Halley Bay – Typical Details of Buildings.* Hand drawing. Undated (archived 1991). British Antarctic Survey Archives, Cambridge.

Secondary Sources:
King, J. C. "Control of Near-Surface Winds over an Antarctic Ice Shelf." *Journal of Geophysical Research* 98, no. D7 (1993): 12949–53.

Halley Bay. "Halley Bay – 1967." 28 January 2021. zfids.org.uk. [online]

1969, Casey Repstat
pp. 712–15

SITE PLAN
Primary Sources:
Commonwealth of Australia, Department of Supply-Antarctic Division. *Casey Site Plan.* Hand-drawn, 19 May 1965.

Google Maps. Accessed 13 February 2021. google.com/maps. [online]

PLAN
Primary Sources:
Incoll, Philip G. *The Influence of Architectural Theory on the Design of Australian Antarctic Stations.* Melbourne: Australian Construction Services, 1990.

SECTION
Primary Sources:
Incoll, Philip G. *Figure 1. Section through Casey Station 1969.* In *The Influence of Architectural Theory on the Design of Australian Antarctic Stations.* Melbourne: Australian Construction Services, 1990.

Secondary Sources:
Antarctic Treaty Visitors Guide. "Site Plan Reconstruction." 2018.

Steverson, Evan. "Initial Environmental Evaluation, Removal of Old Casey Station." Unpublished honours thesis. University of Tasmani. 1991.

Turner, John, Tom A. Lachlan-Cope, and Garet J. Marshall. "An Extreme Wind Event at Casey Station, Antarctica." *Journal of Geophysical Research* 106 no. D7 (2001): 7291–311.

1956/1973, Halley III (III/VI)
pp. 716–21

SITE PLAN
Primary Sources:
British Antarctic Survey. *Halley Bay Base Area January 1974.* Hand-drawn. British Antarctic Survey Archives, Cambridge.

PLAN BELOW SNOW LEVEL
Primary Sources:
British Antarctic Survey. *Antarctic Station Halley (plan internal layout).* Hand-drawn. Undated. British Antarctic Survey Archives, Cambridge.

ELEVATION & SECTION
Primary Sources:
British Antarctic Survey. B.A.S. Halley Bay New Base ARMCO M.P. UNDERPASS 5'-10" SPAN × 7'-8' RISE FOR TUNNEL. Hand-drawn. Undated. British Antarctic Survey Archives, Cambridge.

DETAILS TUNNELS
Primary Sources:
British Antarctic Survey. *B.A.S. New Base Halley Bay, FLOOR DETAILS & GABLE ENDS TO MAIN 'ARMCO' TUNNELS.* Hand-drawn. 21 July 1972. British Antarctic Survey Archives, Cambridge.

Secondary Sources:
King, J.C. "Control of Near-Surface Winds over an Antarctic Ice Shelf." *Journal of Geophysical Research* 98 no. D7 (1993): 12949–53.

1957/1975, Amundsen-Scott South Pole
pp. 722–25

SITE PLAN
Primary Sources:
Department of the Navy, Washington D.C. Naval Facilities Engineering Command. *New South Pole Station Antarctica, PLOT PLAN Civil – Mechanical.* Hand-drawn plan. 19 February 1971. 71.1 × 101.6 cm. United States National Archives Research Center, Washington.

PLAN
Primary Sources:
Department of the Navy, Washington D.C. Naval Facilities Engineering Command. *New Pole Station Antarctica, GEODESIC DOME Floor Plan, Elevations and Details, Architectural.* Hand-drawn. 9 April 1969. 71.1 × 101.6 cm. United States National Archives Research Center, Washington.

ELEVATION & SECTION
Primary Sources:
Department of the Navy, Washington D.C. Naval Facilities Engineering Command. *New Pole Station Antarctica, GEODESIC DOME Floor Plan, Elevations and Details, Architectural.* Hand-drawn. 9 April 1969. 71.1 × 101.6 cm. United States National Archives Research Center, Washington.

DETAIL
Primary Sources:
Department of the Navy, Washington D.C. Naval Facilities Engineering Command. *New South Pole Station Antarctica, GEODESIC DOME Foundation Plan, Structural.* Hand-drawn. 17 February 1970. 71.1 × 101.6 cm. United States National Archives Research Center, Washington.

Secondary Sources:
National Science Foundation. *Draft Environmental Impact Statement: Modernization of the Amundsen-Scott South Pole Station Antarctica.* Arlington, VA: Office of Polar Programs, 1998.

Wikimedia Commons. *Aerial View of the Amundsen-Scott South Pole Scientific Research Station.* Image. 1983.

1975/1980, BASE R > Rothera
pp. 726–29

SITE PLAN
Primary Sources:
Smith, Alan. *New Base Complex at Rothera Point.* Cambridge: British Antarctic Survey 1976, p. 11.

PLAN, SECTION, ELEVATIONS
Primary Sources:
Mosse and Kestaven Architects. *BAS BASE PHASE B Building (all drawings on sheet).* 1:100. July 1977. British Antarctic Survey Archives, Cambridge.

Secondary Sources:
Mackay, D. British Antarctic Survey. British Antarctic Survey Archive (C/1977/R), Cambridge, 1976/77.

Shears, J. R. "The Use of Lichens to Monitor Heavy Metal Levels around Rothera Research Station, Rothera Point, Adelaide Island, Antarctica." In *Biological Methods for use in Monitoring the Arctic,* edited by Kari Viken Olsen. Report. Copenhagen: Directorate for Nature Management (DN) in Norway and The Swedish Environmental Protection Agency, 1995.

1976, Georg Forster
pp. 730–33

SITE PLAN
Google Maps. Accessed 13 February 2021. google.com/maps. [online]

PLAN
Primary Sources:
Alfred Wegener Institute, Helmholtz Centre for Polar and Marine Research (AWI). 1976.

ELEVATION
Original Drawing reconstructed from documents shared by:
Alfred Wegener Institute, Helmholtz Centre for Polar and Marine Research (AWI).

Secondary Sources:
Soni, V. K., M. Sateesh, Anand K. Das, and S. K. Peshin. "Progress in Meteorological Studies around Indian Stations in Antarctica." *Proceedings of the Indian National Science Academy* 83, no. 2 (2017).

Information provided by Dietrich Enss, Polar Engineering Consultant, and Hartwig Gernandt, Alfred Wegener Institute, Helmholtz Centre for Polar and Marine Research (AWI). 2020.

1953/1978, Esperanza
pp. 734–35

SITE PLAN
Primary Sources:
Google Maps. Accessed 13 February 2021. google.com. [online]

Argentine Antarctic Institute. Image (1980) in "Argentine Antarctic Permanent Stations Description and Features." Internal document shared with UNLESS, 25 May 2020.

Secondary Sources:
Van Wessem, Jan Melchior, Carleen H. Reijmer, Willem Jan van de Berg, Michiel R. van den Broeke, Alison J. Cook, Lambertus H. van Ulft, and Erik van Meijgaard. "Temperature and Wind Climate of the Antarctic Peninsula as Simulated by a High-Resolution Regional Atmospheric Climate Model." *Journal of Climate* 28, no. 18 (2015): 7306–26.

1981, Georg von Neumayer
pp. 736–39

SITE PLAN, PLAN, SECTION
Primary Sources:
Archival Project Drawings (dwg) provided courtesy of Dietrich Enss and Hartwig Gernandt, Alfred Wegener Institute, Helmholtz Centre for Polar and Marine Research (AWI). Shared with UNLESS for reproduction in *Antarctic Resolution* in 2020.

Secondary Sources:
Klöwer, Milan, Thomas Jung, Gert König-Langlo, and Tido Semmler. "Aspects of Weather Parameters at Neumayer Station, Antarctica, and Their Represen-

tation in Reanalysis and Climate Model Data." *Meteorologische Zeitschrift* 22, no. 6 (2013): 699–709.

1956/1983, Halley IV
pp. 740–45

SITE PLAN
Primary Sources:
Jamieson Associates Designers & British Antarctic Survey. *New Halley Complex Overall Site Layout: 1:5000 (BAS 110).* Hand drawing. Undated. British Antarctic Survey Archives, Cambridge.

PLANS
Primary Sources:
Jamieson Associates Designers & British Antarctic Survey. *Plan at Upper Floor Level: 1:100 (BAS 108).* Hand drawing. Undated. British Antarctic Survey Archives, Cambridge.

Jamieson Associates Designers & British Antarctic Survey. *Plan at Lower Floor Level: 1:100 (BAS 109).* Hand drawing. Undated. British Antarctic Survey Archives, Cambridge.

ELEVATION
Primary Sources:
Halley IV Plan, *Fig. 2.* Hand drawing. Undated. Elevation Reconstructed by UNLESS for *Antarctic Resolution.* British Antarctic Survey Archives, Cambridge.

SECTION
Primary Sources:
Jamieson Associates Designers & British Antarctic Survey. *Section 2.3, Section 2.2: 1:100 (BAS 106).* Hand drawing. Undated. British Antarctic Survey Archives, Cambridge.

British Antarctic Survey. *Image of Halley IV Cross Section taken in 1991, Antarctica.* British Antarctic Survey Archives, Cambridge.

DETAIL
Primary Sources:
British Antarctic Survey. *Specification of Test Panels and Assembly Details.* Hand drawing. Undated. British Antarctic Survey Archives, Cambridge.

Secondary Sources:
King, J.C. "Control of Near-Surface Winds Over an Antarctic Ice Shelf." *Journal of Geophysical Research* 98 no. D7 (1993): 12949–53.

1969/1984, Base Presidente Eduardo Frei Montalva
pp. 746–51

ALL DRAWINGS
Primary Sources:
Archival project drawings (dwg) provided courtesy of Chilean Antarctic Institute (INACH) 2019 & Chilean Armed Forces, 2005, 2011, 2018. Shared with UNLESS by Polar Lab Chile for reproduction in *Antarctic Resolution* in 2020.

Secondary Sources:
Hernández, Victór. "Medio siglo de la Base Antártica Presidente Eduardo Frei Montalva." *La Prensa Austral,* 27 March 2019.

Peter, Hans-Ulrich, Christina Braun, Susann Janowski, Anja Nordt, Anke Nordt, and Michel Stelter. *The Current Environmental Situation and Proposals for the Management of the Fildes Peninsula Region.* Dessau-Roßlau: Federal Environment Agency, 2013. uba.de. [online]

1986, Mario Zucchelli
pp. 752–55

SITE PLAN
Google Maps. Accessed 13 February 2021. google.com. [online]

PLANS & ELEVATIONS
Primary Sources:
Archival project drawings (dwg) provided courtesy of ENEA, Francesco Pellegrino Italian National Antarctic Research Program (PNRA), 2004. Shared with UNLESS for reproduction in *Antarctic Resolution* in 2020.

Secondary Sources:
Berger, André, Fedor Mesinger, and Djordje Sijacki. *Climate Change: Inferences from Paleoclimate and Regional Aspects.* Vienna: Springer Vienna, 2012.

1969/88, Casey (II/II)
pp. 756–59

SITE PLAN
Primary Sources:
Incoll, Philip G. *Figure 2. New Casey Site Layout, 1980.* In *The Influence of Architectural Theory on the Design of Australian Antarctic Stations,* p. 6. Melbourne: Australian Construction Services, 1990.

PLAN & ELEVATION
Primary Sources:
Incoll, Philip G. *Casey Office.* In *AANBUS: The Creation of a Building System for Antarctica,* p. 44. Canberra: Commonwealth of Australia, Department of Housing & Construction, 1980.

PLAN 1st and 2nd LEVEL
Primary Sources:
Incoll, Philip G. *Figure 3. Casey Domestic Building First Floor Plan* and *Figure 4. Casey Domestic Building Ground Floor Plan.* In *The Influence of Architectural Theory on the Design of Australian Antarctic Stations,* pp. 7, 8. Melbourne: Australian Construction Services, 1990.

Secondary Sources:
Turner, John, Tom A. Lachlan-Cope, and Gareth J. Marshall. "An Extreme Wind Event at Casey Station, Antarctica." *Journal of Geophysical Research* 106 no. D7 (2001): 7291–311.

1989, Zhongshan
pp. 760–61

SITE PLAN
Primary Sources:
Archival Project Drawings (dwg) provided courtesy of Beijing Institute of Architectural Engineering; China Polar Research Institute (1989/2019), Architectural Design and Research Institute of Tsinghua University (renovation project since 2002). Shared with UNLESS by Polar Lab Hong Kong for *Antarctic Resolution* in 2020.

Secondary Sources:
Botao, Shi, and Fu Tao. "China Railway Construction Will Expedite the 16th Expedition to Antarctica." *Capital Construction News,* 27 August 2018.

China Railway Construction Engineering Group (renovation project since 2002).

Council of Managers of National Antarctic Programs (COMNAP). *Antarctic Station Catalogue.* Christchurch: COMNAP, 2017. comnap.aq. [online]

Heng, Wu, and Qian Zhihong. *Contemporary China: China's Antarctic Research Undertakings.* Beijing: Contemporary China Publishing House, 2009.

Wang, Yuting, Lingen Bian, Yongfeng Ma, Jie Tang, Dongqi Zhang, and Xiangdong Zheng. "Surface Ozone Monitoring and Background Characteristics at Zhongshan Station over Antarctica." *Chinese Science Bulletin* 56, no. 10 (2011): 1011–19.

Yi, Zhang, and Zhuang Weimin. "Research on Psychological Demand and Environmental Design in Polar Conditions: Taking the Antarctic Research Station of China as an Example." *Design Community* 4 (2012): 130–35.

1956/1992, Halley V (V/VI)
pp. 762–67

SITE PLAN
Primary Sources:
British Antarctic Survey. *Untitled Drawing.* Hand-drawn. Undated. British Antarctic Survey Archives, Cambridge.

PLAN
Primary Sources:
Christiani & Nielsen for Natural Environment Research Council, British Antarctic Survey. *Halley (5) Surface Station: Fire Fighting in ACB Alarms and Equipments 1:50.* Hand-drawn. 30 November 1987. British Antarctic Survey Archives, Cambridge.

ELEVATION
Primary Sources:
British Antarctic Survey. *Halley V Elevation Windward Side, 1:100.* Hand-drawn. Undated. British Antarctic Survey Archives, Cambridge.

PLAN FOUNDATIONS
Primary Sources:
Christiani & Nielsen for Natural Environment Research Council, British Antarctic Survey. *Halley (5) Surface Station: Platform ACB Cross Beams and Girders 1:50.* Hand-drawn. 30 November 1987. British Antarctic Survey Archives, Cambridge.

SECTION
Primary Sources:
British Antarctic Survey. *Halley V Cross Section, 1:25.* Hand-drawn. Undated. British Antarctic Survey Archives, Cambridge.

DETAIL
Primary Sources:
Christiani & Nielsen for Natural Environment Research Council, British Antarctic Survey. *Halley (5) Surface Station: Platform ACB, Foundation, Plans and Details 1:20.* Hand-drawn. 1987. British Antarctic Survey Archives, Cambridge.

Secondary Sources:
Council of Managers of National Antarctic Programs (COMNAP). *Antarctic Station Catalogue.* Christchurch: COMNAP, 2017. comnap.aq. [online]

King, J. C. "Control of Near-Surface Winds over an Antarctic Ice Shelf." *Journal of Geophysical Research* 98 no. D7 (1993): 12949–53.

Further Information provided by Dietrich Enss, Polar Engineering Consultant.

1981/1992, Neumayer II
pp. 768–69

SITE PLAN
Primary Sources:
Alfred Wegener Institute, Helmholtz Centre for Polar and Marine Research (AWI). 1991.

Secondary Sources:
Council of Managers of National Antarctic Programs (COMNAP). *Antarctic Station Catalogue.* Christchurch: COMNAP, 2017. comnap.aq. [online]

Klöwer, Milan, Thomas Jung, Gert König-Langlo, and Tido Semmler. "Aspects of Weather Parameters at Neumayer Station, Antarctica, and Their Representation in Reanalysis and Climate Model Data." *Meteorologische Zeitschrift* 22, no. 6 (2013): 699–709.

Further information provided by Dietrich Enss, Polar Engineering Consultant.

1957/1993 Syowa
pp. 770–75

SITE PLAN
Primary Sources:
Archival Project Drawings (PDF) provided courtesy of National Institute of Polar Research, Japan. 1993. Shared with UNLESS for reproduction in *Antarctic Resolution* in 2020.

Google Maps. Accessed 13 February 2021. google. com. [online]

PLAN
Primary Sources:
National Academy of Sciences – National Research Council. *Appendix 10020 main hut plan, p. 241.* In *Symposium on Antarctic Logistics.* Boulder, Colorado: University of Colorado. August 13–17, 1962.

ELEVATION
Original Drawing adapted from axonometric: National Academy of Sciences (PNAS) – National Research Council. *Appendix 60000 order of erection,* p. 252. Symposium on Antarctic Logistics. University of Colorado, Boulder. 13–17 August 1962.

SECTION & DETAIL, PREFABRICATED UNIT
Images found on:
Takenaka Corporation. "Project Story Antarctic Research Expedition." Accessed February 2020. [online]

PLAN 1st–3rd LEVELS, SECTION, ELEVATION ADMINISTRATIVE BUILDINGS
Primary Sources:
Archival Project Drawings (PDF) provided courtesy of National Institute of Polar Research, Japan. 1993. Shared with UNLESS for reproduction in *Antarctic Resolution* in 2020.

Secondary Sources:
Antarctic Treaty Secretariat. "Management Plan for Antarctic Specially Protected Area No. 141." *ATCM XXXVII Final Report: Measure 7 (2014) Annex.* env.go.jp. [online]

Council of Managers of National Antarctic Programs (COMNAP). *Antarctic Station Catalogue.* Christchurch: COMNAP, 2017. comnap.aq. [online]

Sato, Kaoru, and Naohiko Hirasawa. "Statistics of Antarctic Surface Meteorology Based on Hourly Data in 1957–2007 at Syowa Station." *Polar Science* 1, no. 1 (2007): 1–15.

Sawaragi, Noi. "Notes on Art and Current Events 68." Accessed 13 February 2021. art-it.asia. [online]

Michael, Gibert, Naziaty Mohd Yaacob, and Zuraini Md Ali. "The Capsule Living Unit Reconsidered a Utopia Transformed Reality." *Pertanika Journal of Social Sciences and Humanities* (2018).

1960/1997, Sanae IV (IV/IV)
pp. 776–79

SITE PLAN
Primary Sources:
Department of Environmental Affairs, Antarctic Division. *Sanae IV, Environmental Health and Safety Impact Assessment (Figure 2).* Hand-drawn. Undated.

Google Maps. Accessed 13 February 2021. google.com. [online]

PLAN
Primary Sources:
Department of Environmental Affairs, Antarctic Division. *Block B.* In *Sanae IV, Environmental Health and Safety Impact Assessment (Figure 5).* Hand drawing. Undated.

LONG ELEVATION
Department of Environmental Affairs, Antarctic Division. *Sanae IV, Environmental Health and Safety Impact Assessment (Figure 4).* Hand drawing. Undated.

SHORT SECTION & ELEVATION
Department of Environmental Affairs, Antarctic Division. *Typical Section & South Elevation.* In *Sanae IV, Environmental Health and Safety Impact Assessment.* Hand drawing. Undated.

Secondary Sources:
Council of Managers of National Antarctic Programs (COMNAP). *Antarctic Station Catalogue.* Christchurch: COMNAP, 2017. comnap.aq. [online]

Hund, Andrew J. (ed). *Antarctica and the Arctic Circle: A Geographic Encyclopedia of the Earth's Polar Regions Volume I: A–I.* Oxford: ABC-CLIO LLC, 2014.

Koegelenberg, Ilana, and Royal Haskoning. "SANAE IV Antarctica Refurbishment: Meticulously Planned, Carefully Executed." 9 January 2019. refrigerationandaircon.co.za. [online]

2005, Concordia
pp. 780–85

ALL DRAWINGS
Primary Sources:
Archival Project Drawings (dwg) provided courtesy of National Antarctic Research Program, Italy (PNRA), 2005. Shared with UNLESS for reproduction in *Antarctic Resolution* in 2020.

Secondary Sources:
Aristidi, E., K. Agabi, M. Azouit, E. Fossat, J. Vernin, T. Travouillon, J. S. Lawrence et al. "An Analysis of Temperatures and Wind Speeds above Dome C, Antarctica." *Astronomy & Astrophysics* 430, no. 2 (2005): 739–46.

Council of Managers of National Antarctic Programs (COMNAP). *Antarctic Station Catalogue.* Christchurch: COMNAP, 2017. comnap.aq. [online]

Cristofanelli, Paolo, Davide Putero, Paolo Bonasoni, Maurizio Busetto, Francescopiero Calzolari, Giuseppe Camporeale, Paolo Grigioni et al. "Analysis of Multi-Year Near-Surface Ozone Observations at the WMO/GAW 'Concordia' Station (75°06'S, 123°20'E, 3280 M A.S.L. – Antarctica)." *Atmospheric Environment* 177 (2018): 54–63.

Institut Polaire. *Concordia: A New Permanent, International Research Support Facility High on the Antarctic Ice Cap; Technical Overview.* Plouzane: French Polar Institute, 2020. latitude.aq. [online]

2007, Princess Elisabeth
pp. 786–89

ALL DRAWINGS
Primary Sources:
Archival project drawings (dwg) provided courtesy of Samyn and Partners Architects & Engineers & PREFALUX, 2007. Shared with UNLESS for reproduction in *Antarctic Resolution* in 2020.

Secondary Sources:
Amin, Nighat F. D. *Princess Elizabeth Antarctica.* Tielt: Lannoo Publishers, 2013.

Council of Managers of National Antarctic Programs (COMNAP). *Antarctic Station Catalogue.* Christchurch: COMNAP, 2017. comnap.aq. [online]

Google Maps. Accessed 13 February 2021. google.com. [online]

International Polar Foundation. "BELARE 2007–2008." Press Release. March 2008. antarcticstation.org. [online]

Pattyn, Frank, Kenichi Matsuoka, and Johan Berte. "Glacio-Meteorological Conditions in the Vicinity of the Belgian Princess Elisabeth Station, Antarctica." *Antarctic Science* 22, no. 1 (2010): 79.

1957/2008, Amundsen-Scott South Pole (III/III)
pp. 790–95

SITE PLAN
Primary Sources:
United States Antarctic Program / Environmental Research and Assessment. *Map 4: ASMA No. 5 – Amundsen-Scott South Pole Station.* (MAP ID: 10069.014.15). 3 April 2017. *Southpolestation.com.* [online]

PLAN
Primary Sources:
Ferraro Choi and Associates for Department of the Navy, Naval Facilities Engineering Command. *South Pole Station Modernisation – Elevated Station: Floor Plan, POD A, Levels 1 and 2; POD B, Levels 1 and 2.* July 1999. 55.9 × 86.4 cm (Size D).

SECTION TOWER UNIT
Primary Sources:
Ferraro Choi and Associates for Department of the Navy, Naval Facilities Engineering Command. *South Pole Station Modernisation – Elevated Station: Vertical Tower - Sections.* July 1999, 55.9 × 86.4 cm (Size D).

ELEVATIONS
Primary Sources:
Ferraro Choi and Associates for Department of the Navy, Naval Facilities Engineering Command. *South Pole Station Modernisation – Exterior Elevation – Pods A and B; Exterior Elevations – Pod A.* July 1999. 55.9 × 86.4 cm (Size D).

SECTION DETAIL
Primary Sources:
Ferraro Choi and Associates for Department of the Navy, Naval Facilities Engineering Command. *South Pole Station Modernisation – Elevated Station – Wall Sections.* July 1999. 55.9 × 86.4 cm (Size D).

Secondary Sources:
"Amundsen-Scott South Pole Station." BBFM Engineers Inc. Accessed 13 February 2021. bbfm.com. [online]

Council of Managers of National Antarctic Programs (COMNAP). *Antarctic Station Catalogue.* Christchurch: COMNAP, 2017. comnap.aq. [online]

Ferraro Choi and Associates Ltd. "Amundsen-Scott South Pole Station." 18 December 2014. architect-magazine.com. [online]

National Science Foundation. *Draft Environmental Impact Statement: Modernization of the Amundsen-Scott South Pole Station Antarctica.* Arlington, VA: Office of Polar Programs, 1998.

West, Peter. "U.S. South Pole Station." National Science Foundation. Accessed 13 February 2021. nsf.gov. [online]

"Amundson-Scott [*sic*] South Pole Station." PDC Engineers. Accessed 13 February 2021. pdceng.com. [online]

Further information shared courtesy of Ferraro Choi and Associates.

1981/2009, Neumayer III (III/III)
pp. 796–803

ALL DRAWINGS
Primary Sources:
Archival Project Drawings (dwg) provided courtesy of Dietrich Enss and Hartwig Gernandt, Alfred Wegener Institute, Helmholtz Centre for Polar and Marine Research (AWI), 2008. Shared with UNLESS for reproduction in *Antarctic Resolution* in 2020.

Secondary Sources:
bof Architekten.

Council of Managers of National Antarctic Programs (COMNAP). *Antarctic Station Catalogue*. Christchurch: COMNAP, 2017. comnap.aq. [online]

Klöwer, Milan, Thomas Jung, Gert König-Langlo, and Tido Semmler. "Aspects of Weather Parameters at Neumayer Station, Antarctica, and Their Representation in Reanalysis and Climate Model Data." *Meteorologische Zeitschrift* 22, no. 6 (2013): 699–709.

2012, Bharati
pp. 804–9

ALL DRAWINGS
Primary Sources:
Archival project drawings (dwg) provided courtesy of bof Architekten, 2010. Shared with UNLESS for reproduction in *Antarctic Resolution* in 2020.

Secondary Sources:
"Projekt 'Indische Forschungsstation Bharati, Antarktis.'" Accessed 13 February 2021. competition-line.com. [online]

Council of Managers of National Antarctic Programs (COMNAP). *Antarctic Station Catalogue*. Christchurch: COMNAP, 2017. comnap.aq. [online]

"KAEFER Competence in Arctic Conditions." Press Release. June 2011. kaefer-sea.com. [online]

1956/2013, Halley VI (VI/VI)
pp. 810–19

ALL DRAWINGS
Primary Sources:
Archival project drawings (dwg) provided courtesy of Hugh Broughton Architects, 2005. Shared with UNLESS for reproduction in *Antarctic Resolution* in 2020.

Secondary Sources:
Council of Managers of National Antarctic Programs (COMNAP). *Antarctic Station Catalogue*. Christchurch: COMNAP, 2017. comnap.aq. [online]

Glancey, Jonathan. "Cold Comfort." *The Guardian*, 19 July 2005.

Hugh Broughton Architects. "Halley VI British Antarctic Research Station." Accessed 13 February 2021. hbarchitects.co.uk. [online]

King, J. C. "Control of Near-Surface Winds over an Antarctic Ice Shelf." *Journal of Geophysical Research* 98 no. D7 (1993): 12949–53.

Further information shared courtesy of Hugh Broughton Architects.

2014, Jang Bogo
pp. 820–23

ALL DRAWINGS
Primary Sources:
Archival project drawings (dwg) provided courtesy of Space Group Architects, 2014. Shared with UNLESS for reproduction in *Antarctic Resolution* in 2020.

Secondary Sources:
Council of Managers of National Antarctic Programs (COMNAP). *Antarctic Station Catalogue*. Christchurch: COMNAP, 2017. comnap.aq. [online]

Wang, Jang-Woon, Jae-Jin Kim, Wonsik Choi, Da-Som Mun, Jung-Eun Kang, Hataek Kwon, Jin-Soo Kim et al. "Effects of Wind Fences on the Wind Environment around Jang Bogo Antarctic Research Station." *Advances in Atmospheric Sciences* 34, no. 12 (2017): 1404–14.

Lee, Sang Leem. *Jangbogo Antarctic Research Station*. Seoul: Space Group, 2011.

2014, Taishan
pp. 824–27

ALL DRAWINGS
Primary Sources:
Archival project drawings (dwg) provided courtesy of Polar Research Institute of China, 1989. Shared with UNLESS by Polar Lab Hong Kong for *Antarctic Resolution* in 2020.

Secondary Sources:
Council of Managers of National Antarctic Programs (COMNAP). *Antarctic Station Catalogue*. Christchurch: COMNAP, 2017. comnap.aq. [online]

"China Opens 4th Antarctic Research Base." *Global Times*, 8 February 2014. globaltimes.cn. [online]

Lei, Zhang, and Dong Fangyu. "Spirit of Adventure Lives on in Antarctic." 19 February 2014. chinadaily.com.cn. [online]

Shen, Jiaxing. "Assembled Structural Member and Joint Design of China Antarctic Taishan Station." *Steel Construction* (2016): 73–76.

Shiping, Liu. "My Country's First Antarctic Under-Snow Building Achieves Five Major Innovations" [translated]. 8 March 2019. xinhuanet.com. [online]

Yuan, Han. "Chief Designer Wang Zhongjun Explains the Functional Area inside Taishan Station" [translated]. 28 March 2014. xinhuanet.com. [online]

1998/2014 Union Glacier
pp. 828–31

SITE PLAN
Adapted from Image:
Antarctic Logistics @Antarctic_ALE. Twitter Post. 10 September 2019, 11:50 p.m. [online]

Antarctic Logistics. "Camp Services: Union Glacier Camp." antarctic-logistics.com. 2020. [online]

Doran, Temujin (StudioCanoe). *A Very Short Guide to Union Glacier Camp*. Video. 3 May 2020. boomers-daily.com. [online]

PLAN & SECTION
Primary Sources:
Bernal, Marcelo, Paul Taylor, and Francisco Valdivia. *Ilaia: Estación Polar Científica Conjunta Glaciar Unión: Antártica, Chile, 2013–2014*. Antarctica/Chile: ARQ, 2014.

Secondary Sources:
Antarctic Logistics & Expeditions. *Union Glacier Camp*. Brochure. 2019. antarctic-logistics.com. [online]

McKay, Christopher, Edward Balaban, Simon Abrahams, and Nick Lewis. "Dry Permafrost over Ice-Cemented Ground at Elephant Head, Ellsworth Land, Antarctica." *Antarctic Science* 31, no. 05 (2019): 1–8.

Ice Trek. "Union Glacier Camp." Accessed 13 February 2021. icetrek.com. [online]

1988/2018, Juan Carlos I (II/II)
pp. 832–37

ALL DRAWINGS
Primary Sources:
Archival project drawings (dwg) provided courtesy of Hugh Broughton Architects, 2008. Shared with UNLESS for reproduction in *Antarctic Resolution* in 2020.

Secondary Sources:
Bañón, Manuel, Ana Justel, David Velázquez, and Antonio Quesada. "Regional Weather Survey on Byers Peninsula, Livingston Island, South Shetland Islands, Antarctica." *Antarctic Science* 25, no. 2 (2013): 146–56.

Council of Managers of National Antarctic Programs (COMNAP). *Antarctic Station Catalogue*.

Christchurch: COMNAP, 2017. comnap.aq. [online]

1984/2020, Comandante Ferraz
pp. 838–43

ALL DRAWINGS
Primary Sources:
Archival project drawings (PDF) provided courtesy of Estudio41, 2013–2015. Shared with UNLESS for reproduction in *Antarctic Resolution* in 2020.

Secondary Sources:
Council of Managers of National Antarctic Programs (COMNAP). *Antarctic Station Catalogue*. Christchurch: COMNAP, 2017. comnap.aq. [online]

McManus, David. "Comandante Ferraz Antarctic Station Building." 2 November 2020. e-architect. com. [online]

"Wind Rose Graphical and Tabelarical Presentation." Accessed 13 February 2020. meis.si. [online]

1988/2020+, St. Kliment Ohridski
pp. 844–47

ALL DRAWINGS
Primary Sources:
Project drawings (dwg) provided courtesy of Penka Stancheva of PS Architects. 2020. Shared with UNLESS for reproduction in *Antarctic Resolution* in 2020.

Secondary Sources:
Council of Managers of National Antarctic Programs (COMNAP). *Antarctic Station Catalogue*. Christchurch: COMNAP, 2017. comnap.aq. [online]

1977/2020+, Henryk Arctowski
pp. 848–51

ALL DRAWINGS
Primary Sources:
Project drawings (dwg) provided courtesy of Kuryłowicz & Associates. 2020. Shared with UNLESS for reproduction in *Antarctic Resolution* in 2020.

Secondary Sources:
Kejna, Marek. "Topoclimatic Conditions in the Vicinity of the Arctowski Station (King George Island, Antarctica) during the Summer Season of 2006/2007." *Polish Polar Research* 29, no. 2 (2008).

Further information provided courtesy of Kuryłowicz & Associates.

1957/2020+ Scott Base
pp. 852–55

ALL DRAWINGS
Primary Sources:
Project Drawings (dwg) provided courtesy of Hugh Broughton Architects. 2020. Shared with UNLESS for reproduction in *Antarctic Resolution* in 2020.

Secondary Sources:
M. R. Sinclair. "Local Topographic Influence on Low-Level Wind at Scott Base, Antarctica." *New Zealand Journal of Geology and Geophysics* 31, no. 2 (1988): 237–45.

"Quantarctica: An Antarctic GIS Package." Accessed 13 February 2021. qgis.org. [online]

Further information provided courtesy of Hugh Broughton Architects.

1975/2020+, Rothera
pp. 856–59

SITE PLAN
Google Maps. Google.com/maps. 2020 [online]

ALL DRAWINGS
Primary Sources:
Project drawings (PDF) provided courtesy of Hugh Broughton Architects and British Antarctic Survey. 2020. Shared with UNLESS for reproduction in *Antarctic Resolution* in 2020.

Secondary Sources:
BAM Nutall Ltd. / BAM International. "New Building at Rothera Breaks Ground." Press Release. 31 January 2020. bam.com. [online]

Shears, J. R. "The Use of Lichens to Monitor Heavy Metal Levels around Rothera Research Station, Rothera Point, Adelaide Island, Antarctica." In *Biological Methods for Use in Monitoring the Arctic*, edited by Kari Viken Olsen. Copenhagen: Nordic Council of Ministers, 1995.

2020+, Ross Sea Station
pp. 860–61

SITE PLAN
Primary Sources:
Polar Research Institute of China and Tongji University (reconstruction). Produced and shared with UNLESS by Polar Lab Hong Kong for *Antarctic Resolution* in 2020.

Secondary Sources:
Adams, Byron, Diana Wall, Ross Virginia, Emma Broos, and Matthew Knox. "Ecological Biogeography of the Terrestrial Nematodes of Victoria Land, Antarctica." *Zookeys* 419 (2014): 29–71.

Yang, Yi. "China Prepares to Build 5th Antarctic Research Station." 11 December 2017. xinhuanet.com. [online]

Tongji University & Polar Research Institute of China. *Proposed Construction and Operation of a New Chinese Research Station, Victoria Land, Antarctica.* Draft Comprehensive Environmental Evaluation, 2018.

Toyobo Engineering Co., Ltd. "Reverse Osmosis Membrane Seawater Concentration System." Accessed 13 February 2021. toyobo-eng.co.jp. [online]

Wang, Zikun. "The Chinese Antarctic Expedition Team Have Once Again Landed on the Unspeakable Island." Translated by UNLESS. 28 January 2019. sohu.com. [online]

1957/2020+ Vostok
pp. 862–65

SITE PLAN
Primary Sources:
Courtesy of Arctic and Antarctic Research Institute (AARI).

PLAN & ELEVATION
Primary Sources:
Project drawings (dwg) provided courtesy of Arctic and Antarctic Research Centre, 2020. Shared with UNLESS for reproduction in *Antarctic Resolution* in 2020.

Secondary Sources:
Olsen, Kari Viken, ed. *Biological Methods for Use in Monitoring the Arctic.* Copenhagen: Nordic Council of Ministers, 1995.

1956/2020+, McMurdo
pp. 866–71

ALL DRAWINGS
Project drawings (dwg) provided courtesy of OZ Architects, 2018. Shared with UNLESS for reproduction in *Antarctic Resolution* in 2020.

Primary Source:
OZ Architecture, 2020.

Secondary Sources:
OZ Architecture. *McMurdo Station Modernization Study: Building Shell & Fenestration Study.* 29 April 2016. usap.gov. [online]

"McMurdo Station: Master Plan 2.1." Future USAP, National Science Foundation. 16 December 2015. usap.gov. [online]

"Quantarctica: An Antarctic GIS Package." Accessed 13 February 2021. qgis.org. [online]

2020+, Dome Fuji
pp. 872–75

SITE PLAN
Recreation based on image from Misawa Homes, 2020.

ELEVATIONS
Primary Sources:
Japan Aerospace Exploration Agency. "A Joint Project of JAXA, NIPR, Misawa Homes, and MHIRD: Demonstration Test of Antarctica Mobile Station Unit." Press release. 26 August 2019. jaxa.jp. [online]

Secondary Sources:
Council of Managers of National Antarctic Programs (COMNAP). *Antarctic Station Catalogue.* Christchurch: COMNAP, 2017. comnap.aq. [online]

Further information provided courtesy of Misawa Homes.

"When you consider that
in 35 years, at the rate we are
going, there will be twice
as many people alive as there
are today, you realize that
science must come up with
solutions for thousands of
new problems. What we have
been learning in Antarctica
is the raw material for some
of these solutions."

Rear Admiral Lloyd Viel Berkner (President of
the International Council of Scientific Unions),
quoted in "What We've Accomplished in Antarctica",
National Geographic, 1959

Image Credits

03
The Governance System of a Citizenless Continent

04
Antarctic Resources. Temptations and Accountability

05
"The Question of Antarctica"

The Antarctic through the Lens of Herbert Ponting

FOUR ELEMENTS

06
The Revolutionary Ocean

07
Twenty-Six Quadrillion Tons of Ice. The Urgencies of Contemporaneity vs. the Truths of "Deep Time"

08

A Continental Archipelago under Ice

09

The White Desert. A Unique Viewing Platform into the Cosmos

The Antarctic through the Lens of Eliot Porter

SURVIVING IN THE CRYOSPHERE

10

Extreme Habitation. Compressed Spaces and Dilated Times

p. 423
J.H. Sutherland, 1963. Reproduced courtesy of the British Antarctic Survey Archives Service. Archives ref. AD8/3/6. Copyright Crown

p. 424
© Mori Art Museum, Tokyo

p. 425 top, 425 middle
Lee Mattis; reproduced courtesy of the US Antarctic Program, National Science Foundation

p. 425 bottom
An-My Lê; © An-My Lê

p. 426 left
Herbert George Ponting; © Scott Polar Research Institute, University of Cambridge

p. 426 middle
© Alexander Turnbull Library, Wellington, New Zealand

p. 426 right
© Scott Polar Research Institute, University of Cambridge

p. 427 left
Kevin Eric W. Walton; reproduced courtesy of the British Antarctic Survey Archives Service. Archives ref. AD6/19/2/E606/4; © Crown (expired)

p. 427 middle
© Commonwealth Copyright – All rights reserved

p. 427 right
Peter Macdonald; © Peter Macdonald

p. 428 left
Alan Smith; reproduced courtesy of the British Antarctic Survey Archives Service. Archives ref. AD11/2Z/4.1/1; © UK Research and Innovation

p. 428 middle
© Commonwealth Copyright – All rights reserved

p. 428 right
Elmer Cranton; © Elmer Cranton

p. 429 left
© Philippe Samyn and Partners

p. 429 middle
Fred Portelli © Subframe.media

pp. 429 right, 430, 431
© Space Group

13

Geography of Science, "Domed Cities", and Mobile Architectures

p. 436 top
Richard J. Wolak; © Richard J. Wolak

pp. 436 bottom, 437
United States Navy; reproduced courtesy of the US Antarctic Program, National Science Foundation

pp. 438, 450 top, 450 middle, 452
© Shaun O'Boyle, courtesy of the artist

p. 439 top left
Robert Smithson; © bpk-Bildagentur

p. 439 bottom
© Anne Noble, courtesy of the artist

p. 442
Scot Jackson; reproduced courtesy of the US Antarctic Program, National Science Foundation

pp. 443, 451
© An-My Lê, courtesy of the artist

p. 444
Chilean Air Force (FACh)

p. 447 top, 447 bottom middle
© Gonzalo Escobar Meza / Pontificia Universidad Catolica de Chile

p. 447 bottom middle, 447 bottom right
Chilean Air Force (FACh)

p. 448
Horacio Villalobos from the Corbis Historical Collection via Getty Images

p. 449
Amancio Williams; © Archivo Williams

p. 450 bottom
© Todd Anderson, courtesy of the artist

pp. 453, 454
OZ Architecture; © OZ Architecture

p. 456
Hugh Broughton Architects; © Hugh Broughton Architects

p. 457
Photograph: Anthony Powell, image: Jasmax / Hugh Broughton © Anthony Powell / Jasmax / Hugh Broughton Architects

p. 458 top
Herbert George Ponting; © Scott Polar Research Institute, University of Cambridge

p. 458 bottom
Tom J. Taylor © Antarctica New Zealand

p. 459
United States Navy

p. 460 top right
Christopher James Gilbert; reproduced courtesy of the British Antarctic Survey Archives Service. Archives ref. AD6/19/4/1/17/27/9; © UK Research and Innovation

p. 460 bottom
Fredrick Ernest Harvey; reproduced courtesy of the British Antarctic Survey Archives Service. Archives ref. AD6/19/3/C/Z47; © UK Research and Innovation

p. 461 top
Peter Tarnas; reproduced courtesy of the British Antarctic Survey Archives Service. Archives ref. AD6/2Z/1983/C1; © UK Research and Innovation

p. 461 bottom
David Boyer; © Nat Geo Image Collection

p. 466 top left
R.I. Walcott, 1957. Reproduced courtesy of the British Antarctic Survey Archives Service. Archives ref. AD6/19/3/C/D3. Copyright Crown

p. 466 top right
Eric James Chinn; reproduced courtesy of the British Antarctic Survey Archives Service. Archives ref. AD6/19/3/C/Z24; © UK Research and Innovation

p. 466 bottom left
Charles Frank Le Feuvre; reproduced courtesy of the British Antarctic Survey Archives Service. Archives ref. AD6/19/3/C/Z5; © Crown (expired)

p. 466 bottom right
Colin Marshall Read; reproduced courtesy of the British Antarctic Survey Archives Service. Archives ref. AD6/19/3/C/Z28; © UK Research and Innovation

p. 467 top left
D.G. Allan, 1982. Reproduced courtesy of the British Antarctic Survey Archives Service. Archives ref. AD6/19/4/1/20/28/61. Copyright Allan, D.G.

p. 467 top right
Vivian Ernest Fuchs; reproduced courtesy of the British Antarctic Survey Archives Service. Archives ref. AD6/19/3/C/Z38; © UK Research and Innovation

p. 467 bottom left
Vivian Ernest Fuchs; reproduced courtesy of the British Antarctic Survey Archives Service. Archives ref. AD6/19/3/C/Z41; © UK Research and Innovation

p. 467 bottom right
Douglas George Allan; reproduced courtesy of the British Antarctic Survey Archives Service. Archives ref. AD6/19/4/1/19/27/76; © Douglas George Allan

p. 468 top
Ewan Scott Hunter; reproduced courtesy of the British Antarctic Survey Archives Service. Archives

ref. AD6/2Z/1989/C3; © UK Research and Innovation

p. 468 bottom
David Maxwell; © David Maxwell/ Royal Engineers

p. 469 top right
Getty format – Photographer / Collection name via Getty Images. Royal Geographic Society via Getty Images

p. 469 middle right
© Expéditions Polaires Françaises

p. 469 bottom right
© unknown; www.bas.ac.uk

p. 469 bottom left
© Herron Archive, Archigram Archival Project

p. 471 top
© James Morris, courtesy of the artist

p. 471 bottom
Pete Bucktrout; courtesy of the British Antarctic Survey

14

From Anthropic Footprints to Disembodied Technologies

pp. 475, 496
"Deep Freeze '65, Task Force 42," Antarctica Cruisebooks, Burdette & Company, Boston, 1965; © Burdette & Company; www.history.navy.mil

p. 476
© Shaun O'Boyle, courtesy of the artist

p. 478 left, 478 top right
D.R.H. Davis; reproduced courtesy of the British Antarctic Survey Archives Service. Archives ref. AD6/19/3/D8; © Crown (expired)

p. 478 bottom right
Beth Healey; © European Space Agency (ESA) / Institut polaire français Paul-Émile-Victor (IPEV)/ Programma Nazionale di Ricerche in Antartide (PNRA) – Beth Healey

p. 479
© Jean de Pomereu, courtesy of the artist

p. 480
Emil Schulthess; © Fotostiftung Schweiz

p. 484
Allen Cull; www.southpolestation.com

p. 485
© James Morris, courtesy of the artist

p. 488 top
Herbert George Ponting; © Harry Pennell Collection, Canterbury Museum

p. 488 bottom
Estúdio 41 Arquitetura; © Estúdio 41 Arquitetura

p. 489 top
Paul Allman Siple, "Man's First Winter at the South Pole," *National Geographic* (January 1958): 468–69, detail

p. 489 bottom
© Nat Geo Image Collection

p. 491
© Novatek, Zapsibgazprom, OZSK

p. 492
© Anne Noble, courtesy of the artist

p. 493, 502 top right
© unknown

pp. 494, 497, 499
© Collection IPF: René Robert, Nighat F.D. Amin, Alain Hubert

p. 498
© International Polar Foundation, Henri Robert

p. 502 bottom left
Steven Brown; © Steven Brown / Australian Antarctic Division

15

The Ideological Use of Relics

16

Antarctica and Outer Space. Laboratories for a Future as Global Commons

The Antarctic through the Lens of Joan Myers

ARCHIVE OF ANTARCTIC ARCHITECTURE

Little America I, 1929
© The Ohio State University, Byrd Polar and Climate Research Center Archival Program, Admiral Richard E. Byrd Papers [image #7803_2]

Little America III, 1940
© unkown

East Base, 941
United States Navy

Base A, 1944
Ivan Mackenzie Lamb; reproduced courtesy of the British Antarctic Survey Archives Service. Archives ref. 1a5_41; © Crown (expired)

Base B, 1944
George Rayner; © Museums Victoria – CC BY-SA 4.0

Base C, 1945
Vivian Ernest Fuchs; reproduced courtesy of the British Antarctic Survey Archives Service. Archives ref. AD6/19/2/E1022/24; © Crown (expired)

Base D > Teniente Ruperto Elichiberty, 1945
© Instituto Antártico Uruguayo

Little America IV > 1946
Reproduced courtesy of the United States Navy Museum of Naval Aviation

Base C2, 1946
Michael W. Sadler; reproduced courtesy of the British Antarctic Survey Archives Service. Archives ref. AD6/19/2/E404/36; © Crown (expired)

Base E, 1946
John Eliot Tonkin; reproduced courtesy of the British Antarctic Survey Archives Service. Archives ref. AD6/19/2/E505/18; © Crown (expired)

Captain Arturo Prat, 1947
© Armada de Chile

Melchior, 1947
Boletin del Instituto Antártico Argentino 2, no. 19 (May 1966): 21; reproduced courtesy of the Instituto Antártico Argentino

Base H > Signy, 1947
Charles J. Skilling; reproduced courtesy of the British Antarctic Survey Archives Service. Archives ref. AD6/19/2/H144; © Crown (expired)

Base F > Vernadsky, 1947
Robert Seddon Moss; reproduced courtesy of the British Antarctic Survey Archives Service. Archives ref. AD6/19/2/BM132; © Crown (expired)

Base G, 1947
Roger Anthony Todd-White; reproduced courtesy of the British Antarctic Survey Archives Service. Archives ref. AD6/19/2/G176; © Crown (expired)

Base Deception, 1948
Boletin del Instituto Antártico Argentino 1, no. 6 (November 1959): 14; reproduced courtesy of the Instituto Antártico Argentino

General Bernardo O'Higgins, 1948
University of Chile

Maudheim, 1949
© Scott Polar Research Institute, University of Cambridge

Port Martin, 1950
© Expéditions Polaires Françaises

Paradise Harbour Shelter, 1950
© unkown

Almirante Brown, 1951
Boletin del Instituto Antártico Argentino 2, no. 18 (November 1965): 9; reproduced courtesy of the Instituto Antártico Argentino.

Gabriel Gonzalez Videla, 1951
© Ministry of Defense Crown

San Martina, 1951
Boletin del Instituto Antártico Argentino 1, no. 6 (November 1959): 15; reproduced courtesy of the Instituto Antártico Argentino

Destacamento Naval Bahia Luna > Teniente Camara, 1953
Boletin del Instituto Antártico Argentino 1, no. 6 (November 1959): 14; reproduced courtesy of the Instituto Antártico Argentino

Esperanza, 1953
Andrew Shiva; © Andrew Shiva – CC BY-SA 4.0

Refugio Potter > Carlini, 1953
© Argentine Ministry of Foreign Affairs, International Trade & Worship

Base V > General Ramon Canas Montalva, 1953
Richard Irving Walcott; reproduced courtesy of the British Antarctic Survey Archives Service. Archives ref. AD6/19/3/C/D3; © Crown (expired)

Mawson, 1954
Phillip Law; © Phillip Law / Australian Antarctic Division

Luis Risopatron, 1954
© Instituto Antártico Chileno (INACH)

General Belgrano I, 1955
Fundación Histarmar

Base N, 1955
Alexander James Rennie; reproduced courtesy of the British Antarctic Survey Archives Service. Archives ref. AD6/19/3/C/N1; © Crown

Base Y, 1955
John Crabbe Cunningham; reproduced courtesy of the British Antarctic Survey Archives ref. AD6/19/2/A669/26; © Crown

Presidente Pedro Aguirre Cerda, 1955
Jorge Valdés R. Jorval

Dumont d'Urville, 1956
© Expéditions Polaires Françaises

Little America V, 1956
Jim Waldon; reproduced courtesy of the US Antarctic Program, National Science Foundation

Hallett, 1956
Report of Charles L. Roberts, 1959; courtesy of the National Oceanic and Atmospheric Administration Photo Library

Halley I, 1956
Reproduced courtesy of the British Antarctic Survey Archives Service. Archives ref. AD6/19/3/C/Z6; © Crown (expired)

Naval Air Facility McMurdo, 1956
United States Navy; reproduced courtesy of the US Antarctic Program, National Science Foundation

Mirny, 1956
© Antarctica New Zealand Pictorial Collection, [ANZPC0115.19], 1956-57

Base O, 1956
Richard Arthur Foster; reproduced courtesy of the British Antarctic Survey Archives ref. AD6/19/2/O3/1; © Crown

Base W, 1946
David Michael Burkitt; reproduced courtesy of the British Antarctic Survey Archives ref. AD9/1/1996/14/27; © UK Research and Innovation

Amundsen-Scott South pole Station, 1957; Byrd, 1957
United States Navy; reproduced courtesy of the US Antarctic Program, National Science Foundation

Charcot, 1957
© Expéditions Polaires Françaises

Davis, 1957
© Australian Antarctic Division Data Centre

Ellsworth, 1957
Ens. William Sumrall, USNR; © Ens. William Sumrall, USNR

Roi Baudouin, 1957
Xavier de Maere; Nighat F. D. Amin, *Princess Elisabeth Antarctica* (Tielt: Lannoo Publishers, 2013)

Scott Base, 1957
Murray Ellis; © The Ellis Family Archives

Syowa, 1957
© unkown

Base J, 1957
Frederick Edward Wooden; reproduced courtesy of the British Antarctic Survey Archives Service. Archives ref. AD6/19/3/C/J3. © Crown (expired)

Vostok, 1957
© Arctic and Antarctic Research Institute Archives

Wilkes, 1957
Alastair Battye; © Alastair Battye / Australian Antarctic Division

Little Rockford, 1958
Larry Johnson; © Larry Johnson

Pole of Inaccessibility, 1958
© Arctic and Antarctic Research Institute Archives

Arrival Height Laboratory, 1959
Dan Smale; © Dan Smale – CC BY-SA 4.0

Base KG > Fossil Bluff, 1961
Reproduced courtesy of the British Antarctic Survey Archives Service. Archives ref. AD6/19/3/C/KG5; © UK Research and Innovation

Base Conjunts Teniente Matienzo, 1961
Jorge Arcadio Martinez; © Jorge Arcadio Martinez – CC BY-SA 4.0

Base T > Teniente Carvajal, 1961
Charles Winthrop Molesworth Swithinbank.; reproduced courtesy of the British Antarctic Survey Archives Service. Archives ref. AD6/19/3/C/T27; © UK Research and Innovation

Novolazarevskaya, 1961; Molodezhnaya, 1962
© Arctic and Antarctic Research Institute Archives

Yelcho, 1962
© unkown

New Byrd, 1962
Elmer Cranton; © Elmer Cranton

Eights, 1963
David de Roo; © David de Roo

Brockton, 1965
Brian Rieusset; © Brian Rieusset / Australian Antarctic Division

Plateau, 1965
© United States Seabee Museum – CC BY-SA 2.0

Old Parlmer, 1965
United States Navy; reproduced courtesy of the US Antarctic Program, National Science Foundation

Halley II, 1967
Colin Marshall Read; reproduced courtesy of the British Antarctic Survey Archives Service. Archives ref. AD6/19/3/C/Z28; © UK Research and Innovation

Bellingshausen, 1968
© Arctic and Antarctic Research Institute Archives

Palmer, 1968
United States Navy; reproduced courtesy of the US Antarctic Program, National Science Foundation

Borga, 1969
© unkown

Casey Repstat, 1964
Brian Rieusset; © Brian Rieusset / Australian Antarctic Division

Centro Metereologico Eduardo Frei > Base Presidente Eduardo Frei Montalva
© INACH

Marambio, 1969
© Fundación IDA, Investigación en Diseño Argentino, Fondo Mario Mariño

Vanda, 1969
Garth Varcoe; © Antarctica New Zealand Pictorial Collection [ANZSC0365.13], 1978–79

Mizuho, 1970
Katsu Kaminuma; © Antarctica New Zealand Pictorial Collection [ANZSC0155.1], 1980–81

Leningradskaya, 1971
Dennis Foster; © Dennis Foster / www.kaibabjournal.com

Damoy point hut, 1975
Micheal Paul Landy; reproduced courtesy of the British Antarctic Survey Archives Service. Archives ref. AD6/19/3/C/DAMOY1; © UK Research and Innovation

Siple, 1973
James R. Logan; © James R. Logan

Halley III, 1973
Vivian Ernest Fuchs; reproduced courtesy of the British Antarctic Survey Archives Service. Archives ref. AD6/19/3/C/Z41; © UK Research and Innovation

Amundsen Scott, 1975
© Connie Samaras / De Soto Gallery, Los Angeles, courtesy of the artist

Base R > Rothera, 1975
Micheal Paul Landy; reproduced courtesy of the British Antarctic Survey Archives Service. Archives ref. 3cr7; © UK Research and Innovation

Georg Forster, 1976
Hartwig Gernandt; reproduced courtesy of the Archiv für deutsche Polarforschung, Alfred Wegener Institute; © Hartwig Gernandt

Giacomo Bove, 1976
Reproduced courtesy of WAP (Worldwide Antarctic Program's website www.waponline.it), with the kind consent of Gianni Varetto, content manager of WAP website

Henryk Arctowksi, 1977
Agata Weydmann; www.eu-polarnet.eu

Primavera, 1977
Grant C.; © Grant C. – CC BY-SA 2.0

Lillie Marleen hut, 1979
© Federal Institute for Geosciences and Natural Resources (BGR), Hannover, Germany

General Belgrano, 1979
© Argentine Ministry of Foreign Affairs, International Trade & Worship

Georg Von Neumayer, 1981
Dietrich Enss; © Dietrich Enss

Filchner, 1982
F. Sitte; © Archiv für deutsche Polarforschung, Alfred Wegener Institute

Julio Ripamonti, 1982
© Instituto Antártico Chileno (INACH)

Gondwana, 1983
Andreas Läufer; © Federal Institute for Geosciences and Natural Resources (BGR), Hannover, Germany

Halley IV, 1983
Douglas George Allan; reproduced courtesy of the British Antarctic Survey Archives Service. Archives ref. AD6/19/4/1/19/27/76; © Douglas George Allan

Artigas, 1984
CC BY-SA 3.0

Comandante Ferraz, 1984
Rolf Weber and Montone Rosalinda, "Gerenciamento Ambiental na Baía do Almirantado, Ilha do Rei George, Antártica," 2006

Dakshin Gingotri, 1984
Prakash Khatkar, Madhya Pradesh; © Prakash Khatkar, Madhya Pradesh – CC BY-SA 4.0

Great Wall, 1985
Chen Gangying; © Chen Gangying

Captain Peter J.Lenie Base, 1985
Wayne Trivelpiece, National Oceanic and Atmospheric Administration (NOAA)

Dresher, 1986
Joachim Plötz; © Archiv für deutsche Polarforschung, Alfred Wegener Institute

Law > Law Racovita, 1986
David Hosken; © David Hosken / Australian Antarctic Division

Mario Zucchelli, 1986
© Programma Nazionale di Ricerche in Antartide (PNRA)

World Park, 1987
© Ricardo Roura, 1992, courtesy of the artist

Aboa, 1988
Polar Research Secretariat; © Polar Research Secretariat – CC BY-SA 3.0

Juan Carlos I, 1988
© unkown

Casey, 1988
© Steven Brown / Australian Antarctic Division

King Sejong, 1988
© Korea Polar Research Institute

Progress, 1988
© Arctic and Antarctic Research Institute

St. Kliment Ohridski, 1988
© unkown

Svea, 1988
Wilfried Bauer; © Wilfried Bauer – CC BY-SA 4.0

Eco – Nelson, 1989
Archiv Václava Stackeho

Gabriel de Castilla, 1989
© Ejercito Tierra

Machu Pichu, 1989
© unkown

Maitri, 1989
Philipp R. Heck; © Philipp R. Heck / Field Museum of Natural History

Wasa, 1989
Cecilia Selberg; www.polar.se

Zhongshan, 1989
Jiansong Zhang; © Jiansong Zhang / Xinhua News Agency Shanghai Branch

Pedro Vincente Maldonado, 1990
Reproduced courtesy of WAP (Worldwide Antarctic Program's website www.waponline.it), with the kind consent of Gianni Varetto, content manager of WAP website

Troll, 1990
Peter Leopold; © Norwegian Polar Institute

Gars Laboratory at O'Higgins, 1991; Dr Guillermo Mann, 1991
© Instituto Antártico Chileno (INACH)

Jinnah, 1991
© unkown

Shirreff, 1991
Dan Costa; © Dan Costa

Halley V, 1992
David Maxwell; © David Maxwell (Royal Engineers)

Neumayer II, 992
Carmen Pia Günther; reproduced courtesy Archiv für deutsche Polarforschung, Alfred Wegener Institute

Ice Station Weddell N.1, 1992
Reproduced courtesy of the US Antarctic Program, National Science Foundation

Tor, 1993
Sebastien Descamps; © Norwegian Polar Institute

Dallmann Laboratory, 1944
Carlos Bellisio; © Instituto Antártico Argentino

Professor Julio Escudero, 1994
© Instituto Antártico Chileno (INACH)

SANAE IV, 1997
Ross Hofmeyr; © Ross Hofmeyr, Wildmedic – CC BY-SA 3.0

Teniente Arturo Parodi, 1999
Paul Taylor and Marcelo Bernal; © Paul Taylor and Marcelo Bernal

Kohnen, 2001
Carmen Pia Günther; © Archiv für deutsche Polarforschung, Alfred Wegener Institute

Concordia, 2005
© Programma Nazionale di Ricerche in Antartide (PNRA)

Collins, 2006
© Instituto Antártico Chileno (INACH)

Johann Gregor Menel, 2007
Kamil Láska; © Kamil Láska

Princess Elisabeth, 2007
© Collection IPF: René Robert, Nighat F.D. Amin, Alain Hubert

Vechernyaya, 2007
© unkown

Amundsen-Scott South Pole, 2008
Jesse Peterson; reproduced courtesy of the US Antarctic Program, National Science Foundation

Kunlun, 2009
Xia Limin; © Xia Limin / Chinese Arctic and Antarctic Administration

Neumayer III, 2009
Christine Wesche; © Christine Wesche / Alfred Wegener Institute Helmholtz Centre for Polar and Marine Research (AWI)

Bharati, 2012
© bof Architekten / Ramboll

Gerritsz laboratory, 2013
Tristan Biggs; © Tristan Biggs / NIOZ Royal Netherlands Institute for Sea Research

Halley IV, 2013
© British Antarctic Survey

Jang Bogo, 2014
© Korea Polar Research Institute (KOPRI)

Taishan, 2014
Tang Zhijian; © Tang Zhijian / China Central Television

Union Glacier Field Camp, 2014
Paul Taylor and Marcelo Bernal; © Pol Taylor and Marcelo Bernal

Juan Carlos I, 2018
ARC Media; © ARC Media

Comandante Ferraz, 2020
Estúdio 41 Arquitetura; © Estúdio 41 Arquitetura

St.Kliment Ohridski, 2020+
PS Architects; © PS Architects

Henryk Arctowki, 2020+
© Kuryłowicz & Associates

Scott Base, 2020+; Rothera, 2020+
Hugh Broughton Architects; © Hugh Broughton Architects

Ross Sea, 2020+
Chinese Arctic and Antarctic Administration

Vostok, 2020+
© Novatek, Zapsibgazprom, OZSK

McMurdo, 2020+
OZ Architecture; © OZ Architecture

Dome Fuji, 2020+
© Misawa Homes Co., Ltd.

p. 614
© Norwegian Polar Institute

p. 618
Roland Bishop & Co.; © Canterbury Museum

p. 622
© The Whaling Museum (Vestfoldmuseene), Sandefjord Norway

p. 624
© Alexander Turnbull Library, Wellington, New Zealand

p. 628
Reproduced courtesy of the National Library of Norway

p. 632
Herbert George Ponting; © Harry Pennell Collection, Canterbury Museum

p. 636
Frank Hurley; reproduced courtesy of the Australian Antarctic Division

p. 640
Frank Hurley; © Royal Geographical Society (with IBG)

p. 644
I.M. Lamb, 1944. Reproduced courtesy of the British Antarctic Survey Archives Service. Archives ref. AD6/19/1/A1/29. Copyright Crown (expired)

p. 648
© Scott Polar Research Institute, University of Cambridge

The Antarctic through the Lens of Jean de Pomereu

APPENDIX

The Antarctic through the Lens of Paolo Pellegrin

The Antarctic through the Lens of Paolo Pellegrin

Photographs taken in 2017 during the National Aeronautics and Space Administration (NASA) Operation IceBridge

"These landscapes are not only extraordinarily beautiful, they create in one a feeling of being in front of a presence that conveys the air of eternity, of being suddenly close to something greater than what is conceivable, the perception of a space so exceeding the human in scale. And yet, I felt compelled to show what a world in danger looks like. Seen from up there, aboard the NASA IceBridge flight, even the ailing ice is beautiful. The crevasses, the ten-storey-high icebergs, the myriad differences in shades of blinding white: everything is fascinating and magical, yet it hides an impending catastrophe. Seemingly devoid of any human history, such a landscape is in fact tarnished by layers upon layers of it. This terrain's decline is macroscopically accelerated by the choices of a civilization that does not appear to know its limits. Perhaps what I photographed is the true horizon of all the other conflicts on this planet."

Paolo Pellegrin (Photographer), in conversation with UNLESS, 2020

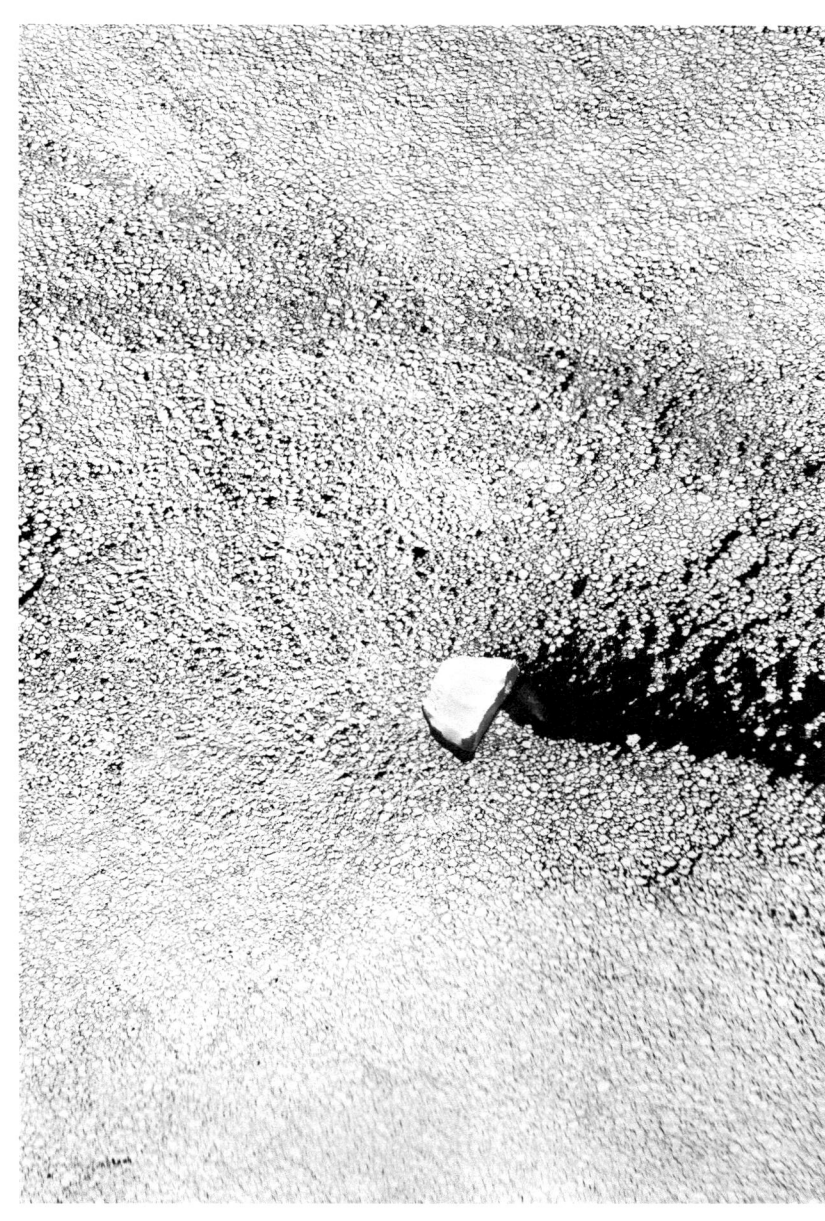

ANTARCTIC RESOLUTION

Editor: Giulia Foscari Widmann Rezzonico / UNLESS
Editorial assistant: Federica Sofia Zambeletti
Editorial coordination: Sabrina Syed
Copyediting: Simon Cowper, Mike Pilewski, James
Roderick O'Donovan
Proofreading: Simon Cowper, Mike Pilewski
Coordination: Federica Sofia Zambeletti, Maya Rüegg
Design: Giulia Foscari Widmann Rezzonico / UNLESS
with Integral Lars Müller
Production: UNLESS, Martina Mullis
Lithography: prints professional, Berlin
Printing and binding: Grafiche Antiga spa,
Crocetta del Montello, Italy
Paper: Sappi Magno Volume, 115 gsm
Font: Everett by Nolan Paparelli

All profits generated with the sale of this publication
will be invested into UNLESS in the pursuit of
Antarctic Resolution.

Lars Müller Publishers is supported by the Swiss
Federal Office of Culture with a structural contribution
for the years 2021–2024.

Lars Müller Publishers
Zurich, Switzerland
www.lars-mueller-publishers.com

ISBN 978-3-03778-640-6

Distributed in North America by ARTBOOK | D.A.P.
www.artbook.com

Printed in Italy

printed
with the sun

FSC
www.fsc.org
MIX
Paper from
responsible sources
FSC® C008309